W9-ANH-306

# REAL AND COMPLEX SINGULARITIES, OSLO 1976

Proceedings of the Nordic Summer School/NAVF
Symposium in Mathematics
Oslo, August 5–25, 1976

# Real and complex singularities, Oslo 1976

Proceedings of the Nordic Summer School/NAVF
Symposium in Mathematics
Oslo, August 5–25, 1976

**P. Holm, editor**

*University of Oslo,*
*Institute of Mathematics,*
*Oslo, Norway*

Sijthoff & Noordhoff International Publishers

1SBN 90 286 0097 3

Set-in-film in The United Kingdom
Printed in The Netherlands

## CONTENTS

## PREFACE

This volume contains the lectures given at the Nordic Summer School in Mathematics 1976, as well as the papers reported on in the adjoint symposium. The Summer School was the 9th in the series sponsored by Nordic Research Courses of the Nordic Cultural Commission. The symposium organized in conjunction with the Summer School was made possible through grants from the Norwegian Research Council for Science and Humanities (NAVF). It is definite that the full project could not have succeeded without the support of both organizations, and I would here like to thank the Nordic Cultural Commission and the Norwegian Research Council for making it possible.

The present book contains nearly all talks given at the meeting with one notable exception: The elegant exposition of John Mather on the genericity and finite type of the topologically stable maps was not written up in time to be included in the book. The loss is partly made up for by two recent expositions in the Springer Lecture Notes (No. 535 and 552), which, however, both differ from the presentation given at the Summer School. It is to be hoped that Mather's proof will appear ultimately.

Finally I take this opportunity to express my gratitude to the staff of the Mathematics Institute, in particular to Mrs Kirkaloff and Mrs Möller.

P. Holm
Oslo, 18 February 1977

Nordic Summer School/NAVF
Symposium in Mathematics
Oslo, August 5–25, 1976

# IRREDUCIBILITY OF THE COMPACTIFIED JACOBIAN

Allen B. Altman[1], Anthony Iarrobino[2], and Steven L. Kleiman[3]

## Introduction

Compactifications of jacobians of integral curves and their compatibility with specialization have been considered by Igusa [11], by Mayer and Mumford [13], and by D'Souza ([5], Bombay thesis, preliminary version, 1973, being improved for *Astérisque*; a copy of this improved manuscript was not available to us at the time of this writing). Seshadri and Oda [14] have studied the compactified jacobian for reducible curves. Nakamura and Namikawa, according to [14], have dealt with compactified jacobians in the complex analytic context.

Igusa worked with a Lefschetz pencil of hyperplane sections on a smooth surface (a general member is a smooth curve and finitely many members have a node as a singularity). He defined the compactification for a singular member as the limit of the jacobians of the non-singular members using Chow coordinates (and Chow's construction of the jacobian). Mayer and Mumford announced an intrinsic characterization of the compactified jacobian as the moduli space of torsion-free sheaves with rank one and Euler characteristic $1-p_a$. They said that it could be constructed for any integral curve using geometric invariant theory and that the jacobian is an open subscheme of it.

D'Souza studied the irreducibility of the compactified jacobian. The irreducibility is equivalent to the denseness of the jacobian because the jacobian is irreducible. He proved the irreducibility for an integral curve $X$ with arithmetic genus $p_a$ and with $\delta_x = 1$ at each singularity $x$ (so the arithmetic and geometric genera differ by the total number of singularities of $X$). Moreover, assuming $X$ has only nodes as singularities, he showed that the completion of the local ring of a singularity of the compactified jacobian

[1] Partially supported by a special grant from Universidad Simón Bolívar.
[2] Supported by CNRS exchange fellowship in France.
[3] Partially supported by the NSF under grant no. MCS74-07275.

is isomorphic to $k[[x_0, \cdots, x_{p_a}]]/x_0x_1$. His proof uses the Abel map from the appropriate Hilbert scheme [see 7] to the compactified jacobian, and he showed the map was smooth when it should be (see (9) below) for a Gorenstein curve. (In his thesis, D'Souza dealt only with the case in which $X$ has only nodes as singularities, but our impression is that he will handle the case in which $X$ is Gorenstein in his article for *Astérisque.*)

Below, we show that the compactified jacobian $P$ is irreducible for an integral curve $X$ lying on a smooth surface $Y$. Since any integral curve with embedding dimension at most 2 at each point can be embedded in a smooth surface [3], this theorem imples D'Souza's result on the irreducibility. Our proof also uses the Abel map to transfer the problem to the Hilbert scheme $B$ of $X$. The Hilbert scheme $B$ is the scheme of zeroes of a section of a locally free sheaf on the Hilbert scheme of $Y$. This gives a lower bound on the dimension of $B$. On the other hand, $B$ has a stratification into sets of low dimension. In addition, the Hilbert scheme of $Y$ is smooth of dimension $2n$ as we prove below in (3). (A different proof is found in [6] for the case that $Y$ is projective and the base is a field or a discrete valuation ring.) The combination of these results yields (5) that $B$ is cut out by a regular section and that it is irreducible. We conclude as corollaries that $P$ is (under the same hypotheses as in (5)) irreducible, reduced and locally a complete intersection. Thus, the jacobian is dense in $P$ and every torsion-free sheaf of rank one on $X$ is a limit of invertible sheaves. It would be interesting to know just when $P$ is a strict complete intersection (that is, when every tangent cone $\mathrm{gr}_{m_p}(\mathcal{O}_{P,p})$ is also a complete intersection) for a curve on a smooth surface; D'Souza showed $P$ is so when the curve has only nodes as singularities.

Our result is sharp. We construct a curve which is integral, smooth except at one point, and a complete intersection in $\mathbb{P}^3$, so Gorenstein in particular, for which the compactified jacobian is reducible, containing a component whose dimension is larger than that of the jacobian. In our example, there are many torsion-free sheaves of rank one which are not limits of invertible sheaves.

In our example, the subset $\mathrm{Sm}^n(X)$ of $\mathrm{Hilb}^n_{(X/S)}$ of smoothable 0-dimensional subschemes of length $n$ is not all of $\mathrm{Hilb}^n_{(X/S)}$; in fact, there are many thick points on $X$ not smoothable in $X$ or even in the ambient $\mathbb{P}^3$. On the other hand, Theorem (8) states that for a reduced curve on a smooth ambient surface, $\mathrm{Sm}^n(X)$ is all of $\mathrm{Hilb}^n_{(X/S)}$.

The general theory of the compactified jacobian of a family $X/S$ can be developed in the spirit of Grothendieck's theory of the Picard scheme, as announced in [2]. The heart of this development is a theory of linear systems, which allows us to represent the Abel map as the twisted family of projective spaces associated to a manageable sheaf. It leads naturally to the surprising conclusion that for a family of non-Gorenstein curves, the theory is more satisfactory with $\mathrm{Quot}_{(\omega/X/S)}$, where $\omega$ is the dualizing sheaf, in place of $\mathrm{Hilb}_{(X/S)}$ as the source of the Abel map. (Note that the two are isomorphic when the family of curves is Gorenstein.) For example, the Abel map from $\mathrm{Quot}^n_{(\omega/X/S)}$ is smooth for $n \geqq 2p_a - 2$ and surjective for $n \geqq p_a - 1$. Since here we deal exclusively with Gorenstein curves except for (10) and (11), we have chosen to work with $\mathrm{Hilb}^n_{(X/S)}$ (but note the awkward proof required for (10)), and we use both [2] and [5] as references.

The authors would like to thank Tor Gulliksen for a fruitful discussion about local complete intersections. We would also like to thank the Mathematics Institutes of the University of Aarhus and the University of Oslo for the generous hospitality extended to us.

Throughout, $S$ will be a locally noetherian (base) scheme.

LEMMA (1). *Let $Z/S$ be a flat quasi-projective family of schemes, and let $n$ be a positive integer.*

(i) *There is an open subscheme of the Hilbert scheme* $\mathrm{Hilb}_{(Z/S)}$ *of $Z/S$ [see 7] parametrizing those subschemes of $Z/S$ that are regularly embedded.*

(ii) *Let $V$ be the open subscheme of* $\mathrm{Hilb}^n_{(Z/S)}$ *(the Hilbert scheme of 0-dimensional subschemes $W$ of $Z/S$ with $\chi(\mathcal{O}_W) = n$) parametrizing those 0-dimensional subschemes which are regularly embedded (see* (i)*), and let $w$ be a geometric point of $V$ representing a subscheme $W$.*

(a) *At the point $w$ of $V$, the scheme* $\mathrm{Hilb}^n_{(Z/S)}$ *is smooth over $S$ with relative dimension*

$$\dim (H^0(W, N))$$

*where $N$ is the normal sheaf of $W$ in the fiber of $Z/S$ containing it.*

(b) *If the fiber of $Z/S$ containing $W$ has dimension $d$ at each point in the support of $W$, then the relative dimension of* $\mathrm{Hilb}^n_{(Z/S)}$ *at $w$ is equal to $dn$.*

PROOF. Assertion (i) follows easily from the fact that the set of points of a base scheme over which a flat and proper subscheme is transversally regularly embedded is open [8; $\mathrm{IV}_4$, 19.2.4]. Assertion (ii(a)) follows easily from the infinitesimal study of the Hilbert scheme [7; 221–23] and the fact that $H^1(W, N)$ is zero since $W$ is zero dimensional. Assertion (ii(b)) follows from (ii(a)), from the fact that $N$ is free with rank $d$, and from the equality $n = h^0(W, \mathcal{O}_W)$.

PROPOSITION (2). *Let $Z/S$ be a quasi-projective family of schemes whose geometric fibers are connected. Then the geometric fibers of* $\mathrm{Hilb}^n_{(Z/S)}$ *are connected.*

PROOF. We may assume $S$ is equal to the spectrum of an algebraically closed field. The norm map (see [7], 221–26),

$$\mathfrak{n}_{(Z/S)} : \mathrm{Hilb}^n_{(Z/S)} \to \mathrm{Symm}^n_S(Z),$$

has connected fibers ([6], Proposition 2.2). It is proper because $Z$ may be embedded in a projective scheme $Y$, the map $\mathfrak{n}_{(Z/S)}$ is the restriction of $\mathfrak{n}_{(Y/S)}$ to $\mathrm{Symm}^n_S(Z)$, and $\mathrm{Hilb}^n_{(Y/S)}$ is proper. Therefore, since $\mathrm{Symm}^n_S(Z)$ is connected (because $Z$ is), $\mathrm{Hilb}^n_{(Z/S)}$ is connected.

PROPOSITION (3). *Let $Y/S$ be a smooth quasi-projective family of surfaces. For each positive integer $n$, $\mathrm{Hilb}^n_{(Y/S)}$ is smooth over $S$ with relative dimension $2n$.*

PROOF. For the smoothness assertion, we use the infinitesimal criterion [8; $\mathrm{IV}_4$, 17.5.4]. Take $S$ to be the spectrum of an artinian local ring, take $S_0$ to be a subscheme of $S$, and take $W_0$ to be a 0-dimensional subscheme of $Y_0 = Y \times_S S_0$ that is flat over $S_0$. We must lift $W_0$ to a flat subscheme $W$ of $Y/S$.

For any open subscheme $U$ of $Y$ that contains $W_0$, it is clearly sufficient to lift $W_0$ to a flat subscheme of $U/S$. Since $W_0$ is a finite set of (closed) points and since $Y$ is quasi-projective, $W_0$ is contained in an affine open subset of $Y$. Hence we may assume $Y$ is affine.

Consider a resolution of $\mathcal{O}_{W_0}$,

$$0 \longrightarrow K \xrightarrow{u_0} \mathcal{O}_{Y_0}^{\oplus(m+1)} \xrightarrow{v_0} \mathcal{O}_{Y_0} \longrightarrow \mathcal{O}_{W_0} \longrightarrow 0. \tag{3.1}$$

Since $W_0$ and $Y_0$ are $S_0$-flat, $K$ is $S_0$-flat. Let $s$ be the point of $S$. Then the restriction of (3.1) to the fiber over $s$ is also exact. Moreover $K(s)$ is locally free because $Y(s)$ is regular [8; $\mathrm{IV}_4$, 17.5.2] with dimension 2. Hence $K$ is locally free. Moreover, the exactness of (3.1) implies that $K$ has rank $m$ off $W_0$, hence it has rank $m$ everywhere. By shrinking $Y$, we may assume $K$ is free.

By a theorem of Burch ([4], p. 944; see [12], p. 148 for an outline of a proof) the map $v_0$ is equal to $a(\bigwedge^m u_0)$ where $a$ is a unit. Since the image and kernel of $a(\bigwedge^m u_0)$ are equal to the image and the kernel of $(\bigwedge^m u_0)$, we may assume $a$ is equal to 1. Now, lift $u_0$ to a map $u : \mathcal{O}_Y^{\oplus m} \to \mathcal{O}_Y^{\oplus(m+1)}$. Then the sequence,

$$0 \longrightarrow \mathcal{O}_Y^{\oplus m} \xrightarrow{u} \mathcal{O}_Y^{\oplus(m+1)} \xrightarrow{\bigwedge^m u} \mathcal{O}_Y \longrightarrow \mathcal{O}_W \longrightarrow 0, \tag{3.2}$$

where $\mathcal{O}_W$ is the cokernel of $\bigwedge^m u$, restricts to (3.1), and its restriction to the fiber is exact. Hence (3.2) is exact and $\mathcal{O}_W$ is $S$-flat (see [8], $\mathrm{IV}_3$, 11.3.7). Thus, there is a lifting $W$ of $W_0$ to a flat subscheme of $Y/S$. So $\mathrm{Hilb}^n_{(Y/S)}$ is smooth over $S$.

To prove the dimension assertion, we may clearly assume that $S$ is the spectrum of a field and that $Y$ is connected. The scheme $\mathrm{Hilb}^n_{(Y/S)}$ is connected (2), so irreducible because $\mathrm{Hilb}^n_{(Y/S)}$ is regular. The subscheme $V$ parametrizing regularly embedded subschemes is open (1, (ii)) and clearly non-empty, so it is dense in $\mathrm{Hilb}^n_{(Y/S)}$. Since $V$ has dimension $2n$ by (1, (ii(b))), $\mathrm{Hilb}^n_{(Y/S)}/S$ has relative dimension $2n$.

PROPOSITION (4). *Let $Y$ be a quasi-projective $S$-scheme, $E$ a locally free $\mathcal{O}_Y$-Module, and $X \subset Y$ the subscheme of zeroes of a section of $E$. Set $A = \mathrm{Hilb}^n_{(Y/S)}$ and $B = \mathrm{Hilb}^n_{(X/S)}$. Let $W \subset Y \times A$ denote the universal 0-dimensional subscheme of $Y/S$ with degree $n$, and let $p: W \to A$ denote the structure morphism. Then $B$ is equal to the subscheme of $A$ of zeroes of a section of the locally free $\mathcal{O}_Y$-Module $p_*(E \mid W)$, whose rank is equal to $nr$ where $r$ is the rank of $E$. Moreover, the formation of $p_*(E \mid W)$ and the formation of the section commute with base change.*

PROOF. Let $I$ denote the Ideal of $X$. Then $B$ is clearly equal to the subscheme of $A$ of zeroes of the natural composition,

$$I_A \to \mathcal{O}_{Y\times A} \to \mathcal{O}_W.$$

The section of $E$ defining $X$ corresponds to a surjection $E^\vee \to I$, so $B$ is also equal to the subscheme of $A$ of zeroes of the induced composition,

$$(E \mid W)^\vee \to I_W \to \mathcal{O}_W. \tag{4.1}$$

Since $E$ is locally free and $p$ is finite, the formation of

$$\underline{\mathrm{Hom}}_W\,((E \mid W)^\vee, \mathcal{O}_W) = E \mid W$$

and the formation of

$$p_*\,\underline{\mathrm{Hom}}_W\,((E \mid W)^\vee, \mathcal{O}_W) = p_*(E \mid W)$$

commute with base change. Consequently, $B$ is equal to the subscheme of $A$ of zeroes of the section of $p_*(E \mid W)$ arising from (4.1). It is clear from the construction that the formation of the section commutes with base change. Finally, since $p$ is finite and flat, with degree $n$, clearly $p_*(E \mid W)$ is locally free with rank $rn$.

THEOREM (5). *Let $X/S$ be a flat family of geometrically reduced curves lying on a smooth quasi-projective family of surfaces $Y/S$. Set $A = \mathrm{Hilb}^n_{(Y/S)}$, and set $B = \mathrm{Hilb}^n_{(X/S)}$. Then $B$ is equal to the subscheme of $A$ of zeroes of a section $\sigma$ of a locally free $\mathcal{O}_A$-Module $F$ with rank $n$, and the formations of $F$ and of $\sigma$ commute with base change; moreover, $\sigma$ is transversally regular over $S$. In addition, $B$ is equal to the set-theoretic closure of $\mathrm{Div}^n_{(U/S)}$ in $A$ where $U$ denotes the smooth locus of $X/S$.*

PROOF. First, $X$ is a relative effective divisor of $Y/S$ because $X/S$ is flat and its fibers are reduced, pure 1-codimensional subschemes of the locally factorial schemes $Y(s)$. Consequently, by (4), $B$ is equal to the subscheme of $A$ of zeroes of a section $\sigma$ of a locally free $\mathcal{O}_A$-Module $F$ with rank $n$, and the formations of $F$ and of $\sigma$ commute with base change.

To show that $\sigma$ is transversally regular, it is enough to check on the geometric fibers of $A/S$ [8; $\mathrm{IV}_3$, 11.3.8]. Moreover, to check that $\mathrm{Div}^n_{(U/S)}$ is dense in $B$, it is clearly enough to check on the geometric fibers. Therefore, we may assume $S$ is the spectrum of an algebraically closed field.

Let $\mathrm{Symm}^n_S(X)$ denote the $n$-fold symmetric product of $X/S$, and let

$$p: X^{\times n} \to \mathrm{Symm}^n_S(X)$$

denote the canonical projection; it is a finite morphism. Let $r$ be an integer satisfying $0 \leqq r \leqq n$, let $\pi$ denote a sequence of integers $(n_1, \cdots, n_r)$ satisfying $n_1 \geqq n_2 \geqq \cdots \geqq n_r$ and $n' = \sum_{i=1}^r n_i \leqq n$, and set $m = n - n'$. Let $U^{(n)}_{(r,\pi)}$ denote the (finite) union

$$U^{(n)}_{(r,\pi)} = \bigcup p\,(\{x_1\}^{n_1} \times \cdots \times \{x_r\}^{n_r} \times U^{\times m}),$$

where $(x_1, \cdots, x_r)$ runs through the set of $r$-tuples of distinct singular points of $X$. Each $U^{(n)}_{(r,\pi)}$ is clearly constructible, and its dimension is $m$ because $U^{\times m}$ has dimension $m$ and $p$ is finite. Clearly the (finite) union of the sets $U^{(n)}_{(r,\pi)}$ over all pairs $(r, \pi)$ covers $\mathrm{Symm}^n_S(X)$.

Let $w_n : B \to \mathrm{Symm}^n_S(X)$ denote the norm map $\mathfrak{n}_{(X/S)}$ [see 7; 221–26], and set $B^{(n)}_{(r,\pi)} = w_n^{-1}(U^{(n)}_{(r,\pi)})$. Clearly $B$ is equal to the finite union of the constructible sets $B^{(n)}_{(r,\pi)}$.

Assume the estimate,

$$\dim (B^{(n)}_{(r,\pi)}) \leqq n - r. \tag{5.1}$$

Since $B$ is equal to the subscheme of $A$ of zeroes of a section of a locally free $\mathcal{O}_A$-Module $F$ with rank $n$, every component of $B$ has codimension less than or equal to $n$ [7; $0_{\mathrm{IV}}$, 16.3.7]. Since $A$ has dimension $2n$ by (2), every component of $B$ has dimension greater than or equal to $n$. Therefore, (5.1) implies that $B$ has pure dimension $n$ and is the closure of $B^n_{(0,\varnothing)}$. Moreover, since $A$ is regular (2), so Cohen-Macaulay, the section of $F$ defining $B$ is regular [7; $0_{\mathrm{IV}}$, 16.5.6]. Finally, $B^{(n)}_{(0,\varnothing)}$ is equal to the open set $\mathrm{Div}^n_{(U/S)}$ because a 0-cycle on a smooth curve is the norm of a unique 0-dimensional subscheme, which is a divisor; so $\mathrm{Div}^n_{(U/S)}$ is dense in $B$.

To complete the proof, we need only establish the estimate (5.1). The fiber of the norm map over a geometric point $n_1x_1 + \cdots + n_rx_r + y_1 + \cdots y_m$ is clearly equal to the product of the fibers of the maps $w_{n'} : B^{(n')}_{(r,\pi)} \to U^{(n')}_{(r,\pi)}$ and $w_m : B^{(m)}_{(0,\varnothing)} \to U^{(m)}_{(0,\varnothing)}$. The dimension of the latter fiber is clearly zero.

Hence the commutativity of the diagram,

$$\begin{array}{ccc} \mathrm{Hilb}^{n'}_{(Y/S)} & \xrightarrow{w} & \mathrm{Symm}^{n'}_S(Y) \\ \uparrow & & \uparrow \\ B^{n'}_{(r,\pi)} & \xrightarrow{w_{n'}} & \mathrm{Symm}^{n'}_S(X) \end{array}$$

and the estimate,

$$\dim(w^{-1}(x)) \leqq n' - r \quad \text{for} \quad x \in \mathrm{Symm}^{n'}_S(Y),$$

([9], Cor. 2, p. 820) imply the estimate,

$$\dim(w_{n'}^{-1}(x)) \leqq n' - r \quad \text{for} \quad x \in U^{(n)}_{(r,\pi)}.$$

Since the dimension of $U^{(n)}_{(r,\pi)}$ is equal to $m$, estimate (5.1) holds by [8; $IV_2$, 5.6.7].

Corollary (6). *The norm map,*

$$\mathfrak{n}_{(X/S)}: \mathrm{Hilb}^n_{(X/S)} \to \mathrm{Symm}^n_S(X)$$

*is birational for all n.*

Proof. By [7; 221–27], the norm map,

$$\mathfrak{n}_{(U/S)}: \mathrm{Hilb}^n_{(U/S)} \to \mathrm{Symm}^n_S(U),$$

is an isomorphism. Obviously $\mathrm{Hilb}^n_{(U/S)}$ is equal to $\mathrm{Div}^n_{(U/S)}$. By (5), $\mathrm{Div}^n_{(U/S)}$ is dense in $\mathrm{Hilb}^n_{(X/S)}$. Since $U$ is dense in $X$, clearly $\mathrm{Symm}^n_S(U)$ is dense in $\mathrm{Symm}^n_S(X)$. Thus $\mathfrak{n}_{(X/S)}$ is birational.

Corollary (7). *The scheme* $\mathrm{Hilb}^n_{(X/S)}$ *is flat over S and each geometric fiber is locally a complete intersection, Cohen-Macaulay, reduced, and n-dimensional. Moreover, if the geometric fibers of X/S are integral (resp. connected), then the geometric fibers of* $\mathrm{Hilb}^n_{(X/S)}$ *are integral (resp. connected).*

Proof. Since the section $\sigma$ is transversally regular, $\mathrm{Hilb}^n_{(X/S)}$ is flat over $S$ [8; $IV_4$, 19.2.1]. Since $\mathrm{Hilb}^n_{(Y/S)}$ is smooth over $S$ with relative dimension $2n$ by (3), and since $F$ is locally free with rank $n$, the geometric fibers are local complete intersections with dimension $n$, so Cohen-Macaulay with dimension $n$ [8; $0_{IV}$, 16.5.6]. Since $\mathfrak{n}_{(X/S)}$ is birational, the geometric fibers of $\mathrm{Hilb}^n_{(X/S)}$ are generically regular, and they are irreducible (resp. connected) if $X$ is irreducible (resp. connected). Since the geometric fibers of $\mathrm{Hilb}^n_{(X/S)}$ are Cohen-Macaulay, they are reduced, and they are integral if $X$ is integral.

Theorem (8). *Let X/S be a flat family of geometrically reduced curves lying on a smooth quasi-projective family of surfaces. Let V be the open subset of* $\mathrm{Symm}^n_S(X)$ *corresponding to 0-cycles* $\sum_{i=1}^n x_i$ *on the fibers of X/S where the* $x_i$

*are distinct smooth points. Then* $(\mathfrak{n}_{(X/S)})^{-1}(V)$ *and* $\mathrm{Div}^n_{(U/S)}$ *are scheme-theoretically dense in* $\mathrm{Hilb}^n_{(X/S)}$.

PROOF. Since the data commute with base change, we may assume $S$ is the spectrum of an algebraically closed field [8; $\mathrm{IV}_3$, 11.10.10]. The set $V$ is obviously dense in $\mathrm{Symm}^n_S(U)$. So, since $\mathfrak{n}_{(U/S)}$ is an isomorphism [7; 221–27] and since $\mathrm{Div}^n_{(U/S)}$ is dense in $\mathrm{Hilb}^n_{(X/S)}$ by (5), the open set $\mathfrak{n}^{-1}_{(X/S)}(V)$ is dense in $\mathrm{Hilb}^n_{(X/S)}$. Finally, since $\mathrm{Hilb}^n_{(X/S)}$ is Cohen-Macaulay, so has no embedded components, $\mathfrak{n}^{-1}_{(X/S)}(V)$ and $\mathrm{Div}^n_{(U/S)}$ are scheme-theoretically dense in $\mathrm{Hilb}^n_{(X/S)}$ [8; $\mathrm{IV}_4$, 20.2.13 (iv)].

THEOREM (9). *Let $X/S$ be a flat projective family of geometrically integral curves with arithmetic genus $p_a$ lying on a smooth quasi-projective family of surfaces parametrized by $S$. Let $P$ denote the compactified jacobian of $X/S$. Then $P/S$ is flat and its geometric fibers are integral, Cohen-Macaulay, local complete intersections with dimension $p_a$. Moreover, $P$ contains the jacobian* $\mathrm{Pic}^0_{(X/S)}$ *as an open scheme-theoretically dense subscheme.*

PROOF. For each integer $m$, the $S$-scheme $P_m$ is defined (see [2], Theorem 2) as the $S$-scheme which parametrizes the isomorphism classes of torsion-free $\mathcal{O}_X$-Modules of rank 1 with Euler characteristic $m$. Clearly $P$ is equal to $P_{(1-p_a)}$. Since $P_m$ is isomorphic to $P_{m+rd}$ for any $r$ where $d$ is the degree of $\mathcal{O}_X(1)$, it suffices to prove the assertions for $P_m$ with $m<3-3p_a$.

Set $n=1-p_a-m$. Then, since $X$ is Gorenstein, and since $n>2p_a-2$ holds, the Abel map,

$$\mathcal{A}_{(X/S)}:\mathrm{Hilb}^n_{(X/S)}\to P_m,$$

is smooth with relative dimension $n-p_a$ and surjective (see [2], Theorem 4 or [5]). So, since $\mathrm{Hilb}^n_{(X/S)}$ is $S$-flat and its geometric fibers are integral and Cohen-Macaulay with dimension $n$ by (7), the scheme $P_m$ is flat over $S$ ($\mathcal{A}_{(X/S)}$ is faithfully flat) and its geometric fibers are integral with dimension $p_a$ and Cohen-Macaulay by [8; $\mathrm{IV}_4$, 17.5.8]. Since $\mathrm{Div}^n_{(X/S)}$ is dense in $\mathrm{Hilb}^n_{(X/S)}$ by (5) and since $P_m$ is Cohen-Macaulay, its image is scheme-theoretically dense in $P_m$.

Finally since the geometric fibers of $\mathrm{Hilb}^n_{(X/S)}$ are local complete intersections (7), the geometric fibers of $P_m$ are local complete intersections by the following general result: If $B/A$ is a formally smooth extension of noetherian local rings, then $A$ is a complete intersection if and only if $B$ is a complete intersection. This result is an easy consequence of the theory of homology of algebras. Alternately, a proof in our case may be based on [8; $\mathrm{IV}_4$, 19.3.10, 19.3.4, and 17.5.3$(d'')$].

LEMMA. (10). *Assume $S$ is a noetherian scheme, and let $X/S$ be a flat projective family of geometrically integral curves with genus $p_a$. Then there* *exists an integer m such that the Abel map,*

$$\mathcal{A}_{(X/S)} : \mathrm{Hilb}^n_{(X/S)} \to P_m \qquad n = 1 - p_a - m,$$

*is surjective and such that P is isomorphic to* $P_m$.

PROOF. Let $(p_1, \cdots, p_t)$ be the generic points of the irreducible components of $P$, and, for each $i$, let $g_i$ be a geometric point with image $p_i$. For each $i$, let $I_i$ be a universal sheaf on the geometric fiber $X(g_i) \times P(g_i)$; one exists ([2], Theorem 3) because $X(g_i)$ has a smooth rational point. By Serre's theorem [8; III, 2.2.2 (iv)], there is an integer $r$ such that for each $i$, the sheaf,

$$\underline{\mathrm{Hom}}_{X(g_i)}(I_i, \mathcal{O}_{X(g_i)})(r) = \underline{\mathrm{Hom}}_{X(g_i)}(I_i(-r), \mathcal{O}_{X(g_i)}), \tag{10.1}$$

is generated by its global sections. Set

$$m = \chi(I_i(-r));$$

clearly $m$ is independent of $i$.

Twisting by $(-r)$ induces an isomorphism $d_m : P \xrightarrow{\sim} P_m$. Set $\bar{p}_i = d_m(p_i)$ and $\bar{g}_i = d_m(g_i)$. Then $d_m$ clearly carries each $I_i(-r)$ to a universal sheaf, $J_i$, on $X(\bar{g}_i) \times P_m(\bar{g}_i)$. Since, for each $i$, the sheaf (10.1) is generated by its global sections, the isomorphic sheaf, $\underline{\mathrm{Hom}}_{X(\bar{g}_i)}(J_i, \mathcal{O}_{X(\bar{g}_i)})$, is also, so it has a non-zero global section $\sigma_i$. The image $\sigma_i(J_i)$ is an Ideal of $\mathcal{O}_{X(\bar{g}_i)}$ and defines a point of $\mathrm{Hilb}^n_{(X/S)}$ in the fiber over $\bar{p}_i$. Since $\mathrm{Hilb}^n_{(X/S)}$ is projective, $\mathcal{A}_{(X/S)}$ is proper, and so its image is closed. Since this image contains each generic point $\bar{p}_i$ of $P_m$, it is therefore all of $P_m$. Thus $\mathcal{A}_{(X/S)}$ is surjective.

PROPOSITION (11). *Let X/S be a flat projective family of geometrically integral curves. Then the compactified jacobian P of X/S has geometrically connected fibers and it is proper over S. If, in addition, S is connected, then P is connected.*

PROOF. First note that the second assertion follows from the first, and for the first we may assume $S$ is noetherian. By (10), there are an integer $n$ and a surjective map (isomorphic to the Abel map),

$$\mathrm{Hilb}^n_{(X/S)} \to P.$$

Since $\mathrm{Hilb}^n_{(X/S)}$ has geometrically connected fibers (2), $P$ has geometrically connected fibers. Moreover, since $\mathrm{Hilb}^n_{(X/S)}$ is projective, $P$ is proper over $S$.

REMARK (12). Theorem (9), Lemma (10) and Proposition (11) hold for each $P_j$ as the proofs in fact really show. The various $P_j$ need not be isomorphic if there is no section of the smooth locus of $X/S$ although they become isomorphic after an étale base change that provides one.

EXAMPLE (13). The result in the theorem is sharp. It fails for a complete intersection $X$ in the projective 3-space $\mathbb{P}^3$ over an algebraically closed field; $X$ is integral with only one singularity.

In fact $P$, the compactified jacobian of $X$, is connected (11) and has a component with dimension strictly greater than $p_a$, while the closure of the jacobian, $\mathrm{Pic}^0_{(X/S)}$, has dimension $p_a$. So, $P$ can be neither irreducible nor Cohen-Macaulay. In particular, $X$ has torsion-free sheaves of rank 1 which are not limits of invertible sheaves.

Let $x$ be a closed point of $\mathbb{P}^3$, let $R$ denote the local ring of $x$, let $\mathfrak{m}$ denote the maximal ideal of $R$, and let $d$ be a positive integer. The homogeneous ideal associated to $\mathfrak{m}^d$ is clearly generated by its homogeneous elements of degree $d$. So by a form of Bertini's Theorem ([3]), there are two surfaces of degree $(d+1)$ whose intersection contains the thick point, $Z = \mathrm{Spec}\,(R/\mathfrak{m}^d)$, and which is smooth except at $x$ and irreducible. Take $X$ to be this intersection.

For simplicity of computation, we take $d = 4e$. We now establish a lower bound,

$$h = \dim\,(\mathrm{Hilb}^n_{(X/S)}) \geqq \frac{d^4}{16} = 16e^4, \tag{13.1}$$

for a certain integer $n$ satisfying,

$$n \leqq \frac{d^3}{3}. \tag{13.2}$$

Each subspace $V$ of $\mathfrak{m}^{d-1}/\mathfrak{m}^d$ is clearly an ideal of $A = R/\mathfrak{m}^d$; the length of $A/V$ is equal to $n = \binom{d+1}{3} + s$ for $s = \dim\,(V)$. It follows that $\mathrm{Hilb}^n_{(Z/S)}$ contains $\mathrm{Grass}_s\,(\mathfrak{m}^{d-1}/\mathfrak{m}^d)$. The former is contained in $\mathrm{Hilb}^n_{(X/S)}$, and the latter has dimension

$$s(\dim\,(\mathfrak{m}^{d-1}/\mathfrak{m}^d) - s) = s(2e(4e+1) - s).$$

Take $s = e(4e+1)$. Then (13.1) and (13.2) follow. (This argument is the one found in ([10], p. 75).)

Note in passing that $\mathrm{Div}^n_{(X/S)}$ is open in $\mathrm{Hilb}^n_{(X/S)}$ and its dimension is equal to $n$ by (1, (ii(b))). Therefore, for large $d$, the scheme $\mathrm{Hilb}^n_{(X/S)}$ is reducible with a component having dimension greater than $n$, and $X$ contains 0-dimensional subschemes of degree $n$ which are not smoothable or equivalently are not limits of divisors.

To verify the assertions about $P$, we first find $p_a$, which (an infinitesimal analysis shows) is equal to the dimension of $\mathrm{Pic}^0_{(X/S)}$. The dualizing sheaf $\omega_X$ of $X$ is given by the formulas,

$$\omega_X \cong (\omega_{\mathbb{P}^3} \mid X) \otimes (\Lambda^2 N(X, \mathbb{P}^3))^{-1} \cong \mathcal{O}_X(2d-2).$$

It follows that $p_a$ is given by the formula,

$$p_a = (d-1)(d+1)^2 - 1. \tag{13.3}$$

Consider the Abel map,

$$\mathcal{A}_{(X/S)} : \mathrm{Hilb}^n_{(X/S)} \to P_m,$$

where $m = 1 - p_a - n$. Let $\delta$ denote the maximal fiber dimension of $\mathcal{A}_{(X/S)}$. Suppose we have proved the estimate

$$\delta \leqq n. \tag{13.4}$$

Then since $P_m$ is isomorphic to $P$, the dimension $h$ of $\mathrm{Hilb}^n_{(X/S)}$ is estimated by

$$h \leqq \dim(P) + n, \qquad [8; \mathrm{IV}_2, 5.6.7]. \tag{13.5}$$

If $P$ were irreducible, then its dimension would be equal to $p_a$ because $\mathrm{Pic}^0_{(X/S)}$ is open in $P$. If $P$ had pure dimension $p_a$, then (13.3) and (13.5) would yield the inequality,

$$h \leqq (d-1)(d+1)^2 - 1 + n.$$

However, for large $d$, this would contradict (13.1) and (13.2). Therefore, for $d$ large, $P$ is not irreducible, and it contains a component whose dimension is greater than $p_a$.

It remains to establish (13.4). Let $D$ represent a point of $\mathrm{Hilb}^n_{(X/S)}$, and let $\delta(D)$ be the dimension of the fiber of the Abel map containing this point. Let $I$ denote the Ideal of $D$. There are formulas,

$$\begin{aligned}\delta(D) &= \dim(\mathrm{Hom}(I \otimes \omega_X, \omega_X)) - 1\\ &= h^1(X, I \otimes \omega_X) - 1;\end{aligned}$$

the first equality holds because $\omega_X$ is invertible and the second holds by duality ([1], VIII, 1.15, p. 167). Now, the exact sequence,

$$0 \to I \otimes \omega_X \to \omega_X \to \mathcal{O}_D \to 0,$$

obtained by tensoring with $\omega_X$ the sequence relating $I$ and $D$, yields the estimate,

$$h^1(X, I \otimes \omega_X) \leqq n + h^1(X, \omega_X).$$

This estimate establishes the bound,

$$\delta(D) + 1 \leqq n + 1.$$

Thus, (13.4) holds.

## BIBLIOGRAPHY

[1] Altman, A. and Kleiman, S., *Introduction to Grothendieck Duality Theory.* Lecture Notes in Mathematics No. 146, Springer-Verlag, (1970).

[2] Altman, A. and Kleiman, S., Compactifying the Jacobian, *Bull. Am. Math. Soc.*, Vol. 82, No. 6 (1976), 947–949.

[3] Altman, A. and Kleiman, S., Bertini Theorems for Hypersurface Sections Containing a Subscheme (to appear).

[4] Burch, L., On Ideals of Finite Homological Dimension in Local Rings. *Proc. Cambridge Phil. Soc.*, Vol. 69 (1968) 941–952.

[5] D'Souza, C., Thesis 'Compactification of Generalised Jacobians', Bombay (1973) (preliminary version).

[6] Fogarty, J., Algebraic Families on an Algebraic Surface. *Am. Jour. Math.*, Vol. 90 (1968) 511–521.

[7] Grothendieck, A., Techniques de Construction et Théorèmes d'Existence en Géométrie Algébrique IV: Les Schémas de Hilbert, Séminaire Bourbaki 13e année (1960/61), n°221.

[8] Grothendieck, A. (with J. Dieudonné), *Éléments de Géométrie Algébrique.* $III_1$, $0_{IV}$, $IV_2$, $IV_3$, $IV_4$, Publ. Math. de I. H. E. S., no. 11, 20, 24, 28, 32 (1961, 1964, 1965, 1966, 1967).

[9] Iarrobino, A., Punctual Hilbert Schemes. *Bull. Am. Math. Soc.*, Vol. 78 (1972) 819–823.

[10] Iarrobino, A., Reducibility of the Families of 0-dimensional Schemes on a Variety. *Inventiones Math.*, Vol. 15 (1972) 72–77.

[11] Igusa, Fiber Systems of Jacobian Varieties I. *Am. Jour. Math.*,Vol. 78 (1956) 177–199.

[12] Kaplansky, I., *Commutative Rings,* The University of Chicago Press, rev. ed. (1974).

[13] Mayer, A. and Mumford, D., 'Further Comments on Boundary Points'. Woods Hole (1964).

[14] Seshadri, C. and Oda, T., Compactifications of the Generalized Jacobian Variety., (preprint).

Allen B. Altman
Universidad Simón Bolívar
Departamento de Matemáticas
Apartado Postal 5354
Caracas, Venezuela

Anthony Iarrobino
University of Texas
Department of Mathematics
Austin, Texas 78712
USA

Steven L. Kleiman
Massachusetts Institute of Technology
Department of Mathematics
Cambridge, Mass. 02139
USA

Nordic Summer School/NAVF
Symposium in Mathematics
Oslo, August 5–25, 1976

# QUASIANALYTIC AND PARAMETRIC SPACES

Aldo Andreotti and Per Holm

The study of hypersurface models obtained by projecting a closed submanifold $X^n \subset P_N(\boldsymbol{C})$ into a projective space $P_{n+1}(\boldsymbol{C})$ were initiated by the classical algebraic geometers. The model obtained is in general no longer a manifold, but acquires singularities, which for a generic projection were called 'ordinary singularities'. Thus manifolds of dimension 1 produced curves with ordinary nodes only, and manifolds of dimension 2 produced surfaces in projective 3-space having a double nodal curve, a finite set of triple points on the curve (triple also for the nodal curve) and a finite set of pinch points (first discovered by Cayley).

In this paper we begin the study of the singularities one obtains by mapping a *real* analytic manifold into a higher dimensional manifold by a generic map, with special attention to questions of reality.

Thus we have been led to discuss in § 1 real-analytic spaces, here considered as reduced complex spaces $X$ with an antiholomorphic involution $\tau$. If $X_\tau$ is the fixed-point set of the antiinvolution, $X_\tau$ has a fundamental system of $\tau$-invariant neighborhoods in $X$ which are Stein. Therefore, following a proof of Grauert, we can always realize $X_\tau$ as a subset of some numerical space $\boldsymbol{C}^N$ by a proper holomorphic imbedding respecting the real structure.

In §2 we introduce the notion of (real) quasianalytic space. Loosely speaking these are germs of complex spaces with real structure along the part $\hat{X} \subset X_\tau$ where $X$ is a faithful (germ-) complexification of $X_\tau$. Also quasianalytic spaces admit proper 'holomorphic' imbeddings preserving real structure into numerical spaces. If we define these spaces by local models, they admit Stein complexifications, as is proved by an argument of Bruhat and Whitney from the manifold case. Moreover proper maps between Stein complexifications, which are surjective and generally one-to-one, preserve the quasianalytic part. Quasianalytic spaces are semianalytic, and this mapping property shows that they avoid a certain amount of pathology present for semianalytic and analytic spaces in general.

In §3 we introduce parametric spaces as pure dimensional quasianalytic spaces whose normalization is a manifold. They therefore present the type of singularities studied by algebraic geometers. The notion of weakly normal parametric space is introduced, and three types of examples are worked out.

In §4 we study transverse regular parametric singularities, i.e. parametric spaces whose normalization map is correct (mostly 1-correct) or even excellent, in the sense of Thom. We prove that a generic real-analytic map from a compact $n$-manifold into a manifold of dimension $n+1$ or $n+e$ with $e \geqq n/2$ produce weakly normal singularities only. In the hypersurface case this theorem is proved by a reduction to the complex case (using a 'subtransversality' result (19.3)) which is settled by a Hartogs type theorem for weak normality, (16.3). In the second case the map is reduced to a normal form whose parametric singularities are seen to fall into the types of examples treated in §3. The normal form is one of two types that we establish in section 24 for maps determined by their 2-jets. This part requires an extensive use of the Malgrange-Weierstrass preparation theorem.

Finally §5 (the appendix) treats maps of finite deficiency, i.e. maps $f$ whose associated local ring $F/f^*\mathfrak{m}$ at any point is finite dimensional. We give detailed proof that a generic map from an $n$-manifold to a $p$-manifold, $n \leqq p$, has deficiency bounded by a function of $n$ and $p$ only.

This covers the preliminary local study necessary to introduce spaces with reasonable singularities. We hope to come back with some global results on the subject.

## Contents

## §1. Real-analytic spaces

### 1. Real-analytic structures

By a *complex space* we understand throughout one which is reduced, Hausdorff and with a countably based topology (i.e. a countable base for open sets).

Let $X$ be a complex space. A *real (analytic) structure* on $X$ is an antiholomorphic map $\tau: X \to X$ such that $\tau^2 = id$ (i.e. $\tau$ is an antiinvolution). A *real point* of $X$ is a fixed point of $\tau$. The set of real points

$$X_\tau = \{x \in X \mid \tau x = x\}$$

is called the *real part* of $X$.

EXAMPLES (1.1). Let $z = (z_1, \cdots, z_n) \in \boldsymbol{C}^n$ and set

$$j(z) = \bar{z} = (\bar{z}_1, \cdots, \bar{z}_n).$$

Then $j$ is an antiinvolution on $\boldsymbol{C}^n$ and $\boldsymbol{C}^n_j = \boldsymbol{R}^n = \{z \in \boldsymbol{C}^n \mid z_i \in \boldsymbol{R};\ 1 \leqq i \leqq n\}$.

(1.2) Set $X = \{(x, y) \in \boldsymbol{C}^2 \mid x^2 + y^2 + 1 = 0\}$. The conjugation $j$ on $\boldsymbol{C}^2$ maps $X$ into itself and so induces an antiinvolution $\tau$ on $X$. Now $X_\tau$ is empty.

Similarly any complex subspace $X \subset \boldsymbol{C}^n$ which is $j$-invariant receives an induced real structure. We write $X_j$ or $X_{\boldsymbol{R}}$ for the real part of $X$. Then $X_{\boldsymbol{R}} = X \cap \boldsymbol{R}^n$.

PROPOSITION (1.3). *Let $X$ be a complex manifold with real structure $\tau$, everywhere of complex dimension $n$. Then $X_\tau$ is a real-analytic submanifold, everywhere of real dimension $n$ (possibly empty[1]).*

PROOF. If $X_\tau \neq \varnothing$, let $x_0 \in X_\tau$. We can choose a coordinate neighborhood $U$ of $x_0$ such that $\tau U = U$, with holomorphic coordinates $z = (z_1, \cdots, z_n)$ vanishing at $x_0$. In this chart $(U, z)$ $\tau$ is given by a set of equations of the form

$$\begin{cases} z_j' = \overline{g_j(z_1, \cdots, z_n)} \\ 1 \leqq j \leqq n \end{cases} \tag{1}$$

where the functions $g_j$ are holomorphic on $U$ and vanish at the origin. As $\tau^2 = id$ we must have in $U$

$$\bar{z} = g(\overline{g(z)}).$$

Therefore if $A = \left(\frac{\partial(g)}{\partial(z)}\right)_0$, we deduce from this identity that

$$A\bar{A} = I.$$

In particular the equations

$$\begin{cases} w_j = g_j(z_1, \cdots, z_n) \\ 1 \leqq j \leqq n \end{cases}$$

have a non-vanishing jacobian at the origin and thus define, if $U$ is sufficiently small, a change of coordinates in $U$. With respect to this new choice of coordinates the equations (1) take the form

$$\begin{cases} w_j' = \bar{w}_j \\ 1 \leqq j \leqq n \end{cases} \tag{2}$$

But this is the conjugation on $\boldsymbol{C}^n$. Thus the proposition is established by example (1.1).

REMARK (1.4). We have proved that if $X$ is a complex manifold with a real structure $\tau$, at a real point $x_0 \in X_\tau$ $(X, x_0)$ is locally isomorphic to $(\boldsymbol{C}^n, 0)$

[1] We use the convention that the empty set can be given the structure of an $n$-manifold for any integer $n$.

with the real structure given by the conjugation, i.e. we have a commutative diagram

$$\begin{array}{ccc} X \supset W & \xrightarrow{h} & \Omega \subset \boldsymbol{C}^n \\ \downarrow \tau & & \downarrow j \\ X \supset W & \xrightarrow{h} & \Omega \subset \boldsymbol{C}^n \end{array}$$

where $n = \dim_{\boldsymbol{C}} (X, x_0)$ and where $W$ and $\Omega$ are suitable neighborhoods of $x_0$ in $X$ and $0$ in $\boldsymbol{C}^n$ respectively.

## 2. Real maps

(2.1) Let $(X, \tau)$, $(Y, \sigma)$ be two complex spaces with real structures $\tau$ and $\sigma$ respectively. A holomorphic map $h: X \to Y$ is called *real* if

$$h \circ \tau = \sigma \circ h \tag{2.1.1}$$

i.e. if it is compatible with the real structures.

In particular if $Y = \boldsymbol{C}^N$ with the real structure $j$ given by the conjugation, a holomorphic map $h: X \to \boldsymbol{C}^N$ is real if $\overline{h(x)} = h(\tau(x))$ for all $x \in X$.

PROPOSITION (2.2). *Let $X$ be a finite dimensional Stein space with a real structure $\tau$. Then there is a real holomorphic map of $X$ into some numerical space $\boldsymbol{C}^N$ which is a closed topological imbedding.*

PROOF. Let $H(X) = \Gamma(X, \mathcal{O})$ be the space of holomorphic functions on $X$. We let $\tau$ operate on $H(X)$ by $(\tilde{\tau} f)(x) = \overline{f(\tau(x))}$. If $V = \{f \in H(X) \mid \tilde{\tau} f(x) = f(x)\}$, then $iV = \{f \in H(X) \mid \tilde{\tau} f(x) = -f(x)\}$ and $H(X)$ decomposes as vector space over $\boldsymbol{R}$ into the direct sum of the eigenspaces corresponding to the eigenvalues $\pm 1$ of $\tilde{\tau}$:

$$H(X) = V \oplus iV$$

Indeed $V \cap iV = \{0\}$, and for every $h \in H(X)$ we have $h(x) = \frac{1}{2}(h(x) + \overline{h(\tau x)}) + \frac{1}{2}(h(x) - \overline{h(\tau x)})$.

Let $f = (f_1, \cdots, f_N)$ be a finite set of holomorphic functions on $X$ such that the map $f: X \to \boldsymbol{C}^N$ they define is proper and injective.

Then the holomorphic map $\varphi: X \to \boldsymbol{C}^{2N}$ defined by

$$\begin{aligned} \varphi_1(x) &= f_1(x) + \overline{f_1(\tau x)} \\ \varphi_2(x) &= i(f_1(x) - \overline{f_1(\tau x)}) \\ \varphi_{2N-1}(x) &= f_N(x) + \overline{f_N(\tau x)} \\ \varphi_{2N}(x) &= i(f_N(x) - \overline{f_N(\tau x)}) \end{aligned}$$

has the following properties

(1) $\varphi(\tau x) = j\varphi(x)$

i.e. $\varphi$ is real. Indeed for every $1 \leqq j \leqq 2N$ $\varphi_j \in V$.

(2) $f$ factorizes continuously through $\varphi$.

For with $\sigma_i = \frac{1}{2}(\varphi_1 - i\varphi_2), \cdots, \sigma_N = \frac{1}{2}(\varphi_{2N-1} - i\varphi_{2N})$, we have $\sigma \circ \varphi = f$. Therefore, as $f$ is injective so $\varphi$ must be injective, and as $f$ is proper, so also must be $\varphi$.

REMARKS (2.3). If $X$ admits a proper holomorphic imbedding in some $\boldsymbol{C}^N$ which is an isomorphism onto its image, then $X$ also admits a proper holomorphic real imbedding in $\boldsymbol{C}^{2N}$ which is an isomorphism onto its image.

We know that this is the case precisely when the imbedding dimension $t(x) = \dim \mathfrak{m}_x/\mathfrak{m}_x^2$ is bounded on $X$. (Here $\mathfrak{m}_x$ denotes the maximal ideal of the local ring $\mathcal{O}_x$ of the structure sheaf.)

(2.4) In particular if $X$ is any complex space with a real structure $\tau$, then at each real point $x_0 \in X_\tau$, $X$ admits a real local imbedding in some numerical space. In other words, there is a neighborhood $U$ of $x_0$ in $X$ which is isomorphic to an analytic set $A$ of some open set $G$ of $\boldsymbol{C}^N$ such that

$$jG = G \quad \text{and} \quad jA = A$$

and moreover $j \mid A$ is the antiinvolution $\tau$ on $U$ (transformed via the imbedding).

We may suppose that the ideal of all holomorphic functions on $G$ vanishing on $A$ is finitely generated and has a set of generators $f_i$ such that

$$\overline{f_i(z)} = f_i(\bar{z}) \qquad 1 \leqq i \leqq k.$$

Therefore $X_\tau \cap U \simeq A_{\boldsymbol{R}} = \{z \in G \cap \boldsymbol{R}^N \mid f_i(z) = 0 \,\forall\, i\}$ is a real-analytic set.

Given the complex space $X$ with real structure $\tau$, we define on $X_\tau$ the subsheaf $\mathscr{A}$ of $\mathcal{O}(X) \mid X_\tau$ by

$$\mathscr{A}_x = \{f \in \mathcal{O}_x \mid \bar{f} = f \circ \tau\}, \qquad x \in X_\tau.$$

The ringed space $(X_\tau, \mathscr{A})$ will be called a *real-analytic space* or simply a *real space*. Morphisms of real spaces, $X_\tau \to Y_\sigma$, may again be called *real holomorphic maps* as they extend uniquely to germs of real holomorphic maps on the complexifications, $(X, X_\tau) \to (Y, Y_\sigma)$.

REMARKS (2.5). When $X = \boldsymbol{C}^n$ with its standard real structure, then $\mathscr{A} = \mathscr{A}(\boldsymbol{R}^n)$ is the sheaf of germs of real analytic functions on $\boldsymbol{R}^n$. In the general case, by the local imbedding of the previous remark, the sheaf $\mathscr{A}$ is locally isomorphic to sheaves of the form

$$\mathscr{A}(\boldsymbol{R}^n \cap G)/\mathscr{A}(\boldsymbol{R}^n \cap G)(g_1, \cdots, g_k), \tag{2.5.1}$$

on $\{g_1 = \cdots = g_k = 0\}$ in $\mathbf{R}^n$, where $g_1, \cdots, g_k$ are the restrictions to $\mathbf{R}^n \cap G$ of the real holomorphic functions $f_1, \cdots, f_k$ on $G \subset \mathbf{C}^n$.

Note that even if the functions $f_i$ are generators for the ideal of holomorphic functions vanishing on the local set

$$A = \{z \in G \mid f_i(z) = \cdots = f_k(z) = 0\},$$

still the $g_i$ need not be generators for the ideal of all real-analytic functions vanishing on

$$A_{\mathbf{R}} = \{z \in \mathbf{R}^n \cap G \mid g_i(z) = \cdots = g_k(z) = 0\}.$$

(2.6) The associated sheaf of germs of real-analytic functions on $A_{\mathbf{R}}$ is canonically a quotient of (2.5.1). As such quotients are respected by real holomorphic maps, there is a corresponding sheaf $\mathscr{A}_{\text{red}}$ on $X_\tau$ which is a quotient of $\mathscr{A}$ and a subsheaf of the sheaf of germs of continuous functions. A morphism of ringed spaces $(X_\tau, \mathscr{A}_{\text{red}})$ is called a *real-analytic* map.

## 3. Global imbeddings

We need the following result on Stein neighborhoods in order to have imbeddings of real spaces in $\mathbf{R}^N$ for suitable $N$. The proof is an adaptation of the proof of Grauert for the case of manifolds ([6]).

THEOREM (3.1). *Let $X$ be a complex space with a real structure $\tau$. There exists a $\tau$-invariant open Stein subset of $X$ containing the real part $X_\tau$.*

As any open subset $U$ of $X$ with $X_\tau \subset U$ and $\tau U = U$, has a real structure $\tau \mid U$ with real part $X_\tau$, the theorem shows that $X_\tau$ has a fundamental system of $\tau$-invariant Stein neighborhoods in $X$.

PROOF. (a) Let us first consider $\mathbf{R}^N \subset \mathbf{C}^N$ and let $z_j = x_j + iy_j$ be holomorphic coordinates in $\mathbf{C}^N$. For every point $a = (a_1, \cdots, a_N) \in \mathbf{R}^N$ we consider the function $v_a : \mathbf{C}^N \to \mathbf{R}$ given by

$$v_a(z) = 2 \sum y_j^2 - \sum (x_j - a_j)^2.$$

If we set $\xi_j = (x_j - a_j) + iy_j$, then

$$v_a(\xi) = -\tfrac{3}{4} \sum (\xi_j^2 + \bar{\xi}_j^2) + \tfrac{1}{2} \sum \xi_j \bar{\xi}_j$$

so that

$$\partial\bar{\partial} v_a(\xi) = \tfrac{1}{2} \sum d\xi_j \, d\bar{\xi}_j > 0.$$

Let

$$K_a = \{z \in \mathbf{C}^N \mid 0 < v_a(z)\}$$

and set

$$w_a(z) \equiv \begin{cases} v_a(z) & \text{for } \ z \in K_a \\ 0 & \text{for } \ z \in \mathbf{C}^N - K_a \end{cases}$$

Then $w_a(z)$ is continuous (and even Lipshitz), and the function

$$p_a(z) \equiv \begin{cases} e^{-(1/w_a(z))} & \text{in } K_a \\ 0 & \text{in } \boldsymbol{C}^N - K_a \end{cases}$$

is a $C^\infty$ function in the whole space. Moreover

$$\partial\bar{\partial} p_a = e^{-(1/w_a)} \left( \frac{1}{w_a^2} \partial\bar{\partial} w_a + \frac{1}{w_a^4} (1 - 2w_a) \partial w_a \bar{\partial} w_a \right).$$

Therefore, if we define

$$V_a = \{z \in \boldsymbol{C}^N \mid 0 < v_a(z) < \tfrac{1}{2}\},$$

we have

$$\partial\bar{\partial} p_a \text{ is } \begin{cases} > 0 & \text{in } V_a \\ 0 & \text{in } \boldsymbol{C}^N - K_a \end{cases}$$

(b) We select a locally finite countable covering $U = \{U_i\}$ of $X_\tau$ by $\tau$-invariant open sets $U_i \subset X$ and isomorphisms

$$\varphi_i : U_i \to A_i, \qquad A_i \subset G_i \subset \boldsymbol{C}^{N_i}$$

where $A_i$ is a conjugation-invariant analytic subset of some conjugation-invariant open set $G_i$ of some numerical space $\boldsymbol{C}^{N_i}$.

Let $d(p, q)$ denote the euclidean distance in $\boldsymbol{C}^N = \boldsymbol{C}^{N_i}$ and let $r(x) > 0$ be a $C^\infty$ function defined for $x \in B_i = G_i \cap \boldsymbol{R}^N$ having the following properties

(i) for $\quad x \in B_i \qquad r(x) < \frac{1}{8} d(x, \partial B_i)^2$

(ii) if we set $G_i' = \{z \in \boldsymbol{C}^N \mid \sum y_j^2 < r(x)\}$, then $U_i' \equiv \varphi_i^{-1} G_i' \subset U_i$.

This is certainly satisfied if $r(x) < d(x, \partial G_i)$.

We then have

$$X_\tau \subset \bigcup_i U_i'.$$

It follows that moreover

(iii) $\forall a \in B_i \qquad \mathrm{pr}(G_i' \cap K_a) \subset\subset B_i$,

where pr is the projection to $\boldsymbol{R}^N$. In fact

$$p \in K_a \Leftrightarrow d(\mathrm{pr}(p), a)^2 < 2d(p, \boldsymbol{R}^N)$$
$$p \in G_i' \Rightarrow d(p, \boldsymbol{R}^N)^2 < r(\mathrm{pr}(p)) < \tfrac{1}{8} d^2(\mathrm{pr}(p), \partial B_i).$$

Thus, setting $\delta = d(\mathrm{pr}(p), \partial B_i)$, we have for $p \in K_a \cap G_i'$

$$d(\mathrm{pr}(p), a) < \tfrac{1}{2}\delta$$

and

$$d(a, \partial B_i) \geqq d(\mathrm{pr}(p), \partial B_i) - d(\mathrm{pr}(p), a) > \tfrac{1}{2}\delta.$$

Therefore, $\forall a \in B_i$ we have

$$\partial G_i' \cap K_a \subset\subset \partial G_i' - (\partial G_i' \cap \boldsymbol{R}^N).$$

Consequently

$$\partial U_i' \cap \varphi_i^{-1}(K_a) \subset\subset \partial U_i' - \partial U_i' \cap X_\tau.$$

(c) For $b \in U_i' \cap X_\tau$ we denote by $p_b^{(i)}$ the function

$$p_b^{(i)} = p_{\varphi_i}(b) \circ \varphi_i$$

which is defined on $U_i$ and is zero outside $\varphi_i^{-1} K_{\varphi_i(b)}$. We extend this function to $X$ by setting it equal to 0 outside $\varphi_i^{-1} K_{\varphi_i(b)}$.

We claim that on a sufficiently small neighborhood $V$ of $X_\tau$ in $X$ the function thus defined is a $C^\infty$ function[2].

Let us choose $\varepsilon_i > 0$ so small that with

$$G_i'' = \{z = x + iy \in G_i' \mid d(x, \partial G_i') > \varepsilon_i\}$$

we do have

$$X_\tau \subset \bigcup \varphi_i^{-1}(G_i'').$$

Set

$$G_i''' = \left( \bigcup_{a \in G_i \cap \boldsymbol{R}^N} K_a \right) \cap G_i'.$$

(Thus $d(a, \partial G_i') > \varepsilon_i$). Then we can find $\sigma_i > 0$ such that

$$z = x + iy \in G_i'' \quad \text{implies} \quad r(x) > \sigma_i.$$

Now, if $V$ is a sufficiently small neighborhood of $X_\tau$ in $X$, we will have

$$\varphi_i(\bar{V}) \subset \{x + iy \in G_i' \mid \|y\|^2 < \sigma_i\}.$$

Therefore the function $p_b^{(i)}$ will be a $C^\infty$ function on $V$ provided that

$$\varphi_i(b) \in G_i'' \cap \boldsymbol{R}^N.$$

(d) If the function $r(x)$ is chosen sufficiently small, we may assume that

(iv) $\forall a \in B_i$ we do have $G' \cap K_a \subset V_a$.

Then on $V$ the function $p_b^{(i)}$ will have the following properties:

$$p_b^{(i)} \text{ is } C^\infty \quad (\text{provided } \varphi_i(b) \subset G_i'' \cap \boldsymbol{R}^N)$$

$$p_b^{(i)} \geqq 0$$

$$\partial\bar{\partial} p_b^{(i)} \geqq 0$$

and

$$\partial\bar{\partial} p_b^{(i)} > 0 \quad \text{on} \quad \varphi_i^{-1}(K_{\varphi_i(b)}) \quad \text{(i.e. is strongly plurisubharmonic).}$$

[2] A function $f: X \to \boldsymbol{R}$ is called $C^\infty$ if for every point $x \in X$ we can find a neighborhood $U(x)$, an isomorphism of $U(x)$ onto an analytic subset $U$ of some open set $G$ of a numerical space and a $C^\infty$ function $F$ on $G$ such that $f \mid U(x)$ corresponds to $F \mid U$ (cf. [2]).

Set

$$W_b^{(i)} = \{q \in \partial V \mid p_b^{(i)}(q) > 0\}.$$

We can find a countable collection of points $b_n$ in $X_\tau$ and for every $n$ an index $i_n$ such that $b_n \in X_\tau \cap \varphi_{i_n}^{-1}(G''_{i_n})$ and such that

(v) $\{W_{b_n}^{(i_n)}\}$ is a locally finite covering of $\partial V$

(vi) $\varphi_{i_n}^{-1}(K_{\varphi_{i_n}(b_n)} \cap \bar{V}$ is a locally finite family of subsets of $\bar{V}$.

Then we can find open subsets $W_{b_n}'^{(i_n)}$ in $\partial V$ such that

$$W_{b_n}'^{(i_n)} \subset\subset W_{b_n}^{(i_n)} \quad \text{and} \quad \partial V \subset \bigcup_n W_{b_n}'^{(i_n)}.$$

For each index $(b_n, i_n)$ we select a constant $c(b_n, i_n) > 0$ such that

$$c(b_n, i_n) p_{b_n}^{(i_n)} > 1 \quad \text{on} \quad W_{b_n}'^{(i_n)}.$$

We can consider now the function

$$h = \sum c(b_n, i_n) p_{b_n}^{(i_n)}.$$

This is well defined and $C^\infty$ on a neighborhood of $\bar{V}$ because of property (vi). Moreover it satisfies the following conditions

(vii) $\partial\bar{\partial}h \geqq 0$ on $V$ (i.e. $h$ is plurisubharmonic)

(viii) $\partial\bar{\partial}h > 0$ on $\partial V$ (i.e. on $\partial V$ $h$ is strongly plurisubharmonic)

(ix) $h > 1$ on $\partial V$

(x) $h = 0$ on $X_\tau$.

(e) If $\rho_i(x, y)$ is a $C^\infty$ function on $G_i$ with

$$\rho_i(x, y) > 0 \quad \text{on} \quad B_i,$$

the function

$$\eta_i = \rho_i(x, y) \sum_1^N y_j^2$$

has the property

$$\partial\bar{\partial}\eta_i \mid \boldsymbol{R}^n = r\partial\bar{\partial} \sum y_j^2 > 0.$$

Therefore, if $\alpha_i = \rho_i \varphi_i$ is a sequence of $C^\infty$ functions on $X$ such that

$$\text{support } \alpha_i \subset\subset U_i'$$
$$\sum \alpha_i(x) > 0 \quad \text{on} \quad X_\tau,$$

then for the function

$$\theta = \sum \eta_i \circ \varphi_i$$

we have

$$\theta \text{ is } C^\infty \text{ on } X; \qquad \partial\bar{\partial}\theta > 0 \text{ on } X_\tau; \qquad \theta \mid X_\tau = 0.$$

Let $g: X \to \boldsymbol{R}$ be a $C^\infty$ function such that

$$\{x \in X \mid g(x) < c\} \subset\subset X \qquad \forall c \in R.$$

We can choose $\rho: X \to \boldsymbol{R}$ $C^\infty$ and growing so rapidly on $X_\tau$ that

$$f = g + \rho\theta$$

has the property

$$\partial\bar{\partial} f > 0 \quad \text{on} \quad X_\tau.$$

Therefore on a neighborhood $W$ of $X_\tau$ we will have

$$\partial\bar{\partial} f > 0 \quad \text{on} \quad W.$$

Now in the previous constructions we may assume $\bar{V} \subset W$. Let us consider the function

$$\varphi = f + (1-h)^{-1} = f + 1 + h + h^2 + \cdots$$

which is well defined on

$$U = \{x \in V \mid h(x) < 1\}.$$

On $U$ we have

$$\partial\bar{\partial}\varphi = \partial\bar{\partial} f + \partial\bar{\partial} h + \partial\bar{\partial} h^2 + \cdots,$$

and therefore $\varphi$ is strongly plurisubharmonic on $U$. Moreover

$$\{x \in U \mid \varphi(x) < c\} \subset\subset U$$

for all $c \in \boldsymbol{R}$. This shows that $U$ is a 0-complete set, and therefore it is a Stein set.

Replacing $U$ with $U \cap \tau U$ we get a $\tau$-invariant Stein neighborhood of $X_\tau$ in $X$. This completes the proof.

Combining theorem (3.1) with proposition (2.2) we get the following

COROLLARY (3.2). *Every real-analytic space admits a closed topological imbedding in some numerical space $\boldsymbol{R}^n$ by a real holomorphic map.*

PROOF. Let $(X, \tau)$ be a complex space with a real structure. By the previous theorem, restricting eventually $X$, we may assume that $X$ is a Stein neighborhood of its real part $X_\tau$. A real proper holomorphic and one to one imbedding $\lambda$ of $X$ in some numerical space $\boldsymbol{C}^N$ gives a map of $X_\tau$ on $\lambda(X_\tau)$, which is the real part of $\lambda(X)$.

REMARK (3.3). If at every point $x \in X_\tau$ we have an estimate, $t(X) = \dim \mathfrak{m}_x/\mathfrak{m}_x^2 \leqq M < \infty$, $\mathfrak{m}_x = \mathfrak{m}_x(\mathscr{A})$, then the imbedding can be chosen to be an isomorphism on its image, cf. remark (2.3).

## 4. Imbedding of real algebraic varieties

Consider the projective space $P_n(\boldsymbol{C})$ as endowed with the real structure $j$ given by conjugation of homogeneous coordinates.

Let $z=(z_0,\cdots,z_n)$ be homogeneous coordinates on $P_n(\boldsymbol{C})$ and let $z=(z_{ij})_{0\leqq i\leqq j\leqq n}$ be homogeneous coordinates on $P_N(\boldsymbol{C})$ where $N=\binom{n+2}{2}-1$. The Segre map $f:P_n(\boldsymbol{C})\to P_N(\boldsymbol{C})$ is defined by the equations

$$\begin{cases} z_{ij}=z_iz_j \\ 0\leqq i\leqq j\leqq n, \end{cases}$$

and it is therefore a real birational and biregular map of $P_n(\boldsymbol{C})$ onto its image $V=f(P_n(\boldsymbol{C}))$.

The real part of $P_n(\boldsymbol{C})$ is $P_n(\boldsymbol{R})$ (canonically imbedded in $P_n(\boldsymbol{C})$), and the hyperplane $L=\{\sum_0^n z_{ii}=0\}$ does not intersect $V_{\boldsymbol{R}}=f(P_n(\boldsymbol{R}))$. Identifying $P_N(\boldsymbol{C})-L$ to $\boldsymbol{C}^N$ we thus obtain a birational and biregular real imbedding of the affine manifold $X=P_n(\boldsymbol{C})-\{\sum z_i^2=0\}$ into $\boldsymbol{C}^N$, which is proper and such that $X_{\boldsymbol{R}}=P_n(\boldsymbol{R})$ is imbedded as a compact algebraic submanifold of $\boldsymbol{R}^N$. Consequently:

(4.1) *Given any $j$-invariant projective algebraic variety $W\subset P_n(\boldsymbol{C})$ we can find*

(a) *an hypersurface section $W\cap L$ which does not intersect the real part $W_{\boldsymbol{R}}$ of $W$.*

(b) *a birational biregular proper imbedding of $W-W\cap L$ into $\boldsymbol{C}^N$ $\left(N=\binom{n+2}{2}-1\right)$ which maps $W_{\boldsymbol{R}}$ onto a compact algebraic subvariety of $\boldsymbol{R}^N$.*

Note that in the case of projective algebraic varieties we are in condition to get a more precise result than that given by the general imbedding theorem. The imbedding can be chosen birational and biregular so that the algebraic structure is also respected.

## 5. Complexifications

Let $X$ be a complex space with a real structure $\tau$, and assume that $X_\tau\neq\emptyset$. Then locally at each point $X_\tau$ is a real-analytic set[3], and, as such, has a dimension $\dim_{\boldsymbol{R}}(X_\tau, a)$. We have always ([11] p. 93)

$$\dim_{\boldsymbol{R}}(X_\tau, a)\leqq\dim_{\boldsymbol{C}}(X, a).$$

[3] If $\Omega$ is an open subset of $\boldsymbol{R}^n$, a reál-analytic set $A\subset\Omega$ is the set of common zeros to a finite set of equations $f_\alpha(x)=0$, $1\leqq\alpha\leqq k$, where $f_\alpha$ are real-analytic functions on $\Omega$.

PROPOSITION (5.1). *If $a \in X_x$ and if*

$$\dim_{\boldsymbol{R}} (X_\tau, a) < \dim_{\boldsymbol{C}} (X, a)$$

*then, locally at a, the real part $X_\tau$ of $X$ sits in the singular set $S(X)$ of $X$.*

This is a straightforward consequence of proposition (1.3).

PROPOSITION (5.2). *If $a \in X_\tau$ and if*

$$\dim_{\boldsymbol{R}} (X_\tau, a) = \dim_{\boldsymbol{C}} (X, a),$$

*then for any open neighborhood $V$ of $a$ in $X$ we have*

$$X_\tau \cap V \not\subset S(X) \cap V.$$

*Moreover, if $a$ is regular on $X_\tau$, then $a$ is regular on $X$.*

PROOF. We know that $S(X)$ is a complex subspace of $X$. The real-analytic structure $\tau$ on $X$ induces a real structure $\tau \mid S(X)$ on $S(X)$ as $S(X)$ must be $\tau$-invariant.

If for some $V$ we have $X_\tau \cap V \subset S(X) \cap V$, then, replacing $X$ by $V$, which can be assumed $\tau$-invariant, we should have $X_\tau \subset S(X)_\tau$.

As $\dim_{\boldsymbol{C}} (S(X), a) \leqq \dim_{\boldsymbol{C}} (X, a) - 1$, we get

$$\dim_{\boldsymbol{R}} (X_\tau, a) \leqq \dim_{\boldsymbol{R}} (S(X)_\tau, a) \leqq \dim_{\boldsymbol{C}} (S(X), a) \leqq \dim_{\boldsymbol{C}} (X, a) - 1.$$

This is a contradiction.

Finally, if $a$ is a regular point of $X_\tau$ and $\dim_{\boldsymbol{R}} (X_\tau, a) = \dim_{\boldsymbol{C}} (X, a)$, then a system of coordinate functions around $a$ in $X_\tau$ is extendable to a system of coordinate functions in $X$.

COROLLARY (5.3). *A point $a \in X_\tau$ where*

$$\dim_{\boldsymbol{R}} (X_\tau, a) = \dim_{\boldsymbol{C}} (X, a)$$

*is an accumulation point of the regular points of $X_\tau$.*

COROLLARY (5.4). *The real part $X_\tau$ of a finite dimensional complex space $X$ with real structure $\tau$ sits as the real part of a $\tau$-invariant complex subspace $Y$ such that* $\dim_{\boldsymbol{C}} Y = \dim_{\boldsymbol{R}} X_\tau$.

PROOF. If $\dim_{\boldsymbol{R}} X_\tau < \dim_{\boldsymbol{C}} X$, then $X_\tau \subset S(X)$ and $X_\tau = S(X)_\tau$. As $\dim_{\boldsymbol{C}} S(X) < \dim_{\boldsymbol{C}} X$ repeated application of this procedure yields a complex subspace $S(\cdots S(X))$ with the desired property.

A complex space $X$ with real structure $\tau$ is called a *complexification* of its real part $X_\tau$ if

$$\dim_{\boldsymbol{R}} X_\tau = \dim_{\boldsymbol{C}} X.$$

The space $X$ is called a *faithful complexification* of $X_\tau$ if for each $x \in X_\tau$ we have

$$\dim_{\mathbf{R}} (X_\tau, a) = \dim_{\mathbf{C}} (X, a).$$

EXAMPLES (5.5). In $C^3$ consider the subvariety $X$ of points $(x, y, z)$ satisfying the equation

$$y^2 - x^2 z = 0 \qquad \text{(The Cayley umbrella).}$$

The real part $X_{\mathbf{R}} \subset \mathbf{R}^3$ is the subset of $\mathbf{R}^3$ given by the same equation. In particular the $z$-axis sits in $X_{\mathbf{R}}$. Thus $X$ is a complexification of $X_{\mathbf{R}}$, but not a faithful one.

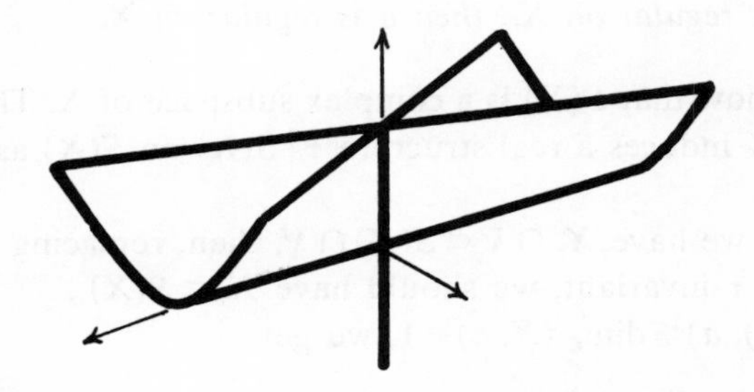

(5.6) Every real-analytic manifold (with countable topology) admits a faithful complexification. This is a theorem due to Whitney and Bruhat.

(5.7) Let $V \subset P_n(R)$ be a real projective manifold. Let $I_V$ be the ideal of real homogeneous polynomials vanishing on $V$. Then $I_V$ equals its radical and the ideal of all homogeneous polynomials vanishing on $V$ is the ideal $I_W = I_V \otimes_{\mathbf{R}} \mathbf{C}$.

The ideal $I_W$ defines a complex projective variety $W$ in $P_n(\mathbf{C})$. Every irreducible component $A$ of $W$ is such that $A \cap V \neq \emptyset$ and contains interior points of $V$. Otherwise $W' = \overline{W - (A \cup jA)}$ will be a $j$-invariant algebraic variety with $W' \subsetneq W$ and $V \subset W'$. This implies that the ideal $I_{W'}$ of homogeneous polynomials vanishing on $W'$ will satisfy $I_{W'} \supsetneq I_W = I_V' \otimes_{\mathbf{R}} \mathbf{C}$. But by the definition of $I_W$ we must have $I_W \supset I_{W'}$, which gives a contradiction.

Therefore, if $V = V_1 \cup \cdots \cup V_k$ is the decomposition of $V$ into its irreducible components, each irreducible component $A$ of $W$ must contain some $V_i$. Hence with obvious notations, we must have $I_A = I_{V_i} \otimes_{\mathbf{R}} \mathbf{C}$ so that $I_V = \bigcap I_{V_i}$ and $I_W = \bigcap I_{V_i} \otimes_{\mathbf{R}} \mathbf{C}$.

The variety $W_i$ of zeros of $I_{W_i} = I_{V_i} \otimes_{\boldsymbol{R}} \boldsymbol{C}$ is irreducible and of the same (complex) dimension as (the real dimension of) $V_i$. Moreover $W_i$ must be non-singular along $V_i \subset W_i$. In conclusion:

(5.7.1) *Each real projective manifold admits a faithful projective complexification.*

(5.8) By the same argument one can show that each real projective variety admits a projective complexification. However, this complexification may fail to be faithful.

## §2. Quasianalytic spaces

### 6. Definitions

Let $X$ be a complex space with real structure $\tau$. Assume that $X_\tau \neq \varnothing$ and that $X$ is a complexification of $X_\tau$. By theorem (5.4) a real part $X_\tau$ can always be presented in this way.

The non-empty set

$$\hat{X}_\tau = \{a \in X_\tau \mid \dim_{\boldsymbol{R}} (X_\tau, a) = \dim_{\boldsymbol{C}} (X, a)\}$$

(briefly $\hat{X}$) is called the *quasiregular part* of $X_\tau$ or the quasiregular real part of $X$. Clearly $\hat{X}_\tau$ is a closed set.

In particular, if $X$ is a $j$-invariant analytic subset of some open set $G$ of $\boldsymbol{C}^N$, we can consider its real part $X_{\boldsymbol{R}}$. If $X$ is a complexification of $X_{\boldsymbol{R}}$, then the non-empty set $\hat{X}_{\boldsymbol{R}}$ is defined. We call $\hat{X}_{\boldsymbol{R}}$ a *quasiregular set.*

REMARKS (6.1). A subset $S$ of a real or complex space $X$ is called semianalytic (in $X$) if locally at any point of $X$ $S$ is a finite union of finite intersections of sets of the form $\{f>0\}$ and $\{g=0\}$, $f$ and $g$ being local real-analytic functions on $X$.

The class of semianalytic subsets of $X$ is closed under the formation of finite unions, finite intersections, and differences. Also the closure of a semianalytic subset is semianalytic [16] p. 76).

If $X$ is a complex space with real structure $\tau$, then $\hat{X}_\tau$ is the closure of $(X - S(X)) \cap X_\tau$. Consequently

(6.1.1) *The quasiregular real part $\hat{X}_\tau$ of a complex space with real structure is a semianalytic subset of $X$ (and hence of $X_\tau$).*

(6.2) By Lojasiewicz, [16], one can triangulate $X$ in such a way that $X_\tau$ and $\hat{X}_\tau$ are the supports of subcomplexes of the triangulation.

EXAMPLES (6.3). For the Cayley umbrella $X$ (example (5.5)) the quasiregular part $\hat{X}_{\boldsymbol{R}}$ is the set defined in $\boldsymbol{R}^3$ by the conditions

$$y^2 - x^2 z = 0 \qquad z \geqq 0.$$

Note that $\hat{X}_{\boldsymbol{R}}$ is given by parametric equations

$$x = u$$
$$y = uv$$
$$z = v^2.$$

(6.4) Consider in $\boldsymbol{C}^3$ the subvariety $X$ defined by the equations

$$y^2 + x^2 + z^2(1-z) = 0.$$

Then $X_{\boldsymbol{R}}$ consists of an isolated point at the origin and a 'bell' above $z = 1$.The quasiregular part $\hat{X}_{\boldsymbol{R}}$ is defined thus by the conditions

$$y^2 + x^2 + z^2(1-z) = 0 \qquad z \geqq 1.$$

Also in this case we have parametric equations for the quasiregular part, namely

$$x = \lambda(1+\lambda^2)\frac{1-\mu^2}{1+\mu^2}$$
$$y = \lambda(1+\lambda^2)\frac{2\mu}{1+\mu^2}$$
$$z = 1+\lambda^2.$$

## 7. Germ complexifications

Let $a \in \boldsymbol{R}^n$ and let $(A, a)$ be a germ at $a$ of a real-analytic set $A$, defined in an open subset of $\boldsymbol{R}^n$ by finitely many real-analytic equations.

The smallest complex germ of complex-analytic sets $(\tilde{A}, a)$ in $\boldsymbol{C}^n$ containing $(A, a)$ is called the *complexification* of the given germ $(A, a)$. For the unsupported statements below, cf. [11], Ch. V, §1.

(7.1) Let $I(A, a)$ be the ideal of germs of real-valued real-analytic functions vanishing on $(A, a)$ and let $I(\tilde{A}, a)$ be its complex analog:

$$I(A, a) = \{g \in \mathscr{A}(\boldsymbol{R}^n)_a \mid g \mid (A, a) = 0\}$$
$$I(\tilde{A}, a) = \{g \in \mathcal{O}(\boldsymbol{C}^n)_a \mid g \mid (\tilde{A}, a) = 0\}.$$

Then we have

$$I(\tilde{A}, a) = I(A, a) \otimes_{\boldsymbol{R}} \boldsymbol{C}. \tag{7.1.1}$$

(7.2) If $(A, a) = \bigcup (A_\nu, a)$ is the decomposition of the germ $(A, a)$ into its irreducible components $(A_\nu, a)$, then the corresponding decomposition of the complexified germ $(\tilde{A}, a)$ is

$$(\tilde{A}, a) = \bigcup (A_\nu, a). \tag{7.2.1}$$

Moreover we have

$$\dim_{\boldsymbol{R}} (A, a) = \dim_{\boldsymbol{C}} (\tilde{A}, a). \tag{7.2.2}$$

Let $(A, a)$ be given in a neighborhood of $a$ in $\boldsymbol{R}^n$ and let $(B, b)$ be another real-analytic germ given in a neighborhood of $b$ in $\boldsymbol{R}^m$. Let $\varphi : (A, a) \rightarrow (B, b)$ be a morphism, i.e. a real-analytic mapgerm. Then $\varphi$ is induced by a morphism $\varphi' : (\boldsymbol{R}^n, a) \rightarrow (\boldsymbol{R}^m, b)$. If $\tilde{\varphi}' : (\boldsymbol{C}^n, a) \rightarrow (\boldsymbol{C}^m, b)$ is the holomorphic extension of $\varphi'$, $\tilde{\varphi}'$ will have the property that

$$g \in I(\tilde{B}, b) \quad \text{implies} \quad g \circ \tilde{\varphi}' \in I(\tilde{A}, a).$$

Hence it induces a map germ $\tilde{\varphi} : (\tilde{A}, a) \rightarrow (\tilde{B}, b)$.

LEMMA (7.3). *A real-analytic mapgerm $\varphi : (A, a) \rightarrow (B, b)$ has a unique holomorphic extension $\tilde{\varphi} : (\tilde{A}, a) \rightarrow (\tilde{B}, b)$.*

PROOF. Let $z_1, \cdots, z_n$ be holomorphic coordinates in $\boldsymbol{C}^n$ and let $u_1, \cdots, u_m$ be holomorphic coordinates in $\boldsymbol{C}^m$. Let

$$\begin{cases} u_i = g_i(z) \\ 1 \leqq i \leqq n \end{cases} \quad \text{and} \quad \begin{cases} u_i = h_i(z) \\ 1 \leqq i \leqq n \end{cases}$$

denote the equations of two holomorphic extensions $\tilde{\varphi}'$, $\tilde{\varphi}''$ of $\varphi'$, $\varphi''$ obtained as above from the mapgerm $\varphi$. We must have

$$g_i(z) - h_i(z) \in I(\tilde{A}, a) \quad \text{for} \quad 1 \leqq i \leqq n$$

as $(g_i - h_i) \mid (A, a) = 0$. Hence $g_i \mid (\tilde{A}, a) = h_i \mid (\tilde{A}, a)$, and therefore the two extensions $\tilde{\varphi}'$ and $\tilde{\varphi}''$ induce the same map from $(\tilde{A}, a)$ to $(\tilde{B}, b)$.

COROLLARY (7.4). *The complexification of a germ of a real analytic set is independent of the particular local presentation of the germ as germ of an analytic set in some numerical space.*

COROLLARY (7.5). *Let $X$ be a complex space with real structure $\tau$ and let $a \in X_\tau$. Then $(X, a)$ is a complexification of $(X_\tau, a)$ if and only if*

(a) *each irreducible component $(X_\nu, a)$ of $(X, a)$ is $\tau$-invariant*

(b) *for each $(X_\nu, a)$ restriction to its real part $(X_{\nu\tau}, a)$ preserves dimension*

$$\dim_{\boldsymbol{R}} (X_{\nu\tau}, a) = \dim_{\boldsymbol{C}} (X_\nu, a).$$

Let $A$ and $B$ be real-analytic sets. A map $f : \hat{A} \rightarrow \hat{B}$ of the corresponding quasiregular sets will be called *real-analytic* if for any $x \in \hat{A}$ there is a neighborhood $U_x$ in $A$ such that $f \mid U_x \cap \hat{A}$ is induced from some real-analytic map $U_x \rightarrow B$. More generally this defines real-analytic maps $\hat{A} \rightarrow \hat{B}$ when $A$ and $B$ are real spaces.

COROLLARY (7.6). *Let X and Y be complex spaces with real structures $\tau$ and $\sigma$. Assume that at each point $a \in \hat{X}_\tau$, the germ $(X, a)$ is the complexification of $(X_\tau, a)$. Then any real-analytic map $\varphi : \hat{X}_\tau \to \hat{Y}_\sigma$ extends uniquely to a real holomorphic mapgerm $\tilde{\varphi} : (X, \hat{X}_\tau) \to (Y, \hat{Y}_\sigma)$.*

PROOF. Choose $a \in \hat{X}_\tau$ and set $\varphi(a) = b$. It suffices to show that the mapgerm $\varphi : (\hat{X}_\tau, a) \to (\hat{Y}_\sigma, b)$ extends *uniquely* to a holomorphic mapgerm $\tilde{\varphi} : (X, a) \to (Y, b)$. Since the situation is local, we can suppose that $X$ and $Y$ are complex analytic sets in small neighborhoods of $a$ and $b$ in certain numerical spaces.

The mapgerm $\varphi$ is induced by restriction from a local real-analytic map

$$\psi : X_\tau \to Y_\sigma,$$

and it suffices to check that $\psi$ is unique (near $a$); the claim then follows by lemma (7.3).

Let $X_{\tau\nu}$ be any irreducible component of $X_\tau$. Then $X_{\tau\nu} = X_{\nu\tau}$ for some irreducible component $X_\nu$ of $X$, and we have $a \in \hat{X}_{\nu\tau}$ ((7.2) and (7.5)). Clearly it suffices to check that $\psi \mid \hat{X}_{\nu\tau}$ is uniquely determined by $\psi \mid (\hat{X}_\tau \cap \hat{X}_{\nu\tau}) = \varphi \mid (\hat{X}_\tau \cap \hat{X}_{\nu\tau})$. This will hold provided $\hat{X}_\tau \cap \hat{X}_{\nu\tau}$ has non-empty interior in $\hat{X}_{\nu\tau}$ (arbitrarily near $a$).

However, we have $\hat{X}_{\nu\tau} - (\hat{X}_{\nu\tau} \cap \hat{X}_\tau) \subset X_{\nu\tau} \cap \bigcup_{\mu \neq \nu} X_{\mu\tau}$, and consequently

$$\dim_{\mathbf{R}} (\hat{X}_{\nu\tau} - (\hat{X}_{\nu\tau} \cap \hat{X}_\tau)) \leqq \dim_{\mathbf{R}} (X_{\nu\tau} \cap \bigcup_{\mu \neq \nu} X_{\mu\tau}) < \dim_{\mathbf{R}} X_{\nu\tau} = \dim_{\mathbf{R}} \hat{X}_{\nu\tau}.$$

Since $\hat{X}_{\nu\tau} \neq \emptyset$, this proves our contention.

## 8. Quasianalytic spaces and their complexifications

Let $A$ be a complex-analytic subset of some open set $G$ of $\mathbf{C}^N$. Assume that

(i) $G$ and $A$ are $j$-invariant, so that $A$ is really imbedded in $G$

(ii) at each point $a$ of the quasiregular part $\hat{A}_{\mathbf{R}}$ of $A_{\mathbf{R}}$ $(A, a)$ is the complexification of $(A_{\mathbf{R}}, a)$.

This is for instance the case if $A$ is locally irreducible at each point of $\tilde{A}_{\mathbf{R}}$, e.g. if $A$ is a normal complex space. The example of the Cayley umbrella (5.5) shows that the above conditions can be realized without assuming $A$ locally irreducible.

Under these conditions the two sheaves $\mathscr{A}$ and $\mathscr{A}_{\text{red}}$ on $A_{\mathbf{R}}$ coincides on all of $\hat{A}_{\mathbf{R}}$ (cf. sec. 2). Consider then the ringed spaces $(\hat{A}_{\mathbf{R}}, \mathscr{A})$ defined in this way and define a ringed space $(X, \mathscr{A})$ to be a *quasianalytic space* if it is

Hausdorff and locally isomorphic to some of the models $(\hat{A}_{\boldsymbol{R}}, \mathscr{A})$ just described. As usual we assume that $X$ has a countable topology.

(8.1) Let $(Y, \mathscr{A})$ be a quasianalytic space. A (*proper*) *complexification* of $Y$ consists of

(a) a complex space $X$ with real structure $\tau$, such that at every point $a \in \hat{X}_\tau$ $(X, a)$ is the complexification of $(X_\tau, a)$

(b) an isomorphism $\varphi$ of $(Y, \mathscr{A})$ onto the ringed space $(\hat{X}_\tau, \mathscr{A})$, this last with its natural structure of a quasianalytic space.

From corollary (7.6) we see that if $(Y, \mathscr{A})$ admits a complexification $(X, \tau; \varphi)$, then the germ of $X$ along $\varphi Y = \hat{X}_\tau$ is determined up to a unique isomorphism.

We have the following result

THEOREM (8.2). *Every quasianalytic space admits complexifications.*

PROOF. The proof is obtained by a careful checking of the proof of Bruhat and Whitney for the case of real analytic manifolds.

(a) Let $(Y, \mathscr{A})$ be a quasianalytic space. We can find three locally finite open coverings of $Y$

$$\{V'_i\}_{i\in I}, \qquad \{U'_i\}_{i\in I}, \qquad \{T'_i\}_{i\in I}$$

over the same set of indices $I$ such that

$$V'_i \subset\subset U'_i \subset\subset T'_i.$$

For each $i$ we can find a local model $T_i \subset \tilde{T}_i$ for $T'_i$, where $\tilde{T}_i$ is a really imbedded complex set in some $\boldsymbol{C}^{N_i}$ and $T_i$ is its quasiregular real part; moreover, at each point $a$ of $T_i$ the germ $(\tilde{T}_i, a)$ is the complexification of $(\tilde{T}_{i\boldsymbol{R}}, a)$. Let $\varphi_i : T'_i \to T_i$ be an isomorphism and form

$$U_i = \varphi_i U'_i, \qquad V_i = \varphi_i V'_i \tag{1}$$

$$U_{ij} = \varphi_i(U'_i \cap U'_j), \qquad V_{ij} = \varphi_i(V'_i \cap V'_j), \qquad T_{ij} = \varphi_i(T'_i \cap T'_j) \tag{2}$$

The isomorphism

$$\varphi_j \circ \varphi_i^{-1} : T_{ij} \to T_{ji} \tag{3}$$

extends to a complex isomorphism $\psi_{ji} : \tilde{T}_{ij} \to \tilde{T}_{ji}$ from a neighbourhood of $T_{ij}$ in $\tilde{T}_i$ to a neighbourhood of $T_{ji}$ in $\tilde{T}_j$; this by virtue of corollary (7.6).

We can assume that

$\tilde{T}_{ij} = \varnothing$ if $T_{ij} = \varnothing$

$T_{ij}$ is the quasiregular real part of $\tilde{T}_{ij}$

$\psi_{ij} = \psi_{ji}^{-1}$.

For every pair $(i, j)$ we can select an open set $\tilde{U}_{ij}$ in $\tilde{T}_{ij}$ such that

$$\tilde{U}_{ij} \subset\subset \tilde{T}_{ij}$$

$$\psi_{ji}\tilde{U}_{ij} = \tilde{U}_{ji}$$

and

$$\begin{cases} \tilde{U}_{ij} \cap T_i = U_{ij} \\ \overline{\tilde{U}_{ij}} \cap T_i = \bar{U}_{ij}. \end{cases} \tag{4}$$

Since $\bar{V}_i \cap \psi_{ij}(\bar{V}_j \cap \bar{U}_{ji})$ is compact and contained in $U_{ij}$, we can choose an open set $\tilde{W}_{ij} \subset \tilde{T}_{ij}$ with

$$\tilde{W}_{ij} \subset\subset \tilde{U}_{ij}$$

$$\tilde{W}_{ij} = \psi_{ij}\tilde{W}_{ji}$$

and

$$\bar{V}_i \cap \psi_{ij}(\bar{V}_j \cap \bar{U}_{ji}) \subset \tilde{W}_{ij}. \tag{5}$$

Moreover the sets

$$\bar{V}_i - \bar{V}_i \cap \tilde{W}_{ij},\ \psi_{ij}(\bar{V}_j \cap \bar{U}_{ji}) - \psi_{ij}(\bar{V}_j \cap \bar{U}_{ji}) \cap \tilde{W}_{ij}$$

are compact and disjoint and are thus contained in disjoint open sets $\tilde{A}_{ij}$ and $\tilde{B}_{ij}$ of $\tilde{T}_i$.

We have

$$\begin{cases} \bar{V}_i \subset \tilde{A}_{ij} \cup \tilde{W}_{ij} \\ \psi_{ij}(\bar{V}_j \cap \bar{U}_{ji}) \subset \tilde{B}_{ij} \cup \tilde{W}_{ij}. \end{cases} \tag{6}$$

Let $\tilde{A}_i$ be open in $\tilde{T}_i$ such that

$$\begin{cases} \tilde{A}_i \cap T_i = V_i \\ \bar{\tilde{A}}_i \cap T_i = \bar{V}_i \end{cases} \tag{7}$$

$$\tilde{A}_i \subset \tilde{A}_{ij} \cup \tilde{W}_{ij} \tag{8}$$

for all indices $j$ such that $T_{ij} \neq \varnothing$ (a finite number only).

We have

$$\overline{\psi_{ji}(\tilde{A}_i \cap \tilde{U}_{ij})} \cap T_j \subset \psi_{ji}(\bar{V}_i \cap \bar{U}_{ij}). \tag{9}$$

Indeed $\bar{\tilde{A}}_i \cap \overline{\tilde{U}_{ij}}$ is compact and contained in $\tilde{T}_{ij}$. Thus $\overline{\psi_{ji}(\tilde{A}_i \cap \tilde{U}_{ij})}$ is contained in $\psi_{ji}(\bar{\tilde{A}}_i \cap \overline{\tilde{U}_{ij}})$. Now

$$\psi_{ji}(\bar{\tilde{A}}_i \cap \overline{\tilde{U}_{ij}}) \cap T_j = \psi_{ji}(\bar{\tilde{A}}_i \cap \overline{\tilde{U}_{ij}} \cap T_i) = \psi_{ji}(\bar{V}_i \cap \bar{U}_{ij})$$

by (4) and (7).

(b) For every point $x \in U_i$ choose an open set $\tilde{U}_{ix}$ containing $x$ in $\tilde{T}_i$ satisfying the following four conditions

$$\forall_j \text{ such that } x \in U_{ij} \text{ we have } \tilde{U}_{ix} \subset \tilde{U}_{ij}. \tag{10}$$

(The number of these $j$'s is finite.)

$$\begin{array}{l}\forall_j \text{ such that } x\in\psi_{ij}(\bar V_j\cap\bar U_{ji}) \text{ we have}\\ \tilde U_{ix}\subset\tilde B_{ij}\cup\tilde W_{ij}\ \text{(cf. (6)).}\end{array}\tag{11}$$

$$\begin{array}{l}\forall_j \text{ such that } \varphi_i^{-1}(x)\notin\overline{V_j'} \text{ we have}\\ \tilde U_{ix}\cap\psi_{ij}(\tilde A_j\cap\tilde U_{ji})=\varnothing.\end{array}\tag{12}$$

If $\tilde U_{ij}=\varnothing$ this is trivially satisfied. Otherwise there are only finitely many such $j$'s. Note that by (9) if $\varphi_i^{-1}(x)\notin\bar V_j'$, then since $x\notin\psi_{ij}(\bar V_j\cap\bar U_{ji})$, we have $x\notin\psi_{ij}(\tilde A_j\cap\tilde U_{ji})$.

$$\begin{array}{l}\forall(j,k)\text{ such that}\\ x\in U_{ij}\cap U_{ik},\quad\text{(i.e. } \varphi_i^{-1}(x)\in U_i'\cap U_j'\cap U_k')\\ \text{we have }\ \tilde U_{ix}\subset\psi_{ij}(\tilde U_{ji}\cap\tilde U_{jk})\cap\psi_{ik}(\tilde U_{ki}\cap\tilde U_{kj}).\end{array}\tag{13}$$

There are finitely many such pairs $(j, k)$.

(c) On $\tilde U_{ix}$ we can assume (since it holds on $U_{ij}\cap U_{ik}$)

$$\psi_{ji}=\psi_{jk}\circ\psi_{ki}.$$

Let $\tilde U_i$ be the union of all $\tilde U_{ix}$ for $x\in U_i$

$$\tilde U_i=\bigcup\tilde U_{ix},\qquad x\in U_i.$$

Let $\tilde V_i$ be a neighborhood of $V_i$ in $\tilde T_i$ contained in $\tilde A_i$ and such that

$$\tilde V_i\subset\subset\tilde U_i.$$

By (7) we have

$$\tilde V_i\cap T_i=V_i,\qquad \overline{\tilde V_i}\cap T_i=\bar V_i.$$

Set

$$\begin{cases}\tilde V_{ij}=\tilde V_i\cap\psi_{ij}(\tilde V_j\cap\tilde U_{ji})\\ \tilde V_{ijk}=\tilde V_{ij}\cap\tilde V_{ik}.\end{cases}\tag{14}$$

We have $\tilde V_{ij}\subset\tilde U_{ij}$ and $\psi_{ji}:\tilde V_{ij}\cong\tilde V_{ji}$. A point $y\in\tilde V_{ijk}$ is contained in some $\tilde U_{ix}$ for some $x\in U_i$, and this set $\tilde U_{ix}$ must meet $\psi_{ij}(\tilde V_j\cap\tilde U_{ji})$ and $\psi_{ik}(\tilde V_k\cap\tilde U_{ki})$ and therefore also $\psi_{ij}(\tilde A_j\cap\tilde U_{ji})$ and $\psi_{ik}(\tilde A_k\cap\tilde U_{ki})$. By (12) this implies that $x\in\tilde U_{ij}\cap\tilde U_{ik}$. Thus by (2)

$$\psi_{ki}(y)\in\tilde V_k\cap\tilde U_{kj}$$

$$\psi_{jk}\circ\psi_{ki}(y)=\psi_{ji}(y).$$

Consequently

$$\begin{aligned}z=\psi_{ji}(y)&\in\tilde V_j\\ &\in\psi_{ji}(\tilde V_i\cap\tilde U_{ij})\\ &\in\psi_{jk}(\tilde V_k\cap\tilde U_{kj})\\ &\in\tilde V_{jik}.\end{aligned}$$

Thus we have $\psi_{ji}\tilde{V}_{ijk} \subset \tilde{V}_{jik}$ and by symmetry $\psi_{ij}\tilde{V}_{jik} \subset \tilde{V}_{ijk}$. This yields

$$\psi_{ji} : \tilde{V}_{ijk} \cong \tilde{V}_{jik}$$
$$\psi_{ij} = \psi_{ji}^{-1}.$$

Moreover

$$\psi_{ji} = \psi_{jk} \circ \psi_{ki}.$$

Hence we have in fact a complex analytic amalgamation system $\{\tilde{V}_i, \psi_{ij}\}$.

Let $\{\tilde{\Omega}, \psi_i\} = \lim \{\tilde{V}_i, \psi_{ij}\}$ be the amalgamated sum. Thus $\tilde{\Omega}$ is obtained from the disjoint sum $\sum \tilde{V}_i$ by dividing out by the equivalence relation

$$x \sim y \Leftrightarrow x \in \tilde{V}_{ij}, \qquad y \in \tilde{V}_{ji} \quad \text{and} \quad y = \psi_{ji}(x),$$

and $\psi_i : \tilde{V}_i \to \tilde{\Omega}$ is the canonical open imbedding $\tilde{V}_i \to \sum \tilde{V}_i \to \tilde{\Omega}$. Then $\tilde{\Omega}$ is a complex space, which a priori need not be separated. Moreover the isomorphisms $\varphi_i$ merge to an isomorphism $\varphi$ of $Y$ onto the quasiregular real part of $\tilde{\Omega}$. We want to show that $\tilde{\Omega}$ is Hausdorff.

(d) We show first that $\tilde{V}_{ij} \subset \tilde{U}_{ij}$, and more precisely that $\tilde{V}_{ij} \subset \tilde{W}_{ij}$. (When $T_{ij} \neq \emptyset$.)

Let $y \in \tilde{V}_{ij}$. Since $\tilde{V}_{ij} \subset \tilde{V}_i \subset \tilde{U}_i$ there exists $x \in U_i$ such that $y \in \tilde{U}_{ix}$.

If $x \notin \psi_{ij}(\bar{V}_j \cap \bar{U}_{ji})$, then $\varphi_i^{-1}(x) \notin V'_j$, thus by (12) $y \notin \psi_{ij}(\tilde{A}_j \cap \tilde{U}_{ji})$ and a fortiori $y \notin \psi_{ij}(\tilde{V}_j \cap \tilde{U}_{ji})$ which contradicts the fact that $y \in \tilde{V}_{ij}$ (cf. (14)).

Thus $x \in \psi_{ij}(\bar{V}_j \cap \bar{U}_{ji})$, thus by (11) $y \in \tilde{W}_{ij} \cup \tilde{B}_{ij}$. Now by (8) $y \in \tilde{V}_i \subset \tilde{A}_i \subset \tilde{A}_{ij} \cup \tilde{W}_{ij}$, and since $\tilde{A}_{ij}$ and $\tilde{B}_{ij}$ are disjoint, we conclude that $y \in \tilde{W}_{ij}$. Hence $\tilde{V}_{ij} \subset \tilde{W}_{ij}$.

Now let $x' \neq y'$ be in $\tilde{\Omega}$ and let $x \in \tilde{V}_i$, $y \in \tilde{V}_j$ be points such that $\psi_i(x) = x'$, $\psi_j(y) = y'$. We show there is a neighborhood $A$ of $x$ in $\tilde{V}_i$ and a neighborhood $B$ of $y$ in $\tilde{V}_j$ such that no point of $A$ is equivalent to a point of $B$.

If this was not true, we could find two sequences $\{x_k\}$, $\{y_k\}$ in $\tilde{T}_i$ and $\tilde{T}_j$ converging to $x$ and $y$ with $x_k \in \tilde{V}_{ij}$, $y_k \in \tilde{V}_{ji}$ and $x_k = \psi_{ij}(y_k)$. Since $\tilde{V}_{ij} \subset \tilde{U}_{ij}$, we have $x \in \tilde{U}_{ij}$, $y \in \tilde{U}_{ji}$, and $x = \psi_{ij}(y)$ by continuity. Thus

$$y \in \tilde{V}_j \cap \tilde{U}_{ji}$$
$$x \in \tilde{V}_i \cap \psi_{ij}(\tilde{V}_j \cap \tilde{U}_{ji}) = \tilde{V}_{ij}.$$

Hence $y \in \tilde{V}_{ji}$ and since $y = \psi_{ji}(x)$, $x$ and $y$ are equivalent, which contradicts the assumption $x' \neq y'$.

Combining the previous theorem with corollary (3.2) we obtain the following

THEOREM (8.3). *Let $Y$ be a quasianalytic space and let $X$ be a complexification of $Y$ with real structure $\tau$.*

*There exist a fundamental system of $\tau$-invariant Stein neighborhoods $U$ of*

*Y in X and proper real holomorphic imbeddings of U in some numerical space.*

*The imbedding can be chosen to be an isomorphism onto its image if there is a finite upper bound for the local imbedding dimension*

$$t(x) = \dim_{\boldsymbol{C}} \mathfrak{m}(\mathcal{O})_x/\mathfrak{m}(\mathcal{O})_x^2, \qquad x \in Y.$$

The last part of the theorem follows from the fact that the function $t(x)$ is upper semicontinuous (i.e. for $x_0 \in X$ there is a neighborhood $U(x_0)$ of $x_0$ in $X$ such that for every $y \in U(x_0)$ we have $t(y) \leqq t(x_0)$). Therefore, if there exist an $M > 0$ such that $t(x) < M$ for $x \in Y$, then the set

$$W = \{x \in X \mid t(x) < M\}$$

is an open neighborhood of $Y$ in $X$.

Note that for $x \in Y$ it is irrelevant whether we compute $t(x)$ in $\mathcal{O}_x$ or $\mathcal{A}_x$ since here $\mathfrak{m}(\mathcal{O})_x \cong \mathfrak{m}(\mathcal{A})_x \otimes \boldsymbol{C}$.

## 9. Hypersurfaces

Under the mild assumption for a quasianalytic space $Y$ that

$$\sup_{x \in Y} \dim_{\boldsymbol{R}} \mathfrak{m}(\mathcal{A})_x/\mathfrak{m}(\mathcal{A})_x^2 < M < \infty$$

we have seen that $Y$ can be identified with the quasiregular part of a $j$-invariant complex-analytic subset $\tilde{Y}$ of a numerical space $\boldsymbol{C}^N$ such that for $a$ in $Y$ $(\tilde{Y}, a)$ is the complexification of $(\tilde{Y}_{\boldsymbol{R}}, a)$.

We add some remarks in the case $\tilde{Y}$ is pure dimensional of codimension 1 everywhere, i.e. when $\tilde{Y}$ is a hypersurface. Then $Y$ is of course also pure dimensional.

PROPOSITION (9.1). *Let $X$ be a complex analytic subset of pure dimension $n$ in $\boldsymbol{C}^{n+1}$. There exist a holomorphic function $f: \boldsymbol{C}^{n+1} \to \boldsymbol{C}$ such that*

(i) $X = \{z \in \boldsymbol{C}^{n+1} \mid f(z) = 0\}$

(ii) $df(z) \neq 0$ *if* $z \in X - S(X)$.

*Moreover if $X$ is conjugation-invariant, then we can choose $f$ real.*

PROOF. Let $\mathcal{I} = \mathcal{I}(X)$ be the sheaf of germs of holomorphic functions vanishing on $X$. This is a coherent sheaf of ideals. Moreover, as the local rings $\mathcal{O}_x$ in $\boldsymbol{C}^{n+1}$ are unique factorization domains, $\mathcal{I}_x$ must be a principal ideal, $\mathcal{I}_x = f_x \mathcal{O}_x$. As $\mathcal{I}$ is coherent, there exist a neighborhood $V(x)$ of $x$ and a representative $f$ of $f_x$ such that $\mathcal{I}_y = f_y \mathcal{O}_y$ for all $y \in V(x)$, $f_y$ denoting the germ of $f$ at $y$. Moreover if $y \in V(x) \cap (X - S(X))$ we must have

$$df(y) \neq 0.$$

In fact in a small neighborhood of $y$, $X$ is a manifold, and therefore $\mathcal{I}_y = u\mathcal{O}_y$ has a generator $u$ with $du(y) \neq 0$. As $f = wu$ with $w$ a unit at $y$, we get $df(y) = w(y)\, du(y) \neq 0$.

Let $\mathfrak{B} = \{V_i\}_{i \in I}$ be an open covering of $\boldsymbol{C}^{n+1}$ by open sets $V_i = V(x_i)$ of the type described before, and let $f_i$ be the corresponding holomorphic functions generating $\mathcal{I} \mid V_i$. Then

$$g_{ij} = f_i/f_j : V_i \cap V_j \to \boldsymbol{C}^*$$

are holomorphic and never vanishing functions with $g_{ij}g_{jk} = g_{ik}$ on $V_i \cap V_j \cap V_k$. They thus define transition functions of a holomorphic line bundle. As $\boldsymbol{C}^{n+1}$ is Stein, we have

$$H^1(\boldsymbol{C}^{n+1}, \mathcal{O}^*) \simeq H^2(\boldsymbol{C}^{n+1}, \boldsymbol{Z}) = 0$$

(use the exact sequence $0 \to \boldsymbol{Z} \to \mathcal{O} \xrightarrow{\exp} \mathcal{O}^* \to 0$). This shows that the holomorphic line bundle is trivial, and therefore (as $H^1(\mathfrak{B}, \mathcal{O}^*) \to H^1(\boldsymbol{C}^{n+1}, \mathcal{O}^*)$ is injective) there exist holomorphic functions

$$\lambda_i : V_i \to \boldsymbol{C}^*, \qquad i \in I,$$

never vanishing, such that $g_{ji} = \lambda_j \lambda_i^{-1}$. Thus

$$f = \lambda_i f_i = \lambda_j f_j \quad \text{in} \quad V_i \cap V_j$$

is a global holomorphic function associated to each $f_i$ in $V_i$ since the functions $\lambda_i$ are units. Conditions (i) and (ii) are thus satisfied by the function $f$.

Now assume that $X$ is $j$-invariant. Then we must have

$$\overline{f(jx)} = \mu(x) f(x)$$

with $\mu(x)$ holomorphic and never vanishing. As $j^2 = id$ we have

$$\begin{aligned} f(x) &= \overline{\mu(jx)}\,\overline{f(jx)} \\ &= \mu(x)\overline{\mu(jx)} f(x). \end{aligned}$$

Therefore

$$\mu(x)\overline{\mu(jx)} = 1$$

for all $x$ such that $f(x) \neq 0$ and thus everywhere (by continuity). Let $\mu^{\frac{1}{2}}$ be one of the square roots of $\mu(x)$. Then

$$\mu^{\frac{1}{2}}(x)\overline{\mu^{\frac{1}{2}}(jx)} = \varepsilon,$$

where $\varepsilon$ is either $+1$ or $-1$. Let $g(x)=\mu^{\frac{1}{2}}(x)f(x)$. We have

$$\begin{aligned} g(jx) &= \mu^{\frac{1}{2}}(jx)f(jx)=\mu^{\frac{1}{2}}(jx)\overline{\mu(x)f(x)} \\ &= (\mu^{\frac{1}{2}}(jx)\overline{\mu^{\frac{1}{2}}(x)})\overline{\mu^{\frac{1}{2}}(x)}\overline{f(x)} \\ &= \varepsilon\overline{g(x)}. \end{aligned}$$

If $\varepsilon=1$, then $g$ is real and associated to $f$. If $\varepsilon=-1$, replace $g$ by $ig$.

REMARKS (9.2). The same argument holds without changes if we replace $\boldsymbol{C}^{n+1}$ by a Stein space $Z$ having the following properties

(i) For $z\in Z$ the local ring $\mathcal{O}_z$ is a unique factorization ring

(ii) $H^2(Z, \boldsymbol{Z})=0$.

For the last part of the statement we have to assume that $Z$ has a real structure $\tau$ and $X$ is $\tau$-invariant.

(9.3) In the hypersurface case (9.1) suppose that $Y=\hat{X}_{\boldsymbol{R}}$ is compact. Then $X_{\boldsymbol{R}}-Y$ is an analytic set in $\boldsymbol{R}^{n+1}-Y$ and of dimension $\leqq n-1$. Thus it does not disconnect the components of $\boldsymbol{R}^{n+1}-Y$, which are open sets. We can choose the defining function $f$ so that $f>0$ on the unique unbounded component of $\boldsymbol{R}^{n+1}-X_{\boldsymbol{R}}$. Then we have

$$Y=\text{bdy}\,\{x\in\boldsymbol{R}^{n+1}\mid f(x)<0\}.$$

In fact $B=\{x\in\boldsymbol{R}^{n+1}\mid f(x)<0\}$ is the union of the bounded connected components of $\boldsymbol{R}^{n+1}-X_{\boldsymbol{R}}$. Each point of $Y$ is a limit of points where $f>0$ and of points where $f<0$, since $f$ changes sign when we cross a non-singular point of $Y$. Therefore $Y\subset\text{bdy}\,B$.

If $b\in\text{bdy}\,B$, then $f(b)=0$, so $b\in X_{\boldsymbol{R}}$. If $b\in X_{\boldsymbol{R}}-Y$, we must have $f<0$ when $U\cap\hat{X}_{\boldsymbol{R}}=\varnothing$, as $X_{\boldsymbol{R}}-Y$ do not disconnect a spherical neighborhood $U$ of $b$, in $U-U\cap X_{\boldsymbol{R}}$. Thus $b\in\text{int}\,\bar{B}$, in particular $b\notin\text{bdy}\,\bar{B}=\text{bdy}\,B$.

## 10. Locally proper holomorphic maps, generally one-to-one

We recall the following two results of Remmert, [12].

(10.1) *Let $X$ and $Y$ be complex spaces and $f: X\to Y$ a holomorphic map*

(i) *If $f$ is proper, then $fX$ is an analytic subset of $Y$.*

(ii) *If $X$ is pure dimensional of dimension $n$, $Y$ is locally irreducible of dimension $p$ and* $\dim f^{-1}f(x)=n-p$, *then $f$ is open.*

Proofs can be found also in [11], Ch. VII.

Let $X$ and $Y$ be topological spaces and $f: X\to Y$ a continuous map. $f$ is called *locally proper* if for each $y$ in $fX$ there is a neighborhood $U$ of $y$ in $Y$ such that the induced map

$$f_U: f^{-1}U\to U$$

is a proper map. This is equivalent to the existence of an open neighborhood $W$ of $fX$ in $Y$ such that the induced map

$$f_W : X \to W$$

is proper.

Assume now that $X$ and $Y$ are complex spaces and that $f$ is a holomorphic map. We say that $f$ is *generally one-to-one* if we can find an analytic subset $A$ of $X$ such that

(a) for each irreducible component $X_\nu$ of $X$

$$A \cap X_\nu \text{ is of codimension } \geqq 1 \text{ in } X_\nu$$

(b) $f : X - A \to Y$ is one-to-one.

PROPOSITION (10.2). *Let $X$ and $Y$ be complex spaces and $f : X \to Y$ a locally proper holomorphic map.*

(i) *If $X$ is Stein, then $f$ has finite fibers and* $\dim (fX, f(x)) = \dim (X, x)$ *for all $x \in X$.*

(ii) *If $X$ is Stein and of pure dimension, and $Y$ is normal and* $\dim Y = \dim X$, *then $f$ is an open map. And if in addition $f$ is generally one-to-one, then $f$ maps $X$ isomorphically onto an open subset of $Y$.*

REMARKS (10.3). Claim (i) shows that one cannot have a locally proper holomorphic map $f : X \to Y$ from a Stein space unless $\dim (Y, f(x)) \geqq \dim (X, x)$ at all points $f(x) \in Y$.

(10.4). Claim (i) and the first part of claim (ii) holds under more general conditions: Instead of $X$ being Stein it suffices that the holomorphic functions on $X$ separate points. Also 'normal' can be weakened to 'locally irreducible'.

PROOF OF PROPOSITION (10.2). (i) $f$ is proper from $X$ into some open subset $W$ of $Y$. By (10.1) $fX$ is an analytic subset of $W$ and so has a well defined dimension at each point $f(x)$.

For $x \in X$, $f^{-1}f(x)$ is a compact analytic subset of $X$ with non-constant holomorphic functions (since $H(X)$ separates points). But then $f^{-1}f(x)$ is a compact discrete set, hence finite. Thus $\dim f^{-1}f(x) = 0$ for all $x \in X$ and so $\dim (fX, f(x)) = \dim (X, x)$.

(ii) Since the fiber dimension is constant, the first part follows by (10.1). As for the last part, notice that if $\Omega \subset X$ is open and relatively compact, so is $f^{-1}f\Omega$. Replacing $X$ by a relatively compact, $f$-saturated, open subset of $X$, we may assume that $X$ is an analytic subset of some numerical space $\boldsymbol{C}^N$.

Set $X' = X - f^{-1}fS(X) - f^{-1}fA$. Then $X'$ is an open dense submanifold of $X$ on which $f : X' \to Y$ is one-to-one. Moreover the map $f : X' - f^{-1}S(Y) \to$

$Y-S(Y)$ is a holomorphic one-to-one map between manifolds of the same dimension. The map is then open, and its jacobian, by a theorem of Bochner and Martin, is never zero. Hence

$$f^{-1}: fX' \cap (Y-S(Y)) \to X' - f^{-1}S(Y)$$

is holomorphic.

If $z_i$, $1 \leqq i \leqq$ , are holomorphic coordinates in $\boldsymbol{C}^N$, then $f^{-1}$ is given by equations

$$\begin{cases} z_i = \varphi_i(y) & y \in fX' \cap (Y-S(Y)) \\ 1 \leqq i \leqq N \end{cases} \tag{1}$$

where $\varphi_i$ are holomorphic functions.

As $f$ is open, we may replace $Y$ by $fX$ which is open in $Y$. We claim that $\varphi_i$ are holomorphic functions on all of $Y$ (i.e. they extend holomorphically to $Y$).

Indeed $B = fS(X) \cup fA \cup S(Y)$ is a proper analytic subset of $Y$ of codimension $\geqq 1$ at each point. In open neighborhoods $V \subset\subset Y$ of any point $y \in Y$ the functions $\varphi_i$ take values in a compact set since $f$ is proper $(Y = fX)$. Thus, since $Y$ is normal, by Riemann's extension theorem the functions $\varphi_i$ extends holomorphically to $V$. Therefore the equations (1) define a holomorphic map $\varphi: Y \to \boldsymbol{C}^N$. We must have Im $\varphi \subset X$. Indeed if $h: \boldsymbol{C}^N \to \boldsymbol{C}$ is any holomorphic function on $\boldsymbol{C}^N$ vanishing on $X$, we have $h \circ \varphi = 0$ on $Y-B$ and therefore on $Y$, by continuity. Also

$$\varphi \circ f = id_X \quad \text{and} \quad f \circ \varphi = id_Y.$$

Indeed these equations are satisfied on $X - f^{-1}(B)$ and $Y-B$ respectively. Therefore, by continuity, they are satisfied everywhere. Altogether $\varphi = f^{-1}$, and so $f$ is an isomorphism.

## 11. Projections of quasianalytic sets

Let $Q$ be a quasianalytic space and $(X, \tau)$ a proper Stein complexification of $Q$. If $Q$ is pure dimensional or irreducible, then $X$ can be chosen pure dimensional or irreducible, respectively.

THEOREM (11.1). *Let $(X, \tau)$ and $(Y, \sigma)$ be Stein complexifications of their quasianalytic real parts $\hat{X}$ and $\hat{Y}$, and let $f: X \to Y$ be a real holomorphic map which is proper and generally one-to-one.*

(i) *If $X$ is pure dimensional with* $\dim X = \dim Y$ *and $Y$ is irreducible, then $f$ is surjective.*

(ii) *If $f$ is surjective, then $f\hat{X} = \hat{Y}$ (invariance of the quasiregular part). If in* (i) *$Y$ is also locally irreducible, then $f$ is also open.*

PROOF. (i) If $\pi: X^* \to X$ and $\omega: Y^* \to Y$ denote normalizations of $X$ and $Y$, then $X^*$ is pure dimensional and Stein, $Y^*$ is connected, $\dim X^* = \dim Y^*$, and the lift $f^*$ of $f$ is proper

$$\begin{array}{ccc} X^* & \xrightarrow{f^*} & Y^* \\ \downarrow \pi & & \downarrow \omega \\ X & \xrightarrow{f} & Y \end{array}$$

Therefore $f^*X^* \subset Y^*$ is open (by proposition (10.2)) as well as closed ($f^*$ is proper), and so $f^*X^* = Y^*$. This shows that $fX = Y$. If $Y$ is locally irreducible, then $\omega$ is open (again by (10.2)). Hence $\omega \circ f^*$ is open, and therefore $f$ is open.

(ii) As $X$ is Stein and $f$ is proper, all fibers $f^{-1}f(x)$ are finite and $f$ is a map that preserves dimension (proposition (10.2)). As $\hat{X}_\tau$ consists of points $p \in X_\tau$ where $\dim_{\boldsymbol{R}}(X_\tau, p) = \dim_{\boldsymbol{C}}(X, p)$, we must have $f\hat{X}_\tau \subset \hat{Y}_\sigma$. Thus we need to show that $\hat{Y}_\sigma \subset f\hat{X}_\tau$.

Let $A \subset X$ be an analytic subset of $X$ of codimension $\geqq 1$ outside of which $f$ is one-to-one. Then $fA$ is of codimension $\geqq 1$ in $Y$.

Let $y_0 \in \hat{Y}_\sigma$. For any open neighborhood $V$ of $y_0$ in $Y$ we have (cf. corollary (5.3))

$$\hat{Y}_\sigma \cap V \not\subset fA.$$

We can then choose a sequence $\{y_\nu\}_{\nu=1,2,\cdots}$ of $Y$-regular points from $\hat{Y}_\sigma - \hat{Y}_\sigma \cap fA$ converging to $y_0$. Since $f$ is surjective, for each $\nu$ there is a *unique* $x_\nu \in X$ such that $f(x_\nu) = y_\nu$. Since $f$ is proper, we may assume that the sequence $\{x_\nu\}$ converges to a point $x_0 \in f^{-1}(y_0)$.

We must have $x_\nu \in X_\tau$ for all $\nu$. In fact $f(x_\nu) = y_\nu = f(\tau x_\nu)$ since $f$ is real. Therefore $x_\nu = \tau x_\nu$, i.e. $x_\nu \in X_\tau$.

If we assume that $S(X) \subset A$ (which we may), then $x_\nu \in X - S(X)$ and therefore $x_\nu \in \hat{X}_\tau$ (proposition (1.3)). Consequently $x_0 \in \hat{X}_\tau$ and $y_0 = f(x_0) \in f\hat{X}_\tau$.

REMARKS (11.2). The parametric equations of the Cayley umbrella (5.5) give a proper map $f$, generally one-to-one, of $X = \boldsymbol{C}^2$ onto the subset $Y$, irreducible and locally irreducible, of $\boldsymbol{C}^3$ defined by the equation $y^2 - x^2 z = 0$. But we have

$$f(X_{\boldsymbol{R}}) \subsetneqq Y_{\boldsymbol{R}}.$$

Thus, even under the assumption in the theorem we have in general $fX_\tau \neq Y_\sigma$.

## 12. Locally proper generally one-to-one maps between quasianalytic spaces

Consider two quasianalytic spaces $Q$, $R$, and an analytic map $f: Q \to R$. If $(X, \tau)$, $(Y, \sigma)$ are sufficiently small Stein complexifications of $Q$ and $R$ then $f$

extends uniquely to a holomorphic map $\tilde{f}:X \to Y$ which respects the real structures $\tau$ and $\sigma$.

We will investigate the following question: Suppose that $f$ is locally proper and generally one-to-one, under which conditions can we assert that for $X$, $Y$ sufficiently small $\tilde{f}$ is also locally proper and generally one-to-one?

PROPOSITION (12.1). *Let $X$, $Y$ be complex spaces and $f:X \to Y$ a locally proper holomorphic map. Then $f$ is generally one-to-one if and only if it is bimeromorphic onto its image.*

PROOF. Replacing $Y$ by a convenient open subset of $Y$ containing $fX$ we may assume that $f$ is proper. Let $Z = fX$. Then $Z$ is an analytic subspace of $Y$. We want to show that $f$ is generally one-to-one if and only if $f:X \to Z$ admits a meromorphic inverse.

Let $G$ be the graph of $f:X \to Z$:

$$G = \{(x, z) \in X \times Z \mid z = f(x)\}.$$

Then $G$ is an analytic subset of $X \times Z$.

Let $A \subset X$ be an analytic subset of $X$ of codimension $\geqq 1$ on each irreducible component of $X$ such that $f:X - A \to Z$ is one-to-one.

Then $B = fA$ is an analytic subset of $Z$ which is of codimension $\geqq 1$ on each irreducible component of $Z$. (The image by a proper holomorphic map of an irreducible analytic space is irreducible.) Then $f^{-1}$ is defined on $Z - B$. We need show that the closure of the graph of $f^{-1}$ in $Z \times X$ is analytic. But the graph of $f^{-1}$ is $tG - tG \cap (B \times X)$ where $t$ is the flip map $X \times Z \to Z \times X$. The closure of this set is $tG$. Therefore $f^{-1}$ is meromorphic as the closure of its graph is analytic, [12].

Conversely if $f^{-1}$ is meromorphic from $X$ to $Z$, this means that its graph $tG$ has the following property:

There exists an analytic set $B \subset Z$ of codimension $\geqq 1$ on each irreducible component of $A$ such that $tG - tG \cap (B \times X)$ in $Z \times X$ is the graph of a holomorphic map $f^{-1}: Z - B \to X$.

Now $tG \cap (B \times X)$ is an analytic subset $\tilde{A}$ in $tG$ of codimension $\geqq 1$ on each irreducible component of $tG$. As $f$ is holomorphic, $G$ (and therefore $tG$) is isomorphic to $X$. Hence $A = \mathrm{pr}_X \tilde{A}$ is an analytic subset of $X$ of codimension $\geqq 1$ on each irreducible component of $X$ and $f:X - A \to Z$ is one-to-one.

EXAMPLES (12.2). Let $f:\boldsymbol{R} \to \boldsymbol{R}$ be $f(\lambda) = \lambda^3$. Then $f$ is proper and generally one-to-one. Its complexification $\tilde{f}:\boldsymbol{C} \to \boldsymbol{C}$ is also proper, but has degree 3 and so cannot have a meromorphic inverse for any choice of neighborhoods of $\boldsymbol{R}$ in $\boldsymbol{C}$.

(12.3) Let $f:\hat{X}\to\hat{Y}$ be an analytic map between quasianalytic spaces and let $\tilde{f}:X\to Y$ be a complexification of $f$ to Stein neighborhoods of $\hat{X}$, $\hat{Y}$ in their complexification spaces. Then $\tilde{f}$, if it is locally proper, must have finite fibers. In particular for any $a\in X$ the ideal

$$\mathcal{O}_a\cdot\tilde{f}^*\mathfrak{m}_{f(a)}\subset\mathcal{O}_a$$

must (by the Nullstellensatz) contain a power of the maximal ideal $\mathfrak{m}_a$

$$\mathcal{O}_a\cdot\tilde{f}^*\mathfrak{m}_b\supset\mathfrak{m}_a^h\qquad b=f(a)\tag{12.3.1}$$

for some integer $h>0$ (depending on $a$).

In particular, if $a\in\hat{X}$ the above condition implies the corresponding real condition

$$\mathcal{A}_a\cdot f^*\mathfrak{m}_b\supset\mathfrak{m}_a^h\qquad b=f(a).\tag{12.3.2}$$

Here $\mathfrak{m}_a$ and $\mathfrak{m}_b$ denote respectively the maximal ideals of $\mathcal{A}_a$ ($\mathcal{A}$ the structure sheaf of $\hat{X}$) and $\mathcal{A}_b$ ($\mathcal{A}$ the structure sheaf of $\hat{Y}$). In fact if $\mathfrak{m}_a^{\boldsymbol{R}}$ is the maximal ideal of $\mathcal{A}_a$, then for $\mathfrak{m}_a$ in $\mathcal{O}_a$ we have

$$\mathfrak{m}_a=\mathfrak{m}_a^{\boldsymbol{R}}\otimes_{\boldsymbol{R}}\boldsymbol{C}.$$

If $x_1,\cdots,x_l$ are generators of $\mathfrak{m}_a^{\boldsymbol{R}}$ over $\mathcal{A}_a$, then $x_1,\cdots,x_l$ are generators of $\mathfrak{m}_a$ over $\mathcal{O}_a$.

By (12.3.1) we have

$$x_{i_1}\cdots x_{i_n}=\sum g_i(x)\tilde{f}_i(x)$$

(the $\tilde{f}_i$ being the components of a real holomorphic map $\boldsymbol{C}^n\to\boldsymbol{C}^k$ representing $\tilde{f}$ at $a$, $b$).

Therefore

$$x_{i_1}\cdots x_{i_n}=\sum\tfrac{1}{2}(g_i(x)+\overline{g_i(\bar{x})})f_i(x)$$

for $x\in\hat{X}$. But $g_i(x)+\overline{g_i(\bar{x})}$ are real valued holomorphic functions on $\hat{X}$, hence they belong to $A_a$. This implies (12.3.2) (with the same $h$).

Conversely if (12.3.2) holds, tensorization by $\boldsymbol{C}$ gives condition (12.3.1) at all points of $\hat{X}$ and therefore at all points of $X$ in a sufficiently small neighborhood of $\hat{X}$ in $X$ by virtue of the following

LEMMA (12.4). *Let $g:X\to Y$ be a holomorphic map between complex spaces. For every $x\in X$ define $\delta(g,x)=\delta(x)$ by*

$$\delta(x)=\dim_{\boldsymbol{C}}\mathcal{O}_x/\mathcal{O}_x g^*\mathfrak{m}_{g(x)}.$$

*Then the function $\delta(x)$ is upper semicontinuous on $X$.*

PROOF. The question being local we may assume $X$ and $Y$ analytic subsets of some open sets $U$, $V$ of $\boldsymbol{C}^n$, $\boldsymbol{C}^p$ respectively. Also we may assume that there exists a finite set of holomorphic functions $h_1, \cdots, h_1$ on $U$ such that $\mathcal{O}(\boldsymbol{C}^n)(h_1, \cdots, h_l)$ coincide with the sheaf of ideals of germs of holomorphic functions vanishing on $X$.

Let

$$\begin{cases} y_i = g_i(x) \\ 1 \leqq i \leqq p \end{cases}$$

be a holomorphic map from $U$ to $V$ inducing the map g.

Then

$$\begin{aligned} \delta(x) &= \dim_{\boldsymbol{C}} \mathcal{O}(X)_x/\mathcal{O}(X)_x(g_1, \cdots, g_p) \\ &= \dim_{\boldsymbol{C}} \mathcal{O}(\boldsymbol{C}^n)_x/\mathcal{O}(\boldsymbol{C}^n)_x(g_1, \cdots, g_p, h_1, \cdots, h_l). \end{aligned}$$

It is enough to prove the lemma for the map $\boldsymbol{C}^n \to \boldsymbol{C}^{p+1}$ given locally near a point $x \in X$ by the equations

$$\begin{cases} y_i = g_i(x) & 1 \leqq i \leqq p \\ z_j = h_j(x) & 1 \leqq j \leqq 1. \end{cases}$$

Hence it suffices to prove the lemma in the case $X$ and $Y$ are complex manifolds. However, in this case the proof is easy.

Proposition (12.5). *Let $f: Q \to R$ be a real-analytic map between quasianalytic spaces. Assume that:*

(i) *$f$ is locally proper.*

(ii) *For each $a \in Q$ there is an integer $h > 0$ such that*

$$\mathcal{A}_a \cdot f^* \mathfrak{m}_b \supset \mathfrak{m}_a^h \qquad b = f(a).$$

*Let $\tilde{f}: X \to Y$ be a holomorphic extension of $f$ to complexifications $X \supset Q$ and $Y \supset R$.*

*Then $Q$ has a fundamental system of invariant Stein neighborhoods $U \subset X$ such that $\tilde{f} \mid U$ is locally proper.*

*Conversely, if a locally proper complexification of f exists, then the condtions* (i) *and* (ii) *are satisfied.*

Proof. The last part of the proposition has already been established. We have to show that conditions (i) and (ii) are sufficient for the existence of a locally proper complexification $\tilde{f}$. We first establish the following

Lemma (12.6). *Let $\Omega$ be open in $\boldsymbol{R}^n$ and let $f: \Omega \to \boldsymbol{R}^{n+k}$ be a real-analytic map. Let $a \in \Omega$ and assume there is an integer $h > 0$ such that*

$$\mathcal{A}_a \cdot f^* \mathfrak{m}_b \supset \mathfrak{m}_a^h \qquad b = f(a).$$

*Then a and b admit fundamental systems of invariant neighborhoods $\{U_\varepsilon\}$ and $\{W_\varepsilon\}$ in $\boldsymbol{C}^n$ and $\boldsymbol{C}^{n+k}$ respectively, such that the complexification $\tilde{f}$ maps $U_\varepsilon$ properly into $W_\varepsilon$.*

PROOF. We may assume $a$ and $b$ at the origin respectively of $\boldsymbol{C}^n$ and $\boldsymbol{C}^{n+k}$. Because of the assumption, if

$$\begin{cases} y_j = f_j(x) \\ 1 \leqq j \leqq n+k \end{cases}$$

are the equations of the map $f$ near the origin, we have in some neighborhood of the origin in $\boldsymbol{R}^n$

$$\begin{cases} x_i^h = \sum_{j=1}^{n+k} g_j^{(i)}(x) f_j(x) \\ 1 \leqq i \leqq n \end{cases} \tag{1}$$

with real analytic $g_j^{(i)}$'s.

These equations still holds if we replace $f_j$ and $g_j^{(i)}$ by their holomorphic extensions in some neighborhood $V$ of $0 \in \boldsymbol{C}^n$, $x_i$ now denoting holomorphic coordinates in $\boldsymbol{C}^n$.

Choose $\varepsilon > 0$ so small that $P_\varepsilon = \{x \in \boldsymbol{C}^n \mid |X_i|^h < \varepsilon\}$ has closure contained in $V$, i.e. $P_\varepsilon \subset\subset V$. On $P_\varepsilon$ the relations (1) give the following estimates

$$\begin{cases} |X_i|^h \leqq c \sup_j |\tilde{f}_j(X)|, \\ 1 \leqq i \leqq n \end{cases} \tag{2}$$

where $c = (n+k) \sup_{x \in P_\varepsilon} |\tilde{g}_j^{(i)}(x)|$. Set $t = (n+k)$ and choose $\sigma$ with $0 < \sigma < \varepsilon/2c$, say $\sigma = \varepsilon/4c$, and set

$$Q_\sigma = \{y \in \boldsymbol{C}^t \mid |y_j| < \sigma\}.$$

From (2) follows that

$$\tilde{f}^{-1} Q_\sigma \cap P_\varepsilon \subset P_{\varepsilon/2}.$$

Therefore $\tilde{f} \mid (\tilde{f}^{-1} Q_\sigma \cap P_\varepsilon)$ is a proper map into $Q_\sigma$; in fact if $K \subset Q_\sigma$ is compact, then $\tilde{f}^{-1} K \cap P_\varepsilon$ is closed and contained in $P_{\varepsilon/2}$, and thus it is compact. We can take $U_\varepsilon = f^{-1} Q_\sigma \cap P_\varepsilon$ and $W_\varepsilon = Q_\sigma$.

COROLLARY (12.7). *Let $A$, $B$ be quasianalytic models and $f: A \to B$ a real-analytic map. Let $a \in A$ and assume that for some integer $h > 0$ we have*

$$\mathscr{A}_a \cdot f^* \mathfrak{m}_b \supset \mathfrak{m}_a^h \qquad b = f(a).$$

*Then there is a fundamental system $\{U_\nu\}$ of a neighborhood of $a$ in the complexification of $A$ and a fundamental system of neighborhoods $\{W_\nu\}$ of $b$ in the complexification of $B$, such that the holomorphic extension of $f$ is defined on $U_\nu$ and yields a proper map $\tilde{f}: U_\nu \to W_\nu$ for every $\nu$.*

The proof is reduced to the previous lemma by the reduction process to manifolds used in the proof of lemma (12.4).

PROOF OF PROPOSITION (12.5) (continued). Eventually restricting $R = \hat{Y}$, we can assume without loss of generality that $f: Q \to R$ is a proper map. Then $Z = fQ$ is a closed subset of $R$. Moreover, $f$ has finite fibers, since in fact any fiber $f^{-1}\{b\}$ is both compact (assumption (i)) and discrete (assumption (ii)).

For $b \in Z$ and $f^{-1}\{b\} = \{a_1, \cdots, a_k\}$, say, choose open neighborhoods $U(a_i) \subset X$ of the points $a_i$ and $W(b) \subset Y$ of $b$ such that $\tilde{f}U(a_i) \subset W(b)$ and $\tilde{f}: U(a_i) \to W(b)$ is a proper map for $1 \leqq i \leqq k$. This is possible by corollary (12.7) and because there are finitely many $a_i$.

Set $F_b = f^{-1}\{b\}$ and $U(F_b) = \bigcup_i U(a_i)$. Then $U(F_b)$ is an open neighborhood in $Y$ of the real fiber $F_b$, and the induced map

$$\tilde{f}: U(F_b) \to W(b)$$

is proper.

Now, let $\{W(b_j)\}_{j=1,2,\ldots}$ be a locally finite family of neighborhoods (of this type) covering $Z$

$$Z \subset \bigcup_j W(b_j).$$

Set $W = \bigcup_j W(b_j)$, $U = \bigcup_j U(F_{b_j})$. Then $U$ and $W$ are open sets in $X$ and $Y$, respectively, with $Q \subset U$, $Z \subset W$, and $\tilde{f}U \subset W$.

We claim that the induced map

$$\tilde{f}: U \to W$$

is proper: Let $K \subset W$ be a compact set. Then $K$ meets only finitely many $W(b_j)$ (since these sets form a locally finite family); say

$$K \cap W(b_j) \neq \varnothing$$

for $j = 1, 2, \cdots, 1$ only.

We can find smaller open neighborhoods in $Y$

$$W'(b_j) \subset\subset W(b_j) \qquad (1 \leqq j \leqq l)$$

such that

$$K \subset \bigcup_{j=1}^{l} W'(b_j).$$

For each $j$ define the corresponding $U'(F_j) \subset U(F_j)$ as

$$U'(F_j) = U(F_j) \cap \tilde{f}^{-1} W'(b_j) \qquad (1 \leqq j \leqq l).$$

Then $\tilde{f}U'(F_j) \subset W'(b_j)$.

Now form

$$K_i = K \cap W'(b_i) \qquad (1 \leqq i \leqq l)$$
$$K_{ij} = K \cap W'(b_i) \cap W(b_j) \qquad (1 \leqq i < j \leqq l)$$
$$\vdots$$
$$K_{12\cdots l} = K \cap W'(b_1) \cap \cdots \cap W'(b_l).$$

Set

$$K_{(2)} = \bigcup K_{ij}$$
$$K_{(3)} = \bigcup K_{ijk}$$
$$\vdots$$

and

$$H_i = \overline{K_i - K_{(2)}} \qquad (1 \leqq i \leqq l)$$
$$H_{ij} = \overline{K_{ij} - K_{(3)}} \qquad (1 \leqq i < j \leqq l)$$
$$\vdots$$
$$H_{12\cdots l} = \bar{K}_{12\cdots l}.$$

Then

$$K = (\bigcup H_i) \cup (\bigcup H_{ij}) \cup \cdots \cup H_{12\cdots l}.$$

This is a decomposition of $K$ into a finite union of compacts $H_I$ with the property that

$$H_I \cap W'(b_j) \neq \varnothing \quad \text{implies} \quad H_I \subset W(b_j). \tag{1}$$

We must check that $U \cap \tilde{f}^{-1}K$ is compact. We have

$$U \cap \tilde{f}^{-1}K = \left(\bigcup_{j=1}^{l} U(F_j)\right) \cap \tilde{f}^{-1}K = \left(\bigcup_{j=1}^{l} U'(F_j)\right) \cap \tilde{f}^{-1}K$$
$$= \bigcup_{j,I} (U(F_j) \cap \tilde{f}^{-1}H_I) = \bigcup_{j,I} (U'(F_j) \cap \tilde{f}^{-1}H_I).$$

Consequently it suffices to check the subsets

$$U(F_j) \cap \tilde{f}^{-1}H_I$$

for those $(j, I)$ such that $U'(F_j) \cap \tilde{f}^{-1}H_I \neq \varnothing$.

Now, $U'(F_j) \cap \tilde{f}^{-1}H_I \neq \varnothing$ implies $W'(b_j) \cap H_I \neq \varnothing$, which again implies $H_I \subset W(b_j)$ by (1). But in that case $U(F_j) \cap \tilde{f}^{-1}H_I$ is compact by the properness of $\tilde{f}: U(F_j) \to W(b_j)$.

This shows that $\tilde{f}$ is locally proper on $U$. By choosing each $W(b)$ and each $U(F_b)$ invariant we can assert that $U$ is invariant, and by shrinking, if necessary, we can even take $U$ to be Stein. This proves proposition (12.5).

EXAMPLE (12.8). Set

$$\hat{X} = \{(x, y; u, v) \in \mathbf{R}^2 \times P_1(\mathbf{R}) \mid xu = yv\}$$

and let $f: \hat{X} \to \mathbf{R}^2$ be the natural projection. The map is proper but no extension $f: X \to \mathbf{C}^2$ to a Stein complexification of $\hat{X}$ is a locally proper map. In fact the condition of proposition (12.5) is not satisfied on $f^{-1}(0)$, which does not consist of isolated points.

## 13. Generic projections of real affine algebraic varieties

Let $X \subset \mathbf{C}^N$ be an irreducible complex algebraic variety of $\dim_{\mathbf{C}} X = n$. For every $l$ with $n < l < N$ we consider the set of linear maps $\mathrm{Hom}^* (\mathbf{C}^N, \mathbf{C}^l)$ of maximal rank $l$, i.e. the $\pi \in \mathrm{Hom}(\mathbf{C}^N, \mathbf{C}^l)$ with $\pi\mathbf{C}^N = \mathbf{C}^l$. If $\pi \in \mathrm{Hom}^* (\mathbf{C}^N, \mathbf{C}^l)$ then Ker $\pi$ is a linear subspace of $\mathbf{C}^N$ of codimension $l$. We have therefore a natural map

$$\mathrm{Hom}^* (\mathbf{C}^N, \mathbf{C}^l) \to G^l(\mathbf{C}^N)$$

where $G^l(\mathbf{C}^N)$ denotes the Grassmann manifold of linear subspaces of codimension $l$ in $\mathbf{C}^N$. This map is holomorphic. In fact the Plücker coordinates of Ker $\pi$ are homogeneous coordinates $[p_{j_1 \cdots j_l}]_{1 \leq j_1 < \cdots < j_l \leq N}$, where the $p_{j_1 \cdots j_l}$ are the coefficients of $\pi_1 \wedge \pi_2 \wedge \cdots \wedge \pi_l$ in the expansion by the standard basis in $\mathbf{C}^{N*} \wedge \cdots \wedge \mathbf{C}^{N*}$,

$$\pi_1 \wedge \cdots \wedge \pi_l = \sum p_{j_1 \cdots j_l} \, dt_{j_1} \wedge \cdots \wedge dt_{j_l}.$$

If $A = \{a_{ij}\}$ is the standard matrix of $\pi$, then $\pi_i = \sum_{j=1}^N a_{ij} \, dt_j$ and so

$$\sum p_{j_1 \cdots j_l} \, dt_{j_1} \wedge \cdots dt_{j_l} = \bigwedge_{i=1}^{l} \sum_{j=1}^{N} a_{ij} \, dt_j,$$

showing that the $p_{j_1 \cdots j_l}$ are the $l$th order subdeterminants of $\{a_{ij}\}$.

Two maps $\pi, \pi' \in \mathrm{Hom}^* (\mathbf{C}^N, \mathbf{C}^l)$ with the same kernel differ by an automorphism of $\mathbf{C}^l$. Hence, up to automorphisms of $\mathbf{C}^l$ we can parametrize the projections $\pi: \mathbf{C}^N \to \mathbf{C}^l$ by the points of $G^l(\mathbf{C}^N)$.

If $\pi$ is real (i.e. $\pi \circ j_{\mathbf{C}^N} = j_{\mathbf{C}^l} \circ \pi$), then it is represented by a point in the real grassmannian $G^l(\mathbf{R}^N)$, which is the real part of the manifold $G^l(\mathbf{C}^N)$.

PROPOSITION (13.1). *There is a proper algebraic subset $A \subsetneq G^l(\mathbf{C}^N)$ such that for $\pi$ in $G^l(\mathbf{C}^N) - A$ we have*

(i) *$\pi \mid X$ is a proper map*

(ii) *$\pi \mid X$ is one-to-one outside a proper algebraic subset of $X$*

(iii) *$A$ is conjugation invariant if $X$ is real, and if $\pi \in G^l(\mathbf{R}^N) - A_{\mathbf{R}}$, then the properties* (i) *and* (ii) *are true for $\pi \mid X_{\mathbf{R}}$.*

PROOF. (a) Consider $P_N(\boldsymbol{C})$ as the projective completion of $\boldsymbol{C}^N$. As $X$ is algebraic, the closure $\bar{X}$ of $X$ in $P_N(\boldsymbol{C})$ is a projective variety that meets $P_{N-1}(\boldsymbol{C})$ in a variety $V=\bar{X}\cap P_{N-1}(\boldsymbol{C})$, $P_{N-1}(\boldsymbol{C})=P_N(\boldsymbol{C})-\boldsymbol{C}^N$. Then $\dim_{\boldsymbol{C}} V=n-1$, and $V$ is pure dimensional. Let $V=V_1\cup\cdots\cup V_K$ be the decomposition of $V$ into irreducible components. For each $i$ we consider in $V_i\times G^l(\boldsymbol{C}^N)$ the set $S_i$ defined by

$$\sum_1^N x_i\, dt_i \wedge \omega = 0$$

where $(x_1,\cdots,x_N)$ are the homogeneous coordinates of a point $x\in V_i$ and $\omega$ represents an element of $G^l(\boldsymbol{C}^N)$. This set is algebraic. Moreover for any $x_0\in V_i$

$$\mathrm{pr}_{V_i}^{-1}(x_0)\cap S_i \simeq G^l(\boldsymbol{C}^{N-1})$$

since it represents the spaces $P_{l-1}$ of $P_{N-1}(\boldsymbol{C})$ which contain $x_0$. Therefore $S_i$ is irreducible and

$$\begin{aligned}\dim_{\boldsymbol{C}} S_i &= (n-1)+\dim_{\boldsymbol{C}} G^l(\boldsymbol{C}^{N-1})\\ &=(n-1)+l(N-l-1)\\ &<l(N-l) \qquad (\text{as } l>n).\end{aligned}$$

Hence $\mathrm{pr}_{G^l(\boldsymbol{C}^N)} S_i$ has for closure an algebraic set $A_i \subsetneqq G^l(\boldsymbol{C}^N)$.

If $\pi\in G^l(\boldsymbol{C}^N)-\bigcup_{i=1}^k A_i$, then $\pi\mid X$ is a proper map. If $X$ is real, $\bigcup_{i=1}^k A_i$ is invariant under the conjugation map in $G^l(\boldsymbol{C}^N)$.

(b) Consider now in $X\times X\times G^l(\boldsymbol{C}^N)$ the set defined by

$$\sum_1^N (a_i-b_i)\, dt_i\wedge\omega=0$$

where $a=(a_1,\cdots,a_N)$, $b=(b_1,\cdots,b_N)$ are two points of $X$. This is also an algebraic set $C$. If $\Delta$ denotes the diagonal of $X\times X$, then $C\supset\Delta\times G^l(\boldsymbol{C}^N)$. Let $S$ be the algebraic set defined by

$$S=\overline{C-\Delta\times G^l(\boldsymbol{C}^N)}.$$

For every $(a,b)\in X\times X-\Delta$ we have

$$\mathrm{pr}_{X\times X}^{-1}(a,b)\cap S\simeq G^l(\boldsymbol{C}^{N-1})$$

since it represents the spaces $\boldsymbol{C}^l$ in $\boldsymbol{C}^N$ which are parallel to the line joining $a$ to $b$.

Therefore $S$ must be irreducible, and

$$\begin{aligned}\dim_{\boldsymbol{C}} S &= 2n+\dim_{\boldsymbol{C}} G^l(\boldsymbol{C}^{N-1})\\ &=2n+l(N-l-1).\end{aligned}$$

We distinguish several cases

CASE 1. $\dim_{\boldsymbol{C}} S < \dim G^l(\boldsymbol{C}^N)$, i.e. $l > 2n$. Then $\mathrm{pr}_{G^l(\boldsymbol{C}^N)} S$ has a closure $B$ in $G^l(\boldsymbol{C}^N)$, which is an algebraic set strictly contained in $G^l(\boldsymbol{C}^N)$. If we take $A = \bigcup_{i=1}^k A_i \cup B$, then for $\pi \notin A$ the map $\pi$ will be proper and one to one.

CASE 2. $\dim_{\boldsymbol{C}} \overline{\mathrm{pr}_{G^l(\boldsymbol{C}^N)} S} < \dim G^l(\boldsymbol{C}^N)$. In this case taking $B = \overline{\mathrm{pr}_{G^l(\boldsymbol{C}^N)} S}$ we conclude as before.

CASE 3. $\dim_{\boldsymbol{C}} \mathrm{pr}_{G^l(\boldsymbol{C}^N)} S = \dim G^l(\boldsymbol{C}^N)$. Then there is an algebraic set $B \subsetneqq G^l(\boldsymbol{C}^N)$ such that for $\pi \notin B$ we have

$$\dim_{\boldsymbol{C}} \mathrm{pr}_{G^l(\boldsymbol{C}^N)}^{-1} \{\pi\} \cap S + \dim_{\boldsymbol{C}} G^l(\boldsymbol{C}^N) = \dim_{\boldsymbol{C}} S$$

i.e.

$$\dim_{\boldsymbol{C}} \mathrm{pr}_{G^l(\boldsymbol{C}^N)}^{-1} \{\pi\} \cap S = 2n - l < n.$$

Then

$$\overline{\mathrm{pr}_X \, \mathrm{pr}_{X \times X} (\mathrm{pr}_{G^l(\boldsymbol{C}^N)}^{-1} \{\pi\} \cap S)} = D$$

is an algebraic subset of $X$ of dimension $<n$ and

$$\pi \mid X - D \text{ is one to one.}$$

Therefore in this case it is enough to take $A = \bigcup_{i=1}^k A_i \cup B$.

Note that if $X$ is real, we can take $B$ to be invariant by the conjugation of $G^l(\boldsymbol{C}^N)$. Then $A_{\boldsymbol{R}} \subsetneqq G^l(\boldsymbol{R}^N)$ is a real algebraic subset of $G^l(\boldsymbol{R}^N)$, since $A_{\boldsymbol{R}} = A \cap G^l(\boldsymbol{R}^N)$. This concludes the proof.

## §3. Parametric spaces

### 14. Definitions

Let $(Q, \mathscr{A})$ be a quasianalytic space and let $X \supset Q$ be a proper Stein complexification of $Q$ with real structure $\tau$. Then the germ of $(X, \tau)$ along $Q = \hat{X}_\tau$ is uniquely determined (cf. (8.1)).

Let $X^*$ be a normalization of $X$, and let $\tau^*$ be the lift of $\tau$ to $X^*$. Then $X^*$ is also a Stein space, and $\tau^*$ is a real structure on $X^*$.

We call $(Q, \mathscr{A})$ a *parametric space* if $Q$ is pure dimensional and $X^*$ is non-singular along its real part $X^*_{\tau^*}$.

REMARKS (14.1). In this case $X^*_{\tau^*}$ is a manifold, and $\dim_{\boldsymbol{R}} (X^*_{\tau^*}, x^*) = \dim_{\boldsymbol{C}} (X^*, x^*)$ at every real point $x^*$ (proposition (1.2)). Hence $X^*_{\tau^*}$ equals its quasiregular part and so is projected onto $\hat{X}_\tau = Q$ by the canonical map $\pi : X^* \to X$ (theorem (11.1)). Consequently $X^*_{\tau^*} = \pi^{-1} Q$, which shows that the germ of $X^*$ along $X^*_{\tau^*}$ is uniquely determined by the parametric space $(Q, \mathscr{A})$.

Set $Q^* = X^*_{\tau^*}$ and let $\pi: Q^* \to Q$ be the induced map. This is an analytic map which is proper and generally one-to-one and which admits a complexification $\pi: X^* \to X$, again proper and generally one-to-one. We call $(Q^*, \pi)$ a *normalization* of $Q$.

(14.2) A real-analytic map $f: Q \to R$ between parametric spaces lifts uniquely to normalizations when it lifts,

$$f^*: Q^* \to R^*.$$

In fact the complexification

$$\tilde{f}: X \to Y,$$

which is unique as germ along $Q$ (corollary (7.6)), lifts uniquely (when it lifts)

$$\tilde{f}^*: X^* \to Y^*,$$

and restriction to the real parts gives $f^*$.

(14.3) The notion of a parametric space is *local*, i.e. can be defined by local models, which we now describe.

Consider an open set $\Omega \subset \boldsymbol{R}^n$ and a real-analytic map into $\boldsymbol{R}^p$, $p \geqq n$,

$$f: \Omega \to \boldsymbol{R}^p.$$

Let $\tilde{\Omega} \subset \boldsymbol{C}^n$ be an open complex neighborhood of $\Omega$, conjugation invariant and connected relative to $\Omega$ (i.e. every connected component of $\tilde{\Omega}$ meets $\Omega$) and such that $\tilde{\Omega} \cap \boldsymbol{R}^n = \Omega$.

For some choice of $\tilde{\Omega}$ $f$ extends to a holomorphic map

$$\tilde{f}: \tilde{\Omega} \to \boldsymbol{C}^p,$$

and because of the assumptions the extension is unique.

Now assume the following

$$\text{For some choice of } \tilde{\Omega} \text{ the complexification } \tilde{f}: \tilde{\Omega} \to \boldsymbol{C}^p \text{ is locally proper and generally one-to-one.} \tag{14.3.1}$$

Once (14.3.1) is satisfied the conclusion actually holds for a fundamental system of Stein neighborhoods $\tilde{\Omega}$ (proposition (12.5)).

If $\tilde{V} = \tilde{f}\tilde{\Omega}$, then $\tilde{V}$ is a complex-analytic set, really imbedded in $\boldsymbol{C}^p$ and of pure complex dimension $n$ ((10.2) or (10.4)). Moreover, $f\Omega$ is the quasianalytic real part of $\tilde{V}$ (by (11.1)).

Set $V = f\Omega$. If $b \in V$, then $f^{-1}\{b\} = \{a_1, \cdots, a_k\}$ is finite, and the germ $(\tilde{V}, b)$ has exactly $k$ irreducible components $(\tilde{V}_i, b) = \tilde{f}(\tilde{\Omega}, a_i)$ and is the complexification of the real-analytic germ $(\tilde{V}_{\boldsymbol{R}}, b)$.

The quasianalytic sets $V$ obtained in this way (with their sheaves of germs of analytic functions, cf. sec 8) are the local models for $n$-dimensional parametric spaces. The map

$$f:\Omega \to \boldsymbol{R}^p \qquad (p \geqq n)$$

will be called a *presentation* of the model $V = f\Omega$. If

$$\begin{cases} y_i = f(x_1, \cdots, x_n) \\ 1 \leqq i \leqq p \end{cases} \tag{14.3.2}$$

are the coordinate expressions of $f$, then $V$ admits these as 'parametric equations' for $(x_1, \cdots, x_n) \in \Omega$.

Since $\Omega$ is the normalization of $V$, the presentation $(\Omega, f)$ is determined up to isomorphism over $V$.

(14.4) A presentation $f:\Omega \to \boldsymbol{R}^n$, i.e. where $p = n$, is an analytic isomorphism onto an open subset $f\Omega$. In fact $\tilde{f}:\tilde{\Omega} \to \boldsymbol{C}^n$ is a holomorphic isomorphism onto an open subset $\tilde{f}\tilde{\Omega} \subset \boldsymbol{C}^n$ for suitable $\tilde{\Omega}$. This follows from proposition (10.2).

## 15. Weakly normal parametric spaces

We shall define weak normality for real parametric spaces (cf. [15] for weakly normal complex spaces).

(15.1) Let $g$ be a real valued function on a local parametric set $V \subset \boldsymbol{R}^{n+k}$ with presentation

$$f:\Omega \to \boldsymbol{R}^{n+k}.$$

We call $g$ *c-analytic* if it is continuous and the pull-back $g \circ f$ is real-analytic on $\Omega$. Every analytic function on $V$ is $c$-analytic.

Consider the real-analytic subset $\Gamma \subset \Omega \times \Omega$ which is the fibered product of $f$ by itself,

$$\Gamma = \{(x, y) \in \Omega \times \Omega \mid f(x) = f(y)\}.$$

Let $\Pi$ be the union of the irreducible components of $\Gamma$ that are not contained in the diagonal $\Delta \subset \Omega \times \Omega$. We call $\Pi$ the *coincidence set* of $f$. It is again a real-analytic subset of $\Omega \times \Omega$ (possibly empty). Let $h:\Omega \to \boldsymbol{R}$ be a real-analytic function. We have the following useful criterion (of immediate verification).

(15.1.1) *A necessary and sufficient condition that $h$ be the pull-back of a continuous (hence c-analytic) function on $V$ is that $H:\Omega \times \Omega \to \boldsymbol{R}$, defined by*

$$H(x, y) \equiv h(x) - h(y),$$

*vanishes on the coincidence set $\Pi$, $H \mid \Pi = 0$.*

(15.2) Pull-back of a $c$-analytic function by an analytic map of local models

$$V \to V'$$

is again $c$-analytic provided $V \to V'$ lifts analytically to $\Omega' \to \Omega$ (remark (14.2)). In particular $c$-analytic functions are preserved under isomorphisms $V \cong V'$.

Let $(Q, \mathscr{A})$ be a parametric space. A real valued function $g$ on $Q$ is again called *$c$-analytic* if for any sufficiently small open set $U \subset Q$ $g \mid U$ is $c$-analytic, i.e. if $g$ is $c$-analytic on the local models of $Q$.

Let $\mathscr{P}(Q)$ be the sheaf of germs of (local) $c$-analytic functions on $Q$. Clearly $\mathscr{P}$ is the sheaf of germs of continuous functions on $Q$ which become analytic when we pull back to the normalization $Q^*$,

$$\pi^* \mathscr{P}(Q) = \mathscr{A}(Q^*) \cap \pi^* \mathscr{C}(Q), \tag{15.2.1}$$

(where $\mathscr{C}(Q)$ denotes the sheaf of germs of continuous functions on $Q$).

We have

$$\mathscr{A}(Q) \subset \mathscr{P}(Q),$$

and we say that the parametric space $(Q, \mathscr{A})$ is *weakly normal* if $\mathscr{A}(Q) = \mathscr{P}(Q)$. Being a local property this holds precisely when the local models of $Q$ are weakly normal.

(15.2.2) *A local parametric set $V = f\Omega$ is weakly normal if and only if every function $h$ on $\Omega$, analytic and such that $h(x) = h(y)$ whenever $f(x) = f(y)$, is the pull-back of an analytic function on $V$.*

This is a consequence of (15.1.1).

Of course both (15.1.1) and (15.2.2) hold equally well for global parametric spaces $V$, in which case the presentation $(\Omega, f)$ should be replaced by a normalization $(V^*, \pi)$.

(15.3) The preceding definitions and constructions can equally well be stated in the complex case. This leads to *$c$-holomorphic* functions and weakly normal complex spaces. On a complex space $X$ a complex valued function $g$ is $c$-holomorphic if and only if it is continuous on all of $X$ and holomorphic on $X - S(X)$. Both (15.1.1) and (15.2.2) continue to hold in the complex case.

(15.4) It is convenient to have the following criterion (of immediate verification).

(15.4.1) *Suppose $\tilde{\Pi} \subset \tilde{\Omega} \times \tilde{\Omega}$ is a germ complexification of $\Pi \subset \Omega \times \Omega$. Then $V$ is weakly normal provided $\tilde{V}$ is weakly normal along $V$.*

EXAMPLES (15.5). The parametric set $V \subset \boldsymbol{R}^2$ presented by $f: \boldsymbol{R} \to \boldsymbol{R}^2$,

$$f \equiv \begin{cases} x = \lambda^2 \\ y = \lambda^3 \end{cases}$$

is not weakly normal: Take the function $g$ on $V$ defined by $g(x, y) = y/x$ and $g(0, 0) = 0$. Then $g$ is continuous, and $g \circ f = \mathrm{id}_{\boldsymbol{R}}$ is analytic. But $g$ is not the restriction of even a differentiable local function $G$ around the origin $0 \in \boldsymbol{R}^2$. For then $G \circ f$ would be differentiable, and we should have

$$rk_0(G \circ f) = 0 \qquad (= rk_0 f)$$

contradicting the fact that $G \circ f = \mathrm{id}_{\boldsymbol{R}}$.

(15.6) Consider the holomorphic map $\tilde{f}: \boldsymbol{C}^2 \to \boldsymbol{C}^3$ given by

$$\tilde{f} \equiv \begin{cases} x = u & (u, v) \in \boldsymbol{C}^2 \\ y = v(v^3 - u) & (x, y, z) \in \boldsymbol{C}^3 \\ z = v^3 \end{cases}$$

This map is proper, since its composition with the projection onto the $(x, z)$-plane is proper, and it is one-to-one outside the curve

$$C = \{(u, v) \in \boldsymbol{C}^2 \mid v^3 - u = 0\}.$$

On $C$ the map is three-to-one, the points $(u_0, v_0)$, $(u_0, e^{i\omega} v_0)$, $(u_0, e^{2i\omega} v_0)$, with $(u_0, v_0) \in C$ and $\omega = 2\pi/3$, having the same image. Thus its real restriction $f: \boldsymbol{R}^2 \to \boldsymbol{R}^3$ defines a parametric model $V = f\boldsymbol{R}^2$.

The real-analytic map $f$ is globally one-to-one, because two out of three of the points above on $C$ are always non-real. Since $f$ is proper, it is therefore a homeomorphism onto its image. Thus every analytic function on $\boldsymbol{R}^2$ factorizes through $f$.

On the other hand a holomorphic function on $\boldsymbol{C}^2$ that separates two of the three curve points $(u_0, v_0)$, $(u_0, e^{i\omega} v_0)$, and $(u_0, e^{2i\omega} v_0)$, for some $(u_0, v_0)$ in $C$, cannot factorize through $\tilde{f}$.

For instance the function $g$ on $V$ defined by

$$g(x, y, z) = z^{\frac{1}{3}},$$

lifts to

$$g \circ f(u, v) = v$$

on $\boldsymbol{R}^2$ and is therefore $c$-analytic. However, its complexification $\widetilde{g \circ f}$ does not factorize through $\tilde{f}$ and so $g$ is not the restriction of a $c$-holomorphic function on $\tilde{V} = \tilde{f}\boldsymbol{C}^2$.

Neither $\tilde{V}$ nor $V$ is weakly normal. However, it is a consequence of a general fact that the local $c$-holomorphic functions on $\tilde{V}$ constitute a complex-analytic structure, which is evidently weakly normal (the weak normalization of $\tilde{V}$, cf. [15]). Even so the corresponding real parametric structure induced on $V$ is not weakly normal, as the consideration above shows. This unpleasant behavior shows up because $\Pi(\tilde{f})$ fails to be a faithful complexification of $\Pi(f)$. In fact $\Pi(f)=\emptyset$, while $\tilde{\Pi}=\Pi(\tilde{f})$ is the union of two irreducible components $\tilde{\Pi}_{(1)}$ and $\tilde{\Pi}_{(2)}$,

$$\tilde{\Pi}_{(1)}=\{((u_0, v_0), (u_0, e^{i\omega}v_0)) \mid (u_0, v_0)\in C\}$$
$$\tilde{\Pi}_{(2)}=\{((u_0, v_0), (u_0, e^{2i\omega}v_0)) \mid (u_0, v_0)\in C\}.$$

## 16. Hartogs theorem for weak normality

We first prove a vanishing lemma.

Let $G$ be an open subset of $\boldsymbol{C}^n$ and $A\subset G$ an analytic subset of $G$. Let $\mathscr{F}$ be a coherent analytic sheaf on $G$ and take a point $x\in A$.

LEMMA (16.1). *There exists a fundamental sequence of polycylinders $P$ in $G$, centered at $x$, such that $H^r(P-A;\mathscr{F})=0$ for $0<r<\mathrm{pf}_x\,\mathscr{F}-\dim_{\boldsymbol{C}}(A,x)-1$.*

PROOF. Set $\dim_{\boldsymbol{C}}(A,x)=k$, $\mathrm{pf}_x\,\mathscr{F}=m$ (pf = 'profondeur' = depth). As $\mathscr{F}$ is defined on an open subset of $\boldsymbol{C}^n$, we have $m\leqq n$. For $m=n$ we have $\mathscr{F}\cong\mathcal{O}^p$ locally at $x$, and in this case the result is proved in [2], p. 225 (where 'profondeur' is called 'dimension homologique'). Let $m<n$ and assume the claim proved for sheaves of depth $>m$ at $x$. In a neighborhood of $x$, we have a free resolution

$$0\to\mathcal{O}^{p_{n-m}}\to\mathcal{O}^{p_{n-m-1}}\to\cdots\to\mathcal{O}^{p_0}\to\mathscr{F}\to 0.$$

Set $\ker\mathcal{O}^{p_0}\to\mathscr{F}=\mathscr{G}$. Then $\mathscr{G}$ has a resolution

$$0\to\mathcal{O}^{p_{n-m}}\to\mathcal{O}^{p_{n-m-1}}\to\cdots\to\mathcal{O}^{p_1}\to\mathscr{G}\to 0$$

hence $\mathrm{pf}_x\,\mathscr{G}\geqq m+1$.

From the exact sequence

$$0\to\mathscr{G}\to\mathcal{O}^{p_0}\to\mathscr{F}\to 0$$

we get

$$H^r(P-A;\mathcal{O}^{p_0})\to H^r(P-A;\mathscr{F})\to H^{r+1}(P-A;\mathscr{G}).$$

By our induction assumption $H^{r+1}(P-A;\mathscr{G})=0$ in the range $0<r+1<(m+1)-k-1$ and $H^r(P-A;\mathcal{O}^{p_0})=0$ in the range $0<r<n-k-1$. This yields the lemma.

REMARKS (16.2). Let $X$ be a reduced complex space and $\mathcal{O}=\mathcal{O}(X)$ the structure sheaf of $X$. Then at a point $x\in X$ one has the following properties

(a) $\mathrm{pf}_x\ \mathcal{O} \leqq \dim_{\boldsymbol{C}}(X, x)$

(b) $\mathrm{pf}_x\ \mathcal{O} = \dim_{\boldsymbol{C}}(X, x)$ if $X$ is a complete intersection at $x$.

For the proofs see [2], p. 200.

(16.3) Let $X$ be an analytic subset of an open set $G \subset \boldsymbol{C}^n$. Let $\mathcal{I} \subset \mathcal{O}$ be the sheaf of germs of holomorphic functions on $G$ vanishing on $X$ and assume that $\mathcal{I} = \mathcal{O}(f_1, \cdots, f_s)$ for suitable functions $f_1, \cdots, f_s$ on $G$. Then at any point $x \in X$ we have

$$\mathrm{pf}_x\ \mathcal{O}(X) = \mathrm{pf}_x\ \mathcal{I} - 1.$$

Let $\mathcal{R}(f_1, \cdots, f_s)$ be the sheaf on $G$ defined by the exact sequence

$$0 \to \mathcal{R}(f_1, \cdots, f_s) \to \mathcal{O}^s \xrightarrow{\alpha} \mathcal{I} \to 0$$

with $\alpha(g_1, \cdots, g_s) = \sum_1^s f_i g_i$. It follows that

$$\mathrm{pf}_x\ \mathcal{R}(f_1, \cdots, f_s) = \mathrm{pr}_x\ \mathcal{I} + 1 = \mathrm{pf}_x\ \mathcal{O}(X) + 2.$$

THEOREM (16.4). *Let $X$ be a complex space and suppose $X$ is weakly normal outside an analytic subset $A \subset S(X)$, satisfying*

$$\dim_{\boldsymbol{C}}(A, a) < \mathrm{pf}_a\ \mathcal{O}(X) - 1$$

*at all points $a \in A$. Then $X$ is weakly normal.*

PROOF. As the weak normality of $X$ is a local question we may assume that $X$ is an analytic subset of some open set $G \subset \boldsymbol{C}^n$ with $A$ of codimension $>2$ in $G$.

Let $x \in A$ and let $P$ be a polycylinder in $G$ at $x$. Let $\mathcal{I}_A$ be the sheaf of ideals on $P$ of germs of holomorphic functions vanishing on $A$. By choosing $P$ small we may assume there is a finite set of holomorphic functions $g_1, \cdots, g_t$ on $P$ such that $\mathcal{I}_A = \mathcal{O}_P(g_1, \cdots, g_t)$. Define $U_i \subset P$ by $z \in u_i \Leftrightarrow g_i(z) \neq 0$, $i = 1, \cdots, t$. Then $U_i$ is a Stein open subset of $P$, and $\mathfrak{U} = \{U_i\}_{1 \leqq l \leqq t}$ is a Stein covering of $P - A$.

Let $g: X \cap P \to \boldsymbol{C}$ be a continuous function which is holomorphic outside $S(X) \cap P$. We want to show that $g$ is holomorphic at $x$.

On $X_i = X \cap U_i$ $g$ is holomorphic by assumption. Thus we can find a holomorphic function $G_i: U_i \to \boldsymbol{C}$ with the property

$$G_i \mid X_i = g \mid X_i \qquad i = 1, \cdots, t.$$

We have

$$(G_i - G_j) \mid X_i \cap X_j = 0.$$

Let $\mathcal{I}_{X \cap P}$ be the sheaf of ideals on $P$ defining $X \cap P$. Again, if $P$ is sufficiently small, we can find a finite set of global generators; $\mathcal{I}_{X \cap P} = \mathcal{O}_p(f_1, \cdots, f_s)$.

We thus have

$$G_i - G_j = \sum_{\alpha=1}^{s} \lambda_{ij}^{\alpha} f_\alpha \quad \text{on} \quad U_i \cap U_j$$

with $\lambda_{ij}^{\alpha}$ holomorphic on $U_i \cap U_j$. Since $(G_i - G_j) + (G_j - G_k) + (G_k - G_i)$ is defined and identically 0 on $U_i \cap U_j \cap U_k$, we obtain

$$\sum_{\alpha=1}^{s} (\lambda_{ij}^{\alpha} + \lambda_{jk}^{\alpha} + \lambda_{ki}^{\alpha}) f_\alpha = 0 \quad \text{on} \quad U_i \cap U_j \cap U_k.$$

On $U_i \cap U_j \cap U_k$ define the holomorphic functions $\mu_{ijk}^{1}, \cdots, \mu_{ijk}^{s}$ by

$$\mu_{ijk}^{\alpha} = \lambda_{ij}^{\alpha} + \lambda_{jk}^{\alpha} + \lambda_{ki}^{\alpha} \qquad \alpha = 1, \cdots, s.$$

It follows that $\mu^{\alpha} = \{\mu_{ijk}^{\alpha}\}_{i,j,k}$ is a Čech 2-cocycle on the covering $\mathfrak{U}$ with values in the sheaf $\mathcal{R}(f_1, \cdots, f_s)$. Hence we get a cohomology class

$$\{\mu^{\alpha}\} \in H^2(\mathfrak{U}; \mathcal{R}(f_1, \cdots, f_s))$$

for each $\alpha$. Furthermore, as $\mathfrak{U}$ is a Stein covering of $P - A$, $H^2(\mathfrak{U}; \mathcal{R}(f_1, \cdots, f_s)) \cong H^2(P - A; \mathcal{R}(f_1, \cdots, f_s))$.

Now, the condition

$$0 < 2 < \mathrm{pf}_x \mathcal{R}(f_1, \cdots, f_s) - \dim_{\mathbf{C}} A - 1$$

is equivalent to

$$2 < \mathrm{pf}_x \mathcal{O}_X + 2 - \dim_{\mathbf{C}} A - 1$$

i.e. to

$$\mathrm{pf}_x \mathcal{O}_X - \dim_{\mathbf{C}} A > 1$$

which is our assumption. Hence we can apply lemma (16.1) to conclude that each $\{\mu_{ijk}^{\alpha}\}_{i,j,k}$ is cohomologous to zero on a suitable small $P$. This means we can find holomorphic functions

$$\sigma_{ij}^{\alpha} : U_i \cap U_j \to \mathbf{C}$$

such that

$$\mu_{ijk}^{\alpha} = \sigma_{ij}^{\alpha} + \sigma_{ik}^{\alpha} + \sigma_{ki}^{\alpha} \tag{1}$$

$$\sum_{\alpha=1}^{s} \sigma_{ij}^{\alpha} f_\alpha = 0. \tag{2}$$

Therefore, if we replace $\lambda_{ij}^{\alpha}$ by $\tau_{ij}^{\alpha} = \lambda_{ij}^{\alpha} - \sigma_{ij}^{\alpha}$ in the expression

$$G_i - G_j = \sum_{\alpha=1}^{s} \lambda_{ij}^{\alpha} f_\alpha,$$

we get a new identity of the same type

$$G_i - G_j = \sum_{\alpha=1}^{s} \tau_{ij}^{\alpha} f_\alpha$$

with the additional property

$$\tau_{ij}^{\alpha}+\tau_{jk}^{\alpha}+\tau_{ki}^{\alpha}=0 \quad \text{on} \quad U_i \cap U_j \cap U_k$$

for $1 \leqq \alpha \leqq s$.

Therefore $\{\tau_{ij}^{\alpha}\}_{i,j}$ is a 1-cocycle on $\mathfrak{U}$ for every $\alpha$ and so represents a cohomology class

$$\{\tau^{\alpha}\} \in H^1(\mathfrak{U}; \mathcal{O}) \cong H^1(P-A; \boldsymbol{C}).$$

Because $\mathrm{pf}_x\, \mathcal{O} = n$ and $\dim_{\boldsymbol{C}} A < n-2$, the condition

$$0<1<\mathrm{pf}_x\, \mathcal{O} - \dim_{\boldsymbol{C}} A - 1$$

of lemma (16.1) is satisfied.

Hence the cohomology group above is also zero, and therefore we can find holomorphic functions on each $U_i$

$$\nu_i^{\alpha}: U_i \to \boldsymbol{C} \qquad \alpha = 1, \cdots, r$$

such that

$$\tau_{ij}^{\alpha} = \nu_i^{\alpha} - \nu_j^{\alpha} \quad \text{on} \quad U_i \cap U_j.$$

Hence

$$G_i - \sum_{\alpha=1}^{s} \nu_i^{\alpha} f_{\alpha} = G_j - \sum_{\alpha=1}^{s} \nu_j^{\alpha} f_{\alpha} \quad \text{on} \quad U_i \cap U_j$$

for arbitrary $i$, $j$. Hence there ˙ a global holomorphic function $g'$ on $P-A$ that coincides with $G_i - \sum \nu_i^{\alpha} f_{\alpha}$ on $U_i$. But $\dim_{\boldsymbol{C}} A < n-2$ and so $g'$ extends to a holomorphic function $g'$ on the polycylinder $P$.

The function $g' \mid X \cap P$ is continuous and equals $g$ outside $A \cap P$. Then by continuity $g' \mid X \cap P = g$. Hence $g$ is holomorphic at $x$. This achieves the proof.

REMARK (16.5). The proof actually yields a local version of the theorem: If $X-A$ is weakly normal near $x \in A$ and $\dim_{\boldsymbol{C}}(A, x) < \mathrm{pf}_x\, \mathcal{O}(X) - 1$, then $X$ is weakly normal at $x$.

We now give three examples of weakly normal singularities. The first two will be referred to later.

EXAMPLES (16.6). Let $E_1, E_2, \cdots, E_k$ be linear subspaces of $\boldsymbol{C}^p$ *in general position*, i.e. such that

$$\operatorname{codim} \bigcap_{i=1}^{k} = \sum_{i=1}^{k} \operatorname{codim} E_i.$$

Set $V = E_1 \cup \cdots \cup E_k$. $V$ is called a *p-cross*. We show that $V$ is weakly normal.

Let $g: U \to \boldsymbol{C}$ be a local function on some open set $U \subset V$, and set $U_j = E_j \cap U$, $j = 1, 2, \cdots, k$. Assume that $g$ is continuous everywhere and holomorphic on each subset $U_j - (U_j \cap \bigcup_{i \neq j} U_i)$. By Riemann's extension theorem, $g$ is then holomorphic on each $U_j$. Let $x$ be a point of the singular locus $U \cap S(V)$. We show that $g$ is holomorphic in a neighborhood of $x$; thus $V$ is weakly normal at $x$.

Without loss of generality we may assume $x \in \cap E_i$ (otherwise $x$ sits in the intersection of a $p$-subcross) and therefore take $x$ to be the origin.

By assumption we can split the $p$ coordinates (after a linear transformation) into $k+1$ blocks $w, x_{(1)}, \cdots, x_{(k)}$, where $w$ spans $\cap E_i$ and

$$E_j = \{x_{(j)} = 0\}, \qquad (1 \leqq j \leqq k).$$

A function $G$ on $\boldsymbol{C}^p$, holomorphic near the origin, will have a power series expansion which (in multi-index notation) looks like

$$G = \sum a_{\beta\alpha_1 \cdots \alpha_k} w^\beta x_{(1)}^{\alpha_1} \cdots x_{(k)}^{\alpha_k},$$

while $g_j = g \mid U_j$ will have a power series expansion

$$g_j = \sum b_{\beta\alpha_1 \cdots 0 \cdots \alpha_k} w^\beta x_{(1)}^{\alpha_1} \cdots x_{(j)}^{0} \cdots x_{(k)}^{\alpha_k}.$$

Existence of a $G$ such that $G \mid U_j = g_j$ requires

$$a_{\beta\alpha_1 \cdots 0 \cdots \alpha_k} = b_{\beta\alpha_1 \cdots 0 \cdots \alpha_k}.$$

As $g_j \mid U_j \cap U_k = g_k \mid U_j \cap U_k$, these conditions are compatible, and thus there exists a local holomorphic function $G$ near the origin such that $G \mid U_j = g \mid U_j$ for all $j$. This completes the proof.

Set $V_l = E_1 \cup \cdots \cup E_l$ for $1 \leqq l \leqq k$. The argument above shows that the local rings of $V_l$, $V_{l-1}$, $E_l$, and $V_{l-1} \cap E_l$ at a point 0 of $\cap E_i$ fit into a short exact sequence

$$0 \to \mathcal{O}_{V_l} \to \mathcal{O}_{V_{l-1}} \oplus \mathcal{O}_{E_l} \to \mathcal{O}_{V_{l-1} \cap E_l} \to 0.$$

This gives $\operatorname{pf} \mathcal{O}_{V_l} = 1 + \operatorname{pf} \mathcal{O}_{V_{l-1} \cap E_l}$. Since $V_{l-1} \cap E_l$ is the cross of a system in general position in $E_l$, we can apply induction. One finds

$$\begin{aligned} \operatorname{pf} \mathcal{O}_{V_k} &= p - \sum \operatorname{codim} E_i + k - 1 \\ &= \dim \cap E_i + k - 1. \end{aligned}$$

Thus $V = V_k$ is never a complete intersection unless every $E_i$ is a hyperplane.

(16.7) Consider $\boldsymbol{C}^n$ $(n \geqq 2)$ with coordinates $x = (t_1, \cdots, t_{n-1}, u)$, $\boldsymbol{C}^{2n-1}$ with coordinates $y = (w_1, \cdots, w_{n-1}, z_1, \cdots, z_n)$, and the map $y = f(x)$ given by

$$\begin{cases} w_1 = t_1 \\ \cdot \\ \cdot \\ \cdot \\ w_{n-1} = t_{n-1} \\ z_1 = u^2 \\ z_2 = t_1 u \\ \cdot \\ \cdot \\ \cdot \\ z_n = t_{n-1} u \end{cases} \qquad (16.7.1)$$

Clearly $f$ is a proper polynomial map, and so $W_n = f\boldsymbol{C}^n$ is an analytic subvariety of $\boldsymbol{C}^{2n-1}$. Since $f$ is one-to-one outside the line

$$\Delta = \{t_1 = \cdots = t_{n-1} = 0\},$$

$W_n$ is of pure dimension $n$. We call $W_n$ a (complex) *Whitney umbrella.* For $n = 2$ we have the Cayley umbrella (5.5).

On the line $\Delta$ $f$ is the folding map

$$z_1 = u^2.$$

Since $S(W_n) = f\Delta$, since $f$ is an immersion outside the origin, and since im $df(a) +$ im $df(-a) = \boldsymbol{C}^{2n-1}$ for any point $a \in \Delta - \{0\}$, $W_n$ is locally isomorphic to a (double) $(2n-1)$-cross along $S(W_n) - \{0\} = f(\Delta - \{0\})$. By the preceding example $W_n$ is therefore weakly normal except possibly at the origin $0 \in \boldsymbol{C}^{2n-1}$. By theorem (16.4) it therefore suffices to verify that $\mathrm{pf}_0\, \mathcal{O}(W_n) > 1$. We will show that $\mathrm{pf}_0\, \mathcal{O}(W_n) = 2$ for all $n \geqq 2$.

For brevity set $\mathcal{O}(W_n)_0 = A$. Then $A = \boldsymbol{C}\{t_1, \cdots, t_{n-1}, u^2, t_1 u, \cdots, t_{n-1} u\}$. Set $R = \boldsymbol{C}\{t_1, \cdots, t_{n-1}, u\}$. Then $R$ is a finitely generated $A$-module,

$$R = A + Au, \qquad (1)$$

and we have a short exact sequence of $A$-modules

$$0 \to A \to R \to R/A \to 0.$$

Therefore, to show pf $A = 2$ it suffices to show pf $R/A = 1$, since evidently pf $R > 1$. We determine the $A$-module $R/A$.

There is an $A$-linear map

$$A \xrightarrow{\cdot u} R$$

which is multiplication by $u$ and whose composite with $R \to R/A$ is surjective (follows from (1))

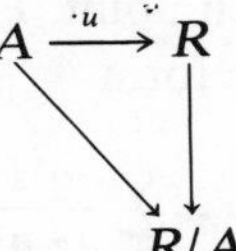

Set $c(R, A) = \ker\{A \to R/A\}$. Then

$$c(R, A) = \{a \in A \mid au \in A\} \tag{2}$$

and

$$A/c(R, A) \cong R/A \tag{3}$$

(always as $A$-modules).

Evidently $c(R, A)$ is an ideal of the ring $A$, but it is even an ideal of the larger ring $R$, since by (1) and (2) we have

$$c(R, A) = \{a \in A \mid Ra \subset A\}. \tag{4}$$

Let $\mathfrak{p} \subset A$ be the prime ideal generated by $t_1, \cdots, t_{n-1}, t_1u, \cdots, t_{n-1}u$

$$\mathfrak{p} = (t_1, \cdots, t_{n-1}, t_1u, \cdots, t_{n-1}u)A.$$

Then $\mathfrak{p}$ is in fact an $R$-ideal since $u\mathfrak{p} \subset \mathfrak{p}$. Evidently $\mathfrak{p} \subset c(R, A)$. We claim that $\mathfrak{p} = c(R, A)$, so that $R/A \cong A/\mathfrak{p} \cong \mathbf{C}\{u^2\}$.

To see this localize at $\mathfrak{p}$. In fact (cf. [3] p. 46)

$$\mathfrak{p} \supset c(R/A) \Leftrightarrow (A/c(R, A))_{\mathfrak{p}} \neq 0$$
$$\Leftrightarrow (R/A)_{\mathfrak{p}} \neq 0,$$

the second equivalence by (3). Now

$$A/\mathfrak{p} \cong \mathbf{C}\{u^2\}, \qquad R/\mathfrak{p} \cong \mathbf{C}\{u\},$$

and so

$$A_{\mathfrak{p}}/\mathfrak{p}A_{\mathfrak{p}} \to R_{\mathfrak{p}}/\mathfrak{p}R_{\mathfrak{p}}$$

is not an isomorphism, confirming that $0 \neq R_{\mathfrak{p}}/A_{\mathfrak{p}} \cong (R/A_{\mathfrak{p}})$. This completes the proof.

Note that since $\mathrm{pf}_0\, \mathcal{O}(W_n) = 2$, $W_n$ is never a complete intersection except in the case $n = 2$ (the Cayley umbrella).

(16.8) Let $F$ be a holomorphic function defined on some open set $G \subset \mathbf{C}^{n+1}$ and set

$$X = \{z \in G \mid F(z) = 0\}.$$

Assume that $F$ has no multiple factors so that

$$S(X) = \left\{ z \in X \,\middle|\, \frac{\partial F}{\partial z_1}(z) = \cdots = \frac{\partial F}{\partial z_{n+1}}(z) = 0 \right\}$$

is of codimension $\geqq 1$ in $X$ at each point. For $z \in S(X)$ consider the Hessean of $F$ at $z$, which is the quadratic form

$$d^2F(z)u = \sum_{i,j} \frac{\partial^2 F}{\partial z_i \partial z_j}(z) u_i u_j.$$

Set

$$S^2(X)=\{z\in S(X)\mid \text{rk } d^2F(z)\leqq 1\}.$$

This is an analytic subset of $S(X)$ and hence of $X$.

(16.8.1) *Suppose* $\dim_{\boldsymbol{C}} S^2(X)\leqq n-2$. *Then $X$ is weakly normal.*

PROOF. Since at any point $x\in X$ $\text{pf}_x\, \mathcal{O}(X)=n$, it suffices to show that $X$ is weakly normal outside $S^2(X)$. So take $x\in S(X)-S^2(X)$. For the same reason it suffices to check the case where $(S(X), x)$ has irreducible components of dimension $n-1$ only. Then $\text{rk } d^2F(x)=2$, and so with suitable local coordinates $z_1, \cdots, z_{n+1}$ we have

$$F(z)=az_1z_2+O(3) \qquad (a\neq 0).$$

This shows that $X$ is locally isomorphic to the $(n+1)$-cross $\{z_1=0\}\cup\{z_2=0\}$ at $x$. The claim now follows from (16.6).

## §4. Transverse singularities

### 17. Whitney's semicanonical form

Here and in the sequel $N$ and $P$ shall be smooth manifolds of dimensions $n$ and $p$, respectively, with $n\leqq p$.

(17.1) We denote by

$$\pi: J(N, P)\to N\times P$$

the canonical smooth vector bundle over $N\times P$ whose fiber at $(x, y)$ is the linear space $\text{Hom}\,(T_xN, T_yP)$. This is a bundle of fiber type $\text{Hom}\,(\boldsymbol{R}^n, \boldsymbol{R}^p)$ and structure group $GL(\boldsymbol{R}^n)\times GL(\boldsymbol{R}^p)$, acting on $\text{Hom}\,(\boldsymbol{R}^n, \boldsymbol{R}^p)$ by conjugation (substitution)

$$(Q, R)A=RAQ^{-1}.$$

Of course, $J(N, P)$ is just the 1. order jet bundle over $N\times P$, usually denoted $J^1(N, P)$. To further simplify the notation we set $\text{Hom}\,(\boldsymbol{R}^n, \boldsymbol{R}^p)=J(n, p)$.

For $i\geqq 0$ let $\Sigma^i=\Sigma^i(n, p)$ be the subset of $J(n, p)$ of linear maps of kernel rank $i$. Then $\Sigma^i$ is a (regular) submanifold of $J(n, p)$, invariant under the action of the structure group. Since $GL(\boldsymbol{R}^n)\times GL(\boldsymbol{R}^p)$ acts transitively on $\Sigma^i$, it is in fact an orbit, so that $J(n, p)$ has orbit structure precisely $\{\Sigma^i\}_{0\leqq i\leqq n}$.

For the submanifold collection $\{\Sigma^i\}$ the following properties are well known and readily verified

(i) codim $\Sigma^i = i(i+e)$, where $e = p - n$ (the *excess*).

(ii) $\overline{\Sigma^i} = \Sigma^i \cup \Sigma^{i+1} \cup \cdots \cup \Sigma^n$.

(iii) $\Sigma^0$ is open and dense.

In addition the following holds, cf. [13],

(iv) Each pair of strata $(\Sigma^i, \Sigma^{i+k})$ satisfies the Whitney regularity conditions.

Again for $i \geqq 0$ let $\Sigma^i = \Sigma^i(N, P)$ be the subset of $J(N, P)$ of homomorphisms of kernel rank $i$. It follows that $\Sigma^i$ is submanifold of $J(N, P)$ and a smooth subbundle over $N \times P$, and that the family $\{\Sigma^i\}_{0 \leqq i \leqq n}$ satisfies (i)–(iv) with respect to $J(N, P)$.

Composing $\pi : J(N, P) \to N \times P$ with the projection maps onto $N$ and $P$, respectively, we obtain smooth fiber bundles

$$\sigma : J(N, P) \to N \quad \text{(source map)}$$

and

$$\tau : J(N, P) \to P \quad \text{(target map)}$$

so that $\pi = \sigma \times \tau$.

If $f : N \to P$ is a smooth map, we denote by $jf : N \to J(N, P)$ the corresponding section of $J(N, P)$ over $N$,

$$jf(x) = T_x f : T_x N \to T_{f(x)} P,$$

and by $\Sigma^i(f) \subset N$ the counterimage of $\Sigma^i \subset J(N, P)$,

$$\Sigma^i(f) = (jf)^{-1} \Sigma^i.$$

If $jf$ is transverse to all $\Sigma^i$ in $J(N, P)$, then $\{\Sigma^i(f)\}$ is a collection of submanifolds of $N$ for which (i)–(iv) is again true (with respect to $N$). We call $f$ 1. *order correct* or briefly 1-*correct* if $jf$ is transverse to all $\Sigma^i$ in $J(N, P)$. By Thoms transversality theorem the 1-correct smooth maps $N \to P$ form an open dense set $E^1(N, P)$ in $E(N, P)$ (the set of all smooth maps) in the Whitney topology.

In the sequel we repeatedly encounter splittings of the coordinate spaces $\boldsymbol{R}^n$ and $\boldsymbol{R}^{n+e}$ of $N$ and $P$, $\boldsymbol{R}^n = \boldsymbol{R}^r \times \boldsymbol{R}^s$ and $\boldsymbol{R}^{n+e} = \boldsymbol{R}^r \times \boldsymbol{R}^{s+e}$. If $x \in \boldsymbol{R}^n$ and $y \in \boldsymbol{R}^{n+e}$ we write $x = (t, u)$ and $y = (w, z)$ with $t \in \boldsymbol{R}^r$, $u \in \boldsymbol{R}^s$, $w \in \boldsymbol{R}^r$, $z \in \boldsymbol{R}^{s+e}$. Thus $x_i = t_i$ for $1 \leqq i \leqq r$ and $x_{j+r} = u_j$ for $1 \leqq j \leqq s$, etc. The following observation is due to Whitney ([13], p. 290).

Lemma (17.2). *If $f : N \to P$ is of rank $r$ and kernel rank $s = n - r$ at $z \in N$, we can select coordinates $x = (t_1, \cdots, t_r, u_1, \cdots, u_s)$ and $y = (w_1, \cdots, w_r, z_1, \cdots, z_{s+e})$ at $a$ and $f(a)$, respectively, such that in a neighborhood of $a$ $f$ is* *given by equations*

$$\begin{cases} w_1 = t_1 \\ \cdot \\ \cdot \\ \cdot \\ w_r = t_r \\ z_1 = \varphi_1(t, u) \\ \cdot \\ \cdot \\ \cdot \\ z_{s+e} = \varphi_{s+e}(t, u) \end{cases} \tag{17.2.1}$$

*with $\varphi_j$ smooth and $\varphi_j(0, 0) = \partial\varphi_j/\partial x_i(0, 0) = 0$ for all i, j.*

*Moreover, jf is transverse to $\Sigma^s$ at a if and only if the entries of the matrix*

$$\frac{\partial \varphi}{\partial u} = \begin{Bmatrix} \dfrac{\partial \varphi_1}{\partial u_1} \cdots \dfrac{\partial \varphi_1}{\partial u_s} \\ \cdot \\ \cdot \\ \cdot \\ \dfrac{\partial \varphi_{s+e}}{\partial u_1} \cdots \dfrac{\partial \varphi_{s+e}}{\partial u_s} \end{Bmatrix} \tag{16.2.2}$$

*are smoothly independent at the origin, hence can be taken in a system of local coordinates* in $\boldsymbol{R}^n$.

REMARKS (17.3). If $a \in \Sigma^s(f)$ and $jf$ is transverse to $\Sigma^s$, then codim $\Sigma^s(f) =$ codim $\Sigma^s = s(s+e)$, which requires $s(s+e) \leqq n$. This agrees with the fact that the $s(s+e)$ functions $\partial\varphi_j/\partial u_k$ are smoothly independent at the origin.

(17.4) From (16.2.1) we have for the jacobian of $f$ (in a neighborhood of $a$)

$$\frac{\partial f}{\partial x} = \begin{Bmatrix} I_r & 0 \\ \dfrac{\partial \varphi}{\partial t} & \dfrac{\partial \varphi}{\partial u} \end{Bmatrix}. \tag{17.4.1}$$

This shows that any sufficiently small neighborhood $U$ of $a$ is stratified by the subsets $\Sigma^s(f)$, $\Sigma^{s-1}(f), \cdots, \Sigma^0(f)$, which are given by determinantal equations,

$$\Sigma^{s-i}(f) \cap U = \left\{ p \in U \mid \mathrm{rk}_{x(p)} \frac{\partial \varphi}{\partial u} = i \right\}.$$

If moreover $jf$ is transverse to all $\Sigma^{s-i}$ on $U$, then the functions $\partial\varphi_j/\partial u_k = v_{jk}$ form part of a local coordinate system, say $\{v_{jk}, v_l\}$, with $1 \leqq l \leqq n - s(s+e)$. Hence in this case $(\Sigma^s(f), \Sigma^{s-i}(f), \cdots, \Sigma^0(f)) \cap U$ is smoothly equivalent to the cylinder $(\Sigma^s, \Sigma^{s-1}, \cdots, \Sigma^0) \times \boldsymbol{R}^{n-s(s+e)}$ over the standard stratification $(\Sigma^s, \Sigma^{s-1}, \cdots, \Sigma^0)$ of $\operatorname{Hom}(\boldsymbol{R}^s, \boldsymbol{R}^{s+e})$. In particular, for $f \in E^1(N, P)$ the stratification of $N$ by the $\Sigma^i(f)$ has no local invariants (except the corank $i$).

## 18. Multitransversality

Given a smooth map $f: N \to P$ we say that $f$ has *k-fold normal crossings*, briefly *k-normal crossings*, if for every set of $l \leqq k$ points $x_1, \cdots, x_l$ in $N$ with $f(x_1) = \cdots = f(x_l)$ the family

$$\{f_* T_{x_j} N\}_{1 \leqq j \leqq l}$$

is in general position in $T_y P$, $y = f(x_i)$, (cf. (16.6)). Equivalently, $f$ has $k$-normal crossings if for $l \leqq k$ the $l$-fold map

$$_l f: {}_l N \to P^l$$

is transversal to the diagonal $\Delta \subset P^l$. Here $_l N \subset N^l$ is the (open) subset of $l$-tuples of distinct points of $N$ and $_l f = f^l \mid {}_l N$.

Let $\Sigma^{i_1}(f), \cdots, \Sigma^{i_l}(f)$ be the corank strata of $f$ through $x_1, \cdots, x_l$, respectively. We say that $f$ is $k$-fold 1-*correct* if $f$ is 1-correct and if for $l \leqq k$ (in the situation above) the family

$$\{f_* T_{x_j} \Sigma^{i_j}(f)\}_{1 \leqq j \leqq l}$$

is in general position in $T_y P$. This can be reformulated as follows.

(18.1) The target map $\tau: J(N, P) \to P$ restricts to a fiber bundle projection $\Sigma^i(N, P) \to P$ for each $i$. Hence, for any $l$-tuple of indices $i_1, \cdots, i_l$ the fibered product $\Sigma^{i_1} \times_P \cdots \times_P \Sigma^{i_l} \subset J(N, P) \times \cdots \times J(N, P)$ is a bundle over $P$ and a submanifold of $J(N, P)^l$. Let $\Sigma^{i_1, \cdots, i_l}$ be the intersection of $\Sigma^{i_1} \times_P \cdots \times_P \Sigma^{i_l}$ with the open subset $_l J(N, P) \subset J(N, P)^l$, where $_l J(N, P)$ is the $l$-fold jet bundle space,

$$_l J(N, P) = (\sigma^l)^{-1} {}_l N.$$

Then we have the diagram

$$\begin{array}{ccccc} \Sigma^{i_1, \cdots, i_l} \subset & _l J(N, P) & \subset & J(N, P)^l \\ & \downarrow {}_l\sigma & & \downarrow \sigma^l \\ & _l N & \subset & N^l \end{array}$$

For smooth $f: N \to P$ denote by

$$_l jf: {}_l N \to {}_l J(N, P)$$

the $l$-fold jet extension of $f$ (the restriction of $(jf)^l : N^l \to J(N, P)^l$) and set

$$\Sigma^{i_1, \cdots, i_l}(f) = ({}_ljf)^{-1}\Sigma^{i_1, \cdots, i_l}.$$

Then $(x_1, \cdots, x_l) \in {}_lN$ is in $\Sigma^{i_1, \cdots, i_l}(f)$ if and only if $x_1 \in \Sigma^{i_1}(f), \cdots, x_l \in \Sigma^{i_l}(f)$ and $f(x_1) = \cdots = f(x_l)$.

Let $x = (x_1, \cdots, x_l)$ be a point of ${}_lN$. The $l$-fold jet extension ${}_ljf$ is seen to be transverse to $\Sigma^{i_1, \cdots, i_l}$ at $x$ if and only if

(1) $jf$ is transverse to $\Sigma^{i_j}$ at $x_j$ $(1 \leqq j \leqq l)$

(2) $f^l \mid \Sigma^{i_1}(f) \times \cdots \times \Sigma^{i_l}(f)$ is transverse to the diagonal of $P^l$ at $x$.

Thus the map $f$ is $k$-fold 1-correct if and only if for $l = 1, 2, \cdots, k$ its $l$-fold (1. order) jet extension is everywhere transverse to the canonical strata $\Sigma^{i_1, \cdots, i_l}$ in ${}_lJ(N, P)$. For $k = 1$ we get back the original 1-correct maps.

The condition of $k$-fold normal crossings give rise to subclasses $E^{0,k}(N, P)$ of smooth maps $N \to P$, and to a descending sequence

$$E(N, P) \supset E^{0,2}(N, P) \supset E^{0,3}(N, P) \supset \cdots.$$

If $n < P$, this sequence stabilizes at stage $p/(p-n)$ (or below), since by transversality a map $N \to P$ with $k$-fold normal crossings can have no more than $p/(p-n)$ elements in a fiber when $k \geqq p/(p-n)$. For consistency set $E(N, P) = E^0(N, P) = E^{0,1}(N, P)$.

Similarly the condition of $k$-fold 1-correctness defines a subclass of 1-correct maps $E^{1,k}(N, P)$, and we get a sequence

$$E^1(N, P) \supset E^{1,2}(N, P) \supset E^{1,3}(N, P) \supset \cdots.$$

We set $E^1(N, P) = E^{1,1}(N, P)$. Then $E^{1,k}(N, P) \subset E^1(N, P) \cap E^{0,k}(N, P)$ for $k \geqq 1$.

Again, if $n < p$, the sequence contains at most $p/(p-n)$ different terms.

(18.2) By induction one defines the collection $E^s(N, P)$ of maps $N \to P$ correct *up to s. order*, briefly *s-correct* maps, for $s \geqq 1$, and the subcollection $E^{s,k}(N, P)$ of *k-fold s*-correct maps. Then

$$E(N, P) \supset E^1(N, P) \supset E^2(N, P) \supset \cdots$$

and

$$E^s(N, P) \supset E^{s,2}(N, P) \supset E^{s,3}(N, P) \supset \cdots.$$

For $n < p$ the first sequence stabilizes at stage $n+1$ and the second, as before, at a stage $\leqq p/(p-n)$. This follows from the transversality requirements (cf. [1] for details).

The members of $B(N, P) = E^{n+1}(N, P)$ are called *correct* maps (or Thom-Boardman maps) and those of $\bigcap_s B^s(N, P) = \bigcap_s E^{n+1,s}(N, P)$ *excellent* maps.

## 19. Subtransversality

If a smooth map $f:N\to P$ is 1-correct, then in fact $f$ has 2-normal crossings for nearby points. That is, there is a neighborhood $W$ of the diagonal $\Delta_N$ in $N\times N$ such that ${}_2f$ is transversal to $\Delta_P\subset P\times P$ on $W-\Delta_N$. We won't give the proof of this fact; instead we investigate what happens along the diagonal $\Delta_N$.

(19.1) Let $X$ be a smooth manifold $a\in X$. $\mathscr{E}(X)$ the sheaf of smooth functions on $X$ and $I\subset\mathscr{E}(X)_a$ an ideal in the local ring $E(X)_a$. We say that $I$ is *regular of codimension* $k$ (or $k$-regular) if either $I=\mathscr{E}(X)_a$ or $I\neq\mathscr{E}(X)_a$ and admits a set of $k$ generators $h_1,\cdots,h_k$ such that $dh_1(a)\wedge\cdots\wedge dh_k(a)\neq 0$. If $I\subsetneq\mathscr{E}(X)_a$ and if it is regular of codimension $k$ then there is a germ of a smooth submanifold $(V,A)$ at $a$, of codimension $k$ such that

$$I=\{f\in\mathscr{E}(X)_a\,|f|\,(V,a)=0\}.$$

Conversely any such ideal is regular of codimension $k$.

Let $X$, $Y$ be smooth manifolds and let $B\subset Y$ be a smooth submanifold of $Y$ of codimension $k$. Let $\mathscr{I}(B)$ denote the sheaf of ideals of germs of smooth functions on $Y$ vanishing on $B$.

Let $f:X\to Y$ be a smooth map. Then $f$ is transversal to $B$ at $a\in X$ if and only if

$$\mathscr{E}(X)_a\cdot f^*\mathscr{I}(B)_b\qquad b=f(a)$$

is a regular ideal of codimension $k$.

Let $X$, $Y$ be smooth manifolds of pure dimensions $n$ and $p$ and let $A\subset X$, $B\subset Y$ be closed submanifolds of $X$ and $Y$ respectively. Consider the set of smooth maps from $X$ to $Y$ that send $A$ into $B$:

$$E(X,A;Y,B)=\{f\in E(X,Y)\mid f(A)\subset B\}.$$

This is a closed subset of $E(X,Y)$ in the Whitney topology.

Let $f\in E(X,A;Y,B)$ and let $a\in A$, $f(a)=b\in B$. We have

$$\mathscr{E}(X)_a\cdot f^*\mathscr{I}(B)_b\subset\mathscr{I}(A)_a$$

with obvious notations. Consider the conductor ideal

$$c_f(\mathscr{I}(A)_a,\mathscr{I}(B)_b)=\{h\in\mathscr{E}(X)_a\mid h\mathscr{I}(A)_a\subset\mathscr{E}(X)_a\cdot f^*(\mathscr{I}(B)_b)\}.$$

We say that $f\in E(X,A;Y,B)$ is *subtransversal* to $B$ at $a\in A$ if $c_f(\mathscr{I}(A)_a,\mathscr{I}(B)_b)$ is a regular ideal of codimension equal to the codimension of $B$ at $b$ $(b=f(a))$. We say that the map $f$ is *strongly subtransversal* to $B$ at $a\in A$ if $c_f(\mathscr{I}(A)_a,\mathscr{I}(B)_b)+\mathscr{I}(A)_a$ is regular of codimension equal to codimension $(B,b)+$codimension $(A,a)$.

For instance if codim $(A, a) = 1$ then $\mathscr{I}(A)_a = \mathscr{E}(X)_a h$ is a principal ideal. If $(B, b)$ is of codimension $k$ at $b = f(a)$ and if $\mathscr{I}(B)_b = \mathscr{E}(Y)_b(\varphi_1, \cdots, \varphi_k)$ then

$$\varphi_i \circ f = h \cdot w_i \qquad 1 \leqq i \leqq k$$

and $c_f(\mathscr{I}(A)_a, \mathscr{I}(B)_b) = \mathscr{E}(X)_a(w_1, \cdots, w_k)$. Therefore $f$ is subtransversal to $B$ at $a$ if $\mathscr{E}(X)_a(w_1, \cdots, w_k)$ is regular of codimension $k$ and strongly subtransversal to $B$ at $a$ if $\mathscr{E}(X)_a(w_1, \cdots, w_k, h)$ is regular of codimension $k+1$.

(19.2) Let $f \in E(X, A; Y, B)$. If $(A, a)$ is of codimension $>1$, we can consider the blow-up of $X$ along $A$. This is a smooth manifold $\tilde{X}$ with a surjective proper smooth map $\sigma : \tilde{X} \to X$ such that

(i) $\sigma \mid \tilde{X} - \sigma^{-1}A \to X - A$ is a diffeomorphism

(ii) $\tilde{A} = \sigma^{-1}A$ is a submanifold of $\tilde{X}$ of codimension 1.

The manifold $\tilde{X}$ is determined up to isomorphisms over $X$. Composition with $\sigma$ gives a natural map $\sigma^* : E(X, A; Y, B) \to E(\tilde{X}, \tilde{A}; Y, B)$.

We say that $f \in E(X, A; Y, B)$ is *$\sigma$-subtransversal* (*strongly $\sigma$-subtransversal*) to $B$ at $a \in A$ if $\sigma^*(f)$ is so at each point $\tilde{a} \in \sigma^{-1}\{a\}$.

We apply these considerations to the following situation. Let $f \in E(N, P)$; then $f \times f \in E(N \times N, \Delta_N;\ P \times P, \Delta_P)$.

THEOREM (19.3). *If* $f \in E(N, P)$ *is* 1-correct *then* $f \times f$ *is strongly $\sigma$-subtransversal to* $\Delta_P$ *at all points of* $\Delta_N$.

PROOF. (a) Since the statement is a local statement on $N \times N$ along $\Delta_N$ we may assume that $N = \boldsymbol{R}^n$, $P = \boldsymbol{R}^p$, and $f : \boldsymbol{R}^n \to \boldsymbol{R}^p$ is such that $f(0) = 0$. Set $p = n + e$ $(e \geqq 0)$. We may assume that $f$ is in Whitney semicanonical form with $0 \in \Sigma^s(f)$. Set $r = n - s$. Then $f$ is given by

$$\begin{cases} y_i = x_i, & (1 \leqq i \leqq r) \\ y_{r+\alpha} = \varphi_\alpha(x) \equiv \displaystyle\sum_{i,j=1}^{n} a_{ij}^{(\alpha)} x_i x_j + O(3), & (1 \leqq \alpha \leqq s + e) \end{cases}$$

with the conditions that the linear forms

$$\left\{ \sum_{j=1}^{n} a_{ij}^{(\alpha)} x_j \qquad \begin{matrix} (r+1 \leqq i \leqq n) \\ (1 \leqq \alpha \leqq s+e) \end{matrix} \right. \qquad (*)$$

are linearly independent (as $f \in E^1(N, P)$).

Let $x$ be coordinates in $\boldsymbol{R}^n$ and $x'$ coordinates in a copy of $\boldsymbol{R}^n$. In $\boldsymbol{R}^n \times \boldsymbol{R}^n$

we can choose coordinates adapted to $\Delta_{\boldsymbol{R}^n}$ by setting

$$u_i = x_i - x'_i \qquad (1 \leqq i \leqq n)$$

$$u'_i = x_i + x'_i \qquad (1 \leqq i \leqq n)$$

so that $\Delta_{\boldsymbol{R}^n}$ has the equations $u_i = 0$, $1 \leqq i \leqq n$.

Therefore $x_i = \frac{1}{2}(u_i + u'_i)$ and $x'_i = \frac{1}{2}(u'_i - u_i)$ for $1 \leqq i \leqq n$ so that

$$x_i x_j - x'_i x'_j = \tfrac{1}{2}(u'_i u_j + u_i u'_j).$$

The ideal $\mathscr{E}(\boldsymbol{R}^n \times \boldsymbol{R}^n)(f \times f)^* \mathscr{I}(\Delta_{\boldsymbol{R}^p})$ is generated over $\mathscr{E}(\boldsymbol{R}^n \times \boldsymbol{R}^n)$ by

$$x_1 - x'_1, \cdots, x_r - x_r, \varphi_1(x) - \varphi_1(x'), \cdots, \varphi_{s+e}(x) - \varphi_{s+e}(x').$$

Therefore in the coordinates $u$, $u'$ this ideal is given by

$$\mathscr{E}(\boldsymbol{R}^n \times \boldsymbol{R}^n)(u_1, \cdots, u_r, \{\Sigma a_{ij}^{(\alpha)} u_i u'_j + O_3(u, u')\}_{1 \leqq \alpha \leqq s+e}).$$

(b) Now we blow up the diagonal $\Delta_{\boldsymbol{R}^n}$ of $\boldsymbol{R}^n \times \boldsymbol{R}^n$. We have

$$\widetilde{\boldsymbol{R}^n \times \boldsymbol{R}^n} = \{(u, u', \tau) \in \boldsymbol{R}^n \times \boldsymbol{R}^n \times P_{n-1} \mid u_i \tau_j = u_j \tau_i,\ 1 \leqq i \leqq n\}$$

where $\tau = (\tau_1, \cdots, \tau_n)$ are the homogeneous coordinates in $P_{n-1}(\boldsymbol{R})$. The manifold $\widetilde{\boldsymbol{R}^n \times \boldsymbol{R}^n}$ is covered with $n$ coordinate patches

$$W_i = \{(u, u', \tau) \in \widetilde{\boldsymbol{R}^n \times \boldsymbol{R}^n} \mid \tau_i \neq 0\}$$

on which $u_i$, $\tau_j/\tau_i$ for $j \neq i$ and $u'$ are the coordinates. With respect to the map $f$, we have to distinguish these into two types, those with $1 \leqq i \leqq r$ and those with $r < i \leqq n$. Typical of the first type is $W_1$, where the coordinates are

$$u_1, \frac{\tau_2}{\tau_1}, \cdots, \frac{\tau_n}{\tau_1}, u'_1, \cdots, u'_n;$$

while $W_n$ is typical of the second type, and on it the coordinates are

$$u_n, \frac{\tau_1}{\tau_n}, \cdots, \frac{\tau_{n-1}}{\tau_n}, u'_1, \cdots, u'_n.$$

If $\sigma: \widetilde{\boldsymbol{R}^n \times \boldsymbol{R}^n} \to \boldsymbol{R}^n \times \boldsymbol{R}^n$, then $\sigma^{-1}\Delta_{\boldsymbol{R}^n} = \tilde{\Delta}$ is given in the coordinate patch $W_i$ by the equation $u_i = 0$. At each point $w \in \tilde{\Delta}$ we have to consider the conductor

$$c(\mathscr{I}(\tilde{\Delta})_w, \mathscr{I}(\Delta_{\boldsymbol{R}^p})_v) \quad \text{for} \quad v = (f \times f \circ \sigma)(w).$$

Claim 1. For $1 \leqq i \leqq r$ and $w \in W_i \cap \tilde{\Delta}$

$$c(\mathscr{I}(\tilde{\Delta})_w, \mathscr{I}(\Delta_{\boldsymbol{R}^p})_v) = \mathscr{E}(\widetilde{\boldsymbol{R}^n \times \boldsymbol{R}^n})_w$$

(i.e. the conductor is the unit ideal).

To fix the ideas let us suppose that $i = 1$. Then the pull-back of the ideal of the diagonal $\Delta_{\boldsymbol{R}^p}$ is generated by

$$u_1, u_1\frac{\tau_2}{\tau_1}, \cdots, u_1\frac{\tau_r}{\tau_1}, \quad \text{and} \quad u_1\left\{\sum_{j=1}^{n}\frac{\tau_j}{\tau_1}\left(\sum a_{jr}^{(\alpha)}u_r' + O_2\right)\right\} \qquad (1 \leqq \alpha \leqq s+e).$$

As $\mathscr{I}(\tilde{\Delta})$ is generated by $u_1$, this is contained in the pull-back of the ideal of the diagonal $\Delta_{\boldsymbol{R}^p}$ and therefore the unit 1 is contained in the conductor.

CLAIM 2. For $r<i\leqq n$ and $w\in W_i\cap\mathscr{C}$ we have for the conductor

$$c=\mathscr{E}(\widetilde{\boldsymbol{R}^n\times\boldsymbol{R}^n})_w\left(\frac{\tau_1}{\tau_i}, \cdots, \frac{\tau_r}{\tau_i}, \left\{\sum_{k=r+1}^{n}\frac{\tau_k}{\tau_i}\left(\sum_{j=1}^{n} a_{jk}^{(\alpha)}u_j' + O(2)\right)\right\} \quad 1\leqq\alpha\leqq s+e\right).$$

Again, to fix the ideas, let us suppose $i=n$. The pull-back of the ideal of the diagonal $\Delta_{\boldsymbol{R}^p}$ is then generated by

$$u_n\frac{\tau_1}{\tau_n}, \cdots, u_n\frac{\tau_r}{\tau_n} \text{ and by the functions}$$

$$u_n\sum_{i=1}^{n}\frac{\tau_i}{\tau_n}\left(\sum a_{ij}^{(\alpha)}u_j' + O(2)\right) \qquad (1\leqq\alpha\leqq s+e)$$

and therefore $c$ is generated by

$$\frac{\tau_1}{\tau_n}, \cdots, \frac{\tau_r}{\tau_n} \text{ and by the functions}$$

$$\sum_{i=1}^{n}\frac{\tau_i}{\tau_n}\left(\sum a_{ij}^{(\alpha)}u_j' + (2)\right) \qquad (1\leqq\alpha\leqq s+e).$$

Now taking into account the first set of $r$ generators, we see that the conductor $c$ is also generated by the functions

$$\frac{\tau_1}{\tau_n}, \cdots, \frac{\tau_r}{\tau_n} \quad \text{and by the functions}$$

$$\sum_{i=r+1}^{n}\frac{\tau_i}{\tau_n}\left(\sum a_{ij}^{(\alpha)}u_j' + O(2)\right) \qquad (1\leqq\alpha\leqq s+e).$$

This proves our second claim.

(c) We went to show that we have strong $\sigma$-subtransversality over $\Delta_{\boldsymbol{R}^n}\subset\boldsymbol{R}^n\times\boldsymbol{R}^n$, when $f\in E^1(N,P)$. At a point $(a,a)\in\Delta_{\boldsymbol{R}^n}$ we may choose local coordinates in source and target space to bring $f$ into Whitney semicanonical form. It is therefore enough to prove this statement, with the notations used up to now, over the point $(0,0)\in\boldsymbol{R}^n\times\boldsymbol{R}^n$.

Thus let us consider one of the covering charts of $\boldsymbol{R}^n\times\boldsymbol{R}^n$ on which $c$ is not the unit ideal. To fix the ideas let $W_n$ be the chart we consider. We can make a change of variables

$$u_n^* = u_n$$
$$\left(\frac{\tau_i}{\tau_n}\right)^* = \frac{\tau_i}{\tau_n} \qquad (1 \leqq i \leqq n-1)$$
$$w_i^{(\alpha)} = \sum a_{ij}^{(\alpha)} u_j' + O_2\left(u_n, \frac{\tau_i}{\tau_n}, u'\right) \qquad \begin{cases} r+1 \leqq i \leqq n \\ 1 \leqq \alpha \leqq s+e \end{cases}$$
$$w_l' = \sum b_{lj} u_j' \qquad (1 \leqq l \leqq n - s(s+e))$$

where the functions $\sum b_{lj} u_j'$ are suitable linear functions of the coordinates $u_j'$. This is possible because of condition (*) (which was a consequence of the assumption $f \in E^1$). This change of variables is permissible all over the region where $u_n = u_1' = \cdots = u_n' = 0$, which is the part of $\widetilde{\boldsymbol{R}^n \times \boldsymbol{R}^n}$ over $(0, 0) \in \boldsymbol{R}^n \times \boldsymbol{R}^n$. In fact the term $O_2$ vanishes of second order when $u_n = u_1' = \cdots = u_n' = 0$ by the way it is obtained:

$$\varphi_\alpha(x) - \varphi_\alpha(x') = \sum_1^n u_i \left( \sum a_{ij}^{(\alpha)} u_j + \mu_i^\alpha(u, u') \right)$$

with $\mu_i^\alpha(u, u')$ of order two so that the term $O_2$ is of the form

$$\mu_i^\alpha\left(u_n \frac{\tau_1}{\tau_n}, \cdots, u_n \frac{\tau_{n-1}}{\tau_n}, u_n, u_1', \cdots, u_n'\right).$$

In these new coordinates the generators of the conductor $c$ take the form

$$\frac{\tau_1}{\tau_n}, \cdots, \frac{\tau_r}{\tau_n}$$

$$w_n^{(\alpha)} + \sum_{i=r+1}^{n-1} w_i^{(\alpha)} \frac{\tau_i}{\tau_n} \qquad (1 \leqq \alpha \leqq s+e).$$

Now at a point $w$ where $u_n = u_1' = \cdots = u_n' = 0$ the coordinates $w_i^{(\alpha)}$ vanish and

$$\left\{ d\left(\frac{\tau_1}{\tau_n}\right) \wedge \cdots \wedge d\left(\frac{\tau_r}{\tau_n}\right) \wedge \bigwedge_{\alpha=1}^{s+e} d\left(w_n^{(\alpha)} + \sum_{i=r+1}^{n-1} w_i^{(\alpha)} \frac{\tau_i}{\tau_n}\right) \wedge du_n \right\}_w \neq 0.$$

This shows that $E(\widetilde{\boldsymbol{R}^n \times \boldsymbol{R}^n})_w(c, u_n)$ is regular of codimension $p+1$, as we wanted to prove.

Remark (19.4). Theorem (19.3) also holds for holomorphic maps $f: N \to P$ between two complex manifolds of complex dimensions $n \leqq p$ respectively, provided $f$ satisfies the corresponding (complex) transversality conditions.

## 20. Generic properties of real-analytic maps

Let $N$ and $P$ be real-analytic manifolds and $A(N, P)$ and $E(N, P)$ the set of analytic and smooth maps from $N$ to $P$, respectively. With the Whitney topology $E(N, P)$ is a Baire space containing $A(N, P)$ as a dense subset; the last claim by a theorem of Grauert, [6]. Henceforth we give $A(N, P)$ the topology induced from $E(N, P)$.

We call a property $P$ of analytic maps $N \to P$ *generic* if it is satisfied for the elements of a dense open subset of $A(N, P)$ (or we say that a generic map has property $P$). More specifically we say that $P$ is generic for *proper* maps or *locally proper* maps if it holds for the elements of a dense open subset of $A_{\mathrm{pr}}(N, P)$ or $A_{\mathrm{lpr}}(N, P)$, respectively (always with the induced Whitney topology). Note that $A_{\mathrm{pr}}(N, P) \subset A_{\mathrm{lpr}}(N, P)$ are *open* subsets of $A(N, P)$. In fact $E_{\mathrm{pr}}(N, P)$ is open in $E(N, P)$ (even in the graph topology) and $E_{\mathrm{lpr}}(N, P) = \bigcup E_{\mathrm{pr}}(N, U)$, where $U$ runs through the open subsets of $P$.

Let $n$ and $p$ be the dimensions of $N$ and $P$. Throughout this section we assume $n < p$.

Proposition (20.1). *Assume $n < p$. Then the following properties are generic*

(i) *$f$ is correct* (cf. (18.2)).

(ii) *$f$ has bounded deficiency, i.e. the function $\delta(x) = \delta(f, x)$ defined by*

$$\delta(x) = \dim_{\mathbf{R}} \mathscr{A}_x / \mathscr{A}_x \cdot f^* \mathfrak{m}_{f(x)}$$

*is bounded on $N$.*

By (i) a generic map $N \to P$ is in particular 1-correct, as was already pointed out in 16. We leave the proof of the sharper fact to the reader; it will not be used in the sequel. Claim (ii) is a consequence of theorem (28.4) of §5.

Proposition (20.2). *Assume $n < p$. The following properties are generic for locally proper maps*

(i) *$f$ is excellent* (cf. (18.2)).

(ii) *$f$ has a locally proper complexification $\tilde{f}: \tilde{N} \to \tilde{P}$ with $\tilde{N}$ a Stein manifold.*

The fact that property (i) holds for a dense subset of $E_{\mathrm{pr}}(N, P)$ (and even for a dense subset of $E(N, P)$) is a consequence of Mather's multitransversality theorem, [7]. However, the fact that (i) holds for an *open* dense subset of $E_{\mathrm{pr}}(N, P)$ requires the techniques applied in Mather's proof of the genericity of topological stability, [9]. We give no details here.

From this fact follows that (i) is a generic property for $A_{\mathrm{pr}}(N, P)$. Since this is true for arbitrary $P$, (i) is generic for $A_{\mathrm{lpr}}(N, P) = \bigcup A_{\mathrm{pr}}(N, U)$.

In (ii) we may assume that $f$ is of bounded deficiency (by the preceding lemma). If $s$ is a bound for $\delta(f, -)$, then $\mathfrak{m}_x^s \subset \mathscr{A}_x \cdot f^* \mathfrak{m}_{f(x)}$ for all $x \in N$ (§5, sec. 26). But then $f$ admits a locally proper complexification $\tilde{f}: \tilde{N} \to \tilde{P}$ with $N$ Stein by proposition (12.5).

REMARK (20.3). By (20.2) a generic locally proper map $N \to P$ has in particular 2-normal crossings. In fact one can show directly that ${}_2f$ is transverse to $\Delta_p$ in $P \times P$ for a dense open set of 1-correct maps $f$ from $E_{\text{pr}}(N, P)$. The restriction to (locally) proper maps cannot be omitted.

Let $f$ and $\tilde{f}$ be as above. We denote by $\tilde{\Gamma}_r = \Gamma_r(\tilde{f})$ the $r$-fold fibered product of $\tilde{N}$ by $\tilde{f}$. Let $\tilde{\Delta}^r = \Delta^r_{\tilde{P}}$ and $\tilde{\mathbf{\Delta}}^r = \mathbf{\Delta}^r_{\tilde{N}}$ be respectively the diagonal and the fat diagonal of $\tilde{P}^r$ and $\tilde{N}^r$. (Thus $\mathbf{\Delta}^r_{\tilde{N}}$ is the set of $r$-tuples of $\tilde{N}^r$ with repetitions, i.e. the complement of ${}_r\tilde{N}$ in $\tilde{N}^r$.) By definition $\tilde{\Gamma}_r = (\tilde{f}^r)^{-1} \tilde{\Delta}^r$ and so is a complex-analytic subset of $\tilde{N}^r$. We define the $r$-fold coincidence set $\tilde{\Pi}_r = \Pi_r(\tilde{f})$ to be the union of all irreducible components of $\tilde{\Gamma}_r$ not contained in the fat diagonal $\tilde{\mathbf{\Delta}}^r$. Then

$$\tilde{\Pi}_r = \overline{\tilde{\Gamma}_r - \tilde{\Gamma}_r \cap \tilde{\mathbf{\Delta}}^r}.$$

$\tilde{\Pi}_r$ is a complex-analytic set with a real part $\tilde{\Pi}_r \cap N^r$. Define $\Gamma_r = \Gamma_r(f)$ and $\Pi_r \subset \Gamma_r$ analogously. Then

$$\overline{\Gamma_r - \Gamma_r \cap \mathbf{\Delta}^r} \subset \Pi_r \subset \tilde{\Pi}_r \cap N^r.$$

Let $\text{pr}: \tilde{N}^r \to \tilde{N}$ be the projection to the first factor. Then for any subset $K \subset \tilde{N}$ we have

$$(\text{pr} \mid \tilde{\Pi}_r)^{-1} K = \tilde{\Pi}_r \cap (K \times \tilde{f}^{-1} \tilde{f} K \times \cdots \times \tilde{f}^{-1} \tilde{f} K).$$

It follows that $\text{pr} \mid \tilde{\Pi}_r$ pulls back compact subsets to compact subsets, and so $\text{pr} \mid \tilde{\Pi}_r$ is proper. Since $\tilde{\Pi}_r$ is Stein (as analytic subset of a Stein space), $\text{pr} \mid \tilde{\Pi}_r$ has finite fibers and preserves dimension (cf. (10.2)).

Set $\tilde{D}_r = D_r(\tilde{f}) = \text{pr}\ \tilde{\pi}_r$. It follows that $\tilde{D}_r$ is a complex-analytic set of the same dimension as $\tilde{\pi}_r$, really imbedded in $\tilde{N}$. Thus we get filtrations of $\tilde{N}$ and $N$ by analytic subsets,

$$\tilde{N} \supset \tilde{D}_2 \supset \tilde{D}_3 \supset \cdots$$

and

$$N \supset \tilde{D}_2 \cap N \supset \tilde{D}_3 \cap N \supset \cdots.$$

We can extend our definitions and notations from section (18.2) to the real-analytic and complex-analytic cases, thus defining subclasses of ($s$-correct, $k$-fold $s$-correct, $\cdots$) analytic maps $A^{s,k}(N, P) \subset A^s(N, P) \subset A(N, P)$ and $H^{s,k}(\tilde{N}, \tilde{P}) \subset H^s(\tilde{N}, \tilde{P}) \subset H(\tilde{N}, \tilde{P})$.

PROPOSITION (20.4). *Let $f: N \to P$ be a real-analytic map of manifolds of dimensions $n < p$, and let $\tilde{f}: \tilde{N} \to \tilde{P}$ be a complexification of $f$. Restricting $\tilde{f}$ to a* *sufficiently small (Stein) neighborhood of N in $\tilde{N}$ we have*

(i) *If f is locally proper and of finite (bounded) deficiency, so is $\tilde{f}$.*

(ii) *If $f \in A^{s,k}$, then $\tilde{f} \in H^{s,k}$.*

(iii) *If f has k-normal crossings, then* $\operatorname{codim}_{\boldsymbol{C}} \tilde{D}_r = (r-1)(p-n)$ *for* $2 \leqq r \leqq k$.

(iv) *If f has 2-normal crossings, then $\tilde{f}$ is generally one-to-one, and if moreover f is 1-correct, then $\tilde{\Pi}_2$ is a germ complexification of $\Pi_2$.*

PROOF. $\tilde{f}$ is locally proper, again by proposition (12.5), and of finite deficiency by the remark preceding lemma (12.4). This proves (i).

Claim (ii) follows immediately from Boardman's characterization of the (repeated) corank strata, cf. [1]. However, one can also proceed inductively. First observe that $j\tilde{f}: \tilde{N} \to J(\tilde{N}, \tilde{P})$ is a holomorphic extension of $jf: N \to J(N, P)$, and that $\Sigma^i(\tilde{N}, \tilde{P}) \subset J(\tilde{N}, \tilde{P})$ are (faithful) complexifications of $\Sigma^i(N, P) \subset J(N, P)$. Then it is immediate that $f \in A^1$ implies $\tilde{f} \in H^1$ (in a sufficiently small neighborhood of $N$). Moreover $\Sigma^i(\tilde{f})$ is a complexification of $\Sigma^i(f)$ in $\tilde{N}$ (for dimensional reasons). Hence we can restrict $f$ and $\tilde{f}$ to $N_2 = \Sigma^i(f)$ and $\tilde{N}_2 = \Sigma^i(\tilde{f})$ and repeat the procedure. It follows that $f \in A^s$ implies $\tilde{f} \in H^s$. For the $k$-fold case we use (2) of (18.1) in addition. This proves (ii).

If $f$ has $k$-normal crossings, so has $\tilde{f}$ by property (ii). Then, for $r \leqq k$, $\tilde{\Pi}_r$ is a manifold outside $\tilde{\Delta}^r$ of dimension $p - r(p-n)$ (eventually empty). Then $\tilde{D}_r$ has dimension $p - r(p-n)$ and therefore codimension $n - p + r(p-n) = (r-1)(p-n)$. This gives (iii).

If $f$ has 2-fold normal crossings, then $\tilde{D}_2$ has codimension $p - n \geqq 1$ by (iii), and therefore $\tilde{f}$ is generally one-to-one (since it is one-to-one outside $\tilde{D}_2$). Moreover, outside the diagonal $\tilde{\Delta}$ in $\tilde{N} \times \tilde{N}$ $\tilde{\Pi}_2$ is manifold which is the complexification of its real part. Finally if $f$ is 1-correct, then by theorem (19.3) every irreducible component of the germ $(\tilde{\Pi}_2, (a, a))$ for $a \in N$ contains a real part of the same dimension. Hence, even along $\Delta \subset N \times N$ $\tilde{\Pi}_2$ is a germ complexification of its real part. This proves (iv).

COROLLARY (20.5). *Let $f: N \to \boldsymbol{R}^p$ be a real-analytic map, locally proper and of finite deficiency. Suppose f is 1-correct and has 2-normal crossings. Let $\tilde{f}: \tilde{N} \to \boldsymbol{C}^p$ be a suitable holomorphic extension of f.*

*Then fN is a parametric set in $\boldsymbol{R}^p$ with complexification $\tilde{f}\tilde{N}$* (cf. (8.1)). *If $\tilde{f}\tilde{N}$ is weakly normal, so is fN.*

PROOF. Only the last part (if any) needs comment. Since the question is local, we can (by restricting to a small neighborhood in $\boldsymbol{C}^p$ and pulling back) assume that $N$ and $\tilde{N}$ are open sets $\Omega \subset \boldsymbol{R}^n$ and $\tilde{\Omega} \subset \boldsymbol{C}^n$, respectively. By proposition (20.4) $\tilde{\Pi}_2$ is a germ complexification of its real part, which then equals the coincidence set of $f$. The claim now follows by (15.4.1).

EXAMPLE (20.6). The map $f: \boldsymbol{R}^2 \to \boldsymbol{R}^3$ of (15.6) has 2-fold normal crossings and is almost everywhere 1-correct; it fails at the origin only. However

$\Pi_2(\tilde{f})$ is a union of two curves
$\Pi_2(\tilde{f})_{\boldsymbol{R}}$ is the origin
$\Pi_2(f)$ is empty.

As an application of the preceding results we show that parametric hypersurfaces with transversal singularities are weakly normal. More precisely

THEOREM (20.7). *Let $f: N \to \boldsymbol{R}^{n+1}$ be a real-analytic map, locally proper and of finite deficiency. Suppose $f$ is 1-correct and has normal crossings. Then $V = fN$ is weakly normal.*

PROOF. Let $\tilde{f}: \tilde{N} \to \boldsymbol{C}^{n+1}$ be a suitable complexification of $f$. Then $\tilde{V} = \tilde{f}\tilde{N}$ is a complex-analytic subset of some open set $\tilde{G} \subset \boldsymbol{C}^{n+1}$. By corollary (20.5) it suffices to show that $\tilde{V}$ is a weakly normal complex space.

Since $\tilde{V}$ is purely of codimension 1 in $\boldsymbol{C}^{n+1}$, we have $\text{pf}_a\, \mathcal{O}(\tilde{V}) = n$ at any point $a \in \tilde{V}$. By Hartog's theorem for weak normality it will therefore be enough to show that $\tilde{V}$ is weakly normal outside some analytic subset $\tilde{A}$ with $\dim \tilde{A} \leqq n-2$.

Take $\tilde{A} = \tilde{f}\Sigma(\tilde{f})$, where $\Sigma(\tilde{f}) = \Sigma^1(\tilde{f}) \cup \Sigma^2(\tilde{f}) \cup \cdots$ is the singular set of $\tilde{f}$. Then $\dim \tilde{A} = \dim \Sigma^1(\tilde{f}) = n-2$ (since $\tilde{f}$ can be assumed 1-correct). At a point $a \in \tilde{V}$ outside $\tilde{A}$ $\tilde{V}$ is at worst an $(n+1)$-cross (since $\tilde{f}$ can be assumed to have normal crossings), hence weakly normal by example (16.6).

REMARKS (20.8). The proof works equally well for maps $f: N \to \boldsymbol{R}^{n+k}$, $k \geqq 1$, provided $\text{pf}_a\, \mathcal{O}(\tilde{V}) \geqq n-k+1$ at all points $a \in \tilde{A}$, this because $\tilde{A} = \tilde{f}\Sigma(\tilde{f})$ has dimension $n-k-1$ in this case. In particular it works if $\tilde{V}$ is locally a complete intersection along $V$ (then $\text{pf}_a\, \mathcal{O}(\tilde{V}) = n$).

(20.9) Theorem (20.7) (and its extension (20.8)) also holds in the complex case.

## 21. The quadratic differential

Let $f: N \to \boldsymbol{R}^{n+e}$ be a smooth map presented in Whitney form at $a \in N$ by suitable coordinate systems $x = (t_1, \cdots, t_r, u_1, \cdots, u_s)$ and $y = (w_1, \cdots, w_r, z_1, \cdots, z_{s+e})$, as in lemma (17.2)

$$\begin{cases} w_i = t_i & (1 \leqq i \leqq r) \\ z_j = \varphi_j(t, x) & (1 \leqq j \leqq s+e). \end{cases}$$

The functions $\varphi_j$ vanish at the origin together with their first derivatives.

Set

$$\varphi_j(t, u) = q_j(u) + \sum t_i a_i^{(j)}(t, u) + O(3)$$

where $q_j(u)$ are quadratic forms in $u_1, u_2, \cdots, u_s$, and $a_i^{(j)}(0, 0) = 0$ for all $i, j$.

Assume that $a \in \Sigma^s(f)$ and $df$ is transverse to $\Sigma^s$ at $a$. Then locally at $a$

$$\Sigma^s(f) = \left\{ x \;\middle|\; \frac{\partial \varphi_j}{\partial u_k}(x) = 0, \qquad 1 \leqq j \leqq s + e, \qquad 1 \leqq k \leqq s \right\}$$

$$\ker df(a) = \text{span} \left\{ \frac{\partial}{\partial u_k}(a) \;\middle|\; 1 \leqq k \leqq s \right\}$$

$$\text{coker } df(a) \cong \text{span} \left\{ \frac{\partial}{\partial z_j}(a) \;\middle|\; 1 \leqq j \leqq s + e \right\}.$$

Let

$$d^2 f(a) : \ker df(a) \to \text{coker } df(a)$$

denote the quadratic differential of $f$ at $a$ (cf. [1]). Then in the coordinate systems above $d^2 f$ is given by equations

$$d^2 z_j = q_j(du_1, \cdots, du_s) \qquad (1 \leqq j \leqq s + e)$$

in a neighborhood of $a$ in $\Sigma^s(f)$.

Since $f$ is 1-correct at $a$, the 2. order corank strata $\Sigma^{(s, t)}(f) \subset \Sigma^s(f)$ are defined in a neighborhood of $a$,

$$\Sigma^{(s, t)}(f) = \Sigma^t(f'),$$

where $f' = f \mid \Sigma^s(f)$. A point $x \in \Sigma^s(f)$ belongs to $\Sigma^{(s, t)}(f)$ precisely when $\dim T_x \Sigma^s(f) \cap \ker df(x) = t$. In particular

$$\Sigma^{(s, 0)}(f) = \{x \in \Sigma^s(f) \mid T_x \Sigma^s(f) \cap \ker df(x) = \{0\}\}.$$

PROPOSITION (21.1). *Assume $df$ is transverse to $\Sigma^s$ at $a$. Then $a \in \Sigma^{(s, 0)}(f)$ if and only if $d^2 f(a)$ has no (non-trivial) double zero.*

PROOF. A *double zero* of a set of real quadratic forms $q_1, q_2, \cdots, q_{s+e}$ (in $s$ variables) is a non-trivial common solution of the equations

$$\frac{\partial q_j}{\partial u_k}(U) = 0 \qquad (\forall j, k). \tag{21.1.1}$$

In the linear coordinate systems from $x$ and $y$ $T_a N$ is identified with $\boldsymbol{R}^n$, $\ker df(a)$ with the subspace of vectors $(0, \cdots, 0, U_1, \cdots, U_s)$ and $T_a \Sigma^s(f) \cap \ker df(a)$ with the smaller subspace of vectors satisfying

$$\sum_{l=1}^{s} \frac{\partial^2 \varphi_j}{\partial u_k \partial u_l}(0, 0) U_l = 0 \qquad (\forall j, k). \tag{21.1.2}$$

But (21.1.2) is equivalent to (21.1.1).

REMARK (21.2). If $U=(U_1,\cdots,U_s)$ is a double zero for the real quadratic forms $q_1,\cdots,q_{s+e}$, then by Euler's theorem on homogeneous functions

$$q_j(U)=0 \qquad (\forall j).$$

If $q=(q_1,\cdots,q_{s+e})$ admits a complex double zero $U$, then $\bar{U}+U$ and $i(\bar{U}-U)$ are also solutions of (21.1.1) and at least one is $\neq 0$. Therefore, if a rea. quadratic map admits a complex double zero, it admits a real double zero.

A real quadratic map $q$ in $s$ variables which has a double zero, can by a linear change of variables be written as a quadratic map in *less* than $s$ variables.

## 22. The span of the quadratic differential

Denote by $\Phi_2(s)$ the vector space of real quadratic forms in $s$ variables. Then $\dim \Phi_2(s)=\frac{1}{2}s(s+1)$.

We continue with the analysis of the map $f: N\to \boldsymbol{R}^{n+e}$ of the preceding section. Let

$$d^2z_j=q_j(du_1,\cdots,du_s)$$

be $f$'s quadratic differential as presented by $f$ in some Whitney form.

We denote by $L(q)$ the linear subspace of $\Phi_2(s)$ spanned by $(q_1,\cdots,q_{s+e})$. It is an element of $G_t(\Phi_2(s))$, the grassmannian of $t$-dimensional linear subspaces of $\Phi_2(s)$, where $t$ is the dimension of $L(q)$. We call $t$ the *quadratic rank* and $\tau=\frac{1}{2}s(s+1)-t$ the quadratic corank of $f$ at $a$, and write $\mathrm{rk}_a^2 f=t$, $\mathrm{cork}_a^2 f=\tau$.

The space $L(q)$ is invariant under linear coordinate changes in the target $\boldsymbol{R}^{s+e}$. On the other hand linear coordinate changes in the source space $\boldsymbol{R}^s$ produce "similar" quadratic forms. Thus $GL(s)=GL(s,\boldsymbol{R})$ acts linearly on $\Phi_2(s)$ and therefore $PGL(s)=GL(s)/\boldsymbol{R}^*\cdot I$ acts on $G_t(\Phi_2(s))$. We denote by $o_a^2(f)$ the orbit of $L(q)$ in $G_t(\Phi_2(s))$.

It is clear that $\mathrm{rk}_a^2 f$, $\mathrm{cork}_a^2 f$, and $o_a^2(f)$ are invariants of $f$ which depend only on the second order jet $j^2f$ (and even on that only up to 2. order equivalence).

(22.1) Given $s>0$ and $t\geqslant 0$ we can ask when the quadratic differential of map with rank $n-s$ and quadratic rank $t$ at a point $a\in N$ has, up to a convenient choice of coordinates in source and target spaces, a finite set of representatives (canonical forms).

Evidently this requires that $PGL(s)$ have a finite number of orbits in $G_t(\Phi_2(s))$. In particular at least one orbit must be open. Since

$$\dim G_t(\Phi_2(s))=t(\tfrac{1}{2}s(s+1)-t)$$

and

$$\dim PGL(s)=s^2-1,$$

we can expect canonical forms for the quadratic differential only if

$$t(\tfrac{1}{2}s(s+1)-t)\leqq s^2-1, \tag{22.1.1}$$

i.e. only if

$$t^2-\varphi_2(s)t+s^2-1\geqq 0,$$

where $\varphi_2(s)=\frac{1}{2}s(s+1)$. Besides we must have

$$0\leqq t\leqq \varphi_2(s).$$

Now the solutions of (22.1.1) for $t$ in this range (given $s$) come in pairs symmetric with respect to $\frac{1}{2}\varphi_2(s)$. Therefore it suffices to check the solutions in the range

$$0\leqq t\leqq \tfrac{1}{2}\varphi_2(s) \tag{22.1.2}$$

To any such there is then a corresponding solution $t'=\varphi_2(s)-t$ in the range $\frac{1}{2}\varphi_2(s)\leqq t'\leqq \varphi_2(s)$.

Within (22.1.2) we find that $t=0$ is a solution for all $s\geqq 1$, that $t=1$ is a solution for all $s\geqq 2$, and that $t\geqq 3$ is not a solution for any $s$. The last part is checked by inspection for $t=3$; for $t\geqq 4$ it follows because the estimate

$$4(\tfrac{1}{2}s(s+1)-t)\leqq s^2-1$$

is inconsistent with (22.1.2).

There remains the exceptional case $t=2$, which is a solution precisely for $s=3$.

Thus we see we can expect canonical forms for the quadratic differential provided the corank is 0 or 1, say. In the following sections we give canonical forms for some smooth mapgerms determined by their second order jet when the quadratic span is of codimension 0 and 1. As an application we show that parametric spaces $X^n\subset \mathbf{R}^{n+e}$ with transversal singularities are weakly normal if $e\geqq n/2$.

## 23. Coordinate changes preserving Whitney form

Coordinate systems in $N$ and $P$ presenting $f$ in Whitney form, as in lemma (17.2), are called *adapted* to $f$. A change of coordinates in $\mathbf{R}^n$, 0 and $\mathbf{R}^p$, 0 is called *admissible* if it transforms adapted systems into adapted systems. We list the standard admissible changes.

(a) $(w, z)$ *is changed to* $(w, \zeta)$ *with*

$$\zeta_j=\zeta_j(w,z), \qquad 1\leqq j\leqq s+e, \quad \textit{and}$$

$$\zeta(0,0)=0$$

$$\det\frac{\partial\zeta}{\partial z}(0,0)\neq 0$$

$$\frac{\partial\zeta}{\partial w}=0.$$

This is a coordinate change in $\mathbf{R}^p$.

(b) $(t, u)$ *is changed to* $(t, \theta)$ *with*

$$\theta_j = \theta_j(t, u), \qquad 1 \leqq j \leqq s, \quad \textit{and}$$

$$\theta(0, 0) = 0$$

$$\det \frac{\partial \theta}{\partial u}(0, 0) \neq 0.$$

This is a coordinate change in $\boldsymbol{R}^n$.

(c) $(w, z)$ *is changed to* $(\eta, z)$ *and* $(t, u)$ *is changed to* $(\tau, u)$ *with*

$$\eta_j = \eta_j(w, z) \quad \textit{and} \quad \tau_j = \tau_j(t, u) = \eta_j(t, \varphi(t, u)), \qquad 1 \leqq j \leqq r,$$

*and such that*

$$\eta(0, 0) = 0$$

$$\det \frac{\partial \eta}{\partial w}(0, 0) \neq 0.$$

This is a simultaneous coordinate change in $\boldsymbol{R}^n$ and $\boldsymbol{R}^p$.

Note that in adapted coordinates the jacobian of $f$ shows the block form

$$J = \begin{Bmatrix} I & 0 \\ J'' & J' \end{Bmatrix}$$

with

$$J'(0) = 0, \qquad J''(0) = 0$$

and that a coordinate change is admissible if it keeps this property. Then it is clear that (a), (b), and (c) are admissible changes.

Morin has proved that every admissible coordinate change is a composite of the standard changes, [10].

## 24. Canonical forms of some $\Sigma^{(s, 0)}$-singularities

We will show that canonical forms exist for mapgerms $f: (\boldsymbol{R}^n, 0) \to \boldsymbol{R}^p$ under suitable transversality conditions on $df$ and $d^2 f$.

THEOREM (24.1). *Let* $f: (\boldsymbol{R}^n, 0) \to \boldsymbol{R}^{n+e}$ *be a smooth mapgerm of corank* $s \geqq 1$ *(at 0). Assume that*

(i) $df$ *is transverse to* $\Sigma^s$ *at* 0

(ii) $d^2 f(0)$ *span the full space of quadratic forms in s variables.*

*Then there are adapted coordinate systems* $(t, u)$ *on* $\boldsymbol{R}^n$ *and* $(w, z)$ *on* $\boldsymbol{R}^{n+e}$, *at* 0 *and $f(0)$, in which $f$ takes the form*

$$\begin{cases} w_1 = t_1 \\ \cdot \\ \cdot \\ \cdot \\ w_r = t_r & (r = n-s) \\ z_1 = u_1^2 + t.u_2 + \cdots + t.u_s \\ z_2 = u_1 u_2 + t.u_1 + \cdots + t.u_s \\ \cdot \\ \cdot \\ \cdot \\ z_q = u_s^2 + t.u_1 + \cdots + t.u_{s-1} & \left(q = \binom{s+1}{2}\right) \\ z_{q+1} = t.u_1 + \cdots + t.u_s \\ \cdot \\ \cdot \\ \cdot \\ z_{s+e} = t.u_1 + \cdots + t.u_s \end{cases}$$

REMARKS. (1) The $z$-component of $f(t, u)$ depends on parameters $u_1, \cdots, u_s$ and $t_1, \cdots, t_{s(s+e)-s}$, as the terms with 'linear' variable $u_j$ are omitted from expressions containing $u_j^2$. The notation $t.$ means that the indices run consecutively (from 1 to $s(s+e)-s$). That $s(s+e)-s \leqq r$ follows from condition (i) and the assumption $0 \in \Sigma^s(f)$ (which ensures that codim $\Sigma^s(f) \leqq n$, i.e. $s(s+e) \leqq n$).

(2) For $r = s(s+e)-s$, i.e. $n = s(s+e)$, all $t$-variables are effective in the expression of $f$. This is the lowest dimension $n$ for which a map $f: \boldsymbol{R}^n \to \boldsymbol{R}^{n+e}$ can display transversally a $\Sigma^s$-singularity. In this case $\Sigma^s(f)$ is a discrete set and so $\Sigma^s(f) = \Sigma^{(s,0)}(f)$.

For $n > s(s+e)$ the canonical form above is the suspension of the canonical form for $n = s(s+e)$, i.e. with $m = n - s(s+e)$ we have $f_{(m)} = f_{(0)} \times \mathrm{id}_{\boldsymbol{R}^m}$ in obvious notation.

PROOF. (a) We may suppose that $f$ is in Whitney form

$$\begin{cases} w_i = t_i & (1 \leqq i \leqq r) \\ z_j = q_j(u) + \sum t_i a_i^j(t, u) + O(3) & (1 \leqq j \leqq s+e) \end{cases}$$

with $a_i^j(0, 0) = 0$. Moreover by the transversality assumption the functions $\partial z_j / \partial u_l$ are part of a coordinate system at the origin of $\boldsymbol{R}^n$.

Let $\omega_{(l,h)}(U)=u_l u_h$ be the standard basis in $\Phi_2(s)$, $1\leqq l\leqq h\leqq s$.

As $L(q)=\Phi_2(s)$, a linear change in the variables $z_1,\cdots,z_{s+e}$ (which is admissible of type (a)) will bring $f$ to the form

$$f\equiv\begin{cases} w_i=t_i & (1\leqq i\leqq r)\\ z_\alpha=\omega_\alpha(u)+\sum_1^r t_i a_i^\alpha(t,u)+O(3) & \\ z_j=\qquad\quad \sum_1^r t_i a_i^j(t,u)+O(3) & \end{cases}$$

where $\alpha=(l,h)$, $1\leqq l\leqq h\leqq s$, and where $\binom{s+1}{2}<j\leqq s+e$.

(b) If we here forget the last set of coordinates $z_j$, we obtain a map $g:(\boldsymbol{R}^n,0)\to(\boldsymbol{R}^\sigma,0)$ where $\sigma=r+\binom{s+1}{2}$. $g$ is the composite $\mathrm{pr}\circ f$ where pr is the natural projection $\mathrm{pr}:\boldsymbol{R}^{n+e}\to\boldsymbol{R}^\sigma$, and has the equations

$$g\equiv\begin{cases} w_i=t_i & (1\leqq i\leqq r)\\ z_\alpha=\omega_\alpha+\sum_1^r t_i a_i^\alpha(t,u)+O(3) & \end{cases}$$

$\alpha=(l,h), 1\leqq l\leqq h\leqq s$.

Let $F=\boldsymbol{R}[[t,u]]$ be the ring of formal power series in the $n$ variables $t_1,\cdots,t_r,u_1,\cdots,u_s$, and let $F'=\boldsymbol{R}[[w,z']]$ be the ring of power series in the $q$ variables $w_1,\cdots,w_r,z_{(1,1)},\cdots,z_{(s,s)}$. Let $\mathfrak{m}'$ denote the maximal ideal in $F'$.

Consider the ideal $I=F\cdot g^*\mathfrak{m}'$ in $F$. We claim that

$$I=F(t_1,\cdots,t_r,\omega_{(1\ 1)}(u),\cdots,\omega_{(s\ s)}(u)).$$

In fact if we set $g_\alpha(t,u)\equiv\omega_\alpha(u)+\sum_1^r t_i a_i^\alpha(t,u)+O(3)$, we need to prove that the ideal generated by the $\omega_\alpha(u)$'s over the ring $\boldsymbol{R}[[u]]$ of formal power series in $u_1,\cdots,u_s$ is the same as the ideal generated by the functions $g_\alpha(0,u)$. Now we can write

$$g_\alpha(0,u)=\omega_\alpha(u)+\sum\omega_\beta(u)h_\beta^\alpha(u)$$

with $h_\beta^\alpha(0)=\theta$. Since the matrix

$$A=(h_\beta^\alpha(u)+\delta_\beta^\alpha)$$

(where $\delta_\beta^\alpha$ is the Kronecker symbol) equals the identity for $u=0$, it is invertible over $\boldsymbol{R}[[u]]$. Consequently the $\omega_\alpha(u)$ can be written as linear combinations of the $g_\alpha(0, u)$. It follows that

$$F/I \cong \boldsymbol{R} + \sum_{i=1}^{s} \boldsymbol{R} u_i.$$

(c) We can now apply Malgrange's preparation theorem which asserts that the ring $E(t, u)$ of germs of smooth functions in $t$ and $u$ at the origin is a finitely generated module over the ring $g^*E(w, z')$ (the pull-back by $g$ of the ring of smooth functions in the $w$'s, and $z_\alpha$'s at the origin), the generators being $1, u_1, \cdots, u_s$.

Hence we obtain relations of the form

$$\omega_\alpha(u) = a^\alpha(g(t, u)) + \sum_{l=1}^{s} a_l^\alpha(g(t, u))u_l$$

$$f_j(t, u) = b^j(g(t, u)) + \sum_{l=1}^{s} b_l^j(g(t, u))u_l$$

for $\alpha = (l, h)$, $1 \leqq l \leqq h \leqq s$, and for $\binom{s+1}{2} < j \leqq s+e$; the $f_j$ being the $j$th component of the map $f$ in the form given at the end of part (a) above.

We claim that the jacobian matrix $\left(\text{of type } \binom{s+1}{2} \times \binom{s+1}{2}\right)$

$$\frac{\partial(a^{(1,1)}, \cdots, a^{(s,s)})}{\partial(z_{(1,1)}, \cdots, z_{(s,s)})}$$

has non-vanishing determinant at the origin.

In fact, applying the operator $\partial^2/\partial u_l \partial u_h|_{(0,0)}$ to the expression for $\omega_\alpha$ above, we get

$$\frac{\partial a^\alpha}{\partial z_\beta}(0, 0) = \begin{cases} 1 & \text{if } \beta = \alpha \\ 0 & \text{if } \beta \neq \alpha \end{cases}$$

since we have

$$\frac{\partial g_\alpha}{\partial u_j}(0, 0) = 0 \quad \text{and} \quad \frac{\partial^2 g_\alpha}{\partial u_l \partial u_h}(0, 0) = \frac{\partial^2 \omega_\alpha}{\partial u_l \partial u_h}.$$

We can therefore perform the following admissible change of coordinates (of type (a)) in the target space

$$\begin{cases} w_i' = w_i & (1 \leqq i \leqq r) \\ z_\alpha' = a^{\alpha'}(w, (z_\beta)) & (\alpha = (l, h), 1 \leqq l \leqq h \leqq s) \\ z_j' = z_j & \left(\binom{s+1}{2} < j \leqq s+e\right). \end{cases}$$

In this new system $f$ is given by the equations (where we drop the primes to simplify the notations)

$$f \equiv \begin{cases} w_i = t_i & (1 \leqq i \leqq r) \\ z'_\alpha = \omega_\alpha(u) - \sum_{l=1}^{s} a_l^\alpha(g(t, u))u_l & (\alpha = (l, h), 1 \leqq l \leqq h \leqq s) \\ z_j = b^j(g(t, u)) + \sum_{l=1}^{s} b_l^j(g(t, u))u_l & \left(\binom{s+1}{2} < j \leqq s+e\right). \end{cases}$$

Note that the functions $a_l^\alpha$, $b^j$, and $b_l^j$ can be expressed as functions of $w$ and the $z_\alpha$'s through the admissible change of coordinates given above.

Since $b^j(w, z)$ is such that $b^j(0, 0) = 0$ for all $j$'s, we can make another admissible change of reference (in the target space)

$$\begin{cases} w'_i = w_i \\ z'_\alpha = z_\alpha \\ z'_j = z_j - b^j(w, z). \end{cases}$$

This brings $f$ to the form (again dropping the primes)

$$f \equiv \begin{cases} w_i = t_i \\ z_\alpha = \omega_\alpha(u) - \sum_{l=1}^{s} a_l^\alpha(g(t, u))u_l \\ z_j = \sum_{l=1}^{s} b_l^j(g(t, u))u_l \end{cases}$$

(the range of the indices being the same as before).

Again the functions $a_l^\alpha$ and $b_l^j$ can be expressed as functions of $w$ and the $z_\alpha$'s.

(d) Next we change coordinates in the source space by setting

$$\begin{cases} t'_i = t_i & (1 \leqq i \leqq r) \\ u'_j = u_j - \frac{1}{2}a_j^{(j,j)}(t, z_\alpha(t, u)) & (1 \leqq j \leqq s). \end{cases}$$

This is an admissible change of coordinates (of type (b)). In fact

$$a_j^{(j,j)}(0, 0) = 0 \quad \text{and} \quad \frac{\partial a_j^{(j,j)}}{\partial u_l}(0, 0) = 0$$

as the functions $z_\alpha(t, u)$ start with second order terms.

Forgetting primes we obtain equations for $f$ of the form

$$\begin{cases} w_i = t_i \\ z_{ll} = u_l^2 + \sum\limits_{j \neq l} a_j^{(l,l)}(t, (z_\alpha))u_j + c^{(l,l)}(t, (z_\alpha)) \\ z_{lh} = u_l u_h + \sum\limits_{j} a_j^{(l,h)}(t, (z_\alpha))u_j + c^{(l,h)}(t, (z_\alpha)) \\ z_j = \sum b_l^j(t, (z_\alpha))u_l + d^j(t, (z_\alpha)) \end{cases}$$

with obvious notations (being understood that for $z_{lh}$ $l \neq h$).

Here the functions $a$ and $b$ are new functions of $t$ and $u$ expressible through functions of $w$ and $z_\alpha$'s and with the same properties as the previous one. The functions $c$ and $d$ are also expressible through functions of the $w$ and $z_\alpha$'s, and vanishing of second order at the origin in the target space.

Therefore the change of coordinates (in the target space)

$$\begin{cases} w_i' = w_i \\ z_\alpha' = z_\alpha - c^\alpha(w, (z_\alpha)) \\ z_j' = z_j - d^j(w, (z_\alpha)) \end{cases}$$

is an admissible change of coordinates, after which $f$ takes the form (dropping primes):

$$\begin{cases} w_i = t_i \\ z_{ll} = u_l^2 + \sum\limits_{j=l} a_j^{(l,l)}(t, (z_\alpha))u_j \\ z_{lh} = u_l u_h + \sum\limits_{j} a_j^{(l,h)}(t, (z_\alpha))u_j \\ z_j = \sum b_l^j(t, (z_\alpha))u_l . \end{cases}$$

(e) At this point we bring in the assumption that $f$ is 1-correct. This tells us that the derivatives of the $z$'s with respect to the $u$'s can be taken among a set of local coordinates at the origin. Equivalently stated, this says that the jacobian matrix of $\partial z/\partial u$ with respect to the variables $t$ and $u$ at the origin must have maximal rank $(= s(s+e))$. This matrix has the following explicit form (where $I$ denotes the identity matrix).

| | $\partial/\partial t_1 \quad \cdots \quad \partial/\partial t_r$ | $\partial/\partial u_1 \quad \cdots \quad \partial/\partial u_s$ |
|---|---|---|
| $\dfrac{\partial z_{(i,i)}}{\partial u_i}$ | $0$ | $2I$ |
| $\dfrac{\partial z_{(i,i)}}{\partial u_j}$ | $\dfrac{\partial a_j^{(i,i)}}{\partial t_1}(0,0)$ <br> $\vdots$ | $0$ |
| $\dfrac{\partial z_{(i,j)}}{\partial u_l}$ | $\dfrac{\partial a_l^{(i,j)}}{\partial t_1}(0,0) \quad \cdots$ <br> $\vdots$ | $\delta^{(i,j)}_{(l,h)}$ |
| $\dfrac{\partial z_j}{\partial u_l}$ | $\dfrac{\partial b_l^j}{\partial t_1}(0,0) \quad \cdots$ <br> $\vdots$ | $0$ |

By subtraction of rows we can kill $\delta^{(i,j)}_{(l,k)}$. Therefore the matrix

$$\frac{\partial(\cdots, a_j^{(i,i)}, \cdots, a_l^{(i,j)}, \cdots, b_l^j, \cdots)}{\partial(t_1, t_2, \cdots, t_r)}$$

has maximal rank at the origin ($= s(s+e)-s =$ the number of $a$'s + the number of $b$'s).

By an admissible change of coordinates (of type (c) we can therefore convert the $a$'s and $b$'s into the first set of $s(s+e)-s$ $t$-parameters. This yields the desired normal form.

We turn to the remaining case where we expect finite canonical forms for $f:(\boldsymbol{R}^n, 0)\to \boldsymbol{R}^{n+e}$; namely when $f$ is of quadratic corank 1 (cf. 22).

Let $P^*(\Phi_2(s))$ be the grassmannien of codimension 1 subspaces of $\Phi_2(s)$. This is a projective space isomorphic to $P(\Phi_2(s))$: If $\{\omega_{ij}\}$ denotes the coordinates of the general element $\omega \in \Phi_2(s)$ relative to the standard basis $\{u_i u_j\}_{1\leqq i\leqq j\leqq s}$ (so that $\omega = \sum \omega_{ij} u_i u_j$), then to the hyperplane with equation

$$\sum_{1\leqq i\leqq j\leqq s} a_{ij}\omega_{ij} = 0$$

corresponds the element of $P(\Phi_2(s))$ with homogeneous coordinates $(a_{ij})$.

We need to know the orbit decomposition of $P^*(\Phi_2(s))$ under the action of $GL(s)$. Now, if $a \in P(\Phi_2(s))$ has homogeneous coordinates $(a_{ij}) = A$ (construed as a matrix) and $B \in GL(s)$, then $a' = Ba$ has homogeneous coordinates $(a'_{ij}) = A'$ with $A' = {}^tBAB$. From Sylvester's theorem on the classification of quadratic forms we therefore obtain

LEMMA (24.2). *Under the action of $GL(s, \boldsymbol{R})$ the space $P^*(\Phi_2(s))$ decomposes into a finite set of orbits; one for each hyperplane $\Pi(p, q)$ with equation*

$$\sum_{i=1}^{p} \omega_{ii} - \sum_{j=p+1}^{p+q} \omega_{jj} = 0,$$

$0 \leqq q \leqq p$, $0 < p+q \leqq s$.

*The orbit of $\Pi(p, q)$ is open precisely when $p+q = s$.*

REMARKS. (1) If $p+q<s$, the orbit is contained in the subvariety of $P^*(\Phi_2(s))$ given by the homogenous equation $\det A = 0$. Thus these orbits are nowhere dense.

(2) The hyperplane

$$\omega_{11} + \varepsilon_2\omega_{22} + \cdots + \varepsilon_l\omega_{ll} = 0$$

where $\varepsilon_i = \pm 1$ $(2 \leqq i \leqq l)$, is the span of the $\frac{1}{2}s(s+1)-1$ quadratic forms

$$\begin{cases} u_1^2 - \varepsilon_2 u_2^2 & \\ \cdot & \\ \cdot & \\ \cdot & \\ u_1^2 - \varepsilon_l u_l^2 & \\ u_i^2 & (l+1 \leqq i \leqq s) \\ u_i u_j & (1 \leqq i < j \leqq s) \end{cases}$$

We will consider a 1-correct mapgerm $f:(\boldsymbol{R}^n, 0) \to \boldsymbol{R}^{n+e}$ such that $\operatorname{cork}_0 f = s$, $\operatorname{cork}_0^2 f = 1$, and whose orbit $o^2(f)$ is open in $P^*(\Phi_2(s))$. For dimensional reasons we have to have $s + e \geqq \frac{1}{2}s(s+1)-1$, or $e \geq \frac{1}{2}s(s-1)-1$. That is, the smallest value of the excess $e$ for which we can have $\operatorname{cork}_0^2 f = 1$ is $e = \frac{1}{2}s(s-1)-1$.

THEOREM (24.3). *Let $f:(\boldsymbol{R}^n, 0) \to \boldsymbol{R}^{n+e}$ be a smooth mapgerm of corank $s \geqq 2$ (at 0). Assume that*

(i) *$df$ is transverse to $\Sigma^s$ at 0.*

(ii) *$d^2f(0)$ has a codimension 1 span in $\Phi_2(s)$ belonging to an open orbit in $P^*(\Phi_2(s))$.*

(iii) $e = \frac{1}{2}s(s-1)-1$.

*Then there are adapted coordinate systems $(t, u)$ on $\boldsymbol{R}^n$ and $(w, z)$ on $\boldsymbol{R}^{n+e}$, at $0$ and $f(0)$, in which $f$ takes the form*

$$\begin{cases} w_1 = t_1 \\ \cdot \\ \cdot \\ \cdot \\ w_r = t_r & (r = n-s) \\ z_1 = u_1^2 - \varepsilon_2 u_2^2 + t.u_3 + t.u_4 + t.u_5 + \cdots \\ z_2 = u_1 u_2 + t.u_1 + t.u_2 + \cdots \\ \cdot \\ \cdot & (\varepsilon_j = \pm 1) \\ \cdot \\ z_s = u_1 u_s + t.u_1 + t.u_2 + \cdots \\ z_{s+1} = u_1^2 - \varepsilon_3 u_3^2 + t.u_1 + t.u_2 + t.u_4 + \cdots \\ \cdot \\ \cdot \\ \cdot \\ z_{s+e} = u_{s-1} u_s + t.u_1 + t.u_2 + \cdots \end{cases}$$

As before the notation $t.$ means that the indices run consecutively (from 1 to $s(s+e)-s$, since the 'linear' variable $u_j$ lacks in the $z$-component with term $u_1^2 - \varepsilon_j u_j^2$ and, in addition, the 'linear' variable $u_1$ lacks in $z_1$).

PROOF. Since the proof goes much like the proof of theorem (24.1), a briefer exposition will suffice here.

(a) We start with $f$ in Whitney form

$$\begin{cases} w_i = t_i & (1 \leqq i \leqq r) \\ z_j = \omega_j(u) + \sum t_i a_i^j(t, u) + O(3) & (1 \leqq j \leqq s+e) \end{cases}$$

where $\omega_1, \cdots, \omega_{s+e}$ are quadratic forms in the variables $u_1, \cdots, u_s$. By assumption $\omega_1, \cdots, \omega_{s+e}$ span a hyperplane in $\Phi_2(s)$. After a linear coordinate change in the $u_k$'s we can assume the hyperplane given by an equation

$$\omega_{11} + \varepsilon_2 \omega_{22} + \cdots + \varepsilon_s \omega_{ss} = 0, \qquad (\varepsilon_i = \pm 1).$$

This hyperplane is also spanned by the quadratic forms

$$\begin{cases} u_1^2 - \varepsilon_j u_j^2 & (2 \leqq j \leqq s) \\ u_i u_j & (1 \leqq i < j \leqq s). \end{cases}$$

Thus, by a linear change of the variables $z_j$ we can make the Whitney form of $f$ look like

$$\begin{cases} w_i = t_i & (1 \leqq i \leqq r) \\ z_j = u_1^2 - \varepsilon_j u_j^2 + \sum_{i=1}^{r} t_i a_i^j(t, u) + O(3) & (2 \leqq j \leqq s) \\ z_{ij} = u_i u_j + \sum_{l=1}^{r} t_l a_l^{ij}(t, u) + O(3) & (1 \leqq i \leqq j \leqq s). \end{cases}$$

(b) To apply Malgrange's preparation theorem we need to check the dimension of the (formal) local ring of $f$. Set $F = \mathbf{R}[[t, u]]$ and $F' = \mathbf{R}[[w, z]]$ and let $I \subset F$ be the pullback by $f$ of the maximal ideal $\mathfrak{m}' \subset F'$,

$$I = F \cdot f^* \mathfrak{m}'.$$

Then we have

$$F/I \cong \mathbf{R} + \sum_{j=1}^{s} \mathbf{R} u_j + \mathbf{R} u_1^2. \tag{3}$$

(As a first step one shows that with $F'' = \mathbf{R}[[u]]$ and $I'' = F''(u_1^2 - \varepsilon_2 u_2^2, \cdots, (u_i u_j)_{i<j})$ the relations

$$\mathfrak{m}''^3 \subset I''$$

and

$$F''/I'' \cong \mathbf{R} + \sum_{j=1}^{s} \mathbf{R} u_j + \mathbf{R} u_1^2$$

holds, $\mathfrak{m}''$ being the maximal ideal in $F''$.)

(c) By Malgrange's preparation theorem and the relation (3) the germ of any smooth function at the origin $0 \in \mathbf{R}^n$ can be written as a linear combination of the generators $1, u_1, u_2, \cdots, u_1^2$, with germs of smooth functions that factorize through $f$ as coefficients.

In particular we have

$$\begin{cases} u_1^2 - \varepsilon_j u_j^2 = a_j \circ f + \sum_{l=1}^{s} (b_l^j \circ f) u_l + (d^j \circ f) u_1^2 \\ u_i u_j = a_{ij} \circ f + \sum_{l=1}^{s} (b_l^{ij} \circ f) u_l + (d^{ij} \circ f) u_1^2 \end{cases} \tag{4}$$

for suitable germs of functions $a_j, a_{ij}, b_l^j, \cdots$, with $2 \leqq j \leqq s$, $1 \leqq i < j \leqq s$.

By differentiation (with respect to the $u_k$'s) one confirms that

$$\begin{cases} b_l^j(0, 0) = 0 \\ b_l^{ij}(0, 0) = 0 \end{cases} \tag{5}$$

that

$$\begin{cases} d^j(0,0)=0 \\ d^{ij}(0,0)=0 \end{cases} \tag{6}$$

and that

$$\det \frac{\partial(a_2,\cdots,a_s,a_{12},\cdots)}{\partial(z_2,\cdots,z_s,z_{12},\cdots)} \neq 0.$$

Therefore, by an admissible change of coordinates we can replace the $z$'s by the $a$'s (and relabel them $z$'s) so that $f$ by virtue of (4) takes the form

$$\begin{cases} w_i = t_i \\ z_j = u_1^2 - \varepsilon_j u_j^2 + \sum (a_l^j \circ f)u_l + (\sigma_j \circ f)u_1^2 \\ z_{ij} = \qquad u_i u_j + \sum (a_l^{ij} \circ f)u_l + (\sigma_{ij} \circ f)u_1^2 \end{cases} \tag{7}$$

Moreover, the conditions (5) and (6) show that

$$\begin{cases} a_l^j(0,0) = \sigma_j(0,0) = 0 \\ a_l^{ij}(0,0) = \sigma_{ij}(0,0) = 0 \end{cases} \tag{8}$$

with the indices in the usual range.

(d) Next we normalize the quadratic part in the $u$'s in (7). We set

$$\begin{cases} q_j = u_1^2 - \varepsilon_j u_j^2 + (\sigma_j \circ f)u_1^2 \\ q_{ij} = \qquad u_i u_j + (\sigma_{ij} \circ f)u_1^2 \end{cases}$$

for $2 \leqq j \leqq s$, $1 \leqq i < j \leqq s$. These forms span a hyperplane in $\Phi_2(s)$ with equation

$$\xi_{11} + \sum_{j=2}^{s} \varepsilon_j(1+\sigma_j)\xi_{jj} - \sum_{1 \leqq i < j \leqq s} \sigma_{ij}\xi_{ij} = 0.$$

For $(w, z)$ near the origin this hyperplane belongs to the same orbit as for $(w, z) = (0, 0)$ (since the orbit is open), which is the orbit of

$$\xi_{11} + \varepsilon_2\xi_{22} + \cdots + \varepsilon_s\xi_{ss} = 0.$$

Therefore, by a linear change of $z$-coordinates $f$ can be brought from (7) into the form

$$\begin{cases} w_i = t_i & (1 \leqq i \leqq r) \\ z_j = u_1^2 - \varepsilon_j u_j^2 + \sum_{l=1}^{s} (b_l^j \circ f)u_l & (2 \leqq j \leqq s) \\ z_{ij} = \qquad u_i u_j + \sum_{l=1}^{s} (b_l^{ij} \circ f)u_l. \end{cases}$$

(e) Next replace $u_i$ by $u_i - e_i$, where

$$2e_1 = b_1^2 \circ f$$
$$2\varepsilon_j e_j = -b_j^j \circ f \qquad (2 \leqq j \leqq s).$$

This brings $f$ to the form

$$\begin{cases} w_i = t_i & (1 \leqq i \leqq r) \\ z_2 = u_1^2 - \varepsilon_2 u_2^2 + \sum_{l \geqq 3} (c_l^2 \circ f) u_l + \nu_2 \circ f & \\ z_j = u_1^2 - \varepsilon_j u_j^2 + \sum_{l \neq j} (c_l^j \circ f) u_l + \nu_j \circ f & (3 \leqq j \leqq s) \\ z_{ij} = \qquad u_i u_j + \sum_{l \geqq 1} (c_l^{ij} \circ f) u_l + \nu_{ij} \circ f & (1 \leqq i < j \leqq s) \end{cases}$$

where all $\nu_j$ and $\nu_{ij}$ vanish of second order at the origin in $\boldsymbol{R}^{n+e}$. Therefore

$$z' = z_j - \nu_j$$
$$z'_{ij} = z_{ij} - \nu_{ij}$$

is an admissible change of coordinates in $\boldsymbol{R}^{n+e}$, which brings $f$ to the form (dropping primes)

$$\begin{cases} w_i = t_i \\ z_2 = u_1^2 - \varepsilon_2 u_2^2 + \sum_{l \geqq 3} (b_l^2 \circ f) u_l \\ z_j = u_1^2 - \varepsilon_j u_j^2 + \sum_{l \neq j} (b_l^j \circ f) u_l \\ z_{ij} = \qquad u_i u_j + \sum_{l \geqq 1} (b_l^{ij} \circ f) u_l. \end{cases}$$

(f) From the fact that the entries of the Whitney matrix $\partial(z \circ f)/\partial u$ form a subset of a set of local coordinates in $\boldsymbol{R}^n$ (condition (i)) we find (by computing the gradients of the entries, as in the proof of theorem (24.1)) that the rank of the matrix

$$\left. \frac{\partial(\cdots, b_l^j \circ f, \cdots, b_l^{ij} \circ f, \cdots)}{\partial(t_1, \cdots, t_r)} \right|_0$$

is maximal. Therefore, by a coordinate change (admissible of type (c)) we can convert the $b$'s into the first $s(s+e)-s$ $t$-parameters. This yields (up to enumeration of the $z$'s) the announced normal form.

Remark. (24.4) Examples have been given by W. A. Adkins which show that none of the requirements in theorem (24.3) can be dropped.

## 25. Weak normality in the metastable range

Consider a 2-correct map $f: N \to \boldsymbol{R}^{n+e}$, where $e \geqq n/2$ ($n = \dim N$). If $a \in N$ is a singular point for $f$, then $a \in \Sigma^1(f)$ as $\Sigma^2(f) = \Sigma^3(f) = \cdots = \varnothing$. (We have $\operatorname{codim} \Sigma^s(f) = s(s+e) \geqq n+4$ for $s \geqq 2$.) Moreover, since $f \mid \Sigma^1(f)$ is 1-correct, we find $a \in \Sigma^{(1,0)}(f) \subset \Sigma^1(f)$ as $\Sigma^{(1,1)}(f) = \Sigma^{(1,2)}(f) = \cdots = \varnothing$ (again for dimensional reasons). Thus, in the range $e \geqq n/2$ a 2-correct map $f$ is in fact correct, and $\Sigma^1(f) = \Sigma^{(1,0)}(f)$ is its total singular set (which is empty if $e \geqq n$).

Since $a \in \Sigma^{(1,0)}(f)$, $d^2f(a)$ has no double zero (proposition (21.1)) and so $d^2f(a) \neq 0$. Then $d^2f(a)$ spans $\Phi_2(1)$ which has dimension 1. By theorem (24.1) adapted coordinate systems exist at $a$ and $f(a)$ presenting $f$ in the form $(w, z) = f(u, t)$

$$\begin{cases} w_1 = t_1 \\ \cdot \\ \cdot \\ \cdot \\ w_{n-1} = t_{n-1} \\ z_1 = u^2 \\ z_2 = t_1 u \\ \cdot \\ \cdot \\ \cdot \\ z_{e+1} = t_e u \end{cases}$$

This result is due to Whitney, [13] (who stated it in the case $e = n-1$).

Next consider the case $e > n/2$ and assume that $f$ is 3-fold 1-correct (in addition to being 2-correct). We denote by $\Delta \subset N$ the closure of the set of (genuine) multiple points of $f$ (the points $x \in N$ for which there are other points $y$ with $f(x) = f(y)$). The extra transversality condition ensures that in fact there will be no multiple points but double points, and that these belong to $\Sigma^0(f)$ (where $f$ is immersive). Thus the map $f$ is excellent.

Any point $a$ in the singular set $\Sigma$ of $f$ belongs to $\Delta$, since at such a point $\Delta$ and $\Sigma$ are given by equations

$$t_1 = \cdots = t_e = 0 \qquad (\text{for } \Delta)$$

and

$$t_1 = \cdots = t_e = u = 0 \qquad (\text{for } \Sigma).$$

This follows from the expression of $f$ above.

Thus $\Delta$ is a closed submanifold of $N$ of codimension $e$ and $\Sigma$ a closed submanifold of $\Delta$ of codimension 1. If $N$ and $f$ are analytic, then $V = fN$ is a parametric subspace of $\boldsymbol{R}^{n+e}$, which is singular along $D=f\Delta$. $D$ is a manifold with boundary $S=f\Sigma$, and $f\mid\Delta$ folds $\Delta$ along $\Sigma$.

It follows that $V$ is locally isomorphic to a (double) $(n+e)$-cross or a suspended Whitney umbrella at a singular point. Thus, by (16.6), (16.7), and (20.5) $V$ is weakly normal.

If $e=n/2$, triple points may occur. However, if $f$ is 4-fold 1-correct (in addition to being 2-correct), no quadruple points occur, while triple points occur isolated only (for dimensional reasons). All multiple points belong to $\Sigma^0(f)$ (so the map is excellent). The singularities of $V$ are triple and double $(n+e)$-crosses and suspended Whitney umbrellas. Therefore $V$ is again weakly normal.

We collect our observations in

THEOREM (25.1). *Let $f:N\to\boldsymbol{R}^{n+e}$, $e\geqq n/2$, be a real-analytic map, locally proper and of finite deficiency. Suppose $f$ is 2-correct and 4-fold 1-correct.*
*Then $V=fN$ is weakly normal.*

This shows that weak normality is a generic property for parametric models with presentations $\Omega\to\boldsymbol{R}^{n+e}$, $\Omega\subset\boldsymbol{R}^n$, $e\geqq n/2$. We would like to suggest that models with *excellent* presentations $\Omega\to\boldsymbol{R}^p$ quite generally are weakly normal. However, at the moment we have no proof of this.

Of course we already know that weak normality is generic for hypersurface models $\Omega\to\boldsymbol{R}^{n+1}$, by theorem (20.7).

Theorem (25.1) has been proved independently and by different methods by W. A. Adkins and J. V. Leahy.

## §5. Appendix: Maps of finite deficiency

### 26. Deficiency of a map

Denote by $J^k(n,p)$ the vector space of $k$-jets of smooth mapgerms $(\boldsymbol{R}^n,0)\to(\boldsymbol{R}^p,0)$ and by $\pi_l^k:J^k(n,p)\to J^l(n,p)$ the canonical projection (truncation), which exists for $k\geqq l\geqq 0$. We assume throughout this § that $n\leqq p$.

Set $J^\infty(n,p)=\varprojlim J^k(n,p)$ and let $\pi_l:J^\infty(n,p)\to J^l(n,p)$ be the canonical projection associated to the inverse limit. Then

$$J^\infty(n,p)=\mathfrak{M}\times\cdots\times\mathfrak{M},\qquad (p\text{ times})$$

where $\mathfrak{M}$ is the maximal ideal in $F_n$, the ring of formal power series over $\boldsymbol{R}$ in $n$ variables. We denote by $\mathfrak{m}$ the maximal ideal in $F_p$, and for brevity we set $F_n=F$. Finally if $\alpha\in J^\infty(n,p)$, we set $I(\alpha)=F\cdot\alpha^*\mathfrak{m}$; then $I(\alpha)=F(f_1,\cdots,f_p)$ for $\alpha=(f_1,\cdots,f_p)$.

(26.1) We define the *deficiency* $\delta(\alpha)$ by

$$\delta(\alpha) = \dim_{\boldsymbol{R}} F/I(\alpha) \tag{26.1.1}$$

for $\alpha \in J^\infty(n, p)$. Similarly, for $k \geqq 0$ we define the $k$-*deficiency* of $\alpha \in J^\infty(n, p)$ by

$$\delta_k(\alpha) = \dim_{\boldsymbol{R}} F/(I(\alpha) + \mathfrak{M}^{k+1}). \tag{26.1.2}$$

Evidently the $k$-deficiency $\delta_k(\alpha)$ depends only on the $k$-jet $\pi_k(\alpha)$ of $\alpha$.

REMARK. (26.2) While $\delta(\alpha)$ may be infinite, $\delta_k(\alpha)$ is always finite. We have

$$\delta_k(\alpha) \leqq \dim F/\mathfrak{M}^{k+1} = \binom{n+k}{k}.$$

Set $A_k = F/(I(\alpha) + \mathfrak{M}^{k+1})$. Then $A_k$ is an $F$-module as well as a real vector space, and $\mathfrak{M}^l A_k = 0$ for $l > k$. Thus $A_k \cong \oplus_{l=0}^k \mathfrak{M}^l A_k / \mathfrak{M}^{l+1} A_k$ (as vector spaces); i.e. we have

$$A_k \cong \bigoplus_{l=0}^{k} \mathfrak{M}^l / \{(I(\alpha) \cap \mathfrak{M}^l) + \mathfrak{M}^{l+1}\} \tag{1}$$

(with $\mathfrak{M}^0 = F$).

Therefore, if we introduce the quantities $\nu_l(\alpha)$,

$$\nu_l(\alpha) = \dim_{\boldsymbol{R}} \mathfrak{M}^l / \{(I(\alpha) \cap \mathfrak{M}^l) + \mathfrak{M}^{l+1}\}$$

for $l \geqq 0$, we have by (1)

LEMMA (26.3) $\delta_k(\alpha) = 1 + \sum_{l=1}^{k} \nu_l(\alpha)$.

LEMMA (26.4) *The following are equivalent:*

(i) $\nu_k(\alpha) = 0$

(ii) $\mathfrak{M}^k \subset I(\alpha)$.

PROOF. $\nu_k(\alpha) = 0$ if and only if $\mathfrak{M}^k = (I(\alpha) \cap \mathfrak{M}^k) + \mathfrak{M}^{k+1}$ which is equivalent to $\mathfrak{M}^k \subset I(\alpha)$ by Nakayama's lemma.

COROLLARY (26.5)

(i) $\nu_k(\alpha) = 0$ implies $\nu_k(\alpha) = \nu_{k+1}(\alpha) = \cdots = 0$

(ii) $\delta_k(\alpha) = \delta_{k+1}(\alpha)$ implies $\delta_k(\alpha) = \delta_{k+1}(\alpha) = \cdots = \delta(\alpha)$

(iii) $\delta_k(\alpha) \leqq k$ implies $\delta(\alpha) \leqq k$.

PROOF. (i) is immediate, (ii) follows from (i) and (26.3), and (iii) from (i), (ii), and (26.3).

Notice that by (ii) we have always

$$\delta(\alpha) = \sup_k \delta_k(\alpha). \tag{26.5.1}$$

REMARK. (26.6) Denote by $E(n, p)$ the vector space of germs of smooth maps $(\boldsymbol{R}^n, 0) \to (\boldsymbol{R}^p, 0)$ and by $\pi_k : E(n, p) \to J^k(n, p)$ the canonical homomorphism which to a germ associates its $k$-jet. Let $\pi : E(n, p) \to J^\infty(n, p)$ be the associated limit map, $\pi = \varprojlim \pi_k$. Finally let $E_n = E$ and $E_p$ be the rings of germs of smooth functions at the origin in $\boldsymbol{R}^n$ and $\boldsymbol{R}_p$, respectively.

Replacing $F_n$, $F_p$, and $J^\infty(n, p)$ with $E_n$, $E_p$, and $E(n, p)$ we can define $\delta(\alpha)$, $\delta_k(\alpha)$, and $\nu_l(\alpha)$ as above for $\alpha \in E(n, p)$. Then all claims and remarks above hold in the new context; in fact the proofs work without change.

Now, $\delta_k(\alpha) = \delta_k(\pi(\alpha))$ since we have $E/(I(\alpha) + \mathfrak{M}^{k+1}) \cong F/(I(\pi(\alpha)) + \mathfrak{M}^{k+1})$ (with a slight abuse of notation). Therefore, by (26.5.1) $\delta(\alpha) = \delta(\pi(\alpha))$. Consequently 'smooth' deficiency coincide with 'formal'.

The same remark applies to real-analytic mapgerms and the corresponding 'analytic' deficiency.

## 27. The Bochnak–Lojasiewicz criterion

Let $\alpha = (f_1, \cdots, f_p)$ be any element of $J^\infty(n, p)$ (so that $f_i \in F$). Set

$$\Omega(s) = \left\{ (\sigma_1, \cdots, \sigma_n) \in \boldsymbol{N}^n \,\middle|\, \sum_1^1 \sigma_i \leqq s \right\}$$

where $\boldsymbol{N} = \{0, 1, \cdots\}$, and write $|\sigma| = \sum \sigma_i$, where $\sigma = (\sigma_1, \cdots, \sigma_n)$. Construct a matrix $A(\alpha, s)$ with entries $u_\gamma^\beta$,

$$A(\alpha, s) = \{u_\gamma^\beta\},$$

where $\beta \in \Omega(s)$ and $\gamma \in \{1, 2, \cdots, p\} \times \Omega(s)$ (i.e. $\gamma$ runs through all pairs $(i, \sigma)$ with $1 \leqq i \leqq p$ and $\sigma \in \Omega(s)$) and where

$$u_{(i,\sigma)}^\beta = \text{coefficient of } x^\sigma \quad \text{in} \quad x^\beta f_i.$$

(Cf. [4] for this construction.)

PROPOSITION (27.1). *For any $\alpha \in J^\infty(n, p)$ the following are equivalent*

(i) $\delta(\alpha) > s$

(ii) $\delta_s(\alpha) > s$

(iii) $\binom{n+s}{s} - \operatorname{rk} A(\alpha, s) > s$.

PROOF. The equivalence (i)$\Leftrightarrow$(ii) follows from (26.5), property (iii).

We check the equivalence (ii)$\Leftrightarrow$(iii). Then we need to compute the number

$$d = \dim I(\alpha)/I(\alpha) \cap \mathfrak{M}^{s+1},$$

since

$$\delta_s(\alpha) = \dim F/I(\alpha) + \mathfrak{M}^{s+1} = \dim F/\mathfrak{M}^{s+1} - \dim I(\alpha)/I(\alpha) \cap \mathfrak{M}^{s+1} = \binom{n+s}{s} - d.$$

Let $\sum \lambda_i(x) f_i(x)$ be the general element of $I(\alpha)$ ($\lambda_i \in F$, $i = 1, \cdots, p$). The part of $I(\alpha)$ not in $\mathfrak{M}^{s+1}$ is spanned by elements

$$\sum \lambda_i(x) f_i(x) = \sum_{i=1}^{p} \sum_{|\beta| \leqq s} \lambda^i_\beta x^\beta f_i(x) = \sum_{i=1}^{p} \sum_{|\beta| \leqq s} \sum_{|\sigma| \leqq s} \lambda^i_\beta u^\beta_{(i,\sigma)} x^\sigma$$

where we have set $\lambda_i(x) = \sum \lambda^i_\beta x^\beta$. Now $x^\sigma$ with $|\sigma| \leqq s$ form a basis for the vector space $F/\mathfrak{M}^{s+1}$. Hence the part of $I(\alpha)$ not contained in $\mathfrak{M}^{s+1}$ is generated by

$$\sum_{i=1}^{p} \sum_{|\beta| \leqq s} \lambda^i_\beta \sum_{|\sigma| \leqq s} x^\sigma u^\beta_{(i,\sigma)} = \sum_{i,\beta} \lambda^i_\beta p^\beta_i$$

for varying coefficients $\lambda^i_\beta (|\beta| \leqq s)$, where $p^\beta_i = \sum_{|\sigma| \leqq s} x^\sigma u^\beta_{(i,\sigma)}$. The dimension of the linear subspace of $F/\mathfrak{M}^{s+1}$ spanned by the vectors $p^\beta_i$ is therefore

$$d = \text{rank of matrix of } p^\beta_i = \text{rank } A(\alpha, s).$$

Thus we have quite generally

$$\delta_s(\alpha) = \binom{n+s}{s} - \text{rk } A(\alpha, s). \tag{27.1.1}$$

From this follows the equivalence (ii)$\Leftrightarrow$(iii).

We have remarked that the $s$-deficiency $\delta_s(\alpha)$ of any $\alpha \in J^\infty(n, p)$ depends on the $s$-jet of $\alpha$ only. Therefore the function $\delta_s$ factorizes through $\pi_s : J^\infty(n, p) \to J^s(n, p)$ and defines the deficiency $\delta$ on $J^s(n, p)$ by $\delta_s = \delta \circ \pi_s$.

Define the $s$-deficiency variety $V^s(n, p) \subset J^s(n, p)$ by

$$V^s(n, p) = \{\alpha \in J^s(n, p) \mid \delta(\alpha) > s\}.$$

Corollary (27.2). *$V^s(n, p)$ is an algebraic subvariety of $J^s(n, p)$, invariant under the jet bundle structure group.*

Proof. The first part is immediate from proposition (27.1). The jet bundle structure group associated to $J^s(n, p)$ is $L^s(n) \times L^s(p)$, where $L^s(n)$ denotes the open subset of $J^s(n, n)$ of invertible jets (a group under composition). This group acts on $J^s(n, p)$ by conjugation (substitution)

$$(\beta, \gamma) \cdot \alpha = \gamma \alpha \beta^{-1}.$$

Since $\beta$ and $\gamma$ are jets of germs of diffeomorphisms, it is clear that $\delta_s(\gamma \alpha \beta^{-1}) = \delta_s(\alpha)$. Thus $V^s(n, p)$ is invariant.

## 28. Maps of bounded deficiency

We will show that the codimension of $V^s(n, p)$ in $J^s(n, p)$ tends to infinity with $s$.

LEMMA (28.1). *For* $l \geqq k$ $\pi_k^l V^l(n, p) \subset V^k(n, p)$. *Moreover, for any* $\alpha_k \in J^k(n, p)$ *there exists* $l \geqq k$ *and* $\alpha_l \in J^l(n, p) - V^l$ *such that* $\pi_k^l(\alpha_l) = \alpha_k$.

PROOF. By proposition (27.1) we find for the subset $\pi_s^{-1} V^s \subset J^\infty(n, p)$

$$\pi_s^{-1} V^s = \{\alpha \in J^\infty(n, p) \mid \delta(\alpha) > s\}.$$

For $l \geqq k$ this gives $\pi_l^{-1} V^l \subset \pi_k^{-1} V^k$, and by projecting down via $V^l$, $\pi_k^l V^l \subset V^k$.

For the last part of the claim it suffices to produce an $\alpha \in J^\infty(n, p)$ with $\delta(\alpha) < \infty$ and $\pi_k(\alpha) = \alpha_k$. Then we can take $l \geqq \delta(\alpha)$ and $\alpha_l = \pi_l(\alpha)$.

Let $\alpha_k$ be given by equations

$$y_i = p_i(x) \qquad (1 \leqq i \leqq p)$$

where $p_i$ are polynomials of degree $\leqq k$. Define $\alpha$ by

$$\alpha \equiv \begin{cases} y_i = p_i(x) + x_i^{k+1} & (1 \leqq i \leqq n) \\ y_{n+j} = p_{n+j}(x) & (1 \leqq j \leqq p-n). \end{cases}$$

We show that $\delta(\alpha) \leqq \dim W$, where $W$ is the space of polynomials of degree $\leqq (k+1)n - 1$. This gives $\delta(\alpha) \leqq \binom{n+(k+1)n-1}{n}$.

In fact every monomial of degree $\geqq (k+1)n$ must contain a power $x_i^{k+1}$ for some $i$ $(1 \leqq i \leqq n)$. Thus for a polynomial $\omega$ of degree $l \geqq (k+1)n$ we have

$$\omega \equiv \sum x_i^{k+1} q_i(x) \qquad (\mathrm{mod}\ W)$$

where the $q_i$ are polynomials of degree $\leqq l-(k+1)$.

From this we get

$$\omega \equiv \sum (p_i(x) + x_i^{k+1}) q_i(x) - \sum p_i(x) q_i(x) \qquad (\mathrm{mod}\ W).$$

Since degree $\sum p_i(x) q_i(x) \leqq l-1$, we can repeat the procedure. This finally gives

$$\omega \equiv \sum (p_i(x) + x_i^{k+1}) h_i(x) \qquad (\mathrm{mod}\ W)$$

and shows that we have

$$F = W + I(\alpha) + \mathfrak{M}^{l+1}$$

for every $l \geqq 0$. Consequently $\delta_l(\alpha) \leqq \dim W$, and so $\delta(\alpha) \leqq \dim W$ by (26.5:1). This completes the proof.

LEMMA (28.2). *The codimension of $V^k(n, p)$ in $J^k(n, p)$ tends to $\infty$ with $k$.*

PROOF. We have $V^l \subset (\pi_k^l)^{-1} V^k$ and we need to show that $V^l$ contains no interior points in $(\pi_k^l)^{-1} V^k$, when $l$ is large. Let $V^k = \bigcup_\nu V_\nu^k$ be the finite decomposition of $V^k$ into irreducible components; then $(\pi_k^l)^{-1} V^k = \bigcup_\nu (\pi_k^l)^{-1} V_\nu^k$ is the corresponding decomposition of $(\pi_k^l)^{-1} V^k$. If $V_\mu^l$ is an irreducible component of $V^l$, then $V_\mu^l \subset (\pi_k^l)^{-1} V_\nu^k$ for some $\nu$, and $V_\mu^l \neq (\pi_k^l)^{-1} V_\mu^k$ if $l$ is sufficiently large, say $l = l(\mu, \nu)$ (lemma (28.1)). Then with $l' = \sup_{\mu,\nu} l(\mu, \nu)$ we must have codim $V^{l'} >$ codim $V^k$.

Given $n$ and $p$ with $1 \leqq n \leqq p$ we denote by $\varphi = \varphi(n, p)$ the smallest integer such that

$$\operatorname{codim} V^\varphi(n, p) > n. \tag{28.2.1}$$

This number exists by lemma (28.2).

Let $N$ and $P$ be real-analytic or smooth manifolds of dimension $n$ and $p$. By corollary (27.2) all jet bundles $J^k(N, P)$ contain closed subbundles $V^k(N, P)$ of fiber type $V^k(n, p)$. Since $V^k(n, p)$ is an invariant algebraic subvariety of $J^k(n, p)$, it admits a finite stratification by invariant semi-algebraic submanifolds of $J^k(n, p)$, [14]. $V^k(n, p)$ has at least one (open) stratum of minimal codimension in $J^k(n, p)$. It follows that $V^k(N, P)$ receives a corresponding finite stratification by real-analytic (respectively smooth) submanifolds of $J^k(N, P)$ of codimension $\geqq$codim $V^k(n, p)$.

Let $f: N \to P$ be a real-analytic (respectively smooth) map and $a \in N$ any point of $N$. If $\varphi$ and $\psi$ are local charts in $N$ and $P$ centered at $a$ and $f(a)$, then the mapgerm

$$\psi f \varphi^{-1}: (\boldsymbol{R}^n, 0) \to (\boldsymbol{R}^p, 0)$$

has a well defined deficiency (cf. (26.6)), independent of $\varphi$ and $\psi$, which we denote $\delta(f, x)$. In the analytic case it coincides with the deficiency as defined in proposition (20.1).

PROPOSITION (28.3). *The functions $\delta_s(x) = \delta_s(f, x)$ and $\delta(x) = \delta(f, x)$ are upper semi-continuous on $N$.*

PROOF. For $\delta_s$ this follows from formula (27.1.1). Then, for $\delta$ it follows from (26.5.1).

THEOREM (28.4) *The smooth maps $f: N \to P$ such that*

$$\sup_{x \in N} \delta(f, x) \leqq \varphi(n, p)$$

*form a dense open set in $E(N, P)$.*

PROOF. If $j^k f: N \to J^k(N, P)$ denotes the $k$-jet extension of $f$, then $\delta(f, x) \leqq k$ if and only if $j^k f(x) \notin V^k(N, P)_x$. Since $V^k(N, P)$ is closed, the set of maps $f$ such that $j^k fN \cap V^k(N, P) = \emptyset$ is open in $E(N, P)$. Now if $k \geqq \varphi(n, p)$, then $j^k fN$ and $V^k(N, P)$ disjoint means that $j^k f$ is transversal to (the strata of) $V^k(N, P)$, by the definition of $\varphi(n, p)$ (cf. 28.2.1)). The claim now follows from the transversality theorem.

This proves the genericity of maps of bounded deficiency.

## REFERENCES


[1] ARNOLD, *Singularities of smooth mappings.* Russian Math. Surveys 23 (1968).

[2] ANDREOTTI and GRAUERT, *Théorèmes de finitude pour la cohomologie des espaces complexes.* Bull. Soc. Math. France 90 (1962).

[3] ATIYAH and MACDONALD, *Introduction to commutative algebra.* Addison-Wesley.

[4] BOCHNAK and LOJASIEWICZ, *Remarks on finitely determined analytic germs.* Proc. Liverpool Sing. Symp., Springer Lecture Notes 192.

[5] BRUHAT and WHITNEY, *Quelques propriétés fondamentales des ensembles analytiques-réels.* Com. Math. Helv. 33 (1959).

[6] GRAUERT, *On Levi's problem and the imbedding of real analytic manifolds.* Annals of Math. 68 (1960).

[7] MATHER, *Stability of $C^\infty$-mappings V. Transversality.* Advances in Mathematics 4 (1970).

[8] MATHER, *Stability of $C^\infty$-mappings VI. The nice dimensions.* Proc. Liverpool Sing. Symp., Springer Lecture Notes 192.

[9] MATHER, *Stratifications and mappings.* In Peixoto: Dynamical systems. Academic Press (1973).

[10] MORIN, *Formes canoniques des singularités d'une application differentiable.* C. R. Acad. Sc. Paris t. 260 (1965) p. 5662.

[11] NARASIMHAN, *Introduction to the theory of analytic spaces.* Springer Lecture Notes 25.

[12] REMMERT, *Holomorphe und meromorphe Abbildungen komplekser Räume.* Math. Annalen 133.

[13] WHITNEY, *Singularities of mappings of Euclidean spaces.* Sym. Int. de Top. Alg., Mexico (1958).

[14] WHITNEY, *Tangents to an analytic variety.* Ann. Math. 81 (1965).

[15] ANDREOTTI and NORGUET, *La convexité holomorphe dans l'espace analytique des cycles d'une varieté algébrique.* Ann. Sc. Norm. Sup. di Pisa, 21 (1967).

[16] LOJASIEWICZ, *Ensembles semianalytiques.* I.H.E.S., Mimeographed notes (1965).



ALDO ANDREOTTI
Università di Pisa
Istituto di Matematica
Via Derna 1, Pisa, Italy

PER HOLM
Oslo Universitet
Matematisk Institutt
P.O. Box 1053, Oslo 3, Norway

Nordic Summer School/NAVF
Symposium in Mathematics
Oslo, August 5–25, 1976

# TOPOLOGIE NORMALER GEWICHTET HOMOGENER FLÄCHEN

Gottfried Barthel

## Abstract

We investigate the topological structure of those normal hypersurfaces in $\mathbb{P}_3(\mathbb{C})$ whose affine part is defined by a weighted homogeneous polynomial. The main results are contained in sections 3 and 4. In section 3, we use singular Poincaré duality (3.2) to relate the local homology groups in the singular points and the topology of the curve at infinity with the global homotopy invariants. Complete results on the cohomology ring with integer coefficients are known in the case that the curve at infinity is irreducible. We get complete results for real coefficients in the general case, together with detailed information on the torsion (3.5). In section 4 we prove that almost all surfaces whose defining equations coincide in what we call 'numerical type', namely weights, degree, and the multiplicities in the factorization of the leading form, are homeomorphic.

## Einleitung

Im Gegensatz zum Fall kompakter komplexer Kurven ist die topologische Klassifikation normaler kompakter komplexer Flächen ein schwieriges Problem. Ist die Fläche einfach zusammenhängend, so ist ihr Homotopietyp durch globale Homologieinvarianten bestimmt. Dabei ist der singuläre Poincaré-Dualitätshomomorphismus, der wiederum eng mit der lokalen Homologie in den Singularitäten verknüpft ist, von zentraler Bedeutung. Allerdings zeigen schon einfache Beispiele, daß homotopieäquivalente Flächen nicht notwendig homöomorph sind.

Eine interessante Klasse einfach zusammenhängender kompakter komplexer Flächen ist durch die Hyperflächen im komplex-projektiven Raum $\mathbb{P}_3(\mathbb{C})$ gegeben. Unter diesen sind die 'gewichtet homogenen' Flächen – die affin durch ein gewichtet homogenes Polynom definiert sind – besonders ausgezeichnet: Diese Flächen sind invariant unter einer Aktion der Gruppe $\mathbb{C}^*$ auf $\mathbb{P}_3(\mathbb{C})$, sie liefern eine Fülle interessanter Beispiele, und durch die

$\mathbb{C}^*$-Aktion sind ihre lokalen Homologiegruppen explizit berechenbar.

In diesem Artikel wird die Untersuchung der Topologie normaler gewichtet homogener Flächen aus [Ba-Ka] und [Ka-Ba$_{1,2}$] weitergeführt. Neben der weiteren Diskussion des Homotopietyps werden dabei zur Untersuchung des topologischen Typs die Ideen aufgegriffen, die ansatzweise bereits in [Ka-Ba$_{1,2}$] enthalen sind: Flächen in einer Familie zusammenzufassen, deren affine Gleichungen in wichtigen Daten – wir nennen diese den 'numerischen Typ' – übereinstimmen, und Faserungsargumente anzuwenden, um auf Homöomorphie schließen zu können.

Der Artikel ist in vier Abschnitte gegliedert:



Jedem Abschnitt ist ein Überblick vorangestellt.

Im Abschnitt 1 sind einige bekannte Definitionen und Ergebnisse kurz zusammengestellt. Der folgende Abschnitt ist – vor allem in (2.3) und (2.4) – recht 'technisch'; dort werden im wesentlichen die möglichen Gleichungen und die auftretenden Singularitäten untersucht. Die Hauptergebnisse sind in den Abschnitten 3 und 4 enthalten. Zur Bestimmung des Homotopietyps sind dabei die Reduktion auf den genzzahligen Poincaré-Homomorphismus (3.3.1) und die Resultate aus (3.5) von besonderer Bedeutung: für den Fall, daß die unendlich ferne Kurve reduzibel ist, kann der Kohomologiering mit reellen Koeffizienten vollständig bestimmt werden, und für ganzzahlige Koeffizienten erhalten wir detaillierte Informationen über die Torsion. Im vierten Abschnitt können wir zeigen, daß fast alle Flächen vom gleichen numerischen Typ – die also in Gewichten, Grad, und Vielfachheiten bei der Faktorisierung der Leitform übereinstimmen – homöomorph sind (4.3.1).

Für die gute und fruchtbare Zusammenarbeit, der auch dieser Artikel sein Entstehen verdankt, möchte der Autor bei dieser Gelegenheit Ludger Kaup herzlich danken. Ebenso sei an dieser Stelle Per Holm und Nils Øvrelid für die Einladung nach Oslo und für die Organisation der Sommerschule Dank gesagt, wie auch der DFG, deren Beihilfe die Teilnahme ermöglichte.

## 1. Gewichtet homogene Polynome

### (1.0) Überblick

In diesem Abschnitt stellen wir einige bekannte Ergebnisse über gewichtet homogene Polynome zusammen. Wir beschränken uns dabei auf den für uns wesentlichen Fall von drei Variablen; diese Einschränkung ist aber nur für (1.3) und (1.4) erforderlich. Eine ausführliche Darstellung findet sich in Orliks Lecture Notes [$\mathrm{Or}_1$].

### (1.1) Definitionen

Ein Polynom $f \in \mathbb{C}[z_1, z_2, z_3]$ heißt *gewichtet homogen*, falls es positive ganze Zahlen $q_1$, $q_2$, $q_3$, $r$ gibt – die ohne Einschränkung teilerfremd angenommen werden können –, so daß

$$f(t^{q_1}z_1, t^{q_2}z_2, t^{q_3}z_3) = t^r f(z_1, z_2, z_3)$$

für beliebige $t \in \mathbb{C}$, $z \in \mathbb{C}^3$ gilt. Die Quotienten $w_i := r/q_i$ heißten die *Gewichte* von $f$, und $f$ ist genau dann gewichtet homogen mit Gewichten $(w_1, w_2, w_3)$, falls für jedes Monom $z_1^{i_1}z_2^{i_2}z_3^{i_3}$ von $f$ gilt: $i_1/w_1 + i_2/w_2 + i_3/w_3 = 1$. (In der Literatur wird $f$ auch oft *quasihomogen* bezüglich der Gewichte – oder Exponenten – $q_1$, $q_2$, $q_3$ (bzw. $1/w_1$, $1/w_2$, $1/w_3$) vom quasihomogenen Grad $r$ (bzw. 1) genannt.) Offenbar sind insbesondere die Brieskorn-Polynome $z_1^{a_1} + z_2^{a_2} + z_3^{a_3}$ gewichtet homogen. Sind alle $w_i < 1$, so hat $f$ eine Singularität im Ursprung. Ist die Singularität isoliert, so treten in der Gleichung die Monome $z_i^{a_i}$ oder $z_i^{a_i}z_j$ sowie $z_j^{c_j}z_k^{c_k}$ (für $\{i, j, k\} = \{1, 2, 3\}$) auf. Ordnet man die Gewichte gemäß $w_1 \leqq w_2 \leqq w_3$ an, so gilt bei isolierten Singularitäten zusätzlich $w_2 \leqq \operatorname{grad} f \leqq w_3$. Die lokalen topologischen Eigenschaften dieser. Singularität (Milnor-Faserung, charakteristische Abbildung, $\cdots$) sind unter anderem von Milnor [$\text{Mi}_2$, §9] und Milnor-Orlik [Mi-Or] untersucht worden.

### (1.2) Die $\mathbb{C}^*$-Aktion

Die Hyperfläche $V=\{f=0\}\subset\mathbb{C}^3$ ist *invariant* unter der $\mathbb{C}^*$-*Aktion*

$$(t, (z_1, z_2, z_3)) \to (t^{q_1} z_1, t^{q_2} z_2, t^{q_3} z_3),$$

und umgekehrt ist eine bei dieser Aktion invariante Hyperfläche des $\mathbb{C}^3$ durch ein gewichtet homogenes Polynom definiert (siehe [$\text{Or-Wa}_1$, Prop. 1.1.2]). Als (topologischer) offener *Kegel* über $K=V\cap S_\varepsilon^5$ ist $V$ *zusammenziehbar.* Der *Umgebungsrand* $K$ ist invariant unter $U(1)\subset\mathbb{C}^*$.

Weiter sind $K$ und $V\backslash\{0\}$ genau dann *glatt,* wenn die *Singularität* im Ursprung *isoliert* ist.

### (1.3) Auflösung und lokale Homologie

Die $U(1)$-Aktion auf $K$ benutzen Orlik und Wagreich in [$\text{Or-Wa}_{1,2,3}$], um die *kanonische äquivariante Auflösung* einer gewichtet homogenen Singularität zu bestimmen. Sie zeigen unter anderem, daß der duale Graph sternförmig ist, wobei nur die zentrale Kurve das Geschlecht $g>0$ haben kann. Für den Fall einer isolierten Singularität in $\mathbb{C}^3$ geben sie explizite Rechenregeln an, um den dualen Graphen aus den Gewichten zu berechnen [$\text{Or-Wa}_1$, 3; $-_2$, 3.6]. (Zur Auflösung für den Brieskorn-Fall siehe auch [Hirzebruch-Jänich, 4], vgl. weiter [Hirzebruch-Neumann-Koh, §9 sowie Appendix, 8]). Insbesondere gestatten diese Untersuchungen die Berechnung der *lokalen Homologiegruppen* $H_i(V, V\backslash\{0\})$; so ergibt sich unter anderem Rang $H_2(V, V\backslash\{0\})=2g$. Die allgemeinen Formeln dafür finden sich bei Orlik [$\text{Or}_2$], eine explizite Liste enthält [$\text{Ka-Ba}_2$, 9] (für den Brieskorn-Fall bereits bei [Neumann]).

### (1.4) Die acht Klassen

Nach einem weiteren wesentlichen Ergebnis von Orlik und Wagreich [$\text{Or-Wa}_1$, Theorem 3.1.4] *hängt der topologische Typ* einer gewichtet homogenen *isolierten* Singularität *nur von den Gewichten ab;* genauer sind die Umgebungsränder äquivariant diffeomorph. In jeder Familie gibt es mindestens eine Fläche, deren Gleichung in der folgenden Liste auftritt (siehe [$\text{Or-Wa}_1$, Proposition 3.1.2] und Arnold [Ar, Proposition 11.1]):

I $\quad Z_1^{a_1}+Z_2^{a_2}+Z_3^{a_3}$

II $\quad Z_1^{a_1}+Z_2^{a_2}+Z_2Z_3^{a_3}, \qquad a_2\geqq 2$

III $\quad Z_1^{a_1}+Z_2^{a_2}Z_3+Z_2Z_3^{a_3}; \qquad a_2,\ a_3\geqq 2$

IV $Z_1^{a_1}+Z_1Z_2^{a_2}+Z_2Z_3^{a_3}$, $a_1 \geqq 2$

V $Z_1^{a_1}Z_2+Z_2^{a_2}Z_3+Z_1Z_3^{a_3}$

VI $Z_1^{a_1}+Z_1Z_2^{a_2}+Z_1Z_3^{a_3}+Z_2^{c_2}Z_3^{c_3}$; $a_1 \geqq 2$, $c_2/w_2+c_3/w_3=1$

VII $Z_1^{a_1}Z_2+Z_1Z_2^{a_2}+Z_1Z_3^{a_3}+Z_2^{c_2}Z_3^{c_3}$; $a_1, a_2 \geqq 2$, $c_2/w_2+c_3/w_3=1$

VIII $Z_1^{a_1}+Z_2Z_3$

Singularitäten der Klasse VI existieren genau dann, wenn $(a_1-1) \mid kgV(a_2, a_3)$ gilt, solche der Klasse VII genau dann, wenn $(a_1-1)(a_2, a_3) \mid (a_2-1)a_3$ gilt (siehe [Ar, Proposition 11.2]). Die Singularitäten der Klasse VIII sind algebraisch äquivalent zu $Z_1^{a_1}+Z_2^2+Z_3^2$ (Klasse I), aber die Gewichte $w_2$ und $w_3$ sind nicht eindeutig bestimmt, so daß verschiedene $\mathbb{C}^*$-Aktionen auf $V=\{f=0\}$ auftreten können: Ist $1<u/v \leqq 2$ rational, $(u, v)=1$, so ist die Gleichung gewichtet homogen bezüglich $r=kgV(a_1, u)$, $q_1=r/a_1$, $q_2=vr/u$, $q_3=(u-v)r/u$, also mit Gewichten $(a_1, u/v, u/(u-v))$. In den anderen Klassen sind die Gewichte – und damit die $\mathbb{C}^*$-Aktion – durch die Gleichung bestimmt. (Die Bezeichnung weicht bei VI–VIII von [Or-Wa$_4$, 3.6] ab).

## 2. Verhalten im Unendlichen

### (2.0) Überblick

In diesem Abschnitt untersuchen wir die Eigenschaften gewichtet homogener Flächen im Unendlichen, wobei vor allem Normalitätsbedingungen betrachtet werden. Dabei wird nach den bekannten Definitionen (2.1) zuerst der Fall, daß das Ausgangspolynom $f$ homogen ist, behandelt (2.2). Im nicht-homogenen Fall ergeben sich Fallunterscheidungen durch die möglichen Anordnungen der Gewichte. In (2.3.1) werden die Leitformen bei isolierter affiner Singularität angegeben, als Folgerung ergeben sich die auftretenden unendlich fernen Teile (2.3.2). In (2.4.3) werden die Gleichungen, die normale Flächen definieren, bestimmt; in (2.4.4) sind die Singularitäten im Unendlichen aufgeführt. In (2.5) wird schließlich die $\mathbb{C}^*$-Aktion im Unendlichen betrachtet – wo in vielen Fällen eine weitere gewichtet homogene Singularität auftritt (2.5.1) – und die Fixpunkte angegeben.

### (2.1) Homogenisierung und projektiver Abschluß

Ist $f \in \mathbb{C}[z_1, z_2, z_3]$ ein Polynom vom Grad $d$, so läßt sich $f$ *eindeutig als Summe* $f=\sum_{j=0}^d f_j$ darstellen, wobei $f_j$ *homogen* vom Grad $j$ ist. Mit $f$ sind auch alle $f_j$ gewichtet homogen (mit den gleichen Gewichten), insbesondere also die *Leitform* $f_d$ und die '*Sub-Leitform*' $f_{d-1}$. Ist $\hat{f}=\sum_{j=0}^d z_0^{d-j} f_j \in \mathbb{C}[z_0, z_1, z_2, z_3]$ die *Homogenisierung* von $f$, so ist $X=\{\hat{f}=0\} \subset \mathbb{P}_3(\mathbb{C})$ der *projektive Abschluß* von $V=\{f=0\}\subset\mathbb{C}^3$. Der *unendlich ferne Teil* oder die *Kurve im Unendlichen* ist dann die durch $\{z_0=f_d=0\}$ definierte ebene Kurve $X_\infty$ in der unendlich fernen Hyperebene $\{z_0=0\}=\mathbb{P}_2(\mathbb{C})$.

Als *gewichtet homogene Fläche* bezeichnen wir den projektiven Abschluß einer Hyperfläche im $\mathbb{C}^3$, deren Gleichung gewichtet homogen ist.

**(2.2) Der homogene Fall**

Ist $f$ *homogen*, so ist $X_\infty$ die ebene projektive Kurve $\{f=0\}$ in $P_2(\mathbb{C})$. Diese ist genau dann glatt, wenn die affine Singularität isoliert ist, und man sieht sofort, daß $X$ dann nur diese eine Singularität besitzt. In der Tat ist $X\backslash\{[1:0:0:0]\}$ der Totalraum des komplexen Geradenbündels, das sich durch Einschränkung des Normalbündels der Hyperebene $\{z_0=0\}$ auf $X_\infty$ ergibt (siehe etwa [Ka-Ba$_1$, §3]).

**(2.3) Leitformen bei isolierter Singularität**

Ist $f$ *nicht homogen*, so ist die *Leitform* – bis auf invertierbare Konstanten und Wahl der Indizes – von der Gestalt $z_1^{m_1}z_2^{m_2}z_3^{m_3}\prod_{j=1}^k(z_1^m+\lambda_j z_2^l z_3^{m-l})$ mit $m_1+m_2+m_3+km=d$ und $1\leqq l<m$, $(m,l)=1$ für $k\neq 0$, weil die auftretenden Exponenten $(i_1, i_2, i_3)$ auf der Schnittgerade der Ebenen $\{i_1+i_2+i_3=d\}$ und $\{i_1/w_1+i_2/w_2+i_3/w_3=1\}$ liegen. Die Bedingung, daß der *affine Teil V normal* ist, schränkt die Leitform wesentlich ein; es gilt nämlich.

(2.3.1) LEMMA. *Hat ein gewichtet homogenes, aber nicht homogenes Polynom f vom Grad d und mit Gewichten $w_1\leqq w_2\leqq w_3$ (höchstens) eine isolierte Singularität, so ist die Leitform $f_d$ von f – bis auf invertierbare Konstanten – durch die Gewichte wie folgt bestimmt:*

| Fall | Gewichtsbedingung | Leitform |
|---|---|---|
| $A$ | $w_1=w_2<d=w_3$ | $z_3^d$ |
| $B$ | $w_1<w_2<d=w_3$ | $z_3^d$ |
| $C$ | $w_1<w_2=d<w_3$ | |
| $C_1$ | $1/w_1+a_3/w_3=1,\ a_3<d-1,\ w_1<d/(d-a_3)$ | $z_2^d$ |
| $C_2$ | $1/w_1+(d-1)/w_3=1$ | $z_2^d+\mu z_1 z_3^{d-1},\ \mu\neq 0$ |
| $D$ | $w_1=w_2<d<w_3$ | $(\lambda z_1+\mu z_2)z_3^{d-1}\ (\neq 0)$ |
| $E$ | $w_1<w_2<d<w_3,\ 1/w_1+(d-1)/w_3\neq 1$ | |
| $E_k$ | $k/w_2+(d-k)/w_3=1$ für ein $k,\ 1\leqq k<d$ | $z_2^k z_3^{d-k}$ |
| $F$ | $w_1<w_2<d<w_3,\ 1/w_1+(d-1)/w_3=1$ | |
| $F_1$ | $k/w_2+(d-k)/w_3\neq 1$ für alle $k,\ 1<k<d$ | $z_1 z_3^{d-1}$ |
| $F_{2,k}$ | $k/w_2+(d-k)/w_3=1$ für ein $k,\ 1<k<d$ | $z_3^{d-k}(z_1 z_3^{k-1}+\mu z_2^k)$ |
| $G$ | $w_1<w_2=w_3$ | $\prod_{j=1}^l(\alpha_j z_2+\beta_j z_3)^{\nu_j}$, $\sum_{j=1}^l\nu_j=d$ |

BEWEIS. Wegen $w_1 < d$ kommen in allen Monomen der Leitform die Variablen $z_2$ oder $z_3$ vor. Da die affine Singularität von $f$ nach Voraussetzung isoliert ist, tritt in $f$ jeweils mindestens ein Monom aus $z_1 z_2^{a_2}$, $z_2^{a_2}$, $z_2^{a_2} z_3$ sowie aus $z_1 z_3^{a_3}$, $z_2 z_3^{a_3}$, $z_3^{a_3}$ auf. Damit ergibt sich die oben aufgeführte Liste bis auf den Fall $E_k$ mit $1 < k < d-1$. Daß dieser Fall ebenfalls auftritt, zeigen etwa die Gleichungen der Klassen VI und VII bei geeigneter Wahl der Exponenten.

Daraus folgt sofort

(2.3.2) KOROLLAR. *Ist f wie in* (2.3.1), *so ergibt sich die unendlich ferne Kurve* $X_\infty$ *wie folgt:*

| Fall | $X_\infty$ |
|---|---|
| $A, B, C_1$ | $\mathbb{P}_1$ |
| $C_2$ | $\{z_0 = z_2^d + z_1 z_3^{d-1} = 0\} =: Q_d$ |
| $D, E, F_1, F_2$ mit $\mu = 0$ | $\mathbb{P}_1 \vee \mathbb{P}_1$ |
| $F_{2,k}$ mit $\mu \neq 0$ | $\mathbb{P}_1 \vee Q_k$ |
| $G$ | $\mathbb{P}_1^{(1)} \vee \cdots \vee \mathbb{P}_1^{(l)}$ |

*Insbesondere ist* $X_\infty$ *stets einfach zusammenhängend.*

### (2.4) Normalität

Die Forderung, daß $X$ *normal* ist, bestimmt die auftretenden *Gewichte* der definierenden Gleichung $f$ über (2.3.1) hinaus; insbesondere ergeben sich *Bedingungen an die Sub-Leitform* $f_{d-1}$ aus der folgenden

(2.4.1) BEMERKUNG. Ist die gewichtet homogene Fläche $X$ normal und hat die Leitform $f_d$ der affinen Gleichung $f$ einen quadratischen Faktor $g$, so ist $g$ kein Teiler der Sub-Leitform $f_{d-1}$ (*insbesondere* $f_{d-1} \neq 0$). Anderenfalls wäre nämlich $z_0 = g = 0$ eine Menge nicht-isolierter Singularitäten.

Aus der Liste (2.3.1) sind nun sofort die Fälle zu entnehmen, in denen die Leitform nicht quadratfrei ist. Aus der Normalitätsbedingung folgt dann

(2.4.2) LEMMA. *Ist die gewichtet homogene Fläche X normal und ist die affine Gleichung f nicht homogen, so genügen die Gewichte* (*zusätzlich zu* 2.3.1) *einer der folgenden Bedingungen, und die Sub-Leitform* $f_{d-1}$ *ist – bis auf invertierbare Konstanten – wie folgt bestimmt:*

| Fall | Gewichte | $f_{d-1}$, Bemerkungen |
|---|---|---|
| $A, D$ | $w_1 = w_2 = d-1$ | $\prod_{j=1}^{d-1}(\alpha_j z_1 + \beta_j z_2)$ quadratfrei |
| $B_1, E_{11}, F_1$ | $w_1 < w_2 = d-1$ | $z_2^{d-1} + z_1 g$, $g(0,1,0) = 0$ |
| $B_2, E_{12}$ | $1/w_1 + (d-2)/w_2 = 1$ | $z_1 z_2^{d-2} + z_1^2 g$ |
| $C_1, E_{d-1}$ | $1/w_1 + (d-2)/w_3 = 1$ | $z_1 z_3^{d-2} + z_1^2 g$ $(w_1 < d/2)$ |
| $C_2$ | | keine Bedingung |
| $E_k$ | $1/w_1 + (d-2)/w_3 = 1$, $w_1 < d/2$, $w_2 = d-1$ | $z_1 z_3^{d-2} + \mu z_2^{d-1}$, $\mu \neq 0$ $(1 < k < d-1)$ |
| $F_{2,d-1}$ | | keine Bedingung, aber $\mu \neq 0$ *in* $f_d = z_3(z_1 z_3^{d-2} + \mu z_2^{d-1})$ |
| $F_{21,k}$ | $w_2 = d-1$ $(1 < k < d-1)$ | wie $B_1$ |
| $F_{22,k}$ | $1/w_1 + (d-2)/w_2 = 1$ $(1 < k < d-1)$ | wie $B_2$, zusätzlich $\mu \neq 0$ in $f_d = z_3^{d-k}(z_1 z_3^{k-1} + \mu z_2^k)$ |
| $G_1$ | (alle $\nu_j = 1$) | keine Bedingung |
| $G_2$ | (ein $\nu_j > 1$) $w_1 = d/2$ | $z_1 p_{d-2}(z_2, z_3)$, $p_{d-2}$ homogen vom Grad $d-2$, $(\nu_j > 1 \Rightarrow (\alpha_j z_2 + \beta_j z_3) \nmid p_{d-2})$ |

BEWEIS. In den Fällen, in denen die Leitform einen quadratischen Faktor enthält, folgen die Bedingungen aus Bemerkung (2.4.1) und aus der Forderung, daß die affine Singularität isoliert ist (vgl. Beweis zu (2.3.1)). Im Fall $F_{2,d-1}$ folgt aus $\mu = 0$ mit (2.4.1), daß $f$ durch $z_1$ teilbar ist; analog ergibt sich $\mu \neq 0$ für die Fälle $F_{22,k}$.

Durch (2.3.1) und das obige Lemma sind die Gewichte in den Fällen $A$, $D$, $G_2$ bestimmt; im Fall $F_{21,k}$ ergibt sich $w = (d-k,\ d-1,\ (d-1)(d-k)/(d-k-1))$ als einzige Möglichkeit. Bis auf die Fälle $E_k$ mit $1 < k < d-1$ und $G_2$, alle $\nu_j \leqq d-2$, können sämtliche auftretenden Fälle durch Gleichungen aus den Klassen I – VII realisiert werden; dabei sind die Klassen VI und VII nur in den Fällen $F_{22,k}$ erforderlich. Im Fall $G_2$, alle $\nu_j \leqq d-2$, können entsprechende Gleichungen leicht angegeben werden, während für $E_k$ mit $1 < k < d-1$ gewisse Teilbarkeitsbedingungen erfüllt sein müssen.
Umgekehrt gilt

(2.4.3) SATZ. *Eine gewichtet homogene Fläche X, deren affine Gleichung f nicht homogen ist, ist genau dann normal, wenn f irreduzibel ist, eine isolierte (affine) Singularität hat und den Bedingungen aus* (2.3.1) *und* (2.4.2) *genügt.*

BEWEIS. Eine isolierte zweidimensionale Hyperflächensingularität ist normal (siehe [Laufer, Theorem 3.1]); es genügt daher zu zeigen, daß die

Singularitäten $S_\infty(X)$ im unendlich fernen Teil $z_0=0$ der Fläche isoliert sind. Wegen $\hat{f}=f_d+z_0f_{d-1}+z_0^2h$ für geeignetes $h$ ist $S_\infty(X)$ durch die Gleichungen $f_d=f_{d-1}=(\partial/\partial z_i)f_d=(\partial/\partial z_j)f_d=0,\ z_k=1$ (mit $\{i, j, k\}=\{1, 2, 3\}$) gegeben. Eine Diskussion der Fälle ergibt dann die folgende Liste und liefert so den Beweis des Satzes:

(2.4.4) LEMMA. *Sind $X$ und $f$ wie in* (2.4.3), *so liegen im unendlich fernen Teil $X_\infty$ die folgenden Singularitäten (wobei $f_d$ und $f_{d-1}$ bis auf invertierbare Konstanten bestimmt sind und $g$ ein geeignetes Polynom bezeichnet):*

| Fall | $f_d$ | $f_{d-1}$ | $S_\infty(X)$ |
|---|---|---|---|
| $A$ | $z_3^d$ | $\prod_{j=1}^{d-1}(\alpha_jz_1+\beta_jz_2)$ quadratfrei | $[0:\beta_j:-\alpha_j:0]$, $j=1,\cdots,d-1$ |
| $B_1$ | $z_3^d$ | | |
| $E_{11}$ | $z_2z_3^{d-1}$ | | |
| $F_1$; $F_{211,k}$ | $z_1z_3^{d-1}$ | $z_2^{d-1}+z_1z_3g$ | $[0:1:0:0]=:o_1$ |
| $F_{212,k}$ | $z_3^{d-k}(z_1z_3^{k-1}+\mu z_2^k)$ | | |
| $B_2$ | $z_3^d$ | | |
| $E_{12}$ | $z_2z_3^{d-1}$ | $z_1z_2^{d-2}+z_1^2z_3g$ | $o_1$, $[0:0:1:0]=:o_2$ |
| $F_{22,k}$ | wie $F_{212,k}$ | | |
| $C_1$ | $z_2^d$ | $z_1z_3^{d-2}+z_1^2z_2g$ | $o_1$, $[0:0:0:1]=:o_3$ |
| $E_{d-1}$ | $z_2^{d-1}z_3$ | | |
| $C_{21}$ | $z_2^d+\mu z_1z_3^{d-1}, \mu\neq 0$ | $z_1^{d-1}$ $(w_1=d-1)$ | $\emptyset$ |
| $C_{22}$ | wie $C_{21}$ | $(w_1<d-1)$ | $o_1$ |
| $D$ | $(\gamma z_1+\delta z_2)z_3^{d-1}$ | wie $A$, $(\gamma z_1+\delta z_2)$ $\nmid f_{d-1}$ | wie $A$; $(\gamma, \delta)$ $\neq(0, 0)$ |
| $E_k$ | $z_2^kz_3^{d-k}$ | $z_1z_3^{d-2}+\mu z_2^{d-1}$, $\mu\neq 0$ | $o_1$, $o_3$ |
| $F_{2,i,d-1}$ | $z_3(z_1z_3^{d-2}+\mu z_2^{d-1})$ | wie $C_{2,i}$ | wie $C_{2,i}$ |
| $G_{1,i}$ | $\prod_{j=1}^d(\alpha_jz_2+\beta_jz_3)$ quadratfrei | wie $C_{2,i}$ | wie $C_{2,i}$ |
| $G_2$ | $\prod_{j=1}^l(\alpha_jz_2+\beta_jz_3)^{\nu_j}$, $\sum\nu_j=d, l<d$ | $z_1\ \prod_{j=1}^{d-2}(\gamma_jz_2+\delta_jz_3)$, $\nu_j>1\Rightarrow(\alpha_jz_2+\beta_jz_3)$ $\nmid f_{d-1}$ | $o_1$ $[0:0:\beta_j:-\alpha_j]$, $\nu_j>1$ |

## (2.5) Die $\mathbb{C}^*$-Aktion

Bei der *Fortsetzung der $\mathbb{C}^*$-Aktion* aus (1.2) auf $\mathbb{P}_3$, die durch $(t, [z_0:z_1:z_2:z_3])\mapsto[z_0:t^{q_1}z_1:t^{q_2}z_2:t^{q_3}z_3]$ $(t\in\mathbb{C}^*, z\in\mathbb{P}_3)$ gegeben ist, ist $X$ invariant. Damit ergibt sich (ohne Normalitätsbedingung)

(2.5.1) LEMMA. *Gilt für die Gewichte $w_i=r/q_i$ eines gewichtet homogenen Polynoms $w_1<w_2\leqq d\leqq w_3$, so ist die affine Hyperfläche $V_1:=X\cap\{z_1=1\}$ ebenfalls gewichtet homogen, und die Gewichte der definierenden Gleichung*

$f_1(z_0, z_2, z_3) = \hat{f}(z_0, 1, z_2, z_3)$ *sind* $(\tilde{w}_0, \tilde{w}_2, \tilde{w}_3) = (d - w_1, (q_1 d - r)/(q_1 - q_2), (q_1 d - r)/(q_1 - q_3))$.

Beweis. Der Ursprung $o_1 := [0:1:0:0]$ der Karte $\{z_1 = 1\}$ liegt auf $X$, da jedes $z_1$ enthaltende Monom von $\hat{f}$ auch $z_0$, $z_2$ oder $z_3$ enthält. Nun gilt

$$\hat{f}(t^{q_1} z_0, z_1, t^{q_1 - q_2} z_2, t^{q_1 - q_3} z_3) = t^{q_1 d} \hat{f}(z_0, t^{-q_1} z_1, t^{-q_2} z_2, t^{-q_3} z_3) = t^{q_1 d - r} \hat{f}(z_0, z_1, z_2, z_3),$$

und daraus folgt die Behauptung.

Insbesondere ist $o_1$ ein Fixpunkt der $\mathbb{C}^*$-Aktion, der für $\tilde{w}_0 > 1$ singulär ist (Fälle $B$, $C$, $E$, $F$, $G$ außer $w_1 = d - 1$). Im normalen Fall sind die weiteren singulären Punkte ebenfalls Fixpunkte. Die Menge *aller* Fixpunkte der $\mathbb{C}^*$-Aktion auf einer normalen gewichtet homogenen Fläche ist leicht anzugeben: Auf der affinen Varietät $V$ ist der Ursprung $o_0$ der einzige Fixpunkt; es bleiben also nur die Fixpunkte im unendlich fernen Teil zu bestimmen. Ersichtlich gilt

(2.5.2) Bemerkung. Ist $f \in \mathbb{C}[z_1, z_2, z_3]$ *homogen*, so besteht die *Kurve im Unendlichen* $X_\infty$ nur aus *Fixpunkten* der $\mathbb{C}^*$-Aktion.

In den übrigen Fällen erhalten wir

(2.5.3) Bemerkung. Ist $f$ *gewichtet homogen*, aber *nicht homogen*, so liegen *im unendlich fernen Teil* die folgenden *Fixpunkte* der $\mathbb{C}^*$-Aktion

| Fall | $X_\infty \cap \mathrm{Fix}(X, \mathbb{C}^*)$ |
|---|---|
| $A$ | $[0:z_1:z_2:0] = \mathbb{P}_1$ |
| $D$ | $[0:z_1:z_2:0]$, $o_3$ |
| $B$ | $o_1, o_2$ |
| $C$ | $o_1, o_3$ |
| $E, F$ | $o_1, o_2, o_3$ |
| $G$ | $o_1, [0:0:\alpha_j:-\beta_j], j = 1, \cdots, l$ |

## 3. Der Homotopietyp

### (3.0) Überblick

In diesem Abschnitt untersuchen wir den Homotopietyp normaler gewichtet homogener Flächen. Die Grundlage liefert der Homotopieklassifikationssatz (3.1), nach dem der Homotopietyp durch die ganzzahligen Homologiegruppen, die Poincaré-Homomorphismen und die Pontrjagin-Quadrate bestimmt ist. In (3.2) werden kurz einige allgemeine Eigenschaften des Poincaré-Homomorphismus erwähnt; in (3.3) sind speziellere

Ergebnisse für Flächen erwähnt. Besonders wichtig sind die Reduktion auf den ganzzahligen Poincaré-Homomorphismus (3.3.1) und das Diagramm (P). In (3.4) wird der Homotopietyp der Flächen, deren affine Gleichung homogen ist, vollständig bestimmt. Der Schwerpunkt der Untersuchungen liegt in (3.5), wo mit der äquivarianten Auflösung der Singularitäten und der Klassifikation glatter algebraischer Flächen mit $\mathbb{C}^*$-Aktion aus den lokalen Homologiegruppen und dem unendlich fernen Teil die Homologiegruppen und Poincaré-Homomorphismen weitgehend, für einen irreduziblen unendlich fernen Teil sogar vollständig, bestimmt werden. In (3.6) zeigen drei explizite Beispiele, daß aus den Informationen in (3.5) diese Größen auch dann in geeigneten Fällen vollständig berechenbar sind, wenn der unendlich ferne Teil zwei irreduzible Komponenten hat.

### (3.1) Der Homotopieklassifikationssatz

Eine Hyperfläche im projektiven Raum $\mathbb{P}_3(\mathbb{C})$ ist *kompakt* und nach dem Homotopie-Lefschetzsatz *einfach zusammenhängend* (siehe etwa [Mi$_1$, 7.4]). Ist die Fläche zusätzlich *irreduzibel*, so kann das folgende Resultat zur Bestimmung des *Homotopietyps* benutzt werden:

(3.1.1) HOMOTOPIEKLASSIFIKATIONSSATZ. *Ist X eine einfach zusammenhängende, irreduzible kompakte komplexe Fläche, so ist ihr Homotopietyp bestimmt durch*

(i) $H_2(X,\mathbb{Z})$ *und* $b_3(X,\mathbb{Z})$,

(ii) *den Äquivalenztyp der Poincaré-Homomorphismen* $P_2(X,\mathbb{Z}/(m))$: $H^2(X,\mathbb{Z}/(m)) \to H_2(X,\mathbb{Z}/(m))$ *für* $m=0$ *und* $m=p_i^{n_i}$, *wobei die* $p_i^{n_i}$ *durch* $H_2(X,\mathbb{Z}) = b_2\mathbb{Z} \oplus \sum_{i=1}^{t} \mathbb{Z}/(p_i^{n_i})$ *bestimmte Primzahlpotenzen sind,*

(iii) *zusätzlich in dem Fall, daß 2-Torsion auftritt: Die Pontrjagin-Quadrate* $\mathfrak{p}_{2^k}$: $H^2(X,\mathbb{Z}/(2^k)) \to H^4(X,\mathbb{Z}/(2^{k+1}))$ *für* $p_i^{m_i} = 2^k$ *aus* (ii).

Der Beweis in [Ba-Ka] beruht auf J. H. C. Whiteheads Homotopieklassifikation für einfach zusammenhängende endliche (topologisch) vierdimensionale Polyeder [Wh] und auf L. Kaups Ergebnissen über den singulären Poincaré-Homomorphismus [Ka$_1$].

Zu (i) und (ii), insbesondere zum Begriff 'Äquivalenztyp' merken wir an, daß die Daten (i) mit den Voraussetzungen an $X$ die ganzzahligen Homologie- und Kohomologiegruppen bestimmen. Wegen der Kronecker-Dualität von $H_2(X,\mathbb{Z}/(m))$ und Hom $(H_2(X,\mathbb{Z}/(m);\mathbb{Z}/(m)) = H^2(X,\mathbb{Z}/(m))$ ist der Poincaré-Homomorphismus $P_2(X): H^2(X) \to H_2(X)$ die Korrelationsabbildung einer Bilinearform auf $H^2(X)$, die – symmetrische – *Cup-Produktform* auf $H^2(X,\mathbb{Z}/(m))$. Die Kronecker-Dualität und die Poincaré-Homomorphismen sind mit Koeffizientenhomomorphismen $\mathbb{Z}/(l) \to \mathbb{Z}(m)$

verträglich, und als '*Äquivalenztyp*' bezeichnen wir den Äquivalenztyp dieses Systems von symmetrischen Bilinearformen und Koeffizientenhomomorphismen. Das *Pontrjagin-Quadrat* schließlich ist eine *quadratische* Abbildung, die das Cup-Produkt als assoziierte symmetrische bilineare Abbildung hat; für 2-Torsionsklassen wird dadurch die Produktstruktur auf der Kohomologie verfeinert.

Der Homotopieklassifikationssatz gilt unter allgemeineren Voraussetzungen; insbesondere kann er auf den reduziblen Fall übertragen werden (siehe [Ka-Ba$_{1,2}$]).

**(3.2) Allgemeines zum Poincaré-Homomorphismus**

Für die Topologie von *Mannigfaltigkeiten* sind die *Dualitätssätze* von Poincaré, Alexander und Lefschetz von herausragender Bedeutung. Solche Sätze können für singuläre komplexe Räume aus einem einfachen Grund nicht gelten: Die komplementären Homologie- und Kohomologiegruppen sind bei gegebenen Koeffizienten im allgemeinen nicht isomorph. In [Ka$_{1,2}$] wurde eine *singuläre Dualitätstheorie* entwickelt und auf die Untersuchung komplexer Räume angewandt. Ein wichtiges Hilfsmittel ist dabei die gemischte Spektralsequenz aus [Bredon, V, 8.4–8.6]. Die Ergebnisse und Methoden sind topologischer Natur; die Dualitätstheorie gilt daher nicht nur für komplexe Räume, sondern allgemeiner für Räume $X$ mit '*topologischer Normalisierung*' $\pi: \tilde{X} \to X$ (siehe [Ka$_1$, Definition 1.1]): Es gibt Poincaré-Homomorphismen $P_j(X): H^j(\tilde{X}) \to H_{t-j}(X)$ für $0 \leqq j \leqq t$ – wobei $t$ die topologische Dimension von $X$ bezeichnet –, und für lokal abgeschlossene Unterräume $A \subset X$ gibt es relative Homomorphismen $P_j(X, A): H^j(\tilde{X}, \tilde{A}) \to H_{t-j}(X \setminus A)$ und $Q_j(A): H^j(\tilde{A}) \to H_{t-j}(X, X \setminus A)$, so daß sich mit den exakten Homologie- und Kohomologiesequenzen eine kommutative 'Leiter' ergibt (siehe [Ka$_1$, Theorem 2.1]). – Falls $X$ nicht topologisch normal ist, nennt man die zusammengesetzte Abbildung $P_j\pi^j: H^j(X) \to H_{t-j}(X)$ kurz ebenfalls Poincaré-Homomorphismus und bezeichnet sie mit $P_j(X)$, falls keine Verwechselungen zu befürchten sind.

Für unsere folgenden Untersuchungen benötigen wir von der allgemeinen Dualitätstheorie mit beliebigen Trägerfamilien, Koeffizientengarben und topologischer Normalisierung nur den Spezialfall konstanter Koeffizienten, kompakter oder abgeschlossener Träger und *normaler Räume*. In dieser Situation ist der absolute *Poincaré-Homomorphismus* – wie im Fall von Mannigfaltigkeiten – durch das *Cap-Produkt mit der Fundamentalklasse* gegeben (siehe [Ka$_1$, Satz 3.3]), und die gemischte Spektralsequenz reduziert sich im absoluten Fall auf die bereits von [Zeeman] definierte, deren detaillierte geometrische Analyse sich in der Dissertation von [McCrory] findet.

Für topologisch *normale* Räume mit *isolierten* (homologischen) *Singularitäten* – also insbesondere für normale komplexe Flächen – treten alle *Poincaré-Homomorphismen* mit den *lokalen Homologiegruppen* in einer *langen exakten Sequenz*

$$\cdots \to H^0(X, H_{t-j+1}) \to H^j(X) \to H_{t-j}(X) \to H^0(X, H_{t-j}) \to \cdots$$

auf (siehe [$Ka_2$, Einleitung]); entsprechend auch die relativen Poincaré-Homomorphismen. Mit der erwähnten 'Leiter' ergibt sich ein großes *kommutatives Diagramm* (siehe [$Ka_1$, §2, insbesondere Korollar 2.4]).

Das Komplement einer 'guten' Umgebung der Singularitätenmenge $S(X)$ ist eine (Homologie-)Mannigfaltigkeit $W$ mit Rand, deren Homologie eng mit der von $X$ zusammenhängt:

(3.2.1) BEMERKUNG [McC, I. 4. D, Prop. 6]. Die lange exakte *Poincaré-Sequenz* ist *isomorph zur Homologiesequenz des Paares* $(W, \partial W)$.

Es sei am Rande bemerkt, daß diese Homologiesequenz (mit einer Dimensionsverschiebung) isomorph zur Homologiesequenz von $(X, X\backslash S(X))$ ist. Die darin auftretenden absoluten Gruppen identifizieren sich mit geeigneten *'Schnitt-Homologiegruppen'*, die von Goresky und MacPherson definiert und untersucht worden sind.

## (3.3) Der Poincaré-Homomorphismus bei Flächen

Wir kehren jetzt zur Betrachtung komplexer Flächen zurück. Für die Homotopieklassifikation nach 3.1.1 ist es im allgemeinen erforderlich, die Poincaré-Homomorphismen $P_2(X, \mathbb{Z}/(m))$ für verschiedene Koeffizientengruppen zu bestimmen. Für die Klassen aus $H^2(X, \mathbb{Z})\otimes\mathbb{Z}/(m)$ genügt wegen des universellen Koeffizientensatzes für Poincaré-Homomorphismen ([$Ka_2$, Satz 4.3]) die Kenntnis von $P_2(X, \mathbb{Z})$; es bleibt somit $P_2(X, \mathbb{Z}/(m))$ auf der Gruppe der 'geborenen' Torsionsklassen Hom (Tors $H_2(X, \mathbb{Z})$; $\mathbb{Z}/(m)$) zu bestimmen. Wir geben eine Bedingung an – die bei normalen gewichtet homogenen Flächen erfüllt ist –, unter der die Poincaré-Homomorphismen auf diesen Klassen verschwinden. Es folgt damit, daß der Äquivalenztyp von $P_2(X, \mathbb{Z}/(m))$ vollständig durch den von $P_2(X, \mathbb{Z})$ bestimmt ist und daß im Fall von 2-Torsion für die Pontrjagin-Quadrate von 'geborenen' 2-Torsionsklassen $2\mathfrak{p} = 0$ gilt.

(3.3.1) SATZ. *Ist die komplexe Fläche X aus* (3.1.1) *zusätzlich normal und ist* Kern $P_2(X, \mathbb{Z})$ *ein direkter Summand von* $H^2(X, \mathbb{Z})$, *so gilt*

$$P_2(X, \mathbb{Z}/(m)) \mid \text{Hom}(\text{Tors } H_2(X, \mathbb{Z}); \mathbb{Z}/(m)) = 0$$

*für alle* $m \in \mathbb{Z}$.

BEWEIS. Wir betrachten das kommutative Diagramm mit exakten Zeilen

$$\begin{array}{ccccccc}
H^0(X, \mathcal{H}_3) & \xrightarrow[\gamma_1]{} & H^2(X) & \xrightarrow[P_2]{} & H_2(X) & \xrightarrow[\alpha_2]{} & H^0(X, \mathcal{H}_2) \\
\uparrow\cong & & \uparrow\cong & & \uparrow\cong & & \uparrow\cong \\
H_2(\partial W) & \longrightarrow & H_2(W) & \xrightarrow[j_2]{} & H_2(W, \partial W) & \xrightarrow[\partial_1]{} & H_1(\partial W) \\
\uparrow\cong & & \uparrow\cong & & & & \\
H^1(\partial W) & \xrightarrow{\delta^1} & H^2(W, \partial W) & & & &
\end{array}$$

Dabei ist $(W, W)$ die kompakte berandete Mannigfaltigkeit aus (3.2.1), und die senkrechten Pfeile sind durch Poincaré-Lefschetz-Dualität, Ausschneidung oder Homotopie gegebene Isomorphismen. Da $H^2(X, \mathbb{Z})$ frei ist, folgt mit der Voraussetzung, daß die Torsionsuntergruppe $T_2 := \operatorname{Tors} H_2(X, \mathbb{Z})$ unter $\alpha_2$ injektiv nach $H^0(X, \mathcal{H}_2(X, \mathbb{Z}))$ abgebildet wird. Aus der entsprechenden Aussage für $\partial_1$ folgt mit dem 'kontravarianten' universellen Koeffizientensatz, daß $\operatorname{Hom}(T_2, \mathbb{Z}/(m))$ im Bild von $\gamma_1$ liegt.

Um $\operatorname{Hom}(T_2, \mathbb{Z}/(m))$ als Untergruppe von $H^2(X, \mathbb{Z}(m))$ zu interpretieren, muß eine Spaltung von $H_2(X, \mathbb{Z})$ in $T_2$ und einen komplementären freien Summanden $\bar{H}_2$ gewählt werden. Bei geeigneter Wahl folgt aus (3.3.1), daß $H^2(X, \mathbb{Z})$ unter $P_2(X)$ in $\bar{H}_2$ abgebildet wird. Weiter können wir $P_2(X)$ auf den *nicht ausgearteten Anteil* reduzieren. Bezeichnen wir Kern $P_2(X, \mathbb{Z})$ kurz mit $I$ und einen zu $I$ komplementären Summanden von $H^2(X, \mathbb{Z})$ mit $M$, so bildet $P_2(X, \mathbb{Z})$ diese Gruppe $M$ injektiv in die duale (bezüglich des Kronecker-Produktes $\langle\ ,\ \rangle$) Gruppe $M^* := I^\perp \cap \bar{H}_2$ ab (dabei ist $I^\perp = \{a \in H_2(X, \mathbb{Z}) : \langle a, I\rangle = 0\}$), und $P_2(X, \mathbb{Z})$ ist durch diese Einschränkung $\bar{P}_2 : M \to M^*$ eindeutig bestimmt. Als Folgerung ergibt sich die direkte Summenzerlegung

$$\text{Kokern } P_2(X, \mathbb{Z}) = I^* \oplus T_2 \oplus \text{Kokern } \bar{P}_2 \quad (\text{mit } I^* := M^\perp \cap \bar{H}_2).$$

Neben diesen Ergebnissen sind für unsere Untersuchungen weitere allgemeine Eigenschaften wichtig. In unserem Fall einer *Hyperfläche* $X$ vom Grad $d$ in $\mathbb{P}_3(\mathbb{C})$ gilt für den *Rang* des ganzzahligen Poincaré-Homomorphismus stets $rg\ P_2(X, \mathbb{Z}) \geqq 1$: Ist $j : X \hookrightarrow \mathbb{P}_3$ die Einbettung und $w \in H^2(\mathbb{P}_3)$ das kanonische Erzeugende, so gilt $j_* P_2(X, \mathbb{Z}) j^* w = d[\mathbb{P}_1]$. Da $j^* w$ unteilbar ist, folgt insbesondere für $b_2(X) = 1$, daß $P_2(X, \mathbb{Z})$ die Multiplikation mit $d$ (modulo Torsion) ist. Diese Aussagen sind in [Ba-Ka, Lemma 3.2 und 3.3] bewiesen. Wir bemerken, daß sie – mit der Konvention von (3.2) – auch für den nicht normalen Fall gelten.

Zu unserer Untersuchung des Homotopietyps normaler gewichtet homogener Flächen benutzen wir das große *kommutative Poincaré-Diagramm* $(P)$

aus [Ba-Ka, Satz 2.1]. Wählen wir darin als Unterraum $A$ die Kurve $X_\infty$, so ist $X\backslash A = V$ zusammenziehbar und hat den Nullpunkt als einzige Singularität. Es gilt also $H^0(V, \mathcal{H}_i(X)) = \mathcal{H}_i(X)_0 =: \mathcal{H}_{i,0}$ (für $i = 2$ und 3). Mit den Bezeichnungen $\Gamma_X\mathcal{H}_i := H^0(X, \mathcal{H}_i(X))$ und $\Gamma_\infty\mathcal{H}_i := H^0(X_\infty, \mathcal{H}_i(X))$ erhalten wir in unserer Situation das Diagramm

(P)

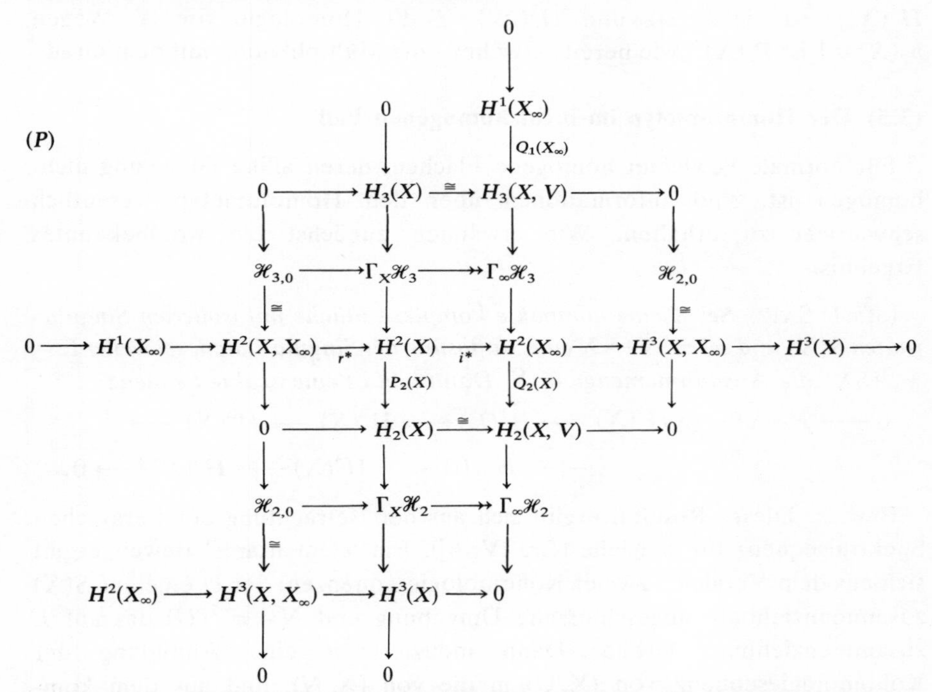

Eine zusätzliche Vereinfachung ergibt sich, wenn die affine Gleichung von $X$ zwar gewichtet homogen, aber *nicht homogen* ist: Nach (2.3.2) ist $X_\infty$ dann einfach zusammenhängend und folglich gilt $H^1(X_\infty) = 0$. Ist eine normale gewichtet homogene Fläche $X$ durch ihre affine Gleichung gegeben, so ist $X_\infty$ – und damit auch die Kohomologie von $X_\infty$ – bestimmt; weiter ist die lokale Homologie $\mathcal{H}_{2,0}$ explizit berechenbar. Die Gruppe $\Gamma_\infty\mathcal{H}_2$ ist die direkte Summe der lokalen Homologien $\mathcal{H}_{2,p}$ für die Punkte $p \in S(X)$; auch diese Gruppen sind aus der lokalen Gleichung berechenbar. Weiter ist wegen der lokalen Poincaré-Dualität (siehe [Ka₂, Korollar 1.2]) die lokale Homologiegarbe $\mathcal{H}_3(X, \mathbb{Z})$ der freie Anteil von $\mathcal{H}_2(X, \mathbb{Z})$.

### (3.4) Der Homotopietyp im homogenen Fall

Als erste Anwendung erhalten wir:

(3.4.1) Satz. *Der Homotopietyp einer normalen Fläche X, deren affine*

*Gleichung homogen vom Grad d ist, ist durch*

$$H_2(X,\mathbb{Z})=\mathbb{Z},\qquad H_3(X,\mathbb{Z})=(d-1)(d-2)\mathbb{Z},\qquad P_2(X,\mathbb{Z})=(d)$$

*– also durch ihren Grad – vollständig bestimmt.*

BEWEIS. Aus (2.2) folgt $\Gamma_\infty\mathcal{H}_2=0$, daher ergibt das Diagramm $(P)$ mit $H^1(X_\infty)=(d-1)(d-2)\mathbb{Z}$ und $H^2(X_\infty)=\mathbb{Z}$ die Homologie von $X$. Wegen $b_2(X)=1$ ist $P_2(X)$ – wie bereits erwähnt – die Multiplikation mit dem Grad.

### (3.5) Der Homotopietyp im nicht homogenen Fall

Für normale gewichtet homogene Flächen, deren affine Gleichung nicht homogen ist, sind Informationen über den Homotopietyp wesentlich schwieriger zu erhalten. Wir erwähnen zunächst ein wohlbekanntes Ergebnis.

(3.5.1) SATZ. *Sei $X$ eine kompakte komplexe Fläche mit isolierten Singularitäten $S(X)$ und sei $\pi:\hat{X}\to X$ eine Auflösung der Singularitäten, sei weiter $E=\pi^{-1}(SX)$ die Ausnahmemenge in $\hat{X}$. Dann gibt es eine exakte Sequenz*

$$0\longrightarrow H^1(X)\xrightarrow[\pi^*]{} H^1(\hat{X})\xrightarrow[i^*]{} H^1(E)\longrightarrow H^2(X)\longrightarrow H^2(\hat{X})\longrightarrow$$
$$\longrightarrow H^2(E)\longrightarrow H^3(X)\longrightarrow H^3(\hat{X})\longrightarrow 0.$$

BEWEIS. Dieses Resultat ergibt sich aus der Betrachtung der Lerayschen Spektralsequenz für $\pi$ (siehe [Br, IV, 6]). Ein 'elementarer' Beweis ergibt sich aus dem Vergleich zweier Kohomologiesequenzen: Sei $U$ eine auf $S(X)$ zusammenziehbare abgeschlossene Umgebung und $N=\pi^{-1}(U)$ das auf $E$ zusammenziehbare Urbild. Dann induziert $\pi$ eine Abbildung der Kohomologiesequenz von (X, U) in die von $(\hat{X},N)$, und aus dem kommutativen Diagramm erhält man die Behauptung. Einen detaillierten Beweis (für reelle Koeffizienten) findet man bei [Brenton, Lemma (1.7)].

Für unsere Untersuchungen ist eine Folgerung besonders wichtig:

(3.5.2) KOROLLAR. *Sind $X$ und $\hat{X}$ wie in* 3.5.1, *so gilt*

$$b_3(X)=b_3(\hat{X}).$$

BEWEIS. In der obigen exakten Sequenz mit ganzzahligen Koeffizienten hat die Abbildung $H^2(\hat{X})\to H^2(E)$ maximalen Rang, denn die duale Homologieabbildung $H_2(E)\to H_2(\hat{X})$ ist injektiv, weil die Schnittform auf dem Bild nach [Mumford, S. 6] und [Grauert, S. 367] negativ definit ist.

Wir bemerken, daß beide Ergebnisse richtig bleiben, wenn $\pi$ auch endlich viele glatte Punkte modifiziert.

Der Bezug zur Diskussion des *Homotopietyps einer normalen gewichtet homogenen Fläche $X$* liegt in der $\mathbb{C}^*$-Aktion, die auf $X$ definiert ist (2.5). Die

Singularitäten von $X$ gehören zu der Fixpunktmenge dieser Aktion. Auf der *minimalen Auflösung* von $X$ gibt es dann nach [Or-Wa$_4$, 3.2, Lemma] genau eine $\mathbb{C}^*$-Aktion, so daß $\pi$ *äquivariant* ist. Aus den Ergebnissen von Orlik und Wagreich über *algebraische Flächen mit* $\mathbb{C}^*$-*Aktion* folgt.

(3.5.3) SATZ. *Ist $X$ eine normale gewichtet homogene Fläche, deren affine Gleichung nicht homogen ist, und ist $\hat{X}$ eine äquivariante Auflösung von $X$, so gilt*

$$b_3(\hat{X}) = b_{3,0} = rg\mathscr{H}_3(X, \mathbb{Z})_0 = b_{3,\infty} = rg\Gamma_\infty\mathscr{H}_3(X, \quad ).$$

BEWEIS. Ohne Einschränkung kann angenommen werden, daß in der glatten Fläche $\hat{X}$ keine *elliptischen* Fixpunkte der $\mathbb{C}^*$-Aktion existieren. Dabei heißt ein Fixpunkt elliptisch, wenn $rs > 0$ gilt, wobei die ganzen Zahlen $r$ und $s$ dadurch eindeutig bestimmt sind, daß in lokalen Koordinaten $(x, y)$ die Aktion durch $t(x, y) = (t^r x, t^s y)$ gegeben ist (siehe [Or-Wa$_4$, 2.3]). Man sieht nämlich sofort, daß durch iterierte quadratische Transformationen in den elliptischen Fixpunkten erreicht werden kann, daß genau eine Komponente der Ausnahmemenge nur aus Fixpunkten besteht – in denen dann in der Notation von oben $rs = 0$ gilt; solche Fixpunkte heißen *parabolisch*-, während sonst auf der Ausnahmemenge nur isolierte Fixpunkte auftreten, in denen entsprechend $rs < 0$ gilt – diese heißen *hyperbolisch*.

Auf $\hat{X}$ kann also der 'topologische' *Klassifikationssatz für glatte Flächen mit* $\mathbb{C}^*$-*Aktion und ohne elliptische Fixpunkte* [Or–Wa$_4$, Theorem 2.5] angewandt werden. Demnach besteht die Fixpunktmenge von $\hat{X}$ aus (isolierten) hyperbolischen Fixpunkten und zwei zueinander isomorphen glatten kompakten Kurven $F^+$ und $F^-$, und $\hat{X}$ entsteht aus $F^+ \times \mathbb{P}_1(\mathbb{C})$ durch Modifikation in Fixpunkten und 'Abblasen' von Ausnahmekurven der ersten Art. Es folgt $b_3(\hat{X}) = b_3(F^+ \times \mathbb{P}_1(\mathbb{C})) = b_1(F^+)$. Nun ist $F^+$ gerade die *zentrale Kurve der kanonischen äquivarianten Auflösung der affinen Singularität* [Or-Wa$_1$, Theorem (2.3.1)]. Ist $K$ der Umgebungsrand, so gilt nach [Or-Wa$_1$, Remark (2.6.2)] $b_1(K) = b_1(F^+) = 2g$, wobei $g$ das Geschlecht von $F^+$ ist. Nun ist $b_{3,0} = b_2(K) = b_1(K)$, und damit ist $b_3(\hat{X}) = b_{3,0}$ gezeigt.

Hat $X$ eine weitere *gewichtet homogene Singularität im Unendlichen* (siehe 2.5.1), so ist $F^-$ die zentrale Kurve der Auflösung dieser Singularität. Unsere Behauptung $b_3(\hat{X}) = b_{3,\infty}$ ist für diesen Fall gezeigt, wenn wir nachgewiesen haben, daß weitere Singularitäten keinen Beitrag zu der lokalen Bettizahl leisten. Sei also $p$ ein solcher Punkt und $E_p = \pi^{-1}(p)$. Dann besteht $E_p$ aus Bahnen der $\mathbb{C}^*$-Aktion und hyperbolischen Fixpunkten, die von $F^+$ und $F^-$ getrennt sind; der duale Graph ist also – da notwendigerweise zusammenhängend (siehe [La, Lemma 4.1]) – linear (als zusammenhängender Teilgraph des Graphen von $(\hat{X}, \mathbb{C}^*)$, der $F^+$ und $F^-$ nicht enthält, siehe

[Or-Wa$_4$, Theorem 2.5, (ii)]), und die Kurven sind isomorph zu $\mathbb{P}_1(\mathbb{C})$. Es folgt sofort mit der üblichen Rechnung (siehe [Mumford]) $b_1K_p = b_1E_p = 0$, und wie oben also $b_{3,p} = 0$.

Es bleibt der Fall zu betrachten, daß $X_\infty$ *keine gewichtet homogene Singularität* enthält. Nach (2.5.1) sind dies die Fälle $A$ und $D$. Die obige Überlegung liefert $b_{3,\infty} = 0$. Nach (2.5.3) besteht die Fixpunktmenge im Unendlichen aus $\mathbb{P}_1$ und zusätzlich in Fall $D$ einem weiteren Punkt. Bei der Auflösung der auf $\mathbb{P}_1$ liegenden singulären Punkte – die durch die $d-1$ Linearfaktoren der Sub-Leitform gegeben sind, siehe (2.4.4) – bleibt als eigentliche Transformierte diese Fixpunktkurve $F^- = \mathbb{P}_1$ erhalten. Nach dem Klassifikationssatz folgt $F^+ = \mathbb{P}_1$ und somit $b_3(\hat{X}) = 0$, was zu zeigen war.

Aus der Gleichheit $b_3(X) = b_{3,0} = b_{3,\infty}$, die also nach (3.5.2) und (3.5.3) gilt, folgt mit dem Poincaré-Diagramm $(P)$.

(3.5.4) Satz. *Ist $X$ eine normale gewichtet homogene Fläche mit nicht homogener affiner Gleichung, so gilt (mit ganzzahligen Koeffizienten)*

(i) $b_2(X) = b_3(X) + b_2(X_\infty)$
(ii) $rgP_2(X) = b_2(X_\infty)$
(iii) Kern $P_2(X) =$ Bild $H^2(X, X_\infty)$
(iv) Kern $P_2(X)$ *ist ein direkter Summand von* $H^2(X)$
(v) Bild $P_2(X) \subset$ Bild $H_2(X_\infty)$.

Beweis. Aus der exakten Zeile

$$0 \to H^2(X, X_\infty) \to H^2(X) \to H^2(X_\infty) \to H^3(X, X_\infty) \to H^3(X) \to 0$$

folgt mit $H^2(X, X_\infty) \cong \mathcal{H}_{3,0}$ und $H^3(X, X_\infty) \cong \mathcal{H}_{2,0}$ sowie mit $b_{2,0} = b_{3,0} = b_3(X)$ die Behauptung (i). Ensprechend ergibt sich (ii) aus der exakten Spalte

$$0 \longrightarrow H^3(X) \longrightarrow \Gamma_X\mathcal{H}_3 \longrightarrow H^2(X) \xrightarrow{P_2} H_2(X)$$

mit (i) und $rg\Gamma_X\mathcal{H}_3 = b_{3,0} + b_{3,\infty} = 2b_3(X)$.

Der Anfang der obigen exakten Zeile

$$0 \to H^2(X, X_\infty) \to H^2(X) \to H^2(X_\infty)$$

besteht aus freien Gruppen; also ist das Bild von $H^2(X, X_\infty)$ ein direkter Summand von $H^2(X)$. Diese Gruppe liegt im Bild von $\Gamma_X\mathcal{H}_3$ und hat den gleichen Rang $b_3(X)$, sie muß also mit Bild $\Gamma_X\mathcal{H}_3 =$ Kern $P_2(X)$ übereinstimmen. Damit sind auch (iii) und (iv) gezeigt.

Die Aussage (v) folgt aus (iii); sie kann aber auch unmittelbar aus dem kommutativen Diagramm

$$\begin{array}{ccc} H^2(X, V) & \xrightarrow{\cong} & H^2(X) \\ \downarrow{\scriptstyle P_2(X,V)} & & \downarrow{\scriptstyle P_2(X)} \\ H_2(X_\infty) & \xrightarrow[i_*]{} & H_2(X) \end{array}$$

– einem Ausschnitt des Poincaré-Diagramms für $(X, V)$ mit *abgeschlossenen* Trägern – abgelesen werden.

Daß in (v) im allgemeinen nicht die Gleichheit gilt, zeigt etwa [Ba-Ka, Satz 3.7, (iv)].

Als Anwendung können wir für den Fall, daß die Kurve $X_\infty$ *irreduzibel* ist, die Homologie und den Poincaré-Homomorphismus vollständig bestimmen. Nach (2.3.2) treten entweder $\mathbb{P}_1$ oder die singuläre Kurve $Q_d := \{z_0 = z_2^d + \mu z_1 z_3^{d-1} = 0\}$, $\mu \neq 0$ auf. Der zweite Fall reduziert sich auf den ersten, indem man $X$ in der lokalen Karte $\{z_1 = 1\}$ betrachtet: Aus der Gewichtsbedingung $C_2$ folgt, daß jedes Monom von kleinerem Grad durch $z_1$ teilbar ist; in der Karte $\{z_1 = 1\}$ ergibt sich als neue Leitform dann $z_2^d$.

(3.5.5) Satz. *Ist X eine im Affinen normale gewichtet homogene Fläche mit unendlich ferner Kurve $\mathbb{P}_1$, so gilt*

(i) $H_2(X; \mathbb{Z}) = \mathbb{Z} \oplus \mathcal{H}_2(X; \mathbb{Z})_0$, $H_3(X, \mathbb{Z}) = \mathcal{H}_3(X, \mathbb{Z})_0$

(ii) $H^2(X, \mathbb{Z}) \cong j^* H^2(\mathbb{P}_3(\mathbb{C}), \mathbb{Z}) \oplus r^* H^2(X, X_\infty; \mathbb{Z})$

(iii) $P_2(X)(j^* w) = d i_* [X_\infty]$.

Die *Beweise* aus [Ba-Ka; Satz 3.5, Bem. 3.6, Satz 3.7] übertragen sich sofort; zum Beweis von [Ba-Ka, Satz 3.5] sei ergänzend angemerkt, daß auch die exakte Sequenz

$$0 \longrightarrow H_2(X_\infty) \xrightarrow[i_*]{} H_2(X) \longrightarrow H_2(X, X_\infty) \longrightarrow 0$$

spaltet und deshalb $H_2(X) = i_* H_2(X_\infty) \oplus H_2(X, X_\infty)$ gilt.

Natürlich gilt im normalen Fall auch das Ergebnis $\Gamma_\infty \mathcal{H}_2 = \mathcal{H}_{2,0} \oplus \mathbb{Z}/(d)$ aus [Ba-Ka, Korollar 3.8, Beweis]: (hier ist nachzutragen, daß die Behauptung

$$\mathcal{H}_2(V, \mathbb{Z})_0 = (s_3 - 1)(s_2 - 1)\mathbb{Z} \oplus (s_3 - 1)\mathbb{Z}/(t_3) \oplus (s_2 - 1)\mathbb{Z}/(t_2) \oplus \mathbb{Z}/(a_3)$$

heißen muß).

Diese Ergebnisse sind bereits in [Ka-Ba$_1$, Proposition 3.2] und [Ka-Ba$_2$, 6] aufgeführt.

In den verbleibenden Fällen haben wir keine allgemeine Aussage zur Verfügung, mit der sich die Torsion von $H_2(X, \mathbb{Z})$ oder der Äquivalenztyp von $P_2(X)$ aus den bekannten Daten (lokale Homologie, Grad, $b_2(X)$) ergibt. Das Diagramm $(P)$ ergibt jedoch einige weitere Informationen. Sei dazu $T_2$ (bzw. $T_{2,0}$ bzw. $T_{2,\infty}$) die Torsionsuntergruppe von $H_2(X, \mathbb{Z})$ (bzw. $\mathcal{H}_{2,0}$ bzw. $\Gamma_\infty \mathcal{H}_2$, Koeffizienten $\mathbb{Z}$). Dann gibt es – wie in [Ka-Ba$_2$, 7] für $b_2(X_\infty) = 2$ bereits gezeigt – *exakte Sequenzen*

$$0 \to T_2 \to T_{2,\infty} \to G_\infty \to 0$$

und

$$0 \to G_0 \to T_{2,0} \to T_2 \to 0,$$

wobei $G_0 := H^2(X_\infty)/\text{Bild}\, H^2(X)$ und $G_\infty$ die Torsionsuntergruppe von Kokern $(H^2(X_\infty) \to H_2(X, V)/T_2)$ ist. Dazu betrachten wir die kurze exakte Sequenz

$$0 \longrightarrow H^2(X_\infty) \xrightarrow[Q_2(X_\infty)]{} H_2(X, V) \longrightarrow \Gamma_\infty \mathcal{H}_2 \longrightarrow 0$$

sowie die exakte Zeile aus $(P)$:

$$H^2(X) \to H^2(X_\infty) \to H^3(X, X_\infty) \to H^3(X) \to 0$$

und beachten, daß $H^3(X) = b_{2,0}\mathbb{Z} \oplus T_2$ gilt.

Zur weiteren Diskussion des Poincaré-Homomorphismus reduzieren wir auf den *nicht-ausgearteten Anteil* $\bar{P}_2$. Mit den Bezeichnungen aus (3.3) ist $M$ isomorph zu der freien abelschen Gruppe $H^2(X)/\text{Bild}\, H^2(X, X_\infty)$ vom Rang $b_2(X_\infty)$ und $M^*$ zu der freien abelschen Gruppe $(r^*H^2(X, X_\infty))^\perp/T_2$, und $\bar{P}_2$ ist die Korrelationsabbildung zum nicht ausgearteten Anteil des Cup-Produktes. Für $\bar{P}_2$ – oder die dadurch repräsentierte symmetrische Bilinearform $\sigma(X)$ auf $M$ – erhalten wir

(3.5.6) Satz. *Ist $X$ eine normale gewichtet homogene Fläche mit nicht homogener affiner Gleichung vom Grad $d$, so gilt mit den oben eingeführten Bezeichnungen*

(i) *Es gibt eine exakte Sequenz*

$$0 \to G_0 \to \text{Kokern}\, \bar{P}_2 \to G_\infty \to 0$$

*insbesondere ist* $|\det \sigma| = |T_{2,0} \oplus T_{2,\infty}|/|T_2|^2$.

(ii) $\sigma(j^*w, j^*w) = d$.

(iii) $\text{sign}\, \sigma = 2 - rg\sigma = 2 - b_2(X_\infty)$.

Beweis. Aus dem Poincaré-Diagramm $(P)$ erhält man das kommutative Diagramm

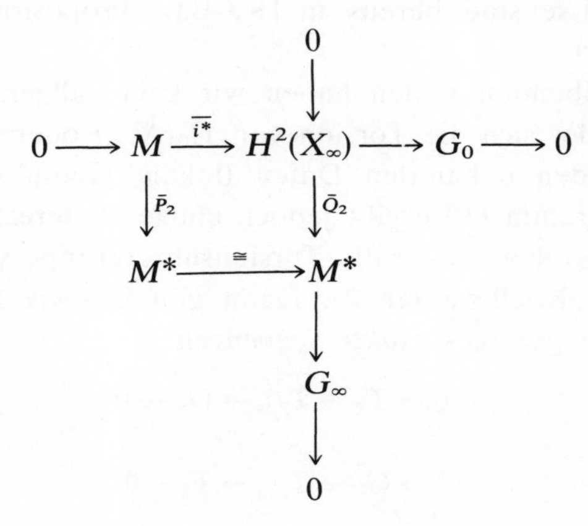

$$\begin{array}{ccccccc} & & & & 0 & & \\ & & & & \downarrow & & \\ 0 \longrightarrow & M & \xrightarrow{\overline{i^*}} & & H^2(X_\infty) & \longrightarrow G_0 \longrightarrow & 0 \\ & \downarrow \bar{P}_2 & & & \downarrow \bar{Q}_2 & & \\ & M^* & \xrightarrow{\cong} & & M^* & & \\ & & & & \downarrow & & \\ & & & & G_\infty & & \\ & & & & \downarrow & & \\ & & & & 0 & & \end{array}$$

mit exakten Zeilen und Spalten. Mit den Bezeichnungen $Q := \text{Bild } \bar{Q}_2$, $P := \text{Bild } \bar{P}_2$ gilt dann $G_\infty = M^*/Q \cong (M^*/P)/(Q/P) = \text{Kokern } \bar{P}_2/(Q/P)$. Nun ist $Q/P$ zu $G_0$ isomorph, und daraus folgt die Behauptung (i). Die Aussage (ii) ist äquivalent zu der in (3.3) erwähnten Gleichung $j_* P_2(X) j^* w = d[\mathbb{P}_1]$; sie besagt insbesondere, daß die Form $\sigma$ nicht negativ definit ist. Es bleibt die Aussage über die Signatur zu beweisen. Sei dazu $\pi : \hat{X} \to X$ eine äquivariante Auflösung und $X$ ohne elliptische Fixpunkte. Die Fläche $F^+ \times \mathbb{P}_1(\mathbb{C})$, aus der $\hat{X}$ durch Auf- und Abblasen entsteht, hat eine unimodulare Cup-Produktform von Rang 2 und der Signatur 0. Es folgt, daß $\hat{X}$ die Signatur $2 - b_2(\hat{X})$ hat. Die Cup-Produktform auf $\hat{X}$ ist unimodular – insbesondere also nicht ausgeartet –, und $\pi$ ist eine Abbildung vom Grad 1 (es gilt $\pi_*[\hat{X}] = [X]$). Daher ist Kern $P_2(X)$ auch der Kern von $\pi^* : H^2(X) \to H^2(\hat{X})$, und $\pi^*$ induziert eine Abbildung, die $M$ als Untergitter in das unimodulare Gitter $H^2(X)$ eingebettet. Da die Form $\sigma$ auf $M$ einen positiven Wert annimmt, ist der positive Eigenraum eindimensional, und daraus folgt die Behauptung.

Ergänzend bemerken wir noch, daß für den Fall torsionsfreier Homologie von $X$ der Kokern von $P_2$ (und $\bar{P}_2$) die Torsionsuntergruppe von $\Gamma_X \mathcal{H}_2$ ist. Dieser Fall tritt beispielsweise für die affinen Gleichungen $z_1^{a_1} + z_2^d + z_3^d$, $1 \leqq a_1 < d$ und $z_1^{a_1} z_2 + z_2^d + z_3^d$, $1 \leqq a_1 < d-1$, auf (siehe [Ba-Ka, Satz 3.11] und [Ka-Ba$_2$, Proposition 8.1] für die Torsionsfreiheit der Homologie).

## (3.6) Beispiele

Durch die Ergebnisse des Satzes (3.5.7) ist die Bestimmung des Poincaré-Homomorphismus auf die Untersuchung indefiniter, nicht ausgearteter ganzzahliger symmetrischer Bilinearformen reduziert worden. Dabei treten häufig *binäre* Formen auf. Solche Formen sind ausgiebig untersucht worden, und unsere Beispiele zeigen, daß in geeigneten Fällen der Poincaré-Homomorphismus explizit bestimmt werden kann.

(3.6.1) Beispiel. Sei $X$ durch die affine Gleichung

$$z_1 z_2 + z_2 z_3^2 + z_1 z_3 = 0$$

definiert; dann gilt

$$H_2(X, \mathbb{Z}) = 2\mathbb{Z},\ H_3(X, \mathbb{Z}) = 0,$$

$$P_2(X, \mathbb{Z}) = \begin{bmatrix} 3 & 0 \\ 0 & -6 \end{bmatrix}.$$

Beweis. Die Fläche hat zwei gewichtet homogene Singularitäten mit $w = (3/2, 3, 3)$ in $o_0$ und $o_1$ und eine weitere in $o_2$ mit der lokalen Gleichung $z_0 z_1 (1 + z_3) + z_3^2 = 0$ (dabei $o_i$: Ursprung in der Karte $\{z_i = 1\}$). Die zweiten lokalen Homologiegruppen sind $C_3$, $C_3$ und $C_2$ (dabei $C_n$: zyklische Gruppe der Ordnung $n$). Somit gilt $b_{2,0} = b_3(X) = 0$ und $b_2(X) = b_2(X_\infty) = 2$. Bezüglich einer Basis von $H^2(X)$, die $j^*w$ enthält, und der dualen Basis von $H_2(X)/T_2$ hat $P_2(X)$ die Matrixdarstellung

$$\begin{bmatrix} 3 & b \\ b & a \end{bmatrix}.$$

Aus $\mathscr{H}_{2,0} = C_3$ folgt $T_2 = C_3$ oder $T_2 = 0$. Die Annahme $T_2 = C_3$ führt auf $\det P_2 = -2$ und damit auf die Bedingung $b^2 = 3a + 2$. Da die Kongruenz $b^2 \equiv 2 \bmod 3$ unlösbar ist, ist also notwendig $T_2 = 0$, und daraus folgt $\det P_2 = -18$. Die Gleichung $b^2 = 3a + 18$ wird gelöst durch $a = 3\alpha$, $b = 3\beta$ mit $\beta^2 = \alpha + 2$. Die Formen

$$\begin{bmatrix} 3 & 3\beta \\ 3\beta & 3\beta^2 - 6 \end{bmatrix} \quad \text{und} \quad \begin{bmatrix} 3 & 3(\beta+1) \\ 3(\beta+1) & 3(\beta+1)^2 - 6 \end{bmatrix}$$

sind äquivalent, und folglich repräsentiert jede die Cup-Produktform. Da die Homologie torsionsfrei ist, ist der Homotopietyp dieser Fläche vollständig bestimmt.

Unser nächstes Beispiel ist eine Fläche vom Grad 4. (Es handelt sich dabei um eine Gleichung, die zum Fall $G_{2,\nu_1} = \nu_2 = 2$ gehört, der (vgl. (2.4.2)) nicht in den Klassen I–VII auftritt.)

(3.6.2) Beispiel. Sei $X$ durch die affine Gleichung

$$z_1^2 + z_1 z_2^2 + (z_2 + z_3)^2 z_3^2$$

definiert; dann gilt

$$H_2(X, \mathbb{Z}) = 4\mathbb{Z} \oplus C_2, \qquad H_3(X, \mathbb{Z}) = 2\mathbb{Z},$$

$$P_2(X, \mathbb{Z}) = \begin{bmatrix} 4 & 2 \\ 2 & 0 \end{bmatrix} \oplus 0.$$

Beweis. Es gilt $\mathscr{H}_{2,0} = 2\mathbb{Z} \oplus C_2$ und $\Gamma_\infty \mathscr{H}_2 = 2\mathbb{Z} \oplus 3C_2$ sowie $b_2(X_\infty) = 2$. Die Form $\bar{P}_2(X)$ hat die Matrixdarstellung $\begin{bmatrix} 4 & b \\ b & a \end{bmatrix}$. Aus $T_{2,0} = C_2$ folgt $T_2 = C_2$ oder $T_2 = 0$. Nun ist $T_{2,\infty} = 3C_2$ eine Gruppe mit drei Erzeugenden. Da $G_\infty$ höchstens zwei Erzeugende hat, folgt $T_2 = C_2 = T_{2,0}$, also $G_0 = 0$, und somit Kokern $\bar{P}_2 = G_\infty = 2C_2$, so daß $\bar{P}_2$ die Elementarteiler $2, 2$ hat. Aus $b^2 = 4a + 4$ ergibt sich mit $b = 2\beta$ dann $a = \beta^2 - 1$. Für gerades $\beta$ ist $a$ ungerade, so daß sich die Elementarteiler $1, 4$ ergeben. Für ungerades $\beta$ sind die resultierenden Formen äquivalent, und für $\beta = 1$ erhalten wir den obigen Repräsentanten.

Eine kleine Variation ist

(3.6.3) BEISPIEL. Sei $X$ durch die affine Gleichung

$$z_1^2 + z_1 z_2^2 + z_2 z_3^3$$

definiert; dann gilt

$$H_2(X, \mathbb{Z}) = 4\mathbb{Z}, \qquad H_3(X, \mathbb{Z}) = 2\mathbb{Z}$$

$$P_2(X, \mathbb{Z}) = \begin{bmatrix} 4 & 2 \\ 2 & -2 \end{bmatrix} \oplus 0.$$

Der Beweis bleibt dem Leser überlassen.

## 4. Der topologische Typ

### (4.0) Überblick

In diesem Abschnitt untersuchen wir den Homöomorphietyp normaler gewichtet homogener Flächen. In (4.1) zeigen wir durch einige Beispiele, daß aus Homotopieäquivalenz oder auch aus der Übereinstimmung der Gewichte, Grade, und Zahl der irreduziblen Faktoren der Leitform der affinen Gleichung nicht allgemein die Homöomorphie folgt. Die Daten der affinen Gleichung, die übereinstimmen müssen, nennen wir den 'numerischen Typ'. In (4.2) führen wir die Familie $Y$ der normalen gewichtet homogenen Flächen vom gegebenen numerischen Typ ein und zeigen, daß der Parameterraum $Z$ eine Untermannigfaltigkeit eines geeigneten $\mathbb{C}^N$ ist. In (4.3) benutzen wir das 'generische Whitneylemma' und das erste Thom-Mathersche Isotopielemma, um zu zeigen, daß fast alle Flächen vom gleichen numerischen Typ homöomorph sind.

### (4.1) Beispiele nicht-homöomorpher Flächen

Beispiele für *homotopieäquivalente*, aber *nicht homöomorphe* normale gewichtet homogene Flächen sind leicht anzugeben, wie etwa die durch die Gleichungen $z_1 + z_2^4 + z_3^5$ und $z_1^3 + z_2^4 + z_3^5$ definierten Flächen (siehe [Ba-Ka, §3.3]).

In (1.4) hatten wir bereits eine Aussage über den topologischen Typ erwähnt: Stimmen die *Gewichte* der affinen Gleichung überein, so sind die affinen Teile homöomorph. Daß daraus nicht die Homöomorphie der projektiven Flächen folgt, zeigt

(4.1.1) BEISPIEL. Seien $X_1$ und $X_2$ durch die Gleichungen

$$f_1 = z_1^2 + z_2^3 + z_1 z_3^2, \qquad f_2 = z_1^2 + z_2^3 + z_3^4$$

vom Gewicht (2, 3, 4) gegeben. Es gilt $b_2(X_1) = b_2(X_2) = 1$ und $P_2(X_1) = (3)$, $P_2(X_2) = (4)$; also sind die Flächen nicht vom gleichen Homotopietyp.

Daß auch bei gleichen Gewichten und gleichem *Grad* der affinen Gleichung die Flächen nicht homöomorph sein müssen, zeigt

(4.1.2) Beispiel. Seien $X_1, X_2, X_3, X_4$ durch die Gleichungen

$$f_1 = z_1^2 + z_1 z_2^2 + z_3^4, \qquad f_2 = z_1^2 + z_1 z_2^2 + z_2 z_3^3,$$
$$f_3 = z_1^2 + z_1 z_2^2 + (z_2^2 + z_3^2) z_3^2, \qquad f_4 = z_1^2 + z_1 z_2^2 + z_2^4 + z_3^4$$

vom Gewicht $(2, 4, 4)$ gegeben. Dann gilt $b_2(X_{j,\infty}) = j$ und somit $b_2(X_j) = 2 + j$; also sind die Flächen nicht vom gleichen Homotopietyp.

Daß schließlich auch die Übereinstimmung der Bettizahlen noch nicht ausreicht, zeigen die Beispiele (3.6.2) und (3.6.3). Die beiden Gleichungen stimmen in Gewicht, Grad, und der *Anzahl der irreduziblen Faktoren* der Leitform, aber nicht in deren Vielfachheiten, überein.

Wir bemerken ergänzend, daß in den Fällen, in denen zwei Gewichte übereinstimmen, auch der Grad bestimmt ist. Mit der Bezeichnung aus (2.3), (2.4) sind dadurch in den Fällen $A$ und $D$ sowie im Fall $G$ mit $w_1 \neq d/2$ die Daten $\Gamma_{S(X)}\mathcal{H}_2$ und $H_*(X_\infty)$ zur Bestimmung des Homotopietyps auch bestimmt. In den Fällen $B$, $C$, $E$ sowie $F_1$, $F_{2,d-1}$ und $F_{22,k}$ genügen Gewicht und Grad. Lediglich für die Fälle $F_{21,k}$ $(1 < k < d-1)$ und $G$ mit $w_1 = d/2$ sind zusätzliche Angaben über die Leitform erforderlich: Bei $F_{21,k}$, ob nicht-lineare Faktoren auftreten, und bei $G_2$ die Angabe der Vielfachheiten der Linearfaktoren.

## (4.2) Der numerische Typ

Aus den in (4.1) erwähnten Beispielen folgt, daß im allgemeinen normale Flächen in den Gewichten und dem Grad der affinen Gleichungen und – in den Fällen $F_{21,k}$ mit $k \neq d-1$ und $G$ mit $w_1 = d/2$ – qualitativ in der Faktorisierung der Leitform übereinstimmen müssen, um homöomorph sein zu können.

Diese Daten nennen wir den '*numerischen Typ*' der Gleichung. Genauer ist also der numerische Typ gegeben durch die Gewichte $w = (w_1, w_2, w_3)$ (wobei ohne Einschränkung $w_1 \leqq w_2 \leqq w_3$ gelte), den Grad $d$, sowie zusätzlich im Fall $F_{21,k}$ mit $1 < k < d-1$ – also für $w = (d-k, d-1, (d-k)(d-1)/(d-k-1))$ – die Zahl $\mu = 0$ oder 1 und im Fall $G$ mit $w = (d/2, d, d)$ das ungeordnete $l$-tupel $\langle \nu \rangle := \langle \nu_1, \cdots, \nu_l \rangle$ ganzer Zahlen mit $\nu_j \geqq 1$ und $\sum_{j=1} \nu_j = d$.

Zu einem gegebenen numerischen Typ seien $m_1, \cdots, m_N \in \mathbb{C}[z_1, z_2, z_3]$ die normierten Monome mit Grad $m_i =: d_i \leqq d$, die gewichtet homogen mit Gewichten $w = (w_1, w_2, w_3)$ sind, lediglich im Fall $F_{21,k}$ mit $\mu = 0$ werde

$z_2^k z_3^{d-k}$ ausgeschlossen. Für jeden Punkt $\alpha = (\alpha_1, \cdots, \alpha_N) \in \mathbb{C}^N \setminus \{0\}$ ist $f_\alpha := \sum_{i=1}^N \alpha_i m_i$ ein gewichtet homogenes Polynom mit Grad $f_\alpha \leqq d$. Sei $X_\alpha$ die durch $f_\alpha$ definierte Fläche. Wir betrachten die Menge $Z$ aller $\alpha \in \mathbb{C}^N \setminus \{0\}$, so daß $f_\alpha$ vom gegebenen numerischen Typ ist und eine normale Fläche $X_\alpha$ definiert; dabei sei im Fall $F_{21,k}$, $\mu = 1$, der Koeffizient von $z_2^k z_3^{d-k}$ stets invertierbar.

Die Menge $Y := \{(z, \alpha) \in \mathbb{P}_3(\mathbb{C}) \times Z : f_\alpha(z) = 0\}$, zusammen mit der natürlichen Projektion $\pi$ auf $Z$, ist dann die *Familie der normalen gewichtet homogenen Flächen vom gegebenen numerischen Typ*, und $Z$ ist der zugehörige *Parameterraum*. Wir bemerken, daß $\pi : Y \to Z$ eine *eigentliche* Abbildung ist.

Wir erhalten das folgende Ergebnis:

(4.2.1) LEMMA. *Der Parameterraum $Z$ für die normalen Flächen vom gegebenen numerischen Typ ist eine Untermannigfaltigkeit des $\mathbb{C}^N$, und zwar außer im Fall $G_2$ sogar eine offene dichte Teilmenge.*

BEWEIS. In den Fällen $A$–$F$ sind die Koeffizienten der Leitform und der Sub-Leitform nach (2.3.1), (2.4.2), und (2.4.4) bis auf invertierbare Konstanten bestimmt. In der Form kleinsten Grades $e$ muß – je nach den Gewichten – mindestens eins der Monome $z_1^e$, $z_1^{e-1} z_2$ oder $z_1^{e-1} z_3$ mit einem invertierbaren Koeffizienten auftreten. Durch diese Bedingungen wird eine echte algebraische Teilmenge des $\mathbb{C}^N$ ausgeschlossen. Als einzige weitere Bedingung bleibt dann noch, daß $f_\alpha$ irreduzibel ist, und das besagt ebenfalls, daß $\alpha$ außerhalb einer echten algebraischen Teilmenge des $\mathbb{C}^N$ liegt. In diesen Fällen ist $Z$ also offen und dicht in $\mathbb{C}^N$.

Im *homogenen* Fall ist $f_\alpha$ eine ternäre Form von Grad $d$, und mit (2.5) ergibt sich $Z$ als Komplement eines echten algebraischen Kegels in $\mathbb{C}^N$.

Im Fall $G$ treten unter den $m_i$ die Monome $z_2^j z_3^{d-j}$ auf, nach Umindizierung sei $\alpha_j$ der Koeffizient von $z_2^j z_3^{d-j}$.

Das Polynom $p_\alpha := \sum_{j=0}^d \alpha_j z_2^j z_3^{d-j}$ hat eine Faktorisierung in $l$ Linearformen mit Vielfachheiten $\langle \nu \rangle = \langle \nu_1, \cdots, \nu_l \rangle$. Der Fall $G_1$ – also $\langle \nu \rangle = \langle 1, \cdots, 1 \rangle$ – liegt genau dann vor, wenn die Diskriminante $D(\alpha_0, \cdots, \alpha_d)$ nicht verschwindet, wenn also $(\alpha_0, \cdots, \alpha_d)$ außerhalb der Diskriminantenmenge $\Delta \subset \mathbb{C}^{d+1}$ liegt. Wie oben ergibt sich $Z$ als offene dichte Teilmenge des $\mathbb{C}^N$.

Es bleibt der Fall $G_2$ zu behandeln. Die Diskriminantenmenge $\Delta = \{D(\alpha_0, \cdots, \alpha_d) = 0\} \setminus \{0\}$ hat eine natürliche Stratifikation in die Mengen $\Delta^{\langle \nu \rangle}$, wo $\langle \nu \rangle$ die Vielfachheit der Linearfaktoren beschreibt. Diese Mengen $\Delta^{\langle \nu \rangle}$ sind Untermannigfaltigkeiten von $\mathbb{C}^{d+1}$, und wie oben ergibt sich $Z$ als offene dichte Teilmenge von $\Delta^{\langle \nu \rangle} \times \mathbb{C}^{N-d-1}$.

**(4.3) Numerischer und topologischer Typ**

Wir wollen das folgende Ergebnis beweisen:

(4.3.1) SATZ. *Fast alle Flächen vom gleichen numerischen Typ sind homöomorph.*

Der Satz ist eine unmittelbare Folgerung aus

(4.3.2) LEMMA. *Die natürliche Projektion* $\pi: Y \to Z$ *aus* (4.2) *ist außerhalb einer echten analytischen Teilmenge von* $Z$ *lokal trivial.*

BEWEIS. Der Beweis beruht auf dem *ersten Isotopielemma* von Thom und Mather und auf dem *'generischen Whitneylemma'* für stratifizierte analytische Räume. Wir haben in $Y$ die natürliche Zerlegung $(Y \setminus S(Y), S(Y))$ in die Mannigfaltigkeit der regulären Punkte und die Singularitätenmenge. Nun gilt

(4.3.3) LEMMA. *Die Singularitätenmenge* $S(Y)$ *ist eine Mannigfaltigkeit, und* $\pi \mid S(Y)$ *ist ein lokaler Isomorphismus.*

Zum *Beweis* benutzt man, daß $S(Y) \cap X_\alpha = S(X_\alpha)$ gilt. Außer in den Fällen $A$, $D$, und $G_2$ hängt aber $S(X_\alpha)$ nach (2.4.4) nicht von $\alpha$ ab, so daß sich $S(Y) = S(X_\alpha) \times Z$ ergibt. In den genannten Fällen hängen die unendlich fernen Singularitäten (siehe (2.4.4)) analytisch von den Koeffizienten der Sub-Leitform (Fall $A$ und $D$) beziehungsweise der Leitform (Fall $G_2$) ab, die affine Singularität ist unabhängig.

Da die Zerlegung endlich ist und die Randbedingung erfüllt, handelt es sich also um eine Stratifizierung von $Y$ im Sinne von Whitney [Wy, §18]. Weiter gilt

(4.3.4) LEMMA. *Die Projektion* $\pi$ *ist auf den Strata eine Submersion.*

Das ist für $S(Y)$ klar, für $Y \setminus S(Y)$ genügt es zu zeigen, daß – in einer lokalen Karte $\{z_i = 1\}$ – der Gradient der Gleichung $\hat{f}_\alpha(z)$ im Punkt $(z, \alpha)$ nicht im Tangentialraum $T_\alpha Z$ liegt. Da aber $z$ nicht in $S(X_\alpha)$ liegt, ist mindestens eine partielle Ableitung $\partial/\partial z_j$ von Null verschieden.

Das 'generische Whitneylemma' [Wy, Lemma 19.3] besagt, daß die *Stratifikation generisch* – außerhalb einer echten analytischen Teilmenge $S_b$ von $S(Y)$ – *regulär* ist, also Whitneys Bedingungen $a$ und $b$ [Wy, §19] genügt. Nach (4.3.3) ist $\pi(S_b)$ eine echte analytische Teilmenge von $Z$. Dann ist $Z_r := Z \setminus \pi(S_b)$ eine zusammenhängende offene dichte Teilmenge der Mannigfaltigkeit $Z$. Ist $Y_r$ das Urbild von $Z_r$, so kann auf $\pi: Y_r \to Z_r$ das erste Thom-Mathersche Isotopielemma [Mather, Proposition 11.1] angewandt werden: *Eine eigentliche Abbildung einer regulär stratifizierten Menge auf eine Mannigfaltigkeit, die auf den Strata submersiv ist, ist topologisch lokal trivial.* Damit ist (4.3.2) bewiesen.

LITERATUR

V. I. ARNOLD
Normal forms of functions in neighborhoods of degenerate critical points. Uspehi Mat. Nauk *29*:2 (1974), 11–49; Russian Math. Surveys *29*:2 (1974), 10–50.

G. BARTHEL, L. KAUP
Homotopieklassifikation einfach zusammenhängender normaler kompakter komplexer Flächen. Math. Annalen 212 (1974), 113–144.

G. BREDON
Sheaf Theory. McGraw-Hill, New York 1967.

L. BRENTON
A Riemann-Roch theorem for singular surfaces. Sonderforschungsbereich 40 Theoretische Mathematik, Bonn 1974.

H. GRAUERT
Über Modifikationen und exzeptionelle analytische Mengen. Math. Annalen 146 (1962), 331–368.

F. HIRZEBRUCH, K. JÄNICH
Involutions and singularities. Tata Sympos. on Alg. Geometry, Oxford University Press 1969, 219–240.

F. HIRZEBRUCH, W. NEUMANN, S. KOH
Differentiable manifolds and quadratic forms. Lecture notes in Pure and Applied Math. 4, Marcel Dekker, New York 1971.

L. KAUP
[1] Poincaré-Dualität für Räume mit Normalisierung. Notas de mat. No. 12, Univ. Nac. La Plata 1970 und Ann. Sc. Norm. Sup. Pisa *26* (1972), 1–31.
[2] Zur Homologie projektiv algebraischer Varietäten. Notas de mat. No. 15, Univ. Nac. La Plata 1970 und Ann. Sc. Norm. Sup. Pisa *26* (1972), 479–513.

L. KAUP, G. BARTHEL
[1] On the homotopy type of weighted homogeneous compact complex surfaces. Notes for the 22nd AMS Summer Institute on Several Complex Variables, Williamstown (Mass.) 1975.
[2] On the Homotopy type of weighted homogeneous normal complex surfaces. Several Complex Variables, Proc. of Symp. in Pure Math. 30: 1 (1977), 263–271.

H. LAUFER
Normal two-dimensional singularities. Annals of Math. Studies 71, Princeton University Press 1971.

J. MATHER
Notes on topological stability. Harvard Univ. 1970.

C. MCCRORY
Poincaré Duality in spaces with singularities. Dissertation, Brandeis Univ. 1972.

J. MILNOR
[1] Morse theory. Annals of Math. Studies 51, Princeton University Press 1963.
[2] Singular Points of complex hypersurfaces. Annals of Math. Studies 61, Princeton Univ. Press 1968.

J. MILNOR, P. ORLIK
Isolated singularities defined by weighted homogeneous polynomials. Topology *9* (1970), 385–393.

D. MUMFORD
The topology of normal singularities of an algebraic surface and a criterion for simplicity. Pub. Math. I.H.E.S. *9*, 5–22, Paris 1961.

W. NEUMANN
$S^1$-Actions and the -invariant of their involutions. Bonner Math. Schriften 44, Bonn 1970.

P. Orlik

[1] Seifert manifolds. Lecture notes in math. 291, Springer 1972.

[2] On the homology of weighted homogeneous manifolds. Proc. Sec. Conf. Comp. Transf. Groups I, 260–269, Lecture notes in math. 298, Springer 1972.

P. Orlik, P. Wagreich

[1] Isolated singularities of algebraic surfaces with $\mathbb{C}^*$-action. Ann. Math. *93* (1971), 205–228.

[2] Singularities of algebraic surfaces with $\mathbb{C}^*$-action. Math. Ann. *193* (1971), 121–135.

[3] Equivariant resolution of singularities with $\mathbb{C}^*$-action. Proc. Sec. Conf. Comp. Transf. Groups I, 270–290, Lecture notes in math. 298, Springer 1972.

[4] Algebraic surfaces with $k^*$-action. Erscheint in Acta math.

E. Spanier

Algebraic topology. McGraw-Hill, New York 1966.

J. H. C. Whitehead

On simply connected, 4-dimensional polyhedra. Comm. Math. Helv. *22* (1949), 48–92.

H. Whitney

Tangents to an analytic variety. Ann. Math. 81 (1965), 496–549.

C. Zeeman

Dihomology III, A generalization of the Poincaré duality for manifolds. Proc. London Math. Soc. *13* (1963), 155–183.

Gottfried Barthel
Universität Konstanz
Fachbereich Mathematik
Postfach 7733
D-7750 Konstanz
Bundesrepublik Deutschland

*Zusatz bei der Korrektur:* Mit Teissiers Äquisingularitätskriterium "$\mu^*$ konstant" (siehe dazu seine Vortragsreihe sowie seine Artikel in "Singularités à Cargèse" und "Algebraic Geometry, Arcata 1974") kann gezeigt werden, daß für einige (vermutlich sogar für alle) Familien aus (4.2) die Stratifizierung aus (4.3) überall regulär ist. Damit kann (4.3.1) verschärft werden. Diese Ergebnisse sollen – mit einer partiellen Umkehrung der topologischen Klassifikation – an anderer Stelle ausgeführt werden.

Nordic Summer School/NAVF
Symposium in Mathematics
Oslo, August 5–25, 1976

# GENERIC EQUIVARIANT MAPS

Edward Bierstone

## 1. Introduction

In this paper we obtain several results concerning generic properties of smooth ($\mathscr{C}^\infty$) maps which preserve a compact Lie group of symmetries. The results center on a theory of general position of jets of equivariant maps – an analog of Thom's transversality theory in the category of equivariant maps.

General position is used to prove that equivariant stability of a smooth proper equivariant map implies equivariant infinitesimal stability. This completes the program begun by Poenaru [13], [14] and Ronga [15] to prove Mather's stability theorem in the presence of symmetries.

Let $G$ be a compact Lie group, and $V$, $W$ linear $G$-spaces. The module $\mathscr{P}(V, W)^G$ of equivariant polynomial maps from $V$ to $W$ is finitely generated over the ring $\mathscr{P}(V)^G$ of invariant polynomials on $V$. Let $F_1, \cdots, F_k$ be a set of generators. Then any map $f$ in the space $\mathscr{C}(V, W)^G$ of smooth equivariant maps from $V$ to $W$ may be written $f(x) = \sum_{i=1}^k h_i(x)F_i(x)$, where $h_1, \cdots, h_k$ are smooth invariant functions on $V$. In other words, $f$ factors as $f = U \circ (I, h)$, where $I$ is the identity map of $V$, $h(x) = (h_1(x), \cdots, h_k(x))$, and $U(x, h) = \sum_{i=1}^k h_iF_i(x)$. In [4] we proposed:

Definition 1.1. $f$ is *in general position with respect to* $0 \in W$ *at* $0 \in V$ if the map $(I, h)$ is transverse to the minimum Whitney stratification of the real affine algebraic subvariety $(U = 0)$ of $V \times \mathbb{R}^k$, at $0 \in V$.

Let $N$, $P$ be smooth $G$-manifolds, and $Q$ an invariant submanifold of $P$. Using the differentiable Slice Theorem, the above local definition provides a notion of general position of a map $f \in \mathscr{C}(N, P)^G$ with respect to $Q$. (M. J. Field gave another definition of '$G$-transversality' in [7], and showed that it is equivalent to Definition 1.1 in [6]). Equivariant maps in general position display a rich variety of singular behavior (see the examples of [4, Section 2]). We showed in [4] that Definition 1.1 is, nevertheless, a natural extension to the category of equivariant maps of Thom's idea of transversality. In particular, if $Q$ is closed, then the set of smooth equivariant maps which are in general position with respect to $Q$ is open and dense in the Whitney topology.

The theory presents two problems, however, which are resolved in this paper.

The first concerns general position of jets of equivariant maps. The actions of $G$ on $N$, $P$ induce a natural action on the $q$-jet bundle $J^q(N, P)$, for each natural number $q$. Suppose $Q$ is an invariant submanifold of $J^q(N, P)$. If $f \in \mathscr{C}(N, P)^G$, one might use the above definition to express the idea of general position of the $q$-jet extension $j^q f \in \mathscr{C}(N, J^q(N, P))^G$ with respect to $Q$. With this definition, however, it is in general *false* that the set of smooth equivariant maps whose $q$-jets are in general position with respect to $Q$ is dense.[1]

EXAMPLE 1.2. Let $S^1$ act on $\mathbb{R}\times\mathbb{C}$ by $e^{i\theta}(t, x) = (t, e^{i\theta}x)$, where $e^{i\theta} \in S^1$, $(t, x) \in \mathbb{R}\times\mathbb{C}$. Identify $J^1(\mathbb{R}\times\mathbb{C}, \mathbb{R})$ with $(\mathbb{R}\times\mathbb{C})\times\mathbb{R}\times(\mathbb{R}\times\mathbb{C})$, so that if $f \in \mathscr{C}(\mathbb{R}\times\mathbb{C}, \mathbb{R})$, then $j^1 f(t, x)$ is identified with $((t, x), f(t, x), Df(t, x))$. The set of smooth invariant functions $f: \mathbb{R}\times\mathbb{C} \to \mathbb{R}$ such that $j^1 f$ is in general position with respect to the invariant submanifold $Q = \mathbb{R}\times\mathbb{C}\times\mathbb{R}\times\mathbb{R}\times\{0\}$ of $J^1(\mathbb{R}\times\mathbb{C}, \mathbb{R})$ is not dense in the space of smooth invariant functions.

Why? Let $\mathbb{C}$ be the last factor in $J^1(\mathbb{R}\times\mathbb{C}, \mathbb{R})$ (i.e. the orthogonal complement of $Q$). Any map $F \in \mathscr{C}(\mathbb{R}\times\mathbb{C}, \mathbb{C})^{S^1}$ may be written $F(t, x) = H_1(t, \bar{x}x)x + H_2(t, \bar{x}x)ix$, where $H_1$, $H_2$ are smooth real-valued functions. $F$ is in general position with respect to 0 (in a neighborhood of the $t$-axis) if and only if for each $t$, $H_1(t, 0)$, $H_2(t, 0)$ are not both zero. But if $f(t, x) = H(t, \bar{x}x)$ is a smooth invariant function, then

$$j^1 f(t, x) = \left((t, x), H(t, \bar{x}x), \left(\frac{\partial H}{\partial t}(t, \bar{x}x), 2\frac{\partial H}{\partial y}(t, \bar{x}x)x\right)\right),$$

so that $j^1 f$ is in general position with respect to $Q$ (in a neighborhood of the $t$-axis) if and only if $(\partial H/\partial y)(t, 0) \neq 0$ for each $t$.

Consider $f(t, x) = H(t, \bar{x}x) = t\bar{x}x$. Since $(\partial H/\partial y)(t, 0) = t$, then $j^1 f$ is not in general position with respect to $Q$. Neither is $j^1 g$ for any smooth invariant function $g$ in a suitable neighborhood of $f$.

A generic notion of general position for jets of equivariant maps is given in section 7, but depends on a more careful study of the local structure of equivariant maps which is made in sections 2–4.

Consider linear $G$-spaces $V$, $W$. Let $p_1, \cdots, p_l$ be a set of generators for the algebra $\mathscr{P}(V)^G$, and $F_1, \cdots, F_k$ a set of generators for the $\mathscr{P}(V)^G$-

[1] Example 1.2 is a counterexample to the theorem announced by M. Field in C.R. Acad. Sci. Paris 282 (1976), A379–A380 (MR 53 #4124).

module $\mathscr{P}(V, W)^G$. By the theorem of G. W. Schwarz [16], a map $f \in \mathscr{C}(V, W)^G$ may be written $f = \sum_{i=1}^{k} (H_i \circ P)F_i$, where $P = (p_1, \cdots, p_l)$, and $H = (H_1, \cdots, H_k)$ is a smooth map. Thus $j^q f$ factors as the composition

$$j^q f = U^q_{P,F} \circ (I, H^q \circ P), \tag{1.3}$$

where $H^q(x) = (H(x), DH(x), \cdots, D^q H(x))$, and $U^q_{P,F}: V \times \mathscr{P}^q(l, k) \to J^q(V, W)$ is an equivariant polynomial map depending on the $q$-jets of $P$ and $F_1, \cdots, F_k$ (section 4.3). Here $\mathscr{P}^q(l, k) = \mathbb{R}^k \times L(\mathbb{R}^l, \mathbb{R}^k) \times L^2_s(\mathbb{R}^l; \mathbb{R}^k) \times \cdots \times L^q_s(\mathbb{R}^l; \mathbb{R}^k)$, where $L^j_s(\mathbb{R}^l; \mathbb{R}^k)$ denotes the space of $j$-multilinear symmetric maps from $\mathbb{R}^l$ to $\mathbb{R}^k$. The polynomial map $U^q_{P,F}$ will play a role, in general position of jets of equivariant maps, analogous to that of $U$ in definition 1.1. Definition 1.1 is recovered when $q = 0$.

Theorems 7.6, 7.7, and 7.8 give the openness, density, and isotopy properties of equivariant general position. We have grouped together in section 5 the main technical lemmas on which the theory depends. These lemmas provide natural transformations of the local representation (1.3) under changes in the local data, and a comparison of the local representations at nearby points. In section 6, we recall some basic facts about transversality to a Whitney stratification, and interpret the invariance lemmas of section 5 in terms of invariance of the local transversality conditions for equivariant general position.

The second problem, which was raised in [4], is illustrated by the following example.

EXAMPLE 1.4. Consider $S^1$ acting on $\mathbb{C} = \mathbb{R}^2$ by the standard action, and on $\mathbb{C}^2$ by $e^{i\theta}(y_1, y_2) = (e^{i\theta} y_1, e^{2i\theta} y_2)$. Any map $f \in \mathscr{C}(\mathbb{C}, \mathbb{C}^2)^{S^1}$ may be written $f(x) = (\alpha(\bar{x}x)x, \beta(\bar{x}x)x^2)$, where $\alpha$, $\beta$ are smooth complex-valued functions. $f$ is in general position with respect to $0 \in \mathbb{C}^2$ at $0 \in \mathbb{C}$ if the map $x \mapsto (x, \alpha(\bar{x}x), \beta(\bar{x}x))$ is transverse at $0 \in \mathbb{C}$ to the affine algebraic variety $\{(x, \alpha, \beta) \in \mathbb{C}^3 \mid \alpha x = 0, \beta x^2 = 0\}$. The map $x \mapsto (x, \alpha(\bar{x}x), \beta(\bar{x}x))$ is transverse to this variety at $x = 0$, provided that $\alpha(0)$, $\beta(0)$ are not both zero. Definition 1.1 does not distinguish maps $f \in \mathscr{C}(\mathbb{C}, \mathbb{C}^2)^{S^1}$ which are non-singular at 0; i.e. maps $f$ such that $\alpha(0) \neq 0$. However, if we consider also the derivative of the local representation $f(x) = (\alpha x, \beta x^2)$, then the $\beta$-axis is obtained as a distinguished subvariety of the above affine variety. An additional requirement of transversality with respect to this subvariety gives the condition $\alpha(0) \neq 0$.

Example 1.4 may be regarded from the following simple point of view. Suppose $N$, $P$ are smooth manifolds, and $Q$ is a submanifold of $P$. If $f \in \mathscr{C}(N, P)$ is transverse to $Q$, then for each $q = 0, 1, \cdots$, the map $j^q f \in \mathscr{C}(N, J^q(N, P))$ is automatically transverse to the submanifold $J^q(N, Q)$ of $J^q(N, P)$ (in fact $j^q f(N) \cap J^q(N, Q) = \varnothing$ for $q > 0$, except in the trivial case dim $Q = \dim P$). The corresponding statement for equivariant general position, however, is false.

This observation leads to a refinement of the theory of equivariant general position which is given in section 8. Let $N$, $P$ be $G$-manifolds, and $Q$ an invariant submanifold of $P$. We will say (definition 8.1) that $f \in \mathscr{C}(N, P)^G$ is in $r$th approximation to general position with respect to $Q$ if, for each $q = 0, \cdots, r$, the map $j^q f$ is in general position with respect to $J^q(N, Q)$. The sequence of approximations (locally) stabilizes (theorem 8.4), giving in some sense an ultimate notion of general position of equivariant maps. The general position theorems extend to this context (theorem 8.6).

In section 10, we prove that stability of a proper smooth equivariant map $f \in \mathscr{C}(N, P)^G$ (i.e. stability under the action of equivariant diffeomorphisms) implies equivariant infinitesimal stability. We show that the stability of $f$ implies a general position condition, using a multijet version of the equivariant general position theorem, which is given in section 9. It follows from the general position condition, by the same method as that used by Mather [11], that $f$ is infinitesimally stable in the space of equivariant maps. Together with Poenaru's result, this gives the equivalence, for proper equivariant maps, of stability, infinitesimal stability, and a general position condition (theorem 10.7).

## 2. Preliminaries

2.1. Throughout the paper, $G$ will denote a compact Lie group, and $e$ the identity element of $G$. Manifolds will always be smooth ($\mathscr{C}^\infty$). If $N$, $P$ are smooth manifolds, then $\mathscr{C}(N, P)$ denotes the space of smooth maps from $N$ to $P$; $\mathscr{C}(N) = \mathscr{C}(N, \mathbb{R})$ is the ring of smooth functions on $N$. If $G$ acts smoothly on $N$, $P$ ($N$, $P$ are '$G$-manifolds'), then there is an induced action on $\mathscr{C}(N, P)$, defined by $(g \cdot f)(x) = g(f(g^{-1}x))$, where $g \in G$, $x \in N$, $f \in \mathscr{C}(N, P)$. The fixed point set $\mathscr{C}(N, P)^G$ of this action is the space of smooth equivariant maps from $N$ to $P$. $G$ will be assumed to act trivially on $\mathbb{R}$, so that $\mathscr{C}(N)^G$ is the ring of smooth invariant functions on $N$. $\mathscr{C}(N, P)^G$ is a Baire space in either the $\mathscr{C}^\infty$ or Whitney topology [4, Proposition 3.1].

Let $V$, $W$ be finite dimensional vector spaces. $\mathscr{P}(V, W)$ denotes the space of polynomial maps from $V$ to $W$; $\mathscr{P}(V) = \mathscr{P}(V, \mathbb{R})$. If $G$ acts linearly on $V$, $W$ ($V$, $W$ are 'linear $G$-spaces'), then there is an induced action on $\mathscr{P}(V, W)$. $\mathscr{P}(V)^G$ is the ring of invariant polynomial functions on $V$, and $\mathscr{P}(V, W)^G$ is the $\mathscr{P}(V)^G$-module of equivariant polynomial maps from $V$ to $W$. Likewise, of course, $\mathscr{C}(V, W)^G$ is a $\mathscr{C}(V)^G$-module.

The symbol $I$ will always denote an identity map.

2.2. Let $N$ be a $G$-manifold. The orbit space of $N$ is denoted $N/G$. If $x \in N$, then $Gx$ denotes the orbit of $x$ under the action of $G$, and $G_x$ is the isotropy subgroup of $x$. If $H=G_x$, then $Gx$ is called an orbit of *type* $(H)$ ($(H)$ denotes the conjugacy class of the closed subgroup $H$ of $G$). Let $N_G(H)$ be the normalizer of $H$ in $G$. Let $N^H=F(H,N)$ be the fixed point submanifold of $H$ in $N$, $N_{(H)}$ the bundle of orbits of type $(H)$ in $N$, and $N_H=N^H\cap N_{(H)}$, the fixed point submanifold of $H$ in $N_{(H)}$. Recall that $N_H$ is a (right) principal bundle over $N_{(H)}/G\cong N_H/N_G(H)$, with structure group $N_G(H)/H$. $N_{(H)}$ is the bundle associated to $N_H$, with fiber $G/H$.

2.3. If $\pi: E\to N$ is a smooth fiber bundle, and $Q$ a smooth submanifold of $N$, then $E\mid Q$ will denote $\pi^{-1}(Q)$.

Let $H$ be a closed subgroup of $G$, and $V$ a linear $H$-space. Then $G\times_H V$ denotes the orbit space of $G\times V$ under the action of $H$ defined by $h(g,x)=(gh^{-1},hx)$; $[g,x]$ denotes the image of $(g,x)$ under the orbit projection. $G\times_H V$ is the $G$-vector bundle with fiber $V$ associated to the principal bundle $G\to G/H$. Since $G$ has the structure of a real algebraic linear group, then $G/H$ and $G\times_H V$ are real affine algebraic varieties, and the projection $G\times_H V\to G/H$ is a morphism. We identify $G/H$ with the image of the zero section (so that the identity coset $eH\in G/H$ is identified with $[e,0]\in G\times_H V$). The inclusion map $i(x)=[e,x]$ of $V$ into $G\times_H V$ identifies $V$ with an $H$-invariant submanifold of $G\times_H V$.

The canonical projection $G\to G/H$ is equivariant with respect to the action of $H$ on $G$ by conjugation, and on $G/H$ by left translation. There is an $H$-invariant open neighborhood $U$ of $eH$ in $G/H$, and an analytic $H$-equivariant local section $\sigma: U\to G$, with $\sigma(eH)=e$. The map $(gH,x)\mapsto[\sigma(gH),x]$ of $U\times V$ onto $G\times_H V\mid U$ defines an analytic $H$-vector bundle equivalence over $U$.

Let $\pi: E\to G\times_H V$ be a smooth $G$-fiber bundle. If $Q$ is an invariant submanifold of $E$, then $Q\mid V=Q\cap\pi^{-1}(V)$ is an $H$-invariant submanifold of $E\mid V$. The inclusion map $E\mid V\to E$ induces a $G$-fiber bundle equivalence $G\times_H(E\mid V)\cong E$, under which $G\times_H(Q\mid V)\cong Q$.

The local structure of a smooth $G$-manifold $N$ is described by the Slice Theorem [5, Chapter VI, Theorem 2.2]. Let $x\in N$, and $H=G_x$. There is a linear $H$-space $V$ and an equivariant diffeomorphism $\eta$ of $G\times_H V$ onto an invariant open tubular neighborhood of $Gx$ in $N$, such that $\eta([e,0])=x$.

## 3. Jets

3.1. Let $N$, $P$ be smooth manifolds, and $q$ a natural number. If $f\in\mathscr{C}(N,P)$, then $j^qf(x)$ or $j^q_xf$ denotes the $q$-jet of $f$ at $x\in N$. $J^q(N,P)$ denotes the bundle of $q$-jets of maps in $\mathscr{C}(N,P)$. Though $J^q(N,P)$ is not a vector bundle over $N\times P$, it has a canonical zero-section which maps $(x,y)\in N\times P$ into the $q$-jet at $x$ of the constant map to $y$.

Let $V$, $W$ be finite-dimensional vector spaces, and $L_s^k(V; W)$ the space of $k$-multilinear symmetric maps from $V$ to $W$. Let $\mathcal{P}^q(V, W) = W \times L(V, W) \times L_s^2(V; W) \times \cdots \times L_s^q(V; W)$. $\mathcal{P}^q(V, W)$ identifies with the space of polynomial maps of degree at most $q$ from $V$ to $W$. We also write $J^q(n, p) = L(\mathbb{R}^n, \mathbb{R}^p) \times L_s^2(\mathbb{R}^n; \mathbb{R}^p) \times \cdots \times L_s^q(\mathbb{R}^n; \mathbb{R}^p)$, and $\mathcal{P}^q(n, p) = \mathcal{P}^q(\mathbb{R}^n, \mathbb{R}^p) = \mathbb{R}^p \times J^q(n, p)$.

If $n = \dim N$, $p = \dim P$, then $J^q(n, p)$ identifies with the typical fiber of $J^q(N, P) \to N \times P$. If $V$, $W$ are linear spaces, we identify $J^q(V, W)$ with $V \times \mathcal{P}^q(V, W)$: under this identification, if $f \in \mathscr{C}(V, W)$, then

$$j^q f(x) = (x, f(x), Df(x), \cdots, D^q f(x)),$$

where $D^k f(x) \in L_s^k(V; W)$ is the $k$th (total) derivative of $f$ at $x$ ($D^0 f = f$, $D^1 f = Df$).

3.2. Let $f \in \mathscr{C}(\mathbb{R}^n, \mathbb{R}^p)$. We write $f^{(k)}(x) = D^k f(x)$, $k = 0, 1, \cdots$, and $f^q(x) = (f^{(0)}(x), \cdots, f^{(q)}(x)) \in \mathcal{P}^q(n, p)$, so that $j^q f(x) = (x, f^q(x))$. We also denote by $H^q = (H^{(0)}, \cdots, H^{(q)})$ a *variable* in $\mathcal{P}^q(n, p)$.

Given $F^q \in \mathcal{P}^q(n, 1)$, $G^q \in \mathcal{P}^q(n, p)$, the *product* $F^q \cdot G^q \in \mathcal{P}^q(n, p)$ is determined by Leibniz's rule. In other words, if $f \in \mathscr{C}(\mathbb{R}^n)$, $g \in \mathscr{C}(\mathbb{R}^n, \mathbb{R}^p)$, then $f^q(x) \cdot g^q(x) = (f \cdot g)^q(x)$. The product is linear in each factor.

Given $F^q \in \mathcal{P}^q(m, n)$, $G^q \in \mathcal{P}^q(n, p)$, the *composition* $c(F^q, G^q) \in \mathcal{P}^q(m, p)$ is determined by Leibniz's rule and the chain rule. If $f \in \mathscr{C}(\mathbb{R}^m, \mathbb{R}^n)$, $g \in \mathscr{C}(\mathbb{R}^n, \mathbb{R}^p)$, then $c(f^q(x), g^q(f(x))) = (g \circ f)^q(x)$. Note that $c$ is a polynomial map which is linear in the second variable.

We will use the following elementary properties of these operations. If $F^q \in \mathcal{P}^q(l, m)$, $G^q \in \mathcal{P}^q(m, n)$, $H^q \in \mathcal{P}^q(n, p)$, then

$$c(F^q, c(G^q, H^q)) = c(c(F^q, G^q), H^q). \tag{3.2.1}$$

If $F^q \in \mathcal{P}^q(m, n)$, $G^q \in \mathcal{P}^q(n, 1)$, $H^q \in \mathcal{P}^q(n, p)$, then

$$c(F^q, G^q \cdot H^q) = c(F^q, G^q) \cdot c(F^q, H^q). \tag{3.2.2}$$

## 3.3. Jets of equivariant maps

If $N, P$ are $G$-manifolds, then $J^q(N, P)$ has an induced $G$-fiber bundle structure. The action of $G$ on $J^q(N, P)$ is defined by $g \cdot (j^q f(x)) = j^q(gfg^{-1})(gx)$. If $f \in \mathscr{C}(N, P)^G$, then the $q$-jet extension $j^q f$ of $f$ is an equivariant section of the projection $J^q(N, P) \to N$. If $x \in N$, then the space $j_x^q(\mathscr{C}(N, P)^G)$ of $q$-jets at $x$ of equivariant maps, is a subspace of the fixed point set of $G_x$ in the fiber over $x$.

Let $H$ be a closed subgroup of $G$, and $V$ a linear $H$-space. Suppose $y \in P^H$. By the Slice Theorem, there is a linear $G_y$-space $V_y$, and an equivariant diffeomorphism $\rho$ from $T = G \times_{G_y} V_y$ onto an invariant open tubular neighborhood of $Gy$ in $P$, such that $\rho([e, 0]) = y$. There is a linear $G_y$-space $W$, and an analytic $G_y$-equivariant isomorphism $\tau$ of $W$ onto a neighborhood of $[e, 0]$ in $T$, such that $\tau(0) = [e, 0]$.

LEMMA 3.4. *$\tau$ determines an analytic isomorphism of $j_0^q(\mathscr{C}(V, W)^H)$ onto an open subset of $j_{eH}^q(\mathscr{C}(G \times_H V, T)^G)$.*

PROOF. The mapping of $j_0^q(\mathscr{C}(V, W)^H)$ into $j_{eH}^q(\mathscr{C}(G \times_H V, T)^G)$ is defined by $j^q f(0) \mapsto j^q \tilde{f}(eH)$, where $f \in \mathscr{C}(V, W)^G$ and $\tilde{f}([g, x]) = g(\tau \circ f)(x)$. Let $\Omega$ be the set of maps $f \in \mathscr{C}(G \times_H V, T)^G$ such that $f(eH) \in \tau(W)$. The image of the above map is the open subset $j_{eH}^q(\Omega)$ of $j_{eH}^q(\mathscr{C}(G \times_H V, T)^G)$. Its inverse, defined on $j_{eH}^q(\Omega)$, is given by the 'restriction' map $j^q f(eH) \mapsto j^q(\tau^{-1} \circ (f \mid \mathscr{V}))(0)$, where $f \in \Omega$ and $\mathscr{V}$ is a sufficiently small $H$-invariant neighborhood of 0 in $V$.

Note that although $j_x^q(\mathscr{C}(N, P)^G)$ is a submanifold of the fiber of $J^q(N, P)$ over $x \in N$, the subspace of $J^q(N, P)$ of jets of equivariant maps is not, in general, a submanifold.

3.5. In the notation of section 3.3, let $U$ be an $H$-invariant open neighborhood of $eH$ in $G/H$, and $\sigma$ an analytic $H$-equivariant local section of $G \to G/H$, defined in $U$, such that $\sigma(eH) = e$ (cf. section 2.3). The analytic section $\sigma$ induces an analytic $H$-equivariant immersion

$$\mathscr{A}_\sigma : J^q(V, T) \to J^q(G \times_H V, T), \tag{3.5.1}$$

defined by $\mathscr{A}_\sigma(j^q f(x)) = j^q \tilde{f}([e, x])$, where

$$\tilde{f}([\sigma(gH), x]) = \sigma(gH) f(x).$$

Let $i : V \to G \times_H V$ be the canonical $H$-equivariant inclusion. Then $i$ induces an $H$-equivariant submersion

$$i^* : J^q(G \times_H V, T) \mid V \to J^q(V, T),$$

and $i^* \circ \mathscr{A}_\sigma = I$.

Note that if $f \in \mathscr{C}(G \times_H V, T)^G$, then $\mathscr{A}_\sigma(j^q(f \mid V)(x)) = j^q f([e, x])$, independently of the choice of $\sigma$. In other words, if $\tau_* : J^q(V, W) \to J^q(V, T)$ denotes the diffeomorphism $\tau_*(j^q f(x)) = j^q(\tau \circ f)(x)$ induced by $\tau$, then $\mathscr{A}_\sigma \circ \tau_*$ restricts to the isomorphism of $j_0^q(\mathscr{C}(V, W)^H)$ into $j_{eH}^q(\mathscr{C}(G \times_H V, T)^G)$ induced by $\tau$.

## 4. Local structure of equivariant maps

### 4.1. Invariant functions

Let $V$ be a linear $G$-space. The $\mathbb{R}$-algebra $\mathscr{P}(V)^G$ of invariant polynomial functions on $V$ is finitely generated [21, Theorem 8.14A]. A finite generating set will be called a *Hilbert basis* of $\mathscr{P}(V)^G$ [10]. A Hilbert basis consisting of homogeneous polynomials will be called *homogeneous*. It will be called *minimal* if no proper subset is still a Hilbert basis.

Let $\mathscr{V}$ be an invariant open subset of $V$, and $\mathscr{U}$ an open subset of $\mathbb{R}^l$. A smooth invariant map $Q:\mathscr{V}\to\mathscr{U}$ induces a ring homomorphism $Q^*:\mathscr{E}(\mathscr{U})\to\mathscr{E}(\mathscr{V})^G$, defined by $Q^*(f)=f\circ Q$. $Q$ will be called an *orbit map* if $Q^*\mathscr{E}(\mathscr{U})$ is dense in $\mathscr{E}(\mathscr{V})^G$ (in the $\mathscr{E}^\infty$ topology). In Mather's terminology [10], $Q$ induces an embedding of the orbit space $\mathscr{V}/G$ into $\mathscr{U}$.

If $Q$ is a proper orbit map, then $Q^*:\mathscr{E}(\mathscr{U})\to\mathscr{E}(\mathscr{V})^G$ is split surjective [10, Theorem 2] (surjectivity was first proven in [16]). Any Hilbert basis $q_1,\cdots,q_l$ of $\mathscr{P}(V)^G$ defines a proper map $Q=(q_1,\cdots,q_l)\colon V\to\mathbb{R}^l$. Since $\mathscr{P}(V)^G$ is dense in $\mathscr{E}(V)^G$, then $Q$ is a proper orbit map.

### 4.2. Equivariant maps

Let $V$, $W$ be linear $G$-spaces, and $\mathscr{V}$ an invariant open neighborhood of 0 in $V$. Then $\mathscr{E}(\mathscr{V}, W)^G$ is a module over the ring $\mathscr{E}(\mathscr{V})^G$, and is generated by finitely many polynomial maps (this was apparently first observed by Malgrange). Let $F_1,\cdots,F_m$ be a finite set of polynomial generators. In the $\mathscr{E}^\infty$ topology, the map $(h_1,\cdots,h_m)\mapsto\sum_{i=1}^m h_iF_i$ of $(\mathscr{E}(\mathscr{V})^G)^m$ onto $\mathscr{E}(\mathscr{V}, W)^G$ is a continuous linear surjection of Fréchet spaces, hence open by the Open Mapping Theorem. In fact it is split surjective [4, Proposition 3.2]. We may, of course, choose homogeneous polynomial generators $F_1,\cdots,F_m$ (replacing any set of polynomial generators by their homogeneous parts). If $F_1,\cdots,F_m$ are homogeneous, then they also generate $\mathscr{P}(V, W)^G$ over $\mathscr{P}(V)^G$. Conversely, any set of generators of $\mathscr{P}(V, W)^G$ over $\mathscr{P}(V)^G$ is a set of generators for $\mathscr{E}(V, W)^G$ over $\mathscr{E}(V)^G$.

Let $A=\mathscr{E}_0(V)^G$, the ring of germs at 0 of functions in $\mathscr{E}(V)^G$, and $M=\mathscr{E}_0(V, W)^G$, the finitely generated $A$-module of germs at 0 of maps in $\mathscr{E}(V, W)^G$. $A$ is a local ring, its unique maximal ideal $\mathfrak{n}$ comprising germs of smooth invariant functions which vanish at $0\in V$. Then $A/\mathfrak{n}\cong\mathbb{R}$, and $M\otimes_A A/\mathfrak{n}$ is a finite dimensional vector space over the field $A/\mathfrak{n}$, of dimension $k$ say.

If $F_1,\cdots,F_m$ generate $\mathscr{E}(\mathscr{V}, W)^G$ over $\mathscr{E}(\mathscr{V})^G$, then $m\geqq k$ and (after reordering) $F_1,\cdots,F_k$ generate $M$ over $A$ ($F_1,\cdots,F_k$ will be called a *minimal* set of generators). If $f=\sum_{i=1}^k h_iF_i\in M$, where $h_i\in A$, $i=1,\cdots,k$, then the $h_i(0)$ are uniquely determined by $f$ and $F_1,\cdots,F_k$. If the generators $F_1,\cdots,F_m$ are homogeneous polynomials, then $F_1,\cdots,F_k$ generate $\mathscr{E}(V, W)^G$ over $\mathscr{E}(V)^G$.

### 4.3. Local representation of jets of equivariant maps

Let $V$, $W$ be linear $G$-spaces, and $\mathscr{V}$ an invariant open neighborhood of 0 in $V$. Let $P:\mathscr{V}\to\mathscr{U}$ be a polynomial orbit map, where $\mathscr{U}$ is an open subset of

$\mathbb{R}^l$. Let $F_1, \cdots, F_k$ be a set of polynomial generators for $\mathscr{C}(\mathscr{V}, W)^G$ over $\mathscr{C}(\mathscr{V})^G$. Any map $f \in \mathscr{C}(\mathscr{V}, W)^G$ can be written $f(x) = \sum_{i=1}^k H_i(P(x))F_i(x)$, where $H_i \in \mathscr{C}(\mathscr{U})$, $i = 1, \cdots, k$. Hence for $q = 0, 1, 2, \cdots$:

$$f^q(x) = \sum_{i=1}^{k} c(P^q(x), H_i^q(P(x))) \cdot F_i^q(x). \tag{4.3.1}$$

For each $q = 0, 1, 2, \cdots$ we define an equivariant polynomial map

$$U^q_{P,F}: \mathscr{V} \times \mathscr{P}^q(l, k) \to J^q(\mathscr{V}, W)$$
$$U^q_{P,F}(x, H^q) = \left(x, \sum_{i=1}^{k} c(P^q(x), H_i^q) \cdot F_i^q(x)\right) \tag{4.3.2}$$

($G$ acts trivially on $\mathscr{P}^q(l, k)$).

The $q$-jet map $j^q f: \mathscr{V} \to J^q(\mathscr{V}, W)$ can then be written as the composition

$$j^q f = U^q_{P,F} \circ (I, H^q \circ P), \tag{4.3.3}$$

where $(I, H^q \circ P): \mathscr{V} \to \mathscr{V} \times \mathscr{P}^q(l, k)$.

Now let $H$ be a closed subgroup of $G$, and $V$ a linear $H$-space. Let $P$ be a $G$-manifold, and $y \in P^H$. By the Slice Theorem, there is a linear $G_y$-space $V_y$, and an equivariant diffeomorphism $\rho$ from $T = G \times_{G_y} V_y$ onto an invariant open tubular neighborhood of $G_y$ in $P$, such that $\rho([e, 0]) = y$. There is a linear $G_y$-space $W$, and an analytic $G_y$-equivariant isomorphism $\tau$ of $W$ onto a neighborhood of $[e, 0]$ in $T$, such that $\tau(0) = [e, 0]$.

Let $P \in \mathscr{P}(V, \mathbb{R}^l)^H$ be an orbit map for $V$, and $F_1, \cdots, F_k$ a set of polynomial generators for $\mathscr{C}(V, W)^H$ over $\mathscr{C}(V)^H$. The following is easily verified.

Lemma 4.4. *The $H$-equivariant composed map $\tau_* \circ U^q_{P,F}: V \times \mathscr{P}^q(l, k) \to J^q(V, T)$ naturally induces an analytic $G$-equivariant map*

$$\mathfrak{U}^q_{P,F}: (G \times_H V) \times \mathscr{P}^q(l, k) \to J^q(G \times_H V, T) \tag{4.4.1}$$

*such that*

$$i^* \circ \mathfrak{U}^q_{P,F} \circ (i \times I) = \tau_* \circ U^q_{P,F}. \tag{4.4.2}$$

(Here $i: V \to G \times_H V$ is the inclusion map, and $I$ the identity map of $\mathscr{P}^q(l, k)$). In fact

$$\mathfrak{U}^q_{P,F} \circ (i \times I) = \mathscr{A}_\sigma \circ \tau_* \circ U^q_{P,F} \tag{4.4.3}$$

for any choice of $\mathscr{A}_\sigma$ as in (3.5.1).

## 5. Invariance lemmas

Lemma 5.1. *Let $u \in \mathscr{P}(\mathbb{R}^n \times \mathbb{R}^k, \mathbb{R}^p)$ be a polynomial map which is linear in $h \in \mathbb{R}^k$, and $U(x, h) = (x, u(x, h))$. Suppose that $\alpha \in \mathscr{C}(\mathbb{R}^n, \mathbb{R}^k)$ is a map such* *that* $U \circ (I, \alpha) = 0$. *Then the map* $A(x, h) = (x, h + \alpha(x))$ *is a diffeomorphism of* $\mathbb{R}^n \times \mathbb{R}^k$ *such that* $U \circ A = U$.

This is completely trivial.

Let $V$, $W$ be linear $G$-spaces. If $P \in \mathcal{P}(V, \mathbb{R}^l)^G$ is an orbit map for $V$, and $F_1, \cdots, F_k$ a set of polynomial generators for $\mathcal{E}(V, W)^G$ over $\mathcal{E}(V)^G$, then the equivariant polynomial map $U^q_{P,F}: V \times \mathcal{P}^q(l, k) \to J^q(V, W)$ is defined for each $q = 0, 1, 2, \cdots$ by (4.3.2).

LEMMA 5.2. *Let* $p_1, \cdots, p_l$ *and* $q_1, \cdots, q_l$ *be minimal homogeneous Hilbert bases of* $\mathcal{P}(V)^G$. *Set* $P = (p_1, \cdots, p_l)$ *and* $Q = (q_1, \cdots, q_l)$. *Then:*

(a) *There is an algebraic isomorphism* $\phi$ *of* $\mathbb{R}^l$ *to itself, such that* $P = \phi \circ Q$.

(b) $\phi$ *induces an equivariant algebraic isomorphism* $\Phi(x, H^q) = (x, \Phi_1(x, H^q))$ *of* $V \times \mathcal{P}^q(l, k)$ *to itself for any* $k$, *such that* $U^q_{P,F} = U^q_{Q,F} \circ \Phi$ *for any set of polynomial generators* $F_1, \cdots, F_k$ *of* $\mathcal{E}(V, W)^G$.

PROOF. (a) Let $d_i = \deg p_i$ and $e_i = \deg q_i$, $i = 1, \cdots, l$. Denote by $(y_1, \cdots, y_l)$ the coordinates of $\mathbb{R}^l$, and assign $y_i$ weight $e_i$. For $i = 1, \cdots, l$, there is a polynomial $\phi_i(y_1, \cdots, y_l)$, weighted homogeneous of degree $d_i$, such that $p_i = \phi_i \circ Q$. Then $\phi = (\phi_1, \cdots, \phi_l)$ is an algebraic isomorphism such that $P = \phi \circ Q$.

(b) Define $\Phi(x, H^q) = (x, \Phi_1(x, H^q))$ by $\Phi_1(x, H^q) = c(\phi^q(Q(x)), H^q)$. Then $\Phi$ is an equivariant algebraic isomorphism, and $U^q_{P,F} = U^q_{Q,F} \circ \Phi$ by (4.3.2), since

$$
\begin{aligned}
c(P^q(x), H^q) &= c((\phi \circ Q)^q(x), H^q) \\
&= c(c(Q^q(x), \phi^q(Q(x))), H^q) \\
&= c(Q^q(x), c(\phi^q(Q(x)), H^q)) \quad \text{by (3.2.1).}
\end{aligned}
$$

LEMMA 5.3. *Let* $p_1, \cdots, p_l$ *be a minimal homogeneous Hilbert basis of* $\mathcal{P}(V)^G$, *and* $P = (p_1, \cdots, p_l)$. *Let* $Q = (q_1, \cdots, q_m)$ *be a polynomial orbit map defined in an invariant open neighborhood of* 0 *in* $V$. *Then:*

(a) *There exist an algebraic immersion* $\psi: \mathcal{U} \to \mathbb{R}^m$, *where* $\mathcal{U}$ *is an open neighborhood of* 0 *in* $\mathbb{R}^l$, *and an analytic submersion* $\phi$ *from an open neighborhood of* $Q(0)$ *to* $\mathbb{R}^l$, *such that* $\phi \circ \psi = I$ *in* $\mathcal{U}$, *and* $Q = \psi \circ P$ *in* $P^{-1}(\mathcal{U})$.

(b) *Let* $\mathcal{V} = P^{-1}(\mathcal{U})$. *There is an equivariant analytic immersion* $\Phi(x, H^q) = (x, \Phi_1(x, H^q))$ *of* $\mathcal{V} \times \mathcal{P}^q(l, k)$ *into* $\mathcal{V} \times \mathcal{P}^q(m, k)$, *and an equivariant algebraic submersion* $\Psi(x, H^q) = (x, \Psi_1(x, H^q))$ *from* $\mathcal{V} \times \mathcal{P}^q(m, k)$ *to* $\mathcal{V} \times \mathcal{P}^q(l, k)$, *for any* $k$, *such that* $\Psi \circ \Phi = I$, *and* $U^q_{Q,F} = U^q_{P,F} \circ \Psi$ *for any set of polynomial generators* $F_1, \cdots, F_k$ *of* $\mathcal{E}(V, W)^G$.

PROOF. (a) The existence of an algebraic immersion $\psi$ such that $Q=\psi\circ P$ in a neighborhood of 0 is proven in [10]. The existence of $\phi$ then follows from the implicit function theorem.

(b) Define $\Phi(x, H^q)=(x, \Phi_1(x, H^q))$ by $\Phi_1(x, H^q)=c(\phi^q(Q(x)), H^q)$, where $(x, H^q)\in\mathscr{V}\times\mathscr{P}^q(l, k)$. Define $\Psi(x, H^q)=(x, \Psi_1(x, H^q))$ by $\Psi_1(x, H^q)=c(\psi^q(P(x)), H^q)$, where $(x, H^q)\in\mathscr{V}\times\mathscr{P}^q(m, k)$. Then $\Phi$ is an equivariant analytic immersion, and $\Psi$ an equivariant algebraic submersion, such that $\Psi\circ\Phi=I$. Since

$$c(Q^q(x), H^q)=c((\psi\circ P)^q(x), H^q)$$
$$=c(P^q(x), c(\psi^q(P(x)), H^q))$$

in $\mathscr{V}\times\mathscr{P}^q(m, k)$, then $U^q_{Q,F}=U^q_{P,F}\circ\Psi$ by (4.3.2).

LEMMA 5.4. *Let $F_1,\cdots,F_k$ and $G_1,\cdots,G_k$ be minimal sets of polynomial generators of $\mathscr{C}(\mathscr{V}, W)^G$ over $\mathscr{C}(\mathscr{V})^G$, where $\mathscr{V}$ is an invariant open neighborhood of 0 in $\mathscr{V}$. Let $P=(p_1,\cdots,p_l)$ be a polynomial orbit map for $\mathscr{V}$. Then there is an invariant open neighborhood $\mathscr{V}'$ of 0 in $\mathscr{V}$, and an equivariant analytic isomorphism $A(x, H^q)=(x, A_1(x, H^q))$ of $\mathscr{V}'\times\mathscr{P}^q(l, k)$ to itself, such that $U^q_{P,F}=U^q_{P,G}\circ A$. Moreover, if $F_1,\cdots,F_k$ and $G_1,\cdots,G_k$ are homogeneous, and $P\in\mathscr{P}(V,\mathbb{R}^l)^G$, then $A$ may be defined throughout $V\times\mathscr{P}^q(l, k)$, and algebraic.*

PROOF. There are invariant analytic functions $\gamma_{ji}(x)$, $i, j=1,\cdots,k$, defined in an invariant open neighborhood $\mathscr{V}'$ of 0 in $\mathscr{V}$, such that the matrix $\gamma(x)=(\gamma_{ij}(x))$ is invertible and

$$F_i(x)=\sum_{j=1}^{k}\gamma_{ji}(x)G_j(x),\qquad i=1,\cdots,k,\tag{5.4.1}$$

for $x\in\mathscr{V}'$ (that the $\gamma_{ij}$ may be chosen analytic, and, in fact, algebraic over the ring of polynomials, follows from implicit function theorems of M. Artin [1], [2]). By [9], there are analytic functions $\Gamma_{ji}$ defined in an open neighborhood of $P(\mathscr{V}')\subset\mathbb{R}^l$, such that $\gamma_{ji}=\Gamma_{ji}\circ P$, $i, j=1,\cdots,k$. Note that if $F_1,\cdots,F_k$ and $G_1,\cdots,G_k$ are homogeneous, then we may choose homogeneous polynomials $\gamma_{ji}$, so that $\gamma(x)$ is invertible for all $x\in V$ ($\mathscr{V}'=V$). In this case we may also choose polynomials $\Gamma_{ji}$ (by [21, Theorem 8.14A]).

Define $A_1:\mathscr{V}'\times\mathscr{P}^q(l, k)\to\mathscr{P}^q(l, k)$, $A_1=(A_{11},\cdots,A_{1k})$, by $A_{1i}(x, H^q)=\sum_{j=1}^k\Gamma^q_{ij}(P(x))\cdot H^q_j$, $i=1,\cdots,k$. Define $A:\mathscr{V}'\times\mathscr{P}^q(l, k)\to\mathscr{V}'\times\mathscr{P}^q(l, k)$ by $A(x, H^q)=(x, A_1(x, H^q))$. Then $A$ is an equivariant analytic isomorphism (algebraic in the case that the $\Gamma_{ji}$ are polynomials). We compute

$$\sum_{i=1}^{k} c(P^q(x), H_i^q) \cdot F_i^q(x)$$

$$= \sum_{i=1}^{k} c(P^q(x), H_i^q) \cdot \sum_{j=1}^{k} c(P^q(x), \Gamma_{ji}^q(P(x))) \cdot G_j^q(x) \quad \text{by (5.4.1)}$$

$$= \sum_{i=1}^{k} \sum_{j=1}^{k} c(P^q(x), \Gamma_{ij}^q(P(x)) \cdot H_j^q) \cdot G_i^q(x) \qquad \text{by (3.2.2).}$$

Hence

$$U_{P,F}^q(x, H^q) = \left(x, \sum_{i=1}^{k} c(P^q(x), H_i^q) \cdot F_i^q(x)\right) \qquad \text{by (4.3.2)}$$

$$= \left(x, \sum_{i=1}^{k} c\left(P^q(x), \sum_{j=1}^{k} \Gamma_{ij}^q(P(x)) \cdot H_j^q\right) \cdot G_i^q(x)\right)$$

$$= U_{P,G}^q \circ A(x, H^q).$$

LEMMA 5.5. *Let $F_1, \cdots, F_m$ be a set of polynomial generators of $\mathscr{E}(\mathscr{V}, W)^G$ over $\mathscr{E}(\mathscr{V})^G$, where $\mathscr{V}$ is an invariant open neighborhood of 0 in V. Let $P = (p_1, \cdots, p_l)$ be a polynomial orbit map for $\mathscr{V}$. Assume that $F_1, \cdots, F_k$ represent a minimal set of generators of the module of germs of equivariant maps at 0. Set $G_i = F_i$, $i = 1, \cdots, k$, and denote by $\pi : \mathscr{P}^q(l, m) \to \mathscr{P}^q(l, k)$ the canonical projection $\pi(H_1^q, \cdots, H_m^q) = (H_1^q, \cdots, H_k^q)$. Then there is an invariant open neighborhood $\mathscr{V}'$ of 0 in $\mathscr{V}$, and an equivariant analytic isomorphism $A(x, H^q) = (x, A_1(x, H^q))$ of $\mathscr{V}' \times \mathscr{P}^q(l, m)$ to itself, such that $U_{P,G}^q \circ (I \times \pi) = U_{P,F}^q \circ A$, where I is the identity map of $\mathscr{V}'$. Moreover, if $F_1, \cdots, F_m$ are homogeneous, and $P \in \mathscr{P}(V, \mathbb{R}^l)^G$, then A may be defined throughout $V \times \mathscr{P}^q(l, m)$, and algebraic.*

PROOF. There are invariant analytic functions $\phi_{ji}(x)$, $j = 1, \cdots, k$, $i = k+1, \cdots, m$, defined in an invariant open neighborhood $\mathscr{V}'$ of 0 in $\mathscr{V}$, such that

$$F_i(x) = \sum_{j=1}^{k} \phi_{ji}(x) F_j(x), \qquad i = k+1, \cdots, m \tag{5.5.1}$$

(by [2]). Then by [9], there are analytic functions $\Phi_{ji}$, defined in an open neighborhood of $P(\mathscr{V}') \subset \mathbb{R}^l$, such that $\phi_{ji} = \Phi_{ji} \circ P$. Again note that if the $F_i$ are homogeneous, then we can choose homogeneous polynomials $\phi_{ji}$ and polynomials $\Phi_{ji}$.

Define $A_1 : \mathscr{V}' \times \mathscr{P}^q(l, m) \to \mathscr{V}' \times \mathscr{P}^q(l, m)$, $A_1 = (A_{11}, \cdots, A_{1m})$, by

$$A_{1i}(x, H^q) = H_i^q - \sum_{j=k+1}^{m} \Phi_{ij}^q(P(x)) \cdot H_j^q, \qquad i = 1, \cdots, k,$$

$$A_{1i}(x, H^q) = H_i^q \qquad i = k+1, \cdots, m.$$

Define $A(x, H^q) = (x, A_1(x, H^q))$. Then $A$ is an equivariant analytic isomorphism (algebraic in the case that the $\Phi_{ji}$ are polynomials). We compute

$$\sum_{i=1}^{m} c(P^q(x), H_i^q) \cdot F_i^q(x)$$

$$= \sum_{i=1}^{k} \left( c(P^q(x), H_i^q) + \sum_{j=k+1}^{m} c(P^q(x), \Phi_{ij}^q(P(x))) \cdot c(P^q(x), H_j^q) \right) \cdot F_i^q(x)$$

by (5.5.1)

$$= \sum_{i=1}^{k} c\left( P^q(x), H_i^q + \sum_{j=k+1}^{m} \Phi_{ij}^q(P(x)) \cdot H_j^q \right) \cdot F_i^q(x) \qquad \text{by (3.2.2).}$$

Hence $U_{P,F}^q = U_{P,G}^q \circ (I \times \pi) \circ A^{-1}$ by (4.3.2).

5.6. Let $N$, $P$ be $G$-manifolds, and $\mathcal{N}$, $\mathcal{P}$ invariant open subsets of $N$, $P$ (respectively). Let $\gamma$ be an equivariant diffeomorphism of $\mathcal{N} \times \mathcal{P}$ into $N \times P$, of the form $\gamma(x, y) = (\alpha(x), \gamma_1(x, y))$, $(x, y) \in \mathcal{N} \times \mathcal{P}$. Note that $\gamma$ induces an isomorphism $\gamma\cdot$ from $\mathscr{C}(\mathcal{N}, \mathcal{P})$ into $\mathscr{C}(\alpha(\mathcal{N}), P)$, defined by the formula

$$(I, \gamma \cdot f) = \gamma \circ (I, f) \circ \alpha^{-1}, \tag{5.6.1}$$

where $f \in \mathscr{C}(\mathcal{N}, \mathcal{P})$; i.e. $(\gamma \cdot f)(x) = \gamma_1(\alpha^{-1}(x), f(\alpha^{-1}(x)))$, $x \in \alpha(\mathcal{N})$. $\gamma\cdot$ restricts to an isomorphism from $\mathscr{C}(\mathcal{N}, \mathcal{P})^G$ into $\mathscr{C}(\alpha(\mathcal{N}), P)^G$. $\gamma$ induces an equivariant bundle map $\gamma_* : J^q(\mathcal{N}, \mathcal{P}) \to J^q(N, P)$ over $\gamma$; in particular $\gamma_*$ is an equivariant diffeomorphism into $J^q(N, P)$, and the following diagram commutes:

$$\begin{array}{ccc} J^q(\mathcal{N}, \mathcal{P}) & \xrightarrow{\gamma_*} & J^q(N, P) \\ \downarrow & & \downarrow \\ \mathcal{N} \times \mathcal{P} & \xrightarrow{\gamma} & N \times P \end{array} \tag{5.6.2}$$

$\gamma_*$ is defined by

$$\gamma_* \circ j^q f = j^q(\gamma \cdot f) \circ \alpha \tag{5.6.3}$$

for $f \in \mathscr{C}(\mathcal{N}, \mathcal{P})$.

In particular, if $\alpha$ is an equivariant diffeomorphism of $\mathcal{N}$ into $N$, we define $\alpha_\# = (\alpha \times I)_*$, where $I$ is the identity map of $P$; i.e. for all $f \in \mathscr{C}(\mathcal{N}, P)$,

$$\alpha_\# \circ j^q f = j^q(f \circ \alpha^{-1}) \circ \alpha. \tag{5.6.4}$$

If $\beta$ is an equivariant diffeomorphism of $\mathcal{N} \times \mathcal{P}$ into $\mathcal{N} \times P$ of the form $\beta(x, y) = (x, \beta_1(x, y))$, $(x, y) \in \mathcal{N} \times \mathcal{P}$, then $\beta$ operates on maps $f \in \mathscr{C}(\mathcal{N}, \mathcal{P})^G$ by the formula

$$(I, \beta \cdot f) = \beta \circ (I, f), \tag{5.6.5}$$

and for all $f \in \mathscr{C}(\mathcal{N}, \mathscr{P})$:

$$\beta_* \circ j^q f = j^q(\beta \cdot f). \tag{5.6.6}$$

In the following two lemmas, $V$ and $W$ are linear $G$-spaces. Fix a minimal homogeneous Hilbert basis $p_1, \cdots, p_l$ of $\mathscr{P}(V)^G$, and a minimal set of homogeneous generators $F_1, \cdots, F_k$ of $\mathscr{P}(V, W)^G$ over $\mathscr{P}(V)^G$. Set $P = (p_1, \cdots, p_l)$. We also denote by $P$ the unique (algebraic) map in $\mathscr{C}(G \times_H V, \mathbb{R}^l)^G$ induced by $P$.

LEMMA 5.7. *Let $\mathscr{V}$ be an invariant open neighborhood of 0 in $V$, and $\alpha$ an equivariant diffeomorphism of $\mathscr{V}$ into $V$ such that $\alpha(0) = 0$. Then there is an invariant open neighborhood $\mathscr{V}'$ of 0 in $\mathscr{V}$, and an equivariant diffeomorphism $A(x, H^q) = (\alpha(x), A_1(x, H^q))$ of $\mathscr{V}' \times \mathscr{P}^q(l, k)$ into $V \times \mathscr{P}^q(l, k)$, such that the following diagram commutes:*

$$\begin{array}{ccc} \mathscr{V}' \times \mathscr{P}^q(l, k) & \xrightarrow{U^q_{P,F}} & J^q(\mathscr{V}', W) \\ \Big\downarrow A & & \Big\downarrow \alpha_\# \\ V \times \mathscr{P}^q(l, k) & \xrightarrow{U^q_{P,F}} & J^q(V, W). \end{array}$$

PROOF. There is an open neighborhood $\mathscr{U}$ of 0 in $\mathbb{R}^l$, and a diffeomorphism $\beta$ of $\mathscr{U}$ into $\mathbb{R}^l$, such that $\beta \circ P = P \circ \alpha$ in $P^{-1}(\mathscr{U})$. We can write

$$F_i(\alpha^{-1}(x)) = \sum_{j=1}^{k} \Gamma_{ji}(P(x)) F_j(x), \qquad i = 1, \cdots, k, \tag{5.7.1}$$

for $x \in \alpha(\mathscr{V})$, where $\Gamma_{ji} \in \mathscr{C}(\mathscr{U})$, and the matrix $(\Gamma_{ij}(0))$ is invertible. Hence there is a neighborhood $\mathscr{U}'$ of 0 in $\mathscr{U}$ such that $(\Gamma_{ij}(y))$ is invertible for $y \in \beta(\mathscr{U}')$. Let $\mathscr{V}' = P^{-1}(\mathscr{U}')$. From (5.7.1):

$$\begin{aligned} (F_i \circ \alpha^{-1})^q(\alpha(x)) &= \sum_{j=1}^{k} (\Gamma_{ji} \circ P)^q(\alpha(x)) \cdot F_j^q(\alpha(x)) \\ &= \sum_{j=1}^{k} c(P^q(\alpha(x)), \Gamma_{ji}^q(P(\alpha(x)))) \cdot F_j^q(\alpha(x)) \end{aligned} \tag{5.7.2}$$

for $x \in \mathscr{V}$.

Define $\gamma = (\gamma_1, \cdots, \gamma_k) : \mathscr{U}' \times \mathscr{P}^q(l, k) \to \mathscr{P}^q(l, k)$ by

$$\gamma_i(y, H^q) = \sum_{j=1}^{k} \Gamma_{ij}^q(\beta(y)) \cdot c((\beta^{-1})^q(\beta(y)), H_j^q). \tag{5.7.3}$$

Define $A(x, H^q) = (\alpha(x), A_1(x, H^q))$ by $A_1(x, H^q) = \gamma(P(x), H^q)$, $x \in \mathscr{V}'$. $A$ is an equivariant diffeomorphism. Note that by the chain rule, (5.6.4) may be written $\alpha_\#(x, f^q(x)) = (\alpha(x), c((\alpha^{-1})^q(\alpha(x)), f^q(x)))$, for $f \in \mathscr{C}(\mathscr{V}, W)$. Then

$$\alpha_\# \circ U^q_{P,F}(x, H^q) = (\alpha(x), c((\alpha^{-1})^q(\alpha(x)), \sum_{i=1}^{k} c(P^q(x), H_i^q) \cdot F_i^q(x)))$$

by (4.3.2). But

$$c((\alpha^{-1})^q(\alpha(x)), \sum_{i=1}^{k} c(P^q(x), H_i^q) \cdot F_i^q(x))$$

$$= \sum_{i=1}^{k} c((\alpha^{-1})^q(\alpha(x)), c(P^q(x), H_i^q)) \cdot (F_i \circ \alpha^{-1})^q(\alpha(x))$$

by (3.2.2) and the chain rule

$$= \sum_{i=1}^{k} c((P \circ \alpha^{-1})^q(\alpha(x)), H_i^q) \cdot \sum_{j=1}^{k} c(P^q(\alpha(x)), \Gamma_{ji}^q(P \circ \alpha(x))) \cdot F_j^q(\alpha(x))$$

by (3.2.1), (5.7.2)

$$= \sum_{i=1}^{k} \sum_{j=1}^{k} c(P^q(\alpha(x)), \Gamma_{ij}^q(P \circ \alpha(x))) \cdot c(P^q(\alpha(x)),$$

$$c((\beta^{-1})^q(P \circ \alpha(x)), H_j^q)) \cdot F_i^q(\alpha(x)) \qquad \text{since } P \circ \alpha^{-1} = \beta^{-1} \circ P$$

$$= \sum_{i=1}^{k} c(P^q(\alpha(x)), \sum_{j=1}^{k} \Gamma_{ij}^q(\beta \circ P(x)) \cdot c((\beta^{-1})^q(\beta \circ P(x)), H_j^q)) \cdot F_i^q(\alpha(x))$$

by (3.2.2)

$$= \sum_{i=1}^{k} c(P^q(\alpha(x)), \gamma_i(P(x), H^q)) \cdot F_i^q(\alpha(x)) \qquad \text{by (5.7.3).}$$

Hence $\alpha_{\#} \circ U^q_{P,F}(x, H^q) = U^q_{P,F} \circ A(x, H^q)$.

LEMMA 5.8. *Let $\mathcal{V}$, $\mathcal{W}$ be invariant open neighborhoods of 0 in $V$, $W$ respectively, and $\beta : \mathcal{V} \times \mathcal{W} \to \mathcal{V} \times W$ an equivariant diffeomorphism of the form $\beta(x, y) = (x, \beta_1(x, y))$, $(x, y) \in \mathcal{V} \times \mathcal{W}$. Let $\mathcal{O} = (U^0_{P,F})^{-1}(\mathcal{V} \times \mathcal{W}) = \{(x, H) \in \mathcal{V} \times \mathbb{R}^k \mid \sum_{i=1}^k H_i F_i(x) \in \mathcal{W}\}$, so that $\mathcal{O} \times J^q(l, k)$ is an invariant open subset of $\mathcal{V} \times \mathcal{P}^q(l, k)$. There is a smooth equivariant map $B(x, H^q) = (x, B_1(x, H^q))$ of $\mathcal{O} \times J^q(l, k)$ into $\mathcal{V} \times \mathcal{P}^q(l, k)$, such that the germ of $B$ at $(x = 0)$ is invertible, and the following diagram commutes:*

$$\begin{array}{ccc} \mathcal{O} \times J^q(l, k) & \xrightarrow{U^q_{P,F}} & J^q(\mathcal{V}, \mathcal{W}) \\ {\scriptstyle B}\downarrow & & \downarrow{\scriptstyle \beta_*} \\ \mathcal{V} \times \mathcal{P}^q(l, k) & \xrightarrow{U^q_{P,F}} & J^q(\mathcal{V}, W). \end{array}$$

PROOF. There is an open neighborhood $\mathcal{U}$ of 0 in $\mathbb{R}^l \times \mathbb{R}^k$, such that for $(x, H) \in \mathcal{O}$, $H = (H_1, \cdots, H_k)$:

$$\beta_1\left(x, \sum_{i=1}^{k} H_i F_i(x)\right) = \sum_{i=1}^{k} \Gamma_i(P(x), H) F_i(x),$$

where $\Gamma = (\Gamma_1, \cdots, \Gamma_k) \in \mathcal{C}(\mathcal{U}, \mathbb{R}^k)$. The germ at $(y = 0)$ of the map $(y, H) \mapsto (y, \Gamma(y, H))$ is invertible because $F_1, \cdots, F_k$ is a minimal set of generators.

Given $f \in \mathscr{C}(\mathscr{V}, \mathscr{W})^G$, $f(x) = \sum_{i=1}^k H_i(P(x))F_i(x)$, we have $\beta_* \circ U^q_{P,F}(x, H^q(P(x))) = \beta_* \circ j^q f(x) = j^q(\beta \cdot f)(x)$ by (4.3.2), (5.6.6). On the other hand, $(\beta \cdot f)(x) = \sum_{i=1}^k \Gamma_i(P(x), H(P(x)))F_i(x)$, so that

$$(\beta \cdot f)^q(x) = \sum_{i=1}^k c(P^q(x), c((I, H)^q(P(x)), \Gamma_i^q(P(x), H(P(x))))) \cdot F_i^q(x). \tag{5.8.1}$$

Define $\gamma : \mathscr{U} \times J^q(l, k) \to \mathscr{P}^q(l, k)$ by

$$\gamma(y, H^q) = c((I^q(y), H^q), \Gamma^q(y, H^0)).$$

Define $B(x, H^q) = (x, B_1(x, H^q))$ by $B_1(x, H^q) = \gamma(P(x), H^q)$. Then the germ of $B$ at $(x = 0)$ is invertible. From (5.8.1):

$$(\beta \cdot f)^q(x) = \sum_{i=1}^k c(P^q(x), \gamma_i(P(x), H^q(P(x)))) \cdot F_i^q(x).$$

Hence $j^q(\beta \cdot f)(x) = U^q_{P,F} \circ B(x, H^q(P(x)))$, so that

$$\beta_* \circ U^q_{P,F} = U^q_{P,F} \circ B.$$

More generally, let $N$, $P$ be $G$-manifolds. Let $x^0 \in N$ and $H = G_{x^0}$. By the Slice Theorem, there is a linear $H$-space $V$, and an equivariant diffeomorphism $\eta$ from $G \times_H V$ onto an invariant open tubular neighborhood of $Gx^0$ in $N$, such that $\eta([e, 0]) = x^0$. Suppose $y^0 \in P^H$. There is a linear $G_{y^0}$-space $V_{y^0}$, and an equivariant diffeomorphism $\rho$ from $T = G \times_{G_{y^0}} V_{y^0}$ onto an invariant open tubular neighborhood of $Gy^0$ in $P$, such that $\rho([e, 0]) = y^0$. Then there is a linear $G_{y^0}$-space $W$, and an analytic $G_{y^0}$-equivariant isomorphism $\tau$ of $W$ onto an open neighborhood of $[e, 0]$ in $T$, such that $\tau(0) = [e, 0]$.

Let $p_1, \cdots, p_l$ be a minimal homogeneous Hilbert basis of $\mathscr{P}(V)^H$, and $F_1, \cdots, F_k$ a minimal set of homogeneous generators of $\mathscr{P}(V, W)^H$ over $\mathscr{P}(V)^H$. Set $P = (p_1, \cdots, p_l)$. Recall Lemma 4.4.

Suppose $\mathscr{V}$, $\mathscr{W}$ are $H$-invariant open neighborhoods of 0 in $V$, $W$ respectively, and $\beta$ is an equivariant diffeomorphism of $(G \times_H \mathscr{V}) \times T$ into itself, such that $\beta(\mathscr{V} \times \tau(\mathscr{W})) \subset \mathscr{V} \times \tau(W)$, and the following diagram commutes:

$$\begin{array}{ccc} (G \times_H \mathscr{V}) \times T & & \\ \downarrow \beta & \searrow & G \times_H \mathscr{V} \\ (G \times_H \mathscr{V}) \times T & \nearrow & \end{array}$$

(each slanted arrow is the projection onto $G \times_H \mathscr{V}$). Let $\mathscr{O} = (U^0_{P,F})^{-1}(\mathscr{V} \times \mathscr{W}) = \{(x, H) \in \mathscr{V} \times \mathbb{R}^k \mid \sum_{i=1}^k H_i F_i(x) \in \mathscr{W}\}$, so that $(G \times_H \mathscr{O}) \times J^q(l, k)$ identifies with an invariant open subset of $(G \times_H \mathscr{V}) \times \mathscr{P}^q(l, k)$. The following is a simple consequence of Lemma 5.8.

COROLLARY 5.9. *There is a smooth equivariant map* $B:(G\times_H \mathcal{O})\times J^q(l,k)\to(G\times_H \mathcal{V})\times\mathcal{P}^q(l,k)$, *commuting with the projections onto* $G\times_H \mathcal{V}$, *such that the germ of* $B$ *at* $G\times_H(x=0)$ *is invertible, and the following diagram commutes:*

$$\begin{array}{ccc}(G\times_H \mathcal{O})\times J^q(l,k) & \xrightarrow{\amalg^q_{P,F}} & J^q(G\times_H \mathcal{V}, T)\\ \downarrow{\scriptstyle B} & & \downarrow{\scriptstyle \beta_*}\\ (G\times_H \mathcal{V})\times\mathcal{P}^q(l,k) & \xrightarrow{\amalg^q_{P,F}} & J^q(G\times_H \mathcal{V}, T)\end{array}$$

## 5.10. Invariance under the choice of slice

We keep the above notation. Let $\eta':G\times_H V\to N$ be another equivariant diffeomorphism onto an invariant open tubular neighborhood of $Gx^0$ in $N$, such that $\eta'([e,0])=x^0$. It will suffice to assume that the image of $\eta'$ is contained in that of $\eta$, so that $\alpha=\eta^{-1}\circ\eta':G\times_H V\to G\times_H V$ is defined. By the uniqueness of invariant tubular neighborhoods [5, Chapter VI, Theorem 2.6], there is a smooth equivariant isotopy $\alpha_t:G\times_H V\to G\times_H V$, $t\in[0,1]$, such that $\alpha_0$ is a $G$-vector bundle equivalence, $\alpha_1=\alpha$, and $\alpha_t\mid G/H=I$ for each $t$.

PROPOSITION 5.11. *There is an* $H$*-invariant open neighborhood* $\mathcal{O}$ *of* $(U^0_{P,F})^{-1}(0,0)$ *in* $V\times\mathbb{R}^k$, *and a smooth equivariant map* $A:(G\times_H \mathcal{O})\times J^q(l,k)\to(G\times_H V)\times\mathcal{P}^q(l,k)$, *such that the germ of* $A$ *at* $G\times_H(x=0)$ *is invertible, and each square in the following diagram commutes:*

$$\begin{array}{ccccc}G\times_H V & \longleftarrow & (G\times_H \mathcal{O})\times J^q(l,k) & \xrightarrow{\amalg^q_{P,F}} & J^q(G\times_H V, T)\\ \downarrow{\scriptstyle \alpha} & & \downarrow{\scriptstyle A} & & \downarrow{\scriptstyle \alpha_\#}\\ G\times_H V & \longleftarrow & (G\times_H V)\times\mathcal{P}^q(l,k) & \xrightarrow{\amalg^q_{P,F}} & J^q(G\times_H V, T).\end{array}$$

PROOF. First we find a convenient representation of $\alpha$. The closed subgroup $H$ of $G$ acts on the right of $G\times V$ by $(g,x)\cdot h=(gh,h^{-1}x)$, where $(g,x)\in G\times V$ and $h\in H$. With respect to this action, the canonical projection $G\times V\to G\times_H V$ is a principal fiber bundle with structure group $H$, on which $G$ operates as a group of bundle maps. Since $\alpha_0$ may be written $\alpha_0([g,x])=[g,\theta(x)]$, where $\theta$ is an $H$-equivariant linear automorphism of $V$, then $\alpha_0$ lifts to a $G$-equivariant principal bundle map $\tilde\alpha_0(g,x)=(g,\theta(x))$. By the equivariant covering homotopy theorem for smooth $G$-fiber bundles [3, Corollary 3.3], the isotopy $\alpha_t$ lifts to a smooth isotopy $\tilde\alpha_t$ of equivariant bundle maps, such that $\tilde\alpha_t\mid G\times\{0\}=I$ for each $t$. Then $\tilde\alpha_1$ is equivariant with respect to the right action of $H$, and the left action of $G$ on $G\times V$. It follows that $\tilde\alpha_1$ may be written $\tilde\alpha_1(g,x)=(g\xi(x),\omega(x))$, where $\omega$ is an $H$-equivariant diffeomorphism of $V$, such that $\omega(0)=0$, and $\xi:V\to G$ is a smooth map, such that $\xi(0)=e$ and $\xi(hx)=h\xi(x)h^{-1}$ for all $x\in V$, $h\in H$. Then $\alpha([g,x])=[g\xi(x),\omega(x)]$.

Hence $\alpha$ can be written as the composition $\alpha=\psi\circ\phi$, where $\phi([g,x])=[g\xi(x),x]$ and $\psi([g,x])=[g,\omega(x)]$. It suffices to prove the proposition separately for $\phi$ and $\psi$.

The statement for $\psi$ follows from Lemma 5.7.

On the other hand, $\phi$ induces an equivariant diffeomorphism $\beta$ of $(G\times_H V)\times T$ into itself, defined by

$$\beta([g,x],y)=([g,x],g\xi^{-1}(x)g^{-1}y),$$

where $([g,x],y)\in(G\times_H V)\times T$. $\beta$ has the property that $f\circ\phi^{-1}=\beta\cdot f$ for any map $f\in\mathscr{C}(G\times_H V,T)^G$, so that $\beta_*\circ j^qf=j^q(\beta\cdot f)=j^q(f\circ\phi^{-1})=\phi_\#\circ j^qf\circ\phi^{-1}$. For a sufficiently small $H$-invariant open neighborhood $\mathcal{O}$ of $(U^0_{P,F})^{-1}(0,0)$ in $V\times\mathbb{R}^k$, there is a smooth equivariant map $B:(G\times_H\mathcal{O})\times J^q(l,k)\to(G\times_H V)\times\mathscr{P}^q(l,k)$ given by Corollary 5.9. If $H\in\mathscr{C}(\mathbb{R}^l,\mathbb{R}^k)$, and the pair $(0,H(0))\in V\times\mathbb{R}^k$ lies in $\mathcal{O}$, then

$$\begin{aligned}\phi_\#\circ\mathfrak{U}^q_{P,F}\circ(I,H^q\circ P)&=\beta_*\circ\mathfrak{U}^q_{P,F}\circ(I,H^q\circ P)\circ\phi\\&=\mathfrak{U}^q_{P,F}\circ B\circ(I,H^q\circ P)\circ\phi\end{aligned}$$

in a neighborhood of the zero section of $G\times_H V$. Define a map $\Phi$ from $(G\times_H\mathcal{O})\times J^q(l,k)$ into $(G\times_H V)\times\mathscr{P}^q(l,k)$ by $\Phi([g,x],H^q)=B(\phi([g,x]),H^q)$. Then $\Phi$ is a smooth equivariant map whose germ at $G\times_H(x=0)$ is invertible, and $\phi_\#\circ\mathfrak{U}^q_{P,F}=\mathfrak{U}^q_{P,F}\circ\Phi$.

5.12. Let $V$, $W$ be linear $G$-spaces. Let $P=(p_1,\cdots,p_l)$ be a polynomial orbit map for $V$, and $F_1,\cdots,F_k$ a set of polynomial generators for $\mathscr{C}(V,W)^G$ over $\mathscr{C}(V)^G$. Suppose $x^0\in V$ and $K=G_{x^0}$. Choose a $K$-invariant linear subspace $T$ of $V$, complementary to the tangent space $T_{x^0}(Gx^0)$ in $V$. Then the image of the map $i(y)=x^0+y$ of some invariant neighborhood $S$ of $0\in T$ into $V$, is a slice for the orbit $Gx^0$ at $x^0$. We identify the corresponding tubular neighborhood of $Gx^0$ with $G\times_K S$; the map $i(y)=x^0+y$ is then identified with the canonical $K$-invariant inclusion $i:S\to G\times_K S$.

Note that $Q(y)=P(x^0+y)$ is a polynomial orbit map for the $K$-space $S$, and that $G_i(y)=F_i(x^0+y)$, $i=1,\cdots,k$, form a set of polynomial generators for $\mathscr{C}(S,W)^K$ over $\mathscr{C}(S)^K$.

Let $\sigma$ be an analytic $K$-equivariant local section of $G\to G/K$, defined on a $K$-invariant open neighborhood $U$ of $eK$, such that $\sigma(eK)=e$. Let $\mathscr{A}_\sigma:J^q(S,W)\to J^q(G\times_K S,W)$ be the immersion (3.5.1) induced by $\sigma$.

LEMMA 5.13. *The following diagram is commutative*

$$\begin{array}{ccc} S\times\mathcal{P}^q(l,k) & \xrightarrow{U^q_{Q,G}} & J^q(S,W) \\ \downarrow{\scriptstyle i\times I} & & \downarrow{\scriptstyle \mathcal{A}_\sigma} \\ (G\times_K S)\times\mathcal{P}^q(l,k) & \xrightarrow{U^q_{P,F}} & J^q(G\times_K S, W). \end{array}$$

PROOF. If $f\in\mathscr{C}(S,W)^K$, $f(y)=\sum_{i=1}^k H_i(Q(y))G_i(y)$, then $j^qf(y)=U^q_{Q,G}(y,H^q(Q(y)))$. But $f$ extends to the map $\tilde{f}(x)=\sum_{i=1}^k H_i(P(x))F_i(x)$ in $\mathscr{C}(G\times_K S,W)^G$. Hence $\mathcal{A}_\sigma\circ j^qf(y)=j^q\tilde{f}(x^0+y)$, or $\mathcal{A}_\sigma\circ U^q_{Q,G}(y,H^q(Q(y)))=U^q_{P,F}(x^0+y,H^q(P(x^0+y)))$. In other words, $\mathcal{A}_\sigma\circ U^q_{Q,G}=U^q_{P,F}\circ(i\times I)$.

## 6. Transversality to a Whitney stratification

6.1. We recall some basic facts about stratified sets. See [12] for details. Let $P$ be a smooth manifold. A *stratification* of a subset $E$ of $P$ is a locally finite partition of $E$ into connected manifolds, called the *strata*, such that the frontier $Cl(X)-X$ of each stratum $X$ is the union of a set of lower dimensional strata ($Cl(X)$ denotes the closure of $X$ in $E$). A *Whitney stratification* of $E$ has the additional property that the following Condition $B$ of Whitney is satisfied by every pair $(X,Y)$ of strata, with $X$ in the frontier of $Y$ (to state Condition $B$, we assume $P$ is an open subset of $\mathbb{R}^p$; this is sufficient since the condition is diffeomorphism invariant):

Condition $B$. Let $x$ be any point of $X$. Let $\{x_i\}$, $\{y_i\}$ be any sequences of points in $X$, $Y$ (respectively), such that each sequence converges to $x$, the line joining $x_i$ and $y_i$ converges (in projective space $\mathbb{P}^{p-1}$) to a line $l$, and $T_{y_i}Y$ converges (in the Grassmannian of (dim $Y$)-planes) to a plane $\tau$. Then $l\subset\tau$.

To a Whitney stratification $\mathcal{S}$ of $E$ is associated a *filtration by dimension*, defined by letting $E_j$ be the union of the strata of dimension at most $j$ [12]. Suppose $\mathcal{S}$ and $\mathcal{S}'$ are two Whitney stratifications of $E$, and $\{E_j\}$, $\{E'_j\}$ the associated filtrations. Say that $\mathcal{S}<\mathcal{S}'$ if there exists an integer $j$ such that $E_j\subset E'_j$, and $E_k=E'_k$ for $k>j$.

A semianalytic subset $E$ of $\mathbb{R}^p$ has a canonical 'minimum' Whitney stratification $\mathcal{S}$, as constructed inductively by Mather [12].or Lojasiewicz [8]. Each stratum of $\mathcal{S}$ is a semianalytic manifold. 'Minimum' means that if $\mathcal{S}'$ is any other Whitney stratification by semianalytic manifolds, then $\mathcal{S}\leqq\mathcal{S}'$. Mather [12] points out that $\mathcal{S}$ is, in fact, minimum among stratifications by smooth manifolds. It follows that if $\alpha$ is a diffeomorphism of $\mathbb{R}^p$, then $\alpha(E)$ has a minimum Whitney stratification by smooth manifolds: the strata are the images of those in $\mathcal{S}$.

6.2. If $N$ is a smooth manifold, then by transversality of a map $f \in \mathscr{C}(N, \mathbb{R}^p)$ to a semianalytic subset $E$ of $\mathbb{R}^p$, we mean transversality of $f$ to each stratum of the minimum Whitney stratification of $E$ ([19], [17, Section II.D]). By Whitney's Condition $A$, if $f$ is transverse to a stratum $X$ of $E$, then $f$ is transverse to each stratum to which $X$ is adherent, in some neighborhood of $X$ (since Condition $A$ is weaker than $B$ [12], we skip its definition here).

If $f$ is an immersion which is transverse to $E$, then $f^{-1}(E)$ is a Whitney stratified subset of $N$. If $E$ is closed, $N$ compact, and $g \in \mathscr{C}(N, \mathbb{R}^p)$ is sufficiently close to $f$, then Thom's First Isotopy Lemma [12, Theorem 8.1] provides a topological ambient isotopy of $g^{-1}(E)$ to $f^{-1}(E)$.

The following simple consequences of the invariance lemmas of section 5 will be used, in section 7, to show that equivariant general position is well-defined. In each of these corollaries, the notation is that of the specified lemma from section 5.

COROLLARY 6.3 (*to lemma* 5.1). *Let $Q$ be an analytic submanifold of $\mathbb{R}^p$, and $h \in \mathscr{C}(\mathbb{R}^n, \mathbb{R}^k)$. Then the map $(I, h): \mathbb{R}^n \to \mathbb{R}^n \times \mathbb{R}^k$ is transverse to $U^{-1}(Q)$ at $x \in \mathbb{R}^n$ if and only if $(I, h+\alpha)$ is transverse to $U^{-1}(Q)$ at $x$.*

COROLLARY 6.4 (*to lemma* 5.3). *Let $Q$ be an invariant analytic submanifold of $J^q(V, W)$. Let $f \in \mathscr{C}(\mathscr{V}, W)^G$, $f(x) = \sum_{i=1}^k H_i(P(x))F_i(x)$, where $H = (H_1, \cdots, H_k) \in \mathscr{C}(\mathscr{U}, \mathbb{R}^k)$. Then $j^q f = U^q_{P,F} \circ (I, H^q \circ P) = U^q_{Q,F} \circ (I, \Phi \cdot (H^q \circ P))$. The map $(I, H^q \circ P): \mathscr{V} \to \mathscr{V} \times \mathscr{P}^q(l, k)$ is transverse to $(U^q_{P,F})^{-1}(Q)$ at $x \in \mathscr{V}$ if and only if $(I, \Phi \cdot (H^q \circ P)) = \Phi \circ (I, H^q \circ P): \mathscr{V} \to \mathscr{V} \times \mathscr{P}^q(m, k)$ is transverse to $(U^q_{Q,F})^{-1}(Q)$ at $x$.*

COROLLARY 6.5 (*to lemma* 5.4). *Let $Q$ be an invariant analytic submanifold of $J^q(\mathscr{V}, W)$. Let $f \in \mathscr{C}(\mathscr{V}', W)^G$, $f(x) = \sum_{i=1}^k H_i(P(x))F_i(x)$. Then $j^q f = U^q_{P,F} \circ (I, H^q \circ P) = U^q_{P,G} \circ (I, A \cdot (H^q \circ P))$. The map $(I, H^q \circ P)$ is transverse to $(U^q_{P,F})^{-1}(Q)$ at $x \in \mathscr{V}'$ if and only if $(I, A \cdot (H^q \circ p))$ is transverse to $(U^q_{P,G})^{-1}(Q)$ at $x$.*

COROLLARY 6.6 (*to lemma* 5.5). *Let $Q$ be an invariant analytic submanifold of $J^q(\mathscr{V}, W)$. Let $f \in \mathscr{C}(\mathscr{V}', W)^G$, $f = \sum_{i=1}^m (H_i \circ P)F_i$. Then $j^q f = U^q_{P,F} \circ (I, H^q \circ P) = U^q_{P,G} \circ (I, \pi \circ (A^{-1} \cdot (H^q \circ P)))$. The map $(I, H^q \circ P)$ is transverse to $(U^q_{P,F})^{-1}(Q)$ at $x \in \mathscr{V}'$ if and only if $(I, \pi \circ A^{-1} \cdot (H^q \circ P)))$ is transverse to $(U^q_{P,G})^{-1}(Q)$ at $x$.*

COROLLARY 6.7 (*to proposition* 5.11). *Let $Q$ be an invariant analytic submanifold of $J^q(G \times_H V, T)$. Suppose $f \in \mathscr{C}(G \times_H V, T)^G$, and $f(V) \subset \tau(W)$, so that $\tau^{-1} \circ f$ is defined. The map $\tau^{-1} \circ f \mid V \in \mathscr{C}(V, W)^H$ may be written $\tau^{-1} \circ f \mid V = \sum_{i=1}^k (H_i \circ P)F_i$, where $H = (H_1, \cdots, H_k) \in \mathscr{C}(\mathbb{R}^l, \mathbb{R}^k)$, so that*

*$j^q f = \mathfrak{U}^q_{P,F} \circ (I, H^q \circ P)$. Then $j^q(f \circ \alpha^{-1}) = \mathfrak{U}^q_{P,F} \circ (I, A \cdot (H^q \circ P))$ in some invariant neighborhood of the zero section of $G \times_H V$. The map $(I, H^q \circ P): G \times_H V \to (G \times_H V) \times \mathscr{P}^q(l, k)$ is transverse to $(\mathfrak{U}^q_{P,F})^{-1}(Q)$ at $[g, 0] \in G \times_H V$ if and only if $(I, A \cdot (H^q \circ P))$ is transverse to $(\mathfrak{U}^q_{P,F})^{-1}(\alpha_\#(Q))$ at $\alpha([g, 0]) = [g, 0]$.*

Remark. It follows from proposition 5.11 and lemma 5.13 that $(\mathfrak{U}^q_{P,F})^{-1}(\alpha_\#(Q))$ has a minimum Whitney stratification.

Proof. We have:

$$\begin{aligned} j^q(f \circ \alpha^{-1}) &= \alpha_\# \circ j^q f \circ \alpha^{-1} && \text{by (5.6.4)} \\ &= \alpha_\# \circ \mathfrak{U}^q_{P,F} \circ (I, H^q \circ P) \circ \alpha^{-1} \\ &= \mathfrak{U}^q_{P,F} \circ A \circ (I, H^q \circ P) \circ \alpha^{-1} && \text{by proposition 5.11} \\ &= \mathfrak{U}^q_{P,F} \circ (I, A \cdot (H^q \circ P)) && \text{by (5.6.1)}, \end{aligned}$$

where the last two equalities hold in some invariant neighborhood of the zero section of $G \times_H V$. The second assertion follows immediately from proposition 5.11.

Corollary 6.8 (*to corollary* 5.9). *Let $Q$ be an invariant analytic submanifold of $J^q(G \times_H V, T)$. Let $f \in \mathscr{C}(G \times_H V, T)^G$, with $f(V) \subset \tau(W)$, and $\tau^{-1} \circ f \mid V = \sum_{i=1}^k (H_i \circ P)F_i$. Then $j^q f = \mathfrak{U}^q_{P,F} \circ (I, H^q \circ P)$, and $j^q(\beta \cdot f) = \mathfrak{U}^q_{P,F} \circ (I, B \cdot (H^q \circ P))$ in some invariant neighborhood of the zero section of $G \times_H V$. The map $(I, H^q \circ P): G \times_H V \to (G \times_H V) \times \mathscr{P}^q(l, k)$ is transverse to $(\mathfrak{U}^q_{P,F})^{-1}(Q)$ at $[g, 0] \in G \times_H V$ if and only if $(I, B \cdot (H^q \circ P))$ is transverse to $(\mathfrak{U}^q_{P,F})^{-1}(\beta_*(Q))$ at $[g, 0]$.*

## 7. General position of jets of equivariant maps

Let $N, P$ be $G$-manifolds, and $Q$ an invariant submanifold of $J^q(N, P)$, for some natural number $q$. Let $(x, y) \in N \times P$, and $H = G_x$. By the Slice Theorem, there is a linear $H$-space $V$, and an equivariant diffeomorphism $\eta$ from $G \times_H V$ onto an invariant open tubular neighborhood of $Gx$ in $N$, such that $\eta([e, 0]) = x$. Also there is a linear $G_y$-space $V_y$, and an equivariant diffeomorphism $\rho$ from $T = G \times_{G_y} V_y$ onto an invariant open tubular neighborhood of $Gy$ in $P$, such that $\rho([e, 0]) = y$. Finally, there is a linear $G_y$-space $W$, and an analytic $G_y$-equivariant isomorphism $\tau$ of $W$ onto an invariant open neighborhood of $[e, 0]$ in $T$, such that $\tau(0) = [e, 0]$. The equivariant diffeomorphism $\eta \times \rho$ of $(G \times_H V) \times T$ into $N \times P$ induces an equivariant diffeomorphism $(\eta \times \rho)^*$ of $J^q(G \times_H V, T)$ into $J^q(N, P)$ (cf. section 5.6).

*We assume that for each $(x, y) \in N \times P$, $\eta$ and $\rho$ may be chosen so that $(\eta \times \rho)_*^{-1}(Q)$ is an analytic submanifold of $J^q(G \times_H V, T)$.*

Let $P=(p_1,\cdots,p_l)$ be a polynomial orbit map for the linear $H$-space $V$, and $F_1,\cdots,F_k$ a set of polynomial generators for $\mathscr{C}(V,W)^H$ over $\mathscr{C}(V)^H$. The unique (algebraic) map in $\mathscr{C}(G\times_H V,\mathbb{R}^l)^G$ induced by $P$ is also denoted $P$. Recall that the $H$-equivariant composed map $\tau_*\circ U^q_{P,F}: V\times\mathscr{P}^q(l,k)\to J^q(V,T)$, where $U^q_{P,F}$ is defined by (4.3.2), induces an analytic $G$-equivariant map $\mathfrak{U}^q_{P,F}:(G\times_H V)\times\mathscr{P}^q(l,k)\to J^q(G\times_H V,T)$ (lemma 4.4).

If $f\in\mathscr{C}(N,P)^G$ and $y=f(x)$, then $\eta,\rho,\tau$ may be chosen so that $f\circ\eta(V)\subset\rho\circ\tau(W)$. In particular $(\rho\circ\tau)^{-1}\circ f\circ\eta\mid V$ is defined. We can write $(\rho\circ\tau)^{-1}\circ f\circ\eta\mid V=\sum_{i=1}^k(H_i\circ P)F_i$, where $H=(H_1,\cdots,H_k)\in\mathscr{C}(\mathbb{R}^l,\mathbb{R}^k)$.

Definition 7.1. $j^qf$ *is in general position with respect to* $Q$ *at* $x\in N$ if either $j^qf(x)\notin Q$, or $j^qf(x)\in Q$ and the following condition is satisfied: The map

$$(I,H^q\circ P): G\times_H V\to(G\times_H V)\times\mathscr{P}^q(l,k)$$

is transverse to the minimum Whitney stratification of $((\eta\times\rho)_*\circ\mathfrak{U}^q_{P,F})^{-1}(Q)$ *at* $[e,0]\in G\times_H V$.

$j^qf$ is *in general position with respect to* $Q$ if it is in general position with respect to $Q$ at each $x\in N$.

Remark. Once we assume that $\eta$ and $\rho$ may be chosen so that $(\eta\times\rho)_*^{-1}(Q)$ is an analytic submanifold of $J^q(G\times_H V,T)$, then it follows from lemmas 5.8, 5.13, and proposition 5.11 that $((\eta\times\rho)_*\circ\mathfrak{U}^q_{P,F})^{-1}(Q)$ has a minimum Whitney stratification for any choice of $\eta$, $\rho$, and $\tau$.

Proposition 7.2. *General position of $j^qf$ with respect to $Q$ is well-defined.*

Proof. It follows easily from corollaries 6.3 to 6.8 of the invariance lemmas of section 5, that definition 7.1 is independent of all choices involved: of $\eta$, $\rho$, and $\tau$, and of $P$, $F_i$, and the representation of $(\rho\circ\tau)^{-1}\circ f\circ\eta\mid V$ in terms of these generators. Note that although corollaries 6.3 to 6.8 are stated for analytic submanifolds of the target, the above remark shows that they apply to any choice of $\eta$ and $\rho$ here.

Remark 7.3. If $G=\{e\}$, then general position = transversality.

Proposition 7.4. *In the above notation, $j^qf\mid\eta(G\times_H V)$ is in general position with respect to $Q$ if and only if the map $(I,H^q\circ P)$ is transverse to $((\eta\times\rho)_*\circ\mathfrak{U}^q_{P,F})^{-1}(Q)$.*

This is an immediate consequence of lemma 5.13.

Corollary 7.5. *If $j^qf$ is in general position with respect to $Q$ at $x\in N$, then $j^qf$ is in general position with respect to $Q$ in some neighborhood of $x$.*

In view of proposition 7.4, corollary 7.5 follows from the corresponding property for transversality to a Whitney stratified set.

THEOREM 7.6. *If $Q$ is a closed invariant submanifold of $J^q(N, P)$, then the set of maps $f \in \mathscr{C}(N, P)^G$, such that $j^q f$ is in general position with respect to $Q$, is open (in the Whitney topology).*

Recall that, in the local context, the map $(H_1, \cdots, H_k) \mapsto \sum_{i=1}^k (H_i \circ P)F_i$ of $\mathscr{C}(\mathbb{R}^l)^k$ onto $\mathscr{C}(V, W)^H$ is open (in the $\mathscr{C}^\infty$ topology), by the Open Mapping Theorem. Theorem 7.6 then follows from proposition 7.4 and corollary 7.5, together with the openness theorem for transversality to a Whitney stratified set (cf. [4, Theorem 1.3]).

THEOREM 7.7. *If $Q$ is an invariant submanifold of $J^q(N, P)$, then the set of maps $f \in \mathscr{C}(N, P)^G$, such that $j^q f$ is in general position with respect to $Q$, is a residual subset of $\mathscr{C}(N, P)^G$ (in either the $\mathscr{C}^\infty$ or Whitney topology).*

In fact we will use a more general multijet version of this theorem, which will be given in section 9. The crux of the theorem is the local lemma 9.3. Theorem 7.7 follows from lemma 9.3 (with $r = 1$) in a standard way (cf. [20, Chapter VII, Theorem 4.2], [4, Theorem 1.4]).

If $Q$ is an invariant submanifold of $J^q(N, P)$ and $j^q f$ is in general position with respect to $Q$, where $f \in \mathscr{C}(N, P)^G$, then $(j^q f)^{-1}(Q)$ is Whitney stratified. The following isotopy theorem is an easy consequence of Thom's First Isotopy Lemma (cf. [4, Theorem 1.5]).

THEOREM 7.8. *Suppose that $N$ is compact, and $Q$ closed in $J^q(N, P)$. Let $S$ be a smooth manifold, with the trivial action of $G$, and $f(s, x) = f_s(x)$ a map in $\mathscr{C}(S \times N, P)^G$. Let $F(s, x) = j^q f_s(x)$, $(s, x) \in S \times N$. If $j^q f_{s^0}$ is in general position with respect to $Q$, then there is an open neighborhood $\mathscr{U}$ of $s^0$ in $S$, and an equivariant homeomorphism $A : \mathscr{U} \times N \to \mathscr{U} \times N$ covering the identity map of $\mathscr{U}$, such that $A \mid \{s^0\} \times N = I$, and*

$$A((F \mid \mathscr{U} \times N)^{-1}(Q)) = \mathscr{U} \times (j^q f_{s^0})^{-1}(Q).$$

## 8. General position of equivariant maps: higher order conditions

Let $N, P$ be $G$-manifolds, and $Q$ an invariant submanifold of $P$. For each $q = 0, 1, \cdots$, $J^q(N, Q)$ canonically identifies with an invariant submanifold of $J^q(N, P)$. For any point $y \in P$, an invariant tubular neighborhood $\rho : T = G \times_{G_y} V_y \to P$ of $Gy$ can be chosen so that $\rho^{-1}(Q)$ is a $G$-vector subbundle of $T$ (hence for any invariant tubular neighborhood $\eta : G \times_H V \to N$ of an orbit in $N$, $(\eta \times \rho)_*^{-1}(Q)$ is an invariant algebraic submanifold of $J^q(G \times_H V, T)$). Let $r$ be a non-negative integer.

DEFINITION 8.1. A map $f \in \mathscr{C}(N, P)^G$ is *in* $r$th *approximation to general position with respect to $Q$ at $x \in N$* if for each $q = 0, 1, \cdots, r$, $j^q f$ is in general position with respect to $J^q(N, Q)$ at $x$. $f$ is *in general position with respect to*

*Q at x* if it is in $r$th approximation to general position, for each non-negative integer $r$. $f$ is *in general position with respect to Q* if it is in general position at each $x \in N$.

We will see, in theorem 8.4 below, that the infinite sequence of transversality conditions involved in the definition locally stabilizes. *The notion of equivariant general position given by* definition 1.1 *is what we now call zeroth approximation to general position.*

It will be convenient to reformulate the definition as a sequence of local regularity conditions, analogous to definition 1.1. Let $V$, $W$ be linear $G$-spaces. For each $q = 0, 1, \cdots$, $V$ is embedded in $J^q(V, W)$ as the image of the zero section. Let $P = (p_1, \cdots, p_l)$ be a polynomial orbit map for $V$, and $F_1, \cdots, F_k$ a set of polynomial generators for $\mathscr{C}(V, W)^G$.

PROPOSITION 8.2. *Let* $f \in \mathscr{C}(V, W)^G$, $f = \sum_{i=1}^k (H_i \circ P)F_i$, *where* $H = (H_1, \cdots, H_k) \in \mathscr{C}(\mathbb{R}^l, \mathbb{R}^k)$. *Then f is in* $r$th *approximation to general position with respect to* $0 \in W$ *at* $0 \in V$ *if and only if, for each* $q = 0, 1, \cdots, r$, *the map* $(I, H^q \circ P): V \to V \times \mathscr{P}^q(l, k)$ *is transverse to the affine algebraic subvariety* $(U^q_{P,F})^{-1}(V)$ *of* $V \times \mathscr{P}^q(l, k)$, *at* $0 \in V$.

Now let $N$, $P$ be $G$-manifolds, and $Q$ an invariant submanifold of $P$. Let $(x, y) \in N \times P$, and $H = G_x$. There is a linear $H$-space $V$, and an $H$-equivariant diffeomorphism $\eta$ from $V$ onto a slice for the orbit $Gx$ at $x$. There is a linear $G_y$-space $W$, and a $G_y$-equivariant diffeomorphism $\tau$ of $W$ onto a $G_y$-invariant open neighborhood of $y$ in $P$, such that $W$ is a $G_y$-direct sum $W = W_1 \oplus W_2$, and $\tau(W_1) = Q \cap \tau(W)$. Let $\pi: W \to W_2$ be the projection. If $f \in \mathscr{C}(N, P)^G$, and $y = f(x)$, then $\eta$ and $\tau$ may be chosen so that $f \circ \eta(V) \subset \tau(W)$.

PROPOSITION 8.3. *f is in* ($r$th *approximation to*) *general position with respect to Q at x if and only if either* $f(x) \notin Q$, *or* $f(x) \in Q$ *and* $(\pi \circ \tau^{-1}) \circ f \circ \eta : V \to W_2$ *is in* ($r$th *approximation to*) *general position with respect to* $0 \in W_2$ *at* $0 \in V$.

THEOREM 8.4. *Let V, W be linear G-spaces. There is a non-negative integer* $r = r(V, W)$, *depending only on V, W, such that if* $f \in \mathscr{C}(V, W)^G$ *is in* $r$th *approximation to general position with respect to* $0 \in W$ *at* $0 \in V$, *then f is in general position with respect to* $0 \in W$ *at* $0 \in V$.

PROOF. Let $p_1, \cdots, p_l$ be a minimal homogeneous Hilbert basis of $\mathscr{P}(V)^G$, and $F_1, \cdots, F_k$ a minimal set of homogeneous generators of $\mathscr{P}(V, W)^G$. Then $f \in \mathscr{C}(V, W)^G$ may be written $f = \sum_{i=1}^k (H_i \circ P)F_i$, where $P = (p_1, \cdots, p_l)$, and $H = (H_1, \cdots, H_k) \in \mathscr{C}(\mathbb{R}^l, \mathbb{R}^k)$.

For each positive integer $q$, denote by $\pi_{q+1}$ the canonical projection of $V\times\mathcal{P}^{q+1}(l,k)$ onto $V\times\mathcal{P}^{q}(l,k)$. The map $(I, H^q\circ P): V\to V\times\mathcal{P}^q(l,k)$ is transverse to the affine algebraic subvariety $(U^q_{P,F})^{-1}(V)$ of $V\times\mathcal{P}^q(l,k)$ at $0\in V$, if and only if $(I, H^{q+1}\circ P)$ is transverse to the subvariety $(U^q_{P,F}\circ\pi_{q+1})^{-1}(V)$ of $V\times\mathcal{P}^{q+1}(l,k)$ at $0\in V$.

By its definition (4.3.2), $(U^{q+1}_{P,F})^{-1}(V)$ is an affine algebraic subvariety of $(U^q_{P,F}\circ\pi_{q+1})^{-1}(V)$. Each irreducible component of $(U^{q+1}_{P,F})^{-1}(V)$ either coincides with the irreducible component of $(U^q_{P,F}\circ\pi_{q+1})^{-1}(V)$ in which it lies, or else has strictly greater codimension in $V\times\mathcal{P}^{q+1}(l,k)$ (for basic properties of real algebraic varieties see [22]). Hence there is a non-negative integer $q$, such that if $(I, H^q\circ P)$ is transverse to the subvariety $(U^q_{P,F})^{-1}(V)$ of $V\times\mathcal{P}^q(l,k)$ at $0\in V$, then $(I, H^t\circ P)$ is transverse to $(U^t_{P,F})^{-1}(V)$ at 0 for each $t\geqq q$. Let $r(V, W)$ be the least such $q$.

REMARK 8.5. For a given compact Lie group $G$, $r(V, W)$ may be arbitrarily large. For example, if $V=\mathbb{C}$, with $S^1$ acting in the standard way, and $W=\mathbb{C}^2$, with $S^1$ acting by $e^{i\theta}(y_1, y_2)=(e^{ni\theta}y_1, e^{(n+1)i\theta}y_2)$, then $r(V, W)=n$. In this example, any map $f\in\mathscr{C}(V, W)^{S^1}$ may be written $f(x)=(\alpha(\bar{x}x)x^n, \beta(\bar{x}x)x^{n+1})$, where $x\in\mathbb{C}$ and $\alpha, \beta$ are smooth $\mathbb{C}$-valued functions. It is easily checked that if $r<n$, then $f$ is in $r$th approximation to general position with respect to $0\in W$ at $0\in V$ provided that $\alpha(0), \beta(0)$ are not both zero. $f$ is in general position with respect to $0\in W$ at $0\in V$ if and only if $\alpha(0)\neq 0$.

If $G=\{e\}$, then $r(V, W)=0$ for any linear spaces $V, W$. In [4, Section 2], $r(V, W)=0$ in examples 2.3, 2.4, and 2.6, but $r(V, W)=1$ in example 2.5.

THEOREM 8.6. *Let $N, P$ be smooth $G$-manifolds, and $Q$ an invariant submanifold of $P$. Then the set of maps $f\in\mathscr{C}(N, P)^G$ which are in general position with respect to $Q$ is a residual subset of $\mathscr{C}(N, P)^G$ (in either the $\mathscr{C}^\infty$ or Whitney topology). If $Q$ is closed in $P$, then the set of maps $f\in\mathscr{C}(N, P)^G$ which are in general position with respect to $Q$ is open in $\mathscr{C}(N, P)^G$, with the Whitney topology.*

PROOF. Consider linear $G$-spaces $V, W$, and $f\in\mathscr{C}(V, W)^G$. Let $P\in\mathcal{P}(V, \mathbb{R}^l)^G$ be an orbit map for $V$, and $F_1, \cdots, F_k$ a set of generators of $\mathcal{P}(V, W)^G$. Write $f=\sum_{i=1}^k(H_i\circ P)F_i$, where $H=(H_1, \cdots, H_k)\in\mathscr{C}(\mathbb{R}^l, \mathbb{R}^k)$. By lemma 5.13, proposition 8.2, and theorem 8.4, $f$ is in general position with respect to $0\in W$ if and only if, for $q=0, 1, \cdots, r(V, W)$, the map $(I, H^q\circ P): V\to V\times\mathcal{P}^q(l,k)$ is transverse to $(U^q_{P,F})^{-1}(V)$. This local result suffices to prove the openness assertion, as in theorem 7.6.

If $f=\sum_{i=1}^k(H_i\circ P)F_i\in\mathscr{C}(V, W)^G$, then, as in lemma 9.3, there exists a polynomial map $B=(B_1, \cdots, B_k)\in\mathcal{P}(\mathbb{R}^l, \mathbb{R}^k)$, of degree at most $r(V, W)$, with arbitrarily small coefficients, such that the equivariant map $g(x) = \sum_{i=1}^{k} (H_i + B_i)(P(x))F_i(x)$ is in general position with respect to the origin of $W$. By proposition 8.3, this local result suffices to prove the density assertion, in the standard way.

REMARKS 8.7. If $G = \{e\}$, then general position = transversality. A map $f \in \mathscr{C}(N, P)^G$ which is in general position with respect to an invariant submanifold $Q$ of $P$, is stratumwise transverse to $Q$; i.e. for each isotropy subgroup $H$ of $N$, the map $f \mid N_H : N_H \to P^H$ is transverse to $Q^H$ [4, Proposition 6.4]. If $f$ is in general position with respect to $Q$, then $f^{-1}(Q)$ is strongly stratified (in the sense of Thom [18]) by invariant submanifolds. Since a map $f \in \mathscr{C}(N, P)^G$ in general position with respect to $Q$ is, in particular, in zeroth approximation to general position, then theorem 7.8 provides an isotopy theorem in the present context.

The definition 7.1 of general position for the $q$-jets of equivariant maps can also be refined, using a sequence of transversality conditions involving the higher jets. The sequence of local conditions stabilizes, as in theorem 8.4. Definition 7.1, though, will suffice for our purposes in this paper.

## 9. Multijet general position

Let $r$ be a positive integer. If $X$ is a space, then ${}_rX$ denotes the subspace of the $r$-fold product $X^r$ comprising points $(x^1, \cdots, x^r)$ such that $x^i \neq x^j$ if $i \neq j$.

Let $N, P$ be smooth $G$-manifolds. Denote by $\pi : N \to N/G$ the orbit map, and by $\pi^r : N^r \to (N/G)^r$ its $r$-fold product. Let ${}_r^GN = (\pi^r)^{-1}({}_r(N/G))$. ${}_r^GN$ is the invariant open submanifold of $N^r$, of $r$-tuples of points with disjoint orbits. Let $q$ be a non-negative integer, and ${}_r^GJ^q(N, P)$ be the subspace of $J^q(N, P)^r$ projecting onto ${}_r^GN \times P^r$. ${}_r^GJ^q(N, P)$ is clearly invariant under the action of $G$ on $J^q(N, P)^r$, and has the structure of a $G$-fiber bundle over ${}_r^GN \times P^r$. A map $f \in \mathscr{C}(N, P)^G$ induces an $r$-fold $q$-jet extension

$$
{}_rj^qf : {}_r^GN \to {}_r^GJ^q(N, P)
$$
$$
{}_rj^qf(x^1, \cdots, x^r) = (j^qf(x^1), \cdots, j^qf(x^r)).
$$

Let $Q$ be an invariant submanifold of ${}_r^GJ^q(N, P)$, and $z = (z^1, \cdots, z^r) \in Q$; say $z^i = j^qf^i(x^i)$, where $f^i \in \mathscr{C}(N, P)^G$ and $y^i = f^i(x^i)$, $i = 1, \cdots, r$. For each $i = 1, \cdots, r$, choose $\eta^i : G \times_{H^i} V^i \to N$, $\rho^i : T^i \to P$, $\tau^i : W^i \to T^i$ as at the beginning of section 7, making sure that the $\eta^i(G \times_{H^i} V^i)$ are disjoint. The equivariant diffeomorphisms $\eta^i \times \rho^i$ of $(G \times_{H^i} V^i) \times T^i$ into $N \times P$, $i = 1, \cdots, r$, induce an equivariant diffeomorphism $\prod_{i=1}^r (\eta^i \times \rho^i)_*$ of $\prod_{i=1}^r J^q(G \times_{H^i} V^i, T^i)$ into ${}_r^GJ^q(N, P)$. We again assume that the $\eta^i, \rho^i$ may be chosen so that $Q$ pulls back to an analytic submanifold of $\prod_{i=1}^r J^q(G\times_{H^i} V^i, T^i)$.

For each $i=1,\cdots,r$, choose an orbit map $P^i\in\mathcal{P}(V^i,\mathbb{R}^{l_i})$ for $V^i$, and a set $F_1^i,\cdots,F_{k_i}^i$ of polynomial generators of $\mathcal{C}(V^i,W^i)^{H^i}$.

If $f\in\mathcal{C}(N,P)^G$, and $y^i=f(x^i)$, $i=1,\cdots,r$, then $\eta^i,\rho^i,\tau^i$ may be chosen so that $f\circ\eta^i(V^i)\subset\rho^i\circ\tau^i(W^i)$. Then write $(\rho^i\circ\tau^i)^{-1}\circ f\circ\eta^i\mid V^i=\sum_{j=1}^{k_i}(H_j^i\circ P^i)F_j^i$, where $H^i=(H_1^i,\cdots,H_{k_i}^i)\in\mathcal{C}(\mathbb{R}^{l_i},\mathbb{R}^{k_i})$.

DEFINITION 9.1. ${}_rj^qf$ is in general position with respect to $Q$ at $x=(x^1,\cdots,x^r)\in{}_r^GN$ if either ${}_rj^qf(x)\notin Q$, or ${}_rj^qf(x)\in Q$ and the following condition is satisfied: The map

$$\prod_{i=1}^r(I^i,(H^i)^q\circ P^i):\prod_{i=1}^r(G\times_{H^i}V^i)\to\prod_{i=1}^r((G\times_{H^i}V^i)\times\mathcal{P}^q(l_i,k_i))$$

is transverse to the minimum Whitney stratification of $(\prod_{i=1}^r(\eta^i\times\rho^i)_*\circ\mathfrak{U}_{P^i,F^i}^q)^{-1}(Q)$ at the point $(\xi^1,\cdots,\xi^r)$, where $\xi^i=[e,0]\in G\times_{H^i}V^i$, $i=1,\cdots,r$.

THEOREM 9.2. *The set of maps $f\in\mathcal{C}(N,P)^G$ such that ${}_rj^qf$ is in general position with respect to $Q$ is a residual subset of $\mathcal{C}(N,P)^G$ (with either the $\mathcal{C}^\infty$ or Whitney topology).*

The main part of the proof is the following local lemma. We keep the above notation, but for simplicity assume that $Q$ is an invariant submanifold of $\prod_{i=1}^r J^q(G\times_{H^i}V^i,T^i)$, so we can suppress $\eta^i,\rho^i$.

LEMMA 9.3. *Let $(f^1,\cdots,f^r)\in\prod_{i=1}^r\mathcal{C}(G\times_{H^i}V^i,T^i)^G$. Suppose $f^i(V^i)\subset\tau^i(W^i)$, and write $(\tau^i)^{-1}\circ f^i\mid V^i=\sum_{j=1}^{k_i}(H_j^i\circ P^i)F_j^i$, $i=1,\cdots,r$. Then there exist polynomial maps $B^i=(B_1^i,\cdots,B_{k_i}^i)\in\mathcal{P}(\mathbb{R}^{l_i},\mathbb{R}^{k_i})$, $i=1,\cdots,r$, of degree at most $q$, with arbitrarily small coefficients, such that the map*

$$\prod_{i=1}^r(I^i,(H^i+B^i)^q\circ P^i):\prod_{i=1}^r(G\times_{H^i}V^i)\to\prod_{i=1}^r((G\times_{H^i}V^i)\times\mathcal{P}^q(l_i,k_i))$$

*is transverse to the minimum Whitney stratification of $(\prod_{i=1}^r\mathfrak{U}_{P^i,F^i}^q)^{-1}(Q)$.*

PROOF. Let $\mathcal{P}$ be the space of $r$-tuples of polynomial maps $(B^1,\cdots,B^r)\in\prod_{i=1}^r\mathcal{P}(\mathbb{R}^{l_i},\mathbb{R}^{k_i})$ of degree at most $q$. $\mathcal{P}$ is a finite dimensional vector space. Consider the smooth equivariant map

$$\Gamma:\prod_{i=1}^r(G\times_{H^i}V^i)\times\mathcal{P}\to\prod_{i=1}^r((G\times_{H^i}V^i)\times\mathcal{P}^q(l_i,k_i))$$

$$((x^1,\cdots,x^r),(B^1,\cdots,B^r))\mapsto\prod_{i=1}^r(x^i,(H^i+B^i)^q\circ P(x^i)).$$

$\Gamma$ is a submersion, so that it satisfies any transversality condition. Hence by a well-known corollary of Sard's theorem due to Thom [20, Chapter VII, Corollary 1.7], one can choose $(B^1, \cdots, B^r)$ arbitrarily small, so that the required transversality condition is satisfied.

Let $f \in \mathscr{C}(N, P)^G$, and $z = {}_rj^qf(x) \in {}^G_rJ^q(N, P)$, where $x = (x^1, \cdots, x^r) \in {}^G_rN$. Denote by $J_x$ the fiber of $J = {}^G_rJ^q(N, P)$ over $x$. Let $\omega : T_z(J) \to T_z(J_x)$ be the projection with respect to the internal direct sum decomposition $T_z(J) = T_z(J_x) \oplus \operatorname{Im} d({}_rj^qf)_x$ ($T_z(\cdot)$ denotes the tangent space at $z$, and $d(\cdot)_x$ the tangent map at $x$). The following implication of equivariant general position will be used in section 10; this proposition is an analog, for multijet general position, of the fact that general position of an equivariant map implies stratumwise transversality.

Proposition 9.4. *Let $f \in \mathscr{C}(N, P)^G$. If ${}_rj^qf$ is in general position with respect to $Q$ at $x$, and $z = {}_rj^qf(x) \in Q$, then*

$$T_z({}_rj^q_x(\mathscr{C}(N, P)^G)) \subset \omega(T_z(Q)).$$

Proposition 9.4 is a consequence of the observation that, in the notation of this section, $\prod_{i=1}^r (\eta^i \times \rho^i)_* \circ \mathfrak{U}^q_{P^i, F^i}(0, \cdot)$ maps $\prod_{i=1}^r \mathscr{P}^q(l_i, k_i)$ submersively onto ${}_rj^q_x(\mathscr{C}(N, P)^G)$.

## 10. Stability of smooth equivariant maps

Let $G$ be a compact Lie group, and $N, P$ smooth $G$-manifolds. Denote by $\mathscr{D}iff(N)$ the group of diffeomorphisms of $N$. $\mathscr{D}iff(N)$ is a subspace of $\mathscr{C}(N, N)$ which is invariant under the action of $G$ on $\mathscr{C}(N, N)$. $\mathscr{D}iff(N)^G$ is the subgroup of $\mathscr{D}iff(N)$ of equivariant diffeomorphisms.

$\mathscr{D}iff(N) \times \mathscr{D}iff(P)$ acts on $\mathscr{C}(N, P)$ by $(g, h) \cdot f = h \circ f \circ g^{-1}$, where $(g, h) \in \mathscr{D}iff(N) \times \mathscr{D}iff(P)$, $f \in \mathscr{C}(N, P)$. This action induces an action of $\mathscr{D}iff(N)^G \times \mathscr{D}iff(P)^G$ on $\mathscr{C}(N, P)^G$. Give $\mathscr{C}(N, P)^G$ the Whitney topology.

Definition 10.1. A map $f \in \mathscr{C}(N, P)^G$ is *stable* if its orbit under $\mathscr{D}iff(N)^G \times \mathscr{D}iff(P)^G$ is open in $\mathscr{C}(N, P)^G$.

If $f \in \mathscr{C}(N, P)$, then $\theta(f)$ denotes the $\mathscr{C}(N)$-module of smooth vector fields along $f$ (i.e. smooth sections of $f^*TP$). If $f \in \mathscr{C}(N, P)^G$, then there is a natural induced action of $G$ on $\theta(f)$; $\theta(f)^G$ is a $\mathscr{C}(N)^G$-module. Let $\theta(N) = \theta(I_N)$, where $I_N$ is the identity map of $N$. Then $\theta(N)^G$ is the $\mathscr{C}(N)^G$-module of smooth equivariant vector fields on $N$. Define maps

$$\alpha_f : \theta(P)^G \to \theta(f)^G \qquad \beta_f : \theta(N)^G \to \theta(f)^G$$
$$\eta \mapsto \eta \circ f \qquad\qquad \xi \mapsto -df \circ \xi.$$

DEFINITION 10.2. A map $f \in \mathcal{C}(N, P)^G$ is *infinitesimally stable* if $\theta(f)^G = \alpha_f(\theta(P)^G) + \beta_f(\theta(N)^G)$.

In theorem 10.7, we will prove that a proper stable equivariant map is infinitesimally stable. Poenaru showed that a proper infinitesimally stable equivariant map is stable [13]. The stability theorem for germs of equivariant maps was proven by Ronga [15]. The proof that stability implies infinitesimal stability for proper equivariant maps will follow that of Mather [11] (see also [20, Chapter X, Section 3]), once we interpret stability of $f \in \mathcal{C}(N, P)^G$ in terms of general position of the multijet extensions ${}_rj^qf$ with respect to orbits of $\mathcal{D}iff(N)^G \times \mathcal{D}iff(P)^G$ on ${}^G_rJ^q(N, P)$. We begin by examining these orbits.

LEMMA 10.3. *Let $x \in N$ and $H = G_x$. Then the orbit of $x$ under the action of $\mathcal{D}iff(N)^G$ on $N$ is a union of components of $N_H$.*

PROOF. By the Slice Theorem, it certainly suffices to show that if $N = G \times_H V$, where $V$ is a linear $H$-space, then the orbit of $[e, 0]$ under $\mathcal{D}iff(N)^G$ is $N_H$. Note that $N_H = F(H, G \times_H V) \cong N_G(H)/H \times V^H$. If $a \in N_G(H)$, then the map $[g, x] \mapsto [ga^{-1}, ax]$, where $[g, x] \in G \times_H V$, is an equivariant diffeomorphism. On the other hand, the orbit of $0 \in V$ under $\mathcal{D}iff(V)^H$ is clearly $V^H$. The result follows easily.

Let $\Delta = \{(y^1, \cdots, y^r) \in P^r \mid y^1 = \cdots = y^r\}$. Denote by ${}^G_r\Delta^q(N, P)$ the subspace of ${}^G_rJ^q(N, P)$ projecting onto ${}^G_rN \times \Delta$. ${}^G_r\Delta^q(N, P)$ has the structure of a $G$-fiber bundle over ${}^G_rN \times \Delta$.

The groups $\mathcal{D}iff(N)^G \times \mathcal{D}iff(P)^G$ and $G \times \mathcal{D}iff(N)^G \times \mathcal{D}iff(P)^G$ act in a natural way on ${}^G_r\Delta^q(N, P)$, and on the subspace of multijets of equivariant maps. Note that an orbit of the latter group is just the saturation by $G$ of an orbit of the former; in particular it is $G$-invariant (hence more convenient for discussion of equivariant general position in the framework of this paper).

Let $z = (z^1, \cdots, z^r) \in {}^G_r\Delta^q(N, P)$; say $z^i = j^qf^i(x^i)$, where $f^i \in \mathcal{C}(N, P)$, $i = 1, \cdots, r$. Let $y = f^i(x^i)$. We denote by $O = O_z$ the orbit of $z$ under $\mathcal{D}iff(N)^G \times \mathcal{D}iff(P)^G$, and by $GO = GO_z$ the orbit of $z$ under $G \times \mathcal{D}iff(N)^G \times \mathcal{D}iff(P)^G$.

LEMMA 10.4. *The orbits $O$ and $GO$ are submanifolds of ${}^G_r\Delta^q(N, P)$.*

PROOF. First we examine the orbits of these groups on the base ${}^G_rN \times \Delta$. Let $H^i = G_{x^i}$, $i = 1, \cdots, r$, and $K = G_y$. Choose disjoint invariant tubular neighborhoods $\eta^i : G \times_{H^i} V^i \to N$ of the orbits $Gx^i$, and an invariant tubular neighborhood $\rho : G \times_K W \to \Delta$ of $Gy$ in $\Delta$ (we may, of course, identify $\Delta$ with $P$).

By lemma 10.3, the orbit of $(x^1, \cdots, x^r; y)$ under $\mathcal{D}iff(N)^G \times \mathcal{D}iff(P)^G$ intersects the open subset $(\prod_{i=1}^r \eta^i(G \times_{H^i} V^i)) \times \rho(G \times_K W)$ of ${}^G_r N \times \Delta$ in the submanifold

$$\left(\prod_{i=1}^r \eta^i(F(H^i, G \times_{H^i} V^i))\right) \times \rho(F(K, G \times_K W)).$$

The isotropy subgroup of any point in the orbit of $(x^1, \cdots, x^r; y)$ is $H^1 \cap \cdots \cap H^r \cap K$. The subgroup $N_G(H^1) \cap \cdots \cap N_G(H^r) \cap N_G(K)$ acts on this orbit, which inherits the structure of a (right) principal bundle with structure group $\bigcap_{i=1}^r N_G(H^i) \cap N_G(K)/\bigcap_{i=1}^r H^i \cap K$. The orbit of $(x^1, \cdots, x^r; y)$ under $G \times \mathcal{D}iff(N)^G \times \mathcal{D}iff(P)^G$ is the bundle associated to this principal bundle, with fiber $G/\bigcap_{i=1}^r H^i \cap K$.

The orbits $O$ and $GO$ fiber over the corresponding orbits of $(x^1, \cdots, x^r; y)$ in ${}^G_r N \times \Delta$. In each case, the fiber over $(x^1, \cdots, x^r; y)$ is $(\mathcal{L}^q_N(x^1) \times \cdots \times \mathcal{L}^q_N(x^r) \times \mathcal{L}^q_P(y)) \cdot z$, where $\mathcal{L}^q_N(x')$ denotes the Lie group of $q$-jets at $x' \in N$ of invertible germs of equivariant maps fixing $x'$ (the action of $q$-jets of diffeomorphisms on $q$-jets of maps is defined in the standard way).

Consider a closed subgroup $H$ of $G$, and a linear $H$-space $V$. Recall (section 2.3) that there is an $H$-invariant open neighborhood $U$ of $eH$ in $G/H$, and an $H$-equivariant local section $\sigma: U \to G$, with $\sigma(eH) = e$ ($H$ acts by conjugation on $G$). The map $(gH, x) \mapsto [\sigma(gH), x]$ of $U \times V$ onto $G \times_H V \mid U$ defines an $H$-vector bundle equivalence over $U$. An equivariant diffeomorphism $\phi$ of $G \times_H V$ fixing $[e, 0]$ is determined by $\phi \mid V$. In a neighborhood of 0, $\phi \mid V$ is the composition of the equivalence $U \times V \to G \times_H V \mid U$ with an $H$-equivariant map $x \mapsto (\xi(x), \omega(x))$ of $V$ into $U \times V$, such that $\omega(x)$ is invertible. Using [11, Lemma 1.5], it is easy to see from this local representation of equivariant diffeomorphisms that the orbit $(\mathcal{L}^q_N(x^1) \times \cdots \times \mathcal{L}^q_N(x^r) \times \mathcal{L}^q_P(y)) \cdot z$ is a submanifold. (It is trivial that the orbit is an immersed submanifold. This will, in fact, suffice for our purposes since we will be concerned only with the density of equivariant maps whose multijets are in general position with respect to an orbit).

REMARK. The above proof shows, moreover, that for any choice of invariant tubular neighborhoods $\eta^i: G \times_{H^i} V^i \to N$ of $Gx^i$, $i = 1, \cdots, r$, and $\rho: G \times_K W \to \Delta$ of $Gy$, the orbits $O$ and $GO$ pull back to analytic submanifolds of $\prod_{i=1}^r J^q(G \times_{H^i} V^i, G \times_K W)$.

NOTATION 10.5. If $A$ is a commutative unitary ring, then $r(A)$ denotes the radical of $A$; i.e. the intersection of the maximal ideals.

If $X$ is a closed subset of a smooth manifold $N$, then $\mathcal{C}_X(N)$ denotes the ring of germs at $X$ of functions in $\mathcal{C}(N)$. Write $\mathfrak{m}_X = r(\mathcal{C}_X(N))$. If $X$ is a closed invariant subset of a $G$-manifold $N$, write $\mathfrak{n}_X = \mathfrak{m}_X \cap \mathscr{C}_X(N)^G = r(\mathscr{C}_X(N)^G)$.

Let $f \in \mathscr{C}(N, P)^G$. Suppose $y \in P$, and $S$ is a closed subset of $f^{-1}(y)$. Let $A_{Gy} = \mathscr{C}_{Gy}(P)^G$, $B_{GS} = \mathscr{C}_{GS}(N)^G$. Note that $A_{Gy}$ is a local ring with maximal ideal $\mathfrak{n}_{Gy}$. In fact if $H = G_y$, and $W$ is a slice for the orbit $Gy$ at $y$, then $A_{Gy} \cong \mathscr{C}_y(W)^H$. Also define

$$\theta_{GS}(f)^G = \theta(f)^G \otimes_{\mathscr{C}(N)^G} B_{GS}$$

$$\theta_{GS}(N)^G = \theta(N)^G \otimes_{\mathscr{C}(N)^G} B_{GS}$$

$$\theta_{Gy}(P)^G = \theta(P)^G \otimes_{\mathscr{C}(P)^G} A_{Gy}.$$

Denote by

$$\alpha_{GS} : \theta_{Gy}(P)^G \to \theta_{GS}(f)^G$$

$$\beta_{GS} : \theta_{GS}(N)^G \to \theta_{GS}(f)^G$$

the maps induced by $\alpha_f$, $\beta_f$.

We will need the following elementary lemma of Ronga [15]:

LEMMA 10.6. *Let $H$ be a closed subgroup of $G$, and $V$ a linear $H$-space. Let $p_1, \cdots, p_l$ be a minimal homogeneous Hilbert basis of $\mathscr{P}(V)^H$. Set $m = \min_i \deg p_i$, $M = \max_i \deg p_i$. Then*

(1) $\mathfrak{n}_{G/H} \subset \mathfrak{m}^m_{G/H}$;

(2) $\mathfrak{m}^{qM}_{G/H} \cap \mathscr{C}_{G/H}(G \times_H V)^G = \mathfrak{m}^{qM}_{G/H} \cap \mathfrak{n}_{G/H} \subset \mathfrak{n}^q_{G/H}$ *for any positive integer $q$.*

PROOF. (1) is trivial. To see (2), let $d_i = \deg p_i$, $i = 1, \cdots, l$, and let $r \geqq M$. If $f \in \mathfrak{m}^r_{G/H} \cap \mathscr{C}_{G/H}(G \times_H V)^G$, then $f \mid V$ may be written $(f \mid V)(x) = \sum_{i=1}^l (h_i(x) + g_i(x)) p_i(x)$, where $h_i \in \mathfrak{m}_0^{r-d_i} \cap \mathscr{C}_0(V)^H$, and $g_i$ is a polynomial of degree at most $r - d_i - 1$, $i = 1, \cdots, l$. In fact, then, $(f \mid V)(x) = \sum_{i=1}^l h_i(x) p_i(x)$, so that

$$\mathfrak{m}^r_{G/H} \cap \mathscr{C}_{G/H}(G \times_H V)^G \subset (\mathfrak{m}^{r-M}_{G/H} \cap \mathscr{C}_{G/H}(G \times_H V)^G) \cdot \mathfrak{n}_{G/H}.$$

The result follows by iterating this statement.

THEOREM 10.7. *Let $f \in \mathscr{C}(N, P)^G$ be a proper map. Then the following are equivalent:*

(1) *$f$ is infinitesimally stable;*

(2) *$f$ is stable;*

(3) *for each $r$ and $q$, ${}_rj^q f$ is in general position with respect to all orbits $GO$ of ${}^G_r\Delta^q(N, P)$.*

REMARK. Assume, moreover, that $N, P$ have only finitely many orbit types. Then $N, P$ have only finitely many inequivalent slice representations.

For each $x \in N$, let $M(x)$ be the maximum degree of the polynomials in a minimal homogeneous Hilbert basis for the invariants of a slice for the orbit $Gx$. For each $y \in P$, let $t(y)$ be the minimum number of generators of $\theta_{Gy}(P)^G$ over $A_{Gy}$. Define

$$t(P) = \max_{y \in P} t(y)$$

$$q(N, P) = \max_{y \in P} \left\{ (t(y)+1) \max_{x \in N,\, G_x \subset G_y} M(x) \right\} - 1 .$$

The following proof will show that conditions (1)–(3) of theorem 10.7 are each equivalent to:

(4) *for each* $r \leqq t(P)+1$, ${}_rj^{q(N,P)}f$ *is in general position with respect to all orbits* $GO$ *of* ${}^G_r\Delta^{q(N,P)}(N, P)$.

Proof of Theorem 10.7. If $f$ is infinitesimally stable, then $f$ is stable by [13]. Suppose $f$ is stable. For any $r$, $q$, the set of $f' \in \mathscr{C}(N, P)^G$ such that ${}_rj^qf'$ is in general position with respect to an orbit $GO_z$ of ${}^G_r\Delta^q(N, P)$ is dense in $\mathscr{C}(N, P)^G$, by theorem 9.2. Since $f$ is stable, and $GO_z$ is invariant under $\mathscr{D}iff(N)^G \times \mathscr{D}iff(P)^G$, then ${}_rj^qf$ is in general position with respect to $GO_z$. Hence (2) implies (3). It remains to show that (3) implies $f$ is infinitesimally stable.

It suffices to prove that the criterion (10.2) for infinitesimal stability is satisfied in some neighborhood of the inverse image $f^{-1}(Gy)$ of each orbit $Gy$ in $P$ (since $f$ is proper, this can be globalized using an invariant partition of unity on $P$). In other words, we must show that for each $y \in P$:

$$\theta_{Gf^{-1}(y)}(f)^G = \alpha_{Gf^{-1}(y)}(\theta_{Gy}(P)^G) + \beta_{Gf^{-1}(y)}(\theta_{Gf^{-1}(y)}(N)^G). \tag{10.7.1}$$

Let $t(y)$ be the minimum number of generators of $\theta_{Gy}(P)^G$ over $A_{Gy}$. We will see that for every non-empty finite subset $S$ of $f^{-1}(y)$:

$$\theta_{GS}(f)^G = \alpha_{GS}(\theta_{Gy}(P)^G) + \beta_{GS}(\theta_{GS}(N)^G) + \mathfrak{n}_{Gy} \cdot \theta_{GS}(f)^G + \mathfrak{n}_{GS}^{t(y)+1} \cdot \theta_{GS}(f)^G. \tag{10.7.2}$$

First, though, we prove (10.7.1) under the assumption that (10.7.2) is satisfied for each $y \in P$, and each finite subset $S = \{x^1, \cdots, x^r\} \subset f^{-1}(y)$ such that $Gx^i \neq Gx^j$ if $i \neq j$, and $r \leqq t(y)+1$. By the Equivariant Preparation Theorem [13], the homomorphism $f^*: A_{Gy} \to B_{GS}$ induced by $f$ is excellent. Since $\mathfrak{n}_{GS}$ is a finitely generated ideal in $B_{GS}$, and $t(y)$ is the minimum number of generators of $\theta_{Gy}(P)^G$ over $A_{Gy}$, then (10.7.2) implies

$$\theta_{GS}(f)^G = \alpha_{GS}(\theta_{Gy}(P)^G) + \beta_{GS}(\theta_{GS}(N)^G) \tag{10.7.3}$$

by [20, Chapter III, Remark 1.6].

Let $\mathscr{S}_y$ be the set of orbits $Gx$ contained in $Gf^{-1}(y)$ such that $\theta_{Gx}(f)^G \supsetneqq \beta_{Gx}(\theta_{Gx}(N)^G)$. Then card $(\mathscr{S}_y) \leqq t(y)$. Otherwise let $S$ be a subset $\{x^1, \cdots, x^{t(y)+1}\}$ of $f^{-1}(y)$ such that $Gx^i \neq Gx^j$ if $i \neq j$, and $Gx^i \in \mathscr{S}_y$, $i = 1, \cdots, t(y)+1$. Then

$$t(S) = \dim_{\mathbb{R}} \left( \frac{\theta_{GS}(f)^G}{\beta_{GS}(\theta_{GS}(N)^G)} \otimes_{A_{Gy}} \frac{A_{Gy}}{\mathfrak{n}_{Gy}} \right) \geqq t(y)+1 .$$

But

$$t(y) = \dim_{\mathbb{R}} \left( \theta_{Gy}(P)^G \otimes_{A_{Gy}} \frac{A_{Gy}}{\mathfrak{n}_{Gy}} \right),$$

so that $t(S) \leqq t(y)$ by (10.7.3); a contradiction.

The proof of (10.7.1) now follows immediately, so that $f$ is infinitesimally stable. It remains to prove (10.7.2).

Let $S = \{x^1, \cdots, x^r\} \subset f^{-1}(y)$, such that $Gx^i \neq Gx^j$ if $i \neq j$. We will show that if $z = (j^q f(x^1), \cdots, j^q f(x^r))$, then general position with respect to $GO_z$ at $x = (x^1, \cdots, x^r)$ implies

$$\theta_{GS}(f)^G = \alpha_{GS}(\theta_{Gy}(P)^G) + \beta_{GS}(\theta_{GS}(N)^G) + (\mathfrak{m}_{GS}^{q+1} \cap B_{GS}) \cdot \theta_{GS}(f)^G . \quad (10.7.4)$$

Let $M(x^i)$ be the maximum degree of the polynomials in a minimal homogeneous Hilbert basis for the invariants of a slice for the orbit $Gx^i$, $i = 1, \cdots, r$. If $q+1 \geqq (t(y)+1)M(x^i)$, $i = 1, \cdots, r$, then (10.7.4) implies (10.7.2) by lemma 10.6.

Let $J_x$ be the fiber of $J = {}_r^G J^q(N, P)$ over $x \in {}_r^G N$, and $O_x$ the fiber of $O = O_z$ over $x$ ($O_x$ is also the fiber of $GO = GO_z$ over $x$). Recall that ${}_r j_x^q(\mathscr{C}(N, P)^G)$ is a submanifold of $J_x$.

There is a natural isomorphism of real vector spaces

$$T_z({}_r j_x^q(\mathscr{C}(N, P)^G)) \cong \frac{\theta_{GS}(f)^G}{(\mathfrak{m}_{GS}^{q+1} \cap B_{GS}) \cdot \theta_{GS}(f)^G} . \quad (10.7.5)$$

In fact let $\zeta \in \theta_{GS}(f)^G$, and let $f_t : N \to P$ be a 1-parameter family of smooth equivariant maps such that $f_0 = f$ and $\partial f_t/\partial t|_{t=0}$ is a representative of $\zeta$. Then

$$\frac{\partial}{\partial t} {}_r j^q f_t(x) \bigg|_{t=0}$$

is an element of $T_z({}_r j_x^q(\mathscr{C}(N, P)^G))$, depending only on $\zeta \bmod (\mathfrak{m}_{GS}^{q+1} \cap B_{GS}) \cdot \theta_{GS}(f)^G$. This gives (10.7.5).

The tangent space $T_z O$ to the orbit $O$ at $z$ is spanned by two types of vectors. First there are vectors of the form

$$V_\xi = \frac{\partial}{\partial t} ({}_r j^q (f \circ g_t^{-1})(g_t(x))) \bigg|_{t=0},$$

where

$$\bar{\xi} \in \frac{\theta_{GS}(N)^G}{(\mathfrak{m}_{GS}^{q+1} \cap B_{GS}) \cdot \theta_{GS}(N)^G},$$

and $\{g_t\}$ is the 1-parameter group generated by a compactly supported representative of $\bar{\xi}$ in $\theta(N)^G$ (it is clear that $V_{\bar{\xi}}$ depends only on $\bar{\xi}$). Moreover, if

$$\bar{\xi} \in \frac{\mathfrak{n}_{GS} \cdot \theta_{GS}(N)^G}{(\mathfrak{m}_{GS}^{q+1} \cap B_{GS}) \cdot \theta_{GS}(N)^G},$$

and $\xi$ is an element of $\mathfrak{n}_{GS} \cdot \theta_{GS}(N)^G$ projecting onto $\bar{\xi}$, then $V_{\bar{\xi}}$ identifies (under (10.7.5)) with $\beta_{GS}(\bar{\xi}) = \beta_{GS}(\xi) \bmod (\mathfrak{m}_{GS}^{q+1} \cap B_{GS}) \cdot \theta_{GS}(f)^G$; $V_{\bar{\xi}}$ is an element of $T_zO_x \subset T_z({}_rj_x^q(\mathscr{C}(N, P)^G))$.

Secondly there are vectors of the form

$$V_{\bar{\eta}} = \frac{\partial}{\partial t}\, {}_rj^q(h_t \circ f)(x)\bigg|_{t=0},$$

where

$$\bar{\eta} \in \frac{\theta_{Gy}(P)^G}{(\mathfrak{m}_{Gy}^{q+1} \cap A_{Gy}) \cdot \theta_{Gy}(P)^G},$$

and $\{h_t\}$ is the 1-parameter group generated by a compactly supported equivariant vector field representing $\bar{\eta}$. Note that $V_{\bar{\eta}} \in T_zO_x$. If $\eta$ is an element of $\theta_{Gy}(P)^G$ projecting onto $\bar{\eta}$, then $V_{\bar{\eta}}$ identifies (under (10.7.5)) with $\alpha_{GS}(\bar{\eta}) = \alpha_{GS}(\eta) \bmod (\mathfrak{m}_{GS}^{q+1} \cap B_{GS}) \cdot \theta_{GS}(f)^G$.

Let $\omega : T_z(J) \to T_z(J_x)$ be the projection with respect to the internal direct sum decomposition

$$T_z(J) = T_z(J_x) \oplus \mathrm{Im}\,(d({}_rj^qf)_x).$$

Since $V_{\bar{\eta}} \in T_z(O_x)$, then $\omega(V_{\bar{\eta}}) = V_{\bar{\eta}} = \alpha_{GS}(\bar{\eta})$. On the other hand,

$$\begin{aligned} V_{\bar{\xi}} &= \frac{\partial}{\partial t}({}_rj^q(f \circ g_t^{-1})(g_t(x)))\bigg|_{t=0} \\ &= \frac{\partial}{\partial t}\, {}_rj^q(f \circ g_t^{-1})(x)\bigg|_{t=0} + d({}_rj^qf)_x\left(\frac{\partial g_t(x)}{\partial t}\bigg|_{t=0}\right) \end{aligned}$$

by the chain rule. The first term in this sum equals $\beta_{GS}(\bar{\xi})$, while the second lies in Ker $\omega$. In terms of the preceding identifications, then

$$\omega(T_zO) = \frac{\alpha_{GS}(\theta_{Gy}(P)^G) + \beta_{GS}(\theta_{GS}(N)^G) + (\mathfrak{m}_{GS}^{q+1} \cap B_{GS}) \cdot \theta_{GS}(f)^G}{(\mathfrak{m}_{GS}^{q+1} \cap B_{GS}) \cdot \theta_{GS}(f)^G}. \quad (10.7.6)$$

If ${}_rj^qf$ is in general position with respect to $GO$ at $x$, then

$$\omega(T_zO) = T_z({}_rj_x^q(\mathscr{C}(N, P)^G))$$

by the preceding paragraph and proposition 9.4. By (10.7.5) and (10.7.6), this condition is equivalent to (10.7.4). This completes the proof.

## REFERENCES


[1] M. Artin, Algebraic approximation of structures over complete local rings, *Inst. Hautes Etudes Sci. Publ. Math.* No. 36 (1969), 23–58.

[2] M. Artin, On the solutions of analytic equations, *Invent. Math.* 5 (1968), 277–291.

[3] E. Bierstone, The equivariant covering homotopy property for differentiable $G$-fibre bundles, *J. Diff. Geom.* 8 (1973), 615–622.

[4] E. Bierstone, General position of equivariant maps, *Trans. Amer. Math. Soc.* (to appear).

[5] G. E. Bredon, *Introduction to Compact Transformation Groups*, Academic Press, New York (1972).

[6] M. J. Field, Stratifications of equivariant varieties (preprint, University of Sydney, 1976).

[7] M. J. Field, Transversality in $G$-manifolds, *Trans. Amer. Math. Soc.* (to appear).

[8] S. Lojasiewicz, *Ensembles Semi-Analytiques*, Inst. Hautes Etudes Sci., Bures-sur-Yvette, France (1964).

[9] D. Luna, Fonctions différentiables invariantes sous l'opération d'un groupe réductif, *Ann. Inst. Fourier (Grenoble)* 26 (1976), 33–49.

[10] J. N. Mather, Differentiable invariants, *Topology* (to appear).

[11] J. N. Mather, Stability of $\mathscr{C}^\infty$ mappings: V, Transversality, *Advances in Math.* 4 (1970), 301–336.

[12] J. N. Mather, Stratifications and mappings, *Dynamical Systems*, Academic Press, New York (1973), 195–232.

[13] V. Poenaru, Singularités $\mathscr{C}^\infty$ en présence de symétrie, *Lecture Notes in Mathematics* No. 510, Springer-Verlag, Berlin (1976).

[14] V. Poenaru, Stability of equivariant smooth maps, *Bull. Amer. Math. Soc.* 81 (1975), 1125–1126.

[15] F. Ronga, Stabilité locale des applications équivariantes, Differential Topology and Geometry, Dijon 1974, *Lecture Notes in Mathematics* No. 484, Springer-Verlag, Berlin (1975), 23–35.

[16] G. W. Schwarz, Smooth functions invariant under the action of a compact Lie group, *Topology* 14 (1975), 63–68.

[17] R. Thom, Ensembles et morphismes stratifiés, *Bull. Amer. Math. Soc.* 75 (1969), 240–284.

[18] R. Thom, Local topological properties of differentiable mappings, *Differential Analysis:* Papers presented at the Bombay Colloquium, 1964, Oxford University Press, London (1964), 191–202.

[19] R. Thom, Propriétés différentielles locales des ensembles analytiques, *Séminaire Bourbaki* No. 281 (1964–65).

[20] J.-Cl. Tougeron, *Idéaux de Fonctions Différentiables*, Springer-Verlag, Berlin (1972).

[21] H. Weyl, *The Classical Groups*, Princeton University Press, Princeton (1946).

[22] H. Whitney, Elementary structure of real algebraic varieties, *Ann. of Math.* 66 (1957), 545–556.



Edward Bierstone

University of Toronto
Department of Mathematics
Toronto, Canada
M5S 1A1

Nordic Summer School/NAVF
Symposium in Mathematics
Oslo, August 5–25, 1976

# COHOMOLOGY OF A TYPE OF AFFINE HYPERSURFACES

Jean-Luc Brylinski

## Introduction

This note should be viewed as a complement to Orlik's lecture notes at this Conference [1]. The main theorem 2.2 of Orlik's chapter IV will be proved by a different, shorter and more 'intrinsic' method.

The question is: given $k$, a field of characteristic 0, put $\mathcal{O} = k[z_1, \cdots, z_n]$, and take a polynomial $f \in \mathcal{O}$, which has an isolated singularity at 0. Assume $f$ is 'weighted homogeneous'. Consider the smooth, affine, algebraic variety $F$ over $k$ defined as $F = f^{-1}(1) \subset \mathbb{A}_n$. For any positive integer $q$, denote by $DR^q(F)$ the $q$th cohomology group of the complex of regular algebraic differential forms globally defined on $F$, and $k$-rational. $DR^q(F)$ is a vector space over $k$.

On the other hand, denote by $\partial f$ the Jacobian ideal of $f$. Then $\mathcal{O}/\partial f$ is a (graded) artinian algebra over $k$.

Now, in [2], Orlik and Solomon describe a family of isomorphisms from $\mathcal{O}/\partial f$ to $DR^{n-1}(F)$. An isomorphism of the family is precisely described by the choice of a homogeneous $k$-vector subspace of $\mathcal{O}$, which is supplementary to $\partial f$.

Our proof is based on the following philosophy: 'If one wants to prove that a graded vector space $E$ is isomorphic to a filtered vector space $V$, one should not try to force a graduation on $V$, but rather one should forget that $E$ is graded, and compare the layers of the filtered vector spaces $E$ and $V$. This tactical retreat allows one to prove a better result'.

In our case, of course, $E = \mathcal{O}/\partial f$ and $V = DR^{n-1}(F)$. From the technical viewpoint, the proof is completely elementary and self-contained. The proof will be written in the case $f$ is homogeneous. I will briefly indicate at the end which (slight) modifications are in order to treat the weighted-homogeneous case. A few related (and open) questions are alluded to as a conclusion.

*Afterthought:* I warn the reader that the emergence of roots of unity in my proof has given me some tremendous headaches, so that he tries not to catch such headaches too.

**Proof of Orlik-Solomon's isomorphism**

I keep the notations of the introduction. The base field $k$ will, for simplicity, contain whatever roots of unity I use in the proof. The polynomial will be homogeneous of degree $m > 1$. As usual, denote by $G_m$ the algebraic group Spec $k[\lambda, 1/\lambda]$, with comultiplication: $\lambda \to \lambda \otimes_k \lambda$.

Let $\psi$ be the map

$$\varphi : \Omega^{n-1}(F) \to \Omega^n(F \times \mathbb{C}_n)$$

$$\psi(\eta) = \eta \Lambda \frac{d\lambda}{\lambda}.$$

By Künneth formula, $\psi$ induces an isomorphism (also called $\psi$) from $DR^{n-1}(F)$ to $DR^n(F \times G_m)$.

Now the map

$$\pi : F \times G_m \to \mathbb{A}^n - F_0$$

$$\pi(\underline{z}, \lambda) = \lambda \cdot \underline{z}$$

is an etale covering with group $\mu_m$ (and $F_0$ is the subscheme $F_0 = f^{-1}(0)$ of $\mathbb{A}^n$).

It follows that $F \times G_m = \operatorname{Spec}(\mathcal{O}[1/f^{1/m}])$.

LEMMA 1. *$\varphi$ carries the restriction of $P \cdot \omega$ to $F$ to the form $P/f^{p+n/m} \cdot (dz_1 \wedge \cdots \wedge dz_n)$ in $\Omega^n(\mathcal{O}[1/f^{1/m}])$ for any homogeneous polynomial $P$ of degree $p$. ($\omega$ is the $n-1$ form $\sum_{i=1}^n (-1)^i z_i \times dz_1 \wedge \cdots \wedge \widehat{dz_i} \wedge \cdots \wedge dz_n$).*

PROOF. Locally* in $\mathbb{A}^n - F_0$, there is an inverse to $\pi$

$$\pi^{-1} : \underline{z} \to (f(\underline{z})^{-1/m} \cdot \underline{z}, f(\underline{z})^{1/m}).$$

So we have:

$$\psi((P \cdot \omega)_{|F})$$

$$P(f^{-1/m}(\underline{z}) \cdot \underline{z}) \cdot \left( \sum_{i=1}^{n} (-1)^i f^{-1/m}(\underline{z}) z_i \cdot d(f^{-1/m}(\underline{z}) \cdot z_1) \right.$$
$$\left. \wedge \cdots \wedge \widehat{i} \wedge d(f^{-1/m}(\underline{z}) \cdot z_n) \right) \wedge \frac{df^{1/m}(\underline{z})}{f^{1/m}(\underline{z})}.$$

Now use that $P$ is homogeneous, and get:

$$\varphi((P \cdot \omega)_{|F}) = \frac{1}{m} \cdot f^{-p/m}(\underline{z}) \cdot f^{-n/m}(\underline{z}) \cdot P(\underline{z}) \cdot \omega(\underline{z}) \wedge \frac{df(\underline{z})}{f(\underline{z})}.$$

Since $f$ is homogeneous of degree $m$, Euler's identity shows that: $\omega \wedge df/f = m \cdot dz_1 \wedge \cdots \wedge dz_n$, whence the result.

* I leave to the vicious reader the pleasure of discovering what 'locally' might signify in this context.

REMARK. If $\mathcal{O}[1/f^{1/m}]$ is given the obvious grading, with deg $(1/f^{1/m}) = -1$, then the image of $\psi$ consists of homogeneous differential forms of degree 0; this, of course, was 'a priori' obvious. Note also that $\lambda$ in $\mu_m$ acts as $\lambda^{p+n}$ on $(P \cdot \omega)_{|F}$ and on $\psi((P \cdot \omega)_{|F})$, the latter in a suitable sense.

Since $\mu$ acts on $\mathcal{O}(1/f^{1/m}]$, it acts on $V = Dr^n(\mathcal{O}[1/f^{1/m}])$ and we decompose it under this action as follows.

For any $0 \leqq i \leqq m-1$, we view $i-n$ as an element of $\mathbb{Z}/m$ and we define $V_{i-n}$ as the subspace of $V$ on which $\lambda$ in $\mu_m$ acts as $\lambda^i$. One has: $V \bigoplus_{j \in (\mathbb{Z}/m)} V_j$. Also put: $C_{i-n} = 1/f^{i/m} \cdot \mathcal{O}[1/f] \subset \mathcal{O}[1/f^{1/m}]$ and $B_{i-n} = \sum_{j=1}^n \partial/\partial z_j(C_{i-n})$. ('$C$' stands for cycles, and '$B$' for boundaries.) These ugly notations are (perhaps) justified (see the above remark).

LEMMA 2. *For any integer i, with $0 \leqq i \leqq m-1$, one has: $V_{i-n} = C_{i-n} \mid B_{i-n}$. The map $C_{i-n} \to V_{i-n}$ is just $g \to g\, dz_1 \wedge \cdots \wedge dz_n$.*

LEMMA 3. *Let P in $\mathcal{O}$ be homogeneous of degree p. Then $\sum_{j=1}^n \partial/\partial z_j(P \cdot z_j/f^{i/m+k}) = (p+n-i-mk) \times P/f^{i/m+k}$. Therefore, if $p \neq i-n+mk$, then $P/f^{i/m+k}$ belongs to $B_{i-n}$.*

PROOF. Write: $\partial/\partial z_j(P \cdot z_j/f^{i/m+k}) = 1/f^{i/k+m+1}(f \times z_j \times (\partial P/\partial z_j) + f \times P - ((i/m)+k)P \cdot z_j \cdot \partial f/\partial z_j)$. Sum up over $j$, remembering Euler's relations: $P = p \times (\sum_1^n z_j \cdot \partial P/\partial z_j)$ and $f = m \times (\sum_1^n z_j \cdot \partial f/\partial z_j)$ to get the equality.

Now we reach the essential step of the proof.

PROPOSITION 4. *Let i and k be two integers, with $0 \leqq i \leqq m-1$, and $k \geqq 0$, and put $p = i-n+mk$. If P is homogeneous of degree p, then $P/f^{i/m+k}$ belongs to $B_{i-n}$ if and only if one can write:*

$$\frac{P}{f^{i/m+k}} = \sum_{j=1}^{n} \frac{\partial}{\partial z_j}\left(\frac{Q_j}{f^{i/m+k-1}}\right)$$

*with $Q_j$ homogeneous of degree $p+1-m$. If $(i/m)+k=1$, then $P=0$.*

PROOF. The statement will follow by an easy induction if we can prove: 'If $P/f^{i/m+k} = \sum_{j=1}^n \partial/\partial z_j(Q_j/f^{i/m+k})$, then there exist polynomials $R_j$ of degree $p+1-m$ such that

$$\frac{P}{f^{i/m+k}} = \sum_{j=1}^{n} \frac{\partial}{\partial z_j}\left(\frac{R_j}{f^{i/m+k-1}}\right).$$

Our assumption then reads:

$$f \cdot P = \sum_{j=1}^{n} \left(f \cdot \frac{\partial Q_j}{\partial z_j} - \left(\frac{i}{m}+k\right) Q_j \cdot \frac{\partial f}{\partial z_j}\right).$$

Now, one has $(i/m)+k>0$; otherwise $i=k=0$ and $p=-n$, which is absurd.

Therefore $\sum_1^n Q_j \cdot \partial f/\partial z_j = h \times f$, with some homogeneous $h$ of degree $p$ which we won't write down. We get:

$$\sum_1^n Q_j \cdot \frac{\partial f}{\partial z_j} = h \times f = m \cdot h \cdot \left(\sum_1^n z_j \cdot \frac{\partial f}{\partial z_j}\right).$$

If we set:

$$Q_j' = Q_j - m \cdot h \cdot z_j, \quad \text{then} \quad \sum_1^n Q_j' \cdot \frac{\partial f}{\partial z_j} = 0.$$

Next, consider the map $\rho$ from $\mathcal{O}^n$ to $B_{i-n}$

$$\rho : (T_1, \cdots, T_n) \to \sum_{j=1}^n \frac{\partial}{\partial z_j}\left(\frac{T_j}{f^{i/m+k}}\right),$$

lemma 3 says that $\rho(hz_1, \cdots, hz_n) = 0$.

Therefore:

$$\frac{P}{f^{i/m+k}} = \rho(Q_1, \cdots, Q_n) = \rho(Q_1', \cdots, Q_n').$$

SUBLEMMA. *Let $\mathcal{R}$ be a ring, $(x_1, \cdots, x_n)$ a regular sequence in $\mathcal{R}$, with $n \geqq 2$. Then the submodule of $\mathcal{R}^n$ consisting of all $(a_1, \cdots, a_n)$ such that $\sum_1^n a_j \times x_j = 0$ is generated, as an $\mathcal{R}$-module, by the $n(n-1)/2$ elements:*

$$(0, \cdots, 0, -x_j, 0, \cdots, 0, x_i, 0, \cdots, 0)$$

$\downarrow$ $i$th place $\quad$ $\downarrow$ $j$th place

PROOF. Freshman algebra.

Since $(\partial f/\partial z_1, \cdots, \partial f/\partial z_n)$ is a regular sequence in $\mathcal{O}$ (see [2]), one can write $(Q_1', \cdots, Q_n')$ as a $k$-linear combination of elements of type $(-\varphi(\partial f/\partial z_2), \varphi(\partial f/\partial z_1), 0, \cdots 0)$ with $\varphi$ in $\mathcal{O}$ homogeneous of degree $p-m+2$. In case $(i/m)+k=1$, then $p=m-n$ and $\varphi$ is of degree $2-n$. Since $n \geqq 2$, one has $n=2$ and $\varphi$ is a constant. Then one sees that $\rho(-(\partial f/\partial z_2), (\partial f/\partial z_1), 0, \cdots, 0) = 0$ and so $P=0$. If $(i/m)+k \neq 1$, one writes:

$$\rho\left(-\varphi\frac{\partial f}{\partial z_2}, \varphi\frac{\partial f}{\partial z_1}, 0, \cdots, 0\right) = -\frac{\partial}{\partial z_1}\left(\frac{\varphi \cdot \dfrac{\partial f}{\partial z_2}}{f^{i/m+k}}\right) + \frac{\partial}{\partial z_2}\left(\frac{\varphi \cdot \dfrac{\partial f}{\partial z_1}}{f^{i/m+k}}\right).$$

The next trick is to consider the function $G = \varphi/f^{i/m+k-1}$, then

$$\frac{\partial G}{\partial z_1} = \frac{\dfrac{\partial \varphi}{\partial z_1}}{f^{i/m+k-1}} - \left(\frac{i}{m} + k - 1\right)\frac{\varphi \cdot \dfrac{\partial f}{\partial z_1}}{f^{i/m+k}}.$$

Consequently:

$$\rho\left(-\varphi\cdot\frac{\partial f}{\partial z_2},\varphi\cdot\frac{\partial f}{\partial z_1},0,\cdots,0\right)$$

$$=\frac{1}{i/m+k-1}\left[-\frac{\partial}{\partial z_1}\left(\frac{\dfrac{\partial\varphi}{\partial z_2}}{f^{i/m+k-1}}\right)+\frac{\partial}{\partial z_2}\left(\frac{\dfrac{\partial\varphi}{\partial z_1}}{f^{i/m+k-1}}\right)\right]$$

Since $P/f^{i/m+k}$ is a $k$-linear combination of elements of that type, we have won.

The situation forces me to the following definitions. In $C_{i-n}$ (resp. $B_{i-n}$), let $\tilde{C}_{i-n}$ (resp. $\tilde{B}_{i-n}$) be the space of homogeneous elements of degree $-n$. By lemmas 2 and 3, one has: $V_{i-n}=C_{i-n}\mid B_{i-n}=\tilde{C}_{i-n}\mid \tilde{B}_{i-n}$. Now filter these spaces as follows:

$$W_{-1}(C_{i-n})=0 \quad\text{and}\quad W_k(C_{i-n})=\frac{1}{f^{i/m+k}}\cdot\mathcal{O}\quad\text{for}\quad k\geqq 0,$$

$$W_{-1}(B_{i-n})=0 \quad\text{and}\quad W_k(B_{i-n})=\sum_1^n\frac{\partial}{\partial z_j}(W_{k-1}(C_{i-n}))\quad\text{for}\quad k\geqq 0.$$

Note: $C_{i-n}=\bigcup_k W_k(C_{i-n})$ and the same for $B_{i-n}$.

Denote also by $W$ the filtrations induced on $\tilde{C}_{i-n}$ and $\tilde{B}_{i-n}$. Proposition 4 is easily translated into:

$$W_k(\tilde{C}_{i-n})\cap\tilde{B}_{i-n}=W_k(\tilde{B}_{i-n}).$$

therefore, we state:

Proposition 5. *The inclusion map* $\tilde{B}_{i-n}\hookrightarrow\tilde{C}_{i-n}$ *is* strictly compatible *with the filtrations* $W$. *So* $V_{i-n}=\tilde{C}_{i-n}\mid\tilde{B}_{i-n}$ *is filtered by* $W_k(V_{i-n})=W_k(\tilde{C}_{i-n})\mid W_k(\tilde{B}_{i-n})$

One has:

$$W_k(V_{i-n})\mid W_{k-1}(V_{i-n})=W_k(\tilde{C}_{i-n})\mid W_{k-1}(\tilde{C}_{i-n})+W_k(\tilde{B}_{i-n})$$

and this is naturally isomorphic to $E_{i-n+mk}$, the homogeneous part of degree $i-n+mk$ in $\mathcal{O}/\partial f$.

We need only prove the very last statement.

We identify $W_k(\tilde{C}_{i-n})$ with the space of homogeneous polynomials of degree $p=i-n+mk$. Then $W_{k-1}(\tilde{C}_{i-n})$ is the space of polynomials which are divisible by $f$ in $\mathcal{O}$. And $W_k(\tilde{B}_{i-n})$ is identified with the sum of $n$ vector spaces, the $j$th space being the space of $f\cdot(\partial Q/\partial z_j)+((i/m)+k-1)\cdot Q\cdot(\partial f/\partial z_j)$ with $Q$ homogeneous of degree $p-m+1$.

As before, if $(i/m)+k=1$, then $k=1$, $i=0$ and $W_0(\tilde{C}_{-n})=W_1(\tilde{B}_{-n})=0$; then $W_1\mid W_0(V_{i-n})$ is identified with the set of polynomials of degree $m-n$; since $n\geqq 2$, $\partial f$ has no elements of that degree, and the statement follows.

If $(i/m)+k\neq 1$, it appears that $W_{k-1}(\tilde{C}_{i-n})$ and $W_k(\tilde{B}_{i-n})$ generate together the homogeneous part of $\partial f$ of degree $p$, Q.E.D.

I will just briefly indicate how theorem 2.2 of Orlik's chapter IV follows from our proposition 5.

First, one chooses in $\mathcal{O}$ a subspace $H$ such that $H$ is homogeneous

$$\left( H = \bigoplus_{l\in\mathbb{Z}} Hl = \bigoplus_{0\leqq i\leqq m-1} H_{(i-n)}, \right.$$

with

$$\left. H_{(i-n)} = \bigoplus_{l=i-n\ (\mathrm{mod.}\ m)} Hl \right) \quad \text{and} \quad \mathcal{O} = H \oplus \partial f.$$

One assigns to $h\in H$ the $n-1$ form on $F$, i.e. $(h\times\omega)_{|F}$.

By $\psi$, one finds a form on $F\times\mathbb{G}_m$, described by lemma 1.

To show that the map from $H_{(i-n)}$ to $V_{i-n}$ thus defined is an isomorphism, one notices that it preserves the filtrations; then proposition 5 shows it is a 'filtered isomorphism'.

Notice that, in this proof, one does not need to assume that $V_{i-n}$ is finite dimensional, since it is forced to be so by the proof itself.

## Extension to the weighted-homogeneous case

Here each variable $z_i$ has integral degree $d_i>0$, and $f$ has degree $m$. The new $\omega$ is $\omega=\sum_1^n(-1)^i d_i\times z_i\, dz_1\wedge\cdots\wedge\widehat{dz_i}\wedge\cdots\wedge dz_n$ and $\omega\wedge df/f=m(dz_1\wedge\cdots\wedge dz_n)$.

Whenever degrees of polynomials are computed, $n$ should be replaced by $\sum_1^n d_i$. Note that $\partial f/\partial z_j$ is homogeneous of degree $m-d_j$. In proposition 4, $Q_j$ has degree $p+d_j-m$. In the proof of that proposition, $\varphi$ has degree $p-m+d_1+d_2$. These are the changes that make the whole proof go through.

## Final remarks

Fixing a field $k$, an integer $n$, and a grading of $\mathcal{O}=k[z_1,\cdots,z_n]$ such that each $z_i$ has positive degree, one constructs or category $\mathscr{C}$ as follows. The objects of $\mathscr{C}$ are weighted-homogeneous maps: $f:\mathbb{A}_k^n\to\mathbb{A}_k^1$, and a map from $f$ to $g$ is an algebraic mapping $\varnothing$ from $\mathbb{A}_k^n$ to itself, preserving the graduation, such that

$$\begin{array}{ccc} \mathbb{A}_k^n & \longrightarrow & \mathbb{A}_k^n \\ & {\scriptstyle f}\searrow \quad \swarrow{\scriptstyle g} & \\ & \mathbb{A}_k^1 & \end{array}$$

commutes.

It is an exercise to check that: $V:(f\to DR^{n-1}(f^{-1}(1))$ and $E:(f\to\mathcal{O}/\partial f)$ are contravariant functors from $\mathscr{C}$ to the category vect of $k$-vector spaces. The category Hom $(\mathscr{C}^0,\text{Vect})$ is an additive category; since it is the category of sheaves of $k$-vector spaces on the simplicial classifying space of $\mathscr{C}$, it is even an abelian category.

It is clear, I hope, that $V$ and $E$ are filtered objects of Hom $(\mathscr{C}^0, \text{Vect})$. Since this category is abelian, it makes good sense to speak of Gr $(V)$ and Gr $(E)$, the associated graded objects. It appears that we have proved that Gr $(V)$ and Gr $(E)$ are isomorphic (for instance, as functors from $\mathscr{C}$ to graded $k$-vectors spaces).

Now, if a linear group $G$ acts on $\mathbb{A}^n_k$, preserving the graduation and preserving $f$, then one gets a functor of inclusion from $G$ (viewed as a category with one object and elements of $G$ as arrows) to $\mathscr{C}$. Therefore $V$ and $E$ restrict to functors from $G^0$ to Vect, i.e. to $G$-modules (or to simplicial sheaves on the classifying space of $G^0$).

So the graded $G$-modules Gr $(V)$ and Gr $(E)$ are isomorphic; and so are their images in $k[G]\otimes_k k[t]$, i.e. their characters. But $V$ and Gr $(V)$ have the same image in this 'graded representation ring'; the same remark holds for $E$ and Gr $(E)$. Therefore the character of $G$, acting on $V$ *or* on $E$, gives the same element in $k[G]\otimes_k k[t]$.

This might explain a little bit the 'graded character formulas' of ([1], chapter IV, §6) and of ([2], yet unpublished part).

A few concluding remarks:

(1) Proposition 4 exhibits an example where the exterior differential is strictly compatible with a filtration of forms by the order of the pole along a divisor with an isolated singular point ([3], [4]).

(2) By putting together the filtrations $W$ on the $V_{i-n}$, one gets on $V = H^{n-1}(F, \mathbb{C})$, a filtration 'by the order of the pole at infinity'. What is the geometrical significance (if any) of this filtration? Find the relation with Deligne's work ([3] and [4], §9.2). What happens for a general $f$?

(3) For a general $f$ (with isolated singularity), try to prove $\mu_a = \mu_t$ by an ingenious examination of the Gauss Manin connection ([5], [6]).

## BIBLIOGRAPHY


[1] P. Orlik. The multiplicity of a holomorphic map at an isolated critical point (this volume).

[2] P. Orlik and L. Solomon. Singularities (I: preprint, II: in preparation).

[3] P. Deligne. Théorie de Hodge II, Publ. Math. I.H.E.S. n° 40 (1971), p. 5–58.

[4] P. Deligne. Théorie de Hodge III, Publ. Math. I.H.E.S. n° 44 (1972), p. 5–77.

[5] N. Katz and T. Oda. On the differentiation of De Rham cohomology classes with respect to parameters, J. Math. Kyoto. Univ. n° 8 (1968), p. 199–213.

[6] J. M. Bony. Séminaire Bourbaki, exposé n° 459 (février 1975).


*Added in proof:* "Question (3) was solved by B. Malgrange in his paper: 'Intégrales asymptotiques et monodromie', Ann. scient. Ec. Norm. Sup.

1974, p. 405–430; see the 'analytical index theorem' page 408. Although the proof of Malgrange is of complex-analytic nature, it could be modified into an algebraic proof".

JEAN-LUC BRYLINSKI
Centre de Mathématiques de l'Ecole Polytechnique
Plateau de Palaiseau.
91128 Palaiseau Cedex, France
"Laboratoire de Recherche associé au C.N.R.S."

Nordic Summer School/NAVF
Symposium in Mathematics
Oslo, August 5–25, 1976

# RESIDUAL INTERSECTIONS AND THE DOUBLE POINT FORMULA

William Fulton[1] and Dan Laksov

## §0. Introduction

Numerical formulas for double points go back at least to Severi's 1902 paper. Double points and singular points of mappings have also been intimately related to the search for higher dimensional invariants, or *canonical classes*, of a variety. The canonical divisor was known in the nineteenth century. Severi introduced canonical zero-cycles in 1932, and then Segre, Todd, and Eger in the 1930's developed a theory of canonical classes in all dimensions on a non-singular variety. In one of his treatments of this theory, Todd [9] showed the relation between canonical classes and double points; his proof used a residual intersection theorem. At about the same time, Stiefel and Whitney were developing their characteristic classes and studying their relations with singularities of mappings [10]. It was some time before the relations between the topology and the algebraic geometry were understood, even to the extent of showing that the canonical classes and the Chern classes were the same (up to sign).

A recent treatment of Todd's double point formula has been given in [4] and [3] for mappings to projective space, and in [5], supplemented by [1], for general mappings. The latter papers also proved and used a residual intersection formula, but in a different way from Todd; the residual intersection formula was derived from an important special case called 'formula clef', which was conjectured by Grothendieck in 1957, and proved by Jouanolou (unpublished) and by Lascu, Mumford, and Scott ([6]).

Johnson [3] found canonical classes for use in the double point formula on a singular variety. These classes generalize to the singular case one of Segre's methods for constructing canonical classes on non-singular varieties [8]. In topology double point cycles on singular spaces are related to

[1] Partially supported by the National Science Foundation.

homology operations corresponding to Steenrod squares (cf. [7]). (A historical discussion emphasizing numerical formulas can be found in Kleiman's lectures.)

We present here a proof of general residual intersection and double point theorems. The essential idea of the proofs follows [5], but the calculations are made simpler by using the intersection theory developed in [2].

## §1. Residual schemes and Segre classes

1.1. All our schemes are quasi-projective over a field.

The Segre class of a closed imbedding $Y \subset X$ is a class in the group $A.Y$ of cycles modulo rational equivalence on $Y$. It is defined by blowing $X$ up along $Y$, so we have a diagram

$$\begin{array}{ccc} \tilde{Y} & \longrightarrow & \tilde{X} \\ {\scriptstyle p}\downarrow & & \downarrow{\scriptstyle q} \\ Y & \longrightarrow & X. \end{array}$$

Let $\tilde{N}$ be the normal bundle to $\tilde{Y}$ in $\tilde{X}$. Then the Segre class of $Y$ in $X$ is

$$s(Y, X) = p_*(c(\tilde{N})^{-1} \frown [\tilde{Y}]).$$

If $Y$ is a local complete intersection in $X$ with normal bundle $N$, then $s(Y, X)$ is the total inverse Chern class of $N$. In case $Y$ is a divisor on $X$, $N$ is the restriction of $\mathcal{O}(Y)$ to $Y$, so $s(Y, X)$ is the alternating sum of the self-intersections $Y$, $Y \cdot Y$, $Y \cdot Y \cdot Y, \cdots$, with these intersections all lying in $A.Y$.

The *Segre class* $s(X)$ of a scheme $X$ is the Segre class of the diagonal imbedding of $X$ in $X \times X$ (cf. [8] and [3]).

1.2. Let $Y$ be a closed subscheme of a scheme $X$. Assume $X$ is purely $n$-dimensional, and $\dim(Y) < n$. If $D$ is an arbitrary closed subscheme of $Y$, there appears to be no canonical way to choose a closed subscheme $Z$ so that $Y = D \cup Z$. If $D$ is a Cartier divisor on $X$, however, the ideal sheaf $I(D)$ is invertible, and we can define such a $Z$ by the equation

$$I(Z) = I(Y)I(D)^{-1}$$

relating their ideal sheaves (cf. [5]). We then call $Z$ the *residual scheme* to $D$ in $Y$.

Since $D$ and $Z$ are contained in $Y$, we may regard the Segre classes $s(D, X)$ and $s(Z, X)$ as elements of $A \cdot Y$.

PROPOSITION 1. *If $D$ is a Cartier divisor on $X$, $D \subset Y \subset X$, and $Z$ is residual to $D$ in $Y$, then*

$$s(Y, X) = s(Z, X) + s(D, X) + \text{terms of dimension} < \dim(Z).$$

PROOF. Let

$$\begin{array}{ccc} \tilde{Y} & \xrightarrow{j} & \tilde{X} \\ {\scriptstyle p}\downarrow & & \downarrow{\scriptstyle q} \\ Y & \xrightarrow[i]{} & X \end{array}$$

be the blow-up diagram of $X$ along $Y$; let $\tilde{D}=q^{-1}(D)$, $\tilde{Z}=q^{-1}(Z)$. We can also regard $\tilde{X}$ as the blow-up of $X$ along $Z$, except that in this case $\tilde{Z}$ becomes the exceptional divisor; this is because the ideals of $Y$ and $Z$ differ by an invertible sheaf.

Let us write, for this proof only, $c_1(D)$ for $i^*c_1\mathcal{O}(D)$, $c_1(\tilde{D})$ for $j^*c_1\mathcal{O}(\tilde{D})$, and $c_1(\tilde{Z})$ for $j^*c_1\mathcal{O}(\tilde{Z})$. We note first the formula

$$c_1(\tilde{Z})\frown[\tilde{D}]=c_1(\tilde{D})\frown[\tilde{Z}] \tag{*}$$

in $A_{n-2}\tilde{Y}$. In fact, if $s$ is the direct sum of the sections of $\mathcal{O}(\tilde{D})$ and $\mathcal{O}(\tilde{Z})$ whose zeros and $\tilde{D}$ and $\tilde{Z}$ respectively, the localized Chern class $c_2^{(s)}(\mathcal{O}(\tilde{D})\oplus\mathcal{O}(\tilde{Z}))$ lives in $A_{n-2}(\tilde{D}\cap\tilde{Z})$ and maps to each of the classes of (*) in $A_{n-2}(\tilde{D})$ and $A_{n-2}(\tilde{Z})$ respectively ([2] Prop. 7.2). A less conceptual proof of (*) is given in [1]. Now

$$\begin{aligned} s(Y,X) &= p_*(c(j^*\mathcal{O}(\tilde{Y}))^{-1}\frown[\tilde{Y}]) \\ &= \sum_{i\geq 0}(-1)^i p_*((c_1(\tilde{Z})+c_1(\tilde{D}))^i\frown([\tilde{Z}]+[\tilde{D}])) \\ &= \sum_{i\geq 0}(-1)^i p_*(c_1(\tilde{Z})^i\frown[\tilde{Z}])+\sum_{i\geq 0}(-1)^i p_*(c_1(\tilde{D})^i\frown[\tilde{D}]) \\ &\quad +\sum_{j>0} n_{ij}p_*(c_1(\tilde{Z})^i c_1(\tilde{D})^j\frown[\tilde{Z}]) \end{aligned}$$

for some integers $n_{ij}$. We have used (*) to replace each occurrence of $c_1(\tilde{Z})\frown[\tilde{D}]$ by $c_1(\tilde{D})\frown[\tilde{Z}]$. Since $\tilde{X}$ is the blow-up of $X$ along $Z$, and since $c_1(\tilde{D})=p^*c_1(D)$ and $p_*[\tilde{D}]=[D]$, this sum is equal to

$$s(Z,X)+s(D,X)+\sum_{j>0} n_{ij}c_1(D)^j p_*(c_1(\tilde{Z})^i\frown[\tilde{Z}]).$$

Now the result is clear since the last terms all have codimension at least one in $Z$.

1.3. Suppose $Y\subset X$ as in §1.2, but now suppose $D$ is an arbitrary closed subscheme of $Y$. Let $\pi:\tilde{X}\to X$ be the blow-up of $X$ along $D$, $\tilde{D}=\pi^{-1}(D)$ the exceptional divisor, $\tilde{Y}=\pi^{-1}(Y)$, $\varphi:\tilde{Y}\to Y$ the map induced by $\pi$. Now we can define the residual scheme $\tilde{Z}$ to $\tilde{D}$ in $\tilde{Y}$ by the procedure of §1.2, i.e. so that $I(\tilde{Y})=I(\tilde{D})I(\tilde{Z})$.

PROPOSITION 2. *With these assumptions,*

$$s(Y, X) = \varphi_* s(\tilde{Z}, \tilde{X}) + s(D, X) + \text{terms of dimension} < \dim(\tilde{Z})$$

PROOF. By proposition 1,

$$s(\tilde{Y}, \tilde{X}) = s(\tilde{Z}, \tilde{X}) + s(\tilde{D}, \tilde{X}) + \text{terms of dimension} < \dim(\tilde{Z}).$$

Now the result follows by the 'birational invariance' of the Segre classes ([2] Proposition 3.2).

## §2. Residual intersections

Let $Y$ be a local complete intersection in $X$ of codimension $d$ with normal bundle $N$. Let $f: X' \to X$ be a morphism from a purely $n$-dimensional scheme $X'$, and let $Y' = f^{-1}(Y)$. Let $D'$ be any closed subscheme of $Y'$. Let $\pi: \tilde{X} \to X'$ be the blow-up of $X'$ along $D'$. Let $\tilde{D} = \pi^{-1}(D')$ be the exceptional divisor, $\tilde{Y} = \pi^{-1}(Y')$. We label the induced maps as in the following diagram:

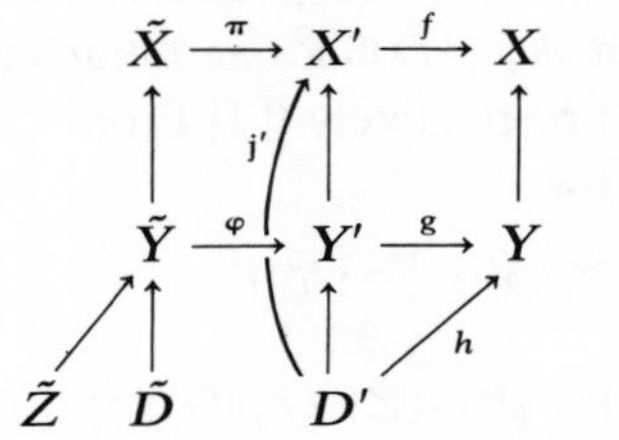

Define $\tilde{Z}$ to be the residual scheme of $\tilde{D}$ in $\tilde{Y}$, so $I(\tilde{Y}) = I(\tilde{D})I(\tilde{Z})$. Note that $\tilde{Z}$ is locally defined by $d$ equations, since the same is true for $\tilde{Y}$, and since $I(\tilde{D})$ is locally principal.

PROPOSITION 3. *Assume $\tilde{Z}$ is generically a local complete intersection of codimension $d$ in $\tilde{X}$. Then the localized intersection class is given by the equation*

$$X' \cdot_f Y = \varphi_*[\tilde{Z}] + \{g^* c(N) \frown s(D', X')\}_{n-d}.$$

*Here $X' \cdot_f Y$ is the class in $A.Y'$ defined in* [2] §1; *$c(N)$ is the total Chern class of $N$; the Segre class $s(D', X')$ is regarded in $A.Y'$ by the inclusion of $D'$ in $Y'$; and $\{\ \}_{n-d}$ denotes the $(n-d)$-dimensional component of the class inside.*

PROOF. $X' \cdot_f Y$ is defined to be $\{g^* c(N) \frown s(Y', X')\}_{n-d}$, which by proposition 2 is

$$\{g^* c(N) \frown \varphi_* s(\tilde{Z}, \tilde{X})\}_{n-d} + \{g^* c(N) \frown s(D', X')\}_{n-d},$$

and the result follows since $s(\tilde{Z}, \tilde{X}) = [\tilde{Z}] + \text{lower dimensional terms}$ when $\tilde{Z}$ is generically a local complete intersection (cf. [2] Remark after Prop. 3.1).

The following theorem then follows from this proposition and from theorem 1 of [2].

RESIDUAL INTERSECTION THEOREM. *Assume that $\tilde{Z}$ is generically a local complete intersection of codimension d, and assume that X and Y are non-singular. Then*

$$f^*[Y] = \pi_*[\tilde{Z}] + j'_*\{h^*c(N) \frown s(D', X')\}_{n-d}$$

*in $A_{n-d}X'$. (See the above diagram for notation.)*

REMARKS 1. In case $D'$ is a Cartier divisor, no blowing up is necessary, so $\tilde{Z}$ is the residual scheme to $D'$ in $Y'$, and $s(D', X')$ is $j'^*c(\mathcal{O}(D'))^{-1} \frown [D']$; the theorem specializes to Todd's residual intersection theorem, as given in ([5]+[1]).

2. If $X'$ is Cohen–Macaulay, the residual scheme will be a local complete intersection (everywhere) provided its codimension is $d$.

3. Even if the residual scheme $\tilde{Z}$ is too big, our procedure will give a formula as in the theorem, but with the term $[\tilde{Z}]$ replaced by a definite class in $A_{n-d}(\tilde{Z})$ involving the Segre classes of $\tilde{Z}$ in $\tilde{X}$ and the restriction of $c_1\mathcal{O}(\tilde{D})$ to $\tilde{Z}$. Such a class may not give a positive cycle on $X'$, however (cf. [1]), and its geometric significance becomes less clear when $\tilde{Z}$ is large. In the other direction, one can give conditions which assure that $\tilde{Z}$ is reduced (cf. [4] and [5]), when $\tilde{Z}$ is a double point scheme.

## §3. Double points

Let $f: X \to Y$ be a morphism from a purely $n$-dimensional scheme $X$ to a non-singular $m$-dimensional scheme $Y$. We shall apply the results of §2 to the morphism

$$f \times f: X \times X \to Y \times Y.$$

The diagonal $\Delta_Y$ is a local complete intersection of codimension $m$ in $Y \times Y$, and the inverse image $(f \times f)^{-1}(\Delta_Y)$ contains the diagonal $\Delta_X$ as a subscheme.

Let $\pi: X \times X \to X \times X$ be the blow-up of $X \times X$ along $\Delta_X$, and let $P(X)$ be the exceptional divisor. The residual scheme to $P(X)$ in $\pi^{-1}(f \times f)^{-1}(\Delta_Y)$ is called the *double point scheme* of $f$, and is denoted $Z(f)$. The name is justified by the following proposition (cf. [5] Remark 14).

PROPOSITION 4. (1) *If $x_1$ and $x_2$ are distinct rational points of X, the point $\pi^{-1}(x_1, x_2)$ is in $Z(f)$ if and only if $f(x_1) = f(x_2)$.*

(2) *If u is a rational point of $P(X)$, projecting to x in X, then u is in $Z(f)$ if and only if $df(u) = 0$.*

In (2) we regard $P(X)$ as a subscheme of Proj (Sym $(\Omega^1_X)$), so $u$ represents a line in the Zariski tangent space to $X$ at $x$. Note, however, that not all lines in the Zariski tangent space are represented by points in $P(X)$ (cf. [3] for a discussion of this distinction).

PROOF. (1) is immediate from the definition. For (2), let $\mathcal{O}_u$ be the local ring of $\widetilde{X \times X}$ at $u$. Then $u$ does not belong to $Z(f)$ if and only if $I(Z)\mathcal{O}_u = \mathcal{O}_u$, i.e. the induced map $I(\Delta_Y)\mathcal{O}_u \to I(\Delta_X)\mathcal{O}_u$ is surjective. By Nakayama's lemma we may restrict this to $P(X)$. Note that $I(X)$ restricts to $\Omega^1_X$ on $\Delta_X$, and similarly for $Y$. Moreover, if $\varphi: P(X) \to X = \Delta_X$ is the projection, and $\mathcal{L}$ is the conormal sheaf of the imbedding of $P(X)$ in $\widetilde{X \times X}$, there is a surjection

$$\varphi^*\Omega^1_X \to \mathcal{L}$$

which may be defined either as the restriction to $P(X)$ of the surjection

$$I(\Delta_X) \otimes_{\mathcal{O}_{X\times X}} \mathcal{O}_{\widetilde{X\times X}} \to I(\Delta_X)\mathcal{O}_{\widetilde{X\times X}}$$

on $\widetilde{X \times X}$, or as the restriction to $P(X)$ of the universal line-bundle quotient on Proj (Sym $(\Omega^1_X)$). The condition that $u \notin Z(f)$ now says that the composite

$$\varphi^* f^* \Omega^1_Y \to \varphi^* \Omega^1_X \to \mathcal{L}$$

is surjective at $u$, which is the desired assertion.

DOUBLE POINT THEOREM. *Assume the double point scheme $Z(f)$ is generically a local complete intersection of codimension $m$ in $\widetilde{X \times X}$. Define the* double point cycle $\mathbb{D}(f)$ *to be* $pr_{1*}\pi_*[Z(f)]$, *where* $pr_1: X \times X \to X$ *is the projection. Then*

$$\mathbb{D}(f) = f^* f_*[X] - \{f * c(Y) \frown s(X)\}_{2n-m}$$

*in* $A_{2n-m}(X)$.

PROOF. By the residual intersection theorem,

$$(f \times f)^*[\Delta_Y] = \pi_*[Z(f)] + j'_*\{h * c(N) \frown s(\Delta_X, X \times X)\}_{2n-m}.$$

Now $N$ is the tangent bundle to $Y$, so $c(N)$ is the total Chern class of $Y$. And $pr_{1*}((f \times f)^*[\Delta_Y]) = f * f_*[X]$ by ([2] §2 Lemma, cf. also [5]). The theorem follows by applying $pr_{1*}$ to the displayed equation.

REMARKS. 1. One may use this formula to define the Segre canonical classes of a variety $X$. We always have $\{s(X)\}_n$ equal to $[X]$. Then take generic projections $X \to \mathbb{P}^{n+k}$ for $k = 1, 2, \cdots, n$ to determine $\{s(X)\}_{n-1}$, $\{s(X)\}_{n-2}, \cdots, \{s(X)\}_0$ successively; no Chern classes are necessary since $c(\mathbb{P}^m) = (1+H)^{m+1}$, $H$ a hyperplane section. Then the theorem says these classes are independent of the choice of projections, and that the formula remains true for other mappings $f: X \to Y$. If $X$ is a non-singular curve, for example, this gives the familiar expression for the canonical class (and therefore the genus) of a plane model with ordinary double points; it then determines the relation between canonical classes and double points for any generically one-to-one mapping from the curve to a non-singular surface.

2. For a mapping of a non-singular curve to a non-singular surface, the double-point cycle is given by the conductor ([1]). For a singular curve, however, this may no longer be the case. For example, if the curve consists of three non-coplanar lines passing through a point in projective 3-space, and the morphism is a generic projection to a plane passing through the point, then the double-point cycle is zero, but the conductor is not trivial.

3. There are interesting relations between the double-point cycle and the corresponding notion of ramification cycle (cf. [3]).

## REFERENCES


[1] W. Fulton, A note on residual intersections and the double point formula, preprint, Aarhus University.

[2] W. Fulton and R. MacPherson, Intersecting cycles on an algebraic variety, these proceedings.

[3] K. Johnson, Immersion and embedding of projective varieties, thesis, Brown University, June 1976.

[4] D. Laksov, Secant bundles and Todd's formula for the double points of maps into $\mathbb{P}^n$. To appear.

[5] D. Laksov, Residual intersections and Todd's formula for the double locus of a morphism, to appear.

[6] A. T. Lascu, D. Mumford and D. B. Scott, The self-intersection formula and the 'formule-clef', *Math. Proc. Camb. Phil. Soc.* 78 (1975) 117–123.

[7] C. McCrory, Geometric homology operations, to appear in *Advances in Math.*

[8] B. Segre, Nuovi metodi e risultati nella geometria sulle varietà algebriche, *Ann. di Mat.* (4) 35, (1953) 1–128.

[9] J. A. Todd, Invariant and covariant systems on an algebraic variety, *Proc. London Math. Soc.* (2) 46, (1940) 199–230.

[10] H. Whitney, On the topology of differentiable manifolds, Lectures in Topology, Univ. of Michigan Press, (1941) 101–141.



William Fulton
Brown University
Dept. of Mathematics
Providence, Rhode Island 02912
USA

Dan Laksov
Oslo Universitet
Mathematisk Inst.
P.O. Box 1053
Oslo 3
Norway

Nordic Summer School/NAVF
Symposium in Mathematics
Oslo, August 5–25, 1976

# INTERSECTING CYCLES ON AN ALGEBRAIC VARIETY

William Fulton[1] and Robert MacPherson[1]

## §0. Introduction

If $C$ and $D$ are subvarieties (or subschemes) of a smooth variety $X$, they determine classes $[C]$, $[D]$ in the group $A.X$ of cycles modulo rational equivalence on $X$. The problem of intersection, in its simplest form, is to define an intersection class $[C]\cdot[D]$ in $A_mX$, where $m=\dim C+\dim D-\dim X$.

The traditional-modern solution to this problem proceeds in two steps: (A) If $C$ and $D$ intersect properly, i.e. each component $V$ of $C\cap D$ is $m$-dimensional, one assigns (cf. [11]) an intersection number $i(V)$ to each $V$; in this case $[C]\cdot[D]=\sum i(V)[V]$ is a well-defined cycle. (B) One proves (cf. [1] or [9]) a moving lemma, at least if $X$ is quasi-projective, which says there are cycles $\alpha$ and $\beta$, rationally equivalent to $[C]$ and $[D]$ respectively, which do intersect properly, and the resulting cycle $\alpha\cdot\beta$ is well-defined up to rational equivalence.

In fact it is only necessary to move one of the cycles $[C]$ or $[D]$. If we let $i: C\to X$, $j: D\to X$ be the inclusions, we get well-defined classes $i^*[D]$ in $A_mC$ and $j^*[C]$ in $A_mD$. For example if $\beta$ is rationally equivalent to $[D]$ and meets $C$ properly, then the cycle $[C]\cdot\beta$ represents $i^*[D]$. This raises the question of whether $[C]\cdot[D]$ can be constructed in $A_m(C\cap D)$. Of course the exact sequence

$$A_m(C\cap D)\to A_mX\to A_m(X-(C\cap D))\to 0$$

shows that there are classes on $C\cap D$ that map to $[C]\cdot[D]$ in $A_mX$, but this does not provide even an existence solution to the following problem.

INTERSECTION PROBLEM. To construct a class $C\cdot D$ in $A_m(C\cap D)$ that maps to $i^*[D]$ in $A_mC$ and to $j^*[C]$ in $A_mD$.

[1] Partially supported by the National Science Foundation.

Before discussing our solution to this problem, we look at some facts that indicated that a solution was possible.

(1) Let $T = \Sigma(-1)^i \operatorname{Tor}_i^{\mathcal{O}_X}(\mathcal{O}_C, \mathcal{O}_D)$ be the virtual sheaf with which Serre harpooned intersection theory [11]; $T$ is an element of the Grothendieck group $K_0(C \cap D)$ of coherent sheaves on $C \cap D$. Then one may show that the Riemann-Roch map ([2])

$$\tau : K_0(C \cap D) \rightarrow A.(C \cap D)_Q$$

maps $T$ to a class whose component in dimension $m$ solves the problem, but with rational coefficients (cf. §6, where we show that our construction agrees with this).

(2) The existence of such a class in ordinary homology $H_{2m}(C \cap D)$ follows easily from the Lefschetz duality isomorphisms $H^{\cdot}(X, X-V) \cong H.V$ for subvarieties $V$ of $X$, and the relative cup product (cf. §7, but note that our construction for algebraic cycles is more explicit than is possible for general cycles).

(3) Known formulas like the self-intersection formula, 'formule clef' (cf. [7]) and Todd's residual intersection formula (cf. [5]) are explicit solutions to this problem.

(4) As Severi emphasized [12], to apply intersection theory to classical problems of geometry, it is not enough to know the intersection of cycles in the ambient space; one needs a solution to the above intersection problem. He gives the following problem as an illustration.

How many conics are there which are tangent to five given (general) conics $C_1, \cdots, C_5$ in the plane? The conics tangent to $C_i$ form a hypersurface $H_i$ of degree six in the projective 5-space of all conics. These five hypersurfaces intersect in points representing the desired solutions, but also in the Veronese surface which consists of all conics which are double lines. The intersection cycle $H_1 \cdot \cdots \cdot H_5$ has $6^5 = 7776$ points, but to solve the problem one needs the correct contribution to each component of $H_1 \cap \cdots \cap H_5$; in this case the Veronese surface contributes 4512 points, leaving 3264 for the true solutions.

In Severi's later work on rational equivalence, he apparently thought of $[C] \cdot [D]$ as a limit of cycles $\alpha_t \cdot \beta_t$, where $\alpha_t$ and $\beta_t$ vary in families of cycles rationally equivalent to $[C]$ and $[D]$ respectively; such a limiting cycle would live on $C \cap D$ (cf. [14] for a discussion of, and objections to, this approach).

Our formula for $C \cdot D$ in $A_m(C \cap D)$ depends only on the normal cone to the diagonal inclusion of $C \cap D$ in $C \times D$, and on the restriction of the tangent bundle of $X$ to $C \cap D$.

In fact, the construction does not need either a theory of intersection numbers or a moving lemma. All that is needed is a theory of Chern classes for vector bundles, which is much more elementary. For example, no projective hypotheses are needed in such a theory, so this gives a direct generalization of the ring of rational equivalence to non-projective varieties. We will develop intersection theory from this point of view in another paper (see §8 for related discussion). Here we work within the already developed theory of rational equivalence (cf. [4] for Chern classes, cap products, etc.). In particular we assume all schemes are quasi-projective over a field.

Here is an outline of our procedure. By reduction to the diagonal (§2), the problem reduces to the following: If $Y \subset X$ is a local complete intersection of codimension $d$, $f: X' \to X$ a morphism from an $n$-dimensional variety $X'$, to construct $f^*[Y]$ in $A_{n-d}(Y')$, $Y' = f^{-1}(Y)$. Such a class is constructed in §1 in terms of the Chern classes of the normal bundle to $Y$ in $X$, and the 'Segre class' of $Y'$ in $X'$. This Segre class (cf. [10]) can be constructed by blowing up $X'$ along $Y'$, and pushing down various self-intersections (which are Chern classes) of the exceptional divisor (§1); it depends only on the normal cone to $Y'$ in $X'$ (§3). In fact, the construction of $f^*[Y]$ can be carried out entirely from the map that $f$ induces on the normal cones (§3).

The proof that the construction solves the problem (§4) is based on deformation to the normal cone, which was a basic construction in the Riemann-Roch theorem [2]. Following Verdier [15], we give a simpler construction of this deformation in §4. In the non-singular case, the total space of this deformation is seen to be the same space used by Mumford to prove the self-intersection formula in 1959 (cf. [7], [6]).

In §5 we give an intersection formula (theorem 3) which simultaneously generalizes the self-intersection formula and formule-clef. The proof is similar in spirit to the rest of the paper, but it is logically independent; it may be regarded as a simplification of the previous approach to such problems.

In §6 we relate this to Serre's intersection theory, as in (1) above. We use the same methods to strengthen Serre's result on the vanishing of the intersection number in case of excess intersection (see theorem 4).

In §7 we use the construction to localize Chern classes. This shows how the construction here relates to the graph construction. A result proved here is useful in proving the residual intersection theorem (cf. [5]).

Finally, in §8, we discuss briefly relations with other homology theories, Gysin maps, and how one can use them to form a general intersection theory.

**Contents**

## §1. The basic construction

Let $Y$ be a closed subscheme of a scheme $X$. For the moment we assume that $Y$ contains no irreducible component of $X$. Construct the blow-up diagram of $X$ along $Y$:

$$\begin{array}{ccc} \tilde{Y} & \xrightarrow{\tilde{\imath}} & \tilde{X} \\ {\scriptstyle p}\downarrow & & \downarrow{\scriptstyle q} \\ Y & \xrightarrow{i} & X \end{array}$$

Let $\tilde{N}$ be the normal bundle of $\tilde{Y}$ in $\tilde{X}$. Define the *Segre class* $s(Y, X)$ of $Y$ in $X$ by the formula

$$s(Y, X) = p_*(c(\tilde{N})^{-1} \frown [\tilde{Y}]).$$

Here $c(\tilde{N}) = 1 + c_1(\tilde{N})$ is the total Chern class of $\tilde{N}$, and $[\tilde{Y}]$ is the fundamental cycle of $\tilde{Y}$. The fundamental cycle of a non-reduced scheme includes the multiplicities of its components in the definition ([4] §1.1); Chern classes for bundles on singular quasi-projective schemes are defined by restriction from non-singular schemes ([4] §3.2). The Segre class lies in the group $A.Y$ of cycles modulo rational equivalence on $Y$. If $Y$ is a local complete intersection in $X$ with normal bundle $N$, then $s(Y, X) = c(N)^{-1} \frown [Y]$. For this and other properties of Segre classes see §3.

For the rest of §1, assume $Y$ is a local complete intersection in $X$ of codimension $d$ with normal bundle $N$. Let $f: X' \to X$ be a morphism, and let $Y' = f^{-1}(Y)$. We have a fiber square.

$$\begin{array}{ccc} X' & \xrightarrow{f} & X \\ {\scriptstyle i'}\uparrow & & \uparrow{\scriptstyle i} \\ Y' & \xrightarrow[g]{} & Y \end{array}$$

Assume $X'$ is purely $n$-dimensional. Then we define the *localized intersection class* $X' \cdot_f Y$ in $A_{n-d}Y'$ by the formula

$$X' \cdot_f Y = \{c(g^*N) \frown s(Y', X')\}_{n-d}$$

where $\{\ \}_{n-d}$ denotes the component of dimension $n-d$ of a class in $A.Y'$. The notation is that used by Serre [11] in case the intersection is proper, i.e. $Y'$ has dimension $n-d$; we will see that our definition agrees with Serre's in that case (§6).

Note that the class $X' \cdot_f Y$ depends only on the normal cone of $Y'$ in $X'$, and on the pull-back of the normal bundle of $Y$ in $X$.

THEOREM 1. *Assume $X$ and $Y$ are non-singular. Then $f^*[Y] = i'_*(X' \cdot_f Y)$ in $A.X'$.*

Here $f^*: A.X \to A.X'$ is the Gysin map that exists because $X$ is non-singular (cf. [4]). The theorem will be proved in §4.

REMARK. If $Y \subset X$ and $Y' \subset X'$ are both local complete intersections of codimensions $d$, $d'$ and normal bundles $N$, $N'$ respectively, then we have an exact sequence

$$0 \to N' \to g^*N \to E \to 0$$

with $E$ a vector bundle of rank $e = d - d'$. Since $s(Y', X') = c(N')^{-1} \frown [Y']$, we have a formula for the localized intersection class:

$$X' \cdot_f Y = c_e(E) \frown [Y'].$$

If in addition $X$ and $Y$ are non-singular, theorem 1 says

$$f^*[Y] = i'_*(c_e(E) \frown [Y']).$$

This will be generalized in §5.

## §2. Solution to the intersection problem

Let $X$ be a non-singular, purely $n$-dimensional scheme, with closed subschemes $C, D$ of pure dimensions $c, d$ respectively. Let $i: C \to X$, $j: D \to X$ be the inclusions. We define the intersection class $C \cdot D$ in $A_m(C \cap D)$, $m = c + d - n$, as follows: The diagonal $\Delta_X$ is a local complete intersection of codimension $n$ in $X \times X$. Let $f: C \times D \to X \times X$ be the product of the inclusions. Then we have a fiber square

$$\begin{array}{ccc} C \times D & \xrightarrow{f} & X \times X \\ \uparrow & & \uparrow \\ C \cap D & \longrightarrow & \Delta_X \end{array}$$

as in §1. We can therefore define

$$C \cdot D = (C \times D) \cdot_f \Delta_X.$$

It is clear from this definition that $D \cdot C = C \cdot D$.

More generally, if $X$ is non-singular and $i: C \to X$, $j: D \to X$ are any proper morphisms, the above process defines $C \cdot D$ in $A_m(C \cap D)$, where $C \cap D$ denotes the fiber product $C \times_X D$.

THEOREM 2. *Assume $X$ is projective. Then the inclusion of $C \cap D$ in $C$ (respectively $D$) maps $C \cdot D$ to $i^* j_* [D]$ in $A_m C$ (respectively to $j^* i_* [C]$ in $A_m D$).*

PROOF. By theorem 1, $C \cdot D$ maps to $f^*[\Delta_X]$ in $A_m(C \times D)$. So it suffices to invoke the following lemma from standard intersection theory.

LEMMA. *Let $p: C \times D \to C$ be the projection. Then $p_* f^*[\Delta_X] = i^* j_*[D]$ in $A_m C$.*

PROOF. Let $q: C \times X \to C$ be the projection. Consider the factorization of $f$:

$$C \times D \xrightarrow{g} C \times X \xrightarrow{h} X \times X.$$

Choose a cycle $\beta$ on $X$ which is rationally equivalent to $j_*[D]$, so that the cycle $i^*\beta$ is defined. Then $i^* j_*[D]$ is represented by the cycle $q_*(([C] \times \beta) \cdot_h \Delta_X)$. And $p_* f^*[\Delta_X]$ is represented by the cycle $p_*(([C] \times [D]) \cdot_f \gamma)$, where $\gamma$ is any cycle on $X \times X$ rationally equivalent to $\Delta_X$ such that $f^*\gamma$ is defined. We may assume also that $\gamma$ meets $[C] \times \beta$ properly. Then $g_*(([C] \times [D]) \cdot_f \gamma) = ([C] \times g_*[D]) \cdot_h \gamma$ ([11] V-30) so $p_*(([C] \times [D]) \cdot_f \gamma) = q_*(([C] \times g_*[D] \cdot_h \gamma)$. Now the cycles $([C] \times \beta) \cdot_h \Delta_X$ and $([C] \times g_*[D]) \cdot_h \gamma$ are both rationally equivalent to $([C] \times \beta) \cdot_h \gamma$ on $C \times X$ (cf. [4] §2.3 or [9]), and the result follows since rational equivalence pushes forward by $q_*$.

## §3. Cones and Segre classes

Let $C$ be a cone on a scheme $Y$, i.e. $C = \operatorname{Spec}(S^\cdot)$, where $S^\cdot$ is a graded sheaf of $\mathcal{O}_Y$-algebras generated by the coherent sheaf $S^1$. Denote the projectivized cone $\operatorname{Proj}(S^\cdot)$ by $P(C)$, with structural morphism $p: P(C) \to Y$, and relatively ample line bundle $\mathcal{O}(1)$ (cf. [3]). Assume $p$ is surjective. Then we define the *Segre class of the cone* $s(C)$ in $A.Y$ by the equation

$$s(C) = p_*(c(\mathcal{O}(-1))^{-1} \frown [P(C)]).$$

We denote by $C \oplus 1$ the cone $\operatorname{Spec}(S^\cdot[z])$, where $z$ is a variable and the $k$th graded piece of $S^\cdot[z]$ is $\sum_{i+j=k} S^i z^j$. Then $P(C \oplus 1)$ is the union of the closed subscheme $P(C)$ and the dense open subscheme $C$. We call $P(C \oplus 1)$ the *completion* of $C$. We have the following 'stability' result.

LEMMA. *If $C$ is a cone on $Y$, and $p: P(C) \to Y$ is surjective, then $s(C \oplus 1) = s(C)$.*

PROOF. The canonical bundle $\mathcal{O}(1)$ on $P(C \oplus 1)$ restricts to the canonical bundle on $P(C)$, and $c_1(\mathcal{O}(1)) \frown [P(C \oplus 1)] = [P(C)]$ since $\mathcal{O}(1)$ has a section vanishing precisely (scheme-theoretically) on $P(C)$. Let $\pi : P(C \oplus 1) \to Y$ be the projection, $j : P(C) \to P(C \oplus 1)$ the inclusion. Then $s(C \oplus 1) = \pi_*(\sum c_1(\mathcal{O}(1))^i \frown [P(C \oplus 1]) = \pi_*([P(C \oplus 1] + \sum_{i>0} j_*(c_1(j^*\mathcal{O}(1)^{i-1} \frown [P(C)]) = s(C)$ since $\pi_*([P(C \oplus 1]) = 0$, this last since none of the fibers of $\pi$ have any isolated points.

If $C$ is a cone for which $p : P(C) \to Y$ is not surjective, we define $s(C)$ to be $s(C \oplus 1)$.

Let $Y$ be a closed subscheme of a scheme $X$, defined by an ideal sheaf $I$. Then the normal cone to $Y$ in $X$ is the cone $C_Y X = \operatorname{Spec}(\sum I^n/I^{n+1})$. The *projectivized normal cone* $P(C)$ may be identified with the exceptional divisor $\tilde{Y}$ of the blow-up $\tilde{X}$ of $X$ along $Y$, and $\mathcal{O}(1)$ is then dual to the normal bundle of $\tilde{Y}$ in $\tilde{X}$. Therefore $c(\tilde{N})^{-1} = c(\mathcal{O}(-1))^{-1}$ so *the Segre class of $Y$ in $X$ is the Segre class of the normal cone to $Y$ in $X$.*

The fact that $s(Y, X) = c(N)^{-1} \frown [Y]$ for a local complete intersection with normal bundle $N$ is a special case of the following simple result, which has been rediscovered many times since Segre.

PROPOSITION 3.1. *Let $E$ be a vector bundle on $Y$. Then $s(E) = c(E)^{-1} \frown [Y]$.*

PROOF. As usual $E$ is identified with the cone Spec (Sym $E^\vee$), where $E$ is the sheaf of sections of $E$. On $P(E)$ we have the universal exact sequence

$$0 \to \mathcal{O}(-1) \to p^*E \to F \to 0.$$

Then $c(E) \frown s(E) = p_*(p^*c(E)c(\mathcal{O}(-1))^{-1} \frown [P(E)]) = p_*(c(F) \frown [P(E)]) = p_*(c_f(F) \frown [P(E)])$, where $f = \operatorname{rank} F$; the other terms vanish for dimension reasons. To show that this last term is $[Y]$ is a local question on $Y$, so we may assume $E$ is trivial. Then $c_f(F) = c_1(\mathcal{O}(1))^f$ is represented on $Y \times \mathbb{P}^f$ by the intersection of $f$ independent hyperplanes, from which the result is clear.

REMARK. The last part of the argument shows that if $C$ is a cone on $Y$ which is generically a vector bundle, then $s(C) = [Y] +$ lower dimensional terms.

The following 'birational invariance' of the Segre class can be useful in computations.

PROPOSITION 3.2. *Let $Y \subset X$, $f : X' \to X$ a birational morphism, $Y' = f^{-1}(Y)$, $g : Y' \to Y$ the induced morphism. Then $g_* s(Y', X') = s(Y, X)$.*

PROOF. In case $f$ is the blow-up of $X$ along $Y$, this is just the definition of $s(Y, X)$. In case $Y$ and $Y'$ are Cartier divisors on $X$ and $X'$ respectively, the proposition follows from the fact that the normal bundle to $Y$ in $X$ pulls back to the normal bundle to $Y'$ in $X'$, and the fact that the cycle $[Y']$ pushes down to the cycle $[Y]$ (cf. [4] §1.5).

For the general case, let $\pi:\tilde{X}\to X$ be the blow-up of $X$ along $Y$, $\pi':\tilde{X}'\to X'$ the blow-up of $X'$ along $Y'$. Then there is a unique morphism $\tilde{f}:\tilde{X}'\to\tilde{X}$ making the diagram

$$\begin{array}{ccc} \tilde{X}' & \xrightarrow{\pi'} & X' \\ {\scriptstyle \tilde{f}}\downarrow & & \downarrow{\scriptstyle f} \\ \tilde{X} & \xrightarrow[\pi]{} & X \end{array}$$

commutative. The above cases prove the assertion for $\pi$, $\pi'$, and $\tilde{f}$, and the result for $f$ follows immediately.

Even if $Y$ should contain a component of $X$, we may define the Segre class of $Y$ in $X$ to be the Segre class of the *completed normal cone* $P(C_YX\oplus 1)$. The completed normal cone may be identified with the exceptional divisor of $X\times\mathbb{A}^1$ blown up along $Y\times\{0\}$. In particular it is purely $n$-dimensional if $X$ is purely $n$-dimensional; and $s(Y,X)=s(Y,X\times\mathbb{A}^1)$.

We often denote the completed normal cone $P(C_YX\oplus 1)$ simply by $\bar{X}$. If $i:Y\to X$ is the given imbedding, let $\bar{i}:Y\to X$ be the imbedding given by the zero-section of $Y$ in $C_YX$ followed by the inclusion of $C_YX$ in $P(C_YX\oplus 1)$. Denote the projection from $\bar{X}\to Y$ by $\pi$. Note that the imbeddings $Y\subset X$ and $Y\subset\bar{X}$ have the same normal cone, and the same Segre classes, but in the latter case we have a retraction $\pi$ from $\bar{X}$ back to $Y$.

These constructions are functorial. Suppose $f:X'\to X$ is a morphism, and set $Y'=f^{-1}(Y)$. Then we have a fiber square

$$\begin{array}{ccc} X' & \xrightarrow{f} & X \\ {\scriptstyle i'}\uparrow & & \uparrow{\scriptstyle i} \\ Y' & \xrightarrow{g} & Y \end{array}$$

Write $\bar{X}=P(C_YX\oplus 1)$, $\bar{X}'=P(C_{Y'}X'\oplus 1)$, and let $\bar{f}:\bar{X}\to\bar{X}'$ be the induced morphism. Then we have commutative diagrams

$$\begin{array}{ccc} \bar{X}' & \xrightarrow{\bar{f}} & \bar{X} \\ {\scriptstyle \bar{i}'}\uparrow & & \uparrow{\scriptstyle \bar{i}} \\ Y' & \xrightarrow{g} & Y \end{array} \qquad \begin{array}{ccc} \bar{X}' & \xrightarrow{\bar{f}} & \bar{X} \\ {\scriptstyle \pi'}\downarrow & & \downarrow{\scriptstyle \pi} \\ Y' & \xrightarrow{g} & Y \end{array}$$

Now assume that $Y$ is a local complete intersection in $X$ of codimension $d$ with normal bundle $N$. Then $\bar{X}$ is the projective completion of $N$. Let

$$0\to\mathcal{O}(-1)\to\pi^*(N\oplus 1)\to\xi\to 0$$

define the universal quotient bundle $\xi$ of rank $d$ on $\bar{X}$.

PROPOSITION 3.3. *In addition to these assumptions, assume $X'$ is purely $n$-dimensional. Then the localized intersection class defined in* §1 *is given by the formula*

$$X' \cdot_f Y = \pi'_*(c_d(\bar{f}^*\xi) \frown [\bar{X}']).$$

PROOF. Since $\bar{f}^*\mathcal{O}(1)$ is the canonical bundle on $\bar{X}'$,

$$s(Y', X') = \pi'_*(c(\bar{f}^*\mathcal{O}(-1))^{-1} \frown [\bar{X}']).$$

Therefore

$$\begin{aligned} c(g^*N) \frown s(Y', X') &= \pi'_*(\pi'^*g^*(c(N))c(\bar{f}^*\mathcal{O}(-1))^{-1} \frown [\bar{X}']) \\ &= \pi'_*(\bar{f}^*(c(\pi^*N)c(\mathcal{O}(-1))^{-1}) \frown [\bar{X}']) \\ &= \pi'_*(\bar{f}^*c(\xi) \frown [\bar{X}']), \end{aligned}$$

the last equality from the exact sequence defining $\xi$. The result follows by equating terms of dimension $n-d$.

REMARK. The theme of this paper is the comparison of the given situation to the normal situation. Note that the normal bundle $N$ extends to a vector bundle $\xi$ on all of $\bar{X}$. The mapping of the second factor in $\pi^*(N \oplus 1) \to \xi$ gives a section of $\xi$ which vanishes precisely on $Y$. So $[Y]$ is represented by the top Chern class $c_d(\xi)$, which explains why it is easier to pull back $[Y]$ in the normal situation; the retraction $\pi'$ localizes the result on $Y'$. Theorem 1 now says that the pullback constructed this way in the normal situation agrees with the pull-back in the given situation.

## §4. Deformation to the normal cone

Let $i: Y \to X$ be a closed imbedding. We will deform this imbedding to the imbedding $\bar{i}: Y \to \bar{X}$ into the completed normal cone $\bar{X} = P(C_Y X \oplus 1)$. Let $\tilde{X}$ be the blow-up of $X$ along $Y$.

Let $\mathbb{P}^1$ be the projective line, covered as usual by the two affine lines $\mathbb{A}^1_0$ and $\mathbb{A}^1_\infty$, neighborhoods of $\{0\}$ and $\{\infty\}$ respectively.

We shall construct a scheme $W$, a flat morphism $\varphi: W \to \mathbb{P}^1$, and a closed imbedding $\psi: Y \times \mathbb{P}^1 \to W$, and an open imbedding of $X \times \mathbb{A}^1_0$ in $W$, with the following properties:

(1) There is a Cartesian diagram

$$\begin{array}{ccccccc} Y \times \{0\} & \longrightarrow & Y \times \mathbb{A}^1_0 & \longrightarrow & Y \times \mathbb{P}^1 & \longleftarrow & Y \times \{\infty\} \\ \downarrow & & \downarrow & & \downarrow{\scriptstyle \psi} & & \downarrow \\ X \times \{0\} & \longrightarrow & X \times \mathbb{A}^1_0 & \longrightarrow & W & \longleftarrow & W_\infty \\ \downarrow & & \downarrow & & \downarrow{\scriptstyle \varphi} & & \downarrow \\ \{0\} & \longrightarrow & \mathbb{A}^1_0 & \longrightarrow & \mathbb{P}^1 & \longleftarrow & \{\infty\} \end{array}$$

Here $W_\infty$ is defined to be the fiber $\varphi^{-1}(\infty)$ of $W$ over $\{\infty\}$. The other maps are the canonical inclusions and projections.

(2) There are closed imbeddings of the completed normal cone $\bar{X}$ and the blow-up $\tilde{X}$ as Cartier divisors on $W$, and

$$W_\infty = \bar{X} + \tilde{X},$$

i.e. the Cartier divisor $W_\infty$ is the sum of the Cartier divisors $\bar{X}$ and $\tilde{X}$. In fact the scheme-theoretic intersection of $\bar{X}$ and $\tilde{X}$ in $W$ may be identified with the projectivized normal cone $P(C_Y X)$ in $\bar{X}$, and with the exceptional divisor $\tilde{Y}$ in the blow-up $\tilde{X}$. The inclusion of $Y = Y \times \{\infty\}$ in $W_\infty$ in the diagram is the inclusion $\bar{i}$ of $Y$ in $\bar{X}$; the image $\bar{i}(Y)$ does not meet $\tilde{X}$.

(3) There is a morphism $\rho: W \to X$ inducing the following maps on the various subschemes described in (1) and (2): The projection $Y \times \mathbb{P}^1 \to Y \subset X$, the projection $X \times \mathbb{A}^1_0 \to X$, the projection $\pi: \bar{X} \to Y \subset X$, and the blowing down map $\tilde{X} \to X$.

(4) If $X$ is purely $n$-dimensional, then $\bar{X}$, $\tilde{X}$ are purely $n$-dimensional, and $W$ is purely $(n+1)$-dimensional.

(5) If $X$ is reduced, complete, or irreducible, respectively, the same is true for $W$.

(6) If $X$ and $Y$ are non-singular, then $W$ is non-singular.

This is a more complete version of the construction used for Riemann-Roch [2]. Following Verdier [15], it may be constructed much more easily, and without taking any closures of subvarieties, as follows:

Define $W$ to be the blow-up of $X \times \mathbb{P}^1$ along the subscheme $Y \times \{\infty\}$. Then $\varphi: W \to \mathbb{P}^1$ is the composition of the blow-down map $W \to X \times \mathbb{P}^1$ followed by the projection to $\mathbb{P}^1$. The other projection gives $\rho: W \to X$.

The completed normal cone $\bar{X}$ is included in $W$ as the exceptional divisor of the map $W \to X \times \mathbb{P}^1$. The inclusion $Y \times \{\infty\} \subset X \times \{\infty\} \subset X \times \mathbb{P}^1$ gives an imbedding of the blow-up $\tilde{X}$ of $X \times \{\infty\}$ along $Y \times \{\infty\}$ in $W$. Similarly the inclusion $Y \times \{0\} \subset Y \times \mathbb{P}^1 \subset X \times \mathbb{P}^1$ gives an imbedding $\psi$ of $Y \times \mathbb{P}^1$ (identified with the blow-up of $Y \times \mathbb{P}^1$ along $Y \times \{0\}$) in $W$.

All the properties (1)–(6) follow easily from this description of $W$, and of standard facts about blowing-up. Perhaps (2) is the least obvious, so we prove this. We may assume $X = \operatorname{Spec}(R)$, and $Y \subset X$ is given by an ideal $I \subset R$. Write $\mathbb{A}^1_\infty = \operatorname{Spec} k[\mu]$. Then, over $X \times \mathbb{A}^1_\infty = \operatorname{Spec} R[\mu]$, $W$ is given by $\operatorname{Proj}(\sum J^n)$, where $J$ is the ideal in $R[\mu]$ generated by $I$ and $\mu$. For $h \in I$, let $U_h = \operatorname{Spec}(\sum J^n_{(h)})$, an affine open subscheme of $W$; here $\sum J^n_{(h)}$ is the subring of the localization consisting of homogeneous elements. On $U_h$, $W_\infty$ is the divisor of $\mu$, and $\bar{X}$ is the divisor of $h$ (regarded as elements in $J^0$) and $\tilde{X}$ is the divisor of $\mu/h$. So $W = \bar{X} + \tilde{X}$ on $U_h$, and since, as $h$ varies in $I$, the $U_h$ cover $W_\infty$, (2) follows. (The flatness of $W$ over $\mathbb{P}^1$ is also clear from this description.)

REMARK 1. The fact that the divisors $\varphi^{-1}(0) = X \times \{0\}$ and $\varphi^{-1}(\infty) = W_\infty = \bar{X} + \tilde{X}$ are linearly equivalent, and the consequence that $[X \times \{0\}] = [\bar{X}] + [\tilde{X}]$ in $A_n W$, will play an important role. (Even in the non-singular case this equation is enough to simplify the proofs in [7] and [6].)

2. When one blows up a variety along a subvariety, the blown-up variety is isomorphic to the original except over the subvariety, but one thinks of the variety being 'deformed' near the subvariety in order to put in the exceptional divisor. Likewise the family of imbeddings $\psi : Y \times \mathbb{P}^1 \to W$ is isomorphic to the trivial family $Y \times \mathbb{A}^1_0 \to X \times \mathbb{A}^1_0$ away from infinity, but one should think of a deformation taking place in order to complete it to the normal imbedding at infinity.

3. The entire construction is functorial. If $f : X' \to X$ is a morphism, $Y' = f^{-1}(Y)$, denote all the corresponding schemes and maps for $Y' \subset X'$ with primes. There is a map $F : W' \to W$ which is compatible with all the maps of (1), (2), (3) above. This follows from the functoriality of blowing up subschemes.

PROOF OF THEOREM 1. We use the above notation. In addition, let $h_0$ be the inclusion of $X = X \times \{0\}$ in $W$, and $h_\infty$ the inclusion of $\bar{X}$ in $W$.

First note that by proposition 3 and the identity $i'\pi' = \rho' h'_\infty$, we have

(i) $i'_*(X' \cdot_f Y) = \rho'_* h'_{\infty*}(c_d(\bar{f}^*\xi) \frown [\bar{X}'])$.

We have seen (§3 remark) that $[Y] = c_d(\xi)$ in $A.\bar{X}$, and so

(ii) $c_d(\bar{f}^*\xi) \frown [\bar{X}'] = \bar{f}^*[Y] \frown [\bar{X}']$ in $A_{n-d}\bar{X}'$.

We also need

(iii) $h^*_\infty[Y \times \mathbb{P}^1] = [Y]$ in $A.(\bar{X})$, as follows e.g. from the fact that $Y \times \mathbb{P}^1$ and $\bar{X}$ intersect transversally in $Y$.

Now we have

(iv) $h'_{\infty*}(\bar{f}^*[Y] \frown [\bar{X}']) = h'_{\infty*}(\bar{f}^* h^*_\infty [Y \times \mathbb{P}^1] \frown [\bar{X}'])$

$= h'_{\infty*}(h'^*_\infty F^*[Y \times \mathbb{P}^1] \frown [\bar{X}'])$ since $h_\infty \bar{f} = F h'_\infty$

$= F^*[Y \times \mathbb{P}^1] \frown [X']$ by the projection formula

$= F^*[Y \times \mathbb{P}^1] \frown ([\bar{X}'] + [X'])$ since $F^{-1}(Y \times \mathbb{P}^1)$ does not meet $\tilde{X}'$,

$= F^*[Y \times \mathbb{P}^1] \frown ([X'] \times \{0\}])$ since $[X' \times \{0\}$ is rationally equivalent to $[\bar{X}'] + [X']$.

Now that we have gone from $\{\infty\}$ to $\{0\}$, we reverse the above steps:

$= h'_{0*}(h'^*_0 F^*[Y \times \mathbb{P}^1] \frown [X' \times \{0\}])$

$= h'_{0*}(f^* h^*_0 [X \times \mathbb{P}^1] \frown [X'])$

$= h'_{0*}(f^*[Y] \frown [X'])$.

Now apply $\rho'_*$, and by (i) and (iv) we get

(v) $i'_*(X' \cdot_f Y) = \rho'_* h'_{0*}(f^*[Y] \frown [X']) = f^*[Y] \frown [X']$, since $\rho' h'_0 = id_{X'}$, and this is the desired formula. (We have used the notation $f^*[Y] \frown [X]$ instead of simply $f^*[Y]$ so that the argument goes over without change to other contexts – cf. §6, §8.)

## §5. An intersection formula

Let $i: Y \to X$ be a closed imbedding of non-singular varieties, of codimension $d$ and with normal bundle $N$. Let $f: X' \to X$ be a morphism from a purely $n$-dimensional scheme $X'$ to $X$, and construct the fiber square.

$$\begin{array}{ccc} X' & \xrightarrow{f} & X \\ \uparrow{\scriptstyle i'} & & \uparrow{\scriptstyle i} \\ Y' & \xrightarrow{g} & Y \end{array}$$

Assume $Y'$ is a local complete intersection of codimension $d'$ in $X'$, with normal bundle $N'$, and define a bundle $E$ of rank $e = d - d'$ on $Y'$ by the exact sequence

$$0 \to N' \to g^*N \to E \to 0.$$

THEOREM 3. *For all* $\alpha \in A.Y$,

$$f^* i_* \alpha = i'_*(c_e(E) \frown g^* \alpha) \quad in \quad A.X'.$$

REMARKS 1. When $X' = Y$, this is the self-intersection formula. When $X'$ is the blow-up of $X$ along $Y$, it is formule clef (cf. [7]). For the corresponding result in topology see [8]. We thank L. Illusie for suggesting the formula to us.

2. The case when $e = 0$ appears in ([4] §2.2(4)), but the hypothesis that the maps are Tor-independent was omitted. We include a proof here.

PROOF. Assume $e = 0$. By moving $\alpha$ we may assume $\alpha = [V]$, where the cycle $g^*[V]$ is defined. Then $i'_* g^* \alpha$ is the cycle determined by the virtual sheaf $\sum (-1)^i \operatorname{Tor}_i^{\mathcal{O}_Y}(\mathcal{O}_V, \mathcal{O}_{Y'})$, and $f^* i_* \alpha$ is the cycle determined by $\sum (-1)^i \operatorname{Tor}_i^{\mathcal{O}_X}(\mathcal{O}_V, \mathcal{O}_{X'})$. Since a regular sequence defining $Y$ (locally) on $X$ remains regular on $X'$, we see that $\operatorname{Tor}_i^{\mathcal{O}_X}(\mathcal{O}_Y, \mathcal{O}_{X'})$ is zero if $i > 0$, and $\mathcal{O}_{Y'}$ if $i = 0$. Therefore the spectral sequence

$$\operatorname{Tor}_p^{\mathcal{O}_Y}(\mathcal{O}_V, \operatorname{Tor}_q^{\mathcal{O}_X}(\mathcal{O}_Y, \mathcal{O}_{X'})) \Rightarrow \operatorname{Tor}_{p+q}^{\mathcal{O}_X}(\mathcal{O}_V, \mathcal{O}_{X'})$$

degenerates to give an isomorphism of the Tors, which proves that $i'_* g^* \alpha = f^* i_* \alpha$ in this case.

For the general case we use the following commutative diagram from the deformation construction in §4:

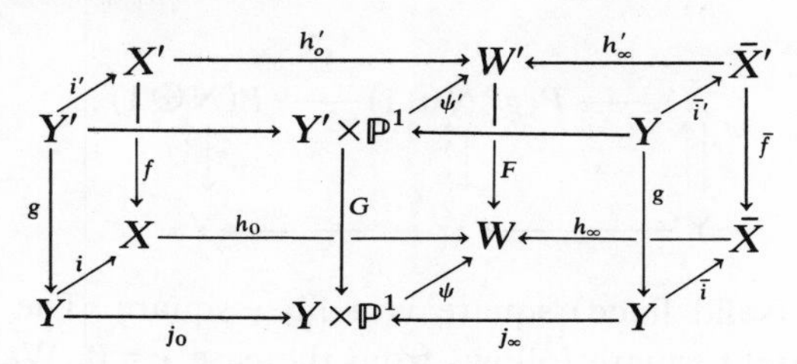

The bottom of this array is from the deformation diagram for $Y \subset X$, with the imbedding at 0 on the left, and that at $\infty$ on the right; we have let $j_0$, $j_\infty: Y \to Y \times \mathbb{P}^1$ be the inclusions at 0 and $\infty$ respectively, while $h_0$ and $h_\infty$ are the inclusions of $X$ (respectively $\bar{X}$) in $W$. The top of the array is the corresponding diagram for $Y' \subset X'$, with primes on corresponding spaces and maps. Let $F: W' \to W$, $G = g \times 1: Y' \times \mathbb{P}^1 \to Y \times \mathbb{P}^1$ be the induced maps.

We will work our way around the diagram from the given situation at 0 to the normal situation at $\infty$. Note that $h'_{0*}$ is injective, since the projection $\rho': W' \to X'$ retracts $W'$ back on $X'$. So the formula to prove is

$$h'_{0*} f^* i_* \alpha = h'_{0*} i'_*(c_e(E) \frown g^* \alpha).$$

Now $h'_{0*} f^* = F^* h_{0*}$ by the $e = 0$ case established above. So

$$\begin{aligned} h'_{0*} f^* i_* \alpha &= F^* h_{0*} i_* \alpha \\ &= F^* \Psi_* j_{0*} \alpha \\ &= F^* \Psi_* j_{\infty*} \alpha \quad \text{since} \quad j_{0*} = j_{\infty*} \\ &= h'_{\infty*} \bar{f}^* \bar{i}_* \alpha \quad \text{by reversing the steps on the right side of the diagram.} \end{aligned}$$

Similarly

$$\begin{aligned} h'_{0*} i'_*(c_e(E) \frown g^* \alpha) &= \Psi'_* j'_{0*}(c_e(E) \frown g^* \alpha) \\ &= \Psi'_* j'_{0*}(j'^*_0 c_e(q'^* E) \frown g^* \alpha) \end{aligned}$$

where $q': Y' \times \mathbb{P}^1 \to Y'$ is the projection

$$\begin{aligned} &= \Psi'_*(c_e(q'^* E) \frown j'_{0*} g^* \alpha) \\ &= \Psi'_*(c_e(q'^* E) \frown G^* j_{0*} \alpha) \quad \text{by the case } e = 0 \\ &= \Psi'_*(c_e(q'^* E) \frown G^* j_{\infty*} \alpha) \quad \text{since } j_{0*} = j_{\infty*} \\ &= h'_{\infty*} \bar{i}'_*(c_e(E) \frown g^* \alpha) \quad \text{by reversing the steps at } \infty. \end{aligned}$$

So we have reduced the problem to the normal case.

We factor the normal map $\bar{f}:\bar{X}'\to\bar{X}=P(N\oplus 1)$, giving a commutative diagram

$$\begin{array}{ccccc} \bar{X}' & \longrightarrow & P(g^*N\oplus 1) & \longrightarrow & P(N\oplus 1) \\ {\scriptstyle\pi'}\downarrow\uparrow & & \downarrow\uparrow & & {\scriptstyle\pi}\downarrow\uparrow{\scriptstyle i} \\ Y' & \longrightarrow & Y' & \longrightarrow & Y \end{array}$$

where the right (solid lined) square is a fiber square. The validity of the formula for the right square follows from the case $e=0$. We leave it to the reader to check that if the formula holds for each of two squares as above, then it holds for the composite square.

So we are left with proving the formula for a situation as in the left square:

$$\begin{array}{ccc} P(N'\oplus 1) & \xrightarrow{f} & P(N\oplus 1) \\ {\scriptstyle\pi'}\downarrow\uparrow{\scriptstyle i'} & & {\scriptstyle\pi}\downarrow\uparrow{\scriptstyle i} \\ Y & = & Y \end{array}$$

where $f$ is induced by an inclusion of vector bundles $N'\subset N$ with quotient bundle $E$. We may assume $\alpha=[V]$, where $V$ a subvariety of $Y$, and then $V=Y$ by restricting the bundle to $V$.

We use the exact sequence

$$0\to\xi'\to f^*\xi\to\pi'^*E\to 0$$

relating the universal quotient bundles, and the fact that

$$i_*[V]=c_d(\xi)\frown[P(N\oplus 1)] \quad \text{(cf. §3)}.$$

So

$$\begin{aligned} f^*i_*\alpha &= c_d(f^*\xi)\frown[P[N'\oplus 1)] \\ &= c_e(\pi'^*E)c_d(\xi')\frown[P(N'\oplus 1)] \\ &= \pi'^*c_e(E)\frown i'_*[V] \\ &= i'_*(c_e(E)\frown[V]) \end{aligned}$$

as desired.

## §6. Tor and Riemann-Roch

Let $i:C\to X$, $j:D\to X$ be morphisms to a non-singular variety $X$. Write $C\cap D$ for the fiber product $C\times_X D$. Let

$$\operatorname{Tor}^X(C,D)=\sum_{i\geq 0}(-1)^i\operatorname{Tor}_i^{\mathcal{O}_X}(\mathcal{O}_C,\mathcal{O}_D)$$

be Serre's virtual sheaf, an element in the Grothendieck group $K_0(C\cap D)$ of coherent sheaves on $C \cap D$. We denote by $F_k K_0(C \cap D)$ the subgroup of $K_0(C \cap D)$ generated by sheaves whose support have dimension $\leqq k$.

Let $\tau : K_0(C \cap D) \to A.(C \cap D)_{\mathbb{Q}}$ be the Riemann-Roch map ([2]) with $\tau_k : K_0(C \cap D) \to A_k(C \cap D)_{\mathbb{Q}}$ the component of dimension $k$.

THEOREM 4. *Assume $X$, $C$, and $D$ are purely $n$, $c$, and $d$ dimensional. Let $m = c + d - n$. Then*

(1) $\mathrm{Tor}^X(C, D) \in F_m K_0(C \cap D)$

(2) $\tau_k(\mathrm{Tor}^X(C, D)) = \begin{cases} 0 & \text{if} \quad k > m \\ C \cdot D & \text{if} \quad k = m. \end{cases}$

Here $C \cdot D$ is the class constructed in §2, regarded in the group of cycles on $C \cap D$ with rational coefficients.

PROOF. By reduction to the diagonal, as in §2 (cf. [11]) we are reduced to proving the following assertion:

'Let $Y \subset X$ be a local complete intersection of codimension $d$, $f : X' \to X$ a morphism from a purely $n$-dimensional scheme $X'$, $Y' = f^{-1}(Y)$. Then

(1) $\mathrm{Tor}^X(X', Y) \in F_{n-d} K_0 Y'$

(2) $\tau(\mathrm{Tor}^X(X', Y)) = X' \cdot_f Y +$ lower dimensional terms'.

Let $\bar{X}$, $\bar{X}'$ be the completed normal cones of $X$ along $Y$, $\bar{X}'$ along $Y'$. We claim first that $\mathrm{Tor}^X(X', Y) = \mathrm{Tor}^{\bar{X}}(\bar{X}', Y)$. This fact was proved by Verdier [15]. In the language we are using here, one can prove it by following the same sequence of steps as in (iv) of the proof of theorem 1 in §4, but replacing each cap product of the form $h^*[T] \frown [S']$, where $h : S' \to S$ is a morphism, $T \subset S$, by the element $\mathrm{Tor}^S(S', T)$. Each of these Tors can be regarded in $K_0(\rho'^{-1}(Y'))$, $\rho' : W' \to X'$ as in §4; for the rational equivalence step, note that if $D_1$ and $D_2$ are linearly equivalent effective Cartier divisors on a scheme $T$ as above, then $\mathrm{Tor}^S(S', D_1) = \mathrm{Tor}^S(S', D_2)$ on $h^{-1}(T)$, since $\mathcal{O}_{D_1} = \mathcal{O}_{D_2}$ in $K_0 T$. We leave the other details to the reader.

On $\bar{X}$ we have a Koszul complex $\Lambda^{\cdot} \xi^{\vee}$ resolving $\mathcal{O}_Y$, so

$$\begin{aligned} \mathrm{Tor}^{\bar{X}}(\bar{X}, Y) &= \sum (-1)^i \bar{f}^* \Lambda^i \xi^{\vee} \otimes \mathcal{O}_{X'} \\ &= \lambda_{-1}(\bar{f}^* \xi^{\vee}) \otimes \mathcal{O}_{\bar{X}}, \quad \text{in} \quad K_0(\bar{X}'). \end{aligned}$$

But $\lambda_{-1}(\bar{f}^* \xi^{\vee})$ is in the $d$th $\lambda$-filtration of the Grothendieck group of vector bundles on $\bar{X}'$, and so the tensor product is in $F_{n-d}(K_0(\bar{X}'))$ (cf. [13]). Finally,

$$\begin{aligned} \tau(\mathrm{Tor}^X(X', Y)) &= \tau(\pi'_*(\bar{f}^* \lambda_{-1}(\xi^{\vee}) \otimes \mathcal{O}_{\bar{X}'})) \\ &= \pi'_*(\bar{f}^* \mathrm{ch}(\lambda_{-1}(\xi^{\vee})) \frown \tau(\mathcal{O}_{\bar{X}'})) \quad \text{by [2]} \\ &= \pi'_*(\bar{f}^* c_d(\xi) \frown [\bar{X}']) + \text{lower terms}, \end{aligned}$$

as desired.

REMARKS. If $\dim C \cap D = l > m$, Serre showed that $\operatorname{Tor}^X(C, D) \in F_{l-1}K_0(C \cap D)$.

If the intersection is proper, i.e. $\dim C \cap D = m$, theorem 4 shows that our cycle agrees with Serre's. The positivity of the intersection number may also be seen geometrically: if $A$ is an $m$-dimensional component of $C \cap D$, then the construction of §2 shows that the intersection number of $A$ in $C \cap D$ is the degree of the morphism from the completed normal cone to $C \cap D$ in $C \times D$ to the restriction to $A$ of the completed tangent bundle $T_X$. One would like a similar description which does not involve reduction to the diagonal.

## §7. Localized Chern classes

Let $E$ be a vector bundle of rank $d$ on a purely $n$-dimensional scheme $X$, $s$ a section of $E$, $Y$ the zero-scheme of $s$. We will define a localized Chern class $c_d^{(s)}(E) \in A_{n-d}Y$ which generalizes the fact that $[Y] = c_d(E) \cap [X]$ if $s$ is suitably transversal to the zero-section.

The section $s$ induces a map from $E^\vee$ to $\mathcal{O}_X$ which maps $E^\vee$ onto the ideal sheaf $I$ of $Y$; this gives a surjection $\operatorname{Sym}(E^\vee) \to \sum I^n$. This map of graded rings gives an imbedding of the blow-up $\tilde{X}$ of $X$ along $Y$ into $P(E)$. Similarly when we restrict to $Y$ we get an imbedding of the normal cone $C_YX$ in $E|_Y$ and an imbedding of $P(C_YX \oplus 1)$ in $P(E|_Y \oplus 1)$. We have a commutative diagram

$$\begin{array}{ccccccc}
\bar{X} = P(C_YX \oplus 1) & \xrightarrow{j} & P(E \oplus 1) & \longleftarrow & P(E) & \longleftarrow & \tilde{X} \\
\downarrow{\scriptstyle \pi} & & \downarrow{\scriptstyle p} & & & & \downarrow{\scriptstyle q} \\
Y & \xrightarrow{i} & X & = & = & = & X
\end{array}$$

let $p^*(E \oplus 1) \to \xi_E \to 0$ be the universal $d$-dimensional quotient bundle.

DEFINITION. The localized Chern class is defined by

$$c_d^{(s)}(E) = \pi_*(c_d(j^*\xi_E) \frown [\bar{X}]).$$

REMARK. At least if $X$ is quasi-projective, there is always a map $f: X \to X^+$, with $X^+$ non-singular, a vector bundle $E^+$, and a section $s^+$ of $E^+$ transversal to the zero section, so that $E = f^*E^+$, $s = s^+ \circ f$. If $Y^+$ is the zero-scheme of $s^+$, then the definition of $c_d^{(s)}(E)$ coincides with the definition of $X \cdot_f Y^+$ in §1. On the other hand it coincides with graph construction applied to the homomorphism $s: 1 \to E$ of vector bundles, but we have localized the top Chern class instead of the Chern character as in [2]; the decomposition of the cycle at infinity of the graph construction (cf. [2] Ch. II 1.1) is just $Z_\infty = [\bar{X}] + [\tilde{X}]$, with $\bar{X}$ and $\tilde{X}$ as above.

PROPOSITION 7.1. *The localized class* $c_d^{(s)}(E)$ *in* $A_{n-d}Y$ *maps to* $c_d(E) \frown [X]$ *in* $A_{n-d}X$.

The proof is the same as the proof of theorem 1 in §4 (or it follows from theorem 1 by the above remark); or use the essentially equivalent proof in ([2] Ch. II 2.1).

Now let $E_1$, $E_2$ be two vector bundles on $X$ of ranks $d_1$, $d_2$, with sections $s_1$, $s_2$ vanishing on subschemes $Y_1$, $Y_2$, and let $i_1: Y_1 \to X$, $i_2: Y_2 \to X$ be the inclusions. Let $E = E_1 \oplus E_2$, $d = d_1 + d_2$, $s = s_1 \oplus s_2$. So $c_d^{(s)}(E) \in A_{n-d}(Y_1 \cap Y_2)$.

PROPOSITION 7.2. $c_d^{(s)}(E)$ *maps to* $i_1^* c_{d_2}(E_2) \frown c_{d_1}^{(s_1)}(E_1)$ *in* $A_{n-d}Y_1$, *and to* $i_2^* c_{d_1}(E_1) \frown c_{d_2}^{(s_2)}(E_2)$ *in* $A_{n-d}Y_2$.

PROOF. The inclusion $E_1 \oplus 1 \to E \oplus 1$ gives a map $\varphi: P(E_1 \oplus 1) \to P(E \oplus 1)$ with $\Phi^*(\xi_E) = \xi_{E_1} \oplus E_2$. It follows that $i_1^* c_{d_2}(E_2) \frown c_{d_1}^{(s_1)}(E_1)$ is equal to $c_d^{(s_1 \oplus 0)}(E)$ in $A_{n-d}Y_1$ (cf. [2] II 2.3). Now consider the family of sections $s_1 \oplus ts_2$ of $E$, parametrized by $t \in \mathbb{A}^1$, all vanishing on the scheme $Y_1$. Then the 'homotopy property' (cf. [2] II 2.5), applied to the Chern class instead of the Chern character, implies that the classes obtained by specializing $t \to 0$ and $t \to 1$ agree.

There are other similar facts about localized Chern classes. We have included proposition 2 here because it is useful in proving the residual intersection theorem (cf. [5]).

## §8. Some related results

8.1. Most of what we have done here works for any homology-cohomology theory (in place of rational equivalence), and one may see that the answers provided by different theories are in agreement. For example on an $n$-dimensional non-singular complex projective variety $X$, we have a Lefschetz duality isomorphism

$$L: H^{2n-i}(X, X - V) \to H_i(V)$$

for any subvariety $V$ of $X$. In particular if $\dim C = c$, $\dim D = d$, we have fundamental classes $L^{-1}[C] \in H^{2n-2c}(X, X - C)$, $L^{-1}[D] \in H^{2n-2d}(X, X - D)$, so $L^{-1}[C] \smile L^{-1}[D] \in H^{4n-2c-2d}(X, X - (C \cap D))$, hence $L(L^{-1}[C] \smile L^{-1}[D] \in H_{2m}(C \cap D)$. To show that the cycle map $A_m(C \cap D) \to H_{2m}(C \cap D)$ takes our intersection class $C \cdot D$ to this one, one can reduce to the diagonal as before, and then argue as in the proof of theorem 1.

8.2. More generally, if $Y \subset X$ is any local complete intersection of codimension $d$ of complex projective varieties, then $Y$ determines a relative cohomology class $\mathrm{cl}(Y) \in H^{2d}(X, X-Y)$ (cf. [2] IV.4). In particular, if $f: X' \to X$ is a morphism, capping with $\mathrm{cl}(Y)$ induces a homomorphism $H_i X' \to H_{i-2d} Y'$, $Y' = f^{-1}(Y)$.

The theory of rational equivalence does not yet have relative groups $A^{\cdot}(X, X-Y)$, but we may define, for a local complete intersection $i: Y \to X$ of codimension $d$, an *operator*

$$\mathrm{cl}(Y): A_k X' \to A_{k-d} Y'$$

for any morphism $f: X' \to X$, $Y' = f^{-1}(Y)$: If $V'$ is a $k$-dimensional variety on $X'$, then the action of $\mathrm{cl}(Y)$ on $[V']$ is defined to be the image of the intersection class $V' \cdot_f Y$ in $A_{k-d} Y'$. The fact that this operation on cycles passes to rational equivalence is proved, much as in ([$V$]), by reducing it to the case of divisors. In fact when $X' = X$, this agrees with the Gysin map $i^*: A_k X \to A_{k-d} Y$ constructed by Verdier. One can prove that these 'operator cohomology classes' have the formal properties one would expect them to have.

8.3. When one has Gysin maps for local complete intersections, it is easy to construct an intersection theory: For any map $f: X' \to X$ from a scheme $X'$ to a non-singular variety $X$, the graph morphism $\gamma_f: X' \to X \times X'$, $\gamma_f(x') = (f(x), x')$, is a local complete intersection of codimension $n = \dim(X)$. Then the external product

$$A_{n-p} X \otimes A_q X' \to A_{n-p+q}(X \times X')$$

followed by the Gysin map

$$\gamma_f^*: A_{n-p+q}(X \times X') \to A_{q-p}(X')$$

defines a *cap product*

$$A^p X \otimes A_q X' \to A_{q-p} X',$$

where $A^p X = A_{n-p} X$. If we let $X' = X$ we get an intersection product on $A^{\cdot} X$. The expected relations among these products follow from facts about Gysin maps. Detailed statements and proofs will be provided in another place, where we will develop foundations of intersection theory based on the principles discussed here.

REFERENCES

[1] C. Chevalley, A. Grothendieck, and J.-P. Serre, *Anneaux de Chow et applications*, Séminaire C. Chevalley, 2e année, Secr. Math., Paris, 1958.

[2] P. Baum, W. Fulton, and R. MacPherson, Riemann-Roch for singular varieties, *Publ. Math. I.H.E.S.*, no. 45 (1975), 101–145.

[3] A. Grothendieck and J. Dieudonné, Eléments de géométrie algébrique II, *Publ. Math. I.H.E.S.*, no. 8 (1961).

[4] W. Fulton, Rational equivalence for singular varieties, *Publ. Math. I.H.E.S.*, no. 45 (1975), 147–167.

[5] W. Fulton and D. Laksov, Residual intersections and the double point formula, these proceedings.

[6] J.-P. Jouanolou, Riemann-Roch sans dénominateurs, *Invent. Math.* 11 (1970), 15–26.

[7] A. T. Lascu, D. Mumford, and D. B. Scott, The self-intersection formula and the 'formule-clef', *Math. Proc. Camb. Phil. Soc.* 78 (1975), 117–123.

[8] D. Quillen, Elementary proofs of some results of cobordism theory using Steenrod operations, *Advances in Math.* 7 (1971), 29–56.

[9] P. Samuel, Séminaire sur l'équivalence rationnelle, *Publ. Math. d'Orsay*, Paris-Orsay (1971), 1–17.

[10] B. Segre, Nuovi metodi e risultati nella geometria sulle varietà algebriche, *Ann. Mat. Pura Appl.* (4), 35 (1953), 1–127.

[11] J.-P. Serre, Algèbre locale, Multiplicités, *Springer Lecture Notes in Mathematics*, 11 (1965).

[12] F. Severi, Sulle interregioni delle varietà algebreche e sopra i loro caratteri e singularità proiettive, *Mem. d. R. Accad. d. Sci. di Torino*, S. II vol. 52 (1902), 61–118.

[13] P. Berthelot, A. Grothendieck, L. Illusie *et al.*, Théorie des intersections et théorème de Riemann-Roch, *Springer Lecture Notes in Mathematics*, 225 (1971).

[14] B. L. van der Waerden, The theory of equivalence systems of cycles on a variety, *Symposia Mathematica* V, Istituto Nazionale di Alta Matematica (1971), 255–262.

[15] J.-L. Verdier, Le Théorème de Riemann-Roch pour les intersections complètes, *Astérisque* 36–37 (1976), 189–228.

*Added in proof:* We have been informed that the intersection formula described here has been discovered independently by H. Gillet.

William Fulton
Robert MacPherson
Brown University
Dept. of Mathematics
Providence, Rhode Island 02912
USA

Nordic Summer School/NAVF
Symposium in Mathematics
Oslo, August 5-25, 1976

# STRATIFICATION AND FLATNESS

Heisuke Hironaka

## 1. Review on semi-analytic sets and subanalytic sets

Let us first recall some elementary notions to fix ideas on terms that we shall speak of. We often speak of a real-analytic space $X$. This means a certain type of $\mathbb{R}$-ringed space, which is more than just a certain type of point-set with topology. For instance, $\mathbb{R}^n$ as real-analytic space is a combination of the two

(1) $\mathbb{R}^n$ as a topological space

(2) structure sheaf (as real-analytic space), denoted by $\mathscr{A}_{\mathbb{R}^n}$ which is the sheaf of germs of real-analytic functions.

In the general case, a real-analytic space $X$ is locally (at least) described as follows: we have an open subset $V$ of some $\mathbb{R}^n$ and a finite system of real-analytic functions $(f_1, \cdots, f_m)$ on $V$ so that

$$|X| = \{x \in V \mid f_1(x) = \cdots = f_m(x) = 0\}$$

and $\mathscr{A}_X = \mathscr{A}_V/(f_1, \cdots, f_m)\mathscr{A}_V$, restricted to its support $|X|$, where $\mathscr{A}_V = \mathscr{A}_{\mathbb{R}^n}/V$. It is useful, when speaking of complexification and transformations (mostly blowing-ups), to have a structure such as the above which is given by a specific ideal sheaf or by a system of equations, rather than simply by point-set with topology.

Let $X$ be a real-analytic space. Given a point $\xi \in X$, we consider a special class of germs of subsets of $|X|$ about the point $\xi$, first of all the class of semi-analytic germs at $\xi$. This is defined to be the smallest class of germs of subsets about $\xi$ such that

(1) it contains germs at $\xi$ of such subsets as $\{x \mid g(x) > 0\}$ where $g$ is any real-analytic function (i.e. section of $\mathscr{A}_X$) in any neighborhood of $\xi$ in $|X|$,

(2) it is closed under elementary set-theoretical operations as follows:

(a) finite intersection

(b) finite union

(c) taking complementary set.

DEFINITION. A subset $A \subset X$ is said to be semi-analytic $\Leftrightarrow$ the germ of $A$ at every point of $X$ is semi-analytic in $X$.

NOTE. The semi-analyticity of the germ of $A$ must be tested at every point of $X$ (in fact, at every point of the closure $\bar{A}$), not only at every point of $A$.

It is convenient to know the following rather explicit local characterization of semi-analyticity:

$A \subset X$ is semi-analytic $\Leftrightarrow$ for every point $\xi \in X$, there can be found an open neighborhood $V$ of $\xi$ in $X$ and a finite system of real-analytic functions $f_{ij}$ and $g_{ik}$ on $V$ such that

$$A \cap V = \bigcup_i \{x \in V \mid f_{ij}(x) = 0, \quad \forall_j, \, g_{ik}(x) > 0, \forall_k\}.$$

When $X$ is given a structure of real-algebraic space (i.e. it is locally described by Zariski open subset $V$ and by polynomial functions on $\mathbb{R}^n$, and globally pieced together by isomorphisms by rational functions), we have a notion of semi-algebraic set in $X$. Take $X = \mathbb{R}^n$ for simplicity, then a subset $A$ of $X$ is said to be semi-algebraic in $X$ $\Leftrightarrow$ (globally) there exists a finite system of polynomial functions $f_{ij}$ and $g_{ik}$ such that

$$A = \bigcup_i \{x \in X \mid f_{ij}(x) = 0, \quad \forall_j, \, g_{ik}(x) > 0, \forall_k\}.$$

For a general real-analytic space $X$ as before, and for a point $\xi \in X$, we define the class of subanalytic germs at $\xi$ to be the smallest class of germs of subsets of $X$ at $\xi$ such that

(1) it contains the germs at $\xi$ of those subsets $A$ of $X$ which are described as follows: There exists a real-analytic map from a real-analytic space into $X$, say $\pi : Y \to X$, and a semi-analytic subset $B$ of $Y$ such that $\pi : \bar{B} \to X$ is proper (i.e. the inverse image of every compact is compact) and $A = \pi(B)$,

(2) it is closed under elementary set-theoretical operations.

DEFINITION. A subset $A$ of $X$ is subanalytic $\Leftrightarrow$ the germ of $A$ at every point of $X$ is subanalytic in $X$.

NOTE. It is clear that the class of subanalytic sets is bigger than that of semi-analytic sets.

To see that the subanalyticity is strictly weaker than the semi-analyticity in general, let us take two sets of examples as follows.

EXAMPLE 1. We take $s$ to be an integer $\geqq 2$, and we find a system $f_1, \cdots, f_r \in \mathbb{R}\{x_1, \cdots, x_s\}$ (which denotes the ring of convergent power series of the $s$ variables) which are analytically independent, i.e. for every non-zero power series $F(y_1, \cdots, y_r)$ of $r$ variables $y_1, \cdots, y_r$, $F(f_1, \cdots, f_r)$ is non-zero as a power series in $x_1, \cdots, x_s$. Such a system $(f_1, \cdots, f_r)$ always exists, for every integer $r$. For instance, take $s=2$ and $r>2$.

NOTE. The existence of such $(f_1, \cdots, f_r)$ can be shown as follows. Since $\mathbb{R}\{x_1\}$ has an infinite transcendence degree over $\mathbb{R}$, we have $h_1, \cdots, h_r \in \mathbb{R}\{x_1\}$ which are algebraically independent over $\mathbb{R}$. For instance, $h_1=1$, $h_2=x_1$, $h_3=\sin x_1, \cdots, h_r=(\sin)^{r-2}(x_1)$. If $h_1, \cdots, h_r$ are algebraically independent over $\mathbb{R}$, then $f_i=x_2h_i(x_i)$, $i=1, 2, \cdots, r$, are analytically independent over $\mathbb{R}$. In fact, if not, there should exist $F(y_1, \cdots, y_r) \in \mathbb{R}\{y_1, \cdots, y_r\}, \neq 0$, such that $F(f_1, \cdots, f_r)$ is zero in $\mathbb{R}\{x_1, x_2\}$. Write $F=F_d+F_{d+1}+\cdots$ where $F_i$ is a homogeneous polynomial of degree $i$ for all $i$ and $F_d \neq 0$. Then we get

$$F(f_1, \cdots, f_r)=x_2^dF_d(h_1, \cdots, h_r)+x_2^{d+1}F_{d+1}(h_1, \cdots, h_r)+\cdots$$

For this to be identically zero, we must have $F_d(h_1, \cdots, h_r)=0$ which contradicts the algebraically independence of $h_1, \cdots, h_r$.

Now, let $D$ be a small disc about the origin in $\mathbb{R}^2$ such that all the $f_i$ are real-analytic in a neighborhood of $D$ in $\mathbb{R}^2$. Let $\pi: \mathbb{R}^2 \to \mathbb{R}^r$ be the real-analytic map defined in a neighborhood of $D$ in $\mathbb{R}^2$. As $\pi \mid D$ is clearly proper, $\pi(D)$ is subanalytic in $\mathbb{R}^r$ by definition. But we claim that $\pi(D)$ is not semi-analytic at the image $\eta=\pi(0)$ in $\mathbb{R}^r$. We shall prove this. First of all, by a translation in $\mathbb{R}^r$, we may assume that $\eta=0$. Suppose that $\pi(D)$ were semi-analytic at 0 in $\mathbb{R}^r$. Then, for a sufficiently small neighborhood $V$ of 0 in $\mathbb{R}^r$, we have an expression

$$\pi(D)\cap V=\bigcup_i \{y \in V \mid f_{ij}(y)=0, \forall_j, g_{ik}(y)>0, \forall_k\}$$

Here, by replacing $V$ by a smaller neighborhood of 0 if necessary, we may assume that for every $i$ we have

$$0 \in \overline{\{y \in V \mid g_{ik}(y)>0, \forall_k\}},$$

i.e. the open subset $U_i=\{y \in V \mid g_{ik}>0, \forall_k\}$ of $V$ has 0 on its boundary. For each $i$, there then should exist at least one index $j(i)$ such that $f_{ij(i)}$ is not identically zero in any neighborhood of 0 in $\mathbb{R}^r$. In fact, if not, $\pi(D)\cap V$, should contain an open subset $\neq \emptyset$ and hence, by a theorem of Sard, we should have a non-critical point of $\pi$ in any neighborhood of $D$ in $\mathbb{R}^2$. This clearly contradicts the assumption that $r>2$. Now let

$$F=\prod_i f_{ij(i)}(y)$$

which is a non-zero power series in $y$ about 0 in $\mathbb{R}^r$ but vanishes identically on $\pi(D)$ about 0. This means $F(f_1(x), \cdots, f_r(x))$ is identically zero in a neighborhood of 0 in $\mathbb{R}^2$, i.e. it is a zero power series in $x_1$, $x_2$. This contradicts the analytical independence of $f_1, \cdots, f_r$.

EXAMPLE 2. Let $A$ be a bounded semianalytic set in $\mathbb{R}^n$. Let $C(A)$ denote the cone over $A$, in $\mathbb{R}^{n+1}$. To be precise, we place $A$ in the translation by 1 of $\mathbb{R}^n$ in $\mathbb{R}\times\mathbb{R}^n = \mathbb{R}^{n+1}$. Then

$$C(A) = \{tu \mid u \in A, t \in \mathbb{R}_+\}(A \subset 1 \times \mathbb{R}^n).$$

We then claim that $C(A)$ is semi-analytic at the vertex 0 in $\mathbb{R}^{n+1} \Leftrightarrow A$ is semi-algebraic in $\mathbb{R}^n$.

Here, the implication ($\Leftarrow$) is clear. To prove ($\Rightarrow$), we make a little more general discussion from which the implication ($\Rightarrow$) follows immediately. Let $B$ be a semi-analytic subset of some $\mathbb{R}^m$ (for instance, $B = C(A)$ and $\mathbb{R}^m = \mathbb{R}^{n+1}$). We define the inner cone $C$ of $B$ at $0 \in \mathbb{R}^m$ as follows. Namely $C$ is characterized by the following properties:

(a) $C$ is a cone at 0, i.e. it is stable by any positive scalar multiplication in $\mathbb{R}^m$,

(b) $0 \in C \Leftrightarrow 0 \in B$

(c) for every open half line $L_v$ issuing from 0 in $\mathbb{R}^m$ (which is parametrized by a point $v$ in the unit sphere $S^{m-1}$ in $\mathbb{R}^m$ about 0), we have

$$L_v \subset C \Leftrightarrow 0 \in \overline{L_v \cap B}.$$

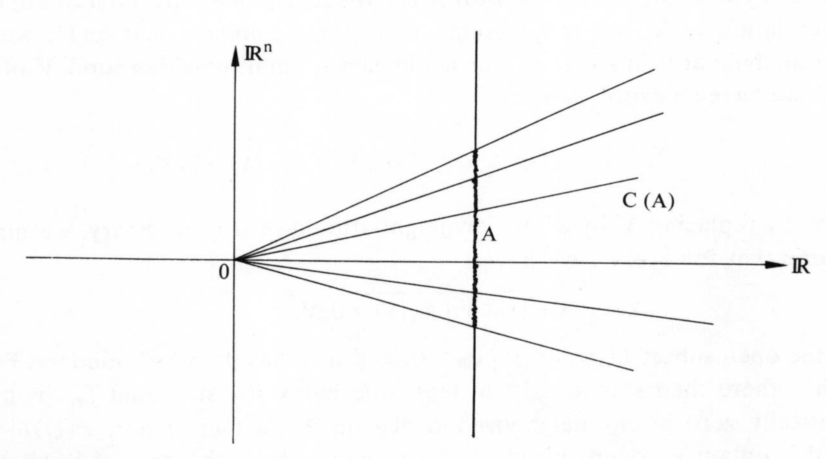


We can then prove

(i) $C$ is semi-algebraic in $\mathbb{R}^m$

(ii) if $B$ is a cone at 0, then $C = B$. Here it is easily seen that these two facts applied to $B = C(A) \subset \mathbb{R}\times\mathbb{R}^n$ imply the desired implication ($\Rightarrow$) of Example 2.

Let us now prove (i) and (ii). (ii) is obvious. As for (i), we shall find an explicit expression for $C$ by which (i) is clear. Namely, we start with an expression

$$B \cap V = \bigcup_i B_i,$$

where $V$ is a small neighborhood of 0 in $\mathbb{R}^m$ and

$$B_i = \{z \in V \mid f_{ij}(z) = 0, \forall_j, g_{ik}(z) > 0, \forall_k\}$$

with finitely many real-analytic functions $f_{ij}$ and $g_{ik}$ in $V$. We let $K$ denote the set of indices $k$. (Without any loss of generality, we may assume that $K$ is independent of $i$.) In what follows, we shall write

$$h = \sum_{a \in \mathbb{Z}_0} h^{(a)}$$

for each real-analytic function $h$ about 0 in $\mathbb{R}^2$ in terms of homogeneous polynomials $h^{(a)}$ of degree $a$ in the coordinate functions $z = (z_1, \cdots, z_m)$ of $\mathbb{R}^m$. Now, by Hilbert's basis theorem, there exists an integer $N(\gg 0)$ such that for every $(i, j, k)$, we have

$$\{f_{ij}^{(a)}, a < N\}\mathbb{R}[z] = \{f_{ij}^{(b)}, b \in \mathbb{Z}_0\}\mathbb{R}[z]$$

and

$$\{g_{ik}^{(a)}, a < N\}\mathbb{R}[z] = \{g_{ik}^{(b)}, b \in \mathbb{Z}_0\}\mathbb{R}[z].$$

In other words, all the $f_{ij}^{(b)}$ are linear combinations (with coefficients in $\mathbb{R}[z]$) of those $f_{ij}^{(a)}$, $a = 0, 1, \cdots, N-1$, and similarly for $g_{ik}$. We can now write

$$C = \bigcup_i \bigcup_{\alpha \in \hat{N}^K} C_{i\alpha},$$

where $\hat{N} = \{0, 1, 2, \cdots, N\}$ and $\hat{N}^K$ denotes the set of all maps $\alpha : K \to N$. Moreover, $C_{i\alpha}$ is defined as follows

$$C_{i\alpha} = f_{ij}^{(a)} = 0, \forall_j, \forall_a \in \hat{N}, g_{ik}^{(b)} = 0, \forall_b < \alpha(k),$$
$$\text{and } g_{ik}^{(\alpha(k))} > 0.$$

This $C$ is clearly semi-algebraic in $\mathbb{R}^m$, and hence it is enough to prove that $C$ is in fact the inner cone of $B$ at 0. The first two properties (a) and (b) for inner cone are trivially true. The last property, (c) can be shown as follows for each $v \in S^{m-1}$,

$0 \in \overline{L_v \cap B}$.

$\Leftrightarrow$ $\exists_i$ such that $0 \in \overline{L_v \cap B_i}$.

$\Leftrightarrow$ $\exists_i$ such that $f_{ij} \mid L_v \equiv 0$ (because of the analyticity of $f_{ij}$ in a neighborhood of 0 in $\mathbb{R}^m$) and $g_{ik} \mid L_v$ is positive near 0, for all $j$ and $k$.

$\Leftrightarrow \exists_i$ such that $f_{ij}^{(a)} \mid L_v \equiv 0$ for all $j$ and all $a \geqq 0$ (where it is enough to have it only for all $a \in \hat{N}$) and for each $k \in K$, $\exists \alpha(k) \in \mathbb{Z}_0$ such that $g_{ik}^{(b)} \mid L_v \equiv 0$ for all $b < \alpha(k)$ and $g_{ik}^{(\alpha(k))} \mid L_v > 0$ (where necessarily we then have $\alpha(k) \in \hat{N}$).

$\Leftrightarrow \exists_i$ such that for some $\alpha \in \hat{N}^K$ $L_v \subset C_{i\alpha}$.

$\Leftrightarrow L_v \subset C = \bigcup_{i,\alpha} C_{i\alpha}$.

The point of the above discussions in Example 2 is that the cone $C(A)$ is not semi-analytic in $\mathbb{R}^{n+1}$ when $A$ is semi-analytic (and bounded) but not semi-algebraic in $\mathbb{R}^n$. On the other hand, $C(A)$ is always subanalytic in $\mathbb{R}^{n+1}$. This can be seen as follows. Let $S^n$ be the unit sphere about 0 in $\mathbb{R}^{n+1}$ and let $p: S^n \times \mathbb{R} \rightarrow \mathbb{R}^{n+1}$ be the map defined by $p(v, t) = tv$ (scalar multiplication). Let $g: 1 \times \mathbb{R}^n \rightarrow S^n$ be the map given by the scalar multiplication by the inverse of the distance from the origin. Note that $g$ is an isomorphism onto a

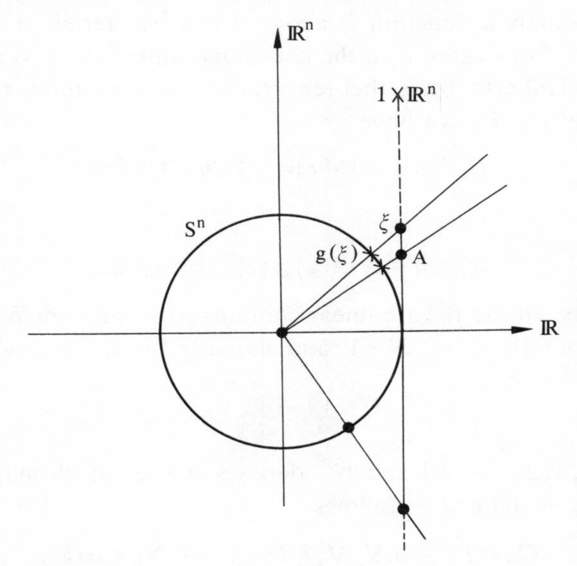


hemisphere on $S^n$. Moreover, if $\mathbb{R}_0 = \{t \in \mathbb{R} \mid t \geqq 0\}$, $p(g(A) \times \mathbb{R}_0) = C(A)$. Clearly $g(A) \times \mathbb{R}_0$ is semi-analytic in $S^n \times \mathbb{R}$ and $p$ is proper. Therefore, by definition, $C(A)$ is subanalytic in $\mathbb{R}^{n+1}$.

We now proceed to review the three basic properties in common for semi-algebraic sets in the algebraic case and for subanalytic sets in the analytic case. Let (*) denote the property of being semi-algebraic when the ambient space is given a structure of an algebraic space, respectively that of being semi-analytic or that of being subanalytic when the ambient space is real-analytic.

I. *Closure property.* If $A$ has the property (*), then its closure $\bar{A}$ has the same. (This is true in all the three cases.)

II. *Stratification property.* If $A$ has the property (*), then $A$ has a stratification with the same property, i.e. there exists a locally finite disjoint decomposition $A = \bigcup_i A_i$ such that each $A_i$ (called stratum) is a smooth connected real-analytic submanifold which has the same property (*). This is again true in all the three cases.

III. *Projection property.* Let $\pi : X \to Y$ be an algebraic map in the algebraic case, and a real-analytic map in the analytic case. If $A \subset X$ has the property (*) and $\pi$ induces a proper map $\bar{A} \to Y$, then $\pi(A)$ has the same property. This is true for semi-algebraicity and subanalyticity, but not in general for semi-analyticity.

REMARK (1). The property II induces the following: If $A \subset X$ has the property (*) then each of its connected components has the same, because a connected component is a locally finite union of strata in a decomposition of $A$ as in II.

REMARK (2). In II, we can obviously require (as we usually do when speaking of stratification) the so called frontier condition: $\bar{A}_i \cup A_j \neq \emptyset \Rightarrow \bar{A}_i \supset A_j$. What is not obvious, but is true, is that we can find a decomposition $A = \bigcup_i A_i$ of II such that every pair $(A_i, A_j)$ with $\bar{A}_i \cap A_j \neq \emptyset$ satisfies Whitney condition.

REMARK (3). It is the projection property that makes the essential difference in dealing with semi-analytic sets on one hand and with subanalytic sets on the other. In fact, it could be said that the subanalyticity is the minimal generalization of semi-analyticity in order to regain the projection property.

REMARK (4). In the algebraic case, the projection property is valid in the notation of III. But this is not an essential change in that case, because an algebraic space can always be compactified and the semi-algebraicity is preserved by any algebraic change of the ambient space.

REMARK (5). Let $A$ be a semi-analytic set in $\mathbb{R}^n$, say bounded for simplicity. Consider various linear projections $\sigma : \mathbb{R}^n \to \mathbb{R}^d$ with a fixed $d < n$. In general $\sigma(A)$ is not semi-analytic, though subanalytic. However, we have

III′ "*Almost all*" *projection property.* For almost all linear projections $\sigma : \mathbb{R}^n \to \mathbb{R}^d$, $\sigma(A)$ is semi-analytic in $\mathbb{R}^d$. Here, almost all means every one in a dense Baire category subset (a countable intersection of open dense subsets) in the space of linear projections.

REMARK (6). The projection property III implies the cone property that if $A \subset \mathbb{R}^n$ has the property (*) and is bounded, then the cone $C(A) \subset \mathbb{R}^{n+1}$ has the same (*). This is therefore true for semi-algebraicity and subanalyticity, but not for semi-analyticity. (See Example 1.)

## 2. Complex-proper image theorem

Let $X$ be a real-analytic space, and let $\tilde{X}$ be a complexification of $X$ with autoconjugation $\sigma$. Let us recall the meanings of these notions. For this, we first take the most simple but typical example as follows.

EXAMPLE 1. Let $X = \mathbb{R}^n$, viewed as a real-analytic space. Then we can take $\tilde{X} = \mathbb{C}^n$ as its complexification. Note that we have the complex conjugation map $\sigma$ of $\mathbb{C}^n$ into itself, defined by

$$\sigma(x_1, x_2, \cdots, x_n) = (\bar{x}_1, \bar{x}_2, \cdots, \bar{x}_n).$$

$\mathbb{C}^n$ as complex space, has its structure sheaf $\mathcal{H}_{\tilde{X}}$ which is the sheaf of germs of holomorphic functions in $\mathbb{C}^n$. $\sigma$ extends to an automorphism of $\tilde{X}$ as ringed space over $\mathbb{R}$ (but not as an automorphism of complex space, i.e. or ringed space over $\mathbb{C}$), as follows. Take any open subset $W$ of $\mathbb{C}^n$. Then the action of $\sigma$ on $\mathcal{H}_{\tilde{X}}$ or the pull-back of functions by $\sigma$ as automorphism of ringed space) is defined by

$$h \in \mathcal{H}_{\tilde{X}}(\sigma(W)) \xrightarrow[\sigma^*]{} h(\bar{x}) \in \mathcal{H}_{\tilde{X}}(W)$$

(Bar denotes the complex conjugation in $\mathbb{C}$). Note that this $\sigma^*$ is an isomorphism which induces the complex conjugation in the constants $\mathbb{C}$. Also note that, in terms of Taylor expansion in the coordinate functions $x_1, \cdots, x_n$ (if it is possible), $\sigma^*$ means simply replacing the coefficients by their complex conjugates. At any rate, $\sigma$ is now defined as an automorphism of $\tilde{X}$ as ringed space over $\mathbb{R}$ with respect to $\mathbb{R} \subset \mathbb{C}$. The pair $(\tilde{X}, \sigma)$ is a complexification of $X = \mathbb{R}^n$.

EXAMPLE 2. The first example can be restricted to an open subset $V$ of $\mathbb{C}^n$, symmetric about the real part $\mathbb{R}^n$ and produces a new example of complexification. Namely, if $\tilde{V} = V \cap \mathbb{R}^n$, then $\tilde{X} \mid \tilde{V}$ with autoconjugation induced by $\sigma$ of Ex. 1, is a complexification of $X \mid V$.

EXAMPLE 3. In the general case, a complexification $(\tilde{X}, \sigma_{\tilde{X}})$ of a real-analytic space $X$ is described (at least within a small neighborhood of a point $\xi \in X$ in $\tilde{X}$) as follows. Suppose (locally) $X$ is given by an open subset $V \subset \mathbb{R}^n$ and by a finite system of real-analytic functions $(f_1, \cdots, f_m)$ which generate

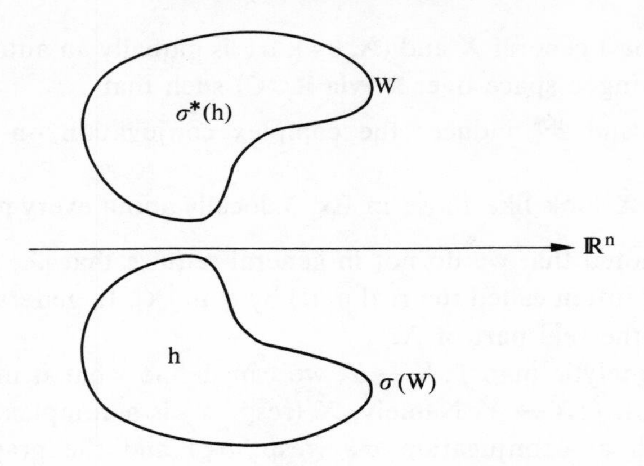


the ideal of $X$ in $V$. That $(\tilde{X}, \sigma_{\tilde{X}})$ within some neighborhood of $|X| \cap V$ will look like the following. Namely, there exists an open subset $\tilde{V}$ of $\mathbb{C}^n$ such that

(a) $V = \tilde{V} \cap \mathbb{R}^n$.
(b) $\tilde{V}$ is symmetric about the real part $\mathbb{R}^n$.

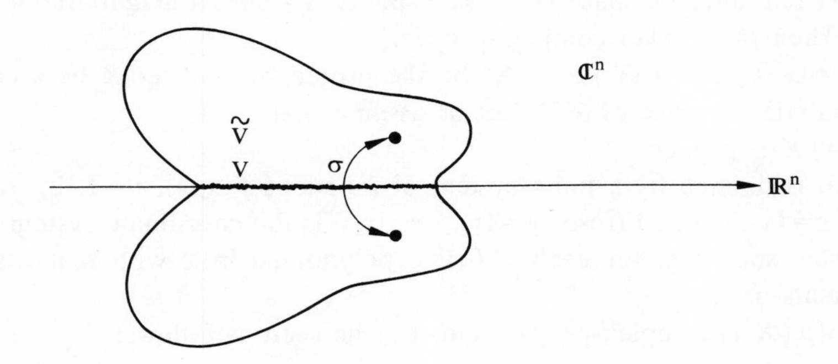


(c) $f_1, \cdots, f_m$ extend to holomorphic functions on $\tilde{V}$ (for simplicity we assume $\tilde{V}$ is connected, so that the extensions are unique. The complex-analytic space $\tilde{X}$ is then defined by

$$|\tilde{X}| = \{x \in \tilde{V} \mid f_1(x) = \cdots = f_m(x) = 0\}$$

and

$$\mathcal{H}_{\tilde{X}} = \mathcal{H}_{\tilde{V}}/(f_1, \cdots, f_m)\mathcal{H}_{\tilde{V}}$$

which is restricted to its support $|\tilde{X}|$), where $\mathcal{H}_{\tilde{V}}$ denotes the sheaf of holomorphic functions on $\tilde{V} \subset \mathbb{C}^n$. Note that $\sigma$ induces $\sigma_{\tilde{X}}$ because $\sigma^*(f_i) = f_i$.

REMARK. For a general $X$ and $(X, \sigma_{\tilde{X}})$, $\sigma_{\tilde{X}}$ is globally an automorphism of $\tilde{X}$ viewed as ringed space over $\mathbb{R}$ (via $\mathbb{R} \subset \mathbb{C}$) such that

(1) $\sigma_{\tilde{X}}^2 = id$ and $\sigma_{\tilde{X}}^*$ induces the complex conjugation on the constant functions $\mathbb{C}$.

(2) $\sigma_{\tilde{X}}$ and $\tilde{X}$ look like those in Ex. 3, locally about every point of $X$.

It should be noted that we do not in general require that $|X|$ is exactly the fixed point-set (often called the real part) by $\sigma$ in $|\tilde{X}|$. In general, $|X|$ is open and closed in the real part of $\tilde{X}$.

For a real-analytic map $f: X \to Y$, we can define what it means to be a complexification $\tilde{f}: \tilde{X} \to \tilde{Y}$. Namely, $\tilde{X}$ (resp. $\tilde{Y}$) is a complexification of $X$ (resp. $Y$) with autoconjugation $\sigma_{\tilde{X}}$ (resp. $\sigma_{\tilde{Y}}$) and the graph of $\tilde{f}$ is a complexification of that of $f$ with respect to $\sigma_{\tilde{X}}$ and $\sigma_{\tilde{Y}}$. In other words, $\tilde{f}$ is a complex-analytic map which induces $f$ and $\sigma_{\tilde{Y}}\tilde{f} = \tilde{f}\sigma_{\tilde{X}}$.

We say that a real-analytic map $f: X \to Y$ is *complex-proper* (or *admits a proper complexification*) if there exists a complexification $\tilde{f}: \tilde{X} \to \tilde{Y}$ of $f$ which is proper. Note that if $f$ is complex-proper, it is proper by itself. But the converse is not true. For instance, take the real-analytic map $f$ defined in Example 1 of the first chapter. Then take any real-analytic map $h$ from a compact real-analytic space (such as a sphere $S^r$) onto a neighborhood of 0 in $\mathbb{R}^n$. Then $fh$ is never complex-proper.

EXAMPLE 1. Let $p: \mathbb{R}^r \times \mathbb{R}^s \to \mathbb{R}^s$ be the projection and let $X$ be a closed real-analytic subspace of $\mathbb{R}^{r+s}$. Let us assume that

(1) $p \mid X$ is proper

(2) $X$ is defined by a finite system of functions $f_i(x, y)$, $i = 1, 2, \cdots, m$, where $x = (x_1, \cdots, x_r)$ (resp. $y = (y_1, \cdots, y_s)$) is the coordinate system in $\mathbb{R}^r$ (resp. $\mathbb{R}^s$), such that for each $i$, $f_i$ is a polynomial in $x$ with real-analytic coefficients in $y$.

Then $p \mid X$ is complex-proper. This can be seen as follows:

$$
\begin{array}{ccc}
\mathbb{R}^r \times \mathbb{R}^s & \xrightarrow{\ f\ } & \mathbb{R}^s \\
\cap & & \cap \\
\mathbb{P}_{\mathbb{R}}^r \times \mathbb{R}^s & \longrightarrow & \mathbb{R}^s \\
\cap & & \cap \\
\mathbb{P}_{\mathbb{C}}^r \times \tilde{V} & \xrightarrow{\ \tilde{f}\ } & \tilde{V}
\end{array}
$$

where $\mathbb{P}_{\mathbb{R}}^r$ is the real projective space which is the real part of the complex projective space $\mathbb{P}_{\mathbb{C}}^r$, $\hat{V}$ is an open symmetric neighborhood of $\mathbb{R}^s$ in $\mathbb{C}^s$ to which all the coefficients of the polynomials $f_i$'s extend as holomorphic functions, and the inclusion $\mathbb{R}^r \subset \mathbb{P}_{\mathbb{R}}^r$ is by $(x_1, \cdots, x_r) \to (1 : x_1 \cdots x_r)$. Let $(z_0, z_1, \cdots, z_r)$ be the homogeneous coordinate system for $\mathbb{P}_{\mathbb{C}}^r$ so that

$x_i = z_i \mid z_0$, $1 \leqq i \leqq r$. Let $N$ be the maximum of the degrees of the $f_i$'s in $x$. Let

$$F_i(z, y) = z_0^N f_i\left(\frac{z}{z_0}, y\right)$$

and let $\tilde{X}$ be the complex-analytic subspace of $\mathbb{P}^r_{\mathbb{C}} \times \tilde{V}$ defined by $F_i(z, y)$, $1 \leqq i \leqq m$. Then $\tilde{X}$ is a complexification of $X$ and $\tilde{p} \mid \tilde{X}$ into $\tilde{V}$ is a complexification of $p \mid X$, which is clearly proper as $\mathbb{P}^r_{\mathbb{C}}$ is compact.

EXAMPLE 2. Let $p: X \to Y$ be a real-analytic map of real-analytic spaces, and let $\tilde{p}: \tilde{X} \to \tilde{Y}$ be a complexification of $p$. Let $\xi \in X$ and $\eta = p(\xi) \in Y$. Then the following conditions are equivalent to one another: (For the proof, use Weierstrass preparation theorem.)

(1) the local ring $\mathcal{A}_{X,\xi}$ as a module over $\mathcal{A}_{Y,\eta}$ is finitely generated.

(2) $\xi$ is an isolated point of $\tilde{p}^{-1}(\eta)$.

(3) there exists an open neighborhood $\tilde{V}$ of $\xi$ in $\tilde{X}$ such that $\tilde{p} \mid \tilde{V}$ is finite-to-one map into $\tilde{Y}$.

Under one (and hence all) of these conditions, we say that $p$ is *complex-finite* at $\xi$. Suppose now that $p: X \to Y$ is complex-finite at $\xi \in X$. Then for a sufficiently small neighborhood $U$ of $\eta = p(\xi)$ in $Y$, there exists an open neighborhood $V$ of $\xi$ in $p^{-1}(U)$ such that

$$p: X \mid V \to Y \mid U$$

is complex-proper. This can be proved as follows. Since the question is local, we assume that $X$ is embedded in $Y \times \mathbb{R}^r$ so that $p$ is induced by the projection. We may also assume that $\xi = \eta \times 0$. Let $(x_1, \cdots, x_r)$ be the coordinate system in $\mathbb{R}^r$. Then the local ring of $Y \times \mathbb{R}^r$ at $\xi$ is

$$\mathcal{A}_{Y,\eta}\{x_1, \cdots, x_r\}.$$

Let $I$ be the ideal in this ring, which defines $X$. Then, by the assumption (1), we have

$$\mathcal{A}_{Y,\eta}\{x\}/I$$

as $\mathcal{A}_{Y,\eta}$-module, is generated by a finite number of monomials in $x$. This means that $I$ is generated by a finite number of polynomial in the variables $x$, with coefficients in $\mathcal{A}_{Y,\eta}$. We can then deduce the assertion in Example 2 from that of Example 1.

If $\tilde{X}$ with $\sigma$ is a complexification of a real-analytic space $X$ and if $\tilde{Z}$ is a complex subspace of $\tilde{X}$ which is invariant by $\sigma$, then $\tilde{Z}$ with the autoconjugation induced by $\sigma$ is a complexification of a real-analytic subspace $Z$ of $\tilde{X}$ where $|Z| = |\tilde{Z}| \cap |X|$. Here the $\sigma$-invariance of $\tilde{Z}$ means not only $\sigma(|\tilde{Z}|) = |\tilde{Z}|$ but also that $\sigma^*$ maps the ideal sheaf of $\tilde{Z}$ in $\mathfrak{X}_{\tilde{X}}$ into itself. The structure sheaf $\mathcal{A}_Z$ of $Z$ is then exactly the $\sigma^*$-invariant part of $\mathfrak{X}_{\tilde{Z}} \mid |Z|$. Call $Z$ the real part of $\tilde{Z}$ in $X$.

THEOREM (*Galbiati*). *Let $f: X \to Y$ be a real-analytic map which admits a proper complexification $\tilde{f}: \tilde{X} \to \tilde{Y}$. Let $\sigma$ denote the autoconjugation of $\tilde{X}$. Let $\tilde{Z}$ be a* closed *complex-analytic subspace of $\tilde{X}$ which is invariant by $\sigma$. Let $Z$ be the real part of $\tilde{Z}$ in $X$. Then $f(X-Z)$ is semi-analytic* (*as subset*) *in $Y$.*

REMARK 1. The question is local in $Y$, we may assume, with no loss of generality, that $\tilde{Y}$ and $\tilde{X}$ are countable at infinity. We do this in what follows.

REMARK 2. For simplicity in notation, we denote the autoconjugation of $\tilde{Y}$ also by $\sigma$. The theorem will be proven (under the assumption of being countable at $\infty$) in such a way that $f(X-Z)$ is *globally* semi-analytic in $Y$ *with respect to the given complexification* $\tilde{Y}$ of $Y$. This means the following:

(*) The image set is a locally finite union of subsets $A_\alpha$ which are of the following form:

$$A_\alpha = \text{a connected component of } X \cap (\tilde{F}_\alpha - \tilde{G}_\alpha)$$

where $\tilde{F}_\alpha$ and $\tilde{G}_\alpha$ are closed complex subspaces of $\tilde{Y}$ which are invariant by $\sigma$.

REMARK 3. This (*) of Remark 2 has the following implication in a special case. Namely, assume that $\tilde{Y}$ is compact algebraic; e.g. $\tilde{Y}$ is a closed complex-analytic subspace of a complex projective space $\mathbb{P}^N_{\mathbb{C}}$. Then a subset of $Y$, which is globally semi-analytic in $Y$ with respect to $\tilde{Y}$, is in fact *semi-algebraic* in $Y$ (with respect to the algebraic structure induced from $\tilde{Y}$). In fact, by *GAGA*, $\tilde{F}_\alpha$ and $\tilde{G}_\alpha$ of (*) are necessarily algebraic in $\tilde{Y}$. Therefore $X \cap (\tilde{F}_\alpha - \tilde{G}_\alpha)$ are semi-algebraic and so are $A_\alpha$. Hence any locally finite (and hence finite) union of such $A_\alpha$ is semi-algebraic.

PROOF OF THE THEOREM (cf. Remarks 1 and 2). To prove the theorem in the form of Remark 2, we may replace $\tilde{Y}$ by the union $\tilde{Y}_1 \cup \sigma(\tilde{Y}_1)$ for every irreducible component $\tilde{Y}_1$ of $\tilde{Y}$. Therefore, without any loss of generality, we may assume that $\tilde{Y}$ has a finite dimension. Similarly we may assume that $\tilde{X}$ has a finite dimension. Let $n = \dim \tilde{X}$ and $m = \dim \tilde{Y}$. The proof will be done by induction on $(n, m)$. We assume that $\tilde{X}$ and $\tilde{Y}$ are reduced, as the nilpotent functions on them have nothing to do with the assertion. Moreover, we may assume that $\tilde{Z}$ is nowhere dense in $\tilde{X}$ as we may replace $\tilde{X}$ by the closure of $\tilde{X} - \tilde{Z}$. We then use the resolution of singularities, applied to $\tilde{X}$, in the following form: There exists

$$\tilde{X}' \xrightarrow{\tilde{\pi}} \tilde{X},$$

such that

(1) $\tilde{\pi}$ is proper and $\tilde{X}'$ is smooth

(2) $\tilde{\pi}$ is isomorphic outside a nowhere dense complex-analytic subset of $\tilde{X}$

(3) the autoconjugation $\sigma$ of $\tilde{X}$ lifts to an autoconjugation $\sigma'$ of $\tilde{X}'$ in a way compatible with $\tilde{\pi}$ (i.e. $\tilde{\pi}\sigma' = \sigma\pi$).

Let $\tilde{S}$ be the set of points of $\tilde{X}$ at which $\tilde{\pi}^{-1}$ is not an isomorphism. It is a nowhere dense closed complex-analytic subspace (with reduced structure) of $\tilde{X}$ and is clearly invariant by $\sigma$. Hence $\tilde{S}$ is a complexification of its real part $S$ in $X$.

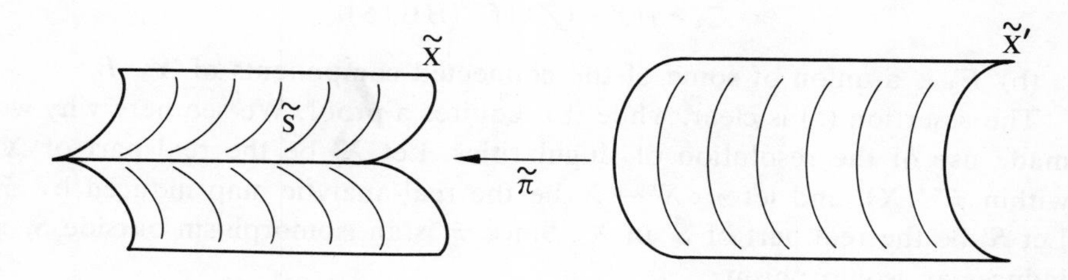


Let $\tilde{B}_1$ be the smallest closed complex-analytic subspace of $\tilde{Y}$ such that the composed map $\tilde{f}\tilde{\pi}: \tilde{X}' \to \tilde{Y}$ is smooth outside $\tilde{B}_1$. Let $\tilde{B}_2 = \operatorname{Sing}(\tilde{Y})$, the singular locus of $\tilde{Y}$ (with reduced structure. Thus $\tilde{f}\tilde{\pi}$ induces a proper and smooth (i.e. submersive) map of complex manifolds (i.e. smooth complex-analytic spaces, not necessarily connected):

$$\tilde{X}' - (\tilde{f}\tilde{\pi})^{-1}(\tilde{B}_1 \cup \tilde{B}_2) \to \tilde{Y} - (\tilde{B}_1 \cup \tilde{B}_2).$$

Next, let $\tilde{Z}' = \tilde{\pi}^{-1}(\tilde{Z})$ and $\tilde{S}' = \tilde{\pi}^{-1}(\tilde{S})$. Let $\tilde{B}_3$ be the smallest closed complex-analytic subspace of $Y$ such that the map

$$\tilde{Z}' \cup \tilde{S}' \to \tilde{Y}$$

induced by $\tilde{f}\tilde{\pi}$ is flat outside $\tilde{B}_3$.

Now, let $\tilde{B} = \tilde{B}_1 \cup \tilde{B}_2 \cup \tilde{B}_3$, It is then clear that $\tilde{B}$ is a closed complex-analytic subspace of $\tilde{Y}$ which is invariant by $\sigma$, so that it is a complexification of its real part $B$ in $Y$. It is a non-trivial fact, however well-known, that $\tilde{B}$ is nowhere dense in $\tilde{Y}$.

Let us summarize

(i) $\tilde{f}\tilde{\pi}$ induces

$$\tilde{h}: \tilde{X}' - (\tilde{f}\tilde{\pi})^{-1}(\tilde{B}) \to \tilde{Y} - \tilde{B}$$

which is a proper and smooth map of complex manifolds,

(ii) $\tilde{h}$ induces a flat map

$$(\tilde{Z}' \cup \tilde{S}') - (\tilde{f}\tilde{\pi})^{-1}(\tilde{B}) \to \tilde{Y} - \tilde{B}.$$

We now claim the following

(a) $f(X-Z)$ is the union of the following three sets:

$$E_1 = f(f^{-1}(B) \cap X - Z),$$
$$E_2 = f(S - Z),$$

and

$$E_3 = f(X - (Z \cup f^{-1}(B) \cup S)).$$

(b) $E_3$ is a union of some of the connected components of $Y-B$.

The assertion (a) is clear, while (b) requires a proof. We see here why we made use of the resolution of singularities. Let $X'$ be the real part of $\tilde{X}'$ within $\tilde{\pi}^{-1}(X)$, and let $\pi: X' \to X$ be the real-analytic map induced by $\tilde{\pi}$. Let $S'$ be the real part of $\tilde{S}'$ in $X'$. Since $\tilde{\pi}$ is an isomorphism outside $\tilde{S}$, $\pi$ induces an isomorphism

$$X' - S' \xrightarrow{\sim} X - S.$$

Therefore, we have

$$E_3 = f\pi(X' - (Z' \cup (f\pi)^{-1}(B) \cup S')),$$

where $Z'$ is the real part of $\tilde{Z}'$ in $X'$. By (i), $\tilde{h}$ (and $\tilde{f}\tilde{\pi}$) induces

$$h: X' - (f\pi)^{-1}(B) \to Y - B$$

which is a proper and smooth (i.e. submersive) real-analytic map of real-analytic manifolds. The smoothness of $h$ implies that $E_3$ is open in $Y-B$. Let us prove that $E_3$ is closed in $Y-B$ (and hence (b) follows).

Take any point $\eta \in Y - B$ which is in the closure of $E_3$. Since $h$ is proper, there exists a point $\xi \in h^{-1}(\eta)$ which belongs to the closure of

$$X' - (Z' \cup (f\pi)^{-1}(B) \cup S').$$

Let $\tilde{F}' = \tilde{h}^{-1}(\eta)$, which is a complexification of $F' = h^{-1}(\eta)$. Let $\tilde{G}' = \tilde{F}' \cap \tilde{Z}' \cup \tilde{S}')$. What we want to show is that

$$F' \not\subset \tilde{G}',$$

which implies

$$\eta \in h(F' - \tilde{G}') \subset E_3$$

and completes the proof of (b). Since $\xi \in F'$, $F'$ is not empty. Since $\tilde{F}'$ is a complexification of a smooth $F'$, any complex subspace of $\tilde{F}'$ containing $F'$ should contain some neighborhood of $F'$ in $\tilde{F}'$. In particular, if we did have

$F' \subset \tilde{G}'$, then $\tilde{G}'$ should contain a component of $\tilde{F}'$. But this is impossible because $\tilde{Z}' \subset \tilde{S}'$ is nowhere dense in $\tilde{X}'$ and $\tilde{h}: \tilde{Z}' \cup \tilde{S}' \to \tilde{Y}$ is flat outside $\tilde{B}$. Thus we have completed the proof of (b).

The proof of the theorem is now quickly completed as follows. To $E_1$, we apply the induction on $m$. Namely, we take $\tilde{f}^{-1}(\tilde{B}) \to \tilde{B}$, which is a complexification of $f^{-1}(B) \to B$, and $\tilde{f}^{-1}(\tilde{B}) \cap \tilde{Z}$ in the place of $\tilde{f}$ and $\tilde{Z}$ of the theorem, where $\dim \tilde{B} < m$. To $E_2$, we apply the induction on $n$. Namely, we take $\tilde{S} \to \tilde{Y}$ and $\tilde{S} \cap \tilde{Z}$, where $\dim \tilde{S} < n$. These, combined with (b), we get the theorem on $\tilde{f}$ and $\tilde{Z}$ as is restated in Remark 2.

COROLLARY 1. *Let $p: \mathbb{R}^r \times \mathbb{R}^s \to \mathbb{R}^s$ be the projection. Let $A$ be a semi-analytic set in $\mathbb{R}^r \times \mathbb{R}^s$ which is* relatively algebraic *with respect to $p$, i.e. every point $\eta \in \mathbb{R}^s$ has an open neighborhood $U$ such that*

$$A \cap (\mathbb{R}^r \times U) = \bigcup_i \{f_{ij} = 0, \forall_j, g_{ik} > 0, \forall_k\},$$

*where $f_{ij}$ and $g_{ik}$ are finitely many polynomials in the coordinate functions in $\mathbb{R}^r$ whose coefficients are real-analytic functions in $U$. Then $p(A)$ is semi-analytic in $\mathbb{R}^s$.*

PROOF. The question is local in $\mathbb{R}^s$ and every point in $\mathbb{R}^s$ has an arbitrarily small neighborhood which is real-analytically isomorphic to $\mathbb{R}^s$. Also $p(\bigcup_i A_i) = \bigcup_i p(A_i)$, and therefore we may assume that

$$A = \{f_j = 0, j \in J, g_k > 0, k \in K\}$$

with finite index sets $J$ and $K$, where $f_j$ and $g_k$ are polynomials in the variables $(x_1, \cdots, x_r)$ for $\mathbb{R}^r$ whose coefficients are real-analytic functions in $\mathbb{R}^s$. Let $(y_1, \cdots, y_s)$ be the coordinate system for $\mathbb{R}^s$. We also let $K = \{1, 2, \cdots, d\}$ and let $(t_1, \cdots, t_d)$ be the coordinate system in $\mathbb{R}^d$. We define the subset $B$ of $\mathbb{R}^r \times \mathbb{R}^d \times \mathbb{R}^s$ defined by

$$\{f_j(x, y) = 0, \forall_j \in J, g_k(x, y) + t_k^2 = 0, \forall_k \in K\}.$$

Let $C = B \cap (\mathbb{R}^r \times (0) \times \mathbb{R}^s)$. If $g: \mathbb{R}^{r+d} \times \mathbb{R}^s \to \mathbb{R}^s$ is the projection then $p(A) = q(B - C)$. For simplicity, let $n = r + d$ and let $x_{r+i} = t_i$, $1 \leqq i \leqq d$. We now take standard embedding $\mathbb{R}^n \subset \mathbb{P}^n_{\mathbb{R}} \subset \mathbb{P}^n_{\mathbb{C}}$ such that if $(z_0, \cdots, z_n)$ is the homogeneous coordinate system of $\mathbb{P}^n_{\mathbb{C}}$ then $z_i/z_0$ induces $x_i$, $1 \leqq i \leqq n$. Take $N \gg 0$ so that

$$F_j(z, y) = z_0^N f_j(z/z_0, y)$$

and $H_k(z, y) = z_0^N g_k(z/z_0, y) + z_0^N (z_{r+k}/z_0)^2$ are homogeneous polynomials in $z$. Now let $X$, resp. $\tilde{X}$, be the real-analytic, resp. complex-analytic, subspace of $\mathbb{P}^n_{\mathbb{R}}$, resp. $\mathbb{P}^n_{\mathbb{C}}$, defined by the ideal generated by $F_j$, $j \in J$, and $H_k$, $k \in K$. Let

$Z$, resp. $\tilde{Z}$, be the subspace of $X$, resp. $\tilde{X}$, defined by the ideal generated by

$$(z_0 z_{r+1}, \cdots, z_0 z_n) = (z_0) \cap (z_{r+1}, \cdots, z_n).$$

Take also the standard complexification $\mathbb{R}^s \subset \mathbb{C}^s$. Let $f: X \to Y = \mathbb{R}^s$, and $\tilde{f}: \tilde{X} \to \tilde{Y} = \mathbb{C}^s$ be the ones induced by the projection $\mathbb{P}^n_{\mathbb{C}} \times \mathbb{C}^s \to \mathbb{C}^s$. Clearly $\tilde{f}$ is a proper complexification of $f$ and $B - C = X - Z$. Thus the theorem can be applied to $f$, $\tilde{f}$ and $\tilde{Z}$ so defined and it implies Corollary 1.

COROLLARY 2. *Let $p: \mathbb{R}^r \to \mathbb{R}^s$ be a polynomial map and let $A$ be a semi-algebraic subset of $\mathbb{R}^r$. Then $p(A)$ is semi-algebraic in $\mathbb{R}^s$.*

PROOF. By the method similar to the one used in the proof of Corollary 1, we are reduced to the following problem: We have a closed real-algebraic subspace $X_0$ of $\mathbb{P}^n_{\mathbb{R}} \times \mathbb{R}^s$ and a closed real-algebraic subspace $Z_0$ of $X_0$. If $q_0: \mathbb{P}^n_{\mathbb{R}} \times \mathbb{R}^s \to \mathbb{R}^s$ is the projection, then $q_0(X_0 - Z_0)$ is semi-algebraic in $\mathbb{R}^s$.

We can then use the inclusion $\mathbb{R}^s \subset \mathbb{P}^s_{\mathbb{R}}$, and reduce the problem as follows: We have a closed real-algebraic subspace $X$ of $\mathbb{P}^n_{\mathbb{R}} \times \mathbb{P}^s_{\mathbb{R}}$ and a closed real-algebraic subspace $Z$ of $X$. If $q: \mathbb{P}^n_{\mathbb{R}} \times \mathbb{P}^s_{\mathbb{R}} \to \mathbb{P}^s_{\mathbb{R}}$ is the projection, then $q(X - Z)$ is semi-algebraic in $\mathbb{P}^s_{\mathbb{R}}$. (Take a compactification $X$ of $X_0$ in $\mathbb{P}^n_{\mathbb{R}} \times \mathbb{P}^s_{\mathbb{R}}$ and let $Z$ be the union of a compactification of $Z_0$ and $X \cap \mathbb{P}^n_{\mathbb{R}} \times (\mathbb{P}^s_{\mathbb{R}} - \mathbb{R}^s)$.) Then we take a complexification $\tilde{X}$ of $X$, $\tilde{Z}$ of $Z$, which are closed complex-algebraic subspaces of $\mathbb{P}^n_{\mathbb{C}} \times \mathbb{P}^s_{\mathbb{C}}$, and then take $\tilde{Y} = \mathbb{P}^s_{\mathbb{C}}$. Now, $\tilde{X} \to \tilde{Y}$ induced by the projection is a complexification of $X \to Y = \mathbb{P}^s_{\mathbb{R}}$. Corollary 2 follows from the theorem as restated in Remark 2 in view of Remark 3.

COROLLARY 3. *In the theorem, let us assume that $\tilde{Y}$ is compact and admits an algebraic structure. (We do not require that $\tilde{X}$ admits any algebraic structure.) Then $f(X - Z)$ is semi-algebraic in $Y$ with respect to the induced algebraic structure from $\tilde{Y}$.*

PROOF. Use Remarks 2 and 3.

## 3. Whitney stratifications

Let $E \subset \mathbb{R}^n$ be a subanalytic set.

DEFINITION 1. A *subanalytic stratification* of $E$ is by definition a decomposition

$$E = \bigcup_{\alpha} E_{\alpha}$$

such that

(1) for each $\alpha$, $E_{\alpha}$ is a connected real-analytic submanifold of $\mathbb{R}^n$, which is subanalytic in $\mathbb{R}^n$.

(2) (frontier condition) If $\bar{E}_{\alpha} \cap E_{\beta} \neq \varnothing$, then $\bar{E}_{\alpha} \supset E_{\beta}$.

(3) $\{E_{\alpha}\}$ are mutually disjoint and locally finite at every point of $\mathbb{R}^n$.

NOTE 1. We call each $E_{\alpha}$ a stratum of the stratification of $E$, or simply a stratum of $E$ when a stratification of $E$ is given.

Let $M$ be a connected real-analytic submanifold of $\mathbb{R}^n$ and let $N$ be any subset of $\bar{M}-M$. Let $\xi \in N$.

DEFINITION 2. We say that the pair $(M, N)$ satisfies the *Whitney condition* at $\xi$, if the following condition is satisfied:

(*) Take any sequence of points $\{x_i\}_{i\in\mathbb{N}}$ in $M$ and $\{y_i\}_{i\in\mathbb{N}}$ such that

(a) $\lim_{i\to\infty} x_i = \lim_{i\to\infty} y_i = \xi$

(b) $\lim l_{x_i y_i} = \tilde{l}$ exists

(c) $\lim_{i\to\infty} T_{M,x_i} = \tilde{T}$ exists.

Then $\tilde{l} \subset \tilde{T}$.

NOTE 2. In this definition, $l_{ab}$ denote the line passing through $a$ and $b$ in $\mathbb{R}^n$. The tangent space of $\mathbb{R}^n$ at every point is identified with $\mathbb{R}^n$ itself in a natural way (with reference to the linear structure of $\mathbb{R}^n$). Hence we view the tangent space $T_{M,x_i}$ as a $d$-plane in $\mathbb{R}^n$, where $d = \dim M$.

NOTE 3. We call $E = \bigcup_\alpha E_\alpha$ of def. 1 a *Whitney stratification* of $E$ if for every pair $(\alpha, \beta)$ such that $\bar{E}_\alpha - E_\alpha \supset E_\beta$, the pair $(E_\alpha, E_\beta)$ satisfies the Whitney condition at every point $E_\beta$.

In this lecture, our main interest is the following question: Say we have a real-analytic map of real-analytic manifolds, $\pi: X \to Y$. For simplicity, assume that $X$ and $Y$ are compact. Then we ask if we can find a Whitney *subanalytic* stratification of $X$ and such of $Y$, respectively, in such a way that $\pi$ induces a surjective submersion (i.e. smooth map) from each of the strata of $X$ to one of the strata of $Y$.

NOTE 4. Here we cannot in general expect that there exist *semi-analytic* stratifications having the required properties. In other words, the *subanalyticity* is the best we can hope for.

The answer to the above question is affirmative. Namely, to be precise, we can prove

THEOREM. *Let $\pi:\mathbb{R}^n \to \mathbb{R}^m$ be a real-analytic map. Let $E \subset \mathbb{R}^n$ and $F \subset \mathbb{R}^m$ be bounded subanalytic sets, such that $\pi(E) \subset F$. Suppose we are given finitely many subanalytic sets $A_i$ in $\mathbb{R}^n$, with $A_i \subset E$, and finitely many subanalytic sets $B_j$ in $\mathbb{R}^m$, $B_j \subset F$. Then there exist Whitney subanalytic stratifications:*

$$E = \bigcup_\alpha E_\alpha \quad \text{in } \mathbb{R}^n, \qquad F = \bigcup_\beta F_\beta \quad \text{in } \mathbb{R}^m,$$

*such that*

(1) *the stratification of $E$ (resp. $F$) is* compatible *with each of the $A_i$ (resp. $B_j$),*

(2) *for every* $\alpha$, *there exists* $\beta$ *such that* $\pi$ *induces* $E_\alpha \to F_\beta$ *which is surjective and* submersive.

NOTE 5. A stratification $E = \bigcup_\alpha E_\alpha$ is said to be *compatible* with $A_i$ if for every $\alpha$ we have either $E_\alpha \subset A_i$ or $E_\alpha \cap A_i = \emptyset$.

NOTE 6. A real-analytic map of real-analytic manifolds $E_\alpha \to F_\beta$ is said to be *submersive*, if locally at every point of $E_\alpha$ the map can be identified with a projection $\mathbb{R}^r \times \mathbb{R}^s \to \mathbb{R}^r$. This is equivalent to saying that the differential of the map at every point of $E_\alpha$ is surjective from the tangent space of $E_\alpha$ at the point to that of $F_\beta$ at the image point.

The key step in the proof of the theorem consists of the following two lemmas, the first of which is useful for other purposes as well.

RECOVERING LEMMA. *Let* $D$ *be a closed subanalytic set in* $\mathbb{R}^n$. *Then there exists a real-analytic map* $h: X \to \mathbb{R}^n$, *from some real-analytic space* $X$, *together with a closed real-analytic subspace* $S$ *of* $X$, *such that*

(1) $h$ *is proper and* $D = h(X) = h(X - S)$

(2) $X - S$ *is* smooth *as a real-analytic space*

(3) *the map of real-analytic manifolds* $X - S \to \mathbb{R}^n$, *induced by* $\pi$, *is* immersive.

NOTE 7. A real-analytic space (with structure sheaf) is said to be *smooth* if it is locally isomorphic to a ball in some $\mathbb{R}^d$ (where $d$ denotes the dimension of the local ring of the real-analytic space at the point in question).

NOTE 8. A real-analytic map, like $X - S \to \mathbb{R}^n$, is said to be *immersive*, if for every point $\xi$ of $X - S$, there exist an open neighborhood $V$ of $\xi$ in $X - S$ and a closed real-analytic subspace, say $Z$, of some open subset of $\mathbb{R}^n$, such that the given map induces an isomorphism $X \mid V \to Z$.

EXAMPLE. Let $D$ be the set

$$\{(x, y, z) \in \mathbb{R}^3 \mid x^2 - zy^2 = 0, z \geqq 0\}.$$

Then, let $X$ be the disjoint union of 3 real-analytic manifolds, $X_0$, $X_1$, $X_2$, where $X_0$ is a point, $X_1 = \mathbb{R}$ and $X_2$ is the normalization of the surface in $\mathbb{R}^3$ defined by the ideal $(x^2 - zy^2)$. Let $h: X \to \mathbb{R}^3$ be defined as follows: $h$ maps $X_0$ to the origin of $\mathbb{R}^3$, $X_1$ to the positive half of the $z$-axis (i.e. $x = y = 0$, $z \leqq 0$) by $t \in \mathbb{R} \to (0, 0, t^2) \in \mathbb{R}^3$, and $h \mid X_2$ is the normalization map. Let $S$ be the union of the inverse image of the $z$-axis in $X_2$ and the origin in $X_1 = \mathbb{R}$. It can be seen that $h: X \to \mathbb{R}^3$ and $S$ have all the properties of the above lemma. (Note that the normalization loses the negative part of the $z$-axis which is included in the surface of $x^2 - zy^2 = 0$.)

NOTE 9. Using the resolution of singularities in the real-analytic case, we can even require that $X$ of the lemma is smooth (but not connected).

GENERIC WHITNEY LEMMA. *Let $M$ be a connected real-analytic submanifold (smooth) of $\mathbb{R}^n$, which is subanalytic in $\mathbb{R}^n$. Let $N \subset \bar{M} - M$, which is subanalytic in $\mathbb{R}^n$. Assume that $N$ is locally closed (for simplicity). Then there exists a closed subset $S$ of $N$ such that*

(1) *$S$ is subanalytic in $\mathbb{R}^n$*

(2) *$N - S$ is dense in $N$*

(3) *$(M, N)$ satisfies the Whitney condition at every point of $N - S$.*

Using these two lemmas, we now want to prove the theorem.

REMARK 1. The Recovering Lemma implies the following fact: Let $D$ be as in the lemma. Then there exists a closed subset $D'$ of $D$, subanalytic in $\mathbb{R}^n$, such that $D - D'$ is dense in $D$ and is smooth (i.e. it coincides with a locally closed real-analytic submanifold, not necessarily connected, of $\mathbb{R}^n$).

PROOF. Let the notation be the same as in the Recovering Lemma. Let us take the fibre product:

$$\begin{array}{ccc} W = X \times_{\mathbb{R}^n} X & \xrightarrow{q} & \mathbb{R}^n \\ {\scriptstyle p_1} \swarrow \quad \searrow {\scriptstyle p_2} & & \\ X \qquad X & & \end{array}$$

where $p_1$ and $p_2$ are projections and $q = hp_1 = hp_2$. Then there exists a closed real-analytic subspace $V$ of $W$ such that

(a) $V \supset (S \times_{\mathbb{R}^n} X \cup X \times_{\mathbb{R}^n} S) = K$, say.

(b) for a point $\zeta \in W - K$, $p_1$ is submersive at $\zeta$ if and only if $\zeta \in W - V$.

We can then prove that $D' = q(V)$ has the required property.

REMARK 2. In the theorem, we can assume that $E$ and $F$ are closed. Take $F$ for instance. Let $F_0 = \bar{F}$, $F_0' = \overline{F_0 - F}$, $F_1 = \overline{F_0 \cap F}$, and inductively

$$F_i = \overline{F_{i-1}' \cap F},$$

$$F_i' = \overline{F_i - F},$$

for all $i \geqq 1$. Then we have $F = \bigcup_i (F_i - F_i')$, a disjoint union. Moreover, $\dim F_{i+1} < \dim F_i'$ until we reach the empty set, so that the set $\{F_i, F_i'\}$ is finite. To prove the theorem as stated, it is enough to prove the theorem after replacing $F$ by $F_0$ and $\{B_j\}$ by $\{B_j, F_i, F_i'\}$. The argument is quite similar for $E$. The point is that any stratification of $F_0$, compatible with all the $F_i$

and $F'_i$, induces a stratification of $F$ (automatically). Let us further remark that we can also assume that all the $A_i$ and $B_j$ are closed, without any loss of generalities.

We first consider the case in which $\pi = id$, $E = F$ and $\{A_i\} = \{B_j\}$.

PROPOSITION 1. *Let $E$ be a compact subanalytic set in $\mathbb{R}^n$. Let $\{A_i\}$ be finitely many subsets of $E$, which are all subanalytic in $\mathbb{R}^n$. Suppose we have a closed subset $E'$ of $E$, subanalytic in $\mathbb{R}^n$, and a Whitney subanalytic stratification of $E - E'$,*

$$E - E' = \bigcup_{\alpha \in \Lambda}, E_\alpha$$

*which is compatible with $A_i \cap (E - E')$ for all $i$. Then there exists a Whitney subanalytic stratification of $E$,*

$$E = \bigcup_{\alpha \in \Lambda} E_\alpha,$$

*where $\Lambda'$ is a subset of $\Lambda$, which is compatible with $A_i$ for all $i$. (Note that the strata of $E - E'$ are also taken as some of the strata of $E$.)*

PROOF. The proof will be done by induction on $r = \dim E'$. Therefore, it is enough to prove the following:

(*) There exists a closed subset $E''$ of $E'$, subanalytic in $\mathbb{R}^n$, such that
(a) $E' - E''$ is dense in $E'$
(b) $E' - E''$ is smooth real-analytic in $\mathbb{R}^n$
(c) if $\{E_\beta\}_{\beta \in \Gamma}$ is the set of connected components of $E' - E''$, then

$$E - E'' = \bigcup_{\alpha \in \Lambda \cup \Gamma} E_\alpha$$

is a Whitney subanalytic stratification compatible with all the $A_i \cap (E - E'')$.

To prove (*), we first take $G_1 \subset E'$, closed and subanalytic, such that $E' - G_1$ has the properties (a) and (b). We shall obtain $E''$ by adding some sets to $G_1$. Let $C_\alpha = \bar{E}_\alpha - E_\alpha$ for each $\alpha \in \Lambda'$. Let us define

$$A_i = (E' - G_1) \cap E' - A_i \cap A_i,$$
$$\tilde{C}_\alpha = (E' - G_1) \cap \overline{E' - C\alpha} \cap C_\alpha.$$

Then $\bigcup_i \tilde{A}_i \bigcup_\alpha \tilde{C}_\alpha, = H$ say, is a nowhere dense closed subanalytic subset of $E' - G_1$. Next, let $N_\alpha$ be a closed nowhere dense subset of $C_\alpha \cap E'$, subanalytic in $\mathbb{R}^n$, such that $(E_\alpha, C_\alpha \cap E')$ satisfies the Whitney condition at every point of $C_\alpha \cap E' - N_\alpha$. Clearly $N_\alpha$ is nowhere dense in $E' - G_1$. Set $E' = G_1 \cup H \cup (\bigcup_{\alpha \in \Lambda}, N_\alpha)$, then $E'$ has all the required properties (a), (b), and (c) of (*).

REMARK 3. Given a stratification $E=\bigcup_\alpha E_\alpha$ as in Def. 1., a *refinement* of it means another stratification (always subanalytic) $E=\bigcup_\beta E'_\beta$, which is compatible with $E_\alpha$ for all $\alpha$. Now, a closed subanalytic subset $E'$ of $E$ being given, let $E=\bigcup_\alpha E_\alpha$ be a Whitney subanalytic stratification of $E$ which is compatible with $E'$. Let

$$E'=\bigcup_{\alpha\in\Lambda'} E_\alpha$$

and

$$E-E'=\bigcup_{\alpha\in\Lambda''} E_\alpha.$$

Now take any refinement of the first one:

$$E'=\bigcup_{\gamma\in\Gamma} E'_\gamma$$

which is Whitney subanalytic. Then the combined decomposition

$$E=(\bigcup_{\alpha\Lambda''} E_\alpha)\cup(\bigcup_{\gamma\in\Gamma} E'_\gamma)$$

is a Whitney subanalytic stratification of $E$.

PROOF OF THE THEOREM. By Remark 2, we assume that $E$ and $F$ are closed and hence compact. The theorem will be proven by induction on $r=\dim F$. We proceed in the following steps:

STEP 1. Take any Whitney subanalytic stratification

$$F=\bigcup_\beta F'_\beta$$

which is compatible with all the $B_j$ and with $\pi(E)$. Given such, we define

$$U'=\bigcup_{\beta\in\Delta'} F'_\beta$$

where $\Delta'=\{\beta \mid \dim F_\beta=r \text{ and } F_\beta\subset\pi(E)\}$.

STEP 2. Take any Whitney subanalytic stratification

$$E=\bigcup_\alpha E'_\alpha$$

which is compatible with all the $A_i$ and also with all the $\pi^{-1}(F'_\beta)$. We then divide the set of the strata $\{F'_\beta\}$ into two parts.

$$\Gamma_0=\{\alpha \mid \pi(E'_\alpha) \text{ contains an open non-empty subset of } U'\}$$

$$\Gamma_1=\{\alpha \mid \pi(E'_\alpha)\cap U' \text{ is nowhere dense in } U'\}$$

for each $\alpha\in\Gamma_0$, let $S_\alpha$ be the critical set in $E'_\alpha$ for the map $E'_\alpha\to U'$, induced by $\pi$. We know that $\pi(S_\alpha)$ is nowhere dense in $U'$. Now, let $D$ be the union of all the sets as follows:

$$\pi(E'_\alpha), \forall \alpha \in \Gamma_1, \quad \text{and} \quad \pi(S_\alpha), \forall \alpha \in \Gamma_0.$$

Then $D$ is a closed subanalytic set in $\mathbb{R}^n$, such that $D \cup U = \pi(E)$ and such that $D \cap U$ is nowhere dense.

STEP 3. We take a Whitney subanalytic refinement

$$F = \bigcup_\gamma F''_\gamma$$

of the stratification of $F$ in Step 1, such that it is compatible with the set $D$ of Step 2. We then define $U''$ for this stratification in the same way as we did $U$ in Step 1. We let $U'' = \bigcup_{\gamma \in \Delta''} F_\gamma$. Clearly $U'' \subset U' - D$. Therefore if we let $\hat{E}_\alpha = E'_\alpha \cap \pi^{-1}(U'')$ for $\alpha \in \Gamma_0$, then

(a) $\hat{E}_\alpha \to U$, induced by $\pi$, is submersive.

(b) $\bigcup_{\alpha \in \Gamma_0} \hat{E}_\alpha = \pi^{-1}(U'')$ and $\pi^{-1}(U'') \to U''$ is proper.

It follows from these two facts, that we have

(c) for every $\alpha \in \Gamma_0$, every connected component of $\hat{E}_\alpha$ is mapped *onto* a connected component of $U''$, i.e. onto a stratum $F''_\gamma$ contained in $U''$.

Let us take the collection of all the connected components of $\hat{E}_\alpha$ for all $\alpha \in \Gamma_0$, and call it $\{E''_\varepsilon\}_{\varepsilon \in \mathscr{C}}$.

Let $F^* = \pi(E) - U''$ and $E^* = \pi^{-1}(F^*)$. Then we have (in view of (a), (b), (c))

(i) $\{E''_\varepsilon\}_{\varepsilon \in \mathscr{C}}$ is a Whitney subanalytic stratification of $E - E^*$, compatible with all the $A_i$,

(ii) $\{F''_\gamma \mid F''_\gamma \cap F^* = \varnothing\}$ is a Whitney subanalytic stratification of $F - E^*$, compatible with all the $B_j$,

(iii) for each $\varepsilon$, there exists $F''_\gamma$ in the family of (ii) such that $E''_\varepsilon \to F''_\gamma$, induced by $\pi$, is surjective and submersive.

STEP 4. (Final) We apply the induction assumption to $\pi : \mathbb{R}^n \to \mathbb{R}^m$, $E^* \subset \mathbb{R}^n$, $F^* \subset \mathbb{R}^m$, together with the families of subsets:

$$\{E'_\alpha, \alpha \in \Gamma_1; \bar{E}''_\alpha \cap E^*, \alpha \in \Gamma_1\} \text{ in } \mathbb{R}^n$$

and

$$\{F''_\gamma \mid F''_\gamma \subset F^*\} \text{ in } \mathbb{R}^m.$$

(Note: $\dim F^* < \dim F = r$.) Simply combining the result of Step 4 with those strata in (i) and (ii), we obtain $\{E_\alpha\}$ and $\{F_\beta\}$, respectively, having the properties required in the theorem. (cf. Remark 3 about combined stratifications.)

Let us now assume that there is given a Whitney subanalytic stratification of a real-analytic map of real-analytic spaces, say $\pi : X \to Y$, i.e. there are given Whitney subanalytic stratifications

$$X = \bigcup_{\alpha} X_\alpha \quad \text{and} \quad Y = \bigcup_{\beta} Y_\beta$$

such that for every $\alpha$ there exists $\beta$ for which $\pi$ induces a surjective submersion $X_\alpha \to Y_\beta$. We then often ask if the stratification satisfies the *$A_\pi$-condition* (or *Thom condition*) with respect to the map $\pi$.

To be precise,

DEFINITION 3. *Let M and N be two real-analytic subspaces, locally closed, in X. Assume that M is smooth, that the differential of $\pi \mid M$ has constant rank on each connected component of M, and that $N \subset \bar{M} - M$. We say that (M, N) satisfies the $A_\pi$-condition at $\eta \in N$ (with respect to the map $\pi$) if the following conditions are satisfied:*

(1) *Within a sufficiently small neighborhood of $\eta$ in N, N is smooth and the differential of $\pi \mid N$ has constant rank.*

(2) *Take any sequence of points $\{x_i\}$ in M such that $\lim_i x_i = \eta$ and such that the limit of the tangent spaces of the fibres*

$$\tilde{T} = \lim_{i \to \infty} T_{M \cap \pi^{-1}(\pi(x_i)),\, x_i}$$

*exists. Then we have $\tilde{T} \supset T_{N \cap \pi^{-1}(\pi(\eta)), \eta}$.*

We say that $(M, N)$ satisfies the $A_\pi$-condition if it does so at every point $\eta \in N$.

DEFINITION 4. *We say that a Whitney subanalytic stratification of $\pi : X \to Y$, say $X = \bigcup_\alpha X_\alpha$ and $Y = \bigcup_\beta Y_\beta$, satisfies the $A_\pi$-condition, if every pair $(X_\alpha, X_{\alpha'})$ with $X_{\alpha'} \subset \bar{X}_\alpha - X_\alpha$ satisfies the $A_\pi$-condition.*

EXAMPLE (a). Let $n \geqq 2$ and let $S$ be the unit sphere with center at the origin in $\mathbb{R}^n$. Let

$$\pi_1 : \mathbb{R} \times S \to \mathbb{R}^n$$

by $(t, v) \to tv$ (scalar multiplication by $t$ of the unit vector $v \in S$). This is a degree 2 proper map of real-analytic manifolds, which is called the *double oriented blowing-up* of $\mathbb{R}^n$ with center $0 \in \mathbb{R}^n$. It has the following two properties:

(1) $\pi_1$ induces a doubly sheeted covering (locally isomorphic)

$$\mathbb{R} \times S - \pi_1^{-1}(0) \to \mathbb{R}^n - 0.$$

(The source of this map has two connected components, $\mathbb{R}_+ \times S$ and $\mathbb{R}_- \times S$, where $\mathbb{R}_+ = \{t \in \mathbb{R} \mid t > 0\}$ and $\mathbb{R}_- = t \in \mathbb{R} \mid t < 0\}$.)

(2) $\pi_1^{-1}(0) = 0 \times S = S$

The map $\pi_1$ induces

$$\pi_2 : \mathbb{R}_0 \times S \to \mathbb{R}^n$$

with $\mathbb{R}_0 = \{t \in \mathbb{R} \mid t \geqq 0\}$, which is called the *simple oriented blowing-up* (or the *oriented blowing-up*). It also induces

$$\pi_3 : X = \mathbb{R} \times S/\mathbb{Z}_2 \to \mathbb{R}^n$$

where $\mathbb{Z}_2 = \mathbb{Z}/(\ )\mathbb{Z} = \{\pm 1\}$ (the last is the multiplicative group) acts on $\mathbb{R} \times S$ by

$$(t, v) \mapsto (-t, -v).$$

This $\pi_3$ is called the *non-oriented blowing-up* (or, simply, the *blowing-up*) of $\mathbb{R}^n$ with center $0 \in \mathbb{R}^n$. Note that $X$ is a real-analytic manifold and $\pi_3$ has the following property:

(1*) $\pi_3$ induces an isomorphism

$$X - \pi_3^{-1}(0) \xrightarrow{\sim} \mathbb{R}^3 - 0$$

(2*) $\pi_3^{-1}(0) = \mathbb{P}_{\mathbb{R}}^{n-1}$ (= the real-projective space obtained from $S$ by identifying the antipodal pairs of points). Either one of the maps $\pi_1$ or $\pi_3$ has the following local picture. The lines through 0 in $\mathbb{R}^n$ are all separated to form a family of lines parametrized by $S$ (resp. $\mathbb{R}$).

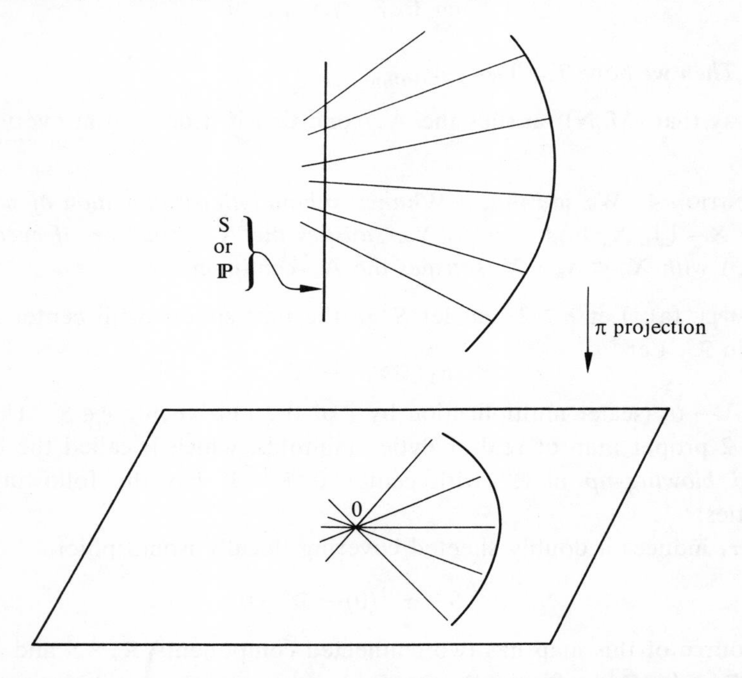


Remark (a). For $\pi = \pi_1$ or $\pi_3$, we can easily prove that every Whitney stratification of $\pi$ fails to satisfy the $A_\pi$-condition. Namely, we can find a pair of strata $M$, $N$ in the source $X$ of the map $\pi$, such that $N \subset \bar{M}$, $M \subset \pi^{-1}(\mathbb{R}^n - 0)$ and $N \subset \pi^{-1}(0)$. Note that the fibres in $M$ with respect to $\pi$

are single points while the fibre in $N$ (which is $N$ itself) is $(n-1)$-dimensional. It is obvious that any limit of tangents to the first cannot contain the tangent space to the second so long as $n \geqq 2$.

EXAMPLE (b). Let $Y=\mathbb{C}^n$, $n \geqq 2$. We have a canonical map $\mathbb{C}^n-0 \rightarrow (\mathbb{C}^n-0)/\mathbb{C}^* = \mathbb{P}_{\mathbb{C}}^{n-1}$. We get a complex manifold $X$ in $\mathbb{C}^n \times \mathbb{P}_{\mathbb{C}}^{n-1}$ by taking the closure of the graph of the map. (The projection $X \rightarrow \mathbb{P}_{\mathbb{C}}^{n-1}$ is in fact a complex line bundle, dual to the one associated with hyperplanes in $\mathbb{P}_{\mathbb{C}}^{n-1}$.) The projection $\pi: X \rightarrow \mathbb{C}^n$ is called the *blowing-up* (or the *complex blowing-up*) of $\mathbb{C}^n$ with center 0. Analogous to Example (b), we have

(1) $\pi$ induces an isomorphism

$$X-\pi^{-1}(0) \longrightarrow \mathbb{C}^n-0.$$

(2) $\pi^{-1}(0)=\mathbb{P}_{\mathbb{C}}^{n-1}$.

REMARK (b). Every Whitney stratification of $\pi$ fails to satisfy $A_\pi$ for Example (b). Let us consider the following situation: Suppose $n \geqq 3$, and let $g: \mathbb{C}^n \rightarrow {}^{n-1}$ be the projection, say to the first $(n-1)$-factor. Take the map

$$f=g \circ \pi: X \rightarrow \mathbb{C}^{n-1}$$

with the $\pi$ of Example (b). This time, we let $Y=\mathbb{C}^{n-1}$. Note that $\dim_{\mathbb{C}} f^{-1}(y)=1$ got $y \in Y-0$ but $\dim_{\mathbb{C}} f^{-1}(0)=n-1$. ($f^{-1}(0)$ is the union of a complex line, $\approx \mathbb{C}$, and $\mathbb{P}_{\mathbb{C}}^{n-1}=\pi^{-1}(0)$, the two being connected at a point.) It can easily be proven that $f$ does not admit any Whitney stratification satisfying $A_f$. But now let us take the base change

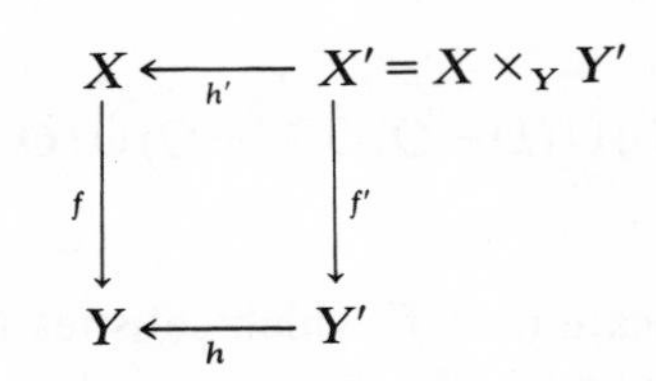

where $C_{\eta'}$ is the complex line in $T_{Y,\eta}=\mathbb{C}^{n-1}=Y$, which corresponds to admits a Whitney stratification satisfying $A_{f'}$. To be precise, let $\eta=0 \in Y$ and let $\xi_0=0 \in \mathbb{C}^n$. Then $\pi^{-1}(\xi_0)=\mathbb{P}_{\mathbb{C}}^{n-1}, =\mathbb{P}$ for simplicity. Let $Z=h^{-1}(\eta)=\mathbb{P}_{\mathbb{C}}^{n-2}, =S$ for simplicity. Then $X'$ is the union of two submanifolds

$$\mathbb{P}'=\mathbb{P} \times S=(h')^{-1}(\mathbb{P})$$

and $X''=$ (the closure of $X'-\mathbb{P}'$ in $X'$). Moreover, $X'' \cap \mathbb{P}'=Q$, say, is a locally trivial $\mathbb{P}_{\mathbb{C}}^1$-bundle over $S$. The points in $S$ correspond to the complex lines in $\mathbb{C}^{n-1}=Y$ through the origin $\eta$, and the fibre in $Q$ above each point

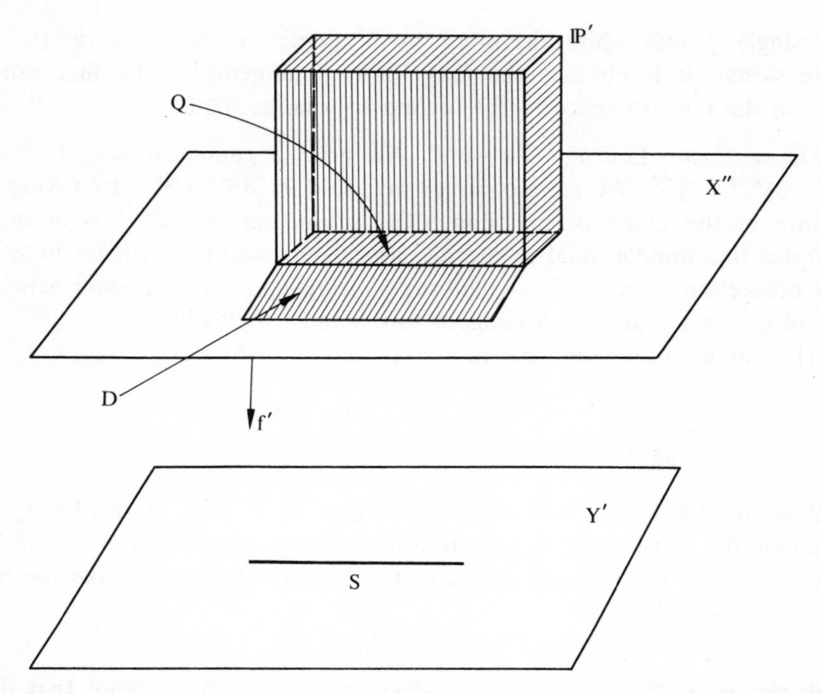


$\eta' \in S$ is the associated complex projective line of the vector space ($\approx \mathbb{C}^2$):

$$(dg)^{-1}(C_{\eta'}),$$

where $C_{\eta'}$ is the complex line in $T_{Y,\eta} = \mathbb{C}^{n-1} = Y$, which corresponds to $\eta' \in S$. Moreover, $(f')^{-1}(S) = \mathbb{P}' \cup D$ where $D = \mathbb{C} \times S$. Now, take the following stratifications:

$$X' = (X' - (f')^{-1}(S)) \cup (D - Q) \cup (\mathbb{P}' - Q) \cup (Q - D) \cup (Q \cap D),$$
$$Y' = (Y' - S) \cup S.$$

This is a Whitney stratification of $f'$ which satisfies the $A_{f'}$-condition. (The smooth lines $(f')^{-1}(y')$, $y' \in Y' - S$, degenerate into conics with ordinary double points $(f')^{-1}(\eta') \cap X''$, $\nu' \in S$.)

In the following two sections, we shall discuss the kind of base changes as the one used in this last remark in connection with $A_f$-conditions.

## 4. Flattening in the complex-analytic case

We shall be mostly interested in a complex-analytic map of complex-analytic spaces $f: X \to Y$. When $Y$ is smooth at a point $y$, i.e. the local ring $A = \mathfrak{X}_{Y,y}$ is regular, $f$ is flat at a point $x \in f^{-1}(y)$ if and only if a regular system of parameters $(t_1, \cdots, t_m)$ of $A$ (or the local coordinate system $Y$, having $y$ as its origin, with reference to an identification of a neighborhood of $y$ in $Y$ with an open ball in $\mathbb{C}^m$, $m = \dim_{\mathbb{C}} Y$ at $y$) is a regular sequence for $B = \mathfrak{X}_{X,x}$. This implies, for instance, that if $Y$ is smooth and $X$ is Cohen-Macauley (or, in particular, smooth, then the flatness of $f$ at $x \in X$ is equivalent to saying that, within a sufficiently small neighborhood of $x$ in $X$, the fibres of $f$ have the same dimension.

REMARK 1. Let $f: X \to Y$ be a general complex-analytic map, except that we assume $Y$ to be reduced (i.e. the structure sheaf $\mathfrak{X}_Y$ has no nilpotent elements, i.e. no local section of $\mathfrak{X}_Y$, $\neq 0$, represents a function taking the value 0 identically). In this case, $f$ is flat if and only if for every complex-analytic $h: Y' \to Y$, with the derived diagram:

$$\begin{array}{ccc} X & \xleftarrow{h'} & X' = X \times_Y Y' \\ {\scriptstyle f}\downarrow & & \downarrow{\scriptstyle f'} \\ Y & \xleftarrow{h} & Y' \end{array}$$

the map $f'$ is flat. In this criterion, what is more interesting is that we may take only those $h$ with $Y' = \mathbb{D}$ (the unit disc in $\mathbb{C}$). Moreover, when $Y = \mathbb{D}$, the flatness of $f$ is equivalent to saying that no irreducible components (even the imbedded ones) are mapped to a point in $\mathbb{D}$. In any case, this is so after reducing the structure sheaf of $X$ by killing nilpotents in it. It is easy to deduce from this fact, that

(*) If $f: X \to Y$ is any complex-analytic map of complex-analytic spaces (with no added assumptions), which is flat, than for every $h: Y' \to Y$ as above the derived map $f'$ has the following topological property: For every $y' \in Y'$ with $y' \in \overline{Y' - y'}$, we have

$$(f')^{-1}(y') \subset \overline{X' - (f')^{-1}(y')}$$

where bar denotes the closure (in the sense of topological space) and the inclusion is set-theoretical.

In the actual point-set geometry, this consequence (*) of the flatness is often the property that is most important. This property (for all $h$) is often referred to as the *universal openness* of $f$. In other words, as a consequence of the flatness, we get

(**) (*universal openness*) For every $h: Y' \to Y$ as above, the derived map $f': X' \to Y'$ is an open map (i.e. maps open subsets into open subsets) when $f'$ is viewed only as a map of topological spaces.

It should be noted that this consequence (**) of the flatness holds true only in the *complex-analytic* case, but not in the *real-analytic* case (mainly due to the lack of a Nullstellen Satz).

EXAMPLE 1. If $f$ is the map defined in the last remark (Remark (b) of §3), then the conclusion of (*) fails for $h:\mathbb{C}\to Y$ which is the inclusion of a complex line through the origin. In the real case, the projection of the unit sphere in $\mathbb{R}^{n+1}$ to the unit disc in $\mathbb{R}^n$ is not an open map, but it is flat as a real-analytic map of real-analytic manifolds.

REMARK 2. Let $f:X\to Y$ be a surjective complex-analytic map of connected complex-analytic manifolds. In this case, if $f$ admits a Whitney stratification satisfying the $A_f$-condition, then $f$ is flat. In fact, since both $Y$ and $X$ are smooth (all we need for $X$ is the Cohen-Macauley property), the flatness of $f$ is equivalent to the universal openness (or the property of (*), Remark 1). In particular, if $f$ is not flat, then there exists a point $\xi\in X$ such that in any small neighborhood $V$ of $\xi$ in $X$, there exists a point $\xi'\in V$ such that the dimension of $f^{-1}(f(\xi))$ at $\xi$ is bigger than that of $f^{-1}(f(\xi'))$ at $\xi'$. It then follows that the $A_f$-condition fails at $\xi'$ in any Whitney stratification of $f$.

REMARK 3. Let $f:X\to Y$ be the same as in Remark 2. If there exist triangulations of $X$ and $Y$, respectively, so as to make $f$ a simplical map, then $f$ is again flat. Let $n$, $m$ be the complex dimensions of $X$, $Y$, respectively. Let $S$ be the set

$$\{\xi\in X \mid \dim_{\xi} f^{-1}(f(\xi))>n-m\}.$$

Then we see that $S$ is a closed complex-analytic subset of $X$ and that if $f$ is not flat then $S\neq\emptyset$. Assume that $f$ is not flat and hence $S\neq\emptyset$. Taking $S$ as a reduced complex-analytic subspace of $X$, we can find a connected non-empty open subset $V$ of $S$ such that $S\mid V$ is smooth and the differential $df\mid V$ has constant rank. Suppose we had triangulations of $X$ and $Y$ for which $f$ is simplical. Then the triangulation of $X$ must be compatible with $S$, so that there should exist a simplex $\sigma$ of dimension $=2\dim_C(S\mid V)$, such that $\sigma\subset S$ and $\sigma\cap V\neq\emptyset$. Then pick any simplex $\tau$ of $\dim=2n$ such that $\sigma\subset\partial\tau$. By the definition of $S$,

$$\dim_{\mathbb{R}} f(\sigma)<\dim_{\mathbb{R}}\sigma-2(n-m)$$

while

$$\dim_{\mathbb{R}} f(\tau)=2m=\dim_{\mathbb{R}}\tau-2(n-m).$$

This is absurd. As a matter of fact, we can find a two-simplex, which is either equal to $\tau$ or is in $\partial\tau$, whose image by $f$ should look like:

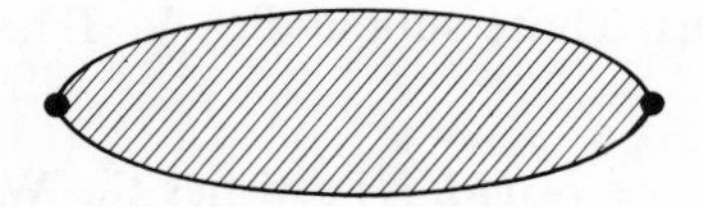

(one of the two end points is the image of some side in the two-simplex).

REMARK 4. When we are dealing with complex-analytic spaces, we can use complex tangent spaces and complex lines connecting points in the definition of Whitney condition and Thom condition. The complex definition is in fact equivalent to the real one, when complex-analytic spaces are viewed as real-analytic spaces (with real dimensions doubling complex dimensions).

We say that a complex-analytic subspace $M$ of a complex-analytic space $X$ has $\mathbb{C}$-*boundary*, if $M$ can be expressed as $M = M_1 - M_2$ where $M_1$, $M_2$ are two closed complex-analytic subspaces of $X$. For instance, a complex submanifold $M$ of $X$ has $\mathbb{C}$-boundary if and only if both $\bar{M}$ and $\bar{M} - M$ are closed complex-analytic subsets of $X$.

*The generic Whitney lemma* in the complex-analytic case can be stated as follows.

LEMMA 1. *Let $M$ be a connected complex-analytic submanifold of a complex-analytic space $X$. Let $N$ be a closed complex-analytic subspace of $X$, contained in $\bar{M} - M$. Then there exists a closed complex-analytic subspace $S$ of $N$ such that*

*(1) $S$ is nowhere dense in $N$*

*(2) $(M, N)$ satisfies Whitney condition in some open neighborhood of $y \in N$ in $N$ if and only if $y \in N - S$.*

PROOF. Clearly we may replace $X$ by $\bar{M}$ with reduced complex structure. Moreover, since $S$ is globally defined by (2) as a subset of $N$, the other assertions ((1) and the complex-analyticity of $S$) are of local nature. Therefore, we may assume that $X$ is embedded as a locally closed complex-analytic subspace in some $\mathbb{C}^m$. Let $\sigma$ be the map

$$M \times N \to \mathbb{P}_{\mathbb{C}}^{m-1} \times \text{Grass}\,(m, n),$$

defined by $(x, y) \to (L_{xy}, T_{M,x})$ where $L_{xy}$ is the translation through the origin of the complex line connecting $x$ and $y$ in $\mathbb{C}^N$ and $T_{M,x}$ is the tangent space of $M$ at $x$ which is naturally identified as a complex vector subspace of $\mathbb{C}^m$, where $n = \dim_{\mathbb{C}} M$. Let $W$ be the closure of the graph of $\sigma$ in $X \times N \times \mathbb{P}_{\mathbb{C}}^{m-1} \times \text{Grass}\,(m, n)$. Then we obtain two complex vector subbundles $V$ and $T$ of the trivial $\mathbb{C}^m$-bundle on $W$ as follows. For each point $w \in W$, the fibre $V_w$ of $V$ at $w$ is the complex line in $\mathbb{C}^m$ corresponding to the projection of $w$ to $\mathbb{P}_{\mathbb{C}}^{m-1}$, while $T_w$ is the complex $n$-plane in $\mathbb{C}^m$, corresponding to the projection of $w$ to Grass $(m, n)$. Those subbundles $V$, $T$ have the following virtue for our problem:

NOTE (a). For a point $y \in N$, $(M, N)$ satisfies the Whitney condition at $y$ if and only if

$$T_w \supset V_w, \forall w \in q^{-1}(y, y),$$

where $q: W \to X \times N$ denotes the projection.

In fact, every sequence of points in $M \times N$ which converges to $y \times y$, has a subsequence, say $\{(x_i, y_i)\}_{i \in \mathbb{N}}$ which is the image by $q$ of a convergent sequence $\{w_i\}_{i \in \mathbb{N}}$ in $W$. We then see that if $w = \lim_{i \to \infty} w_i$, then

$$\lim_{i \to \infty} T_{M,x_i} \supset \lim_{i \to \infty} L_{x_i y_i} \Leftrightarrow T_w \supset V_w$$

This proves Note (a). Now, clearly

$$B = \{w \in W \mid T_w \supset V_w\}$$

is a closed complex-analytic subset of $W$. By Note (a) the set $S$ (defined by (2) of Lemma 1) is obtained as the set of those $y \in N$ such that there exists an open neighborhood $E$ of $y$ in $N$ for which $q^{-1}(y', y') \subset B$ for all $y' \in E$. Since $q$ is proper, $S$ is a closed complex-analytic subset of $N$. By the generic Whitney lemma in the real-analytic (or subanalytic) case, $N - S$ is dense in $N$. (This can also be proven directly without using the corresponding lemma in the real case, in fact, with a proof analogous and considerably simpler.)

Once we have this lemma, the following theorem can be proven in essentially the same way as the corresponding theorem in the real case which we proved in §3.

THEOREM 1. *Let $f: X \to Y$ be a proper complex-analytic map of complex-analytic spaces. Then there exist Whitney stratifications*

$$X = \bigcup_{\alpha} X_\alpha \quad \text{and} \quad Y = \bigcup_{\beta} Y_\beta$$

*such that*

(1) *$X_\alpha$ (resp. $Y_\beta$) is a connected complex-analytic submanifold of $X$ (resp. $Y$) which has $\mathbb{C}$-analytic boundary in $X$ (resp. $Y$).*

(2) *for every $\alpha$, there exists $\beta$ for which $f$ induces a surjective submersion $X_\alpha \to Y_\beta$.*

TERMINOLOGY. The pair of stratifications having all the properties of this theorem will be called a Whitney $\mathbb{C}$-analytic stratification of $f$.

In the next section, we shall consider the question of the $A_f$-condition for such a stratification of $f: X \to Y$ (which means $X = \bigcup_\alpha X_\alpha$ and $Y = \bigcup_\beta Y_\beta$ having the properties of Theorem 1). For the proof of the theorems concerning the $A_f$-condition in §5, we need the technique of complex-analytic *flattening* which we shall now discuss.

Let us examine, once again, the following local situation discussed in the Appendix on flatness

$$\begin{array}{ccc} \xi = \eta \times 0 \in X & \xrightarrow[j]{} & Y \times \mathbb{C}^n \\ {\scriptstyle f}\downarrow & \swarrow {\scriptstyle \text{projection}} & \\ \eta = f(\xi) \in Y & & \end{array}$$

where $j$ is a locally closed imbedding and $f$ is induced by the projection. As was shown in the Appendix on flatness, with a suitable coordinate system $(t) = (t_1, \cdots, t_n)$ in $\mathbb{C}^n$, we have a natural homomorphism

$$\sum_{i=0}^{n} \sum_{\alpha \in F_i} t^\alpha A\{t(i)\} \xrightarrow{\kappa} B.$$

where $A = \mathfrak{X}_{Y,\eta}$, $B = \mathfrak{X}_{X,\xi} = A\{t\}/I$ with the ideal $I$ of $X$ in the local ring $A\{t\}$ of $Y \times \mathbb{C}^n$ at $\xi$, and $t(i) = (t_{i+1}, \cdots, t_n)$. This $\kappa$ has the properties listed in Theorem 1 of the Appendix. We then defined a complex-analytic subspace $Z$ of $Y$ within a sufficiently small neighborhood of $\eta$, whose ideal $D(\kappa)$ at $\eta$ in $A$ was obtained from Ker $(\kappa)$ as follows: Take any $g \in \operatorname{Ker}(\kappa)$ and write it as

$$g = \sum_{i=0}^{n} \sum_{\alpha \in F_i} t^\alpha \sum_{\gamma \in \mathbb{Z}_0^{n-1}} g_{i\alpha\gamma} t(i)^\gamma$$

with $g_{i\alpha\gamma} \in A$. The ideal $D(\kappa)$ is then generated by all the $g_{i\alpha\gamma}$ for all $g \in \operatorname{Ker}(\kappa)$. We have seen an intrinsic characterization of $Z$ with respect to $f: X \to Y$ and $\xi \in X$, which was stated as Theorem 2 in the Appendix. In view of this, we give the following

DEFINITION 1. Let $f: X \to Y$ be a complex-analytic map of complex-analytic spaces. Let $\xi \in X$, and $\eta = f(\xi) \in Y$. Then a locally closed complex-analytic space $Z$ of $Y$ in some neighborhood of $\eta$ is said to be a *flattener* of $f$ with respect to the given point $\xi$, if the following condition is satisfied: For any complex-analytic map $h: Y' \to Y$ with $\eta' \in h^{-1}(\eta)$, the map $f'$ in the diagram

$$\begin{array}{ccc} X & \xleftarrow[h']{} & X' = X \times_Y Y' \\ {\scriptstyle f}\downarrow & & \downarrow{\scriptstyle f'} \\ Y & \xleftarrow{h} & Y' \end{array}$$

is flat at $\xi' = \xi \times \eta'$ if and only if the map of germs $h:(Y', \eta') \to (Y, \eta)$ factors through $(Z, \eta)$.

NOTE. The germ $(Z, \eta)$ is uniquely determined by $f$ and $\xi$, and from time to time we write $Z(f, \xi)$ for $Z$ to indicate its reference to $f$ and $\xi$.

The explicit construction of the ideal $D(\kappa)$ of $Z(f, \xi)$, obtained after localizing $X$ about $\xi$ and choosing an imbedding $j: X \to Y \times \mathbb{C}^n$ as before, has another important application to the problem of flattening.

Namely, let $f: X \to Y$, $\xi \in X$ and $Z = Z(f, \xi)$ be the same as in Definition 1. Then, by restricting $Y$ to any open neighborhood of $\eta = f(\xi)$ in which $Z$ is closed, we take the blowing-up with center $Z$, say

$$q: Y_1 \to Y.$$

Consider the cartesian diagram:

$$\begin{array}{ccc} X & \xleftarrow{q_1} & X_1 = X \times_Y Y_1 \\ {\scriptstyle f}\downarrow & & \downarrow{\scriptstyle f_1} \\ Y & \xleftarrow{q} & Y_1 \end{array}$$

Then let $X^*$ be the 'smallest' *closed* complex-analytic subspace of $X_1$ (i.e. the one with the 'largest' ideal sheaf) such that

$$X^* \mid V = X_1 \mid V$$

for the open subset $V = f_1^{-1}(Y_1 - q^{-1}(Z))$. For each $\eta_1 \in q^{-1}(\eta)$, we denote

$$(X^*)_{\eta_1} = (f_1)^{-1}(\eta_1) \cap X^*$$

which is the fibre in $X^*$ for $f_1$ above $\eta_1$. The corresponding fibre in $X_1$,

$$(X_1)_{\eta_1} = (f_1)^{-1}(\eta_1)$$

can be canonically (by $q_1$) identified with the fibre $X_\eta = f^{-1}(\eta)$ because the diagram is obtained by the fibre product. Thus, via $q_1$, we see it as

$$(X^*)_{\eta_1} \subset X_\eta$$

(closed imbedding). Now we can state

THEOREM 2. *For every point $\eta_1 \in q^{-1}(\eta)$, the fibre $(X^*)_{\eta_1}$ is* strictly *smaller than the corresponding fibre $X_\eta$ at $\xi_1 = \xi \times \eta_1$.*

PROOF. The question being local at $\xi$ in $X$, we take an imbedding $j: X \to Y \times \mathbb{C}^n$ and define a homomorphism $\kappa$ as before. We shall use the same notation such as $A$, $B$ etc. as before. Let $A_1$ (resp. $B_1$) *be the local ring of* $Y_1$ (resp. $X_1$) at $\eta_1$ (resp. $\xi_1$). We have a surjection $\kappa_1$, derived from $\kappa$, as follows:

$$\sum_{i=0}^{n}\sum_{\alpha\in F_i} t^{\alpha}A_1\{t(i)\}\xrightarrow{\kappa_1} B_1.$$

By Proposition 12 of the Appendix on flatness, we have $D(\kappa_1)=D(\kappa)A_1$. Since $q$ is the blowing-up of the ideal $D(\kappa)$ (i.e. blowing-up whose center $Z$ is defined by $D(\kappa)$), $D(\kappa)A_1$ is invertible (i.e. free of rank 1) as $A_1$-module. In other words, $D(\kappa_1)$ is generated by a single element, say $e$, which is not a zero-divisor in $A_1$. In view of the way in which $D(\kappa_1)$ is obtained from the elements in Ker $(\kappa_1)$, it follows that there exists $g\in \text{Ker}(\kappa_1)$ such that if we write

$$g=\sum_{i=0}^{n}\sum_{\alpha\in F_i}\sum_{\gamma\in Z_0^{n-1}} t^{\alpha}g_{i\alpha\gamma}t(i)^{\gamma}$$

with $g_{i\alpha\gamma}\in A_i$, then $(e)A_1=(g_{i\alpha\gamma})A_1$ for at least one $(i, \alpha, \gamma)$. This means that we have $g'$ in

$$\sum_{i=0}^{n}\sum_{\alpha\in F_i} t^{\alpha}A_1\{t(i)\}=\Delta_1, \text{ say,}$$

such that $g=eg'$ and $g'_{i\alpha\gamma}$ is a unit in $A_1$ for at least one $(i, \alpha, \gamma)$. In other words, $\kappa_1(g')\in B_1$ is annihilated by $e$ and is not contained in $m_1B_1$ with $m_1=\max(A_1)$. (This last assertion is due to the fact that $\kappa_1$ induces an isomorphism modulo $m_1$; see (6) of Theorem 1 of the Appendix on flatness.) Now, $e$ generates the ideal of $q^{-1}(Z)$ in $Y_1$ locally about $\eta_1$, and hence the pullback of $e$ to $X_1$ induces a non-vanishing (or invertible) function on $V=(f_1)^{-1}(Y_1-q^{-1}(Z))$. Hence, by the definition of $X^*\subset X_1$, all the annihilators of $e$ in $B_1$ should induce zero in the local ring, say $B^*$, of $X^*$ at $\xi_1$. In particular, $\kappa_1(g')$ for the above $g'$ is mapped to zero in $B^*$. On the other hand, as $\kappa_1(g')\not\equiv 0$ modulo $m_1B_1$, the natural homomorphism (surjective)

$$B_1/m_1B_1\to B^*/m_1B^*$$

has a non-trivial kernel (containing the image of $\kappa_1(g')$). This means $(X^*)_{\eta_1}$ at $\xi_1$ is strictly smaller than $(X_1)_{\eta_1}$, i.e. strictly smaller than $X_\eta$ at $\xi=q_1(\xi_1)$. Theorem 2 is now established.

REMARK 5. In the above Theorem 2, it is quite possible that $Z$ contains an open neighborhood of $\eta$ in $Y$ (point-set theoretically). In this case, $q^{-1}(\eta)$ is empty and Theorem 2 is trivially true. If $Y$ is reduced, however, this can happen only when $f$ is flat at the given point $\xi\in X$. Thus no flattening problem exists at least about $\xi$ in $X$.

With Theorem 2, we are ready to state and prove a flattening theorem (local version) as follows.

STEP 0. Start with any complex-analytic map $f: X \to Y$, a point $\eta \in Y$ and a compact subset $K(\neq \emptyset)$ in $f^{-1}(\eta)$. We assume that $Y$ is reduced. We want to 'flatten' $f$ around $K$. (If $f$ is proper, we get the best result by taking $K = f^{-1}(\eta)$.)

STEP 1. Pick any $\xi \in K$ at which $f$ is not flat, if it exists at all. Then pick a sufficiently small open neighborhood $V$ of $\eta$ in $Y$ such that we have $Z = Z(f, \xi)$ as a closed complex-analytic subspace of $Y \mid V$. If $Z$ contains any non-empty open subset of $V$, then divide the set of irreducible components of $Y \mid V$ into two, one consisting of those contained in $Z$ and the other consisting of the rest. Thus we can write $Y \mid V = Y_1 \cup Y_2$ with closed reduced complex-analytic subspaces $Y_i$ of $Y \mid V$, $i = 1, 2$, such that $Y_1 \subset Z$ and such that $Z \cap Y_2$ is nowhere dense in both $Y_1$ and $Y_2$. Let $h^{(1)}: Y^{(1)} \to Y$ be the blowing-up of $Y \mid V$ with center $Z \cap Y_2$, composed with the inclusion $Y \mid V \subset Y$. Then take the diagram

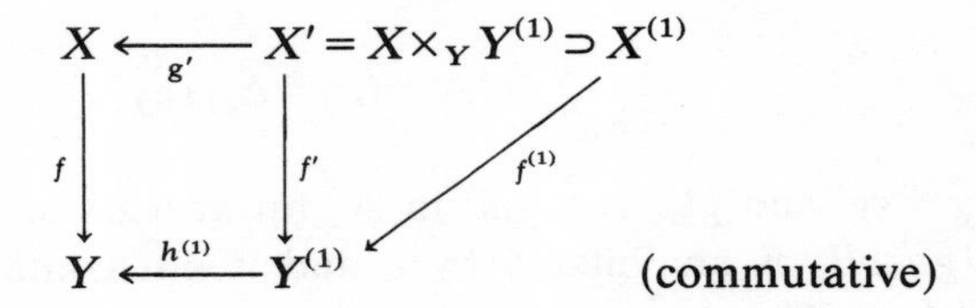


where the left square is the cartesian diagram and $X^{(1)}$ is the smallest *closed* complex-analytic subspace of $X'$ such that $X'$ and $X^{(1)}$ have the same restriction to the open dense subset

$$(f')^{-1}(Y^{(1)} - (h^{(1)})^{-1}(Z \cap Y_2)).$$

We keep the resulting diagram

$$\begin{array}{ccc} X & \xleftarrow{g^{(1)}} & X^{(1)} \\ {\scriptstyle f}\downarrow & & \downarrow{\scriptstyle f^{(1)}} \\ Y & \xleftarrow{h^{(1)}} & Y^{(1)} \end{array}$$

NOTE 1. Let $Y_1' \to Y_1$ be the blowing-up with center $Y_1 \cap Y_2 = Y_1 \cap (Z \cap Y_2)$. Let $Y_2' \to Y_2$ be the blowing-up with center $Z \cap Y_2$. Then $Y^{(1)}$ is a *disjoint* union of $Y_1'$ and $Y_2'$, and $h^{(1)}$ induces those two blowing-ups. Let $X_i' = (f')^{-1}(Y_i')$. Then $X'$ is a disjoint union of $X_1'$ and $X_2'$, and $f'$ induces $f_i': X_i' \to Y_i'$, $i = 1, 2$. Since $Z \supset Y_1$, $f$ induced to $Y_1$ is flat at $\xi$. Since the flatness is preserved by any base change, $f_1'$ is flat at every point of

$X_1' \cap (g')^{-1}(\xi)$. Hence $X^{(1)} \to X'$ is isomorphic around any point in $X_1' \cap (g')^{-1}(\xi)$. Write $X^{(1)}$ as a disjoint union on $X_i^{(1)}$, $i=1, 2$, where $X_i^{(1)}$ is the inverse image of $X_i'$. Then we have

(a) $f^{(1)}$ is flat at every point of

$$(g^{(1)})^{-1}(\xi) \cap X_1^{(1)}.$$

NOTE 2. Let us now look at the other part $X_2^{(1)}$. Let $X_2 = f^{-1}(Y_2)$ and let $f_2: X_2 \to Y_2$ be the map induced by $f$. Then in view of the intrinsic characterization of $Z$, it is easy to see that

$$Z \cap Y_2 = Z(f_2, \xi).$$

Hence the blowing-up $Y_2' \to Y_2$ is the one considered in Theorem 2 with respect to $f_2$. Hence, by the theorem, we have

(b) For every point $\xi'$ of

$$(g^{(1)})^{-1}(\xi) \cap X_2^{(1)}$$

the fibre of $f^{(1)}$ through $\xi'$ is strictly smaller than the corresponding fibre of $f$ through $\xi$, even locally at $\xi'$.

NOTE 3. It is important for later applications, that the center of the blowing-up $h^{(1)}: Y^{(1)} \to Y$ is nowhere dense in $Y$.

STEP 2. Pick any point $\eta^{(1)} \in (h^{(1)})^{-1}(\eta)$ and let

$$K^{(1)} = (g^{(1)})^{-1}(K) \cap (f^{(1)})^{-1}(\eta^{(1)})$$

which is compact. We then pick any $\xi^{(1)} \in K^{(1)}$ at which $f^{(1)}$ is not flat, if it exists at all. We then apply the same process to

$$\{f^{(1)}, \eta^{(1)}, K^{(1)}, \xi^{(1)}\}$$

as we did in Step 1 to

$$\{f, \eta, K, \xi\}$$

so that we obtain a diagram

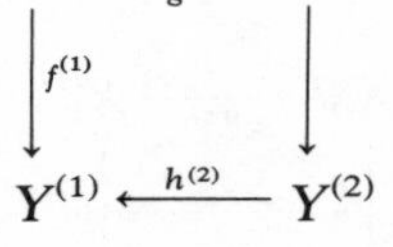

$$\begin{array}{ccc} X^{(1)} & \xleftarrow{\;g^{(2)}\;} & X^{(2)} \\ \downarrow{\scriptstyle f^{(1)}} & & \downarrow \\ Y^{(1)} & \xleftarrow{\;h^{(2)}\;} & Y^{(2)} \end{array}$$

(which corresponds to the second diagram in Step 1).

General Step. Repeat Step 2 (inductively) so as to produce

$$\begin{array}{ccc} X^{(j)} & \xleftarrow{g^{(j+1)}} & X^{(j+1)} \\ \downarrow{\scriptstyle f^{(j)}} & & \downarrow{\scriptstyle f^{(j+1)}} \\ Y^{(j)} & \xleftarrow{h^{(j+1)}} & Y^{(j+1)} \end{array}$$

so long as we have a point $\xi^{(j)} \in K^{(j)}$ at which $f^{(j)}$ is not flat.

Theorem 3. *However we choose* $\{\eta^{(j)}\}_{j=1,2,\cdots}$, *the process terminates after a finite number of steps, i.e. for some j we find that* $f^{(j)}$ *is flat at every point of the compact set* $K^{(j)}$.

Proof. Suppose we had an infinite sequence of steps with some $\{\eta^{(j)}\}_{j=1,2,\cdots}$. Then we obtain an infinite sequence of strictly decreasing closed complex-analytic subspaces

$$F^{-1}(\eta) \supsetneq F^{(1)} \supsetneq F^{(2)} \supsetneq \cdots,$$

where $F^{(j)} = (f^{(j)})^{-1}(\eta^{(j)})$ is identified with its isomorphic image in $f^{-1}(\eta)$ with respect to the maps $g^{(i)}$, $1 \leqq i \leqq j$, for each $j$. But this is impossible within a sufficiently small neighborhood of a compact $K$.

Remark 6. The advantage of Theorem 3 is that the process (and the proof) are so simple, while the disadvantage is its *strictly local* nature. More on the side of advantage is that $f$ is not assumed to be *proper*, so that the theorem applies to a complexification $f: X \to Y$ of a real-analytic map of real-analytic spaces, say $f_0: X_0 \to Y_0$. Assuming that $f_0$ is proper, we can take $K = f_0^{-1}(\eta)$ for a given point $\eta \in Y_0$. It is important to note then, that for every $\xi \in K$ (real point) $Z(f, \xi)$ can be chosen to be invariant under the autoconjugation of $Y$, where $Y$ denotes a complexification of a real-analytic subspace of $Y_0$. It then follows that $Y_1$ and $Y_2$ of Step 1 are both invariant under the autoconjugation. Thus the center $Z \cap Y_2$ of the blowing-up $h^{(1)}$ is also invariant under the auto-conjugation, which implies that the autoconjugation of $Y$ lifts to an autoconjugation of $Y^{(1)}$. Likewise, the autoconjugation of $X$ lifts to a such of $X'$. Further – in view of the definition of $X^{(1)}$, the autoconjugation of $X'$ induces a such of $X^{(1)}$. By repeating this, we get a sequence of diagrams of real-analytic maps

$$\begin{array}{ccc} X_0^{(j)} & \xleftarrow{g^{(j+1)}} & X_0^{(j+1)} \\ \downarrow{\scriptstyle f_0^{(j)}} & & \downarrow{\scriptstyle f_0^{(j+1)}} \\ Y_0^{(j)} & \xleftarrow{h_0^{(j+h)}} & Y_0^{(j+1)} \end{array}$$

of which the diagram in the General Step in the complex case is exactly the complexification.

REMARK 7. When $f: X \to Y$ is a *proper* complex-analytic map, where $Y$ is countable at $\infty$, we have the *global flattening theorem* as is proven in the paper: 'Flattening theorem in complex-analytic geometry', Am. J. Math., Vol. 97, No. 2, pp. 503–547. The statement of the theorem is as follows: Let $B$ be a closed complex-analytic subspace of $Y$. We assume that $X$, $Y$, and $B$ are all reduced and that every irreducible component of $X$ is mapped onto $Y$. We want to turn

$$f^{-1}(B) \to B$$

into a flat map by applying a locally finite (but, possibly, globally infinite) sequence of blowing-ups on $Y$. To be precise, we have a well ordered index set $\Lambda$ with a minimal element $0 \in \Lambda$. Let $f^{(0)} = f$, $X^{(0)} = X$, $Y^{(0)} = Y$ and $B^{(0)} = B$. For each $\alpha \in \Lambda$ with the next largest element $\alpha + 1 \in \Lambda$, we have

$$\begin{array}{ccc} X^{(\alpha)} & \xleftarrow{g^{(\alpha+1)}} & X^{(\alpha+1)} \\ \downarrow{\scriptstyle f^{(\alpha)}} & & \downarrow{\scriptstyle f^{(\alpha+1)}} \\ Y^{(\alpha)} & \xleftarrow{h^{(\alpha+1)}} & Y^{(\alpha+1)} \end{array}$$

such that

(1) the center $D^{(\alpha)}$ of the blowing-up $h^{(\alpha+1)}$ is a closed complex-analytic subspace of $Y^{(\alpha)}$; $f^{(\alpha)}$ induces a flat map

$$(f^{(\alpha)})^{-1}(D^{(\alpha)}) \to D^{(\alpha)}$$

and $g^{(\alpha+1)}$ is the blowing-up with center $(f^{(\alpha)})^{-1}(D^{(\alpha)})$.

(2) We have $D^{(\alpha)} \subset B^{(\alpha)}$, and $D^{(\alpha)}$ is nowhere dense in $B^{(\alpha)}$.

(3) $B^{(\alpha+1)}$ is the strict transform of $B^{(\alpha)}$, i.e. it is the smallest closed complex-analytic subspace of $Y^{(\alpha+1)}$ which induces an isomorphism

$$B^{(\alpha+1)} \to (h^{(\alpha+1)})^{-1}(D^{(\alpha)}) \xrightarrow{\sim} B^{(\alpha)} - D^{(\alpha)}$$

(so that $h^{(\alpha+1)}$ induces the blowing-up $B^{(\alpha+1)} \to B^{(\alpha)}$ with center $D^{(\alpha)}$).

(4) $f^{(\alpha+1)}$ is the unique map that makes the diagram commutative.

Moreover, we have

(5) for $\alpha$ with no $\alpha' \in \Lambda$ such that $\alpha = \alpha' + 1$, $X^{(\alpha)} = \varprojlim_{\beta<\alpha} X^{(\beta)}$, $Y^{(\alpha)} = \varprojlim_{\beta<\alpha} X^{(\beta)}$ and $f^{(\alpha)} = \varprojlim_{\beta<\alpha} f^{(\beta)}$

(6) the images of the $D^{(\alpha)}$ back into $Y$ are locally finite in $Y$ (so that the sequence is in effect finite over any relatively compact subset of $Y$)

(7) if we let $\tilde{X} = \varprojlim_\alpha X^{(\alpha)}$, $\tilde{Y} = \varprojlim_\alpha Y^{(\alpha)}$, $\tilde{f} = \varprojlim_\alpha f^{(\alpha)}$ and $\tilde{B}^{(\alpha)} = \varprojlim_\alpha B^{(\alpha)}$,

then the map $\tilde{f}$ induces a flat map

$$\tilde{f}^{-1}(\tilde{B}) \rightarrow \tilde{B}.$$

NOTE. Combining the technique of resolution of singularities and that of global flattening, we can even require that all the centers $D^{(\alpha)}$ in the above process are smooth.

REMARK 8. Let $\tilde{f}: \tilde{X} \rightarrow \tilde{Y}$ denote the last map of the process of Theorem 3 where the process terminates, and denote by

$$\begin{array}{ccc} X & \xleftarrow{\tilde{g}} & \tilde{X} \\ \downarrow{\scriptstyle f} & & \downarrow{\scriptstyle \tilde{f}} \\ Y & \xleftarrow{\tilde{h}} & \tilde{Y} \end{array}$$

the diagram obtained by taking $\tilde{g}$ to be the composition of those $g^{(i)}$ and $\tilde{h}$ to be that of those $h^{(i)}$. This final diagram depends upon

(1) $\{\eta^{(j)}\}$ and the open neighborhoods $V^{(j)}$ of $\eta^{(j)}$ (which are analogous to $V$ for $\eta$ in Step 1).

(2) $\{\xi^{(j)}\}$.

The dependence on the choice of (2) is not serious in the sense that both maps $h^{(j+1)}$ and $g^{(j+1)}$ are proper and surjective in the following restricted diagram

$$\begin{array}{ccc} X^{(j)} \mid (f^{(j)})^{-1}(V^{(j)}) & \xleftarrow{g^{(j+1)}} & X^{(j+1)} \\ \downarrow{\scriptstyle f^{(j)}} & & \downarrow \\ Y^{(j)} \mid V^{(j)} & \xleftarrow{h^{(j+1)}} & Y^{(j+1)} \end{array}$$

(independent of the choice of $\xi^{(j)}$ so long as $V^{(j)}$ remains the same). However, the choice of (1) is more serious with reference to the strongly *local* nature of the result in Theorem 3. Due to this local nature,

(A) the image $\tilde{h}(\tilde{Y})$ (as a point-set) is *not in general* a neighborhood of the given point $\eta$ in $Y$.

By selecting $\{\eta^{(j)}\}$, however, we can easily verify that

(B) for every germ $(\Gamma, \eta)$ of a complex *curve* $\Gamma$, locally irreducible at $\eta$, in $Y$, we can choose $\{\eta^{(j)}\}$ of Theorem 3 so that, however $V^{(j)}$ and $\xi^{(j)}$ may be chosen, the image $(\tilde{h}(\tilde{Y}), \eta)$ contains $(\Gamma, \eta)$.

In fact, we let $(\Gamma^{(0)}, \eta^{(0)}) = (\Gamma, \eta)$ and at every stage, we pick $\eta^{(i+1)}$ in such a way that we have a locally irreducible curve $\Gamma^{(i+1)}$ at $\eta^{(i+1)}$ in $Y^{(i+1)}$ such that $h^{(i+1)}$ maps $(\Gamma^{(i+1)}, \eta^{(i+1)}$ onto $(\Gamma^{(i)}, \eta^{(i)})$ (as map of germs). (Note: When $Z(f^{(i)}, \xi^{(i)})$ does not contain $(\Gamma^{(i)}; \eta^{(i)})$, then $(\Gamma^{(i+1)}, \eta^{(i+1)}$ is unique in this process.)

It is more difficult, but possible, to prove that for given $(f, \eta, K)$ we can choose *finitely many* flattening diagrams as in Theorem 3, say

$$\begin{array}{ccc} X & \xleftarrow{\tilde{g}_i} & \tilde{X}_i \\ {\scriptstyle f}\downarrow & & \downarrow{\scriptstyle f_i} \\ Y & \xleftarrow{\tilde{h}_i} & \tilde{Y}_i, \quad i=1,2,\cdots,r \end{array}$$

such that $U_i\tilde{h}_i(\tilde{Y}_i)$ is a neighborhood of the given point $\eta$ in $Y$.

For this, we need the technique of "*La voûte etoilée*" as is shown in the paper: '*Introduction to real-analytic sets and real-analytic maps*', Publ. Inst. Mat. "L. Tonelli", Pisa, Italy, 1973.

## 5. Thom $A_f$-condition and flattening

Let $f: X \to Y$ be a complex-analytic map. We are interested in finding (if possible at all) Whitney stratification ($\mathbb{C}$-analytic) of $f$ which satisfies Thom's $A_f$-condition.

We can perceive some deep (but not exact) relationship between the two notions:

(I) Flatness.

(II) Existence of an $A_f$-Whitney stratification.

The relationship is not so obvious, however, and it can be easily seen by numerous examples that the flatness of $f$ does not imply (II) nor does (II) imply the flatness of $f$, even if we add such reasonable assumptions as: $Y$ is reduced and irreducible while $X$ is reduced and every irreducible component of $X$ is mapped onto $Y$ by $f$.

The difficulties in (II)$\Rightarrow$(I) are mainly due to the rather delicate *algebraic* nature of flatness, for instance those related to possible imbedded components (in the sense of Lasker-Noether primary decompositions of ideals) in the inverse images by $f$ of various sections or strata of $Y$. Imbedded components do not in general have much geometric (point-set topological) significance, unless we look into jet-bundles or such. Conversely, the difficulty in (I)$\Rightarrow$(II) is seen in the general phenomena that even if $f$ itself is flat, the induced maps, say in the critical locus of $f$ or say in some canonically associated strata of $X$, are not always flat onto their images in $Y$. In other words, a flat map $f$ may contain much hidden non-flatness.

Let us first examine the following situation:

$M$: connected complex submanifold with $\mathbb{C}$-analytic boundary in $X$

$N$: connected complex submanifold with $\mathbb{C}$-analytic boundary in $X$

such that

(1) the differential of $f \mid M$, resp. $f \mid N$, has constant rank.

(2) $N \subset \partial M = \bar{M} - M$.

We then ask if $(M, N)$ satisfies the $A_f$-condition or not, say within an open dense subset of $N$.

REMARK 1. If the differentials of $f \mid M$ and $f \mid N$ have the same rank, the answer to this question is rather simple. Namely, by the Generic Whitney Lemma, we have an open dense subset $N_0$ of $N$, which has also $\mathbb{C}$-analytic boundary in $X$, such that $(M, N)$ satisfies Whitney condition at every point of $N_0$. It then follows that $(M, N)$ satisfies the $A_f$-condition at every point of $N_0$. In fact, take any point $y \in N_0$ and any sequence of points $\{x_i\}_{i=1,2,\dots}$ in $M$ such that

$$\lim_{i \to \infty} x_i = y$$

and

$$\exists : \lim_{i \to \infty} T_{M \cap f^{-1}(f(x_i)), x_i} = \tilde{T}.$$

By replacing the sequence by a suitable subsequence, we may also assume that

$$\exists : \lim_{i \to \infty} T_{M, x_i} = \bar{T}.$$

Whitney condition implies that $\bar{T} \supset T_{N,y}$. If $Z$ is the complex submanifold of $Y$ within a sufficiently small neighborhood of $f(y)$ in $Y$ such that $f \mid M$ and $f \mid N$ are submersive locally about the $x_i$ ($i \gg 0$) and about $y$ onto $Z$, then we have

$$T_{M \cap f^{-1}(f(x_i)), x_i} = \operatorname{Ker}(T_{M, x_i} \xrightarrow[\alpha_i]{} T_{Z, f(x_i)})$$

and

$$\tilde{T} \subset \operatorname{Ker}(\bar{T} \xrightarrow[\alpha]{} T_{Z, f(y)}),$$

Here $\alpha_i$ are surjective and so is $\alpha$ because $T_{N,y} \to T_{Z,f(y)}$ is so. Hence, by comparing the dimensions,

$$\tilde{T} = \operatorname{Ker}(\bar{T} \xrightarrow[\alpha]{} T_{Z, f(y)}),$$

which contains

$$T_{N \cap f^{-1}(f(y)), y} = \operatorname{Ker}(T_{N,y} \to T_{Z, f(y)}).$$

Thus the $A_f$-condition is verified at $y$.

REMARK 2. In view of Remark 1, we have only to consider the case in which the rank of the differential of $f \mid N$ is strictly less than that of $f \mid M$.

REMARK 3. The condition of having $\mathbb{C}$-analytic boundary implies (by Cartan) that the closure is complex-analytic everywhere. Therefore, we may replace $X$ by the closure of $M$, given the reduced complex structure. Namely,

(3) $X$ is reduced and irreducible, and $M$ is open dense in $X$. Furthermore, we may assume that

(4) $Y$ is the smallest closed complex-analytic subspace of itself which contains the set $f(M)$, and, in particular, $Y$ is reduced irreducible.

REMARK 4. For a complex-analytic space $X$, quite generally, we have the *sheaf of differentials* $\Omega_X$ defined as follows. Let $J$ be the ideal sheaf of the diagonal in $X \times X$. Then

$$\Omega_X = (p_1)_*(J/J^2) \quad \text{with} \quad p_1 : X \times X \to X.$$

For the smooth part $X_0$ of $X$, $\Omega_X \mid X_0$ is in a canonical way the dual of the sheaf of holom. sections of the tangent bundle $T_{X_0}$. For a given $f : X \to Y$, we define the *sheaf of relative differentials* $\Omega_f$. This is defined to be the cokernel of the natural homomorphism

$$f^*\Omega_Y \to \Omega_X.$$

We shall sometimes write $\Omega_{X/Y}$ for $\Omega_f$. If $X_1$ is the open subset of $X$ consisting of those points in $X_0$ at which $f$ is smooth, then $\Omega_f \mid X_1$ is in a natural way the dual of the sheaf of holomorphic sections of

$$T_{X_1/Y} = T_{f \mid X_1} \quad \text{(notations)}$$

which denotes the vector bundle formed by the tangent vectors to the fibres of $f$ in $X_1$.

REMARK 5. Let us assume (1)–(3) for simplicity. We then have a canonical complex-analytic map

$$p' : X' \to X \quad (X' : \text{reduced})$$

called (df)-*modification* of $X$, which is uniquely determined by the following properties:

(a) $p'$ is a *bimeromorphic modification* of $X$, i.e. there exists an open dense subset $U$ of $X$ such that $(p')^{-1}(U)$ is open dense in $X'$ and $p'$ induces an isomorphism

$$X' \mid (p')^{-1}(U) \xrightarrow{\sim} X \mid U.$$

(b) There exists a surjective homomorphism of sheaves

$$(p')^*\Omega_f \xrightarrow{r} \Omega'_f$$

where $\Omega'_f$ is locally free and $\mathrm{Ker}\,(r)$ has nowhere dense support in $X'$.

(c) If $q:Z\to X$ is any map having the properties (a) and (b) (with reduced $Z$), then there exists a unique complex-analytic map $h:Z\to X'$ such that $q=p'h$.

Moreover, this $p'$ has the following additional properties:

(d) $p'$ is proper and surjective.

(e) If $X_1$ is the open subspace of $X$ consisting of those points at which $f$ is smooth, then $p'$ induces an isomorphism

$$(p')^{-1}(X_1)\xrightarrow{\sim} X_1.$$

(f) $(p')^{-1}(X_1)$ is dense in $X'$, and hence there is no *closed* complex-analytic subspace $X_1'$ of $X'$, $\neq X'$, which contains $(p')^{-1}(X_1)$.

REMARK 6. In view of the duality between sheaf of differentials and tangent space, the map $p':X'\to X$ has the following *local* but rather explicit description. Take any point $y\in X$ and an open neighborhood $V$ of $y$ in $X$ such that we have an imbedding $j$ with the commutative diagram

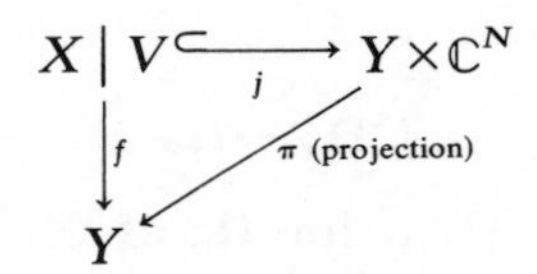


Then, to each point $x\in M\cap V$, there is associated with it an $m$-plane

$$T_{M\cap f^{-1}(f(x)),x}\subset \mathbb{C}^N$$

where $m$ is the dimension of the fibres of $f$ in $M$ and $\mathbb{C}^N$ is canonically identified with

$$T_{\pi^{-1}(f(x)),j(x)}.$$

In other words, we have a map

$$\sigma: M\mid V\to \mathrm{Grass}\,(m,\mathbb{C}^N)$$

(the largest being the Grassmannian of $m$-planes in $\mathbb{C}^N$). Then $X'\mid (p')^{-1}(V)$ is canonically identified with the closure (with reduced complex structure) of the graph of $\sigma$ in the product

$$(X\mid V)\times \mathrm{Grass}\,(m,N)$$

so that $p'$ is induced by the projection to $X\mid V$.

REMARK 7. The (df)-modification $p':X'\to X$ together with the locally free sheaf $\Omega_f'$ has the following important property. Let us define (a sheaf on $N'=(p')^{-1}(N)$)

$$\Omega'_{f|N}=(p')^*\Omega_{f|N}.$$

(Recall that $\Omega_{f|N}$ is the cokernel of

$$(f \mid N)^*\Omega_Y \to \Omega_N$$

as $\Omega_f$ was defined in Remark 4.)

We have the canonical *surjective* homomorphisms

$$(p')^*\Omega_X \mid N' \begin{array}{l} \xrightarrow{\varepsilon} \Omega'_f \mid N' \\ \xrightarrow[\delta]{} \Omega'_f \mid N \end{array}$$

Let us denote by $(y')$ the fibre at $y'$. (Note: For a coherent sheaf $\mathscr{F}$ on $X'$, the *fibre* $\mathscr{F}(y')$ of $\mathscr{F}$ at $y' \in X'$ means

$$\mathscr{F}_{y'}/m_{y'}\mathscr{F}_{y'},$$

where $m_{y'}$ is the maximal ideal of the local ring of $X'$ at $y'$.)

(*) $(M, N)$ satisfies $A_f$-condition at $y \in N$ if and only if for every $y' \in (p')^{-1}(y)$, the kernel of

$$\epsilon(y') : ((p')^*\Omega_X)(y') \to \Omega'_f(y')$$

is contained in the kernel of

$$\delta(y') : ((p')^*\Omega_X)(y') \to \Omega'_{f|N}(y').$$

PROOF. It may be easier to understand the maps $\varepsilon$ and $\delta$ by taking the dual maps. Namely, localizing $X$ about $y$ and take the locally closed imbedding $j$ of Remark 6. Let $\mathscr{C}$ be the trivial $\mathbb{C}^N$-bundle

$$T_{Y\times\mathbb{C}^N/Y} \mid V$$

on the complex-analytic space $X \mid V$. Let $\mathscr{C}' = (p')^*\mathscr{C}$. Let $\mathscr{T}'$ be the vector bundle associated with the dual of $\Omega'_f$, and $\mathscr{S}'$ the one associated with the dual of $\Omega'_{f|N}$. Then $\mathscr{T}'$ is a vector subbundle of $\mathscr{C}'$ and $\mathscr{S}'$ is a vector subbundle of $\mathscr{E}' \mid N'$ with $N' = (p')^{-1}(N)$. Note that, via $p'$,

(i) For every $x' \in X'$ with $x = p'(x') \in M$,

$$\mathscr{T}'_{x'} = T_{M\cap f^{-1}(f(x)),x}.$$

(ii) For every $y' \in N'$ with $y = p'_{(y')} \in N$,

$$\mathscr{S}'_{y'} = T_{N\cap f^{-1}(f(y)),y}.$$

Moreover, (*) is equivalent to

(**) $(M, N)$ satisfies $A_f$-condition at $y \in N$ if and only if

$$\mathscr{S}'_{y'} \subset \mathscr{T}'_{y'}$$

for all $y' \in (p')^{-1}(y)$.

This assertion is clear by (i) and (ii) and by the fact that $\mathscr{T}'$ is a vector subbundle extension of $(p')^*T_{M/Y}$. By taking the duals of (**), we get (*).

Now we want to prove the following

THEOREM 1. *Let* $f: X \to Y$ *be the same as before with* (1)–(4) *of Remark* 3. *We also assume Remark* 2. *Let* $p': X' \to X$ *be the* (*df*)-*modification. Let* $S$ *be a closed complex-analytic set in* $\bar{N}$ *such that*

(1) *every irreducible component of* $(p')^{-1}(\bar{N})$ *is either mapped onto* $\bar{N}$ *or mapped into* $S$,

(2) $S \supset \bar{N} - N$.

(3) *For every* $y \in N - S$, *the composed map* $fp': X' \to Y$ *is* universally open *at every point of* $(p')^{-1}(y)$.

*Then* $(M, N)$ *satisfies the* $A_f$-*condition at every point of* $N - S$.

NOTE. Quite generally, a complex-analytic map, say $g: X' \to Y$ is said to be *universally* open at $x' \in X'$ if there exists an open neighborhood $V'$ of $x'$ in $X'$ such that for every complex-analytic map $h: Y' \to Y$ with the cartesian diagram

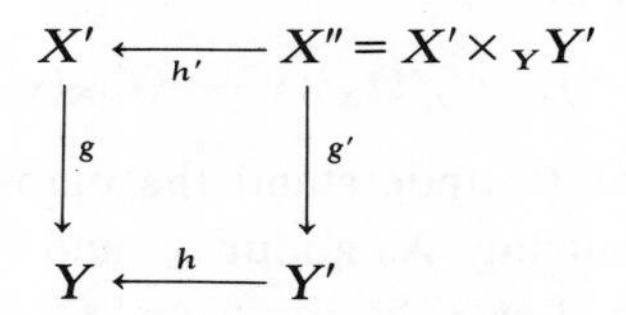


the map $g'$ induces $(h')^{-1}(V') \to Y'$ which is an open map of topological spaces. It is known that if $g$ is *flat* at $x'$, it is universally open at $x'$.

PROOF OF THE THEOREM. Take any point $y \in N - S$. To prove the assertion at $y$, we may restrict $X$ to any open neighborhood of $y$ whenever necessary. For instance, with no loss of generality, we may assume

(4) $N = \bar{N}$ and $fp': X' \to Y$ is universally open everywhere.

(5) there is given an imbedding $j$ as in Remark 6, where $X = X \mid V$.

(6) $f(N)$ is a closed connected complex submanifold of dimension $d$ in $Y$, where $d$ is the rank (which is constant by the assumption (1)) of the differential of $f \mid N$. (For this, we restrict $Y$ to a suitably small neighborhood of $f(y)$ in $Y$.)

(7) if $s = \dim_{\mathbb{C}} Y - d$, then there exists a system of $s$ holomorphic functions $(b_1, \cdots, b_s)$ on $Y$ such that $f(N)$ *as point-set* is defined by $b_1 = \cdots = b_s = 0$ in $Y$. (Note that this is always possible locally at the given point $f(y)$ in $Y$.)

Let us now consider the following commutative diagram.

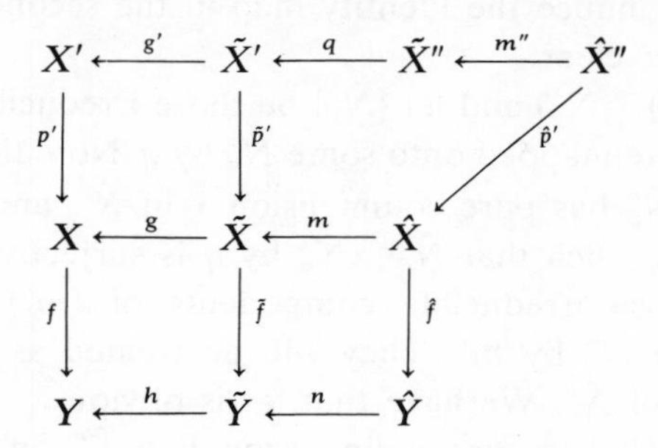


Here we let:

(i) If $s>1$, then $h$ (resp. $g$, resp. $g'$) is the blowing-up of the ideal sheaf $(b)\,\mathfrak{X}_Y$ (resp. $(b)\,\mathfrak{X}_X$, resp. $(b)\,\mathfrak{X}_{X'}$). If $s=1$, then they are all identities. ($s\geqq 1$ by Remark 2.)

(ii) $\tilde{f}$ and $\tilde{p}'$ are the unique maps which make the diagram commutative.

(iii) $q$ is the blowing-up whose center is

$$(i'g')^{-1}(N).$$

(iv) $n$, $m$, and $m''$ are the normalization maps of the respective spaces, while $\hat{f}$ and $\hat{p}'$ are the unique maps which make the diagram commutative.

Let $\{N'_\alpha\}$ be those irreducible components of $(p')^{-1}(N)$ which are mapped onto $N$. (Recall that $p'$ is proper and surjective.) Let us first prove

(8) $h^{-1}(f(N))$, $g^{-1}(N)$, $(g')^{-1}(N'_\alpha)$ are all irreducible, and for each $\alpha$ the two maps

$$(g')^{-1}(N'_\alpha)\rightarrow g^{-1}(N) \quad \text{by } \tilde{p}'$$

and

$$g^{-1}(N)\rightarrow h^{-1}(f(N)) \quad \text{by } \tilde{f}$$

are both *surjective.*

In fact, the ideal $(b)\,\mathfrak{X}_Y$ is generated by $s$ elements and defines a complex-analytic subset $N$ of pure codimension $s$. Since $fp'$ is universally open, it can be deduced from this that $(b)\,\mathfrak{X}_{X'}$ also defines a complex-analytic subset of pure codimension $s$ in $X'$. Since $p'$ is a proper modification, the complex-analytic subset of $X$ defined by $(b)\,\mathfrak{X}_X$ cannot have codimension $<3$ anywhere in $X$. Then, by Krull, $(b)\,\mathfrak{X}_X$ defines a complex-analytic subset of pure codimension $s$ in $X$. It follows that

$$h^{-1}(f(N))=f(N)\times\mathbb{P}_{\mathbb{C}}^{s-1},$$
$$g^{-1}(N)=N\times\mathbb{P}_{\mathbb{C}}^{s-1},$$
$$(g')^{-1}(N'_\alpha)=N'_\alpha\times\mathbb{P}_{\mathbb{C}}^{s-1},\quad \forall\alpha,$$

where $f(N)$, $N$, and $N'_\alpha$ are given the canonical *reduced* complex structure. Moreover, $\tilde{p}'$ and $\tilde{f}$ induce the identity map in the second factor $\mathbb{P}^{s-1}_{\mathbb{C}}$. The assertion (8) is hence clear.

Next, let $\tilde{N}'_\alpha \doteq (g')^{-1}(N'_\alpha)$ and let $\{\tilde{N}''_\beta\}$ be those irreducible components of the $q^{-1}(\tilde{N}'_\alpha)$ which are mapped onto some $\tilde{N}'_\alpha$ by $q$. Note that, as is obvious,

(9) for each $\beta$, $\tilde{N}''_\beta$ has pure codimension 1 in $\tilde{X}''$, and for each $\alpha$ there exists at least one $\beta$ such that $\tilde{N}''_\beta \to \tilde{N}'_\alpha$ by $q$ is surjective.

Let $\{\hat{N}''_\gamma\}$ be those irreducible components of $(m'')^{-1}(\tilde{N}''_\beta)$, which are mapped onto some $\tilde{N}''_\beta$ by $m''$. They will be treated as reduced complex-analytic subspaces of $\hat{X}''$. We have that, as is obvious,

(10) for each $\gamma$, $\hat{N}''_\gamma$ has pure codimension 1 in $\hat{X}''$, and for each $\beta$ there exists at least one $\gamma$ such that $\hat{N}''_\gamma \to \tilde{N}''_\beta$ by $m''$ is surjective.

Now, let $\hat{N}_\gamma$ be the image (with reduced complex structure) of $\hat{N}''_\gamma$ by $\hat{p}'$ for each $\gamma$. Let $\tilde{N} = g^{-1}(N)$, endowed with reduced complex structure. It is irreducible by (8). We have

(11) the map $m$ induces a surjective map

$$\hat{N}_\gamma \to \tilde{N}$$

for every $\gamma$.

This is clear by (8) and by the above definition of $\{\tilde{N}''_\beta\}$ and $\{\hat{N}''_\gamma\}$.

Now, let $L_\gamma$ be the closure of the image of $\hat{N}_\gamma$ by $\hat{f}$. Then we claim that

(12) $L_\gamma$ is a closed complex-analytic set of pure codimension 1 in $\hat{Y}$ and $L_\gamma \to h^{-1}(f(N))$ is surjective by $n$.

In fact, by (8), $\tilde{N} \to h^{-1}(f(N))$ is surjective. Hence, by (11), $L_\gamma \to h^{-1}(f(N))$ is surjective. Since $f(N)$ is a closed complex-analytic set in $Y$ by (6), so is $h^{-1}(f(N))$. Hence $n^{-1}(h^{-1}(f(N)))$ is closed complex-analytic in $\hat{Y}$. Since $n$ is finite and $\hat{f}(\hat{N}_\gamma) \to h^{-1}(f(N))$ is surjective, the closure $L_\gamma$ of $\hat{f}(\hat{N}_\gamma)$ is complex-analytic in $\hat{Y}$. Since $h$ is the blowing-up whose center coincides with $f(N)$ at least point-set-theoretically. Therefore $h^{-1}(f(N))$ is pure codimension 1 in $\tilde{Y}$. It follows that $L_\gamma$ has pure codimension 1 in $\hat{Y}$.

Now we are ready to work for the $A_f$-condition of $(M, N)$ using the above commutative diagram. With respect to the imbedding $j$ of (5) as in Remark 6, we define the trivial $\mathbb{C}^N$-bundle $\mathscr{C}'$ on $X'$, a vector subbundle $\mathscr{T}'$ of $\mathscr{C}'$ on $X'$ and a vector subbundle $\mathscr{S}'$ of the restriction $\mathscr{C}' \mid (p')^{-1}(N)$ as we did in Remark 7 in connection with the assertion (*) and (**). By (**) of Remark 7, it suffices to prove.

(A) $\mathscr{S}'_{y'} \subset \mathscr{T}'_{y'}$ for every $y' \in (p')^{-1}(y)$.

By (1), it is enough to prove

(B) $\mathscr{S}' \mid N'_\alpha \subset \mathscr{T}' \mid N'_\alpha$ for all $\alpha$.

In fact, the irreducible components of $(p')^{-1}(N)$ other than those $N'_\alpha$ are all mapped into $S$. Since $y \in N - S$, $(B) \Rightarrow (A)$. Now $\tilde{N}'_\alpha \to N_\alpha$ is surjective because g′ is so. Hence, by (9) and (10), for each $\alpha$ there exists at least one $\gamma$ such that $\hat{N}''$ is mapped onto $N_\alpha$ by the composition $\pi = g'qm'' : \hat{X}'' \to X'$. Therefore, if we let

$$\hat{\mathscr{C}} = \pi^*\mathscr{S}', \qquad \hat{\mathscr{T}} = \pi^*\mathscr{T}' \quad \text{and} \quad \mathscr{S}'' = \pi^*\mathscr{S}'$$

then it is enough to prove

(C) for every $\gamma$, we have

$$\hat{\mathscr{S}} \mid \hat{N}''_\gamma \subset \hat{\mathscr{T}} \mid \hat{N}''_\gamma.$$

For this end, it is enough to prove

(D) for each $\gamma$, there exists a non-empty open subset $U$ of $\hat{N}''_\gamma$ such that

$$\hat{\mathscr{S}}'' \mid U \subset \hat{\mathscr{T}}'' \mid U.$$

The implication (D)⇒(C) is by the following lemma (which is easy to prove).

LEMMA. *Let K be any reduced and irreducible complex-analytic space. Let E be a complex vector bundle on K. Let S and T be two complex vector subbundles of E on K. If there exists a non-empty open subset U of K such that* $S \mid U \subset T \mid U$, *then* $S \subset T$ *(entirely).*

*We now want to prove* (D). *Let*

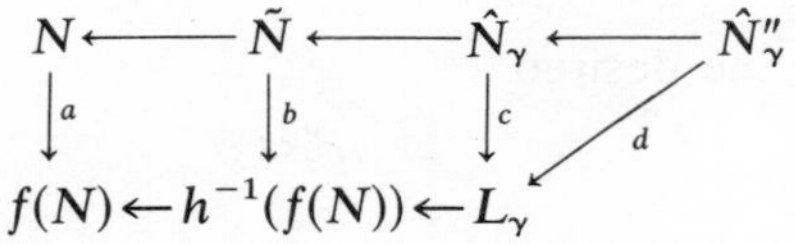

$$\begin{array}{ccccccc} N & \longleftarrow & \tilde{N} & \longleftarrow & \hat{N}_\gamma & \longleftarrow & \hat{N}''_\gamma \\ \downarrow a & & \downarrow b & & \downarrow c & \swarrow d & \\ f(N) & \longleftarrow & h^{-1}(f(N)) & \longleftarrow & L_\gamma & & \end{array}$$

*induced by the earlier commutative diagram.*

By the argument that follows (8), it is clear that every fibre of $b$ is mapped onto the corresponding fibre of $a$. Since $m$ (which induces $\tilde{N}_\gamma \to \tilde{N}$) is the normalization, we have a non-empty open subset $\tilde{U}$ of $\tilde{N}$ such that if $\hat{U}$ is its inverse image in $\hat{N}_\gamma$ then

$$\hat{N}_\gamma \mid \hat{U} \to \tilde{N} \mid \tilde{U}$$

is a locally isomorphic covering. Here we need the result (11). As $\hat{N}''_\gamma \to \hat{N}_\gamma$ is surjective by definition, we can find a non-empty open subset $U_0$ of $\hat{N}''_\gamma$ which is mapped into $\hat{U}$ and

$$\hat{N}''_\gamma \mid U_0 \to \hat{N}_\gamma \mid \hat{U}$$

is smooth. We then have a non-empty open subset $U_1$ of $U_0$ such that

$$\hat{N}''_\gamma \mid U_1 \xrightarrow{d} L_\gamma$$

is smooth. (Here we see the fact that the image of $d$ contains an open dense subset of $L_\gamma$, as was seen below (12).) Now a normal complex-analytic space, such as $\hat{X}''$ and $\hat{Y}$, has its singular locus of codimension $\geqq 2$. Therefore we can choose a non-empty open subset $U$ of $U_1$ such that both $\hat{N}''_\gamma$ and $\hat{X}''$ are smooth at every point of $U$ and such that both $L_\gamma$ and $\hat{Y}$ are smooth at every point of $d(U)$. We shall now claim that this $U$ has the required property in the assertion (D). Pick any point $z \in U$. By the selection of $U \subset U_0$ and $\tilde{U}$, the composition $gm\hat{p}'$ induces a submersion from the fibre of $d$ in $U$ through $z$ to the corresponding fibre of $a$ (through the image point of $z$). Hence we have a canonical *surjective* map

$$T_{N_\gamma'' \cap r^{-1}(r(z)), z} \to \hat{\mathscr{S}}''_z \quad \text{where} \quad r = \hat{f}\hat{p}'.$$

Let $\hat{M}''$ be the inverse image of $M \subset X$ all the way back to $\hat{X}''$. Then for every point $w \in \hat{M}''$, where $\hat{f}\hat{p}'$ is smooth, we have a canonical homomorphism (induced by $p'g'qm''$)

$$T_{r^{-1}(r(w)), w} \to \hat{\mathscr{T}}''_w.$$

Therefore, by taking a sequence of such $w$, say $\{w_i\}$, with

$$\lim_{i \to \infty} w_i = z$$

it is enough for the desired

$$\hat{\mathscr{S}}''_z \subset \hat{\mathscr{T}}''_z$$

to prove that

$$T_{\hat{N}_\gamma'' \cap r^{-1}(r(z)), z} \subset \lim_{i \to \infty} T_{r^{-1}(r(w_i)), w_i}.$$

But this is easy to prove by the following facts. $\hat{X}''$ is smooth at $z$, $\hat{N}''_\gamma$ has codimention 1 in $\hat{X}''$, $\hat{Y}$ is smooth at $r(z)$ and the image $L_\gamma$ of $\hat{N}''_\gamma$ has codimension 1 in $\hat{Y}$. Moreover $\hat{N}''_\gamma$ is smooth at $z$ and $L_\gamma$ is smooth at $r(z)$. Furthermore, $\hat{N}''_\gamma \to L_\gamma$ is smooth at $z$. In such a situation, we can choose a local coordinate system $(u, t_1, \cdots, t_e)$ of $\hat{Y}$ at $r(z)$ and a local coordinate system

$$(u', t_1', \cdots, t_e', v_1, \cdots, v_d)$$

of $\hat{X}''$ at $z$ so that the map $r$ is locally given by

$$\begin{cases} t_i = t_i', & 1 \leqq i \leqq e \\ u = (u')^b \end{cases}$$

where $b$ is the multiplicity with which $\hat{N}''_\gamma$ appears in the fibre $r^{-1}(L_\gamma)$.

Q.E.D.

REMARK 8. From a technical point of view, it is often more convenient to work with some complex-analytic map which *dominates* the (df)-modification, rather than the (df)-modification itself. To be precise,

DEFINITION 1. Let $f:X\to Y$ be a complex-analytic map where $X$ is a reduced complex-analytic space. Then a complex-analytic map $q:\tilde{X}\to X$ is said to dominate the (df)-modification if the following condition is satisfied: For each irreducible component (reduced) $X_i$ of $X$, there is an open subset $V_i$ of $\tilde{X}$, contained in $q^{-1}(X_i)$, and there is a factorization of maps (or commutative diagram) with $q_i$ induced by $q$:

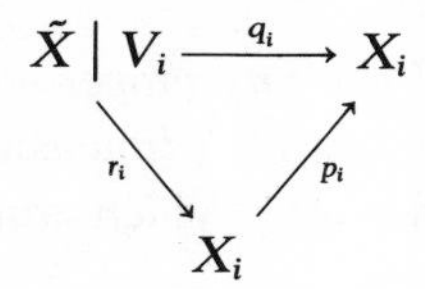


such that

(1) $p_i'$ is the $(\mathrm{df}_i)$-modification where $f_i:X_i\to Y$ (or the map into the smallest closed complex-analytic subspace of $Y$ that contains $f(X_i)$) is induced by $f$.

(2) $r_i$ is surjective.

NOTE 1. For instance, pick any surjective map $r_i:Z_i\to X_i'$ with reference to the $(\mathrm{df}_i)$-modification $p_i'$. Then let $\tilde{X}$ be the disjoint union of those $Z_i$ for all $i$ and let $q:\tilde{X}\to X$ be the map defined by $\{p_i'r_i\}$. Then $q$ dominates the (df)-modification.

REMARK 9. Given $f:X\to Y$ with a reduced $X$, let us suppose that we have a map $q:\tilde{X}\to X$ such that

(1) $q$ dominates the (df)-modification

(2) $fq$ is universally open.

Then, for each irreducible component $X_i$ of $X$, the composition $fp_i'$ is universally open where $p_i':X_i'\to X_i$ is the $(\mathrm{df}_i)$-modification of $f_i:X_i\to Y$ induced by $f$.

In fact, using the notation of Definition 1, it is clear that $fq_i$ is universally open. Since $r_i$ is surjective, it follows that $fp_i'$ is universally open.

THEOREM 2. *Let $f:X\to Y$ be a complex-analytic map where $X$ and $Y$ are reduced. Assume that there exists a complex-analytic map $q:\tilde{X}\to X$ such that*

*(1) $q$ dominates (df)-modification*

*and*

*(2) $fq$ is universally open.*

*Now take any connected open complex sub-manifold $M$ of $X$, with $\mathbb{C}$-analytic boundary and also take any closed reduced complex-analytic subspace $N$ of $X$*

*which is contained in $\bar{M}-M$. Then there exists a nowhere dense closed complex-analytic subset $S$ of $N$ such that $(M, N-S)$ satisfies $A_f$-condition (at every point of $N-S$.)*

PROOF. We apply Theorem 1 to the map $\bar{M}\to Y$ induced by $f$. In this case, by Remark 9, the map $fp'$ of the Theorem 1 is universally open (everywhere). The set $S$ in Theorem 1 can be chosen to be nowhere dense in $N$. ($S$ is the union of the images by $p'$ of those irreducible components of $(p')^{-1}(N)$ whose images in $N$ are nowhere dense in $N$. Here we add to $S$ the set of points of $N$ at which either $N$ is singular or the differential of $f\,|\,N$ does not have constant rank.)

COROLLARY 1. *Let $f: X\to Y$ be any proper complex-analytic map where $Y$ is a smooth complex-analytic curve (dimension one). Then there exists a $\mathbb{C}$-analytic Whitney stratification of $f$ which satisfies the $A_f$-condition.*

PROOF. This is an easy consequence of Theorem 2 in view of the following two facts:

(1) Remark 1 (together with *Generic Whitney Lemma*)

(2) Every reduced irreducible complex-analytic space mapped into a smooth curve, but not to a single point, is flat everywhere over the curve.

## 6. Appendix: Flatness in analytic maps

We are mainly interested in the question of flatness for a complex- (resp. real-) analytic map $\pi: X\to Y$. Let $\xi\in X$ and $\pi(\xi)=\eta\in Y$. Let $\mathcal{M}$ be a coherent sheaf on $X$, and let $M=\mathcal{M}_\xi$. Our questions are how we decide whether $M$ is $A$-flat or not, and if not, then how far it is from being $A$-flat. Before we get into these questions, we recall basics on flatness in general commutative algebra.

Thus, quite abstractly, let $A$ be any commutative ring with unity and let $M$ be any $A$-module.

PROPOSITION–DEFINITION 1. *$M$ is said to be flat over $A$ (or $A$-flat) if one of the following equivalent conditions is satisfied:*

(1) *If $0\to N'\to N''\to N\to 0$ is an exact sequence of $A$-modules, then the derived sequence*

$$0\to N'\otimes_A M\to N''\otimes_A M\to N\otimes_A M\to 0$$

*is exact.*

(2) *If $N'\to N''$ is an injective homomorphism of $A$-modules, then $N'\otimes_A M\to N''\otimes_A M$ is injective.*

(3) *For every $A$-module $N$, $\mathrm{Tor}_1^A(N, M)=(0)$.*

(4) *For every $A$-module $N$, $\mathrm{Tor}_i^A(N, M)=(0)$ for all $i>0$.*

PROOF. (1)$\Leftrightarrow$(2) is by the right exactness of $\otimes_A M$. (1)$\Leftrightarrow$(3) by the long exact sequence of $\otimes_A M$ and $\mathrm{Tor}_1^A(N, M)$, associated with each sequence $0 \to N' \to N'' \to N \to 0$. (3)$\Leftrightarrow$(4) because if we take an exact sequence

$$0 \to N' \to L \to N \to 0$$

with a free $A$-module $L$, then

$$\mathrm{Tor}_{i+1}^A(N, M) = \mathrm{Tor}_i^A(N', M)$$

for all $i > 0$.

REMARK 1. Note that the first two equivalences in this proof are for each $0 \to N' \to N'' \to N \to 0$, while the last are for "all" $N$.

REMARK 2. In the definition above, it is in fact enough to check any one of the conditions (1)–(4) only for *finitely generated* $A$-modules $N'$, $N''$, $N$.

PROOF. Each element of a tensor product is written by finitely many elements in the given module. Moreover, whether such an element is zero or not can be checked by means of finitely many (possibly other) elements. These principles apply to the conditions (1) and (2). The claim for (3) and (4) are proven by (4)$\Rightarrow$(3)$\Leftrightarrow$(1) for each sequence of finitely generated $A$-modules.

In the following proposition, we assume that $A$, $B$ are noetherian local rings, that the homomorphism $A \to B$ (by which $B$ is an $A$-algebra) is local, i.e. $\max(A)B \subset \max(B)$, and that $M$ is a finite $B$-module. $M$ is viewed as $A$-module via $A \to B$.

PROPOSITION 2. *$M$ is $A$-flat*
$\Leftrightarrow$(1) $\mathrm{Tor}_1^A(M, k) = 0$ *with* $k = A/\max(A)$.
$\Leftrightarrow$(2) *the natural homomorphism*

$$\mathrm{gr}_m(A) \otimes_k M/mM \xrightarrow{\theta} \mathrm{gr}_m(M)$$

*is an isomorphism, where* $m = \max(A)$.

NOTE. $\mathrm{gr}_m(A) = \bigoplus_{\nu=0}^{\infty} m^\nu/m^{\nu+1}$, which is a graded $k$-algebra with multiplication induced by that in $A$.

$$\mathrm{gr}_m(M) = \bigoplus_{\nu=0}^{\infty} m^\nu M/m^{\nu+1}M$$

which is a graded $\mathrm{gr}_m(A)$-module (in fact, it is a graded $\mathrm{gr}_{mB}(B)$-module). The map $\theta$ of (2) is obtained as follows. The action of $A$ in $M$ gives

$$m^\nu \otimes_A M \to m^\nu M$$

which induces

$$(m^\nu/m^{\nu+1}) \otimes_k M/mM \xrightarrow{\theta_\nu} m^\nu M/m^{\nu+1}M.$$

The map $\theta$ is the collection of those $\theta_\nu$ for $\nu \in \mathbb{Z}_0$. (2) means that $\theta_\nu$ is bijective for all $\nu \in \mathbb{Z}_0$.

PROOF OF PROP. 2. $M$ is $A$-flat$\Leftrightarrow \mathrm{Tor}_1^A(M, N) = (0)$ for all finitely generated $A$-modules $N$. So the $A$-flatness of $\Rightarrow 1$. Conversely, assume (1). Take any Artin $A$-module $N$ (i.e. an $A$-module of finite length). Then we have an exact sequence

$$0 \to k \to N \to N' \to 0$$

with length $(N') = \text{length}(N) - 1$. It gives

$$\mathrm{Tor}_1^A(M, k) \to \mathrm{Tor}_1^A(M, N) \to \mathrm{Tor}_1^A(M, N')$$

(exact). Hence, by induction on length $(N)$, we deduce $\mathrm{Tor}_1^A(M, N) = (0)$ for any Artin $A$-module $N$. For a general (finitely generated as always) $N$, define

$$\mathrm{Supp}(N) = A/\{a \in A \mid aM = (0)\}$$

(the support of $N$). We use induction on

$$r = \dim(\mathrm{Supp}(A)).$$

Here $r = 0 \Leftrightarrow N$ is Artin, in which case we already have $\mathrm{Tor}_1^A(M, N) = (0)$. Say $r > 0$. Then we have an element $z \in m$ such that

$$\dim(\mathrm{Supp}(N)/z\,\mathrm{Supp}(N)) = s < r.$$

We have an integer $m(\gg 0)$ such that $z^m N$ has no $z$-torsion, i.e. if $N' = z^m N$ then

$$0 \to N' \to N' \to C \to 0$$

is exact, where $z$ denotes the multiplication by $z$. Moreover, the support of $C$ has dimension $\leqq s$ and hence by the induction assumption, $\mathrm{Tor}_1^A(M, C) = (0)$. Hence the multiplication

$$\mathrm{Tor}_1^A(M, N') \xrightarrow{z} \mathrm{Tor}_1^A(M, N')$$

is surjective. Clearly, $\mathrm{Tor}_1^A(M, N')$ is a finite $B$-module (as it is definable by any free resolution of $N'$ as $A$-module). Since $zB \subset \max(B)$ as $A \to B$ is local, Nakayama's lemma asserts that $\mathrm{Tor}_1^A(M, N') = (0)$. Now we have an exact sequence

$$0 \to K \to N \xrightarrow{z^m} N' \to 0$$

which gives

$$\mathrm{Tor}_1^A(M, K) \to \mathrm{Tor}_1^A(M, N) \to \mathrm{Tor}_1^A(M, N').$$

As $K$ has support of dimension $\leqq s$, this implies $\operatorname{Tor}_1^A(M, N) = (0)$. Next, the implication: $A$-flatness of $M \Rightarrow (2)$, can be seen as follows. We have an exact sequence

$$0 \to m^\nu/m^{\nu+1} \to A^{(\nu)} \to A^{(\nu-1)} \to 0,$$

where $A^{(i)} = A/m^{i+1}$ for every $i \in \mathbb{Z}_0$. Hence

$$\begin{array}{ccccccc} 0 \to (m^\nu/m^{\nu+1}) \otimes_A M & \to & A^{(\nu)} \otimes_A M & \to & A^{(\nu-1)} \otimes_A M \to 0 \\ \| & & \| & & \| \\ (m^\nu/m^{\nu+1}) \otimes_k M/mM & \to & M/m^{\nu+1}M & \xrightarrow{\alpha} & M/m^\nu M \end{array}$$

is exact. But $\operatorname{Ker}(\alpha) = m^\nu M/m^{\nu+1}M$. Hence it shows the bijectivity of $\theta_\nu$ (where $\theta_\nu$ is defined in the above Note). Conversely, we now want to prove $(2) \Rightarrow (1)$. By

$$0 \to m \to A \to k \to 0$$

we get

$$\begin{aligned} \operatorname{Tor}_1^A(M, k) &= \operatorname{Ker}(m \otimes_A M \to A \otimes_A M), \\ &= \operatorname{Ker}(m \otimes_A M \to mM), \end{aligned}$$

because $\operatorname{Tor}_1^A(M, A) = (0)$. (Tor-sequence.) So what we want is the injectivity of

$$m \otimes_A M \xrightarrow{\alpha_1} mM.$$

By

$$\begin{array}{ccc} m/m^2 \otimes_A M & \xrightarrow{\theta_1} & mM/m^2M \\ \uparrow & & \uparrow \\ m \otimes_A M & \xrightarrow{\alpha_1} & mM \\ \uparrow & & \uparrow \\ m^2 \otimes_A M & \xrightarrow{\alpha_2} & m^2M \end{array}$$

the assumption (2) implies

$$\operatorname{Ker}(\alpha_1) \subset \operatorname{Im}(\beta_1)$$

and, since $\gamma_1$ is injective, we get

$$\operatorname{Ker}(\alpha_1) \subset \beta_1(\operatorname{Ker}(\alpha_2)).$$

By the analogous diagram for $\theta_2$, we get

$$\operatorname{Ker}(\alpha_2) \subset \beta_2(\operatorname{Ker}(\alpha_3)),$$

where

$$\begin{array}{ccc} m^\nu \otimes_A M & \xrightarrow{\alpha_\nu} & m^\nu M \\ \uparrow \beta_\nu & & \uparrow \gamma_\nu \\ m^{\nu+1} \otimes_A M & \xrightarrow{\alpha_{\nu+1}} & m^{\nu+1} M \end{array}$$

for $\nu \in \mathbb{Z}_0$. In fact, $\operatorname{Ker}(\alpha_\nu) \subset \beta_\nu(\operatorname{Ker}(\alpha_{\nu+1}))$ by the injectivity of $\theta_\nu$ for all $\nu \in \mathbb{Z}_0$. By composing these inclusions, we get

$$\operatorname{Ker}(\alpha_1) \subset \beta_{\nu 1}(\operatorname{Ker}(\alpha_\nu)),$$

where $\beta_{\nu 1} : m^\nu \otimes_A M \to m \otimes_A M$, for all $\nu \geqq 2$. Hence

$$\operatorname{Ker}(\alpha_1) \subset \operatorname{Im}(\beta_{\nu 1}) = m^{\nu-1}(m \otimes_A M)$$

for all $\nu \geqq 2$. Since $m \otimes_A M$ is a finite $B$-module, where $B$ is a noetherian local ring, the Artin-Rees theorem asserts that the inclusion for $\nu \gg 0$ implies

$$\operatorname{Ker}(\alpha_1) \subset m \operatorname{Ker}(\alpha_1)$$

since $mB \subset \max(B)$ as $A \to B$ is local, Nakayama lemma asserts $\operatorname{Ker}(\alpha_1) = (0)$.

REMARK 3. (Geometric interpretation of Prop. 2) Let us first recall the notion of the normal cone. Let $X$ be a complex- (resp. real-) analytic space and $F$ a closed complex- (resp. real-) analytic subspace of $X$. Then we have a coherent ideal sheaf $I$ in the structure sheaf $\mathcal{O}_X$ of $X$ which defines $F$, i.e. $\mathcal{O}_X/I$ (restricted to its support) is the structure sheaf $\mathcal{O}_F$ of $F$. We then obtain a sheaf of graded $\mathcal{O}_F$-algebra

$$\mathcal{O}_F \oplus I/I^2 \oplus I^2/I^3 \oplus \cdots$$

which we denote by $\operatorname{gr}_F(\mathcal{O}_X)$. It is locally of finite presentation, i.e. when restricted to a small neighborhood of every point, it can be presented as

$$\operatorname{gr}_F(\mathcal{O}_X) = \mathcal{O}_F[t_1, \cdots, t_n]/\mathcal{J}$$

with some indeterminates $(t_1, \cdots, t_n)$ and a homogeneous (with respect to $t$) ideal $\mathcal{J}$ which is finitely generated. Then the complex-(resp. real-) analytic space associated with $\operatorname{gr}_F(\mathcal{O}_X)$,

$$p : C(X, F) \to F, \qquad C(X,F) = \operatorname{Specan}(\operatorname{gr}_F(\mathcal{O}_X))$$

is called the *normal cone* of $F$ in $X$. The meaning of this notation 'Specan' is defined as follows. Locally, assuming that $\mathrm{gr}_F(\mathcal{O}_X)$ is presented as above, $C(X, F)$ is a closed complex- (resp. real-) analytic subspace of $F \times k^n$ defined by the ideal sheaf generated by $\mathcal{J}$, where $k = \mathbb{C}$ (resp. $\mathbb{R}$) and $(t_1, \cdots, t_n)$ is identified as the coordinate system of $k^n$. The map $p$ is induced by the projection $F \times k^n \rightarrow F$. Globally, such locally presented normal cones are pieced together in a natural fashion so as to produce the total $p: C(X, F) \rightarrow F$. Now, as a special case of normal cone, if $F$ is a single point $\xi \in X$, then $C(X, \xi)$ is called the *tangent cone* of $X$ at $\xi$. Let us next consider a complex- (resp. real-) analytic map $\pi: X \rightarrow Y$ as in the beginning of Appendix. Let $\eta \in Y$. Then we have a canonical map

$$C(X, F) \xrightarrow[q_F]{} C(Y, \eta)$$

for every closed complex- (resp. real-) analytic subspace $F$ of the fibre $\pi^{-1}(\eta)$ in $X$. We shall call $q_F$ the *differential* of $\pi$ with respect to $F$. In particular, if $F = \pi^{-1}(\eta)$, we call $q_F$ the *differential* of $\pi$ at $\eta \in Y$. If $F = \xi \in \pi^{-1}(\eta)$, then we call $q_F$ the *differential* of $\pi$ at $\xi \in X$. These two will be denoted by $(\partial\pi)_\eta$ and $(\partial\pi)_\xi$ respectively. We have a commutative diagram of canonical maps associated with $F_1 \subset F_2 \subset \pi^{-1}(\eta)$:

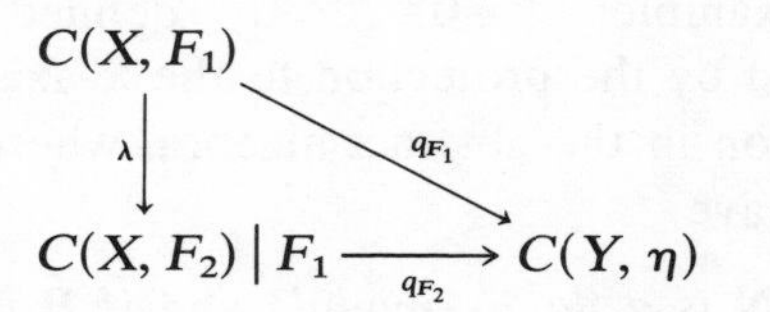


where $C(X, F_2) \mid F_1$ means the inverse image of $F_1$ by the projection $C(X, F_2) \xrightarrow{p} F_2$. In particular,

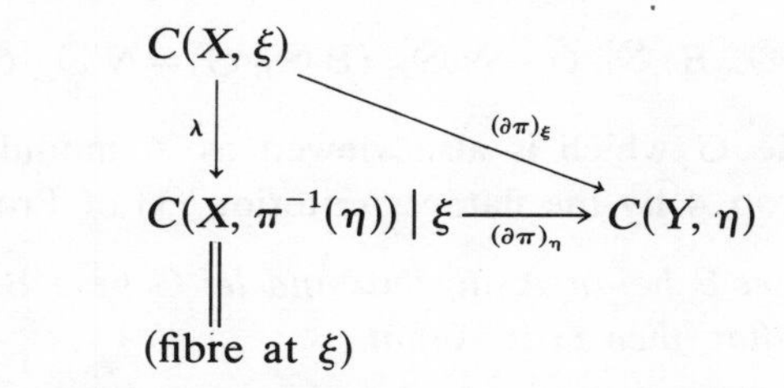


This $\lambda$ is the standard map from a tangent cone to the fibre of the normal cone, where $C(\pi^{-1}(\eta), \xi)$ is mapped into $\lambda^{-1}(0)$ as is well-known.

Now, the following is a direct translation of Prop. 2.

PROPOSITION 3. *Let $\pi: X \rightarrow Y$ and $\eta = \pi(\xi)$ be as above. Then $\pi$ is flat at $\xi$ if and only if the differential $(\partial\pi)_\eta: C(X, \pi^{-1}(\eta)) \rightarrow C(Y, \eta)$ is locally trivial* *at $\xi$, i.e. when $X$ is restricted to a sufficiently small neighborhood of $\xi$ in $X$,*

$$\begin{array}{ccc} C(X, \pi^{-1}(\eta)) & \xrightarrow{\sim} & \pi^{-1}(\eta)\times C(Y, \eta) \\ {\scriptstyle (\partial\pi)_\eta}\downarrow & \swarrow{\scriptstyle \text{projection}} & \\ C(Y, \eta) & & \end{array}$$

(a commutative diagram with the top arrow an isomorphism).

REMARK 4. When $\pi: X \to Y$ is flat at $\xi \in X$, $\pi$ itself is in general far from being locally trivial as we know well in the theory of deformations of a singular point $\xi \in \pi^{-1}(\eta)$. (Here I am viewing $\pi$ as the family of $\pi^{-1}(t)$, $t \in Y$.) Recall that, by taking normal and tangent cones, we lose a great deal of delicate analytic structures as the trivialization in Prop. 3 tells us. In this trivialization, we are keeping the analytic structure of the fibre $\pi^{-1}(\eta)$ as it is, but greatly reducing the analytic structure in the direction normal to the fibre, so to speak.

REMARK 5. When we take $(\partial\pi)_\xi$ (instead of $(\partial\pi)_\eta$), we do not obtain a characterization of the flatness as in Prop. 3. But we can prove that: If the tangential map

$$(\partial\pi)_\xi : C(X, \xi) \to C(Y, \eta)$$

is flat, then $\pi$ is flat at $\xi$. But the converse is not true, as we see by the following simple example: $\xi = 0 \in X \subset \mathbb{C}^2$ defined by $y^3 + x^2 = 0$ and $\pi : X \to Y = \mathbb{C}$ induced by the projection to the $x$-axis.

Back to the situation in the abstract algebra, where $A$ is a commutative ring with unity, we have

PROPOSITION 4. *If $N$ is a flat $A$-module and if $B$ is any $A$-algebra, then $N \otimes_A B$ is flat as $B$-module.*

PROOF. We have the associativity:

$$(N \otimes_A B) \otimes_B G = N \otimes_A (B \otimes_B G) = N \otimes_A G$$

for every $B$-module $G$ which is also viewed as $A$-module. Taking $G$ as variable, we get Prop. 4 by the flatness criterion (1) of Prop-Def. 1.

PROPOSITION 5. *Let $B$ be an $A$-algebra and let $G$ be a $B$-module. If $B$ is $A$-flat and $G$ is $B$-flat, then $G$ is $A$-flat.*

PROOF. Use the same associativity as in the proof of Prop. 4, though this time $G$ is fixed and $N$ is variable.

PROPOSITION 6. *Let $A \to B$ be as in the paragraph preceeding Prop. 2. Let $N$ be a finitely generated $A$-module. Assume that $B$ is $A$-flat. Then $N$ in $A$-flat $\Leftrightarrow N \otimes_A B$ Is $B$-flat.*

PROOF. (⇒) is by Prop. 4. To prove (⇐), take any injective homomorphism $H' \to H''$ of finitely generated $A$-modules. We want to show that the kernel of

$$K = \mathrm{Ker}\,(h' \otimes_A N \to H'' \otimes_A N)$$

is zero. Since $B$ is $A$-flat, we have an exact

$$0 \to K \otimes_A B \to H' \otimes_A N \otimes_A B \to H'' \otimes_A N \otimes_A B$$

is exact. But this last map is injective, because $N \otimes_A B$ is $B$-flat and hence is $A$-flat by Prop. 5. Therefore $K \otimes_A B = (0)$. Since tensor product is right-exact,

$$K \otimes_A B \to K \otimes_A B/mB$$

is surjective. Hence this last module is zero. But it is nothing but

$$(K/mK) \otimes_k (B/mB).$$

Since $A \to B$ is local, $B/mB$ is a non-zero vector space over the field $k$. Therefore it follows that $K/mK = (0)$. By Nakayama, $K = (0)$.

PROPOSITION 7. *Let $A \to B$ and $M$ be the same as those in the paragraph preceeding Prop. 2. Let $(x) = (x_1, \cdots, x_n)$ be a regular $A$-sequence with $x_i \in m$ for all $i$. Then $M$ is $A$-flat if and only if we have*

(1) *$M/(x)M$ is $A/(x)A$-flat*

(2) *$(x)$ is a regular $M$-sequence.*

PROOF. By a simple induction on $n$, we can reduce the problem to the case of $n = 1$. We shall consider only this case. Let $\bar{M} = M/xM$ and $\bar{A} = A/xA$. If $M$ is $A$-flat, then we get (1) by Prop. 4 and then (2) by taking $\otimes_A M$ of the injection $A \xrightarrow{x} A$. Conversely, let us assume (1) and (2). Take any $\bar{A}$-module $N$ and an exact sequence of $\bar{A}$-modules

$$0 \to K \to L \to N \to 0,$$

where $L$ is a finitely generated free $\bar{A}$-module. Let us look at the exact sequence

$$\begin{array}{ccccccc} M \otimes_A K & \xrightarrow{\alpha} & M \otimes_A L & \to & M \otimes_A N & \to & 0 \\ \| & & \| & & \| & & \\ \bar{M} \otimes_{\bar{A}} K & \xrightarrow{\alpha} & \bar{M} \otimes_{\bar{A}} L & \to & \bar{M} \otimes_{\bar{A}} N & \to & 0 \end{array}$$

for the upper sequences, we have

$$\mathrm{Ker}\,(\alpha) = \mathrm{Coker}\,(\mathrm{Tor}_1^A\,(M, L) \to \mathrm{Tor}_1^A\,(M, N)),$$

for the lower sequence, we have

$$\operatorname{Ker}(\alpha) = \operatorname{Tor}_1^{\bar{A}}(\bar{M}, N),$$

because $L$ is $\bar{A}$-free. Here $\operatorname{Tor}_1^A(\bar{M}, N) = (0)$ by (1). Moreover, $\operatorname{Tor}_1^A(M, N)$ is a direct sum of copies of $\operatorname{Tor}_1^A(M, \bar{A})$. By applying $\otimes_A M$ to the sequence

$$0 \to A \xrightarrow{x} A \to \bar{A} \to 0$$

we get $\operatorname{Tor}_1^A(M, \bar{A}) = \operatorname{Ker}(M \xrightarrow{x} M)$ which is zero by (2). Thus we conclude $\operatorname{Tor}_1^A(M, N) = (0)$. By taking $N = k$, we obtain the $A$-flatness of $M$ by Prop. 2.

COROLLARY 1. *In the Prop. 7, let us assume that $A$ is regular and $(x_1, \cdots, x_n)$ is a minimal base of $m$ (i.e. a regular system of parameters). Then $M$ is $A$-flat $\Leftrightarrow (x)$ is a regular $M$-sequence.*

REMARK 6. Let us now go back to the geometric situation with $\pi: X \to Y$ and $\eta = \pi(\xi)$. We let $A = \mathcal{O}_{Y,\eta}$ and $B = \mathcal{O}_{X,\xi}$. Let $M = \mathcal{M}_\xi$ with a coherent analytic sheaf $\mathcal{M}$ on $X$. Let $(t_1, \cdots, t_n)$ be the coordinate system of $k^n$ with $k = \mathbb{C}$ (resp. $=\mathbb{R}$). We write $A\{t_1, \cdots, t_n\}$ for the local ring of $Y \times k^n$ at $\eta \times 0$. Let $id: k^n \to k^n$ be the identity map, and let

$$\pi' = \pi \times id: X \times k^n \to Y \times k^n.$$

Let $\mathcal{M}'$ be the pull-back of $\mathcal{M}$ by the projection $X \times k^n \to X$. Let $\xi' = \xi \times 0$ and $\eta' = \eta \times 0$. Let $M' = \mathcal{M}'_{\xi'}$. Let $B' = B\{t_1, \cdots, t_n\}$ and $A' = A\{t_1, \cdots, t_n\}$. Then

(*) $M'$ is $A'$-flat $\Leftrightarrow M$ is $A$-flat.

In other words, $\mathcal{M}'$ is flat over $Y'$ at $\xi'$ if and only if $\mathcal{M}$ is flat over $Y$ at $\xi$. This is an easy consequence of Prop. 7. Namely, $(t_1, \cdots, t_n)$ is then clearly a regular $M'$-sequence. Taking this as $(x)$ of Prop. 7, (1) of Prop. 7 means the $A$-flatness of $M$. Thus we have (*).

REMARK 7. Take any complex- (resp. real-) analytic map $h: (T, t) \to (Y, h)$. This makes a commutative diagram (cartesian)

$$\begin{array}{ccc} X & \xleftarrow{h''} & X_T = X \times_Y T \\ \downarrow{\scriptstyle \pi} & & \downarrow{\scriptstyle \pi''} \\ Y & \xleftarrow{h} & T \end{array}$$

Let $\xi'' = \xi \times t \in X_T$ and let $\mathcal{M}''$ be the pull-back of $\mathcal{M}$ by $h'$. In this situation, we have

PROPOSITION 8. *The flatness is preserved by any base change. Namely, in the diagram as above, if $\mathcal{M}$ is $\pi$-flat at $\xi$ then $\mathcal{M}''$ is $\pi''$-flat at $\xi''$.*

PROOF. By localizing $T$ about the given point $t$, we may assume that

$$\begin{array}{ccc} t=\eta\times 0\in T & \overset{j}{\hookrightarrow} & Y\times k^n \\ h\downarrow & \swarrow_{\text{projection}} & \\ Y & & \end{array}$$

where $j$ is an imbedding and $h$ is induced by the projection. Thus we get

$$\begin{array}{ccc} X\times k^n & \overset{i}{\hookleftarrow} & X_T \\ \downarrow \pi' & & \downarrow \pi'' \\ Y\times k^n & \underset{j}{\hookleftarrow} & T. \end{array}$$

By Remark 7, if $\mathscr{m}$ *is* $\pi$-flat at $\xi$ then $\mathscr{m}'$ is $\pi'$-flat at $\xi'$. If $M''=\mathscr{m}''_{\xi''}$ and $A''$ is the local ring $\mathcal{O}_{T,t}$, then we have $M''=M'\otimes_{A'}A''$ because $j$ is an imbedding. Thus, by Prop. 4, if $M'$ is $A'$-flat, then $M''$ is $A''$-flat. This means $\mathscr{m}''$ is $\pi''$-flat at $\xi''$.

So far, we had only those facts deduced from the known results in general algebra. We now look into some deeper facts about the flatness in analytic geometries. Let us fix once and for all the following diagram with which we shall work.

$$\begin{array}{ccc} \xi=\eta\times 0\in X & \rightarrow & Y\times k^n \quad \text{(locally closed imbedding)} \\ \pi\downarrow & \swarrow_{\text{projection}} & \\ \eta\in Y & & \end{array}$$

Hence we have an ideal $I$ in $A\{t_1,\cdots,t_n\}$ such that $A\{t\}/I=B$. Our objective is

(**) when $\pi$ is flat at $\xi$, we want to find some sort of "trivialization" of $X$ about $\xi$ with respect to $\pi$, i.e. a trivialization of $B$ as $A$-module.

For this purpose, we must disregard some portion of the algebra structure of $B$ to be trivialized as $A$-module, as the local analytic structure of $X$ obviously varies in the $\pi$-fibres. The technique most suitable for this purpose is Weierstrass Preparation Theorem. So let us first recall this theorem.

Let $\mathbb{C}\{z\}=\mathbb{C}\{z_1,\cdots,z_m\}$ and let $L:\mathbb{R}^m\rightarrow\mathbb{R}$ be a positive linear function, i.e.

$$L(\alpha)=\sum_{i=1}^{m}a_i\alpha_i$$

with $a_i\in\mathbb{R}_+$ for all $i$. Then we get a valuation $v_L$ in $\mathbb{C}\{z\}$ which is defined as follows. Given $f \in \mathbb{C}\{z\}$, we write

$$f = \sum_{\alpha \in \mathbb{Z}_0^m} c_\alpha z^\alpha \quad \text{with} \quad c_\alpha \in \mathbb{C}$$

and define

$$v_L(f) = \min\{L(\alpha) \mid c_\alpha \neq 0\}.$$

For a finite system $A_1, \cdots, A_b \in \mathbb{Z}_0^m$, we define

$$[A_1, \cdots, A_b] = \bigcup_{i=1}^{b} A_i + \mathbb{Z}_0^m.$$

For a subset $E$ of $\mathbb{Z}_0^m$, we write

$$\mathbb{C}\{z\}^E = \{f \in \mathbb{C}\{z\} \mid c_\alpha = 0, \quad \forall \alpha \in E\}.$$

WEIERSTRASS PREPARATION THEOREM. *Suppose we are given an ideal $J$ in $\mathbb{C}\{z\}$ and $g_i \in J$, $i = 1, 2, \cdots, b$, such that*

$$v_L(g_i - c_i z^{A_i}) > v_L(c_i z^{A_i}) = L(A_i)$$

*with $A_i \in \mathbb{Z}_0^m$ and $c_i \in \mathbb{C} - \{0\}$, for every $i$. Let $E = [A_1, \cdots, A_b]$. Then the natural map*

$$\mathbb{C}\{z\}^E \to \mathbb{C}\{z\}/J$$

*is surjective.*

Let us go back to our $B = A\{t_1, \cdots, t_n\}/I$ defined above. Let $\bar{I}$ be the ideal induced by $I$ in $A\{t\}/mA\{t\}$ which we identify with $\mathbb{C}\{t\}$. Each element $h \in \bar{I}$ is written in the form

$$h = h_d + h_{d+1} + \cdots$$

where $h_i$ is a homogeneous polynomial of degree $i$ for all $i$ and $h_d \neq 0$, unless of course $h = 0$. We let $h_d = \text{in}(h)$ ($\text{in}(h) = 0$ if $h = 0$) which we call the *initial form* of $h$. We also write

$$\text{in}(h) = \sum_\alpha e_\alpha z^\alpha \quad \text{with} \quad e_\alpha \in \mathbb{C}$$

and let $\text{lex}_t(h)$ denote the lexicographically maximal element in the finite set

$$\{\alpha \in \mathbb{Z}_0 \mid E_\alpha \neq 0\}.$$

We define

$$\text{Lex}_t(\bar{I}) = \{\alpha \in \mathbb{Z}_0 \mid \exists h \in \bar{I}, \quad \alpha = \text{lex}_t(h)\}.$$

We write $E$ for $\text{Lex}_t(\bar{I})$ for simplicity. Then this $E$ is an *ideal* in the semi-group $\mathbb{Z}_0^n$ in the sense that $\alpha \in E$ and $\beta \in \mathbb{Z}_0^n \Rightarrow \alpha + \beta \in E$. As before, we let

$$A\{t\}^E = \left\{ \sum_{\alpha \in \mathbb{Z}_0{}^n} a_\alpha t^\alpha \in A\{t\} \mid a_\alpha \neq 0, \quad \forall \alpha \in E \right\},$$

where $a_\alpha$ are elements of $A$. Now Weierstrass Preparation Theorem implies

PROPOSITION 9. *The natural map*

$$A\{t\}^E \xrightarrow{k} B = A\{t\}/I$$

*is surjective.*

We can prove (Weierstrass $\Rightarrow$ Prop. 9) as follows. We first write $A = \mathbb{C}\{x_1, \cdots, x_r\}/K$ with an ideal $K$ in the convergent power series in $r$ variables. Let

$$(z_1, \cdots, z_m) = (t_1, \cdots, t_n, x_1, \cdots, x_r).$$

Let us then write

$$E = \bigcup_{i=1}^{b} \alpha_i + \mathbb{Z}_0^n.$$

NOTE. It is a general fact, easily proven, that every ideal $E$ in the semi-group $\mathbb{Z}_o^n$ can be written as above with a finite number of $\alpha_1, \cdots, \alpha_b \in E$.

Next we let $A_i = \alpha_i \times (0)^r \in \mathbb{Z}_0^m$. For each $i$, pick one $f_i \in I$ such that if $\bar{f}_i$ is the class of $f_i$ in $\bar{I}$ then $\text{lex}_t\,(\bar{f}_i) = \alpha_i$. We then pick a representative $g_i$ of $f_i$ in $\mathbb{C}\{z\}$ as we identify $A\{t\}$ with $\mathbb{C}\{z\}/K\mathbb{C}\{z\}$. We can then choose a positive linear function $L:\mathbb{R}^n \to \mathbb{R}$ such that

$$v_L(g_i - c_i z^{A_i}) > L(A_i)$$

with suitable constants $c_i \in \mathbb{C} - \{0\}$ for all $i$. For instance, let $N$ be the maximum of all the $|\alpha_i|$ (which denotes the sum of the $n$ components of $\alpha_i \in \mathbb{Z}_0^n$). Then define $L$ by

$$L(\alpha) = \sum_{i=1}^{n} \left(1 + \frac{i-1}{Nn}\right)\alpha_i + (N+1)\left(\sum_{i=n+1}^{m} \alpha_i\right).$$

Proposition 9 is then immediate from Weierstrass.

REMARK 8. It is important to note that the homomorphism $\kappa$ of Prop. 9 induces bijection modulo $m = \max(A)$. Namely

$$\mathbb{C}\{t\}^E \xrightarrow{\bar{\kappa}} \mathbb{C}\{t\}/\bar{I}$$

is bijective.

PROOF. Take any $f \in \mathbb{C}\{t\}^E$ with $\bar{\kappa}(f) = 0$. Then we have $f \in \bar{I}$. But then, unless $f = 0$, $\text{lex}_t\,(f) \in E$ which contradicts $f \in \mathbb{C}\{t\}^E$.

We could say that Prop. 9 and Remark 8 give us a clearer picture of the $A$-*module* structure in $B$ (although the multiplicative structure of $B$ is lost to a great extent in $A\{t\}^E$). We are not quite satisfied, however, unless we know the structure of $E$ more precisely. Instead of analyzing $E$ in the general case, we find it convenient to take a suitable linear transformation in $t$ and simplify the structure of $E$.

To be precise, let $k$ be any infinite subfield of $\mathbb{C}$ (especially, $k=\mathbb{R}$).

REMARK 9. For almost all $\sigma \in GL(n, k)$ (i.e. for all $\sigma$ in a dense Zariski open subset of $GL(n, k)$), if we let $t'=\sigma(t)$, then

$$E' = \mathrm{Lex}_t\,(\bar{I}) \quad \text{is } \textit{monotonous} \text{ in } \mathbb{Z}_0^n,$$

i.e. (by definition) $E'$ has the property:

$$\forall i, \quad 1 \leqq i < n, \quad \forall \alpha = (\alpha_1, \cdots, \alpha_n) \in E'$$
$$\Rightarrow (\alpha_1, \cdots, \alpha_{i-1}, \alpha_i + \alpha_{i+1}, 0, \alpha_{i+2}, \cdots, \alpha_n) \in E'.$$

(Namely, for each $i$, $E'$ is closed under the operation of adding the $(i+1)$th component to the $i$th component while replacing the first by zero.)

We leave the detail of the proof of this assertion to the reader, except that we make a few comments useful for the proof.

(1) The problem on $\mathrm{Lex}_t$ is only about the homogeneous ideal in $\mathbb{C}[t]$ which is generated by $\mathrm{in}\,(h)$, $h \in \bar{I}$.

(2) For each one $\alpha \in E$ and $i$, $1 \leqq i < n$, we can find $\sigma_0 \in GL(n, k)$ of the form: $\sigma_0(t)_j = t_j$ for $j \neq i+1$ and $\sigma_0(t)_{i+1} + c\sigma_0(t)_i = t_i$ with $c \in K$, such that

$$(\alpha_1, \cdots, \alpha_{i-1}, \alpha_i + \alpha_{i+1}, 0, \alpha_{i+2}, \cdots, \alpha_n) \in E_0',$$

where $E_0' = \mathrm{Lex}_{t'}\,(\bar{I})$ with $t' = \sigma_0(t)$.

(3) If $\sigma$ is a generic $n \times n$-matrix (with indeterminates as its entries), then $\mathrm{Lex}_{\sigma(t)}\,(\bar{I})$ makes sense (for instance by (1)) and for every $\sigma_0 \in GL(n, k)$, the composition $\sigma_0\sigma$ is still generic.

(4) For a dense Zariski open subset $U$ of $GL(n, k)$, $\mathrm{Lex}_{\sigma(t)}\,(\bar{I})$ is independent of $\sigma \in U$.

REMARK 10. If an ideal $E$ in $\mathbb{Z}_0^n$ is non-empty and monotonous, then we have a finite subset $F_i \subset \mathbb{Z}_0^i$ for each $i$, $1 \leqq i \leqq n$, such that

$$\mathbb{Z}_0^n - E = \bigcup_{i=1}^{n} F_i \times \mathbb{Z}_0^{n-i}$$

which is a disjoint union.

PROOF. The detail is left to the readers. We point out that

$$F_i = (\mathrm{pr}_{i-1}\, E) \times \mathbb{Z}_0 - \mathrm{pr}_i\, E$$

for $1 \leqq i \leqq n$, where $(\mathrm{pr}_0\, E) \times \mathbb{Z}_0 = \mathbb{Z}_0$ and for $i \geqq 1$,

$$\mathrm{pr}_i : \mathbb{Z}_0^n \to \mathbb{Z}_0^i$$

is the projection to the first $i$ components. The key is that $F_i$ is finite for all $i$.

REMARK 11. Let $E$ and $F_i$ be as in Remark 10. Then we have a direct sum decomposition:

$$A\{t\}^E = \sum_{i=1}^{n} \sum_{\alpha \in F_i} t^\alpha A\{t(i)\}$$

where $t(i) = (t_{i+1}, \cdots, t_n)$ and $t^\alpha$ means $t^A$ with $A = \alpha \times (0)^{n-1}$ for $\alpha \in \mathbb{Z}_0^i$.

REMARK 12. If an ideal $E \subset \mathbb{Z}_0^n$ is monotonous, then $A\{t\}^E$ is $A$-flat.

PROOF. $A\{t(i)\}$ is $A$-flat (for instance, by (2) of Prop. 2). Hence the assertion follows from the direct sum decomposition of Remark 11.

We summarize the results up to here as follows.

THEOREM 1. *Assume that we are given a commutative diagram of complex-analytic maps as follows:*

$$\begin{array}{ccc} \xi = \eta \times 0 \in X & \xrightarrow{j} & Y \times \mathbb{C}^n \\ \downarrow \pi & \swarrow \text{projection} & \\ \eta \in Y & & \end{array}$$

*where $j$ is a locally closed imbedding. Let $A$ (resp. $B$) be the local ring of $Y$ (resp. $X$) at $\eta$ (resp. $\xi$). Then, after a suitable $\mathbb{R}$-linear transformation in $\mathbb{C}^n$, we have a homomorphism*

$$\sum_{i=0}^{n} \sum_{\alpha \in F_i} t^\alpha A\{t(i)\} \xrightarrow{\kappa} B,$$

*where*

(1) *$(t_1, \cdots, t_n)$ is the coordinate system of $C^n$, and $t(i) = (t_{i+1}, \cdots, t_n)$, $0 \leqq i \leqq n$,*

(2) *$F_i$ is a finite subset of $\mathbb{Z}_0^i$ and for $\alpha = (\alpha_1, \cdots, \alpha_i) \in F_i$,*

$$t^\alpha = t_1^{\alpha_1} t_2^{\alpha_2} \cdots t_i^{\alpha_i}.$$

(3) *The sum in the source of $\kappa$ is direct (in fact, the subset $F_i \times \mathbb{Z}_0^{n-i}$ of $\mathbb{Z}_0^n$ are mutually disjoint).*

(4) *The same sum is an $A$-submodule of $A\{t\}$ and $\kappa$ is induced by the natural homomorphism $A\{t\} \to B$.*

(5) *$\kappa$ is surjective.*

(6) *$\kappa$ induces a bijection modulo $m$, i.e. the induced map*

$$\sum_{i=0}^{n} \sum_{\alpha \in F_i} t^\alpha \mathbb{C}\{t(i)\} \xrightarrow{\bar{\kappa}} B/mB$$

*is bijective, where $m = \max(A)$ and $B/mB$ is the local ring of the fibre $\pi^{-1}(\eta)$ at $\xi$.*

NOTE 1. $\mathbb{Z}_0^0$ consists of a single element, say *, and $t^*$ means $1 \in \mathbb{C}$. In the theorem, if $F_0$ is not empty, then all the other $F_i$ must be empty, in which case the sum of the theorem is nothing but $A\{t\}$.

NOTE 2. The reason we speak of $\mathbb{R}$-linear transformation (instead of $\mathbb{C}$-linear) in the theorem, is that when $(\pi, j)$ are given as complexification of a real-analytic diagram we want to have the transformation to be compatible with autoconjugations.

EXERCISE. Generalize the theorem to a coherent analytic sheaf $\mathcal{m}$ on $X$, so that the above theorem becomes a special case in which $\mathcal{m} = \mathfrak{X}_X$ (the sheaf of holomorphic functions on $X$). To be more precise, take a presentation (locally around $\xi$)

$$0 \to K \to L \to j * \mathcal{m} \to 0$$

with a finite *free* sheaf $L$ on $Y \times \mathbb{C}^n$. Pick any free base $(b_1, b_2, \cdots, b_s)$ of $L$. Then $\kappa$ of the generalized theorem is of the form

$$\sum_{i=0}^{n} \sum_{(a,\alpha) \in F_i} b_a t^\alpha A\{t(i)\} \xrightarrow{\kappa} M = \mathcal{m}_\xi,$$

where $F_i$ is a subset of $\{1, 2, \cdots, \mathrm{b}\} \times \mathbb{Z}_0^i$. Of course, all the 6 conditions analogous to those of the theorem must be satisfied. It is useful to note that in this type of presentation the nature of 'extension' for a module by another will be completely lost. Namely, if we have an exact sequence of coherent sheaves

$$0 \to \mathcal{m}' \to \mathcal{m} \to \mathcal{m}'' \to 0$$

(for instance, $\mathcal{m}'$ is the subsheaf of $\mathcal{m}$ generated by the images of $b_1, \cdots, b_r$, $r < s$, and $\mathcal{m}'' = \mathcal{m}/\mathcal{m}'$), then the presentation of the theorem for $\mathcal{m}$ is obtained by taking the direct sum of those for $\mathcal{m}'$ and $\mathcal{m}''$. (Note that the above $\kappa$ has a canonical lifting to a map into $L_\xi$ due to its presentation.)

NOTE. This excercise can be "generalized" to include most of the statements below about the flatness of complex- (resp. real-) analytic maps.

Now, going back to the situation of the theorem, we have

PROPOSITION 10. *In the theorem, $\pi$ is flat at $\xi$ (i.e. $B$ is $A$-flat) if and only if $\kappa$ is bijective.*

PROOF. In view of Remark 12, this assertion is immediate from the following

LEMMA. *Let $A$ be a noetherian local ring with maximal ideal $m$. Let $B_i$ be an $A$-algebra which is a noetherian local ring, such that $A \to B_i$ is local. Let*

*$N_i$ be a finitely generated $B_i$-module, and let $N = \oplus_{i=1}^{n} N_i$ (a direct sum of finitely many $N_i$). Let $K$ be an $A$-submodule of $N$ and let $N' = N/K$. Let $k = A/m$. Assume*

(1) *$N \otimes_A k \to N' \otimes_A k$ is bijective.*

(2) *$N'$ is $A$-flat.*

*Then $K = (0)$.*

PROOF. By (2), $\mathrm{Tor}_1^A(N', k) = (0)$. Hence

$$0 \to K \otimes_A k \to N \otimes_A k \to N' \otimes_A k \to 0$$

is exact. Hence by (2), we must have $K \otimes_A k = (0)$, i.e. $K = mK$. Hence $K = m^\nu K$ for all $\nu > 0$. In particular, $K \subset \bigcap_\nu m^\nu N = \oplus_{i=1}^{n} (\bigcap_\nu m^\nu N_i) = (0)$.

NOTE. This lemma is applied to $B_i = A\{t(i)\}$ and $N_i = \sum_{\alpha \in F_i} t^\alpha A\{t(i)\}$ for $0 \leqq i \leqq n$.

Let the situation be the same as in the theorem. Let us take an arbitrary complex-analytic map $h: Y' \to Y$ with a point $\eta'$ in $h^{-1}(\eta)$. Let $A' = \mathfrak{X}_{Y',\eta'}$. By $h$, we take the so-called base change

$$\begin{array}{ccc} X & \xleftarrow{h'} & X' = X \times_Y Y' \ni \xi' = \xi \times \eta' \\ \downarrow \pi & & \downarrow \pi' \\ Y & \xleftarrow{h} & Y' \end{array}$$

Let $B' = \mathfrak{X}_{X',\xi'}$. Note that the embedding $j$ induces a new embedding $j'$:

$$\begin{array}{ccc} X' & \xrightarrow{j'} & Y' \times \mathbb{C}^n \\ \downarrow \pi' & \swarrow \text{projection} & \\ Y' & & \end{array}$$

In view of the proof of Theorem 1, the following assertion is easily verified.

PROPOSITION 11. *With the same $(t_1, \cdots, t_n)$ and $F_i$, $0 \leqq i \leqq n$, the homomorphism $\kappa$ induces*

$$\sum_{i=0}^{n} \sum_{\alpha \in F_i} t^\alpha A'\{t(i)\} \xrightarrow{\kappa'} B'$$

*with the same properties (1)–(6) of the theorem with respect to the new diagram of $\pi'$ and $j'$. Moreover, if $h$ is finite at $\eta'$, i.e. $A'$ is finitely generated as $A$-module, then $\mathrm{Ker}(\kappa')$ is equal to the canonical image of $\mathrm{Ker}(\kappa) \otimes_A A'$ into the above direct sum.*

NOTE. As for the last statement, we first note that if

$$0 \to I \to A\{t\} \to B \to 0$$

is the canonical exact sequence, then the derived sequence

$$I \otimes_A A' \to A'\{t\} \to B' \to 0$$

is exact, because of the finiteness assumption on $h$ at $\xi'$. Since $\operatorname{Ker}(\kappa)$ consists of the remainders of the elements of $I$ when we apply Weierstrass theorem and so is $\operatorname{Ker}(\kappa')$ for the image $IA'$ of $I \otimes_A A'$, we see that $\operatorname{Ker}(\kappa')$ is the image $\operatorname{Ker}(\kappa)A'$ of $\operatorname{Ker}(\kappa) \otimes_A A'$. (For its details, review the proof of the theorem.)

Using $\kappa$ of the theorem, we define an ideal $D(\kappa)$ in $A$ as follows. Pick any $g \in \operatorname{Ker}(\kappa)$. Write (with $a_{i\alpha\beta}(g) \in A$)

$$g = \sum_{i=0}^{n} \sum_{\alpha \in F_i} t^\alpha \sum_{\beta \in Z_0^{n-1}} a_{i\alpha\beta}(g) t(i)^\beta.$$

Then $D(\kappa)$ is generated by $a_{i\alpha\beta}(g)$ for all $(i, \alpha, \beta)$ and for all $g \in \operatorname{Ker}(\kappa)$.

PROPOSITION 12. *Take an arbitrary base change with $h$ as in the paragraph preceeding Prop.* 11. *Then we always have*

$$D(\kappa') = D(\kappa)A'$$

*with respect to the canonical map $A \to A'$.*

PROOF. The equality is clear in the case in which $h$ is finite at $\eta'$, by the second half of Prop. 11. Now, for the general case, we first note that to prove the equality of Prop. 12, it is enough to have

$$D(\kappa') \equiv D(\kappa)A' \bmod (m')^{\nu+1}$$

for all $\nu$, where $m' = \max(A')$. This means that, letting $Y'_\nu$ be the complex-analytic subspace of $Y'$ defined by $(m')^{\nu+1}$ and letting $h_\nu : Y'_\nu \to Y$ be the map induced by $h$, it is enough to prove Prop. 12 for all $h_\nu$. But these maps are obviously all finite.

Let $Z$ be the complex-analytic subspace of $Y$ (within a sufficiently small neighborhood of $\eta$ in $Y$), whose ideal sheaf has the stalk $D(\kappa)$ at $\eta$. Then

THEOREM 2. *$Z$ has the following property: For every map $h:(Y', \eta') \to (Y, \eta)$, the map $\pi': X' \to Y'$ (in the paragraph preceeding Prop.* 11) *is flat at $\xi' = \xi \times \eta'$ if and only if $h$ factors through $Z$ as*

$$(Y', \eta') \to (Z, \eta) \xrightarrow[\text{inclusion}]{} (Y, \eta)$$

*(when $Y'$ is restricted to a sufficiently small neighborhood of $\eta'$).*

PROOF. By Prop. 12, $h$ factors through $Z$ if and only if $D(\kappa')=(0)$. This is so if and only if $\kappa'$ is an isomorphism. Then, Prop. 10 implies Theorem 2.

COROLLARY. *Assume that $Y$ is reduced. Then $\pi$ is flat at $\xi$ if and only if for every $h:(Y', \eta)\to(Y, \eta)$ with $Y=\mathbb{D}$ (which denotes the 1-dimensional complex disc), $\pi'$ is flat at $\xi'$.*

PROOF. Since $Y$ is reduced, $Y=Z$ (locally at $\eta$) if and only if there exists no $h$ which does not factor through $Z$.

REMARK 13. The germ of $Z$ of Theorem 2 at the point $\eta$ is uniquely determined by the property described by Theorem 2. (This is obvious.) Moreover, if $\pi:X\to Y$ is given as a complexification of a real-analytic of real-analytic spaces, then the germ of $Z$ at $\eta$ is invariant by the autoconjugation $\sigma$ of $Y$. (Check that $\sigma(Z)$ has the same property as $Z$ and hence, by the uniqueness, $\sigma(Z)=Z$ at $\eta$.)

HEISUKE HIRONAKA
Harvard University
Dept. of Mathematics
Cambridge, Mass. 02138
USA

Nordic Summer School/NAVF
Symposium in Mathematics
Oslo, August 5–25, 1976

# EXAKTE SEQUENZEN FÜR GLOBALE UND LOKALE POINCARÉ-HOMOMORPHISMEN

Ludger Kaup

Die Poincaré-Homomorphismen $P_j^\varphi(X, A)$, welche die Poincaré-Dualität beschreiben, bieten nicht nur im Falle orientierbarer Mannigfaltigkeiten, sondern auch auf singulären Räumen ein nützliches Mittel zur Analyse der topologischen Daten. Während sie jedoch für $X$ Mannigfaltigkeit Isomorphismen sind für alle $j$, trifft dies bei Vorliegen topologischer Singularitäten in $X$ nur noch für gewisse $j$ zu, in Abhängigkeit vom Ausmaß und vom Charakter des singulären Teiles. Will man für die restlichen $j$ mehr Informationen als nur die Existenz und die Natürlichkeit der $P_j^\varphi(X, A)$ ausnutzen, so kommt es entscheidend darauf an, Kerne und Kokerne in einer zugänglichen Weise darzustellen.

Im Falle isolierter Singularitäten ist das einfach: Bezeichnet (in dieser Einleitung) $X$ einen lokal irreduziblen rein $n$-dimensionalen komplexen Raum, $A$ eine offene oder abgeschlossene Teilmenge von $X$ und $\mathscr{H}_*(X)$ die lokale Homologie von $X$ (mit Koeffizienten in einem festen Hauptidealring $L$), so ist $P_0^\varphi(X, A)$ ein Isomorphismus und für $j \neq 0$ existiert für isolierte Singularitäten eine exakte Sequenz

$$\cdots \longrightarrow H_\varphi^j(X, A) \xrightarrow{P_j^\varphi(X,A)} H_{2n-j}^{\varphi|X\setminus A}(X\setminus A)$$
$$\longrightarrow H_\varphi^0(X, A; \mathscr{H}_{2n-j}(X)) \longrightarrow H_\varphi^{j+1}(X, A) \xrightarrow{P_{j+1}} \cdots,$$

in der alle restlichen $P_j(X, A)$ auftreten (vgl. [7]).

Ein Ziel der vorliegenden Arbeit besteht darin, auch für nichtisolierte Singularitäten exakte 'Poincaré-Sequenzen' aufzustellen, in denen möglichst viele $P_j$ auftreten, wobei die Störgruppen als Homologiegruppen bzw. Kohomologiegruppen mit unveränderten Koeffizienten dargestellt sind und nicht etwa mit Koeffizienten in lokalen Homologiegarben, da deren hinreichend genaue Bestimmung für nichtisolierte Singularitäten oft mit erheblichem Aufwand verbunden ist.

Dies führte zu einer Reihe von Beispielen, wo derartige Sequenzen sozusagen a priori existieren, als Konsequenz von Verschwindungssätzen recht allgemeiner Art. Als Exempel führen wir hier an, daß schon aus der Tatsache, daß die topologische Singularitätenmenge eines lokal irreduziblen Raumes höchstens $(n-2)$-dimensional ist, sich ergibt: Es gibt eine analytische Teilmenge $T$ von $X$, so daß für jedes offene $U \subset X$ exakte Sequenzen existieren:

$$0 \longrightarrow H^1_\varphi(X, U) \xrightarrow{P_1} H^{\varphi|X\setminus U}_{2n-1}(X \setminus U) \longrightarrow H^2_\varphi(X, X \setminus T \setminus U) \longrightarrow$$

$$H^2_\varphi(X, U) \xrightarrow{P_2} H^{\varphi|X\setminus U}_{2n-2}(X \setminus U) \longrightarrow H^3_\varphi(X, X \setminus T \setminus U) \longrightarrow$$

$$H^3_\varphi(X, U) \xrightarrow{P_3} H^{\varphi|X\setminus U}_{2n-3}(X \setminus U) \longrightarrow H^4_\varphi(X, X \setminus U \setminus T)$$

sowie

$$H^{2n-3}_\varphi(X, U) \xrightarrow{P_{2n-3}} H^{\varphi|X\setminus U}_3(X \setminus U) \longrightarrow H^{\varphi|X\setminus U}_3(X \setminus U, X \setminus U \setminus T) \longrightarrow$$

$$H^{2n-2}_\varphi(X, U) \xrightarrow{P_{2n-2}} H^{\varphi|X\setminus U}_2(X \setminus U) \longrightarrow H^{\varphi|X\setminus U}_2(X \setminus U, X \setminus U \setminus T) \longrightarrow$$

$$H^{2n-1}_\varphi(X, U) \xrightarrow{P_{2n-1}} H^{\varphi|X\setminus U}_1(X \setminus U) \longrightarrow 0.$$

Die allgemeine Situation ist in den sehr technischen Lemmata 2.2 und 2.7 dargelegt; die anschließenden Bemerkungen zeigen, wie allgemein die Ergebnisse tatsächlich sind. Insbesondere folgt, daß für $2 \cdot \dim S(X) \leqq n-1$ *stets zwei solche exakte Sequenzen existieren, mit denen alle $P_j$ erfaßt werden.*

Für die lokale Homologie isolierter Singularitäten wurde für $j \neq 2n$ in [6] folgender Dualitätssatz bewiesen:

$$\mathcal{H}_j(X)_x \cong \mathcal{H}^{2n-j+1}_x(X).$$

Im ersten Paragraphen der vorliegenden Arbeit wird dieses Resultat auf den allgemeinen Fall übertragen: Ist $T \subset X$ abgeschlossen, so existiert zunächst ein Garbenhomomorphismus

$$\mathcal{P}_{2n-j}(T) : \mathcal{H}^{2n-j}_T(X) \to \mathcal{H}_j(T),$$

wobei $\mathcal{H}^*_T(X)$ die Kohomologiegarbe von $X$ mit Trägern in $T$ bezeichnet. Sind nun für alle $j$ mit $1 \leqq j \leqq k$ die Garben $\mathcal{H}_{2n-j}(X \setminus T) = 0$, so existiert eine exakte Garbensequenz

$$0 \longrightarrow \mathcal{H}^1_T(X) \xrightarrow{\mathcal{P}_1} \mathcal{H}_{2n-1}(T) \longrightarrow \mathcal{H}_{2n-1}(X) \longrightarrow \mathcal{H}^2_T(X) \xrightarrow{\mathcal{P}_1} \mathcal{H}_{2n-2}(T) \longrightarrow$$

$$\cdots \longrightarrow \mathcal{H}^{k+1}_T(X) \xrightarrow{\mathcal{P}_{k+1}} \mathcal{H}_{2n-k-1}(T) \longrightarrow \mathcal{H}_{2n-k-1}(X).$$

Der Spezialfall einer nulldimensionalen Singularitätenmenge $S(X)$ gibt mit $T = S(X)$ ersichtlich den obigen lokalen Dualitätssatz.

Ein zu diesem Ergebnis gewissermaßen duales lautet wie folgt: Ist $T$ in $X$ abgeschlossen, $L$ ein Körper und

$$\mathcal{H}_{2n-k}(X \backslash T) = 0 = \mathcal{H}_a^k(T)$$

für einen $a \in T$ und alle $k$ mit $1 \leqq k \leqq j$, dann existiert ein Homomorphismus

$$\mathcal{H}_{k+1}(X)_a \to \mathcal{H}_T^{2n-k}(X)_a,$$

der bijektiv ist für $1 \leqq k \leqq j-1$ und surjektiv für $k = j$. Da die Poincaré-Homomorphismen nicht bijektiv sind, ist ihr Rang nicht durch die entsprechenden Bettizahlen gegeben. Man erhält jedoch Rangtheoreme folgenden Typs: Ist $b_{2n-j}^{\varphi|T}(T) = 0 = b_{2n-j-1}^{\varphi|T}(T)$ und $Q_j^{\varphi}(X \backslash T)$ bijektiv, dann ist

$$\text{rang}\, P_j^{\varphi}(X) + \text{rang}\, P_{j+1}^{\varphi}(X) = b_{2n-j}^{\varphi}(X) + b_{\varphi}^{j+1}(X) - b_{\varphi}^{j+1}(X, X \backslash T)).$$

Als Anwendung bringen wir Sätze vom Lefschetztyp. Lokale Sätze dieser Art wurden in [4] mit Methoden der Morsetheorie für relative Homotopiemengen bewiesen, unter der Voraussetzung, daß die jeweilige Differenzmenge singularitätenfrei ist (man vergleiche dazu auch [10] und [11]). Unsere homologischen Resultate lassen sich bislang nur in gewissem Umfang auf die Homotopie übertragen, dafür verallgemeinern sie aber in der Homologie [4] auf den Fall singulärer Differenzmengen. Es ergeben sich hier zwei verschiedene Typen von Resultaten für abgeschlossenes $A \subset X$; wobei das erste wesentlich dimensionsabhängig, das zweite von der Anzahl der beschreibenden Gleichungen abhängig ist:

(i) $\mathcal{H}_A^j(X)_a = 0$ *falls* $j+1 \leqq \min\,[n - \text{tab}_a X - \dim_a S(X), 2n - \dim_a^{\mathbb{R}} A]$.

(ii) $\mathcal{H}_{n-\text{tab}(X \backslash A)-r}(X, A)_a = 0$, *falls* $A$ *nahe* $a$ *in* $X$ *als Nullstellenmenge von höchstens* $r$ *holomorphen Gleichungen beschreibbar ist und falls* $\dim S(X \backslash A) \leqq r$.

Diese lokalen Lefschetzsätze implizieren globale Lefschetzsätze, etwa:

*Ist* $A \subset X$ *abgeschlossen und* $\dim^{\mathbb{R}} A \leqq 2n-j-1$, *existiert eine Umgebung* $U$ *von* $A$ *in* $X$ *mit*

$$\dim^{\mathbb{R}} \text{supp}\, \mathcal{H}_{q+j}(U) \leqq q-1 \quad \textit{für} \quad 0 \leqq q \leqq \min\,(2 \dim S(U), 2n-j-1),$$

*dann ist der Beschränkungshomomorphismus*

$$H_{\varphi}^i(X) \to H_{\varphi \cap (X \backslash A)}^i(X \backslash A)$$

*bijektiv für* $i \leqq j-1$ *und injektiv für* $i = j$. – Von den Anwendungen auf die Homotopie, etwa in Zusammenhang mit §3 von [9], haben wir hier nur jeweils einen Satz für offene und für abgeschlossene projektiv algebraische Varietäten gebracht.

Es sei nur erwähnt, daß die Theorie der ersten beiden Paragraphen auch für orientierbare reellanalytische Räume mit mindestens 2-codimensionaler Singularitätenmenge, allgemeiner für Räume mit $L$-$p$ Normalisierung im Sinne von [6] mutatis mutandis zutrifft; der wesentliche Unterschied liegt darin, daß man in der allgemeinen Theorie die durch tab gelieferten Verschwindungssätze nicht zur Verfügung hat. – Herrn Jean-Pierre Ramis, in dessen Seminar ich aufmerksame Zuhörer für eine erste Version der vorliegenden Theorie fand, weiß ich mich zu besonderem Dank verpflichtet.

## 0. Notationen und Vorbereitungen

Mit $X$ werde in der ganzen Arbeit ein komplexer Raum der Dimension $n$ bezeichnet; obwohl die ganze Theorie für nicht reindimensionales $X$ entwickelbar ist (analog [7]), haben wir außer in 3.4 – 3.7 stets Reindimensionalität vorausgesetzt, um die Resultate nicht noch technischer zu machen. Unter Dimension wird stets die komplexe Dimension verstanden, andernfalls schreiben wir $\varphi-\dim$ oder $\dim^{\mathbb{R}}$. Grundlegend für alles folgende sind [6] und [7], die wir im wesentlichen als bekannt voraussetzen. Auch in den Notationen halten wir uns weitgehend daran. Insbesondere bezeichnet

$$\pi : \tilde{X} \to X$$

die (ggf. topologische) Normalisierungsabbildung und $\sim$ bedeutet stets 'topologisches Urbild bezüglich $\pi$'. $L$ bezeichnet einen Hauptidealring (ohne Nullteiler, mit 1), alle Koeffizientengarben $\mathcal{F}$ werden als Garben von $L$-Moduln betrachtet. Wie üblich sei $\mathcal{F}$ lokalkonstant und lokal endlich erzeugt, es sei denn, die Teilmenge $A \subset X$ (lokal abgeschlossen und $\varphi$-taut) ist abgeschlossen und $\varphi$ parakompaktifizierend.
Weiter ist für $x \in X$ und den Raumkeim $X_x$:

$\operatorname{tab}_x(X) := \min\{d, X_x$ ist homöomorph zu einem analytischen Mengenkeim $Y_0 \subset \mathbb{C}_0^t$, der sich durch $t-n+d$ holomorphe $f_j \in \mathcal{O}_0^t$ beschreiben läßt$\}$.

$$\operatorname{tab}(X) := \max_{x \in X} \operatorname{tab}_x(X).$$

$\operatorname{Tab}_\varphi(X, A; \mathcal{F}) := \min\{n, q-p\}$, wobei alle $p$ und alle $q \neq 2n$ zu betrachten sind, für die $H^p(X, A; \mathcal{H}_q(X, \mathcal{F})) \neq 0$.

Es gilt damit:

$$P^{\varphi}_{2n-\operatorname{Tab}_\varphi(X,A;\mathcal{F})+j}(X, A; \mathcal{F}) \quad \text{ist} \quad \begin{cases} \text{surjektiv für } j=1 \\ \text{bijektiv für } j \geqq 2. \end{cases} \tag{0.1}$$

Eine entsprechende Aussage gilt für $Q_*^{\varphi}(X\backslash A, \mathscr{F})$.

$$\mathrm{Tab}_\varphi\,(X, A; \mathscr{F}) \geqq n - \mathrm{tab}\,(X\backslash B) - \dim S(X\backslash B) \tag{0.2}$$

falls $B \subset A$ eine abgeschlossene Teilmenge ist, und $\varphi$ parakompaktifizierend, oder $\mathscr{F}$ konstant und endlich erzeugt über einem vollreflexiven Hauptidealring oder einem Körper $L$. Beweis: Für (0.2) ist

$$H^j_\varphi(X, A; \mathscr{H}_{n-\mathrm{tab}(X\backslash B)+i}(X, \mathscr{F})) = H^j_\varphi(X, A; \mathscr{H}_{n-\mathrm{tab}(X\backslash B)+i}(X, \mathscr{F})_{X\backslash B})$$

zu untersuchen. Im folgenden Lemma zeigen wir, daß wegen Satz 3.9 aus [6] für $L$ vollreflexiv (insbesondere für $L = \mathbb{Z}$) oder $L$ Körper, und sowieso für $\varphi$ parakompaktifizierend, $j = 2i$ das maximale $j$ ist, für welches möglicherweise diese Kohomologie nicht verschwindet; damit ist $\mathrm{Tab}_\varphi\,(X, A; \mathscr{F}) \geqq n - \mathrm{tab}\,(X\backslash B) - i \geqq n - \mathrm{tab}\,(X\backslash B) - \dim S(X\backslash B)$, da es genügt, $i \leqq \dim S(X\backslash B)$ zu betrachten.

LEMMA 0.1. *Die Garbe* $\mathscr{H}_{n-\mathrm{tab}(X)+i}(X, \mathscr{F})$ *hat einen Träger mit komplexer Dimension höchstens* $i$.

BEWEIS. Für $i \leqq -1$ ist dies ein Spezialfall aus 1.4 Satz in [7], da dann $\mathscr{H}_{n-\mathrm{tab}(X)+i}(X, \mathscr{F}) = 0$. Dabei verstehen wir unter der komplexen Dimension einer Menge $B \subset X$ die kleinstmögliche Dimension einer analytischen Menge $A$ mit $B \subset A \subset X$. Für allgemeines $i$ genügt es, wegen des universellen Koeffiziententheorems der lokalen Homologie

$$0 \to \mathscr{H}_j(X, \mathbb{Z}) \otimes \mathscr{F} \to \mathscr{H}_j(X, \mathscr{F}) \to \mathscr{H}_{j-1}(X, \mathbb{Z}) * \mathscr{F} \to 0$$

ersichtlich, $\mathscr{H}_*(X) = \mathscr{H}_*(X, \mathbb{Z})$ zu untersuchen. In der holomorphen Whitneystratifikation von $X$ ist jedes Stratum $Y$ eine komplexe Untermannigfaltigkeit konstanter Dimension, deren abgeschlossene Hülle eine analytische Teilmenge von $X$ ist. Außerdem ist $X\backslash AS(X)$, die Menge aller Mannigfaltigkeitspunkte von $X$, ein Stratum. Für jedes Stratum $Y$ der Whitneystratifikation gilt nach [5] 5.2 Korollar

$$\mathscr{H}_j(X) \mid Y = 0,$$

falls $j \leqq (n - \mathrm{tab}\,(X) - 1 - \dim Y) + 2 \dim Y$. Also folgt aus $\mathscr{H}_{n-\mathrm{tab}(X)+i}(X) \mid Y \neq 0$, daß $n - \mathrm{tab}\,(X) + \dim Y \leqq n - \mathrm{tab}\,(X) + i$, also $\dim Y \leqq i$.

BEMERKUNG 0.2. Wie der Beweis zeigt, kann (0.2) zu folgender Ungleichung verschärft werden

$$\mathrm{Tab}_\varphi\,(X, A; \mathscr{F}) \geqq n - s, \tag{0.3}$$

wobei $s$ wie folgt definiert ist: Für $B \subset A$ abgeschlossen in $X$ und jede natürliche Zahl $t$ sei $A_t$ die kleinste analytische Menge in $X\backslash B$ mit

$$\{x \in S(X\backslash B), \operatorname{tab}_x (X) \geqq t\} \subset A_t.$$

Dann ist

$$s := \max_{t \geqq 0} t + \dim A_t.$$

Dabei ist zu berücksichtigen, daß $\dim \emptyset = -\infty$.

Weiter gilt (0.4): Ist der Grundring $L$ ein Körper $K$ und $U$ eine gut offene Umgebung eines Punktes $x$ (etwa ein offener Simplexstern um $x$), so gilt

$$\mathscr{H}_{2n-\operatorname{Tab}_c(U,K)+j}(X, \mathscr{F})_x = 0 \quad \text{für} \quad 2 \leqq j \leqq \operatorname{Tab}_c (U, K) - 1.$$

Beweis. Nach (0.1) ist

$$P^c_{2n-\operatorname{Tab}_c(U,K)+j}(U, K)\pi^{2n-\operatorname{Tab}_c(U,K)+j}$$

bijektiv für $j \geqq 2$. Nun existiert eine exakte Sequenz mit Koeffizienten in $K$

$$0 \longrightarrow H^1(\tilde{U}) \xrightarrow{P_1} H^{cld}_{2n-1}(U) \longrightarrow H^0(U, \mathscr{H}_{2n-1}(U)) \longrightarrow H^2(\tilde{U}) \xrightarrow{P_2} H^{cld}_{2n-2}(U).$$

Es sei ohne Einschränkung $\operatorname{Tab}_c (U, K) \geqq 3$, dann ist $P^c_{2n-1}(U, K)\pi^{2n-1}$ injektiv, woraus folgt, daß $P^{cld}_1(U)\pi^1$ und damit $P^{cld}_1(U)$ surjektiv ist. Wenn $U$ eine gute Umgebung von $x$ ist, dann ist

$$H^j(\tilde{U}, K) \cong (\mathscr{R}^j\pi K)_x,$$

wobei $\mathscr{R}^j\pi K$ die $j$te Bildgarbe der konstanten Garbe $K$ auf $\tilde{X}$ unter $\pi$ bezeichnet. $\pi$ ist endlich, also ist für $j = 1$

$$0 = (\mathscr{R}^1\pi K)_x = H^{cld}_{2n-1}(U, K) = \mathscr{H}_{2n-1}(X, K)_x.$$

Mit dem universellen Koeffiziententheorem folgt $\mathscr{H}_{2n-1}(X, \mathscr{F})_x = 0$ für jede Garbe von $K$-Vektorräumen $\mathscr{F}$. – Falls $\operatorname{Tab}_c (U, K) \geqq 4$, ermöglicht $\mathscr{H}_{2n-1}(X, \mathscr{F})_x = 0$ nun einen Induktionsbeweis.

Korollar 0.3. *Ist $U$ eine gute offene Umgebung von $x \in X$, so ist*

$$\mathscr{H}_{2n-\operatorname{Tab}_c(U,L)+j}(X, \mathscr{F})_x = 0 \quad \text{für} \quad 2 \leqq j \leqq \operatorname{Tab}_c (U, L) - 1.$$

Beweis. Da $L$ ein Hauptidealbereich ist, genügt es, das Verschwinden der lokalen Homologie für $\mathscr{F} = L/m$ nachzuweisen, wobei $m$ alle maximalen Ideale von $L$ durchläuft. Damit können wir (0.4) anwenden, falls gezeigt ist

$$\operatorname{Tab}_c (U, L) \leqq \operatorname{Tab}_c (U, L/m) \quad \text{für alle } m. \tag{0.5}$$

Nun existiert zunächst lokal eine exakte spaltende Garbensequenz, da $L$ Hauptidealbereich ist

$$0 \to \mathscr{T}\!ors_L\, \mathscr{H}_j(U, L) \to \mathscr{H}_j(U, L) \to \mathscr{H}_j(U, L)/\mathscr{T}\!ors_L\, \mathscr{H}_j(U, L) \to 0,$$

andererseits existiert ein exaktes Diagramm für alle $i, j$:

$$\begin{array}{ccccc}
H_c^i(U, \mathscr{H}_j(U, L)) \otimes L/\mathfrak{m} & & & & H_c^i(U, \mathscr{T}or\, \mathscr{H}_{j-1}(U, L)) \otimes L/\mathfrak{m} \\
\downarrow & & & & \downarrow \\
H_c^i(U, \mathscr{H}_j(U, L) \otimes L/\mathfrak{m}) & \to & H_c^i(U, \mathscr{H}_j(U, L/\mathfrak{m})) & \to & H_c^i(U, \mathscr{H}_{j-1}(U, L) * L/\mathfrak{m}) \\
\downarrow & & & & \downarrow \\
H_c^{i+1}(U, \mathscr{H}_j(U, L)) * L/\mathfrak{m} & & & & H_c^{i+1}(U, \mathscr{H}_{j-1}(U, L)) * L/\mathfrak{m}.
\end{array}$$

Es sei $j-i<\mathrm{Tab}_c\,(U, L)$, so ist

$$\begin{aligned} H_c^i(U, \mathscr{H}_j(U, L)) &= 0 = H_c^{i+1}(U, \mathscr{H}_{j-1}(U, L)) \\ &= H_c^{i+1}(U, \mathscr{H}_j(U, L)) = H_c^i(U, \mathscr{T}or\, \mathscr{H}_{j-1}(U, L)), \end{aligned}$$

woraus sich ergibt $H_c^i(U, \mathscr{H}_j(U, L/\mathfrak{m}))=0$ und folglich (0.5).

Lemma 0.4. Ist $B$ eine abgeschlossene Teilmenge von $A$ mit $a:=\dim S(X\backslash B) \leqq n-2$, bezeichnet $t$ eine natürliche Zahl $\geqq \mathrm{tab}\,(X\backslash B)$, dann existiert eine exakte Sequenz

$$H_\Psi^{n+t+a}(X, A; \mathscr{F}) \xrightarrow{P_{n+t+a}^\Psi(X,A;\mathscr{F})} H_{n-t-a}^{\Psi|X\backslash A}(X\backslash A; \mathscr{F}) \longrightarrow E_3^{2a,t-n-a} \longrightarrow$$

$$H_\Psi^{n+t+a+1}(X, A; \mathscr{F}) \xrightarrow{P_{n+t+a+1}^\Psi(X,A;\mathscr{F})} H_{n-t-a-1}^{\Psi|X\backslash A}(X\backslash A, \mathscr{F}) \longrightarrow 0,$$

wobei $E_3^{2a,t-n-a}$ ein Faktormodul von $H^{2a}(X, A;\ \mathscr{H}_{n-t+a}(X, \mathscr{F}))$ nach einem gewissen Bild von $H^{2a-2}(X, A; \mathscr{H}_{n-t+a+1}(X, \mathscr{F}))$ ist.

Beweis. Es ist zu zeigen, daß

$$E_3^{n+t+a,-2n} \to H_{n-t-a} \to E_3^{2a,t-n-a} \to E_3^{n+t+a+1,-2n} \to H_{n-t-a-1} \to 0$$

eine exakte Sequenz ist. Es ist leicht zu sehen, daß alle betrachteten $E_3^{**}$-Terme gleich den entsprechenden $E_2^{**}$-Terme sind, außer $E_3^{2a,t-n-a}$, das offensichtlich die Form wie in 0.4 hat, da $2a \geqq \dim_{\mathbb{R}} AS(X\backslash B)$ und folglich $H_\Psi^{2a+2}(X, A; \mathscr{H}_{n+a-t-1}(X, \mathscr{F}))=0$.

Exaktheit bei $H_{n-t-a-1}$: Es ist zu zeigen, daß $P_{n+t+a+1}$ surjektiv ist. Dies wurde in (0.1) bewiesen, falls $\mathrm{Tab}_\Psi\,(X, A; \mathscr{F}) \geqq n-t-a$, was (0.2) zeigt. Weiter ist die natürliche Folge

$$H_\Psi^{n+t+a+i}(X, A; \mathscr{F}) \to H_{\tilde\Psi}^{n+t+a+i}(\tilde X, \tilde A;\ \tilde{\mathscr{F}})$$

ein Isomorphismus für jedes $i \geqq 0$, da $\dim_{\mathbb{R}} \mathrm{supp}\, \mathscr{Q}(\mathscr{F}) \leqq 2a$ und $n+t+a+i \geqq 2a+1$.

Um nun die Exaktheit bei den anderen Moduln der gegebenen Sequenz zu zeigen, wenden wir die Ergebnisse von [3] an. Dafür müssen wir nachweisen, daß gewisse Moduln $E_r^{u,v}$ verschwinden. Zur Erleichterung setzen wir $v = n + t - i$. Dann genügt es nach 0.1 Lemma stets zu beweisen, daß $i = n + t$ ausdrücklich ausgeschlossen ist, und daß für alle $i$ mit $u \leqq 2i$ gilt, daß entweder diese $u$ nicht auftreten können, oder die Verschwindungsbedingung aus der Annahme folgt.

Exaktheit bei $E_3^{n+t+a+1}$: Wir wenden Proposition 5.9a, Kapitel XV an mit $p = n + t + 1 + a$, $q = -2n$, $s = n + t + 1 - a \geqq 3$, da $a \leqq n - 2$. Dann haben wir für $u = a + i + 2 \geqq n + t + a + 3$ stets $a + i + 2 \geqq 2i + 1$ und $i = n + t$ ist ausgeschlossen. Für $u = a + i + 1$ mit $2a - 2 \leqq a + i + 1 \neq n + t + 1 + a$ ist $i = n + t$ ausgeschlossen und stets $a + i + 1 \geqq 2i + 1$. Für $u = a + i$ mit $2a \neq a + i \leqq n + t + a - 1$ sind $i = a$ und $i = n + t$ ausgeschlossen und für $i \leqq a - 1$ haben wir $a + i \geqq 2i + 1$.

Exaktheit bei $E_3^{2a, t-n-a}$: Wir wenden Proposition 5.9, Kapitel XV mit $p = 2a$, $q = t - n - a$, $s = n + t - a + 1$ an, das mindestens 3 ist. Für $u = a + i - 1 \leqq 2a - 3$ sind $i = a, a - 1, n + t$ ausgeschlossen und für $i \leqq a - 3$ setzen wir $u \geqq 2i + 1$. Für $u = a + i \leqq n + t + a - 2$ mit $a + i \neq 2a$ sind $i = a$, $n + t$ ausgeschlossen und für $i \leqq a - 1$ setzen wir $a + i \geqq 2i + 1$. Für $u = a + i + 1 \neq n + t + a + 1$ mit $u \geqq 2a + 3$ haben wir $i \leqq n + t$ und $u \geqq 2i + 1$.

Exaktheit bei $H_{n-t-a}$: Wir wenden Proposition 5.7 in Kapitel XV mit $\nu$ anstatt $n$ an in der Voraussetzung von [3], da $n$ in unserem Falle auf andere Weise schon festgelegt ist: $p = n + t + a$, $\nu = n + t + a$, $k = n + t - a$. Für $u = a + i \neq n + t + a$, $a + i \neq 2a$ sind $i = n + t, a$ ausgeschlossen und für $i \leqq a - 1$ haben wir $u \leqq 2i + 1$. Für $u = a + i + 1 \geqq n + t + a + 3$ sind $i = n + t$ und $u \leqq 2i$ ausgeschlossen. Für $u = a + i - 1 \leqq 2a - 3$ haben wir $i \leqq a - 2$, also $u \geqq 2i + 1$.

## 1. Lokale Poincaré-Homomorphismen

Es sei $T$ in diesem Abschnitt stets eine *abgeschlossene* Teilmenge von $X$. Dann definiert für $U \subset X$ offen das System

$$\begin{array}{ccc} H^{2n-j}(\tilde{X}, \tilde{X} \setminus \tilde{T}) & \xrightarrow{P_{2n-j}(X, X \setminus T)} & H_j^{cld}(T) \\ \downarrow & & \downarrow \\ H^{2n-j}(\tilde{U}, \tilde{U} \setminus \tilde{T}) & \xrightarrow{P_{2n-j}(U, U \setminus T)} & H_j^{cld}(U \cap T) \end{array}$$

mit den kanonischen Beschränkungshomomorphismen und Koeffizienten in einer *lokal endlich erzeugten* und *lokal konstanten Garbe* $\mathscr{F}$ einen Poincaré-Homomorphismus von Garben

$$\pi_0 \mathscr{H}_{\tilde{T}}^{2n-j}(\tilde{X}, \tilde{\mathscr{F}}) \xrightarrow{\mathscr{P}_{2n-j}(T, \mathscr{F})} \mathscr{H}_j(T, \mathscr{F}),$$

wobei $\mathscr{H}_T^*(X, \mathscr{F})$ die Garbe der Kohomologie von $X$ mit Koeffizienten in $\mathscr{F}$ und Trägern in $T$ bezeichnet.

$\mathscr{P}_{2n-j}(T,\mathscr{F})$ ist ein Garbenisomorphismus auf $T$, falls $T$ die topologische Singularitätenmenge von $X$ nicht berührt. Hier soll jedoch der allgemeine Fall untersucht werden.

SATZ 1.1. *Ist $j \geqq 0$ und $\mathscr{H}_{2n-k}(X \setminus T, \mathscr{F}) = 0$ für $1 \leqq k \leqq j$, so existiert eine exakte Garbensequenz*

$$0 \longrightarrow \pi_0 \mathscr{H}_{\tilde{T}}^1(\tilde{X}, \tilde{\mathscr{F}}) \xrightarrow{\mathscr{P}_1} \mathscr{H}_{2n-1}(T, \mathscr{F}) \longrightarrow \mathscr{H}_{2n-1}(X, \mathscr{F}) \longrightarrow \pi_0 \mathscr{H}_{\tilde{T}}^2(\tilde{X}, \tilde{\mathscr{F}}) \xrightarrow{\mathscr{P}_2}$$

$$\cdots \longrightarrow \mathscr{H}_{2n-j-1}(X, \mathscr{F}).$$

Der Beweis ergibt sich leicht aus dem folgenden

LEMMA 1.2. *Für $j \neq 2n$ existiert eine exakte Garbensequenz*

$$\mathscr{H}_j(T,\mathscr{F}) \longrightarrow \mathscr{H}_j(X,\mathscr{F}) \longrightarrow \pi_0 \mathscr{H}_{\tilde{T}}^{2n-j+1}(\tilde{X}, \tilde{\mathscr{F}}) \xrightarrow{\mathscr{P}} \mathscr{H}_{j-1}(T,\mathscr{F}) \longrightarrow \mathscr{H}_{j-1}(X,\mathscr{F})$$

*falls für eine hinreichend kleine Umgebung $U$ von $T$ in $X$ mit $d := \dim_{\mathbb{R}} S(U \setminus T)$ je eine der Bedingungen aus ($\alpha$) und ($\beta$) erfüllt ist:*

($\alpha$) (i) $\mathscr{H}_{\tilde{T}}^{2n-j+1}(\tilde{X}, \tilde{\mathscr{F}}) = 0$,

(ii) $\mathscr{H}_{j+1}(U \setminus T, \mathscr{F}) = 0$,

$\mathscr{H}_T^{q+2}(U, \mathscr{H}_{q+j+1}(X, \mathscr{F})) = 0$,

*für* $\max(1, 2(n - \operatorname{tab}(U \setminus T) - j)) \leqq q \leqq \min(d-2, 2n-j-2)$

$\mathscr{H}_T^{q+1}(U, \mathscr{H}_{q+j+1}(X, \mathscr{F})) = 0$,

*für* $\max(1, 2(n - \operatorname{tab}(U \setminus T) - j) - 1) \leqq q \leqq \min(d-1, 2n-j-2)$,

(iii) $j = 2n-1$.

($\beta$) (i) $\mathscr{H}_j(X, T; \mathscr{F}) = 0$,

(ii) $\mathscr{H}_j(X \setminus T, \mathscr{F}) = 0$ *und* $\mathscr{H}_T^{q+1}(X, \mathscr{H}_{q+j}(X, \mathscr{F})) = 0$

*für* $\max(1, 2(n - j - \operatorname{tab}(U \setminus T)) + 1) \leqq q \leqq \min(d-1, 2n-j-1)$,

(iii) *$X \setminus T$ ist lokal irreduzibel und $j = 1$.*

BEWEIS. Wir haben formal folgendes exakte kommutative Diagramm zu betrachten:

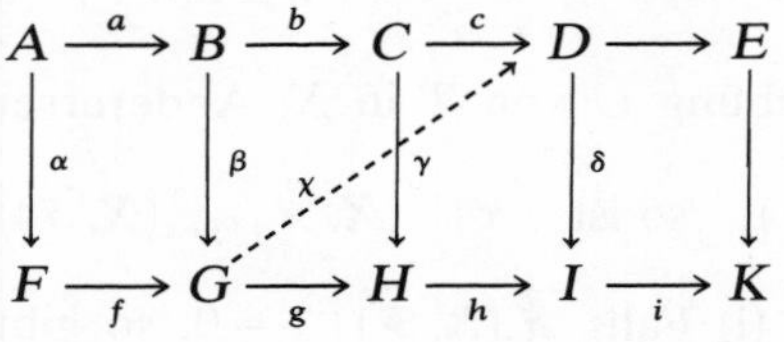

$$\begin{array}{ccccccccc} A & \xrightarrow{a} & B & \xrightarrow{b} & C & \xrightarrow{c} & D & \longrightarrow & E \\ \downarrow{\scriptstyle\alpha} & & \downarrow{\scriptstyle\beta} & \overset{\chi}{\nearrow} & \downarrow{\scriptstyle\gamma} & & \downarrow{\scriptstyle\delta} & & \downarrow \\ F & \xrightarrow[f]{} & G & \xrightarrow[g]{} & H & \xrightarrow[h]{} & I & \xrightarrow[i]{} & K \end{array} \tag{1.1}$$

in dem ein Homomorphismus $\chi$ definiert werden soll, der eine exakte Sequenz

$$F \xrightarrow{f} G \xrightarrow{\chi} D \xrightarrow{\delta} I \xrightarrow{i} K \tag{1.2}$$

liefert. Es sei $\operatorname{Ker} \gamma \subset b(B)$ und $g(G) \subset \gamma(C)$.

Dann ist $\chi := c\gamma^{-1}g$ wohldefiniert. Ist $E=0$, so ist (1.2) ein Komplex. Ist $B=0$, so ist $\chi^{-1}(0) \subset f(F)$; ist $E=0$, so ist $\delta^{-1}(0) \subset \chi(G)$; ist $\gamma$ surjektiv, so ist $i^{-1}(0) \subset \delta(D)$. Wenden wir dies an: Für jedes $x \in X$ erhalten wir wegen $\mathcal{H}^{2n-j}(X, \mathcal{F}) = 0$ für $j \neq 2n$ ein Diagramm vom Typ (1.1), in dem $j: \tilde{X} \setminus \tilde{T} \to \tilde{X}$ die Inklusion bezeichnet:

$$\begin{array}{ccccccccc} & & \pi_0 \mathcal{R}^{2n-j} j_*(\tilde{\mathcal{F}} \mid \tilde{X} \setminus \tilde{T}) & \cong & \pi_0 \mathcal{H}_{\tilde{T}}^{2n-j+1}(\tilde{X}, \tilde{\mathcal{F}}) & & \\ & \overset{x}{\nearrow} & \downarrow Q & & \downarrow \mathcal{P}_{2n-j-1}(T,\mathcal{F}) & & \\ \mathcal{H}_j(T, \mathcal{F}) \to \mathcal{H}_j(X, \mathcal{F}) & \to & \mathcal{H}_j(X, T; \mathcal{F}) & \to & \mathcal{H}_{j-1}(T, \mathcal{F}) & \to & \mathcal{H}_{j-1}(X, \mathcal{F}) \end{array}$$

wobei die relative lokale Homologiegarbe $\mathcal{H}_j(X, T; \mathcal{F})$ definiert wird durch die Prägarbe $U \rightsquigarrow H_p^{cld}(U, U \cap T; \mathcal{F})$, in dem folglich $Q := \lim_{U \to x} Q_{2n-j}^{cld}(U \setminus T, \mathcal{F})$ bijektiv sein soll:

($\alpha$) $Q$ ist injektiv, falls $x \notin T$ oder falls für $x \in T$ gilt

(i) $\mathcal{H}_{\tilde{T}}^{2n-j+1}(\tilde{X}, \tilde{\mathcal{F}}) = 0$, oder falls gilt

(ii) $\lim_{U \to x} H^{q+1}(X \setminus T, \mathcal{H}_{q+j+1}(X, \mathcal{F})) = 0 = \lim_{U \to x} H^q(X \setminus T, \mathcal{H}_{q+j+1}(X, \mathcal{F}))$
$= 0$ für $0 \leqq q \leqq 2n-j-2$, oder falls

(iii) $j = 2n-1$.

Für $q=0$ bedeutet die zweite Gleichung in (ii), daß $\mathcal{H}_{j+1}(X \setminus T, \mathcal{F}) = 0$; dies braucht jedoch ersichtlich nur für eine Umgebung von $T$ in $X$ zu gelten.

($\beta$) $Q$ ist surjektiv, falls

(i) $\lim_{U \to x} H_j^{cld}(U, T \cap U; \mathcal{F}) = 0$, oder

(ii) $\lim_{U \to x} H^q(U \setminus T, \mathcal{H}_{q+j}(X, \mathcal{F})) = 0$, $0 \leqq q \leqq 2n-j-1$ oder falls

(iii) gilt.

Es bleibt schließlich noch zu zeigen, daß gewisse Garben $\mathcal{H}_T^p(X, \mathcal{H}_q)$ stets verschwinden: Es ist $\dim \operatorname{supp} \mathcal{H}_{q+j+1}(U \setminus T, \mathcal{F}) \leqq 2(q+j+1+\operatorname{tab}(U \setminus T) - n)$ und folglich nach Lemma 1.4

$$\mathcal{H}_T^{q+1}(X, \mathcal{H}_{q+j+1}(X, \mathcal{F})) = 0 \quad \text{für} \quad q \leqq 2(n-j-\operatorname{tab}(U \setminus T) - 1)$$

für eine kleine Umgebung $U$ von $T$ in $X$. Andererseits, wenn

$$d := \dim_{\mathbb{R}} S(U \setminus T), \quad \text{so ist} \quad \mathcal{H}_T^{q+1}(X, \mathcal{H}_{q+j+1}(X, \mathcal{F})) = 0 \quad \text{für} \quad q \geqq d.$$

Bemerkungen 1.3. (i) Falls $\mathcal{H}_j(X, \mathcal{F}) \mid T = 0$, so gibt der Beweis von 1.2 falls eine der Bedingungen ($\alpha$) erfüllt ist, eine Inklusion

$$\mathcal{P}: \pi_0 \mathcal{H}_{\tilde{T}}^{2n-j+1}(\tilde{X}, \tilde{\mathcal{F}}) \hookrightarrow \mathcal{H}_{j-1}(T, \mathcal{F}).$$

Ist $\mathcal{H}_{j-1}(T, \mathcal{F}) = 0$ und eine der Bedingungen ($\beta$) erfüllt, so ist $\mathcal{P}_{2n-j+1}(T, \mathcal{F})$ surjektiv.

(ii) $\mathscr{P}_0(T, \mathscr{F})$ ist stets bijektiv, $\mathscr{P}_1(T, \mathscr{F})$ injektiv, $\mathscr{P}_{2n}(T, \mathscr{F})$ stets surjektiv; für nahe $T$ lokal irreduzibles $X$ ist $\mathscr{P}_{2n}(T, \mathscr{F})$ bijektiv und $\mathscr{P}_{2n-1}(T, \mathscr{F})$ surjektiv.

(iii) Hat $X \setminus T$ isolierte Singularitäten, so ist für festes $j \neq 2n$ die Sequenz

$$\mathscr{H}_j(T, \mathscr{F}) \to \mathscr{H}_j(X, \mathscr{F}) \to \pi_0 \mathscr{H}_{\tilde{T}}^{2n-j+1}(\tilde{X}, \tilde{\mathscr{F}}) \to \mathscr{H}_{j-1}(T, \mathscr{F}) \to \mathscr{H}_{j-1}(X, \mathscr{F})$$

auf $T \cup [X \setminus \operatorname{supp} \mathscr{H}_j(X \setminus T, \mathscr{F})]$ exakt.

(iv) Hat $S(X)$ eine komplexe Dimension $\leqq k$, so ist

$$\mathscr{H}_j(T, \mathscr{F}) \to \mathscr{H}_j(X, \mathscr{F}) \to \pi_0 \mathscr{H}_{\tilde{T}}^{2n-j+1}(\tilde{X}, \tilde{\mathscr{F}}) \to \mathscr{H}_{j-1}(T, \mathscr{F}) \to \mathscr{H}_{j-1}(X, \mathscr{F})$$

exakt, falls $\mathscr{H}_j(X \setminus T, \mathscr{F}) = 0$ und $j \leqq n - \operatorname{tab}(U \setminus T) - k - 1$, oder $n + \operatorname{tab}(U \setminus T) + k + 1 \leqq j \leqq 2n - 1$.

(v) Ist $\operatorname{tab}(U \setminus T) + k \leqq n - 3$, $\dim_{\mathbb{R}} T \leqq 2n - 3$ und $X$ in $T$ lokal irreduzibel, so gilt (iv) auch für $j = n - \operatorname{tab}(U \setminus T) + k$.

(vi) Ist $\varphi$ eine Trägerfamilie auf $X$, so liefert Satz 1.1 eine exakte Sequenz

$$0 \to H^1_{\tilde{\varphi} | \tilde{T}}(\tilde{X}, \tilde{\mathscr{F}}) \to H^0(X, \mathscr{H}_{2n-1}(T, \mathscr{F})) \to H^0(X, \mathscr{H}_{2n-1}(X, \mathscr{F})).$$

Sie läßt sich entsprechend für $j > 1$ beweisen, wenn zu den Voraussetzungen von Satz 1.1 noch $\mathscr{H}^j_{\tilde{T}}(\tilde{X}, \tilde{\mathscr{F}}) = 0$ für $j < j_0$ erfüllt ist.

(vii) Nach isolierten Singularitäten ist der nächsteinfache Fall der, daß $X \setminus T$ lokal vollständiger Durchschnitt ist und die Singularitätenmenge $S(X \setminus T)$ die (komplexe) Dimension 1 hat. Dann ist die Dualitätssequenz aus Lemma 1.2 exakt für alle $j \neq n-1, n, n+1, n+2, 2n$.

Wir haben bemerkt, daß $\mathscr{H}_r(X, \mathscr{F})$ die Voraussetzung des folgenden Resultates erfüllt:

LEMMA 1.4. *$X$ sei lokal trianguliert, $\mathscr{G}$ eine Garbe auf $X$, die im 'Inneren' eines jeden Simplex konstant ist, $T$ sei eine abgeschlossene Teilmenge von $X$. Dann gilt für alle*

$$q : \geqq \dim_{\mathbb{R}} \operatorname{supp} \mathscr{G} \mid X \setminus T : \mathscr{H}_T^{q+1}(X, \mathscr{F}) = 0.$$

BEWEIS. Da es sich um ein lokales Problem handelt, dürfen wir annehmen, daß $X = \operatorname{supp} \mathscr{G}$ ein abgeschlossener Simplexstern um einen Punkt $t \in T$ ist. Bezeichnet $X_{q-1}$ das $q-1$ Gerüst von $X$, so existiert mit $U := X \setminus X_{q-1}$ eine exakte Garbensequenz

$$0 \to \mathscr{G}_U \to \mathscr{G} \to \mathscr{G}_{X_{q-1}} \to 0.$$

Die zugehörige exakte Kohomologiesequenz gibt $H^q(X \setminus T, \mathscr{G}_U) \to H^q(X \setminus T, \mathscr{G}) \to H^q(X \setminus T, \mathscr{G}_{X_{q-1}}) = H^q(X_{q-1} \setminus T, \mathscr{G}) = 0$, so daß $H^q(X \setminus T, \mathscr{G}_U) = 0$ für den Beweis ausreicht. Durchläuft $\Delta$ die abgeschlossenen $q$-Simplizes von $X$, so ist $U = \cup \mathring{\Delta}$ und $\mathscr{G}_U = \oplus \mathscr{G}_{\mathring{\Delta}}$, also genügt $H^q(\Delta \setminus T, M_{\mathring{\Delta}}) = 0$ zu zeigen, für $M:=H^0(\mathring{\Delta},\mathscr{G})$. Nun ist $H^q(\Delta\setminus T, M_{\mathring{\Delta}})=H^q(\Delta\setminus T, S^{q-1}\setminus T; M)=H^{q-1}(S^{q-1}\setminus T; M)$, weil etwa $H^c_q(\Delta\setminus T, M)\cong H^0(\Delta, T;\mathscr{H}_q(\Delta, M))\hookrightarrow H^0(\Delta,\mathscr{H}_q(\Delta, M))=H^0(\Delta, S^{q-1}; M)=0$. Andererseits ist $H^{q-1}(S^{q-1}\setminus T, M)\cong H_0(S^{q-1}, T; M)=0$.

KOROLLAR 1.5. *Ist T abgeschlossen mit* $\dim_{\mathbb{R}} T\leqq 2n-j-1$ *und existiert eine Umgebung U von T mit* $\dim_{\mathbb{R}}\operatorname{supp}\mathscr{H}_{q+j+1}(U\setminus T,\mathscr{F})\leqq q$ *für* $-1\leqq q\leqq\min(\dim_{\mathbb{R}} S(U)-1,\ 2n-j-2)$, *so ist* $\mathscr{H}^j_{\tilde{T}}(\tilde{X},\tilde{\mathscr{F}})=0$.

BEWEIS. Man benutze Lemma 1.2 und 1.4.

Während 1.5 vor allem für kleine $j$ interessant ist, gilt für $j=2n$: Ist $X$ nahe $T$ lokal irreduzibel und $T$ ohne isolierte Punkte, so ist $\mathscr{H}^{2n}_T(X,\mathscr{F})=0$. – Dies trifft nicht zu für reduzibles $X$, wie schon das Beispiel zweier in einem Punkt $T$ verhefteter $\mathbb{C}^n$ zeigt.

## 2. Exakte Sequenzen für globale Poincaré-Homomorphismen

Es bezeichne $\mathscr{F}$ eine Garbe von $L$-Moduln auf $X$, $X\setminus T$ eine lokal abgeschlossene Teilmenge von $X$, $\varphi$ eine Trägerfamilie auf $X$, so daß $X\setminus T$ $\varphi$-taut ist. Falls $T$ nicht offen oder $\varphi$ nicht parakompaktifizierend ist, sei $\mathscr{F}$ *lokalkonstant* mit *endlich erzeugten* Halmen. – Um exakte Sequenzen für nichtinjektive und nichtsurjektive Poincaré-Homomorphismen zu erhalten, in denen die Hindernismoduln durch Kohomologiemoduln mit Werten in der Ausgangsgarbe $\mathscr{F}$ (und nicht etwa in lokalen Kohomologiegarben $\mathscr{H}_*(X,\mathscr{F})$) gegeben sind, betrachten wir die Homomorphismen

$$Q_j:=Q_j^{\varphi\cap(X\setminus T)}(X\setminus T,\mathscr{F}): H^j_{\varphi\cap(X\setminus T)}(X\setminus T,\mathscr{F})\to H^{\varphi}_{2n-j}(X, T;\mathscr{F}).$$

Grundlegend ist das folgende technische

LEMMA 2.1. (i) *Ist*

$$\operatorname{Kern} Q_j\cap\operatorname{Bild}[H^j_{\tilde{\varphi}}(\tilde{X},\tilde{\mathscr{F}})\to H^j_{\tilde{\varphi}\cap(\tilde{X}\setminus\tilde{T})}(\tilde{X}\setminus\tilde{T},\tilde{\mathscr{F}})]=0$$

*und*

$$\rho: H^{j-1}_{\tilde{\varphi}\cap(\tilde{X}\setminus\tilde{T})}(\tilde{X}\setminus\tilde{T},\tilde{\mathscr{F}})\xrightarrow{Q_{j-1}} H^{\varphi}_{2n-j+1}(X, T;\mathscr{F})/\operatorname{Bild} H^{\varphi}_{2n-j+1}(X,\mathscr{F})$$

*surjektiv, so existiert eine exakte Sequenz mit Koeffizienten in* $\mathscr{F}$:

$$\operatorname{Kern}\rho\longrightarrow H^j_{\tilde{\varphi}}(\tilde{X},\tilde{X}\setminus\tilde{T})\xrightarrow{\Psi_j} H^{\varphi|T}_{2n-j}(T)\oplus H^j_{\tilde{\varphi}}(\tilde{X})\xrightarrow{\Phi_j} H^{\varphi}_{2n-j}(X). \tag{2.1}$$

(ii) *Ist*

$$\operatorname{Bild}[H^{\varphi}_{2n-j}(X,\mathscr{F})\to H^{\varphi}_{2n-j}(X, T;\mathscr{F})]\subset\operatorname{Bild} Q_j$$

*und*

$$\operatorname{Kern} Q_j\subset\operatorname{Bild}[H^j_{\tilde{\varphi}}(\tilde{X},\tilde{\mathscr{F}})\to H^j_{\tilde{\varphi}\cap(\tilde{X}\setminus\tilde{T})}(\tilde{X}\setminus\tilde{T},\tilde{\mathscr{F}})],$$

*so existiert eine exakte Sequenz mit Koeffizienten in* $\mathscr{F}$:

$$H^{\varphi|T}_{2n-j}(T) \oplus H^j_{\tilde\varphi}(\tilde X) \xrightarrow{\Phi_j} H^{\varphi}_{2n-j}(X) \xrightarrow{\Delta_j} H^{j+1}_{\varphi}(\tilde X, \tilde X \setminus \tilde T) \xrightarrow{\Psi_{j+1}} H^{\varphi|T}_{2n-j-1}(T) \oplus H^{j+1}_{\tilde\varphi}(\tilde X) \quad (2.2)$$

(iii) *Beide Sequenzen sind natürlich in* $\mathscr{F}$.

(iv) *Ist* $f: X \to Y$ *eine holomorphe Abbildung* $n$*-dimensionaler Räume,* $T \subset X$ *und* $S \subset Y$ *abgeschlossen mit* $f(T) \subset S$, $f(X \setminus T) \subset X \setminus S$, $f_*[X] = a[Y]$ *für ein* $a \neq 0$, $\varphi$ *und* $\psi$ *parakompaktifizierende Familien auf* $X$ *bzw.* $Y$ *mit* $f^{-1}\psi \subset \varphi$, $f\varphi \subset \psi$, *und* $f \mid K$ *eigentlich für alle* $K \in \varphi$, *so gilt für Koeffizienten in lokalkonstantem lokal endlich erzeugtem* $\mathscr{F}$:

($\alpha$) *Erfüllt* $(Y, S)$ *mutatis mutandis ebenfalls die Voraussetzungen von* (i) *und ist* $H^{\varphi\setminus T}_{2n-j}(T) \xrightarrow{f^1_{2n-j}} H^{\psi\setminus S}_{2n-j}(S)$ *die Nullabbildung, so induziert* $f$ *ein Diagramm mit den beiden Sequenzen zu* (2.1), *das kommutativ ist bis auf Multiplikation mit* $a$.

($\beta$) *Erfüllt* $(Y, S)$ *mutatis mutandis ebenfalls die Voraussetzungen von* (ii) *und sind* $f^T_{2n-j}$ *und* $f^T_{2n-j-1}$ *die Nullabbildungen, so induziert* $f$ *ein Diagramm mit den beiden Sequenzen zu* (2.2), *das kommutativ ist bis auf Multiplikation mit* $a$.

Beweis. Wir benutzen für den Beweis von (i) folgende algebraische Aussage (vgl. [9], Lemma 1.1): Es sei ein exaktes kommutatives Diagramm von $L$-Moduln gegeben

$$\begin{array}{ccccccc} \tilde C & \xrightarrow{\tilde\gamma} & \tilde A & \xrightarrow{\tilde\alpha} & R \oplus \tilde B & \xrightarrow{\tilde\beta} & \tilde C_1 \\ \downarrow{\scriptstyle\tau} & & \downarrow{\scriptstyle\mu} & & \downarrow{\scriptstyle 1_R \oplus \sigma} & & \downarrow{\scriptstyle\tau_1} \\ C & \xrightarrow{\gamma} & A & \xrightarrow{\alpha} & R \oplus B & \longrightarrow & C_1 \end{array}$$

mit $\tau_1 \mid \tilde\beta(\tilde B)$ ist injektiv und $\rho: \tilde C \to C/\mathrm{Ker}\,\gamma$ ist surjektiv. Dann ist mit

$$\Psi := (\tilde\alpha, \mu) \qquad \Phi := \pi_B \alpha - \sigma$$

folgende Sequenz exakt:

$$\mathrm{Kern}\,\rho \xrightarrow{\tilde\gamma} \tilde A \xrightarrow{\Psi} \tilde B \oplus A \xrightarrow{\Phi} B. \quad (2.3)$$

Wenden wir dies an mit $R = 0$ auf das kommutative Diagramm mit Koeffizienten in $\mathscr{F}$

$$\begin{array}{ccccccc} H^{j-1}_{\tilde\varphi \cap (\tilde X\setminus \tilde T)}(\tilde X \setminus \tilde T) & \to & H^j_{\tilde\varphi}(\tilde X, \tilde X \setminus \tilde T) & \to & H^j_{\tilde\varphi}(\tilde X) & \to & H^j_{\tilde\varphi \cap (\tilde X\setminus \tilde T)}(\tilde X \setminus \tilde T) \\ \downarrow{\scriptstyle Q^{\varphi}_{j-1}(X\setminus T)} & & \downarrow{\scriptstyle P^{\varphi}_j(X, X\setminus T)} & & \downarrow{\scriptstyle P^{\varphi}_j(X)} & & \downarrow{\scriptstyle Q^{\varphi}_j(X\setminus T)} \\ H^{\varphi}_{2n-j+1}(X, T) & \to & H^{\varphi|T}_{2n-j}(T) & \to & H^{\varphi}_{2n-j}(X) & \to & H^{\varphi}_{2n-j}(X, T), \end{array}$$

so erhalten wir (2.1).

Ist weiter ein kommutatives exaktes Diagramm von $L$-Moduln gegeben

$$\begin{array}{ccccccc} & & R\oplus\tilde{B} & \xrightarrow{\tilde{\beta}} & \tilde{C}_1 & \xrightarrow{\tilde{\lambda}} & \tilde{A}_1 & \longrightarrow & \tilde{B}_1 \\ & & \downarrow{\scriptstyle 1_R\oplus\sigma} & & \downarrow{\scriptstyle \tau_1} & & \downarrow & & \\ A & \longrightarrow & R\oplus B & \xrightarrow{\beta} & C_1 & \longrightarrow & A_1 & & \end{array}$$

mit $(\beta\,|\,B)(B)\subset\tau_1(\tilde{C}_1)$ und Kern $\tau_1\subset$ Bild $\tilde{\beta}$, so existiert eine exakte Sequenz mit $\Delta:=\tilde{\lambda}\tau_1^{-1}\beta\,|\,B$

$$\tilde{B}\oplus A\xrightarrow{\Phi}B\xrightarrow{\Delta}\tilde{A}_1\xrightarrow{\psi}A_1\oplus\tilde{B}_1. \tag{2.4}$$

Dies wenden wir an auf das kommutative Diagramm mit Koeffizienten in $\mathscr{F}$, und $R=0$:

$$\begin{array}{ccccccc} H^j_{\tilde{\varphi}}(\tilde{X}) & \to & H^j_{\tilde{\varphi}\cap(X\setminus T)}(\tilde{X}\setminus\tilde{T}) & \to & H^{j+1}_{\tilde{\varphi}}(\tilde{X},\tilde{X}\setminus\tilde{T}) & \to & H^{j+1}_{\tilde{\varphi}}(\tilde{X}) \\ \downarrow{\scriptstyle P^{\varphi}_j(X)} & & \downarrow{\scriptstyle Q^{\varphi}_j(X\setminus T)} & & \downarrow{\scriptstyle P^{\varphi}_{j+1}(X,X\setminus T)} & & \\ H^{\tilde{\varphi}|T}_{2n-j}(T) & \to & H^{\tilde{\varphi}}_{2n-j}(X) & \to & H^{\tilde{\varphi}}_{2n-j}(X,T) & \to & H^{\tilde{\varphi}|T}_{2n-j-1}(T), \end{array}$$

woraus die exakte Sequenz (2.2) folgt.

Die Natürlichkeit von (2.1) und (2.2) in $\mathscr{F}$ folgt mit Theorem 2.1 aus [6] wegen der rein algebraischen Konstruktion von (2.3) und (2.4).

Die Voraussetzungen an $f$, $\varphi$, und $\psi$ in (iv) garantieren, daß $f$ Homomorphismen aller zu betrachtenden Homologie- und Kohomologiegruppen induziert. Wenn $f^T_{2n-j}$ bzw. zusätzlich $f^T_{2n-j-1}$ die Nullabbildungen sind, dann sind damit ersichtlich Diagramme definiert, in denen wir uns bei den Doppelpfeilen $\downarrow\!\uparrow$ nur um den einen zu kümmern haben. Als Beispiel für die Kommutativität betrachten wir

$$\begin{array}{ccc} H^{\varphi}_{2n-j}(X) & \xrightarrow{\Delta} & H^{j+1}_{\tilde{\varphi}}(\tilde{X},\tilde{X}\setminus\tilde{T}) \\ \downarrow & & \uparrow \\ H^{\psi}_{2n-j}(Y) & \xrightarrow{\Delta} & H^{j+1}_{\tilde{\varphi}}(\tilde{Y},\tilde{Y}\setminus\tilde{S}). \end{array}$$

Zunächst wird wegen $a\neq 0$ keine der irreduziblen Komponenten des (reindimensionalen) Raumes $X$ auf ein niederdimensionales Bild geworfen. Daher kann man $f$ eindeutig zu einer holomorphen Abbildung $\tilde{f}:\tilde{X}\to\tilde{Y}$ auf die Normalisierungen liften. Gemäß Definition von $\Delta$ zerfällt das zu betrachtende Rechteck in

$$\begin{array}{ccccccc} H^{\varphi}_{2n-j}(X) & \longrightarrow & H^{\varphi}_{2n-j}(X,T) & \xleftarrow{Q_j} & H^j_{\tilde{\varphi}\cap(\tilde{X}\setminus\tilde{T})}(\tilde{X}\setminus\tilde{T}) & \longrightarrow & H^{j+1}_{\tilde{\varphi}}(\tilde{X},\tilde{X}\setminus\tilde{T}) \\ \downarrow{\scriptstyle f_*} & & \downarrow{\scriptstyle f_*} & & \uparrow{\scriptstyle f^*} & & \uparrow{\scriptstyle f^*} \\ H^{\psi}_{2n-j}(Y) & \longrightarrow & H^{\psi}_{2n-j}(Y,S) & \xleftarrow{Q_i} & H^j_{\tilde{\psi}\cap(\tilde{Y}\setminus\tilde{S})}(\tilde{Y}\setminus\tilde{S}) & \longrightarrow & H^{j+1}_{\tilde{\varphi}}(\tilde{Y},\tilde{Y}\setminus\tilde{S}). \end{array}$$

Bis auf das mittlere Rechteck sind offensichtlich alle Teilrechtecke kommutativ. Weil $T$ und $S$ abgeschlossen sind, gilt

$$H_*^{\varphi}(X, T) \cong H_*^{\varphi \cap (X \setminus T)}(X \setminus T)$$
$$H_*^{\psi}(Y, S) \cong H_*^{\psi \cap (Y \setminus S)}(Y \setminus S)$$

für die betrachteten Koeffizientengarben. Also bleibt

$$\begin{array}{ccc} H^j_{\tilde{\varphi} \cap (\tilde{X} \setminus \tilde{T})}(\tilde{X} \setminus \tilde{T}) & \xrightarrow{P_j(X \setminus T)} & H_{2n-j}^{\varphi \cap (X \setminus T)}(X \setminus T) \\ \uparrow f^* & & \downarrow f_* \\ H^j_{\tilde{\psi} \cap (\tilde{Y} \setminus \tilde{S})}(\tilde{Y} \setminus \tilde{S}) & \xrightarrow{P_j(Y \setminus S)} & H_{2n-j}^{\psi \cap (Y \setminus S)}(Y \setminus S) \end{array}$$

zu untersuchen: $\varphi \cap (X \setminus T)$ und $\psi \cap (Y \setminus S)$ sind parakompaktifizierend, folglich werden die Poincaré-Homomorphismen durch das Cupprodukt mit der Fundamentalklasse dargestellt; für

$$b \in H^j_{\tilde{\psi} \cap (\tilde{Y} \setminus \tilde{S})}(\tilde{Y} \setminus \tilde{S}) \quad \text{gilt} \quad f_*([\tilde{X} \setminus \tilde{T}] \cap f^* b) = f_*[\tilde{X} \setminus \tilde{T}] \cap b = a[\tilde{Y} \setminus \tilde{S}] \cap b,$$

woraus der Rest leicht folgt.

Als wichtigste Anwendung ergibt sich eine exakte Sequenz für Poincaré-Homomorphismen im Falle niederdimensionaler $T$, in der die Hindernisgruppen globale Kohomologiegruppen sind:

Satz 2.2. *Es sei* $(\varphi \mid T) - \dim T \leqq 2n - j - 1$ *für ein festes* $j$ *und* $\mathscr{H}_{2n-k}(X, \mathscr{F}) \mid X \setminus T = 0$ *für alle* $k$ *mit* $1 \leqq k \leqq j$. *Dann existiert eine exakte Sequenz:*

$$0 = H^1_{\tilde{\varphi}}(\tilde{X}, \tilde{X} \setminus \tilde{T}; \tilde{\mathscr{F}}) \longrightarrow H^1_{\tilde{\varphi}}(\tilde{X}, \tilde{\mathscr{F}}) \xrightarrow{P_1} H^{\varphi}_{2n-1}(X, \mathscr{F}) \longrightarrow$$
$$H^2_{\tilde{\varphi}}(\tilde{X}, \tilde{X} \setminus \tilde{T}; \tilde{\mathscr{F}}) \longrightarrow \cdots \longrightarrow H^{j+1}_{\tilde{\varphi}}(\tilde{X}, \tilde{X} \setminus \tilde{T}; \tilde{\mathscr{F}}) \xrightarrow{\Psi}$$
$$H^{\varphi|T}_{2n-j-1}(T, \mathscr{F}) \oplus H^{j+1}_{\tilde{\varphi}}(\tilde{X}, \tilde{\mathscr{F}}) \xrightarrow{\Psi} H^{\varphi}_{2n-j-1}(X, \mathscr{F}).$$

Beweis. Wegen $\mathscr{H}_{2n-k}(X, \mathscr{F}) \mid X \setminus T = 0$ ist $Q_k^{\varphi \cap (X \setminus T)}(X \setminus T, \mathscr{F})$ bijektiv für $0 \leqq k \leqq j$ nach [6] Satz 2.2 und Satz 2.3. Damit folgt die Behauptung aus 2.1 Lemma.

Korollar 2.3. *Ist* $U \subset X$ *offen und* $\mathscr{F}$ *lokalkonstant und lokal endlich erzeugt,* $[\varphi \mid (T \setminus U)] - \dim T \leqq 2n - j - 1$ *und* $\mathscr{H}_{2n-k}(X, \mathscr{F}) \mid X \setminus U \setminus T = 0$ *für alle* $k$ *mit* $1 \leqq k \leqq j$, *so existiert eine exakte Sequenz*

$$0 = H^1_{\tilde{\varphi}}(\tilde{X}, \tilde{X} \setminus \tilde{U} \setminus \tilde{T}; \tilde{\mathscr{F}}) \longrightarrow H^1_{\tilde{\varphi}}(\tilde{X}, \tilde{U}; \tilde{\mathscr{F}}) \xrightarrow{P_1} H^{\varphi|X \setminus U}_{2n-1}(X \setminus U, \mathscr{F}) \longrightarrow$$
$$H^2_{\tilde{\varphi}}(\tilde{X}, U \setminus \tilde{T}; \tilde{\mathscr{F}}) \longrightarrow \cdots \longrightarrow H^{j+1}_{\tilde{\varphi}}(\tilde{X}, \tilde{X} \setminus \tilde{U} \setminus \tilde{T}; \mathscr{F}).$$

BEWEIS. Man wähle $\psi := \varphi \mid X \setminus U$, dann ist für lokal konstantes $\mathscr{F}$

$$H^*_{\tilde\psi}(\tilde X, \tilde{\mathscr{F}}) \cong H^*_{\tilde\varphi}(\tilde X, \tilde U; \tilde{\mathscr{F}})$$
$$H^{\psi}_*(X, \mathscr{F}) \cong H^{\varphi}_*(X \setminus U, \mathscr{F})$$
$$H^*_{\tilde\psi}(\tilde X, \tilde X \setminus \tilde T; \tilde{\mathscr{F}}) \cong H^*_{(\tilde\varphi|\tilde X\setminus \tilde U)|\tilde T}(\tilde X, \tilde{\mathscr{F}}) \cong H^*_{\tilde\varphi|\tilde X \setminus \tilde T \setminus \tilde U}(\tilde X, \tilde{\mathscr{F}}) \cong H^*_{\tilde\varphi}(\tilde X, \tilde X \setminus \tilde T \setminus \tilde U, \mathscr{F}).$$

Damit läßt sich Satz 2.2 anwenden.

Andererseits ergibt sich für $T$ mit verschwindender niedrigdimensionaler Homologie

SATZ 2.4. *Es sei ein* $j$ *mit* $2n - j \leqq \mathrm{Tab}_{\varphi \cap (X \setminus T)}(X \setminus T, \mathscr{F}) + 2$ *gegeben mit* $H^{\varphi|T}_{2n-j}(T, \mathscr{F}) = 0 = H^{\varphi|T}_{2n-j-1}(T, \mathscr{F})$. *Dann existiert eine exakte Sequenz*

$$H^j_{\tilde\varphi}(\tilde X, \tilde X \setminus \tilde T; \tilde{\mathscr{F}}) \longrightarrow H^j_{\tilde\varphi}(\tilde X, \tilde{\mathscr{F}}) \xrightarrow{P_j} H^{\varphi}_{2n-j}(X, \mathscr{F}) \longrightarrow$$
$$H^{j+1}_{\tilde\varphi}(\tilde X, X \setminus \tilde T; \tilde{\mathscr{F}}) \longrightarrow \cdots \xrightarrow{P_{j+1}} H^{\varphi}_{2n-j-1}(X, \mathscr{F}) \longrightarrow H^{j+2}_{\tilde\varphi}(\tilde X, \tilde X \setminus \tilde T; \tilde{\mathscr{F}}).$$

BEWEIS. Wegen $2n - j \leqq \mathrm{Tab}_{\varphi \cap (X \setminus T)}(X \setminus T, \mathscr{F}) + 2$ sind $Q^{\varphi \cap (X \setminus T)}_j(X \setminus T, \mathscr{F})$ und $Q^{\varphi \cap (X \setminus T)}_{j+1}(X \setminus T, \mathscr{F})$ bijektiv, so daß man aus den exakten Sequenzen (2.1) und (2.2) für $j$ bzw. $j+1$ die Behauptung erhält. In gleicher Weise wie 2.3 aus 2.2 erhält man aus 2.4:

KOROLLAR 2.5. *Ist* $U \subset X$ *offen,* $\mathscr{F}$ *lokalkonstant mit endlich erzeugten Halmen, ferner ein* $j$ *gegeben mit* $2n - j \leqq \mathrm{Tab}_{\varphi \cap (X \setminus T \setminus U)}(X \setminus T \setminus U, \mathscr{F}) + 2$ *und mit* $H^{\varphi|T \setminus U}_{2n-j}(T \setminus U, \mathscr{F}) = H^{\varphi|T \setminus U}_{2n-j-1}(T \setminus U, \mathscr{F}) = 0$, *so existiert eine exakte Sequenz mit Koeffizienten in* $\mathscr{F}$:

$$H^j_{\tilde\varphi}(\tilde X, \tilde X \setminus \tilde T \setminus \tilde U) \longrightarrow H^j_{\tilde\varphi}(\tilde X, \tilde U) \xrightarrow{P_j} H^{\varphi|X \setminus U}_{2n-j}(X \setminus U) \longrightarrow$$
$$H^{j+1}_{\tilde\varphi}(\tilde X, \tilde X \setminus \tilde T \setminus \tilde U) \longrightarrow \cdots \longrightarrow H^{j+2}_{\tilde\varphi}(\tilde X, \tilde X \setminus \tilde T \setminus \tilde U).$$

BEMERKUNGEN 2.6. (i) Ist $(\varphi \mid T)-\dim T \leqq 2n-1$, dann ist $H^0_{\varphi}(X, X \setminus T; \mathscr{F}) = 0$. Dies ist evident für lokal irreduzibles $X$; im allgemeinen Fall ist $H^0_{\varphi}(X, X \setminus T; \mathscr{F})$ ein Untermodul von $H^0_{\tilde\varphi}(\tilde X, \tilde X \setminus \tilde T; \tilde{\mathscr{F}})$, woraus die Behauptung folgt. Nach [6] Satz 3.1 gilt dies für beliebige Garben von $L$-Moduln $\mathscr{F}$ auf $X$.

Falls $(\varphi \mid T)-\dim T \leqq 2n-2$, dann ist nach 2.2

$$H^1_{\tilde\varphi}(\tilde X, \tilde X \setminus \tilde T; \tilde{\mathscr{F}}) = 0.$$

(ii) Ist $X \setminus T$ lokal irreduzibel, so existiert eine exakte Sequenz

$$H^{2n}_{\varphi}(X, X \setminus T; \mathscr{F}) \longrightarrow H^{2n}_{\varphi}(X, \mathscr{F}) \xrightarrow{P_{2n}} H^{\varphi}_0(X, \mathscr{F}) \longrightarrow 0,$$

falls $H^{2n}_{\varphi}(X, X \setminus T; \mathscr{F}) \to H^{\varphi|T}_0(T, \mathscr{F}) \to H^{\varphi}_0(X, \mathscr{F})$ die Nullabbildung ist.

(iii) Für lokal irreduzibles $X$ kann man $T$ so wählen, daß $(\varphi \mid T)-\dim T \leqq 2n-4$ und daß $\mathcal{H}_{2n-k}(X, \mathcal{F}) \mid X \backslash T = 0$ für $1 \leqq k \leqq 3$ (etwa $T \supset AS(X)$). Dann existiert nach 2.2 eine exakte Sequenz mit Koeffizienten in $\mathcal{F}$:

$$0 \to H^1_\varphi(X) \to H^\varphi_{2n-1}(X) \to H^2_\varphi(X, X\backslash T) \to H^2_\varphi(X) \to H^\varphi_{2n-2}(X) \to \cdots \to H^4_\varphi(X, X\backslash T).$$

Nach 1.1 Satz ist dann $\mathcal{H}_{2n-k}(X, \mathcal{F}) \cong \mathcal{H}^{k+1}_T(X, \mathcal{F})$ für $k = 1, 2$. Man hat einen kanonischen Homomorphismus mit Koeffizienten in einer beliebigen Garbe $\mathcal{G}$:

$$H^p_\varphi(X, X\backslash T) \to H^0_\varphi(X, \mathcal{H}^p_T(X)) \tag{2.5}$$

wie folgt: Ist $Z \subset X$ abgeschlossen, so wähle man

$$H^p_Z(X, X\backslash T) = H^p_{Z\cap T}(X) \to H^0(X, \mathcal{H}^p_{Z\cap T}(X)) \to H^0_Z(X, \mathcal{H}^p_T(X)).$$

Dann wird für allgemeines $\varphi$ der Homomorphismus (2.5) als induktiver Limes über alle $Z \in \varphi$ definiert. – In der vorliegenden Situation resultieren aus Satz 1.1 mit (2.5) Homomorphismen

$$\tau_k : H^{k+1}(X, X\backslash T) \to H^0_\varphi(X, \mathcal{H}_{2n-k}(X)),$$

wobei $\tau_1$ bijektiv ist (mit einem Spektralsequenzargument für $\varphi = cld(X)$). Daher kann die exakte Sequenz aus (iii) als eine Fortsetzung der exakten Sequenz aus Diagramm 3.3 in [6] angesehen werden.

(iv) Ist $S(X\backslash T)$ isoliert und $H^0_{\varphi\cap(X\backslash T)}(X\backslash T, \mathcal{H}_{2n-j}(X, \mathcal{F})) = 0$ für ein festes $j \neq 2n$ und $H^{\varphi|T}_{2n-j}(T, \mathcal{F}) = 0 = H^{\varphi|T}_{2n-j-1}(T, \mathcal{F})$, so ist exakt

$$H^j_{\tilde\varphi}(\tilde X, \tilde{\mathcal{F}}) \to H^\varphi_{2n-j}(X, \mathcal{F}) \to H^{j+1}_{\tilde\varphi}(\tilde X, \tilde X\backslash \tilde T; \tilde{\mathcal{F}}) \to H^{j+1}_{\tilde\varphi}(\tilde X, \tilde{\mathcal{F}}) \to H^\varphi_{2n-j-1}(X, \mathcal{F}),$$

auch wenn $H^0_{\varphi\cap(X\backslash T)}(X\backslash T, \mathcal{H}_{2n-j+1}(X, \mathcal{F})) \neq 0$.

(v) Ist $\dim T = 0$, $\dim S(X\backslash T) = 0$, $H^0_\varphi(X\backslash T, \mathcal{H}_{2n-j}(X, \mathcal{F})) = 0$ für ein $2n \neq j \neq 2n-1$, so ist analog (iii) mit 1.1 Satz für abgeschlossenes $T$ folgende Sequenz mit Koeffizienten in $\mathcal{F}$ exakt:

$$H^j_{\tilde\varphi}(\tilde X) \to H^\varphi_{2n-j}(X) \to H^0_{\tilde\varphi}(\tilde X, \mathcal{H}^{j+1}_{\tilde T}(\tilde X)) \to H^{j+1}_{\tilde\varphi}(\tilde X) \to H^\varphi_{2n-j-1}(X).$$

(vi) Es sei $X\backslash T$ topologisch lokal vollständiger Durchschnitt mit $\dim_{\mathbb{C}} S(X\backslash T) \leqq 1$, $\varphi$ sei parakompaktifizierend (oder allgemeiner $[\varphi \mid S(X\backslash T)]-\dim S(X\backslash T) \leqq 2$). Dann gilt

(2.1) ist exakt, falls $j < n-2$ oder $j \geqq n+3$

(2.2) ist exakt, falls $j \leqq n-3$ oder $j \geqq n+3$

(vii) Wenn $S(X\backslash T)$ und $T$ sich nicht berühren, dann ist

$$H^*_\varphi(X, \mathcal{H}_*(X, \mathcal{F})) \cong H^*_{\varphi|X\backslash T}(X\backslash T, \mathcal{H}_*(X, \mathcal{F})) \oplus H^*_{\varphi|T}(\bar T, \mathcal{H}_*(X, \mathcal{F})).$$

Daher ist dann in (ii) von 2.1 Lemma stets Kern $Q_j$ enthalten in

$$\text{Bild}\,[H^j_{\tilde\varphi}(\tilde X, \tilde{\mathscr F}) \to H^j_{\tilde\varphi \cap (\tilde X \setminus \tilde T)}(\tilde X \setminus \tilde T, \tilde{\mathscr F})].$$

(viii) $T$ sei Vereinigung von $r$ offenen (bzw. von $r$ abgeschlossenen) Steinschen Teilmengen von $X$, $B$ eine (abgeschlossene) analytische Teilmenge von $T$, $\mathscr F \mid T\setminus B$ sei konstant und $\varphi \mid T = c(T) \cap (T\setminus B)$. Dann existiert eine exakte Sequenz mit Koeffizienten in $\mathscr F$, falls $\mathscr H_{2n-k}(X, \mathscr F) \mid X\setminus T = 0$ für alle $k$ mit $1 \leqq k \leqq r-1$:

$$0 \longrightarrow H^1_{\tilde\varphi}(\tilde X) \xrightarrow{P_1} H^{\varphi}_{2n-1}(X) \longrightarrow H^2_{\tilde\varphi}(\tilde X, \tilde X \setminus \tilde T) \longrightarrow H^2_{\tilde\varphi}(\tilde X) \xrightarrow{P_2}$$
$$\cdots \quad \cdots \longrightarrow H^{n-r}_{\tilde\varphi}(\tilde X) \xrightarrow{P_{n-r}} H^{\varphi}_{n+r}(X).$$

Zum Nachweis der Voraussetzungen von 2.1 benutzt man, daß für einen komplexen Raum $T$, der sich als Vereinigung von $r$ offenen Steinschen Teilmengen schreiben läßt, $H^c_{n+r}(T, B; M) = H^{c(T)\cap(T\setminus B)}_{n+r}(T\setminus B, M) = 0$ für beliebige konstante Koeffizienten $M$ (vgl. [8]).

(ix) Es sei $B \subset T$ eine abgeschlossene Teilmenge, $t \geqq \operatorname{tab}(X\setminus B)$ und $s := \dim S(X\setminus B) \leqq n-2$. Dann ist in 2.1 Lemma (ii) mit $j = n+t+s$ die Voraussetzung

$$\text{Bild}\,[H^{\varphi}_{n-t-s}(X, \mathscr F) \to H^{\varphi}_{n-t-s}(X, T; \mathscr F)] \subset \text{Bild}\, Q_{n+t+s}$$

erfüllt, falls $H^{2s}_{\varphi}(X, T; \mathscr H_{n-t+s}(X, \mathscr F)) \to H^{2s}_{\varphi}(X, \mathscr H_{n-t+s}(X, \mathscr F))$ die Nullabbildung ist; weiter ist die Voraussetzung

$$\text{Kern}\, Q_{n+t+s} \subset \text{Bild}\,[H^{n+t+s}_{\tilde\varphi}(\tilde X, \tilde{\mathscr F}) \to H^{n+t+s}_{\tilde\varphi \cap (\tilde X \setminus \tilde T)}(\tilde X \setminus \tilde T, \tilde{\mathscr F})]$$

erfüllt, falls $t \geqq \operatorname{tab}(X\setminus B) + 1$ und

$$H^{2s}_{\varphi}(X, \mathscr H_{n-t-1+s}(X, \mathscr F)) \to H^{2s}_{\varphi \cap (X\setminus T)}(X\setminus T, \mathscr H_{n-t-1-s}(X, \mathscr F))$$

surjektiv ist.

Der Beweis wird mit Hilfe von Lemma 0.4 geführt.

(x) Für $s := \dim S(X)$ gilt, daß mit $T = S(X)$ für vernünftige Trägerfamilien $\varphi$ (etwa parakompaktifizierend) die Sequenzen aus 2.2 Satz und 2.3 Korollar exakt sind bis $H^{2(n-s)}_{\tilde\varphi}(\tilde X, \tilde X \setminus \tilde T; \tilde{\mathscr F})$ bzw. $H^{2(n-s)}_{\tilde\varphi}(\tilde X, \tilde X \setminus \tilde U \setminus \tilde T; \mathscr F)$ einschließlich. Es sei schon hier vermerkt, daß die Sequenz aus 2.2d Satz (und die aus dem nicht explizit formulierten Korollar 2.3d) exakt ist ab $H^{2s+1}(\tilde X, \tilde{\mathscr F})$; diese beiden Sequenzen beinhalten für $s \leqq (n-1/2)$ folglich alle $P^{\varphi}_j(X, \mathscr F)$. – Falls $\mathscr F$ konstant ist und $S(X)$ keine irreduzible kompakte Komponente der Dimension $s = \dim S(X)$ hat, so gilt: Ist $\varphi \mid T = c(T)$, so setzt sich die Sequenz aus 2.2 Satz fort bis

$$\cdots \longrightarrow H^{2(n-s)}_{\tilde\varphi}(\tilde X, \tilde X \setminus \tilde T; \tilde{\mathscr F}) \longrightarrow H^{2(n-s)}_{\tilde\varphi}(\tilde X, \tilde{\mathscr F}) \xrightarrow{P}$$
$$H^{\varphi}_{2s}(X, \mathscr F) \longrightarrow H^{2(n-s)+1}_{\tilde\varphi}(\tilde X, \tilde X \setminus \tilde T; \tilde{\mathscr F}).$$

Ist $\varphi \cap T = cld\, T$, so beginnt die Sequenz aus 2.2d Satz bereits mit

$$H^{2s}_{\tilde\varphi}(\tilde X, \tilde{\mathscr F}) \xrightarrow{P_{2s}} H^{\varphi}_{2n-2s}(X, \mathscr F) \longrightarrow H^{\varphi}_{2n-2s}(X, X \setminus T; \mathscr F) \longrightarrow H^{2s+1}_{\tilde\varphi}(\tilde X, \tilde{\mathscr F}) \longrightarrow \cdots.$$

Also ergibt sich in diesen Fällen auch für $s \leqq n/2$ eine fast vollständige Überdeckung aller $P^c_j(X)$. Das analoge Resultat gilt für $P^c_j(X, U)$.

(xi) *Rangtheorem für* $P^{\varphi}_*(X)$: Für Rangberechnungen der Poincaré-Homomorphismen genügt es ersichtlich, $L$ durch den Körper der Quotienten $Q(L)$ zu ersetzen. Daher setzen wir in (xi) voraus, daß $L$ bereits ein Körper ist. Dann gilt mit

$$b^j_{\psi}(Y, A) := \dim_L H^j_{\psi}(Y, A; L) \quad \text{und} \quad b^{\psi}_j(Y, A) := \dim_L H^{\psi}_j(Y, A; L):$$

*Ist* $Q^{\varphi}_j(X \setminus T)$ *bijektiv und für* $i = j, j+1$

$$\text{Bild}\, H^{\varphi|T}_{2n-i}(T) \subset \text{Bild}\, P^{\varphi}_i(X),$$

*so gilt*

$$\begin{aligned}\text{rang}\, P^{\varphi}_j(X) + \text{rang}\, P^{\varphi}_{j+1}(X) = b^{\varphi}_{2n-j}(X) + b^{j+1}_{\tilde\varphi}(\tilde X)\\ + b^{\varphi|T}_{2n-j-1}(T) - b^{j+1}_{\tilde\varphi}(\tilde X, \tilde X \setminus \tilde T).\end{aligned}$$

Beweis. Wir wollen die exakten Sequenzen (2.3) sowie (2.4) ausnutzen, wobei wir $R = H^j_{\tilde\varphi}(\tilde X)/\text{Kern}\, P^{\varphi}_j(X) \cong \text{Bild}\, P^{\varphi}_j(X)$ wählen, $\tilde B$ sei Kern $P^{\varphi}_j(X)$; weil $L$ nach Voraussetzung ein Körper ist, erhalten wir eine direkte Summenzerlegung

$$H^j_{\tilde\varphi}(\tilde X) \cong H^j_{\tilde\varphi}(\tilde X)/\text{Kern}\, P_j \oplus \text{Kern}\, P_j \to \text{Bild}\, P_j \oplus H^{\varphi}_{2n-j}(X)/\text{Bild}\, P_j \cong H^{\varphi}_{2n-j}(X).$$

Der Beweis für (2.3) und (2.4) zeigt, daß wegen $Q^{\varphi}_j(X \setminus L)$ bijektiv eine exakte Sequenz resultiert

$$H^{\varphi|T}_{2n-j}(T) \oplus \text{Kern}\, P^{\varphi}_j \to H^{\varphi}_{2n-j}(X)/\text{Bild}\, P_j \to H^{j+1}_{\tilde\varphi}(\tilde X, \tilde X \setminus \tilde T) \to H^{\varphi|T}_{2n-j-1}(T) \oplus H^{j+1}_{\tilde\varphi}(\tilde X) \to H^{\varphi}_{2n-j-1}(X).$$

Wegen der Voraussetzung Bild $H^{\varphi|T}_{2n-i}(T) \subset \text{Bild}\, P^{\varphi}_i(X)$ induziert diese Sequenz nach Konstruktion eine exakte Sequenz

$$0 \to H^{\varphi}_{2n-j}(X)/\text{Bild}\, P_j \to H^{j+1}_{\tilde\varphi}(\tilde X, \tilde X \setminus \tilde T) \to H^{\varphi|T}_{2n-j-1}(T) \oplus H^{j+1}_{\tilde\varphi}(\tilde X) \to \text{Bild}\, P^{\varphi}_{j+1}(X) \to 0,$$

woraus die Behauptung folgt.

Die Dualitätstheorie ist nicht 'symmetrisch' in Homologie und Kohomologie, da einerseits nur in der Kohomologie die Normalisierung zu betrachten ist, andererseits ist bei Unterräumen $A \subset X$ in der Kohomologie $\varphi \cap A$, in der Homologie $\varphi \mid A$ anzusetzen. Daher bleibt es uns nicht erspart, eine 'duale' Situation zu 2.1 zu betrachten. – Es sei dazu $A \subset X$ lokal abgeschlossen und $\varphi$-taut; wir setzen abkürzend

$$\psi := \varphi \mid X \backslash A.$$

$\mathscr{F}$ darf beliebig sein, wenn $A$ abgeschlossen und $\varphi$ parakompaktifizierend. Dann gilt zunächst:

LEMMA 2.7. ($\alpha$) *Es existiert eine exakte Sequenz mit Koeffizienten in* $\mathscr{F}$

$$\text{Kern}\, \rho \longrightarrow H^j_{\tilde{\varphi}}(\tilde{X}) \xrightarrow{\psi_j} H^{\varphi}_{2n-j}(X) \oplus H^j_{\tilde{\varphi} \cap \tilde{A}}(\tilde{A}) \xrightarrow{\Phi_j} H^{\varphi}_{2n-j}(X, X \backslash A) \quad (2.6)$$

*falls*

$$P^{\varphi}_{j+1}(X, A) \mid \text{Bild}\, H^j_{\tilde{\varphi} \cap \tilde{A}}(\tilde{A})$$

*ist injektiv*

$$\rho := \bar{P}^{\varphi}_j(X, A) : H^j_{\tilde{\varphi}}(\tilde{X}, \tilde{A}) \to H^{\psi}_{2n-j}(X \backslash A)/\text{Bild}\, H^{\varphi}_{2n-j+1}(X, X \backslash A)$$

*ist auf.*

($\beta$) *Es existiert eine exakte Sequenz mit Koeffizienten in* $\mathscr{F}$

$$H^{\varphi}_{2n-j}(X) \oplus H^j_{\tilde{\varphi} \cap \tilde{A}}(\tilde{A}) \xrightarrow{\Phi_j} H^{\varphi}_{2n-j}(X, X \backslash A) \xrightarrow{\Delta} H^{j+1}_{\tilde{\varphi}}(\tilde{X}) \xrightarrow{\psi_{j+1}}$$

$$H^{\varphi}_{2n-j-1}(X) \oplus H^{j+1}_{\tilde{\varphi} \cap \tilde{A}}(\tilde{A}) \quad (2.7)$$

*falls*

$$\text{Bild}\, H^{\varphi}_{2n-j}(X, X \backslash A) \subset P^{\varphi}_{j+1}(X, A) H^{j+1}_{\tilde{\varphi}}(\tilde{X}, \tilde{A})$$
$$\text{Kern}\, P^{\varphi}_{j+1}(X, A) \subset \text{Bild}[H^j_{\tilde{\varphi} \cap \tilde{A}}(\tilde{A}) \to H^{j+1}_{\tilde{\varphi}}(\tilde{X}, \tilde{A})]$$

($\gamma$) *Beide Sequenzen sind natürlich in* $\mathscr{F}$.

($\delta$) *Ist* $f: X \to Y$ *eine holomorphe Abbildung, sind* $A \subset X$ *und* $B \subset Y$ *abgeschlossene Teilräume, so daß mutatis mutandis die Voraussetzungen von Lemma* 2.1, (iv) *erfüllt sind, dann existieren entsprechende kommutative Diagramme.*

Der Beweis ergibt sich aus (2.3) und (2.4) mit ähnlichen Schlüssen wie Lemma 2.1. Wir werden daher auf Einzelheiten verzichten und statt dessen die Voraussetzungen von Lemma 2.7 erläutern; dabei bezeichne $B$ eine abgeschlossene Teilmenge von $X$ mit $B \subset A$:

(I) $P^{\varphi}_{j+1}(X, A; \mathscr{F}) \mid \text{Bild}\, H^j_{\tilde{\varphi} \cap \tilde{A}}(\tilde{A}, \tilde{\mathscr{F}})$ ist injektiv, falls eine der folgenden Bedingungen erfüllt ist:

(i) $H^j_{\tilde{\varphi} \cap \tilde{A}}(\tilde{A}, \tilde{\mathscr{F}}) = 0$,

(ii) $j \geqq 2n - \text{Tab}_{\varphi}\, (X, A; \mathscr{F}) + 1$,

(iii) jedes $x \in X \backslash B$ hat eine gute offene Umgebung $U_x \subset X \backslash B$, und $j \leqq \min \text{Tab}_c\, (U_x, L) - 2$.

(iv) $j = -1, 0, 2n$; oder $X \backslash B$ lokal irreduzibel und $j = 2n - 1$.

(II) $\bar{P}_j(X, A; \mathscr{F}) : H^j_{\tilde{\varphi}}(\tilde{X}, \tilde{A}; \tilde{\mathscr{F}}) \to H^{\psi}_{2n-j}(X \backslash A, \mathscr{F})/\text{Bild}\, H^{\varphi}_{2n-j+1}(X, X \backslash A; \mathscr{F})$

ist surjektiv, falls eine der folgenden Bedingungen gilt:

(i) $H^{\psi}_{2n-j}(X\backslash A, \mathscr{F}) = 0$ oder $H^{\varphi}_{2n-j}(X, \mathscr{F}) = 0$,

(ii) $j \geqq 2n - \mathrm{Tab}_{\varphi}(X, A; \mathscr{F}) + 1$,

(iii) $j \leqq \min_{x \in X\backslash B} \mathrm{Tab}_c(U_x, L) - 2$ mit $U_x$ wie in (I),

(iv) $j = 0, 2n$; oder $X\backslash B$ lokal irreduzibel und $j = 2n-1$,

(v) $H^q_{\varphi|X\backslash B}(X\backslash B, \mathscr{H}_{2n+q-j}(X, \mathscr{F})) = 0$ für

$$\max(0, j - \mathrm{tab}(X\backslash B) - n) \leqq q$$
$$\leqq \min(j-1, (\varphi \mid X - B) - \dim S(X\backslash B) \cap \operatorname{supp} \mathscr{F})$$

(III) Bild $H^{\varphi}_{2n-j}(X, X\backslash A; \mathscr{F}) \subset P^{\varphi}_{j+1}(X, A; \mathscr{F}) H^{j+1}_{\varphi}(\tilde{X}, \tilde{A}; \tilde{\mathscr{F}})$, falls eine der folgenden Bedingungen gilt:

(i) $H^{\varphi}_{2n-j}(X, X\backslash A; \mathscr{F}) = 0$, oder $H_{2n-j-1}(X\backslash A, \mathscr{F}) = 0$,

(ii) $P^{\varphi}_{j+1}(X, A; \mathscr{F})$ ist surjektiv (dazu vgl. Bedingungen (II) außer (i)),

(iii) $H^q_{\varphi|X\backslash B}(X\backslash B, \mathscr{H}_{2n-j-1+q}(X, \mathscr{F})) = 0$ für

$$\max(1, j - \mathrm{tab}(X\backslash B) - n) \leqq q \leqq \min(j, (\varphi \mid X\backslash B) - \dim S(X\backslash B) \cap \operatorname{supp} \mathscr{F}).$$

(IV) Kern $P^{\varphi}_{j+1}(X, A; \mathscr{F}) \subset$ Bild $H^j_{\tilde{\varphi} \cap \tilde{A}}(\tilde{A}, \tilde{\mathscr{F}})$ gilt, falls eine der folgenden Bedingungen erfüllt ist:

(i) $H^{j+1}_{\tilde{\varphi}}(\tilde{X}, \tilde{A}; \tilde{\mathscr{F}}) = 0$ oder $H^{j+1}_{\tilde{\varphi}}(\tilde{X}, \tilde{\mathscr{F}}) = 0$,

(ii) $P^{\varphi}_{j+1}(X, A; \mathscr{F})$ ist injektiv (dazu vgl. (I) außer (i)),

(iii) $H^{q+1}(X, A; \mathscr{H}_{2n+q-j}(X, \mathscr{F})) = 0$ für $1 \leqq q \leqq j-1$ und

$$H^q(X, A; \mathscr{H}_{2n+q-j}(X, \mathscr{F})) = 0 \quad \text{für} \quad 0 \leqq q \leqq j-1.$$

Zum Beweis gehen wir nur auf (III) (iii) ein. Man erhält ein Diagramm

$$\begin{array}{ccc}
H^{j+1}_{\tilde{\varphi}}(\tilde{X}, \tilde{A}; \tilde{\mathscr{F}}) & \longrightarrow & H^{j+1}_{\tilde{\varphi}}(\tilde{X}, \tilde{\mathscr{F}}) \\
\downarrow {\scriptstyle P^{\varphi}_{j+1}(X,A)} & & \downarrow {\scriptstyle P^{\varphi}_{j+1}(X)} \\
H^{\psi}_{2n-j-1}(X\backslash A, \mathscr{F}) & \xrightarrow{\vartheta} & H^{\varphi}_{2n-j-1}(X, \mathscr{F}) \\
\downarrow & & \downarrow \\
H^0_{\varphi}(X, A; \mathscr{H}_{2n-j-1}(X, \mathscr{F})) & \hookrightarrow & H^0(X, \mathscr{H}_{2n-j-1}(X, \mathscr{F})),
\end{array}$$

welches nach [6] Satz 2.2 kommutativ und exakt ist. Dann folgt Kern $\vartheta \subset$ Bild $P^{\varphi}_{j+1}(X, A)$.

Es ist einleuchtend, daß Lemma 2.7 eine lange exakte 'Poincaré-Sequenz' liefert, wenn $\tilde{\mathscr{F}} \mid \tilde{A}$ $\tilde{\varphi} \cap \tilde{A}$-azyklisch und $X\backslash A$ eine $L$-Homologiemannigfaltigkeit ist. Das ist in der Regel jedoch eine viel zu starke Voraussetzung, wenn man die exakten 'Poincaré-Sequenzen' nur für gewisse Dimensionen benötigt.

Satz 2.2d. *Es sei* $(\varphi \cap A) - \dim A \leqq j-1$ *für ein festes* $j$ *und* $\mathscr{H}_{2n-k}(X, \mathscr{F}) \mid X\backslash A = 0$ *für alle* $k$ *mit* $1 \leqq k \leqq l$ *für ein* $l \geqq j$*. Dann existiert eine* *exakte Sequenz mit Koeffizienten in $\mathscr{F}$:*

$$H^j_{\tilde\varphi}(\tilde X) \xrightarrow{p_j} H^\varphi_{2n-j}(X) \longrightarrow H^\varphi_{2n-j}(X, X\backslash A) \xrightarrow{\chi}$$
$$H^{j+1}_{\tilde\varphi}(\tilde X) \xrightarrow{p_{j+1}} \cdots \longrightarrow H^{l+1}_{\tilde\varphi}(\tilde X) \xrightarrow{p_{l+1}} H^\varphi_{2n-l-1}(X).$$

Beweis. Es ist $(\varphi \cap A) - \dim A = (\tilde\varphi \cap \tilde A) - \dim \tilde A$, so daß die Voraussetzungen von $(\alpha)$ und $(\beta)$ aus 2.7 Lemma erfüllt sind, wie man leicht sieht.

Entsprechend dualisiert sich Korollar 2.3 zu einer relativen exakten Sequenz für $P^\varphi_j(X, U; \mathscr{F})$.

Satz 2.4d. *Für ein* $j \geqq 2n - \mathrm{Tab}_\varphi\,(X, A; \mathscr{F}) + 1$ *existiert eine exakte Sequenz mit Koeffizienten in $\mathscr{F}$:*

$$H^j_{\tilde\varphi}(\tilde X) \xrightarrow{p_j} H^\varphi_{2n-j}(X) \oplus H^j_{\tilde\varphi\cap\tilde A}(\tilde A) \longrightarrow H^\varphi_{2n-j}\, X, X\backslash A) \longrightarrow$$
$$H^{j+1}_{\tilde\varphi}(\tilde X) \xrightarrow{p_{j+1}} H^\varphi_{2n-j-1}(X) \oplus H^{j+1}_{\tilde\varphi\cap\tilde A}(\tilde A) \longrightarrow H^\varphi_{2n-j-1}(X, X\backslash A) \longrightarrow H^{j+2}_{\tilde\varphi}((\tilde X).$$

Der Beweis wird mit 2.7 Lemma geführt, ebenso wie die relative Form 2.5d Korollar, die wir dem Leser überlassen.

Es ergeben sich wieder eine Reihe an Bemerkungen:

Bemerkungen 2.8. (i) Falls $\mathscr{H}_{2n-1}(X, L)$ in eine Umgebung von $\tilde A$ verschwindet und $\dim_{\mathbb{R}} A \leqq 2n-2$, so ist

$$H^1_\varphi(X, X\backslash A, M) = 0$$

für $\varphi = c(X)$ und $\varphi = cld(X)$, $M$ beliebige $L$-Modul (2.4d und universelle Koeffiziententheoreme der Kohomologie).

(ii) Falls $(\varphi \cap A) - \dim A \leqq 2n-2$ und $\mathrm{Tab}_\varphi\,(X, A; \mathscr{F}) \geqq 2$ (insbesondere also, wenn $X$ in $X\backslash A$ lokal irreduzibel ist), so existiert eine exakte Sequenz mit Koeffizienten in $\mathscr{F}$

$$H^{2n-1}_{\tilde\varphi}(\tilde X) \xrightarrow{p_{2n-1}} H^\varphi_1(X) \longrightarrow H^\varphi_1(X, X\backslash A) \longrightarrow$$
$$H^{2n}_\varphi(X) \xrightarrow{p_{2n}} H^\varphi_0(X) \longrightarrow 0.$$

(iii) Für lokal irreduzibles $X$ kann man $A$ so wählen, daß $(\varphi \cap A) - \dim A \leqq 2n-4$ und $\mathrm{Tab}_\varphi\,(X, A; \mathscr{F}) \geqq 4$, insbesondere natürlich, falls $A \supset S(X)$. Dann existiert eine exakte Sequenz mit Koeffizienten in $\mathscr{F}$:

$$H^{2n-3}_\varphi(X) \xrightarrow{p_{2n-3}} H^\varphi_3(X) \longrightarrow H^\varphi_3(X, X\backslash A) \longrightarrow H^{2n-2}(X) \longrightarrow$$
$$\cdots \longrightarrow H^{2n}_\varphi(X) \xrightarrow{p_{2n}} H^\varphi_0(X) \longrightarrow 0.$$

(iv) Ist $(X\backslash A)\cap S(X)$ isoliert und $H^0_\varphi(X, A; \mathcal{H}_{2n-j}(X, \mathcal{F}))=0$ für festes $j\neq 0$ sowie $H^j_{\tilde\varphi\cap\tilde A}(\tilde A, \tilde{\mathcal{F}})=0$, so existiert eine exakte Sequenz mit Koeffizienten in $\mathcal{F}$:

$$H^\varphi_{2n-j-1}(X, X\backslash A)\longrightarrow H^j_\varphi(\tilde X)\xrightarrow{P_j} H^\varphi_{2n-j}(X)\longrightarrow$$
$$H^\varphi_{2n-j}(X, X\backslash A)\longrightarrow H^{j+1}_{\tilde\varphi}(\tilde X)\longrightarrow H^\varphi_{2n-j-1}(X)\oplus H^{j+1}_{\tilde\varphi\cap\tilde A}(\tilde A),$$

weil $(\beta)$ erfüllt ist für $j+1$ und weil man aus der exakten 'Poincaré-Sequenz' für isolierte Singularitäten ableitet, daß $\chi_{j-1}$ definierbar ist. Falls zusätzlich $H^{j+1}_{\tilde\varphi\cap\tilde A}(\tilde A, \tilde{\mathcal{F}})=0$, so existiert eine exakte Fortsetzung der Sequenz bis $H^\varphi_{2n-j-1}(X, \mathcal{F})$.

(v) Es sei $X\backslash A$ topologisch lokal vollständiger Durchschnitt und $\dim_c S(X\backslash A)\leqq 1$, $\varphi$ parakompaktifizierend. Es sei $H^j_{\tilde\varphi\cap\tilde A}(\tilde A, \tilde{\mathcal{F}})=0$. Dann ist (2.6) exakt, falls $j\leqq n-3$ oder $j\geqq n+2$; (2.7) ist exakt falls $j\leqq n-2$ oder $j\geqq n+2$.

(vi) Ist $P^\varphi_j(X, \mathcal{F})$ surjektiv und $H^{\varphi|X\backslash A}_{2n-j}(X\backslash A; \mathcal{F})=0=H^j_{\tilde\varphi\cap\tilde A}(\tilde A, \tilde{\mathcal{F}})$, so ist $H^\varphi_{2n-j}(X)=0$ nach (2.6). Beispiel: Falls $A$ Durchschnitt von $r$ Hyperflächenschnitten in einer projektiv algebraischen Varietät $X$ ist und $\mathcal{F}$ auf jedem dieser Hyperflächenschnitte konstant, sowie $\varphi=c(X)$, so ist $H^c_{2n-j}(X\backslash A)=0$ für $j\leqq n-r$. Falls $P_{n-r}(X)$ surjektiv und $H_{n+r}(X)\neq 0$, so ist auch $H^{n-r}(\tilde A)\neq 0$ (schwache Form des Lefschetzsatzes für Hyperflächenschnitte).

(vii) $A$ lasse sich durch $r=n-\mathrm{Tab}_\varphi(X, A; \mathcal{F})+1$ offene (bzw. $r$ abgeschlossene) Steinsche Teilmengen von $X$ überdecken, wobei $\varphi= cld(A)\mid X\backslash B$ für eine analytische Teilmenge $B$ von $A$, $\mathcal{F}$ sei konstant auf jeder der überdeckenden Teilmengen. Dann existiert eine exakte Sequenz mit Koeffizienten in $\mathcal{F}$:

$$H^{n+r}_{\tilde\varphi}(\tilde X)\xrightarrow{P_{n+r}} H^\varphi_{n-r}(X)\longrightarrow H^\varphi_{n-r}(X, X\backslash A)\longrightarrow H^{n+r+1}_{\tilde\varphi}(\tilde X)\longrightarrow$$
$$\cdots\longrightarrow H^{2n}_\varphi(X)\longrightarrow H^\varphi_0(X)\longrightarrow 0$$

falls $H^{\tilde\varphi}_{n+r-1}(\tilde A, \tilde{\mathcal{F}})$ endlich erzeugt ist oder $L$ ein Körper ist oder $\mathcal{F}$ endliche Halme hat (sonst beginnt die Sequenz erst bei $n+r+1$).

Beispiel. Ist $X$ Steinsch und der Einfachheit halber $A\supset S(X)$, $\dim_{\mathbb{C}} A=:a$, $\mathcal{F}$ konstant, so ist

$$H^{cld}_j(X, \mathcal{F})\cong H^{cld}_j(X, X\backslash A; \mathcal{F})\quad\text{für}\quad\begin{cases} j\leqq 2n-a-2 & \text{stets}\\ j=2n-a-1, & \text{falls } H^c_a(A, L)\\ & \text{endlich erzeugt.}\end{cases}$$

Ist $\tilde A$ zusätzlich Mannigfaltigkeit, so ist

$$H^c_{2n-j}(X, X\backslash A, \mathcal{F})\to H^{j+1}_c(\tilde X, \tilde{\mathcal{F}})$$

bijektiv für $j \leqq a-2$ und injektiv für $j = a-1$. Ist darüber hinaus $A$ global in $X$ durch $s$ holomorphe Gleichungen definierbar, so ist

$$H^j_c(\tilde{X}, \tilde{\mathscr{F}}) = 0 \quad \text{für} \quad j \leqq n-s-1\,.$$

(viii) Ist $A$ in (vii) abgeschlossen, $X$ in $A$ lokal irreduzibel und $\mathscr{F}$ in einer Umgebung von $A$ eine kohärente $\mathcal{O}(X)$-Garbe, so existiert nach Theorem $B$ eine exakte Sequenz mit Koeffizienten in $\mathscr{F}$ wie in (vii), darüber hinaus

$$H^r_{\tilde{\varphi}}(\tilde{X}) \to H^{\varphi}_{2n-r}(X) \to H^{\varphi}_{2n-r}(X, X \backslash A) \to H^{r+1}_{\tilde{\varphi}}(\tilde{X}) \to \cdots \to H^{\mathrm{Tab}_{\varphi}(X,A)-1}_{\tilde{\varphi}}(\tilde{X}) \to H^{\varphi}_{2n-\mathrm{Tab}+1}(X)\,.$$

(ix) Ist $A$ abgeschlossen und $x \in A$, so gilt für konstantes $\mathscr{F}$ und eine geeignete kleine Umgebung $U$ von $x$ in $X$ (etwa für den offenen Simplexstern um $x$ mit $x$ als neu eingeführtem Eckpunkt):

$$H^c_{2n-j}(U \backslash A, \mathscr{F}) \cong \mathscr{H}^j_x(X, \mathscr{F})$$

falls $j \geqq \dim_R S(U \backslash A)$, $H^j_c(\tilde{U} \cap \tilde{A}, \tilde{\mathscr{F}}) = 0 = H^{j+1}_c(\tilde{U} \cap \tilde{A}, \tilde{\mathscr{F}})$ und

$$\dim_{\mathbb{R}} \operatorname{supp} \mathscr{H}_{2n-j+q}(XA, \mathscr{F}) \leqq q-1$$

für

$$\max(1, j - \operatorname{tab}(X \backslash A) - n) \leqq q \leqq (j-1, \dim_{\mathbb{C}} S(U \backslash A))\,.$$

(Der Beweis benutzt die Tatsache, daß $U$ so klein wählbar ist, daß $U \backslash A$ keinen Eckpunkt einer Triangulierung enthält, so daß

$$H^0_c(X \backslash A, \mathscr{H}_{2n-j}(X, \mathscr{F})) = 0\,.).$$

Für $L$ ein Körper gilt ein duales Resultat zu Paragraph 1:

Satz 2.9. *Ist $A$ abgeschlossen in $X$ und $\mathscr{F}$ eine lokalkonstante Garbe von $L$-Vektorräumen, ferner $\mathscr{H}_{2n-k}(X \backslash A, L) = 0$ für alle $k$ mit $1 \leqq k \leqq j$, so existiert eine exakte Garbensequenz auf $X$ mit Koeffizienten in $\mathscr{F}$:*

$$\pi_0 \mathscr{H}_{j+1}(\tilde{A}) \to \pi_0 \mathscr{H}_{j+1}(\tilde{X}) \to \mathscr{H}^{2n-j}_{A}(X) \to \pi_0 \mathscr{H}_j(\tilde{A}) \to \cdots \to \pi_0 \mathscr{H}_1(\tilde{A}) \to 0\,.$$

Beweis. Wie in Lemma 1.2 lassen sich Aussagen mit erheblich schwächeren Voraussetzungen ableiten, darauf wollen wir aber hier verzichten. Nach 2.7 Lemma erhalten wir für jedes offene $U \subset X$ mit Koeffizienten in $\mathscr{F}$ eine exakte Sequenz

$$H^1_c(\tilde{U}) \to H^c_{2n-1}(U) \oplus H^1_c(\tilde{U} \cap \tilde{A}) \to H^c_{2n-1}(U, U \backslash A) \to H^2_c(\tilde{U}) \to \cdots \to H^c_{2n-j-1}(U) \oplus H^{j+1}_c(\tilde{U} \cap \tilde{A})\,.$$

Um die exakte Sequenz von Satz 2.9 zu etablieren, genügt es, $U$ eine Basis offener Mengen durchlaufen zu lassen. Daher sei ohne Einschränkung $U$ zusammenziehbar. Von der dann entstehenden exakten Sequenz erhält man für so kleine $U$, daß $\mathscr{F} \mid U$ konstant ist, durch Transposition mit den universellen Koeffiziententheoremen, weil $L$ ein Körper ist, eine exakte Sequenz mit Koeffizienten in $\mathscr{F}$

$$H_1^{\mathrm{cld}}(\tilde{U}) \leftarrow H_1^{\mathrm{cld}}(\tilde{U} \cap \tilde{A}) \leftarrow H^{2n-1}(U, U \backslash A) \leftarrow H_2^{\mathrm{cld}}(\tilde{U}) \leftarrow \cdots.$$

Weil die Konstruktion funktoriell ist, liefert sie auf $X$ eine exakte Garbensequenz des obigen Typs. $\tilde{X}$ ist lokal irreduzibel, daher endet sie mit $\pi_0 \mathscr{H}_1(\tilde{X}) = 0$.

Wie im ersten Paragraphen, so liefert auch Satz 2.9 eine Reihe von Anwendungen; diese Dualisierung bleibt dem Leser überlassen. Statt dessen wollen wir noch ein Rangtheorem beweisen:

SATZ 2.10. *Ist $P_{j+1}^{\varphi}(X, A; Q(L))$ bijektiv, $\bar{P}_j^{\varphi}(X, A; Q(L))$ surjektiv und sind die Beschränkungen* Kern $P_i^{\varphi}(X, Q(L)) \to H^i_{\bar{\varphi} \cap \tilde{A}}(\tilde{A}, Q(L))$ *der kanonischen Homomorphismen für $i = j, j+1$ die Nullabbildungen, so gilt*

$$\operatorname{rang}_L P_j^{\varphi}(X) + \operatorname{rang}_L P_{j+1}^{\varphi}(X) = b_{2n-j}^{\varphi}(X) + b_{\varphi}^{j+1}(\tilde{X}) + b^j_{\varphi \cap \tilde{A}}(\tilde{A}) - b_{2n-j}^{\varphi}(X, X \backslash A).$$

BEWEIS. Man ersetzt das Diagramm, das zur exakten Sequenz (2.4) führt, durch ein geeignetes Diagramm

$$\begin{array}{ccccccc} & & \tilde{B} & \longrightarrow & \tilde{C}_1 & \longrightarrow & S \oplus \tilde{A}_1 & \longrightarrow & \tilde{B}_1 \\ & & \downarrow & & \downarrow & & \downarrow{\scriptstyle 1 \oplus 0} & & \\ A & \longrightarrow & B & \longrightarrow & C_1 & \longrightarrow & S \oplus A_1, & & \end{array}$$

mit $S = H_{\bar{\varphi}}^{j+1}(\tilde{X}, Q(L))/\text{Kern}\, P_{j+1}^{\varphi}(X, Q(L))$, woraus mit (2.3) eine exakte Sequenz mit Koeffizienten in $Q(L)$ folgt:

$$H^j_{\bar{\varphi}}(\tilde{X}) \xrightarrow{\psi_j} H_{2n-j}^{\varphi}(X) \oplus H^j_{\bar{\varphi} \cap \tilde{A}}(\tilde{A}) \longrightarrow H_{2n-j}^{\varphi}(X, X \backslash A) \xrightarrow{\Delta_j} \text{Kern}\, P_{j+1}^{\varphi}(X, A) \longrightarrow H^{j+1}_{\bar{\varphi} \cap \tilde{A}}(\tilde{A}).$$

$\Delta_j$ ist nach Voraussetzung surjektiv, ebenso ist Kern $\psi_j =$ Kern $P_j^{\varphi}(X, A)$, womit die Behauptung folgt.

## 3. Lokale und globale Lefschetzsätze für Teilräume

Sei $A \subset X$ eine analytische Teilmenge, $a \in A$ ein gegebener Punkt und $\mathscr{G}$ eine lokalkonstante Garbe mit Halm abelsche Gruppe $G$. Wir setzen zur Abkürzung: $R(X) := \{X \in X, X_x \text{ ist reduzibel}\}$. Weiter sagen wir, daß $A$ (bzw. der Keim $A_a$) topologisch durch $r$ Gleichungen beschrieben werden kann, wenn ein Homöomorphismus $h: X \to Y$ auf einen komplexen Raum $Y$ (bzw. Raumkeim $Y_b$) existiert, mit $h(X)$ bzw. $h(X_a)$ läßt sich im Bild durch $r$ holomorphe Gleichungen beschreiben.

**3.1. Lokaler Lefschetzsatz der Kohomologie**

*Es sei $j \in \mathbb{N}_0$ fest gegeben. Dann ist*

($\alpha$) $\mathcal{H}^j_{\tilde{A}}(\tilde{X}, \mathcal{G})_a = 0$, *falls*

$$j \leqq \min[\dim_a X - \operatorname{tab}_a X - \dim_a S(X), 2(\dim_a X - \dim_a A)] - 1$$

($\beta$) $\mathcal{H}^j_A(X, \mathcal{G})_a = 0$, *falls für j aus* ($\alpha$) *noch gilt* $j \geqq \dim^{\mathbb{R}}_a \overline{R(X \backslash A)} + 2$.

($\gamma$) $\mathcal{H}^{2n-k}_A(X, \mathcal{G})_a = 0$, *falls*

$$k \leqq \min[\dim_a X - \operatorname{tab}_a (X) - \dim_a S(X) - 1, \dim_a A - \operatorname{tab}_a A] - 1.$$

($\delta$) $\mathcal{H}^{n+r+1}_A(X, \mathcal{G})_a = 0$, *falls $A_a$ topologisch durch r holomorphe Gleichungen beschrieben werden kann.*

Beweis. Für $1 \leqq k \leqq \dim^{\mathbb{C}}_a X - \operatorname{tab} X - \dim^{\mathbb{C}}_a S(X) - 1$ gilt $\mathcal{H}_{2n-k+1}(X, \mathcal{G})_a = 0$ mit $n := \dim_a X$, also existiert nach Satz 1.1 eine exakte Sequenz im Punkt $a$ für jede endlich erzeugte abelsche Gruppe $G$ (da $\mathcal{H}^j_{\tilde{A}}(\tilde{X}, \mathbb{Z})$ endlich erzeugt ist, erhält man daraus mit dem universellen Koeffiziententheorem der Kohomologie und dem der lokalen Homologie die Endaussage für beliebige lokalkonstante Garbe $\mathcal{G}$)

$$0 = \mathcal{H}_{2n-k+1}(X, G) \rightarrow \pi_0 \mathcal{H}^k_{\tilde{A}}(\tilde{X}, G) \rightarrow \mathcal{H}_{2n-k}(A, G).$$

Für $2n - k \geqq 2 \dim^{\mathbb{C}}_a A + 1$ ist $\mathcal{H}_{2n-k}(A, G) = 0$, also auch $\mathcal{H}^k_{\tilde{A}}(\tilde{X}, G)$. Dies beweist ($\alpha$). – Nun ist $\pi_0 \mathcal{H}^k_{\tilde{A}}(\tilde{X}, G) \cong \mathcal{H}^k_A(X, \mathcal{H}_{2n}(X, G))$, wie man leicht nachrechnet. Es existiert eine Garbe $Q$ auf $X$ mit Träger $R(X)$, so daß $0 \rightarrow \mathcal{G} \rightarrow \mathcal{H}_{2n}(X, \mathcal{G}) \rightarrow Q \rightarrow 0$ eine exakte Garbensequenz ist. Dies liefert die exakte Sequenz

$$\cdots \rightarrow \mathcal{H}^{k-1}_A(X, Q) \rightarrow \mathcal{H}^k_A(X, G) \rightarrow \pi_0 \mathcal{H}^k_{\tilde{A}}(\tilde{X}, \mathcal{G}) \rightarrow \mathcal{H}^k_A(X, Q) \rightarrow \cdots.$$

$Q$ erfüllt die Voraussetzungen von Lemma 1.4, also ist $\mathcal{H}^{j-1}_A(X, Q)_a = 0$, falls $j \geqq 2 + \dim_{\mathbb{R}} \operatorname{supp} Q$. Dies beweist ($\beta$). – Weiter ist für $k \leqq \dim_a X - \operatorname{tab}_a X - \dim_a S(X) - 2$ und beliebiges $\mathcal{F}$

$$\mathcal{H}_{2n-k}(X, \mathcal{F}) = 0.$$

Daraus folgt $\mathcal{H}_{k+1}(\tilde{X}, G) = 0$, denn man hat eine exakte Sequenz für kleine offene $U \ni a$:

$$0 = H^0_c(U, \mathcal{H}_{2n-k}(U, G)) \rightarrow H^{k+1}_c(\tilde{U}, G) \rightarrow H^c_{2n-k-1}(U, G) = 0$$

für alle $G$, woraus mit den universellen Koeffiziententheoremen der Kohomologie und der lokalen Homologie die Behauptung folgt. Satz 2.9 liefert eine Surjektion

$$\pi_0 \mathcal{H}_{k+1}(\tilde{X}, G)_a \rightarrow \mathcal{H}^{2n-k}_A(X, G)_a,$$

aus der die Behauptung ($\gamma$) folgt. – Für ($\delta$) sei $(X_a, A_a)$ homöomorph zu einem Paar von komplexen Raumkeimen $(Y_b, B_b)$, in dem $B_b$ durch höchstens $r$ holomorphe Funktionen gegeben ist. Dann ist nach dem Verschwindungssatz für die Kohomologie Steinscher Räume [8] für jeden kleinen Steinschen Repräsentanten $Y$ von $Y_b$, der ohne Einschränkung endlich erzeugte Homologie habe, $H^{n+j}(Y, G)=0=H^{n+r+j}(Y\backslash B, G)$ für alle $j \geqq 0$. Daher ist $\mathscr{H}_A^{n+r+1}(X, G)=0$.

**3.2. Globaler Lefschetzsatz der Kohomologie**

*Es gelte für eine Umgebung $U$ von $A$ in $X$*

$$j \leqq \min_{a \in A}[\dim_a U - \operatorname{tab}_a U - \dim_a S(U),\, 2(\dim_a U - \dim_a A)] - 1,$$

*dann ist die kanonische Abbildung*

$$H^i_\varphi(\tilde{X}, \mathscr{G}) \to H^i_{\varphi \cap (\tilde{X}\backslash \tilde{A})}(\tilde{X}\backslash \tilde{A}, \mathscr{G})$$

*bijektiv für $i \leqq j-1$ und injektiv für $i=j$; falls zusätzlich $j \geqq 2 + \max_{a \in A} \dim_a^{\mathbb{R}} R(U\backslash A)$, so gilt die analoge Aussage für*

$$H^i_\varphi(X, \mathscr{G}) \to H^i_{\varphi \cap (X\backslash A)}(X\backslash A, \mathscr{G}).$$

Beweis. Dies folgt mit der Spektralsequenz $E_2^{p,q} = H^p_{\tilde{\varphi}}(\tilde{X}, \mathscr{H}^q_{\tilde{A}}(\tilde{X}, \mathscr{G})) \to H^{p+q}_\varphi(\tilde{X}, \tilde{X}\backslash \tilde{A}; \mathscr{G})$ weil $A$ abgeschlossen ist, aus dem lokalen Lefschetzsatz der Kohomologie, Teil $(\alpha)$ und $(\beta)$. Die Spektralsequenz ergibt sich wie die für $\varphi = cld(X)$ übliche mit gleicher Beweismethode, da welke Garben $\varphi$-azyklisch sind und $\Gamma_\varphi \Gamma_A(\mathscr{F}) = \Gamma_{\varphi|A}(\mathscr{F})$, $R^*\Gamma_{\varphi|A}(\mathscr{F}) = H^*_\varphi(X, X\backslash A; \mathscr{F})$.

**3.3. Lokaler Lefschetzsatz der Homologie**

*Ist $A \subset X$ in $a \in A$ topologisch durch höchstens $r$ holomorphe Gleichungen gegeben und $r \geqq \operatorname{tab}(U\backslash A)$, so ist $\mathscr{H}_{n-r}(X, A; \mathscr{G})_a = 0$, falls*

$$\dim_{\mathbb{R}} \operatorname{supp} \mathscr{H}_{q+n-r}(U\backslash T, \mathscr{G}) \leqq q \quad \text{für} \quad 1 + 2(r - \operatorname{tab}(U\backslash T)) \leqq q \leqq \min(\dim_{\mathbb{R}} S(U\backslash T),\, n+r) - 1$$

*für eine kleine Umgebung $U$ von $a$ in $X$.*

Beweis. Nach Teil $(\delta)$ des lokalen Lefschetzsatzes der Kohomologie ist $\mathscr{H}_A^{n+r+1}(X, \mathscr{G})_a = 0$. Da die Normalisierung eines Steinschen Raumes wieder Steinsch ist, gilt damit auch $\mathscr{H}_{\tilde{A}}^{n+r+1}(\tilde{X}, \tilde{\mathscr{G}}) = 0$, damit auch $\pi_0 \mathscr{R}^{n+r} j_*(\tilde{\mathscr{G}} \mid \tilde{X}\backslash \tilde{A})_a = 0$ für die kanonische Inklusion $j: \tilde{X}\backslash \tilde{A} \to \tilde{X}$. Es bleibt zu zeigen, daß $\lim_{U \to x} Q^{cld}_{n+r}(U\ A, \mathscr{G}): \pi_0 \mathscr{R}^{n+r} j_*(\tilde{G} \mid \tilde{X}\ \tilde{A})_a \to \mathscr{H}_{n-r}(X, A; \mathscr{G})_a$ surjektiv ist. Dazu geht man vor wie im Beweis von 1.2 Lemma, wobei man $U$ so klein wählt, daß $\mathscr{H}_{n-r}(U\backslash T, \mathscr{G}) = 0$.

Wir haben bislang der Übersichtlichkeit halber alles für reindimensionale $X$ durchgeführt. Es erschien zweckmäßig, diese Einschränkung im folgenden Resultat fallen zu lassen. Für nicht reindimensionales $U$ bezeichne im folgenden $U_j := \{x \in U, \dim^x_{\mathbb{C}} U \leqq j\}$.

3.4. Korollar *Es sei $U$ eine offene Umgebung von $a \in A$ in $X$, $A_a$ werde in $X_a$ durch höchstens $r \geqq \max_j \dim_{\mathbb{C}} S(U_j \backslash A)$ holomorphe Gleichungen topologisch beschrieben. Dann ist*

$$\mathcal{H}_{n-\mathrm{tab}(U\backslash A)-r}(X, A; \mathcal{G})_a = 0.$$

Beweis. Wegen $r \geqq 1$ ist für $\mathrm{tab}\,(U\backslash A) \geqq n-1$ nichts zu zeigen. Also sei $\mathrm{tab}\,(U\backslash A) \leqq n-2$ und somit $U\backslash A$ lokal irreduzibel. Dann ist $U_j \cap U_k \subset A$ für $j \neq k$, alle $U_j \backslash A$ sind reindimensional und

$$\mathcal{H}_*(X, A; \mathcal{G}) = \bigoplus_j \mathcal{H}_j(X_j, A \cap X_j; \mathcal{G}).$$

Weiter ist $n-j-\mathrm{tab}\,(U_j\backslash A) \geqq n-\mathrm{tab}\,(U\backslash A)$, so daß es genügt, $\mathcal{H}_{n-\mathrm{tab}(U\backslash A)-r}(X_j, A \cap X_j; \mathcal{G})$ zu untersuchen. Dazu sei ohne Einschränkung $j = n$. Wie in 3.3 ist als Hindernis

$$\mathcal{H}^{q+1}_{A}(X, \mathcal{H}_{q+n-\mathrm{tab}(U\backslash A)-r}(X, \mathcal{G}))_a$$

zu untersuchen für $2r+1 \leqq q \leqq 2 \dim_{\mathbb{C}} S(U\backslash A)-1$. Nach Voraussetzung über $r$ existieren aber solche $q$ nicht.

3.5. Bemerkung Ist $X\backslash A$ singularitätenfrei, so verschärft 3.1 ($\beta$) den homologischen Teil des mit Hilfe der Morsetheorie bewiesenen Satzes 2.19 aus [4]. In 3.4 ist dann $\mathrm{tab}\,(U\backslash A)$ die (komplexe) Dimension der niedrigstdimensionalen irreduziblen Komponente von $U\backslash A$, die $a$ berührt. Damit ist 3.4 eine gewisse Verschärfung des homologischen Teiles von Satz 2.9 in [4]. – Entsprechende Lefschetz-Aussagen für die relative Homotopie erhält man in der vorliegenden Situation insbesondere, wenn man weiß, daß das relative $\pi_1$ verschwindet und daß der kleinere Raum einfach zusammenhängend ist:

**3.6. Lefschetzsatz der Homotopie für Zariskioffene Teilräume**

*Es sei $A \subset X$ eine analytische Menge mit $X\backslash A$ lokal irreduzibel nahe $A$ und*

$$\min_{a \in A} [\dim_a A - \mathrm{tab}_a\,(A) - \dim_a S(A)] \geqq 3.$$

*Dann ist $\pi_j(X, X\backslash A) = 0$ für*

$$j \leqq \min_{a \in A} [\dim_a X - \mathrm{tab}_a X - \dim_a S(X), 2(\dim_a X - \dim_a A)] - 1,$$

*falls $\pi_1(X\backslash A)$ $(=\pi_1(X))$ trivial auf allen $\pi_i(X, X\backslash A)$ mit $3 \leqq i \leqq j$ operiert (insbesondere also für einfach zusammenhängendes $X$).*

Beweis. Zunächst ist für die angegebene $j$ nach 3.1 ($\beta$) für jede abelsche Gruppe $G$ die relative Kohomologie $H^j(X, X\backslash A; G) = 0$, so daß nach dem universellen Koeffiziententheorem der Kohomologie folgt $H^c_j(X, X\backslash A; \mathbb{Z}) = 0$.

Nach dem relativen Satz von Hurewicz genügt es zu zeigen, daß $\pi_i(X, X\setminus A) = 0$, $i = 1, 2$. Dies folgt aus Korollar 5.2 und Lemma 2.2 in [5].

Als Anwendung auf projektiv algebraische Varietäten folgt aus den bisherigen Ergebnissen unter Verwendung des bekannten Diagramms

$$\begin{array}{ccccccccc} 0 & \longrightarrow & \mathbb{C}^* & \longrightarrow & \varphi^{-1}(A) & \longrightarrow & A & \longrightarrow & 0 \\ & & \| & & \downarrow & & \downarrow & & \\ 0 & \longrightarrow & \mathbb{C}^* & \longrightarrow & \varphi^{-1}(X) & \longrightarrow & X & \longrightarrow & 0 \\ & & \| & & \downarrow & & \downarrow & & \\ 0 & \longrightarrow & \mathbb{C}^* & \longrightarrow & (\mathbb{C}^{m+1})^* & \xrightarrow{\varphi} & \mathbb{P}_m & \longrightarrow & 0 \end{array}$$

von $\mathbb{C}^*$-Bündeln über den projektiv algebraischen Varietäten $A \hookrightarrow X \hookrightarrow \mathbb{P}_m$:

### 3.7. Lefschetzsatz der Homotopie für abgeschlossene Teilräume

*Für die projektiv algebraischen Varietäten $A \hookrightarrow X$ gelte unter Verwendung obiger Bezeichnungen in $0 \in \mathbb{C}^{m+1}$:* (i) $\operatorname{tab}_0 \varphi^{-1}(A) \leqq \dim_0 \varphi^{-1}(A) - 3$, (ii) $\operatorname{tab}_0 \varphi^{-1}(X) \leqq \dim_0 \varphi^{-1}(X) - 3$. *Dann ist $\pi_j(X, A) = 0$ für $j \leqq \dim X - \operatorname{tab}(X\setminus A) - r$, falls $A$ sich durch höchstens $r$ holomorphe Gleichungen beschreiben läßt und falls jede irreduzible Komponente von $X\setminus A$ eine höchstens $r$-dimensionale Singularitätenmenge hat.*

BEWEIS. Die obige Inklusion von Faserbündeln liefert ein Diagramm von $S^1$-Bündeln

$$\begin{array}{ccccccccc} 0 & \to & S^1 & \to & \Sigma_A & \to & A & \to & 0 \\ & & \| & & \downarrow & & \downarrow & & \\ 0 & \to & S^1 & \to & \Sigma_X & \to & X & \to & 0, \end{array}$$

wobei $\Sigma_X$ (entsprechend $\Sigma_A$) den Schnitt einer hinreichend kleinen $(2n+1)$ -Sphäre um 0 in $\mathbb{C}^{m+1}$ mit $\varphi^{-1}(X)$ bezeichnet. Wegen (i) und (ii) sind $\Sigma_A$ und $\Sigma_X$ 1-zusammenhängend, insbesondere damit auch $A$ und $X$. Damit ist $\pi_j(\Sigma_X, \Sigma_A) = \mathscr{H}_{j+1}(\varphi^{-1}(X), \varphi^{-1}(A); \mathbb{Z})_0 = 0$ für $j \leqq \dim_0 \varphi^{-1}(X) - \operatorname{tab}_0(\varphi^{-1}(U\setminus A)) - r - 1 = \dim X - \operatorname{tab}(U\setminus A) - r$ nach 3.4. Es bleibt noch zu zeigen, daß $\pi_j(\Sigma_X, \Sigma_A) \to \pi_j(X, A)$ surjektiv ist. Das ist einfach zu verifizieren.

LITERATUR

G. BARTHEL, L. KAUP

[1] Homotopieklassifikation einfach zusammenhängender normaler kompakter komplexer Flächen. Math. Ann. *212*, 113–144 (1974), Springer-Verlag 1974.

G. BREDON

[2] Sheaf theory. New York: McGraw-Hill 1967.

H. Cartan, S. Eilenberg

[3] Homological Algebra. Princeton 1956.

H. Hamm

[4] Lokale topologische Eigenschaften komplexer Räume. Math. Ann. *191*, 235–252 (1971).

M. Kato

[5] Partial Poincaré duality for $k$-regular spaces and complex algebraic sets. Sonderforschungsbereich 40, Theoretische Mathematik, Bonn 1974.

L. Kaup

[6] Poincaré Dualität für Räume mit Normalisierung. Ann. della Sc. Norm. Sup. di Pisa, Vol. XXVI, Fasc. I (1972).

[7] Zur Homologie projektiv algebraischer Varietäten. Ann. della Sc. Norm. Sup. di Pisa, Vol. XXVI, Fasc. II (1972).

[8] Eine topologische Eigenschaft Steinscher Räume. Nachrichten der Akademie der Wissenschaften, Göttingen, Jahrg. 1966, Nr. 8.

L. Kaup, G. Weidner

[9] Mayer-Vietoris Sequenzen und Lefschetzsätze für mehrfache Hyperflächenschnitte in der Homotopie. Math. Z. *142*, 243–269 (1975), Springer-Verlag 1975.

Lê Dũng Tráng

[10] Singularités isolées des intersections complètes. Sem. Top. singularités var. alg. 1969/70, I.H.E.S.

J. Milnor

[11] Singular points of complex hypersurfaces. Ann. Math. Studies 61, Princeton 1968.

Ludger Kaup
Universität Konstanz
Fachbereich Mathematik
Postfach 7733
D-7750 Konstanz
Bundesrepublik Deutschland

Nordic Summer School/NAVF
Symposium in Mathematics
Oslo, August 5–25, 1976

*But a mathematical theory cannot thrive indefinitely on greater and greater generality. A proper balance must ultimately be maintained between the generality and the concreteness of the structure studied.*[1]

# THE ENUMERATIVE THEORY OF SINGULARITIES

Steven L. Kleiman[2]

## Table of Contents



[1] Oscar Zariski ([120]; p. xiii).

[2] Partially supported by the NSF under grant no. MCS 74-07275. Guest during June and July 1976 at the Mathematics Institute of the University of Aarhus, Denmark, guest from Dec. 17, 1976 to Jan. 14, 1977 of the Mathematics Department of the California Institute of Technology, Pasadena, CA; many thanks are due for the generous hospitality extended.

# I. Introduction

## A. Preface

These notes are aimed at building a bridge between differential topology and algebraic geometry. The two fields lie close together at some points, they have enjoyed parallel developments at some points, and they have been bridged before at some points. Yet at times each field has suffered because its workers were unaware of similar work done in the opposite field or in their own. Both fields can only benefit when their workers learn more about the accomplishments and activities in each other's field and in their own.

In these notes algebraic geometry is done, but differential topology provides the perspective. The proofs belong to algebraic geometry, the methods stem from differential topology. Most results stem from classical algebraic geometry, most formulations stem from modern differential topology.

Differential topology inspires elegant proofs. And elegance is a joy for its beauty. But elegance also is indicative of the right point of view. And indeed, on a cliff at the edge of the land of differential topology we can stand above many scattered results in algebraic geometry. We can see more clearly what is behind them and between them. Our understanding is deepened, the results are unified. At the same time we cannot fail to admire the ingenuity used in discovering these results. More importantly, from this vantage point we can see the general lay of the land and some of its concrete features beyond.

Below we enter some lively areas, where new developments are expected soon. We pass by other areas, where important developments have taken place, contenting ourselves with a glance and a few references at most. But the purpose of these notes is not to cover our subject completely, which is impossible anyway; the purpose of these notes is to convey the spirit of our subject and to suggest its interest and its significance.

The mathematics and philosophy in these notes has evolved over the years through the formal writings of a great many authors. (These works will be cited in due course.) These notes also owe a great deal to informal discussions with William Fulton (particularly about multiple-point formulas), Dan Laksov, Ragni Piene, and Israel Vainsencher. Finally, special thanks go to Viola Wiley for doing with remarkable speed her usual fine job in preparing the typescript and to Ragni Piene for voluntarily and carefully proofreading chapters I–IV.

The main ground covered in these notes is contained in three overlapping subject areas. Here is a brief description of each one:

1. *Thom polynomials.* They express invariants of the singularities of an appropriately generic mapping $f:X \to Y$ in terms of invariants of $X$ and invariants of $Y$. Once we have the local structure of the singularities of $f$ suitably in hand, we can enumerate the singularities of each type using an appropriate Thom polynomial.

2. *Classical enumerative geometry.* The object is to find the number of geometric figures satisfying given geometric conditions in terms of invariants of the figures and the conditions. D. Hilbert's 15th problem (see [45]) calls for making rigorous the impressive work of the classical enumerative geometers, particularly Schubert.

3. *Multiple-point formulas.* A point $x$ of $X$ is called an *$n$-fold point* of a mapping $f:X \to Y$ if there exist $n-1$ other points with the same image; some of the points may lie infinitely near others, that is, be ramification points. If $f$ is appropriately generic, an $n$-fold-point formula will enumerate the $n$-fold points in terms of invariants of $X$ and $Y$.

In each of these three areas there are situations in which the enumerative formulas can be turned around and used to determine the invariants of $X$. Historically such extrinsic relations were used as the first definitions of the invariants. History is repeating itself now. At the beginning geometers were interested in enumeration. Next they sought invariants. Now with the theory of invariants secured by means of Chern class we are again interested in enumeration and interpretation.

Throughout these notes the ground field is the field of complex numbers. The ground field never enters explicitly, and any other algebraic closed field of characteristic 0 works as well. In fact, so does one of positive characteristic, or nearly so. The qualifications necessary in positive characteristic will be indicated within parentheses. (Other remarks will be made parenthetically also.)

In these notes the term 'variety' implies reduced and irreducible and quasi-projective. Varieties are usually assumed to be smooth and often complete (compact). An '$n$-fold' is a variety of dimension $n$. 'Point' means a closed point, never a generic point in the technical sense. Always $f:X \to Y$ denotes a mapping of primary interest in which $X$ is a smooth $n$-fold and $Y$ is a smooth $m$-fold. Rarely is $f$ assumed to be generic, usually only appropriately generic. 'Generic' has come to mean in particular that certain loci under consideration are smooth of the right dimension. 'Appropriately generic' will be used to mean in particular that these loci simply have the right dimension; they may have singularities and even nilpotents.

The rest of this chapter is intended to introduce in more detail the subject of these notes. The material is divided into sections according to the three main areas described above. The rest of the notes is organized into chapters according to technique.

## B. Thom polynomials

One of the oldest and most famous, simplest, and nearly ideal examples of a Thom polynomial appears in the Riemann–Hurwitz formula. (H. G. Zeuthen's name is sometimes associated with it, and properly so. Zeuthen obtained it (and a more general formula) by geometric means in 1870. A. Hurwitz obtained it topologically in 1891 and by using abelian integrals in 1892. B. Riemann obtained the special case in which the target is $\mathbb{P}^1$ in 1857. (See [119; Footnote 150, p. 298], [8; Footnote 190, p. 373].)) The formula deals with a surjective mapping $f: X \to Y$ between compact Riemann surfaces, or projective, smooth algebraic curves. In this context, '$f$ is generic' means that $f$ is a simple covering, that is, the critical points are nondegenerate and at most one appears in each fiber. However, no additional hypotheses are necessary for the validity of the formula; every $f$ is appropriately generic because the ramification locus automatically has the right dimension, 0. (In positive characteristic, $f$ must be separable.)

The formula is

$$\boxed{\sum_{x \in X} [e_x - 1] = [2g_X - 2] - \deg(f)[2g_Y - 2]} \quad . \tag{I,1}$$

Here $e_x$ is the *branching number* or *ramification index* at $x \in X$. Topologically $e_x$ is the number of sheets that come together at $x$. Algebraically $e_x$ appears as the exponent in the relation,

$$t = s^{e_x} u, \qquad u(x) \neq 0, \tag{I,2}$$

between a local uniformizing parameter $t$ at $f(x)$ and one, $s$, at $x$. For instance, we have $e_x = 2$ if and only if $x$ is a nondegenerate critical point. In

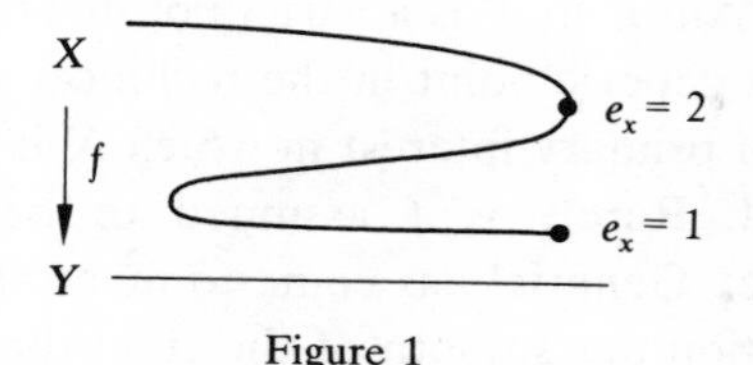


Figure 1

the formula, $g_X$ and $g_Y$ are the *genera* of $X$ and $Y$. Topologically $g_X$ is the number of handles on $X$ or half the first Betti number, and $[2g_X-2]$ is the negative of the Euler characteristic. Algebro–geometrically $g_X$ is the number of linearly independent holomorphic differentials $\omega$ on $X$, and $[2g_X-2]$ is the degree of any $\omega$.

The formula can be proved topologically by making compatible triangularizations of $X$ and $Y$. The formula can be proved algebro-geometrically from the point of view of the theory of Thom polynomials as follows. First of all, the left hand side is viewed differently; it is viewed as the sum of the numbers,

$$\boxed{\mu_x = \dim\,(O_X/F^0\Omega^1_{X/Y})_x}\,, \tag{I,3}$$

where the term in the denominator is the $O$th Fitting ideal of the sheaf of relative differentials.

The first job is to establish the relation,

$$\boxed{\mu_x = e_x - 1}\,. \tag{I,4}$$

Consider the standard exact sequence on $X$ of sheaves of differentials,

$$f^*\Omega^1_Y \to \Omega^1_X \to \Omega^1_{X/Y} \to 0. \tag{I,5}$$

Since the two terms on the left are locally free sheaves, the sequence can be used to compute $F^0\Omega^1_{X/Y}$. Differentiating relation (I,2) yields

$$dt = e_x s^{e_x-1} u\, ds + s^{e_x}\, du. \tag{I,6}$$

However $ds$ and $dt$ generate the stalks $\Omega^1_{X,x}$ and $\Omega^1_{Y,f(x)}$. Hence by (I,5) the stalk $(F^0\Omega^1_{X/Y})_x$ is generated by $s^{e_x-1}$ because $u$ is a unit and $e_x$ is non-zero. Therefore (I,3) yields (I,4). (In positive characteristic $e_x \cdot 1$ can be zero and then (I,4) does not hold; however Formula (I,1) remains valid if the left hand side is replaced by $\sum \mu_x$.)

Globally (and without any local computation) Sequence (I,5) yields

$$F^0\Omega^1_{X/Y} = f^*\Omega^1_Y \otimes (\Omega^1_X)^{-1}. \tag{I,7}$$

Extracting the degrees of both sides and using some general properties of this operation, we get

$$\begin{aligned} -\sum \mu_x &= \deg\,(f^*\Omega^1_Y) + \deg\,(\Omega^1_X)^{-1} \\ &= \deg\,(f)[2g_Y-2] - [2g_X-2]. \end{aligned} \tag{I,8}$$

Therefore (I,4) yields the Formula (I,1).

Formula (I,1) is a numerical relation. However, our proof (unlike the topological one) involves a finer relation, (I,7), and the additional information in it is lost in extracting degrees. There is a nice way to express all the information in (I,7). First we replace the numerical sum $\sum \mu_x$ by the formal sum,

$$\boxed{[R]=\sum_{x\in X}\mu_x[x]}\ , \tag{I,9}$$

where $[x]$ is a symbol associated to the point $x$. This formal sum is called the *ramification cycle*, or *ramification divisor, of f*. It is the fundamental cycle of the ramification locus $R$, which is the subscheme of $X$ defined by the ideal $F^0\Omega^1_{X/Y}$.

Two other divisors are involved in the finer expression. They are called canonical divisors. A canonical divisor on $X$, say, is the divisor $(\omega)$ of a meromorphic 1-form $\omega$ on $X$. It is defined as the formal sum,

$$\boxed{(\omega)=\sum_{x\in X}v_x(\omega)[x]}\ . \tag{I,10}$$

where the coefficients $v_x(\omega)$ are determined as follows: Let as before $s$ be a local uniformizing parameter at $x$. Then we have $\omega = g\,ds$ for a suitable meromorphic function $g$ on $X$. The order of vanishing of $g$ at $x$ (or the negative of its growth rate) is the coefficient $v_x(\omega)$.

Formula (I,1) can now be replaced by a finer relation among divisors, namely,

$$\boxed{[R]=(\omega)-f^*(\eta)}\ , \tag{I,11}$$

where $\omega$ and $\eta$ are meromorphic 1-forms on $X$ and $Y$. Here $f^*(\eta)$ denotes the pullback of $(\eta)$. This is a divisor on $X$ defined as follows. Let $x$ be a point of $X$ and set $y=f(x)$. Say that $(\eta)$ is defined at $y$ by a meromorphic function $h$; that is, the order of vanishing of $h$ at $y$ is the coefficient of $[y]$ in $(\eta)$. Then $f^*(\eta)$ is defined at $x$ by the pullback $f^*h=h\circ f$; that is, the order of vanishing of $f^*h$ at $x$ is the coefficient of $[x]$ in $f^*(\eta)$.

Precisely the assertion is that given $\eta$ there exists an $\omega$ such that (I,11) holds. This assertion is logically equivalent to the validity of (I,7). It yields (I,7); indeed, the sheaves in (I,7) are the inverses of the ones associated to the divisors in (I,11). It follows from (I,7); indeed, $[R]+f^*(\eta)$ is a canonical divisor because its associated sheaf is

$$(F^0\Omega^1_{X/Y})^{-1}\otimes f^*\Omega^1_Y, \tag{I,12}$$

which is equal to $\Omega^1_X$ by (I,7).

The preceding assertion involves the choice of a meromorphic 1-form $\eta$ on $Y$. Let $\eta'$ be a second choice. Then we have $\eta' = h\eta$ for a suitable meromorphic function $h$ on $Y$. Hence we have

$$(\eta') = (h) + (\eta) \quad \text{with} \quad (h) = \sum v_x(h)[x], \tag{I,13}$$

where $v_x(h)$ denotes the order of vanishing of $h$ at $x$. The two divisors $(\eta')$ and $(\eta)$ are said to be *linearly*, or *rationally*, *equivalent* because they differ by the divisor of zeros and poles of a meromorphic function, $h$. Linear equivalence of divisors is obviously an equivalence relation. Now, the equivalence class of $(\eta)$ does not depend on the choice of $\eta$. So it is an invariant of the sheaf $\Omega^1_Y$. It is called the *first Chern class* of $\Omega^1_Y$ and denoted $c_1(\Omega^1_Y)$. In these terms the preceding assertion says we have

$$\boxed{[R] = c_1(\Omega^1_X) - f^*c_1(\Omega^1_Y)}\ , \tag{I,14}$$

where by abuse of notation $[R]$ also denotes the linear equivalence class of the divisor $[R]$.

Finally there is another common form of Formula (I,14). It is

$$\boxed{[R] = f^*c_1(Y) - c_1(X)}\ . \tag{I,15}$$

Here $c_1(X)$ and $c_1(Y)$ are invariants of $X$ and $Y$, called their *first Chern classes*. By definition $c_1(X)$ and $c_1(Y)$ are the first Chern classes of the tangent sheaves $\tau_X$ and $\tau_Y$, which are the duals of the sheaves $\Omega^1_X$ and $\Omega^1_Y$. Since dualizing a sheaf changes the sign of its first Chern class, (I,14) and (I,15) are essentially equivalent.

More generally consider a surjective mapping $f: X \to Y$ between smooth varieties of any dimension $n$. (In positive characteristic $f$ must be separable.) The above discussion of (I,11), (I,14), and (I,15) carries over with minor changes and yields exactly similar formulas. Divisors (more correctly they should be called divisorial cycles) now are finite formal linear combinations of closed subvarieties with dimension $n-1$. It makes sense to talk about the order of vanishing of a meromorphic function along a closed subvariety with dimension $n-1$, and so the divisor of a meromorphic function, the divisor of a differential $n$-form, and the pullback of a divisor can all be defined virtually as before. The ramification divisor is defined by the Fitting ideal $F^0\Omega^1_{X/Y}$ as before, but now the sequence of sheaves of 1-forms similar to (I,5) yields

$$F^0\Omega^1_{X/Y} = f^*\Omega^n_Y \otimes (\Omega^n_X)^{-1}, \tag{I,16}$$

generalizing (I,7), because the sheaves of $n$-forms $\Omega^n_X$ and $\Omega^n_Y$ are the $n$-fold exterior powers of the sheaves of 1-forms $\Omega^1_X$ and $\Omega^1_Y$. It is now a simple matter to obtain a formula like (I,11) but with $\omega$ and $\eta$ meromorphic $n$-forms on $X$ and $Y$. The first Chern classes $c_1(\Omega_X^n)$ and $c_1(\Omega_Y^n)$ can be defined as the linear equivalence class of the divisors $(\omega)$ and $(\eta)$. The first Chern classes $c_1(\Omega_X^1)$ and $c_1(\Omega_Y^1)$ can be taken as these same classes, and $c_1(X)$ and $c_1(Y)$ as their negatives. Then Formulas (I,14) and (I,15) generalize without even a change in notation.

Classically one case of Formula (I,11) for $n$-folds held special importance. It is the case in which $Y$ is the projective $n$-space $\mathbb{P}^n$. In this case the assertion is that the divisor,

$$[R]-(n+1)f^*[H], \qquad H \text{ a hyperplane}, \tag{I,17}$$

is a canonical divisor on $X$; for, $-(n+1)[H]$ is a canonical divisor of $\mathbb{P}^n$. The assertion came in to play when $X$ was studied by viewing it as a 'multiple hyperspace', that is, a several sheeted covering of $\mathbb{P}^n$. Such a presentation $f:X\to\mathbb{P}^n$ was obtained by embedding $X$ into a projective space $\mathbb{P}^N$ with $N \geqq n$ and then projecting the image from a general linear space of codimension $n+1$.

Classically another important way of studying an $n$-fold $X$ was to map $X$ birationally onto a hypersurface in $\mathbb{P}^{n+1}$. Such a presentation $f:X\to\mathbb{P}^{n+1}$ was obtained via a projection from a center in general position. The singularities appearing on the hypersurface $f(X)$ were termed 'ordinary', and today a certain amount is known about their structure, as well as about the structure of the singularities of $f$ (see [58], [65], [66], [88], [91], [93], [27], [28]). For example, when $X$ is a curve, then the singularities of $f(X)$ are double points with distinct tangents, and $f$ has no singular (or critical) points.

For a second example, suppose $X$ is a surface. Then $f$ has a finite number of nondegenerate critical points. (They may be degenerate but only in characteristic 2, see [90; §5].) Classically the number was denoted by $\nu_2$ and called the 'type' of $X$, or of the surface $f(X)$ in $\mathbb{P}^3$. The images of the critical points were termed the *pinch points* of $X$ or of $f(X)$ or of the double curve of $f(X)$. (The term is Cayley's (1868), see [7; p. 158].) At a pinch point, $f(X)$ has two sheets with coincident tangent planes.

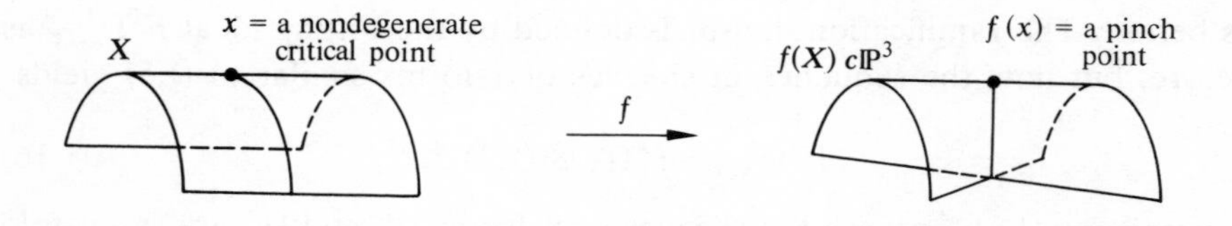


Figure 2

In modern terms, $\nu_2$ is given by a Thom polynomial as follows:

$$\boxed{\nu_2 = 6\int h^2 - 4\int h \cdot c_1(X) + \int c_1(X)^2 - \int c_2(X)} \,. \tag{I,18}$$

Here $h$ denotes the rational equivalence class of the pullback of a hyperplane, and $\int h^2$ denotes the number of points common to the pullbacks of two hyperplanes. (The integral sign, which is common enough in the present usage, is supposed to be reminiscent of integrating de Rham classes.) If the points are counted with multiplicity, then any two hyperplanes will do, provided their pullbacks intersect in a finite number of points. The number $\int h^2$ is an invariant of the class $h$. Similarly $\int h \cdot c_1(X)$ is the weighted number of points common to the pullback of a hyperplane and to the negative of a canonical divisor, whenever the number is finite. Similarly $\int c_1(X)^2$ is found by intersecting two suitable canonical divisors. However, the number $\int c_2(X)$ is something different. It is the degree of the *second Chern class*, another invariant of $X$. It is equal to the (topological) Euler characteristic, and it differs by 4 from the (geometric) Zeuthen–Segre invariant.

In general, as intimated above, a smooth projective $n$-fold $X$ possesses an intersection ring $A^{\cdot}X$ which behaves functorially in $X$ and carries Chern classes. The ring is the set of residue classes of cycles modulo a suitable equivalence relation. Cycles are linear combinations of closed subvarieties with any codimension. The equivalence relation can be homological equivalence or any one of several others. Rational equivalence is the least wasteful of information and so the one most commonly used; it is a generalization of rational equivalence of divisors. Every class $z$ has a well-defined degree $\int z$, which is found as follows: represent $z$ by a cycle and sum the coefficients of the components with dimension 0. The product of classes is induced by the intersection product of cycles. The intersection ring is graded by codimension. The $i$th graded piece $A^iX$ contains an invariant $c_i(X)$, the $i$th *Chern class* of $X$. The sum $c(X) = \sum c_i(X)$ is called the *total Chern class* of $X$. (For more details, see chapter II.)

For example, consider $\mathbb{P}^n$. Let $u$ denote the class of hyperplane. Then $A^{\cdot}\mathbb{P}^n$ is the simple ring $\mathbb{Z}[u]$ with the relation $u^{n+1} = 0$. The total Chern class is given by the formula,

$$\boxed{c(\mathbb{P}^n) = (1+u)^{n+1}} \,. \tag{I,19}$$

One interesting consequence of the simple structure of $A^{\cdot}\mathbb{P}^n$ is Bezout's theorem. It asserts that if two varieties with complementary dimensions intersect in a finite number of points, then the weighted number of points is equal to the product of their degrees. In fact, more generally, the degree of an intersection is equal to the product of the degrees of the intersectors; for the class in $A^{\cdot}\mathbb{P}^n$ of a variety is a multiple of a power of $u$, and the multiplier is the degree of the variety.

Consider a mapping $f: X \to Y$ between smooth varieties of dimension $n$ and $m$. Assume $n \leqq m$ and consider the *critical locus*, or *ramification locus*, $R$ of $f$. It is the subscheme of $X$ defined by the Fitting ideal $F^0\Omega^1_{X/Y}$. Assume that $f$ is appropriately generic in the sense that each irreducible component of $R$ has the appropriate codimension, $m-n+1$. Then the class $[R]$ in $A^{\cdot}X$ is given by the following formula (see chapter III, section A):

$$\boxed{[R] = c_{m-n+1}(\nu)}\ . \tag{I,20}$$

Here $\nu$ is the *virtual normal sheaf*, or *bundle*, of $f$. It is defined as the formal difference,

$$\nu = f^*\tau_Y - \tau_X, \tag{I,21}$$

where $\tau_Y$ and $\tau_X$ are the tangent bundles. Its total Chern class is defined by the formula,

$$\boxed{c(\nu) = f^*c(Y)/c(X)}\ . \tag{I,22}$$

For example, Formula (I,18) and Formula (I,15) as generalized to $n$-folds both come from Formulas (I,20), (I,22), and (I,19) by straightforward computation.

For another example, consider the *Steiner surface*. It is obtained as follows. First embed $\mathbb{P}^2$ in $\mathbb{P}^5$ by the Veronese embedding (it is given by the 6 quadratic monomials in the coordinate functions). Then project into $\mathbb{P}^3$ from a line in general position. The image of $\mathbb{P}^2$ in $\mathbb{P}^3$ is the Steiner surface. Let $u$ denote the hyperplane class of $\mathbb{P}^2$. Then the pullback $h$ of the hyperplane class of $\mathbb{P}^3$ is equal to $2u$. Hence (I,22) and (I,19) yield

$$\boxed{c(\nu) = 1 + 5u + 6u^2}\ . \tag{I,23}$$

Therefore by (I,20) the number $\nu_2$ of pinch points is 6 (which checks with [7; Ex. 5, p. 164]). (In characteristic 2 only, each pinch point appears with multiplicity 2, see [90; §5].)

## C. Classical enumerative geometry

Some of the oldest and best known enumerative formulas are the Plücker formulas for a (reduced, irreducible) plane curve $Z$ that is smooth except for $\delta$ (simple) nodes and $\kappa$ (simple) cusps. Let $d$ denote the *degree*, or order, of

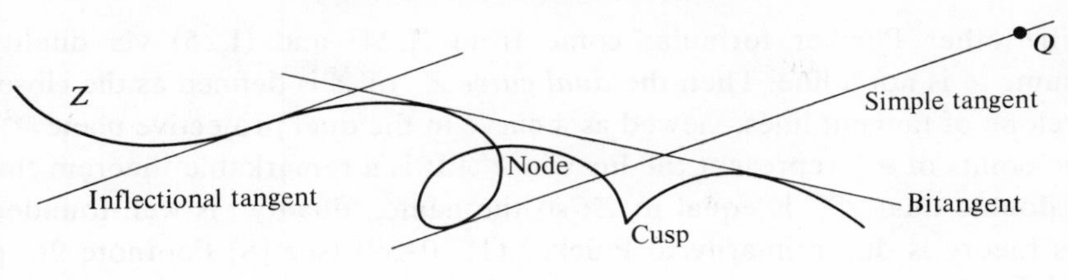


Figure 3

$Z$, that is, the number of (nonsingular) points of $Z$ that lie on a line in $\mathbb{P}^2$ in general position. Let $d^\vee$ denote the *class* of $Z$, that is, the number of nonsingular points of $Z$ at which the tangent line passes through a fixed point $Q$ of $\mathbb{P}^2$ in general position. Let $g$ denote the (geometric) *genus* of $Z$, that is, the genus of the desingularization $X$ of $Z$. The basic two formulas are

$$\boxed{d^\vee = d(d-1) - 2\delta - 3\kappa}, \tag{I,24}$$

$$\boxed{d^\vee = 2d + (2g-2) - \kappa}. \tag{I,25}$$

For example, (I,24) yields this: a smooth cubic has class 6; a nodal cubic 4; and a cuspidal cubic 3. Then (I,25) yields this: a smooth cubic has genus 1; a nodal cubic 0; and a cuspidal cubic 0.

Formula (I,24) reduces to $d^\vee = d(d-1)$ when $Z$ is smooth. This case was first discussed in an anonymous treatise on curves in 1756 (according to [8; Footnote 28, p. 325]). The idea of correction terms for singularities seems due to J. V. Poncelet (1817–22). He asserted that a node diminished the class by 2 and that a cusp diminished it by some number at least 2 (according to [82; p. 55]). J. Plücker (1834–39) (who is equally known for experimental work in physics) obtained Formula (I,24) and added several others to it (see [8; §8, pp. 342–345]). Of course (I,25) was not among Plücker's formulas; B. Riemann did not introduce the concept of genus until 1857. It was A. Clebsch (1864) who first gave the formula,

$$\boxed{g = (d-1)(d-2)/2 - (\delta + \kappa)}, \tag{I,26}$$

which is equivalent to (I,25) in the presence of (I,24) (see [8; p. 330]). Interestingly (see [8; Footnote 39, p. 329], [121; top p. 120]) the genus was called the 'deficiency' by A. Cayley (1865) and later writers in English because, in view of (I,26), the genus $g$ is the amount that the nodes and cusps, $\delta + \kappa$, falls short of the maximum possible, $(d-1)(d-2)/2$.

The other Plücker formulas come from (I,24) and (I,25) via duality. Assume $Z$ is not a line. Then the *dual curve* $Z^\vee$ of $Z$ is defined as the closed envelope of tangent lines, viewed as a curve in the dual projective plane $\mathbb{P}^{2\vee}$. (The points of $\mathbb{P}^{2\vee}$ represent the lines in $\mathbb{P}^2$.) It is a remarkable theorem that the double dual $Z^{\vee\vee}$ is equal to $Z$; so the name, 'duality', is well-founded. This theory is due primarily to Plücker (1830–39) (see [8; Footnote 26, p. 325], [9; Footnote 147, p. 289]).

Here is a simple proof of the 'biduality' theorem. Represent $Z$ with affine coordinates $(x, y)$ which depend on a parameter $t$, and represent $Z^\vee$ with the corresponding parametric affine coordinates $(m, b)$. Here $m$ is the slope of tangent line and it is given by the defining relation,

$$m = y'/x' \tag{I,27}$$

where $y'$ and $x'$ are the derivatives with respect to $t$. However, $b$ is not taken to be the $y$-intercept but its negative; so $b$'s defining relation, $y = mx - b$, will have the more symmetric form,

$$y + b = mx. \tag{I,28}$$

Clearly $Z^{\vee\vee}$ has parametric coordinates,

$$(b'/m', (b'/m')m - b). \tag{I,29}$$

Now, differentiating (I,28) with respect to $t$ yields

$$y' + b' = m'x + mx'. \tag{I,28$'$}$$

Together (I,28)$'$ and (I,27) yield

$$m'x - b' = y' - mx' = 0. \tag{I,30}$$

Using (I,30) and (I,28) we find (I,29) simplifies to $(x, y)$. Thus $Z^{\vee\vee}$ is equal to $Z$.

A *simple tangent* of $Z$ is a line that is tangent to $Z$ at one and only one point; at it, the multiplicity of intersection, or order of contact, must be 2 precisely, and at it $Z$ must be smooth. All but a finite number of tangents are simple. This fact is a corollary of the proof of the biduality theorem. Indeed, it is not hard to see by computing with power series that a branch $B$ of $Z$ determines and is determined by a branch $B^\vee$ of $Z^\vee$ and that the order of contact of $B$ with its tangent diminished by the multiplicity of $B$ is equal to the multiplicity of $B^\vee$. Consequently, if a smooth point $z$ of $Z$ corresponds to a smooth point of $Z^\vee$, then the tangent at $z$ is simple.

It follows that the tangents to $Z$ through a fixed but general point $Q$ of $\mathbb{P}^2$ are all simple; their number is, by definition, the class $d^\vee$. It also follows that $d^\vee$ is equal to the degree of $Z^\vee$; for the lines through a general point $Q$ of $\mathbb{P}^2$

correspond to the points on a general line in $\mathbb{P}^{2\vee}$. For example, the class of a conic is 2 by (I,24), so the dual is also a conic. (Seven interpretations of the class are given in [81; p. 22]; most are from differential geometry and integral geometry.) By duality, the class of $Z^\vee$ is equal to the degree of $Z$.

Assume now (as Plücker did) that $Z^\vee$ is smooth except for $\delta^\vee$ (simple) nodes and $\kappa^\vee$ (simple) cusps. Then applying (I,24) and (I,25) to $Z^\vee$ gives

$$\boxed{d = d^\vee(d^\vee - 1) - 2\delta^\vee - 3\kappa^\vee}, \qquad \text{(I,24)}^\vee$$

$$\boxed{d = 2d^\vee + (2g - 2) - \kappa^\vee}. \qquad \text{(I,25)}^\vee$$

Note as a corollary of biduality that $Z$ and $Z^\vee$ have the same desingularization $X$, so the same geometric genus $g$. Now, it follows from the statement about branches $B$ and $B^\vee$ in the second paragraph above that a (simple) node of $Z^\vee$ corresponds to a (simple) bitangent of $Z$. By the same token, a (*simple*) *bitangent* is a line that is tangent to $Z$ at exactly 2 distinct points; at each, the contact is 2 and $Z$ is smooth. A (simple) cusp of $Z^\vee$ corresponds to a (*simple*) *inflectional tangent* of $Z$. The latter is a line that is tangent to $Z$ at one and only one point; at it, the contact is 3 and $Z$ is smooth. So $\delta^\vee$ is the number of bitangents of $Z$, and $\kappa^\vee$ is the number of inflectional tangents. Thus (I,24)$^\vee$ and (I,25)$^\vee$ can be interpreted without reference to $Z^\vee$.

The four Formulas (I,24), (I,25), (I,24)$^\vee$, and (I,25)$^\vee$ are not independent; any three imply the fourth. However, any three are independent. The formulas yield other interesting relations. For example, eliminating $g$ and $d^\vee$, respectively $g$ and $d$, yields

$$\boxed{\kappa^\vee = 3d(d-2) - 6\delta - 8\kappa}, \qquad \text{(I,31)}$$

$$\boxed{\kappa = 3d^\vee(d^\vee - 2) - 6\delta^\vee - 8\kappa^\vee}. \qquad \text{(I,31)}^\vee$$

For instance, the dual of a cubic is smooth except for (simple) cusps by virtue of the relation between branches $B$ and $B^\vee$ given above because a cubic can have no bitangents and no tangent with contact above 3. Hence (I,31) yields this: a smooth cubic has 9 (simple) inflectional tangents; a nodal cubic 3; and a cuspidal cubic 1. Formulas (I,31) and (I,31)$^\vee$ were given by Plücker (1839) instead of (I,25) and (I,25)$^\vee$.

(In positive characteristic, the Plücker formulas and the biduality theorem do not always hold. For example, in characteristic 2 the tangents to a conic all pass through a certain point, called the conic's *strange point*; so the dual curve is a line. In characteristic 3 there are smooth cubics with only 3 inflectional tangents, and ones with only 1. The root of the problem is that the 'tangent mapping' $f^\vee: X \to Z^\vee$ may be inseparable; there is a minor additional difficulty in characteristic 2.

If the class $d^\vee$ is interpreted properly, then Formulas (I,24) and (I,25) hold in every characteristic except 2; in characteristic 2, the formulas must be replaced by

$$d^\vee = d(d-1) - 2\delta - 4\kappa, \qquad (\mathrm{I},24)_2$$

$$d^\vee = 2d + (2g-2) - 2\kappa. \qquad (\mathrm{I},25)_2$$

Interpreted properly, $d^\vee$ is the weighted number of tangent lines through a general point; each line is counted a number of times equal to the degree of $f^\vee$. In other words, $d^\vee$ is equal to the degree of $Z^\vee$ multiplied by the degree of $f^\vee$.

Suppose $f^\vee$ is separable. Then it actually has degree 1. In fact, biduality holds and can be proved virtually the same way as above. (Just choose $t$ to be a separating transcendental for the function field of $Z^\vee$. Then (I,30) implies $m' \neq 0$.) Finally, notice that if $d^\vee$ is computed using one of the formulas and turns out not to be a multiple of the characteristic, then $f^\vee$ must be separable, and so biduality, the naive interpretation of $d^\vee$, and all the Plücker formulas hold, although the formulas must be modified suitably in characteristic 2. (Warning: the relation between branches $B$ and $B^\vee$ given above can also fail. For instance, one of its consequences fails in characteristic 2. The dual of a smooth cubic has a higher order cusp. In particular, the dual formulas of $(\mathrm{I},24)_2$ and $(\mathrm{I},25)_2$ do not apply.)

The Plücker formulas were generalized from the start (see [82; p. 55], [83], [8; §8, pp. 342–345, §15, pp. 373–77, §18, pp. 381–382], [102; pp. 884–85], [119; §3, pp. 262–65], [7; pp. 158, 163]). Poncelet (1822) erroneously asserted that an ordinary $r$-fold point diminished the class by $r$; Plücker (1839) corrected the $r$ to $r(r-1)$. Cayley (1866) introduced the colorful idea that an arbitrary singularity was equivalent to so many nodes and cusps in an effort to make the formulas applicable to any curve. General formulas were obtained in the 1870's by G.-H. Halphen, H. J. S. Smith, and M. Noether. The formulas were generalized in terms of osculating spaces for curves in 3-space by Cayley (1845) and for curves in $n$-space by Veronese (1882). Various work on these was done by C. Segre (1894), H. and J. Weyl (1943), W. F. Pohl (1962), P. Griffiths (1974) and R. Piene (1976). Alongside these generalizations are formulas of G. Salmon, Cayley (1869), Zeuthen (1871, 76) and Noether (1871, 75) relating tangential and singularity invariants of a surface with mild singularities in 3-space. Some similar formulas were obtained by Roth (1933) [99] for 3-folds in 4-space.

Pohl [80], [81] is the first one to have looked at the Plücker formulas from the point of view of the theory of singularities of mappings. He started with a mapping $f: X \to \mathbb{P}^m$ such that $X$ is smooth and $f$ is an immersion on a dense open subset. When $X$ is a curve, he constructed out of the osculating $p$-planes a mapping $f^{(p)}$ from $X$ into the grassmannian of $p$-planes in $\mathbb{P}^m$ and he established one of the series of Cayley–Veronese formulas. When $X$ is an $n$-fold, he considered the mapping $f^\vee$ determined by the tangent $n$-planes. Although $f^\vee$ need not be defined everywhere on $X$, Pohl was able to obtain a new generalization of Formula (I,25) (see chapter IV, section C).

Piene [82], [83] brought Pohl's approach significantly further in three directions. First, Piene generalized Formula (I,24) to the case in which $Z$ is a hypersurface of any dimension with arbitrary singularities, recovering Teissier's formula for the case of isolated singularities (see chapter IV, section C). A major difference between Pohl's formula and Piene's, when they both apply, is this: Pohl's formula involves the singularities of the mapping $f: X \to \mathbb{P}^m$, in other words, the cuspidal singularities of $Z = f(X)$; Piene's involves all the singularities of $Z$. Second, Piene gave the first modern derivations of some of the formulas of Salmon, Cayley, Zeuthen, and Noether for surfaces in $\mathbb{P}^3$. (The classical extrinsic invariants involved here are expressed in terms of Chern classes in [68]. These expressions are useful to have, but they are derived assuming the classical formulas.) Third, Piene established, for the first time, all the Cayley–Veronese formulas for curves in $n$-space.

The class of a plane curve $Z$ has a significant reinterpretation. Defined as the number of nonsingular points of $Z$ at which the tangent line passes through a general point $Q$ of $\mathbb{P}^2$, it can be reinterpreted as the number of nonsingular points $z$ of $Z$ at which the line $\overline{zQ}$ is tangent to $Z$ at $z$. This reinterpretation underlies some important generalizations of the Plücker formulas to the case of a smooth $n$-fold $X$ embedded in $\mathbb{P}^m$ (see chapter IV, section D). The class $d^\vee$ of $X$ is defined as the number of points $x$ of $X$ such that the hyperplane $\overline{xQ}$ spanned by $x$ and a general $(m-2)$-plane $Q$ is tangent to $X$ at $x$. There are 'Plücker' formulas for $d^\vee$. The various hyperplanes tangent to $X$ form a variety $X^\vee$ in $\mathbb{P}^{m\vee}$, ordinarily a hypersurface, called the *dual variety*. There is a biduality theorem, and so forth. (In positive characteristic, there are the same sort of reservations as for curves.)

Consider the system of sections of $X$ by the hyperplanes through the general $(m-2)$-plane $Q$. The singular sections are precisely those cut out by the hyperplanes tangent to $X$. The locus of singular points of sections is called a *polar locus* of $X$. It is a finite set and has $d^\vee$ points. The notion of polar locus was generalized classically, under the name of *jacobian locus*, to arbitrary linear systems of divisors. There are corresponding generalized 'Plücker' formulas. These formulas played the key role in the development in algebraic geometry of the theory of canonical classes, before Chern's fundamental work. (For more details, see chapter III, section D and chapter IV, section B.)

## D. Multiple-point formulas

One of the oldest and most familiar multiple-point formulas concerns a generic mapping $f: X \to \mathbb{P}^2$; here $X$ is a smooth, complete curve, $f$ maps $X$ birationally onto its image $Z = f(X)$, and $Z$ is smooth except for $\delta$ (simple) nodes. The formula is

$$\boxed{\delta = \tfrac{1}{2}(d-1)(d-2) - g}, \tag{I,32}$$

where $d$ denotes the degree of $Z$ and $g$ denotes the genus of $X$. It is a special case of Clebsch's formula (I,26).

Each node of $Z$ is the image of 2 points of $X$, each of which is a double-point of $f$. So the number of double-points of $f$ is equal to $2\delta$. Hence (I,32) may be rewritten in the equivalent form,

$$\int [M_2] = d^2 - 3d + (2 - 2g), \tag{I,33}$$

where $M_2$ denotes the double-point locus of $f$ and $\int [M_2]$ denotes the degree of the cycle $[M_2]$.

Formulas (I,32) and (I,33) are numerical double-point formulas. They are consequences of the following cycle-theoretic assertion: let $H$ be a hyperplane of $\mathbb{P}^2$; then there exists a meromorphic differential $\omega$ on $X$ satisfying the equation,

$$\boxed{[M_2] = (d-3)f^*[H] - (\omega)}. \tag{I,34}$$

Equivalently, this assertion says that the following double-point formula holds in the rational equivalence ring:

$$\boxed{[M_2] = f^*f_*[X] - f^*c_1(\mathbb{P}^2) + c_1(X)}. \tag{I,35}$$

The two formulations are equivalent because $Z = f(X)$ is linearly equivalent to $dH$ and because of Formula (I,19).

Equation (I,34) is essentially part of the classical theory of adjoint curves. An adjoint curve of $Z$ is a plane curve $C$ (possibly reducible and with multiple components) of degree $(d-3)$ that pass through all the nodes of $Z$. It was known that $f^*[C] - [M_2]$ is a positive canonical divisor, that is, the divisor of a holomorphic differential, and that all of them have this form.

This statement is clearly equivalent to the validity of (I,34). The theory of adjoint curves grew out of Riemann's explicit determination (1857) of the abelian integrals of the first kind on $X$, the integrals of the holomorphic differentials. He proved (according to [11; (V,15), p. 56]) that these are precisely the integrals of the form,

$$\int \frac{\Phi(x, y)}{\partial F/\partial y}\, dx, \tag{I,36}$$

where $F$ is a polynomial defining $Z$ and $\Phi$ is a polynomial of degree at most $d-3$ that vanishes at all the nodes of $Z$; the nodes are assumed to be at finite distance. The theory of adjoint curves was generalized classically to the case that $Z$ is a curve with arbitrary singularities by Clebsch and Gordon (1866), Brill and Noether (1873), and Noether (1883) in particular (according to [8; p. 374, p. 377]) and to the case that $X$ is a smooth $n$-fold mapping nicely into $\mathbb{P}^{n+1}$ by Noether (1869, 70, 75) in particular (according to [102; p. 926]).

During the past year (1975–76), tremendous progress was made in double-point theory by D. Laksov especially. The following double-point theorem is now proved (see chapter V, section C): Let $f: X \to Y$ be an appropriately generic proper mapping from a smooth $n$-fold into a smooth $m$-fold with $m \geqq n$. Then the double-point set $M_2$ of $f$ is the support of a positive cycle whose rational equivalence class $[M_2]$ is given by the formula,

$$\boxed{[M_2] = f^* f_* [X] - c_{m-n}(\nu)}\;, \tag{I,37}$$

where $\nu$ is the virtual normal bundle (see (I,21) and (I,22)). 'Appropriately generic' is a more subtle concept in the case of this formula. Fulton [14; §2.5] points out with an example that it is not enough to require that each irreducible component of $M_2$ have the right codimension, $m-n$. What is required for the proof is that a certain more fundamental locus $Z_2$ have pure codimension $m-n$.

Formula (I,37) yields Formula (I,35) as a special case. In fact, it shows (I,35) is valid without any restriction whatsoever on the kinds of singularities on $Z = f(X)$. In fact, it yields a very similar formula valid for any finite mapping $f: X \to Y$ such that $X$ is a smooth $n$-fold, $Y$ is a smooth $(n+1)$-fold, and $X \to f(X)$ is birational (see fig. 2 above). In this case (see chapter V, section A) the conductor makes the set $M_2$ into a scheme whose fundamental cycle $[M_2]$ satisfies (I,37). In general, it is not known whether there is such a scheme structure on $M_2$; however, Fulton [14; §2.4] points out that the conductor does not always do the trick when it might.

For an example, consider the Steiner surface. We considered it at the end of section B, and we now use the same notation. Since the surface has degree 4, Formulas (I,37) and (I,23) yield

$$\boxed{[M_2]=3u}\ . \tag{I,38}$$

The double curve of the Steiner surface is the image of $M_2$, and its degree was denoted by $\varepsilon_0$ classically. Since $M_2$ is a 2-to-1 (ramified) covering of its image, $\varepsilon_0$ is equal to $\frac{1}{2}\int[M_2]h$. Since $h$ is equal to $2u$, Formula (I,38) yields $\varepsilon_0=3$ (which checks with [7; Ex. 5, p. 164]).

The present interest among algebraic geometers in the double-point Formula (I,37) stems from independent pioneering work (see chapter V, section B) of A. Holme [29] and of C. A. M. Peters and J. Simonis [84], which was available in 1974 in preprint form. Holme was interested in numerical embedding obstructions for a mapping $f: X \rightarrow \mathbb{P}^m$ induced by a projection from a center in general position. Holme's obstructions were found by direct hard work, but they can be obtained from (I,37) by requiring $[M_2]$ to be 0. Holme's work shows that $f$ is appropriately generic, which is expected but not obvious; hence (I,37) does apply. In the case that $X$ has dimension $n=m-1$, Vainsencher [114] strengthened Holme's result and gave a short direct proof; a single inequality suffices to ensure that $f: X \rightarrow \mathbb{P}^{n+1}$ is an embedding and $X$ may be allowed arbitrary singularities. Holme [30] himself went on (independently) to obtain suitable embedding obstructions for the general singular case.

Peters and Simonis considered a smooth $n$-fold $X$ in $\mathbb{P}^{2n+1}$. They obtained a formula for the number of secant lines passing through a general point $Q$ of $\mathbb{P}^{2n+1}$. They observed (implicitly) that this number is 0 if and only if the projection with center $Q$ induces an embedding $f: X \rightarrow \mathbb{P}^{2n}$. They applied this observation to reprove some interesting results: 1. (Van de Ven, 1973) An abelian variety of dimension $n$ at least 3 cannot be embedded in $\mathbb{P}^{2n}$. (Holme also discussed this example.) 2. (Feder, 1965) $\mathbb{P}^n$ can be embedded in $\mathbb{P}^{2n}$ only as a subvariety of degree 1 or $2^n$. Of course, the end points on $X$ of the secants through $Q$ form the double-point set $M_2$, and so Peters and Simonis' formula is also a special case of (I,37). However, again, (I,37) does not apply unless $f$ is appropriately generic; their work shows it is.

The first general double-point formula is due to Severi (1902) (according to [102; p. 925]). He considered the case of a smooth $n$-fold $X$ under projection from a general center, and he gave a rather nice formula (V,37) for the degree of the locus $M_2$. His formula yields Peters and Simonis' formula and Holme's formula for his embedding obstructions. The first

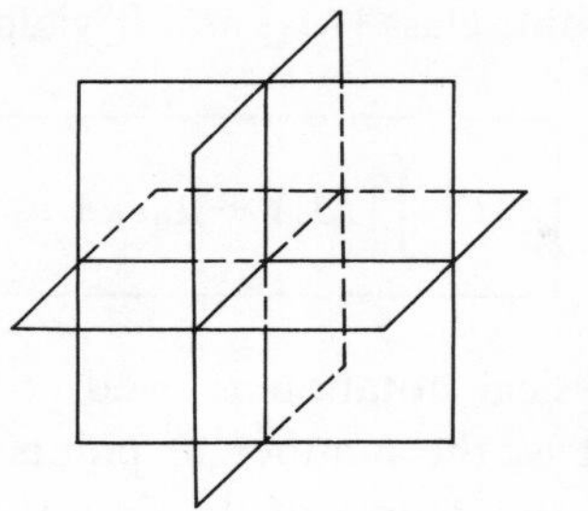

$f(X)$ locally at an ordinary triple point

Figure 4

general statement of a compact formula like (I,37) is due to Todd (1940) (see [111; bottom of p. 32]); he worked modulo rational equivalence and did not require $f$ to be induced by a projection nor $Y$ to be a projective space. A formula like (I,37) was given by B. Segre (1953) as part of his theory of canonical classes. Appropriately it is Segre's classes that come naturally into play in today's proof of (I,37). Indeed, they are necessary for the generalization of (I,37) to the case in which $X$ is singular as carried out by Johnson [40] for projections and in Johnson's spirit by Fulton and Laksov [16] for the general case. On the topological side, recent work on the double-point formula was done by Ronga [96] for differentiable varieties, by McCrory [64] for *PL*-varieties and by McCrory and Hardt [67] for subanalytic varieties.

Higher-order multiple-point theory is at present only beginning to be developed. In many cases however we do have a constructive procedure that in principle will yield a multiple-point formula, and moreover in some cases the formula can easily be worked out (see chapter V, section D). On the topological side, multiple-point formulas were obtained notably by Lashof and Smale [60] (this work is well-known to be wrong), by Haefliger [24] (who does not attempt to enumerate multiple points on $X$) and by Herbert [23]; however, all of this work is done for the case of an immersion.

One case in which a triple-point formula is established is the case of a mapping $f: X \to \mathbb{P}^3$ induced by a projection from a general center, where $X$ is a smooth surface. This case is important because it is classical (it complements Piene's work on the formulas of Salmon, Cayley, Zeuthen, and Noether) and because it seems to be one of only 2 cases in which higher-order multiple-point formulas have already been obtained in the presence of ramification (the other case is that of a general projection of a smooth 3-fold into $\mathbb{P}^4$; it was investigated by Roth [99]). In the present case, there is a formula for the triple-point class $[M_3]$ and it yields the following numerical formula:

$$\boxed{3t = \int [M_2]^2 - \mu_0\varepsilon_0 + \nu_2}\ . \tag{I,40}$$

Here the following classical notation is used: $t$ for the number of triple points of $Z = f(X)$, that is, the number of points in $f(M_3)$; and $\mu_0$ for the degree of $Z$; and $\varepsilon_0$ for the degree of the double curve $f(M_2)$ of $Z$; and $\nu_2$ for the number of pinch points. For example, for the Steiner surface, from what we already know (Formulas (I,23) and (I,38), etc.) and from Formula (I,39) or Formula (I,40), we obtain $t = 1$ (which checks with [7; Ex. 5, p. 164]).

## II. A résumé of intersection theory

### A. Basics

(See [43; §2, §3] for a further outline; see [10], [105], [106], [107; Ch. V], [13], [54; Ch. VIII], [6], [36], [89], [122], for details.) Here we work exclusively with smooth, projective (irreducible) ambient varieties, $X$, $Y$, etc.

The group of *k-cycles*, or *cycles of dimension k*, on an $n$-fold $X$ is defined as the free abelian group on symbols $[V]$, one for each $k$-dimensional, closed (irreducible) subvariety $V$ of $X$. The group is denoted by $Z_kX$. It is ordered; a cycle $\sum a_i[V_i]$ is called *positive* if each coefficient $a_i$ is positive. Each subscheme $W$ of $X$ determines a cycle $[W]$, called the *fundamental cycle* of $W$ as follows: let $W_1, \cdots, W_r$ be the irreducible components of $W$ and consider them as subvarieties of $X$, determining generators $[W_i]$ of $Z_kX$; define $[W]$ by

$$[W] = \sum a_i[W_i] \quad \text{with} \quad a_i = \text{length}\ (O_{W,W_i}). \tag{II,1}$$

Note that if $W$ is reduced and irreducible, the fundamental cycle $[W]$ is equal to the generator $[W]$.

The group of $k$-cycles behaves covariantly; a map $f: X \to Y$ induces a linear map,

$$f_*: Z_kX \to Z_kY. \tag{II,2}$$

The induced map $f_*$ is defined on a generator $[V]$ by

$$f_*[V] = \begin{cases} \text{degree}\ (f \mid V) \cdot [f(V)] \\ 0 \quad \text{if} \quad \dim(F(V)) < k. \end{cases} \tag{II,3}$$

Sometimes it is more convenient to work by codimension. The group of cycles of codimension $k$ is denoted by $Z^k$ or $Z^kX$. So by definition we have

$$Z^kX = Z_{n-k}X \qquad (n = \dim(X)). \tag{II,4}$$

There exists a linear map from $Z^kX$ into $H^{2k}(X, \mathbb{Q})$, the cohomology group. The map is a natural transformation; a map $f: X \to Y$ induces a commutative diagram

$$\begin{array}{ccc} Z^kX & \longrightarrow & H^{2k}(X, \mathbb{Q}) \\ \downarrow f_* & & \downarrow f_* \\ Z^{m-n+k}Y & \longrightarrow & H^{2(m-n+k)}(Y, \mathbb{Q}) \end{array} \qquad \begin{array}{l} (n = \dim(X)) \\ \\ (m = \dim(Y)). \end{array} \tag{II,5}$$

The map $f_*$ on the right is the *Gysin homomorphism*; it is the dual, under Poincaré duality, of the canonical map $f^*$ on cohomology. The algebraic cohomology classes, the images of the algebraic cycles, form as $k$ varies a graded ring under cup-product. This ring behaves functorially under the contravariant, multiplicative map $f^*$ on cohomology. The Gysin homomorphism $f_*$ induces a linear map. The two maps are related by the rather useful *projection formula*,

$$\boxed{f_*(f^*y \cdot x) = y \cdot f_*x}\ . \tag{II,6}$$

The formula holds in fact for any cohomology classes $x$ and $y$; it holds simply because $f^*$ is the dual of $f_*$. (In arbitrary characteristic, there is an analogous theory using étale cohomology; see [122].)

The ring of algebraic cohomology classes is a model intersection ring. It could have been used exclusively in these notes. Historically it was available first; it was developed by Lefschetz (1924, 1926) from some ideas of Poincaré and Kronecker. However it is wasteful of information and it is not constructed directly out of cycles. We now discuss two other intersection rings with the same formal properties.

In making an intersection ring out of the cycles on $X$, the first task is to define whenever possible the product of two cycles. By linearity it is enough to consider the case of two generators, $[V]$ and $[W]$. Suppose the subvarieties $V$ and $W$ intersect 'properly'; that is, suppose each irreducible component $Z_i$ of $V \cap W$ satisfies the relation,

$$\operatorname{codim}(Z_i, X) = \operatorname{codim}(V, X) + \operatorname{codim}(W, X). \tag{II,7}$$

(The right hand side is the maximum value for the codimension of $Z_i$ and it is the 'generic' value.) Define the *intersection product*, or *intersection cycle*, $[V]\cdot[W]$ by

$$[V]\cdot[W] = \sum_i m_i[Z_i] \tag{II,8}$$

with $m_i = \sum_j (-1)^j$ length $(\operatorname{tor}_j^{O_{X,Z_i}}(O_{V,Z_i}, O_{W,Z_i}))$. This definition of the multiplicities $m_i$ is known as 'Serre's tor formula'; it was introduced in [107;

Ch. V] and has become popular because it is convenient to work with. The number $m_i$ itself measures the degree of contact of $V$ and $W$ along $Z_i$. That is a fairly sophisticated notation in full generality, and it has taken a century or two and a lot of work to master it.

The intersection product of two cycles is not always defined, because two subvarieties do not always intersect properly. However, given two cycles $v$ and $w$, it is always possible to deform $v$ in a family of cycles parametrized by the projective line into a cycle $v'$ such that the product $v' \cdot w$ is defined; this result is known as the 'moving lemma'. The idea of the proof is this. We may assume $v = [V]$. Embed $X$ in a projective space $\mathbb{P}^m$, choose a general linear space $A$ of codimension $n+1$ with $n = \dim(X)$, and form the cone $C$ over $V$ with vertex $A$. It can be proved that $[C] \cdot [X]$ is defined on $\mathbb{P}^m$ and equal to $v + v''$ where $v''$ is a cycle intersecting $w$ properly on $X$ or more nearly so than $v$. Moreover, $C$ can be deformed by translation into a variety $C'$ such that $v''' = [C'] \cdot [X]$ is defined on $\mathbb{P}^m$ and $v''' \cdot w$ is defined on $X$. Then $v''' - v'$ is a deformation of $v$ that intersects $w$ properly or more nearly so.

Two cycles are said to be *rationally equivalent*, or *linearly equivalent*, if they are both members of the same family of cycles parametrized by the projective line $\mathbb{P}^1$. Such a family consists of the fibers over $\mathbb{P}^1$ of a cycle on $X \times \mathbb{P}^1$. It is not hard to see that rational equivalence is an equivalence relation and that it respects addition. The equivalence class of a cycle $v$ will be denoted by $[v]$; if the cycle is the fundamental cycle $[V]$ of a subscheme $V$, the class will be denoted by $[V]$ also.

The rational equivalence classes form a ring under the intersection product by virtue of the moving lemma. It is graded by codimension. It has a unit, namely, $1 = [X]$. This ring is denoted by $A^{\cdot}X$. Functorial behavior in both directions can be established. A map $f: X \to Y$ induces both a homomorphism of graded rings, $f^*: A^{\cdot}Y \to A^{\cdot}X$, and a homogeneous linear map, $f_*: A^{\cdot}X \to A^{\cdot}Y$, that drops degrees by $m - n$ (with $m = \dim(Y)$ and $n = \dim(X)$). The two maps, $f^*$ and $f_*$, are related by a projection formula like (II,6).

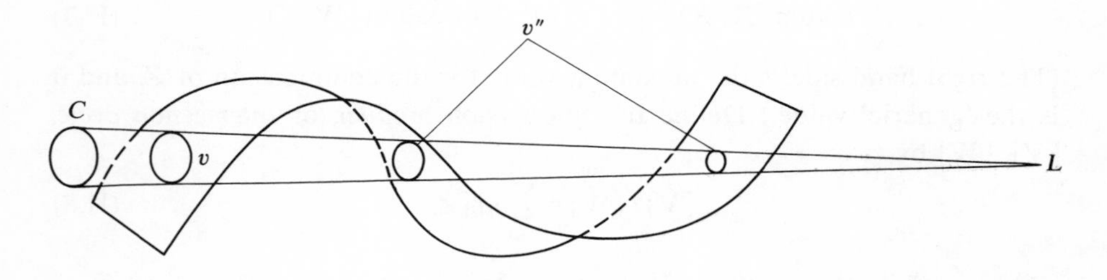


Figure 5

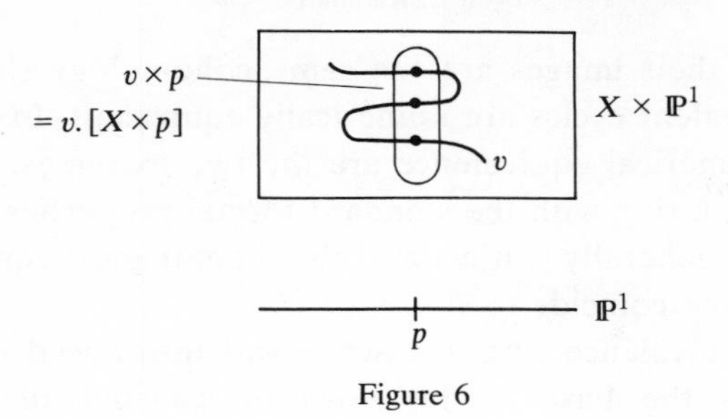


Figure 6

The map $f_*$ is induced by the corresponding map (II,2) on cycles. The map $f^*$ is given on a generator $[W]$ by the formula,

$$f^*[W] = pr_{1*}([\Gamma_f] \cdot [X \times W]), \tag{II,9}$$

where $\Gamma_f$ denotes the graph of $f$ (it is a subvariety of $X \times Y$). If each irreducible component of $f^{-1}W$ has the appropriate codimension in $X$, namely, the same as that of $W$ in $Y$, then $f^*[W]$ is given by a modified form of the tor formula. If in addition $W$ is Cohen–Macaulay, then the higher tor's vanish and we have $f^*[W] = [f^{-1}W]$ where $f^{-1}W$ means the scheme-theoretic preimage.

The *degree* of a class $x$ in $A^{\cdot}X$ is defined as follows. Let $P$ be a point, and let $g: X \to P$ be the constant map. We clearly have an isomorphism,

$$\alpha: \mathbb{Z} \xrightarrow{\sim} A^{\cdot}P \text{ by } \alpha(r) = r[P]. \tag{II,10}$$

The degree of $x$ is defined as the integer, $\alpha^{-1}g_*x$. It is denoted variously by $|x|$, by $\deg(x)$, by $\langle\alpha\rangle$, and by $\int x$. Note that given a mapping $f: X \to Y$ we have

$$\boxed{\int x = \int f_* x}\,. \tag{II,11}$$

The degree of a cycle is defined as the degree of its associated class; clearly it is equal to the sum of coefficients of the components with dimension 0 (or codimension $n$ with $n = \dim(X)$). The *order* of a class $x$ in $A^{n-k}X$ is defined as the number $\int x \cdot h^k$, where $h$ is the class of a hyperplane section of $X$. The order of a subvariety of $X$ is defined as the order of its associated class; the order of a subvariety is often also called the degree of that subvariety.

Two classes $x$, $y$ in $A^{\cdot}X$ are called *numerically equivalent* if the intersection numbers, $\int x \cdot z$ and $\int y \cdot z$, are equal for every $z$ in $A^{\cdot}X$. Two cycles are called numerically equivalent if their classes in $A^{\cdot}X$ are. Modulo numerical equivalence, the cycles clearly form a ring with the same formal properties as $A^{\cdot}X$. It can be proved that rationally equivalent cycles are homologically equivalent (that is, their images are the same cohomology class) and that homologically equivalent cycles are numerically equivalent. In fact, rational equivalence and numerical equivalence are the two extremes; any equivalence relation giving a ring with the standard formal properties lies between them. However it is generally conjectured that homological equivalence and numerical equivalence coincide.

The numerical equivalence ring in essence was introduced in the 1870's by H. Schubert as the basis of his powerful calculus of enumerative geometry. Schubert was inspired by some remarks made in 1873 by G. Halphen on M. Chasles's enumerative method (1864) for conics. From about 1910 to 1940 F. Severi and B. van der Waerden, striving to interpret and justify Schubert's calculus, began the modern development of intersection theory. In the late 1940's and the 1950's tremendous progress took place, particularly in the hands of A. Weil, C. Chevalley, P. Samuel, W. Chow, J.-P. Serre, and A. Grothendieck; the basic theory was secured. In the 1960's and the 1970's there have been three major developments: (1) various analogs of the topological cohomology ring (étale cohomology, crystalline cohomology and some others); (2) higher algebraic $K$-theory and its tie-in with rational equivalence; (3) intersection theory on singular and non-projective varieties. Intersection theory is pervasive, and most every algebraic geometer has contributed something to its advancement. Intersection theory is rich and remains a lively topic of research.

## B. Chern classes

(See [21], [22].) Here again we work with smooth, projective ambient varieties, $X, Y, \cdots$. A theory of Chern classes consists in giving, for each locally free sheaf $E$ on an $n$-fold $X$, a class,

$$c_i(E) \in A^i X, \quad \text{for each } i, \tag{II,12}$$

called the $i$th *Chern class* of $E$. The sum,

$$c(E) = \sum c_i(E), \tag{II,13}$$

is called the *total Chern class* of $E$. Let $e$ denote the rank of $E$. The following properties hold.

1. *Normalization*

We have $c_0(E) = 1$ and $c_i(E) = 0$ for $i > e$. If $E$ is invertible (that is, its rank is 1), then the only nontrivial Chern class $c_1(E)$ is equal to the class $[D]$

of any divisor $D$ whose associated sheaf $O_X(D)$ is equal to $E$; in short, we have

$$\boxed{c(E) = 1 + [D]} \quad \text{for} \quad E = O_X(D). \tag{II,14}$$

2. *Pullback*

For any map $g: Z \to X$, we have

$$\boxed{c(g^*E) = g^*c(E)}\,. \tag{II,15}$$

3. *Additivity*

For each short exact sequence $0 \to E' \to E \to E'' \to 0$ of locally free sheaves on $X$, we have

$$c(E) = c(E')c(E''). \tag{II,16}$$

This property implies that the formation of Chern classes extends over the Grothendieck group $K(X)$, the free abelian group on the locally free sheaves modulo the relations coming from the short exact sequences. Thus for example we have

$$\begin{aligned} c(E - E') &= c(E)/c(E') \quad \text{by definition} \\ &= c(E'') \quad \text{if} \quad 0 \to E' \to E \to E'' \to 0. \end{aligned} \tag{II,17}$$

4. *Duality*

Letting $E^\vee$ denote the dual of $E$, we have

$$\boxed{c_i(E^\vee) = (-1)^i c_i(E)} \quad \text{for each } i. \tag{II,18}$$

5. *Projective bundle*

Set $P = \mathbb{P}(E)$; this is the bundle of projective spaces associated to $E$ and it is defined as Proj (Sym $(E)$). Denote by $p: P \to X$ the structure map and by

$$0 \to K \to p^*E \to O_P(1) \to 0 \tag{II,19}$$

the tautological exact sequence. Set $t = c_1(O_P(1))$. Then we have

$$\boxed{t^e - p^*c_1(E)t^{e-1} + \cdots + (-1)^e p^*c_e(E) = 0}\,. \tag{II,20}$$

Indeed, the left side is, up to sign, equal to the $e$th Chern class of $K$ by (II,16) and (II,15); so it is equal to 0 because $K$ has rank $e - 1$.

Equation (II,20) determines the Chern classes of $E$ because the powers of $t$, it can be proved, are independent over $A^{\cdot}X$. In fact the powers of $t$

generate $A^{\cdot}P$; so we have an isomorphism of $A^{\cdot}X$-algebras,

$$\begin{aligned} A^{\cdot}X[T]/(T^e - c_1(E)T + \cdots) &\xrightarrow{\sim} A^{\cdot}P \\ T &\longrightarrow t, \end{aligned} \tag{II,21}$$

where $T$ is an indeterminate. For example $P$ is the projective $(e-1)$-space and $t$ is the hyperplane class if $X$ is a point; in this case we conclude that $A^{\cdot}P$ is the simple ring $\mathbb{Z}[t]$ with the relation $t^e = 0$.

Grothendieck [21] turned equation (II,20) into the definition of the $c_i(E)$ and brought the theory of Chern classes definitively into algebraic geometry. Just before, Hirzebruch [32] had axiomatized the theory and Chow (unpublished) had algebrized the theory using grassmannians.

6. *Invertible product*

Let $L$ be an invertible sheaf, let $r$ be an integer. Then we have

$$\boxed{c_r(E \otimes L) = \sum_i \binom{e-r+i}{i} c_{r-i}(E) c_1(L)^i} \, . \tag{II,22}$$

Indeed, if $E$ is invertible $(e = 1)$, then (II,22) follows from (II,14). Now, in the notation of property 5, it is enough to verify (II,22) for $p^*E$ and $p^*L$ because $A^{\cdot}P$ is free over $A^{\cdot}X$. In the tautological sequence (II,19), the rank of $K$ is $e-1$. So we may apply induction, and (II,22) follows from (II,16) via a straightforward computation. This method of proof is known as the *splitting principle.*

7. *Determinant*

The splitting principle easily yields

$$\boxed{c_1(E) = c_1(\Lambda^e E)} \, . \tag{II,23}$$

8. *Varieties*

The *Chern classes* $c_i(X)$ and the *total Chern class* $c(X)$ are defined as those of the tangent sheaf $\tau_X$, the dual of the sheaf of 1-forms $\Omega_X^1$. (These sheaves are locally free of rank $n$ because $X$ is a smooth $n$-fold.) It is convenient to let $c(X)^*$ denote the alternating sum of the $c_i(X)$, which by (II,18) is equal to the total Chern class $c(\Omega_X^1)$.

The Chern classes $c_i(X)$ are said to be generalizations of the topological Euler characteristic $\chi(X)$ because of the formula,

$$\boxed{\chi(X) = \int c_n(X)} \, . \tag{II,24}$$

For example, if $X$ is a curve, then (II,24) yields $\chi(X)=-\deg(K)$ where $K$ is a canonical divisor. Formula (II,24) follows from three general results. By the Lefschetz fix-point formula, $\chi(X)$ is equal to the self-intersection number $\int[\Delta]^2$ of the diagonal $\Delta$ in $X\times X$. By the self-intersection formula, this number is equal to the degree $\int c_n(\nu)$ of the highest Chern class of the normal bundle $\nu$ of $\Delta$. Finally, $\nu$ is equal to $\tau_X$.

For example, consider $P=\mathbb{P}(E)$. Its Chern classes can be computed using the usual two exact sequences,

$$0\to\tau_{P/X}\to\tau_P\to\tau_X\to 0, \tag{II,25}$$

$$0\to O_P\to(p^*E^\vee)(1)\to\tau_{P/X}\to 0, \tag{II,26}$$

where $p:P\to X$ denotes the structure map. Formulas (II,16) and (II,14) with $D=0$ yield

$$\boxed{c(P)=p^*c(X)c((p^*E^\vee)(1))}\,. \tag{II,27}$$

For instance, if $X$ is a point, then $P$ is a projective space and (II,27) yields (I,19).

There is a useful formula for the pushout of the total Chern class of a smooth divisor $Z$ on $X$. Known as the *adjunction formula*, it is

$$\boxed{\begin{aligned}&\alpha_*c(Z)=\frac{[Z]}{1+[Z]}\,c(X),\quad\text{or}\\&\alpha_*c_i(Z)=\sum_j(-1)^jc_{i-j}(X)\cdot[Z]^{j+1}\end{aligned}}\quad\text{for all } i, \tag{II,28}$$

where $\alpha$ denotes the inclusion map. The formula comes formally from the tangent sheaves–normal sheaf exact sequence,

$$0\to\tau_Z\to\alpha^*\tau_X\to\nu\to 0. \tag{II,29}$$

Since $Z$ is a divisor, $\nu$ is equal to $O_Z(Z)$, or $\alpha^*O_X(Z)$. By formulas (II,14), (II,15), and (II,16) we obtain from (II,29) the relation,

$$c(Z)\alpha^*(1+[Z])=\alpha^*c(X). \tag{II,30}$$

Applying $\alpha_*$, then using the projection formula (II,6) on each side (with $x=1$ on the right hand side), and finally solving for $\alpha_*c(Z)$, we obtain (II,28).

For example suppose $Z$ is a smooth hypersurface of degree $d$ in $\mathbb{P}^{n+1}$. Then (II,28), (II,24), (II,11), and (I,19) yield

$$\chi(Z)=\sum_{j=0}^{n}(-1)^j\binom{n+2}{n-j}d^{j+1}. \tag{II,31}$$

For instance if $Z$ is a smooth plane curve, then (II,31) yields

$$g = \tfrac{1}{2}(d-1)(d-2). \qquad \text{(II,32)}$$

9. *Segre classes*

For each $i$, the *ith Segre class* of $E$ is defined in the notation of property 5 by the formula,

$$\boxed{s_i(E) = p_* t^{e-1+i} \in A^i X}\ . \qquad \text{(II,33)}$$

It is a well-known and basic fact that we have

$$s_0(E) = 1. \qquad \text{(II,34)}$$

The sum $s(E) = \sum s_i(E)$ is called the *total Segre class* of $E$. It is related to the total Chern class of the dual sheaf $E^\vee$ by the formula,

$$\boxed{c(E^\vee)s(E) = 1}\ . \qquad \text{(II,35)}$$

Indeed, formulas (II,20) and (II,18) yield

$$p_* t^{i-1}(t^e + p^* c_1(E^\vee)t^{e-1} + \cdots + p^* c_e(E^\vee)) = 0, \quad \text{for} \quad i \geqq 1. \qquad \text{(II,36)}$$

Applying the projection formula and the defining equation (II,33), we obtain

$$s_i(E) + c_1(E^\vee)s_{i-1}(E) + \cdots + c_e(E^\vee)s_{i-e}(E) = 0. \qquad \text{(II,37)}$$

Finally, (II,37) and (II,34) imply (II,35).

The Segre classes are sometimes called *inverse Chern classes* because of (II,35). Using (II,35) we can easily determine the Segre classes by inverting $c(E^\vee)$ and applying formula (II,18); for example, we find

$$\begin{aligned} s_1(E) &= c_1(E), \\ s_2(E) &= c_1(E)^2 - c_2(E), \\ s_3(E) &= c_1(E)^3 - 2c_1(E)c_2(E) + c_3(E). \end{aligned} \qquad \text{(II,38)}$$

Another consequence of (II,35) is additivity of the Segre classes; namely, given a short exact sequence, $0 \to E' \to E \to E'' \to 0$, we have $s(E) = s(E')s(E'')$. Therefore the formation of Segre classes too extends over the Grothendieck group $K(X)$. We have

$$\begin{aligned} s(E-E') &= s(E)/s(E') \quad \text{by definition} \\ &= s(E'') \quad \text{if} \quad 0 \to E' \to E \to E'' \to 0. \\ &= c(E'^\vee)/c(E^\vee) \quad \text{by (II,35).} \end{aligned} \qquad \text{(II,39)}$$

Obviously the formation of Segre classes commutes with pullback.

The classes are named after B. Segre. While it does not seem that he introduced them explicitly, they are closely related to the covariants discussed below, which he did introduce in effect. The classes are useful and they were introduced by a number of authors, apparently independently, sometimes under Segre's name and sometimes not.

## C. General ambient varieties

(See [13], [122].) The theory in sections A and B extends without difficulty to the case of quasi-projective ambient varieties, ones that need not be complete (compact), so long as they are smooth; virtually the only qualification necessary is that, for the covariant map $f_*$ to be defined and behave right, the map $f$ must be proper. However, even when smooth ambient varieties hold our primary interest, we may be led to consider singular ones. So, it was an important advance when recently Fulton generalized the theory appropriately to arbitrary quasi-projective ambient schemes $X$. (While nilpotents are allowed, their presence is little in evidence.) We now discuss the salient points of this generalization.

Some matters are virtually the same as before. The group $Z_kX$ of $k$-cycles is defined as the free abelian group on the symbols $[V]$, with $V$ closed, *reduced*, irreducible, and $k$-dimensional. Also as before are the definition of the map $f_*: Z_kX \to Z_kY$ associated to a map $f: X \to Y$, and the definition of rational equivalence (there are several useful, equivalent definitions of rational equivalence though). As before, the group of $k$-cycles modulo rational equivalence, which is denoted $A_kX$, behaves covariantly under proper maps. However, the sum $A.X = \sum A_kX$ is not a ring if $X$ is singular.

A new ring $A^{\cdot}X$ with the desired formal properties can be constructed as follows. If $X$ is smooth, $A^{\cdot}X$ is as before. If $X$ is singular, $A^{\cdot}X$ is the (direct) limit over all mappings $f: X \to Y$ with $Y$ smooth, of the rings $A^{\cdot}Y$. The ring $A^{\cdot}X$ is graded. It behaves contravariantly. There is a *cap product*,

$$A^iX \otimes A_jX \to A_{j-i}X, \tag{II,40}$$

and a projection formula like (II,6). The ring $A^{\cdot}X$ carries Chern and Segre classes that behave appropriately.

## D. Segre covariants

Let $X$ be a quasi-projective scheme and let $V$ be a closed subscheme of $X$ defined by an ideal $I$. Consider the projectivized normal cone of $V$ in $X$. It is the scheme.

$$D = \text{Proj}\,(O_X/I \oplus I/I^2 \oplus I^2/I^3 \oplus \cdots). \tag{II,41}$$

(It is the exceptional divisor of the blowup $X'$ of $X$ along $V$.) Assume that $X$ is irreducible of dimension $n$ or that each irreducible component is of dimension $n$. Then each irreducible component of $D$ has dimension $n-1$. For each $i$, the *ith Segre covariant class* of $V$ in $X$ is defined as the class,

$$\boxed{s^i(V,X) = p_*(c_1(O_D(1))^{n-1-i})} \in A_iX, \tag{II,42}$$

where $p: D \to X$ is the structure map. The sum $s(V,X) = \sum s^i(V,X)$ is called the *total Segre covariant class* of $V$ in $X$. It is convenient to let $s(V,X)^*$ denote the alternating sum $\sum (-1)^{n+i} s^i(V,X)$.

Let $q: X' \to X$ denote the structure mapping and let $\alpha: V \to X$ and $\beta: D \to X'$ denote the inclusion mappings. Set $I' = IO_{X'}$. Then we have

$$\boxed{\alpha_* s^i(V,X) = -q_* c_1(I')^{n-i}} \quad \text{for} \quad i = 0, \cdots, n-1. \tag{II,43}$$

Indeed, by the projection formula we have

$$\beta_* c_1(I' \mid D)^{n-1-i} = c_1(I')^{n-1-i} \cdot [D]. \tag{II,44}$$

Now, applying $q_*$ to the left hand side we get the left hand side of (II,43) because of the definition (II,42) and the basic relation $I' \mid D = O_D(1)$. On the other hand applying $q_*$ to the right hand side of (II,44) we get the right hand side of (II,43) because $I'$ is the ideal of $D$ and so we have $c_1(I') = -[D]$.

Suppose $V$ is locally a complete intersection in $X$; for example, this is the case when $X$ and $V$ are both smooth. Then the conormal sheaf $N = I/I^2$ is locally free and we have $D = \mathbb{P}(N)$. Hence the theory of Segre classes yields

$$\begin{aligned} s(V,X) &= s(N) \quad \text{(by (II,33))} \\ s(V,X)^* &= s(\nu) \\ c(\nu)s(V,X) &= 1 \qquad \text{(by (II,35))}, \end{aligned} \tag{II,45}$$

where $\nu$ is the normal sheaf, the dual of $N$.

The *Segre covariant canonical classes* $s^i(X)$ are defined as the Segre covariant classes of the diagonal $\Delta$ in $X \times X$, and we set

$$\begin{aligned} s(X) &= \sum s^i(X) \\ s(X)^* &= \sum (-1)^{n+i} s^i(X). \end{aligned} \tag{II,46}$$

If $X$ is smooth, then the conormal sheaf of $\Delta$ is $\Omega^1_X$ and (II,45) yields

$$\begin{aligned} s(X) &= s(\Omega^1_X), \\ s(X)^* &= s(\tau_X), \\ c(X)s(X) &= 1. \end{aligned} \tag{II,47}$$

B. Segre (according to [111; §§3, 4, 5]) in 1953 introduced these covariants in terms of residual intersections of sections of $X$ by hypersurfaces containing $V$. In 1954 he related them to the blowup of $X$ along $V$. Relations (II,47) were established by Vesentini in 1954 also.

Let $V_1, \cdots, V_r$ denote the irreducible components of $V$, each equipped with its reduced structure. Each generator $[V_i]$ appears in $s(V, X)$, of course, but with an interesting weighting coefficient. It is the multiplicity $e_i$ of $V$ in $X$ at $V_i$. In other words, we have

$$\boxed{s(V, X) = \sum e_i[V_i] + v}, \tag{II,48}$$

where $v$ is the class of a cycle involving no $[V_i]$. Indeed, let $A$ denote the local ring of $X$ at $V_i$. Set $J = IA$; it is the ideal of $V$ in $A$. Then $J$ is primary for the maximal ideal of $A$, and the multiplicity $e_i$ appears as the leading coefficient in the Hilbert–Samuel polynomial,

$$\text{length}_A\,(J^k/J^{k+1}) = e_i\binom{k+c-1}{c-1} + \cdots \qquad (k \gg 0), \tag{II,49}$$

where $c$ is the dimension of $A$, the codimension of $V_i$ in $X$. However, (II,49) is also the Hilbert polynomial of the scheme Proj $(\oplus J^k/J^{k+1})$, which is the fiber $F$ of $D$ over the generic point of $V_i$. Hence $e_i$ is equal to the degree $\int c_1(O_F(1))^{c-1}$ of the projective scheme $F$. Therefore $s^{c-1}(V, X)$ contains $[V_i]$ with weighting $e_i$. Thus (II,49) holds. (This proof is essentially found in [100].)

For example, consider a hypersurface $X$ in $\mathbb{P}^{n+1}$ with isolated singular points $x_1, \cdots, x_r$. Take $V$ to be the singular locus of $X$. It is the subscheme defined by the jacobian ideal $I$. (If $X$ is given in affine coordinates by the vanishing of a polynomial $\Phi$, then $I$ is generated by the partial derivatives of $\Phi$.) In this case, (II,48) yields

$$\boxed{s(V, X) = \sum e_i[x_i]}. \tag{II,50}$$

Here $e_i$ is the multiplicity on $X$ at $x_i$ of the singular locus, or equivalently, of the jacobian ideal. In this case the $e_i$ are given in terms of generalized Milnor numbers by Teissier's formula [112; Cor. 1.5, p. 320],

$$\boxed{e_i = \mu_{x_i}^{(n+1)}(X) + \mu_{x_i}^{(n)}(X)}. \tag{II,51}$$

The first summand is the ordinary Milnor number at $x_i$ of $X$ and the second is that of a section of $X$ by a general hyperplane through $x_i$.

*Relative Segre covariant canonical classes*, $s^i(X/Y)$ or $s^i(f)$, can, of course, be defined for any mapping $f: X \to Y$ into an arbitrary scheme $Y$; they are, by definition, the Segre covariant classes of the diagonal $\Delta$ in the fibered product $X \times_Y X$. If $Y$ is smooth, then the formula,

$$\boxed{s(X/Y) = f^* c(Y) \cdot s(X)}, \tag{II,52}$$

makes sense, and it may be interesting to know where it holds.

Formula (II,52) holds, for example, if (1) $X$ is smooth along a dense open subset of the ramification locus of $f$ and if (2) each irreducible component of $Q = \mathbb{P}(\Omega^1_{X/Y})$ has dimension $n - m - 1$. Indeed, let $T$ and $T'$ denote the projectivized normal cones of $\Delta$ in $X \times X$ and $X \times_Y X$. Condition (2) implies that the cycles $[Q]$ and $[T']$ are equal on $T$. Now, $Q$ is equal to the zero locus of the natural map $\Omega^1_Y \mid T \to O_T(1)$, and Condition (1) implies that $T$ is smooth, so Cohen–Macaulay, along a dense open subset of $Q$. Hence $[Q]$ and so $[T']$ represent $c_m((\tau_Y \mid T)(1))$. Formula (II,22) and the projection formula now yield formula (II,52). Condition (2) holds, for instance, if $Y$ is equal to $\mathbb{P}^m$ and $f$ is a projection from a general center (see chapter VI, section B).

## III. Ramification

### A. Porteous' formula

Let $f: X \to Y$ be a mapping from a smooth $n$-fold $X$ into a smooth $m$-fold $Y$. We seek to measure the degree to which $f$ differs from being a submersion if $n \geqq m$, and from being an immersion if $n \leqq m$. The points at which $f$ is not a submersion, respectively not an immersion, are called the *singular points* of $f$.

The singular points of $f$ are classified according to the amount by which the jacobian map $\partial f$ drops rank. (This is the Thom–Boardman classification, see [17; Ch. VI].) More precisely, for each $i \geqq 0$ set

$$\boxed{j = \min(n, m) - i} \tag{III,1}$$

and consider the closed set

$$\boxed{\begin{aligned} \bar{S}_i &= \{x \in X \mid \operatorname{rank}(\partial f)(x) \leqq j\} \\ &= \{x \in X \mid \dim(\Omega^1_{X/Y}(x)) \geqq n - j\} \end{aligned}}\ . \tag{III,2}$$

The two descriptions of $\bar{S}$ are obviously equivalent in view of the basic exact sequence,

$$f^*\Omega^1_Y \xrightarrow{\partial f} \Omega^1_X \longrightarrow \Omega^1_{X/Y} \longrightarrow 0. \tag{III,3}$$

Now, there is a natural subscheme structure on $\bar{S}_i$; it is defined by the

Fitting ideal,

$$F^{n-j-1}(\Omega^1_{X/Y}), \qquad \text{(III,4)}$$

which is generated by the $(j+1)$-minors of $\partial f$.

For example, for $i=0$ we have $\bar{S}_0 = X$. For $i=1$, we have $\bar{S}_1 = R$, where $R$ is the *ramification locus* of $f$. It is also called the *singular locus.* The term 'ramification' might better be reserved for maps with finite fibers, in keeping with the usage of 'unramified' (which means immersive), but we shall use it in general. The ramification locus is the most important locus of singular points. If it is empty, then $f$ is (by definition) a submersion or an immersion.

When we classify the singularities of $f$, we usually work with the differences $S_i = \bar{S}_1 - \bar{S}_{i+1}$. We can characterize $S_i$ as the largest subscheme of $X$ on which the restriction of $\Omega^1_{X/Y}$ is locally free of rank $n-j$. The classification into the $S_i$ is known as the first-order classification. If the $S_i$ are smooth, as happens when $f$ is 'generic', then the classification process can be repeated for the restrictions $S_i \to Y$ of $f$, etc., and in this way we obtain the higher-order classifications of $f$. Alternately the higher-order classifications can be obtained through consideration of the higher derivatives of $f$; for example, the second-order classification comes from consideration of the Hessian $\partial^2 f$. In algebraic geometry, the study of the higher-order singularities is not well-developed, but it has been begun in the abstract by Mount and Villamayor [74], [76], for generic projections by Roberts [88], [91], [92], [93] and by Mather [65], [66], and in connection with linear systems by Lascoux [63], by Roberts [94], and by Vainsencher [116; §6].

For the enumerative theory of singularities, the closed subschemes $\bar{S}_i$ are the right ones to consider. First of all, for each irreducible component $\Sigma$ of $\bar{S}_i$, we have

$$\boxed{\operatorname{codim}(\Sigma, X) \leqq (m-j)(n-j)}\,. \qquad \text{(III,5)}$$

Now, suppose $f$ is *appropriately generic* for a given $i$ in the following sense: $\bar{S}_i$ has codimension at least $(m-j)(n-j)$; equivalently, equality holds in (III,5) for every $\Sigma$, but possibly $\bar{S}_i$ is empty. Then the fundamental class $[\bar{S}_i]$ in the rational equivalence ring $A^{\cdot}X$ is given by the following formula, known as *Porteous' formula:*

$$\boxed{[\bar{S}_i] = \left.\begin{vmatrix} c_{n-j} & c_{n-j+1} & \cdots & c_{n+m-2j-1} \\ c_{n-j-1} & c_{n-j} & & \vdots \\ \vdots & & \ddots & \vdots \\ c_{n-m} & c_{n-m+1} & \cdots & c_{n-j} \end{vmatrix}\right\} m-j}$$

$$\begin{aligned} \text{with} \quad c_k &= c_k(\Omega^1_X - f^*\Omega^1_Y) \\ &= k\text{th term of } [c(\Omega^1_X)/c(f^*\Omega^1_Y)]. \end{aligned} \qquad \text{(III,6)}$$

The second expression for $c_k$ is the definition of the first.

A similar but more general pair of results, the second also known as *Porteous' formula*, holds for any map $u: E \to F$ from a locally free sheaf $E$ of rank $m$ into one $F$ of rank $n$ on a smooth ambient variety $Z$; in fact, $Z$ need only be Cohen–Macaulay or just satisfy an appropriate depth condition. Here $\bar{S}_i$ is replaced by the subscheme of $Z$ defined by the $(n-j-1)$st Fitting ideal of Coker $(u)$. The original case is the special case with $\partial f$ for $u$. Another important special case is that with $E = O_X$; here, $u$ is essentially a section of $F$. In this case the result is already familiar; it says that the scheme of zeros of $u$ represents the highest Chern class $c_n(F)$ if $u$ is *regular*, that is, if each component of the scheme of zeros has codimension $n$.

Porteous' formula was first stated and proved by Porteous [85]; he dealt with a generic map $u: E \to F$. The genericity hypothesis was eliminated (in algebraic geometry) by Kempf and Laksov [42], who further generalized the formula to include Giambelli's determinantal formula from classical Schubert calculus. (Lascoux [62] later gave a powerful method for determining such general first-order formulas explicitly.) Porteous' formula (III,6) gives the $i$th first-order Thom polynomial explicitly. Thom (see [26]) suggested an abstract existence proof for polynomials that work for singularities of arbitrary order. Ronga [97] gave an algorithm for calculating all the polynomials, and he [96] determined explicitly some second order polynomials. Menn [70], Sergeraert [109], Lascoux [15–48], and Roberts [94] determined explicitly some important $n$th order polynomials. Only the last two did algebraic geometry, but the polynomials are the same in abstract algebraic geometry as in complex differential topology.

The ramification locus $R = \bar{S}_1$ is the only locus of singularities we shall work with from now on (although the other $\bar{S}_i$ and Porteous' formula for them have been used effectively in algebraic geometry, for example, to establish the existence of special divisors on curves (see [41], [48], [49])). Note that according to (III,4) the subscheme $R$ is defined by the $r$th Fitting ideal of $\Omega^1_{X/Y}$ where $r$ is the standard rank for $\Omega^1_{X/Y}$, namely, $n-m$ if $n \geqq m$ and 0 if $n \leqq m$. Now, by (III,5), for each irreducible component $\sum$ of $R$, we have

$$\boxed{\operatorname{codim}\left(\textstyle\sum, X\right) \leqq \begin{cases} n-m+1 & \text{for} \quad n \geqq m \\ m-n+1 & \text{for} \quad n \leqq m \end{cases}}. \tag{III,7}$$

Suppose $f$ is appropriately generic in the sense that $R$ has codimension at least $|n-m|+1$ or equivalently that equality holds in (III,7) for every $\sum$. Then we have

$$\boxed{[R] = \begin{cases} s_{n-m+1}(\tau) & \text{with} \quad \tau = f^*\tau_Y - \tau_X \quad \text{for} \quad n \geqq m \\ c_{n-m+1}(\nu) & \text{with} \quad \nu = f^*\tau_Y - \tau_X \quad \text{for} \quad n \leqq m \end{cases}}. \tag{III,8}$$

The formal difference $f^*\tau_Y - \tau_X$ is an element of $K(X)$. It is denoted by $\tau$ and called the *virtual tangent sheaf* (or *bundle*) of $f$ if $n \geqq m$; it is equal in $K(X)$ to the tangent sheaf $\tau_{X/Y}$ of $f$ if $f$ is a submersion. It is denoted by $\nu$ and called the *virtual normal sheaf* (or *bundle*) of $f$ if $n \leqq m$; it is equal in $K(X)$ to the normal sheaf $\nu_{X/Y}$ of $f$ if $f$ is an immersion. Note that (in view of sections C and D of chapter II) we have

$$s(\tau) = c(X)^*/f^*c(Y)^* = c(X)^* \cdot f^*s(Y)^* \tag{III,9}$$

$$c(\nu) = f^*c(Y)/c(X) = f^*c(Y) \cdot s(X). \tag{III,10}$$

Formula (III,8) is a corollary of Porteous' formula. Indeed, for $n \geqq m$ we have

$$\begin{aligned} [R] &= c_{n-m+1}(\Omega_X^1 - f^*\Omega_Y^1) \quad \text{by (III,6)} \\ &= s_{n-m+1}(\tau) \qquad\qquad\quad \text{by (III,9).} \end{aligned} \tag{III,11}$$

For $n \leqq m$, we apply Porteous' formula not to $\partial f$ but to its transpose, the differential of $f$,

$$df = (\partial f)^\vee : \tau_X \to f^*\tau_Y, \tag{III,12}$$

noting that a matrix and its transpose have the same set of $(j+1)$-minors.

## B. Ruled surfaces (see fig. 2)

Today, the term *ruled surface of genus p* usually refers to the bundle $X = \mathbb{P}(E)$ of projective lines over a smooth projective curve $C$ of genus $p$ associated to a locally free sheaf $E$ of rank 2 on $C$. Classically, the term usually referred to the image of $X$ in $Y = \mathbb{P}^3$ under a mapping $f: X \to Y$ with the following two properties: the restriction $X \to f(X)$ is birational and $f$ carries the fibers of $X$ onto lines in $Y$.

$$\begin{array}{ccc} X = \mathbb{P}(E) & \xrightarrow{f} & Y = \mathbb{P}^3 \\ \big\downarrow g & & \\ C \text{ genus } p & & \end{array} \tag{III,13}$$

Such an $f$ can be obtained as follows: first embed $X$ in a high dimensional projective space so that the fibers of $X$ becomes lines; do this by taking as the ample sheaf the tensor product of the tautological sheaf $O_X(1)$ and the pullback of a high tensor power of some ample sheaf on $C$; finally project $X$ into $Y$ from a center in general position. No matter how $f$ comes about, $f^*O_Y(1)$ necessarily has the form $O_X(1) \otimes g^*N$ for some invertible sheaf $N$ on $C$. Replacing $E$ by $E \otimes N$, we may assume

$$O_X(1) = f^*O_Y(1). \tag{III,14}$$

Let $h$ denote the pullback to $X$ of the hyperplane class $c_1(O_Y(1))$ of $Y$. Then by (III,14) we have

$$h = f^*c_1(O_Y(1)) = c_1(O_X(1)). \tag{III,15}$$

For convenience, set

$$e = g^*c_1(E) \quad \text{and} \quad \gamma = g^*c_1(C). \tag{III,16}$$

Since $g^*$ is multiplicative and since $C$ is a curve (so $A^2C = 0$), we have

$$e^2 = 0, \qquad e\gamma = 0, \quad \text{and} \quad \gamma^2 = 0. \tag{III,17}$$

Moreover the basic relation (II,20) becomes

$$h^2 - eh = 0. \tag{III,18}$$

Now, we have

$$\begin{aligned} c(X) &= g^*c(C)c((p^*E^\vee)(1)) && \text{by (II,27)} \\ &= (1+\gamma)(1+[-e+2h]+[0-e\cdot h+h^2]) && \text{by (II,22)} \\ &= 1+(\gamma-e+2h)+2h\gamma. && \text{(III,19)} \end{aligned}$$

On the other hand by (I,19) we have

$$f^*c(Y) = (1+h)^4. \tag{III,20}$$

Dividing (III,20) by (III,19) and simplifying, we obtain

$$\boxed{c(\nu) = 1+(2h-\gamma+e)+(2h^2-2h\gamma)}\;. \tag{III,21}$$

Therefore by (III,8) we have

$$\boxed{[R] = 2h^2 - 2h\gamma}\;, \tag{III,22}$$

assuming the ramification locus $R$ is a finite set of points.

A classical numerical relation comes from (III,22) by extracting degrees. The degree of $[R]$ is what was called the number of *pinch points* and denoted by $\nu_2$. By the projection formula with respect to $f$ and by (II,11), the degree $\int h^2$ is equal to the degree $\int f(X)\cdot c_1(O_Y(1))^2$; the latter number is by definition the degree of $f(X)$ and it was denoted by $\mu_0$ classically. By the projection formula with respect to $g$, the degree $\int h\gamma$ is equal to the degree $\int c_1(C)$, or $2-2p$. Putting it all together, we obtain the relation,

$$\boxed{\nu_2 = 2\mu_0 + 4(p-1)}\;. \tag{III,23}$$

It was obtained classically by semi-empirical means (see [7; p. 166]).

**C. Algebraic functions**

Let $f:X\to Y$ be a mapping from a smooth $n$-fold onto a smooth *curve* $Y$. We may think of $f$ as an algebraic function defined everywhere on $X$. Assume $f$ is appropriately generic; that is, the ramification locus $R$ has codimension $n-1+1=n$, or in other words, $f$ has isolated critical points.

The ramification locus $R$ is the scheme defined by the ideal, denoted $F$, of 1-minors of the jacobian map $\partial f$. Let $x$ be a point of $X$ and choose local coordinates $x_1,\cdots,x_n$ about $x$ and $t$ about $f(x)$. Then obviously $F$ is generated by the partial derivatives, $\partial t/\partial x_1,\cdots,\partial t/\partial x_n$. Hence the number,

$$\text{length}\,(O_{R,x})=\dim\left(O_{X,x}\Big/\left(\frac{\partial t}{\partial x_i}\right)\right), \tag{III,24}$$

is the Milnor number $\mu_x(f^{-1}fx)$ of the fiber through $x$. Therefore the fundamental cycle $[R]$ is given by the formula,

$$\boxed{[R]=\sum_{x\in X}\mu_x(f^{-1}fx)[x]}\ . \tag{III,25}$$

Now, the rational equivalence class of $[R]$, which is also denoted $[R]$, is equal to the Segre class $s_n(\tau)$ by (III,8). By (III,9) we have

$$s(\tau)=(1+f^*c_1(Y))(1-c_1(X)+\cdots+(-1)^nc_n(X)). \tag{III,26}$$

Therefore we have

$$\boxed{[R]=(-1)^n(c_n(X)-f^*c_1(Y)\cdot c_{n-1}(X))}\ . \tag{III,27}$$

Note that because $Y$ is a curve, here only the following special case of Porteous' formula was used: the scheme of zeros of a regular section of a locally free sheaf represents its highest Chern class.

An important numerical relation comes from (III,27) and (III,25) by extracting degrees in the case that $X$ and $Y$ are complete. Before extracting degrees it is convenient to reduce (III,27) modulo numerical equivalence, (homological equivalence would work as well.) Modulo numerical equivalence, we have

$$c_1(Y)\equiv\chi(Y)[\eta] \tag{III,28}$$

for any point $\eta$ of $Y$, because $\chi(Y)$ is equal to the degree $\int c_1(Y)$ by (II,24). Pulling (III,28) back by $f^*$, we obtain

$$f^*c_1(Y)\equiv\chi(Y)[Z], \tag{III,29}$$

where $Z$ denotes the fiber $f^{-1}\eta$. Take $\eta$ general enough so that $Z$ is smooth.

Then the adjunction formula (II,28) applies and it yields

$$\alpha_* c_{n-1}(Z) = c_{n-1}(X)[Z], \tag{III,30}$$

where $\alpha : Z \to X$ is the inclusion map; note that the powers $[Z]^i$ for $i \geqq 2$ are zero because $f^*$ is multiplicative and $Y$ is a curve. Putting (III,25), (III,27), (III,29), and (III,30) together we obtain modulo numerical equivalence

$$\sum \mu_x(f^{-1}fx)[x] \equiv (-1)^n (c_n(X) - \chi(Y)\alpha_* c_{n-1}(Z)). \tag{III,31}$$

Extracting the degrees of both sides and rewriting the result using (II,24) and (II,11), we obtain the important relation,

$$\boxed{\chi(X) = \chi(Y)\chi(Z) + (-1)^n \sum \mu_x(f^{-1}fx)}\,. \tag{III,32}$$

For example, if $X$ is a curve, then (III,32) is just a form of the Riemann–Hurwitz formula (I,1) and in fact the derivation here is virtually the same as the one in chapter I.

Classically they knew another case of relation (III,32). It concerned a pencil, or 1-parameter family, of curves on a smooth surface $X$. The pencil was assumed to be free of base points; these are the common points of all the members. For example, the most important case was that of an irrational pencil, one parametrized by a curve $Y$ of positive genus. Then the member curves appear as the fibers of a mapping $f : X \to Y$. It was further assumed that the singularities of the member curves are all ordinary double points; thus $f$ has nondegenerate isolated critical points, and a simple direct computation shows that the corresponding Milnor numbers are all 2 (provided the characteristic is not 2). Denote by $\delta$ the total number of singularities, by $\rho$ the genus of a smooth member curve $Z$, and by $\rho'$ the genus of the parameter curve $Y$. Finally let I denote the Zeuthen–Segre invariant of $X$, which is 4 less than the Euler characteristic $\chi(X) = \int c_2(X)$. Then (III,32) becomes

$$I + 4 = (2 - 2\rho')(2 - 2\rho) + \delta. \tag{III,33}$$

This relation is used in [7; p. 190] to define $I$.

The proof above of (III,32) was given independently by Iversen [35] and by Deligne [104; Exposé XVI, 2, pp. 204–206]. Iversen goes on to treat the case of multiple fibers. An algebro-topological proof under the assumption of nondegenerate isolated critical points is found in [1; §5, p. 9] and a similar étale-topological proof under the same assumption is found in [104; Exposé XVIII, §3.2, pp. 268–271]; both of these proofs are given in the case that $Y$ is $\mathbb{P}^1$ but they extend to the general case with virtually no change.

**D. Jacobian cycles**

(Compare [63; §2], [94; §1].) Let $X$ be a smooth $n$-fold. Let $L$ be an invertible sheaf on $X$, let $V$ be a finite dimensional vector space, and let $v: V \to H^0(X, L)$ be a linear map. They define a linear system, or linear family, of divisors on $X$ as follows. Set $Y = \mathbb{P}(V^\vee)$. This projective space parametrizes the elements of $V$ up to scalar multiple. (It is necessary to use the dual $V^\vee$ because $\mathbb{P}(W)$ is the projective space of 1-codimensional linear subspaces of a vector space $W$.) The tautological map $u: O_Y(-1) \to V_Y$ is the universal 'element' of $V$ up to scalar multiple; its fiber over a point $y$ of $Y$ is isomorphic to the map $O_Y \to V$ defined by an element of $V$ corresponding to $y$. Similarly the composition,

$$q^* O_Y(-1) \xrightarrow{q^* u} V_{X \times Y} \xrightarrow{p^* v^\#} p^* L, \tag{III,34}$$

is the universal 'section' of $L$ coming from $V$ up to scalar multiple, where $p$

$$\begin{array}{ccccc} & X \times Y & \xleftarrow{\alpha} & D & \\ \swarrow p & & \searrow q & \downarrow g & \\ X & & & Y = \mathbb{P}(V^\vee) & \end{array} \tag{III,35}$$

and $q$ are the projections and where $v^\#: V_X \to L$ is the adjoint of $v$.

Twisting the composition (III,34) yields a 'section',

$$O_{X \times Y} \to p^* L \otimes q^* O_Y(1). \tag{III,36}$$

It defines a divisor $D$ on $X \times Y$ and so a family of divisors on $X$ parametrized by $Y$, the linear system in question. The members of the system are the fibers of $D$; they are just the divisors on $X$ defined by the sections of $L$ contained in $V$. The system is called linear not simply because $Y$ is a projective space but because it is a linear condition on the members that they pass through a given point; that is, those that do are parametrized by a linear subspace of $Y$. These linear subspaces of $Y$ form an algebraic family parametrized by $X$. The total space is $D$. In fact $D$ can be identified with the linear subscheme $\mathbb{P}(\text{Coker}(v^{\#\vee}))$ of $\mathbb{P}(V_X^\vee) = X \times Y$.

Let $g: D \to Y$ denote the restriction of $q$ and consider its ramification locus $R$. By definition $R$ is the scheme defined by the $(n-1)$st Fitting ideal of $\Omega^1_{D/Y}$, and we are going to find the class $[R]$ using Porteous' formula. However instead of resolving $\Omega^1_{D/Y}$ using $g^*\Omega^1_Y \xrightarrow{\partial g} \Omega^1_D$ we now use $N \to \Omega^1_{X \times Y/Y} | D$, where $N$ is the conormal sheaf. First we assume $R$ has the appropriate codimension in $D$, namely, $n$. Then Porteous' formula yields

$$[R] = \text{the } n\text{th term of } \{c(\Omega^1_{X \times Y/Y} | D)/c(N)\}. \tag{III,37}$$

Let $\alpha: D \to X \times Y$ denote the inclusion map. Then $N$ is equal to $\alpha^* O_{X\times Y}(-D)$. Finally we obtain

$$\boxed{\begin{aligned}\alpha_*[R] &= \text{the } n\text{th term of } \left\{\frac{[D]}{1-[D]}\, c(\Omega^1_{X\times Y/Y})\right\} \\ &= \sum c_{n-i}(\Omega^1_{X\times Y/Y})[D]^{i+1}\end{aligned}} \quad . \tag{III,38}$$

by applying $\alpha_*$ to both sides of (III,37) and using the projection formula (taking $f = \alpha$ and $x = 1$ in the equivalent of (II,6) for $A^{\cdot}X$).

In most applications $D$ is smooth. However, even if $D$ is singular, the derivation of (III,38) goes through virtually as is, using Fulton's intersection theory (see ch. II, sect. C). In deriving (III,38) we have not used the hypothesis that $D$ is the total space of a linear system; $D$ could have been the total space of an arbitrary algebraic system of divisors parametrized by an arbitrary variety $Y$. However in the case at hand we have

$$[D] = p^* c_1(L) + q^* c_1(O_Y(1)), \tag{III,39}$$

because $D$ is defined by the 'section' (III,36).

Set $r = \dim(Y)$ (so $r+1$ is the dimension of $V$) and assume $r \leqq n$. Then the *jacobian locus* $J$ of our linear system is defined as the locus of singular points of members; thus $J$ is the image of $R$. Using local coordinates $x_1, \cdots, x_n$ for $X$ we can describe the locus explicitly by equations in the following way: the singular points of a member are given by the simultaneous vanishing of a local equation and its derivatives; hence picking a basis $f_0, \cdots, f_r$ for the local equations of all the members (which amounts to picking a basis for $V$), we can express the condition that a point of $X$ be a singular point of some member by the vanishing of the maximal minors of the $(r+1)\times n$ augmented jacobian matrix,

$$\begin{bmatrix} f_0 & f_1 & \cdots & f_r \\ \dfrac{\partial f_0}{\partial x_1} & \dfrac{\partial f_1}{\partial x_1} & \cdots & \dfrac{\partial f_r}{\partial x_n} \\ \vdots & \vdots & & \vdots \\ \dfrac{\partial f_0}{\partial x_n} & \dfrac{\partial f_1}{\partial x_n} & & \dfrac{\partial f_r}{\partial x_n} \end{bmatrix} . \tag{III,40}$$

From this description of $J$ we can see that it has a natural scheme structure and that each of its irreducible components has dimension at least $r-1$. We shall formalize and pursue this approach at the end of section B, chapter IV.

Here we define the *jacobian cycle* $[J]$ of our linear system as the direct image,

$$[J]=p_*\alpha_*[R]. \tag{III,41}$$

It is an $(r-1)$-cycle on $X$. Its class, which is also denoted by $[J]$, is easily determined from (III,38) and (III,39); we find

$$\boxed{[J]=\sum_{i\geqq r-1}\binom{i+1}{r}c_{n-i}(\Omega^1_X)c_1(L)^{i+1-r}}\,. \tag{III,42}$$

Indeed, note that $\Omega^1_{X\times Y/Y}$ is equal to $p^*\Omega^1_X$ and that $p_*q^*c_1(O_Y(1))^i$ is equal to 1 for $i=r$ and to 0 for $i\neq r$. Then use the projection formula.

From now on assume $r=1$. This is an important case, the case of a linear pencil. Let $B$ denote the *base locus*, that is, the intersection $\cap D_y$ of all the members $D_y$ of the pencil, or equivalently, the intersection $D_{y_1}\cap D_{y_2}$ of two distinct members. Then $p$ induces an isomorphism,

$$p_B:(D-p^{-1}B)\xrightarrow{\sim}(X-B), \tag{III,43}$$

because the fibers of $p_B$ are 0-dimensional linear subspaces of $Y$. So $q\circ p_B^{-1}$ is a well-defined mapping from $X-B$ to $Y$, and $D-p^{-1}B$ is its graph. Note that this mapping is equal to the one usually associated to the linear system (or to the map $V_X\to L$, which is surjective off $B$) if we make the canonical identification of the projective line $Y=\mathbb{P}(V^\vee)$ with the dual projective line $Y^\vee=\mathbb{P}(V)$.

Assume the base locus $B$ has no irreducible components of codimension 1; in other words, our pencil has no *fixed components.* Then $D$ is the closure of $D-p^{-1}B$. Indeed, since $D$ is a divisor on $X\times Y$, each of its irreducible components must have dimension at least $n$. On the other hand, $p^{-1}B$ has dimension at most $1+\dim(B)$, so at most $n-1$; hence it can contain no irreducible component of $D$. In fact, $D$ is the scheme-theoretic closure of its open subscheme $D-p^{-1}B$ because $D$ has no embedded components, being a divisor. Moreover, $D-p^{-1}B$ is the graph of a map from $X-B$ to $Y$ with the following property: locally on $X$ there exist two regular functions $f_0$, $f_1$ such that the map is given by $y=(f_0(x),f_1(x))$ off $B$ and such that $f_0$ and $f_1$ generate the ideal of the scheme $B$. Therefore $D$ is the blowup of $X$ along $B$, by one of the definitions of blowing-up (or the Hopf $\sigma$-process).

The appropriate dimension for $R$ is 0 because $r=1$. So if there are just finitely many pairs $(x,D_y)$ where $x$ is a point of $X$ and $D_y$ is a pencil member with a singular point at $x$, then Formula (III,42) for $[J]$ applies. If in addition $B$ is empty, then this formula reduces term by term to Formula (III,27) in which we take $q\circ(p\mid D)^{-1}$ for $f:X\to Y$ (that $f$ is well-defined we noted above). Indeed, the first term of (III,42) is obviously equal to that of

(III,27). The second term involves the factor $2c_1(L)$ and this factor is equal to $f^*c_1(Y)$. For, $Y$ is the projective line and so we have $c_1(Y)=2[y]$ for any point $y$ of $Y$. Furthermore $c_1(L)$ is represented by anyone of the fibers $f^{-1}y$ because these fibers are the members of our pencil. The other terms of (III,42), which each involve a factor of $c_1(L)^i$ with $i \geqq 2$, are all 0 because the fibers $f^{-1}y$, which each represent $c_1(L)$, are disjoint. Thus these terms constitute the correction for the presence of base points. It is not surprising a priori that (III,42) reduces to (III,27) in the absence of base points because then the resolution $N \to \Omega^1_{X\times Y/Y} \mid D$ of $\Omega^1_{D/Y}$ is isomorphic under $p \mid D$ to the resolution $f^*\Omega^1_Y \to \Omega^1_X$ of $\Omega^1_{X/Y}$.

Assume that the singular points of our pencil's members are finite in total number and that they all lie off the base locus $B$; in other words, the jacobian locus $J$ is finite and disjoint from $B$. Then it is obvious that $[J]$ is equal to the ramification cycle of the mapping $f = q \circ p_B^{-1}$ from $X-B$ to $Y$. Therefore by (III,25) we have

$$[J] = \sum_{y\in Y, x\in D_y} \mu_x(D_y)[x]. \tag{III,44}$$

Moreover Formula (III,42) for $[J]$ applies. Assume from now on that $X$ is complete. Combining Formulas (III,42) and (III,44) and extracting degrees we obtain the following numerical relations:

$$\boxed{\begin{aligned}\sum_{y,x} \mu_x(D_y) &= \sum (i+1)\int c_{n-i}(\Omega^1_X)c_1(L)^i \\ &= (-1)^n \int \frac{c(X)}{(1+c_1(L))^2}\end{aligned}}\;. \tag{III,45}$$

The second line comes from the first formally.

Let $D_\eta$ be a smooth member of our pencil. If the base locus $B$ is smooth, then Formula (III,45) may be rewritten as

$$\boxed{\sum_{y,x} \mu_x(D_y) = (-1)^n[\chi(X) - 2\chi(D_\eta) + \chi(B)]}\;. \tag{III,46}$$

Indeed (compare [104; Exposé XVII, proof of Prop. (5.7.2) p. 250]), since $D_\eta$ represents $c_1(L)$, the adjunction formula (II,28) and the invariance of degree under pushout (II,11) and Formula (II,24) yield

$$\chi(D_\eta) = \int \frac{c_1(L)}{1+c_1(L)}\, c(X). \tag{III,47}$$

Similarly we obtain

$$\chi(B) = \int_{D_\eta} \frac{c_1(L \mid D_\eta)}{1 + c_1(L \mid D_\eta)} c(D_\eta). \qquad \text{(III,48)}$$

Then the adjunction formula, the projection formula, and the invariance formula yield

$$\chi(B) = \int_X \left[\frac{c_1(L)}{1 + c_1(L)}\right]^2 c(X). \qquad \text{(III,49)}$$

Finally a straightforward computation yields (III,46). Note that we may allow $D_\eta$ and $B$ to have several components if we interpret their Euler characteristics as the sum of the Euler characteristics of their components.

For example suppose $X$ is a curve and consider a base-point-free linear pencil on $X$. Write the pencil's members out in the form,

$$D_y = \sum_{x \in X} n_{y,x}[x] \qquad n_{y,x} \in \mathbb{Z},\ y \in Y. \qquad \text{(III,50)}$$

The members $D_y$ are the fibers of the mapping $f = q \circ (p \mid D)^{-1}$ from the curve $X$ onto the projective line $Y$, and the jacobian cycle $[J]$ is the ramification cycle of $f$. Hence (see the discussion about Formula (I,4)) we have

$$[J] = \sum_{n_{y,x} > 1} (n_{y,x} - 1)[x]. \qquad \text{(III,51)}$$

(In positive characteristic this can fail.) Now, Formula (III, 42) implies that for any particular $D_y$ the difference $J - 2D_y$ is a canonical divisor. Formulas (III,45) and (III,46) both become

$$\sum_{n_{y,x} > 1} (n_{y,x} - 1) = 2d + 2g - 2, \qquad \text{(III,52)}$$

where $d$ is the degree $\sum_x n_{y,x}$ of any $D_y$ and where $g$ is the genus of $X$. Formula (III,52) is just that case of the Riemann–Hurwitz formula (I,1) in which $Y$ is the projective line, the case considered by Riemann himself. Note that if $d \geqq 2$ then the right hand side of (III,52) is strictly positive, so some member of the pencil must be singular; in other words every nontrivial covering of the projective line must ramify, or the projective line is simply connected.

For a second example suppose $X$ is a surface and consider a linear pencil without fixed components, such that the singular points of its members $D_y$ are finite in total number and all lie off the base locus. Let $b_1, \cdots, b_\sigma$ denote the base points; repetition is allowed so that $\sum [b_i]$ is the intersection cycle $D_{y_1} \cdot D_{y_2}$ of any two distinct members. Let $x_1, \cdots, x_\delta$ denote the singular points; each $x_i$ is repeated a certain number of times, namely, the Milnor

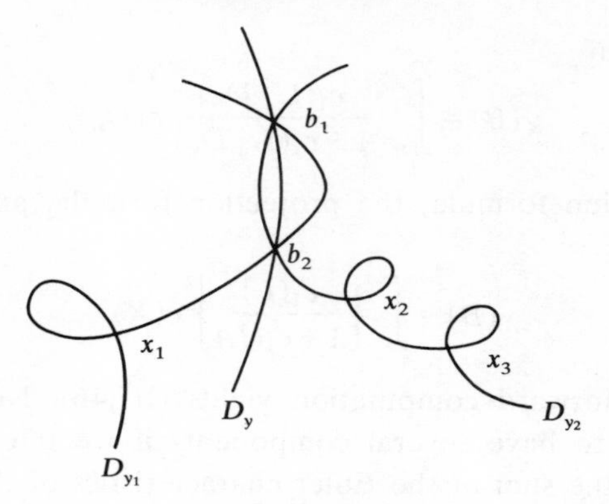


Figure 7

number $\mu_{x_i}(D_y)$ where $D_y$ contains $x_i$. (Note $D_y$ is unique because $x_i$ is not a base point.) Since the Milnor number is equal to 2 if and only if the singularity is an ordinary, or nondegenerate, double point (except in characteristic 2), it may be thought of as the number of ordinary double points concentrated at the singular point, and $\delta$ may be thought of as the total number of ordinary double points. Let $g$ denote the genus of any smooth member $D_\eta$, and let $I$ denote the Zeuthen–Segre invariant of $X$, which is equal to $\chi(X)-4$. In these terms, Formula (III,46) becomes

$$\boxed{\delta = I + 4g + \sigma} \quad . \tag{III,53}$$

While Formula (III,46) was derived under the assumption that the base locus $B$ is smooth, the derivation goes through for the case at hand, and is even easier than the general case, provided $\chi(B)$ is interpreted as $\sigma$. On the other hand combining Formulas (III,42) and (III,44) without extracting degrees yields the following relation among rational equivalence classes:

$$\boxed{\sum [x_i] = c_2(X) - \alpha_* c_1(D_\eta) + \sum [b_i]} \quad , \tag{III,54}$$

where $\alpha : D \to X$ denotes the inclusion map.

Historically the jacobian cycle played the leading role in the discovery and study of the higher canonical classes of an $n$-fold $X$ for $n \geqq 2$ (see [119; pp. 300–301], [102; p. 926], [19; p. 62], and [111; §2, pp. 27–33]). (The canonical divisor class by contrast has been tied in with the $n$-forms, or $n$-fold integrals, from the beginning; this theory was begun for surfaces by A. Clebsch in a note at the end of 1868 and developed for $n$-folds by M. Noether in a series of articles over the next 6 years.) In 1896, C. Segre introduced a numerical character $\Pi_n$. The relation $\Pi_n = (-1)^n \chi(X)$ with the topological Euler characteristic was found by J. W. Alexander in 1914. Segre defined the $\Pi_n$ inductively via a relation amounting to Formula (III,46). Earlier H. G. Zeuthen in 1871 had introduced essentially this character for surfaces and Segre in 1894 had placed in evidence its invariance from pencil to pencil on the same surface; virtually from then on, it has been known as the Zeuthen–Segre invariant of a surface.

The next step was taken in 1932 by F. Severi. In the very same article in which he initiated the general theory of rational equivalence, he introduced the rational equivalence class $c_2(X)$ for $X$ a surface by a formula like (III,54). Two years later, B. Segre (C. Segre's nephew) discovered the rational equivalence class $c_2(X)$ for $X$ a 3-fold by considering the jacobian curve of a general net (or 2-parameter family) of surfaces on $X$. The generalization of this work to all the rational equivalence classes $c_i(X)$ for $X$ an $n$-fold was carried out between 1937 and 1943 by J. A. Todd and M. Eger. Notice that using the adjunction formula in conjunction with Formula (III,42) we can obtain expressions for the $c_i(X)$ in terms of corresponding classes of lower dimensional subvarieties and of jacobians; such expressions were used by Todd and Eger to define the canonical classes, $(-1)^i c_i(X)$. These rational equivalence classes of Todd and Eger were finally related to the topological Chern classes in the 1950's; more about how the connection was made is found in chapter IV, section B.

## IV. The embedded tangent space

### A. Principal parts

(See [12; §16], [82; §2], [83; Appendix]). In this section we prepare the way for the next sections by reviewing some general theory. Let $X$ be a variety, not necessarily smooth. Consider the diagram,

$$\begin{array}{ccc} & \Delta\subset X\times X & \\ \swarrow^{p} & & \searrow^{q} \\ X & & X \end{array} \qquad \text{(IV,1)}$$

where $\Delta$ is the diagonal and $p$ and $q$ are the projections. Let $I$ denote the ideal of $\Delta$. Then we have

$$\Omega^1_X = I/I^2, \qquad \text{(IV,2)}$$

confirming our geometric intuition that the sheaf of 1-forms should be the conormal sheaf of the diagonal. (This relation is used in [12; (16.3.1)] to define $\Omega^1_X$.)

Let $F$ be an $O_X$-Module and $k$ a nonnegative integer. The sheaf $P^k(F)$ of $k$th *order principal parts* of $F$ is defined by

$$\boxed{P^k(F) = p_*(q^*F/I^{k+1}q^*F)}\ . \qquad \text{(IV,3)}$$

For example we obviously have

$$P^0(F) = F. \tag{IV,4}$$

The sheaf $P^k(F)$ is also called the sheaf of $k$-jets of sections of $F$.

For $k \geqq j$ consider the natural surjection,

$$(q^*F/I^{k+1}q^*F) \twoheadrightarrow (q^*F/I^{j+1}q^*F). \tag{IV,5}$$

Applying $p_*$ to it we obtain a surjection,

$$a_{k,j} : P^k(F) \twoheadrightarrow P^j(F), \tag{IV,6}$$

because the terms of (IV, 5) have set-theoretic support in $\Delta$ and $p \mid \Delta$ is a homeomorphism.

Consider the exact sequence,

$$0 \to I/I^2 \to O_{X \times X}/I^2 \to O_\Delta \to O. \tag{IV,7}$$

Tensoring it with $q^*F$ yields a sequence (IV,7) $\otimes\, q^*F$, which is exact because $O_\Delta$ is flat over $O_X$ through $q$. Applying $p_*$ to (IV,7) $\otimes\, q^*F$ yields an important exact sequence,

$$\boxed{0 \longrightarrow \Omega^1_X \otimes F \longrightarrow P^1(F) \xrightarrow{a_{1,0}} F \longrightarrow 0}\,. \tag{IV,8}$$

The left hand term is correct because of (IV,2) and because $I/I^2$ is an $O_\Delta$-Module. Note that by virtue of (IV,8) if $X$ is smooth of dimension $n$ and if $F$ is locally free of rank $f$, then $P^1(F)$ is locally free of rank $(n+1)f$.

A global section of $F$ yields one of $q^*F$, so one of $P^k(F)$. Thus there exists a natural map of $O_X$-Modules,

$$b_k : H^0(X, F)_X \to P^k(F). \tag{IV,9}$$

(Since $p$ is flat, the natural map from $H^0(X, F)_X$ to $p_*q^*F$ is an isomorphism; hence $b_k$ is naturally isomorphic to the pushdown under $p$ of the natural map from $q^*F$ to $q^*F/I^{k+1}q^*F$.) For $k \geqq j$ we obviously have $b_j = a_{k,j} \circ b_k$.

Functoriality in $F$ of the above constructions is obvious. Functoriality in $X$ also holds but its meaning and validity require a little explanation. Let $f : X \to Y$ be a mapping and $G$ a sheaf on $Y$. Then there exists a natural map,

$$c_k : f^*P^k(G) \to P^k(f^*G). \tag{IV,10}$$

It exists basically because, letting $J$ denote the ideal of the diagonal of $Y$, we have $JO_{X\times X}\subset I$. It is evident that the $c_k$ are functorial in $G$ and that they are compatible with the $a_{k,j}$ and the $b_k$. Also, it is evident that we have a commutative diagram,

$$\begin{array}{ccccccccc} 0 & \longrightarrow & f^*\Omega^1_Y\otimes f^*G & \longrightarrow & f^*P^1(G) & \longrightarrow & f^*G & \longrightarrow & 0 \\ & & \Big\downarrow{\scriptstyle \partial f\otimes f^*G} & & \Big\downarrow{\scriptstyle c_1} & & \Big\downarrow{\scriptstyle c_0\,=} & & \\ 0 & \longrightarrow & \Omega^1_X\otimes f^*G & \longrightarrow & P^1(f^*G) & \longrightarrow & f^*G & \longrightarrow & 0. \end{array} \qquad \text{(IV,11)}$$

Exactness holds in both lines except possibly along the top at $f^*\Omega^1_Y\otimes f^*G$; there exactness does hold if $G$ is locally free because the top line is the pullback of an exact sequence that then is locally split.

Let $f{:}X\rightarrow Y=\mathbb{P}(V)$ be a mapping of $X$ into a projective $m$-space; so $V$ is an $(m+1)$-dimensional vector space. Let $\alpha: V_Y\rightarrow O_Y(1)$ denote the tautological surjection and set $L=f^*O_X(1)$. Then we have a commutative diagram of natural maps,

$$\begin{array}{ccccccc} f^*\alpha: V_X & \xrightarrow[\simeq]{\alpha'} & H^0(Y, O_Y(1))_X & \longrightarrow & H^0(X, L)_X & \xrightarrow{b_0} & L \\ & & \Big\downarrow{\scriptstyle f^*b_k} & & \Big\downarrow{\scriptstyle b_k} & & \Big\downarrow{\scriptstyle =} \\ & & f^*P^k(O_Y(1)) & \xrightarrow{c_k} & P^k(L) & \xrightarrow{a_{k,0}} & L. \end{array} \qquad \text{(IV,12)}$$

Here $\alpha'$ is the pullback of the natural isomorphism $V\xrightarrow{\sim} H^0(Y, O_Y(1))$ of Serre. The left square expresses the compatibility of $c_k$ and $b_k$. The composition,

$$\boxed{a_k: V_X\rightarrow P^k(L)}\,, \qquad \text{(IV,13)}$$

is an important map for the extrinsic geometry of $X$ in $Y$.

Here is a local analysis of the map $a_1: V_X\rightarrow P^1(L)$ at a smooth point $x$ of $X$. Let $s_1,\cdots,s_n$ be a system of regular parameters at $x$. For each $i$, the difference $1\otimes s_i-s_i\otimes 1$ is an element of the stalk $I_x$, whose class modulo $I_x^2$ is equal to the element $ds_i$ of $\Omega^1_{X,x}$ under the identification (IV,2). Together the $ds_i$ form a basis of $\Omega^1_{X,x}$ over $O_{X,x}$. Hence $1, ds_1,\cdots, ds_n$ is a basis for $P^1(O_X)_x$ over $O_{X,x}$ by virtue of the exact sequence (IV,8) with $F=O_X$. Let $v_0,\cdots,v_m$ be a basis of $V$ such that the image $(f^*\alpha)v_0$ generates $L$ at $x$. Then the map $O_X\rightarrow L$ sending 1 to $(f^*\alpha)v_0$ is an isomorphism at $x$, and so the induced map $P^1(O_X)\rightarrow P^1(L)$ is also an isomorphism at $x$. Take the latter's inverse and form the composition,

$$V_{X,x}\xrightarrow{a_{1,x}} P^1(L)_x\xrightarrow{\sim} P^1(O_X)_x. \qquad \text{(IV,14)}$$

This composition is isomorphic to the map $a_{1,x}$ we want to analyze, and we have bases for its source and target.

It is easy to see that, in the bases we picked, the composition (IV,14) is represented by the matrix,

$$\begin{bmatrix} 1 & t_1 & \cdots & t_m \\ & \frac{\partial t_1}{\partial s_1} & \cdots & \cdot \\ \bigcirc & \cdot & & \cdot \\ & \vdots & & \cdot \\ & & \cdots & \frac{\partial t_m}{\partial s_n} \end{bmatrix}, \qquad \text{(IV,15)}$$

where $t_i$ is the pullback to $X$ of the affine coordinate function $v_i/v_0$. Note that this matrix is essentially the jacobian matrix of $f$ augmented by $f$. Hence if $f$ is an embedding, or more generally an immersion, at $x$, then the rows of (IV,15) evaluated at $x$ span an $n$-plane $T$ in $Y$. Moreover $T$ contains $f(x)$, and its tangent space $T_{T,x}$ in $T_{Y,x}$ contains the image $f_*(T_{X,x})$ of the tangent space of $X$ at $x$. We call $T$ the *embedded tangent space* of $X$ at $x$. Conversely, if the rows of (IV,15) evaluated at $x$ span an $n$-plane $T$, then $f$ is an immersion at $x$ and $T$ is the embedded tangent space. We conclude that if $X$ is smooth, then $f$ is an immersion everywhere if and only if $a_1: V_X \to P^1(L)$ is surjective and that if $X$ is smooth and $a_1$ is surjective, then $\mathbb{P}(P^1(L))$

$$\mathbb{P}(P^1(L)) \xhookrightarrow{\mathbb{P}(a_1)} \mathbb{P}(V_X) = X \times Y \qquad \text{(IV,16)}$$

is the total space of the family of embedded tangent spaces.

Take $X$ to be $Y$. Then $a_1$ is an isomorphism, being a surjective map between locally free sheaves of the same rank. So we obtain the following isomorphism of sequences, between the tautological sequence of $Y$ and the sequence (IV,8) with $F = O_Y(1)$:

$$\begin{array}{ccccccccc} 0 \longrightarrow & K & \longrightarrow & V_Y & \longrightarrow & O_Y(1) & \longrightarrow 0 \\ & \downarrow \simeq & & a_1 \downarrow \simeq & & \downarrow \simeq & \\ 0 \longrightarrow & \Omega^1_Y(1) & \longrightarrow & P^1(O_Y(1)) & \longrightarrow & O_Y(1) & \longrightarrow 0. \end{array} \qquad \text{(IV,17)}$$

The right hand square is commutative by virtue of the commutativity of the right hand square of (IV,12). The left hand vertical map exists and is an isomorphism by the 5-lemma.

Let $X$ be arbitrary (possibly singular) again, but assume $f$ is an immersion, that is $\partial f$ is surjective. Let $N$ denote the conormal sheaf of $X$ in $Y$. Consider Diagram (IV,11) with $G = O_X(1)$. Since $N \otimes L$ maps naturally onto Ker $(\partial f \otimes L)$, it maps naturally onto Ker $(c_1)$ since $\partial f$ is surjective, $c_1$ is also. Now, in view of the preceding paragraph and of Diagram (IV,12), it is easy to see that there is a commutative triangle with an isomorphism at the top,

$$\begin{array}{ccc} V_X & \xrightarrow[\simeq]{f^*a_1} & f^*P^1(O_Y(1)) \\ & \searrow^{a_1} & \downarrow c_1 \\ & & P^1(L). \end{array} \qquad \text{(IV,18)}$$

Therefore $N\otimes L$ maps naturally onto the Kernel of $a_1$ as well, and $a_1$ is surjective. In other words, there exists a natural short sequence,

$$\boxed{0 \longrightarrow N\otimes L \longrightarrow V_X \xrightarrow{a_1} P^1(L) \longrightarrow 0}, \qquad \text{(IV,19)}$$

which is exact except possibly at $N\otimes L$. At $N\otimes L$ exactness holds if $X$ is smooth, because then the three natural surjections just considered are isomorphisms. (Note that the preceding analysis also shows that $f$ is an immersion if and only if $a_1$ is surjective.)

If $X$ is smooth, then $N$ can be defined as $\operatorname{Ker}(\partial f)$. If $f$ is a closed embedding, then $N$ can be defined as $f^*(I/I^2)$, where $I$ denotes the ideal of $f(X)$ in $Y$. We shall need no other case than these two, but in general $N$ can be defined in either of the following two ways. (1) Let $I$ denote the kernel of the natural map $f^{-1}O_X \to O_Y$, where $f^{-1}$ denotes the pullback in the category of abelian sheaves. Define $N$ as $I/I^2$. (Note $f^{-1}O_X \to O_Y$ is surjective because $f$ is an immersion; hence $I/I^2$ is an $O_Y$-Module.) (2) $f$ factors locally on $X$ into a closed embedding followed by an étale mapping. Define $N$ as the conormal sheaf of the embedding.

The preceding theory was initiated some time before 1955. Then, for $X$ smooth and $f$ an embedding, Sequence (IV,8) with $F=L$ and its tie-in with the embedded tangent space were discovered in an equivalent form. According to M. Atiyah [5; Footnote 5, p. 199] they are due to J.-P. Serre (unpublished) and to S. Nakano [78]. Shortly afterwards Atiyah [5] developed a general theory of the functor $P^1(F)$, which he denoted $D(F)$ and defined differently from us. On the other hand, for $F$ locally free and $X$ smooth, $P^k(F)$ is equal to the sheaf of $k$-jets of sections of the vector bundle $V(F^*)$. The theory of $k$-jets goes back to C. Ehresmann's work around 1951. Recently an algebraic theory of jets was proposed by K. Mount and O. Villamayor [74].

In 1962, Pohl studied the $P^k(F)$ for $F$ locally free and $X$ smooth. For $f$ an immersion, he related the $P^k(L)$ to the osculating spaces and he used this relation to obtain enumerative formulas. It seems however that Pohl did not realize he was dealing with jets or principal parts in their dual form. This observation was made in 1971 by A. Kato [50], who related the exceptional osculating hyperplanes of an elliptic curve to its division points. Recently R.

Piene [82]; [83] developed this point of view further to establish certain generalized Plücker formulas of Cayley and Veronese for a curve in $n$-space.

There are other algebro-geometric works on enumerative problems, using higher order principal parts more or less explicitly. Ogawa [79], Mount and Villamayor [75], and Lax [51] studied Weierstrass points. (These were also treated by Mattuck [72] in an interesting way involving computations in the intersection ring of the symmetric product.) Mattuck [73], Lascoux [63], and Vainsencher [116] advanced the theory of de Jonquières formulas, which enumerates the divisors in a family that have prescribed kinds of singularities. The latter works involve the secant sheaves of Schwarzenberger [110]. These sheaves generalize the sheaves of principal parts; they do for the secant spaces what the sheaves of principal parts do for the osculating spaces.

## B. Polar loci

Let $f: X \to Y$ be an immersion of a smooth $n$-fold $X$ into a projective $m$-space $Y = \mathbb{P}(V)$. (So $V$ is a $(m+1)$-dimensional vector space.) Let $A$ be a linear subspace of $Y$ of codimension $w$; say we have $A = \mathbb{P}(Q)$ where $Q$ is a quotient of $V$ by a subspace $W$ of codimension $w$.

Assume $1 \leqq w \leqq n+1$. Let $\Gamma_w$ denote the locus of points $x$ of $X$ such that the embedded tangent space $T_x$ of $X$ at $x$ intersects $A$ in a linear space of dimension at least $n-w+1$, that is, dimension 1 more than it should have. We call $\Gamma_w$ a *Polar locus* (The use of term 'polar' has evolved over the years from its use by J. V. Poncelet (1822) at the beginning of the golden age of projective geometry. Our usage goes back at least to J. A. Todd (1937).) Here is a second description of $\Gamma_w$: it is the locus of points $x$ of $X$ at which the composition,

$$a: W_X \xrightarrow{\beta_x} V_X \xrightarrow{a_1} P^1(L) \qquad \text{(IV,20)}$$

has rank at most $w-1$, that is, at which the fiber $a(x)$ is not injective, where $\beta$ is the inclusion map of $W$ into $V$ and where $a_1$ is the map defined at Display (IV,13). The two descriptions of $\Gamma_w$ are equivalent because a nonzero element of $V$ lies in $W \cap \ker\,(a_1(x))$ if and only if its zero-locus is a hyperplane containing both $A$ and $T_x$; for $T_x$ is (see the discussion at (IV,16)) equal to $\mathbb{P}(P^1(L)(x))$.

The second description of $\Gamma_w$ shows that $\Gamma_w$ has a natural scheme structure, namely, the one defined by the $w$-minors of $a$. Moreover (see the middle of section A, chapter III) each irreducible component of $\Gamma_w$ has codimension at most $n-w+2$, and if $\Gamma_w$ has codimension at least $n-w+2$, then by Porteous' formula we have

$$\boxed{[\Gamma_w] = c_{n-w+2}(P^1(L))} \qquad (1 \leqq w \leqq n+1) \tag{IV,21}$$

because the Chern classes of $W_X$ are trivial.

Suppose $A$ is in general position. Then each irreducible component of $\Gamma_w$ does have codimension $n-w+2$ and so Formula (IV,21) holds. Indeed consider the tangent mapping $f^\vee$ of $X$ into the grassmannian of $n$-planes in $Y$,

$$f^\vee : X \to G = \mathrm{Grass}_n\,(Y). \tag{IV,22}$$

It sends a point $x$ of $X$ to the point representing the embedded tangent space $T_x$. The mapping is algebraic because it corresponds to the map $a_1$. It is also called the Gauss mapping, the (first) associated mapping, and if $m = n+1$, the dual mapping. Now, clearly $\Gamma_w$ is equal to the pullback $f^{\vee -1}S$ of the Schubert variety $S$ of all $n$-planes meeting $A$ in an $(n-w+1)$-plane at least. Moreover the general linear group acts transitively on $G$, and with respect to this action, $S$ is in general position because $A$ is so by hypothesis. It follows by a short, elementary argument [44; (2,i) p. 290] that each irreducible component of $\Gamma_w = f^{\vee -1}S$ has the same codimension in $X$ as $S$ has in $G$. Furthermore, since $S$ is normal and Cohen–Macaulay, it follows that $\Gamma_w$ is also. (In positive characteristic, the normality of $\Gamma_w$ can fail.)

Similarly, for $n+1 \leqq w \leqq m$ we have a polar locus $\Gamma_w$. It consists of the points $x$ of $X$ at which the intersection $T_x \cap A$ is nonempty, or equivalently, at which the map $a: W_X \to P^1(L)$ has rank at most $n$. The $(n+1)$-minors of $a$ define a scheme structure on $\Gamma_w$. Each irreducible component of $\Gamma_w$ has codimension at most $w-n$, and if $\Gamma_w$ has codimension at least $w-n$, then we have

$$\boxed{[\Gamma_w] = s_{w-n}(P^1(L))} \qquad (n+1 \leqq w \leqq m), \tag{IV,23}$$

by Porteous' formula applied to the dual map $a^\vee : P^1(L)^\vee \to W_X^\vee$. Formula (IV,23) is equivalent to

$$\boxed{[\Gamma_w] = c_{w-n}(\nu \otimes L^{-1})} \qquad (n+1 \leqq w \leqq m), \tag{IV,24}$$

where $\nu$ is the normal sheaf of $X$ in $Y$, by virtue of Sequence (IV,19). Suppose $A$ is in general position. Then $\Gamma_w$ has codimension at least $w-n$; in particular, it is empty if $w \geqq 2n+1$. Also, then $\Gamma_w$ is normal and Cohen–Macaulay. (Normality can fail in positive characteristics. However, except in characteristic 2 with $w = 2n$, if $X$ is suitably reembedded in projective space, then the $\Gamma_w$ are normal because they are smooth in codimension 1 by [93; Cor. 1.2].) Finally, then Formulas (IV,23) and (IV,24) hold.

Formulas (IV,21) and (IV,23) show (in view of (II,35) and (II,18)) that each of the $n$ classes $[\Gamma_w]$ for $2 \leqq w \leqq n+1$ can be expressed as a polynomial in the $n$ classes $[\Gamma_w]$ for $n+1 \leqq w \leqq 2n$, and vice versa. Now, there are natural generalizations of these loci. They are defined by imposing general Schubert conditions on the embedded tangent spaces $T_x$. More precisely, given a nested sequence of linear spaces $A_0 \subseteq \cdots \subseteq A_n$ let $\Gamma$ denote the locus of points $x$ of $X$ at which the intersection $T_x \cap A_i$ has dimension at least $i$ for each $i$. Then $\Gamma$ has a natural scheme structure, each component has at most a certain codimension, and if $\Gamma$ has at least that codimension, then $[\Gamma]$ may be expressed as a certain determinant in the $[\Gamma_w]$ for $2 \leqq w \leqq n+1$, or for $n+1 \leqq w \leqq 2n$; this fact follows from [42; Th. 10]. Schubert initiated the general study of conditions of this sort on linear spaces in an important series of three articles, which appeared in 1886. In a 1902 article Severi [123] brought out the basic importance of the loci $\Gamma$ as projective characters.

Consider the projection $\pi$ with center $A$. Choose a homogeneous coordinate system so that $A$ is given by the vanishing of the first $w$ coordinate functions, in other words, so that these functions form a basis of $W$; then $\pi$ is given by

$$\pi(y_0, \cdots, y_{w-1}, \cdots, y_m) = (y_0, \cdots, y_{w-1}). \qquad \text{(IV,25)}$$

In coordinate-free terms $\pi$ is the mapping,

$$\pi = \mathbb{P}(\beta) : (Y - A) \to Z = \mathbb{P}(W) \qquad (\beta : W \to V). \qquad \text{(IV,26)}$$

Assume $A \cap f(X)$ is empty. Then we have a mapping,

$$\boxed{p = \pi \circ f : X \to Z}. \qquad \text{(IV,27)}$$

The assumption holds automatically in the range $n+1 \leqq w \leqq m$ if $A$ is in general position; otherwise we replace $X$ by $X - f^{-1}A$.

Consider the ramification locus $R$ of $p$. It is defined by the ideal $F$ of $(j+1)$-minors of the jacobian map $\partial p$, where we have set $j = \min(n, w-1) - 1$. Obviously $F$ is also the ideal of $(j+1)$-minors of the twisted map $\partial p \otimes L$. Note that $L = f^*O_Y(1)$ is equal to $p^*O_Z(1)$ because $\pi^*O_Z(1)$ is equal to $O_Y(1) \mid (Y - A)$ since $\pi$ is defined as $\mathbb{P}(\beta)$. Consider Diagram (IV,11) with $p$ for $f$ and $O_Z(1)$ for $G$. It shows that $J$ is also the ideal of $(j+2)$-minors of the map $c_1$. The map $c_1$ also appears in the following diagram:

$$\begin{array}{ccc} W_X & \xrightarrow{\sim} & p^* P^1(O_Z(1)) \\ {\scriptstyle \beta_X}\downarrow & \searrow^{\underline{a}} & \downarrow{\scriptstyle c_1} \\ V_X & \xrightarrow{a_1} & P^1(L). \end{array} \qquad \text{(IV,28)}$$

Here the lower triangle commutes by the definition $a$. Another look at the local analysis of $a_1$ in section A shows readily that $a$ is the analogous map for $p$. Hence the upper triangle is just Triangle (IV,18) for $p$, so it commutes. Thus $c_1$ is isomorphic to $a$. Therefore $F$ is also the ideal of $(j+2)$-minors of $a$. Consequently we have $R=\Gamma_w$ as schemes. Thus the ramification locus consists of the points at which the embedded tangent space meets the center of projection, confirming our intuition.

Consider the range $n+1 \leqq w \leqq m$. Assume $R=\Gamma_w$ has codimension at least $w-n$. Then, for $2n+1 \leqq w \leqq m$, the mapping $p$ is nowhere ramified. (This is an old result, going back at least to L. Kronecker (1869).) For example, suppose $X$ is a curve and $Z$ is the plane; then $p$ is unramified, so $p(X)$ has no cusps. For $n+1 \leqq w \leqq 2n$, ramification is to be expected although it need not occur, and the class of the ramification locus is given by Formulas (IV,23) and (IV,24). (The latter formula was obtained by Roberts [90; Th. 2, p. 159].) For example (see [90; Prop. 6, Example, p. 163]) for $w=2n$, the ramification locus is finite, and if $X$ is complete, then the number of ramification points, each counted with suitable multiplicity, is equal to the degree $\int c(\nu \otimes L^{-1})$; the multiplicities are all equal to 1 if $A$ is in general position because then the ramification locus is normal. (Only in characteristic 2 can the multiplicities be greater than 1. They are so, for example, when $X$ is a plane conic and $Z$ is the line, and when $X$ is the plane, $Z$ is 3-space and $p(X)$ is the Steiner surface.)

For $1 \leqq w \leqq n+1$, the locus $\Gamma_w$ is also a jacobian locus. (The latter is defined in (III,D).) It is the jacobian locus of the linear system of sections $f^{-1}H$ of $X$ by the hyperplanes $H$ through $A$. (This system is the one associated to the map $V \to H^0(X, L)$ whose adjoint is the map $V_X \to L$, the pullback of the tautological map on $Y$.) Indeed it is easy to see by looking at differentials that the section $f^{-1}H$ of $X$ by any hyperplane $H$ is singular at a point $x$ of $X$ if and only if $H$ contains the embedded tangent space $T_x$ of $X$ at $x$. So if a hyperplane $H$ contains $A$ and if the section $f^{-1}H$ of $X$ is singular at $x \in X$, then $T_x$ and $A$ both lie in $H$, and so they intersect in a linear space with dimension at least $n-w+1$. Conversely if a point $x$ of $X$ is such that $T_x$ intersects $A$ in a linear space of dimension at least $n-w+1$, then there exists a hyperplane $H$ containing both $T_x$ and $A$, and so $f^{-1}H$ contains $x$ and is singular there.

Assume $A$ is in general position. We claim the following formula:

$$\boxed{c_i(\Omega_X^1)=(-1)^i\binom{n+1}{i}c_1(L)^i+\sum_{j=0}^{i-1}(-1)^j\binom{n+1-i+j}{j}[\Gamma_{n-i+j+2}]c_1(L)^j}$$

$$\text{for} \quad 1 \leqq i \leqq n. \qquad \text{(IV,29)}$$

Indeed, take Sequence (IV,8) with $L$ for $F$, twist it by $L^{-1}$ and use the additivity of Chern classes (II,16) to obtain

$$c_i(\Omega_X^1) = c_i(P^1(L) \otimes L^{-1}) \quad \text{for each } i. \tag{IV,30}$$

Finally to obtain (IV,28) use Formula (II,22) for the Chern classes of a twist and use Formula (IV,21), which applies because $A$ is in general position.

For example, for $i = 1$, Formula (IV,29) becomes

$$\boxed{c_1(\Omega_X^1) = [\Gamma_{n+1}] - (n+1)c_1(L)}\,, \tag{IV,31}$$

a simple formula for the canonical divisor class in terms of the hyperplane class and the jacobian class of a general $n$-parameter linear system of hyperplane sections. While derived differently, the formula is, as it should be, a special case of the general formula (III,42) for the jacobian class. On the other hand $\Gamma_{n+1}$ is also the ramification locus of the mapping induced by projecting from $A$, and Formula (IV,31) is, as it should be, equivalent to Formula (I,17).

In general, for $1 \leqq i \leqq n$, Formula (IV,29) can be derived formally from Formula (III,42) by solving the system of $n$ equations obtained by taking $[\Gamma_w]$ for $[J]$ for $2 \leqq w \leqq n+1$. However Formula (III,42) was established under a hypothesis on the dimension of the ramification locus $R$ of the mapping from the total space to the parameter space of the linear system. In the present case $R$ represents the pairs $(x, f^{-1}H)$ where $x$ is a point of $X$ and $f^{-1}H$ is a section of $X$ by a hyperplane $H$ containing $A$ and $T_x$ (recall that $H$ contains $T_x$ if and only if $f^{-1}H$ has a singular point at $x$). To check the dimension of $R$, fix $x$ and consider the corresponding hyperplanes $H$. They are parametrized by a projective space of dimension $i-1$, where $i$ is the codimension in $Y$ of the span $T_x * A$. Filter $\Gamma_w$ by the closed sets $F_i$ of points $x$ where $T_x * A$ has codimension at most $i$, or what is the same, where $T_x \cap A$ has dimension at least $m - w + i$. Since $A$ is in general position, $F_i$ has codimension at least $(n+1-w+i)i$. Hence the preimage of $F_i - F_{i-1}$ in $R$ has dimension at most

$$d = (i-1) + n - (n+1-w+i)i \leqq w-2. \tag{IV,32}$$

Hence $R$ has dimension at most $w-2$, q.e.d.

We may conclude more. Since $d$ achieves its minimum, $w-2$, for $i=1$ and for no other value of $i$ and since each irreducible component of $R$ has dimension at least $w-2$, it follows that these components have dimension $w-2$ exactly and that the natural map $p: R \to \Gamma_w$ restricts to a bijection between dense open subsets, namely $F_1 - F_2$ and its preimage. In fact this bijection is an isomorphism between dense open subschemes, and so $[J] = p_*[R]$ is equal to $[\Gamma_w]$. It is an isomorphism between subschemes because $R$

is equal as a scheme to $\mathbb{P}(\text{coker } (a^\vee))$, where $a^\vee$ denotes the dual of the map $a$ in (IV,20). The latter holds whether or not $A$ is general position, and its proof requires a finer analysis. A key step in this analysis is the observation that $R$ is the scheme of zeros of a natural section of $P^1(M)$, where $M$ is the invertible sheaf associated to the total space $D$ of our linear system.

Todd (1937) originally defined the canonical classes in terms of those of lower dimensional subvarieties and of jacobian loci. At the same time he observed that the polar loci $\Gamma_w$ for $2 \leqq w \leqq n+1$ are jacobian loci of linear systems of hyperplane sections and that the canonical classes are given by an expression amounting to the right hand side of Formula (IV,29). This observation proved instrumental around 1954 in the complete establishment of the connection between the canonical classes of Todd (and Eger) and the Chern classes. To establish this connection, Nakano [78] constructed a sequence like (IV,8) with $F = L$, and from it he obtained a formula like (IV,29). The same thing was done by Kodaira and Serre (unpublished), according to Pohl [81; p. 23]. It is less well-known that independently G. Galbura [18] made the connection; he noted it followed from Todd's observation and from a formula like (IV,29) announced by Gamkrelidze [20]. It seems that Hodge [25] was the first one to explicitly conjecture the connection, and he established it without involving polar loci in two special cases. Although Todd worked modulo rational equivalence (even before the theory had been secured), the other authors mentioned used homological equivalence; for a theory of Chern classes modulo rational equivalence was not yet available. More recent treatments of the connection were given by Porteous [86], who did not involve polar varieties, and by Lascoux [62; pp. 177, 178], who did.

Much of the preceding treatment of jacobian loci of linear systems of hyperplane sections using the tool of first order principal parts can be abstracted with profit for arbitrary linear systems although the interpretation with embedded tangent spaces is no longer available. The resulting treatment of jacobian loci is a formalization of the discussion in chapter III, section D. (Porteous [86] is probably the first to have done something like this. Lascoux [63] also did a little of this in effect.) Let $L$ be an invertible sheaf on $X$ and let $\beta: W \to H^0(X, L)$ be a map of vector spaces. Let $w$ denote the dimension of $W$ and assume $w \leqq n+1$. Consider the composition,

$$a: W_X \xrightarrow{\beta_X} H^0(X, L)_X \xrightarrow{b_1} P^1(L). \tag{IV,33}$$

The locus where $a$ has rank at most $w-1$ is the jacobian locus $J$ of the linear system defined by $L$ and $\beta$; indeed, a local analysis like that of $a_1$ in section A shows that given $s \in W$ and $x \in X$ we have $a(s)(x) = 0$ if and only if the divisor defined by $s$ has a singular point at $x$.

Consequently the jacobian locus $J$ has a natural scheme structure, each irreducible component has codimension at most $n-w+2$, and if $J$ has codimension at least $n-w+2$, then by virtue of Porteous' formula and Sequence (IV,8) we have

$$[J]=c_{n-w+2}(P^1(L)), \tag{IV,34}$$

and

$$\boxed{[J]=\sum_{i\geqq w-2}\binom{i+1}{w-1}c_{n-i}(\Omega_X^1)c_1(L)^{i-w+2}}\,. \tag{IV,35}$$

Note that (IV,35) is the same as (III,42) with $r=w-1$, as it should be. Note also that the class $[J]$ is independent of the choice of $\beta$; in other words, two different choices give rise to linearly equivalent jacobian cycles.

For example, suppose $w-1=n$. Then $J$ is defined by an ideal whose dual is isomorphic to $\bigwedge^{n+1}P^1(L)$ or $\Omega_X^n\otimes L^{\otimes n+1}$. Locally $J$ is cut out by the determinant of an $(n+1)$-square matrix like (IV,15) with $m=n$ and like (III,40) with $r+1=n$. This case is easily handled by direct elementary means (see Galbura [19; §2]). For a second example, suppose $w-1=1$; this is the case of a pencil. Suppose $J$ is finite. Then Formula (IV,35) holds. However if one of the points of $J$ is a base point, that is, if some point of $X$ is a singular point of every member of the pencil, then the derivation of Formula (III,42) fails because the locus $R$ is not finite. So, using Formula (IV,35) and taking care at the points of $J$ that are base points, we can generalize Formula (III,45) in particular.

In general, consider the ramification locus $R$ of the mapping from the total space to the parameter space of our system. As a scheme, $R$ is equal to $\mathbb{P}(\text{Coker }(a^\vee))$. Hence if each component of $R$ has the appropriate dimension, $w-2$, then, since each component of $J$ has dimension at least $w-2$, the mapping $R\to J$ restricts to an isomorphism from an open subscheme of $R$, conceivable not dense, onto a dense open subscheme of $J$; consequently, the pushdown of $[R]$ is equal to $[J]$.

The jacobian $J$ has the appropriate dimension, $w-2$, or is empty if $W$ is a general subspace of a vector space $V$ equipped with a map $\Phi: V\to H^0(X,L)$ such that the composition $b_1\Phi: V_x\to P^1(L)$ is surjective. This fact can be proved by abstracting the proof for systems of hyperplane sections. However this time the apparent gain in generality is illusory. Indeed, the composition $a_{0,1}b_1\Phi: V_X\to L$ is also surjective. Hence it corresponds to a mapping $f$ of $X$ into $Y=\mathbb{P}(V)$ such that it is the pullback of the tautological surjection $\alpha: V_Y\to O_Y(1)$, and $f$ is unramified because $b_1\Phi$ is surjective.

Yet another way to treat the jacobian locus of a linear system was introduced by Ingleton and Scott [38] and cleaned up nicely by Vainsencher

[115]. This approach is also in the spirit of these notes, especially that of chapter V, section C.

**C. Plücker formulas**

Let $f:X\to Y$ be a mapping from a smooth $n$-fold $X$ into a projective $m$-space $Y=\mathbb{P}(V)$. (We may allow $X$ to have several components.) If $f$ is an immersion, then Formula (IV,31) holds and it generalizes the Plücker formula (I,25) for plane curves without cusps; indeed, the degree of $c_1(\Omega^1_X)$ is $2g-2$, where $g$ is the genus of $X$, the degree of $[\Gamma_2]$ is the number of (embedded) tangent lines passing through a general point of $Y$, and the degree of $c_1(L)$ is the degree of $X$. This observation was made by Pohl [82], [83], who went on to give a generalization of (IV,31) that yields (I,25) for general plane curves. We now discuss this generalization.

Assume that the mapping $f:X\to Y$ is an immersion on a dense open subset $U$ of $X$. Then the tangent mapping into the grassmannian of $n$-planes in $Y$ is defined on $U$,

$$f^\vee: U\to \mathrm{Grass}_n(Y). \tag{IV,36}$$

The mapping $f^\vee$ sends a point of $U$ to the point representing the embedded tangent space of $X$ at $x$. Algebraically $f^\vee$ is defined so that the map $a_1: V_X\to P^1(L)$ when restricted to $U$, where it is surjective, is the pullback of the tautological surjection. (The map $a_1$ is discussed in section A.)

The mapping $f^\vee$ extends (uniquely) from $U$ to a larger open subset $W$, whose complement in $X$ has codimension at least 2, because $X$ is smooth. In other words, at the points of $W$ there is a well-defined limiting position of the embedded tangent spaces. Algebraically the extension corresponds to a surjection $a: V_W\to P$, the pullback of the tautological surjection. Here $P$ is a locally free sheaf with rank $n+1$. We claim $P$ is canonically isomorphic to the image of $a_1\mid W$. Indeed, $a_1(\mathrm{Ker}\,(a))$ is a subsheaf of $P^1(L)$, which is zero on $U$; so it is identically zero because $P^1(L)$ is locally free. Hence $a_1\mid W$ factors into $a$ followed by a surjection from $P$ to $\mathrm{Im}\,(a_1\mid W)$. The latter is also an injection because it is on $U$ and $P$ is locally free. Alternatively, we could prove directly that $\mathrm{Im}\,(a_1)$ is locally free on a set like $W$ and then extend $f^\vee$.

The *cuspidal edge* $\kappa$ of $f$ is the divisor defined as follows. Let $F$ denote the 0th Fitting Ideal of Coker $(a_1)$. It can be computed on $W$ by using $P$ and $P^1(L)\mid W$. We obtain

$$F\mid W=\bigwedge^{n+1} P\otimes\left(\bigwedge^{n+1} P^1(L)\mid W\right)^{-1}. \tag{IV,37}$$

Thus $F\mid W$ is invertible. Hence it defines a divisor on $W$, so one $\kappa$ on all of $X$ because $W$ intersects every subvariety of $X$ with codimension 1. This divisor $\kappa$ on $X$ is called the cuspidal edge.

It is easy to see, by identifying $V_Y$ with $P^1(O_Y(1))$ and using Diagram (IV,11) with $G=O_Y(1)$, that $F$ is equal to the 0th Fitting ideal of $\partial f \otimes L$, so to that of $\partial f$. In other words, on $W$ the singular locus of $f$ is equal to the divisor $\kappa$. So $\kappa$ may be described concretely as the following divisor. Its components $[V]$ come from the irreducible components $V$ with codimension 1 of the singular locus of $f$. Each component $[V]$ appears with the coefficient $k$ determined as follows. The various $n \times n$-minors of the jacobian $\partial f$ all vanish along $V$. The minimum order of vanishing is the coefficient $k$ in question.

For example, suppose that $X$ is a curve and that $f$ is given at a point $x$ in terms of local coordinates by the equation

$$f(t)=(y_1(t), \cdots, y_m(t)) \quad \text{with} \quad y_i(t)=t^{l_i}u_i(t) \quad \text{and} \quad u_i(0)\neq 0 \quad \text{for} \quad i=1, \cdots, m. \tag{IV,38}$$

Differentiating $y_i(t)$ with respect to $t$ we obtain

$$y_i'(t)=l_i t^{l_i-1}u_i(t)+t^{l_i}u_i'(t). \tag{IV,39}$$

Set $l=\min(l_i)$. Then $l-1$ is obviously the minimum order of vanishing of the $y_i'(t)$ at $x$ or $t=0$; hence $l-1$ is the coefficient of $[x]$ in the cuspidal edge $\kappa$. Note on the other hand that $l$ is the multiplicity of the branch corresponding to $x$ of the curve $f(X)$. Thus *the coefficient is the branch multiplicity diminished by* 1. (In positive characteristic $p$, we have to replace $l$ by the minimum of those $l_i$ not divisible by $p$ to obtain the coefficient of $[x]$, while $l$ remains the multiplicity of the branch.) For instance if the branch has a simple cusp, then the coefficient of $[x]$ is 1, for the multiplicity of the branch is 2 (except that in characteristic 2 the coefficient is 2).

Finally, let $\Gamma_w$ for $1 \leqq w \leqq n+1$ denote the closure on $X$ of the locus of points of $U$ at which the embedded tangent space meets a given general linear subspace $A$ of codimension $w$ in a linear space of dimension at least $n-w+1$. It is not hard to see that $\Gamma_w \mid W$ represents $c_{n-w+2}(P)$. Hence extracting first Chern classes of both sides of (IV,37) and using Sequence (IV,8) and Formula (II,23), we obtain the restriction to $W$ of the formula,

$$\boxed{c_1(\Omega_X^1)=[\Gamma_{n+1}]-(n+1)c_1(L)+[\kappa]}\,. \tag{IV,40}$$

Since the complement of $W$ in $X$ has codimension 2 or more, it follows (using [13; Prop., p. 154] or [19; §4]) that Formula (IV,40) holds as stated.

Formula (IV,40) is the sought generalization of (IV,31). For, $\kappa=\sum k_j[V_j]$ is the divisorial part of the singular locus of $f$. Moreover if $X$ is a complete curve of genus $g$, of class $d^\vee$ and of order $d$, then extracting degrees from both sides of (IV,40) we get

$$\boxed{d^\vee = 2d + (2g-2) - \sum k_j}\,. \tag{IV,41}$$

This formula is one of the generalized Plücker formulas for curves in $n$-space of Cayley and Veronese (see [83; Cor. (3.3)]). In particular (I,25) is a special case because there all the $k_j$ are 1 by virtue of the example just above and so $\sum k_j$ is simply the total number of cusps. (Similarly $(I,25)_2$ is a special case.) It would be good to have a similar generalization of (IV,29) for $i = 2, \cdots, n$.

Pohl chose a slightly different route to Formula (IV,40). He worked entirely within the category of complete, smooth varieties. (Note however that for us $X$ need not be complete.) Instead of working directly on $W$ with the extension of $f^\vee$, Pohl took its graph and resolved the singularities (Hironaka had just proved this is possible) to obtain a smooth, complete variety $X^\vee$ over which $f^\vee$ extends entirely. Then he got a formula on $X^\vee$ and pushed it down to $X$. Resolution of singularities is a deep matter (and unproved for $n \geqq 4$ in positive characteristic). On the other hand Pohl's use of blowing up $X$ to extend $f^\vee$ led Piene [82] to prove a generalization of the other basic Plücker formula (I,24), which we discuss next.

The setup is different now. We consider a $n$-dimensional hypersurface $Z$ of degree $d$ in a projective $(n+1)$-space $Y = \mathbb{P}(V)$. (We can allow $Z$ to have several components, but it must be reduced.) Let $J$ denote the singular locus of $Z$. It is the scheme defined by the jacobian ideal $F$, which is by definition the Fitting ideal $F^n(\Omega^1_Z)$. Set $L = O_Z(1)$. Then $F$ is also the Fitting ideal $F^{n+1}(P^1(L))$ by virtue of Sequence (IV,8) with $F = L$. Hence $F$ may be computed using Sequence (IV,19) and so there is a natural surjection,

$$V^\vee_Z \otimes (N \otimes L) \twoheadrightarrow \quad F, \tag{IV,42}$$

where $N$ is the conormal sheaf of $Z$ in $Y$. Here we have $N = L^{-d}$.

Let $q: Z' \to Z$ denote the blowing-up of $Z$ along $J$. Then the ideal $F' = FO_{Z'}$ is invertible. Set

$$M = F' \otimes q^* L^{d-1}. \tag{IV,43}$$

Then $M$ is invertible and Surjection (IV,42) yields a surjection $V^\vee_{Z'} \twoheadrightarrow M$. The latter's kernel is locally free with rank $n+1$; let $P$ denote its dual. Then we have a commutative diagram,

$$\begin{array}{ccccccccc} & & q^*L^{1-d} & \longrightarrow & V_{Z'} & \longrightarrow & q^*P^1(L) & \longrightarrow & 0 \\ & & \downarrow & & \downarrow{=} & & \downarrow & & \\ 0 & \longrightarrow & M^\vee & \longrightarrow & V_{Z'} & \longrightarrow & P & \longrightarrow & 0 \end{array} \tag{IV,44}$$

The left vertical map comes from the inclusion of $F'$ in $O_{Z'}$. The right vertical map is induced by the rest of the diagram.

The tangent mapping $f^{\vee}$ into $\mathrm{Grass}_n\,(Y)$ is defined on the smooth locus, $U=Z-J$. Algebraically it corresponds to the surjection $a_1: V_Z \to P^1(L)$ restricted to $U$, where $P^1(L)$ is locally free with rank $n+1$. Hence in view of Diagram (IV,44) it is clear that $f\vee^\circ\ q$ extends everywhere on $Z'$, the extension $f'$ corresponding to the surjection $V_{Z'} \to P$. In other words, at each point of $Z'$ there is a well-defined limiting position of the embedded tangent spaces $T_z$ for $z \in q^{-1}U$. (Note that the restriction $q^{-1}U \to U$ is an isomorphism.) Note that $\mathrm{Grass}_n\,(Y)$ is equal to $Y^{\vee}=\mathbb{P}(V^{\vee})$, the dual projective space, and that we have $M=f'^{*}O_{Y^{\vee}}(1)$.

Set $d^{\vee}=\int c_1(M)^n$. This number has the following geometric interpretation. Consider the locus $\Gamma'_2$ on $Z'$ where the embedded tangent hyperplanes or their limits contain a given $(n-1)$-plane $A$. Then $\Gamma'_2$ is the pullback under $f'$ of a certain Schubert variety, which in this case is obviously just a line in $Y^{\vee}$. It follows that if $\Gamma'_2$ consists of a finite number of points, then the number of points, counted with multiplicities, is equal to $d^{\vee}$. Moreover if $A$ is in general position, then $\Gamma'_2$ does consist of a finite number of points, the points all lie in $q^{-1}U$ (for the codimension of $\Gamma'_2 \cap q^{-1}J$ in $q^{-1}J$ cannot be less than $n$), and the multiplicities are all 1; thus $d^{\vee}$ is the exact number of smooth points of $Z$ at which the embedded tangent space contains $A$. (In positive characteristic, the number may not be exact, for the multiplicities may be a power of the characteristic.)

The generalized Plücker formula for $d^{\vee}$ is easy to derive as follows. In view of the definition above of $d^{\vee}$ and the definition (IV,43) of $M$, we have

$$d^{\vee}=\int[(d-1)q^{*}c_1(L)+c_1(F')]^n. \tag{IV,45}$$

Expanding the right hand side and using Formula (II,11) and the projection formula, we obtain

$$d^{\vee}=\sum_{i=0}^{n}\binom{n}{i}(d-1)^i\int c_1(L)^i q_{*}c_1(F')^{n-i}. \tag{IV,46}$$

Using Formula (II,43) and the fact $d=\int c_1(L)^n$, we arrive at the sought formula,

$$\boxed{d^{\vee}=(d-1)^n d-\sum_{i=0}^{n-1}\binom{n}{i}(d-1)^i\int c_1(L)^i\alpha_{*}s^i(J,Z)}\,, \tag{IV,47}$$

where $\alpha$ denotes the inclusion mapping of $J$ into $Z$. The second term on the right hand side is a weighted sum of the orders of the Segre covariants of the singular locus. The term constitutes the correction to the number $(d-1)^n d$ due to the presence of singularities.

In deriving Formula (IV,47) we could have refrained from taking degrees all along and we would have obtained a similar formula for the polar class $[\Gamma_2]=q_*[\Gamma_2']$ itself. In fact, virtually in the same way we can obtain the following formula:

$$\boxed{[\Gamma_w]=\{(d-1)c_1(L)\}^{n-w+2}-\sum_{i=0}^{n-w+1}\binom{n-w+2}{i}\{(d-1)c_1(L)\}^i\alpha_*s^{i-w+2}(J,Z)}\,. \tag{IV,48}$$

Here, for $2\leqq w\leqq n+1$, the polar locus $\Gamma_w$ is the closure of the locus of points of $U$ at which the embedded tangent space meets a given general linear subspace $A$ of codimension $w$ in a linear space of dimension at least $n-w+1$. It is not hard to see that $\Gamma_w$ represents $q_*c_{n-w+2}(P)$.

Suppose $Z$ has isolated singular points $z_1,\cdots,z_r$. Then the classes $s^i(J,Z)$ for $i>0$ vanish and Formula (IV,47) reduces by virtue of Formula (II,50) to

$$\boxed{d=(d-1)^n d-\sum e_j}\,, \tag{IV,49}$$

where $e_j$ is the multiplicity at $z_j$ of the jacobian ideal $F$. Note that $e_j$ can be computed using a local analytic equation $\Phi=0$ for $Z$ at $z_j$. Indeed, $F$ is defined as the Fitting ideal $F^n(\Omega^1_Z)$ and so may be computed using the map $N\to\Omega^1_Y\,|\,Z$. We find that the completion of $F$ at $z_j$ is the ideal generated by the partial derivatives of $\Phi$. Hence the multiplicity of the latter ideal is $e_j$.

For example suppose $Z$ is a plane curve. Suppose $z_j$ is a simple node. At it $Z$ is given analytically by the equation $xy=0$. The ideal of partial derivatives is $(x,y)$, which is the maximal ideal. So $e_j$, the multiplicity of this ideal, is 2. Suppose $z_j$ is a simple cusp. At it $Z$ is given by an equation of the form

$$y^2-x^3+yg(x,y)+h(x,y)=0 \tag{IV,50}$$

where $g$ is a homogeneous polynomial of degree 2 and each term of $h$ has degree at least 4. The ideal of partial derivatives is

$$(-3x^2+yg_x+h_x,\,2y+g+yg_y+h_y). \tag{IV,51}$$

To compute its multiplicity we describe $Z$ at $z_j$ in terms of a parameter $t$ by power series

$$x=t^2+\cdots,\ y=t^3+\cdots. \tag{IV,52}$$

The ideal becomes $(-3t^4+\cdots,2t^3+\cdots)$. Its multiplicity $e_j$, is the smallest exponent of $t$ appearing, namely 3. (In characteristic 2 it is 4.) Hence Formula (IV,49) yields Formula (I,24) (as well as $(\text{I},24)_2$) as a special case. More general, similar computations show this: if $z_j$ is an ordinary $r$-fold point, then $e_j = r(r-1)$ holds; if $z_j$ is a cusp at which $Z$ is given parametrically by power series,

$$x = t^a + \cdots, \; y = t^b + \cdots \quad \text{with} \quad a < b \quad \text{and} \quad (a, b) = 1, \qquad \text{(IV,53)}$$

then $e_j = (a-1)b$ holds (except that if the characteristic divides $b$, then $e_j = a(b-1)$ holds). It is possible to give a (more complicated) formula even when $(a, b) \neq 1$.

Formula (IV,49) may be rewritten using Teissier's formula (II,51) as

$$d^\vee = (d-1)^n d - \sum (\mu_{z_j}^{(n+1)}(Z) + \mu_{z_j}^{(n)}(Z)). \qquad \text{(IV,54)}$$

If $Z$ is a plane curve, then Teissier's formula (II,51) becomes

$$e_j = \mu_{z_j}(Z) + m_{z_j}(Z) - 1, \qquad \text{(IV,55)}$$

where $\mu_{z_j}(Z)$ is the (ordinary) Milnor number and $m_{z_j}(Z)$ is the multiplicity (see [112; (1.6.1), p. 300]), and (IV,49) may be rewritten accordingly. Formula (IV,49) and the above variations are due to Teissier [113; App. II]. Teissier's proof is based on his theory of polar curves; the work is set over $\mathbb{C}$ but is algebraic in character. (Teissier [113; II.1, p. 45] says he was led to the matter by a question from G. Laumon.) Laumon [59] gave his own proof of (IV,54), using étale topology.

Suppose again $Z$ is a plane curve. Denote by $X$ its desingularization, by $f: X \to Y = \mathbb{P}^2$ the corresponding mapping, and by $\kappa = \sum k_i[x_i]$ the cuspidal edge of $f$ (the ramification divisor). There is an interesting relation among the coefficients $k_i$, the multiplicities $e_j$ of the jacobian ideal of $Z$, and the discrepancy $\delta(Z)$ between the arithmetic genus $(d-1)(d-2)/2$ of $Z$ and the geometric genus $g$ of $Z$ (the genus of $X$). The relation is

$$\boxed{\sum e_j = 2\delta(Z) + \sum k_i} \qquad \text{(IV,56)}$$

and it comes from (IV,41) and (IV,49) by straightforward manipulation. Conversely given (IV,56) we can turn either of (IV,41) and (IV,49) into the other.

Relation (IV,56) has a sharper, local form; it is

$$e_j = 2\delta_j + \sum_{f(x_i) = z_i} k_i \quad \text{for each } j. \qquad \text{(IV,57)}$$

Here $\delta_j$ is the local contribution at $z_j$ to the discrepancy between the genera. (See [108; Ch. IV] for details about $\delta_j$.) Informally $\delta_j$ is sometimes thought of as the number of double points concentrated at $z_j$ because at a double point the contribution is 1. Precisely $\delta_j$ is defined as the codimension of the local ring of $Z$ at $z_j$ in its normalization. It is a theorem (due to

Gorenstein and to Samuel) that $2\delta_j$ is equal to the codimension of the conductor in the normalization. Piene [82; (3.13) p. 44–46] established (IV,57) by proving a finer relation among ideals, which is

$$F^1(\Omega^1_Z)O_{X'} = C \cdot F^0(\Omega^1_{X/Z}) \qquad \text{(IV,58)}$$

where $C$ is the conductor. The proof involves comparing in diagrams the various sheaves of principal parts. The conductor comes in to play via a formula (due to Rosenlicht) relating it, the differentials on $X$, those on the plane $Y$, and the conormal sheaf of $Z$ in $Y$. (The formula is (V,7) with $n = 1$.)

The local relation (IV,57) may be rewritten as

$$\boxed{e_j = 2\delta_j + m_j - r_j}\ , \qquad \text{(IV,59)}$$

where $m_j$ is the multiplicity of $Z$ at $z_j$ and $r_j$ is the number of branches of $Z$ at $z_j$; for, as we have seen above, $k_i$ is the multiplicity of the branch corresponding to $x_i$ diminished by 1 (in positive characteristic $p$, we must assume $p$ does not divide $k_j + 1$), and the sum of the multiplicities of the branches at $z_j$ is equal to $m_j$. Comparing (IV,59) and (IV,55) (which is established algebraically over $\mathbb{C}$) we find the relation

$$\boxed{2\delta_j = \mu_j + r_j - 1}\ , \qquad \text{(IV,60)}$$

which is due to Milnor [71; Th. 10.5, p. 85], who gave a topological proof.[3]

## D. Duality

For varieties, there is a fairly well-developed and pretty duality theory, going back to J. V. Poncelet, J. D. Gergonne, and J. Plücker from about 1822 to 1839. The nonenumerative part is elementary (see [118], [77; §1], [104; Exposé XVII, pp. 212–253], [82; Ch. 4]). Here we concentrate on the degree of the dual variety. Formulas for it are generalized Plücker formulas.

We consider a closed $n$-dimensional subvariety $Z$ of an $m$-dimensional projective space $Y = \mathbb{P}(V)$. The *dual variety* $Z^\vee$ in $Y^\vee = \mathbb{P}(V^\vee)$ is defined as the closed envelope of tangent hyperplanes. At a smooth point $z$ of $Z$ the notion of tangency is straightforward; a hyperplane is tangent to $Z$ at $z$ if

[3] J.-J. Risler [Bull. Soc. Math. France, 99 (1971) 305–311] gave an algebraic proof. J. Lipman, reviewing it [MR 46-5334], pointed out that Jung gave an algebraic proof antedating Milnor's proof.

and only if it contains the embedded tangent space of $Z$ at $z$, or equivalently, if and only if its intersection with $Z$ has a singular point at $z$. At a singular point $z$, a hyperplane is considered tangent if it is a limit of hyperplanes tangent at smooth points approaching $z$. A main theorem asserts that the dual of $Z^\vee$ is equal to $Z$. (In positive characteristic this biduality fails on occasion.) (A curve has another natural dual variety; it is the curve of osculating hyperplanes and their limits (see [83; §5]).)

Assume for the time being that $Z$ is smooth. This is an important case; it lies at the foundation of Lefschetz theory. In this case it is useful to approach the dual variety via the linear system of hyperplane sections. The total space of the system is a divisor $D$ in $Z \times Y^\vee$. Consider the projection,

$$q: D \to Y^\vee. \tag{IV,61}$$

Let $R$ denote its ramification locus. Obviously we have $Z^\vee = q(R)$. Now, $R$ is irreducible and has dimension $m-1$ because, with respect to the projection $R \to Z$, the fiber over a point $z$ of $Z$ is the $(m-n-1)$-plane in $Y^\vee$ parametrizing the hyperplanes containing the embedded tangent space at $z$. A finer analysis, involving the exact sequence (IV,19), shows that $R$ and $\mathbb{P}((N(1))^\vee)$ are equal as subschemes of $X \times Y^\vee = \mathbb{P}(V_X^\vee)$. So $R$ is smooth as well.

The dual variety $Z^\vee$ tends to have dimension $m-1$ because $Z$ is smooth. Here however are some examples (quoted from [52]) in which $Z^\vee$ has lower dimension. (If singularities are allowed, examples are cheap. Simply take the dual of a low dimensional subvariety of $Y^\vee$. By biduality, its dual is that low dimensional subvariety. Moreover every example is of this type.) (1) If $Z$ is an $n$-plane, then (obviously) $Z^\vee$ is an $(m-n-1)$-plane. (2) If $Z$ is such that each point lies in an $r$-plane contained in $Z$ with $2r > n$, then $Z^\vee$ has dimension at most $(m-1)-(2r-n)$; specific examples of such $Z$ are the Segre variety $\mathbb{P}^r \times \mathbb{P}^s$ with $r > s$ (it is embedded naturally in $\mathbb{P}^m$ with $m+1 = (r+1)(s+1)$) and more generally any $\mathbb{P}^r$-bundle over a projective, smooth $s$-fold $X$ with $r > s$, embedded by a very ample tautological sheaf. (3) If $Z$ is the grassmannian of lines in a $k$-dimensional projective space, embedded in $\mathbb{P}^m$ with $m+1 = \binom{k+1}{2}$ using Plücker coordinates, then $Z^\vee$ has dimension $m-3$ if $k$ is even and at least 4 (but $Z^\vee$ has dimension $m-1$ otherwise).

One reason why examples (2) and (3) are interesting is this: in them $q: D \to Y^\vee$ is a flat, surjective mapping between smooth varieties, but the set of critical values, $Z^\vee$, while nonempty, does not have codimension 1 in the target, $Y^\vee$. By contrast, for a flat, surjective mapping between varieties of the same dimension, the set of critical values, if nonempty, has pure codimension 1; this is a famous result, known as 'the purity of the branch locus'. (The third example and its connection with 'purity' were suggested to the author first by D. Mumford in conversation.) It would be good to have a better idea of when $Z$ has an $(m-1)$-dimensional dual and when, a lower dimensional one. Below we shall see that $Z^\vee$ is $(m-1)$-dimensional when $Z$ is a curve, a surface, a hypersurface, or a complete intersection, while not a linear space, and when $Z$ is suitably re-embedded.

Consider the $(m-1)$-cycle $\Delta = q_*[R]$ on $Y^\vee$. Since $q(R) = Z^\vee$ holds, by Definition (II,3) we have

$$\Delta = \begin{cases} \text{degree}\,(\gamma: R \to Z^\vee)\,[Z^\vee], & \text{or} \\ 0 \quad \text{if} \quad \dim(Z^\vee) < m-1. \end{cases} \tag{IV,62}$$

Now, it is a theorem that, if $Z^\vee$ has dimension $m-1$, then $\gamma$ is birational. Hence we have $\Delta = [Z^\vee]$ if $Z^\vee$ has dimension $m-1$ and we have $\Delta = 0$ if $Z^\vee$ has smaller dimension. (In positive characteristic $p$, the theorem fails on occasion, but when it does, $\gamma$ must be inseparable and so we have $\Delta = p^r[Z^\vee]$ for some $r$. By the same token, whatever the dimension of $Z^\vee$, when biduality fails, $\gamma$ must be inseparable. Wallice [118] shows there are plane curves for which $\gamma$ has arbitrary separable degree.)

The class $[\Delta]$ is a multiple of the hyperplane class $c_1(O_{Y^\vee}(1))$. Let $d^\vee$ denote the multiplier. By the above, we have

$$\boxed{d^\vee = \begin{cases} \text{degree}\,(Z^\vee) \\ 0 \end{cases}} \quad \begin{matrix} \text{if} \quad \dim(Z^\vee) = m-1 \\ \text{if} \quad \dim(Z^\vee) < m-1. \end{matrix} \tag{IV,63}$$

(In positive characteristic, $d^\vee$ can be a non-trivial multiple of degree $(Z^\vee)$, but then the multiplier must be a multiple of the characteristic.) Using Formulas (III,38) and (III,39), we obtain the following generalized Plücker formula for $d^\vee$:

$$\boxed{\begin{aligned} d^\vee &= \sum_{i=0}^{n} (i+1) \int c_{n-i}(\Omega_Z^1) c_1(O_Z(1))^i \\ &= \int \frac{c(\Omega_Z^1)}{(1-c_1(O_Z(1)))^2} \end{aligned}} \;. \tag{IV,64}$$

For example, if $Z$ is a curve of genus $g$ and of degree $d$, then Formula (IV,64) becomes

$$d^\vee = 2g - 2 + 2d. \tag{IV,65}$$

If also $Z$ is not a line, so $d > 1$, then we find $d^\vee \neq 0$ and so $Z$ is a hypersurface. In view of (IV,64) it is also obvious that, whatever $n$ is, there exists an $r$ (in fact all but $n$ values of $r$ work) such that re-embedding $Z$ using $O_Z(r)$ we obtain $d^\vee \neq 0$ and so $Z^\vee$ is a hypersurface then.

Here is another derivation of Formula (IV,64). (This is essentially the one in [104; Exposé XVII, §5].) First, nearly by definition, we have

$$d^\vee = \int [\Delta] \cdot c_1(O_{Y^\vee}(1))^{m-1}. \tag{IV,66}$$

Applying the projection formula and Formula (II,11) we obtain

$$d^\vee = \int (q \mid R)^* c_1(O_{Y^\vee}(1))^{m-1}. \tag{IV,67}$$

Now, under the identification of $R$ with $\mathbb{P}((N(1))^\vee)$, it is evident that $(q \mid R)^* O_{Y^\vee}(1)$ is identified with $O_R(1)$; hence we have

$$d^\vee = \int c_1(O_R(1))^{m-1} \qquad \text{by (IV,67)} \tag{IV,68}$$

$$d^\vee = \int s_n((N(1))^\vee) \qquad \text{by (II,33)} \tag{IV,69}$$

$$\boxed{d^\vee = \int \frac{1}{c(N(1))}} \qquad \text{by (II,35)} \tag{IV,70}$$

$$d^\vee = \int \frac{(1 - c_1(O_X(1)))^{m-1}}{c(N)} \qquad \text{by (II,22)} \tag{IV,71}$$

Finally, using the conormal sheaf-cotangent sheaves exact sequence and using Formula (I,19) for $c(Y^\vee)$, we obtain Formula (IV,64).

Formula (IV,70) is convenient for computing $d^\vee$ when $Z$ is a hypersurface. We easily find $d^\vee = d(d-1)^n$, where $d$ denotes the degree of $Z$; for we have $N = O_Z(-d)$. Hence if $Z$ is not a hyperplane ($d \neq 1$), we have $d^\vee \neq 0$ and so $Z^\vee$ is a hypersurface. More generally, when $Z$ is a nonlinear complete intersection, we find $d^\vee \neq 0$; indeed, using (IV,70) it is easy to see that $d^\vee$ is a sum of certain positive terms.

Formal manipulations exactly like those used to pass from (III,45) to (III,46) yield the formula,

$$\boxed{d^\vee = (-1)^n[\chi(Z) - 2\chi(Z \cap H) + \chi(Z \cap H \cap H')]}\ , \tag{IV,72}$$

where $H$ and $H'$ are hyperplanes in $Y$ chosen in general position so that $Z \cap H$ and $Z \cap H \cap H'$ are smooth (but possibly reducible). Combining formula (IV,72) with basic Picard–Lefschetz theory (valid in positive characteristic for étale cohomology), Landman [53] obtained some remarkable results. For example, he proved that $d^\vee$ is even if $n$ is odd (this holds ultimately because 'odd Betti numbers are even'). He established, for nonlinear $Z$ of dimension $n$ at least 2, the lower bound

$$\boxed{\dim (Z^\vee) \geqq m-n+1} \quad . \tag{IV,73}$$

For instance, (IV,73) implies that if $Z$ is a nonlinear surface, then $Z$ is a hypersurface. He also proved that, for the specific $Z$ in Example (2) above, $Z^\vee$ has the maximum dimension, $(m-1)-(2r-n)$.

Consider a general pencil of hyperplane sections of $Z$. It is parametrized by a line $T$ in $Y^\vee$ in general position, and it has total space $q^{-1}T$. Let $R_T$ denote the ramification locus of the mapping $q^{-1}T \to T$. We have

$$R_T = R \cap q^{-1}T = (q \mid R)^{-1}T \tag{IV,74}$$

because the formation of sheaves of relative differentials and the formation of Fitting ideals commute with base change. Since $R$ has dimension $m-1$ and since $T$ has codimension $m-1$ and is in general position, $R_T$ therefore has dimension 0. The 0-cycle $[R_T]$ has degree $d^\vee$ by virtue of (IV,67) because $T$ represents $c_1(O_{Y^\vee}(1))^{m-1}$. Alternately we may reason like this: since $T$ is in general position, $d^\vee$ is equal to the exact number of points in $Z^\vee \cap T$ by (IV,63) and $\gamma : R \to Z^\vee$ is an isomorphism over each point of $Z^\vee \cap T$ because $\gamma$ is birational; hence $d^\vee$ is the exact number of points in $R_T$. (However in positive characteristic this alternate reasoning fails.) Consequently $d^\vee$ is also the degree of the pencil's jacobian cycle $[J]$, the pushdown of $[R_T]$ to $X$. Thus it is no accident that the right hand side of (III,45) is virtually identical with that of (IV,64).

A hyperplane $H$ is said to be a *simple tangent* of $Z$ if it is tangent to $Z$ at a unique point $z$ and if the intersection $H \cap Z$ has a nondegenerate quadratic singularity at $z$; that is, $H \cap Z$ is given on $Z$ at $z$ by the vanishing of a regular function whose hessian matrix of second partials is nondegenerate. It is a theorem (valid in every characteristic) that a hyperplane $H$ is a simple tangent of $Z$ with unique point of tangency $z$ if and only if the pair $(z, H)$ corresponds to a point of $R$ at which the mapping $\gamma : R \to Z$ is an isomorphism. (The locus in $R$ where $\gamma$ is not an isomorphism is the locus of second-order singularities of $q : D \to Y^\vee$. Each irreducible component has dimension $m-2$ (in positive characteristic we must assume $\gamma$ is birational) and Roberts [94; §4] found a formula for its class and a formula for the class of its pushdown to $Y^\vee$ as a cycle.)

Consider the pencil of hyperplanes parametrized by $T$; it induces our pencil of hyperplane sections of $Z$. Since $T$ is in general position, it follows from the theorem stated just above that the hyperplanes in the pencil that are tangent to $Z$ are simple tangents. Since $d^\vee$ is the exact number of points in $R_T$, we conclude that $d^\vee$ is exactly the number of simple tangent hyperplanes in the pencil. The hyperplanes in the pencil can also be described as the hyperplanes through a certain general $(m-2)$-plane $A$, namely, the one dual to $T$. Therefore $d^\vee$ is also the exact number of points $z$ of $Z$ such that the hyperplane spanned by $z$ and a fixed general $(m-2)$-plane $A$ is tangent at $z$; moreover these tangent hyperplanes are simple tangents. (In positive characteristic $p$, the last several statements fail on occasion, but they hold if $p$ does not divide $d^\vee$.)

One dual variety played an important role in the development of classical enumerative geometry (see [119; especially §20, pp. 302–304, and §23, pp. 307–308], [45; especially p. 469], [46; especially pp. 1–6]). This dual variety is the hypersurface parametrizing the plane curves of degree $r$ that are tangent to a given plane curve $Z$ of degree $s$. The hypersurface is easily seen to be the dual variety of $Z$ re-embedded in $\mathbb{P}^N$ with $N+1=\binom{r+2}{2}$ by the $r$-fold Veronese embedding; the Veronese embedding extends to the ambient $\mathbb{P}^2=\mathbb{P}(V)$, and the hyperplanes of $\mathbb{P}^N=\mathbb{P}(S^rV)$ cut out on $\mathbb{P}^2$ the curves of degree $r$. The degree $d^\vee$ of the hypersurface can be computed using Formula (IV,65); we find

$$\boxed{d^\vee = s(s+2r-3)}\ . \qquad \text{(IV,75)}$$

Indeed, $g$ is the genus of $Z$, so $2g-2$ is equal to $s^2-3s$ by (II,32). The ample sheaf of $Z$ in $\mathbb{P}^N$ is $\mathcal{O}_Z(r)$, so the degree $d$ of $Z$ is equal to $rs$.

Formula (IV,75) was asserted without proof by J. Steiner (1854). It was given a proof by J. Bischoff (1859) and became associated with his name. Steiner and Bischoff drew remarkable conclusions. For instance, by (IV,75) the variety of conics tangent to a given conic is a hypersurface of degree 6 in $\mathbb{P}^5$. Hence there are $6^5=7776$ conics tangent to 5 general conics by Bezout's theorem; for the conics in question are represented by the points common to 5 hypersurfaces of degree 6. The number 7776 is wrong! L. Cremona (1864) explained why: the degenerate conics consisting of two coincident lines are parametrized by a surface (the Veronese surface) lying in each of the 5 hypersurfaces; so Bezout's theorem does not apply. M. Chasles (1864) was the first to publish the correct answer, 3264. (This answer is correct in every characteristic except 2, where (see [117]) the correct answer is 51.) Chasles's work led to Schubert's and the blossoming of enumerative geometry.

Finally, consider the case of a singular $Z$. We can still define $d^\vee$ as the degree of $Z^\vee$ if $Z^\vee$ is a hypersurface and as 0 if not, and we can still interpret $d^\vee$ as the number of smooth points $z$ of $Z$ such that the hyperplane spanned by $z$ and a fixed general $(m-2)$-plane $A$ is tangent to $Z$ at $z$, etc. (with the same qualifications in positive characteristic as before). Moreover we have formulas for $d^\vee$ in some cases. If $Z$ is a curve, then $d^\vee$ is given by Formula (IV,41) because the embedded tangent line at a smooth point $z$ of $Z$ meets a given $(m-2)$-plane $A$ if and only if the hyperplane spanned by $z$

and $A$ is tangent to $Z$ at $z$. For example it follows that Bischoff's formula (IV,75) has to be replaced by

$$\boxed{d^\vee = s^\vee + s(2r-2)} \tag{IV,76}$$

where $s^\vee$ denotes the class of the plane curve $Z$. (This formula is a special case of a more sophisticated one, known classically; see [101; Beispiel 4, p. 15].)

If $Z$ is a hypersurface, it is evident that $d^\vee$ is given by Formula (IV.47), The derivation above (due to Piene) goes through more generally if $Z$ is a singular hypersurface section of a smooth variety of any dimension and it yields an appropriately more complicated formula. In general, Piene [82; Ch. 2] showed the following: let $g: Z' \to Z$ denote the blowing-up of the jacobian ideal and let $U$ denote the open subset on which $Z$ is smooth; then there exists a locally free sheaf $M$ on $Z'$ such that $M \mid g^{-1}U$ is equal to $g^*((N(1))^\vee \mid U)$, where $N$ is the conormal sheaf of $Z$ in $Y$, and such that there exists a natural map $f': \mathbb{P}(M) \to Z^\vee$. If we could get a hold of the Segre classes of $M$, then we could get a formula for $d^\vee$. In fact then we could get a formula for the class $[\Gamma_w]$ of the polar locus for $2 \leqq w \leqq n+1$, generalizing Formula (IV,48), because it can be proved that $[\Gamma_w]$ is equal to $g_* c_{n-w+2}(P)$ with $P = V_{Z'}/M^\vee$.

## V. The double-point formula

### A. Adjunction

(See [47].) Let $f: X \to Y$ be a finite mapping from a smooth $n$-fold $X$ into a smooth $(n+1)$-fold $Y$ such that the induced mapping $g: X \to f(X)$ is birational. We are going to give a simple, direct proof of the double-point formula (I,37) in this case.

Let $Z$ denote the image $f(X)$; it is a closed subvariety of $Y$. Let $C$ denote the *conductor* on $Z$ of the finite, birational mapping $g: X \to Z$; it is the ideal of $O_Z$ given by the formula,

$$C = \underline{\mathrm{Hom}}_{O_Z}(g_* O_X, O_Z). \tag{V,1}$$

The mapping $g: X \to Z$ is affine; so each quasi-coherent $g_* O_X$-Module $F$ has an associated $O_X$-Module $\tilde{F}$, which is characterized by the relation $g_* \tilde{F} = F$. The associated $O_X$-Module $\tilde{C}$ is the conductor on $X$ of $g$. It is an ideal of $O_X$ and so defines a closed subscheme of $X$. The subscheme's support is the double-point set, because its complement is equal to the preimage $g^{-1}U$ of the largest open subset $U$ of $Z$ such that the restriction $g^{-1}U \to U$ is an isomorphism. Denote this subscheme by $M_2$.

Since $Z$ is an $n$-fold and $Y$ is a smooth $(n+1)$-fold, $Z$ is defined locally on $Y$ by one element. So the normal sheaf $\nu_{Z,Y}$ is invertible; in fact, we have the formula,

$$\nu_{Z,Y} = O_Y(Z) \mid Z. \tag{V,2}$$

Moreover, the invertible $O_Z$-Module,

$$\omega_Z = \nu_{Z,Y} \otimes_{O_Z} (\Omega_Y^{n+1} \mid Z), \tag{V,3}$$

is the dualizing sheaf of $Z$, and so we have the formula,

$$g_*\Omega_X^n = \underline{\mathrm{Hom}}_{O_Z}(g_*O_X, \omega_Z). \tag{V,4}$$

These formulas are part of general Grothendieck duality theory. (See [31; Ch. III, §2, §6] and [2; Prop. (2.4), p. 7, Th. (4.6), p. 14, Th. (5.4), p. 79]. A relatively minor modification of the proof of [2; Th. (5.4), p. 79] will make it yield Formula (V,4).)

Since $\omega_Z$ is invertible, it may be pulled through the $\underline{\mathrm{Hom}}$ in Formula (V,4). Hence, in view of Formula (V,1), we get the formula,

$$g_*\Omega_X^n = C \otimes_{O_Z} \omega_Z. \tag{V,5}$$

Taking the associated $O_X$-Modules of both sides of Formula (V,5), we get the formula,

$$\Omega_X^n = \tilde{C} \otimes_{O_X} g^*\omega_Z. \tag{V,6}$$

Combining Formulas (V,6), (V,3), and (V,2), we get the formula,

$$\boxed{\Omega_X^n = \tilde{C} \otimes_{O_X} f^*O_Y(Z) \otimes f^*\Omega_Y^{n+1}}\ . \tag{V,7}$$

This formula expresses in modern terms the essence of the classical theory of adjoints.

The double-point scheme $M_2$ is a divisor on $X$ because its ideal, the conductor $\tilde{C}$ of $g$, is invertible by virtue of Formula (V,7). Moreover, extracting first Chern classes from both sides of Formula (V,7), we get the relation,

$$-c_1(X) = -[M_2] + f^*[Z] - f^*c_1(Y). \tag{V,8}$$

Now, we have $[Z] = f_*[X]$ because $g: X \to Z$ is birational. The difference $f^*c_1(Y) - c_1(X)$ is equal to $c_1(\nu)$ where $\nu$ is the virtual normal sheaf of $f$, nearly by definition. Thus we obtain the formula,

$$\boxed{[M_2] = f^*f_*[X] - c_1(\nu)}\ , \tag{V,9}$$

which is the double-point formula in the present case.

For example, assume $X$ is a surface and $Y$ is $\mathbb{P}^3$. Let $u$ denote the hyperplane class $c_1(O_Y(1))$, and let $h$ denote its pullback to $X$. As was done classically, let $\mu_0$ denote the degree of the image $Z$ of $X$ in $Y$. Then, since

$g: X \to Z$ is birational, we have $f_*[X] = \mu_0 u$. Hence, we have $\int h^2 = \mu_0$ by the projection formula and Formula (II,11), and we obviously have $f^* f_*[X] = \mu_0 h$. Finally, we have $c_1(Y) = 4u$ by Formula (I,19). Therefore we get the formula,

$$\boxed{[M_2] = \mu_0 h - 4h + c_1(X)}\ , \tag{V,10}$$

from Formula (V,9).

Assume in addition that $Z$ is smooth except for an ordinary double curve $D$ (possibly reducible). As was done classically, let $\varepsilon_0$ denote the degree of $D$. Then we have $2\varepsilon_0 = \int [M_2] \cdot h$ by the projection formula and Formula (II,11) because $M_2$ is (reduced and) a 2-to-1 covering of $D$. Hence we get the formula,

$$\boxed{2\varepsilon_0 = (\mu_0 - 4)\mu_0 + \int c_1(X) \cdot h}\ , \tag{V,11}$$

from Formula (V,10).

For instance, suppose $X$ (or $Z$) is a ruled surface of genus $p$; see chapter III, section B for the definition. We have

$$\begin{aligned} \int c_1(X) \cdot h &= \int \gamma \cdot h + \int (2h^2 - eh) && \text{by (III,19)} \\ &= 2 - 2p + \mu_0 && \text{by (III,16) and (III,18).} \end{aligned} \tag{V,12}$$

Therefore Formula (V,11) becomes in this case

$$\boxed{2\varepsilon_0 = (\mu_0 - 1)(\mu_0 - 2) - 2p}\ , \tag{V,13}$$

which is a classical formula (see [7; p. 165]).

Finally, we put Formula (V,11) in an equivalent classical form; we replace the last term by an expression involving the rank $\mu_1$ of $X$ or $Z$. The rank $\mu_1$ is defined as the degree of the polar locus $\Gamma_3$, which consists of the points $x$ of $X$ such that the embedded tangent plane of $X$ at $x$ passes through a given general point $Q$ of $Y$. Equivalently, $\mu_1$ may be defined as the class of the plane curve $Z_1$ that is obtained by slicing $Z$ by a plane $H$ of $Y$ in general position. Indeed, it is intuitively clear that the set of smooth points $z$ of $Z_1$ such that the tangent line of $Z_1$ at $z$ passes through a general point $Q$ of $H$ is equal to the set $f(\Gamma_3) \cap H$, where $\Gamma_3$ is the polar locus of $X$ with respect to this point $Q$. It is not hard to back up our intuition with a scheme-theoretic analysis involving principal parts.

By hypothesis, $f: X \to Y$ has a finite set of ramification points. So its cuspidal edge is the divisor 0. Hence, since $\mu_1$ is equal to $\int [\Gamma_3] \cdot h$ by definition, $\int c_1(X) \cdot h$ is equal to $3\mu_0 - \mu_1$ by Formula (IV,40). Therefore, Formula (V,11) is equivalent to the following formula:

$$\boxed{2\varepsilon_0 = (\mu_0 - 1)\mu_0 - \mu_1}\ , \qquad (V,14)$$

This is one of the formulas of Salmon, Cayley, Zeuthen, and Noether (see [7; p. 159]).

### B. Projections

(See [47], [55].) Just as linear systems of hyperplane sections are more tangible but are not untypical of arbitrary ones, projections are more tangible but are not untypical of arbitrary mappings. We now explore the double-point theory of projections from centers in general position. We compare and contrast two approaches, originated by Holme [29] and by Peters and Simonis [84]. Johnson in his lovely thesis [40] took the same approach as Peters and Simonis, and he developed the refinements in interpretation and technique necessary for projections with singular sources.

The heart of the theory is the following massive diagram, which is explained in the next few paragraphs:

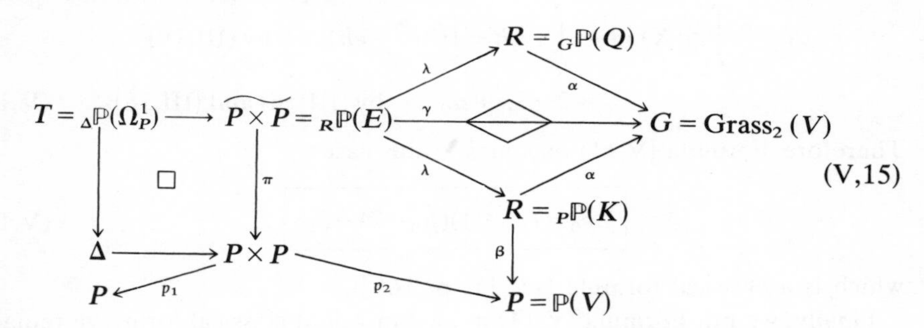

(V,15)

Here $P$ is given as the projective space $\mathbb{P}(V)$. We define $\widetilde{P \times P}$ as the blowup of $P \times P$ along the diagonal $\Delta$, and $T$ as the exceptional divisor. We have $T = \mathbb{P}(\Omega_P^1)$ because $\Omega_P^1$ is the conormal sheaf of $\Delta$ in $P \times P$.

We define $R$ as $\mathbb{P}(K)$, where $K$ is the subbundle in the tautological sequence of $P$,

$$\boxed{0 \to K \to V_P \to L \to 0}\ . \qquad (V,16)$$

The mapping $\lambda : \widetilde{P \times P} \to R$ arises as follows. Off $T$, it is equal to the projection $\phi$ of $\mathbb{P}(V_P) = P \times P$ from the center $\mathbb{P}(L) = \Delta$; the target is $R = \mathbb{P}(K)$ because of the exactness of Sequence (V,16). We may think of $\phi$

as the total mapping of a family of mappings parametrized by $P$; the mapping $\phi_Q$ indexed by a point $Q$ of $P$, is just the projection of $P$ with center $Q$. Now, the blowup $\widetilde{P\times P}$ is the closure in $(P\times P)\times_P R$ of the graph of $\phi$, and the mappings $\pi:\widetilde{P\times P}\to P\times P$ and $\lambda:\widetilde{P\times P}\to R$ are the projections to the two factors.

Intuitively, $P\times P$ parametrizes the pairs $(Q_1, Q_2)$ of points of $P$ but with the pairs of 'infinitely near' points separated. The pairs of infinitely near points fill out the exceptional locus $T=\mathbb{P}(\Omega_P^1)$. A pair of infinitely near points may also be thought of as a tangent vector up to scalar multiple. Now, the fiber of $\beta:R\to P$ over a point $Q$ of $P$ is equal to the target of the projection of $P$ with center $Q$, which is the projective space of lines through $Q$. So $R$ parametrizes the pairs $(Q, \Lambda)$ such that $Q$ is a point of $P$ and $\Lambda$ is a line of $P$ through $Q$.

In the above terms, the mapping $\lambda:\widetilde{P\times P}\to R$ carries a pair $(Q_1, Q_2)$ to the pair $(Q_2, \Lambda)$, where $\Lambda$ is the line determined by $Q_1$ and $Q_2$. This statement is obvious if $Q_1$ is distant from $Q_2$, and it follows by continuity if $Q_1$ is infinitely near $Q_2$. Moreover, the mapping $\lambda$ induces an isomorphism $T\to R$, because a pair $(Q_1, Q_2)$ of infinitely near points determines and is determined by a pair $(Q_2, \Lambda)$ consisting of a point $Q_2$ and a line $\Lambda$ through $Q_2$. (While this reasoning appears only intuitive, it is in fact rigorous if 'point' is interpreted as 'scheme-valued point'.) However, the isomorphism $T\xrightarrow{\sim} R$ does not preserve the tautological sheaves because $K$ is not equal to $\Omega_P^1$ but to $\Omega_P^1\otimes L$.

Consider the following pushout diagram of sheaves on $R$:

$$\begin{array}{ccccccccc} 0 & \to & K_R & \to & V_R & \to & L_R & \to & 0 \\ & & \downarrow & & \downarrow & & \| & & \\ 0 & \to & N & \to & E & \to & L_R & \to & 0 \end{array} \qquad \text{(V,17)}$$

The top line is the pullback of Sequence (V,16), and $N$ denotes the tautological invertible sheaf on $R$. The diagram serves to define the sheaf $E$ and the map $V_R\twoheadrightarrow E$. It is not hard to see that the subscheme $\mathbb{P}(E)$ of $\mathbb{P}(V_R)=P\times R$ is the total space of the family of lines of $P$ parametrized by $R$; in other words, $\mathbb{P}(E)$ parametrizes the triples $(Q_1, Q_2, \Lambda)$ consisting of a pair of points of $P$ and a line containing both of them. It follows that $\mathbb{P}(E)$ is equal to $\widetilde{P\times P}$.

There exists a mapping $\alpha:R\to G=\mathrm{Grass}_2(V)$ such that the relation $E=\alpha^*Q$ holds, where $Q$ denotes the tautological quotient of $V_G$, because $E$ is a rank-2-quotient of $V_R$. Of course, $G$ is also the grassmannian of lines in $P$. Moreover, the mapping $(\beta, \alpha):R\to P\times G$ carries $R$ isomorphically onto the incidence correspondence because, as we observed above, $R$ parametrizes the pairs $(Q, \Lambda)$ consisting of a point $Q$ of $P$ and a line $\Lambda$ through $Q$. Similarly, the mapping $(\lambda, \lambda): \widetilde{P \times P} \to R \times_G R$ is an isomorphism, because $\widetilde{P \times P}$ parametrizes the triples $(Q_1, Q_2, \Lambda)$ consisting of a pair of points of $P$ and a line containing both of them. Finally, we set $\gamma = \alpha \circ \lambda$.

Let $X$ be a smooth, $n$-dimensional, closed subvariety of $P = \mathbb{P}(V)$. Let $W$ be a vector subspace of $V$ with dimension $m+1$. Set $Y = \mathbb{P}(W)$ and $A = \mathbb{P}(V/W)$. Assume that $A$ is in general position with respect to $X$ and that its codimension, $m+1$, is at least $n+1$. Let $f: X \to Y$ denote the projection from $A$. We are going to explore the double-point theory of $f$.

Let $M_2$ denote the double-point set of $f$, whose complement is equal to the preimage $f^{-1}U$ of the largest open subset $U$ of $Y$ such that the restriction $f^{-1}U \to U$ is an embedding. We now prove that a point $x$ of $X$ lies in $M_2$ if and only if there exists a point $y$ of $X$, possibly infinitely near $x$, such that the line $\overline{xy}$, the one determined by $x$ and $y$, meets $A$; in other words, $x$ lies in $M_2$ if and only if $x$ is an end point of a secant, possibly a tangent, meeting $A$.

Indeed, suppose such a point $y$ exists. If $y$ is distant from $x$, then $x$ is a double-point of $f$ because clearly $f(y)$ is equal to $f(x)$. If $y$ is infinitely near $x$, then $\overline{xy}$ is the tangent line to $X$ at $x$ and hence $x$ is a ramification point of $f$ (see chapter IV, section B); so again $x$ is a double-point of $f$. Conversely, suppose $x$ is a double-point of $f$. If $x$ is a ramification point, then the tangent line of $X$ at $x$ meets $A$. If $x$ is not a ramification point, then $f$ is an immersion at $x$ but the scheme-theoretic fiber $f^{-1}f(x)$ is nontrivial; so there exists a point $y$ distant from $x$ such that the secant line $\overline{xy}$ meets $A$.

Let $\widetilde{X \times X}$ denote the proper (or strict) transform of $X \times X$ with respect to $\pi$, that is, the closure in $\widetilde{P \times P}$ of $\pi^{-1}(X \times X - \Delta)$. Viewed in $(P \times P) \times_P R$, clearly $\pi^{-1}(X \times X - \Delta)$ is equal to the graph of the restriction $(X \times X - \Delta) \to \beta^{-1}X$ of the projection $\phi$ of $P \times P$ with center $\Delta$. Hence $\widetilde{X \times X}$ is equal to the blowup of $X \times X$ along $X \times X \cap \Delta$, which is the diagonal $\Delta_X$ of $X \times X$. Thus $\widetilde{X \times X}$ parametrizes the pairs of points of $X$ but with the pairs of infinitely near points separated.

There are two equivalent natural ways to give $M_2$ the structure of a positive cycle $[M_2]$. One way involves the *secant cycle* $\sigma$ on $P \times P$, which is defined by the formula,

$$\boxed{\sigma = \pi_* \lambda^* \lambda_* [\widetilde{X \times X}]}. \tag{V,18}$$

Consider the underlying sets. The mapping $\lambda$ carries a pair $(y, x)$ in $\widetilde{X \times X}$ to the pair $(x, \overline{xy})$. So $\lambda^{-1}\lambda(\widetilde{X \times X})$ consists of the pairs $(Q, x)$ in $\widetilde{P \times P}$ such that $x$ lies in $X$ and $\overline{Qx}$ is a secant of $x$.

The set $\widetilde{X \times X}$ is irreducible and $2n$-dimensional, being the closure of an irreducible, $2n$-dimensional set. So the image $\lambda(\widetilde{X \times X})$ is irreducible. Suppose $X$ is not a linear space. Then some secant of $X$ does not lie entirely in $X$. Hence $\lambda(\widetilde{X \times X})$ is $2n$-dimensional, for some fiber of the mapping from $\widetilde{X \times X}$ to it is finite. Since $\lambda$ is the structure map of the bundle $\mathbb{P}(E)$ of projective lines, $\lambda^{-1}\lambda(\widetilde{X \times X})$ is irreducible and $(2n+1)$-dimensional. Hence $\pi\lambda^{-1}\lambda(\widetilde{X \times X})$ is irreducible, and it is $(2n+1)$-dimensional, because $\lambda^{-1}\lambda(\widetilde{X \times X})$ does not lie entirely within the exceptional locus $T$. Thus $\sigma$ is a positive $(2n+1)$-cycle, whose underlying set is irreducible and consists of the pairs $(Q, x)$ in $P \times X$ such that there exists a secant of $X$ containing both $Q$ and $x$. However, if $X$ is a linear space, then $\sigma$ is the cycle 0.

Define the double-point cycle $[M_2]$ by the formula,

$$\boxed{[M_2] = p_{2*}([A \times P] \cdot \sigma)}. \tag{V,19}$$

If $X$ is a linear space, then $[M_2]$ is 0, but the double-point set $M_2$ is obviously empty.

Suppose $X$ is not a linear space. Then $[A \times P] \cdot \sigma$ is a well-defined $(2n-m)$-cycle because $A$ is in general position. Hence $[M_2]$ is well-defined. Now, no component of $[A \times P] \cdot \sigma$ lies in the diagonal $\Delta$ of $P \times P$; indeed, $A \times P$ intersects any given proper closed subset of the underlying set of $\sigma$ in a set with dimension at most $(2n-m-1)$ because $A$ is in general position. Moreover, $p_2$ induces off $\Delta$ a finite-to-1 mapping on the underlying set of $[A \times P] \cdot \sigma$; in other words, for each point $x$ of $X$ there are at most finitely many points $Q$ of $A$ such that the line $\overline{Qx}$ meets $X$ in a point distant from $x$. Indeed, the linear space $\overline{Ax}$ meets $X$ in a finite set as $A$ does not meet $X$. Hence $p$ carries the underlying set of $[A \times P] \cdot \sigma$ onto the underlying set of $[M_2]$; no components of $[A \times P] \cdot \sigma$ are mapped to 0 under $p_*$. Therefore the underlying set of $[M_2]$ is the set of end points of secants meeting $A$; that is, it is the double-point set $M_2$.

This way of defining the double-point cycle $[M_2]$ is virtually in Holme's fundamental article [29]. Furthermore, Holme computed what amounts to its degree. He did so to get a numerical criterion for $M_2$ to be empty, that is, for $f$ to be an embedding. With this definition of $[M_2]$, however, it is not obvious a priori that its degree – Holme's embedding obstruction – can be expressed as a polynomial in the Chern classes of $X$ and the class, $c_1(O_X(1))$. Moreover, Holme used a local-patching description of $R$, $\beta$, and $\lambda$. (The identification of $R$ with $\mathbb{P}(K)$ and of $\widetilde{P \times P}$ with $\mathbb{P}(E)$ was pointed out in [3; (D9)] and independently in [57].) So Holme's work is all the more impressive. In recent joint work (see [94]), Holme and Roberts investigated the transversality of the intersection $[A \times P] \cdot \sigma$ at a point $(Q, x)$; they proved that if the secant $\overline{Qx}$ is not a tangent and if it meets $X$ in only one other point $y$ besides $x$, then transversality holds if and only if $P$ is spanned by $A$, by the tangent space of $X$ at $x$, and by that at $y$.

The other way of defining the double-point cycle $[M_2]$ involves a certain Schubert variety $\Sigma$ on the grassmannian $G$ of lines in $P$. Namely, $\Sigma$ is the Schubert variety of lines meeting $A$, and $[M_2]$ is defined by the formula,

$$\boxed{[M_2] = p_{2*}\pi_*(\gamma^*[\Sigma] \cdot [\widetilde{X \times X}])}. \tag{V,20}$$

The underlying set $Z_2$ of $\gamma^*[\Sigma] \cdot [\widetilde{X \times X}]$ is given by the formula,

$$\boxed{Z_2 = \gamma^{-1}\Sigma \cap \widetilde{X \times X}}; \tag{V,21}$$

it consists of the pairs $(y, x)$ in $\widetilde{X \times X}$ such that the secant line $\overline{yx}$ meets $A$. The intersection in (V,21) is proper because $\Sigma$ is in general position as $A$ is. Hence $[M_2]$ is well-defined. Note in passing that the set $Z_2$ has a natural scheme structure and that we have

$$\boxed{[Z_2] = \gamma^*[\Sigma] \cdot [\widetilde{X \times X}]} \tag{V,22}$$

because $\gamma$ is flat and because $\Sigma$ is smooth along $\gamma(\widetilde{X \times X})$, so locally a complete intersection there. (In fact, $\Sigma$ is Cohen–Macaulay everywhere.)

No component of $Z_2$ lies in the exceptional locus $T$; in fact, the intersection $\gamma^{-1}\Sigma \cap T$ is proper because $\Sigma$ is in general position. Now, $p_2 \circ \pi$ induces a finite-to-1 mapping on $(Z_2 - T)$; in other words, for each point $x$ of $X$ there are at most finitely many pairs $(y, x)$ in $(\widetilde{X \times X} - T)$ such that the secant line $\overline{yx}$ meets $A$. Indeed, the linear space $\overline{Ax}$ meets $X$ in a finite set as $A$ does not meet $X$. Hence $p_2 \circ \pi(Z_2)$ is equal to the underlying set of $[M_2]$. Therefore the latter is the set of end points of secants meeting $A$; that is, it is the double-point set $M_2$.

The two definitions of $[M_2]$ are equivalent. We now prove this by approaching $P$ via $R$. Using the relation $\beta \circ \lambda = p_2 \circ \pi$ and applying the projection formula with respect to $\pi$ and then to $\lambda$, respectively with respect to $\lambda$, we obtain

$$[M_2] = \beta_*(\lambda_*\pi^*[A \times P] \cdot \lambda_*[\widetilde{X \times X}]) \tag{V,23}$$

$$[M_2] = \beta_*(\alpha^*[\Sigma] \cdot \lambda_*[\widetilde{X \times X}]) \tag{V,24}$$

from (V,19) and (V,20) respectively. It remains to establish the relation,

$$\lambda_*\pi^*[A \times P] = \alpha^*[\Sigma]. \tag{V,25}$$

Since $\alpha$ is flat, $\alpha^*[\Sigma]$ is equal to $[\alpha^{-1}\Sigma]$. Since $\alpha$ is smooth, $\alpha^{-1}\Sigma$ is reduced. On the other hand, $\pi$ is an isomorphism off the diagonal $\Delta$, and $\pi^{-1}(A\times P)$ is clearly equal to the closure of $\pi^{-1}(A\times P-\Delta)$. Hence $\pi^*[A\times P]$ is equal to $[\pi^{-1}(A\times P)]$. Now, $\pi^{-1}(A\times P)$ consists of the pairs $(Q_1, Q_2)$ in $\widetilde{P\times P}$ with $Q_1$ in $A$; while $\alpha^{-1}\Sigma$ consists of the pairs $(Q_2, \Lambda)$ in $R$ such that the line $\Lambda$ contains $Q_2$ and meets $A$. Hence $\lambda$ carries $\pi^{-1}(A\times P)$ onto $\alpha^{-1}\Sigma$, and it induces an isomorphism off $\beta^{-1}A$. Thus (V,25) holds.

The second definition (V,20) of the cycle $[M_2]$ leads naturally to an expression for the cycle's rational equivalence class, denoted $[M_2]$ also. Indeed, we have $[\Sigma]=s_n(Q)$ by Porteous' formula, because $\Sigma$ is the locus on $G$ where the composition $W_G\to V_G\to Q$ has rank at most 1. So, we get the formula,

$$[M_2]=p'_{2*}\pi'_*s_m(\gamma^*Q\,|\,\widetilde{X\times X}), \tag{V,26}$$

where we have let $\pi':\widetilde{X\times X}\to X\times X$ and $p'_i:X\times X\to X$ for $i=1,2$ denote the restrictions of $\pi$ and $p_i$.

Pulling back the bottom row of (V,17), we get the exact sequence,

$$\boxed{0\to I\otimes L_1\to \gamma^*Q\,|\,\widetilde{X\times X}\to L_2\to 0}, \tag{V,27}$$

where $L_i$ denotes the pullback of $O_X(1)$ under $p'_i\circ\pi'$ and $I$ denotes the ideal of the exceptional divisor $T\cap\widetilde{X\times X}$. Indeed, $I$ is the tautological sheaf of $\pi'$. However, the natural embedding of $\widetilde{X\times X}$ in $(P\times P)\times_P R$ is not compatible with the tautological sheaves relative to $P\times P$. Rather, the isomorphic embedding of $\widetilde{X\times X}$ in $\mathbb{P}(F)$ with $F=p_2^*K\otimes p_1^*L^{-1}$ is compatible with them, because it is the natural map $F\to O_{P\times P}$ that has as its image the ideal of the diagonal $\Delta$, (see [3; §5], [124; Prop. (3.2), p. 109]). Therefore, $\lambda^*N\,|\,\widetilde{X\times X}$ is equal to $I\otimes L_1$. Thus the first term in (V,27) is explained.

From Formula (V,26), Sequence (V,27), and the projection formula, we get the formula,

$$\begin{aligned}[M_2]=\text{ the }&(2-m)\text{-dimensional component}\\ &\text{of } s(L\,|\,X)\cdot p'_{2*}\pi'_*s(I\otimes L_1).\end{aligned} \tag{V,28}$$

To compute the direct image of $s(I\otimes L_1)$, we use the formula,

$$s(I\otimes L_1)=s(L_1)\sum_{i=0}^{2n}s(L_1)^ic_1(I)^i; \tag{V,29}$$

it holds because it amounts to the identity,

$$\frac{1}{1-c_1(I)-c_1(L_1)}=\frac{1}{1-c_1(L_1)}\cdot\frac{1}{1-[c_1(I)/(1-c_1(L_1))]}. \tag{V,30}$$

Applying Formula (II,43) and the projection formula, we obtain from (V,29) the formula,

$$p'_{2*}\pi'^{*}s(I \otimes L_1) = d[X] - \sum_{i=1}^{2n} s(L \mid X)^{i+1} s^{2n-i}(X), \qquad \text{(V,31)}$$

where $d$ denotes the degree of $X$. Set $h = c_1(L \mid X)$. Then from Formulas (V,31) and (V,28), we obtain the formula,

$$\boxed{[M_2] = dh^{m-n} - \sum_{i=m}^{m} \binom{m+1}{m-i} h^{m-i} s^{2n-i}(X)} \qquad \text{(V,32)}$$

This formula is, in view of Formulas (I,19) and (II,47), an expanded version of the double-point Formula (I,37) in the present case.

The preceding derivation of Formula (V,32) via Definition (V,20) of $[M_2]$, is nearly in Peters and Simonis' article [84]. In fact, Peters and Simonis considered only the case with $m = 2n$ and $\dim(P) = 2n + 1$, and they computed only the degree of $[M_2]$, which is now a 0-cycle. They noted that $[M_2]$ has no multiple components and that its degree is equal to the (exact) number of secants of $X$ meeting $A$, both holding because of the classical result that each such secant meets $X$ in 2 points and no more. (The result was known only in characteristic 0 at the time [84] was written, but since then Laksov [55; Lemma 15] refined the proof for positive characteristic. Peters and Simonis do not use their hypothesis of characteristic 0 anywhere else.) Also, Peters and Simonis do not go from $G$ to $\widetilde{P \times P}$ via $R$, but establish (V,27) directly on $\widetilde{X \times X}$.

The preceding derivation of Formula (V,32) in full generality is essentially in Johnson's thesis [40]. In fact, Johnson allows $X$ to be singular. The derivation goes through virtually without change in the singular case. However, the result must be interpreted with care. The secants through a fixed point $x$ and all its infinitely near points $y$ – that is, the secants determined by the points on $T \cap \widetilde{X \times X}$ over $x$ – form a cone, which Johnson called the *tangent star*. The tangent star contains the Zariski tangent cone, and it is contained in the Zariski tangent space, but it need not be equal to either one. In particular, the set $M_2 = p_2\pi(Z_2)$, which comes into play, contains every point $x$ such that the fiber $f^{-1}f(x)$ consists of at least two points, but $M_2$ may omit a point $x$ such that the fiber $f^{-1}f(x)$ is nonreduced. (Johnson explained these matters and others, giving examples.) On the other hand, because of this phenomenon, to obtain embedding obstructions in the singular case, Holme [30] worked with both $\widetilde{X \times X}$ and $\mathbb{P}(\Omega_X^1)$.

Consider the ramification locus $R$ of our mapping $f: X \to Y$. From Formulas (I,20), (I,22), (I,19), and (II,47) we obtain the formula,

$$\boxed{[R]=\sum_{i=n}^{m+1}\binom{m+1}{m-i+1}h^{m-i+1}s^{2n-i}(X)}\ . \tag{V,33}$$

However, we can also derive this formula much the same way we derived Formula (V,32). Indeed, $R$ is equal to the polar locus $\Gamma_{m+1}$ of points of $X$ at which the embedded tangent space meets A (see chapter IV, section B). Hence $R$ is set-theoretically equal to the image under $p_2 \circ \pi$ of $T \cap Z_2$. In fact, we have the cycle-theoretic formula,

$$[R]=p_{2*}\pi_*[T\cap Z_2], \tag{V,34}$$

and Formula (V,33) follows easily from it and from what we already know. Formula (V,34) results from the following two observations: (1) each irreducible component of $T\cap Z_2$ has dimension $2n-m-1$, which holds because $A$ is in general position; (2) the fibers of $\widetilde{X\times X}$ and $\gamma^{-1}\Sigma$ over a point on $\Delta\cap X\times X$ are linear subspaces of the fiber of $T=\mathbb{P}(\Omega_P^1)$, which is not hard to see (in fact, $T\cap\widetilde{X\times X}$ is equal to $\mathbb{P}(\Omega_X^1)$, and $T\cap\gamma^{-1}$ is equal to $\mathbb{P}(K/K\cap W_P)$).

The second way of deriving Formula (V,33) is basically due to Johnson [40]. However, while he does observe that the underlying set of the image under $p_{2*}\pi_*$ of $[T\cap Z_2]$ is equal to $R$, he is unconcerned about the definition of the cycle $[R]$ via $F^0\Omega^1_{X/Y}$. On the other hand, Johnson allows $X$ to be singular. Then the image under $p_2\circ\pi$ of $T\cap Z_2$ consists of the points of $X$ at which the tangent star meets $A$. This set can omit points of $X$ at which the Zariski tangent space meets $A$, so it corresponds to a new notion of ramification and immersion. Unfortunately, the set's components need not have the appropriate dimension, $2n-m$. The set itself need not be the underlying set of the cycle $[R]$ defined by (V,34). Johnson gives a simple example in which the set is nonempty but $[R]$ is 0. While Formula (V,33) holds in such a case, it is not so useful.

Finally, some remarkable results come from considering various values of $m$ simultaneously. To avoid confusion, let $f_m: X\to\mathbb{P}^m$ denote the mapping, let $[M_{2,m}]$ denote the double-point cycle, and let $[R_m]$ denote the ramification cycle. From Formulas (V,33) and (V,32), we obtain formally the following interesting relation:

$$\boxed{h\cdot[M_{2,m}]-[M_{2,m+1}]=[R_m]}\ . \tag{V,35}$$

We will obtain two consequences.

The first consequence of Formula (V,35) is this: if (1) $f_{2n}: X\to\mathbb{P}^{2n}$ is an embedding and (2) $f_m: X\to\mathbb{P}^m$ is an immersion for an $m$ at most $2n$, then $f_m$

is in fact an embedding. Indeed, (2) clearly implies that $f_m, \cdots, f_{2n}$ are all immersions. Hence we have

$$[R_m]=0, \cdots, [R_{2n}]=0, [M_{2,2n}]=0. \tag{V,36}$$

Therefore, Formula (V,35) implies that $h^{2n-m}[M_{2,m}]$ is equal to 0. Consequently, $M_{2,m}$ is empty because $X$ is projective, and so $f_m$ is an immersion.

It seems unknown whether Hypothesis (1) is necessary, whether $f_m$ for $m<2n$ is an embedding whenever it is an immersion. The above consequence of Formula (V,35), as well as the formula itself, is a central theme of Johnson's thesis [40]. Johnson, of course, allows $X$ to be singular, observing that the consequence holds for 'embedding' and 'immersion' in the traditional sense and in the sense of the tangent star; in fact, the latter case is used to obtain the former.

Another consequence of Formula (V,35), together with Formula (V,32) for $m=n$, is the following formula:

$$\boxed{[M_{2,m}]=(d-1)h^{m-n}-\sum_{i=1}^{m-n} h^{i-1}[R_{m-i}]} \;. \tag{V,37}$$

Multiplying both sides by $h^{2n-m}$ and extracting degrees, we obtain the following 1902 result of Severi (up to notation, as quoted in [102; p. 925]):

$$\boxed{2\varepsilon_{0,m}=\mu_0(\mu_0-1)-\sum_{j=1}^{m-n}\mu_j} \;. \tag{V,38}$$

Here $\varepsilon_{0,m}$ denotes half the degree of $h^{2n-m}[M_{2,m}]$. Here $\mu_0$ denotes the degree $d$ of $X$, and $\mu_j$ for $j \geqq 1$ denotes the degree of the $j$-codimensional polar locus $\Gamma_{n+j}$, which parametrizes the points at which the tangent space meets an $(n+j)$-codimensional linear space in general position. For example, Formula (V,38) with $n=2$ and $m=3$ reduces to Formula (V,14) for $f_3$.

Alternately, $\varepsilon_{0,m}$ can be interpreted as the degree of the double locus $f_m(M_{2,m})$ of $Z=f_m(X)$. (This interpretation could conceivably fail in positive characteristic because it is not known whether then $Z_2$ and so $[M_2]$ have multiple components, although at any rate $M_2$ is set-theoretically a 2-to-1 cover of its image over a dense open subset.) Alternately $\mu_j$ can be interpreted as an extrinsic character of the $j$-dimensional slice $Z_j$ of $Z$ by an $(n-j)$-codimensional linear space $H$ of $\mathbb{P}^m$ in general position; $\mu_j$ is equal to the number of smooth points $z$ of $Z_j$ such that the tangent space to $Z_j$ at $z$ meets a general $2j$-codimensional linear space in $H$. (In positive characteristic $p$, the points $z$ may have to be counted with multiplicity $p^r$ for some $r$.)

## C. General theory

We are going to follow Laksov's path (see [56], [57], [13]) from the double-point theory of projections to general double-point theory, cleaning it up a little perhaps. The first steps were inspired by Schwarzenberger's theory [101] of the 'secant bundle'.

Let $X$ be a smooth $n$-fold. Let $p_i: X \times X \to X$ for $i = 1, 2$ denote the projections. Let $\Delta_X$ denote the diagonal of $X \times X$. Let $\pi: \widetilde{X \times X} \to X \times X$ denote the blowing-up along $\Delta_X$. Let $T$ denote the exceptional divisor and $I$ its ideal.

We have thought informally of $\widetilde{X \times X}$ as parametrizing the pairs of points of $X$ but with the pairs of infinitely near points separated. We now formalize this view by constructing an appropriate flat family of length-2 subschemes of $X$ parametrized by $\widetilde{X \times X}$. Let $B_i$ for $i = 1, 2$ denote the graph of the composition $p_i \circ \pi$. Let $B$ denote the scheme-theoretic union of $B_1$ and $B_2$ in $X \times \widetilde{X \times X}$. Then $B$ is a natural candidate for the total space of our family; we have only to prove it is flat over $\widetilde{X \times X}$.

Let $q_1$ and $q_2$ denote the projections of $X \times \widetilde{X \times X}$ onto $X$ and $\widetilde{X \times X}$. Under $q_2$, each $B_i$ obviously maps isomorphically onto $\widetilde{X \times X}$. Moreover, the subscheme $B_1 \cap B_2$ of $B_1$ (or of $B_2$) maps isomorphically onto the subscheme $T$ of $\widetilde{X \times X}$. The second statement is obvious on the level of points. In fact, it is obvious on the level of scheme-valued points; hence it holds scheme-theoretically. Therefore, the ideal of $B_1 \cap B_2$ in $B_1$ is equal to the pullback $I_{B_1}$.

There is a basic short exact sequence on $B$,

$$0 \to I_{B_1} \to O_B \to O_{B_2} \to 0. \tag{V,39}$$

It comes from identifying the ideal $J_2$ of $B_2$ in $B$. Indeed, let $K_i$ denote the ideal of $B_i$ in $X \times \widetilde{X \times X}$. Then we have

$$K_2/(K_1 \cap K_2) = (K_1 + K_2)/K_1. \tag{V,40}$$

In other words, $J_2$ is equal to the ideal of $B_1 \cap B_2$ in $B_1$. So $J_2$ is equal to $I_{B_1}$.

The flatness of $B$ over $\widetilde{X \times X}$ follows because of Sequence (V,39). Indeed, $B_1$ and $B_2$ are trivially flat, and $I$ is locally free (of rank 1).

Suppose $X$ is a subvariety of a projective space $P = \mathbb{P}(V)$. Let $L$ denote $O_X(1)$, and let $L_i$ denote $\pi^* p_i^* L$. The tautological surjection $V_X \to L$ induces a surjection $V_B \to (q_1^* L) \mid B$. Consider the latter's adjoint,

$$V_{\widetilde{X \times X}} \to E = q_{2*}((q_1^* L) \mid B). \tag{V,41}$$

Its kernel is a sheaf whose fibers are subspaces of $V$; at a point $\zeta$ of $\widetilde{X \times X}$, the subspace consists of the linear forms that vanish on the subscheme $B(\zeta)$

of $X$, viewed in $P$. Thus the subspace $\mathbb{P}(E)$ of $P\times\widetilde{X\times X}$ is the total space of the family of secant lines of $X$. In other words, the map (V,41) corresponds to the natural mapping of $\widetilde{X\times X}$ into $G$, the grassmannian of lines in $P$. Finally, consider the sequence,

$$\boxed{0\to I\otimes L_1\to E\to L_2\to 0}, \tag{V,42}$$

obtained by tensoring Sequence (V,39) with $q_1^*L$ and applying $q_{2*}$; it is not hard to see that it is equal to Sequence (V,27).

Let $W$ be a general subspace of $V$, set $m=\dim(W)-1$, and assume $m\geqq n$. Set $Y=\mathbb{P}(W)$ and $A=\mathbb{P}(V/W)$. Let $f:X\to Y$ denote the projection from $A$. Consider the composite map,

$$w: W_{\widetilde{X\times X}}\to V_{\widetilde{X\times X}}\to E. \tag{V,43}$$

Clearly, the locus $Z_2$ on $\widetilde{X\times X}$ of points at which $w$ has rank at most 1 parametrizes the secants meeting $A$; in other words, $Z_2$ is equal to the subscheme defined by Formula (V,21). By Porteous' formula, we have $[Z_2]=s_m(E)$, and using Sequence (V,42) and computing formally as in the preceding section, we arrive again at Formula (V,32) for the double-point class $[M_2]=p_{2*}\pi_*[Z_2]$.

At this point Laksov noticed that $P=\mathbb{P}(V)$ is an unnecessary part of the present approach. All that is needed is a suitable proper mapping $f$ from $X$ into $Y=\mathbb{P}(W)$; it need not be a projection. We can define $L$ as $f^*O_Y(1)$ and we can define $w: W_B\to E$ as the adjoint of the map $W_B\to q_1^*L\mid B$ induced by the tautological surjection $W_X\to L$. If $f$ is a projection, then these $L$ and $w$ are equal to the ones defined before. In general, we can define $Z_2$ on $\widetilde{X\times X}$ as the locus of points at which $w$ has rank at most 1. Assuming each component of $Z_2$ has dimension $2n-m$, we can compute formally as before and obtain a formula exactly like (V,32) for the class $[M_2]=p_{2*}\pi_*[Z_2]$.

Next, Laksov gave an important new description of $Z_2$. It comes from the following commutative diagram:

$$\begin{array}{ccccccc} 0\to & \pi^*p_2^*K & \to & W_{\widetilde{X\times X}} & \to & L_2 & \to 0\\ & \downarrow{\scriptstyle k} & & \downarrow{\scriptstyle w} & & =\downarrow{\scriptstyle id} & \\ 0\to & I\otimes L_1 & \to & E & \to & L_2 & \to 0. \end{array} \tag{V,44}$$

The top line is the pullback under $p_2\circ\pi$ of the tautological sequence on $Y$. The bottom line is Sequence (V,42). The right hand square obviously commutes. The map $k$ is defined by the diagram. Now, obviously $w$ has rank at most 1 at a point if and only if $k$ has rank at most 0 there. Hence $Z_2$ is the zero locus of $k$. Set

$$F=L_1\otimes\pi^*p_2^*K^\vee. \tag{V,45}$$

Then $Z_2$ is the zero locus of a certain section $s$ of $I \otimes F$.

The sheaf $F$ is the pullback under $(f \times f) \circ \pi$ of the sheaf

$$\Phi = r_1^* O_Y(1) \otimes r_2^* K^\vee \tag{V,46}$$

on $Y \times Y$, where $r_1$ and $r_2$ are the first and second projections of $Y \times Y$. The sheaf $\Phi$ has a natural section $\sigma$, corresponding to the natural composite map,

$$u : r_2^* K \to W_{Y \times Y} \to r_1^* O_Y(1). \tag{V,47}$$

The pullback $(f \times f)^* \pi^* u$ is clearly equal to the composition of $k$ and the inclusion map of $I \otimes L_1$ into $L_1$. Hence the section $(f \times f)^* \pi^* \sigma$ of $F$ arises from a section of $I \otimes F$, namely, $s$.

The zero locus of the section $\sigma$ of $\Phi$ is obviously equal to the zero locus of the map $u$. The latter locus is easily seen, using scheme-valued points, to be equal to the diagonal $\Delta_Y$ of $Y \times Y$. Therefore, the normal sheaf of $\Delta_Y$ is equal to the restriction $\Phi | \Delta_Y$. Hence we obtain $\tau_Y = K^\vee(1)$, a relation we already know from Diagram (IV,17) in the theory of principal parts. (The proof here works for grassmannians, and it was presented for them by Porteous [85; Prop. 02, p. 292].)

Assume each component of $Z_2$ has dimension $2n - m$. Then $[Z_2]$ is equal to $c_m(I \otimes F)$. Hence by formula (II,22) we get

$$[Z_2] = \sum_{i=0}^{m} c_{m-i}(F) c_1(I)^i. \tag{V,48}$$

Let $\alpha : \Delta_X \to X \times X$ denote the inclusion mapping. By Formula (II,43) and the projection formula, we get

$$\pi_*[Z_2] = (f \times f)^* c_m(\Phi) - \sum_{i=1}^{m} \alpha_*[c_{m-i}((f \times f)^* \Phi | \Delta_X) s^{2n-i}(\Delta_X, X \times X)]. \tag{V,49}$$

Since we have $\Phi | \Delta_Y = \Omega_Y^1$, setting $[M_2] = p_{2*} \pi_* [Z_2]$ we get

$$[M_2] = p_{2*}(f \times f)^* c_m(\Phi) - \sum_{i=n}^{m} f^* c_{m-i}(Y) s^{2n-i}(X). \tag{V,50}$$

This formula is much like the general double-point Formula (I,37). We have to identify the first term on the right. Using Formulas (V,46) and (II,22) and the projection formula, we find that the term is equal to $df^* c_{m-n}(K^\vee)$. Using the tautological sequence on $Y$, we finally get $dh^{m-n}$.

Laksov observed that the theory now generalizes naturally as follows. Let $f : X \to Y$ be a proper mapping into any smooth $m$-fold $Y$ such that the diagonal of $Y \times Y$ is equal to the zero locus of a section $\sigma$ of a locally free sheaf $\Phi$ with rank $m$. Let $F$ denote the pullback of $\Phi$ under $(f\times f)\circ\pi$. Then the pullback of $\sigma$ is a section of $F$, which vanishes along $T=\pi^{-1}\Delta_X$; so it arises from a section $s$ of $I\otimes F$. Define $Z_2$ as the zero locus of $s$. Assuming that each component has dimension $2n-m$ and proceeding as above, we arrive formally at a formula exactly like (V,50). Again, we have to identify the first term on the right.

We have $c_m(\Phi)=[\Delta_Y]$. So we have to prove the formula,

$$\boxed{p_{2*}(f\times f)^*[\Delta_Y]=f^*f_*[X]}\ . \tag{V,51}$$

Let $\Gamma$ denote the graph of $f$, and consider the following diagram:

$$\begin{array}{ccccc} X\times X & \xrightarrow{1\times f} & X\times Y & \xrightarrow{f\times 1} & Y\times Y \\ \downarrow{\scriptstyle p_2} & \square & \downarrow{\scriptstyle p_Y} & & \\ X & \xrightarrow{f} & Y. & & \end{array} \tag{V,52}$$

Since the projection $p_Y$ is flat, and the square is cartesian, the following relation holds (see the proof of Lemma (4) in [13; §2.2]):

$$p_{2*}(1\times f)^*[\Gamma]=f^*p_{Y*}[\Gamma]. \tag{V,53}$$

Now, we have $\Gamma=(f\times 1)^{-1}\Delta_Y$. Hence we get $[\Gamma]=(f\times 1)^*[\Delta_Y]$ because $f\times 1$ is flat. Therefore the left sides of (V,53) and (V,51) are equal. Finally, the restriction $\Gamma\to Y$ of $p_Y$ is equal to $f:X\to Y$. Hence the right sides of (V,53) and (V,51) are equal too. Of course, this proof (found in [57]) does not involve $\Phi$; it establishes Formula (V,51) for any smooth $Y$.

It now seems certain that the general double-point theorem holds, and Laksov saw how to get it about as follows. Let, from now on, $f:X\to Y$ be a proper mapping into any smooth $m$-fold $Y$. The inclusion mapping of subschemes of $X\times X$,

$$\Delta_X\to(f\times f)^{-1}\Delta_Y=X\times_Y X, \tag{V,54}$$

lifts back to an inclusion mapping of subschemes of $\widetilde{X\times X}$,

$$T=\pi^{-1}\Delta_X\to\pi^{-1}(f\times f)^{-1}\Delta_Y. \tag{V,55}$$

In other words, letting $J$ denote the ideal of $\pi^{-1}(f\times f)^{-1}\Delta_Y$, we have $J\subset I$. Hence $J\otimes I^{-1}$ is an ideal. Define $Z_2$ as the corresponding subscheme of $\widetilde{X\times X}$. Beware, $Z_2$ need not be the closure of $(X\times_Y X)-T$. It will have components in $T$ if the ramification locus $R$ of $f$ has dimension $2n-m$ or greater, because of Relation (V,57) below; for example, it will if $f$ arises by desingularizing a plane curve with cusps. This definition of $Z_2$ clearly generalizes the preceding one and so all the earlier ones.

Clearly $Z_2$ is locally defined by $m$ equations because $\Delta_Y$ is. Hence each irreducible component of $Z_2$ has dimension at least $2n-m$. Suppose each has just dimension $2n-m$. Then, to obtain a formula for $[M_2]=p_{2*}\pi_*[Z_2]$, we use the following formula in place of Formula (V,48):

$$\boxed{[Z_2]=\pi^*(f\times f)^*[\Delta_Y]-\alpha'_*\sum_{i=1}^{m} h^*c_{m-i}(\tau_Y)c_1(I\mid T)^{i-1}}\;, \qquad \text{(V,56)}$$

where $\alpha':T\to\widetilde{X\times X}$ denotes the inclusion mapping and $h:T\to Y$ denotes the mapping induced by $\pi$ and $f$. Applying Formulas (V,51), (II,42), and (II,47) and the projection formula, we obtain a formula for $[M_2]$ exactly like the double-point formula (I,37).

A 'residual-intersection' formula like (V,56) holds whenever we deal with a mapping into a smooth variety and with a smooth subvariety whose pullback decomposes into two pieces, one of the appropriate dimension and the other a divisor. (The term $\tau_Y$ in (V,56) is replaced by the normal bundle of the subvariety, and the other terms by their direct analogues.) Such a formula was obtained by Todd (1940) together with the double-point formula, each one from the other by induction on the dimension of the source. Laksov [57] proved the formula by covering the given mapping by the mapping between the blowups of the given source and target and by using the so-called 'formule clef' (which is Grothendieck's (1957) key to describing the intersection ring of a blowup). Laksov ran into technical difficulties and had to assume that the divisorial part contains no component of the other part. However, these difficulties were overcome by Fulton [14] (see also [16]).

Consider the set $\pi(Z_2)$ in $X\times X$. Clearly it contains a pair $(x, y)$ with $x\neq y$ if and only if we have $f(x)=f(y)$. We now prove (following [57; §4]) that it contains a pair $(x, x)$ if and only if $x$ lies in the ramification locus $R$ of $f$; in other words, the following set-theoretic relation holds:

$$\boxed{p_2\pi(Z_2\cap T)=R}\;. \qquad \text{(V,57)}$$

Let $\zeta$ be a point of $T$. Nearly by definition, $\zeta$ lies in $Z_2$ if and only if the inclusion map $u:J\to I$ is not surjective at $\zeta$. Hence, by Nakayama's lemma, $\zeta$ lies in $Z_2$ if and only if the restriction $u\mid T$ is not surjective at $\zeta$. Therefore, it remains to prove that $x$ lies in $R$ if and only if there exists a point $\zeta$ of $T$ lying over $x$ at which $u\mid T$ is not surjective.

Consider the natural diagram on $\widetilde{X \times X}$,

$$\begin{array}{ccc} \pi^*(f\times f)^* J_Y & \longrightarrow & \pi^* I_X \\ \downarrow & & \downarrow \\ J & \xrightarrow{u} & I, \end{array} \tag{V,58}$$

in which $J_Y$ and $I_X$ denote the ideals of $\Delta_Y$ and $\Delta_X$. Pulling the diagram back to $T$ and identifying the terms, we get the diagram,

$$\begin{array}{ccc} p^*f^*\Omega^1_Y & \xrightarrow{p^*\partial f} & p^*\Omega^1_X \\ \downarrow & & \downarrow \\ J \mid T & \xrightarrow{u \mid T} & O_T(1), \end{array} \tag{V,59}$$

in which $p: T \to X$ denotes the natural mapping. Since the right vertical map is the tautological surjection, it follows that $x$ lies in $R$, that is, $\partial f$ is not surjective at $x$, if and only if there exists a point $\zeta$ of $T$ lying over $x$ at which $u \mid T$ is not surjective.

A closer look reveals that $Z_2 \cap T$ is the scheme-theoretic zero locus of the map $p^*f^*\Omega^1_Y \to O_Y(1)$. Hence, on the one hand, $Z_2 \cap T$ is the scheme-theoretic zero locus of a section of $(p^*f^*\tau_Y)(1)$, and on the other hand, since $\Omega^1_{X/Y}$ is the cokernel of $\partial f$, we have the following scheme-theoretic relation:

$$\boxed{Z_2 \cap Y = \mathbb{P}(\Omega^1_{X/Y})}\,. \tag{V,60}$$

Suppose that each irreducible component of $Z_2$ has dimension $2n-m$ and that the intersection $Z_2 \cap T$ is proper. Then (V,60) implies that each irreducible component of $R$ has dimension $2n-m-1$ and that we have the following cycle-theoretic formula:

$$\boxed{p_{2*}\pi_*[Z_2 \cap T] = [R]}\,. \tag{V,61}$$

Then also, we have the following class-theoretic formula:

$$[Z_2 \cap T] = c_m((p^*f^*\tau_Y)(1)) = \sum_{i=0}^{m} c_{m-i}(Y)c_1(O_T(1))^i. \tag{V,62}$$

Formula (V,62) can also be derived from Formula (V,56). Formulas (V,62) and (V,61) yield Formula (I,20) for the class $[R]$. While this derivation of Formula (I,20) involves more stringent hypotheses than the one in chapter III, section A using Porteous' formula, it has the advantage of generalizing à la Johnson [40] to the case that $X$ is singular.

Each irreducible component of $\pi(Z_2)$ has dimension at least $2n-m$. This conclusion is obvious for those components not contained in $\Delta_X$. For those that are, it now follows from a nontrivial result of Artin and Nagata [4; Th. (5.1), p. 318] (as Fulton [14; (2.3)] pointed out).

The set $p_{2*}\pi_*(Z_2)$ consists of the ramification points of $f$ plus the points $x$ for which there exists a point $y$ distinct (or distant) from $x$ satisfying $f(y)=f(x)$; in other words, it is the double-point set $M_2$. Therefore, each irreducible component of $M_2$ has dimension at least $2n-m$. This conclusion holds for the components contained in the ramification locus $R$ because $p_2$ carries $\Delta_X$ isomorphically onto $X$. It holds for the components not contained in $R$ for the following reason: We may assume $Z=f(X)$ has dimension $n$; for otherwise every fiber of $f$ has dimension at least 1 and then $M_2$ is all of $X$. Let $x$ and $y$ be distinct (or distant) points of $X$ with the same image $z$ under $f$. Then correspondingly $Z$ has two analytic components $Z_x$ and $Z_y$ at $z$, and each component has dimension $n$. Hence $f^{-1}Z_y$ is an analytic subscheme of $X$ at $x$ with codimension at least $m-n$. Clearly $f^{-1}Z_y$ lies in $M_2$.

If each irreducible component of $Z_2$ has dimension $2n-m$, it follows that each irreducible component of $M_2$ has dimension $2n-m$ and that the underlying set of $p_{2*}\pi_*[Z_2]$ is indeed $M_2$. Thus, the following double-point theorem has been proved: if $f$ is appropriately generic in the sense that each irreducible component of $Z_2$ has dimension $2n-m$, or equivalently, in the sense that $Z_2$ has dimension at most $2n-m$, then the double-point set $M_2$ is the support of a positive cycle, namely, $p_{2*}\pi_*[Z_2]$ whose rational equivalence class $[M_2]$ is given by the double-point formula (I,37).

It is unknown whether the double-point set $M_2$ has a natural scheme structure such that the associated cycle is equal to $p_{2*}\pi_*[Z_2]$. If $f$ is a general projection and if $m \geqq n+1$, then $p_{2*}\pi_*[Z_2]$ has no multiple components (see [56; Prop. 19], [94]) (in positive characteristic, this statement is not proved in all cases); hence $M_2$ may be given the reduced structure. In the case studied in section A ($f$ finite, $X \to f(X)$ birational, $m=n+1$) it is morally certain that the conductor provides such a scheme structure (it is easy to see that $Z_2$ has dimension at most $2n-m$). Fulton [14; Th. 3] showed that the conductor works if $n=1$, but no other case is proved.

The complement of the double-point set $M_2$ is equal to the preimage $f^{-1}U$ of the largest open subset $U$ of $Y$ such that the restriction $f^{-1}U \to U$ is an embedding. ($X-M_2$ is saturated and $f \mid X-M_2$ is a proper, injective immersion; so it is finite and has trivial scheme-theoretic fibers.) Hence, by virtue of the double-point theorem, if $f$ is appropriately generic, then a necessary and sufficient condition for $f$ to be an embedding is that we have

$$\boxed{f^*f_*[X]=c_{m-n}(\nu)}\,, \tag{V,62}$$

where $\nu$ is the virtual normal bundle of $f$. If also $X$ is projective, then Formula (V,62) may be replaced by the weaker numerical relation obtained by multiplying both sides of (V,62) by $h^{2n-m}$ and extracting degrees, where $h$ denotes the class of a hyperplane section of $X$ for any embedding $X$ into a projective space.

Suppose $f$ is an embedding. Then Formula (V,62) holds and is known as the *self-intersection formula.* It was first established modulo torsion by Grothendieck (1957) and in general (according to [40; p. 15]) by Mumford (1958, unpublished). (It is proved, along with the formule clef, in [69] and [125].) Double-point theory provides the following converse: if $f$ is appropriately generic – for example, if $f$ arises by projecting from a center in general position, or if $f$ is finite, $X \to f(X)$ is birational, and $m = n+1$ – and if the self-intersection formula (V,62) holds – the weaker numerical form is sufficient if $f$ is projective – then $f$ is an embedding.

## D. Higher-order theory

Higher-order multiple-point theory is not yet well-developed, but we are going to see how it looks at present.[4] We freely use notation and results from section C. Fix a proper mapping $f: X \to Y$ from a smooth $n$-fold $X$ into a smooth $m$-fold $Y$.

Consider the following composition, in which $\theta$ is the inclusion mapping:

$$\boxed{f_1: Z_2 \xrightarrow{\theta} \widetilde{X \times X} \xrightarrow{p_2 \circ \pi} X}. \tag{V,63}$$

Since $p_2$ is proper because $f$ is, $f_1$ is proper. Now, a point of $Z_2$ is a pair $(x_2, x_1)$ of points of $X$, with possibly $x_2$ infinitely near $x_1$, satisfying $f(x_2) = f(x_1)$. The pair $(x_2, x_1)$ is a double-point of the mapping $f_1$ if and only if there is a second pair $(x_3, x_4)$ of $Z_2$, possibly infinitely near $(x_2, x_1)$, satisfying $x_4 = x_1$. So $(x_2, x_1)$ is a double-point of $f_1$ if and only if there is a point $x_3$ of $X$, possibly infinitely near $x_1$ or $x_2$ or both, such that $x_3$, as well as $x_2$, lies in the fiber $f^{-1}f(x_1)$. With this intuitive picture in mind, we define the *triple-point set* $M_3$ of $f$ as the image under $f_1$ of the double-point set of $f_1$.

The dimension of $M_3$ is at least $3n-2m$, the appropriate figure, at a point $x_1$ for which there are two other points $x_3$, $x_2$ distinct (or distant) from $x_1$ and from each other such that $f(x_3)$ and $f(x_2)$ are equal to $z = f(x_1)$. Indeed, we may assume that $Z = f(X)$ has dimension $n$; for, otherwise, every fiber of $f$ has dimension at least 1, and so $M_3$ is all of $X$. Then, corresponding to $x_i$ there is an analytic component $Z^i$ of $Z$ at $z$; the three components are

[4] More progress has now been made by the author, see [126].

distinct and each has dimension $n$. Then $Z^3 \cap Z^2$ has codimension in $Y$ at most $2(m-n)$. Hence its preimage under $f$ has codimension at most $2(m-n)$. However, the preimage obviously lies in $M_3$. Thus $M_3$ has dimension at least $3n-2m$ at $x_1$. Conceivably, each irreducible component of $M_3$ always has dimension at least $3n-2m$.

To get a triple-point theorem, we are going to apply the double-point theorem to $f_1$ and push down to $X$. So suppose $Z_2$ is $(2n-m)$-dimensional and reduced. If $Z_2$ is smooth and irreducible, we may apply the theory of section C as is. If $Z_2$ is smooth but reducible, we must first note that the hypothesis of irreducibility in section C was unnecessary. If $Z_2$ is singular and irreducible or not, we must observe (with Fulton and Laksov [16]) that the theory extends to this case virtually without change in the fashion of Johnson [40].

We have to compute the $(m-n)$th Chern class of the virtual normal sheaf $\nu_1$ of $f_1$. The first step is to find the normal sheaf $\nu_\theta$ of $Z_2$ in $\widetilde{X \times X}$. Suppose first that the diagonal of $Y \times Y$ is the zero locus of a section of a locally free sheaf $\Phi$ with rank $m$, and let $F$ denote the pullback of $\Phi$ under $(f \times f) \circ \pi$. Then $Z_2$ is the zero locus of a section of $I \otimes F$. Hence $\nu_\theta$ is equal to the restriction $I \otimes F \mid Z_2$. Now, $F \mid Z_2$ is obviously equal to $f_1^* f^*(\Phi \mid \Delta_Y)$. Moreover, $\phi \mid \Delta_Y$ is equal to the normal bundle of $\Delta_Y$, and the latter is equal to the tangent sheaf $\tau_Y$. Hence the formula,

$$\boxed{\nu_\theta = I \otimes f_1^* f^* \tau_Y}\ , \tag{V,64}$$

follows directly.

Formula (V,64) holds in general. Indeed, there is a natural map from $\nu_\theta$ into $f_1^* f^* \tau_Y$; it is the dual of the map induced by the natural map from the pullback of the ideal of $\Delta_Y$ into the ideal of $Z_2$. So it suffices to check Formula (V,64) locally. Now, $\Delta_Y$ is locally a complete intersection; so locally on $Y \times Y$ it is the zero locus of a section of such a sheaf $\Phi$. Although a (Zariski) open set of $Y \times Y$ need not contain a product of two open sets of $Y$, the preceding analysis goes through yielding the formula locally.

If $Z_2$ is smooth, then the conormal sheaf-cotangent sheaves sequence yields the formula,

$$s(Z_2) = c(\nu_\theta)\theta^* s(\widetilde{X \times X}). \tag{V,65}$$

If $Z_2$ is singular, the formula still holds; indeed, a general such formula for the Segre classes of a local complete intersection in a smooth variety was established by Fulton and Johnson (see [15]).[5]

[5] This is false! (V,65) can fail, and Fulton and Johnson never claimed otherwise. However, (V,80) can still be obtained this way via a double-point formula, different from Johnson's and valid for local complete intersection maps, such as $f_1$. See [126].

Hence the formula,

$$\boxed{c(\nu_1)=f_1^*c(X)\cdot c(\nu_\theta)\cdot \theta^*s(\widetilde{X\times X})}\,, \tag{V,66}$$

follows from the Definition (I,22) and Formula (II,47).

Since $\widetilde{X\times X}$ is the blowup of $X\times X$ along $\Delta_X$, there is an exact sequence,

$$0\to \tau_{\widetilde{X\times X}}\to \pi^*\tau_{X\times X}\to \alpha'_*\tau\to 0, \tag{V,67}$$

in which $\tau$ is the sheaf on $T=\mathbb{P}(\Omega_X^1)$ that appears in the following form of the tautological sequence (see [85; Th. 1, p. 122]):

$$0\to O_T(-1)\to p^*\tau_X\to\tau\to 0. \tag{V,68}$$

Since $\tau_{X\times X}$ is equal to the direct sum of $p_1^*\tau_X$ and $p_2^*\tau_X$, Sequence (V,67) yields the following formula:

$$\boxed{s(\widetilde{X\times X})=c(\alpha'_*\tau)\cdot\pi^*p_1^*s(X)\cdot\pi^*p_2^*s(X)}\,. \tag{V,69}$$

The term $c(\alpha'_*\tau)$ can be evaluated using the Riemann–Roch theorem (see [103; XIV, 3]).

Consider the case $m=n+1$. Since $\tau$ is locally free with rank $n-1$ in view of Sequence (V,68), it follows that we have

$$c_1(\alpha'_*\tau)=(n-1)[T]. \tag{V,70}$$

From Formulas (V,64) and (II,22) we get

$$c_1(\nu_\theta)=f_1^*f^*c_1(Y)-m\theta^*[T]. \tag{V,71}$$

From Formula (V,66) and the last three formulas we get

$$c_1(\nu_1)=f_1^*f^*c_1(Y)-2\theta^*[T]-\theta^*\pi^*p_1^*c_1(X); \tag{V,72}$$

for, we have $f_1=p_2\pi\theta$ and $c(X)s(X)=1$ by Definition (V,63) and by Formula (II,45), and so the potential first and last terms cancel.

We have to compute $f_{1*}c_1(\nu_1)$. Obviously $f_{1*}[Z_2]$ is equal to $[M_2]$. Hence the projection formula yields

$$f_{1*}f_1^*f^*c_1(Y)=[M_2]\cdot f^*c_1(Y). \tag{V,73}$$

By the same token we get the following formula, which we also need:

$$f_{1*}f_1^*f_{1*}[Z_2]=[M_2]^2. \tag{V,74}$$

Now, $\theta_*\theta^*[T]$ is equal to $[Z_2]\cdot[T]$ by the projection formula. Hence, if the intersection $Z_2\cap T$ is proper, then Formula (V,62) implies that $f_{1*}\theta^*[T]$ is equal to $[R]$, and so by Formula (I,20) we have

$$f_{1*}\theta^*[T]=c_2(\nu) \tag{V,75}$$

where $\nu$ is the virtual normal sheaf of $f$. This formula holds, however, whether or not the intersection $Z_2 \cap T$ is proper; it can be established by using the excess-intersection formula (V,56) for $[Z_2]$.

By the projection formula we have

$$f_{1*}\theta^*\pi^*p_1^*c_1(X) = p_{2*}\pi_*([Z_2]\cdot\pi^*p_1^*c_1(X)), \qquad \text{(V,76)}$$

and the right hand side can be evaluated using Formula (V,56) for $[Z_2]$. We get the difference of two terms. The second comes by straightforward manipulation; we find

$$p_{2*}\pi_*\left(\left\{\alpha'_*\sum_{i=1}^{m} h^*c_{m-i}(\tau_Y)c_1(I\,|T)^{i-1}\right\}\cdot\pi^*p_1^*c_1(X)\right)$$
$$= (c_{m-n}(Y)+c_{m-n-1}(Y)s^{n-1}(X))c_1(X) = c_1(\nu)c_1(X). \qquad \text{(V,77)}$$

As to the first term, we find

$$p_{2*}\pi_*(\pi^*(f\times f)^*[\Delta_Y]\cdot\pi^*p_1^*c_1(X)) = f^*f_*c_1(X); \qquad \text{(V,78)}$$

in fact, we get a similar formula with $c_1(X)$ replaced by any class on $X$. Indeed, we may cancel the $\pi_*$ with the $\pi^*$'s by the projection formula. Then we proceed virtually the same way we proved Formula (V,53).

Assume $f_1$ is appropriately generic; that is, letting $Z_3$ denote the scheme associated to $f_1$ analogous to $Z_2$, we assume $Z_3$ has the appropriate dimension, namely, $n-2$. Then the double-point theorem may be applied to $f_1$. Let $[M_3]$ denote the direct image under $f_1$ of the double-point cycle of $f_1$; call it the *triple-point cycle* of $f$. It is easy to see that the underlying set of $[M_3]$ is the triple-point set $M_3$ if each irreducible component of $M_3$ contains a point $x_1$ not on any other component, for which there are two other points $x_3$, $x_2$ distinct (or distant) from $x_1$ and from each other such that $f(x_3)$ and $f(x_2)$ are equal to $f(x_1)$. Conceivably, the underlying set is $M_3$ always, $f_1$ being appropriately generic.

Let $[M_3]$ also denote the corresponding class; call it the *triple-point class* of $f$. From the double-point theorem applied to $f_1$, we get

$$[M_3] = f_{1*}f_1^*f_{1*}[Z_2] - f_{1*}c_1(\nu_1). \qquad \text{(V,79)}$$

Putting together the last eight formulas, we get the formula,

$$\boxed{[M_3] = [M_2]^2 - [M_2]f^*c_1(Y) + 2c_2(\nu) + f^*f_*c_1(X) - c_1(\nu)c_1(X)}\ , \qquad \text{(V,80)}$$

which is therefore a *triple-point formula* for the case $m = n+1$.

Using the double-point formula for $f$ and performing formal manipulations, we can put Formula (V,80) in the following form:

$$[M_3] = (2[M_2]^2 - [M_2]f^*f_*[X] + 2c_2(\nu))$$
$$+ (f^*f_*c_1(X) - f^*f_*[X]\cdot c_1(X)). \qquad \text{(V,81)}$$

Let $u$ be a class on $Y$ and set $h=f^*u$. Then Formula (V,81) yields the following numerical formula:

$$\boxed{\frac{1}{2}\int[M_3]h^{n-2}=\int[M_2]^2h^{n-2}-\frac{1}{2}\int f_*[M_2]f_*[X]u^{n-2}+\int c_2(\nu)h^{n-2}} \quad . \qquad (V,82)$$

Indeed, the projection formula yields the middle term on the right directly, and it shows that the potential last terms cancel. We expect that the integer $\int f_*[M_2]f_*[X]u^{n-2}$ is divisible by 2 because we expect that $[M_2]$ is reduced and that $M_2$ is a double cover of $f(M_2)$. Similarly we expect that the integer $\int[M_3]$ is divisible by 2.

For example, suppose that $Y$ is equal to $\mathbb{P}^{n+1}$ and that $f$ is a projection from a general center $A$. Then Formulas (V,80) and (V,81) hold and the underlying set of the triple-point cycle is the triple-point set. Indeed, $Z_2$ is smooth of dimension $n-1$ in view of Formula (V,21), because $\Sigma$ is in general position since $A$ is; this conclusion follows from a general result about transitive group actions (see [44; Th. 2, p. 209]). There is a description of a dense, open subset of $Z_3$ like Formula (V,21) for $Z_2$; it will be discussed later. It follows that $Z_3$ is reduced and has dimension $n-2$ and that each irreducible component of $Z_3$ contains a triple $(x_3, x_2, x_1)$ of distinct points, a triple lying in no other component.

(In positive characteristic, the result about group actions no longer applies to yield that $Z_2$ is smooth. However, Roberts [93; Lemma 7.3, Th. 7.6] proved that if $X$ is suitably reembedded and then projected into $Y$, then $Z_2-T$ is smooth. Hence $Z_2$ is reduced because it is locally a complete intersection and because $Z_2-T$ is dense. In fact, except in characteristic 2 if $n=2$, then $Z_2$ is normal because it is smooth in codimension 1 by virtue of (V,60) and [93; Cor. 1.2]. It follows that then $Z_2$ is smooth by virtue of the theorem of purity of the branch locus and the reasoning used to prove the result about group actions. Some similar qualifications apply to $Z_3$.)

For instance, suppose $X$ is a surface ($n=2$). Then Formula (V,82) with $u$ taken to be the hyperplane class is easily reduced to Formula (I,39). Moreover, using the techniques of chapter IV to study the tangent planes to $Z=f(X)$ at points of its double curve $f(M_2)$ (see [82; §7]) and letting $\rho$ denote the number of such tangent planes passing through a general point, we can derive from Formula (I,39) the following formula of Salmon, Cayley, Zeuthen, and Noether (see [7; p. 159]):

$$\boxed{\varepsilon_0(\mu_0-2)=\rho+3t} \quad . \qquad (V,83)$$

A specific case is furnished by a ruled surface of degree $\mu_0$ and genus $p$. From Formula (I,39) and various others ((III,19), (III,16), (III,18), (III,23), (V,10) and (V,13)), we get the formula,

$$\boxed{6t = (\mu_0 - 4)\{(\mu_0 - 2)(\mu_0 - 3) - 6p\}}, \qquad \text{(V,84)}$$

which is a classical formula (see [7; p. 166]).

It is evident à priori that with an appropriate amount of work the above procedure can effectively be carried out and a triple-point formula obtained for any given $n$ and $m$. Unfortunately there is no evident pattern to aim for, like the compact form (I,37) of the double-point formula. It is also evident that if $f_1 : Z_2 \to X$ is such that we have a triple-point formula for it, then we can effectively get a quadruple-point formula for $f$; etc.[6] This attractive way of obtaining a higher-order formula, namely, by pushing down a lower-order one, was suggested to the author by H. Salomonsen in a conversation (July 30, 1976). While Salomonsen had no suggestion for explicitly evaluating the direct image of a lower-order formula, the author had just obtained Formula (V,80) and its corollaries by another approach. The computational methods are easily adapted, and this adaptation was carried out above for the case $m = n + 1$.

The other approach to higher-order theory has two principal advantages. They concern the higher-order analogues $Z_r$ of $Z_2$. First, if $Z_r$ has the appropriate dimension $(r+1)n - rm$, then it is possible in principle to effectively compute its class and the class's image on $X$, this producing an $r$th order multiple-point formula. There is no need to assume anything special about $Z_2, \cdots, Z_{r-1}$. Second, in the case of a projection, the approach leads to an analysis of $Z_r$.

Here is a brief look at the approach. We define a smooth ambient scheme $X_r$ for $Z_r$ inductively as follows. Set $X_1 = X$, set $X_2 = \widetilde{X \times X}$ and set

[6] The general triple-point formula is

$$[M_3] = f^*f_*[M_2] - 2c_{m-n}(\nu)[M_2] + \sum_{i=0}^{m-n-1} 2^{m-n-i} c_i(\nu) c_{2(m-n)-i}(\nu).$$

For $m = n+1$, the quadruple-point formula is

$$[M_4] = f^*f_*[M_3] - 3c_1(\nu)[M_3] + 6c_2(\nu)[M_2] - 6c_1(\nu)c_2(\nu) - 12c_3(\nu).$$

For a $\mathbb{P}^2$-ruled 3-fold embedded by generic projection in $\mathbb{P}^4$ with degree $n$ and genus $p$, the latter formula implies that the number of quadruple-points $\chi$ satisfies

$$24\chi = (6t - 18\varepsilon_0 + 60n - 210)n + (6\varepsilon_0 - 36n + 174)(2 - 2p)$$

where $\varepsilon_0$ is the degree of the double surface and $t$ the degree of the triple curve, both expressable in terms of $n$ and $p$ by formulas like (V,13) and (V,84). This result checks with Roth's Formula (9) p. 121 in Proc. Lond. Math. Soc. (33)2 (1931–2).

$p_{2,i}=p_i \circ \pi$ for $i=1, 2$. For $r \geqq 2$ define $X_{r+1}$ as the blowup along the diagonal $\Delta_r$ of the fibered self-product $X_r \times_{X_{r-1}} X_r$ with $p_{r,2}$ as the structure mapping for $X_r$. For $i=1, 2$ define $p_{r+1,i}$ as the mapping from $X_{r+1}$ into $X_r$ arising from the $i$th projection. Since $\Delta_r$ is smooth over $X_r$ (in fact, isomorphic to it), $p_{r+1,i}$ is smooth; hence $X_{r+1}$ is smooth.

In Salomonsen's approach, $Z_{r+1}$ is defined as the scheme $Z_2$ of the mapping $Z_r \to Z_{r-1}$ analogous to $f_1$. Now, it is easy to see that $Z_{r+1}$ is equal to the residual intersection in $X_{r+1}$ with respect to the exceptional divisor, of the preimage of the self-product $Z_r \times_{X_{r-1}} Z_r$. Hence, $Z_{r+1}$ is equal to the residual intersection in $X_{r+1}$ with respect to a certain divisor, of the preimage of the diagonal of $Y^{\times r}$. Therefore, the class of $Z_{r+1}$ is given by the residual intersection formula. Pushing the formula down to $X$ under $p_{r+1,2} \cdots p_{2,2}$ we get an $(r+1)$th order multiple-point formula.

There is a natural flat family of length-$r$ subschemes of $X$ parametrized by $X_r$. Its total space is the union of the graph of $p_{r,2} \cdots p_{2,2}$ and the pullback under $p_{r,1}$ of the corresponding total space over $X_{r-1}$. It can be proved that $Z_r$ is equal to the subscheme of $X_r$ parametrizing those members of the family contained in the fibers of $f$.

Suppose that $X$ is embedded in $\mathbb{P}^N$, that $Y$ is $\mathbb{P}^m$, and that $f$ is the projection from a general center $A$. We may assume that $X$ spans $\mathbb{P}^N$. Then for $r \leqq N+1$ most $r$-tuples of points of $X$ span an $(r-1)$-plane. Let $U$ denote the open subset of $X_r$ formed by the points representing length-$r$ subschemes that span an $(r-1)$-plane. Let $g$ denote the natural mapping from $U$ into the grassmannian of $(r-1)$-planes. Clearly, a length-$r$ subscheme of $X$ lies in a fiber of $f$ if and only if it lies in an $(r-1)$-plane meeting $A$ in an $(r-2)$-plane. Hence $U \cap Z_r$ is the preimage under $g$ of the Schubert variety of $(r-1)$-planes meeting $A$ in an $(r-2)$-plane.

It seems likely that $U \cap Z_r$ is dense in $Z_r$. It is dense for $r=3$ because, clearly, if a length-3 subscheme spans a line, then every plane through that line is a limit of planes through triples of distinct points approaching the subscheme. Suppose $U \cap Z_r$ is dense. It follows that $Z_r$ is reduced and has the appropriate dimension, $(r+1)n-rm$, (in positive characteristic, conceivably $Z_r$ is not reduced) and that under $p_{r,2} \cdots p_{2,2}$ the direct image $[M_r]$ of $[Z_r]$ has $M_r$, the image of $Z_r$, as underlying set. In particular, for $n/(m-n) < r \leqq N+1$, then $Z_r$ is empty; hence $Z_r$ is empty for all $r > n/(m-n)$.

There exists a new embedding of $X$ such that, for all $s$ satisfying $s(m-n) \leqq 2m-n$, the embedded tangent space of $X$ at any $s$ distinct points of $X$ are in general position with respect to each other; this fact was proved by Roberts [93; Lemma 7.3]. It follows that, if $X$ is so reembedded, then $U$ is all of $X_r$ for any $r$ satisfying $r(m-n) \leqq 2m-n$, and hence moreover $Z_r$ is smooth (except possibly in characteristic 2 if $m=2n-1$).

There is another way to define an $r$th order multiple-point cycle. It involves the Hilbert scheme $\mathrm{Hilb}^r_{X/Y}$, which is the natural parameter space of the family of length-$r$ subschemes of $X$ lying in the fibers of $f$. Let $Z'_r$ denote the total space of the family. Then the direct image of $[Z'_r]$ on $X$ is a canonical choice for an $r$th order multiple-point cycle. However, at present there is no way to get a formula for it.

There is a natural mapping from $Z_r$ into $Z'_r$. It carries a point $\zeta$ of $Z_r$ to the pair $(s, x)$ where $s$ represents the same length-$r$ subscheme of $X$ that $\zeta$ does and $x$ is the image of $\zeta$ in $X$ under $p_{r,2} \cdots p_{2,2}$. Conceivably, this mapping is not surjective. After all, the family of length-$r$ subschemes of $X$ parametrized by $X_r$ does not include every length-$r$ subscheme of $X$ if $n \geqq 3$ and $r \gg 0$; Iarrobino [33] showed that the length $r$-subschemes of $X$ form a reducible family with a component of dimension $> rn$ then. On the other hand, it is likely that the mapping $Z_r \to Z'_r$ has pure degree $(r-1)!$ and hence that the cycle $[M_r]$ is divisible by $(r-1)!$. Yet, the right hand side of (V,80) does not appear to be divisible by 2!.

## BIBLIOGRAPHY


[1] ANDREOTTI, ALDO and FRANKEL, THEODORE, 'The second Lefschetz theorem on hyperplane sections', *Global analysis.* Papers in honor of K. Kodaira, edited by D. C. SPENCER and S. IYANAGA, Princeton Math. Series No. 29, Univ. of Tokyo Press, Princeton Univ. Press (1969).

[2] ALTMAN, ALLEN B. and KLEIMAN, STEVEN L., *Introduction to Grothendieck duality theory*, Springer lecture notes in math. [146 (1970)].

[3] ALTMAN, A. B. and KLEIMAN, S. L., 'Joins of schemes, linear projections', Compositio Math. [31 (1975), 309–343].

[4] ARTIN, MICHAEL and NAGATA, MASAYOSHI, 'Residual intersections in Cohen–Macaulay rings', J. Math. Kyoto Univ. [12 (1972), 307–323].

[5] ATIYAH, MICHAEL F., 'Complex analytic connections in fiber bundles', Trans. American Math. Soc. [85 (1957), 181–207].

[6] BOREL, ARMAND and HAEFLIEGER, ANDRÉ, 'La classe d'homologie fundamentale d'un espace analytique', Bull. Soc. Math. France [89 (1961), 461–513].

[7] BAKER, HENRY F., *Principles of Geometry.* Volume VI, *Introduction to the theory of algebraic surfaces and higher loci,* Cambridge University Press (1933).

[8] BERZOLARI, LUIGI, 'Allgemeine Theorie der höheren ebenen algebraischen Kurven', *Enzyklopädie der Math. Wissenschaften* [III, $2^1$] Teubner, Leipzig (1906).

[9] BERZOLARI, L., 'Théorie générale des courbes planes algébriques', *Encyclopédie des sciences mathematiques* [III, 19], Teubner, Leipzig (1915).

[10] CHEVALLEY, CLAUDE, 'Les classes d'équivalences rationelles (I, II)', in *Anneaux de Chow et applications,* Séminaire C. Chevalley, 2ᵉ année, Secr. Math. Paris (1958).

[11] DIEUDONNÉ, JEAN, *Cours de géométrie algébrique 1/Aperçu historique sur le développement de la géométrie algébrique,* Collection sup, Presses Univ. de France (1974).

[12] GROTHENDIECK, ALEXANDER, 'Eléments de géométrie algébrique IV (Quatrième partie)', rédigés avec la collaboration de J. Dieudonné, Publ. Math. I.H.E.S. N° 32, Presses Univ. de France, Vendôme (1967).

[13] FULTON, WILLIAM, 'Rational equivalence for singular varieties', appendix to 'Riemann–Roch for Singular Varieties' by P. BAUM, W. FULTON, and R. MACPHERSON, Publ. Math. I.H.E.S. N° 45, Presses Univ. de France, Vendôme (1975).

[14] FULTON, W., 'A note on residual intersections and the double point formula', preprint.

[15] FULTON, W., 'Canonical classes on algebraic varieties: an introduction', Lecture at Aarhus University, June 23, 1976.

[16] FULTON, WILLIAM and LAKSOV, DAN, 'Residual intersections and the double point formula', this volume pp. 171–178.

[17] GOLUBITSKY, MARTIN and GUILLEMIN, VICTOR, *Stable mappings and their singularities*, Springer graduate text in math. [14 (1973)].

[18] GALBUBRA, G., 'Despre varietatile canonice si ciclurile caracteristice ale unei varietati algebrice', Acad. Repub. Pop. Romine Bul. Sti. Sect. Mat. Fiz. [6 (1954), 61–64] (Roumanian).

[19] GALBUBRA, G., 'Le faisceau jacobien d'un système de faisceaux inversibles', Rev. Roumaine Math. Pures Appl. [14 (1969), 785–792].

[20] GAMKRELIDZE, R. V., 'Computation of the Chern classes of algebraic manifolds', C. R. (Doklady) Acad. Sci. SSSR. (N.S.) [90 (1953), 719–722] (Russian).

[21] GROTHENDIECK, ALEXANDER, 'Théorie des classes de Chern', Bull. Soc. Math. France [86 (1958), 137-154].

[22] GROTHENDIECK, A., 'Sur propriétés fondamentales en théorie des intersections', Séminaire C. Chevalley, 2e année, Secr. Math. Paris (1958).

[23] HERBERT, RALPH J., *Multiple points of immersed manifolds*, Doctoral thesis, Univ. of Minnesota, June (1976).

[24] HAEFLIGER, ANDRÉ, 'Points multiples d'une application et produit cyclique réduit', Bull. Soc. Math. France [87 (1959), 351–359].

[25] HODGE, WILLIAM V. D., 'The characteristic classes on algebraic varieties', Proceedings London Math. Soc. [3 (1951), 138–151].

[26] HAEFLIGER, ANDRÉ and KOSINSKI, A., 'Un théorème de Thom sur les singularités des applications différentiables', Séminarie H. Cartan E.N.S. 1956–1957, exposé no. 8.

[27] HOLME, AUDUN, 'Formal embedding and projection theorems', American Journal Math. [93 (1971), 527–571].

[28] HOLME, A., 'Projection of non-singular projective varieties', Journal Math. Kyoto University [13 (1973), 301–322].

[29] HOLME, A., 'Embedding-obstructions for algebraic varieties I', University of Bergen, Norway, preprint (1974), to appear in Adv. Math.

[30] HOLME, A., 'Embedding-obstructions for singular algebraic varieties in $\mathbb{P}^N$', Acta Math. [135 (1975), 155–185].

[31] HARTSHORNE, ROBIN, *Residues and Duality*, Springer lecture notes in math. [20 (1966)].

[32] HIRZEBRUCH, FREDERICH, *Topological methods in algebraic geometry*, Grundlehren der math. Wissenschaften, Vol. 131, Third enlarged edition, Springer-Verlag (1966).

[33] IARROBINO, ANTONI, 'Reducibility of the families of 0-dimensional schemes on a variety', Inventiones math. [15 (1972), 72–77].

[34] IVERSEN, BIRGER, 'Numerical invariants and multiple planes', American Journal Math. [92 (1970), 968–996].

[35] IVERSEN, B., 'Critical Points of an Algebraic Function', Inventiones math. [12 (1971), 210–224].

[36] IVERSEN, B., 'On the Exceptional Tangents to a Smooth, Plane Curve', Math. Z. [125 (1972), 359–360].

[37] IVERSEN, B., *Sheaf Theory*, a forthcoming book.

[38] INGLETON, A. S. and SCOTT, D. B., 'The tangent direction bundle of an algebraic variety', Ann. di Mat. [(4) 56 (1961), 359–374].

[39] JOUANOLOU, J.-P., 'Riemann–Roch sans dénominateurs', Inventiones math. [11 (1970), 15–26].

[40] JOHNSON, KENT W., *Immersion and embedding of projective varieties*, Doctoral thesis, Brown University, Providence, RI, June (1976).

[41] KEMPF, GEORGE, 'Schubert methods with an application to algebraic curves', Publication of the Mathematisch Centrum, Amsterdam (1971).

[42] KEMPF, GEORGE and LAKSOV, DAN, 'The determinantal formula of Schubert calculus', Acta Math. [132 (1974), 153–162].

[43] KLEIMAN, STEVEN L., 'Motives', in *Algebraic geometry, Oslo* 1970, edited by F. OORT, Wolters–Noordhoff Publ., Groningen (1972).

[44] KLEIMAN, S. L., 'The transversality of a general translate', Compositio Math. [38 (1974), 287–297].

[45] KLEIMAN, S. L., 'Problem 15. Rigorous foundation of Schubert's enumerative calculus', in Proceedings of Symposia in Pure Mathematics, Vol. 28, A.M.S., Providence (1976).

[46] KLEIMAN, S. L., 'Chasles's enumerative theory of conics. A historical introduction', Aarhus University Preprint Series 1975/76 No. 32, Aarhus Denmark, to appear in an M.A.A. volume on algebraic geometry.

[47] KLEIMAN, S. L., 'Todd's double-point formula', Lecture at Aarhus University, June 28, 1976.

[48] KLEIMAN, S. L. and LAKSOV, D., 'On the existence of special divisors', American Journal Math. [93 (1972), 431–436].

[49] KLEIMAN, S. L. and LAKSOV, D., 'Another proof of the existence of special divisors', Acta Math. [132 (1974), 163–176].

[50] KATO, AKIKUNI, 'Singularities of projective embedding (points of order $n$ on an elliptic curve)', Nagoya Math. Journal [45 (1971), 97–107].

[51] LAX, ROBERT F., 'Weierstrass points of the universal curve', Math. Ann. [216 (1975), 35–42].

[52] LANDMAN, ALAN, 'Examples of varieties with 'small' dual varieties', lecture at Aarhus University, June 24, 1976.

[53] LANDMAN, A., 'Picard–Lefschetz theory and dual varieties', lecture at Aarhus University, June 30, 1976.

[54] LEFSCHETZ, SOLOMON, *Topology*, 2nd Ed., Chelsea Publ. Co., New York (1956).

[55] LAKSOV, DAN, 'Some enumerative properties of secants to nonsingular projective schemes', preprint, M.I.T., November (1975).

[56] LAKSOV, D., 'Secant bundles and Todd's formula for the double points of maps into $\mathbb{P}^n$', preprint, M.I.T., April (1976).

[57] LAKSOV, D., 'Residual intersections and Todd's formula for the double locus of a morphism', preprint, M.I.T., June (1976).

[58] LLUIS, E., 'De las singularidades que aparecen al proyectar variedades algebraicas', Bol. Soc. Mat. Mexicana [1 (1956), 1–9].

[59] LAUMON, G., 'Degré de la variété duale d'une hypersurface à singularités isolées , Bull. Soc. Math. France [104 (1976), 51–63].

[60] LASHOF, R. and SMALE, S., 'Self-intersections of immersed manifolds', J. Math. Mech. [8 (1959), 143–157].

[61] LASCOUX, ALAIN, 'Calcul de certains polynômes de Thom', Comptes Rendus Acad. Sci. Paris [278 (1974), 889–891].

[62] LASCOUX, A., 'Puissance extérieures, déterminants et cycles de Schubert', Bulletin de la Société Mathématique de France [102 (1974), 161–179].

[63] LASCOUX, A., 'Sistemi lineari de divisori sulle curve e sulle superficie', preprint, to appear in Ann. di Mat.

[64] MCCRORY, CLINT, an article to appear.

[65] MATHER, JOHN N., 'Stable map-germs and algebraic geometry', *Manifolds-Amsterdam* 1970, Lecture notes in math. [197 (1971), 176–193].

[66] MATHER, J. N., 'Generic projections', Annals Math. [98 (1973), 226–245].

[67] MCCRORY, C. and HARDT, an article to appear.

[68] MUKHERJIA, KALYAN K., 'Chern classes and projective geometry', J. Differential Geometry [7 (1972), 473–478].

[69] LASCU, ALEX T., MUMFORD, DAVID, and SCOTT, D. B., 'The self-intersection formula and the 'formule-clef', Math. Proc. Camb. Phil. Soc. [78 (1975), 117–123].

[70] MENN, MICHAEL, 'Singular Manifolds', Journal diff. geometry [5 (1971), 523–542].

[71] MILNOR, JOHN, *Singular points of complex hypersurfaces,* Annals of Math. Study 61, Princeton Univ. Press, Princeton (1968).

[72] MATTUCK, ARTHUR P., 'On symmetric products of curves', Proceedings of the American Mathematical Society [13 (1962), 82–87].

[73] MATTUCK, A. P., 'Secant bundles on symmetric products', American Journal of Math. [87 (1965), 779–797].

[74] MOUNT, KENNETH R., and VILLAMAYOR, O. E., 'An algebraic construction of the generic singularities of Boardman–Thom', Inst. Hautes Etudes Sci. Publ. Math. [43 (1974), 205–244].

[75] MOUNT, K. R., and VILLAMAYOR, O. E., 'Weierstrass points as singularities of maps in arbitrary characteristic', J. Alg. [31 (1974), 343–353].

[76] MOUNT, K. R. and VILLAMAYOR, O. E., 'Special frames for Thom–Boardman singularities', preprint.

[77] MOISEZON, BORIS G., 'Algebraic homology classes on algebraic varieties', Izv. Akad. Nauk. SSSR Ser. Mat. [31 (1967)] = Math USSR-Izvestija [1 (1967), 209–251].

[78] NAKANO, S., 'Tangent vector bundles and Todd canonical systems of an algebraic variety', Mem. Coll. Sci. Kyoto [(A) 29 (1955), 145–149].

[79] OGAWA, ROY H., 'On the points of Weierstrass in dimensions greater than one', Transactions of the American Mathematical Society [184 (1973), 401–417].

[80] POHL, WILLIAM F., 'Differential geometry of higher order', Topology [1 (1962), 169–211].

[81] POHL, W. F., 'Extrinsic Complex Projective Geometry', *Proceedings of the Conference on Complex Analysis,* Minneapolis 1964, Springer-Verlag (1965), 18–29.

[82] PIENE, RAGNI, *Plücker Formulas,* Doctoral thesis, M.I.T., Cambridge, MA, May (1976).

[83] PIENE, R., 'Numerical characters of a curve in projective $n$-space', preprint of an article prepared for this volume.

[84] PETERS, C. A. M. and SIMONIS, J., 'A secant formula', Quart. J. Math. Oxford [(2), 27 (1976), 181–189].

[85] PORTEOUS, Ian R., 'Simple singularities of maps', *Liverpool singularities symposium* I, Lecture notes in math., Springer-Verlag [192 (1971), 286–307].

[86] PORTEOUS, I. R., 'Todd's canonical classes', *Liverpool singularities symposium* I, Lecture notes in math., Springer-Verlag [192 (1971), 308–312].

[87] PORTEOUS, I. R., 'Blowing-up Chern classes', Proc. Cambridge Phil. Soc. [56 (1960), 118–124].

[88] ROBERTS, JOEL, 'Generic projections of algebraic varieties', American Journal Math. [93 (1971), 191–215].

[89] ROBERTS, J., 'Chow's moving lemma', in *Algebraic geometry, Oslo* 1970, edited by F. OORT, Wolters–Noordhoff Publ., Groningen (1972).

[90] ROBERTS, J., 'The variation of critical cycles in algebraic families', Transactions of the American Mathematical Society [168 (1972), 153–164].

[91] ROBERTS, J., 'Singularity subschemes and generic projections' (research announcement), Bull. Amer. Math. Soc. [78 (1972), 706–708].

[92] ROBERTS, J., 'Generic coverings of $\mathbb{P}^r$ when char $(k)>0$', Notices Amer. Math. Soc. [20 (1973), abstract 704–A5].

[93] ROBERTS, J., 'Singularity subschemes and generic projections', Trans. Amer. Math. Soc. [212 (1975), 229–268].

[94] ROBERTS, J., 'A stratification of the dual variety (Summary of results with indications of proof)', preprint (1976).

[95] ROBERTS, J., 'Embedding obstructions and multiple point cycles', lecture at Aarhus University, June 24, 1976.

[96] RONGA, FÉLIX, 'Les classes duales aux singularités de Boardman d'ordre deux', C. R. Acad. Sci. Paris [270 (1970), 582–584].

[97] RONGA, F., 'Le calcul des classes duales aux singularités de Boardman d'ordre deux', Commentarii math. helvetici [47 (1972), 15–35].

[98] RONGA, F., 'La classe duale aux points doubles d'une application', Compositio Math. [27 (1973), 223–232].

[99] ROTH, LEONARD, 'Some formulae for primals in four dimensions', Proc. Lond. Math. Soc. [(2) 35 (1933), 540–550].

[100] RAMANUJAM, C. P., 'On a geometric interpretation of multiplicity', Inventiones Math. [22 (1973/74), 63–67].

[101] SCHUBERT, HERMANN CÄSER HANNIBAL, *Kalkul der abzählenden Geometrie*, Teubner, Leipzig (1879).

[102] SEGRE, CORRADO, 'Mehrdimensionale Räume', *Encyclopädie der Mathematischen Wissenschaften mit einschluss ihrer anwendungen* [III, 2, 2, C7; 669–972], (Abgeschlossen Ende 1912), B. G. Teubner, Leipzig (1921–1934).

[103] BERTHELOT, PIERRE *et al.*, *Théorie des intersections et théorème de Riemann–Roch*, Springer lecture notes in math. [225 (1971)].

[104] DELIGNE, PIERRE and KATZ, NICHOLAS, *Groupes de monodromie en géométrie algébrique*, Springer lecture notes in math. [340 (1973)].

[105] SAMUEL, PIERRE, 'Relations d'équivalence en géométrie algébrique', Proc. Internat. Congress of Math., Cambridge, England (1958).

[106] SAMUEL, P., 'L'équivalence rationelle et le 'moving lemma' ', preprint (1971).

[107] SERRE, JEAN-PIERRE, *Algèbre locale, Multiplicités*, Lecture notes in math., Springer-Verlag [11 (1965)].

[108] SERRE, J.-P., *Groups algébriques et corps de classes*, Publ. Math. Univ. Nancago VII, Hermann, Paris (1959).

[109] SERGERAERT, FRANCIS, 'Expression explicite de certains polynômes de Thom', Comptes Rendus Acad. Sci. Paris [276 (1973), 1661–1663].

[110] SCHWARZENBERGER, R. L. E., 'The secant bundle of a projective variety', Proc. London Math. Soc. [(3) 14 (1964), 369–384].

[111] TODD, JOHN ARTHUR, 'Canonical systems on algebraic varieties', Bol. Soc. Mat. Mexicana [2 (1957), 26–44].

[112] TEISSIER, BERNARD, 'Cycles évanescents, sections planes, et conditions de Whitney', Astérisque [7 et 8 (1973), 285–362].

[113] TEISSIER, B., 'Sur diverses conditions numériques d'équisingularité des familles de courbes (et un principe de specialisation de la dépendance intégrale)', preprint.

[114] VAINSENCHER, ISRAEL, 'A numerical criterion for hypersurfaces', Advances in Math. [19 (1976), 289–291].

[115] VAINSENCHER, I., 'On a formula of Ingleton and Scott', preprint, to appear in Ann. di Mat.

[116] VAINSENCHER, I., *On the formula of de Jonquières for multiple contacts,* Doctoral thesis, M.I.T., Nov. (1976).

[117] VAINSENCHER, I., 'Conics in characteristic 2', preprint, to appear in Compositio Math.

[118] WALLACE, ANDREW H., 'Tangency and duality over arbitrary fields', Proc. London Math. Soc. [(3) 6 (1956), 321–342].

[119] ZEUTHEN, HIERONYMUS G. and PIERI, M., 'Géométrie énumérative', *Encyclopédie des sciences mathématiques* [III, 2, 260–331] Teubner, Leipzig (1915).

[120] ZARISKI, OSCAR, *Collected papers. Volume* I: *Foundations of algebraic geometry and resolution of singularities,* edited by H. HIRONAKA and D. MUMFORD, M.I.T. Press (1972).

[121] COOLIDGE, JULIEN LOWEL, *A treatise on algebraic plane curves,* Dover Publ., New York (1959).

[122] DOUADY, ADRIEN and VERDIER, JEAN-LOUIS, 'Séminaire de géométrie analytique', Astérisque [36–37 (1976)].

[123] SEVERI, FRANCESCO, 'Sulle interregioni delle varietà algebreche e sopra i loro caratteri e singularità proiettive', Memorie della R.Acc. della Scienze di Torino [52 (1902), 61–118].

[124] KLEIMAN, S. L. and LANDOLFI, JOHN, 'Geometry and deformation of special Schubert varieties', *Algebraic geometry. Oslo* 1970, F. Oort, editor, Wolters–Noordhoff Publ., Groningen (1972).

[125] FULTON, W. and MACPHERSON, ROBERT, 'Intersecting cycles on an algebraic variety', Aarhus Univ. Preprint, Series 1976/77 No. 14, this volume pp. 179–198.

[126] KLEIMAN, S. L., 'Multiple-point formulas', in preparation.

STEVEN L. KLEIMAN
Massachusetts Institute of Technology
Dept. of Mathematics
Cambridge, Mass. 02139, USA

Nordic Summer School/NAVF
Symposium in Mathematics
Oslo, August 5–25, 1976

# SOME REMARKS ON RELATIVE MONODROMY

Lê Dũng Tráng

In this talk we are going to give some means to study the topology of analytic sets even when their singularities are not isolated. Some of the results cited here extend results of J. Milnor in [14], H. Hamm in [3], Lê Dũng Tráng in [9].

## 1. The fibration theorem

In [14], J. Milnor has given a fibration theorem related to the situation of an analytic function $f$ defined in the neighborhood of a critical point of $f$ in $\mathbb{C}^n$.

We first show that this theorem extends to the situation where we have an analytic function on an analytic subset $X$ of an open subset $U$ of $\mathbb{C}^N$ (cf. [10]).

THEOREM (1.1). *Let $X \subset U \subset \mathbb{C}^N$ be an analytic subset of an open set $U$ of $\mathbb{C}^N$. Let $f: X \to \mathbb{C}$ be an analytic function. Let $x \in X$ and suppose that $f(x) = 0$. Then if $\varepsilon > 0$ is small enough and $\eta > 0$, $\varepsilon \gg \eta$, the mapping induced by $f$:*

$$\Psi_{\varepsilon,\eta}: B_\varepsilon \cap X \cap f^{-1}(\mathring{D}_\eta - \{0\}) \to \mathring{D}_\eta - \{0\}$$

*– where $B_\varepsilon$ is the closed real ball in $\mathbb{C}^N$ of center $x$ and radius $\varepsilon > 0$, $\mathring{D}_\eta$ is the open disc of $\mathbb{C}$ centered at 0 and with radius $\eta > 0$, is a topological fibration.*

REMARK (1.2). In [3], H. Hamm, using an argument similar to the one of J. Milnor in [14], proves that this fibration is smooth, i.e. $C^\infty$, when $X - f^{-1}(0)$ is non-singular. In [4] H. Hamm obtains a result when the restriction of $f$ to the critical locus of $f$ is finite on $\mathbb{C}$.

Actually in [1], P. Deligne indicates that such a theorem is true, but the hint of proof he proposes works only if one gets a proper morphism $\varphi$ from an analytic space $X$ onto $\mathbb{C}$. In that case if $x \in X$ and $\varphi(x) = 0$, there is $\eta > 0$ small enough such that $\varphi$ induces a topological fibration $X \cap \varphi^{-1}(\mathring{D}_\eta - \{0\}) \to \mathring{D}_\eta - \{0\}$.

Proof of the Theorem (1.1). Let us stratify such that the stratification of $X$ satisfies Whitney conditions. This is possible because of [19] (cf. [6] too). Thus in a neighborhood of $x$ sufficiently small there are only a finite number of strata. Let $B_\varepsilon$ a real ball of $\mathbb{C}^N$ of sufficiently small radius centered at $x$. Then $S_\varepsilon$, boundary of $B_\varepsilon$, cuts transversally the strata of $X$. Let $S_1, \cdots, S_k$ be the strata of $X$ such that $S_i \cap X \cap B_\varepsilon \neq \emptyset$. The strata $S_1 \cap X \cap \mathring{B}_\varepsilon$ and $S_i \cap X \cap S_\varepsilon$ satisfy Whitney conditions.

Now let $\mathring{D}_\eta$ be the open disc of radius $\eta > 0$ with center at 0. If $\varepsilon \gg \eta$, then the restriction of $f$ to $f^{-1}(\mathring{D}_\eta - \{0\}) \cap X \cap \mathring{B}_\varepsilon$ induces a mapping

$$f^{-1}(\mathring{D}_\eta - \{0\}) \cap X \cap \mathring{B}_\varepsilon \to \mathring{D}_\eta - \{0\}$$

which is a submersion on each stratum. But as it is not proper, we cannot apply the first Thom-Mather isotopy theorem (cf. [17] and [13]). In order to obtain the topological fibration $\Psi_{\varepsilon,\eta}$ we need to be sure that the fibers of the mapping:

$$f^{-1}(\mathring{D}_\eta - \{0\}) \cap X \to \mathring{D}_\eta - \{0\}$$

induced by $f$, cut transversally the strata of $X \cap S_\varepsilon$.

For this reason we need a stratification of $X$ which moreover satisfies Thom condition (also called $A_f$ condition) (cf. [17] and [6]). In [5] we have called stratifications with this property *good stratifications.*

From [6] we find that $X$ can be stratified with Whitney conditions and Thom condition such that $f^{-1}(0)$ is a union of strata.

Using Thom-Mather first isotopy theorem, because $f$ induces a mapping from $f^{-1}(\mathring{D}_\eta - \{0\}) \cap X \cap B_\varepsilon$ onto $\mathring{D}_\eta - \{0\}$ which is a submersion on each stratum, the theorem (1.1) is proved.

Remark (1.3). (a) We may avoid using Thom-Mather isotopy theorem by using Verdier's result in [18].

(b) Moreover in the theorem (1.1) instead of considering a ball $B_\varepsilon$ we might be interested to consider a polydisc $P$. In this purpose we may define sufficiently small privileged polydiscs, neighborhoods of $x$ in $\mathbb{C}^N$, with respect to $X$ and $f^{-1}(0)$ in a similar way as it has been done in [9] such that the theorem (1.1) remains true if we replace $B_\varepsilon$ by these polydiscs.

Finally from Mather's proof of Thom-Mather first isotopy theorem in [13] or from Verdier's result in [18], if $\eta_0$, $0 < \eta_0 < \eta$, $\Psi_{\varepsilon,\eta}$ induces a topological fibration

$$\varphi_{\varepsilon,\eta_0} : f^{-1}(\partial D_{\eta_0}) \cap X \cap B_\varepsilon \to \partial D_{\eta_0}$$

such that the unit vector field of $\partial D_{\eta_0}$ is lifted in a continuous vector field of $f^{-1}(\partial D_{\eta_0}) \cap X \cap B_\varepsilon$ the restriction of which to each stratum is smooth and the integration of which gives a characteristic homeomorphism of the fibration $\varphi_{\varepsilon,\eta_0}$.

## 2. The generic hyperplane section

We suppose that $X$ has a stratification which satisfies Whitney conditions and Thom condition relatively to $f: X \to \mathbb{C}$.

Now let us consider an affine form, say $l$ of $\mathbb{C}^N$ onto $\mathbb{C}$, such that $l(x) = 0$.

Then we have an analytic morphism

$$\Phi: X \to \mathbb{C}^2$$

– the components of which are $f$ and $l$.

We are going to prove the following theorem:

THEOREM (2.1). *There is an open Zariski dense set $\Omega$ in the space of affine hyperplanes of $\mathbb{C}^N$ at $x$ such that if $l = 0$ belongs to $\Omega$, there exists an analytic curve $\Delta_0 \subset V \subset \mathbb{C}^2$ in an open neighborhood $V$ of $0 \in \mathbb{C}^2$, such that for any $\varepsilon > 0$ small enough and $\varepsilon' > 0$, $\varepsilon \gg \varepsilon'$, the mapping induced by $\Phi$:*

$$X \cap B_\varepsilon \cap \Phi^{-1}(\mathring{\mathscr{B}}_{\varepsilon'} - \Delta_0) \to \mathring{\mathscr{B}}_{\varepsilon'} - \Delta_0$$

*where $\mathring{\mathscr{B}}_{\varepsilon'}$ is the open ball of $\mathbb{C}^2$ centered at 0 with radius $\varepsilon' > 0$ is a topological fibration.*

PROOF OF THE THEOREM (2.1). We are going to use again Thom-Mather first isotopy theorem or Verdier's result of [18]. For this purpose if $S_i$ is a stratum of $X$, we must get rid of the points where the restriction of $\Phi$ to $S_i$ has not the maximal rank. In this purpose we have:

LEMMA (2.2). *Let $V$ be an analytic set, $V \subset U \subset \mathbb{C}^N$, and $W \subset V$ an analytic subset such that $V - W$ is smooth. Let $x \in W$ and $f: V \to \mathbb{C}$ be an analytic function such that $f(x) = 0$. Then there is a Zariski open dense set $\tilde{\Omega}$ of affine hyperplanes of $\mathbb{C}^N$ passing through $x$ such that if $l = 0$ belongs to $\tilde{\Omega}$ the critical space $\Gamma$ of the restriction of $\Phi$ to $V - (W \cup f^{-1}(0))$ is void or a smooth curve in a neighborhood of $x$.*

The proof of this lemma we are going to omit here is based on Bertini theorem.

Notice that if the dimension of $V - W$ is one and it is not contained in $f^{-1}(0)$ we have $V - W = \Gamma$.

Then consider $\Omega_1, \cdots, \Omega_k$ the Zariski dense open sets obtained from lemma (2.2) and associated to $S_1, \cdots, S_k$ where the $S_i$ $(i = 1, \cdots, k)$ are the strata of $X$ adherent to $x$ and not contained in $f^{-1}(0)$.

Then with $\Omega_0 = \Omega_1 \cap \cdots \cap \Omega_k$ and $\{l = 0\} \in \Omega_0$ we have $\Gamma = \Gamma_1 \cup \cdots \cup \Gamma_k$ defined in some neighborhood $U_0$ of $x$. Moreover the restriction of $\Phi$ to $\Gamma$ is finite. Let us call $\Delta = \Phi(\Gamma)$ and $\Delta_0 = \Delta \cup \{\lambda = 0\}$ where $\lambda$ is the coordinate of $\mathbb{C}^2$ value of $f$.

Finally let us suppose that $l = 0$ belongs to an open dense $\Omega'_0$ of affine hyperplane through $x$ such that $l = 0$ is transverse to all strata of $f^{-1}(0)$ except may be $\{x\}$ in a neighborhood of $x$. We may choose as $\Omega'_0$ the Zariski dense open set of affine hyperplanes at $x$ such that $L \in \Omega'_0$ does not contain the limit of tangents to the strata of $f^{-1}(0)$ adherent to $x$.

Consider $\Omega = \Omega_0 \cap \Omega'_0$ and $\{l = 0\} \in \Omega$.

The stratification of $X$ obtained from the previous one by taking the trace of the previous strata on $X - \{l = 0\}$ and the intersection of the previous strata with $l = 0$ satisfies Thom condition with respect to $\Phi: X \to \mathbb{C}^2$ in a neighborhood $U_1$ of $x$. This fact is left to the reader to be proved, but this has been implicitly done when $X = \mathbb{C}^n$ in [5].

Now choose $\varepsilon > 0$ small enough such that $B_\varepsilon \subset U_0 \cap U_1$ and $S_\varepsilon$ is transverse to all the strata of $f^{-1}(0) \cap l^{-1}(0)$. Because the new stratification on $X$ satisfies Thom condition with respect to $\Phi$ in a neighborhood $U_1$ of $x$, we can choose $\varepsilon' > 0$, $\varepsilon \gg \varepsilon'$ such that the restrictions of $\Phi$ to the strata of $X$ outside $f^{-1}(0) \cap l^{-1}(0)$ has their fibers over any point of $\mathring{\mathscr{B}}_{\varepsilon'} - \{0\}$ transverse to $S_\varepsilon$. Let us call $\Sigma_1, \cdots, \Sigma_m$ the strata of $X$ outside $f^{-1}(0) \cap l^{-1}(0)$. Now stratify $B_\varepsilon \cap X \cap f^{-1}(\mathring{\mathscr{B}}_{\varepsilon'}) - f^{-1}(0) \cap l^{-1}(0)$ with $\Sigma_1 \cap \mathring{B}_\varepsilon \cap f^{-1}(\mathring{\mathscr{B}}_{\varepsilon'}), \cdots, \Sigma_m \cap B_\varepsilon \cap f^{-1}(\mathring{\mathscr{B}}_{\varepsilon'})$ and $\Sigma_1 \cap S_\varepsilon \cap f^{-1}(\mathring{\mathscr{B}}_{\varepsilon'}), \cdots, \Sigma_m \cap S_\varepsilon \cap f^{-1}(\mathring{\mathscr{B}}_{\varepsilon'})$. It is clear that over $\mathring{\mathscr{B}}_{\varepsilon'} - \Delta_0$, the restrictions of $\Phi$ to these strata have the maximal rank. Thus from Thom-Mather first isotopy theorem or using Verdier's result we prove the theorem (2.1).

Remarks (2.3). We call $\Gamma$ the *polar curve* of $f$ relatively to $l = 0$ and the stratification of $X$ (cf. [11]). The curve $\Delta$ is called the *Cerf diagramm* of $f$ relatively to $l = 0$ and the stratification of $X$. Recall that $\Delta_0 = \Delta \cup \{\lambda = 0\}$.

Let us call $u$ the second coordinate of $\mathbb{C}^2$ value of $l$. As $l = 0$ cuts all the strata transversally in a neighborhood of $x$, we have that $\Delta$ does not contain $\{u = 0\}$.

As in [9], as we already said in the remark (1.3), we may suppose that $l$ is the first coordinate of $\mathbb{C}^N$ and we have privileged polydiscs $P$ of the form $D_1 \times P'$. Actually we have.

Theorem (2.4). *Let us suppose* $\{z_1 = 0\} \in \Omega$. *There is a fundamental system of privileged polydiscs* $P_\alpha = D_\alpha \times P'_\alpha$ $(\alpha \in A)$ *relatively to the stratification of* $X$ *and* $f = 0$, *centered at* $x$, *and* $Q_\alpha = D_\alpha \times D'_\alpha$ $(\alpha \in A)$ *centered at* $0$ *in* $\mathbb{C}^2$ *such that* $\Phi$ *induces:*

$$\Phi_\alpha : P_\alpha \cap X \cap f^{-1}(Q_\alpha - \Delta_0) \to Q_\alpha - \Delta_0$$

*which is a topological fibration. Moreover* $f$ *induces*

$$\varphi_\alpha : P_\alpha \cap X \cap f^{-1}(Q_\alpha - D_\alpha \times \{0\}) \to D'_\alpha - \{0\}$$

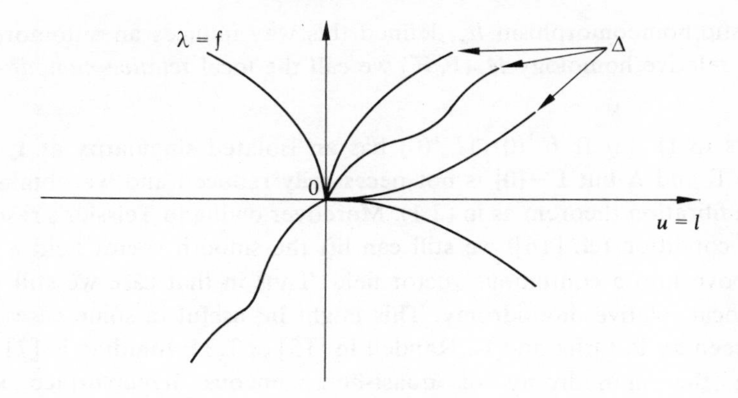


*which is fiber homeomorphic to the fibration of the theorem* (1.1) *and* $\varphi'_\alpha : P_\alpha \cap X \cap f^{-1}(Q_\alpha - D_\alpha \times \{0\}) \cap \{z_1 = 0\} \to D'_\alpha - \{0\}$ *which is fiber homeomorphic to the fibration of the theorem* (1.1) *applied to the restriction of f to* $X \cap \{z_1 = 0\}$.

## 3. Relative monodromy

Let us fix $\alpha \in A$ such that the theorem (2.4) holds. As in [9] we can build a *smooth* vector field $\xi$ on $D_\alpha \times \partial D'_\alpha$ such that:

(1) It is tangent to $\Delta \cap (D_\alpha \times \partial D'_\alpha)$;

(2) It is tangent to $\{\delta = \theta\} \cap (D_\alpha \times \partial D'_\alpha)$ for sufficiently small $\theta$, where $\delta = 0$ is an equation of $\Delta$ at 0;

(3) Its restriction to $\{0\} \times \partial D'_\alpha$ is the unit vector field of $\{0\} \times \partial D'_\alpha$;

(4) It lifts the unit vector field of $\partial D'_\alpha$ by the projection $D_\alpha \times \partial D'_\alpha \to \partial D'_\alpha$.

The construction of $\xi$ as in [9] is very technical. Note that when $\xi$ is integrated we may have orbits closed after one turn around $\partial D'_\alpha$ than $\{0\} \times \partial D'_\alpha$. This last fact depends on the components of $\Delta$ of multiplicity one, i.e. smooth, which are not tangent to $\lambda = 0$. We omit the construction of $\xi$ in this talk to avoid a too long speech.

Now such a smooth vector field $\xi$ can be lifted to a continuous vector field $w$ of $P_\alpha \cap X \cap f^{-1}(D_\alpha \cap \partial D'_\alpha)$ for reasons similar to the ones of [9] and Verdier's results of [18] or Teissier's (c) condition in [16]. Integrating $w$ we obtain a characteristic homeomorphism of the fibration induced by $f$:

$$\Psi_\alpha : P_\alpha \cap X \cap f^{-1}(D_\alpha \times \partial D'_\alpha) \to \partial D'_\alpha$$

and

$$\Psi'_\alpha : P_\alpha \cap X \cap f^{-1}(\{0\} \times \partial D'_\alpha) \to \partial D'_\alpha.$$

Let $F$ be a fiber of $\Psi_\alpha$ and $F' = F \cap \{z_1 = 0\}$ a fiber of $\Psi'_\alpha$. Then the characteristic homeomorphism $h_\alpha$ defined this way induces an automorphism of the relative homology $H_*(F, F')$ we call the local *relative monodromy* of $f$ at $x$.

REMARKS (3.1). (a) If $f^{-1}(0) \cap l^{-1}(0)$ has an isolated singularity at $x$, we can define $\Gamma$ and $\Delta$ but $\Gamma - \{0\}$ is not necessarily reduced and we obtain in that case a fibration theorem as in (2.1). Moreover owing to Teissier's results on the (c) condition (cf. [16]) we still can lift the smooth vector field $\xi$ we defined above into a continuous vector field. Thus in that case we still can define a local relative monodromy. This might be useful in some cases, as has been seen by P. Orlik and R. Randell in [15] or I. N. Iomdine in [7] for computing the monodromy of quasi-homogeneous hypersurface singularities. For instance in the case of $f = \sum_{i=1}^n z_i^{a_i}$ we may choose $z_1 = l = 0$.

(b) If $X = \mathbb{C}^n$ then $\Delta$ is tangent to $\{\lambda = 0\}$ when it is not void. In that case we can construct a characteristic homeomorphism of Milnor's fibration without fixed points (cf. [9]). When $X \neq \mathbb{C}^n$, the methods shown above lead to examples when on the contrary the Lefschetz number of the local monodromy is not zero.

(c) The local relative monodromy is used to prove that local monodromy is quasi unipotent (cf. [12]) without using the resolution of singularities as in [2] or [8].

(d) The main interest of the relative local monodromy is that it precises the use of Lefschetz method using hyperplane sections to study the topology of analytic sets.

## BIBLIOGRAPHY


[1] P. DELIGNE, Equations différentielles à points singuliers réguliers, Springer Lecture Notes n° 163.

[2] A. GROTHENDIECK, Séminaire de géométrie algébrique 7, à l'I.H.E.S., Springer Lecture Notes n° 288.

[3] H. HAMM, Lokale topologische Eigenschaften komplexer Räume, Math. Ann. *191*, 235–252 (1971).

[4] H. HAMM, Zur analytischen und algebraischen Beschreibung der Picard-Lefschetz Monodromie, Habilitationsschrift, Göttingen 1974.

[5] H. HAMM, LÊ DŨNG TRÁNG, Un théorème de Zariski du type de Lefschetz, Ann. Ec. Norm. Sup. 1973.

[6] H. HIRONAKA, Lectures at Nordic Summer School at Oslo 1976, this volume pp. 199–266.

[7] I. N. IOMDINE, Calcul de la monodromie relative d'une hypersurface complexe, Func. Analiz i ievo priloj. 9 (1975), 67–68.

[8] A. LANDMAN, On the Picard-Lefschetz transformation for algebraic manifolds acquiring general singularities, Trans. Amer. Math. Soc. 181 (1973), 89–126.

[9] LÊ DŨNG TRÁNG, La monodromie n'a pas de points fixes, J. of Fac. Sc. Univ. Tokyo, Sec. 1A, Vol. 22, 409–427, 1975.

[10] LÊ DŨNG TRÁNG, Vanishing cycles on analytic sets, lecture à Kyoto University, July 1975, unpublished.

[11] LÊ DŨNG TRÁNG, Topological use of polar curves, Proc. of the AMS meeting on Algebraic Geometry at Arcata 1974, AMS Pub.

[12] LÊ DŨNG TRÁNG, The geometry of the monodromy theorem, to be published in the volume dedicated to C. P. Ramanujam at the Tata Institute.

[13] J. MATHER, Notes on topological stability, Preprint, Harvard Univ.

[14] J. MILNOR, Singular points of complex hypersurfaces, Ann. of Math. Stud. 61, 1968.

[15] P. ORLIK, R. RANDELL, The classification and monodromy of weighted homogeneous singularities, Preprint.

[16] B. TEISSIER, Lectures at Nordic Summer School at Oslo 1976, this volume pp. 565–677.

[17] R. THOM, Ensembles et morphismes stratifiés, Bull. A.M.S. *75* (1969) 240–284.

[18] J.-L. VERDIER, Stratifications de Whitney et théorème de Bertini-Sard, Inv. Math. 36, 1976, 295–312.

[19] H. WHITNEY, Tangents to an analytic variety, Ann. Math. *81* (1965) 496–549.

LÊ DŨNG TRÁNG
Centre de Mathématiques de l'Ecole Polytechnique
Plateau de Palaiseau
91128 Palaiseau Cedex, France
"Laboratoire de Recherche associé au C.N.R.S.
and UER de Mathématiques, Université de Paris VII

Nordic Summer School/NAVF
Symposium in Mathematics
Oslo, August 5–25, 1976

# THE MULTIPLICITY OF A HOLOMORPHIC MAP AT AN ISOLATED CRITICAL POINT

Peter Orlik*

Dedicated to Olaf Hauge and my other friends in Norway.

## Contents



* Partially supported by NSF.

## Introduction

The purpose of these notes is to review recent developments in the local properties of a holomorphic map germ at an isolated critical point. In particular we shall be interested in the various ways an integer 'multiplicity' may be associated with the critical point and prove the well known fact that all these multiplicities are equal. I hope the reader will share some of my pleasure in discovering the strong ties among algebra, function theory, topology, complex analysis and algebraic geometry demonstrated by this excursion.

Let $f:\mathbb{C}^n \to \mathbb{C}$ be a holomorphic map with holomorphic partial derivatives $f_j = \partial f/\partial z_j$, $j = 1, \cdots, n$. A point $z \in \mathbb{C}^n$ is a *critical* point of $f$ if $f_1(z) = \cdots = f_n(z) = 0$. It is called *isolated* if it is the only critical point of $f$ in a sufficiently small neighborhood of $z$. The complex number $f(z)$ is called the corresponding critical value. By a change of coordinates in $\mathbb{C}^n$ and $\mathbb{C}$ we may assume that an isolated critical point is $0 \in \mathbb{C}^n$ with critical value $f(0) = 0$. Define the gradient of $f$, $\partial f:\mathbb{C}^n \to \mathbb{C}^n$ by

$$\partial f(z_1, \cdots, z_n) = (f_1(z), \cdots, f_n(z)).$$

In chapter I we discuss some algebraic aspects of the theory. Let $\mathcal{O}$ be the local ring of germs of holomorphic functions at $0 \in \mathbb{C}^n$ and $(\partial f) = (\hat{f}_1, \cdots, \hat{f}_n)$ the ideal generated by the germs of the partials of $f$. An application of the analytic Nullstellensatz shows that $\mathcal{O}/(\partial f)$ is a finite dimensional vector space over $\mathbb{C}$. Define the *algebraic multiplicity* of the critical point 0 of $f$ by

$$\mu_a = \dim_{\mathbb{C}} \mathcal{O}/(\partial f).$$

Next we prove a theorem of Mather [1] which states that there is a holomorphic change of coordinates, $h:\mathbb{C}^n \to \mathbb{C}^n$, so that $f \circ h$ is a polynomial mapping. We may therefore assume that $f$ is given by a polynomial, and $\partial f$ has polynomial components. We will note that $\partial f$ is an analytic branched cover, so for $y$ in a dense open subset of a suitable neighborhood of 0, the number of points in $(\partial f)^{-1}(y)$ is finite and independent of $y$. Define the *covering multiplicity* of the critical point 0 of $f$ by

$$\mu_c = \#\{(\partial f)^{-1}(y)\}.$$

We prove a result due to Palamodov [1] and Grauert.

(I.5.13) THEOREM. $\mu_a = \mu_c$.

A critical point $z$ is *non-degenerate* if the Hessian matrix $(\partial^2 f/\partial z_i \partial z_j)$ has non-zero determinant at $z$. A function is called *generic* if all of its critical points are non-degenerate and the corresponding critical values are distinct. A *deformation* of $f$ is a map $F:\mathbb{C}^n \times \mathbb{C}^k \to \mathbb{C}$ so that $F(z_1, \cdots, z_n, 0, \cdots, 0) = f(z)$. For fixed coordinates $t_1, \cdots, t_k$, call $F_t(z) = F(z, t):\mathbb{C}^n \to \mathbb{C}$ an *approximation* of $f$. It is a generic approximation if $F_t(z)$ is a generic map. We shall see that the number of critical points in a generic approximation does not depend on $F$ or $t$, so it is an invariant of the critical point 0 of $f$. Call it the *deformation multiplicity:*

$$\mu_d = \text{number of critical points in a generic approximation of } f.$$

(I.6.10) THEOREM. $\mu_d = \mu_c$.

Chapter I concludes with a discussion of current research.
Chapter II is a review of the topological theory. The gradient map, $\partial f$ has no zeros in a deleted neighborhood of 0, so for some small $\varepsilon > 0$, we may define $\varphi : S_\varepsilon^{2n-1} \to S_1^{2n-1}$ by $\varphi(z) = \partial f(z)/|\partial f(z)|$. Define the *gradient multiplicity* of the critical point 0 of $f$ by

$$\mu_g = \deg \varphi.$$

We prove a result of Milnor [3, Appendix].

(II.1.5) THEOREM. $\mu_g = \mu_c$.

Next we prove a version of the fibration theorem of Milnor [3]. Let $B_\varepsilon^{2n}$ be a small closed ball at 0 so that $f$ has no other critical point in $B_\varepsilon^{2n}$. Choose $\delta \ll \varepsilon$, $D = \{t \mid |t| \leqq \delta\}$, so that for $t \in D$, $F_t = f^{-1}(t)$ intersects $S_\varepsilon^{2n-1}$ transversely. Let $D_0 = D - \{0\}$, $E_0 = f^{-1}(D_0) \cap B_\varepsilon^{2n}$ and $T = f^{-1}(D) \cap S_\varepsilon^{2n-1}$. Then

(i) $f: E_0 \to D_0$ is a locally trivial fiber bundle with fiber a smooth compact manifold $M^{2n-2}$ with boundary $K^{2n-3}$, and

(ii) $f: T \to D$ is a trivial fiber bundle with fiber $K$.
It follows from (ii) that $K \cong F_0 \cap S_\varepsilon^{2n-1}$.

We show that if $t$ is a regular value of $f$, then the nonsingular variety $F_t$ is diffeomorphic to the interior of $M$, and use Morse theory to prove that $F = F_t$, $t \neq 0$ has the homotopy type of a wedge of $(n-1)$-spheres. Let their number be the *topological multiplicity*, $\mu_t$ of the critical point 0 of $f$; $F \sim \bigvee_{\mu_t} S^{n-1}$. In particular

$$\mu_t = \dim_{\mathbb{C}} H^{n-1}(F; \mathbb{C}).$$

Milnor [3] proves that $\mu_t = \mu_g$. We give an argument due to Lamotke [1] and Brieskorn [4]:

(II.4.1) THEOREM. $\mu_t = \mu_c$.

This proof constructs the vanishing cycles of $F$ and we define the Picard-Lefschetz monodromy associated with the critical point. The chapter concludes with a brief discussion of various results on the monodromy.

In chapter III we discuss the analytic and algebraic de Rham cohomologies of the non-singular affine complex algebraic variety $F = F_t = f^{-1}(t)$, $t \neq 0$. The main aim is to prove a theorem of Grothendieck [1] which asserts that the cohomology groups of $F$ (with complex coefficients) are isomorphic to the cohomology groups of the complex of rational differential forms which are everywhere regular on $F$. We follow the exposition of Atiyah, Bott, and Gårding [1] suitably amended to include the definitions of the relevant concepts and the proofs of the more elementary statements.

First we define sheaves and the de Rham complex of $F$ in the analytic category, i.e. viewing $F$ as a complex analytic submanifold of $\mathbb{C}^n$. The corresponding de Rham theorem is obtained using the notion of hypercohomology. The latter leads to two spectral sequences which converge to the same groups. The de Rham theorem is obtained by showing the appropriate vanishing theorems so that both spectral sequences collapse, one yielding the cohomology of $F$ in the constant sheaf $\mathbb{C}$, i.e. usual complex cohomology, the other the de Rham cohomology.

In the algebraic setting the approach is similar, but the tools are more sophisticated. After defining the Zariski topology we need the vanishing theorems of Serre [1] and Grothendieck. Next we discuss the comparison results of Serre [2] between the cohomology of a projective algebraic variety $X$ with coefficients in a coherent sheaf $\mathscr{S}$, and the cohomology of the corresponding analytic variety $X^h$ with coefficients in the coherent analytic sheaf $\mathscr{S}^h$. Grothendieck's theorem is deduced from these, and the resolution theorem of Hironaka [1].

Let $\Omega_F^*$ denote the sheaf of everywhere regular differential forms on $F$.

Consider the complex of global sections:

$$0 \to \Gamma(F, \Omega_F^0) \to \Gamma(F, \Omega_F^1) \to \cdots \to \Gamma(F, \Omega_F^{n-1}) \to 0$$

Call the cohomology of this complex the algebraic de Rham cohomology of $F$, $DR^*(\Omega_F^*)$.

The assertion of Grothendieck [1] is with this notation.

(III.9.4) THEOREM. *There are isomorphisms*

$$H^q(F^h, \mathbb{C}) \simeq DR^q(\Omega_F^*)$$

*for all* $q \geqq 0$.

Combining this with the results of chapters I and II we conclude that the finite dimensional complex vector spaces $\mathcal{O}/(\partial f)$ and $DR^{n-1}(\Omega_F^*)$ are of the same dimension.

Chapter IV considers the problem of constructing an explicit isomorphism between these two vector spaces. Using Orlik and Solomon [1] we provide an isomorphism if $f$ is given by a homogeneous polynomial. Let $A = \mathbb{C}[z_1, \cdots, z_n]$ be the polynomial ring and $I = (f_1, \cdots, f_n)$ the ideal of $A$ generated by the partial derivatives of $f$. Then clearly $\mathcal{O}/(\partial f) \simeq A/I$. Let $H$ be a homogeneous subspace of $A$ so that $A = H \oplus I$ and let $P = \mathbb{C}[f_1, \cdots, f_n]$. Then it can be shown that $A \cong P \otimes_{\mathbb{C}} H$. Let $B = A/(f-1)$, the coordinate ring of $F$. We define the complex $\Omega_B$ of Kähler differentials on $B$. It is known that $\Omega_B = \Gamma(F, \Omega_F^*)$. Thus we want an isomorphism $A/I \to H^{n-1}(\Omega_B)$. Define $\omega = \sum_{i=1}^n (-1)^{i-1} z_i \, dz_1 \wedge \cdots \wedge \widehat{dz_i} \wedge \cdots \wedge dz_n$, a global $(n-1)$-form. Let $\pi : \Omega_A \to \Omega_B$ and for a cocycle $\alpha$ let $[\alpha]$ denote its cohomology class. Then we have:

(IV.2.2) THEOREM. *The map* $\theta : A \to H^{n-1}(\Omega_B)$ *defined by*

$$\theta(p \otimes h) = [p(0)\pi(h\omega)]$$

*induces an isomorphism between* $A/I$ *and* $H^{n-1}(\Omega_B)$.

If $G \subset GL(n, \mathbb{C})$ leaves $f$ invariant, then $G$ has a representation both on $A/I$ and on $H^{n-1}(\Omega_B)$. The map $\theta$ is not equivariant, but we have for $g \in G$ and $u \in A/I$

$$\theta(gu) = (\det g) g(\theta u).$$

Using the divisor notation of Milnor and Orlik [1] we compute the representation of $G$ on $H^{n-1}(\Omega_B)$.

The contents of the notes may be given schematically by the following diagram.

$$\begin{array}{ccc} \#\{\partial f^{-1}(y)\} = \mu_c & \xleftrightarrow{\text{Chapter II}} & \mu_t = \dim_{\mathbb{C}} H^{n-1}(F^h; \mathbb{C}) \\ \Big\updownarrow \text{Chapter I} & & H^{n-1}(F^h; \mathbb{C}) \\ \dim_{\mathbb{C}} \mathcal{O}/(\partial f) = \mu_a & \approx & \Big\uparrow \text{Chapter III} \\ \mathcal{O}/(\partial f) & \xrightarrow[\approx]{\text{Chapter IV}} & DR^{n-1}(\Omega_F^*) \end{array}$$

I have tried to present the material assuming only the basics in algebra and topology, so the notes could serve as a primer for the forthcoming book by Brieskorn [7]. The belaboring of certain points is probably caused by my original unfamiliarity with much of the material.

I wish to thank the participants of a seminar at the University of Wisconsin, where the first draft was presented, for their questions and comments; Joe Lipman, Alex Nagel, and Phil Wagreich for helpful discussions, and Louis Solomon for a most enjoyable collaboration.

Finally, I am grateful to Per Holm and the Nordic Summer School at Oslo for providing an excuse for writing these notes, and the participants, in particular G. Barthel and L. Kaup, for their helpful comments on the circulated version.

Oslo, August 1976 and Madison, October 1976.

## Chapter I

## Germs of holomorphic maps

First recall some standard facts from complex analysis, see Gunning and Rossi [1, Ch. I, II].

### 1. Holomorphic maps

Let $D \subset \mathbb{C}^n$ be an open set. The map $f: D \to \mathbb{C}$ is called *holomorphic* in $D$ if each point $w \in D$ has an open neighborhood $U$, $w \in U \subset D$, such that the function $f$ has a power series expansion

$$f(z) = \sum_{\nu_1, \cdots, \nu_n = 0}^{\infty} a_{\nu_1}, \cdots, {}_{\nu_n}(z_1 - w_1)^{\nu_1} \cdots (z_n - w_n)^{\nu_n}$$

which *converges* for all $z \in U$. The set of all functions holomorphic in $D$ is denoted $\mathcal{O}_D$. The *Cauchy-Riemann* criterion states that a complex valued function $f$ defined in an open set $D \subset \mathbb{C}^n$ and continuously differentiable in the underlying real coordinates of $\mathbb{C}^n$ is holomorphic in $D$ if and only if

$$\frac{\partial f}{\partial \bar{z}_j}(z) = 0, \qquad j = 1, \cdots, n.$$

This may be used to show that $\mathcal{O}_D$ is a ring under pointwise addition and multiplication. For open sets $D \subset \mathbb{C}^n$, $D' \subset \mathbb{C}^m$, the map $F: D \to D'$ $F(z) = (f_1(z), \cdots, f_m(z))$ is called holomorphic if each of the $f_i(z)$ is holomorphic.

The Jacobian matrix of $F$ at $w \in D$ is

$$J_F(w) = \left(\frac{\partial f_i}{\partial z_j}\right)_{(w)}.$$

Call $w$ a *simple* point of $F$ if the rank of $J_F(w)$ is maximal, i.e. rank $J_F(w) = \min(m, n)$. Otherwise $w$ is a *critical* point of $F$. A critical point $w$ is *isolated* if in some neighborhood of $w$ all other points are simple. The image of a critical point is called a *critical value.* The complement of the set of critical values in the range is called the set of *regular values.* Note that all points in the inverse image of a regular value are simple, but not all points in the inverse image of a critical value are necessarily critical points. A consequence of Sard's theorem (Milnor [2]) may be stated as follows.

(1.1) The set of regular values of a holomorphic map is everywhere dense.

The holomorphic version of the inverse mapping theorem states:

(1.2) Let $F$ be a holomorphic mapping of a neighborhood $D$ of 0 in $\mathbb{C}^n$ into $\mathbb{C}^n$. Assume that $F(0) = 0$ and 0 is a simple point of $F$. Then there exists a neighborhood $\Delta$ of 0 in which $F$ has a holomorphic inverse $G$ so that $w = F(z)$ if and only if $z = G(w)$.

A subset $M$ of $\mathbb{C}^n$ is a *complex submanifold* of $\mathbb{C}^n$ if to every $p \in M$ corresponds a neighborhood $U$ and a mapping $F: U \to \mathbb{C}^m$ $(n \geqq m)$ so that $p$ is a simple point of $F$ and

$$M \cap U = \{z \in U \mid F(z) = F(p)\}.$$

It follows from (1.1) that there exist coordinate functions $x_1, \cdots, x_n$ at $p$ so that in some neighborhood $V$ of $p$

$$M \cap V = \{z \in V \mid x_1(z) = \cdots = x_m(z) = 0\}.$$

## 2. Local rings

Let $w \in \mathbb{C}^n$ be a fixed point with neighborhoods $U$, $V$, $W$, etc. Call two functions $f: U \to \mathbb{C}$ and $g: V \to \mathbb{C}$ *equivalent* at $w$ if there exists a neighborhood $W$ of $w$, $W \subset U \cap V$ so that $f|_W = g|_W$. The equivalence class of such a function $f$ is called the *germ*, $\hat{f}$ of $f$ at $w$. Note that equivalent functions have the same value at $w$ but this condition is not sufficient for equivalence, e.g. $z$ and $z^2$ are not equivalent at 0.

Pointwise addition and multiplication gives a ring structure to germs of functions. We shall denote by $\mathcal{O}_w$ the ring of germs of holomorphic functions at $w$. When $w$ is the origin 0, we write $\mathcal{O}$ for $\mathcal{O}_0$.

(2.1) THEOREM. *The ring $\mathcal{O}_w$ is isomorphic to the ring of convergent power series centered at $w$.*

By a linear change of coordinates $\mathcal{O}_w$ is isomorphic to $\mathcal{O}$, which can be identified with $\mathbb{C}\{z_1, \cdots, z_n\}$ by (2.1). Note that $\mathcal{O}$ is an integral domain. Its quotient field, $\mathcal{M}$ is called the field of germs of meromorphic functions.

A *unit* $\hat{f} \in \mathcal{O}$ is a germ represented by a function $f$ with $f(0) \neq 0$. It follows that $1/f$ is holomorphic in some neighborhood of 0, so $1/\hat{f} \in \mathcal{O}$. The non-units of $\mathcal{O}$ are therefore all germs vanishing at 0. These form an ideal, $\mathfrak{m}$ which is maximal by the above remark, and therefore $\mathcal{O}$ is a local ring.

It can be shown that $\mathcal{O}$ is a unique factorization domain and a Noetherian ring.

## 3. Varieties and Nullstellensatz

A subset $V \subset \mathbb{C}^n$ is called a *variety* if for every $z \in \mathbb{C}^n$ we can find a neighborhood $U_z$ and functions $f_1, \cdots, f_t$ holomorphic in $U_z$ such that

$$V \cap U_z = \{x \in U_z \mid f_1(x) = \cdots = f_t(x) = 0\} = V(f_1, \cdots, f_t).$$

Now we want to define the germ of a variety. First define the germ of a set as follows. Two subsets $X$ and $Y$ of $\mathbb{C}^n$ are equivalent if there is a neighborhood $U$ of 0 so that $X \cap U = Y \cap U$. Denote the equivalence class of $X$ by $\hat{X}$.

Let $\hat{f} \in \mathcal{O}$. Define $V(f)$ to be the germ of the sets $\{z \in U \mid f(z) = 0\}$, where $f$ is a representative of the germ $\hat{f}$. Note that we may define the following operations on germs in the obvious way: $\hat{X}_1 \cup \hat{X}_2$, $\hat{X}_1 \cap \hat{X}_2$, $\hat{X}_1 - \hat{X}_2$ and that $\hat{X}_1 \subset \hat{X}_2$ is a well defined relation.

Let $\hat{f} \in \mathcal{O}$ and let $\hat{X}$ be the germ of a set. If $\hat{X} \subset \widehat{V(f)}$ then we say that $\hat{f}$ *vanishes* on $\hat{X}$. A germ $\hat{X}$ is the *germ of a variety* if there are elements $\hat{f}_1, \cdots, \hat{f}_t \in \mathcal{O}$ such that

$$\hat{X} = \widehat{V(f_1)} \cap \cdots \cap \widehat{V(f_t)} = \hat{V}(f_1, \cdots, f_t).$$

Let $\hat{V}$ be the germ of a variety. Define the *ideal* of $\hat{V}$ by

$$\mathcal{J}(\hat{V}) = \{\hat{f} \in \mathcal{O} \mid \hat{f} \text{ vanishes on } \hat{V}\}.$$

Let $\mathfrak{A} \subset \mathcal{O}$ be a subset. Define the *locus* of $\mathfrak{A}$ by

$$\mathcal{L}(\mathfrak{A}) = \bigcap_{f \in \mathfrak{A}} \widehat{V(f)}.$$

It is clear that $\mathscr{I}(\hat{V})$ is indeed an ideal in $\mathcal{O}$.

The ideal generated by $\mathfrak{A}$ is finitely generated because $\mathcal{O}$ is Noetherian. It is clear that the locus of a generating set of germs is a variety equal to $\mathscr{L}(\mathfrak{A})$.

Let $\mathscr{F}$ be an ideal in a ring $\mathscr{R}$. Define the *radical* of $\mathscr{F}$ to be the ideal

$$\text{Rad}\,\mathscr{F} = \{x \in \mathscr{R} \mid \exists n > 0,\, x^n \in \mathscr{F}\}.$$

We are now ready to state the analytic Nullstellensatz; Gunning and Rossi [1, p. 90].

(3.1) THEOREM. *Let $\mathfrak{A}$ be an ideal in $\mathcal{O}$. Then*

$$\mathscr{I}\mathscr{L}(\mathfrak{A}) = \text{Rad}\,\mathfrak{A}.$$

The inclusion $\mathscr{I}\mathscr{L}(\mathfrak{A}) \supset \text{Rad}\,\mathfrak{A}$ follows, since clearly $\mathscr{I}\mathscr{L}(\mathfrak{A}) = \mathscr{I}\mathscr{L}(\text{Rad}\,\mathfrak{A})$ and $\mathscr{I}\mathscr{L}(\text{Rad}\,\mathfrak{A}) \supset \text{Rad}\,\mathfrak{A}$. The converse is a deep result and we shall not attempt to prove it here. Our interest is in the following immediate consequence.

(3.2) THEOREM. *Let $f: \mathbb{C}^n \to \mathbb{C}$ be a holomorphic map with an isolated critical point at the origin. Let $f_i = \partial f/\partial z_i$ be the holomorphic partials of $f$ and $(\partial f) = (\hat{f}_1, \cdots, \hat{f}_n)$ the ideal generated by their germs in $\mathcal{O}$. Then $\mathcal{O}/(\partial f)$ is a finite dimensional vector space over $\mathbb{C}$.*

PROOF. Since the only solution to $f_1(z) = \cdots = f_n(z) = 0$ near 0 is 0, we have $\mathscr{L}(\partial f) = \hat{V}(f_1, \cdots, f_n) = 0$ so $\mathscr{I}\mathscr{L}(\partial f) = \mathfrak{m}$ where $\mathfrak{m}$ is the maximal ideal of all functions vanishing at 0. By (3.1) $\mathscr{I}\mathscr{L}(\partial f) = \text{Rad}\,(\partial f)$ so some power of every coordinate function belongs to $(\partial f)$. Thus $\mathfrak{m}/(\partial f)$ is finite dimensional and the dimensions of $\mathcal{O}/(\partial f)$ and $\mathfrak{m}/(\partial f)$ only differ by one, the constants.

(3.3) DEFINITION. The *algebraic multiplicity* of the isolated critical point 0 of $f$ is

$$\mu_a = \dim_{\mathbb{C}} \mathcal{O}/(\partial f).$$

(3.4) REMARK. The definition

$$\text{codim}\,\hat{f} = \dim_{\mathbb{C}} \mathfrak{m}/(\partial f) = \mu_a - 1$$

is also used and has a geometric interpretation, see (7.6).

(3.5) EXAMPLE. Let $f: \mathbb{C}^2 \to \mathbb{C}$ be the map $f(z_1, z_2) = z_1^3 + z_2^3$. Then $f_1 = 3z_1^2$, $f_2 = 3z_2^2$, and $\partial f = (3z_1^2, 3z_2^2)$. Consider the monomials $z_1^{\alpha_1} z_2^{\alpha_2}$, $\alpha_1 \geqq 0$, $\alpha_2 \geqq 0$ schematically represented on page 414.

The monomials in the ideal $(\partial f)$ are below the solid lines. Thus $\mu_a = 4$ and the cosets of the elements $\{1, z_1, z_2, z_1 z_2\}$ may be chosen as a $\mathbb{C}$-basis for $\mathcal{O}/(\partial f)$.

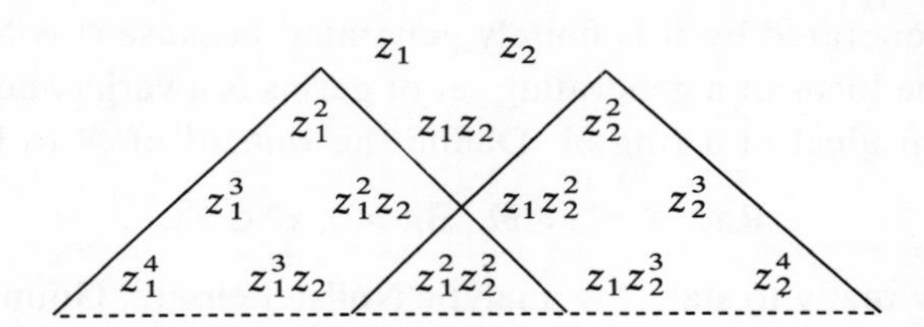


## 4. Finitely determined maps

In this section we shall prove a theorem of Mather [1] which says that provided $f:\mathbb{C}^n \to \mathbb{C}$ is a holomorphic map germ with an isolated critical point at 0, there exists a holomorphic change of coordinates, $h:\mathbb{C}^n \to \mathbb{C}^n$ so that $f \circ h$ is a polynomial map. The theorem is equally valid in the $C^\infty$ and real analytic categories but we shall only prove the version relevant for us.

First let us recall Nakayama's lemma.

(4.1) LEMMA. *Let $R$ be a commutative ring with identity, $\mathfrak{m}$ an ideal, $L$ an $R$-module, $M$ and $N$ submodules of $L$. Suppose*
(a) $(1+x)^{-1}$ *exists in $R$ for every $x \in \mathfrak{m}$,*
(b) *$M$ is finitely generated,*
(c) $M \subset N + \mathfrak{m}M$,
*then $M \subset N$.*

PROOF. Let $e_1, \cdots, e_n$ generate $M$. By (c) we have $f_i \in N$, $\alpha_{ij} \in \mathfrak{m}$ such that

$$e_i = f_i + \sum_{j=1}^{n} \alpha_{ij} e_j \quad \text{for} \quad i = 1, \cdots, n.$$

Using $\vec{e}$ for the column vector $(e_1, \cdots, e_n)$, $\vec{f}$ for the column vector $(f_1, \cdots, f_n)$ and $A = (\alpha_{ij})$ we have $(I-A)\vec{e} = \vec{f}$. Since $\det(I-A) = 1+x$, $x \in \mathfrak{m}$ and $1+x$ is invertible in $R$ by (a), $(I-A)^{-1}$ exists, so $\vec{e} = (I-A)^{-1}\vec{f}$.

In order to include the dimension of the space in our notation, let $\mathcal{O}_n$ be the local ring of holomorphic map germs at 0 in $\mathbb{C}^n$ and $\mathfrak{m}_n$ the maximal ideal of germs that vanish at 0.

(4.2) DEFINITION. Two holomorphic map germs at $0 \in \mathbb{C}^n$, $f:U \to \mathbb{C}$, $g:V \to \mathbb{C}$ are called $k$-jet *equivalent* at 0 if $f(0) = g(0)$ and all their partial derivatives of order $\leqq k$ agree at 0. Call $j_0^k(f)$ the equivalence class of $f$ and $J^k$ the set of $k$-jets at 0. It is clearly a finite dimensional vector space over $\mathbb{C}$, isomorphic to the vector space of all polynomials in $z_1, \cdots, z_n$ of degree $\leqq k$.

Let $f_{(k)}$ denote the power series expansion of $f$ at 0 truncated at degree $k$.

Then clearly $f$ and $g$ are $k$-jet equivalent if and only if $f_{(k)}=g_{(k)}$.

(4.3) DEFINITION. Two germs, $\hat{f}, \hat{g} \in m_n$ are called *right equivalent* and denoted $\hat{f} \underset{R}{\sim} \hat{g}$ if there exists a germ of holomorphic coordinate change $h:\mathbb{C}^n \to \mathbb{C}^n$ so that $f=g \circ h$.

(4.4) DEFINITION. A germ $\hat{f} \in m_n$ is called (right) *k-determined* if for any $\hat{g} \in m_n$

$$f_{(k)}=g_{(k)} \Rightarrow \hat{f} \underset{R}{\sim} \hat{g}.$$

In particular $\hat{f}$ is equivalent to the germ of a polynomial. We may state Mather's result:

(4.5) THEOREM. *If* $\hat{f} \in m_n$ *and*

$$m_n^{k+1} \subset m_n^2(\partial f)+m_n^{k+2}$$

*then* $\hat{f}$ *is* $k$-*determined.*

PROOF. Let $\hat{f}$, $\hat{g} \in m_n$ and let $f$ and $g$ be their representatives with $f_{(k)}=g_{(k)}$. Our aim is to construct a holomorphic germ $h:(\mathbb{C}^n, 0) \to (\mathbb{C}^n, 0)$ such that $f=g \circ h$. Define a family of holomorphic germs $F_t(z)=F(z, t)$: $(\mathbb{C}^n \times \mathbb{R}, 0 \times \mathbb{R}) \to (\mathbb{C}, 0)$ by

(i) $F(z, t)=(1-t)f(z)+tg(z)$.

Note that $F_0=f$, $F_1=g$, and $(F_t)_{(k)}=f_{(k)}$. The theorem follows from the next assertion, whose proof takes up the rest of the argument.

(*) For each $t_0 \in \mathbb{R}$ there exists $\varepsilon>0$ such that $F_t \underset{R}{\sim} F_{t_0}$ for all $t$ provided $|t-t_0|<\varepsilon$.

For every $t_0 \in \mathbb{R}$ we shall construct a local $C^\infty$ family of germs of invertible holomorphic maps $H_t(z)=H(z, t):(\mathbb{C}^n \times U, 0 \times U) \to (\mathbb{C}^n, 0)$ where $U$ is a small neighborhood of $t_0$ in $\mathbb{R}$, such that

(ii) $F_t \circ H_t(z)=f(z)$.

Differentiate (ii) with respect to $t$ to obtain

(iii) $\partial F_t(H_t(z)) \cdot \dot{H}_t(z)+g(H_t(z))-f(H_t(z))=0$

where we let

$$\partial F_t=\left(\frac{\partial F}{\partial z_1}, \cdots, \frac{\partial F}{\partial z_n}\right) \quad \text{and} \quad \dot{H}_t=\left(\frac{\partial H^1}{\partial t}, \cdots, \frac{\partial H^n}{\partial t}\right).$$

Note that by assumption $g \circ H_t - f \circ H_t \in m_n^{k+1}$.

Since the constructed $H_t$ will be invertible, we may define a local family of map germs $\vec{\xi}_t:(\mathbb{C}^n \times U, 0 \times U) \to (\mathbb{C}^n, 0)$ by

(iv) $\vec{\xi}_t(H_t(z))=\dot{H}_t(z)$.

We may write (iii) using (iv) as

$$F_t(H_t(z)) \cdot \vec{\xi}_t(H_t(z)) + g(H_t(z)) - f(H_t(z)) = 0.$$

Using the assumed invertibility of $H_t$ again, solving (iii) is reduced to (iv) and

(v) $F_t(z) \cdot \vec{\xi}_t(z) + g(z) - f(z) = 0.$

It remains to show that we can solve (iv) and (v). First consider (v).

(**) Let $m_n^{k+1} \subset m_n^2(\partial f) + m_n^{k+2}$. Then for all $t_0 \in \mathbb{R}$ there exists a local family of map germs $\vec{\xi}_t(z): (\mathbb{C}^n \times U, 0 \times U) \to (\mathbb{C}^n, 0)$ where $U$ is a neighborhood of $t_0$ in $\mathbb{R}$, such that

(vi) $\vec{\xi}_t(0) = 0$, and

(vii) $\partial F_t(z) \cdot \vec{\xi}_t(z) + g(z) - f(z) = 0$, for all $(z, t)$ near $(0, t_0)$.

To prove (**) let $\mathscr{E}$ be the ring of germs of $C^\infty$ maps at $(0, t_0)$ from $\mathbb{C}^n \times \mathbb{R}$ to $\mathbb{C}$ and $\mathcal{N}$ its maximal ideal. Let $\mathscr{I}^* = \mathscr{E}\left(\frac{\partial F}{\partial z_1}, \ldots, \frac{\partial F}{\partial z_n}\right)$. Since $g - f \in m_n^{k+1}$, we can solve (v) provided

$$m_n^{k+1} \subset \mathscr{I}^* \mathscr{E} m_n$$

where $m_n$ is the maximal ideal of the $t$-invariant functions. From (i)

$$\frac{\partial F}{\partial z_i} = \frac{\partial f}{\partial z_i}(1-t) + \frac{\partial g}{\partial z_i} t$$

$$\frac{\partial f}{\partial z_i} = \frac{\partial F}{\partial z_i} - t \frac{\partial}{\partial z_i}(g - f).$$

So we have

$$(\partial f) \subseteq \mathscr{I}^* + \mathscr{E} m_n^k.$$

Using the assumption that $m_n^{k+1} \subset m_n^2(\partial f) + m_n^{k+2}$ we find

$$\begin{aligned} \mathscr{E} m_n^{k+1} &\subseteq \mathscr{E} m_n^2(\partial f) + \mathscr{E} m_n^{k+2} \subseteq m_n^2 \mathscr{I}^* + \mathscr{E} m_n^{k+2} \\ &\subseteq m_n^2 \mathscr{I}^* + \mathcal{N} \mathscr{E} m_n^{k+1} \end{aligned}$$

and Nakayama's lemma (4.1) completes (**).

Next we turn to (iv). Given $t_0 \in \mathbb{R}$, by the existence theorem for ordinary differential equations, there exist local $C^\infty$ families of map germs $H_t(z): (\mathbb{C}^n \times U, 0 \times U) \to (\mathbb{C}^n, 0)$ satisfying (iv) and the initial condition $H_{t_0} =$ id. Thus there exists $\varepsilon > 0$ such that $H_t$ is a diffeomorphism for all $t$ provided $|t - t_0| < \varepsilon$. The proof of (*) is completed by noting that

$$\frac{d}{dt}(F_t \circ H_t(z)) = F_t(H_t(z)) \cdot \dot{H}_t(z) + g(H_t(z)) - f(H_t(z)) = 0$$

by (**) and therefore

$$F_{t_0} = F_{t_0} H_{t_0} = F_t \circ H_t \quad \text{for all } t \text{ with } \quad |t - t_0| < \varepsilon.$$

In order to convert the $C^\infty$ statement to holomorphic, use standard arguments of Coddington and Levinson [1, Sec. 8].

(4.6) Corollary. *Suppose $\hat{f} \in \mathcal{O}$ is represented by a map $f$ with an isolated critical point at 0. Then there exists a holomorphic germ of change of coordinates $h$ so that $f \circ h$ is a polynomial map.*

Proof. Combine (3.2) with (4.5).

## 5. Covering multiplicity

In view of (4.6) we may assume in the remainder of these notes that $f$ is in fact given by a polynomial map with an isolated critical point at 0.

(5.1) Definition. A mapping $g: X \to Y$ is *light* if $g^{-1}(y)$ consists of only a discrete set of points for all $y \in Y$. It is *proper* if for every compact subset $K \subset Y$, $f^{-1}(K)$ is compact in $X$.

(5.2) Definition. Let $D$ be a domain in $\mathbb{C}^n$. A subset $X \subset D$ is called *thin* if for every point $z \in D$ there are an open polydisc $\Delta(z; r) \subset D$ and a function $g$ holomorphic and not identically zero in $\Delta(z; r)$, such that $g$ vanishes identically on $X \cap \Delta(z; r)$.

Recall here that a polydisc with center $w \in \mathbb{C}^n$ and radius $r = (r_1, \cdots, r_n) \in \mathbb{R}^n$, $(r_j > 0)$ is

$$\begin{aligned}\Delta(w; r) &= \Delta(w_1, \cdots, w_n; r_1, \cdots, r_n) \\ &= \{z \in \mathbb{C}^n \mid |z_j - w_j| < r_j, 1 \leqq j \leqq n\}.\end{aligned}$$

The next definition is not identical to the one usually given, but sufficient for us, Gunning and Rossi [1, p. 101, Def. 3; see also Remark after Def. 4].

(5.3) Definition. An *analytic cover* is a triple $(X, \pi, Y)$ such that
(i) $X$ is a locally compact Hausdorff space;
(ii) $Y$ is a domain in $\mathbb{C}^n$
(iii) $\pi$ is a proper, light, continuous mapping of $X$ onto $Y$;
(iv) there are a thin set $\Delta \subset Y$, and an integer $\lambda$, such that $\pi$ is a $\lambda$-sheeted covering map from $X - \pi^{-1}(\Delta)$ onto $Y - \Delta$;
(v) $X - \pi^{-1}(\Delta)$ is dense in $X$.

(5.4) Proposition. *Let $f: \mathbb{C}^n \to \mathbb{C}$ be a polynomial germ with isolated critical point. Then for some open neighborhood of 0, $X \subset \mathbb{C}^n$ and its image $\partial f(X) = Y \subset \mathbb{C}^n$ the triple $(X, \partial f, Y)$ is an analytic cover.*

Proof. See Gunning and Rossi [1, p. 108].

(5.5) Definition. Let $(X, \pi, Y)$ be an analytic cover. For $x \in X$, define the *branching order* of $x$, as the integer $\lambda$ such that $x$ has a neighborhood basis of $\lambda$-sheeted covers.

(5.6) Definition. Choose $X$ and $Y$ as in (5.4) so $(X, \partial f, Y)$ is an analytic cover. Define the *covering multiplicity* of $f$ at 0 by

$$\mu_c = \text{branching order of } 0 \in X.$$

Note that $\partial f$ is a $\mu_c$-sheeted covering in the complement of the branch locus $\Delta$, so

$$\mu_c = \#\{(\partial f)^{-1}(y)\}, \quad y \in Y - \Delta.$$

(5.7) Example. As in (3.5), let $f(z_1, z_2) = z_1^3 + z_1^3$. Then $f(z_1, z_2) = (3z_1^2, 3z_2^2)$ is an analytic cover with branch locus $\Delta = \{y_1 = 0\} \cup \{y_2 = 0\}$. For $\varepsilon \neq 0$, $y = (3\varepsilon^2, 3\varepsilon^2) \in Y - \Delta$ has clearly $\mu_c = 4$.

The main aim of this section is to prove the result of Palamodov [1] and Grauert that $\mu_c = \mu_a$. This seems to be well known to algebraic geometers, but I have not been able to locate a proof which reveals the main ideas. I am indebted to J. Lipman for help in collecting the pieces of the following argument.

Let us identify the local ring of the domain at 0 with $S = \mathbb{C}\{z_1, \cdots, z_n\}$ and the local ring of the range at 0 with $\mathbb{C}\{y_1, \cdots, y_n\}$. Since $(\partial f)^*$ is a monomorphism, the ring $R = \{f_1, \cdots, f_n\} \subset S$ is isomorphic to $\mathbb{C}\{y_1, \cdots, y_n\}$.

(5.8) Proposition. *$S$ is a finitely generated module over $R$ and there exists a generating set consisting of $\mu_a$ elements.*

Proof. Let $\{\varphi_i\}_{i=1}^{\mu_a}$ generate $\mathcal{O}/(\partial f) = S/(f_1, \cdots, f_n)$ as a $\mathbb{C}$-vector space. Then the Malgrange preparation theorem implies that $S = \sum \varphi_i R \quad i = 1, \cdots, \mu_a$ see Bröcker and Lander [1, 6.7].

In order to prove that $S$ is a *free* $R$-module, we shall need some facts from commutative algebra, see Nagata [1].

(5.9) A ring $A$ is of *altitude* $r$ if there is a chain of distinct prime ideals $\mathcal{P}_i$ such that

$$\mathcal{P}_0 \supset \mathcal{P}_1 \supset \cdots \supset \mathcal{P}_r$$

but there is no such chain with more terms.

If $(A, \mathfrak{m})$ is a local ring of altitude $r$, then there are $r$ elements $a_1, \cdots, a_r$ of $\mathfrak{m}$ which generate a primary ideal belonging to $\mathfrak{m}$, but there is no primary ideal belonging to $\mathfrak{m}$ which is generated by $r-1$ elements.

The set of such elements $a_i$ above is called a *system of parameters* of $A$. A system of parameters $x_1, \cdots, x_r$ of a local ring $A$ is called a *regular system of parameters* if it generates the maximal ideal of $A$. A local ring with such a regular system of parameters is called a *regular local ring*.

A sequence of elements $a_1, \cdots, a_r$ of $A$ is called an *$A$-regular sequence* if for any $x \in A$

$$xa_i \in (a_1, \cdots, a_{i-1}) \Rightarrow x \in (a_1, \cdots, a_{i-1})$$

for $i = 1, \cdots, r$. Here $(a_1, \cdots, a_k)$ is the ideal generated by $a_1, \cdots, a_k$. The crucial fact in translating the topological information that $f$ has an isolated critical point into algebra is the following result.

(5.10) THEOREM. *If $f: \mathbb{C}^n \to \mathbb{C}$ has an isolated critical point at the origin, then the partials $f_1, \cdots, f_n$ form an $S$-regular sequence.*

PROOF. Serre [3, III B, Prop. 6] shows that $f_1, \cdots, f_n$ form a system of parameters in $S$. The conclusion follows by Nagata [1, Th. 25.4].

The corresponding result for polynomial rings is in Bourbaki [1, p. 137, Ex. 5].

(5.11) THEOREM. *$S$ is a free $R$-module of rank $\mu_a$.*

PROOF. We may argue at least three different ways, all amounting to the same.

(i) $S$ and $R$ are regular local rings of dimension $n$. Use Nagata [1, 25.14 and 25.16] to derive the conclusion.

(ii) Use (5.10) to conclude that the homological codimension (depth) of the $R$-module $S$ is $n$, so its homological dimension is $n - n = 0$ and by Nagata [1, 26.4] $S$ is a free $R$-module.

(iii) As in Serre [3, IV A paragraph 4] prove acyclicity of the Koszul complex based on the sequence $f_1, \cdots, f_n$. This is the argument used by Palamodov [1].

(5.12) COROLLARY. *Let $F_R$ and $F_S$ be the respective quotient fields and $d = [F_S : F_R]$ the degree of the field extension. Then $d = \mu_a$.*

(5.13) THEOREM. $\mu_c = \mu_a$.

PROOF. According to the theorem on the primitive element, for some linear combination $\xi = \sum c_i z_i$, $F_S = F_R(\xi)$. By a change of coordinates we may assume that $\xi = z_1$. The minimal equation of $z_1$ over $F_R$ has the form $p(z_1, f_1, \cdots, f_n) = 0$, where

$$p(z_1, f_1, \cdots, f_n) = z_1^d + A_1(f_1, \cdots, f_n) z_1^{d-1} + \cdots + A_d(f_1, \cdots, f_n).$$

Here the $A_i$ are power series convergent in a neighborhood of 0, $U \subset Y$, with $A_i(0)=0$.

Let $V_0 \subset U \times \mathbb{C}$ be the hypersurface

$$V_0 = \{(f_1, \cdots, f_n, z_1) \mid p(z_1, f_1, \cdots, f_n) = 0\}.$$

Let $\pi: V_0 \to U$ be the projection. The $z_1$-discriminant of $p$ is

$$\Delta = \left\{(f_1, \cdots, f_n) \in U \mid \exists z_1, p(z_1, f_1, \cdots, f_n) = 0, \frac{dp}{dz_1}(z_1, f_1, \cdots, f_n) = 0\right\}.$$

Note that $\Delta$ is a nowhere dense $\operatorname{codim}_{\mathbb{R}} 2$ subset of $U$ and with $B_0 = \pi^{-1}(\Delta)$

$$\pi: V_0 - B_0 \to U - \Delta$$

is a $d$-sheeted covering map. The argument is completed by considering the map

$$\sigma: X \to V_0$$
$$\sigma(z_1, \cdots, z_n) = (f_1(z), \cdots, f_n(z), z_1).$$

Let $B = \sigma^{-1}(B_0)$, $V = \sigma^{-1}(V_0)$. After shrinking $U$ and $V$ if necessary as in Gunning [1, p. 28], $\sigma$ maps $V - B$ onto $V_0 - B_0$ *homeomorphically*. Thus the composition $\pi \circ \sigma = \partial f$ is a $d$-sheeted cover on a dense open set near $O$, so $\mu_c = d = \mu_a$.

There is a more sophisticated argument called to my attention by Alex Nagel in Gunning [2, Th. 7]. This proves that if $\varphi: X \to Y$ is a finite analytic branched cover of branching order $d$ and $\mathfrak{A}_x = \varphi^* \mathfrak{m}_{\varphi_x}(Y) \subset \mathcal{O}_{X,x'}$ then $\dim_{\mathbb{C}} \mathcal{O}_{X,x}/\mathfrak{A}_x \geqq d$ and equality holds at $x = 0$ if and only if $\mathcal{O}_{X,0}$ is a free $\mathcal{O}_{Y,0}$ module. The proof depends on the coherence theorem for direct images under finite maps. (For the definition of a coherent sheaf see III.1.11.)

(5.14) The finite $A$-module $M$ is called *flat* if for every exact sequence of finite $A$-modules

$$0 \to N' \to N \to N'' \to 0$$

we have an exact sequence

$$0 \to N' \otimes_A M \to N \otimes_A M \to N'' \otimes_A M \to 0.$$

A map of rings $\pi: A \to B$ is called *flat* if the ring $B$ as an $A$-module is flat.

Thus we may summarize the result of (5.11) by saying that if $f$ has an isolated critical point at the origin, then the induced map $(\partial f)^*: \mathcal{O}_Y \to \mathcal{O}_X$ is a flat map.

## 6. Deformation

In this section we shall discuss small perturbations of the function $f$.

(6.1) DEFINITION. The critical point $z_0$ of $f$ is called *non-degenerate* if the Hessian matrix

$$\left(\frac{\partial^2 f}{\partial z_i \partial z_j}\right)$$

is non-singular at $z_0$.

(6.2) LEMMA. *The critical point* 0 *of* $f$ *is non-degenerate if and only if* $\mu_a = 1$.

PROOF. By (6.1) if 0 is a non-degenerate critical point of $f$, then 0 is a simple point of $\partial f$. The inverse function theorem (1.2) asserts that in some neighborhood of 0, $\partial f$ has a holomorphic inverse $\Phi = (\varphi_1, \cdots, \varphi_n)$ so that $y = \partial f(z)$ if and only if $z = \Phi(y)$. Now $y_i = f_i(z)$ so $z_i = \varphi_i(f_1(z), \cdots, f_n(z))$ is valid in a neighborhood of 0. Thus the ideal

$$(f_1, \cdots, f_n) = (\partial f) \supset m = (z_1, \cdots, z_n)$$

and $\dim_{\mathbb{C}} \mathcal{O}/(\partial f) = \dim_{\mathbb{C}} \mathcal{O}/m = 1$.

If $\mu_a = 1$, then in some neighborhood of 0 we have after suitable change of coordinates

$$f = z_1^2 + \cdots + z_n^2 + \text{higher terms}$$

and the Hessian is non-singular at 0.

Notice the connection of this with the Morse lemma, Milnor [1].

(6.3) DEFINITION. A *deformation* of $f$ is a holomorphic map $F: U \times W \to \mathbb{C}$ so that $F(z, 0) = f(z)$. Here $W$ is an open subset of $\mathbb{C}^k$ with coordinates $t_1, \cdots, t_k$ which we shall refer to as the parameters of the deformation.

(6.4) DEFINITION. A deformation $F$ of $f$ is *infinitesimally versal* if

$$\mathcal{O}_n = (\partial f) + \mathbb{C}[\varphi_1, \cdots, \varphi_k]$$

where $\mathcal{O}_n$ and $(\partial f)$ are as before and $\mathbb{C}[\varphi_1, \cdots, \varphi_k]$ is the $\mathbb{C}$-vector space spanned by the functions $\varphi_i = \left(\frac{\partial F}{\partial t_i}\right)_{t=0}$, $i = 1, \cdots, k$.

(6.5) EXAMPLE. The polynomial

$$F(z_1, z_2; t_1, t_2, t_3, t_4) = z_1^3 + z_2^3 + t_1 + t_2 z_1 + t_3 z_2 + t_4 z_1 z_2$$

is an infinitesimally versal deformation of the polynomial $f(z_1, z_2) = z_1^3 + z_2^3$.

(6.6) Proposition. *Let $f:\mathbb{C}^n \to \mathbb{C}$ be a polynomial map with an isolated critical point at* 0. *Then*

$$F(z;t)=f(z)+\sum_{j=1}^{\mu_a} t_j\varphi_j(z)$$

*is an infinitesimally versal deformation of f for any basis* $\{\varphi_j(z)\}_{j=1}^{\mu_a}$ *of* $\mathcal{O}/(\partial f)$.

(6.7) Definition. For a fixed value of the parameter $t=(t_1,\cdots,t_k)$, call the map $F_t(z)=F(z,t)$ an *approximation* of $f(z)$.

(6.8) Definition. An approximation $F_t$ of $f$ is called *generic* if $F_t$ has only non-degenerate critical points and all distinct critical values.

It is a consequence of Sard's theorem (1.1) that generic approximations of $f$ exist for arbitrarily small $t$; and for some $\delta>0$, the number of critical points near 0 of every generic approximation $F_t$, $|t|<\delta$ is the same.

(6.9) Definition. The *deformation multiplicity* of the critical point 0 of $f$ is

$\mu_d$ = number of critical points in a generic approximation $F_t$ of $f$, $|t|<\delta$.

The next result was observed by Milnor [3, p. 113] and Lamotke [1].

(6.10) Theorem. $\mu_d=\mu_c$.

Proof. Consider the deformation of $f$ given by

$$F(z,t)=f(z)-\sum_{j=1}^{n} t_jz_j.$$

$\partial F/\partial z_j=f_j(z)-t_j$, so the critical points of $F(z,t)$ are, for a fixed $t$, the solutions of $\partial f(z)=t$. If $\Delta$ is the discriminant of (5.13) and $t\in Y-\Delta$, then there are $\mu_c$ distinct solutions. The set of values for which some of the critical values coincide is algebraic and hence at least $\text{codimension}_{\mathbb{R}}$ 2. Since at each of these $z$ with $\partial f(z)=t$, $z$ has branching order 1, $\partial f$ is invertible, so the Hessian in non-trivial, hence $z$ is a non-degenerate critical point of $F_t(z)$ by (6.2).

## 7. Current research

A large and important body of literature deals with the problem of classification and deformation of map germs. A good introduction is Bröcker and Lander [1], where the elementary catastrophes of Thom are also described.

(7.1) The vector space topology on the jet spaces $J^k$ of (4.2) gives $\mathcal{O}$ a topology with induced topology on $\mathfrak{m}$. The group $H$ of biholomorphic coordinate transformations acts on $\mathfrak{m}$ and the orbits of $H$ are the equivalence classes of germs in the sense of (4.3). A germ $\hat{f} \in \mathfrak{m}$ is called *simple* if there exists an open neighborhood $U$ of $\hat{f}$ in $\mathfrak{m}$ such that $U$ intersects only finitely many orbits of $H$. A well known theorem of Arnold [3] states:

(7.2) THEOREM. *The germ $\hat{f}$ is simple if and only if $\hat{f}$ is (right) equivalent to one of the maps below:*

$$
\begin{array}{lll}
A_k: & z_1^{k+1}+z_2^2+\cdots+z_n^2, \quad k \geqq 1, & \mu_a = k \\
D_k: & z_1^2 z_2 + z_2^{k-1}+z_3^2+\cdots+z_n^2, \quad k \geqq 4, & \mu_a = k \\
E_6: & z_1^3+z_2^4+z_3^2+\cdots+z_n^2, & \mu_a = 6 \\
E_7: & z_1^3+z_1 z_2^3+z_3^2+\cdots+z_n^2, & \mu_a = 7 \\
E_8: & z_1^3+z_2^5+z_3^2+\cdots+z_n^2, & \mu_a = 8.
\end{array}
$$

The structure of the orbit space $\mathfrak{m}^* = \mathfrak{m}/H$ has been studied by Arnold [3, 4, 5, 6, 7, 8] and his students.

(7.3) A germ $\hat{g}$ is called *adjacent* to $\hat{f}$ if in any neighborhood of $\hat{f}$ there are germs of the orbit of $\hat{g}$. The adjacency problem has been considered by several authors, Siersma [1], Saito [3]. Here is an interesting result.

(7.4) THEOREM. *If $\hat{f}$ is a simple germ and $\hat{g}$ is adjacent to $\hat{f}$, then $\hat{g}$ is also simple.*

(7.5) An *unfolding* of the holomorphic map germ $f:(\mathbb{C}^n, 0) \to (\mathbb{C}, 0)$ is a holomorphic map germ $F:(\mathbb{C}^n \times \mathbb{C}^p, 0) \to (\mathbb{C} \times \mathbb{C}^p, 0)$ of the form $(F(z, t), t)$ with $F(z, 0) = f(z)$.

This notion with the appropriate definition for morphisms of unfoldings and their stability properties was studied by Thom, Mather, and others. A good description is in Wassermann [1].

The connection between deformations and unfoldings is discussed in the survey article of Brieskorn [7].

(7.6) We may identify the tangent space to the orbit of $H$ at a germ $\hat{f} \in \mathfrak{m}^2$ considered as a subspace of $J^k$ with $\mathfrak{m}(\partial f) \bmod \mathfrak{m}^{k+1}$. This gives rise to the definition

$$\operatorname{codim} \hat{f} = \dim_{\mathbb{C}} \frac{\mathfrak{m}^2}{\mathfrak{m}(\partial f)} = \dim_{\mathbb{C}} \frac{\mathfrak{m}}{(\partial f)}.$$

see Bröcker and Lander [1, p. 99].

Let $\hat{\delta}$ be a germ, $\hat{\delta}:(\mathbb{C}^n \times \mathbb{R}, 0 \times \mathbb{R}) \to (\mathbb{C}^n, 0)$ with $\hat{\delta}(z, 0) = z$, representing a one-parameter family of maps at the identity in $H$. It induces a germ of a path defined by $t \to \hat{f} \circ \hat{\delta}_t$. The tangent vector of this path at $t = 0$ is an element of the tangent space of the orbit of $H$ at $\hat{f}$.

Write $\delta(z, t) = z + \varepsilon(z, t)$, where $\varepsilon : \mathbb{C}^n \times \mathbb{R} \to \mathbb{C}^n$ is restricted by $\varepsilon(z, 0) = 0$ and $\varepsilon(0, t) = 0$.

The following vectors reduced modulo $m^{k+1}$ give the tangent vectors:

$$\frac{\partial}{\partial t}(f(z+\varepsilon(z, t)))|_{t=0} = \sum_{i=1}^{n} \frac{\partial f}{\partial z_i} \cdot \frac{\partial \varepsilon_i}{\partial t}\bigg|_{t=0}.$$

Since $\varepsilon$ satisfies only the restrictions above, $\dfrac{\partial \varepsilon_i}{\partial t}\bigg|_{t=0}$ is an arbitrary element of $m$, so the tangent space is $m(\partial f)$ as asserted.

(7.7) For the algebraic theory of deformations see e.g. Tjurina [1], Kas and Schlessinger [1], Pinkham [1], Brieskorn [7], Saito [3].

(7.8) For the connections between the classification of singularities and rapidly oscillating integrals, wave fronts, caustics, phase transitions and other physical applications see Pham [2] and Arnold [2, 3, 4, 8].

# Chapter II

# Fibration and monodromy

## 1. Gradient multiplicity

This section represents a link between the analytic theory preceding it and the topological considerations which occupy the rest of this chapter.

(1.1) Definition. Let $\alpha : M^n \to N^n$ be a continuous map between closed oriented manifolds of the same dimension. Let $[M] \in H_n(M; \mathbb{Z})$ and $[N] \in H_n(N; \mathbb{Z})$ be the respective orientation classes. Define the degree of $\alpha$, $\deg \alpha$ by the equation $\alpha_*[M] = \deg \alpha\, [N]$.

Next consider the polynomial $f : \mathbb{C}^n \to \mathbb{C}$ with isolated critical point at 0. Let $S_\varepsilon^{2n-1}$ be a small sphere at 0 so that 0 is the only critical point in the closed ball $B_\varepsilon^{2n}$. Define the map

$$\varphi : S_\varepsilon^{2n-1} \to S_1^{2n-1}$$

from $S_\varepsilon^{2n-1}$ to the unit sphere in $\mathbb{C}^n$ by $z \in S_\varepsilon^{2n-1}$

$$\varphi(z) = \partial f(z)/|\partial f(z)|.$$

$\mathbb{C}^n$ has a natural orientation and so does $B_\varepsilon^{2n}$. Orient $S_\varepsilon^{2n-1}$ as its boundary so the outward normal followed by the orientation of the boundary gives the positive frame in $\mathbb{C}^n$.

(1.2) Definition. The *gradient multiplicity* of $f$ at 0 is defined by

$$\mu_g = \deg \varphi.$$

We shall follow Milnor [3, Appendix B] to establish $\mu_g = \mu_c$. First we need a lemma related to (I.6.2).

(1.3) LEMMA. *If the Jacobian of $\partial f$, i.e. the Hessian of $f$ is non-singular at a critical point $z^0$, then $\mu_g = 1$.*

PROOF. Consider the Taylor expansion with remainder

$$\partial f(z) = L(z - z^0) + r(z)$$

where the linear transformation $L$ is non-singular by hypothesis, and where $|r(z)|/|z - z^0|$ tends to zero as $z \to z^0$. Choose $\varepsilon$ small enough so that $|r(z)| < |L(z - z^0)|$ whenever $|z - z^0| = \varepsilon$. Then the one parameter family of mappings

$$h_t(z) = (L(z - z^0) + \mathrm{tr}\,(z))/|L(z - z^0) + \mathrm{tr}\,(z)|$$

from $S_\varepsilon(z^0)$ to the unit sphere demonstrates that the degree $\mu_g$ of $h_1$ is equal to the degree of the mapping $L/|L|$ on $S_\varepsilon(z^0)$.

Now deform $L$ continuously to the identity within the group $GL(n, \mathbb{C})$, consisting of all non-singular linear transformations. This is possible since the Lie group $GL(n, \mathbb{C})$ is connected. It follows easily that the degree of the mapping $L/|L|$ on $S_\varepsilon(z^0)$ is +1.

Milnor calls the fact that the degree of $(L + r)/|L + r|$ on $S_\varepsilon$ is equal to the degree of $L/|L|$ whenever $|r| < |L|$ throughout $S_\varepsilon$ 'Rouché's Principle'.

Next consider a compact region $D$ with smooth boundary in $\mathbb{C}^n$. Assume that $\partial f$ has only finitely many zeros in $D$, and no zeros on the boundary.

(1.4) LEMMA. *The number of zeros of $\partial f$ within $D$, each counted with its appropriate multiplicity, is equal to the degree of the mapping $z \to \partial f(z)/|\partial f(z)|$ from $\partial D$ to the unit sphere.*

PROOF. Remove a small open disc about each zero of $\partial f$ from the region $D$. Then the function $\partial f/|\partial f|$ is defined and continuous throughout the remaining region $D_0$. Since $\partial D$ is homologous to the sum of the small boundary spheres within $D_0$, it follows that the degree of $\partial f/|\partial f|$ on $\partial D$ is equal to the sum of the degrees on the small spheres.

NOTE. (1.3) and (1.4) are valid for arbitrary holomorphic maps $g: \mathbb{C}^n \to \mathbb{C}^n$ with isolated zeros.

(1.5) THEOREM. $\mu_g = \mu_c$.

PROOF. Let $f: \mathbb{C}^n \to \mathbb{C}$ be a polynomial with an isolated critical point at 0 so its gradient $\partial f: \mathbb{C}^n \to \mathbb{C}^n$ has its only zero at 0 in some small disc $D_\varepsilon$ at 0. We have to show that if the gradient multiplicity equals $\mu_g$, then for almost all points $y \in \mathbb{C}^n$ sufficiently close to the origin, the equation $\partial f(z) = y$ has precisely $\mu_g$ solutions $z$ within $D_\varepsilon$. By Sard's theorem (I.1.1), almost every point $y \in \mathbb{C}^n$ is a regular value of $\partial f$. In other words, for all $y$ not belonging to some set of Lebesgue measure zero, the matrix $(\partial f_j/\partial z_k)$ is non-singular at every point $z$ in the inverse image $(\partial f)^{-1}(y)$.

Given any such regular value $y$, note that the solutions $z$ of the system of analytic equations $\partial f(z) - y = 0$ are all isolated, with multiplicity $\mu_g = 1$ by (1.3).

Choose any regular value $y$ of $\partial f$ which is close enough to the origin so that $|y| < |\partial f(z)|$ for all $z \in \partial D_\varepsilon$. Then according to (1.4) the number of solutions of the equation $\partial f(z) - y = 0$ within $D_\varepsilon$ is equal to the degree of the map $(\partial f - y)/|\partial f - y|$ on $\partial D_\varepsilon$. (Each solution must be counted with a certain multiplicity, but we have just seen that these multiplicities are all $+1$.) By Rouché's Principle the degree of this mapping $(\partial f - y)/|\partial f - y|$ is equal to the degree of $\partial f/|\partial f|$, which is $\mu_g$.

*Added in proof:* A direct argument for $\mu_\alpha = \mu_g$ is due to A. G. Kouchnirenko (private communication November 1976).

## 2. Fibration theorem

An important result which led to numerous recent investigations is due to Milnor [3]. Let $f: \mathbb{C}^n \to \mathbb{C}$ be a polynomial map with a critical point at 0. Let $S_\varepsilon^{2n-1}$ be a sufficiently small sphere about 0, $F_t = \{z \in \mathbb{C}^n \mid f(z) = t\}$ and $K = S_\varepsilon^{2n-1} \cap F_0$.

(2.1) THEOREM. *The mapping $f/|f| : S_\varepsilon^{2n-1} - K \to S^1$ is the projection of a smooth fiber bundle.*

Note that 0 is not assumed to be an isolated critical point. With this additional hypothesis the proof is considerably simplified and we shall only present this weaker version.

Let $f: \mathbb{C}^n \to \mathbb{C}$ be a holomorphic map germ with only isolated critical points in a neighborhood of 0, $S_\varepsilon^{2n-1}$ a small sphere at 0 and $B_\varepsilon^{2n}$ the corresponding closed ball. Let $D_\delta$ be a small disc in $\mathbb{C}$ with boundary $S_\delta$. It is a consequence of Sard's theorem (I.1.1) that we may choose $\varepsilon > 0$ and $0 < \delta \ll \varepsilon$ sufficiently small so that

(i) $f$ has no critical points on $S_\varepsilon$,
(ii) $f$ has no critical values on $S_\delta$,
(iii) $F_t = f^{-1}(t)$ is transverse regular to $S_\varepsilon$ for all $t \in D_\delta$.

(Recall that two smooth submanifolds $M_1^{m_1}$ and $M_2^{m_2}$ of the smooth manifold $N^n$ are said to be transverse regular at $p \in M_1 \cap M_2$ if the following holds for the respective tangent spaces at $p: T_pM_1 + T_pM_2 = T_pN$. If this condition holds for every $p \in M_1 \cap M_2$, $M_1$ and $M_2$ are said to be transverse regular submanifolds.)

Let $\Sigma = \{t_1, \cdots, t_r\} \subset D_\delta$ be the set of critical values of $f$ in $D_\delta$. Let $D_0 = D_\delta - \Sigma$ and

$$E = f^{-1}(D_\delta) \cap B_\varepsilon$$
$$E_0 = f^{-1}(D_0) \cap B_\varepsilon$$
$$T = f^{-1}(D_\delta) \cap S_\varepsilon$$
$$C = f^{-1}(S_\delta) \cap B_\varepsilon.$$

Note that $\partial E = C \cup T$.

Figure (2.2) may help in locating these sets.

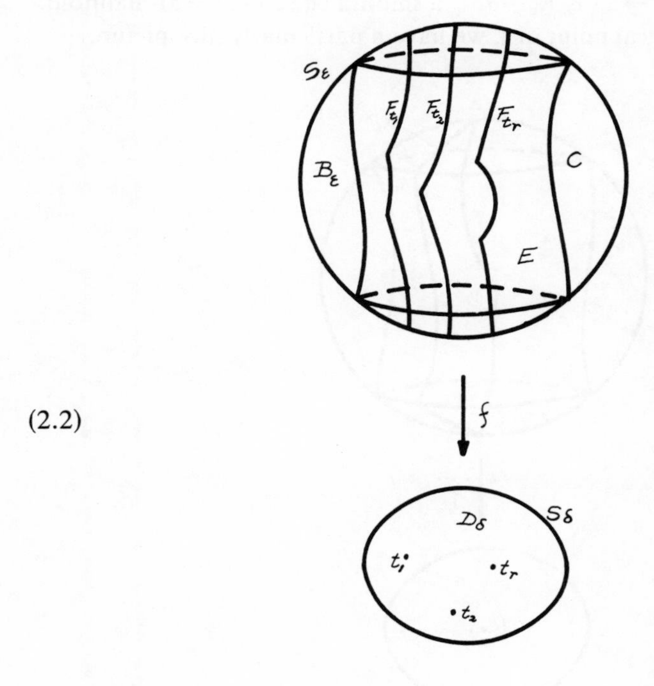


(2.2)

NOTE. $E_0 = E - \bigcup_{i=1}^{r} F_{t_i}$.

(2.3) THEOREM. *The map*

$$f : E_0 \to D_0$$

*is the projection map of a smooth fiber bundle and*

$$f: T \to D_\delta$$

*is the projection map of a trivial fiber bundle.*

PROOF. The fibration theorem of Ehresmann [1] states: Let $E$ and $D$ be smooth manifolds with boundary, $D$ connected and $p: E \to D$ a smooth proper surjective mapping, with the property that for all $x \in D$, the rank of the differential of $p$ at $x$ equals the dimension of $D$. Then $p: E \to D$ is a smooth fiber bundle and all fibers, $p^{-1}(x)$ are diffeomorphic.

Since $\partial f$ has maximal rank at every point of $E_0$ and by the transversality also on $T$, and $D_\delta$ is contractible, the conclusion follows.

(2.4) The typical fiber of $f: E_0 \to D_0$ is a smooth compact $(2n-2)$-manifold with boundary, $M_c = f^{-1}(c) \cap B_\varepsilon$ for some regular value $c \in D_0$. The typical fiber of $f: T \to D$ is $K_c = \partial M_c$ a smooth closed $(2n-3)$-manifold. In case 0 is the only critical point of $f$, we have a particularly nice picture.

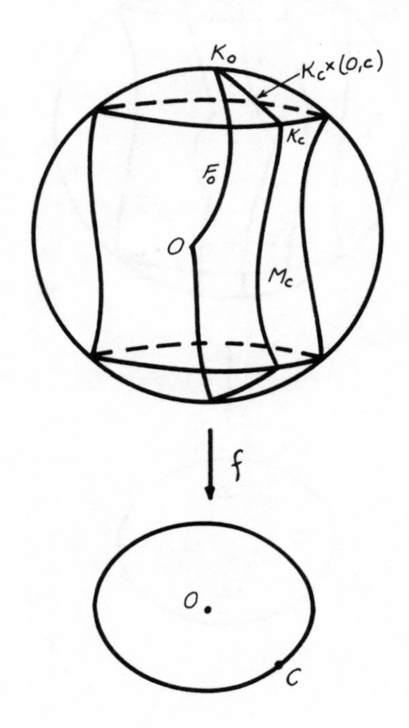


The connection with Milnor's fibration (2.1) is the following. For $n \geqq 3$ the contractible manifold $E$ is a ball with boundary $\partial E = T \cup C$. Then $\partial E - K_0$ is obviously fibered over $S^1$ by open manifolds diffeomorphic to int $M_c$ since $K_c \times (0, c)$ is just an open collar on $\partial M_c = K_c$.

## 3. Homotopy type of the fiber

The next object of study is the typical fiber of the fibration associated with the polynomial map $f:\mathbb{C}^n \to \mathbb{C}$ with an isolated critical point at 0.

(3.1) PROPOSITION. *The interior of the typical fiber $M_c$ is diffeomorphic with the non-singular algebraic variety $F_c = f^{-1}(c)$.*

PROOF. Since $M_c = F_c \cap B_\varepsilon$, it is sufficient to prove that for some Morse function on $F_c$, all critical points lie inside $B_\varepsilon$. Note that it is enough to show this for some suitable value $c$. Let $\varphi(z) = |z - z^0|^2$ denote the distance function from a given point $z^0$. Since $f$ is given by polynomial equations, the set of critical points of the restriction of $\varphi(z)$ to $F$ is finite. By choosing $|c|$ sufficiently small, the critical points will all lie in $B_\varepsilon$.

Let $F$ be the non-singular algebraic variety $F_c$ for any regular value $c$. The next result is due to Milnor [3].

(3.2) LEMMA. *The algebraic variety $F$ has the homotopy type of a finite CW complex of dimension $\leqq n-1$.*

PROOF. We follow Andreotti and Frankel [1] to show that the index (= number of negative characteristic values) of the Morse function $\varphi(z) = |z - z^0|^2$ is $\leqq n-1$ at every critical point of $F$. It follows from (I.1.1) that we may choose a point $z^0 \notin F$ so that $\varphi(z)$ has only non-degenerate critical points. Note that $\varphi(z)$ is bounded from below and the subsets $\{z \in F \mid \varphi(z) \leqq \text{constant}\}$ are compact.

Let $P$ be a (non-degenerate) critical point of $\varphi$. Introduce $z_\beta = x_{2\beta-1} + ix_{2\beta}$, $\beta = 1, \cdots, n$ and choose local coordinates so that $P$ is at the origin and the tangent space to $F$ at $P$ has equation $z_n = 0$. Then the coordinates of $z^0$ will satisfy $z_1^0 = \cdots = z_{n-1}^0 = 0$ and we may assume $z_n^0 = a > 0$, so $x_{2n-1}^0 = a$ and $x_i^0 = 0$ for $i \neq 2n-1$.

By the implicit function theorem we may describe $F$ in a neighborhood of $P$ by the equation

$$z_n = \tfrac{1}{2} \sum_{\beta,\gamma=1}^{n-1} B_{\beta\gamma} z_\beta z_\gamma + \text{higher order terms}$$

so we have

$$x_{2n-1} = \operatorname{Re} z_n = \tfrac{1}{2} \sum_{i,j=1}^{2n-2} b_{ij} x_i x_j + \text{higher order terms}.$$

We may also assume that by a suitable change of coordinates the symmetric form $b_{ij}$ is diagonalized and we have

(i) $x_{2n-1} = \tfrac{1}{2} \sum_{i=1}^{2n-2} b_i x_i^2 + \text{higher order terms.}$

On the other hand

(ii) $\varphi(z)=\sum_{i=1}^{n-1}|z_i|^2+|z_n-a|^2=\sum_{i=1}^{2n-2}x_i^2+(x_{2n-1}-a)^2+x_{2n}^2.$

Substitution of (i) into (ii) shows that the index of $\varphi$ is the number of $b_i$ such that $b_i>\frac{1}{a}>0$. However, the $b_i-s$ are characteristic values of the real part of a complex quadratic form. These occur in pairs with opposite sign, so no more than $(n-1)$ of them can be positive.

The last assertion is shown as follows. Let $Q(z,z)={}^t zBz$ be a complex quadratic form. Put $z=x+iy$, $B=C+iD$. Then

$$\operatorname{Re} Q(z,z)={}^t xCx-{}^t xDy-{}^t yDx-{}^t yCy,$$

where $t$ means transposed. Thus the matrix for $\operatorname{Re} Q(z,z)$ is given by

$$b=\begin{bmatrix} C & -D \\ -D & -C \end{bmatrix}.$$

Observe that if $\begin{pmatrix} x \\ y \end{pmatrix}$ is a characteristic vector of $b$ with characteristic value $\lambda$, then $\begin{pmatrix} -y \\ x \end{pmatrix}$ is a characteristic vector of $b$ with characteristic value $-\lambda$.

(3.3) Lemma. *$H_i(F)=0$ for $i\neq 0$, $n-1$.*

Proof. Consider the smooth sphere $\partial E$ of (2.4) and its submanifold

$$A=K_0\cup K_c\times(0,c)\cup M_c.$$

Note that $A$ is diffeomorphic to $M_c$. Also the complement of $A$ in $\partial E$, $\partial E-A$ is fibered by $K_a\times(0,a)\cup M_a\approx \operatorname{int} M_a\approx F$ over the contractible space $S^1-\{c\}$, so $\partial E-A\sim F$ is a homotopy equivalence.

By Alexander duality, see Spanier [1, p. 296]

$$\tilde{H}_i(\partial E-A)\approx H^{2n-2-i}(A)$$

and the latter group is zero for $2n-2-i>n-1$ by (3.2).

(3.4) Theorem. *$F$ has the homotopy type of a (finite) wedge of $(n-1)$-spheres.*

Proof. By (3.2) and (3.3) $F$ is a finite $CW$-complex whose homology is that of a wedge of spheres. It suffices to prove that $F$ is simply connected and invoke Whitehead's theorem, Spanier [1, p. 399].

Using the Morse function $\varphi$ above, we see that $M_c\approx A$ is built up from a $(2n-2)$ disc $D_0$, by adjoining handles of index $\leqq n-1$. All these handles may be attached inside the $(2n-1)$-sphere $\partial E$. The complement, $\partial E-D_0$ is clearly simply connected and the adjunction of handles of index $\leqq \dim(\partial E)-3=2n-4$ cannot change the fundamental group of the complementary set. So it follows inductively that the complement $\partial E-A$ is also simply connected, provided that $n-1\leqq 2n-4$.

(3.5) A similar argument shows that the closed smooth $(2n-3)$-manifold $K=\partial M$ is $(n-3)$-connected. Thus for $n=3$, $K$ is connected and for $n\geqq 4$, $K$ is simply connected with homology concentrated in dimensions $0$, $n-2$, $n-1$, $2n-3$.

## 4. Topological multiplicity

Define the *topological multiplicity* of the polynomial map $f:\mathbb{C}^n\to\mathbb{C}$ with an isolated critical point at 0, $\mu_t$ as the number of $(n-1)$-spheres in the homotopy type of the typical fiber provided by (3.4):

$$F\sim\bigvee_{\mu_t} S^{n-1}.$$

Milnor [3, Th. 7.2] proves that $\mu_t=\mu_g$. We shall follow Brieskorn [4, Appendix] to show the following equivalent statement, see also Lamotke [1].

(4.1) THEOREM. $\mu_t=\mu_c$.

PROOF. Let $\partial f:\mathbb{C}^n\to\mathbb{C}^n$ be the gradient of $f$. In order to simplify notation let $\mu_c=r$ for this proof. Choose a regular value of $\partial f$, $a=(a_1,\cdots,a_n)$ sufficiently close to 0 so that $(\partial f)^{-1}(a)$ consists of exactly $r$ simple points: $x_1,\cdots,x_r$. The approximation of $f$ defined by

$$f_a(z)=f(z)-\sum_{i=1}^{n} a_i z_i$$

has $r$ distinct non-degenerate critical points $x_1,\cdots,x_r$, see (I.6.10). Thus for each $x_j$ we may choose local coordinates $y_1,\cdots,y_n$ so that

(i) $f_a(y_1,\cdots,y_n)=f_a(x_j)+y_1^2+\cdots+y_n^2$.

As in (2.2) let

$$\begin{aligned}B&=\{z\in\mathbb{C}^n\mid |z|\leqq\varepsilon\}\\ D&=\{t\in\mathbb{C}\mid |t|\leqq\delta\}\\ E&=\{z\in B\mid f(z)\in D\}\\ E_a&=\{z\in B\mid f_a(z)\in D\}.\end{aligned}$$

Choose $a$ sufficiently small so that $E$ and $E_a$ are homeomorphic and $f_a$ is a generic approximation of $f$ with all its critical points $x_1,\cdots,x_r\in E_a$ and distinct critical values $t_1, \cdots, t_r \in D$, and the transversality condition is satisfied by $\varepsilon$ and $\delta$ for $f$ and $f_a$.

Choose small disjoint balls of radius $\varepsilon' > 0$

$$B_j = \{z \in E_a \mid |z - x_j| \leqq \varepsilon'\}, \quad j = 1, \cdots, r$$

so that each $B_j$ is contained in a neighborhood where $f_a$ has the local representation (i). Choose corresponding small discs of radius $\delta'$

$$D_j = \{t \in D \mid |t - t_j| \leqq \delta'\} \subset \operatorname{int} D$$

so that the restriction $f_a : B_j \to D_j$ satisfies the conditions for the fibration theorem (2.3), for $j = 1, \cdots, r$.

Let $\tau_j \in \partial D_j$ be a boundary point for each $j$ and $\tau_0 \in \partial D$. Choose paths $\gamma_j$ from $\tau_0$ to $\tau_j$ so that $(\bigcup_{j=1}^r D_j) \cup (\bigcup_{j=1}^r \gamma_j)$ is a deformation retract of $D$.

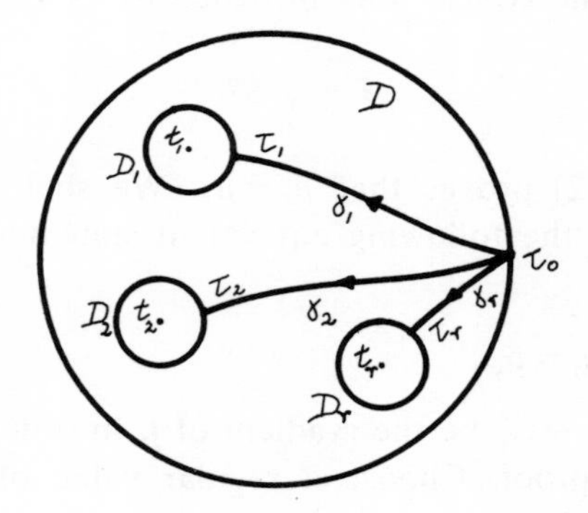

Consider the following subsets of $E_a$:

$$E_j = f_a^{-1}(D_j)$$

$$F_j = f_a^{-1}(\tau_j) \quad \text{and} \quad F_0 = f_a^{-1}(\tau_0)$$

$$E^1 = f_a^{-1}\left(\bigcup_{j=1}^r D_j\right) = \bigcup_{j=1}^r E_j$$

$$E^2 = f_a^{-1}\left(\bigcup_{j=1}^r \gamma_j\right).$$

The retraction of $D$ to $(\bigcup_{j=1}^r D_j) \cup (\bigcup_{j=1}^r \gamma_j)$ shows that $E^1 \cup E^2$ is homotopy equivalent to $E_a$, which in turn is homeomorphic to $E$, which is contractible. Thus the Mayer-Victoris sequence for $(E^1 \cup E^2, E^1, E^2)$ yields for $p > 0$

$$H_p(E^1) + H_p(E^2) \approx H_p(E^1 \cap E^2).$$

Now $E^2$ is a trivial fibration over the contractible base space $\bigcup_{j=1}^{r} \gamma_j$ with fiber $F_0$, and therefore homotopy equivalent to $F_0$. Also, $E^1 \cap E^2 = F_1 \cup \cdots \cup F_r$, where each fiber $F_j$ is diffeomorphic to $F_0$ by (2.3). Thus

$$\text{(ii)} \quad \sum_{j=1}^{r} H_p(E_j) + H_p(F_0) = \sum_{j=1}^{r} H_p(F_j).$$

Let $F$ be the typical fiber of the fibration associated to $f$. By transversality, $F$ is diffeomorphic to the typical fiber $F_0$ so we have using (3.3):

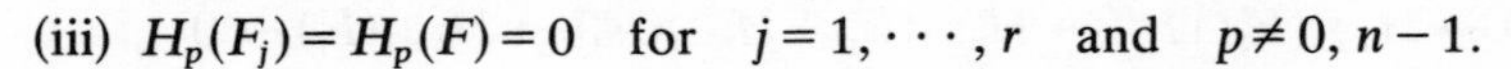

$$\text{(iii)} \quad H_p(F_j) = H_p(F) = 0 \quad \text{for} \quad j = 1, \cdots, r \quad \text{and} \quad p \neq 0, n-1.$$

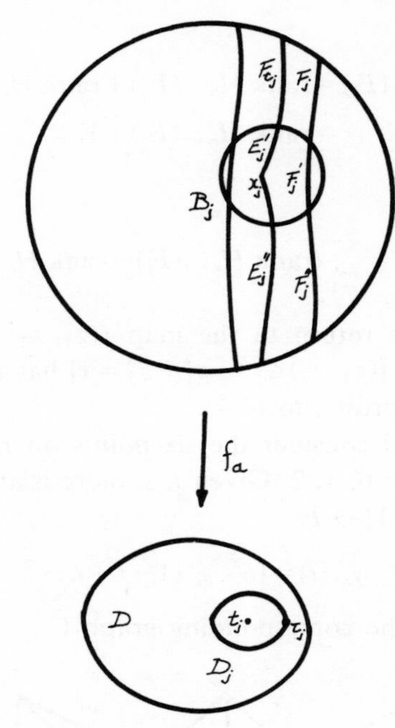


From (ii) and (iii) we conclude that $H_p(E_j) = 0$ for $j = 1, \cdots, r$ and $p \neq 0$, $n-1$.

Determine $H_{n-1}(E_j)$ as follows. Let

$$E_j' = E_j \cap B_j, \qquad E_j'' = E_j - \text{int } B_j,$$
$$F_j' = F_j \cap B_j, \qquad F_j'' = F_j'' - \text{int } B_j.$$

Note that $E_j'$ is contractible, $E_j' \cap E_j'' \approx F_j' \cap F''_j \times D^2$ and

$$F_j'' \to E_j'' \to D_j$$

is a trivial fibration since $f_a$ has no critical points in $E_j''$. Thus the inclusion $F_j'' \to E_j''$ is a homotopy equivalence.

Now consider the restriction of $f_a$ to $E'_j \to D_j$. This map has an isolated non-degenerate critical point at $x_j$ and by (i) $F'_j$ retracts to a real $(n-1)$-sphere. The Mayer-Victoris sequences for $(E_j, E'_j, E''_j)$ and $(F_j, F'_j, F''_j)$ now give:

$$\begin{array}{ccccccc} 0 \to H_{n-1}(E'_j \cap E''_j) & \to & H_{n-1}(E'_j) & + & H_{n-1}(E''_j) & \to H_{n-1}(E_j) \to 0 \\ & & \| & & \| & \\ & & 0 & & H_{n-1}(F''_j) & \end{array}$$

$$0 \to H_{n-1}(F'_j \cap F''_j) \to H_{n-1}(F'_j) + H_{n-1}(F''_j) \to H_{n-1}(F_j) \to 0$$

so we conclude that

$$\begin{aligned} \text{rank } H_{n-1}(F_j) &= \text{rank } H_{n-1}(F''_j) + \text{rank } H_{n-1}(F'_j) \\ &= \text{rank } H_{n-1}(E_j) + 1. \end{aligned}$$

Thus from (ii)

$$\mu_t = \text{rank } H_{n-1}(F_0) = \sum_{j=1}^{r} (\text{rank } H_{n-1}(F_j) - \text{rank } H_{n-1}(E_j)) = r = \mu_c.$$

(4.2) Example. Let us return to the map $f(z_1, z_2) = z_1^3 + z_2^3$. We saw in (I.5.7) that $\mu_c = 4$ so $F = \{(z_1, z_2) \in \mathbb{C}^2 \mid z_1^3 + z_2^3 = 1\}$ has the homotopy type of a wedge of 4 circles according to (4.1).

Let $\xi = \exp(2\pi i/3)$ and consider the six points on $F$ given by $P_r = (\xi^r, 0)$ $r = 0, 1, 2$, $Q_s = (0, \xi^s)$ $s = 0, 1, 2$. Given $r$, $s$, there is an arc on $F$ between $P_r$ and $Q_s$ given by $\gamma_{r,s} : [0, 1] \to F$

$$\gamma_{r,s}(t) = (\xi^r t, \xi^s (1 - t^3)^{\frac{1}{3}}).$$

The homotopy type of the corresponding graph $\Gamma$

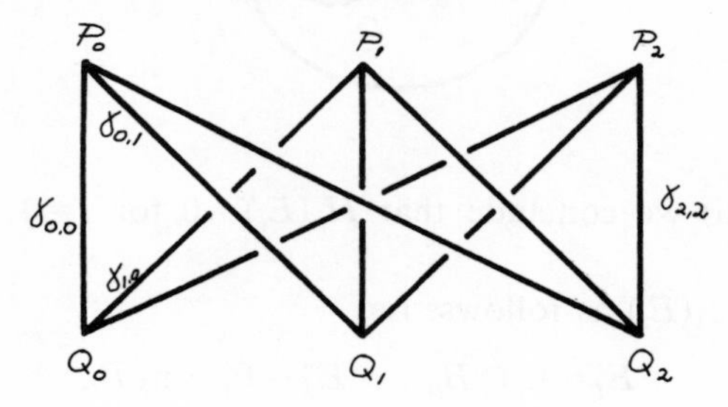


is a wedge of 4 circles. We may retract $F$ onto $\Gamma$ as follows.

(i) Move $z_1$ and $z_2$ continuously in $F$ so that $z_1^3$ and $z_2^3$ move parallel to the imaginary axis to their respective real parts.

(ii) If either $z_1^3$ or $z_2^3$ is negative, move it continuously to zero (and the other to 1). This is clearly in $\Gamma$.

(iii) If both $z_1^3$ and $z_2^3$ are positive with sum 1, the points are obviously in $\Gamma$.

This retraction can be done in general for Brieskorn polynomials, $f(z) = z_1^{a_1} + z_2^{a_2} + \cdots + z_n^{a_n}$, see Pham [1], Milnor [3], and Hirzebruch and Mayer [1].

## 5. Picard-Lefschetz monodromy

The fibration theorem of (2.3) in the case of an isolated critical point as in (2.4) shows that $\partial E - K$ is fibered over $S^1$ with fiber diffeomorphic to $F$. Thus $\partial E - K$ is obtained from $F \times [0, 2\pi]$ by identifying $F \times 0$ and $F \times 2\pi$ via a diffeomorphism $h: F \to F$ called the characteristic map of the fibration. It follows from (3.4) that in homology the only interesting map is

$$h_*^A: H_{n-1}(F; A) \to H_{n-1}(F; A).$$

This isomorphism is called the *Picard-Lefschetz monodromy*. The complete knowledge of this map with $A = \mathbb{Z}$ is sufficient to determine a great deal about the diffeomorphism type of the manifold $K$ and the knot type of the pair $(\partial E, K)$. The problem of computing $h_*^A$ is unsolved in general. Partial results will be quoted in the next section.

Consider the exact sequence of the pair $(\partial E - K, F)$:

$$\to H_i(F) \to H_i(\partial E - K) \to H_i(\partial E - K, F) \to H_{i-1}(F) \to.$$

Recall that $\partial E \approx S^{2n-1}$ so we have for $i \geqq 1$, $n \geqq 3$, $H_i(\partial E - K) \approx H^{2n-i-2}(K)$ by Alexander duality and $H^{2n-i-2}(K) \approx H_{i-1}(K)$ by Poincaré duality. Moreover, we have the isomorphism

$$H_j(F \times [0, 2\pi], F \times 0 \cup F \times 2\pi) \xrightarrow{\approx} H_j(\partial E - K, F)$$

and the left hand group is isomorphic to $H_{j-1}(F)$. Thus for $n \geqq 3$ we obtain the short exact sequence with $I$ the identity map, see Milnor [3, §8]:

$$(5.1)\quad 0 \to H_{n-1}(K) \to H_{n-1}(F) \xrightarrow{h_* - I_*} H_{n-1}(F) \to H_{n-2}(K) \to 0.$$

We obtain immediately in view of (3.5):

(5.2) Lemma. *For $n \geqq 3$, $K$ is a homology sphere and for $n \geqq 4$ $K$ is a homotopy sphere if and only if $h_* - I_*$ is an isomorphism.*

In fact it is sufficient to consider the characteristic polynomial of $h_* - I_*$, $\Delta(t) = \det(tI_* - h_*)$. Then we have:

(5.3) Lemma. *For $n \geqq 4$, $K$ is a homotopy sphere if and only if $\Delta(1) = \pm 1$.*

Next consider the cycles represented by $F_j'$ in (4.1). These are called the *vanishing cycles* because they disappear when moved into the singular fiber

$F_0$. Along the lines of Picard and Lefschetz, Lamotke [1] proves the following.

Let $s_j \in H_{n-1}(F_j')$ be the canonical generator. Let $u_j = \partial D_j$ be the oriented path around the disc $D_j$. For each path $\gamma \in D - \bigcup_{j=1}^r D_j^0$, from $\tau_0$ to $\tau_j$ we have an element $\gamma(s_j) \in H_{n-1}(F)$. The path $\gamma$ also induces an automorphism of $H_{n-1}F$ defined for $x \in H_{n-1}(F)$ by $\sigma_\gamma(x) = \gamma u_j \gamma^{-1}(x)$. The Picard-Lefschetz formula states that

(5.4) $\sigma_\gamma(x) = x - (-1)^{n(n+1)/2}\langle \gamma(s_j), x\rangle \gamma(s_j)$

where $\langle\ ,\ \rangle$ denotes intersection pairing of homology classes.

The paths $\gamma u_j \gamma^{-1}$ generate $\pi_1(D - \bigcup_{j=1}^r D_j^0)$ so we have in fact defined a map

$$\Phi : \pi_1\left(D - \bigcup_{j=1}^{r} D_j^0\right) \to \mathrm{Aut}\,(H_{n-1}(F)).$$

The image of this map is called the *monodromy group*, $W_f$ of the singularity. It does not depend on the particular choice of generic approximation for $f$, see Looijenga [1].

(5.5) It follows from (5.4) that $W_f$ is generated by reflections. If the paths $\gamma_1, \cdots, \gamma_r$ of (4.1) are chosen so that they intersect only in $\tau_0$ and have no self-intersections, then the vanishing cycles $\gamma_1(s_1), \cdots, \gamma_r(s_r)$ form a (weak distinguished) basis for $H_{n-1}(F)$ and the corresponding reflections $\sigma_1, \cdots, \sigma_r$ generate $W_f$. Moreover, this basis may be chosen so that the Coxeter element of $W_f$, $\sigma = \sigma_r \cdot \sigma_{r-1} \cdots \sigma_1$ is the Picard-Lefschetz monodromy. In this case the basis is called *distinguished*.

## 6. Current research

Much of the renewed interest in the topology of the neighborhood boundary of a critical point originated in the discovery by Brieskorn [2], that for example the map $f: \mathbb{C}^5 \to \mathbb{C}$ given by $f(\underline{z}) = z_1^5 + z_2^3 + z_3^2 + z_4^2 + z_5^2$ yields an exotic sphere. More precisely, the intersection $K_0^7 = S_\varepsilon^9 \cap f^{-1}(0)$ is a smooth 7-manifold homeomorphic but not diffeomorphic to $S^7$. In fact it represents a generator of the group of 7-dimensional homotopy spheres.

(6.1) Considerable literature is centered around the diffeomorphism classification of the neighborhood boundary, $K$. See Brieskorn [2], Hirzebruch and Mayer [1], Durfee [1].

(6.2) Many of these results may be extended to the case of complete intersections with an isolated singularity, Hamm [2], Greuel [1], Brieskorn and Greuel [1].

(6.3) If $f:\mathbb{C}^{2k+1}\to\mathbb{C}$, then the typical fiber of the fibration (2.3) is a $4k$-dimensional manifold and it supports a symmetric bilinear form given by intersections

$$H_{2k}F\otimes H_{2k}F\to\mathbb{R}.$$

(If $f:\mathbb{C}^{2k}\to\mathbb{C}$ then it is customary to consider the map $g:\mathbb{C}^{2k+1}\to\mathbb{C}$ given by $g(z)=f(z_1,\cdots,z_{2k})+z_{2k+1}^2$.)

The connection between the monodromy and this intersection form is discussed by Lamotke [1]. Choosing a distinguished basis as in (5.5) the form has special properties, see Durfee [2]. Explicit computations are given by Gabrielov [1, 2, 3], Hefez and Lazzeri [1], Gusein-Zade [1, 2].

(6.4) The computation of the monodromy with complex coefficients and its special properties were studied by Lê [2, 3, 4, 5, 6], Hamm and Lê [1], A'Campo [1, 2, 3, 4, 5], Brieskorn [4], Milnor and Orlik [1].

The integral monodromy is discussed in Orlik [2], Randell [3], Orlik and Randell [1, 2].

(6.5) The differentiable manifolds $K$ have also been studied from the point of view of exotic group actions, knot cobordism, foliations, etc. See Bredon [1, 2], Durfee and Lawson [1], Giffen [1], and Kaufmann [1].

(6.6) The computation of the full monodromy group $W_f$ is in general a difficult problem. Several papers by Gabrielov [1, 2, 3] give explicit results in particular cases.

The general theory connects the monodromy group with versal deformations as follows. Consider the (infinitesimally) versal deformation

$$F(z,t)=f(z)+\sum_{i=1}^{\mu}\varphi_i(z)t_i$$

where $\varphi_i(z)$ form a basis of $\mathcal{O}/(\partial f)$.

Consider the projection map $\mathbb{C}^n\times\mathbb{C}^\mu\to\mathbb{C}^\mu$ restricted to the hypersurface $F(z,t)=0$. The discriminant locus $\Delta\subset\mathbb{C}^\mu$ consists of those values of $(t_1,\cdots,t_\mu)$ where $F(z,t)=0$ has a critical point. It can be shown that the fiber of the projection map over $\mathbb{C}^\mu-\Delta$ is diffeomorphic to the fiber associated with $f$ and therefore $\pi_1(\mathbb{C}^\mu-\Delta)$ operates on the homology of the fiber. The image of $\pi_1(\mathbb{C}^\mu-\Delta)$ in $\operatorname{Aut}(H_{n-1}F)$ is the intrinsic description of the monodromy group, see Looijenga [1], Arnold [8], Brieskorn [5, 7].

(6.7) The investigation of the neighborhood boundary $K$ of (3.5) has been extended to multiple intersections and non-isolated singularities, and to the relative case $(K,L)$ by Hamm [1, 3] under the condition that $K-L$ is non-singular. The case when there are singularities in $K-L$ is treated by L. Kaup: Exakte Sequenzen für globale und lokale Poincaré-homomorphismen, in these Proceedings.

# Chapter III

## Algebraic de Rham cohomology

In the first sections we recall some standard facts from the theory of sheaves and differential forms see Godement [1], Bredon [3], Gunning and Rossi [1], and Wells [1].

### 1. Sheaves

Let $X$ be a topological space.

(1.1) A *presheaf* of abelian groups $\mathcal{F}$ consists of:

(i) an abelian group $\mathcal{F}(U)$ associated to each open set $U \subset X$,

(ii) a homomorphism $\rho_U^V : \mathcal{F}(V) \to \mathcal{F}(U)$ for each inclusion $U \subset V$ so that if $U \subset V \subset W$, then $\rho_U^W = \rho_U^V \circ \rho_V^W$.

The map $\rho_U^V$ is called the restriction.

Given presheaves $\mathcal{F}'$ and $\mathcal{F}''$ over $X$, a map $\theta : \mathcal{F}' \to \mathcal{F}''$ is a homomorphism if for each open set $U \subset X$ we have a homomorphism $\theta(U) : \mathcal{F}'(U) \to \mathcal{F}''(U)$ so that for $U \subset V$ the diagram commutes:

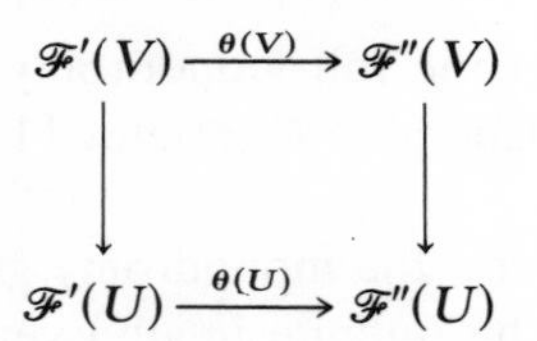

(1.2) Definition. A *sheaf* of abelian groups over $X$ is a topological space $\mathcal{S}$ and a surjective mapping $\pi : \mathcal{S} \to X$ satisfying:

(i) $\pi$ is a local homeomorphism,

(ii) For each $x \in X$, $\pi^{-1}(x) = \mathcal{S}_x$ has the structure of an abelian group, and

(iii) the group operations are continuous in the topology of $\mathcal{S}$.

If $U \subset X$, a section of $\mathcal{S}$ over $U$ is a continuous map $s : U \to \mathcal{S}$ so that $\pi \circ s = id$. Using the pointwise addition, the set of sections over $U$, $\Gamma(U, \mathcal{S})$ forms an abelian group. The assignment $U \to \Gamma(U, \mathcal{S})$ gives a presheaf, and morphisms of sheaves are defined in terms of the corresponding presheaves.

(1.3) Given a presheaf $\mathcal{F}$ on $X$, we may associate a sheaf with it using the direct limit construction. Let $\mathfrak{U}_x$ be a partially ordered set of neighborhoods at $x$ and define

$$\mathcal{S}_x = \operatorname*{dir\,lim}_{U_x \in \mathfrak{U}_x} \mathcal{F}(U_x).$$

Let $\mathcal{S} = \bigcup_x \mathcal{S}_x$ and $\pi : \mathcal{S} \to X$ the obvious projection. For $U \subset \mathfrak{U}_x$ we have the natural map $\rho_x^U:\mathcal{F}(U)\to\mathcal{S}_x$, which is clearly a homomorphism. For $f\in\mathcal{F}(U)$ let $\rho^U(f)=\bigcup_{x\in U}\rho_x^U(f)\subset\mathcal{S}$. Topologize $\mathcal{S}$ by letting the $\rho^U(f)$ be a basis of its topology. It is a routine task to check that $\mathcal{S}$ is indeed a sheaf. By this construction we have a natural map

$$i:\mathcal{F}(U)\to\Gamma(U,\mathcal{S}).$$

(1.4) LEMMA. *Let $\mathcal{F}$ be a presheaf on $X$ and $\mathcal{S}$ the associated sheaf. Consider the following properties:*

(i) *if $\{U_i\}$ are open subsets of $X$ with $U=\bigcup U_i$, and $f, g\in\mathcal{F}(U)$ have the property that $\rho_{U_i}^U(f)=\rho_{U_i}^U(g)$ for all $i$, then $f=g$,*

(ii) *if $\{U_i\}$ are open subsets of $X$ with $U=\bigcup U_i$, and if $f_i\in\mathcal{F}(U_i)$ are such that $\rho_{U_i\cap U_j}^{U_i}(f_i)=\rho_{U_i\cap U_j}^{U_j}(f_j)$ for all pairs $U_i$, $U_j$, then there is an element $f\in\mathcal{F}(U)$ so that $f_i=\rho_{U_i}^U(f)$ for all $i$.*

*The above map $i:\mathcal{F}(U)\to\Gamma(U,\mathcal{S})$ is a monomorphism if and only if $\mathcal{F}$ satisfies* (i) *and epimorphism if and only if $\mathcal{F}$ satisfies* (ii).

*It is common (abuse of language) to call a presheaf $\mathcal{F}$ that satisfies* (i) *and* (ii) *a sheaf.*

The above constructions may be carried out with rings instead of abelian groups, in which case we have a (pre) sheaf of rings.

(1.5) EXAMPLES.

(i) If $R$ is a ring and $X$ a topological space, then the assignment $U\to R$ for each $U\subset X$ gives a presheaf which is clearly a sheaf, called the constant sheaf.

(ii) If $X$ is a smooth manifold, assign to each $U\subset X$, $\mathcal{E}(U)=\{f:U\to R\mid f$ smooth map$\}$. This gives a sheaf called the *structure sheaf* $\mathcal{E}_X$.

(iii) If $X$ is a complex manifold, the assignment $U\subset X$, $\mathcal{O}(U)=\{f:U\to\mathbb{C}\mid f$ holomorphic$\}$ gives the structure sheaf $\mathcal{O}_X$. The stalk of this sheaf at $x\in X$ is the ring of germs of holomorphic functions $\mathcal{O}_x$ as defined in (I.2).

(1.6) DEFINITION. Let $\mathcal{R}$ be a presheaf of commutative rings and $\mathcal{M}$ a presheaf of abelian groups, both over $X$. Suppose that for every open $U\subset X$, $\mathcal{M}(U)$ can be given the structure of an $\mathcal{R}(U)$ module such that if $\alpha\in\mathcal{R}(U)$ and $f\in\mathcal{M}(U)$, then

$$r_V^U(\alpha f)=\rho_V^U(\alpha)r_V^U(f)$$

for $V\subset U$ where $r_V^U$ is the restriction map for $\mathcal{M}$ and $\rho_V^U$ for $\mathcal{R}$. Then $\mathcal{M}$ is called a *presheaf of $\mathcal{R}$-modules.* If $\mathcal{M}$ is a sheaf, it is called a sheaf of $\mathcal{R}$-modules.

(1.7) EXAMPLES. If $X$ is a complex manifold, sheaves of $\mathcal{O}_X$ modules are called *analytic sheaves* over $X$.

(i) If $m$ is a sheaf of $\mathcal{R}$-modules and $m \simeq \mathcal{R}^p$, where $\mathcal{R}^p(U) = \mathcal{R}(U) \oplus \cdots \oplus \mathcal{R}(U)$, we call $m$ a *free* sheaf of modules.

(ii) If $m$ is a sheaf of $\mathcal{R}$-modules so that each $x \in X$ has a neighborhood $U$ such that $m\,|_U$ is free, then $m$ is called *locally free.*

(1.8) Let $\mathcal{S}' \xrightarrow{\alpha} \mathcal{S} \xrightarrow{\beta} \mathcal{S}''$ be maps of sheaves over $X$. The sequence is called *exact* at $\mathcal{S}$ if for each $x \in X$ the induced sequence $\mathcal{S}'_x \xrightarrow{\alpha_x} \mathcal{S}_x \xrightarrow{\beta_x} \mathcal{S}''_x$ is exact.

(1.9) Define restrictions of a sheaf, subsheaves, quotient sheaves, direct sum of sheaves, tensor product of sheaves in the usual way.

(1.10) Let $f: X \to Y$ be a continuous map. Given a sheaf on $X$, its *direct image* sheaf is defined by the presheaf: for an open subset $U$ of $Y$ associate $\mathcal{S}(f^{-1}(U))$. Call the direct image sheaf $f_*\mathcal{S}$.

Given a sheaf $\mathcal{F}$ on $Y$, its *inverse image* sheaf (pull-back) on $X$ called $f^{-1}\mathcal{F}$ is defined by

$$f^{-1}\mathcal{F} = \{(x, f) \in X \times \mathcal{F} \mid f(x) = \pi(f)\},$$

where $\pi: \mathcal{F} \to Y$ is the projection map.

(1.11) The analytic sheaf $\mathcal{F}$ over the complex manifold $X$ is called *coherent* if for each $x \in X$ there is a neighborhood $U$ such that for some $p$ and $q$

$$\mathcal{O}^p|_U \to \mathcal{O}^q|_U \to \mathcal{F}|_U \to 0$$

is exact.

## 2. Differential forms

Let $X$ be a complex manifold of dimension $n$. Letting $T_x(X)$ be the derivations of $\mathcal{O}_x$ at $x$, we find the *holomorphic tangent space*, and their union the holomorphic tangent bundle with fibers isomorphic to $\mathbb{C}^n$ and local basis $(\partial/\partial z_1, \cdots, \partial/\partial z_n)$.

Consider $X$ as a real manifold $X'$ of real dimension $2n$ and its real tangent bundle $T(X')$ and cotangent bundle $T^*(X')$. Let

$$T(X)_c = T(X') \otimes_{\mathbf{R}} \mathbb{C}$$
$$T^*(X)_c = T^*(X') \otimes_{\mathbf{R}} \mathbb{C}$$

be their complexifications. These are $\mathbb{C}^{2n}$ bundles over $X$. Let $\Lambda^r T^*(X)_c$ be the exterior algebra bundle over $X$ and $\mathcal{E}^r(X)$ the presheaf of complex valued smooth sections into $\Lambda^r T^*(X)_c$. This is in fact a sheaf, the sheaf of complex valued differential forms of total degree $r$. We have the usual exterior derivative and complex

$$(2.1)\quad \mathscr{E}^0(X) \xrightarrow{d} \mathscr{E}^1(X) \longrightarrow \cdots \longrightarrow \mathscr{E}^{2n}(X) \to 0 \quad \text{with} \quad d^2 = 0.$$

Using the complex conjugation automorphism in the fiber, we can decompose $T^*(X)_c = T^*(X)^{1,0} \oplus T^*(X)^{0,1}$ and

$$\Lambda T^*(X)_c = \bigoplus_{p,q} \Lambda^{p,q} T^*(X)_c$$

where locally $\Lambda^{p,q}T^*(X)_c$ is generated by elements of the form $u \wedge v$, $u \in \Lambda^p T^*(X)^{1,0}$, $v \in \Lambda^q T^*(X)^{0,1}$.

(2.2) Let $\mathscr{E}^{p,q}(X)$ be the presheaf of complex valued smooth sections into $\Lambda^{p,q}T^*(X)_c$. Then

$$\mathscr{E}^r(X) = \sum_{p+q=r} \mathscr{E}^{p,q}(X)$$

and an $r$-form $u \in \mathscr{E}^{p,q}(X)$ is called a $(p, q)$-form. Local representation shows that

$$u = \sum_{\substack{|I|=p \\ |J|=q}} a_{IJ} w^I \wedge \bar{w}^J$$

where $I = \{i_1 < i_2 < \cdots < i_p\}$, $J = \{j_1 < j_2 < \cdots < j_q\}$ $w^I = dz_{i_1} \wedge \cdots \wedge dz_{i_p}$ and $\bar{w}^J = d\bar{z}_{j_1} \wedge \cdots \wedge d\bar{z}_{j_q}$, and the $a_{IJ}$ are smooth complex valued functions.

(2.3) Define

$$\partial : \mathscr{E}^{p,q}(X) \to \mathscr{E}^{p+1,q}(X)$$
$$\bar{\partial} : \mathscr{E}^{p,q}(X) \to \mathscr{E}^{p,q+1}(X)$$

by setting $\partial = \pi_{p+1,q} \circ d$, $\bar{\partial} = \pi_{p,q+1} \circ d$ where $\pi_{p,q} : \mathscr{E}^r(X) \to \mathscr{E}^{p,q}(X)$ is the projection map.

Since we started with a complex manifold $X$, direct computation shows that $d = \partial + \bar{\partial}$ and in local coordinates $\partial = \sum_{j=1}^n (\partial/\partial z_j)\, dz_j$ and $\bar{\partial} = \sum_{j=1}^n (\partial/\partial \bar{z}_j)\, d\bar{z}_j$.

(2.4) Corresponding to the $\bar{\partial}$ operator we have a complex

$$0 \to \Omega^p \xrightarrow{i} \mathscr{E}^{p,0} \xrightarrow{\bar{\partial}} \mathscr{E}^{p,1} \longrightarrow \cdots \longrightarrow \mathscr{E}^{p,n} \longrightarrow 0$$

where $\Omega^p = \ker \bar{\partial}$ is the sheaf of *holomorphic p-forms* on $X$, i.e. a local representation of $u \in \Omega^p$ is given by

$$u = \sum_{|I|=p} a_I w^I, \quad \text{where the } a_I \text{ are holomorphic.}$$

(2.5) We also have the complex of sheaves

$$0 \longrightarrow \mathbb{C} \xrightarrow{i} \Omega^0 \xrightarrow{\partial} \Omega^1 \xrightarrow{\partial} \cdots \longrightarrow \Omega^n \longrightarrow 0$$

with the $\Omega^i$ defined above, called the complex of *holomorphic differential forms.*

## 3. Sheaf cohomology

Given a complex of sheaves over $X$

$$\mathcal{S}^0 \xrightarrow{d} \mathcal{S}^1 \xrightarrow{d} \cdots \mathcal{S}^i \longrightarrow \mathcal{S}^{i+1} \longrightarrow \cdots \longrightarrow \mathcal{S}^n$$

define

$$\mathcal{L}^i = \ker (\mathcal{S}^i \xrightarrow{d} \mathcal{S}^{i+1})$$
$$\mathcal{B}^i = \operatorname{im} (\mathcal{S}^{i-1} \xrightarrow{d} \mathcal{S}^i)$$

and

$$\mathfrak{H}^i = \mathcal{L}^i / \mathcal{B}^i$$

the $i$th derived sheaf.

(3.1) Given a sheaf $\mathcal{S}$ over $X$, a *resolution* of $\mathcal{S}$ is an exact sequence of sheaves over $X$

$$0 \longrightarrow \mathcal{S} \xrightarrow{j} \mathcal{F}^0 \xrightarrow{d} \mathcal{F}^1 \longrightarrow \cdots \longrightarrow \mathcal{F}^n \longrightarrow \cdots.$$

Clearly, in this sequence $\mathfrak{H}^0 = \mathcal{S}$ and $\mathfrak{H}^i = 0$, $i > 0$.

Every sheaf has a *canonical* resolution, by taking $\mathcal{F}^0$ as the set of all not necessarily continuous sections of $\mathcal{S}$, and iterating. Note that each $\mathcal{F}^i$ here is *flabby*: for all open sets $U$, $\Gamma(X, \mathcal{F}^i) \to \Gamma(U, \mathcal{F}^i)$ is surjective.

(3.2) Definition. Let $\mathcal{S}$ be a sheaf on $X$ and

$$0 \to \mathcal{S} \to \mathcal{F}^0 \to \mathcal{F}^1 \to \cdots$$

its canonical resolution. Consider the derived complex of abelian groups

$$0 \longrightarrow \Gamma(X; \mathcal{S}) \xrightarrow{d^0} \Gamma(X; \mathcal{F}^0) \xrightarrow{d^1} \cdots.$$

Define

$$H^i(X; \mathcal{S}) = \ker d^i / \operatorname{im} d^{i-1}, \quad i = 0, 1, \cdots.$$

(3.3) Sheaf cohomology as defined here has the usual functorial properties. In particular if

$$0 \to \mathcal{S}' \to \mathcal{S} \to \mathcal{S}'' \to 0$$

is a short exact sequence of sheaves, then we have an induced long exact sequence

$$\cdots \to H^i(X, \mathcal{S}') \to H^i(X, \mathcal{S}) \to H^i(X, \mathcal{S}'') \to H^{i+1}(X, \mathcal{S}') \to \cdots.$$

(3.3) Given a sheaf $\mathcal{S}$ on $X$ we may define the *Čech cohomology groups* as follows. If $\mathfrak{U} = \{U_i\}$ is an open cover of $X$, we form the cochain complex $C^*(\mathfrak{U}, \mathcal{S})$, where $C^q(\mathfrak{U}, \mathcal{S})$ consists of alternating functions

$$(i_0, \cdots, i_q) \to f(i_0, \cdots, i_q) \in \Gamma(U_{i_0} \cap \cdots \cap U_{i_q}, \mathcal{S}).$$

Define

$$\check{H}^q(X, \mathcal{S}) = \text{dir lim}_{\mathfrak{U}} H^q(C^*(\mathfrak{U}, \mathcal{S})).$$

The next result is in Godement [1, p. 228].

(3.4) THEOREM. *If $X$ is paracompact and Hausdorff, then*

$$\check{H}^q(X, \mathcal{S}) = H^q(X, \mathcal{S}).$$

Moreover, if $\mathfrak{U}$ has the property, that for all $q$ and every subset $(i_0, \cdots, i_q)$ we have

(3.5) $H^p(U_{i_0} \cap \cdots \cap U_{i_q}, \mathcal{S}) = 0 \quad \text{for} \quad p \geqq 1,$

then the limit is unnecessary and

$$H^q(X, \mathcal{S}) = H^q(C^*(\mathfrak{U}, \mathcal{S})).$$

## 4. Hypercohomology

Assume that $X$ is a smooth manifold (real or complex, open or closed). We shall outline the proof of the de Rham theorem in the appropriate cases, using the Čech definition of hypercohomology.

Given an open covering $\mathfrak{U}$ of $X$ and the sheaf of differential forms (2.1), $\mathcal{E}^*$, the module of Čech cochains with values in $\mathcal{E}^*$, $K^{p,q} = C^p(\mathfrak{U}, \mathcal{E}^q)$ has a double grading with differentials

(4.1) $d': C^p(\mathfrak{U}, \mathcal{E}^q) \to C^{p+1}(\mathfrak{U}, \mathcal{E}^q)$

(4.2) $d'': C^p(\mathfrak{U}, \mathcal{E}^q) \to C^p(\mathfrak{U}, \mathcal{E}^{q+1})$

satisfying $d'd'' + d''d' = 0$. In the usual way, Godement [1, p. 86], we form the total complex

$$K^n = \sum_{p+q=n} K^{p,q}$$

with differential $d = d' + d''$, and define the *hypercohomology* groups

$$\mathcal{H}^r(X, \mathscr{E}^*) = \text{dir} \lim_{\mathfrak{U}} H^r(K^*).$$

If $\mathfrak{U}$ satisfies (3.5) for each $\mathscr{E}^q$, the limit is unnecessary.

(4.3) The complex $K$ may be given two filtrations with two corresponding spectral sequences. The first is by $K_p = \sum_{i \geq p} K^{i,j}$. Call the terms of the corresponding spectral sequence $E_r^{pq}$, where

$$E_1^p = H(K_p/K_{p+1}).$$

Now $K_p/K_{p+1}$ is naturally isomorphic to $K^{p^*} = \sum_j K^{p,j}$ and by this identification the differential $d_0$ becomes $d''$. Thus we have $E_1^{p,q} = h^q(K^{p^*})$ where $h$ indicates that the differential is $d''$. A further calculation shows, Godement [1, p. 87] that

$$E_2^{p,q} = H^p(h^q(K))$$
$$E_2^{p,q} = H^p(X, h^q(\mathscr{E}^*))$$

where $H$ is cohomology with respect to $d'$.

(4.4) A second spectral sequence is obtained by interchanging the roles of $d'$ and $d''$. Calling this $F_r^{p,q}$ to distinguish it from the first, we have

$$F_1^{p,q} = H^q(K^{*p}) = H^q(X, \mathscr{E}^p)$$

and the differential $d_1$ of this spectral sequence is induced by the sheaf homomorphism $\mathscr{E}^p \to \mathscr{E}^{p+1}$. So for $q = 0$ we have the complex, recalling that $H^0(X, \mathscr{E}^p) = \Gamma(X, \mathscr{E}^p)$:

$$0 \to \Gamma(X, \mathscr{E}^0) \to \Gamma(X, \mathscr{E}^1) \to \cdots \to \Gamma(X, \mathscr{E}^n) \to 0.$$

Now $\Gamma(X, \mathscr{E}^p)$, the group of global sections of $X$ into the sheaf $\mathscr{E}^p$ is by definition the group of differential $p$-forms on $X$, so the complex above is what is usually called the *de Rham complex* of $X$. Thus $F_2^{p,0}$ is the cohomology of this complex, and we shall denote these groups by

$$F_2^{p,0} = DR^p(\mathscr{E}^*)$$

the $p$th de Rham group of the complex $\mathscr{E}^*$.

There is a natural homomorphism $DR^p \to \mathcal{H}^p$ from the spectral sequence map $F_2^{p,0} \to F_\infty^{p,0} = \mathcal{H}_p^p/\mathcal{H}_{p+1}^p$, but $\mathcal{H}_{p+1}^p = 0$.

## 5. Analytic de Rham theorem

We may now prove the de Rham theorem for complex manifolds. Here $\mathscr{E}^*$ is the complex of smooth differential forms on $X$, as defined in (2.2). The following is a classical result, see Hu [1, p. 86].

(5.1) POINCARÉ LEMMA. *Let $U$ be a star-shaped domain in $\mathbb{R}^n$. If $f \in \mathscr{E}^p(U)$, $p > 0$, is such that $df = 0$, then there exists a $g \in \mathscr{E}^{p-1}(U)$ so that $dg = f$. Thus*

$$h^q(\mathscr{E}^*) = 0 \quad \textit{for} \quad q \geqq 1.$$

Thus the first spectral sequence, $E^{p,q}$ above is trivial. Since $h^0(\mathscr{E}) = \mathbb{C}$, we have

(5.2) $H^p(X, \mathbb{C}) = E_2^{p,0} \simeq \mathscr{H}^p(X, \mathscr{E}^*)$.

In the second spectral sequence we need the following, see Wells [1, p. 54].

(5.3) DEFINITION. A sheaf of abelian groups $\mathscr{F}$ over a paracompact Hausdorff space $X$ is *fine* if for any locally finite open cover $\{U_i\}$ of $X$ there exists a family of sheaf morphisms $\{\eta_i : \mathscr{F} \to \mathscr{F}\}$ such that

(i) $\sum \eta_i = 1$

(ii) $\eta_i(\mathscr{F}_x) = 0$ for all $x$ in some neighborhood of the complement of $U_i$.

The family $\{\eta_i\}$ is called a *partition of unity* of $\mathscr{F}$ subordinate to the covering $\{U_i\}$.

(5.4) The sheaves $\mathscr{E}^r$ are fine sheaves for $X$ a paracompact complex manifold, since multiplication by a globally defined smooth map defines a sheaf homomorphism and $X$ has a smooth partition of unity.

(5.5) THEOREM. *If $\mathscr{F}$ is a fine sheaf on $X$, then*

$$H^q(X, \mathscr{F}) = 0 \quad \textit{for} \quad q \geqq 1.$$

PROOF. Combine Prop. 3.5 and Theorem 3.11 in Wells [1, p. 54–56]. Thus in the spectral sequence $F^{p,q}$, we have

(5.6) $F_1^{p,q} = H^q(X, \mathscr{E}^p) = 0 \quad$ for $\quad q \geqq 1 \quad$ and all $p$,

so it is also trivial and

(5.7) $DR^p(\mathscr{E}^*) = F_2^{p,0} \simeq \mathscr{H}^p(X, \mathscr{E}^*)$.

We combine (5.2) with (5.7) to obtain the *smooth* de Rham theorem:

(5.8) THEOREM. *If $X$ is a paracompact smooth manifold and $\mathscr{E}^*$ its sheaf of complex valued smooth differential forms, then*

$$H^p(X, \mathbb{C}) \simeq DR^p(\mathscr{E}^*) \quad \textit{for all } p.$$

(5.9) If $X$ is an affine submanifold of $\mathbb{C}^n$, called a Stein manifold, a holomorphic version of the de Rham theorem may be proved. Consider the sheaf of holomorphic differential forms, $\Omega^*$ defined in (2.5). Using the hypercohomology with this sheaf, we have the analog of (5.1) called the holomorphic Poincaré lemma, Gunning and Rossi [1, p. 27 and Remark on p. 28]. Thus

(5.10) $H^p(X;\mathbb{C}) = E_2^{p,0} \simeq \mathscr{H}^p(X,\Omega^*)$.

Since there are no holomorphic partitions of unity, a different result must be used to show that the second spectral sequence also collapses. The sheaves $\Omega^q$ are locally free $\mathcal{O}_X$-modules and therefore coherent. Cartan's theorem B then says, see Gunning and Rossi [1, p. 243]:

(5.11) THEOREM. *Let $X$ be a Stein space with structure sheaf $\mathcal{O}_X$. Let $\mathscr{S}$ be a coherent sheaf on $X$. Then $H^q(X,\mathscr{S})=0$ for all $q \geqq 1$.*

Using (5.11) in place of (5.5) we obtain:

(5.12) THEOREM. *If $X$ is an affine submanifold of $\mathbb{C}^n$ and $\Omega^*$ its sheaf of holomorphic differential forms, then*

$$H^p(X;\mathbb{C}) \simeq DR^p(\Omega^*) \quad \textit{for all } p.$$

Note here that we are still in the analytic category. In order to distinguish (5.12) from the algebraic result we are going to obtain in the next sections, we call it the *analytic* de Rham theorem. Thus every singular cohomology class is represented by a global section into the sheaf of *holomorphic* differential forms.

In the case of an affine algebraic subvariety of $\mathbb{C}^n$ the question of which holomorphic differential forms have algebraic representatives is very old. The problem of integrals of the second kind is related to it and dates back to Picard and Poincaré, see Atiyah and Hodge [1].

Our motivation is the following. We have now shown in (I.5.10), (II.4.1), and (5.12) that

$$\dim_{\mathbb{C}} \mathcal{O}/(\partial f) = \mu_a = \mu_c = \mu_t = \dim_{\mathbb{C}} DR^{n-1}(\Omega^*).$$

The vector space $\mathcal{O}/(\partial f)$ has a polynomial base. The natural question arises, whether $DR^{n-1}(\Omega^*)$ has a rational (or better, polynomial) base and whether we could exhibit an isomorphism of the vector spaces $\mathcal{O}/(\partial f)$ and $DR^{n-1}(\Omega^*)$.

The answer to the first question is given by Grothendieck [1] and we shall outline the proof following Atiyah, Bott, and Gårding [1] in the remaining sections of this chapter.

Chapter 4 is devoted to the construction of an explicit isomorphism when $f$ is a homogeneous polynomial, following Orlik and Solomon [1].

Next we need some background material in algebraic geometry, see Shafarevich [1, Ch. I, §4] or Serre [1, Nos 31, 32].

## 6. Algebraic geometry

(6.1) The *Zariski topology* of $\mathbb{C}^n$ is defined by calling a set *closed* if it is the union of the common zeros of a family of polynomials, $P^\alpha \in \mathbb{C}[Z_1, \cdots, Z_n]$. Since the polynomial ring is Noetherian, $\mathbb{C}^n$ satisfies the decreasing chain condition on closed sets: if

$$F_1 \supset F_2 \supset \cdots \supset F_m \supset \cdots$$

are closed sets, there exists $n$ such that for all $m \geqq n$, $F_m = F_n$.

Let $(z_1, \cdots, z_n) \in \mathbb{C}^n$ and define $\mathcal{O}_z$ as the *local ring* at $z$, the subring of the field of fractions $\mathbb{C}(Z_1, \cdots, Z_n)$ consisting of the rational functions $R = P/Q$, where $P$ and $Q$ are polynomials and $Q(z) \neq 0$.

Such a fraction is called *regular* at all $z$ where $Q(z) \neq 0$. The function $z \to P(z)/Q(z)$ is defined on the set of regular values and takes values in $\mathbb{C}$. It is continuous if $\mathbb{C}$ is also given the Zariski topology.

In order to distinguish $\mathcal{O}_z$ defined here from the local ring of *holomorphic* functions defined in (I.2), we shall denote the latter $\mathcal{O}_z^h$. We have a natural inclusion $\mathcal{O}_z \to \mathcal{O}_z^h$.

(6.2) The sheaf of regular functions on $\mathbb{C}^n$, whose stalks are the $\mathcal{O}_z$ is denoted $\mathcal{O}$ (and the sheaf of germs of holomorphic functions $\mathcal{O}^h$). Clearly, $\mathcal{O}$ is a sheaf of rings.

Let $U$ be an open subset of $\mathbb{C}^n$, $F$ a closed subset of $\mathbb{C}^n$ and $Y = U \cap F$, a locally closed subset of $\mathbb{C}^n$. If $z \in Y$ we may define the ring $\mathcal{O}_{z,Y}$ as the quotient of $\mathcal{O}_z$ by the ideal $I(F) \cdot \mathcal{O}_z$, where $I(F)$ is the ideal of polynomials vanishing on $F$ in $\mathbb{C}[Z_1, \cdots, Z_n]$. The factor sheaf of $\mathcal{O}$ defined by the $\mathcal{O}_{z,Y}$ is called $\mathcal{O}_Y$, the structure sheaf of $Y$.

If $X$ is an algebraic subvariety of $\mathbb{C}^n$, Serre [1, §37, Prop. 1] shows that $\mathcal{O}_X$ is a coherent sheaf of rings. A sheaf of $\mathcal{O}_X$-modules on $X$ is called an *algebraic* sheaf. It is *coherent* if it is locally the cokernel of a map $\mathcal{O}_X^q \to \mathcal{O}_X^p$, where

$$\mathcal{O}_X^p = \mathcal{O}_X \oplus \mathcal{O}_X \oplus \cdots \oplus \mathcal{O}_X \quad p \text{ times.}$$

We call $X$ an *affine* algebraic variety. A section of $\mathcal{O}_X$ on an open set $U$ of $X$ corresponds to a map $f: U \to \mathbb{C}$ which, in a neighborhood of each point $x \in U$, equals the restriction to $X$ of a rational function, regular at $x$. Call such a function *regular* on $U$.

Let $X$ be an affine algebraic variety and $f$ a regular function of $X$. Let $X_f$ denote the open subset of $X$ where $f(x) \neq 0$. Serre [1, §42] proves that the $X_f$ are affine open sets as defined in (6.1) and they form a basis for the (Zariski) topology of $X$.

(6.3) Let $Y=\mathbb{C}^{n+1}-\{0\}$. $\mathbb{C}^*$ operates on $Y$ by $\lambda \in \mathbb{C}^*$

$$\lambda(z_0, \cdots, z_n)=(\lambda z_0, \cdots, \lambda z_n).$$

As usual, we define *complex projective* space $\mathbb{C}P^n$ by identifying in $Y$ two points $z$ and $z'$ if and only if there is a $\lambda \in \mathbb{C}^*$ such that $z=\lambda z'$. Let $\pi: Y \to \mathbb{C}P^n$ be the canonical projection. Let $t_i$ be the $i$th coordinate function of $\mathbb{C}^{n+1}$, $t_i(z)=z_i$.

Let $V_i=\{z \in \mathbb{C}^{n+1} \mid z_i \neq 0\}$ be open sets, $i=0, \cdots, n$, and $U_i=\pi(V_i)$. Then $\{U_i\}$ form a cover of $\mathbb{C}P^n$. For $i \neq j$, the function $t_j/t_i$ is regular on $V_i$ and invariant under $\mathbb{C}^*$, thus a function defined on $U_i$, which we shall again call $t_j/t_i$. For fixed $i$ the functions $t_j/t_i$, $j \neq i$ define a bijection $\psi_i: U_i \to \mathbb{C}^n$.

Let $\mathbb{C}^{n+1}$ have the Zariski topology defined above and $Y=\mathbb{C}^{n+1}-\{0\}$ have the induced topology. Note that $\{0\}=\bigcap_{i=0}^n X_i$, $X_i=\{z \mid t_i(z)=0\}$ so $Y$ is affine open.

The topology of $\mathbb{C}P^n$ is the quotient topology obtained from $\pi: Y \to \mathbb{C}P^n$, i.e. $Z$ is closed in $\mathbb{C}P^n$ if and only if $\pi^{-1}(Z)$ is closed in $Y$.

Let $U$ be open in $\mathbb{C}P^n$ and consider the ring of regular functions on $\pi^{-1}(U)$, $A_U=\Gamma(\pi^{-1}U, \mathcal{O}_Y)$. Let $A_U^0$ be the subring of functions invariant under $\mathbb{C}^*$, i.e. regular functions which are homogeneous of degree 0. If $V \supset U$, we have the restriction maps $\varphi_U^V: A_V^0 \to A_U^0$ and therefore we have a presheaf on $\mathbb{C}P^n$. This presheaf is in fact a sheaf called the structure sheaf, $\mathcal{O}_{\mathbb{C}P^n}$ of $\mathbb{C}P^n$. A function $f$, defined in a neighborhood of a point $x \in \mathbb{C}P^n$ belongs to the stalk $\mathcal{O}_x$ if and only if it agrees locally with a fraction $P/Q$, where $P$ and $Q$ are homogeneous polynomials of the same degree in $t_0, \cdots, t_n$ and $Q(y) \neq 0$ for $y \in \pi^{-1}(x)$.

An algebraic variety is called *projective* if it is isomorphic to a closed subvariety of $\mathbb{C}P^n$.

## 7. Vanishing theorems

Serre [1] defines the cohomology of an algebraic variety with coefficients in an algebraic sheaf using the Čech method. Although we no longer have paracompact spaces and therefore cannot invoke (3.4), a result of Grothendieck asserts that the injective resolution definition is equivalent. Again if (3.5) holds, we need not take limits. Serre [1, no. 46] proves the following result.

(7.1) THEOREM. *Let $X$ be an affine variety and $\mathcal{F}$ a coherent sheaf on $X$. Then*

$$H^q(X, \mathcal{F})=0 \quad \textit{for all} \quad q \geqq 1.$$

We shall not repeat the proof, but the heart of the matter is contained in the next proposition of Serre [1, no. 43, Prop. 5].

(7.2) PROPOSITION. *Let $X$ be an irreducible algebraic variety, $Q$ a regular function on $X$, and $P$ a regular function on $X_Q$. For all sufficiently large $n$, the rational function $Q^nP$ is regular on all of $X$.*

PROOF. We may assume that $X$ is a closed subvariety of $\mathbb{C}^n$ and $Q \in A = \mathbb{C}[Z_1, \cdots, Z_n]/I(X)$, where $A$ is an integral domain. Since $P$ is regular on $X_Q = \{z \in X \mid Q(z) \neq 0\}$, we may write for every $x \in X_Q$, $P = P_x/Q_x$ with $P_x$, $Q_x \in A$ and $Q_x(x) \neq 0$. Let $\mathfrak{A}$ denote the ideal generated by the $Q_x$ for all $x \in X_Q$. The locus of zeros of $\mathfrak{A}$ is by hypothesis contained in the locus of zeros of $Q$. By Hilbert's Nullstellensatz (for polynomials) for some $n$, $Q^n \in \mathfrak{A}$. Thus

$$Q^n = \sum R_x Q_x, \quad \text{with} \quad R_x \in A$$

and

$$Q^n P = \sum R_x P_x$$

thus $Q^nP$ is regular on all of $X$.

Next we turn to some results of Serre [1] on projective varieties.

(7.3) THEOREM. *Let $X$ be a projective variety in $\mathbb{C}P^n$ and $\mathcal{F}$ a coherent sheaf on $X$. The groups $H^q(X, \mathcal{F})$ are finite dimensional vector spaces over $\mathbb{C}$.*

If $Y$ is a closed subvariety of $X$ given locally by one equation, denote by $J$ its sheaf of ideals. The sheaf $\mathcal{O}(nY) = \mathfrak{Hom}\mathcal{O}(J^n, \mathcal{O})$ is the sheaf of functions on $X$ with poles of order $\leqq n$ on $Y$. For any coherent sheaf $\mathcal{S}$ we define $\mathcal{S}(n) = \mathcal{S}(nY) = \mathcal{S} \otimes_{\mathcal{O}} \mathcal{O}(nY)$.

We may interpret Serre [1, no. 66, Th. 2] to imply

(7.4) THEOREM. *If $X$ is projective and $Y$ is a hyperplane section of $X$, then for all $q \geqq 1$,*

$$H^q(X, \mathcal{S}(nY)) = 0 \quad \textit{for all large } n.$$

The analogous analytic theorem with $X^h$ replacing $X$ is the Kodaira vanishing theorem.

Atiyah, Bott, and Gårding [1] then prove the following vanishing theorem.

(7.5) GROTHENDIECK VANISHING THEOREM. *If $X$ is projective and $X - Y$ is affine, then for $q \geqq 1$ and any integer $n$ there exists $m > n$ such that*

$$H^q(X, \mathcal{S}(n)) \to H^q(X, \mathcal{S}(m))$$

*is zero.*

PROOF. In view of (7.3) it is sufficient to show that dir lim $H^q(X,\mathcal{S}(n))=0$ for all $q \geqq 1$. For an affine open set $U$ of $X$ in which $Y$ is defined by $f=0$, we set $U_f = U - U \cap Y$ as in (7.2). The argument of (7.2) then shows that for sufficiently large $n$ we have $\Gamma(U,\mathcal{S}(n)) = \Gamma(U_f, \mathcal{S} \mid U_f)$. Hence for any finite covering $\mathfrak{U}$ of $X$ by affine open sets we have

$$\text{dir lim } H^q(C^*(\mathfrak{U},\mathcal{S}(n)) \simeq H^q(\text{dir lim } C^*(\mathfrak{U},\mathcal{S}(n))) \simeq H^q(C^*(\mathcal{V},\mathcal{S} \mid X-Y))$$

where $\mathcal{V}$ is the affine covering of $X-Y$ induced by $\mathfrak{U}$.

According to Serre [1], since $\mathcal{V}$ is a covering of the affine set $X-Y$ by affine open sets and $\mathcal{S} \mid X-Y$ is coherent, we have

$$H^q(C^*(\mathcal{V},\mathcal{S} \mid X-Y)) = H^q(X-Y,\mathcal{S} \mid X-Y)$$

and (7.1) asserts that this is zero for $q \geqq 1$.

## 8. Comparison theorems

Here we list two results of Serre [2] together with a comparison for hypercohomologies.

Using the natural map $\mathcal{O} \to \mathcal{O}^h$ defined in (6.1), we may define for any $\mathcal{O}$-module $\mathcal{S}$ the module $\mathcal{S}^h = \mathcal{S} \otimes_{\mathcal{O}} \mathcal{O}^h$. Then we have

(8.1) THEOREM. *If $\mathcal{S}$ is a coherent algebraic sheaf on $X$, then $\mathcal{S}^h$ is a coherent analytic sheaf on $X^h$.*

(8.2) THEOREM. *If $X$ is projective algebraic and $\mathcal{S}$ is a coherent algebraic sheaf on $X$, then the natural homomorphism*

$$H^q(X,\mathcal{S}) \to H^q(X^h,\mathcal{S}^h)$$

*is an isomorphism for all $q$.*

The natural homomorphism in (8.2) is obtained as follows. The 'identity' map $i: X^h \to X$ is continuous by construction. Let $\mathcal{S}'$ be the pull back of $\mathcal{S}$ by this map. Then $\mathcal{S}^h = \mathcal{S}' \otimes_{\mathcal{O}} \mathcal{O}^h$ and the map above is the composition

$$H^q(X,\mathcal{S}) \to H^q(X^h,\mathcal{S}') \to H^q(X^h,\mathcal{S}^h).$$

(8.3) HYPERCOHOMOLOGY COMPARISON THEOREM. *Let $\varphi: \Omega^* \to \tilde{\Omega}^*$ be a homomorphism of complexes of sheaves which induces an isomorphism on cohomology sheaves, $h^q(\Omega^*) \simeq h^q(\tilde{\Omega}^*)$ for all $q$. Then $\varphi$ induces an isomorphism of hypercohomology*

$$\mathcal{H}^q(X,\Omega^*) \simeq \mathcal{H}^q(X,\tilde{\Omega}^*).$$

*In particular, if $H^q(X,\Omega^p) = H^q(X,\tilde{\Omega}^p) = 0$ for $q \geqq 1$ and all $p$, then $\varphi$ induces an isomorphism of the de Rham groups*

$$DR^q(\Omega^*) \simeq DR^q(\tilde{\Omega}^*).$$

PROOF. Using the notation of section 4, $\varphi$ induces a homomorphism of spectral sequences $E_r \to \tilde{E}_r$. For $r=2$ the assumption on $\varphi$ and (4.3) show that it is an isomorphism. Hence $E_r \simeq \tilde{E}_r$ for all $r$, and so the end terms of the spectral sequences, namely the hypercohomology groups, are also isomorphic.

If moreover, $H^q(X, \Omega^p)=0$ for $q \geqq 1$, then the second spectral sequence $F_r^{p,q}$ is trivial for $r \geqq 2$ and so $\mathcal{H}^q(X, \Omega^*) \simeq DR^q(\Omega^*)$. Similarly for $\tilde{\Omega}^*$ and so $\varphi$ induces an isomorphism

$$DR^q(\Omega^*) \to DR^q(\tilde{\Omega}^*)$$

as required.

This will be applied in the next section to deduce the algebraic de Rham theorem from the smooth theorem (5.8). We shall introduce an increasing sequence of sheaves

$$\Omega^*(k) \subset \Omega^*(k+1) \subset \cdots$$

of forms on a projective variety $X$ with poles of order determined by $k$ on a codimension one subvariety $Y$. These will satisfy the conditions of the next lemma, which in turn puts bounds into the spectral sequence of (8.3).

(8.4) LEMMA. *Let $\Omega^*(k) \subset \Omega^*(k+1) \subset \cdots$ be an increasing sequence of complexes of sheaves $(k \geqq 0)$ on $X$. Assume that*

(i) *for all $k$, $\Omega^*(k) \to \Omega^*(k+1)$ induces an isomorphism of cohomology sheaves,*

(ii) *there is a function $k \to f(k)$ such that*

$$H^q(X, \Omega^p(k)) \to H^q(X, \Omega^p(f(k)))$$

*is zero for $q \geqq 1$ and all $p$,*

(iii) *$H^q(X, \Omega^p(k))=0$ for $q>n$ and all $p$, $k$, and $\Omega^p(k)=0$ for $p>n$ and all $k$.*

*Then*

(a) *the maps of hypercohomology*

$$\mathcal{H}^q(X, \Omega^*(k)) \to \mathcal{H}^q(X, \Omega^*(k+1))$$

*are all isomorphisms,*

(b) *the natural homomorphism*

$$\varphi_N^q : DR^q(\Omega^*(N)) \to \mathcal{H}^q = \mathcal{H}^q(X, \Omega^*(N))$$

*is surjective provided $N \geqq f^n(0)$, where $f^n = f \circ \cdots \circ f$,*

(c) *the kernel of $\varphi_N^q$ coincides with the kernel of*

$$DR^q(\Omega^*(N)) \to DR^q(\Omega^*(M))$$

*provided $M \geqq f^{n-1}(N)$.*

PROOF. (a) follows from (i) and (4.3) as

$$E_2^{p,q}(k) = H^p(X, h^q\Omega^*(k)) \cong H^p(X, h^q\Omega^*(k+1)) = E_2^{p,q}(k+1).$$

Now consider the spectral sequence $F_r^{p,q}(k) \Rightarrow \mathcal{H}$. We have using (iii) and $F_1^{p,q}(k) = H^q(X, \Omega^p(k))$ that

$$F_{n+1}^{p,q}(k) = F_\infty^{p,q}(k) = \mathcal{H}_p^{p+q}(k)/\mathcal{H}_{p+1}^{p+q}(k)$$

where $\mathcal{H}^s = \mathcal{H}^s(k)$ is filtered

$$\mathcal{H}^s = \mathcal{H}_0^s(k) \supset \mathcal{H}_1^s(k) \supset \cdots \supset \mathcal{H}_s^s(k) \supset \mathcal{H}_{s+1}^s(k) = 0.$$

By (ii) $F_1^{p,q}(k) \to F_1^{p,q}(f(k))$ is zero for $q \geqq 1$, and hence the same is true for $F_\infty^{p,q}$. Thus

$$\mathcal{H}_p^{p+q}(k) \subset \mathcal{H}_{p+1}^{p+q}(f(k)) \quad \text{for} \quad q \geqq 1.$$

Iterating this we see that

$$\mathcal{H}^s = \mathcal{H}_0^s(k) \subset \mathcal{H}_1^s(f(k)) \subset \cdots \subset \mathcal{H}_s^s(f^s(k)) = \mathcal{H}^s$$

and so $\mathcal{H}_s^s(f^s(k)) = \mathcal{H}^s$. In other words the filtration on $\mathcal{H}^s(f^s(k))$ consists of just one term, the last. The same is then true for all integers $> f^s(k)$. Thus for all $N > f^n(0)$ we have a surjection

$$DR^p(\Omega^*(N)) = F_2^{p,0}(N) \to F_\infty^{p,0}(N) = \mathcal{H}_p^p(N) = \mathcal{H}^p(N)$$

proving (b).

The proof of (c) is similar.

## 9. Algebraic de Rham theorem

Atiyah, Bott, and Gårding [1] complete the proof as follows.

Let $Y \subset \mathbb{C}^n$ be given by $z_1 z_2 \cdots z_r = 0$ and let $\Omega^q(k) = \Omega_{r,n}^q(k)$ denote the germ at 0 of meromorphic $q$-forms on $\mathbb{C}^n$ which have poles on $Y$ of total order $\leqq q+k$, i.e.

$$\varphi = \sum_{|\nu| \leqq q+k} z_1^{-\nu_1} z_2^{-\nu_2} \cdots z_r^{-\nu_r} \varphi_\nu$$

where $\varphi_\nu$ is a holomorphic $q$-form, $\nu = (\nu_1, \cdots, \nu_r)$ and $|\nu| = \sum \nu_i$. Clearly $d\Omega^q(k) \subset \Omega^{q+1}(k)$ so that we have a complex

$$\Omega^*(k): \Omega^0(k) \xrightarrow{d} \Omega^1(k) \longrightarrow \cdots \longrightarrow \Omega^n(k).$$

(9.1) LEMMA. *$H^*(\Omega(k))$ is an exterior algebra generated by $dz_i/z_i$, $i = 1, \cdots, r$, and it is isomorphic to $H^*(\mathbb{C}^n - Y)$.*

PROOF. Let $\varphi \in \Omega^q(k)$ with $d\varphi = 0$. If $q = 0$ then $\varphi = \text{const}$. If $q \geqq 1$, writing $\varphi$ in the form $\varphi = dz_1\alpha + \beta$ and expanding in terms of $z_1$,

$$\alpha = \alpha_0 + \alpha_1 z_1^{-1} + \cdots + \alpha_r z_1^{-r}$$
$$\beta = \beta_0 + \beta_1 z_1^{-1} + \cdots + \beta_r z_1^{-r}$$

where $\alpha_0$, $\beta_0$ are holomorphic in $z_1$, we have

$$d\alpha_1 = d\alpha_2 + \beta_1 = d\alpha_3 + 2\beta_2 = \cdots = r\beta_r = 0$$

for the coefficients of $z_1^{-i}\, dz_1$. Differentiating gives

$$d\beta_1 = d\beta_2 = \cdots = d\beta_r = 0.$$

Also for $\varphi_0 = dz_1\alpha_0 + \beta_0$ we have $d\varphi_0 = 0$. Put

$$\theta = -\alpha_2 z_1^{-1} - \alpha_3(2z_1^2)^{-1} - \cdots - \alpha_r[(r-1)z_1^{r-1}]^{-1} \in \Omega^{q-1}(k).$$

Then

$$\varphi = z_1^{-1}\, dz_1\alpha_1 + \varphi_0 + d\theta.$$

Here $\alpha_1 \in \Omega^{q-1}_{n-1,r-1}(k)$ is independent of $z_1$ and $\varphi_0 \in \Omega^q_{n,r-1}(k)$ has no pole on $z_1 = 0$. Since $d\alpha_1 = d\varphi_0 = 0$, induction and the usual Poincaré lemma proves that

$$\varphi = \sum c_{i_1\cdots i_q}(z_{i_1})^{-1}\, dz_{i_1}\cdots (z_{i_q})^{-1}\, dz_{i_q} + d\psi$$

where $1 \leqq i_1 < \cdots < i_q \leqq r$, $c_{i_1\cdots i_q} \in \mathbb{C}$ and $\psi \in \Omega^{q-1}(k)$.

This proves that $H^*(\Omega(k))$ is generated by the $dz_i/z_i$, $i = 1, \cdots, r$. On the other hand for any small ball $B$ around the origin,

$$H^*(\mathbb{C}^n - Y) = H^*(B - B \cap Y)$$

is an exterior algebra on $r$ generators in $H^1(\mathbb{C}^n - Y)$ dual to the $r$ generators of $H_1(\mathbb{C}^n - Y)$ given by small circles $\gamma_k$ around the hyperplanes $z_k = 0$, $k = 1, \cdots, r$. Indeed, $\mathbb{C}^n - Y$ retracts to the torus which is the cartesian product $\gamma_1 \times \cdots \times \gamma_r$. Since

$$\int_{\gamma_k} z_j^{-1}\, dz_j = 2\pi i \delta_{jk}$$

this proves that

$$H^*(\Omega^*(k)) \to H^*(\mathbb{C}^n - Y)$$

is an isomorphism.

(9.2) THEOREM. *Let $X$ be a non-singular projective variety, $Y$ a closed subvariety of codimension one which is a finite union of non-singular subvarieties $Y_i$ with normal crossings, and such that $X - Y$ is affine. Then*

*$H^*(X^h - Y^h)$ is isomorphic to the de Rham group of the complex of rational differential forms on $X$ with poles on $Y$.*

PROOF. Denote by $\Omega^p(k)$ the algebraic sheaf of $p$-forms on $X$ with pole of total order $\leqq p+k$ on $Y$. It is easy to see that these are coherent sheaves and we have $d\Omega^p(k) \subset \Omega^{p+1}(k)$.

Since $X - Y$ is affine, the Grothendieck vanishing theorem (7.5) implies that there is a function $k \to f(k)$ such that

$$H^q(X, \Omega^p(k)) \to H^q(X, \Omega^p(f(k)))$$

is zero for $q \geqq 1$ and all $p$, $k$.

By the Serre comparison theorem (8.2) we may replace $X, \Omega^p(k)$ by $X^h, \Omega^p(k)^h$, their holomorphic counterparts. Applying (9.1) and (8.4) to the sequence of complexes $\Omega^*(k)$ we deduce

$$\operatorname{Im}[DR^q(\Omega^*(N)^h) \to DR^q(\Omega^*(M)^h)] \to \mathcal{H}^q(X^h, \Omega^*(0)^h)$$

is an isomorphism for $N \geqq f^n(0)$, $M \geqq f^{n-1}(N)$.

By Serre's theorem (8.2) applied to $H^0$ we see that the superscript $h$ can be removed from the de Rham groups and hence it remains to prove that

$$\mathcal{H}^q(X^h, \Omega^*(0)^h) \simeq H^q(X^h - Y^h).$$

Let $\tilde{\Omega}^q$ be the sheaf of smooth complex valued $q$-forms on $X^h - Y^h$ and $i: X^h - Y^h \to X^h$ the inclusion map. Let $i_*(\tilde{\Omega}^q)$ be the direct image sheaf on $X^h$, see (1.10). Then by the smooth de Rham theorem (5.8) on $X^h - Y^h$, $H^q(X^h - Y^h) \simeq DR^q(\tilde{\Omega}^*)$, and by the definition of the de Rham groups (4.4)

$$\simeq H^q(\Gamma(X^h - Y^h, \tilde{\Omega}^*))$$

by the functorial properties of direct image sheaves

$$\simeq H^q(\Gamma(X^h, i_*\tilde{\Omega}^*)).$$

The sheaves $i_*\tilde{\Omega}^*$ are fine sheaves, so (5.5) holds and using (5.7) we have

$$\simeq \mathcal{H}^q(X^h, i_*\tilde{\Omega}^*)$$

and using (9.1) together with (8.3) we finally have

$$\simeq \mathcal{H}^q(X^h, \Omega^*(0)^h)$$

which completes the proof of (9.2).

To obtain the general case from (9.2) we must appeal to the resolution theorem of Hironaka [1].

(9.3) RESOLUTION THEOREM. *Let $X$ be an algebraic variety over $\mathbb{C}$, $Y$ a closed subvariety with $X - Y$ non-singular. Then there exists a proper morph-*

*ism* $\pi: X' \to X$ *with* $X'$ *non-singular such that*

(i) $Y' = \pi^{-1}(Y)$ *is a finite union of non-singular submanifolds of codimension one with normal crossings*

(ii) $X' - Y' \to X - Y$ *is an isomorphism.*

Given a non-singular affine variety $F \subset \mathbb{C}^n$, let $\bar{F}$ be its closure in $\mathbb{C}P^n$, $Y = \bar{F} - F$ and apply the resolution theorem. We get $\pi: \bar{F}' \to \bar{F}$ proper with $\bar{F}'$ non-singular, $Y' = \pi^{-1}(Y)$ a finite union of non-singular subvarieties of codimension one with normal crossings, and $\bar{F}' - Y' = F$. Applying (9.2) to $\bar{F}'$, $Y'$ we obtain the result of Grothendieck [1].

Let $\Omega_F^*$ be the sheaf of everywhere regular differential forms on $F$, i.e. forms with poles at infinity.

(9.4) ALGEBRAIC DE RHAM THEOREM. *For any non-singular affine variety* $F$, *there are isomorphisms for all* $q \geqq 0$,

$$H^q(F^h; \mathbb{C}) \simeq DR^q(\Omega_F^*).$$

(9.5) REMARK. The argument given in Atiyah, Bott, and Gårding [1] is stronger than the one here because it obtains bounds on the order of poles at infinity.

## 10. Related research

There are two main areas unfortunately not familiar to me.

(10.1) There is an extensive literature on the description of the monodromy operator in terms of the cohomology groups introduced above, see Bloom [1], Brieskorn [4], and the Springer Lecture Notes of Deligne and Katz [1], where the characteristic $p > 0$ case is discussed and further references are also found.

(10.2) The theory of residues and integrals on complex varieties are closely connected, see Atiyah and Hodge [1], Pham [2], Lefschetz [1].

# Chapter IV

# Isomorphism and group actions

## 1. Kähler differentials

In this section we shall describe a purely algebraic object called the Kähler differentials, see Cartier [1] and Grothendieck [1]. In case of the non-singular algebraic variety $F$, the module of global sections $\Gamma(F, \Omega_F^*)$ into the sheaf of everywhere regular differential forms coincides with the Kähler differentials of the coordinate ring of $F$. The exposition follows Orlik and Solomon [1].

(1.1) Let $A$ be a commutative $\mathbb{C}$-algebra. Give $A \otimes_{\mathbb{C}} A$ the structure of $A$-module by defining

$$a(a' \otimes a'') = aa' \otimes a''.$$

Let $\Omega^1_A$ be the quotient of $A \otimes_{\mathbb{C}} A$ by the submodule $R$ generated by all elements of the form

$$1 \otimes aa' - a \otimes a' - a' \otimes a.$$

Define a map $d_A : A \to \Omega^1_A$ by

$$d_A a = (1 \otimes a) + R.$$

The map $d_A$ is a $\mathbb{C}$-derivation of $A$ into $\Omega^1_A$. The pair $(\Omega^1_A, d_A)$ is called the module of *Kähler differentials* of $A$. It is uniquely defined by the universality property: if $M$ is an $A$-module and $d : A \to M$ is a $\mathbb{C}$-derivation, then there exists a unique $A$-module homomorphism $h$ completing the diagram

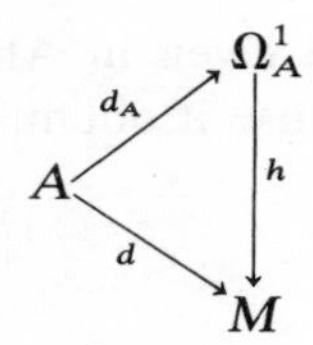

(1.2) Let $\Omega_A = \Lambda(\Omega^1_A)$ be the exterior algebra of $\Omega^1_A$ over $A$. Let $\Omega^p_A = \Lambda^p(\Omega^1_A)$ and $\Omega^0_A = A$. The map $d_A$ extends to a derivation of $\Omega_A$, which we also write as $d_A$. Then for $\alpha \in \Omega^p_A$, $\alpha' \in \Omega^q_A$ we have

$$d_A(\alpha \wedge \alpha') = d_A\alpha \wedge \alpha' + (-1)^p \alpha \wedge d_A\alpha'$$

and $d^2_A = 0$. Let $d^p_A$ be the restriction of $d_A$ to $\Omega^p_A$. Define the cohomology groups

$$H^p(\Omega_A) = \ker d^p_A / \operatorname{im} d^{p-1}_A$$

of the cochain complex $(\Omega_A, d_A)$.

(1.3) Let $\pi^0 : A \to B$ be a homomorphism of rings. We view all $B$-modules as $A$-modules using $\pi^0$. From the definition of $\Omega^1_A$ it follows that there is a unique $A$-module homomorphism $\pi^1 : \Omega^1_A \to \Omega^1_B$ such that $\pi^1 d_A = d_B \pi^0$.

It follows from the universality of the exterior algebra $\Omega_A$, that there is for $p \geqq 1$ an $A$-module homomorphism $\pi^p : \Omega^p_A \to \Omega^p_B$ such that

$$\pi^p(\alpha_1 \wedge \cdots \wedge \alpha_p) = \pi^1\alpha_1 \wedge \cdots \wedge \pi^1\alpha_p$$

for all $\alpha_i \in \Omega^1_A$. The diagram

$$\begin{array}{ccc} \Omega_A^p & \xrightarrow{d_A} & \Omega_A^{p+1} \\ \downarrow{\scriptstyle \pi^p} & & \downarrow{\scriptstyle \pi^{p+1}} \\ \Omega_B^p & \xrightarrow{d_B} & \Omega_B^{p+1} \end{array}$$

commutes for $p \geqq 0$.

Let $\pi : \Omega_A \to \Omega_B$ be the $A$-module homomorphism whose restriction to $\Omega_A^p$ is $\pi^p$.

(1.4) LEMMA. *Let $J = \ker \pi^0$. Then* $\ker \pi$ *is the ideal $L$ of the algebra $\Omega_A$ generated by $J$ and $dJ$.*

PROOF. The inclusion $L \subset \ker \pi$ is clear. To prove the opposite inclusion it suffices to define a $\mathbb{C}$-linear map $\varphi : \Omega_B \to \Omega_A/L$ with the property $\varphi\pi(\alpha) = \alpha + L$, for all $\alpha \in \Omega_A$.

There is a well defined $B$-module structure on $\Omega_A/L$ such that $(\pi a) \times (\alpha + L) = a\alpha + L$ for $a \in A$, and a well defined $\mathbb{C}$-linear map $\theta : B \to \Omega_A/L$ such that $\theta(\pi a) = d_A a + L$. The map $\theta$ is a $\mathbb{C}$-derivation of $B$ into the $B$-module $\Omega_A/L$. By the universality property of $\Omega_B^1$, there exists a $B$-module homomorphism $\varphi : \Omega_B^1 \to \Omega_A/L$ satisfying $\varphi(\pi\alpha) = \alpha + L$ for $\alpha \in \Omega_A^1$. We may extend this map to a $\mathbb{C}$-algebra map $\varphi : \Omega_B \to \Omega_A/L$ such that $\varphi(\pi\alpha) = \alpha + L$ for all $\alpha \in \Omega_A$.

(1.5) Let $F$ be the non-singular affine algebraic variety defined by the polynomial equation $f(z) = 1$. Let $\Omega_F^*$ be the sheaf of everywhere regular differential forms on $F$, and $\Gamma(F, \Omega_F^*)$ the global sections of this sheaf.

Let $A = \mathbb{C}[z_1, \cdots, z_n]$ be the polynomial ring and $B = A/(f-1)$ the coordinate ring of $F$. Then $\pi^0 : A \to B$ is surjective and we have the homomorphism $\pi : \Omega_A \to \Omega_B$.

(1.6) THEOREM. *For the non-singular variety $F$ we have isomorphisms*

$$\Gamma(F, \Omega_F^q) \simeq \Omega_B^q \quad \textit{for} \quad q \geqq 0.$$

PROOF. It is sufficient to show this for $q = 1$, and an elementary proof is in Shafarevich [1, p. 160], see also Grothendieck [1, p. 97, footnote 5].

(1.7) Let $f$ be given by a homogeneous polynomial and $I = (f_1, \cdots, f_n)$ the ideal of partial derivatives of $f$ in the ring $A$. Then clearly $\mathcal{O}/(\partial f) \simeq A/I$ so our aim is to find an isomorphism between $A/I$ and $H^{n-1}(\Omega_B)$.

## 2. The homogeneous case

Let us assume that $f$ is a homogeneous polynomial of degree $m$ with an isolated critical point at the origin. Let $P = \mathbb{C}[f_1, \cdots, f_n]$ be the $\mathbb{C}$-subalgebra of $A$ generated by the partial derivatives of $f$. Let $A_r$ be the submodule of polynomials of degree $r$. Clearly both the ideal $I=(f_1,\cdots,f_n)$ and the ring $P$ are graded.

Let $H$ be a homogeneous subspace of $A$ such that $A=H\oplus I$. Since 0 is the unique common zero of $f_1,\cdots,f_n$, we have the following exercise of Bourbaki [1, Ch. V, Ex. 5.5].

(2.1) LEMMA. *The map from* $P\otimes_{\mathbb{C}}H$ *to* $A$ *defined by* $p\otimes h\to ph$ *is an isomorphism of graded vector spaces.*

Note the similarity between this statement for polynomial rings and the one for local rings given in (I.5.11).

Let us identify $A$ with $P\otimes_{\mathbb{C}}H$. Every element $\beta\in\Omega_B^{n-1}$ is a cocycle. Let $[\beta]$ denote its cohomology class in $H^{n-1}(\Omega_B)$. For $\alpha,\alpha'\in\Omega_A^{n-1}$, write $\alpha\equiv\alpha'$ if $\pi\alpha=\pi\alpha'$ and for cocycles $\alpha,\alpha'$ write $\alpha\sim\alpha'$ if $[\pi\alpha]=[\pi\alpha']$.

Define the $(n-1)$-form

$$\omega=\sum_{j=1}^{n}(-1)^{j-1}z_j\,dz_1\wedge\cdots\wedge\widehat{dz_j}\wedge\cdots\wedge dz_n.$$

(2.2) THEOREM. *The mapping*

$$\theta:p\otimes h\to[p(0)\pi(h\omega)]$$

*is a* $\mathbb{C}$*-linear map of* $A$ *onto* $H^{n-1}(\Omega_B)$ *with kernel* $I$.

The proof of this theorem will occupy the rest of this section. Before giving the argument let us return to an earlier example.

(2.3) EXAMPLE. Let $f=z_1^3+z_2^3$. Then $f_1=3z_1^2$, $f_2=3z_2^2$ and we may choose $\{1,z_1,z_2,z_1z_2\}$ as a $\mathbb{C}$-basis for $H$. We have $\omega=z_1\,dz_2-z_2\,dz_1$ and the theorem asserts that $\pi\omega$, $\pi(z_1\omega)$, $\pi(z_2\omega)$ and $\pi(z_1z_2\omega)$ form a $\mathbb{C}$-basis for $H^1(\Omega_B)\simeq H^1(F^h;\mathbb{C})$. Intuitively this is perhaps as satisfactory an answer as we could expect.

(2.4) It will be notationally convenient to prove a few lemmas for an arbitrary index set $J=\{j_1,\cdots,j_p\}$ of length $|J|=p$. Let $\sigma_J=dz_{j_1}\wedge\cdots\wedge dz_{j_p}$, and

$$\omega_J=\sum_{k=1}^{p}(-1)^{k-1}z_{j_k}\,dz_{j_1}\wedge\cdots\wedge\widehat{dz_{j_k}}\wedge\cdots\wedge dz_{j_p}.$$

In this notation $\omega=\omega_{1,2\cdots,n}$. The symbol $\omega_J$ is skew-symmetric in its indices. Since $\Omega_A^p$ is a free $A$-module with basis consisting of the elements $\sigma_J$ with $|J|=p$ and $j_1<\cdots<j_p$, and $\sigma_J$ is also skew-symmetric in its indices, we may define an $A$-linear map $\delta:\Omega_A^p\to\Omega_A^{p-1}$ for $p\geqq 1$ by $\delta(\sigma_J)=\omega_J$ for all $J$. Computation shows that $\delta^2=0$ and in fact the complex $(\Omega_A,\delta)$ is the Koszul complex based on $z_1,\cdots,z_n$, see Serre [3].

Note that $d\omega_J = p\sigma_J$. Define the index set $(j, J) = (j, j_1, \cdots, j_p)$.

(2.5) LEMMA. *Let $a \in A_r$ and $a_j = \partial a/\partial z_j$. Then we have the identities*

(2.6) $\delta d(a\sigma_J) = \sum_{j=1}^{n} a_j \omega_{(j,J)}$

(2.7) $\delta d(a\sigma_J) = ra\sigma_J - da \wedge \omega_J$

(2.8) $d\delta(a\sigma_J) = pa\sigma_J + da \wedge \omega_J$

(2.9) $(d\delta + \delta d)a\sigma_J = (p+r)a\sigma_J$.

PROOF.

(2.6) $\delta d(a\sigma_J) = \delta\left(\sum_{j=1}^{n} a_j \sigma_{(j,J)}\right) = \sum_{j=1}^{n} a_j \omega_{(j,J)}$

(2.7) $\delta d(a\sigma_J) = \delta\left(\sum_{j=1}^{n} a_j \sigma_{(j,J)}\right)$

$$= \sum_{j=1}^{n} a_j z_j \sigma_J + \left(\sum_{j=1}^{n} a_j\, dz_j\right) \wedge \left(\sum_{k=1}^{p} (-1)^k z_{j_k}\, dz_{j_1} \wedge \cdots \wedge \widehat{dz_{j_k}} \wedge \cdots \wedge dz_{j_p}\right)$$

$$= ra\sigma_J - da \wedge \omega_J$$

(2.8) $d\delta(a\sigma_J) = d(a\omega_J) = da \wedge \omega_J + a d\omega_J$

$$= pa\sigma_J + da \wedge \omega_J \quad \text{since} \quad d\omega_J = p\sigma_J.$$

(2.9) is obtained by adding (2.7) and (2.8). Note that this is the Poincaré lemma for the complex $\Omega_A$.

(2.10) LEMMA. *Let $a \in A_r$ and $J = (j_1, \cdots, j_p)$. Then*

$$d(\delta a\sigma_J) \equiv \frac{p+r}{m} \sum_{j=1}^{n} af_j\, \omega_{(j,J)} - \sum_{j=1}^{n} a_j \omega_{(j,J)}.$$

PROOF. From (2.6) and (2.7) we have

$$ra\sigma_J - da \wedge \omega_J = \sum_{j=1}^{n} a_j \omega_{(j,J)}.$$

Substitute $a = f$ and $r = m$ in this formula. Since $f \equiv 1$ and $df \equiv 0$, we obtain

(2.11) $m\sigma_J \equiv \sum_{j=1}^{n} f_j \omega_{(j,J)}$.

From (2.9) and (2.6) we have

(2.12) $d\delta(a\sigma_J) = (p+r)a\sigma_J - \sum_{j=1}^{n} a_j \omega_{(j,J)}$.

Now multiply the congruence (2.11) by $a/m$ and substitute in (2.12) to obtain the conclusion.

(2.13) REDUCTION FORMULA. Let $a \in A_r$. Then

$$(n-1+r)af_j\omega \sim ma_j\omega.$$

PROOF. Choose $J$ so that $(j, J)$ is an even permutation of $1, \cdots, n$ and apply (2.10).

(2.14) LEMMA. $\Omega_B^{n-1} = \pi(H\omega) + d_B\Omega_B^{n-2}$. *Thus every cohomology class of* $H^{n-1}(\Omega_B)$ *has the form* $[\pi(h\omega)]$, $h \in H$.

PROOF. Since $\Omega_A^{n-1}$ is generated by $\{\sigma_{1,\cdots,\hat{j},\cdots n}\}$, $j=1,\cdots,n$ as an $A$-module, $\{\pi\sigma_{1,\cdots,\hat{j},\cdots,n}\}$, $j=1,\cdots,n$ generate $\Omega_B^{n-1}$ as a $B$-module. It follows from (2.11) that we may assume that any given cohomology class is of the form $[\pi(a\omega)]$ for some $a \in A$. Thus it suffices to show that given $a \in A$ there exists $h \in H$ so that $a\omega \sim h\omega$. We may assume that $a$ is homogeneous, say of degree $r$, and argue by induction on $r$. If $r=0$, then the assertion is clear because $1 \in H$. Since $A = H \oplus I$ and $I = Af_1 + \cdots + Af_n$ we may assume that $a \in Af_i$. The assertion follows from the reduction formula (2.13) and the induction hypothesis.

We may now complete the argument of (2.2) as follows. On the one hand it is easy to see that $I \subseteq \ker\theta$. For let $a \in I$. Let $h_1, \cdots, h_\mu$ be a $\mathbb{C}$-basis for $H$. Write $a = \sum_{\nu=1}^{\mu} p_\nu h_\nu$, where $p_\nu \in P$. Let $P_+ = \{p \in P \mid p(0) = 0\}$. Write $p_\nu = p_\nu(0) + p'_\nu$, where $p'_\nu \in P_+ \subset I$. Then $\sum_{\nu=1}^{\mu} p_\nu(0)h_\nu = a - \sum_{\nu=1}^{\mu} p'_\nu h_\nu$ is in $H \cap I = 0$. Thus $p_\nu(0) = 0$ and $\theta(a) = \sum_{\nu=1}^{\mu}[p_\nu(0)\pi(h_\nu\omega)] = 0$.

On the other hand (2.14) shows that $H^{n-1}(\Omega_B) = \Omega_B^{n-1}/d_B\Omega_B^{n-2} = \operatorname{im}\theta \simeq A/\ker\theta$ is isomorphic to a quotient space of $A/I$. Since $\mu_a = \dim_{\mathbb{C}} A/I = \dim_{\mathbb{C}} H^{n-1}(\Omega_B) = \mu_t$, we conclude that $\theta$ is an isomorphism between the finite dimensional vector spaces $A/I$ and $H^{n-1}(\Omega_B)$.

(2.15) REMARKS. The argument given in Orlik and Solomon [1] is stronger than the one presented here in several respects.

(i) One may work over an arbitrary field $K$ of characteristic 0 provided $f$ has an isolated critical point in the algebraic closure of $K$.

(ii) The polynomial $f$ may be chosen *weighted homogeneous* (quasi-homogeneous) of degree $m$. This means that there exists a grading of $A$ by $\deg z_j = q_j$ so that for every non-zero complex number $t$ we have

$$f(t^{q_1}z_1, \cdots, t^{q_n}z_n) = t^m f(z_1, \cdots, z_n).$$

The positive rational numbers $w_j = m/q_j$ are called the *weights* of $f$.

(iii) It is possible to compute $H^*(\Omega_B)$ purely algebraically. The result is of course that $H^i(\Omega_B) = 0$ for $1 \leqq i \leqq n-2$, and $H^{n-1}(\Omega_B) \simeq A/I$, but in the latter isomorphism it is not necessary to assume that $\mu_a = \mu_\tau$. A direct proof is available to show that $\ker \theta \subseteq I$.

*Added in proof:* J.-L. Brylinski has shown by a different argument that $\theta$ induces an isomorphism, see his paper in these proceedings.

## 3. Residue map

This brief description of the residue theory of Leray is from Pham [2, Ch. III.]. Let $X$ be a complex analytic variety and $S$ a closed subvariety of complex codimension 1 given locally in a neighborhood of $y$ by a single equation $s_y(z)=0$.

(3.1) LEMMA. *If $\alpha_y$ is a regular differential form in some neighborhood of $y$ with the property that $ds_y \wedge \alpha_y = 0$, then there exists a regular form $\beta_y$ near $y$ so that $\alpha_y = ds_y \wedge \beta_y$. Call the restriction $\beta_y \mid S = (\alpha_y/ds_y)$. It only depends on $\alpha_y$ and $s_y$ and it is a holomorphic form if $\alpha$ is holomorphic.*

There is a global version of (3.1) if $S$ is given by a global equation. This is often called de Rham's lemma and there is an equivalent purely algebraic statement in terms of Kähler differentials, see de Rham [1].

(3.2) LEMMA. *Suppose $\varphi$ is a regular form closed on $X-S$ and has a pole of order 1 on $S$, i.e. for all $y \in S$, $s_y\varphi$ is the restriction to $X-S$ of a form $\alpha_y$, which is regular in a neighborhood of $y$. Then there exists in a neighborhood of $y$ closed regular forms $\psi_y$ and $\theta_y$ so that*

$$\varphi = \frac{ds_y}{s_y} \wedge \psi_y + \theta_y.$$

*The form $\psi_y \mid S$ is closed and depends only on $\varphi$. Call it the residue of $\varphi$,*

$$\operatorname{res}[\varphi] = \left.\frac{s_y\varphi}{ds_y}\right|_S.$$

(3.3) Now return to the case $X=\mathbb{C}^n$, $S=F$ with the global equation $f-1=0$, where $f$ is a homogeneous polynomial of degree $m$. Using the Euler identity $\sum_{j=1}^n z_j f_j = mf$ we obtain

$$df \wedge \omega = mf\tau$$

where $\tau = dz_1 \wedge dz_2 \wedge \cdots \wedge dz_n$. Thus

$$\operatorname{res}\left[\frac{\tau}{f-1}\right] = \left.\frac{\tau}{df}\right|_F = \frac{1}{m}\,\omega|_F = \frac{1}{m}\,\pi\omega$$

so the map $\theta$ of theorem (2.2) may be written

$$\theta(p \otimes h) = m \operatorname{res}\left[\frac{p(0)h\tau}{f-1}\right].$$

(3.4) Now assume that $f$ is an arbitrary polynomial. We may try to establish an isomorphism between $\mathcal{O}/(\partial f)$ and $H^{n-1}(\Omega_F^*)$. Let $\mathcal{P} = \{f_1, \cdots, f_n\}$ and $\mathfrak{H}$ a $\mathbb{C}$-vector space so that $\mathcal{O} \cong (\partial f) \oplus \mathfrak{H}$. Then along the lines of (I.5.11) one can prove that $\mathcal{O} \simeq \mathcal{P} \otimes_{\mathbb{C}} \mathfrak{H}$.

(3.5) CONJECTURE. The map

$$\Theta: \mathcal{P} \otimes \mathfrak{H} \to H^{n-1}(\Omega_F^*)$$

defined by

$$\Theta(p \otimes h) = \operatorname{res}\left[\frac{p(0)h\tau}{f-1}\right]$$

induces an isomorphism between $\mathcal{O}/(\partial f)$ and $H^{n-1}(\Omega_F^*)$.

(3.6) Using $\dim_{\mathbb{C}} \mathcal{O}/(\partial f) = \dim_{\mathbb{C}} H^{n-1}(\Omega_F^*)$ it would be sufficient to establish that the map is onto, i.e. a reduction formula similar to (2.13).

## 4. Integrals

In this section the reduction formula (2.13) will be used to obtain certain product formulas for Γ-functions using Dirichlet integrals. These formulas appear to be consequences of Gauss' product formula, Edwards [1, §874, p. 65].

Let $f = z_1^m + z_2^m + \cdots + z_n^m$ and define the vanishing cell of Pham [2, p. 85] on the hypersurface $f^{-1}(1)$ by

$$\gamma: (t_1, \cdots, t_{n-1}) \to (t_1, \cdots, t_n), \qquad 0 \leqq t_i \leqq 1$$

(4.1) $t_n = (1 - t_1^m - \cdots - t_{n-1}^m)^{1/m}$

Choose an element $a \in A_r$ which vanishes on $\partial\gamma$ and an index set $J$ with $|J| = n-1$. By Stokes' theorem

$$\int_\gamma d(a\omega_J) = \int_{\partial\gamma} a\omega_J = 0$$

so (2.13) gives the formula

(4.2) $(n-1+r) \int_\gamma z_j^{m-1}\, a\omega = \int_\gamma a_j \omega.$

For suitable choice of $a$, iteration of (4.2) leads to a product formula for certain definite integrals calculated in Edwards [1, §977, p. 167]:

$$(4.3)\quad \int_0^1 \cdots \int_0^1 t_1^{i_1-1} \cdots t_{n-1}^{i_{n-1}-1}\, dt_1 \cdots dt_{n-1} = m^{1-n} \frac{\Gamma\left(\frac{i_1}{m}\right) \cdots \Gamma\left(\frac{i_n}{m}\right)}{\Gamma\left(\frac{i_1 + \cdots + i_n}{m}\right)},$$

where $t_n$ is given in (4.1).

For simplicity let $n = 2$, $m = 3$ be the previously considered example, $f = z_1^3 + z_2^3$. With $a = z_1 z_2^{k-3}$ and $j = 1$ (4.2) gives the line integrals

$$(4.4)\quad (k-1) \int_\gamma z_1^3 z_2^{k-3} \omega = \int_\gamma z_2^{k-3} \omega.$$

Let

$$I_k = \int_\gamma z_2^k \omega.$$

Then setting $z_1^3 = 1 - z_2^3$ we get from (4.4)

$$I_k = \frac{k-2}{k-1} I_{k-3}.$$

Iteration gives

$$I_{3k} = \frac{3k-2}{3k-1} \cdot \frac{3k-5}{3k-4} \cdots \frac{4}{5} \cdot \frac{1}{2} I_0$$

$$I_{3k+1} = \frac{3k-1}{3k} \cdot \frac{3k-4}{3k-3} \cdots \frac{5}{6} \cdot \frac{2}{3} I_1$$

$$I_{3k+2} = \frac{3k}{3k+1} \cdot \frac{3k-3}{3k-2} \cdots \frac{6}{7} \cdot \frac{3}{4} I_2.$$

We compute the definite integrals $I_0$, $I_1$, $I_2$ from (4.3) with suitable orientation of $\gamma$ as

$$I_0 = \tfrac{1}{3}\Gamma(\tfrac{1}{3})^2 \Gamma(\tfrac{2}{3})^{-1}$$

$$I_1 = \tfrac{1}{3}\Gamma(\tfrac{1}{3})\Gamma(\tfrac{2}{3})$$

$$I_2 = 1.$$

Thus for any choice of integers $\xi_0$, $\xi_1$, $\xi_2$ with $\xi_0 + \xi_1 + \xi_2 = 0$ we have

$$\lim_{k\to\infty} I_{3k}^{\xi_0} I_{3k+1}^{\xi_1} I_{3k+2}^{\xi_2} = 1.$$

Choosing $\xi_0 = \xi_1 = 1$ and $\xi_2 = -2$ we get the following product formula for $\Gamma(\frac{1}{3})$:

$$\Gamma(\tfrac{1}{3})^3 = 9 \lim_{k\to\infty} \left(\frac{3}{1} \cdot \frac{6}{4} \cdots \frac{3k}{3k-2}\right)^3 \frac{1}{(3k+1)^2}.$$

## 5. Computation of $\mu$

We follow Milnor and Orlik [1] to obtain a formula for $\mu$ in case $f$ is weighted homogeneous. Having shown that the different definitions of $\mu$ agree, we shall argue on $\mu_g$, the local degree of the gradient mapping (II, 1.2):

$$\partial f : \mathbb{C}^n \to \mathbb{C}^n$$
$$(\partial f)(z_1, \cdots, z_n) = (\partial f/\partial z_1, \cdots, \partial f/\partial z_n).$$

(5.1) LEMMA. *If a polynomial mapping $G:\mathbb{C}^n \to \mathbb{C}^n$ is such that the ith component of $G(z)$ is a homogeneous polynomial of degree $d_i$, and if $G^{-1}(0)=0$ then the local degree of G at 0 is equal to $d_1 \cdots d_n$.*

PROOF. The hypersurfaces $G_i^{-1}(0) \subset \mathbb{C}^n$ extend to projective hypersurfaces $H_i \subset \mathbb{C}P^n$ of the same degrees. The total intersection multiplicity of $H_1, \cdots, H_n$ is equal to the product $d_1 \cdots d_n$ of degrees. But the intersection $H_1 \cap \cdots \cap H_n$ clearly consists of the single point $0 \in \mathbb{C}^n \quad \mathbb{C}P^n$.

(5.2) THEOREM. *Let f be a weighted homogeneous polynomial of degree m with respect to the grading $q_1, \cdots, q_n$, having an isolated critical point at the origin. Then*

$$\mu = \left(\frac{m}{q_1} - 1\right)\left(\frac{m}{q_2} - 1\right) \cdots \left(\frac{m}{q_n} - 1\right).$$

PROOF. Introduce the auxiliary map

$$G(z_1, \cdots, z_n) = (z_1^{q_1}, \cdots, z_n^{q_n})$$

so that $f \circ G$ is homogeneous of degree $m$. Then the $i$th component of $(\partial f) \circ G$ is a homogeneous polynomial of degree $m - q_i$. So the local degree of $(\partial f) \circ G$ at 0 is $(m - q_1) \cdots (m - q_n)$. Dividing by the local degree $q_1 \cdots q_n$ of $G$ at 0, we see that the local degree of $\partial f$ is equal to

$$\frac{m - q_1}{q_1} \cdot \frac{m - q_2}{q_2} \cdots \frac{m - q_n}{q_n}.$$

(5.3) COROLLARY. *If f is homogeneous of degree m, then $\mu = (m-1)^n$.*

The fact that $\mu_a = \dim_{\mathbb{C}} A/I = (m-1)^n$ follows from Macaulay [1, §58] and $\mu_t = \dim_{\mathbb{C}} H^{n-1}(F, \mathbb{C}) = (m-1)^n$ from Fáry [1], thus in the special case of a homogeneous polynomial $\mu_a = \mu_t$ has been known for some time.

## 6. Divisor formula

Let $G \in GL(n; \mathbb{C})$ be a linear group which leaves the homogeneous polynomial $f$ invariant. The purpose of this section is to obtain a formula which expresses the eigenvalues of the action of an element $g \in G$ on $H^{n-1}(F; \mathbb{C})$ in terms of the eigenvalues of $g$ on $\mathbb{C}^n$.

(6.1) We first introduce the notion of a divisor following Milnor and Orlik [1]. Let $\mathbb{C}^*$ be the multiplicative group of non-zero complex numbers. To each monic polynomial

$$(t-\alpha_1)\cdots(t-\alpha_k)$$

with $\alpha_1, \cdots, \alpha_k \in \mathbb{C}^*$ assign the element

$$\delta((t-\alpha_1)\cdots(t-\alpha_k)) = \langle\alpha_1\rangle + \cdots + \langle\alpha_k\rangle$$

in the integral group ring $\mathbb{Z}\mathbb{C}^*$, called its divisor. If $g \in G$ and $M$ is a $G$-module, we shall call $\delta(g)$ the divisor of the characteristic polynomial $\Delta(g; t) = \det(tI_* - g_*)$.

Let $\xi = \exp(2\pi i/k)$ and let

$$\Lambda_k = \delta(t^k - 1) = \langle 1\rangle + \langle\xi\rangle + \cdots + \langle\xi^{k-1}\rangle.$$

We note the multiplication rule $\Lambda_a\Lambda_b = (a, b)\Lambda_{[a,b]}$ where $(a, b)$ is the greatest common divisor and $[a, b]$ the least common multiple. Let $E_k = k^{-1}\Lambda_k$ in $Q\mathbb{C}^*$ be idempotent with $E_aE_b = E_{[a,b]}$.

(6.2) THEOREM. *Let $g: \mathbb{C}^n \to \mathbb{C}^n$ be a linear isomorphism with eigenvalues $\lambda_1, \cdots, \lambda_n$ of finite orders $u_1, \cdots, u_n$. Suppose that the homogeneous polynomial $f$ of degree $m$ is invariant under $g$ and the polynomial mapping $f: \mathbb{C}^n \to \mathbb{C}$ has an isolated critical point at the origin. Let $F = f^{-1}(1)$ and $M = H^{n-1}(F; \mathbb{C})$. Then*

$$\delta(g) = (mE_{u_1} - 1)(mE_{u_2} - 1)\cdots(mE_{u_n} - 1).$$

PROOF. Since $\delta(g) = \delta(g^{-1})$ we may argue on homology. We have

$$g(z_1, \cdots, z_n) = (\lambda_1 z_1, \cdots, \lambda_n z_n).$$

The characteristic polynomial of $g$ on $F$ is computed by Milnor [3, §9.5 and 9.6] as follows. Let $g^v: F \to F$ be the composition of $g$ be itself $v$ times. Denote by $\chi_v$ the euler number of the fixed point manifold $F(v)$ of $g^v$. Define a sequence of integers $s_1, s_2, s_3, \cdots$ by the formula $\chi_v = \sum_{j|v} s_j$. The numbers $s_j$ are almost all zero, and $s_j$ is divisible by $j$. Milnor [3, §9.6] shows that the euler numbers $\chi_v$ are related to the characteristic polynomial $\Delta(t)$ of $g_*: H_{n-1}(F; \mathbb{C}) \to H_{n-1}(F; \mathbb{C})$ by the formula

$$\Delta(t)^{(-1)^n} = (t-1)/\prod_j (t^j - 1)^{s_j/j}$$

where $\chi_v = \sum_{j|v} s_j$.

In terms of divisors we have

$$(6.3)\quad \delta(g)=(-1)^n\Big(1-\sum (s_j/j)\Lambda_j\Big)=(-1)^n\Big(1-\sum s_jE_j\Big).$$

We may compute the $\chi_v$ and $s_j$ explicitly as follows. The fixed point set $F(v)$ of $g^v$ is defined by the equations

$$\begin{aligned} f(z_1,\cdots,z_n)&=1\\ \lambda_1^v z_1&=z_1\\ &\vdots\\ \lambda_n^v z_n&=z_n. \end{aligned}$$

Renumbering the variables, if necessary, we may assume

$$\begin{aligned} v&\equiv 0 \bmod u_i \quad \text{for} \quad i\leqq k\\ v&\not\equiv 0 \bmod u_i \quad \text{for} \quad i>k. \end{aligned}$$

Then the last equations become $z_{k+1}=\cdots=z_n=0$. Note that the polynomial $f(z_1,\cdots,z_k,0,\cdots,0)$ is homogeneous of degree $m$. It has no critical points other than the origin. For by hypothesis, the gradient of $f(z_1,\cdots,z_n)$ at any point $(z_1,\cdots,z_k,0,\cdots,0)\neq 0$ is non-zero; and this gradient vector, being invariant under the linear transformation

$$g^v(z_1,\cdots,z_n)=(z_1,\cdots,z_k,\lambda_{k+1}^v z_{k+1},\cdots,\lambda_n^v z_n)$$

must be tangent to the plane $z_{k+1}=\cdots=z_n=0$. Thus, thinking of $f(z_1,\cdots,z_k,0,\cdots,0)$ as a homogeneous polynomial in $k$ variables, the associated fiber in $\mathbb{C}^k$ is precisely $F(v)$.

By (5.3) we have $\mu(F(v))=(m-1)^k$ and hence

$$\chi(F(v))=1+(-1)^{k-1}(m-1)^k=1-(1-m)^k.$$

(6.4) Define a sequence of numbers $\chi_1,\chi_2,\cdots$ by the formula

$$1-\chi_v=\prod(1-m)^k$$

where $k$ is the number of $u_i$ such that $v\equiv 0 \pmod{u_i}$. It follows from above that $\chi_v=\chi(F(v))$. Define the numbers $s_1,s_2,\cdots$ by the formula

$$\chi_v=\sum_{j|v} s_j.$$

Then we have reduced the proof of Theorem (6.2) to establishing the formal identity:

$$(6.5)\quad (-1)^n\Big(1-\sum s_jE_j\Big)=(mE_{u_1}-1)\cdots(mE_{u_n}-1).$$

For $n=1$ formula (6.5) is correct, for in that case $\chi_v = m$ or 0 according as $v \equiv 0 \bmod u_1$ or $\not\equiv 0 \bmod u_1$. Thus

$$s_j = m \quad \text{if} \quad j = u_1$$
$$s_j = 0 \quad \text{otherwise.}$$

The general case follows from a multiplicative property of the divisor function proved in the next lemma.

(6.6) LEMMA. *Define the numbers* $\chi_v$ *and* $s_j$ *as in* (6.4). *Let*

$$\delta(u_1, \cdots, u_n) = (-1)^n\left(1 - \sum s_j E_j\right) \in Q\mathbb{C}^*.$$

*Then the identity*

$$\delta(u_1, \cdots, u_k)\delta(u_{k+1}, \cdots, u_n) = \delta(u_1, \cdots, u_n)$$

*is valid in* $Q\mathbb{C}^*$.

PROOF. Let $\chi_j$, $s_j$ be the sequences associated with $u_1, \cdots, u_k$; let $\chi'_j$, $s'_j$ be associated with $u_{k+1}, \cdots, u_n$; and let $\chi''_j$, $s''_j$ be associated with the composite $n$-tuple.

The formula $(1-\chi''_j) = (1-\chi_j)(1-\chi'_j)$ or equivalently

$$\chi''_v = \chi_v + \chi'_v - \chi_v \chi'_v$$

follows from the definition. Substituting $\chi_v = \sum_{j|v} s_j$ this becomes

$$\text{(6.7)} \quad \sum_{j|v} s''_j = \sum_{j|v} s_j + \sum_{j|v} s'_j - \sum_{a|v} \sum_{b|v} s_a s'_b.$$

Next the formula

$$\text{(6.8)} \quad s''_j = s_j + s'_j - \sum_{[a,b]=j} s_a s'_b$$

will be proved by induction on $j$. It is clearly true for $j=1$. Adding the alleged equations (6.8) over all divisors $j$ of $v$ we get the true equation (6.7), proving (6.8) for all $j$.

Now multiply (6.8) by $E_j$ and sum over $j$. Since $(s_a E_a)(s'_b E_b) = s_a s'_b E_{[a,b]}$, this yields

$$\sum s''_j E_j = \sum s_j E_j + \sum s'_j E_j - \left(\sum s_a E_a\right)\left(\sum s'_b E_b\right)$$

or in other words

$$1 - \sum s''_n E_n = \left(1 - \sum s_a E_a\right)\left(1 - \sum s'_b E_b\right).$$

Multiplying the two factors on the right by $(-1)^k$ and $(-1)^{n-k}$ respectively, proves lemma (6.6).

The proof of theorem (6.2) is completed by noting that we have already shown that $\delta(u_1)=(mE_{u_1}-1)$. Induction on the number of variables using (6.6) establishes formula (6.5).

(6.9) REMARK. In Orlik and Solomon [2] a similar formula is obtained by an entirely different method for the divisor of the characteristic polynomial of g acting on $A/I$. Since the latter is a graded vector space, a bit more information is obtained.

(6.10) THEOREM. *Let $g:\mathbb{C}^n\to\mathbb{C}^n$ be a linear isomorphism of finite order, leaving the homogeneous polynomial f of degree m invariant. Suppose that f has an isolated critical point at the origin and the eigenvalues of g on $\mathbb{C}^n$ are $\lambda_1,\cdots,\lambda_n$ of orders $u_1,\cdots,u_n$. Then g acts on $A=\mathbb{C}[z_1,\cdots,z_n]$ by $g(z_i)=\lambda_i^{-1}z_i$. Consider the induced action of g on $A/I$ and let $\Delta(s)=\det(I_*-g_*s)$. Using the indeterminate t for the grading in $A/I$ we have*

$$\delta(g;t)=\prod_{i=1}^{n}\langle\lambda_i\rangle\left[\frac{t^m-1}{t^{u_i}-1}(t^{u_i-1}+\langle\lambda_i\rangle t^{u_i-2}+\cdots+\langle\lambda_i\rangle^{u_i-1})-t^{m-1}\right].$$

*This formula may be specialized as follows.*

(i) *Letting $t\to1$ we obtain the (ungraded) divisor of the action on $A/I$.*

$$\delta(g)=\prod_{i=1}^{n}\langle\lambda_i\rangle\left[\frac{m}{u_i}\Lambda_{u_i}-1\right]=\prod_{i=1}^{n}\langle\lambda_i\rangle(mE_{u_i}-1).$$

*Comparing this with the map $\theta$ of (2.2) which has the property $\theta(gu)=(\det g)g(\theta u)$ and noting that $\prod_{i=1}^{n}\langle\lambda_i\rangle=\delta(\det g)$, we obtain theorem (6.2) as a corollary.*

(ii) *The augmentation map induced by sending $\langle\lambda\rangle$ to the complex number $\lambda$ gives a formula for the trace of the representation on $A/I$.*

(iii) *Finally, the trivializing map $\langle\lambda\rangle\to1$ gives the Poincaré polynomial of $A/I$,*

$$\mathcal{P}(t)=(t^{m-1}+t^{m-2}+\cdots+t+1)^n.$$

(6.11) EXAMPLE. Let us return to the example $f=z_1^3+z_2^3$. The group $G=\mathbb{Z}_3\oplus\mathbb{Z}_3$ acts on $\mathbb{C}^2$ by complex multiplication. Let $\lambda=\exp(2\pi i/3)$ and consider the element

$$g=\begin{bmatrix}1&0\\0&\lambda\end{bmatrix}.$$

We may choose the monomial basis $1, z_1, z_2, z_1z_2$ for $A/I$, and we have $g(1)=1$, $g(z_1)=z_1$, $g(z_2)=\lambda^{-1}z_2$, $g(z_1z_2)=\lambda^{-1}z_1z_2$. This agrees with the formula of (6.10)

$$\delta(g;t)=\langle 1\rangle\left[\frac{t^3-1}{t-1}-t^2\right]\langle\lambda\rangle\left[\frac{t^3-1}{t^3-1}(t^2+\langle\lambda\rangle t+\langle\lambda\rangle^2)-t^2\right]$$
$$=\langle\lambda\rangle(t+1)(\langle\lambda\rangle t+\langle\lambda\rangle^2)=1+(1+\langle\lambda\rangle^2)t+\langle\lambda\rangle^2t^2$$

asserting that the eigenvalue is 1 on the degree 0 generator, there is one eigenvalue 1 and one eigenvalue $\lambda^2=\lambda^{-1}$ in degree one and in degree two the eigenvalue is $\lambda^2=\lambda^{-1}$.

Letting $t\to 1$ and dividing by $\delta(\det g)=\langle\lambda\rangle$ we get the divisor on cohomology,

$$2\cdot(\langle\lambda\rangle+\langle\lambda\rangle^2)=2\langle\lambda\rangle+2\langle\lambda\rangle^2.$$

Since $u_1=1$, $u_2=3$ this agrees with (6.2)

$$(3E_1-1)(3E_3-1)=2(\Lambda_3-1).$$

## REFERENCES


N. A'Campo

[1] Le nombre de Lefschetz d'une monodromie, Indag. Math. 35 (1973), 113–118.

[2] Sur la monodromie des singularités isolées d'hypersurfaces complexes, Invent. math. 20 (1973), 147–169.

[3] On Monodromy Maps of Hypersurface Singularities, Proc. Intern. Conf. on Manifolds, Tokyo (1973), 151–152.

[4] Le groupe de monodromie du déploiement des singularités isolées de courbes planes, Math. Ann. 213 (1975), 1–32.

[5] La fonction zeta d'une monodromie, Comm. Math. Helv. 50 (1975), 233–248.

A. Andreotti and T. Frankel

[1] The Lefschetz theorem on hyperplane sections, Annals of Math. 69 (1959), 713–717.

V. I. Arnold

[1] Singularities of smooth mappings, Russian Math. Surveys 23, no. 1 (1968), 1–43.– Uspehi Mat. Nauk 23, no. 1 (1968), 3–44.

[2] Integrals of rapidly oscillating functions and singularities of projections of Lagrangian manifolds, Funct. Anal. Appl. 6(1972), 222–224 – Funkc. Anal. i Prilož. 6, no. 3 (1972), 61–62.

[3] Normal forms for functions near degenerate critical points, the Weyl groups $A_k$, $D_k$, $E_k$ and Lagrangian singularities, Funct. Anal. Appl. 6 (1972), 254–272 – Funkc. Anal. i Prilož. 6, no. 4 (1972), 3–25.

[4] Some remarks on the stationary phase method and Coxeter numbers, Russian Math. Surveys 28, no. 5 (1973), 19–48 – Uspehi Mat. Nauk 28, no. 5 (1973), 17–44.

[5] Classification of unimodal critical points of functions, Funct. Anal. Appl. 7 (1973), 230–231 – Funkc. Anal. i Prilož. 7, no. 3 (1973), 75–76.

[6] Normal forms of functions near degenerate critical points, Russian Math. Surveys 29, no. 2 (1974), 10–50 – Uspehi Mat. Nauk 29, no. 2 (1974), 11–49.

[7] Critical points of smooth functions and their normal forms, Russian Math. Surveys 30, no. 5 (1975) – Uspehi Math. Nauk 30, no. 5 (1975), 3–65.

[8] Critical points of smooth functions, Proc. Int. Congress Math. (1974), vol. 1, 19–39.

M. Artin

[1] On isolated rational singularities of surfaces, Amer. J. Math. 88 (1966), 129–136.

M. F. Atiyah and W. V. D. Hodge
[1] Integrals of the second kind on an algebraic variety, Annals of Math. 62 (1955), 56–91.
M. F. Atiyah, R. Bott, and L. Gårding
[1] Lacunas for hyperbolic differential operators with constant coefficients II, Acta Math. 131 (1973), 145–206.
T. Bloom
[1] Singularités des hypersurfaces I, II (D'après Brieskorn et Milnor), Séminaire de Géométrie Analytique (1968/69), Exp. 6, 7. Fac. Sci. Univ. Paris, Orsay, 1969.
N. Bourbaki
[1] Groupes et algèbres de Lie, Chapitre V, Hermann, Paris, 1968.
G. E. Bredon
[1] Introduction to Compact Transformation Groups, Academic Press, New York, 1972.
[2] Exotic actions on spheres, Proc. Conf. on Transformation Groups, Springer Verlag, 1968, pp. 47–76.
[3] Sheaf Theory, McGraw-Hill, New York, 1967.
E. Brieskorn
[1] Über die Auflösung gewisser Singularitäten von Holomorphen Abbildungen, Math. Ann. 166 (1966), 76–102.
[2] Beispiele zur Differentialtopologie von Singularitäten, Invent. math. 2 (1966), 1–14.
[3] Rationale Singularitäten komplexer Flächen, Invent. math. 4 (1968), 336–358.
[4] Die Monodromie der isolierten Singularitäten von Hyperflächen, Manuscripta math. 2 (1970), 103–161.
[5] Singular elements of semi-simple algebraic groups, Actes, Congrés Intern. Math. 2 (1970), 279–284.
[6] Vue d'ensemble sur les problèmes de monodromie, Astérisque 7/8 (1973), 393–413.
[7] Special Singularities – Resolution, Deformation and Monodromy, Notes for Summer Institute of AMS, Arcata 1974.
E. Brieskorn and G.-M. Greuel
[1] Singularities of complete intersections, Proc. Intern. Conf. on Manifolds, Tokyo 1973, 123–129.
Th. Bröcker and L. Lander
[1] Differentiable Germs and Catastrophes, London Math. Soc. Lecture Notes 17, Cambridge Univ. Press, 1975.
H. Cartan and S. Eilenberg
[1] Homological Algebra, Princeton Univ. Press, Princeton, 1956.
P. Cartier
[1] Dérivations dans les corps, Séminaire H. Cartan et C. Chevalley, Géométrie Algébrique, École Normale Supérieure, Paris, 1956.
C. H. Clemens
[1] Picard-Lefschetz theorem for families of non-singular algebraic varieties acquiring ordinary singularities, Trans. Amer. Math. Soc. 136 (1969), 93–108.
E. Coddington and N. Levinson
[1] Theory of Ordinary Differential Equations, McGraw-Hill, New York, 1955.
P. Deligne
[1] Equations Différentielles à Points Singuliers Réguliers, Lecture Notes 163, Springer Verlag, 1970.
P. Deligne and N. Katz
[1] Groupes de monodromie en géométrie algébrique, SGA 7 II, Lecture Notes 340, Springer Verlag 1973.

I. V. DOLGACHEV

[1] Automorphic forms and quasihomogeneous singularities, Funct. Anal. Appl. 9 (1975), 149–151 – Funkc. Anal. i Prilož. 9, no. 2 (1975), 67–68.

A. DURFEE

[1] Diffeomorphism classification of isolated hypersurface singularities, Thesis, Cornell U. 1971.

[2] Fibered knots and algebraic singularities, Topology 13 (1974), 47–60.

A. DURFEE and H. B. LAWSON

[1] Fibered knots and foliations of highly connected manifolds, Invent. math. 17 (1972), 203–214.

J. EDWARDS

[1] A treatise on the integral calculus, II. Macmillan, London, 1922.

C. EHRESMANN

[1] Sur l'espaces fibrés differentiables, C. R. Acad. Sci. Paris 224 (1974), 1611–1612.

I. FÁRY

[1] Cohomologie des variétés algébriques, Annals of Math. 65 (1957), 21–73.

A. M. GABRIELOV

[1] Intersection matrices for certain singularities, Funct. Anal. Appl. 7 (1973), 182–193 – Funkc. Anal. i Prilož. 7, no. 3 (1973), 18–32.

[2] Bifurcations, Dynkin diagrams and modality of isolated singularities, Funct. Anal. Appl. 8 (1974), 94–98 – Funkc. Anal. i Prilož. 8, no. 2 (1974), 7–12.

[3] Dynkin diagrams for unimodal singularities, Funct. Anal. Appl. 8 (1974), 192–196 – Funkc. Anal. i Prilož, 8, no. 3 (1974), 1–6.

A. M. GABRIELOV and A. G. KOUCHNIRENKO

[1] Description of deformations with constant Milnor number for homogeneous functions, Funct. Anal. Appl. 9 (1975) 329–331 – Funkc. Anal. i Prilož. 9, no. 4 (1975), 67–68.

C. H. GIFFEN

[1] Smooth homotopy projective spaces, Bull. Amer. Math. Soc. 75 (1969), 509–513.

R. GODEMENT

[1] Topologie algébrique et théorie des faisceaux, 3rd edition, Hermann, Paris, 1973.

G.-M. GREUEL

[1] Der Gauss-Manin Zusammenhang isolierter Singularitäten von vollständigen Durchschnitten, Math. Ann. 214 (1975), 235–266.

A. GROTHENDIECK

[1] On the de Rham cohomology of algebraic varieties, Publ. Math. I.H.E.S. 29 (1966), 95–103.

R. C. GUNNING

[1] Lectures on Complex Analytic Varieties, The Local Parametrization Theorem, Princeton U. Press, Princeton, 1970.

[2] Lectures on Complex Analytic Varieties, Finite Analytic Mappings, Princeton U. Press, Princeton, 1974.

R. C. GUNNING and H. ROSSI

[1] Analytic Functions of Several Complex Variables, Prentice-Hall, 1965.

S. M. GUSEIN-ZADE

[1] Intersection matrices of some singularities of functions of two variables, Funct. Anal. Appl. 8 (1974), 11–13 – Funkc. Anal. i Prilož. 8, no. 1 (1974), 11–15.

[2] Dynkin diagrams for singularities of functions of two variables, Funct. Anal. Appl. 8 (1974), 295–300 – Funkc. Anal. i Prilož. 8, no. 4 (1974), 23–30.

H. Hamm
[1] Lokal topologische Eigenschaften complexer Räume, Math. Ann. 191 (1971), 235–252.
[2] Exotische Sphären als Umgebungsränder in speziellen komplexen Räumen, Math. Ann. 197 (1972), 44–56.
[3] Ein Beispiel zur Berechnung der Picard-Lefschetz Monodromie für nichtisolierte Hyperflächen-singularitäten, Math. Ann. 214 (1975), 221–234.

H. Hamm and Lê Dũng Tráng
[1] Un théorème de Zariski du type de Lefschetz, Ann. Sci., l'École Nor. Sup. 6 (1973), 317–366.

A. Hefez and F. Lazzeri
[1] The intersection matrix of Brieskorn singularities, Invent. math. 25 (1974), 143–158.

H. Hironaka
[1] Resolution of singularities of an algebraic variety over a field of characteristic zero I, II, Annals of Math. 79 (1964), 109–326.

F. Hirzebruch and K. H. Mayer
[1] Differenzierbare 0(n)-Mannigfaltigkeiten, exotische Sphären, und Singularitäten, Lecture Notes 57, Springer Verlag, 1968.

S. T. Hu
[1] Differentiable Manifolds, Holt, Rinehart and Winston, 1969.

I. N. Iomdine
[1] Calculation of the relative local monodromy for singularities of complex hypersurfaces, Funct. Anal. Appl. 9 (1975), 63–64 – Funkc. Anal. i Prilož. 9, no. 1 (1975), 67–68.

A. Kas and M. Schlessinger
[1] On the Versal Deformation of a Complex Space with an Isolated Singularity, Math. Ann. 196 (1972), 23–29.

M. Kato and Y. Matsumoto
[1] On the Connectivity of the Milnor Fiber of a Holomorphic Function at a Critical Point, Proc. Intern. Conf. on Manifolds, Tokyo 1973, 131–136.

L. Kauffman
[1] Branched coverings, open books and knot periodicity, Topology 13 (1974), 143–160.

A. G. Kouchnirenko
[1] Polyèdres de Newton et nombres de Milnor, Invent. math. 32 (1976), 1–31.

K. Lamotke
[1] Die Homologie isolierter Singularitäten, Math. Z. 143 (1975), 27–44.

F. Lazzeri
[1] A theorem on the monodromy of isolated singularities, Astérisque 7/8 (1973), 267–276.

Lê Dũng Tráng
[1] Sur les noeuds algébriques, Comp. Math. 25 (1972), 281–321.
[2] Topologie des singularités des hypersurfaces complexes, Astérisque 7/8 (1973), 171–182.
[3] Les Théorèmes de Zariski de type de Lefschetz, Centre de Math. de l'École Poly. Paris, 1971.
[4] Calcul du nombre de cycles évanouissants d'une hypersurface complexe, Ann. Inst. Fourier Grenoble 23 (1973), 261–270.
[5] Calculation of the Milnor number of isolated singularity of complete intersection, Funct. Anal. Appl. 8 (1974), 127–131 – Funkc. Anal. i Prilož. 8, no. 2 (1974), 45–49.
[6] La monodromie n'a pas de points fixes, Journ. Fac. Sci., Univ. of Tokyo 22 (1975), 409–427.

Lê Dũng Tráng and C. P. Ramanujam
[1] The invariance of Milnor's number implies the invariance of the topological type, Amer. J. Math. 98 (1976), 67–78.

S. LEFSCHETZ
[1] Applications of Algebraic Topology, Applied Math. Sciences vol. 16, Springer Verlag, 1975.

E. LOOIJENGA
[1] The complement of the bifurcation variety of a simple singularity, Invent. math. 23 (1974), 105–116.

F. S. MACAULAY
[1] The algebraic theory of modular systems, Cambridge Univ. Press, 1916.

J. MATHER
[1] Stability of $C^\infty$-Mappings III, Finitely Determined Map-Germs, Publ. Math. I.H.E.S. 35 (1968), 127–156.

J. MILNOR
[1] Morse Theory, Annals of Math. Study 51, Princeton Univ. Press, 1963.
[2] Topology from the Differentiable Viewpoint, Univ. of Virginia Press, 1965.
[3] Singular points of complex hypersurfaces, Annals of Math. Studies 61, Princeton Univ. Press, 1968.

J. MILNOR and P. ORLIK
[1] Isolated singularities defined by weighted homogeneous polynomials, Topology 9 (1970), 385–393.

D. MUMFORD
[1] Topology of normal singularities of an algebraic surface and a criterion for simplicity, Publ. Math. I.H.E.S. 9 (1961).

M. NAGATA
[1] Local Rings, Interscience, 1962.

M. OKA
[1] On the homotopy types of hypersurfaces defined by weighted homogeneous polynomials, Topology 12 (1973), 19–32.
[2] On the Cohomology Structure of Projective Varieties, Proc. Intern. Conf. on Manifolds, Tokyo 1973, 137–143.

P. ORLIK
[1] Weighted homogeneous polynomials and fundamental groups, Topology 9 (1970), 267–273.
[2] On the homology of weighted homogeneous manifolds, Proc. Second Conf. Transformation Groups I, Lecture Notes 298, Springer Verlag 1972, 260–269.

P. ORLIK and R. RANDELL
[1] The structure of weighted homogeneous polynomials, in Several Complex Variables, AMS Proc. of Symp. in Pure Math. XXX, 1977, pp. 57–64.
[2] The monodromy of weighted homogeneous singularities, Invent. Math., to appear.

P. ORLIK and L. SOLOMON
[1] Singularities I; Hypersurfaces with an isolated singularity, Advances in Math., to appear.
[2] Singularities II; Automorphisms of forms, preprint.

P. ORLIK and P. WAGREICH
[1] Isolated singularities of algebraic surfaces with $\mathbb{C}^*$-action, Annals of Math. 93 (1971), 205–228.

V. P. PALAMODOV
[1] Multiplicity of holomorphic mappings, Funct. Anal. Appl. 1 (1967), 218–266 – Funkc. Anal. i Prilož 1, no. 3 (1967), 54–65.

F. PHAM
[1] Formules de Picard-Lefschetz généralisées et ramification des intégrales, Bull. Soc. Math. France 93 (1965), 333–367.

[2] Introduction à l'étude topologique des singularités de Landau, Mémorial des Sciences Math. 164, Paris 1967.

H. PINKHAM

[1] Deformations of algebraic varieties with $G_m$ action, Astérisque 20 (1974).

R. RANDELL

[1] Index invariants of orbit spaces, Math. Scand. 36 (1975), 263–270.

[2] Generalized Brieskorn manifolds, Bull. Amer. Math. Soc. 80 (1974), 111–115.

[3] The homology of generalized Brieskorn manifolds, Topology 14 (1975), 347–355.

G. de RHAM

[1] Sur la division de formes et de courants par une forme linéaire, Comm. Math. Helv. 28 (1954), 346–352.

K. SAITO

[1] Quasihomogene isolierte Singularitäten von Hyperflächen, Invent. math. 14 (1971), 123–142.

[2] Calcul algébrique de la monodromie, Astérisque 7/8 (1973), 195–211.

[3] Einfach-elliptische Singularitäten, Invent. math. 23 (1974), 289–325.

K. SAKAMOTO

[1] Milnor Fiberings and their Characteristic Maps, Proc. Intern. Conf. on Manifolds, Tokyo 1973, 145–150.

M. SEBASTIANI

[1] Preuve d'une conjecture de Brieskorn, Manuscripta Math. 2 (1970), 301–308.

M. SEBASTIANI and R. THOM

[1] Un résultat sur la monodromie, Invent. math. 13 (1971), 90–96.

J.-P. SERRE

[1] Faisceaux algébriques cohérents, Annals of Math. 61.

[2] Géométrie algébrique et géométrie analytique, Ann. Inst. Fourier, 6 (1956), 1–42.

[3] Algèbre Locale - Multiplicités, Lecture Notes 11, Springer Verlag, 1965.

I. R. SHAFAREVICH

[1] Basic Algebraic Geometry, Springer Verlag, 1974.

D. SIERSMA

[1] Classification and Deformation of Singularities, Thesis, Amsterdam, 1974.

E. SPANIER

[1] Algebraic Topology, McGraw-Hill, 1966.

G. N. TJURINA

[1] Locally semiuniversal flat deformation of isolated singularities of complex spaces (in Russian), Izv. Akad. Nauk 33 (1969), 1026–1058.

P. WAGREICH

[1] Elliptic singularities of surfaces, Amer. J. Math. 92 (1970), 419–454.

[2] Singularities of complex surfaces with solvable local fundamental group, Topology 11 (1972), 51–72.

G. WASSERMANN

[1] Stability of unfoldings, Lecture Notes 393, Springer Verlag, 1974.

R. O. WELLS JR.

[1] Differential Analysis on Complex Manifolds, Prentice-Hall, 1973.

E. C. ZEEMAN and D. J. A. TROTMAN

[1] The classification of elementary catastrophes of codimension $\leqq 5$, Mathematics Institute, University of Warwick, Coventry, 1974.

PETER ORLIK
University of Wisconsin
Dept. of Mathematics
213 Van Vleck Hall Madison, Wis. 53706, USA

Nordic Summer School/NAVF
Symposium in Mathematics
Oslo, August 5–25, 1976

# NUMERICAL CHARACTERS OF A CURVE IN PROJECTIVE $n$-SPACE

Ragni Piene[1]

## Contents



## §1. Introduction

The enumerative properties of a curve embedded in projective $n$-space can be studied by considering curves in lower-dimensional projective spaces, derived from the given curve in a natural way. Such was the point of view of Veronese ([14]), who used the 'principle of projections and sections' to establish relations among the numerical characters of a curve in $n$-space by applying Plücker's formulas to $n-1$ plane curves derived from the given one. This method was first used by Cayley ([3]) in the case of a curve in 3-space.

The numerical characters we shall consider are the *ranks* – the number of osculating spaces to the curve which satisfy a given Schubert condition – and the *stationary indices* – the number of *hyperosculating* spaces (i.e. osculating spaces which have 'too high' order of contact with the curve). The points of hyperosculation were called singular points of higher order by Pohl ([13]); in fact, he represented the osculating spaces by vector bundles (dual to the jet bundles) and got the points of hyperosculation as the singularities of maps of bundles. In this paper we show how (the dual of) his representation can be used to prove the geometric statements about projections and sections made by Veronese. Another application of Pohl's viewpoint was given by Kato

[1] An earlier version of this paper forms part of the author's dissertation, written under the direction of Professor S. L. Kleiman at MIT. While a graduate student at MIT (1974–76), the author was supported by the University of Oslo and the Norwegian Research Council for Science and the Humanities.

([9]), who described the higher order singularities of an elliptic curve of degree $n+1$ in $\mathbb{P}^n$ as the points of order $n$ of the curve considered as abelian variety.

We shall assume, for simplicity, that the base field $k$ is algebraically closed, rather than just infinite. By a *curve* we mean a purely 1-dimensional, reduced, proper, algebraic scheme. Any embedding $j: X_0 \hookrightarrow P$ of a curve $X_0$ in projective $n$-space is assumed to be such that $X_0$ spans $P$. Given an embedding $j$, we let $f: X \to P$ denote the finite map obtained by composing $j$ with the normalization $X \to X_0$. In §2 we define, for each integer $m$, $0 \leqq m \leqq n$, the osculating $m$-space $S(x, m)$ to the branch of $X_0$ at $f(x)$ corresponding to a point $x \in X$. We show that these spaces, for *ordinary* (non-hyperosculating) points $x$, are the projective fibers of the bundle $\mathcal{P}^m_X(\mathcal{L})$ of principal parts (or jets) of order $m$ of the canonical line bundle $\mathcal{L} = f^*\mathcal{O}_P(1)$. A modification $\mathcal{P}^m$ of this bundle, called the osculating bundle of order $m$, gives (if char $k = 0$ or char $k$ is sufficiently large) also the hyperosculating $m$-spaces.

The numerical characters of the curve – the ranks and the stationary indices – are defined and interpreted in §3, and the formulas relating them (and the genus of the curve) are deduced (3.2). We then proceed to treat the principle of projections and sections, in §4. *A $q$-projection of $f$* is defined as a generic linear projection of $f$ to $\mathbb{P}^q$, while an *$m$-section* $X_m \to \mathbb{P}^{n-m}$ of $f$ is defined to be the intersection of a (general) subspace $\mathbb{P}^{n-m}$ of $P$ with the $m$th *osculating developable* of $X_0$, i.e. with the ruled $(m+1)$-dimensional variety generated over $X_0$ by the $m$-spaces $S(x, m)$. We prove associativity (4.2, 4.4) and commutativity (4.6) of these two operations and relate the numerical characters of the derived curves to those of $X_0$ (4.7). Finally, in §5, Weyl's proof of duality for analytic curves is adapted. Our key result is an isomorphism of exact sequences (5.2) which shows that all the natural duality statements for a curve and its derived curves hold. As an application we deduce the Cayley-Plücker equations for a curve in 3-space.

Three examples are carried through; the first is the rational $n$-ic $R_n$ in $n$-space. This curve is particularly simple: it has no points of hyperosculation and it is self-dual. The second example is a linearly normal, elliptic curve of degree $d \geqq 3$. The third curve, $\Gamma_d$, was discussed by Dye ([4], [5]). It is the complete intersection of $n-1$ hypersurfaces of degree $d$ which are in special position.

In the appendix (§6) we recall the definition of the functor of principal parts. We show the existence, as well as some general properties, of a natural transformation of functors which will be used to define the osculating bundles.

As far as the characteristic of $k$ is concerned, let us remark that in small, positive characteristic the osculating bundles $\mathcal{P}^m$ (though well defined) may not represent the *hyper*osculating spaces. For $r_m = \deg \mathcal{P}^m$ still to hold the osculating spaces could be *defined* as the fibers of $\mathcal{P}^m$. This definition also makes sense for any finite map (not necessarily birational onto its image) $f: X \to P$ such that $a^n: V_X \to \mathcal{P}^n_X(\mathcal{L})$ (where $V = H^0(P, \mathcal{O}_P(1))$, see §2) is generically surjective. If $f': X' \to f(X)$ denotes the normalization and $g: X \to X'$ the induced map one can show that in this case one has $r_m(f) = (\deg g) r_m(f')$; the restriction to the case of an embedded curve ($\deg g = 1$) is made in order to simplify notations.

## §2. The osculating bundles

Let $X_0 \hookrightarrow P = \mathbb{P}(V)$ be a curve embedded in, and spanning, projective $n$-space. Let $X \to X_0$ denote the normalization and $f: X \to P$ the induced map. For each point $x \in X$ we let $B(x)$ denote the corresponding branch at $f(x)$ of $X_0$. For each integer $m$, $0 \leqq m \leqq n$, we define the *osculating $m$-space* $S(x, m)$ to $B(x)$ to be the linear $m$-dimensional subspace of $P$ which has the highest order of contact with $B(x)$ at $f(x)$. To be more precise, put $A = \mathcal{O}_{X,x}$ and let $t \in A$ denote a uniformizing parameter. There is a parametrization

$$x_i = a_i t^{l_i + i} + (\text{higher order terms}), \qquad (*)$$

$i = 0, 1, \cdots, n$, with $0 = l_0 \leqq l_1 \leqq \cdots \leqq l_n$ (where the $x_i$'s are considered as elements of the completion $\hat{A}$ of $A$), and $a_i \in k^*$. With this choice of coordinates in $P$, $S(x, m)$ is the space defined by $X_{m+1} = X_{m+2} = \cdots = X_n = 0$ and has $(l_{m+1} + m + 1)$ – order contact with $B(x)$ at $f(x)$.

If $l_n = 0$ (and hence $l_i = 0$, $0 \leqq i \leqq n$) holds, we say that $x$ is an *ordinary point.* The *points of hyperosculation* are points with $l_{m+1} > 0$ for some $m$. Pohl called such a point singular of order $m$ ([13]). Classically, a hyperosculating space was also called *stationary*, e.g. a cusp is a stationary point, at a flex there is a stationary tangent, a stall is a stationary osculating hyperplane.

The osculating spaces $S(x, m)$ are the (projective) fibers of an $(m+1)$-bundle on $X$, which we will call the osculating bundle of order $m$; the formulation we use is dual to Pohl's ([13]). Let $\mathcal{P}^m_X(-)$ denote the functor of principal parts (or jets) of order $m$ (§6). Put $\mathcal{L} = f^* \mathcal{O}_P(1)$. Since $X$ is a smooth curve, $\mathcal{P}^m_X(\mathcal{L})$ is an $(m+1)$-bundle – in fact, there are exact sequences, for $m \geqq 1$,

$$0 \longrightarrow \Omega_X^{\otimes m} \otimes \mathcal{L} \longrightarrow \mathcal{P}^m_X(\mathcal{L}) \xrightarrow{b^m} \mathcal{P}^{m-1}_X(\mathcal{L}) \longrightarrow 0$$

([7], 16.10.1, 16.7.3). Let $g$: $x \to \operatorname{spec}(k)$ denote the structure map. There are functorial maps (§6)

$$\alpha^m(\mathcal{L}): g^* g_* \mathcal{L} \to \mathcal{P}^m_X(\mathcal{L})$$

compatible with the maps $b^m$. Let $a^m: V_X \to \mathcal{P}^m_X(\mathcal{L})$ denote the composition of $\alpha^m(\mathcal{L})$ with the map $V_X = g^*H^0(P, \mathcal{O}_P(1)) \to g^*H^0(X, \mathcal{L}) = g^*g_*\mathcal{L}$.

PROPOSITION (2.1). *Let $\mathcal{P}^m \subseteq \mathcal{P}^m_X(\mathcal{L})$ denote the image of the homomorphism $a^m$. For each ordinary point $x \in X$, the embedded linear space $\mathbb{P}(\mathcal{P}^m(x)) \xrightarrow{\mathbb{P}(a^m(x))} \mathbb{P}(V_X(x)) = P$ is equal to the osculating $m$-space $S(x, m)$; if char $k$ is 0 or sufficiently big, then this holds for* all *points of $X$.*

PROOF. Let $x \in X$ and choose a parametrization (*) of $B(x)$. Locally, the map $a^m$ consists of taking the $m$-jets (or the Taylor series development up to order $m$) of the coordinate functions (§6): Consider the basis $\{1, dt, \cdots, (dt)^m\}$ for the free $A$-module $\mathcal{P}^m_X(\mathcal{L}) \otimes A \cong P^m_A$; let $\{d^0, d^1, \cdots, d^m\}$ denote the dual basis for $(P^m_A)^\vee$. Then

$$a^m \otimes A : V_X \otimes A \cong A^{n+1} \to P^m_A \cong A^{m+1}$$

is given by the matrix

$$w^m_x = \begin{pmatrix} x_0 & x_1 & \cdot & \cdot & x_n \\ d^1x_0 & d^1x_1 & \cdot & \cdot & d^1x_n \\ \cdot & \cdot & \cdot & \cdot & \cdot \\ \cdot & \cdot & \cdot & \cdot & \cdot \\ \cdot & \cdot & \cdot & \cdot & \cdot \\ d^mx_0 & \cdot & \cdot & \cdot & d^mx_n \end{pmatrix}$$

Since $d^i$ is a truncated Taylor series, we have $d^i(t^j) = \binom{j}{i}t^{j-i}$, for all $i, j$. Hence we get (considering the entries as elements of $\hat{A}$),

$$w^m_x = \begin{bmatrix} 1+\cdots & a_1t^{l_1+1}+\cdots & \cdots & a_nt^{l_n+n}+\cdots \\ \cdot & (l_1+1)a_1t^{l_1}+\cdots & \cdots & (l_n+n)a_nt^{l_n+n-1}+\cdots \\ \cdot & \cdot & \cdot & \\ \cdot & & \cdot & \\ \cdot & & & \cdot \\ * & \binom{l_m+m}{m}a_mt^{l_m}+\cdots & & \binom{l_n+n}{m}a_nt^{l_n+n-m}+\cdots \end{bmatrix}$$

(where '$+\cdots$' indicates terms of higher powers of $t$, and $*$ indicates elements in the maximal ideal of $\hat{A}$, see also ([9], p. 101)).

CASE 1. $x$ is an ordinary point, i.e. $l_0 = l_1 = \cdots = l_n = 0$. Then $a^m$ is surjective at $x$, so that $\mathcal{P}^m(x) = \mathcal{P}^m_X(\mathcal{L})(x)$, and $a^m(x)$ is given by the matrix

$$\begin{pmatrix} 1 & 0 & 0 & \cdot & \cdot & \cdot & \cdot & 0 \\ 0 & 1 & 0 & & & & & \\ \cdot & & & & & & & \\ \cdot & & \cdot & & & & & \\ \cdot & & & \cdot & & & & \\ & & & & \cdot & & & \\ 0 & \cdot & \cdot & \cdot & 1 & \cdots & & 0 \end{pmatrix},$$

hence it defines ([6], §4) the subspace of $P$ given by $X_{m+1} = X_{m+2} = \cdots = X_n = 0$.

CASE 2. $l_m > 0$ and $\binom{l_i + i}{i} \not\equiv 0$ (char $k$) for $i = 1, \cdots, n$. The inclusion $\mathcal{P}^m \hookrightarrow \mathcal{P}^m_x(\mathcal{L})$ is given, locally at $x$, by the matrix

$$\begin{pmatrix} 1 & & & & \\ & t^{l_1} & & & 0 \\ & & \cdot & & \\ & & & \cdot & \\ & & & & \cdot \\ & 0 & & & t^{l_m} \end{pmatrix},$$

and $a^m(x)$ is given by

$$\begin{pmatrix} 1 & 0 \cdot & \cdot & \cdot 0 & \cdot & \cdot & \cdot & 0 \\ 0 & \binom{l_1+1}{1} & & & & & & \\ \cdot & & & \cdot & & & & \\ \cdot & & & \cdot & & & & \\ \cdot & & & \cdot & & & & \\ 0 & \cdot & \cdot & \binom{l_m+m}{m} & \cdot & \cdot & \cdot & 0 \end{pmatrix}.$$

Hence it defines the same linear $m$-space as in case 1.

The bundle $\mathcal{P}^m = \text{Im}\,(a^m)$ will be referred to as the *osculating bundle of order $m$ of the curve $f: X \to P$. From now on* we assume char$k$ to be such that $\mathcal{P}^m$ represents the osculating $m$-spaces at all points of $X$.

EXAMPLE 1. The rational $n$-ic in $n$-space.

The curve $R_n \subset P$ of degree $n$ and genus 0 has only ordinary points. Every point $x \in R_n$ has a parametrization $(1, t, t^2, \cdots, t^n)$, hence $a^m : V_X \to \mathcal{P}^m_{R_n}(\mathcal{L})$ is everywhere surjective and the osculating bundle of order $m$ is equal to $\mathcal{P}^m_{R_n}(\mathcal{L})$ for $m = 0, 1, \cdots, n$.

EXAMPLE 2. A linearly normal, elliptic curve.

Let $E_d \subset P$ be a curve of genus 1 and degree $d \geqq 3$, which is linearly normal. From Riemann-Roch it follows that at any point $x$, the parametrization (*) satisfies

$$l_i = 0, \qquad i = 0, \cdots, n-1,$$
$$l_n \leqq 1.$$

Hence the curve is non-singular up to order $n-1$.

EXAMPLE 3. Dye's special curve ([4], [5]).

Suppose $\Gamma_d \subseteq P$ is the complete intersection of $n-1$ smooth surfaces $S_i$ of degree $d$ with a common self-polar simplex $S$, i.e., assume $S_i$ has equation $\sum_{j=0}^n a_j^{(i)} X_j^m = 0$, where all $(n-1)$-minors of the matrix $(a_j^{(i)})$ are non-zero. Note that $\Gamma_d$ has degree $d^{n-1}$ and its genus $g$ is given by

$$2g-2 = d^{n-1}((d-1)(n-1)-2).$$

The points of hyperosculation on $\Gamma_d$ are its $(n+1)d^{n-1}$ intersections with the faces of the simplex $S$; the osculating $m$-spaces at these points have order of contact $l_{m+1}+m+1 = md$ and so are hyperosculating (except when $m=1$ and $d=2$).

## §3. Numerical characters

Given a curve in 3-space, we define its *rank* to be the number of tangents that meet a given line. The number of osculating planes that pass through a given point is called the *class* of the curve. In general, for a curve $X_0 \hookrightarrow P = \mathbb{P}(V)$ in $n$-space, we define its *m-rank* to be the number $r_m$ of its osculating $m$-spaces that meet a given (general) sub-$(n-m-1)$-space of $P$. As in §2, let $X \to X_0$ denote the normalization and $f: X \to P$ the induced map. We defined (§2) the osculating bundle of order $m$ of $f$ as an $(m+1)$-bundle $\mathcal{P}^m$ on $X$ together with a surjection $a^m: V_X \to \mathcal{P}^m$, which represents the osculating $m$-spaces of $X_0$. The $(m+1)$-quotient $a^m: V_X \to \mathcal{P}^m$ also defines the $m$th *osculating developable* of $X_0$ as the image of the map $f_{(m)}: \mathbb{P}(\mathcal{P}^m) \to P$ obtained by composing $\mathbb{P}(a^m): \mathbb{P}(\mathcal{P}^m) \hookrightarrow \mathbb{P}(V_X) = X \times P$ with the projection to $P$. Moreover, the 1-quotient $\Lambda^{m+1} V_X \to \Lambda^{m+1}\mathcal{P}^m$ defines *the* $m$th *associated curve* $f^{(m)}: X \to \mathbb{P}(\Lambda^{m+1} V)$. The corresponding interpretations of the ranks are given in the next proposition.

PROPOSITION (3.1). *The following numbers are equal (for* $0 \leqq m \leqq n-1$):

(i) *the m-rank* $r_m$.

(ii) *the degree* $\deg(\mathcal{P}^m)$ *of the osculating bundle of order m.*

(iii) *the degree of the* $m$th *osculating developable* $Y_m = f_{(m)}(\mathbb{P}(\mathcal{P}^m)) \subset P$.

(iv) *the degree of the* $m$th *associated curve* $f^{(m)}: X \to \mathbb{P}(\Lambda^{m+1} V)$.

PROOF. Since the quotient $a^m: V_X \to \mathcal{P}^m$ represents the osculating $m$-spaces of the curve, we see that the $m$-rank $r_m$ is equal to the degree of the pullback to $X$ of the Schubert cycle $\sigma \subseteq \mathrm{Grass}_{m+1}(V)$ of $m$-spaces of $P = \mathbb{P}(V)$ which meet a given $(n-m-1)$-space, via the map $G(a^m): X \to \mathrm{Grass}_{m+1}(V)$ defined by $a^m$. Consider the Plücker embedding, $\mathrm{Grass}_{m+1}(V) \overset{j}{\hookrightarrow} \mathbb{P}(\Lambda^{m+1}V)$; then $\sigma$ is a hyperplane section, and the $m$th associated map $f^{(m)}$ is the composition of $G(a^m)$ with $j$. Hence we get $r_m = \deg(f^{(m)*}\mathcal{O}_{\mathbb{P}(\Lambda^{m+1}V)}(1)) = \deg(\Lambda^{m+1}\mathcal{P}^m) = \deg(\mathcal{P}^m)$. This proves (i) = (ii) = (iv).

We have $\deg(Y_m) = \langle c_1(\mathcal{O}_P(1))^{m+1} \cdot Y_m \rangle_P$, and since $X \to X_0$ is birational, so is $\mathbb{P}(\mathcal{P}^m) \to Y$, therefore we have $\deg(Y_m) = \langle c_1(f^*_{(m)}\mathcal{O}_P(1))^{m+1} \rangle_{\mathbb{P}(\mathcal{P}^m)} = \langle c_1(\mathcal{O}_{\mathbb{P}(\mathcal{P}^m)}(1))^{m+1} \rangle_{\mathbb{P}(\mathcal{P}^m)}$. By Chern class theory this is equal to $\langle c_1(\mathcal{P}^m) \cdot c_1(\mathcal{O}_{\mathbb{P}(\mathcal{P}^m)}(1))^m \rangle_{\mathbb{P}(\mathcal{P}^m)}$, which, by the projection formula, equals $\langle c_1(\mathcal{P}^m) \rangle_X = \deg(\mathcal{P}^m)$.

EXAMPLE 1 ($n=3$). The twisted cubic $R_3 \subseteq \mathbb{P}^3$. We can write $R_3$ as the intersection of three quadrics $Q_1$, $Q_2$, $Q_3$, with its tangent developable defined by $Q_2^2 - 4Q_1Q_3 = 0$. Hence $R_3$ has 1-rank $r_1 = 4$.

Let $x \in X$ and consider, as in §2, a parametrization (*) $x_i = a_i t^{l_i + i} + \cdots$, $a_i \not\equiv 0(t)$, $i = 0, \cdots, n$, $0 = l_0 \leqq l_1 \leqq \cdots \leqq l_n$. The integer $l_{m+1} - l_m$ is called the $m$th stationary index of the curve $f: X \to P$ at $x$, and we define the $m$th *stationary index of* $f$ to be $k_m = \sum_{x \in X}(l_{m+1} - l_m)$. Hence, in the case that $f$ is non-singular of order $m-1$, $k_m$ is the number (counted properly) of hyperosculating (or stationary) $m$-spaces.

THEOREM (3.2). *There are formulas*

$$r_m = (m+1)(r_0 + m(g-1)) - \sum_{i=0}^{m-1} (m-i)k_i,$$

*for* $m = 0, 1, \cdots, n-1$, *and*

$$\sum_{i=0}^{n-1} (n-i)k_i = (n+1)(r_0 + n(g-1)),$$

*where* $g$ *denotes the genus of* $X$ (see [2], p. 200).

PROOF. Put $\mathcal{Q}_m = \mathrm{Coker}(a^m)$. The exact sequences $0 \to \mathcal{P}^m \to \mathcal{P}^m_X(\mathcal{L}) \to \mathcal{Q}_m \to 0$ gives $r_m = \deg(\mathcal{P}^m) = \deg(\mathcal{P}^m_X(\mathcal{L})) - \deg(\mathcal{I}_m^{-1})$, where $\mathcal{I}_m$ is the 0th Fitting ideal $F^0(\mathcal{Q}_m)$ of $\mathcal{Q}_m$, i.e. $\mathcal{I}_m$ is the image of the homomorphism $(\Lambda^{m+1}\mathcal{P}^m) \otimes (\Lambda^{m+1}\mathcal{P}^m_X(\mathcal{L}))^{-1} \to \mathcal{O}_X$. Locally at a point $x \in X$, in terms of a parametrization (*), $\mathcal{I}_m$ is generated by an element of the form $\Pi_{i=0}^m (a_i \binom{l_i+i}{i} t^{l_i} + \cdots)$. Hence we get $\deg(\mathcal{I}_m^{-1}) = \sum_{x \in X} l_i = \sum_{i=0}^{m-1}(m-i)k_i$. The exact sequences (§2)

$$0 \to S^m \Omega_X \otimes \mathcal{L} \to \mathcal{P}^m_X(\mathcal{L}) \to \mathcal{P}^{m-1}_X(\mathcal{L}) \to 0$$

give

$$\deg(\mathcal{P}^m_X(\mathcal{L})) = \sum_{i=1}^{m} \deg(S^i\Omega_X \otimes \mathcal{L}) + \deg \mathcal{L}.$$

Hence we get

$$r_m = \frac{m(m+1)}{2} \deg \Omega_X + (m+1) \deg \mathcal{L} - \sum_{i=0}^{m-1} (m-i)k_i$$
$$= (m+1)(r_0 + m(g-1)) - \sum_{i=0}^{m-1} (m-i)k_i.$$

(Since $r_0 = \deg \mathcal{L}$ and $\deg \Omega_X = 2(g-1)$.) The last equation follows from the fact that $r_n = 0$ holds: the surjection $a^n : V_X \to \mathcal{P}^n$ is necessarily an isomorphism, since the bundles have the same rank.

We note that the map $a^1 : V_X \to \mathcal{P}^1_X(\mathcal{L})$ is canonically isomorphic to the map $f^*\mathcal{P}^1_P(1) \to \mathcal{P}^1_X(\mathcal{L})$ (6.4). Hence we get $\mathrm{Coker}(a^1) = \Omega_{X/P}(\mathcal{L})$ ([7], 16.4.18), and $k_0$ is the degree of the ramification divisor; we have $k_0$ = the number of cusps of $f(X)$, if char$k \neq 2$ (if char$k = 2$, then $k_0 = 2\#$cusps).

COROLLARY (3.3). There are equalities
(i) $(r_{m+1} - r_m) - (r_m - r_{m-1}) = 2(g-1) - k_m$
(ii) $(r_{m+1} + k_m) - (r_{m-2} + k_{m-1}) = 3(r_m - r_{m-1})$
(iii) $(r_{n-2} + k_{n-1}) - (r_1 + k_0) = 2(r_{n-1} - r_0)$.

The formulas (i) are usually referred to as the generalized Plücker formulas for curves (see [13], p. 207; [8], 4.26; [15], p. 43). In the case that the curve is plane ($n = 2$), it is the classical formula

$$r_1 = r_0 + 2g - 2 - k_0 \tag{I}$$

giving the class in terms of the degree, genus, and number of cusps. The other Plücker formula for the class of a plane curve can be written

$$r_1 = r_0(r_0 - 1) - e \tag{II}$$

where $e = \deg(\mathcal{J}^{-1})$ and $\mathcal{J}$ denotes the pullback to $X$ of the Jacobian ideal of $f(X)$. (This formula follows from the fact that, for a plane curve, Ker $(a^1)$ is equal to $f^*\mathcal{N} \otimes \mathcal{J}^{-1}$, where $\mathcal{N}$ is the conormal bundle of $f(X)$ in $P$ ([12]).) One shows $e = 2\delta + k_0$, where $2\delta$ denotes the degree of the conductor of $X$ in $f(X)$. In particular, if $f(X)$ has no other singularities than $D$ (ordinary) double points and $K$ (simple) cusps, we get $e = 2D + 3K$, if char$k \neq 2, 3$.

EXAMPLE 1. $R_n$ has no hyperosculating points. We get

$$r_m = (m+1)(n-m) = r_{n-m-1},$$

hence the degree of the $m$th associated curve $f^{(m)}: X \to \mathrm{Grass}_{m+1}(V) \hookrightarrow \mathbb{P}(\Lambda^{m+1} V)$ is equal to the dimension of $\mathrm{Grass}_{m+1}(V)$.

EXAMPLE 2. We deduce, from the Riemann-Roch theorem, $k_m = 0$, $m = 0, \cdots, n-2$. Hence we get $r_m = (m+1)d$, $m = 0, \cdots, n-1$ and $k_{n-1} = (n+1)d$, so that the curve $E_d$ has exactly $(n+1)d$ hyperosculating hyperplanes.

EXAMPLE 3. For the curve $\Gamma_d$ we have $k_0 = 0$, $k_i = (n+1)d^{n-1}(id - i - 1)$ for $i \geqq 1$, from which we can compute the formulas for $r_m$.

Let us consider the case $d = 2$, $n = 4$. Then $r_0 = 8$, $g = 5$, and $k_0 = 0$, $k_1 = 0$, $k_2 = 40$, $k_3 = 40$. Hence we get $r_1 = 24$, $r_2 = 48$, and $r_3 = 40$. The number of hyperosculating points on $\Gamma = \Gamma_2$ is 40, but their 'weighted number' $\sum_{i=0}^{3} (4-i)k_i$ is 120. In fact the curve $\Gamma \hookrightarrow \mathbb{P}^4$ is *canonical*, i.e. the embedding is given by the canonical bundle $\Omega_\Gamma$. So the 40 hyperosculating points are the Weierstrass points of $\Gamma$; the above remark about their number shows that these points are not ordinary Weierstrass points.

## §4. Projections and sections

If we project a curve in 3-space from a (general) point, we obtain a plane curve, of the same degree, whose class is equal to the rank of the space curve. Another way of obtaining a plane curve from the space curve is to take the intersection of its tangent developable with a (general) plane. We expect this plane curve to have the same class as the space curve, while its degree is that of the tangent developable, hence equal to the rank of the space curve. Veronese ([14]) used the principle of projections and sections to obtain $n-1$ plane curves from a given curve in $n$-space; he then applied the formulas (I), (II) (and their duals) of §3 to these curves, deducing relations among the numerical characters of the given curve (interpreting these in terms of the characters of the plane curves).

Assume from now on that $X$ is a smooth curve, given with a finite map $f: X \to P = \mathbb{P}(V)$ such that $a^n: V_X \to \mathcal{P}^n_X(\mathcal{L})$ is generically surjective (§1, §2). A $(q+1)$-dimensional subspace $V' \subseteq V$ such that $f(X) \cap (V/V') = \varnothing$ (i.e., such that the induced map $V'_X \to \mathcal{L} = f^*\mathcal{O}_P(1)$ is surjective) defines a projection $X \to P' = \mathbb{P}(V')$. If, in addition, the induced map $V'_X \to \mathcal{P}^{q-1}$ is surjective (here $\mathcal{P}^{q-1}$ denotes the osculating bundle of order $q-1$ of $f$, as in §2), we say that $p_q(f) = p_q: X \to \mathbb{P}(V')$ is *a $q$-projection of $f$*. In fact, a *general* subspace $V' \subseteq V$ has this property, as the following lemma shows.

LEMMA (4.1). *If $Y$ is a $d$-dimensional variety and $V_Y \to \mathscr{E}$ is an $r$-quotient on $Y$, then a general $(r+d)$-dimensional subspace $V'$ of $V$ has the property that the induced map $V'_Y \to \mathscr{E}$ is surjective.*

PROOF. Let $\phi: Y \to \mathrm{Grass}_r(V)$ be the morphism defined by the quotient. We want to show ([10], 2.6) that for a general $V' \subseteq V$, we have $\phi^{-1}(\sigma_{d+1}(V')) = \varnothing$, where $\sigma_{d+1}(V') \subseteq \mathrm{Grass}_r(V)$ denotes the $(d+1)$th special Schubert cycle defined by $V'$. Note that $\sigma_{d+1}(V')$ has codimension $d+1$.

Let $G$ denote the general linear group acting on $V$. Given $V' \subseteq V$ there is an open dense subset $U$ of $G$ such that $\phi^{-1}(\sigma_{d+1}(gV')) = \varnothing$ for all $g \in U$ ([11], Cor. 4). There is a surjective map $\psi: G \to G' = \mathrm{Grass}_{n+1-r-d}(V)$ and (by the theorem of generic flatness) there is an open dense subset $U'$ of $G'$ such that $\psi^{-1}(U') \subseteq U$. Hence $gV'$ satisfies the condition of the lemma, for all $g \in U'$.

PROPOSITION (4.2). *The osculating bundles of $p_q: X \to P'$ are equal to the osculating bundles (of order $\leqq q-1$) of $f: X \to P$, and a $q'$-projection of $p_q$ is also a $q'$-projection of $f$.*

PROOF. From the general properties of the maps $a^m$ (6.1) it follows that $a^m(p_q): V'_X \to \mathcal{P}^m_X(\mathcal{L})$ is equal to the composition of $a^m(f): V_X \to \mathcal{P}^m_X(\mathcal{L})$ with the inclusion $V'_X \hookrightarrow V_X$. Therefore the image of $a^m(p_q)$ is equal to $\mathcal{P}^m$, since this is true for $m = q-1$ and since the canonical surjections $b^m: \mathcal{P}^m_X(\mathcal{L}) \to \mathcal{P}^{m-1}_X(\mathcal{L})$ induce surjections $\mathcal{P}^m \to \mathcal{P}^{m-1}$. To prove the last assertion, we observe that if $W' \hookrightarrow V'$ defines a $q'$-projection of $p_q$, then $W' \hookrightarrow V$ defines a $q'$-projection of $f$.

Assume now that $V' \subseteq V$ is an $m$-dimensional subspace. Applying lemma (4.1) to the $(n-m)$-quotient $V^\vee_X \to (\mathrm{Ker}\,(a^m))^\vee$ we see that for a general such subspace $V'$, the induced map $V'_X \to \mathcal{P}^m$ is locally split. As in §2, let $f_{(m)}: \mathbb{P}(\mathcal{P}^m) \to P$ denote the $m$th osculating developable of $f$. Suppose $V' \subseteq V$ is general and put $W = V/V'$. We set $X_m = f^{-1}_{(m)}(\mathbb{P}(W))$ and call $f_m = f_{(m)} \mid X_m: X_m \to \mathbb{P}(W)$ an *$m$-section of $f$*. We let $\mathcal{L}_m$ denote the line bundle $\mathrm{Coker}\,(V'_X \to \mathcal{P}^m)$.

PROPOSITION (4.3). *With the above notations, the map $\gamma_m: X_m \to X$ induced by $\mathbb{P}(\mathcal{P}^m) \to X$ is equal to the canonical isomorphism $\mathbb{P}(\mathcal{L}_m) \tilde{\to} X$.*

PROOF. Put $\mathcal{K}_m = \mathrm{Ker}\,(a^m)$. From the commutative diagram of bundles on $X$

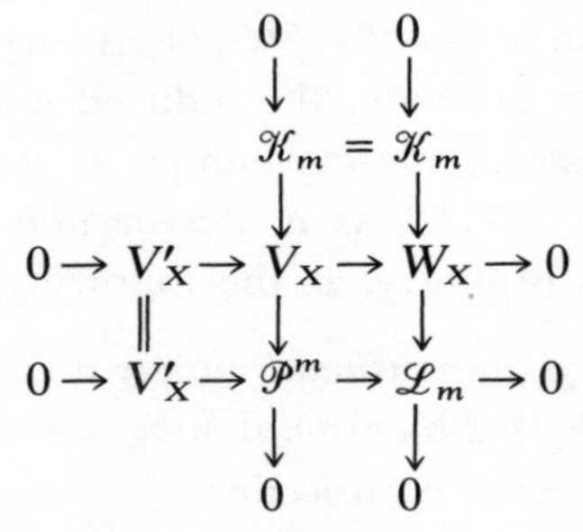
$$\begin{array}{ccccccccc} & & & & 0 & & 0 & & \\ & & & & \downarrow & & \downarrow & & \\ & & & & \mathcal{K}_m & = & \mathcal{K}_m & & \\ & & & & \downarrow & & \downarrow & & \\ 0 & \to & V'_X & \to & V_X & \to & W_X & \to & 0 \\ & & \| & & \downarrow & & \downarrow & & \\ 0 & \to & V'_X & \to & \mathcal{P}^m & \to & \mathcal{L}_m & \to & 0 \\ & & & & \downarrow & & \downarrow & & \\ & & & & 0 & & 0 & & \end{array}$$

it follows that the curves $\mathbb{P}(\mathcal{L}_m)\subseteq\mathbb{P}(\mathcal{P}^m)$ and $X_m=f_{(m)}^{-1}(\mathbb{P}(W))\subseteq\mathbb{P}(\mathcal{P}^m)$ are equal.

The next result we will prove is that an $m'$-section of an $m$-section of $f$ is an $(m+m')$-section of $f$.

THEOREM (4.4). *Suppose $f_m:X_m\to\mathbb{P}(W)$ is an $m$-section of $f$ such that $a^{m'}(f_m)$ is generically surjective, let $\mathcal{P}_m^{m'}$ denote the osculating bundle of order $m'$ of $f_m$, and let $f_{m,(m')}:\mathbb{P}(\mathcal{P}_m^{m'})\to\mathbb{P}(W)$ denote the $m'$th osculating developable of $f_m$. Suppose*

$$f_{m,m'}:X_{m,m'}=(f_{m,(m')})^{-1}(\mathbb{P}(W'))\to\mathbb{P}(W)$$

*is an $m'$-section of $f_m$. Then*

$$f_{m+m'}:X_{m+m'}=(f_{(m+m')})^{-1}(\mathbb{P}(W'))\to P$$

*is an $(m+m')$-section of $f$, and there is a cartesian diagram*

$$\begin{array}{ccc} X_{m,m'} & \xrightarrow{u} & X_{m+m'}\times_X X_m \\ \downarrow & & \downarrow \\ \mathbb{P}(\mathcal{P}_m^{m'}) & \xrightarrow{v} & \mathbb{P}(\mathcal{P}^{m+m'})\times_X X_m \end{array}$$

*where $u$ is an isomorphism and $v$ is an embedding.*

PROOF. We will show that there is a canonical exact sequence

$$0\to V'_{X_m}\to\gamma^*\mathcal{P}^{m+m'}\to\mathcal{P}_m^{m'}\to 0, \tag{4.5}$$

where $V'=\operatorname{Ker}(V\to W)$, and $\gamma=\gamma_m:X_m\to X$ is the isomorphism of (4.3). Once (4.5) is established, the theorem follows: Put $\mathcal{K}_{m+m'}=\operatorname{Ker}(a^{m+m'})$. We have $\operatorname{Ker}(W_{X_m}\to\mathcal{P}_m^{m'})=\operatorname{Ker}(V_{X_m}\to\gamma^*\mathcal{P}^{m+m'})=\gamma^*\mathcal{K}_{m+m'}$. Put $W''=\operatorname{Ker}(W\to W')$ and $\mathcal{L}'=\operatorname{Coker}(W''_{X_m}\to\mathcal{P}_m^{m'})$. Then $\mathcal{L}'$ is invertible (since $W''$ defines an $m'$-section of $f_m$) and there is an induced surjection $W'_{X_m}\to\mathcal{L}'$ with kernel $\gamma^*\mathcal{K}_{m+m'}$. Hence we get a commutative diagram of bundles on $X$, where all arrows are surjective and the three vertical arrows all have kernel equal to $\gamma^*\mathcal{K}_{m+m'}$,

$$\begin{array}{ccccc} V_{X_m} & \to & W_{X_m} & \to & W'_{X_m} \\ \downarrow & & \downarrow & & \downarrow \\ \gamma^*\mathcal{P}^{m+m'} & \to & \mathcal{P}_m^{m'} & \to & \mathcal{L}'. \end{array}$$

Put $V''=\operatorname{Ker}(V\to W')$. Then $V''_X$ is equal to $\operatorname{Ker}(\mathcal{P}^{m+m'}\to\gamma_*\mathcal{L}')$, hence is a sub-$(m+m')$-bundle of $\mathcal{P}^{m+m'}$. This means that $V''$ defines an $(m+m')$-section $f_{m+m'}:X_{m+m'}\to\mathbb{P}(W')$ of $f$. The existence of the cartesian diagram follows from the existence of the above commutative diagram of bundles.

It remains to establish (4.5).

STEP 1. Put $\mathscr{E} = \operatorname{Coker}(\mathscr{K}_{m+m'} \xrightarrow{c} W_X)$. Then $\mathscr{E}$ is an $(m+1)$-bundle.

This follows since $c$ is equal to the composition of the locally split map $\mathscr{K}_{m+m'} \hookrightarrow \mathscr{K}_m$ (induced by the surjection $\mathscr{P}^{m+m'} \to \mathscr{P}^m$) with the map $\mathscr{K}_m \to W_X$ (which is locally split because $W$ defines an $m$-section of $f$).

STEP 2. The map $a: W_X \to \mathscr{P}_X^{m'}(\mathscr{L}_m)$, obtained by composing $\beta^{m'}: W_X \to \mathscr{P}_m^{m'}(W_X)$ with $\mathscr{P}_m^{m'}(a^0): \mathscr{P}_X^{m'}(W_X) \to \mathscr{P}_X^{m'}(\mathscr{L}_m)$ (§6) factors through $b: W_X \to \mathscr{E}$.

Consider the following commutative diagram of exact rows (since $X$ is smooth, $\mathscr{P}_X^{m'}(-)$ is an exact functor ([7], 16.7.3)),

$$
\begin{array}{ccccccccc}
0 \to & V'_X & \to & V_X & \to & W_X & \to 0 \\
 & \downarrow & & \downarrow & \searrow^{e} & \downarrow & \searrow^{a} \\
0 \to & \mathscr{P}_X^{m'}(V'_X) & \to & \mathscr{P}_X^{m'}(V_X) & \to & \mathscr{P}_X^{m'}(W_X) & \to 0 \\
 & \| & & \downarrow{\scriptstyle \mathscr{P}_X^{m'}(a^m)} & & \downarrow & \\
0 \to & \mathscr{P}_X^{m'}(V'_X) & \to & \mathscr{P}_X^{m'}(\mathscr{P}^m) & \to & \mathscr{P}_X^{m'}(\mathscr{L}_m) & \to 0.
\end{array}
$$

The injection $\mathscr{P}^m \hookrightarrow \mathscr{P}_X^m(\mathscr{L})$ gives an injection $j: \mathscr{P}_X^{m'}(\mathscr{P}^m) \hookrightarrow \mathscr{P}_X^{m'}(\mathscr{P}_X^m(\mathscr{L}))$. The map $e': V_X \to \mathscr{P}_X^{m'}(\mathscr{P}_X^m(\mathscr{L}))$, obtained by composing $e$ with $j$, factors through $a^{m+m'}: V_X \to \mathscr{P}_X^{m+m'}(\mathscr{L})$ via $\delta^{m,m'}: \mathscr{P}_X^{m+m'}(\mathscr{L}) \to \mathscr{P}_X^{m'}(\mathscr{P}_X^m(\mathscr{L}))$ (6.2(iv)). Thus we get an equality $e'(V_X) = e(V_X)$ in $\mathscr{P}_X^{m'}(\mathscr{P}^m)$ (since $j$ is injective), also we see that $a^{m+m'}(V_X) = \mathscr{P}^{m+m'}$. Therefore the map $\delta^{m,m'}$ induces a map $\delta: \mathscr{P}^{m+m'} \to e(V_X)$ of $(m+m'+1)$-quotients of $V_X$. It follows that there is a commutative diagram with exact rows

$$
\begin{array}{ccccccccc}
0 \longrightarrow & V'_X & \longrightarrow & V_X & \longrightarrow & W_X & \longrightarrow 0 \\
 & \| & & \downarrow{\scriptstyle a^{m+m'}} & \searrow^{e} & \downarrow{\scriptstyle b} & \searrow^{a} \\
0 \longrightarrow & V'_X & \longrightarrow & \mathscr{P}^{m+m'} & \longrightarrow & \mathscr{E} & \longrightarrow 0 \\
 & \downarrow{\scriptstyle \beta} & & \downarrow{\scriptstyle \delta} & & \downarrow{\scriptstyle \exists} & \\
0 \longrightarrow & \mathscr{P}_X^{m'}(V'_X) & \longrightarrow & \mathscr{P}_X^{m'}(\mathscr{P}^m) & \longrightarrow & \mathscr{P}_X^{m'}(\mathscr{L}_m) & \longrightarrow 0.
\end{array}
$$

STEP 3. The map $\gamma^* a: W_{X_m} \to \gamma^* \mathscr{P}_X^{m'}(\mathscr{L}_m)$ is equal to the map $a^{m'}(f_m)$.

Observe first that proposition (4.3) is a special case $(m'=0)$ of the lemma. Therefore we know that the canonical 1-quotient $a^0(f_m)$ on $X_m$ is equal to $\gamma^* a^0: W_{X_m} \to \gamma^* \mathscr{L}_m$. Applying (6.2(iii)) to $\gamma: X_m \tilde{\to} X$ then gives a commutative diagram

$$
\begin{array}{ccc}
W_{X_m} & \xrightarrow{a^{m'}(f_m)} & \mathscr{P}_{X_m}^{m'}(\gamma^* \mathscr{L}_m) \\
\| & & \uparrow \wr \\
W_{X_m} & \xrightarrow{\gamma^* a} & \gamma^* \mathscr{P}_X^{m'}(\mathscr{L}_m).
\end{array}
$$

This shows that (4.5) exists: Since $a^{m'}(f_m)$ is generically surjective, the quotients $W_X \to \gamma^* \mathscr{E}$ and $W_X \to \mathscr{P}_m^{m'}$ are equal. This completes the proof of theorem (4.4).

THEOREM (4.6). *A $q$-projection of an $m$-section of $f: X \to P$ is the same as an $m$-section of a $(q+m)$-projection of $f$.*

PROOF. Let $f_m: X_m \to \mathbb{P}(V'')$ be an $m$-section of $f$, defined by $0 \to V' \to V \to V'' \to 0$. Let $p_q^m: X_m \to \mathbb{P}(W')$ be a $q$-projection of $f_m$, defined by $0 \to W' \to V'' \to W'' \to 0$. Put $W = \mathrm{Ker}\,(V \to W'')$ (note that $W$ has dimension $m+q+1$). We want to show that $0 \to W \to V \to W'' \to 0$ defines a $(q+m)$-projection of $f$, i.e., that the induced map $c: W_X \to \mathcal{P}^{q+m-1}$ is surjective.

Since $p_q^m$ is a q-projection of $f_m$, we have a commutative diagram

$$\begin{array}{ccc} W'_X & \longrightarrow & V''_X \\ & \searrow \quad \swarrow & \\ & \gamma_* \mathcal{P}_m^{q-1} & \end{array}$$

which fits into the following commutative diagram (with exact rows)

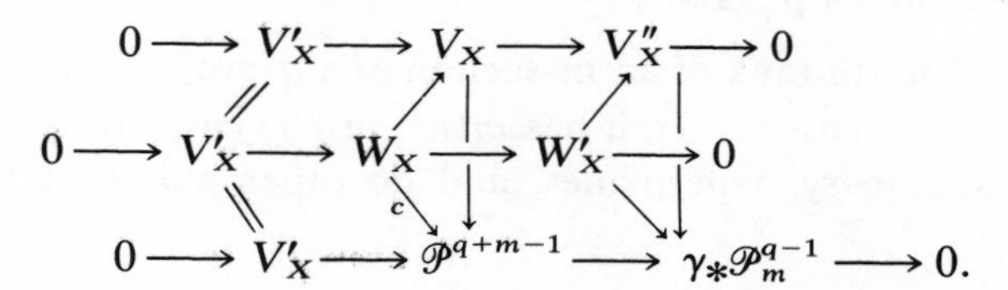

An easy diagram chase shows that $c$ is surjective.

The exact sequence $0 \to V' \to W \to W' \to 0$ defines an $m$-section of the $(q+m)$-projection given above; in fact, we get a commutative diagram

$$\begin{array}{ccccccccc} 0 & \to & V'_X & \to & W_X & \to & W'_X & \to & 0 \\ & & \| & & \downarrow & & \downarrow & & \\ 0 & \to & V'_X & \to & \mathcal{P}^m & \to & \gamma_* \mathcal{P}_m^0 & \to & 0 \end{array}$$

which also shows that this $m$-section is equal to $p_q^m$.

If we suppose, on the other hand, that we are given an $m$-section of a $(q+m)$-projection of $f$, a similar argument shows that this curve is also a $q$-projection of an $m$-section of $f$.

If we put this last result together with (4.2) and (4.4), we see that any curve $f': X' \to P'$ which is obtained from $f$ by projections and sections can be viewed as a $q$-projection of an $m$-section of $f$, $p_q^m$, where $q = \dim P'$ and $m$ is uniquely determined. In particular, from the given curve $f: X \to P$ in $n$-space, we obtain by these methods $n-1$ (and not more) plane curves $p_2^m$, $m = 0, 1, \cdots, n-2$.

PROPOSITION (4.7). *Let $(r_0, r_1, \cdots, r_{n-1})$ and $(k_0, \cdots, k_{n-1})$ denote the ranks and stationary indices of the curve $f: X \to P$. The ranks and stationary* *indices of a q-projection of an m-section* $p_q^m: X_m \to P'$ *of f are given by:*

$$\begin{aligned} r_i(p_q^m) &= r_{m+i}, & i &= 0, \cdots, q-1, \\ k_0(p_q^m) &= r_{m-1} + k_m & &(\text{if } q>1) \\ k_i(p_q^m) &= k_{m+i}, & 1 &\leq i < q-1, \\ k_{q-1}(p_q^m) &= r_{m+q} + k_{m+q-1}. \end{aligned}$$

PROOF. The osculating bundle of order $i$, $\mathcal{P}_m^i$, of an $m$-section $f_m$ fits into an exact sequence (see (4.5))

$$0 \to V_{X_m} \to \gamma^* \mathcal{P}^{m+i} \to \mathcal{P}_m^i \to 0.$$

Hence we get $r_i(f_m) = \deg(\mathcal{P}_m^i) = \deg(\mathcal{P}^{m+i}) = r_{m+i}$ (3.1). From (4.2) it follows that the osculating bundle of $p_q^m$ is equal to that of $f_m$, hence we get also $r_i(p_q^m) = r_{m+i}$.

The formulas for $k_i(p_q^m)$ are straightforward applications of the formula (3.3) (ii) to the curves $p_q^m$ and $f$.

EXAMPLE 1. The $i$th rank of an $m$-section of a $q$-projection of the rational $n$-ic $R_n$ is $(m+i+1)(n-m-i)$; it has $m(n-m+1)$ cusps (if $q>1$), $(m+q+1)(n-m-q)$ stationary hyperplanes, and no other points of hyperosculation.

EXAMPLE 2. For the curve $p_q^m(E_d)$ we get $r_i = (m+i+1)d$, $k_0 = md$ (if $q>1$), $k_i = 0$ for $0<i<q-1$ (if $m+i \leq n-2$), and $k_{q-1} = (m+q+1)d$. In particular, if we take $q=2$, $m=n-2$, we get a plane curve of degree $(n-1)d$, class $nd$, with $(n-2)d$ cusps and $2(n+1)d$ flexes.

EXAMPLE 3. Starting with the curve $\Gamma_2 \subset \mathbb{P}^4$, we obtain 3 plane curves $p_2^i$, $i=0, 1, 2$, with the following numerical characters:

| | $r_0$(degree) | $r_1$(class) | $k_0$(cusps) | $k_1$(flexes) |
|---|---|---|---|---|
| $p_2^0$ | 8 | 24 | 0 | 48 |
| $p_2^1$ | 24 | 48 | 8 | 80 |
| $p_2^2$ | 48 | 40 | 64 | 40. |

## §5. Duality

Suppose $f: X \to P$ is finite and $a^n: V_X \to \mathcal{P}_X^n(\mathcal{L})$ is generically surjective. Assume also that we have either char $k=0$ or char $k>n$. The dual curve

$\check{f}: X \to P^\vee = \mathbb{P}(V^\vee)$ of $f: X \to P$ is defined by the 1-quotient $V_X^\vee \to \mathcal{K}_{n-1}^\vee$, where we have put $\mathcal{K}_{n-1} = \operatorname{Ker}(a^{n-1}: V_X \to \mathcal{P}^{n-1})$. Hence a point $\check{f}(x)$ of the dual curve is equal to the osculating hyperplane to $f$ at $x$, considered as a point in the dual projective space.

THEOREM (5.1). (i) *The $m$-rank of $\check{f}$ is equal to the $(n-1-m)$-rank of $f$, i.e., $r_m(\check{f}) = r_{n-1-m}(f)$, for $m = 0, 1, \cdots, n-1$, and similarly for the stationary indices: $k_m(\check{f}) = k_{n-1-m}(f)$ for $m = 0, 1, \cdots, n-1$.*

(ii) *The dual curve of $\check{f}$ is equal to $f$ (we write $(\check{f})^\vee = f$).*

(iii) *The dual of a $q$-projection of an $m$-section of $f$ is equal to a $q$-projection of an $(n-m-q)$-section of $\check{f}$ (we write $p_q^m(f)^\vee = p_q^{n-m-q}(\check{f})$).*

PROOF. Let $\mathcal{P}_*^m$ denote the $m$th osculating bundle of $\check{f}$ and $a_*^m: V_X^\vee \to \mathcal{P}_*^m$ the canonical quotient. Put $\mathcal{K}_*^m = \operatorname{Ker}(a_*^m)$ and $\mathcal{K}_{n-1-m} = \operatorname{Ker}(a^{n-1-m})$. The key result is the following:

LEMMA (5.2). *There are canonical isomorphisms of exact sequences (for $m = 0, \cdots, n-1$)*

$$\begin{array}{ccccccccc} 0 \to & \mathcal{K}_*^m & \to & V_X^\vee & \to & \mathcal{P}_*^m & \to 0 \\ & \uparrow\wr & & \| & & \uparrow\wr & \\ 0 \to & (\mathcal{P}^{n-1-m})^\vee & \to & V_X^\vee & \to & (\mathcal{K}_{n-m-1})^\vee & \to 0. \end{array}$$

Assuming for a moment that this holds, let us see how it implies the theorem. First of all, (i) follows:

$$\begin{aligned} r_m(\check{f}) &= \deg \mathcal{P}_*^m \text{ (by (3.1))} = \deg(\mathcal{K}_{n-1-m}^\vee) \\ &= -\deg \mathcal{K}_{n-1-m} = \deg \mathcal{P}^{n-1-m} = r_{n-1-m}(f). \end{aligned}$$

The curve $(\check{f})^\vee$ is defined by the 1-quotient $V_X^\vee \to \mathcal{K}_*^{n-1\vee}$, i.e. by $V_X \to \mathcal{P}^0 = \mathcal{L}$, therefore (ii) holds.

Suppose $p_q^m(f)$ is defined by an $(n+1-m)$-quotient $V \to V''$ and a $(q+1)$-subspace $W' \hookrightarrow V''$. We obtain (as in the proof of (4.6)) a commutative diagram

$$\begin{array}{ccccccc} & 0 & & 0 & & & \\ & \downarrow & & \downarrow & & & \\ 0 \to & \mathcal{M} & \to & \mathcal{K}_{m+q-1} & \to & W_X'' \to 0 \\ & \downarrow & & \downarrow & & \| \\ 0 \to & W_X' & \to & V_X'' & \to & W_X'' \to 0 \\ & \downarrow & & \downarrow & & & \\ & \mathcal{P}_m^{q-1} & = & \mathcal{P}_m^{q-1} & & & \\ & \downarrow & & \downarrow & & & \\ & 0 & & 0. & & & \end{array}$$

The dual curve $p_q^m(f)^\vee$ is defined by the 1-quotient $(W'_X)^\vee \to \mathcal{M}^\vee$. Put $W = \mathrm{Ker}\,(V \to W'')$. We have an induced surjection $W \to W'$, and we note $\dim W = m+1+q$. The quotient $V^\vee \to W^\vee$ defines an $(n-m-q)$-section of $f$; we get an exact sequence $0 \to \mathcal{K}_*^{n-m-q} \to W_X^\vee \to \mathcal{M}' \to 0$ defining this section. The isomorphism (5.2) then shows that the 1-quotient $W_X'^\vee \to \mathcal{M}'$ defining the projection $p_q^{n-m-q}(\check{f})$ corresponding to the subspace $W'^\vee \hookrightarrow W^\vee$ is canonically isomorphic to the 1-quotient $W_X'^\vee \to \mathcal{M}^\vee$. This proves (iii).

PROOF OF (5.2). (The proof is essentially the one given by Weyl ([15], p. 47) for analytic curves.) We will first show that the map between the sequences exists, by showing that the composition $a_*^m \circ (a^{n-1-m})^\vee$ is 0. Then we show that $\mathcal{P}_*^m$ has rank $m+1$, for $m = 0, 1, \cdots, n$ (so that the dual curve spans $\check{P}$). From these two facts it then follows that the generically surjective map $a^{n-1-m}(\check{f})$ factors through the quotient $V_X^\vee \to \mathcal{K}_{n-1-m}^\vee$, hence its image $\mathcal{P}_*^m$ is isomorphic to this quotient.

It suffices to show that $a_*^m \circ (a^{n-1-m})^\vee = 0$ holds locally on $X$. Let $x \in X$ and put $A = \mathcal{O}_{X,x}$. With the notations of the proof of (2.1), the map $(a^{n-1-m})^\vee$ is given by $(w_x^{n-1-m})^t : A^{m+1} \to A^{n+1}$. Let $(y): A \to A^{n+1}$ be the $(n+1, n)$-matrix corresponding to the (locally split) map $\mathcal{K}_{n-1} \to V_X$ at $x$, so that the 1-quotient $V_X^\vee \to \mathcal{L}_* = (\mathcal{K}_{n-1})^\vee$ defining $\check{f}$ is given, locally at $x$, by $(y)^t$. It follows (2.1) that the map $a_*^m$ is given by the matrix $w_y^m$. The entries of the matrix $w_y^m \cdot (w_x^{n-1-m})^t$ are $\sum_{\nu=0}^n d^i y_\nu d^j x_\nu$. Hence we want to show:

$$D^{i,j} = \sum_{\nu=0}^n d^i y_\nu d^j x_\nu = 0 \quad \text{holds for} \quad 0 \leqq i \leqq m,\ 0 \leqq j \leqq n-1-m, \qquad (*)$$

$$0 \leqq m \leqq n-1.$$

By definition of $(y)$, $D^{0,j} = 0$ holds for $j = 0, 1, \cdots, n-1$. Assume $D^{i,j} = 0$ holds for some $i$, and for $j = 0, \cdots, n-1-i$. Then we can apply the differential operator $d^1$:

$$0 = d^1(D^{i,j}) = \sum_{\nu=0}^n [(i+1)d^{i+1}y_\nu d^j x_\nu + (j+1)d^i y_\nu d^{j+1} x_\nu]$$
$$= (i+1)D^{i+1,j} + (j+1)D^{i,j+1}$$

Hence we get $D^{i+1,j} = 0$ for $j = 0, \cdots, n-1-(i+1)$, and we can prove (*) by induction.

It remains to prove that $a^m(\check{f})$ is generically surjective, for $0 \leqq m \leqq n$; since we have surjections $\mathcal{P}_*^m \to \mathcal{P}_*^{m-1}$, it is enough to show that $a^n(\check{f})$ is generically surjective. Hence we may assume that $x \in X$ is an ordinary point (see §2), so that $w(x) = \det(w_x^n)$ is a unit in $A = \mathcal{O}_{X,x}$. Suppose we can prove the equality

$$w(x)w(y) = (-1)^{\frac{1}{2}n(n+1)} w(x)^{n+1} \prod_{i=0}^n \binom{n}{i}, \qquad (**)$$

where we have put $w(y) = \det(w_y^n)$. It would follow that $w(y)$ is a unit in $A$, hence $a^n(\check{f})$ is surjective at $x \in X$, as desired.

Consider the matrix $w_y^n \cdot (w_x^n)^t$. Since $D^{i,j} = 0$ holds for $0 \leqq i + j \leqq n-1$, this matrix is lower right triangular; its anti-diagonal entries are $D^{i,n-i}$. In order to prove (**), it thus suffices to show

$$D^{i,n-i} = (-1)\binom{n}{i} w(x). \qquad (***)_i$$

Using the definition of $(y)$ (and linear algebra) we see that $(***)_0$ holds. Assume $(***)_i$ holds and apply $d^1$ to $D^{i,n-1-i}$. From this we obtain $D^{i+1,n-(i+1)} = 0$ and the proof can be completed by induction.

Together with the results of §4, theorem (5.1) shows that from a curve $f: X \to P$ in $n$-space we obtain, by the method of projections, sections and dual curves, $n-1$ (and not more) plane curves $p_2^m(f)$ and their duals. Applying formulas (I) and (II) of §3 to these $2(n-1)$ plane curves we obtain the formulas of Veronese ([14]), if we (similarly to what he does) interpret the character $e$ of (II) for each curve $p_2^m(f)$, $p_2^m(f)^\vee$ in terms of (actual and apparent) double osculating spaces of the curve $f$. We will not discuss this interpretation here; let us only, as an application of (5.1), give the Cayley-Plücker formulas for a curve in 3-space ([2], p. 191).

Corollary (5.3). *Suppose* $f: X \to P = \mathbb{P}^3$ *is a curve such that* $a^3: V_X \to \mathcal{P}_X^3(\mathcal{L})$ *is generically surjective. Suppose* char $k \neq 2, 3$. *Let* $(r_0, r_1, r_2)$ *and* $(k_0, k_1, k_2)$ *denote the ranks and stationary indices of* $f$, *and let* $e_i$ *(resp.* $\check{e}_i$*) denote the degree of the jacobian ideal of the plane curve* $p_2^i(f)$ *(resp.* $p_2^i(f)^\vee$*), for* $i = 0, 1$. *There are formulas*

(1) $k_1 = 2r_1 - (r_0 + r_2) + 2g - 2$
(2) $k_0 = 2r_0 - r_1 + 2g - 2$
(3) $k_2 = 2r_2 - r_1 + 2g - 2$
(4) $e_1 - 3(r_0 + k_1) = (r_1 - 1)(r_1 - 6) + 2r_2 - 6g$
(5) $\check{e}_0 - 3(r_2 + k_1) = (r_1 - 1)(r_1 - 6) + 2r_0 - 6g$
(6) $e_0 - 3k_0 = (r_0 - 1)(r_0 - 6) + 2r_1 - 6g$
(7) $\check{e}_1 - 3k_2 = (r_2 - 1)(r_2 - 6) + 2r_1 - 6g$.

Example 1. The rational $n$-ic is selfdual: its dual curve is also a rational $n$-ic.

Example 2. $E_d \subset \mathbb{P}^3$. We have:

$$e_0 = d(d-3), \qquad \check{e}_0 = d(4d-3)$$
$$e_1 = d(4d-5), \qquad \check{e}_1 = d(9d-5).$$

EXAMPLE 3. $\Gamma_2 \subseteq \mathbb{P}^3$ is an elliptic curve of degree 4, with no cusps and no flexes. It has rank 8 and class 12. The number of apparent double points is $\frac{1}{2}e_0 = 2$, and the degree of the double curve of its tangent developable is $\frac{1}{2}(e_1 - 3(r_0 + k_1)) = 16$. The corresponding characters for the dual curve are $\frac{1}{2}(\check{e}_1 - 3k_2) = 38$ and $\frac{1}{2}(\check{e}_0 - 3(r_2 + k_1)) = 8$.

## §6. Appendix. The functor of principal parts

Let $f: X \to S$ be a morphism of schemes. For each integer $m \geqq 0$, consider the functor of relative principal parts (or jets) of order $m$

$$\mathcal{P}^m_{X/S} = p^{(m)}_{1*} p^{(m)*}_2 : (\mathcal{O}_X - \text{mod}) \to (\mathcal{O}_X - \text{mod}).$$

Here $p_i^{(m)}$ denotes the composition of the $i$th projection $p_i : X \times_S X \to X$ with the $m$th infinitesimal neighborhood $h_m : X^{(m)} \to X \times_S X$ of the diagonal ([7], 16.7.1).

Let $\rho(f): 1 \to f_* f^*$ and $\sigma(f): f^* f_* \to 1$ denote the canonical natural transformations. There is a natural transformation

$$p_{1*}\rho(h_m)p_2^* : p_{1*}p_2^* \to p_{1*}h_{m*}h_m^* p_2^* \cong \mathcal{P}^m_{X/S}.$$

Define $\alpha^m = \alpha^m(f): f^* f_* \to \mathcal{P}^m_{X/S}$ to be the natural transformation obtained by composing the one above with the base change transformation

$$(f_* \rho(p_2))^\# : f^* f_* \to p_{1*} p_2^*$$

(where $\#$ denotes the isomorphism $\operatorname{Hom}_S(-, f_* -) \rightrightarrows \operatorname{Hom}_X(f^* -, -)$). Define also a natural transformation of functors from ($\mathcal{O}_S$-mod) to ($\mathcal{O}_X$-mod)

$$\beta^m = \beta^m(f): f^* \to \mathcal{P}^m_{X/S} \circ f^*$$

by composing $\rho(p_1^{(m)})f^* : f^* \to p^{(m)}_{1*} p_1^{(m)*} f^*$ with the isomorphism $p^{(m)}_{1*} p_1^{(m)*} f^* \cong p^{(m)}_{1*} p_2^{(m)*} f^*$ (which exists since we have $f \circ p_1^{(m)} = f \circ p_2^{(m)}$).

Note that

$$\beta^m(\mathcal{O}_S): \mathcal{O}_X \to \mathcal{P}^m_{X/S}(\mathcal{O}_X),$$

is the homomorphism which defines the (left) $\mathcal{O}_X$-algebra structure on $\mathcal{P}^m_{X/S}(\mathcal{O}_X)$.

PROPOSITION (6.1). *There is a factorization* $\alpha^m = (\mathcal{P}^m_{X/S}\sigma(f)) \circ (\beta^m f_*)$.

PROOF. By the adjoint property ([6]$0_I$, 3.5.3.4) we know that $\alpha^m$ factors through $\beta^m f_*$,

$$\alpha^m = (p^{(m)}_{1*} \alpha^{m\#}) \circ (\beta^m f_*)$$

(here $\#$ denotes the isomorphism $\operatorname{Hom}_X(-, p^{(m)}_{1*} -) \rightrightarrows \operatorname{Hom}_{X^{(m)}}(p^{(m)*} -, -)$). Thus it suffices to show that we have $\mathcal{P}^m_{X/S}\sigma(f) = p^{(m)}_{1*}\alpha^{m\#}$, or, equivalently, that $\alpha^{m\#} = p_2^{(m)*}\sigma(f)$. The proof of this equality is straightforward.

The following proposition lists various properties of $\alpha^m$. The proofs are straightforward (using the naturality of the transformations involved) and will therefore be omitted.

PROPOSITION (6.2).

(i) *We have* $\alpha^0 = \sigma : f^*f_* \to 1$.

(ii) *If* $m' \leqq m$, *the canonical diagram*

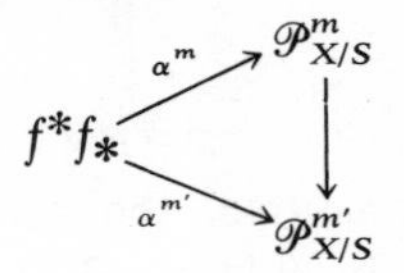


*commutes.*

(iii) *If* $g: Y \to X$ *is an* $S$-*morphism, there is a canonical commutative diagram*

$$\begin{array}{ccc} g^*f^*f_* & \xrightarrow{g^*\alpha^m} & g^*\mathcal{P}^m_{X/S} \\ \downarrow{\scriptstyle g^*f^*f_*\rho(g)} & & \downarrow{\scriptstyle u_g} \\ g^*f^*f_*g_*g^* & \xrightarrow{\alpha^m g^*} & \mathcal{P}^m_{Y/S}g^*. \end{array}$$

(iv) *Let* $\delta^{m,m'} : \mathcal{P}^{m+m'}_{X/S} \to \mathcal{P}^{m'}_{X/S}\mathcal{P}^m_{X/S}$ *denote the natural transformation defined in* ([7] 16.8.8.2). *The following diagram commutes*

$$\begin{array}{ccc} f^*f_* & \xrightarrow{\beta^{m'}f_*} & \mathcal{P}^{m'}_{X/S}f^*f_* \\ \downarrow{\scriptstyle \beta^{m+m'}f_*} & & \downarrow{\scriptstyle \mathcal{P}^{m'}_{X/S}\beta_m} \\ \mathcal{P}^{m+m'}_{X/S}f^*f_* & \xrightarrow{\delta^{m,m'}f^*f_*} & \mathcal{P}^{m'}_{X/S}\mathcal{P}^m_{X/S}f^*f_* \\ \downarrow{\scriptstyle \mathcal{P}^{m+m'}_{X/S}\alpha^0} & & \downarrow{\scriptstyle \mathcal{P}^{m'}_{X/S}\mathcal{P}^m_{X/S}\alpha^0} \\ \mathcal{P}^{m+m'}_{X/S} & \xrightarrow{\delta^{m,m'}} & \mathcal{P}^{m'}_{X/S}\mathcal{P}^m_{X/S}. \end{array}$$

PROPOSITION (6.3). *Let* $\mathscr{E}$ *be a bundle on* $S$. *Put* $P = \mathbb{P}(\mathscr{E})$ *and* $\mathscr{L} = \mathcal{O}_P(1)$. *The map* $\alpha^1(\mathscr{L}) : \mathscr{E}_P \to \mathcal{P}^1_P(\mathscr{L})$ *is an isomorphism.*

PROOF. It suffices to show that $\alpha^1(\mathscr{L})$ is surjective (since it is a map between bundles of the same rank). The question is local on $S$, so we may assume $S = \operatorname{Spec}(A)$ and $E = \bigoplus_{i=0}^n Ae_i$. Write $\operatorname{Sym}(E) = A[T_0, \cdots, T_n]$ and let $U \subseteq P$ denote the open affine scheme defined by $T_j \neq 0$. Then $B = A[T_0/T_j, \cdots, T_n/T_j]$ is the ring of $U$. The map $\alpha^0(\mathscr{L}) : \oplus Be_i \to BT_j$ is the map which sends $e_i$ to $(T_i/T_j)T_j$. Therefore it follows from the factorization $\alpha^1(\mathscr{L}) = \mathcal{P}^1_P(\alpha^0(\mathscr{L})) \circ \beta^1(f_*\mathscr{L})$ (6.1) that $\alpha^1(\mathscr{L}) : \oplus Be_i \to \mathcal{P}^1_P(\mathscr{L}) \mid U = (B \otimes B)/I^2 \otimes BT_j$ is the map which sends $e_i$ to $1 \otimes (T_i/T_j)T_j$ (here we have put $I=\{b\otimes 1-1\otimes b\}B\otimes B)$. Since these elements generate $(B\otimes B)/I^2\otimes BT_j$, we have shown that $\alpha^1(\mathscr{L})$ is surjective.

REMARK (6.4). Put $\mathscr{K}=\operatorname{Ker}(\alpha^1(\mathscr{L}))$. The lemma shows that there is a canonical isomorphism of exact sequences

$$\begin{array}{ccccccccc} 0 & \longrightarrow & \mathscr{K} & \longrightarrow & \mathscr{E}_P & \longrightarrow & \mathscr{L} & \longrightarrow & 0 \\ & & \alpha\downarrow\wr & & \alpha^1\downarrow\wr & & \| & & \\ 0 & \longrightarrow & \Omega_P(\mathscr{L}) & \longrightarrow & \mathscr{P}^1_P(\mathscr{L}) & \xrightarrow{v} & \mathscr{L} & \longrightarrow & 0. \end{array}$$

The isomorphism $\alpha$ is, of course, the canonical isomorphism called the second fundamental form in ([1], I, 3.1). (The isomorphism $\alpha$ can be described as follows: We see that $\mathscr{K}\mid U$ is the free B-module generated by $\{e_i'\}_{i\neq j}$, where $e_i'=(T_i/T_j)e_j-e_i$. Hence

$$\begin{aligned} \alpha(e_i') &= \alpha^1(\mathscr{L})((T_i/T_j)e_j-e_i) \\ &= (T_i/T_j\otimes 1)(1\otimes T_j)-(1\otimes (T_i/T_j)T_j) \\ &= (T_i/T_j\otimes 1-1\otimes T_i/T_j)(1\otimes T_j) \qquad ([1], \text{I}, 3.1). \end{aligned}$$

Note that the map $v\mid U:(B\otimes B)/I^2\otimes BT_j\to BT_j$ is given by

$$v((1\otimes (T_i/T_j))(1\otimes T_j))=v(1\otimes (T_i/T_j)T_j)=(T_i/T_j)T_j.)$$

We shall use $\alpha^1(\mathscr{L})$ to identify $\mathscr{P}^1_P(\mathscr{L})$ (as a left $\mathscr{O}_P$-module) with $\mathscr{E}_P$, and we will identify $\Omega_P(\mathscr{L})$ with $\mathscr{K}$ via $\alpha$.

## BIBLIOGRAPHY


[1] A. ALTMAN, S. L. KLEIMAN, *Introduction to Grothendieck duality theory*, Lecture Notes in Mathematics, Springer-Verlag, 1970.

[2] H. F. BAKER, *Principles of Geometry*, Vol. V, Analytical principles of the theory of curves. Cambridge Univ. Press, 1933.

[3] A. CAYLEY, 'Mémoire sur les courbes à double courbure et les surfaces développables', J. de Math. Pures et Appliquées, (Liouville), t. X (1845), 245–250; Cayley, Papers, I, p. 207.

[4] R. H. DYE, 'Osculating primes to curves of intersection in 4-space, and to certain curves in $n$-space', Proc. Edinburgh Math. Soc. (II), 18 (1973), 325–338.

[5] R. H. DYE, 'The hyperosculating spaces to certain curves in $[n]$', Proc. Edinburgh Math. Soc. (2), 19 (1974/75), 301–309.

[6] A. GROTHENDIECK, J. A. DIEUDONNÉ, *Éléments de Géométrie Algébrique*, I, Springer-Verlag, 1971.

[7] ———, IV, Publ. Math. IHES, no. 32.

[8] P. GRIFFITHS, 'On Cartan's method of Lie groups and moving frames as applied to uniqueness and existence questions in differential geometry'. Duke Math. J., 41 (1974), 775–814.

[9] A. KATO, 'Singularities of projective embedding (point of order $n$ on an elliptic curve)'. Nagoya Math. J., Vol. 45 (1971), 97–107.

[10] S. L. KLEIMAN, 'Geometry on Grassmannians and applications to splitting bundles and smoothing cycles'. Publ. Math. IHES, no. 36 (1969), 291–297.

[11] ———, 'The transversality of a general translate'. Comp. Math., Vol. 28 (1974), 287–297.

[12] R. PIENE, 'Plücker formulas', Ph.D. Thesis, M.I.T., 1976.

[13] W. F. POHL, 'Differential geometry of higher order', Topology 1 (1962), 169–211.

[14] G. VERONESE, 'Behandlung der projectivischen Verhältnisse der Räume von verschiedenen Dimensionen durch das Princip des Projicirens und Schneidens'. Math. Annal. XIX (1882), 161–234.

[15] H. WEYL, J. WEYL, *Meromorphic functions and analytic curves.* Ann. Math. Stud. no. 12, Princeton Univ. Press, 1943.

RAGNI PIENE
Massachusetts Institute of Technology
Department of Mathematics
Cambridge, Mass. 02139, USA

Nordic Summer School/NAVF
Symposium in Mathematics
Oslo, August 5–25, 1976

# PERIODICITIES IN ARNOLD'S LISTS OF SINGULARITIES

Dirk Siersma

**Abstract**

V. I. Arnold discovered experimentally periodicities in the classification of singularities. These periodicities are explained for functions from $\boldsymbol{C}^2$ to $\boldsymbol{C}$, using the blowing up construction.

Moreover the singularities of multiplicity 5 are classified.

In his paper 'Local forms of functions' Arnold [2] gives a list of normal forms of functions in the neighborhood of critical points (the classification of all singularities with number of modules $m = 0, 1, 2$ or with multiplicity $\mu \leqq 16$ included). In his introduction he mentions a *periodicity* in the decomposition of singularities into $\mu$-equivalence classes. According to Arnold the phenomenon of periodicity is only partially explained and for quasi-homogeneous singularities only. The explanation is based upon some root technique for the quasihomogeneous Lie algebra, related to work of Enriques and Demazure [6].

The aim of this paper is to explain the periodicity for all isolated singularities of corank $\leqq 2$ and some singularities of corank 3 (including all singularities in Arnold's lists), using the theory of resolutions. We also give a list of singularities of corank 2 with multiplicity equal to 5.

This paper is an elaborated version of the lecture I gave at the Institut des Hautes Études Scientifiques (Bûres-sur-Yvette, France) in May 1975 about Arnold's paper: 'Local forms of functions'.

I thank the I.H.E.S. for their hospitality.

*Added in proof:*

Arnold communicated to me that:

1° Some of his students have also classified the singularities of corank 2 with multiplicity 5 (unpublished).

2° His remark about periodicities also concerns the actual computations that occur.

AMS(MOS) subject classification scheme (1970): 32 C 40, 58 C 25

## Contents



## 1. Introduction

For results and definitions, mentioned in this introduction see Arnold [2] and Siersma [10].

(1.1) The group $\mathscr{D}$ of germs (or jets) of biholomorphic mappings $(\boldsymbol{C}^n, O) \to (\boldsymbol{C}^n, O)$ acts on $\mathscr{E}$, the set of germs at $O$ (or jets) of holomorphic functions $\boldsymbol{C}^n \to \boldsymbol{C}$ by right-multiplication. A *singularity class* is a subset of $\mathscr{E}$, invariant under this action. Each orbit is a singularity class. Two germs (or jets) belonging to the same orbit are called *right-equivalent.*

(1.2) Let $m = \{f \in \mathscr{E} \mid f(O) = 0\}$ and let $\Delta(f) = (\partial_1 f, \cdots, \partial_n f)$. The codimension of $f \in m^2$ is defined by $\operatorname{codim}(f) = \dim_{\boldsymbol{C}} (m^2 / m\Delta(f))$ and the Milnor number of $f$ is defined by $\mu(f) = \dim_{\boldsymbol{C}} (\mathscr{E}/\Delta(f))$. These are related by: $\mu(f) = 1 + \dim (m/\Delta(f)) = 1 + \operatorname{codim}(f)$.

(1.3) A germ $f \in m$ is called *$k$-determined* (or $k$-sufficient) if for any $g \in m : j^k(f) = j^k(g) \Rightarrow f$ is right-equivalent with $g$. It is well-known that the following are equivalent:
1° $\mu(f) < \infty$.
2° $O$ is an isolated critical point of $f$.
3° there exists $k \in \boldsymbol{N}$ such that $f$ is $k$-determined.

(1.4) Two germs $f$ and $g$ are called:
(1) *Right-left-equivalent* if there is a $\phi \in \mathscr{D}_n$ and a $\psi \in \mathscr{D}_1$ such that $\psi g = f\phi$.
(2) *Contact-equivalent* if there is a $\phi \in \mathscr{D}_n$ and for every $x$ near $O$ a $\psi_x \in \mathscr{D}_1$ (analytically depending on $x$) such that $\psi_x g = f\phi$.

(3) *Topological-equivalent* if there exist homeomorphisms $\phi: \boldsymbol{C}^n \to \boldsymbol{C}^n$ and $\psi: \boldsymbol{C} \to \boldsymbol{C}$ such that $\psi g = f\phi$.
(4) $\mu$-*equivalent* if there exists a family $f_t$ with $\mu(f_t) = \mu$ (constant) and $f_0 = f$ and $f_1 = g$.
If $n \neq 3$: $\mu$-equivalence$\Leftrightarrow$top. equivalence (cf. Lê Dũng Tráng and Ramanujam [8] and Teissier [12]).
A $\mu$-equivalence class will also be called $\mu$-*class*.

REMARK. A $\mu$-class can consist of several orbits; examples are:

$$x^4 + tx^2y^2 + y^4$$
$$x^5 + y^5 + ux^2y^3 + vx^3y^2 + wx^3y^3$$
$$x^3 + y^3 + z^3 + uxyz.$$

(1.5) The *modality* $m(f)$ of $f \in \mathscr{m}$ is the smallest number $k$ such that some neighborhood of (a sufficient $k$-jet of) $f$ at $O$ is covered by a finite number of no more than $m$-parametrized families of orbits of $\mathscr{D}$ in $\mathscr{m}$.

Another characterization of modality is as follows.

Let $(\phi_1, \cdots, \phi_{\mu-1})$ be a basis of $\mathscr{m}/\Delta(f)$ and let $F_u(x) = f(x) + \Sigma u_i\phi_i$ (versal deformation).

Define $S = \{\mu \in \boldsymbol{C}^{\mu-1} \mid F_u$ is $\mu$-equivalent with $f\}$.

Gabrielov [7] proved that the modality of $f$ is equal to the dimension of $S$. So $m(f) = \mu(f) - c(f) - 1$ where $c(f) = \text{codim}_f \{g \in \mathscr{m} \mid \mu(g) = \mu(f)\}$.

(1.6) SPLITTING LEMMA. *Let* $f \in \mathscr{m}^2$ *then:* $f(x_1, \cdots, x_n) \sim g(x_1, \cdots, x_\rho) + Q(x_{\rho+1}, \cdots, x_n)$ *with* $g \in \mathscr{m}^3$ *and* $Q$ *a non-degenerate quadratic form. (Here* $\sim$ *means Right-equivalent.)*

The number $\rho$ is called the *corank* of $f$.

Codimension, sufficiency, and modality of $f$ are equal to those of $g$. Moreover the classification of $f$ follows from the classification of $g$ by adding $Q(x_{\rho+1}, \cdots, x_n)$.

(1.7) Arnold [2] classified in 105 theorems:
1° all singularities with Milnor-number $\mu(f) \leqq 16$.
2° all singularities with modality $m(f) \leqq 2$.

The classification of singularities follows:
1° increasing corank of $f$.
2° increasing multiplicity $\nu$ of $f$; ($f$ has multiplicity $\nu$ if $f \in \mathscr{m}^\nu \backslash \mathscr{m}^{\nu+1}$).
3° different factorizations of $j^\nu(f)$ (decreasing the number of factors).
In this way one finds (see also Siersma [10], [11]).

CORANK 0,1. type $A_n: f = x^{n+1} (n \geqq 0)$.

CORANK 2:

(a) multiplicity $3 \rightarrow j^2(f) \equiv 0$, $j^3(f) \not\equiv 0$. Factorizations of $j^3(f)$:

(a1) three linear factors

$x^2y + y^3$

(a2) two linear factors

$x^2y$

} $f = x^2y + y^k$ type $D_k$ $(k \geqq 4)$

(a3) one linear factor

$x^3$

$f = x^3 + y^4 (E_6)$
$f = x^3 + xy^3 (E_7)$
$f = x^3 + y^5 (E_8)$
$f = x^3 + \alpha xy^4 + \beta y^6 (J_{10})$, etc.

(b) multiplicity 4: 5 different factorizations, etc.

If $\mu$ increases the classification becomes more complicated. To work more systematically and to reduce some of the computations we propose, in the case of corank 2, the use of the blowing up construction.

## 2. The blowing up construction

We consider $f: \boldsymbol{C}^2 \rightarrow \boldsymbol{C}$, a holomorphic mapping.

(2.1) Replace $O \in \boldsymbol{C}^2$ by the set of all its tangent-directions (isomorphic to $\boldsymbol{P}^1(\boldsymbol{C})$). We get a manifold $M$ that can be covered by two charts $(x_1, y_1)$ and $(x_2, y_2)$, together with a projection $\pi: M \rightarrow \boldsymbol{C}^2$.
The charts and the projection are related by the formulae:

$$\pi(x_1, y_1) = (x_1 y_1, y_1) = (x, y)$$
$$\pi(x_2, y_2) = (x_2, x_2 y_2) = (x, y).$$

$C = \pi^{-1}(O) = \boldsymbol{P}^1(\boldsymbol{C})$ is given by $y_1 = 0 \vee x_2 = 0$ and is called the *exceptional divisor.*
$f: \boldsymbol{C}^2 \rightarrow \boldsymbol{C}$ extends in a natural way to a map $\bar{f}: M \rightarrow \boldsymbol{C}$:

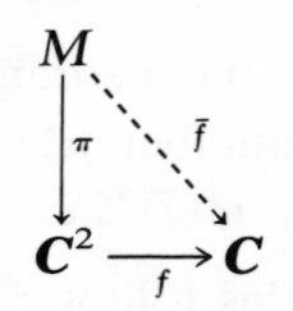


If $f$ has multiplicity $k$ we have:

$$\bar{f}(x_1, y_1) = f(x_1 y_1, y_1) = y_1{}^k g_1(x_1, y_1)$$

and

$$\bar{f}(x_2, y_2) = f(x_2, x_2 y_2) = x_2{}^k g_2(x_2, y_2)$$

for certain $g_1$ and $g_2$ defined in a neighborhood of $C$.

(2.2) The set $\bar{f}^{-1}(O)$ consists of $C$ and some other branches, intersecting $C$ in points $w_1, \cdots, w_s$, corresponding to the tangent directions of $f^{-1}(O)$ in $O$. Those branches are given by $g_1(x_1, y_1)=0$ or $g_2(x_2, y_2)=0$.

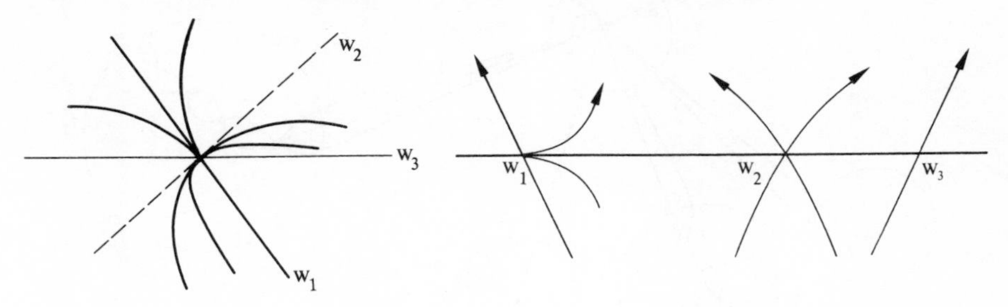


In the neighborhoods of the points $w_1, \cdots, w_s$ we can choose local coordinates $(\xi, \eta)$ such that $g_i(\xi, \eta)=\xi^n+\eta q(\xi, \eta)(i=1,2)$ where $C$ is given by $\eta=0$.

If an intersection point is covered by 2 charts then $g_2(x_2, y_2)= {x_1}^k g_1(x_1, y_1)$, so $g_1$ and $g_2$ are $\mu$-equivalent near that intersection point.

(2.3) In the following paragraphs we shall obtain the following *results*:
1. The classification of $\mu$-classes of $f$ can be done by local investigations around every intersection point.
2. Different intersection points are treated independently.
3. In each intersection point there is a periodicity in the classification of $f$.
First we compare the Milnor number of $f$ with the Milnor numbers at the intersection points. The following proposition is due to Pham [9].

(2.4) PROPOSITION. *Let $f$ have an isolated critical point at $O \in \boldsymbol{C}^2$. Let $f^{-1}(O)$ have $s$ different tangent directions $w_1, \cdots, w_s$ in $O \in \boldsymbol{C}^2$. Let $M$ be constructed from $\boldsymbol{C}^2$ by blowing up $O$ and let $\bar{f}: M \to \boldsymbol{C}$ be the natural extension of $f$. Denote by $f_i$ $(i=1, \cdots, s)$ the restriction of $g_1$ or $g_2$ (defined above) to neighborhoods of the points $w_1, \cdots, w_s \in M$.*
*Then*: $\mu(f)=r_0(r_0-1)+\sum_{i=1}^{s} \mu(f_i)-s+1$ *where $r_0$ is the multiplicity of $f$.*

PROOF. We consider the real morsification $\tilde{f}$ of $f$, as constructed by A'Campo [1]. Let $\tilde{C}$ be the set of zeros of $\tilde{f}$, intersected by a small real 2-disc $D_\varepsilon$. The curve $\tilde{C}$ has as its only singularities multiple normal crossings of branches. Let $p_1, \cdots, p_l$ be the points of crossing of $\tilde{C}$, and $r_k$ the number of branches of $\tilde{C}$ around the points $p_k$.

A'Campo proved: $\mu=\sum_{k=1}^{l} r_k(r_k-1)-q+1$ where $q$ is the number of branches of $f$.

The morsifications of $f_i$ $(i=1, \cdots, s)$ are constructed from $\tilde{f}$ by restricting

$\tilde{f}$, as shown in the following figure:

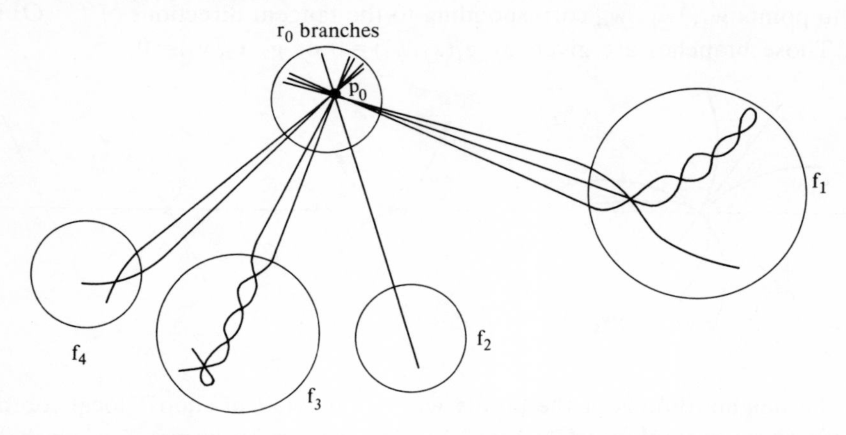


In this way the set $\{p_k \mid k=0,\cdots,l\}$ is subdivided. We write now $p_{i,j}$ and $r_{i,j}$ in the obvious way. Let $p_0$ and $r_0$ correspond to the exceptional divisor $C$. Let $q_i$ the number of branches of $f_i$.

Then:

$$\begin{aligned}\mu(f) &= \sum_{k=0}^{l} r_k(r_k-1)-q+1 = r_0(r_0-1)+\sum_{i=1}^{s}\sum_j r_{i,j}(r_{i,j}-1)-q+1\\ &= r_0(r_0-1)+\sum_{i=1}^{s}\Big(\sum_j r_{i,j}(r_{i,j}-1)-q_i+1\Big)+\sum_{i=1}^{s}(q_i-1)-q+1\\ &= r_0(r_0-1)+\sum_{i=1}^{s}\mu(f_i)-s+1.\end{aligned}$$

## 3. Periodicity of $\mu$-classes in $x^n+m^{n+1}$

(3.1) Let $f(x,y)=x^n+p(x,y)$ with multiplicity $\nu(p)\geqq n+1$. We blow up $O\in \boldsymbol{C}^2$.

Then:

$$\bar{f}(x,y)=f(x_1y_1,y_1)=x_1^ny_1^n+p(x_1y_1,y_1)=y_1^n[x_1^n+y_1q_1(x_1,y_1)]$$
$$\bar{f}(x,y)=f(x_2,x_2y_2)=x_2^n+p(x_2,x_2y_2)=x_2^n[1+x_2q_2(x_2,y_2)]$$

where $q_1(x_1,y_1)=y_1^{-n-1}p(x_1y_1,y_1)$ and $q_2(x_2,y_2)=x_2^{-n-1}p(x_2,x_2y_2)$.
We consider $x_1^n+y_1q_1(x_1,y_1)=0$ and $1+x_2q_2(x_2,y_2)=0$.
The solutions have only one intersection point with the exceptional divisor, namely $(x_1,y_1)=(0,0)$. We next consider the map-germ:
$f_1(x_1,y_1)=y_1^{-n}f(x_1y_1,y_1)=x_1^n+y_1q_1(x_1,y_1)$ in a neighborhood of $(x_1,y_1)=(0,0)$.

(3.2) In order to classify $f$ up to $R$-equivalence it now seems enough to classify $f_1$ up to $R$-equivalence. However we have to take into account the following:

1° Diffeomorphisms of $\boldsymbol{C}^2$ lift only to diffeomorphisms of $M$ with a special form.

2° The polynomial $q_1(x_1, y_1)$ is not arbitrary since $q_1(x_1, y_1) = y_1^{-n-1} p(x_1 y_1, y_1)$.

Since $\mu(f) = n(n-1) + \mu(f_1)$, we see that $\mu(f)$ depends only on $\mu(f_1)$.

Moreover the $\mu$-class of $f$ follows from the configuration of its resolution. If we apply diffeomorphisms to $f_1$ such that this resolution is still the same, then we stay in the same $\mu$-class.

The resolution of $f$ depends only on the different possibilities for tangencies of $f_1^{-1}(O)$ to the exceptional divisor $C$. This can give a subdivision of every $\mu$-class of $f_1$, each subclass giving a $\mu$-class for $f$.

(3.3) We now return to $f_1(x_1, y_1) = x_1^n + y_1 q_1(x_1, y_1)$.

1° Let the multiplicity $\nu(y_1 q_1) \leqq n$.

In this case a detailed study is necessary to find the possibilities for $f_1$ and the corresponding classes for $f$. In some sense the singularity $f_1$ is less complicated than $f$ and is already treated in an earlier part of the classification.

As an example consider $n = 3: x_1^3 + y_1 q_1(x_1, y_1)$.

This singularity must be of type $A$ or $D$ if $\nu(y_1 q_1) \leqq 3$.

A detailed study (cf. §5) gives just the following possibilities for $f_1$:

$$\underset{\bullet}{A_0}\ \underset{\bullet}{A_1}\ \underset{\bullet}{A_2}\quad \times\quad \underset{\bullet}{D_4}\ \underset{\bullet}{D_5}\ \underset{\bullet}{D_6}\ \underset{\bullet}{D_7}\ \underset{\bullet}{D_8}\qquad \underset{\bullet}{D_k}$$

and the following corresponding classes for $f$:

$$\underset{\bullet}{E_6}\ \underset{\bullet}{E_7}\ \underset{\bullet}{E_8}\quad \times\quad \underset{\bullet}{J_{2.0}}\ \underset{\bullet}{J_{2.1}}\ \underset{\bullet}{J_{2.2}}\ \underset{\bullet}{J_{2.3}}\ \underset{\bullet}{J_{2.4}}\qquad \underset{\bullet}{J_{2,k-4}}$$

(Here and in the following we use the same notations as in Arnold [2]).

2° Let the multiplicity $\nu(y_1 q_1) \geqq n+1$.

We now have more or less the same situation as before with $f(x, y) = x^n + p(x, y)$. So we blow up a second time and get: $f_1^1(x_1^1, y_1^1) = (x_1^1)^3 + y_1^1 q_1^1(x_1^1, y_1^1)$. Now we can omit the detailed study, mentioned above, because of the following lemma:

(3.4) Periodicity Lemma. *There is a* 1-1-*correspondence between* $\mu$-*classes of* $f$, *defined by* $f_1$ *and* $\mu$-*classes of* $f$ *defined by* $f_1^1$.

Proof. Let $f(x, y) = x^n + p(x, y)$ with $\nu(p) \geqq 4$.

1° Arrange the normal form:

$$\begin{aligned} f(x, y) &= x^n + x^{n-2} A_{n-2}(y) + \cdots + x y^n A_1(y) + y^{n+1} A_0(y) \\ &= x^n + x^{n-2} y^3 \sum_{k=0}^{\infty} a_{n-2,k} y^k + \cdots + x y^n \sum_{k=0}^{\infty} a_{1,k} y^k + y^{n+1} \sum_{k=0}^{\infty} a_{0,k} y^k. \end{aligned}$$

2° After blowing up:

$$f_1(x_1, y_1) = x_1^n + x_1^{n-2} y_1 \sum_{k=0}^{\infty} a_{n-2,k} y_1^k + \cdots + x_1 y_1 \sum_{k=0}^{\infty} a_{1,k} y_1^k + \\ + y_1 \sum_{k=0}^{\infty} a_{0,k} y_1^k.$$

So

$$\nu(y_1 q_1) \geqq n+1 \Leftrightarrow \begin{cases} a_{n-2,0} = a_{n-2,1} = 0 \\ a_{n-3,0} = a_{n-3,1} = a_{n-3,2} = 0 \\ a_{0,0} = \cdots\cdots\cdots = a_{0,n-1} = 0 \end{cases}$$

If $\nu(y_1 q_1) \geqq n+1$ then:

$$f_1(x_1, y_1) =$$

$$x_1^n + x_1^{n-2} y_1^3 \sum_{k=0}^{\infty} a_{n-2,k+2} y_1^k + \cdots + x_1 y_1^n \sum_{k=0}^{\infty} a_{1,n-1+k} y_1^k + y_1^{n+1} \sum_{k=0}^{\infty} a_{0,n+k} y_1^k.$$

The normal forms of $f$ and $f_1$ differ only by a shift of indices

$$\begin{cases} a_{n-2,k+2} \rightarrow a_{n-2,k} \\ \quad \vdots \\ a_{1,n-1+k} \rightarrow a_{1,k} \\ a_{0,n+k} \quad \rightarrow a_{0,k} \end{cases}$$

So if we blow up our $f$ a second time we have to consider for $f_1^1(x_1^1, y_1^1)$ the same set of germs as before with $f_1(x_1, y_1)$. However the induced action of diffeomorphisms has become even more complicated. But also in this case the $\mu$-class of $f$ depends only on the possibilities for tangencies of $(f_1^1)^{-1}(O)$ to the exceptional divisors $C$ and $C'$:

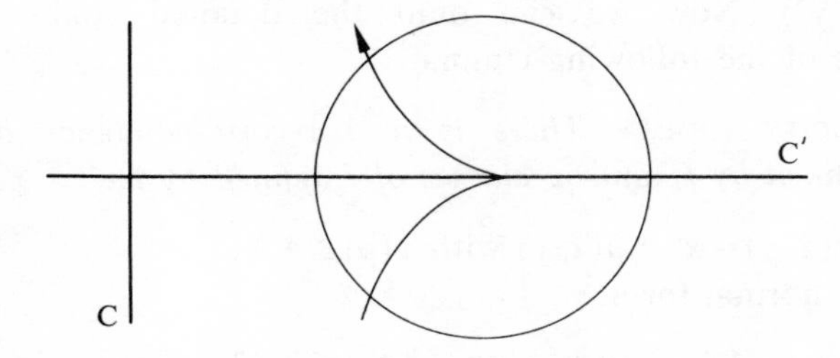


Since $(f_1^1)^{-1}(O)$ and $C$ intersect $C'$ in different points, every $\mu$-class of $f_1^1$ is subdivided in the same classes as before, when we blow up once.

(3.5) From this lemma follows the periodicity in the classification of $\mu$-classes in $x^n + m^{n+1}$. In the case $n=3$ we get the following pattern:

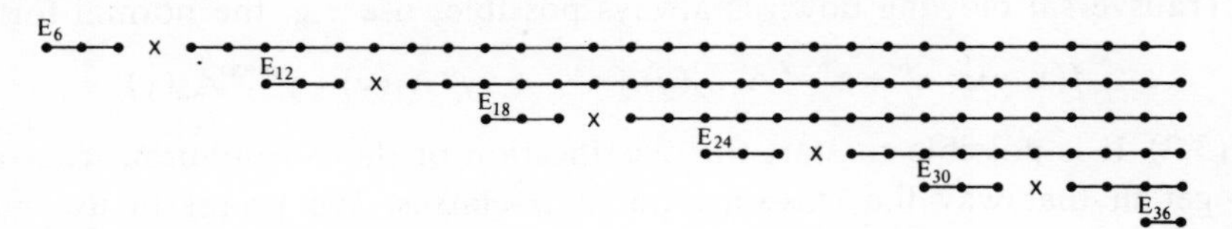


(3.6) Following Zariski's definition of equisingularity, all information about the topological type is contained in the 'resolution-tree', which can be constructed as follows:

Write down the multiplicities $(m_1, \cdots, m_s)$ of the irreducible components of $f^{-1}(O)$, blow up once and do the same for the multiplicities of each $f_i$. Write this as follows:

$$(m_1, \cdots, m_s)$$

$$(m_1^1, \cdots, m_{s_1}^1)(m_1^2, \cdots, m_{s_2}^2) \cdots (\cdots)$$

Repeat this until you get everywhere one branch of multiplicity one.

EXAMPLES

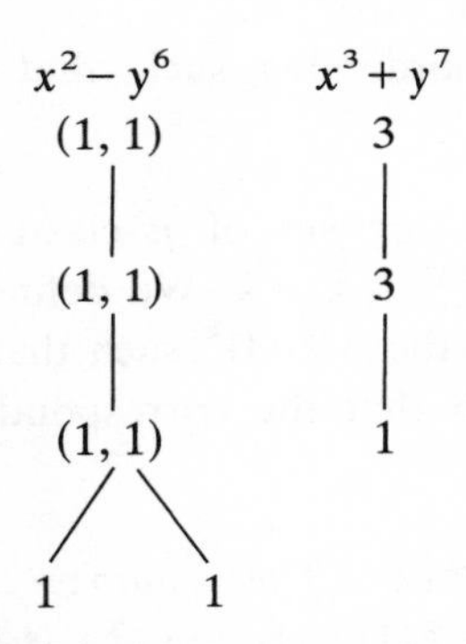


This construction and the following alternative proof of the periodicity-lemma were communicated to me by P. Slodowy.

PROOF. There is a bijective map between trees of class I and trees of class II:

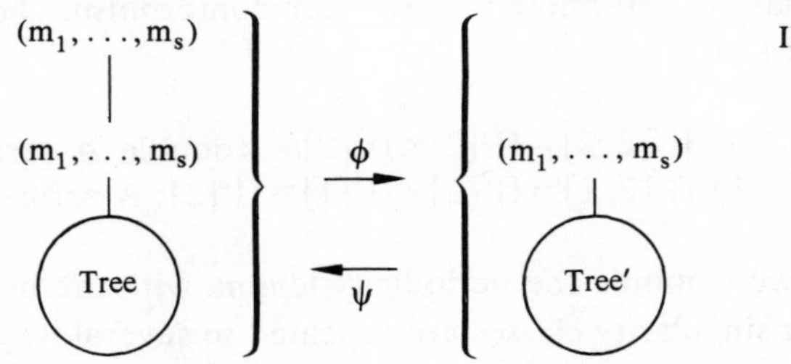

$\phi$ is defined by blowing up and $\psi$ by transversal blowing down. The multiplicities of the new branches are the same as before.

Transversal blowing down is always possible; use e.g. the normal form:

$$f(x, y) = x^n + x^{n-2}A_{n-2}(y) + \cdots + xy^n A_1(y) + y^{n+1}A_0(y)$$

(3.7) It is possible to start the classification of those resolution-trees and to get in that way the classification of $\mu$-classes. We prefer to use map-germs and to get the classification as a chain of semi-algebraic sets in some jet-space, since it is then possible to compare the modality of $f$ and $f_1$ (cf. (5.6)).

## 4. Independence

(4.1) Let $f$ have multiplicity $k$ and let $f$ have $s$ different tangent directions. The $k$-jet of $f$ can be given by a homogeneous polynomial $j^k(f)$ of degree $k$, which factors into linear forms as follows:

$$j^k(f) = (\alpha_1 x + \beta_1 y)^{k_1} \cdot \cdots \cdot (\alpha_s x + \beta_s y)^{k_s}.$$

From Hensel's lemma it follows that we can factor $f$ (at least over $\mathbf{C}[[x_1, x_2]]$): $f = f_1 \cdot \cdots \cdot f_s$ where each $f_i$ is given by $f_i(x, y) = (\alpha_i x + \beta_i y)^{k_i} + p(x, y)$, where $\nu(p) > k_i$.

So we can change coordinates such that $f_i(\xi, \eta) = \xi^{k_i} + p(\xi, \eta)$ with $\nu(p) > k_i$.

Definitions. Let $\Pi^k$ be the set of $\mu$-classes with multiplicity $k$. Let $k_1 \geqq k_2 \geqq \cdots \geqq k_s \geqq 1$ with $\sum_{i=1}^s k_i = k$. We define $\Pi^k[k_1, \cdots, k_s]$ to be the subset of $\Pi^k$, consisting of those $f \in \Pi^k$ such that $f$ has $s$ different tangent-directions $w_1, \cdots, w_s$ such that the corresponding $f_1, \cdots, f_s$ have multiplicities $k_1, \cdots, k_s$.

(4.2) Independence-Lemma. *The map* $\Psi : \Pi^k[k_1, \cdots, k_s] \to \Pi^{k_1}[k_1] \times \cdots \times \Pi^{k_s}[k_s]/\sim$ *defined by* $\Psi(f) = (f_1, \cdots, f_s)$ *is an isomorphism. Here*:

$$(f_1, \cdots, f_i, \cdots, f_j, \cdots, f_s) \sim (f_1, \cdots, f_j, \cdots, f_i, \cdots, f_s) \Leftrightarrow k_i = k_j.$$

Proof. The surjectivity of $\Psi$ is clear. The injectivity follows from the fact that a $\mu$-class is determined by the homeomorphism class of its resolution (cf. Zariski [13]).

Example. (a) $\Pi^4[2, 2] = \Pi^2[2] \times \Pi^2[2]/\sim$ : double $A$-series.
(b) $\Pi^3[2, 1] = \Pi^2[2] \times \Pi^1[1] \cong \Pi^2[2]$ : $A$-series.

(4.3) If we combine the periodicity lemma with the independence lemma we see that singularity classes are repeated in several ways. Moreover in the detailed study of the 'period' in $\Pi^n[n]$ we meet a 1-1-correspondence with earlier treated cases.

This happens e.g. if $f_1 = x_1^n + y_1 q_1(x_1, y_1)$ and $\nu(y_1 q_1) = n$.

We first apply a coordinate transformation of the type $\begin{cases} x_1 = x_1 + \alpha y_1 \\ y_1 = y_1 \end{cases}$ such that the coefficient of $x_1^{n-1} y_1$ is equal to zero. If the $n$-jet of $f_1$ was given by $(x_1 + \beta y_1)^n$ then, after this transformation, $\nu(y_1 q_1) > n$ and the corresponding $\mu$-classes can be studied in the usual way by the periodicity-lemma. In the remaining cases $\nu(y_1 q_1) = n$ and $f_1$ has two or more tangent directions. The possible $\mu$-classes are already treated in an earlier part of the classification, since in each tangent direction the corresponding singularity has multiplicity less than $n$. Moreover it is not difficult to show that all these possibilities can occur.

Since $f_1 = x_1^n + y_1 q_1(x_1, y_1)$ and $\nu(y_1 q_1) \geqq n$, all tangent directions are transversal to the exceptional divisor. Consequently there is no subdivision, but a 1-1-correspondence between $\mu$-classes of $f_1$ and $\mu$-classes of $f$.

So the full pattern of singularities of multiplicity $n$ and with two or more tangent directions is repeated with a jump in $\mu$ of $n(n-1)$.

This pattern becomes a part of the 'period' of $\Pi^n[n]$ and shall be repeated by the periodicity lemma.

(4.4) EXAMPLES.

$n = 2 : \Pi^2[2]$ : Period: 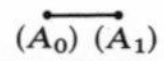 $(A_0)\ (A_1)$

$(A_1)$ has multiplicity 2 and 2 tangent directions.

$n = 3 : \Pi^3[3]$ : Period:

$(A_0)(A_1)(A_2) \times (D_4)(D_5)(D_6) \ldots (D_k) \ldots$

The $D$-series has multiplicity 3 and $\geqq 2$ tangent directions.

$n = 4 : \Pi^4[4]$ see §5.

## 5. Classification of singularities of corank $\leqq 2$ and multiplicity $\nu \leqq 4$

(5.1) As an application of the independence lemma and the periodicity lemma we treat the classification of the singularities, mentioned in the title of this paragraph.

We shall only indicate results, since the classification itself is well-known (cf. Arnold [2], [3], [4], Siersma [11], for $\nu = 3$ see also Briancon [5]). We have to consider the following cases:

$\nu = 1$: The function is regular at $O$, and can be given by $f(x,y) = x$.

$\nu = 2$: (a) 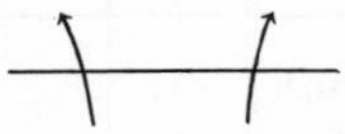 $\Pi^2[1, 1]$

(b) 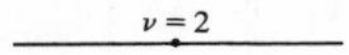 $\nu = 2$ $\Pi^2[2]$

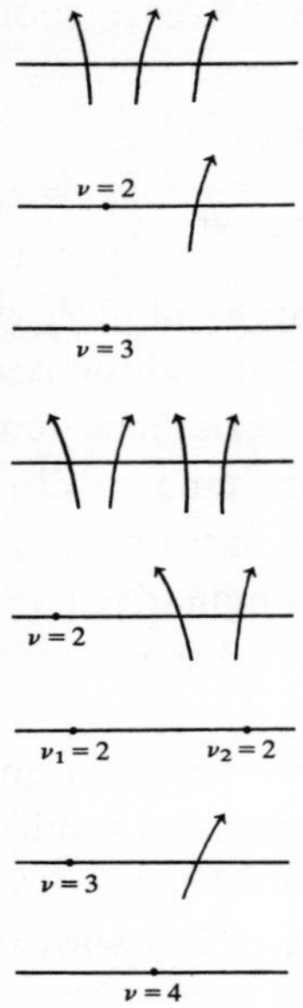

$\nu=3$: (a) $\Pi^3[1, 1, 1]$

(b) $\nu=2$ $\Pi^3[2, 1]$

(c) $\nu=3$ $\Pi^3[3]$

$\nu=4$: (a) $\Pi^4[1, 1, 1, 1]$

(b) $\nu=2$ $\Pi^4[2, 2]$

(c) $\nu_1=2$ $\nu_2=2$ $\Pi^4[2, 2]$

(d) $\nu=3$ $\Pi^4[3, 1]$

(e) $\nu=4$ $\Pi^4[4]$

In the above pictures the transform of $f^{-1}(0)$ is given after one blowing up.

(5.2) $\Pi^2[1, 1]$

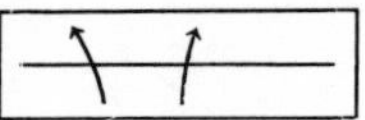

$\Pi^2[1, 1]=\Pi^1[1]\times\Pi^1[1]$ only one $\mu$-class, type $A_1$, given by $f(xy)=xy$ or equivalently by $f(x, y)=x^2+y^2$.

(5.3) $\Pi^2[2]$ $\nu=2$

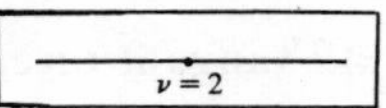

Let $f(x, y)=x^2+p(x, y)$ with multiplicity $\nu(p)\geqq 3$.
We blow up once and get: $f_1(x_1, y_1)=x_1^2+y_1q_1(x_1, y_1)$.
We study the case, that multiplicity $\nu(y_1q_1)\leqq 2$, so:
$f_1(x_1, y_1)=x_1^2+ay_1+bx_1y_1+cy_1^2+\cdots$ and
$f(x, y)=x^2+ay^3+bxy^2+cy^4+\cdots$.

The singularities of $f_1$ have been classified before, so they must be of type $A_0$ or $A_1$. We get the following detailed study:

| $\mu$ | | generic $f_1(x_1, y_1)$ | | | | generic $f(x, y)$ |
|---|---|---|---|---|---|---|
| 2 | | $x_1^2+y_1$ |  | $(A_0)$ | $A_2$ | $x^2+y^3$ |
| 3 | $a_1=0$ | $x_1^2+y_1^2$ |  | $(A_1)$ | $A_3$ | $x^2+y^4$ |

If $b^2-4c=0$, then $x_1^2+bx_1y_1+(b^2/4)y_1^2+\cdots \sim x_1^2+\cdots$ (diffeo respects exc. divisor).

So the detailed study gives the 'period':

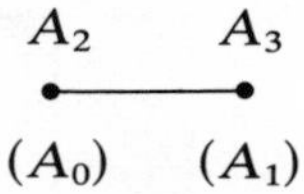

$A_2$ — $A_3$

$(A_0)$ $(A_1)$

The periodicity lemma gives the classes $A_4, A_5, \cdots, A_{2k}, A_{2k+1}, \cdots$.
So in the $\Pi^2[2]$ we get the following singularity classes:

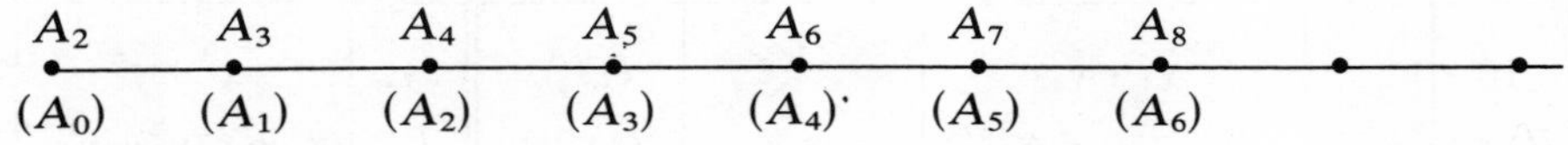

$A_2$ $A_3$ $A_4$ $A_5$ $A_6$ $A_7$ $A_8$

$(A_0)$ $(A_1)$ $(A_2)$ $(A_3)$ $(A_4)$ $(A_5)$ $(A_6)$

$A_k$ is given by $f(x, y)=x^2+y^{k+1}$.

| (5.4) | $\Pi^3[1, 1, 1]$ | 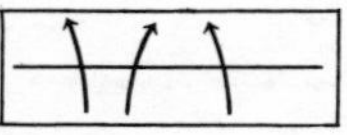 |
|---|---|---|

$\Pi^3[1, 1, 1]=\Pi^1[1]\times\Pi^1[1]\times\Pi^1[1]$: only one $\mu$-class, $D_4$, given by $f(x, y)=x^3+y^3$, or equivalently by $f(x, y)=x^2y+y^3$.

| (5.5) | $\Pi^3[2, 1]$ |  $\nu=2$ |
|---|---|---|

$\Pi^3[2, 1]\cong\Pi^2[2]\times\Pi^1[1]\cong\Pi^2[2]=A$-series $D_{k+5}\leftrightarrow(A_k, A_0)\leftrightarrow(A_k)$.
We get the types:

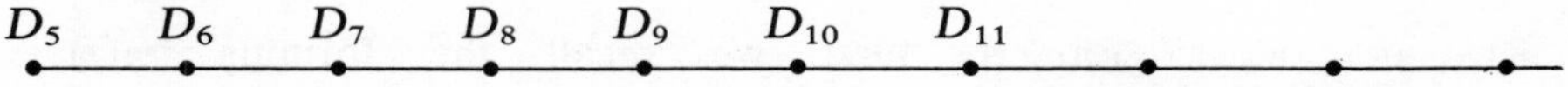

$D_5$ $D_6$ $D_7$ $D_8$ $D_9$ $D_{10}$ $D_{11}$

A class of type $D_k (k\geqq 4)$ can be given by: $f(x, y)=y(x^2+y^{k-1})$.

| (5.6) | $\Pi^3[3]$ |  $\nu=3$ |
|---|---|---|

Let $f(x, y)=x^3+p(x, y)$ with multiplicity $\nu(p)\geqq 4$.
Blowing up once we get: $f_1(x_1, y_1)=x_1^3+y_1q_1(x_1, y_1)$.
We study the case, that multiplicity $\nu(y_1q_1)\leqq 3$, so:

$$f_1(x_1, y_1)=x_1^3+a_1y_1+b_1x_1y_1+b_2y_1^2+c_1x_1^2y_1+c_2x_1y_1^2+c_3y_1^3+\cdots$$

and:

$$f(x, y)=x^3+a_1y^4+b_1xy^3+b_2y^5+c_1x^2y^2+c_2xy^4+c_3y^6+\cdots.$$

The singularities of $f_1$ have been classified earlier, so they must be of type

$A$ or $D$. We get the following detailed study:

| $\mu$ | | generic $f_1(x_1, y_1)$ | | | generic $f(x, y)$ | |
|---|---|---|---|---|---|---|
| 6 | | $x_1^3+y_1$ | | $(A_0)$ | $E_6$ | $x^3+y^4$ |
| 7 | $a_1=0$ | $x_1^3+x_1y_1$ | | $(A_1)$ | $E_7$ | $x^3+xy^3$ |
| 8 | $b_1=0$ | $x_1^3+y_1^3$ | | $(A_2)$ | $E_8$ | $x^3+y^5$ |
| $\times$ | $b_2=0$ | The $A_k$-singularities with $k\geqq 3$ don't occur | | | | |
| 10 | | $x_1^3+y_1^2$ | | $(D_4)$ | $J_{2,0}$ | $x^3+y^6$ |
| $10+i$ | | $x_1^3+x_1^2y_1+y_1^{3+i}$ | | $(D_{i+4})$ | $J_{2,i}$ | $x^3+x^2y^2+y^{6+i}$ |

So the detailed study gives the 'period':

$E_6$ $E_7$ $E_8$ $\times$ $J_{2,0}$ $J_{2,1}$ $J_{2,2}$ $J_{2,3}$ $\cdots$
$(A_0)(A_1)(A_2)$ $(D_4)(D_5)(D_6)(D_7)$ $\cdots$

The periodicity lemma gives classes $E_{12}$, $E_{13}$, $E_{14}$, $J_{3,0}, \cdots, J_{3,i}, \cdots, \cdots,$ $E_{6k}, E_{6k+1}, E_{6k+2}, J_{k+1,0}, \cdots, J_{k+1,i}, \cdots$.

Thus in $\Pi^3[3]$ we get the following pattern of singularity classes:

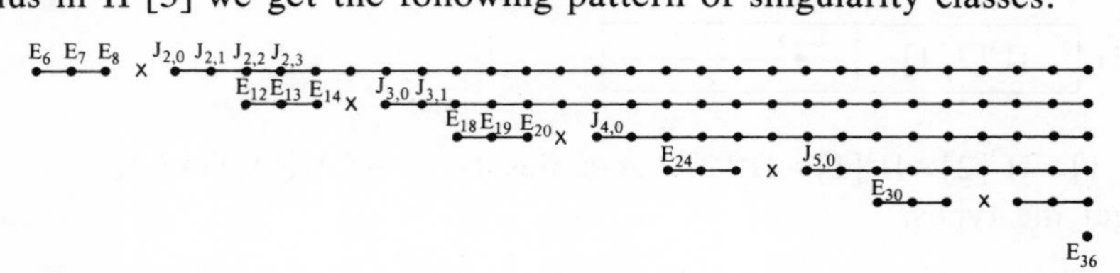


REMARK ABOUT MODALITY. First we recall the formula $m(f)=\mu(f)-c(f)-1$. Consider again the case $\Pi^3[3]$. It is known that $\mu(E_6)=6$; $c(E_6)=5$ and $\mu(E_7)=7$; $c(E_7)=6$.

The semi-algebraic set $A_2$ can be constructed from the semi-algebraic set $A_1$ by one defining equation (of course one has also to change the defining inequalities). After blowing up we see that also $E_8$ can be constructed from $E_7$ by one defining equation.

Next we try to do the same for $A_2 \to A_3$. Since the $(A_3)$-type doesn't occur we see, after blowing up, that the corresponding defining equation for $(A_3)$ must imply the equation for $(D_4)$.

So $c(J_{2,0})=c(E_8)+1$ but $\mu(J_{2,0})=\mu(E_8)+2$.

So $m(J_{2,0})=m(E_8)+1$.

In the same way one shows that $m(J_{2,i})=1$, $m(E_{12})=m(E_{13})=m(E_{14})=1$. But $m(J_{3,0})=2$, etc.

In general: $m(J_{k+1,i}) = k$

$$m(E_{6k}) = m(E_{6k+1}) = m(E_{6k+2}) = k-1.$$

This way of computation of the modality works not only in this example but can be used in general.

(5.7) $\Pi^4[1, 1, 1, 1]$

We get $\Pi^4[1, 1, 1, 1] = \Pi^1[1] \times \Pi^1[1] \times \Pi^1[1]\Pi^1[1]$: one $\mu$-class $X_9 = X_{1,0}$ given by $f(x, y) = x^4 + y^4$.

(5.8) $\Pi^4[2, 1, 1]$ $\nu = 2$

We get $\Pi^4[2, 1, 1] = \Pi^2[2] \times \Pi^1[1] \times \Pi^1[1] \cong \Pi^2[2] \cong A$-series where $X_{i+9} = X_{1,i} \leftrightarrow (A_i, A_0, A_0) \leftrightarrow (A_i)$.
We get the types:

$X_{1,1}\ X_{1,2}\ X_{1,3}\ X_{1,4}\ X_{1,5}$ $\quad\dots\quad$ $X_{1,k}$

$(A_1)(A_2)(A_3)(A_4)(A_5)$ $\quad\dots\quad$ $A_k$

A class of type $X_{1,i}$ can be given by $f(x, y) = x^4 + x^2y^2 + y^{4+i}$.

(5.9) $\Pi^4[2, 2]$ $\nu_1 = 2$ $\nu_2 = 2$

We get $\Pi^4[2, 2] = \Pi^2[2] \times \Pi^2[2]/\sim$.
The type $Y_{p,q}$ corresponds to $(A_{p-1}, A_{q-1})$.
We have a double-$A$-series:

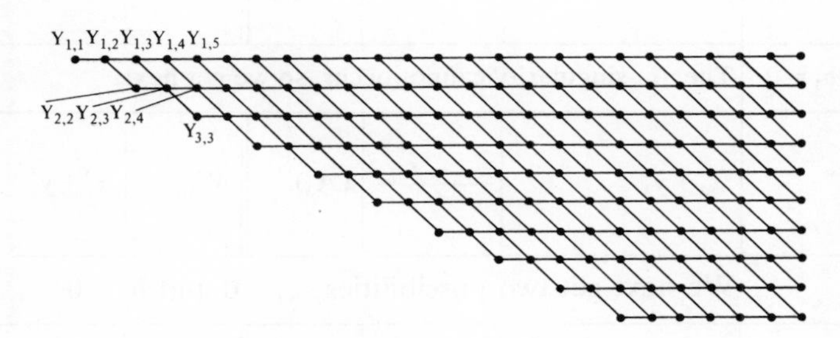


The class $Y_{p,q}$ can be given by $f(x, y) = x^{p+4} + x^2y^2 + y^{q+4}$.

(5.10) $\Pi^4[3, 1]$ $\nu = 3$

We get $\Pi^4[3, 1] = \Pi^3[3] \times \Pi^1[1] \cong \Pi^3[3]$.

The type $Z_{i+5}$ corresponds to $E_i$ and $Z_{k-1,i}$ corresponds to $J_{k,i}$.

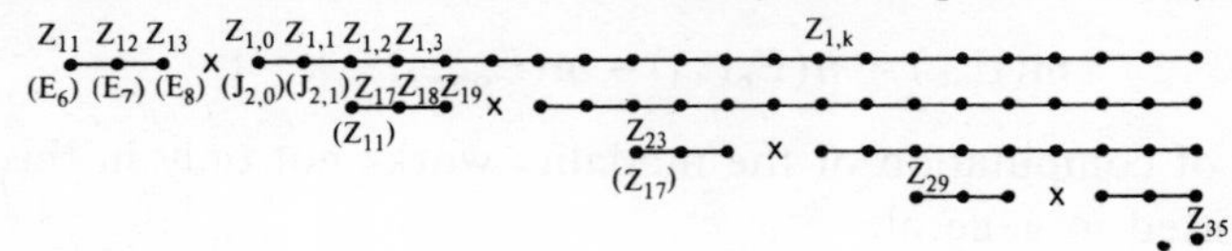


(5.11) $\Pi^4[4]$ $\nu = 4$

Let $f(x, y) = x^4 + p(x, y)$ with multiplicity $\nu(p) \geqq 5$.
Blowing up once we get: $f_1(x_1, y_1) = x_1^4 + y_1 q_1(x_1, y_1)$.
We study the case, that multiplicity $\nu(y_1 q_1) \leqq 4$, so:

$$f_1(x_1, y_1) = x_1^4 + a_1 y_1 + b_1 x_1 y_1 + b_2 y_1^2 + c_1 x_1^2 y_1 + c_2 x_1 y_1^2 + \\ + c_3 y_1^3 + d_1 x_1^3 y_1 + d_2 x_1^2 y_1^2 + d_3 x_1 y_1^3 + d_4 y_1^4 + \cdots$$

and:

$$f(x, y) = x^4 + a_1 y^5 + b_1 x y^4 + b_2 y^6 + c_1 x^2 y^3 + c_2 x y^5 + c_3 y^7 + \\ + d_1 x^3 y^2 + d_2 x^2 y^4 + d_3 x y^6 + d_4 y^8 + \cdots.$$

The singularities of $f_1$ have been classified before; so must be of type $A, D, E, J, X, Y$ or $Z$.
We get the following detailed study:

| $\mu$ | | generic $f_1(x_1, y_1)$ | | | | generic $f(x, y)$ |
|---|---|---|---|---|---|---|
| 12 | | $x_1^4 + y_1$ | | $(A_0)$ | $W_{12}$ | $x^4 + y^5$ |
| 13 | $a_1 = 0$ | $x_1^4 + x_1 y_1$ | | $(A_1)$ | $W_{13}$ | $x^4 + xy^4$ |
| | $b_1 = 0$ | The $A_2$-singularity cannot occur, so we get next: | | | | |
| 15 | | $x_1^4 + y_1^2$ | | $(A_3)$ | $W_{1,0}$ | $x^4 + y^6$ |
| | | We next get two possibilities: $c_1 = 0$ and $b_2 = 0$ | | | | |
| 16 | $4b_2 = c_1^2$ | $(y_1 + x_1^2)^2 + x_1 y_1^2$ | | $(A_4)$ | $W_{1,1}^{\#}$ | $(x^2 + y^3)^2 + xy^5$ |
| 17 | $c_2 = 0$ | $(y_1 + x_1^2)^2 + x_1^2 y_1^2$ | | $(A_5)$ | $W_{1,2}^{\#}$ | $(x^2 + y^3)^2 + x^2 y^4$ |
| $2q + 14$ | | $(y_1 + x_1^2) + x_1 y_1^{q+1}$ | | $(A_{2q+2})$ | $W_{1,2q-1}^{\#}$ | $(x^2 + y^3)^2 + xy^{q+4}$ |
| $2q + 15$ | | $(y_1 + x_1^2) + x_1^2 y_1^{q+1}$ | | $(A_{2q+3})$ | $W_{1,2q}^{\#}$ | $(x^2 + y^3)^2 + x^2 y^{q+3}$ |

| | | | | | | |
|---|---|---|---|---|---|---|
| 16 | $b_2=0$ | $x_1^4+x_1^2y_1+y_1^3$ | | $(D_4)$ | $W_{1,1}$ | $x^4+x^2y^3+y^7$ |
| 17 | | $x_1^4+x_1^2y_1+y_1^4$ | | $(D_5)$ | $W_{1,2}$ | $x^4+x^2y^3+y^8$ |
| $k+15$ | | $x_1^4+x_1^2y_1+y_1^{2+k}$ | | $(D_{k+3})$ | $W_{1,k}$ | $x^4+x^2y^3+y^{6+k}$ |
| 17 | | $x_1^4+x_1y_1^2$ | | $(D_5)$ | $W_{17}$ | $x^4+xy^5$ |
| 18 | | $x_1^4+y_1^3$ | | $(E_6)$ | $W_{18}$ | $x^4+y^7$ |

The classes $E_7$ and $E_8$ don't occur, because of $x_1^4$.
Next $\nu(y_1q_1)=4$ and because of (4.3) we find all singularities of $\Pi^4-\Pi^4[4]$, so the classes $X_1$, $Y$ and $Z$.

Hence the 'period' is:

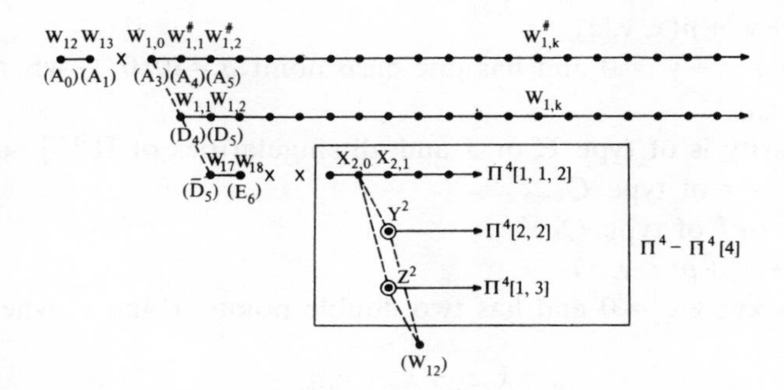


The periodicity lemma gives the classes $W_k$, $X_k$, $Y_k$, $Z_k$, etc.

## 6. Corank 3

(6.1) In the case of corank 3 we need 3 coordinates. The blowing up process now gives $P^2(\boldsymbol{C})$ as exceptional divisor. The intersections of the branches with the exceptional divisor are now not necessarily isolated but form a 1-dimensional algebraic variety $X$. So the induced functions $f_\sigma\,(\sigma \in X)$ can have non-isolated singularities.

If the 3-jet of $f$ has no multiple factor, then $f_\sigma$ has only singularities in the multiple points of $X$; in the other points $f_\sigma$ is of type $A_0$.

In the cases we consider most $f_\sigma$ has a singularity-type that has been studied before.

We list the topological classes of $f_\sigma$ that can occur. Next we assume that different intersection points can be treated independently and compare the results with Arnold's list. The lists are identical and this means that we have found, in this way, all the $\mu$-classes for $f$.

We cannot prove this directly since ($\mu$ constant $\Leftrightarrow$ resolution constant) is not true if $n \geqq 3$.

(6.2) We consider $f(x, y, z)$ with the multiplicity of $f$ equal to 3. The corresponding variety $X \subset P^2(\boldsymbol{C})$ is given by $j^3(f) = 0$.
We have the following possibilities:
(a) $f = xyz + x^3 + y^3 + z^3 + p(x, y, z)$
$X$ is given by $xyz + x^3 + y^3 + z^3 = 0$ and is an elliptic curve without multiple points.
The type of this singularity is called $P_8$.
(b) $f = xyz + x^3 + y^3 + p(x, y, z)$
$X$ is given by $xyz + x^3 + y^3 = 0$ and has one double point $\sigma$, where $f_\sigma = x_\sigma y_\sigma + x_\sigma^3 + y_\sigma^3 + z_\sigma q_\sigma$.
We get a singularity of type $\Pi^2[2]$, which all occur.
$f_\sigma$ of type $A_k \leftrightarrow f$ of type $P_{k+2}$ $(k \geqq 1)$.
(c) $f = x^2 z + y^3 + p(x, y, z)$
$X$ is given by $x^2 z + y^3 = 0$ and has one cusp point $\sigma = (0\!:\!0\!:\!1)$ where $f_\sigma = x_\sigma^2 + y_\sigma^3 + z_\sigma q_\sigma$.
This singularity is of type $E$ or $J$ and all singularities of $\Pi^3[3]$ occur.
$f_\sigma$ of type $E_k \leftrightarrow f$ of type $Q_{k+4}$
$f_\sigma$ of type $J_{k,i} \leftrightarrow f$ of type $Q_{k,i}$.
(d) $f = xyz + x^3 + p(x, y, z)$
$X$ is given by $xyz + x^3 = 0$ and has two double points $\sigma$ and $\tau$, where:

$$f_\sigma = x_\sigma z_\sigma + x_\sigma^3 + y_\sigma q_\sigma$$
$$f_\tau = x_\tau y_\tau + x_\tau^3 + z_\tau q_\tau$$

Both are singularities of type $A$, and all possible combinations can be found.

$$\left.\begin{matrix} f_\sigma \text{ of type } A_{p-4} \\ f_\tau \text{ of type } A_{q-4} \end{matrix}\right\} \leftrightarrow f \text{ of type } R_{p,q} = T_{3,p,q}.$$

This singularity class is isomorphic to $\Pi^2[2] \times \Pi^2[2]/\sim \;\cong \Pi^4[2, 2]$.
(e) $f = x^2 z + yz^2 + p(x, y, z)$
$X$ is given by $x^2 z + yz^2 = 0$ and has one multiple point $\sigma$, where

$$f_\sigma = x_\sigma^2 z_\sigma + z_\sigma^2 + y_\sigma q_\sigma \sim z_\sigma^2 + x_\sigma^4 + y_\sigma q_\sigma.$$

We get the singularities of $\Pi^4[4]$, which all occur.

$$f_\sigma \text{ of type } W_p \leftrightarrow f \text{ of type } S_{p-1}$$
$$f_\sigma \text{ of type } W_{k,i} \leftrightarrow f \text{ of type } S_{k,i}$$
$$f_\sigma \text{ of type } W_{k,i}^\# \leftrightarrow f \text{ of type } S_{k,i}^\#$$

etc., in general:

$$f_\sigma \text{ of type } W_k \leftrightarrow f \text{ of type } S_k$$
$$f_\sigma \text{ of type } X_k \leftrightarrow f \text{ of type } SP_k$$
$$f_\sigma \text{ of type } Y_k \leftrightarrow f \text{ of type } SR_k$$
$$f_\sigma \text{ of type } Z_k \leftrightarrow f \text{ of type } SQ_k.$$

(f) $f = xyz + p(x, y, z)$
$X$ is given by $xyz = 0$ and has three multiple points $\sigma$, $\tau$ and $\mu$, where

$$f_\sigma = y_\sigma z_\sigma + x_\sigma q_\sigma$$
$$f_\tau = x_\tau z_\tau + y_\tau q_\tau$$
$$f_\mu = x_\mu y_\mu + z_\mu q_\mu$$

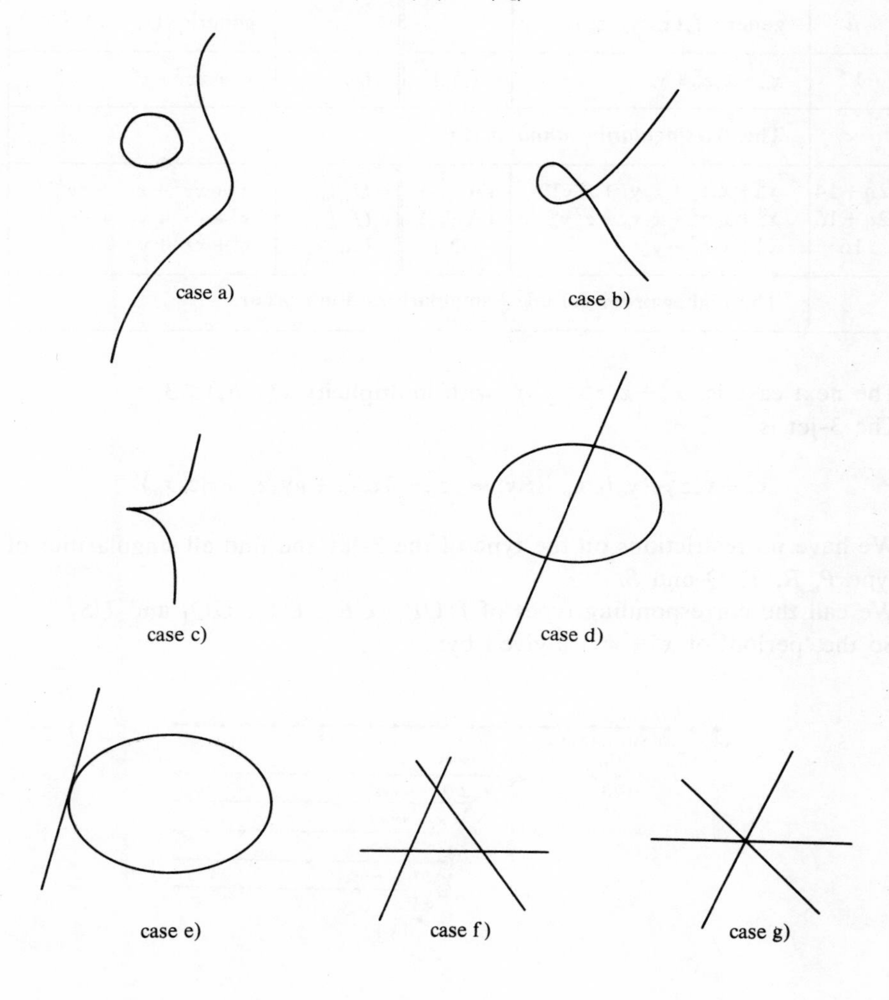

We get three singularities of type $A$ and all combinations can be obtained:

$$\left.\begin{matrix} f_\sigma \text{ of type } A_{p-4} \\ f_\tau \text{ of type } A_{q-4} \\ f_\mu \text{ of type } A_{r-4} \end{matrix}\right\} \leftrightarrow f \text{ of type } T_{p,q,r}.$$

(g) $f = x^3 + xz^2 + p(x, y, z)$
$X$ is given by $x^3 + xz^2 = 0$ and has one multiple point, where

$$f_\sigma = x_\sigma^3 + x_\sigma z_\sigma^2 + y_\sigma q_\sigma.$$

The following possibilities can occur:

| $\mu$ | generic $f_\sigma(x_\sigma, y_\sigma, z_\sigma)$ | | | generic $f(x, y, z)$ |
|---|---|---|---|---|
| 12 | $x_\sigma^3 + x_\sigma z_\sigma^2 + y_\sigma$ | $(A_0)$ | $U_{12}$ | $x^3 + xz^2 + y^4$ |
| $\times$ | The $A_1$-*singularity cannot occur* | | | |
| $2q+14$<br>$2q+15$<br>16 | $x_\sigma^3 + x_\sigma z_\sigma^2 + x_\sigma y_\sigma + z_\sigma y_\sigma^{q+1}$<br>$x_\sigma^3 + x_\sigma z_\sigma^2 + x_\sigma y_\sigma + z_\sigma^2 y_\sigma^{q+1}$<br>$x_\sigma^3 + x_\sigma z_\sigma^2 + y_\sigma^2$ | $(A_{2q+2})$<br>$(A_{2q+3})$<br>$(D_4)$ | $U_{1,2q}$<br>$U_{1,2q+1}$<br>$U_{16}$ | $x^3 + xz^2 + xy^3 + zy^{3+q}$<br>$x^3 + xz^2 + xy^3 + z^2y^{2+q}$<br>$x^3 + xz^2 + y^5$ |
| | The higher order corank 2 singularities don't occur. | | | |

The next case is: $x_\sigma^3 + x_\sigma z_\sigma^2 + y_\sigma q_\sigma$ with multiplicity $\nu(y_\sigma q_\sigma) \leqq 3$.
The 3-jet is

$$x_\sigma^3 + x_\sigma z_\sigma^2 + y_\sigma(\alpha x_\sigma^2 + \beta y_\sigma^2 + \gamma z_\sigma^2 + \delta x_\sigma y_\sigma + \varepsilon y_\sigma z_\sigma + \eta z_\sigma x_\sigma).$$

We have no restrictions on the type of the 3-jet and find all singularities of type $P$, $R$, $T$, $Q$ and $S$.
We call the corresponding types of $f$: $UP_1$, $UR_1$, $UT_1$, $UQ_1$ and $US_1$.
So the 'period' of $x^3 + xz^2$ is given by:

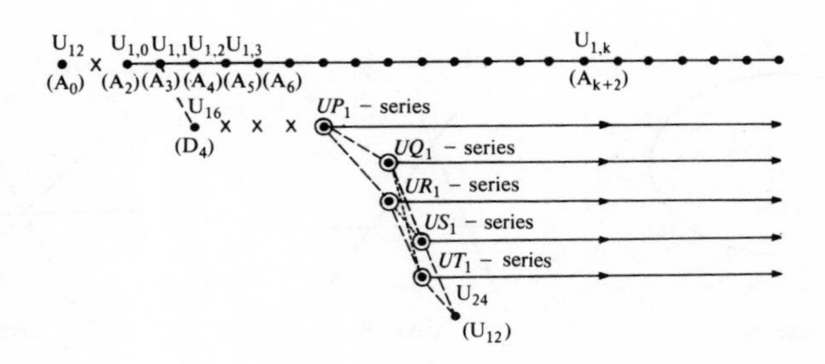

## 7. Classification of corank 2 and multiplicity $\nu = 5$

$$\Pi^5 = \Pi^5[1, 1, 1, 1, 1] \cup \Pi^5[2, 1, 1, 1] \cup \Pi^5[3, 1, 1] \cup \Pi^5[2, 2, 1] \cup \Pi^5[4, 1] \cup \Pi^5[3, 2] \cup \Pi^5[5].$$

(7.1) (a) $\Pi^5[1, 1, 1, 1, 1] \cong \Pi^1[1]$: one $\mu$-class.
(b) $\Pi^5[2, 1, 1, 1] \cong \Pi^2[2]$: $A$-series.
(c) $\Pi^5[3, 1, 1] \cong \Pi^3[3]$: $E/J$-series.
(d) $\Pi^5[2, 2, 1] \cong \Pi^2[2] \times \Pi^2[2]/\sim$: double $A$-series.
(e) $\Pi^5[4, 1] \cong \Pi^4[4]$.
(f) $\Pi^5[3, 2] \cong \Pi^3[3] \times \Pi^2[2]$: $E/J \times A$ series.
(g) $\Pi^5[5]$ shall be described next:

(7.2) $\Pi^5[5]$ —— $\nu = 5$

Let $f(x, y) = x^5 + p(x, y)$ with $\nu(p) \geqq 5$. Blowing up once we get: $f_1(x_1, y_1) = x_1^5 + y_1 q_1(x_1, y_1)$. We first study the case $\nu(y_1 q_1) \leqq 5$ so:

$$\begin{aligned} f_1(x_1, y_1) = x_1^5 &+ a_1 y_1 + b_1 x_1 y_1 + b_2 y_1^2 + c_1 x_1^2 y_1 + c_2 x_1 y_1^2 \\ &+ c_3 y_1^3 + d_1 x_1^3 y_1 + d_2 x_1^2 y_1^2 + d_3 x_1 y_1^3 + d_4 y_1^4 \\ &+ e_1 x_1^4 y_1 + e_2 x_1^3 y_1^2 + e_3 x_1^2 y_1^3 + e_4 x_1 y_1^4 + e_5 y_1^5 + \cdots \end{aligned}$$

and

$$\begin{aligned} f(x, y) = x^5 &+ a_1 y^6 + b_1 x y^5 + b_2 y^7 + c_1 x^2 y^4 + c_2 x y^6 + c_3 y^8 \\ &+ d_1 x^3 y^3 + d_2 x^2 y^5 + d_3 x y^7 + d_4 y^9 + e_1 x^4 y^2 \\ &+ e_2 x^3 y^4 + e_3 x^2 y^6 + e_4 x y^8 + e_5 y^{10} + \cdots \end{aligned}$$

The singularities of $f_1$ must be classified before.
We get the following detailed study:

| $\mu$ | | generic $f_1(x_1, y_1)$ | | | | generic $f(x, y)$ |
|---|---|---|---|---|---|---|
| 20 | | $x_1^5 + y_1$ |  | $(A_0)$ | | $x^5 + y^6$ |
| 21 | $a_1 = 0$ | $x_1^5 + x_1 y_1$ | 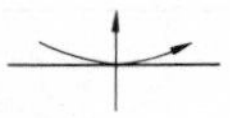 | $(A_1)$ | | $x^5 + xy^5$ |
| × | $b_1 = 0$ | The $A_2$-singularity cannot occur, so we have: | | | | |
| 23 | | $x_1^5 + y_1^2 + x_1^2 y_1$ |  | $(A_3)$ | | $x^5 + y^7 + x^2 y^4$ |

We next get two possibilities: $c_1=0$ and $b_2=0$

| | | | | | |
|---|---|---|---|---|---|
| 24 | $c_1=0$ | $x_1^5+y_1^2$ |  | $(A_4)$ | $x^5+y^7$ |
| – | | Other singularities of type $A_k$ don't occur, because of $x_1^5$. | | | |
| 24 | $b_2=0$ | $x_1^5+x_1^2y_1+y_1^3$ |  | $(D_4)$ | $x^5+x^2y^4+y^8$ |
| 25 | | $x_1^5+x_1^2y_1+y_1^4$ |  | $(D_5)$ | $x^5+x^2y^4+y^9$ |
| $20+k$ | | $x_1^5+x_1^2y_1+y_1^{k-1}$ |  | $(D_k)$ | $x^5+x^2y^4+y^{k+4}$ |
| 26 | | $x_1^5+x_1y_1^2$ |  | $(\bar{D}_6)$ | $x^5+xy^6$ |
| 27 | | $x_1(y_1-x_1^2)^2+y_1x_1^4$ |  | $(\bar{D}_7)$ | $x(y^3-x^2)^2+x^4y^2$ |
| $20+2n$<br>$21+2r$ | | $x_1(y_1-x_1^2)^2+y_1^{n-1}x_1(n\geqq 4)$<br>$x_1(y_1-x_1^2)^2+y_1^n(n\geqq 4)$ | | $(\bar{D}_{2n})$<br>$(\bar{D}_{2n+1})$ | $x(y^3-x^2)^2+xy^{n+3}$<br>$x(y^3-x^2)^2+y^{n+5}$ |
| $x$ | | The $E_6$ singularity cannot occur | | | |
| 27 | | $x_1^5+x_1^3y_1+y_1^3$ |  | $(E_7)$ | $x^5+x^3y^3+y^8$ |
| 28 | | $x_1^5+y_1^3$ |  | $(E_8)$ | $x^5+y^8$ |
| – | | Other singularities of type $J$ or $E$ cannot occur | | | |
| 29 | | $x_1^5+x_1^3y_1+y_1^4$ |  | $(X_9)$ | $x^5+x^3y^3+y^9$ |
| 30 | | $x_1^5+x_1^2y_1^2+y_1^4$ | 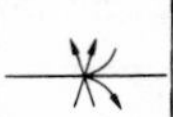 | $(X_{10})$ | $x^5+x^2y^5+y^9$ |
| 30 | | $y_1(x_1-y_1)^2(x_1-2y_1)+x_1^5$ |  | $(\bar{X}_{10})$ | $x^5+(x-y^2)^2$<br>$\times(x-2y^2)y^3$ |

| | | | | | |
|---|---|---|---|---|---|
| – | | Other singularities of type $X$ cannot occur. | | | |
| 31 | | $x_1^5+x_1^2y_1^2+y_1^5$ | | $(Y_{5,5})$ | $x^5+x^2y^5+y^{10}$ |
| $26+k$ | | $x_1^5+x_1^2y_1^2+y_1^k$ | | $(Y_{5,k})$ | $x^5+x^2y^5+y^{5+k}$ |
| – | | Other singularities of type $Y$ and other tangentdirections don't occur. | | | |
| 31 | | $x_1^5+x_1y_1^3$ | | $(\bar{Z}_{11})$ | $x^5+xy^7$ |
| 31 | | $x_1^5+x_1^3y_1+y_1^5$ | | $(Z_{11})$ | $x^5+x^3y^3+y^{10}$ |
| 32 | | $x_1^5+x_1^3y_1+x_1y_1^4$ | | $(Z_{12})$ | $x^5+x^3y^3+xy^8$ |
| 33 | | $x_1^5+x_1^3y_1+y_1^6$ | | $(Z_{13})$ | $x^5+x^3y^3+y^{11}$ |
| | | Also all other singularities of type $Z$ occur. | | | |
| 32 | | $x_1^5+y_1^4$ | | $(W_{12})$ | $x^5+y^9$ |
| – | | Other singularities of type $W$ don't occur. | | | |

Next $\nu(y_1q_1)=5$ and because of (4.3) we find all singularities of $\Pi^5-\Pi^5[5]$. So the 'period' is:

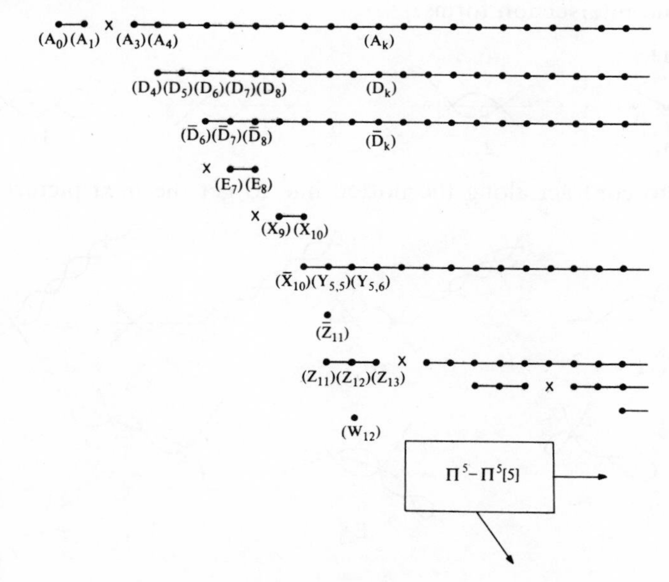

## 8. Morsifications and intersection forms

The general reference for this paragraph is A'Campo [1].

(8.1) A *morsification* $h$ of $f$ is a nearby germ $h$, with only non-degenerate critical points.

A'Campo constructed for every $\mu$-class of $f:\mathbf{C}^2 \to \mathbf{C}$ a morsification with real critical points, using the blowing up construction.

The intersection $h^{-1}(O) \cap \mathbf{R}^2$ consists of a curve $C$ with only double points and:

$$(\text{number of double points}) + (\text{number of regions}) = \mu.$$

The vanishing cycles and their intersections are determined by the curve $C$.

EXAMPLE. ($E_8$)

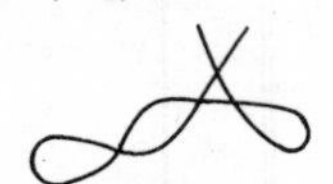

We simplify this picture to

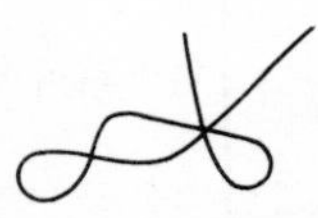

In the following pictures (and those of table III) one has to replace the multiple intersecting lines by lines in general position, as follows:

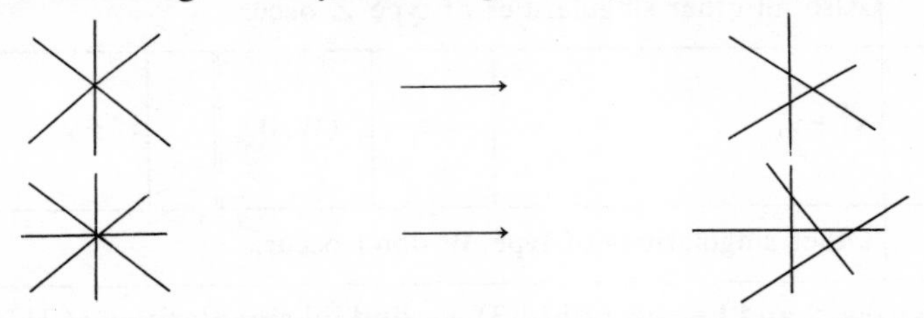

(8.2) The periodicity of the classification induces a periodicity of morsifications and intersection forms.

EXAMPLES

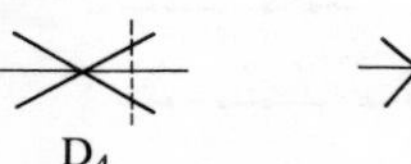

$D_4$

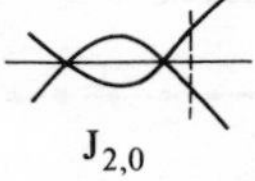

$J_{2,0}$

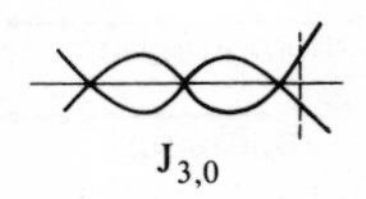

$J_{3,0}$

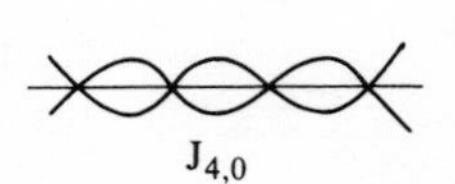

$J_{4,0}$

(one has to contract along the dotted line to get the next picture).

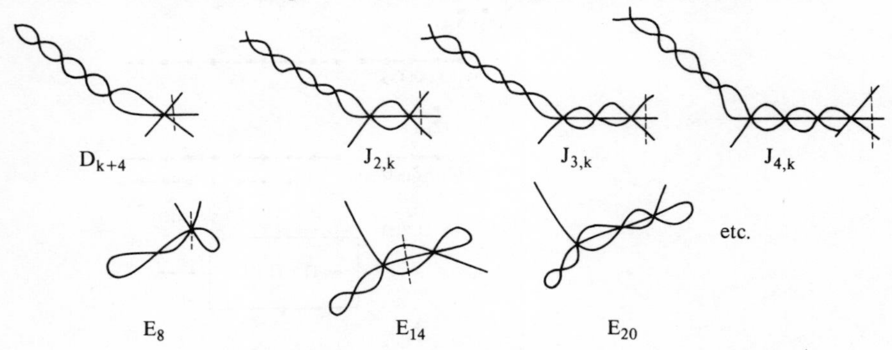

TABLE I. Classification of corank $\leqq 2$ and multiplicity $\nu \leqq 4$.

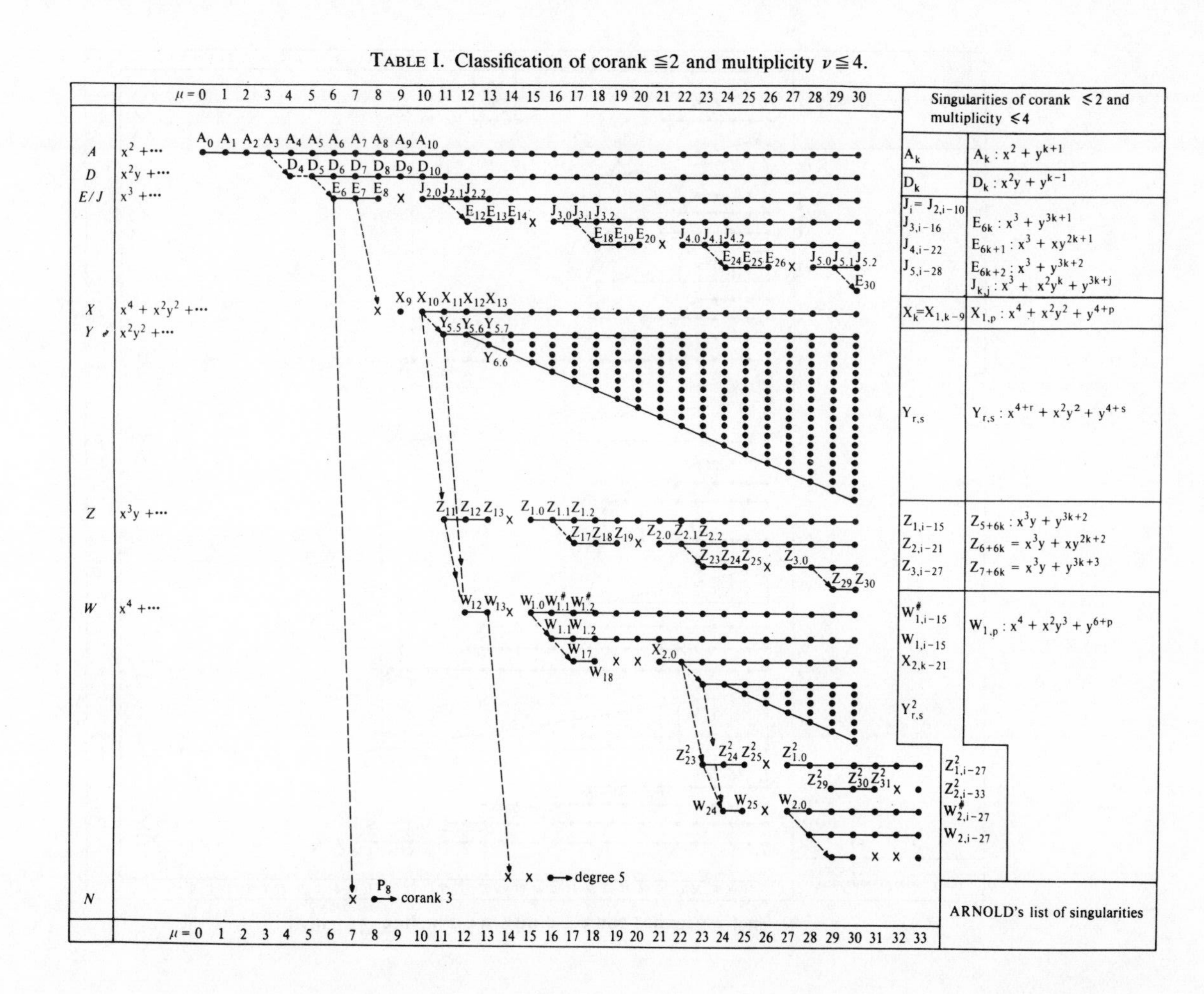

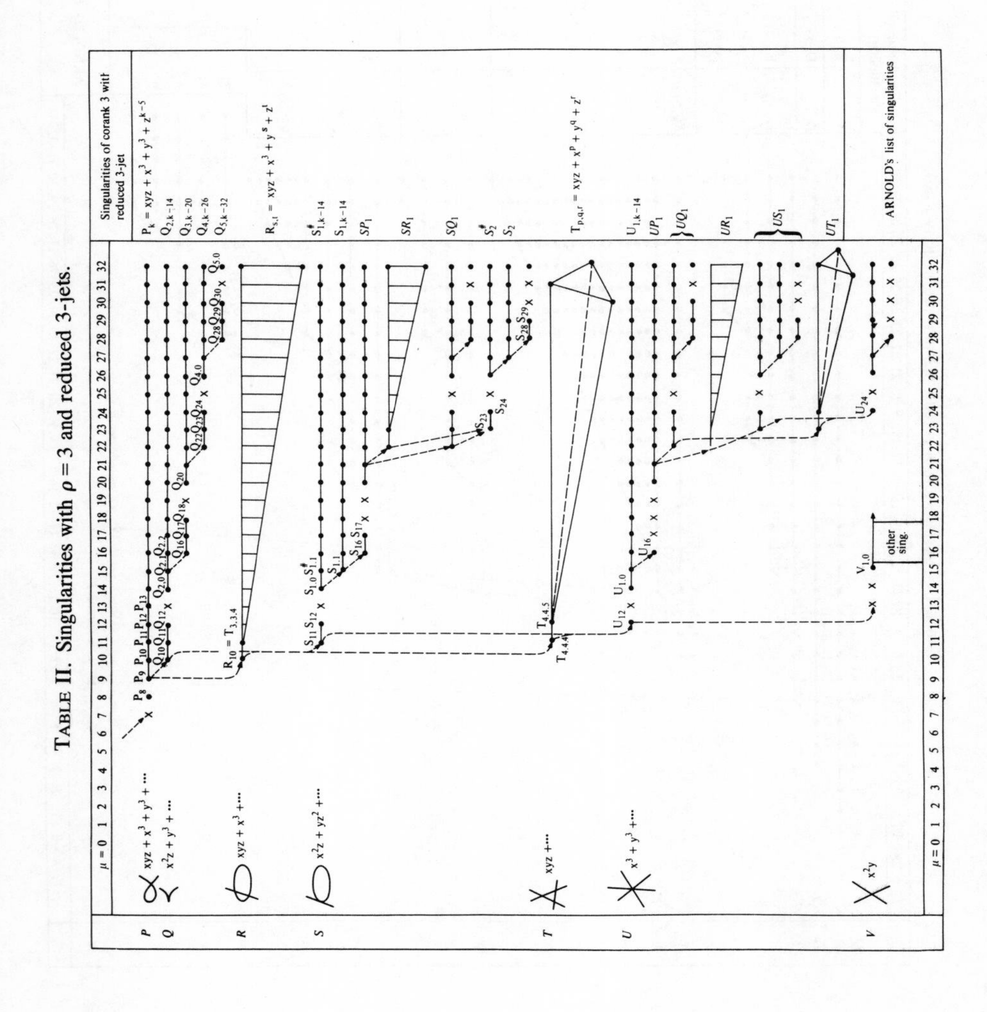


TABLE II. Singularities with $\rho = 3$ and reduced 3-jets.

TABLE III. List of morsifications.

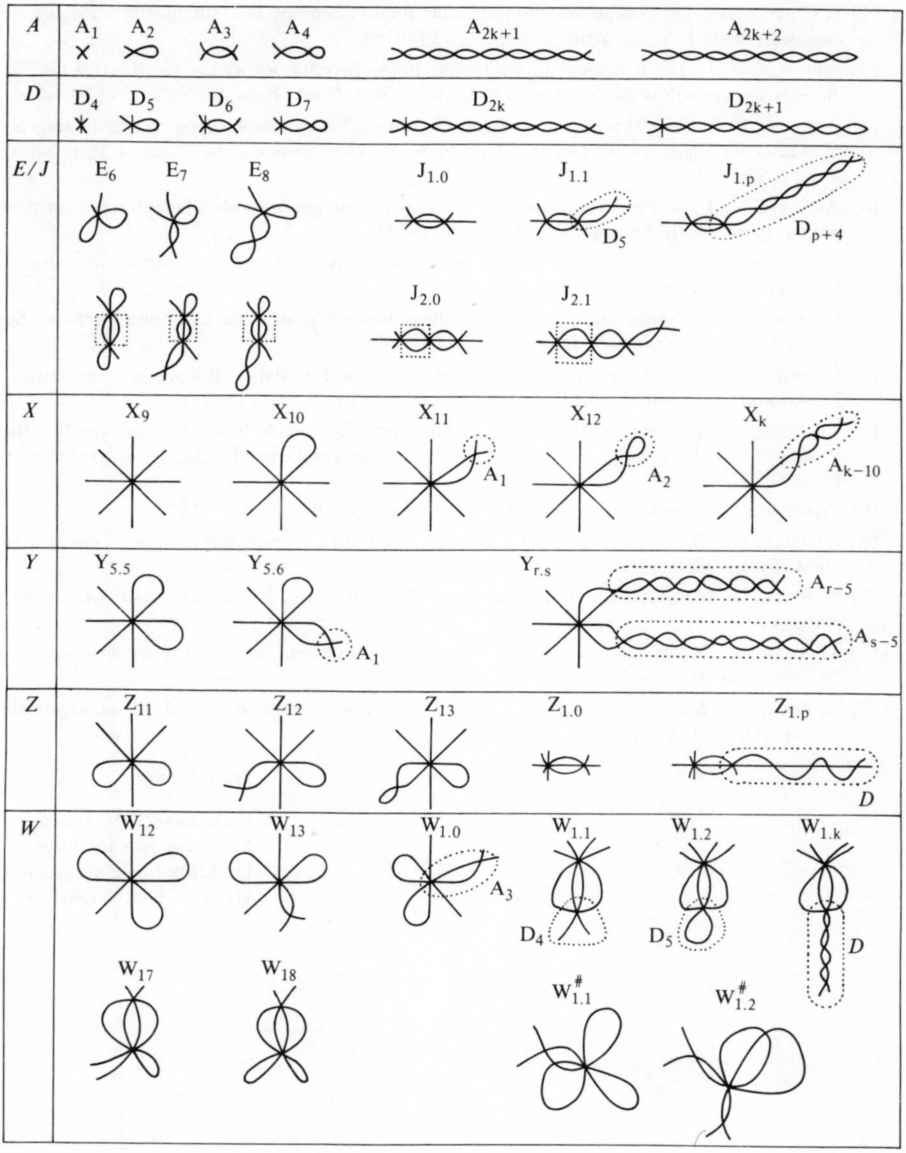

| | |
|---|---|
| *A* | $A_1$ $A_2$ $A_3$ $A_4$ $A_{2k+1}$ $A_{2k+2}$ |
| *D* | $D_4$ $D_5$ $D_6$ $D_7$ $D_{2k}$ $D_{2k+1}$ |
| *E/J* | $E_6$ $E_7$ $E_8$ $J_{1.0}$ $J_{1.1}$ $D_5$ $J_{1.p}$ $D_{p+4}$ <br> $J_{2.0}$ $J_{2.1}$ |
| *X* | $X_9$ $X_{10}$ $X_{11}$ $A_1$ $X_{12}$ $A_2$ $X_k$ $A_{k-10}$ |
| *Y* | $Y_{5.5}$ $Y_{5.6}$ $A_1$ $Y_{r.s}$ $A_{r-5}$ $A_{s-5}$ |
| *Z* | $Z_{11}$ $Z_{12}$ $Z_{13}$ $Z_{1.0}$ $Z_{1.p}$ *D* |
| *W* | $W_{12}$ $W_{13}$ $W_{1.0}$ $A_3$ $W_{1.1}$ $D_4$ $W_{1.2}$ $D_5$ $W_{1.k}$ *D* <br> $W_{17}$ $W_{18}$ $W^{\#}_{1.1}$ $W^{\#}_{1.2}$ |

## REFERENCES

[1] A'Campo, N., Le groupe de monodromie du déploiment des singularité's isolées de courbes planes I. Math. Ann. 213, p. 1–32 (1975).

[2] Arnold, V. I., Local normal forms of functions. Inventiones math. 35, 87–109 (1976). The Russian version is also contained in Uspekhi Mat. Nauk, tome XXX, 5 (1975) 3–65.

[3] Arnold, V. I., Critical points of smooth functions. The Vancouver International Congress of Mathematicians 1974. The Russian version is also contained in Uspekhi Mat. Nauk, tome XXX, 5 (1975) 3–65.

[4] Arnold, V. I., Normal forms of functions in the neighborhoods of degenerate critical points. Russ. Math. Surveys, *29:* 2, 11–49 (1974).

[5] Briançon, J., Quelques calculs inutiles sur les germes de courbes planes d'ordre 3. Université de Nice, dec. 1973.

[6] Demazure, M., Sousgroupes algébrique maximum du groupe de Cremona. Annales Sc. de l'E.N.S., 4–e serie, t. 3, fasc. *4*, 507–588 (1970).

[7] Gabrielov, A. M., Bifurcations, Dynkin diagrams and modality of isolated singularities. Funktional'nyi Analiz i Ego Prilozheniya, Vol. 8, No. 2, 7–12 (1974).

[8] Lê Dũng Tráng, Ramanujam, C. P., The invariance of Milnor's number implies the invariance of the topological type. Centre de Mathématique de l'École Polytechnique, Paris 1973.

[9] Pham, F., Singularités planes d'ordre 3. Astérisque 7–8, 363–391 (1973).

[10] Siersma, D., The singularities of $C^\infty$-functions of right-codimension smaller or equal than eight. Indag. Math. 35, No 1, p. 31–37, (1973).

[11] Siersma, D., Classification and deformation of singularities. Thesis. University of Amsterdam, (1974).

[12] Teissier, B., Cycles evanescent, sections plane et conditions de Whitney. Astérisque 7–8, 285–362 (1973).

[13] Zariski, O., Studies in equisingularity I. Equivalent singularities of plane algebroid curves. Amer. J. Math. *87* (1965) 507–536.

Dirk Siersma

Rijksuniversiteit Utrecht
Mathematisch Instituut
De Uithof, Budapestlaan 6
Utrecht, The Netherlands

Nordic Summer School/NAVF
Symposium in Mathematics
Oslo, August 5–25, 1976

# MIXED HODGE STRUCTURE ON THE VANISHING COHOMOLOGY

J. H. M. Steenbrink*

## Abstract

We construct a mixed Hodge structure on the cohomology of the Milnor fiber associated to an isolated singularity of a complex hypersurface. We determine the relations it has with monodromy, intersection form and local cohomology.

AMS (MOS) subject classification scheme (1970): 14 C 30, 14 D 05, 14 H 20, 57 D 45.

*Keywords:* mixed Hodge structure, vanishing cycles, monodromy, De Rham cohomology, Hodge filtration, weight filtration, intersection form.

## Introduction

Let $P \in \mathbb{C}[z_0, \cdots, z_n]$ with $P(0)=0$. Assume that $0 \in \mathbb{C}^{n+1}$ is a critical point of $P$. Denote $B$ the open ball in $\mathbb{C}^{n+1}$ with center 0 and radius $\varepsilon > 0$. There exists $\eta > 0$ such that $0 < |t| < \eta$ implies that $B_t = P^{-1}(t) \cap B$ is a complex manifold. In this paper we follow a suggestion of Deligne and construct a mixed Hodge structure on the cohomology of $B_t$ (the vanishing cohomology) in the case that $P$ has an isolated critical point at 0.

Let $S$ be the disk with center 0 and radius $\eta$. Denote $X' = P^{-1}(S) \cap B$. Let $\rho : X \to X'$ be a resolution of $P$, i.e. a proper map which is an isomorphism outside $\rho^{-1}(0)$ such that $(P\rho)^{-1}(0)$ is a union of smooth divisors on $X$ with normal crossings. Let $e$ be the least common multiple of the multiplicities occurring in the fiber $(P\rho)^{-1}(0)$. Let $\tilde{S}$ be the disk with radius $\eta^{1/e}$ and define $\sigma : \tilde{S} \to S$ by $\sigma(t) = t^e$. Let $\tilde{X}$ be the normalization of the fiber product $X \times_S \tilde{S}$ and let $\pi : \tilde{X} \to X$ be the natural map. Let $D = (P\rho\pi)^{-1}(0)$ and denote $D_0, \cdots, D_m$ its irreducible components. The cohomology groups of the $D_i$

* Supported by the Netherlands Organization for the Advancement of Pure Research Z.W.O.

and their multiple intersections are the constituents of the mixed Hodge structure on $H^*(B_t)$.

The spaces $\tilde{X}$ and $D_i$ have singularities, but still their cohomology carries a pure Hodge structure because they only have quotient singularities.

In chapter 1 we develop a De Rham cohomology for projective varieties with quotient singularities. In chapter 2 we use this to get hold of the limit Hodge structure of a one parameter family of projective varieties over the punctured disk, whose monodromy is not unipotent. We use the resulting construction to put a mixed Hodge structure on $H^*(B_t)$ in chapter 3, after we have computed the mixed Hodge structure for a projective variety with only one singular point. Chapter 4 contains the study of the relations with problems concerning finiteness of monodromy, intersection form and local cohomology. Finally in chapter 5 we list some open problems.

We thank the I.H.E.S. for its hospitality during the preparation of this work and P. Deligne and N. A'Campo for their continuous stimulation.

## 1. Projective V-manifolds

(1.1) Definition. A $V$-manifold of dimension $n$ is a complex analytic space which admits an open covering $\{U_i\}$ such that each $U_i$ is analytically isomorphic to $Z_i/G_i$ where $Z_i \subset \mathbb{C}^n$ is an open ball and $G_i$ is a finite subgroup of $GL(n, \mathbb{C})$.

$V$-manifolds have been classified locally by D. Prill [15].

To state the result we need the following.

(1.2) Definition. A finite subgroup $G$ of $GL(n, \mathbb{C})$ is called *small* if no element of $G$ has 1 as an eigenvalue of multiplicity precisely $n-1$. In other words: $G$ contains no rotations around hyperplanes other than the identity.

For every finite subgroup $G$ of $GL(n, \mathbb{C})$ denote $G_{\text{big}}$ the normal subgroup of $G$ which is generated by all rotations around hyperplanes. Then the $G_{\text{big}}$-invariant polynomials form a polynomial algebra and hence $\mathbb{C}^n/G_{\text{big}}$ is isomorphic to $\mathbb{C}^n$.

The group $G/G_{\text{big}}$ maps isomorphically to a small subgroup of $GL(n, \mathbb{C})$, once a basis of invariant polynomials has been chosen. Hence local classification of $V$-manifolds reduces to the classification of actions of small subgroups of $GL(n, \mathbb{C})$.

(1.3) Theorem. *Let $G_1$ and $G_2$ be small subgroups of $GL(n, \mathbb{C})$. Then $\mathbb{C}^n/G_1 \cong \mathbb{C}^n/G_2$ if and only if $G_1$ and $G_2$ are conjugate subgroups.* Cf. [15].

We are interested in the Hodge theory of projective $V$-manifolds. The following proposition shows that we can expect an analogous situation as in the smooth projective case:

(1.4) PROPOSITION. *Every V-manifold is a rational homology manifold.*

PROOF. In view of the local classification we need only compute the local cohomology groups of $0 \in \mathbb{C}^n/G$ where $G \subset GL(n, \mathbb{C})$ is a small subgroup. Choose a $G$-invariant metric on $\mathbb{C}^n$. Let $D$ be the ball with radius 1 and let $S$ be its boundary. Then $H^k_{\{0\}}(\mathbb{C}^n/G, \mathbb{Q}) \cong H^k(\mathbb{C}^n/G, \mathbb{C}^n/G\text{-}\{0\}, \mathbb{Q}) \cong H^k(D/G, D/G\text{-}\{0\}, \mathbb{Q}) \cong \tilde{H}^{k-1}(D/G\text{-}\{0\}, \mathbb{Q}) \cong \tilde{H}^{k-1}(D\text{-}\{0\}, \mathbb{Q})^G \cong \tilde{H}^{k-1}(S, \mathbb{Q})^G = 0$ for $k \neq 2n$ and $= \mathbb{Q}$ for $k = 2n$. This follows from the fact that $S$ is an oriented $S^{2n-1}$ and that all elements of $G$ preserve its orientation.

(1.5) COROLLARY. *If $X$ is a complete algebraic V-manifold, then the canonical Hodge structure on $H^k(X)$ is purely of weight $k$ for all $k \geqq 0$. If $p: \tilde{X} \to X$ is a resolution of singularities for $X$, then the map $p^*: H^k(X) \to H^k(\tilde{X})$ is injective for all $k \geqq 0$.* Cf. [5], Th. (8.2.4).

(1.6) In the following sections we will construct a complex $\tilde{\Omega}^{\cdot}_X$ on the projective $V$-manifold $X$ with the following properties:

(i) $\tilde{\Omega}^p_X$ is a coherent analytic sheaf on $X$ for every integer $p$; $\tilde{\Omega}^p_X \neq 0$ if and only if $0 \leqq p \leqq \dim X$; the maps $d: \tilde{\Omega}^p_X \to \tilde{\Omega}^{p+1}_X$ are $\mathbb{C}$-linear;

(ii) $\tilde{\Omega}^{\cdot}_X$ is a resolution of the constant sheaf $\mathbb{C}$ on $X$;

(iii) the spectral sequence of hypercohomology

$$E_1^{pq} = H^q(X, \tilde{\Omega}^p_X) \Rightarrow \boldsymbol{H}^{p+q}(X, \tilde{\Omega}^{\cdot}_X) = H^{p+q}(X, \mathbb{C})$$

degenerates at $E_1 (E_1 = E_\infty)$ and the induced filtration on $H^{p+q}(X, \mathbb{C})$ coincides with the canonical Hodge filtration.

(1.7) DEFINITION. Let $X$ be a $V$-manifold and denote $\Sigma$ its singular locus. Denote $j: X - \Sigma \to X$ the inclusion map.

Then we define $\tilde{\Omega}^{\cdot}_X = j_* \tilde{\Omega}^{\cdot}_{X-\Sigma}$.

(1.8) LEMMA. *Let $D$ be an open ball with center 0 in $\mathbb{C}^n$. Let $G$ be a small subgroup of $GL(n, \mathbb{C})$ which leaves $D$ invariant and let $U = D/G$. Denote $\rho: D \to U$ the quotient map. Then $\Gamma(U, \tilde{\Omega}^p_U) = \Gamma(D, \Omega^p_D)^G$ for all $p \geqq 0$, i.e. $\tilde{\Omega}^{\cdot}_U \cong (\rho_* \Omega^{\cdot}_D)^G$.*

PROOF. Denote $\Sigma = \operatorname{Sing}(U)$ and $N = \rho^{-1}(\Sigma)$. Then $\Gamma(U, \tilde{\Omega}^p_U) = \Gamma(U - \Sigma, \Omega^p_{U-\Sigma}) = \Gamma(D - N, \Omega^p_D)^G$ because $\rho: D - N \to U - \Sigma$ is smooth (use that $G$ is small). Moreover $\Gamma(D - N, \Omega^p_D) = \Gamma(D, \Omega^p_D)$ because $D$ is smooth, $\Omega^p_D$ is locally free on $D$ and $N$ has codimension at least two in $D$.

(1.9) COROLLARY. *$\tilde{\Omega}^{\cdot}_X$ is a resolution of the constant sheaf $\mathbb{C}_X$ for every V-manifold $X$.*

PROOF. Take an open subset $U = D/G$ as in (1.8). The sequence

$$0 \to \mathbb{C} \to \Gamma(D, \mathcal{O}_D) \xrightarrow{d} \Gamma(D, \Omega^1_D) \to \cdots$$

is a $G$-equivariant exact sequence of $\mathbb{C}$-vector spaces, hence the sequence of $G$-invariants is also exact.

(1.10) COROLLARY. *$\tilde{\Omega}^p_X$ is coherent for all $p \geqq 0$.*

PROOF. The local quotient maps $\rho: D \to U$ are proper.

(1.11) LEMMA. *Let $\pi: \tilde{X} \to X$ be a resolution of singularities for the $V$-manifold $X$. Then $\tilde{\Omega}^{\cdot}_X \cong \pi_* \Omega^{\cdot}_{\tilde{X}}$.*

PROOF. Denote $\Sigma = \mathrm{Sing}(X)$, $D = \pi^{-1}(\Sigma)$. Then $D$ is a divisor with normal crossings on $\tilde{X}$. Denote $j: X - \Sigma \to X$ and $i: \tilde{X} - D \to \tilde{X}$ the inclusion maps. Then the map $\Omega^{\cdot}_{\tilde{X}} \to i_* \Omega^{\cdot}_{\tilde{X}-D}$ induces an injective map $\pi_* \Omega^{\cdot}_{\tilde{X}} \to \pi_* i_* \Omega^{\cdot}_{\tilde{X}-D} = j_* \pi_* \Omega^{\cdot}_{\tilde{X}-D} = \tilde{\Omega}^{\cdot}_X$. We have to show that this map is surjective. Let $p$ be a non-negative integer. Assume $X = Z/G$ with $Z$ an open ball in $\mathbb{C}^n$ and $G \subset GL(n, \mathbb{C})$ a small subgroup. The quotient sheaf $\tilde{\Omega}^p_X / \pi_* \Omega^p_{\tilde{X}}$ has support on $\Sigma$, so for every holomorphic function $f$ on $X$ which vanishes on $\Sigma$ there exists an integer $k$ such that $f^k \tilde{\Omega}^p_X \subset \pi_* \Omega^p_{\tilde{X}}$. Let $\omega$ be a holomorphic $p$-form on $\tilde{X} - D$ and let $x$ be a smooth point of $D$. We show that $\omega$ can be extended to a holomorphic $p$-form on a neighborhood of $x$ in $\tilde{X}$. First observe that $\omega$ is meromorphic along $D$: if $f$ is a holomorphic function on $X$ such that $f \tilde{\Omega}^p_X \subset \pi_* \Omega^p_{\tilde{X}}$ then $\pi^* f \cdot \omega$ extends to a holomorphic $p$-form on $\tilde{X}$.

Let $z_1, \cdots, z_n$ be holomorphic coordinates on a neighborhood $W$ of $x$ in $\tilde{X}$, centered at $x$, such that $D$ is given locally by the equation $z_1 = 0$. Write

$$\omega = \sum_{1 \leqq i_1 < \cdots < i_p \leqq n} a_{i_1 \cdots i_p} \, dz_{i_1} \wedge \cdots \wedge dz_{i_p}$$

with $a_{i_1 \cdots i_p}$ meromorphic along $D$. We show that $a_{i_1 \cdots i_p}$ is in fact holomorphic on $W$.

We may assume that $x$ is not contained in the support of the divisor of zeroes of $a_{i_1 \cdots i_p}$, because this divisor intersects $D$ in a set of codimension at least 2 in $\tilde{X}$ and it is sufficient to extend $\omega$ to the complement of a set of codimension two in $\tilde{X}$. So we may even suppose that $a_{i_1 \cdots i_p}$ does not vanish in any point of $W$. Then one may write $a_{i_1 \cdots i_p} = z_1^m \cdot b_{i_1 \cdots i_p}$ for some integer $m$, depending on $i_1, \cdots, i_p$ and some holomorphic function $b_{i_1 \cdots i_p}$ which is invertible on $W$.

If $i_1 = 1$, consider the $2p$-chain $\Gamma$ on $W$ given by the equations $z_j = 0 (j \neq i_1, \cdots, i_p)$. Let $C = \rho^{-1} \pi(\Gamma)$ and let $\eta$ be the holomorphic $G$-invariant

$p$-form on $Z$ corresponding to $\omega$. Then if $g=|G|$:

$$\int |z_1|^{2m}|b_{i_1}\cdots{}_{i_p}|^2\, dz_{i_1}\wedge\cdots\wedge dz_{i_p}\wedge d\bar{z}_{i_1}\wedge\cdots\wedge d\bar{z}_{i_p}$$

$$=\int_\Gamma \omega\wedge\bar{\omega}=\frac{1}{g}\int_C \eta\wedge\bar{\eta}<\infty$$

hence $m\geqq 0$, so $a_{i_1}\cdots{}_{i_p}$ is holomorphic on $W$. If $i_1\neq 1$, consider for all $t\in\mathbb{C}$ the $2p$-chain $\Gamma(t)$ on $W$ given by the equations $z_1=t$, $z_j=0$ $(j\neq 1, i_1,\cdots, i_p)$. Let $C(t)=\rho^{-1}\pi(\Gamma(t))$ and for $t\neq 0$ define

$$\alpha(t)=\int_{\Gamma(t)}\omega\wedge\bar{\omega}=\frac{1}{g}\int_{C(t)}\eta\wedge\bar{\eta}.$$

Then $\lim_{t\to 0}\alpha(t)=1/g\int_{C(0)}\eta\wedge\bar{\eta}$ exists. On the other hand $\alpha(t)\sim c|t|^{2m}$ for some $c>0$. This implies that $m\geqq 0$.

(1.12) THEOREM. *Let $X$ be a projective $V$-manifold. Then the spectral sequence of hypercohomology*

$$E_1^{pq}=H^q(X,\tilde{\Omega}_X^p)\Rightarrow H^{p+q}(X,\mathbb{C})$$

*degenerates at $E_1$ and the induced filtration on $H^{p+q}(X,\mathbb{C})$ coincides with the Hodge filtration.*

PROOF. It follows from [18], proposition IV.12 or from [12], remark 2.3 that every $V$-manifold is Cohen-Macaulay. One concludes from [9], section 3.2 that $\tilde{\Omega}_X^n$ $(n=\dim(X))$ is the canonical dualizing sheaf on $X$. The cup product $\tilde{\Omega}_X^p\otimes\tilde{\Omega}_X^{n-p}\to\tilde{\Omega}_X^n$ defines an injective sheaf homomorphism $\tau:\tilde{\Omega}_X^p\to \underline{\mathrm{Hom}}_{\mathcal{O}_X}(\tilde{\Omega}_X^{n-p},\tilde{\Omega}_X^n)$ for every $p\geqq 0$, given by $(\tau(\omega_1))(\omega_2)=\omega_1\wedge\omega_2$. We first show that $\tau$ is surjective for all $p$. Let $x\in X$ and let $U=Z/G$ be a neighborhood of $x$ in $X$ as in the proof of lemma (1.11). Denote $A=\mathcal{O}_{X,x}$, $B=\mathcal{O}_{Z,0}$, $N=\Omega_{Z,0}^{n-p}$, $M=\Omega_{Z,0}^n$. We have to show that $\mathrm{Hom}_B(N,M)^G=\mathrm{Hom}_A(N^G,M^G)$.

One has $\mathrm{Hom}_A(N^G,M^G)=\mathrm{Hom}_A(N^G,M)^G=\mathrm{Hom}_B(BN^G,M)^G$. The natural map $BN^G\to N$ induces a map $\mathrm{Hom}_B(N,M)\to\mathrm{Hom}_B(BN^G,M)$. It is sufficient to show that this map is an isomorphism. Let $R=N/BN^G$. Because $\rho:Z\to U$ is étale outside $S=\rho^{-1}(\Sigma)$ where $\Sigma=\mathrm{Sing}(X)$, $R$ has support contained in $S$. This implies that $\dim(R)\leqq n-2$. Because $M=A$ is Cohen-Macaulay, depth $(M_p)\geqq 2$ for every prime ideal $p\in\mathrm{Spec}(A)$ with $I(S)\subset p$. Hence $\mathrm{Ext}_A^i(R,M)=0$ for $i<2$ (cf. [10], Proposition 3.7). Therefore $\mathrm{Hom}_B(N,M)$ and $\mathrm{Hom}_B(BN^G,M)$ are isomorphic.

By Grothendieck duality the pairing

$$H^q(X,\tilde{\Omega}_X^p)\otimes H^{n-q}(X,\tilde{\Omega}_X^{n-p})\to H^n(X,\tilde{\Omega}_X^n)=\mathbb{C}$$

is non-singular for every $p,q\geqq 0$.

Let $\pi : \tilde{X} \to X$ be a resolution of singularities for $X$. Then there is a morphism of spectral sequences

$$\begin{array}{ccc} E_1^{pq} = H^q(X, \tilde{\Omega}_X^p) & \Rightarrow & H^{p+q}(X, \mathbb{C}) \\ \downarrow \pi^* & & \downarrow \pi^* \\ E_1^{pq} = H^q(\tilde{X}, \Omega_{\tilde{X}}^p) & \Rightarrow & H^{p+q}(\tilde{X}, \mathbb{C}) \end{array}$$

If $\omega \in H^q(X, \tilde{\Omega}_X^p)$, there exists $\eta \in H^{n-q}(X, \tilde{\Omega}_X^{n-p})$ with $\omega \wedge \eta \neq 0$. Then $\pi^*(\omega) \wedge \pi^*(\eta) = \pi^*(\omega \wedge \eta) \neq 0$ so $\pi^*(\omega) \neq 0$. Hence $\pi^*$ is injective on $E_1$. Because the spectral sequence for $\tilde{X}$ degenerates at $E_1$, so does the spectral sequence for $X$. Finally the induced filtration on $H^{p+q}(X, \mathbb{C})$ has to be the same as the Hodge filtration, because we get it by intersection of the Hodge filtration on $H^{p+q}(\tilde{X}, \mathbb{C})$ with $H^{p+q}(X, \mathbb{C})$ and the Hodge filtration is functorial.

We conclude this chapter with a generalization of the structure theorems for the cohomology of smooth projective varieties to the case of projective $V$-manifolds.

(1.13) THEOREM. *Let $X$ be a projective $V$-manifold of dimension $n$. Let $L \in H^2(X, \mathbb{Z})$ be the cohomology class of an ample divisor on $X$. Then for all $q \in \mathbb{N}$ the map $\omega \mapsto L^q \wedge \omega$ induces an isomorphism between $H^{n-q}(X, \mathbb{C})$ and $H^{n+q}(X, \mathbb{C})$.*

PROOF. Let $\pi : \tilde{X} \to X$ be a resolution of singularities for $X$. Then $\pi$ is a projective map and there exists a divisor $D$ on $\tilde{X}$ which is very ample for $\pi$ and such that, if $[D]$ is the cohomology class of $D$, then $[D] \wedge \pi^*(\eta) = 0$ for all $\eta \in H^i(X, \mathbb{C})$ and all $i$, i.e. $\pi_*[D] = 0$. It follows from [11], prop. (4.4.10)(ii) that for $k$ sufficiently large $C = [D] + k\pi^*L$ is the cohomology class of a very ample divisor on $\tilde{X}$. Hence for all $q \in \mathbb{N}$ the map $\omega \mapsto C^q \wedge \omega$ induces an isomorphism between $H^{n-q}(\tilde{X}, \mathbb{C})$ and $H^{n+q}(\tilde{X}, \mathbb{C})$. For $\eta \in H^{n-q}(X, \mathbb{C})$ one has $C \wedge \pi^*(\eta) = k\pi^*L \wedge \pi^*(\eta) = k\pi^*(L \wedge \eta)$ so $C^q \wedge \pi^*\eta = k^q\pi^*(L^q \wedge \eta)$. Because $\pi^*$ is injective, the map $\omega \mapsto L^q \wedge \omega$ is injective on $H^{n-q}(X, \mathbb{C})$. By Poincaré duality $H^{n-q}(X, \mathbb{C})$ and $H^{n+q}(X, \mathbb{C})$ have equal dimensions, hence $L^q$ is an isomorphism.

(1.14) COROLLARY. *Define the primitive cohomology groups $P^k(X, \mathbb{C})(k \in \mathbb{Z})$ by $P^k(X, \mathbb{C}) = \operatorname{Ker}(L^{n-k+1} : H^k(X, \mathbb{C}) \to H^{2n-k+2}(X, \mathbb{C}))$. Then for all $q \geqq 0$ one has the primitive decomposition*

$$H^q(X, \mathbb{C}) = \bigoplus_{r \geqq 0} L^r P^{q-2r}(X, \mathbb{C}).$$

(1.15) REMARK. Obviously the primitive decomposition depends on the choice of the ample divisor $L$ on $X$. However if one chooses ample divisors

$L$ and $C$ on $X$ resp. $\tilde{X}$ as in the proof of (1.13), the map $\pi^*$ will preserve the primitive decompositions. In particular one concludes that the hermitian form $Q$ on $H^{n-q}(X, \mathbb{C})$ defined by

$$Q(x, y) = (-1)^{(n-q)(n-q-1)/2} Cx \wedge L^q \bar{y}[X]$$

where $C$ is Weil's operator and the bar denotes complex conjugation with respect to $H^{n-q}(X, \mathbb{R})$, induces a positive definite hermitian form on $P^{n-q}(X, \mathbb{C})$.

The following is a generalization of [4] to the case of algebraic $V$-manifolds.

(1.16) Definition. Let $X$ be a $V$-manifold. A divisor $Y$ on $X$ is called a divisor with $V$-normal crossings if locally on $X$ one has $(X, Y) = (Z, D)/G$ with $Z \subset \mathbb{C}^n$ an open domain, $G \subset GL(n, \mathbb{C})$ a small subgroup acting on $Z$ and $D \subset Z$ a $G$-invariant divisor with normal crossings.

(1.17) Definition. Let $X$ be a $V$-manifold and $Y$ a divisor with $V$-normal crossings on $X$. Define the complex $\tilde{\Omega}^{\cdot}_X(\log Y)$ on $X$ by

$$\tilde{\Omega}^{\cdot}_X(\log Y) = j_* \tilde{\Omega}^{\cdot}_{X-\Sigma}(\log (Y - \Sigma))$$

where $\Sigma = \operatorname{Sing}(X)$ and $j : X - \Sigma \to X$ is the inclusion map. One checks like before that if $(X, Y) = (Z, D)/G$ as in (1.16) and if $\rho : Z \to X$ is the quotient map, then

$$\tilde{\Omega}^{\cdot}_X(\log Y) = (\rho_* \Omega^{\cdot}_Z(\log D))^G.$$

If $\pi : \tilde{X} \to X$ is a resolution of singularities for $X$ such that the total transform $\tilde{Y}$ of $Y$ is a divisor with normal crossings on $\tilde{X}$, then

$$\tilde{\Omega}^{\cdot}_X(\log Y) = \pi_* \Omega^{\cdot}_{\tilde{X}}(\log \tilde{Y}).$$

(1.18) Definition. The weight filtration $W$ on $\tilde{\Omega}^{\cdot}_X$ $(\log Y)$ is defined by

$$W_k \tilde{\Omega}^p_X(\log Y) = \tilde{\Omega}^k_X(\log Y) \wedge \tilde{\Omega}^{p-k}_X \qquad (k \in \mathbb{Z})$$

and the Hodge filtration $F$ is given by

$$F^k \tilde{\Omega}^p_X(\log Y) = \tilde{\Omega}^p_X(\log Y) \quad \text{if} \quad p \geqq k;$$
$$F^k \tilde{\Omega}^p_X(\log Y) = 0 \qquad \text{if} \quad p < k.$$

Thus $W$ is increasing and $F$ is decreasing.

Assume that $Y$ is a union of irreducible components $Y_1, \cdots, Y_m$ without self-intersection. Denote $\tilde{Y}^{(p)}$ the disjoint union of all $p$-fold intersections $Y_{i_1} \cap \cdots \cap Y_{i_p}$ for $1 \leqq i_1 < \cdots < i_p \leqq m$.

Denote $a_p : \tilde{Y}^{(p)} \to X$ the natural map. Analogous to the smooth case one has a residue map

$$R : W_k \tilde{\Omega}^p_X(\log Y) \to (a_k)_* \tilde{\Omega}^{p-k}_{\tilde{Y}^{(k)}} \qquad (p, k \geqq 0).$$

(Remark that $\tilde{Y}^{(k)}$ is a $V$-manifold for every $k \geqq 0$).

(1.19) LEMMA. *$R$ induces for every $k \geqq 0$ an isomorphism of complexes*

$$Gr^W_k \tilde{\Omega}^{\cdot}_X(\log Y) \to (a_{k*}) \tilde{\Omega}^{\cdot}_{\tilde{Y}^{(k)}}[-k].$$

PROOF. Analogous to corollary (1.9).

This lemma together with theorem (1.12) show that one may use the bifiltered complex $(\tilde{\Omega}^{\cdot}_X(\log Y), F, W)$ to compute the canonical mixed Hodge structure on $H^k(X-Y)(k \geqq 0)$. In particular one has the spectral sequence

$$E_1^{-n,k+n} = H^{k-n}(\tilde{Y}^{(n)}, \mathbb{Q})(-n) \Rightarrow H^k(X-Y, \mathbb{Q}).$$

With $E_2^{-n,k+n} = E_\infty^{-n,k+n} = Gr^W_{n+k} H^k(X-Y, \mathbb{Q})$. Cf. [5], where the notion of cohomological mixed Hodge complex is used.

(1.20) EXAMPLE. Let $f(z_0, \cdots, z_n)$ be a quasi-homogeneous polynomial. This means that there exist positive rational numbers $w_0, \cdots, w_n$ such that for all $t \in \mathbb{C}$ one has

$$f(t^{w_0} z_0, \cdots, t^{w_n} z_n) = t f(z_0, \cdots, z_n).$$

One may compute the mixed Hodge structure on the affine variety $X \subset \mathbb{C}^{n+1}$ with equation $f(z) = 1$ in the following way, provided $f$ has an isolated critical point at 0.

Write $w_i = u_i / v_i$ with $(u_i, v_i) = 1$. Let $d = lcm(v_0, \cdots, v_n)$ and define $b_i = dw_i$, $i = 0, \cdots, n$.

Define $(n+1) \times (n+1)$-matrices $g^{(0)}, \cdots, g^{(n)}$ by $g^{(i)}_{jk} = 0$ if $j \neq k$, $g^{(i)}_{kk} = 1$ if $i \neq k$ and $g^{(k)}_{kk} = \exp(2\pi i / b_k)$.

Let $G$ be the subgroup of $PGL(n+2, \mathbb{C})$ generated by the elements

$$\begin{bmatrix} g^{(k)} & 0 \\ 0 & 1 \end{bmatrix} \qquad (k = 0, \cdots, n).$$

Then $G$ acts on $\mathbb{P}^{n+1}(\mathbb{C})$. Let $M$ be the quotient $\mathbb{P}^{n+1}/G$.

Denote $h(y_0, \cdots, y_n) = f(y_0^{b_0}, \cdots, y_n^{b_n})$. Let $Z \subset \mathbb{C}^{n+1}$ be given by the equation $h(y) = 1$. Denote $\bar{Z} \subset \mathbb{P}^{n+1}$ its projective closure. Then $G$ leaves the pair $(Z, \bar{Z})$ invariant. Moreover $X = Z/G$. If $f$ has an isolated singularity at 0 then $\bar{X} = \bar{Z}/G$ and $\bar{X} - X$ are $V$-manifolds (though $\bar{Z}$ need not be smooth).

The exact sequence

$$0 \to P^n(\bar{X}) \to H^n(X) \to P^{n-1}(\bar{X}-X) \to 0$$

shows that $Gr_k^W H^n(X) = 0$ for $k \neq n,\ n+1$.

See [20] for a more detailed description.

## 2. Limits of Hodge structures

The purpose of this chapter is: to give an explicit description for the limit Hodge structure associated to a projective one parameter family over the punctured disk. This has been done in [19] for families with unipotent monodromy.

(2.1) NOTATIONS. $S$ is the unit disk in the complex plane, $S^* = S - \{0\}$ and $H = \{z \in \mathbb{C} \mid \mathrm{Im}\,(z) > 0\}$ is the universal covering of $S^*$ by the map $z \to \exp(2\pi i z)$.

$X$ is a smooth, closed, connected subvariety of dimension $n+1$ of $\mathbb{P}^r \times S$ for some $r > 0$. We denote $f: X \to S$ the projection on $S$. We assume that $f$ is surjective, smooth in every point $x \in X$ with $f(x) \neq 0$ and that $f^{-1}(0)$ is a union $E_0 \cup \cdots \cup E_m$ of smooth divisors on $X$ which cross normally. Denote $e_i$ the multiplicity of $E_i$ and let $e = lcm(e_0, \cdots, e_m)$.

Let $\tilde{S}$ be another copy of the unit disk and define $\sigma: \tilde{S} \to S$ by $\sigma(t) = t^e$. Denote $\tilde{X}$ the normalization of $X \times_S \tilde{S}$ and let $\pi: \tilde{X} \to X$ and $\tilde{f}: \tilde{X} \to \tilde{S}$ be the natural maps.

Denote $D_i = \pi^{-1}(E_i)$, $i = 0, \cdots, m$. Let $D = \bigcup_{i=0}^m D_i$. Let $X_\infty = X \times_S H$

$$\begin{array}{ccccc} D_i & \longrightarrow & \tilde{X} & \xrightarrow{\tilde{f}} & \tilde{S} \\ \downarrow{\scriptstyle \pi} & & \downarrow{\scriptstyle \pi} & & \downarrow{\scriptstyle \sigma} \\ E_i & \longrightarrow & X & \xrightarrow{f} & S \end{array}$$

(2.2) LEMMA. *$\tilde{X}$ is a V-manifold and $\tilde{f}^{-1}(0) = D$ is a reduced divisor with V-normal crossings on $\tilde{X}$.*

PROOF. Cover $X$ with coordinate neighborhoods $A$ with coordinates $z_0, \cdots, z_n$ such that there exist integers $\nu \geqq 0$, $d_0, \cdots, d_\nu \geqq 1$ with $f(z_0, \cdots, z_n) = z_0 \cdots z_\nu$. Fix one such $A$.

Let $\tilde{A}$ be the normalization of $A \times_S \tilde{S}$. Then $\tilde{A}$ is isomorphic to an open analytic subset of the normalization of

$$\{(w, z_0, \cdots, z_n) \in \mathbb{C}^{n+2} \mid z_0^{d_0} \cdots z_\nu^{d_\nu} = w^e\}.$$

Let $d = gcd(d_0, \cdots, d_\nu)$, $e' = e/d$, $d_i' = d_i/d$ $(i = 0, \cdots, \nu)$ and $\zeta = \exp(2\pi i/d)$.

Then the normalization $\tilde{R}$ of the ring

$$R = \mathbb{C}[z_0, \cdots, z_n, w]/z_0^{d_0} \cdots z_\nu^{d_\nu} - w^e)$$

is isomorphic to the direct product of $d$ isomorphic copies of the normalization of the ring

$$\mathbb{C}[z_0, \cdots, z_n, w]/(z_0^{d_0'} \cdots z_\nu^{d_\nu'} - w^{e'})$$

so to compute it we only need to consider the case $d = 1$.

Denote $c_i = e/d_i$ for $i = 0, \cdots, \nu$. Denote $R_1 = \mathbb{C}[y_0, \cdots, y_n]$; consider $R$ as a subring of $R_1$ by putting $z_i = y_i^{c_i}$ if $0 \leqq i < \nu$, $z_i = y_i$ for $i > \nu$ and $w = y_0 \cdots y_\nu$.

The group $G = \mathbb{Z}/(c_0) \times \cdots \times \mathbb{Z}/(c_\nu)$ acts on $R_1$ by

$$(a_0, \cdots, a_\nu) \cdot y_j = \exp(2\pi i a_j/c_j) \cdot y_j \quad \text{if} \quad j \leqq \nu;$$

$$(a_0, \cdots, a_\nu) \cdot y_j = y_j \quad \text{if} \quad j > \nu.$$

The ring of $G$-invariants $R_1^G$ is just $\mathbb{C}[z_0, \cdots, z_n]$. Let $G' = \{g \in G \mid gw = w\}$. Then $R$ is isomorphic to $R_1^{G'}$, because $R_1$ integrally closed and $G' = \text{Gal}(K_1/K)$ where $K_1$ and $K$ are the fields of fractions of $R_1$ and $R$ respectively. This shows that $\text{Spec}(\tilde{R})$ and hence $\tilde{A}$ and $\tilde{X}$ are $V$-manifolds. The action of $G'$ on $R_1$ identifies $G'$ with a small subgroup of $GL(n+1, \mathbb{C})$, because no element of $G$ of the form $(0, \cdots, 0, a_i, 0, \cdots, 0)$ with $a_i \neq 0$ leaves $w$ invariant.

Locally $D$ is the quotient under $G'$ of the reduced divisor $y_0 \cdots y_\nu = 0$ and every generic point of this divisor has a trivial isotropy group. Hence $D$ is reduced.

(2.3) Remark. With the same notations a basis for $\tilde{R}$ as a free module over $\mathbb{C}[z_0, \cdots, z_n]$ is obtained as follows. Write $k = q_{ik}c_i + r_{ik}$ $(k = 0, \cdots, e-1;\ i = 0, \cdots, \nu)$ with $q_{ik} \in \mathbb{Z}$ and $0 \leqq r_{ik} < c_i$. Define $x_k = \prod_{i=0}^{\nu} y_i^{r_{ik}} = w^k \prod_{i=0}^{\nu} z_i^{-q_{ik}}$. Then $x_0, \cdots, x_{e-1}$ form a basis for $\tilde{R}$ as a $\mathbb{C}[z_0, \cdots, z_n]$-module. The group $G/G' = \mathbb{Z}(e)$ acts on $\tilde{X}$. This action coincides with the action of $\mathbb{Z}/(e)$ on $\tilde{X}$ which is induced by multiplication with $e$th roots of unity on $\tilde{S}$. One gets $X$ back as the quotient of $\tilde{X}$ under this action.

(2.4) Example. Define $Y \subset P^2(\mathbb{C}) \times S$ by the equation $x^2z - y^3 = tz^3$ where $(x, y, z)$ are homogeneous coordinates on $\mathbb{P}^2(\mathbb{C})$. If one blows up a point three times one obtains a manifold $X$ with a projection $f: X \to S$ such that $f^{-1}(0) = E_0 + 2E_1 + 3E_2 + 6E_3$, the $E_i$ intersecting like in fig. 1 and all $E_i$ non-singular rational curves.

In this case $X$ is smooth and $D_0$, $D_1$, $D_2$ are the disjoint union of 1,2 resp. 3 non-singular rational curves, while $D_3$ is an elliptic curve. See fig. 2.

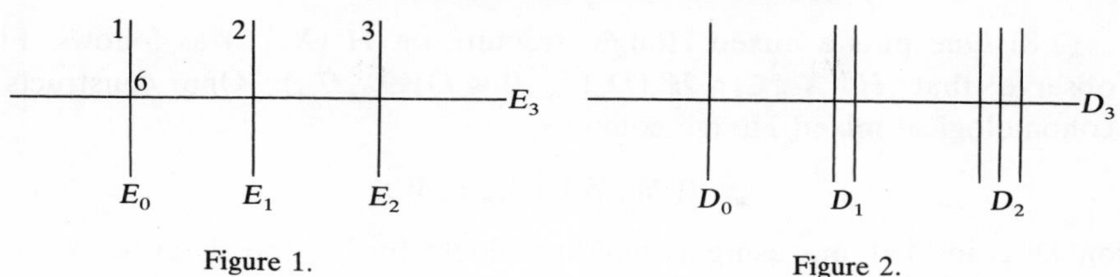


Figure 1. Figure 2.

(2.5) REMARK. From (2.2) one can conclude that for every $p \geqq 0$ the sheaf $\tilde{\Omega}^p_{\tilde{X}}(\log D)$ is locally free of rank $\binom{n+1}{p}$ on $\tilde{X}$. In fact, on an open subset $\tilde{A}$ as above, $\tilde{\Omega}^1_{\tilde{X}}(\log D)$ is the free $\mathcal{O}_{\tilde{X}}$-module on the generators $dz_i/z_i (i = 0, \cdots, \nu)$ and $dz_j (j > \nu)$ and $\tilde{\Omega}^p_{\tilde{X}}(\log D) = \Lambda^p_{\mathcal{O}_{\tilde{X}}} \Omega^1_{\tilde{X}}(\log D)$.

If $w$ is a coordinate on $\tilde{S}$ with $w^e = z_0^{d_0} \cdots z_\nu^{d_\nu}$ on $\tilde{A}$, then $dw/w = \sum_{i=0}^{\nu} (d_i/e)\, dz_i/z_i = \sum_{i=0}^{\nu} dy_i/y_i$. This shows that $\tilde{f}^* \Omega^1_{\tilde{S}}(\log 0)$ is locally a direct factor of $\tilde{\Omega}^1_{\tilde{X}}(\log D)$.

(2.6) DEFINITION.

$$\tilde{\Omega}^p_{\tilde{X}/\tilde{S}}(\log D) = \tilde{\Omega}^p_{\tilde{X}}(\log D)/\tilde{f}^* \Omega^1_{\tilde{S}}(\log 0) \wedge \tilde{\Omega}^{p-1}_{\tilde{X}}(\log D).$$

Then one has a complex $\tilde{\Omega}^{\cdot}_{\tilde{X}/\tilde{S}}(\log D)$ of locally free sheaves of finite rank on $\tilde{X}$ and the differentials in this complex are $f^{\cdot}\mathcal{O}_{\tilde{S}}$-linear. Moreover

$$0 \longrightarrow \tilde{\Omega}^{\cdot}_{\tilde{X}/\tilde{S}}(\log D)[-1] \xrightarrow{\wedge dw/w} \tilde{\Omega}^{\cdot}_{\tilde{X}}(\log D) \longrightarrow \tilde{\Omega}^{\cdot}_{\tilde{X}/\tilde{S}}(\log D) \longrightarrow 0$$

is an exact sequence of complexes on $\tilde{X}$.

Now we have all ingredients to describe the limit Hodge structure of the family $\tilde{f}: \tilde{X} \to \tilde{S}$ (or equivalently: of the family $f: X \to S$). We state the results without proofs, because these are quite the same as for the unipotent case. We refer to [5], for definitions of concepts in Hodge theory.

(2.7) THEOREM. *For every $p \geqq 0$ the sheaf $\boldsymbol{R}^p \tilde{f}_* \tilde{\Omega}^{\cdot}_{\tilde{X}/\tilde{S}}(\log D)$ is locally free of finite rank on $\tilde{S}$.*

The choice of a parameter $w$ on $\tilde{S}$ determines an isomorphism

$$\psi_w : \boldsymbol{H}^p(D, \tilde{\Omega}^{\cdot}_{\tilde{X}/\tilde{S}}(\log D) \otimes_{\mathcal{O}_X} \mathcal{O}_D) \xrightarrow{\sim} H^p(X_\infty, \mathbb{C}).$$

If $w'$ is another parameter on $S$ with $a = (w'w^{-1})(0)$ then $\psi_{w'}^{-1}\psi_w = \exp(-2\pi i \log(a) \operatorname{Res}_0(\tilde{\nabla}))$ where

$$\tilde{\nabla} : \boldsymbol{R}^p \tilde{f}_* \tilde{\Omega}^{\cdot}_{\tilde{X}/\tilde{S}}(\log D) \to \Omega^1_S(\log 0) \otimes_{\mathcal{O}_S} \boldsymbol{R}^p \tilde{f}_* \tilde{\Omega}^{\cdot}_{\tilde{X}/\tilde{S}} \log D)$$

is the Gauss-Manin connection; its residue $\operatorname{Res}_0(\tilde{\nabla})$ is nilpotent.

(2.8) One puts a mixed Hodge structure on $H^p(X_\infty, \mathbb{C})$ as follows. First observe that $H^p(X_\infty, \mathbb{C}) = \boldsymbol{H}^p(D, \tilde{\Omega}^{\cdot}_{\tilde{X}/\tilde{S}}(\log D) \otimes_{\mathcal{O}_X} \mathcal{O}_D)$. One constructs a cohomological mixed Hodge complex

$$((A^{\cdot}_{\mathbb{Q}}, W), (A^{\cdot}_{\mathbb{C}}, F, W))$$

on $D$ as in [19], but using as building blocks for $A^{\cdot}_{\mathbb{C}}$ the sheaves

$$A^{pq} = \tilde{\Omega}^{p+q+1}_{\tilde{X}}(\log D)/W_q \tilde{\Omega}^{p+q+1}_{\tilde{X}}(\log D).$$

This becomes a cohomological mixed Hodge complex by theorem (1.12). It satisfies

$$Gr^W_r A^{\cdot}_{\mathbb{Q}} = \bigoplus_{k \geqq 0, -r} (a_{r+2k+1})_* \mathbb{Q}_{\tilde{D}^{(r+2k+1)}}(-r-k)[-r-2k];$$

$$Gr^W_r A^{\cdot}_{\mathbb{C}} = \bigoplus_{k \geqq 0, -r} (a_{r+2k+1})_* \tilde{\Omega}_{\tilde{D}^{(r+2k+1)}}[-r-2k].$$

(2.9) Corollary. *The spectral sequence of hypercohomology for the filtered complex* $(A^{\cdot}_{\mathbb{Q}}, W)$:

$$E_1^{-r, q+r} = \bigoplus_{k \geqq 0, -r} H^{q-r-2k}(\tilde{D}^{(2k+r+1)}, \mathbb{Q})(-r-k) \Rightarrow H^q(X_\infty, \mathbb{Q})$$

*degenerates at* $E_2$; *the spectral sequence of hypercohomology for the filtered complex* $(A^{\cdot}_{\mathbb{C}}, F)$:

$$E_1^{pq} = H^q(D, \tilde{\Omega}^p_{\tilde{X}/\tilde{S}}(\log D) \otimes_{\mathcal{O}_{\tilde{X}}} \mathcal{O}_D) \Rightarrow H^{p+q}(X, \mathbb{C})$$

*degenerates at* $E_1$ *i.e.* $E_1^{pq} = Gr^p_F H^{p+q}(X_\infty, \mathbb{C})$.

(2.10) Corollary. *For every* $p, q \geqq 0$ *the sheaf* $R^q \tilde{f}_* \tilde{\Omega}^p_{\tilde{X}/\tilde{S}}(\log D)$ *is locally free on* $\tilde{S}$.

Proof. For $w \in \tilde{S}$ with $w \neq 0$ one has $R^q \tilde{f}_* \tilde{\Omega}^p_{\tilde{X}/\tilde{S}}(\log D)(w) \cong H^q(\tilde{X}_w, \Omega^p_{\tilde{X}_w})$. Because $f$ is smooth and proper over the punctured disk, the sheaf $R^q \tilde{f}_* \tilde{\Omega}^p_{\tilde{X}/\tilde{S}}(\log D)$ is locally free on $\tilde{S} - \{0\}$. By semi-continuity one has

$$\dim_{\mathbb{C}} H^q(D, \tilde{\Omega}^p_{\tilde{X}/\tilde{S}}(\log D) \otimes_{\mathcal{O}_X} \mathcal{O}_D) \geqq \dim_{\mathbb{C}} H^q(\tilde{X}_w, \Omega^p_{\tilde{X}_w}).$$

Because of (2.7) and (2.9) one has

$$\sum_{p+q=r} \dim_{\mathbb{C}} H^q(D, \tilde{\Omega}^p_{\tilde{X}/\tilde{S}}(\log D) \otimes_{\mathcal{O}_{\tilde{X}}} \mathcal{O}_D) = \sum_{p+q=r} \dim_{\mathbb{C}} H^q(\tilde{X}_w, \Omega^p_{\tilde{X}_w})$$

for all $r \geqq 0$, both sides being equal to the rank of the locally free sheaf $\boldsymbol{R}^r \tilde{f}_* \tilde{\Omega}^{\cdot}_{\tilde{X}/\tilde{S}}(\log D)$. Thus we must have continuity of the dimensions. Because $\tilde{f}$ is flat we may conclude by [14], Corollary 2 of p. 50.

(2.11) THEOREM. For all $p, q \geqq 0$ the sheaves $R^q f_* \Omega^p_{X/S}(\log E)$ are locally free of finite rank on $S$.

PROOF. With the local computations of (2.2) one shows that $\Omega^p_{X/S}(\log E)$ is equal to the subsheaf of $\mathbb{Z}/(e)$-invariants of $\pi_* \tilde{\Omega}^p_{\tilde{X}/\tilde{S}}(\log D)$ and hence is a direct factor of it. Therefore it is sufficient to show that $R^q f_*(\pi_* \tilde{\Omega}^p_{\tilde{X}/\tilde{S}}(\log D))$ is locally free of finite rank on $S$. One has

$$R^q f_*(\pi_* \tilde{\Omega}^p_{\tilde{X}/\tilde{S}}(\log D)) = R^q(f\pi)_* \tilde{\Omega}^p_{\tilde{X}/\tilde{S}}(\log D) = R^q(\sigma \tilde{f})_* \tilde{\Omega}^p_{\tilde{X}/\tilde{S}}(\log D)$$
$$= \sigma_* R^q \tilde{f}_* \tilde{\Omega}^p_{\tilde{X}/\tilde{S}}(\log D)$$

because $\pi$ and $\sigma$ are finite maps. The last term is locally free on $S$ by corollary (2.10).

(2.12) The maps $\tilde{f}: \tilde{X} - D \to \tilde{S} - \{0\}$ and $f: X - E \to S - \{0\}$ are $C^\infty$-fibrations. For $w \in \tilde{S}$, $w \neq 0$ the fundamental groups $\pi_1(\tilde{S} - \{0\}, w)$ and $\pi_1(S - \{0\}, w^e)$ act on the cohomology of the fiber $\tilde{X}_w = X_{w^e}$. The actions of the positive generators of these groups extend to automorphisms $\tilde{T}$ resp. $T$ of the sheaves $\boldsymbol{R}^q \tilde{f}_* \tilde{\Omega}^{\cdot}_{\tilde{X}/\tilde{S}}(\log D)$ resp. $\boldsymbol{R}^q f_* \Omega^{\cdot}_{X/S}(\log E)$. Denote $\tilde{T}_0$ resp. $T_0$ their fibers over $0 \in \tilde{S}$ resp. $0 \in S$. Denoting $\tilde{\nabla}$ and $\nabla$ the corresponding Gauss-Manin connections, one has:

$$T_0 = \exp(-2\pi i \operatorname{Res}_0(\nabla)); \qquad \tilde{T}_0 = \exp(-2\pi i \operatorname{Res}_0(\tilde{\nabla})).$$

We fix parameters $t, \tau$ on $S, \tilde{S}$ with $t = \tau^e$. These choices determine isomorphisms

$$\boldsymbol{H}^q(D, \tilde{\Omega}^{\cdot}_{\tilde{X}/\tilde{S}}(\log D) \otimes_{\mathcal{O}_{\tilde{X}}} \mathcal{O}_D) \to H^q(X_\infty, \mathbb{C}) \leftarrow \boldsymbol{H}^q(E, \Omega^{\cdot}_{X/S}(\log E) \otimes_{\mathcal{O}_X} \mathcal{O}_E).$$

We consider $T_0$, $\tilde{T}_0$, $\operatorname{Res}_0(\nabla)$ and $\operatorname{Res}_0(\tilde{\nabla})$ as endomorphisms of $H^q(X_\infty)$ by means of these isomorphisms. Then $T_0^e = \tilde{T}_0$ is unipotent. Denote $N = \log(\tilde{T}_0) = \sum_{k=0}^{n+1}(-1)^{k+1}(T_0 - I)^k/k$ and denote $\gamma_s$ the semisimple part of $T_0$.

(2.13) THEOREM. *Let $q \geqq 0$. Then*

(1) *$N$ is a morphism of Hodge structures of type $(-1, -1)$ on $H^q(X_\infty)$, i.e. $N(W_k H^q(X_\infty)) \subset W_{k-2} H^q(X_\infty)$ and $N(F^p H^q(X_\infty)) \subset F^{p-1} H^q(X_\infty)$ for all $p$, $k \geqq 0$.*

(2) *For every $r \geqq 0$ the map*

$$N^r : Gr^W_{q+r} H^q(X_\infty) \to Gr^W_{q-r} H^q(X_\infty)(-r)$$

*is an isomorphism of Hodge structures.*

(3) *$\gamma_s$ is an isomorphism of mixed Hodge structures.*

PROOF. One proves (1) and (2) in the same way as [19], theorem (5.9). To prove (3), consider the map $\lambda: \tilde{X} \to \tilde{X}$ induced by the map $\tau \to \exp(2\pi i \tau/e)$ on $\tilde{S}$. It induces an automorphism of $D_{i_1} \cap \cdots \cap D_{i_r}$ for every $r$-tuple

$(i_1, \cdots, i_r)$ with $0 \leqq i_1 < \cdots < i_r \leqq m$. Therefore $\lambda^*$ acts as an automorphism on each of the spectral sequences of (2.9). We will show that $\gamma_s = \lambda^*$. For this we need $\lambda^* T_0 = T_0 \lambda^*$ and $T_0(\lambda^*)^{-1}$ unipotent. Take $t_0 \in S - \{0\}$, $\tau_0 \in \tilde{S}$ with $\tau_0^e = t_0$. Let $\zeta = \exp(2\pi i/e)$. Because $X_\infty \to H$ is a $C^\infty$ fiber bundle, there exists a continuous family of diffeomorphisms

$$h_\alpha : X_{t_0} \to X_{t_0 \exp(2\pi i \alpha)} \qquad (\alpha \in \mathbb{R})$$

with $h_0 = id$ and $h_{\alpha+\alpha'} = h_\alpha \circ h_{\alpha'}$ for $\alpha, \alpha' \in \mathbb{R}$. Each $h_\alpha$ is uniquely determined up to homotopy. Hence the diagram

$$\begin{array}{ccc} \tilde{X}_{\tau_0} & \xrightarrow{h_1} & \tilde{X}_{\tau_0\zeta} \\ \downarrow \lambda^{-1} & & \downarrow \lambda^{-1} \\ \tilde{X}_{\tau_0\bar{\zeta}} & \xrightarrow{h_1} & \tilde{X}_{\tau_0} \end{array}$$

is commutative up to homotopy (recall that $\tilde{X}_\tau = X_{\tau^e}$ for $\tau \in \tilde{S} - \{0\}$). This implies that on $H^q(\tilde{X}_{\tau_0})$ one has the relation $h_1^*(\lambda^{-1})^* = (\lambda^{-1})^* h_1^*$. Taking limits for $t \to 0$ one gets the relation $T_0(\lambda^*)^{-1} = (\lambda^*)^{-1} T_0$ on $H^q(X_\infty)$, hence also $T_0 \lambda^* = \lambda^* T_0$.

Both $T_0$ and $\lambda^*$ act on the spectral sequence

$$E_2^{pq} = H^p(E, i^{\cdot} R^q k_* \mathbb{C}_{X_\infty}) \Rightarrow H^{p+q}(X_\infty, \mathbb{C})$$

(cf. [19], (2.4) and (2.5)), where $k : X_\infty \to X$ is the natural map and $i : E \to X$ is the inclusion. In fact, denoting $\tilde{i} : D \to \tilde{X}$ and $\tilde{k} : X_\infty \to \tilde{X}$ the analogous maps for $X$, one has

$$i^{\cdot} R^q k_* \mathbb{C}_{X_\infty} = \pi_* \tilde{i}^{\cdot} R^q \tilde{k}_* \mathbb{C}_{X_\infty}$$

because $\pi$ is finite; hence $\lambda^*$ acts on the sheaf $i^{\cdot} R^q k_* \mathbb{C}_{X_\infty}$. Let $Q \in E$. Let $z_0, \cdots, z_n$ be coordinates on a neighborhood $V$ of $Q$ in $X$, centered at $Q$, such that $f(z_0, \cdots, z_n) = z_0^{d_0} \cdots z_\nu^{d_\nu}$ for some integers $\nu \geqq 0$, $d_0, \cdots, d_\nu \geqq 1$. For $\varepsilon > 0$, $0 < \eta \ll \varepsilon$ denote $V_{\varepsilon,\eta}$ the set $\{(z_0, \cdots, z_n) \in V \mid \sum_{i=0}^n |z_i|^2 < \varepsilon$ and $|f(z)| < \eta\}$. For $\varepsilon$, $\eta$ sufficiently small one has (cf. [19], (2.8))

$$(i^{\cdot} R^q k_* \mathbb{C}_{X_\infty})_Q = H^q(k^{-1} V_{\varepsilon,\eta}, \mathbb{C}),$$

and $k^{-1} V_{\varepsilon,\eta}$ consists of $d = gcd(d_0, \cdots, d_\nu)$ components, each of which has the homotopy type of a $\nu$-dimensional torus $S^1 \times \cdots \times S^1$. Hence its cohomology ring is an exterior algebra on $H^0(k^{-1} V_{\varepsilon,\eta}, \mathbb{C})$ which is isomorphic to $\mathbb{C}[\tau]/(\tau^d - 1)$. Both $T_0$ and $\lambda^*$ act on this algebra by cyclic permutation of the components, i.e. by the substitution $\tau \to \exp(2\pi i/d)\tau$ leaving a set of generators for the exterior algebra fixed. This implies that $T_0(\lambda^*)^{-1}$

acts as the identity on $E_2^{pq} = H^p(E, i^{\cdot} R^q k_* \mathbb{C}_{X_\infty})$ for all $p, q \geqq 0$, so $T_0(\lambda^*)^{-1}$ is unipotent on $H^{p+q}(X_\infty, \mathbb{C})$.

(2.14) EXAMPLE. Consider again example (2.4). One can show that $H^1(X_\infty) = H^1(D_3)$. Because $D_3$ is an elliptic curve with an automorphism $\lambda$ of order 6, the Hodge filtration on $H^1(X_\infty, \mathbb{C})$ is given by

$$F^1 H^1(X_\infty, \mathbb{C}) = \mathbb{C}.(e_1 + \rho e_2) \qquad (\rho = \exp(2\pi i/6))$$

if $e_1, e_2$ is a basis for $H^1(X_\infty, \mathbb{Z})$.

(2.15) EXAMPLE (ordinary double point). Suppose $g: Y \to S$ is a family of projective varieties with $Y$ smooth, $g$ smooth outside a point $y_0 \in Y_0$ and $y_0$ a non-degenerate critical point of $g$. By blowing up $y_0$ in $Y$ one obtains a family $f: X \to S$ satisfying the hypotheses of (2.1) with $e = 2$, $E = E_0 \cup E_1$; $E_0$ is a desingularization of $Y_0$, $E_1 = \mathbb{P}^{n+1}$ has multiplicity 2 and $E_0 \cap E_1$ is a non-singular quadric in $E_1$. Here $\tilde{X}$ is non-singular, $D_0 \cong E_0$ and $D_1$ is isomorphic to a non-singular quadric of dimension $n+1$ which is a 2-fold covering of $\mathbb{P}^{n+1} = E_1$, ramified along $E_0 \cap E_1$.

One obtains for $q \geqq 0$:

$$Gr_{q-1}^W H^q(X_\infty) = \mathrm{Coker}\,(H^{q-1}(D_0) \oplus H^{q-1}(D_1) \to H^{q-1}(D_0 \cap D_1));$$
$$Gr_q^W H^q(X_\infty) = H(H^{q-2}(D_0 \cap D_1)(-1) \to H^q(D_0) \oplus H^q(D_1) \to H^q(D_0 \cap D_1));$$
$$Gr_{q+1}^W H^q(X_\infty) = \mathrm{Ker}\,(H^{q-1}(D_0 \cap D_1)(-1) \to H^{q+1}(D_0) \oplus H^{q+1}(D_1));$$
$$Gr_r^W H^q(X_\infty) = 0 \quad \text{if} \quad r \neq q-1, q, q+1.$$

If $n$ is even, $H^q(X_\infty)$ is purely of weight $q$ for all $q \geqq 0$, the monodromy is trivial on $H^q(X_\infty)$ for $q \neq n$ and has order 2 with $-1$ as an eigenvalue of multiplicity one if $q = n$.

If $n$ is odd, $H^q(X_\infty)$ is purely of weight $q$ with trivial monodromy if $q \neq n$ and for $H^n(X_\infty)$ one gets: $Gr_{n-1}^W H^n(X_\infty) = \mathbb{Q}(-(n-1)/2)$, $Gr_{n+1}^W H^n(X_\infty) = \mathbb{Q}(-(n+1)/2)$, $Gr_n^W H^n(X_\infty) = H^n(D_0)$. The monodromy is unipotent, $(T-I)^2 = 0$ and $T - I$: $Gr_{n+1}^W H^n(X_\infty) \xrightarrow{\sim} Gr_{n-1}^W H^n(X_\infty)$.

(2.16) REMARK. One can deduce from theorems (2.10) and (2.13) that the mixed Hodge structure on $H^q(X_\infty)$, constructed in (2.8), coincides with the one, constructed by W. Schmid [16]. See appendix.

(2.17) The relation between the cup product and the mixed Hodge structure on $H^*(X_\infty)$ has been investigated by Schmid [16], §6. For future use we recall the result.

Because $\tilde{T}_0$ satisfies $\tilde{T}_0(x \wedge y) = \tilde{T}_0(x) \wedge \tilde{T}_0(y)$ for all $x, y \in H^*(X_\infty)$, i.e. $\tilde{T}_0$ is an isometry, $N$ is an infinitesimal isometry, i.e. $N(x) \wedge y + x \wedge N(y) = 0$ for all $x, y \in H^*(X_\infty)$.

Recall that all $X_t$ are embedded simultaneously in $\mathbb{P}^r$ for some $r>0$. Let $L$ be the cohomology class of a hyperplane section of $X_t(t\neq 0)$, considered as an element of $H^2(X_\infty)$ by means of the natural map $H^2(X_t)\to H^2(X_\infty)$ (unique after the choice of $\alpha\in H$ with $t=\exp(2\pi i\alpha)$). Then $L$ does not depend on the choice of $t$ or $\alpha$.

Because the map $H^*(X_t)\xrightarrow{\psi_t} H^*(X_\infty)$ is a ring homomorphism for cup product, one has for all $k\geqq 0$:

$$L^k: H^{n-k}(X_\infty)\to H^{n+k}(X_\infty),$$

where $L^k(\omega)=L^k\wedge\omega$ for $\omega\in H^{n-k}(X_\infty)$. Denote $P^q(X_\infty)$ the kernel of the map $L^{n-q+1}: H^q(X_\infty)\to H^{2n-q+2}(X_\infty)$ if $0\leqq q\leqq n$ and $P^q(X_\infty)=0$ for $q>n$ or $q<0$. Then for all $q\geqq 0$ one has a decomposition

$$H^q(X_\infty)=\bigoplus_{k\geqq 0} L^kP^{q-2k}(X_\infty).$$

Elements of $P^q(X_\infty)$ are called primitive classes. Consider on $P^q(X_\infty)$ the bilinear form

$$Q(x, y)=\int_{X_t}(-1)^{q(q-1)/2}L^{n-q}\psi_t^{-1}(x\wedge y).$$

Then $Q$ does not depend on the choice of $t$, because all elements $(\psi_t)_*[X_t]\in H_{2n}(X_\infty)$ are the same.

Because $L$ is a $T_0$-invariant class, $P^q(X_\infty)\subset H^q(X_\infty)$ also carries a mixed Hodge structure, and for all $r\geqq 0$ one has

$$N^r: Gr^W_{q+r}P^q(X_\infty)\xrightarrow{\sim} Gr^W_{q-r}P^q(X_\infty).$$

Denote

$$P_{q,r}(X_\infty)=\operatorname{Ker}(N^{r+1}: Gr^W_{q+r}P^q(X_\infty)\to Gr^W_{q-r-2}P^q(X_\infty)).$$

Then $P_{q,r}(X_\infty)$ carries a Hodge structure of weight $q+r$. Let

$$P_{q,r}(X_\infty)=\bigoplus_{a+b=q+r} P^{a,b}_{q,r}(X_\infty)$$

be its Hodge decomposition. Denote $Q_r$ the bilinear form on $P_{q,r}(X_\infty)$ defined by $Q_r(x, y)=Q(\tilde{x}, N^r\tilde{y})$, where $x, y\in P_{q,r}(X_\infty)$ and $\tilde{x}, \tilde{y}$ are elements of $W_{q+r}P^q(X_\infty)$ whose classes mod $W_{q+r-1}$ are $x$ resp. $y$. The fact that $N$ is an infinitesimal isometry implies that $Q_r(x, y)$ is well-defined.

(2.18) THEOREM. *With notations as in* (2.17) *one has:*
(i) $Q_r(x, y)=0$ *if* $x\in P^{a,b}_{q,r}$, $y\in P^{c,d}_{q,r}$ *and* $(a, b)\neq(d, c)$;
(ii) $i^{a-b}Q_r(x, \bar{x})>0$ *if* $x\in P^{a,b}_{q,r}$, $x\neq 0$.

This theorem describes completely the connection between cup product and mixed Hodge structure, for for all $q$, $r>0$ one has a double decomposition

$$Gr^W_{q+r}H^q(X_\infty)=\bigoplus_{k,j\geqq 0} L^kN^jP_{q-2k,r+2j}$$

## Appendix

We compare the constructions of [16] and (2.8). We keep the notations of the preceeding section. The weight filtration on $H^q(X_\infty)$ is the same in both cases because it is characterized by the property that the map

$$N^r: Gr^W_{q+r}H^q(X_\infty)\to Gr^W_{q-r}H^q(X_\infty)$$

is an isomorphism for all $r\geqq 0$.

Let us recall the construction by Schmid of the Hodge filtration. For $\alpha\in H$ denote $e(\alpha)=\exp(2\pi i\alpha)\in\tilde{S}$ and denote $i_\alpha: X_{e(\alpha)}\to X_\infty$ the natural inclusion. The corresponding map

$$i^*_\alpha: H^q(X_\infty)\to H^q(X_{e(\alpha)})$$

is an isomorphism. Denote $F_\alpha$ the filtration on $H^q(X_\infty)$ obtained by pulling back the Hodge filtration on $H^q(X_{e(\alpha)})$ by means of $i^*_\alpha$. Then $\exp(-\alpha N)F_\alpha=\exp(-(\alpha+1)N)F_{\alpha+1}$ so for $w\in\tilde{S}$, $w\neq 0$ we may define a filtration $\tilde{F}_w$ on $H^q(X_\infty)$ by $\tilde{F}_w=\exp(-N\log(w))F_{\log(w)}$. The filtrations $\tilde{F}_w$, considered as points of a suitable product of Grassmann varieties, tend to a limit $F_\infty$ as $w$ goes to 0. We have to show that $F_\infty=$ our Hodge filtration $F$.

Put $A=H^0(\tilde{S},\mathcal{O}_{\tilde{S}})$, $M=H^0(\tilde{S},\boldsymbol{R}^q\tilde{f}_*\tilde{\Omega}^{\cdot}_{\tilde{X}/\tilde{S}}(\log D))$, $M_w=\boldsymbol{R}^q\tilde{f}_*\tilde{\Omega}^{\cdot}_{\tilde{X}/\tilde{S}}(\log D)(w)$ for $w\in\tilde{S}$ and $p_w: M\to M_w$ the canonical map (evaluation at $w$). Start with a basis $\boldsymbol{e}(0)$ of $M_0$. Claim: there exists a basis $\boldsymbol{e}$ of $M$ over $A$, such that $p_0(\boldsymbol{e})=\boldsymbol{e}(0)$ and such that the Gauss-Manin connection $\tilde{\nabla}$ satisfies $\tilde{\nabla}\boldsymbol{e}=N\boldsymbol{e}\otimes d\tau/\tau$ with $N$ a constant nilpotent matrix. The $\mathbb{C}$-vector space $V\subset M$ generated by the $e_i$ is characterized by $V=\{m\in M\,|(\tilde{\nabla}_{\tau d/d\tau})^k m=0$ for $k$ sufficiently large$\}$. The proof is straightforward. One uses that $\tilde{\nabla}$ has a nilpotent residue at 0. A multivalued horizontal section of $\boldsymbol{R}^q\tilde{f}_*\tilde{\Omega}^{\cdot}_{\tilde{X}/\tilde{S}}(\log D)$ over $\tilde{S}$ is an element $s=\sum_{i=0}^n m_i(\log\tau)^i\in M\otimes_A A[\log\tau]$, which satisfies the relation

$$\sum_{i=0}^n\tilde{\nabla}(m_i)(\log\tau)^i+\sum_{i=0}^n im_i(\log\tau)^{i-1}\,d\tau/\tau=0.$$

The space of all multivalued horizontal sections $\tilde{V}$ is isomorphic with $V$ by the map $y_0: m\to\exp(2\pi iN\log\tau)m(m\in\tilde{V})$.

The computations in [16], §2 show that $\tilde{V}$ and $H^q(X_\infty)$ are canonically isomorphic. The map $y_0^{-1}(p_0)^{-1}: M_0\to H^q(X_\infty)$ is the same as the map, denoted by $\psi_t$ in [19], (4.24). This leads to the following abstract setting.

For $G$ a filtration on $M$ by direct summands one may construct filtrations $G_\infty$ and $G'_\infty$ on $\tilde{V}$ in the following way.

For $\alpha \in H$ denote $y_\alpha : \tilde{V} \to M_{e(\alpha)}$ the isomorphism given by $y_\alpha(s) = \exp(2\pi i \alpha N) p_{e(\alpha)}(s)$. Then for $s \in V$ one has the relation

$$y_\alpha(\exp(-2\pi i N \log \tau)s) = p_{e(\alpha)}(s).$$

Let $G_\alpha$ be the filtration $y_\alpha^{-1} p_{e(\alpha)} G$ of $\tilde{V}$. Then the filtrations $(G_\alpha)_{\alpha \in H}$ have a limit $G_\infty$ uniformly on every vertical strip $R_{a,b} = \{z \in H \mid a < \mathrm{Re}(z) < b\}$. (This is Schmid's construction). We define $G'_\infty = y_0^{-1} p_0 G$. We have to show that $G'_\infty = G_\infty$.

Assume that $G$ is a decreasing filtration with $G^0 = M$. Let $a_j = \text{rank of } G^j$. Choose an $A$-basis $e_1, \cdots, e_{a_0}$ for $M$ such that $G^j = Ae_1 + \cdots + Ae_{a_j}$ for $j \geqq 0$. Let $f_1, \cdots, f_{a_0}$ be the uniquely determined basis of $V$ with $p_0(f_i) = p_0(e_i)$ for $i = 1, \cdots, a_0$. Write $e_i = Cf_i$ with $C \in \mathrm{Aut}_A(M)$; then $C(0) = I$.

Let $g_i = \exp(-2\pi i N \log \tau) f_i \in \tilde{V}$. For $\alpha \in H$ let $h_i(\alpha) = y_\alpha^{-1} p_{e(\alpha)}(e_i)$. Choose norms on $M_w (w \in \tilde{S})$ and $\tilde{V}$ such that the $p_w(e_i)$ (resp. $g_i$) form an orthonormal basis. Then for $\alpha \in H$ with $w = e(\alpha)$ one has:

$$\begin{aligned} h_i(\alpha) &= y_\alpha^{-1} p_w(e_i) \\ &= y_\alpha^{-1} p_w(f_i + (C - I) f_i) \\ &= g_i + y_\alpha^{-1} p_w (C - I) f_i \\ &= g_i + y_\alpha^{-1} (C(w) - I) p_w(f_i). \end{aligned}$$

So

$$\|h_i(\alpha) - g_i\| \leqq \|y_\alpha^{-1}\| \cdot \|C(w) - I\| \cdot \|p_w(f_i)\|.$$

On a vertical strip $R_{a,b}$, $\|y_\alpha^{-1}\|$ and $\|p_w(f_i)\|$ grow at most as a power of $\mathrm{Im}(\alpha)$. Moreover $\|C(w) - I\|$ goes to zero at least as fast as $|w| = \exp(-2\pi \, \mathrm{Im}(\alpha))$, so $\lim_{\mathrm{Im}\,\alpha \to \infty} \|h_i(\alpha) - g_i\| = 0$ uniformly on $R_{a,b}$. This shows in particular that the filtration $G'_\infty$, which is determined by the basis $f_i$, coincides with the filtration $G_\infty$.

## 3. Isolated singularities of hypersurfaces

(3.1) Let $P \in \mathbb{C}[z_0, \cdots, z_n]$ be a polynomial with $P(0) = 0$, such that 0 is an isolated critical point of $P$. This means that 0 is an isolated point of the set $\{Q \in \mathbb{C}^{n+1} \mid \partial P / \partial z_i(Q) = 0 \text{ for all } i\}$. We are interested in the map germ $P : (\mathbb{C}^{n+1}, 0) \to (\mathbb{C}, 0)$. We will construct a mixed Hodge structure on the cohomology of $B_\eta = \{z \in \mathbb{C}^{n+1} \mid P(z) = \eta \text{ and } |z| < \varepsilon\}$ ($\varepsilon$, $\eta$ small enough, $0 < \eta \ll \varepsilon$).

(3.2) One can embed the map germ $P$ in a projective family as follows. Brieskorn ([3], §1.1) has shown that there exists a homogeneous polynomial

$\tilde{P} \in \mathbb{C}[z_0, \cdots, z_{n+1}]$ such that the hypersurface $Y_0 \subset \mathbb{P}^{n+1}$ given by the equation $\tilde{P}=0$ has a unique singular point $y_0=(0, \cdots, 0, 1)$ which is analytically equivalent to the singularity of the zero set of $P$ at 0. One constructs such a $\tilde{P}$ by adding to $P$ a homogeneous polynomial $R(z_0, \cdots, z_n)$ of sufficient generality and high degree and then homogenizing. Let $N=\deg(\tilde{P})$. Then for all $t$ sufficiently close to 0, $t \neq 0$, the zero set of $\tilde{P}-tz_{n+1}^N$ in $\mathbb{P}^{n+1}$ is smooth. Choose $\eta>0$ such that this is the case for $0<|t|<\eta$. Define $Y=\{(z, t) \in \mathbb{P}^{n+1} \times \mathbb{C} \mid \tilde{P}(z)-tz_{n+1}^N=0 \quad \text{and} \quad |t|<\eta\}$. Let $f_1: Y \to S = \{t \in \mathbb{C} \mid |t|<\eta\}$ be the projection on the second factor. Denote $Y_t=f_1^{-1}(t)$ for $t \in S$. Then $y_0$ is the only critical point of $f_1$. Let $\rho: X \to Y$ be a resolution of singularities for the map $f_1$. This means that $\rho$ is proper, induces an isomorphism between $X-\rho^{-1}(y_0)$ and $Y-\{y_0\}$ and $(f_1\rho)^{-1}(0)$ is a divisor with normal crossings on $X$. Put $f=f_1\rho$ and $f^{-1}(0)=E_0 \cup \cdots \cup E_m$ where the restriction of $\rho$ to $E_0$ makes $E_0$ into a resolution of the singularity of $Y_0$.

(3.3) An exact sequence

Let $B$ be a small ball in $Y$ with center $y_0$ and radius $\varepsilon>0$ so small that all spheres with center $y_0$ and radius less than $\varepsilon$ intersect $Y_0$ transversally. Choose $\eta'>0$ such that all fibers $Y_t$ with $|t|<\eta'$ intersect $B$ transversally. Define $B_t=Y_t \cap B$ for $|t|<\eta'$. Then $B_t$ is diffeomorphic with the Milnor fiber of $P$ (cf. [13], Theorem 5.11) if $t \neq 0$. One can construct a homeomorphism between $Y_0$ and $Y_t/B_t$ $(t \neq 0)$, hence $\tilde{H}^i(Y_0) \cong H^i(Y_t, B_t)$ for all $i$. The exact sequence of relative cohomology gives $0 \to \tilde{H}^n(Y_0) \to H^n(Y_t) \to \dot{H}^n(B_t) \to H^{n+1}(Y_0) \to H^{n+1}(Y_t) \to 0$ because $H^i(B_t)=0$ for $i \neq 0, n$ (cf. [13], Theorem 6.5).

Define $B_\infty$ analogously to $X_\infty$ (cf. (2.1)). Then passing to the limit as $t$ tends to 0 one obtains an exact sequence $0 \to \tilde{H}^n(Y_0) \to H^n(X_\infty) \to H^n(B_\infty \to H^{n+1}(Y_0) \to H^{n+1}(X_\infty) \to 0$ which is $T_0$-equivariant because one may take the geometric monodromy $h: Y_t \to Y_t$ of the family $Y$ to be the identity outside $B_t$.

We will put a mixed Hodge structure on $H^n(B_\infty)$ such that the above sequence becomes an exact sequence of mixed Hodge structures.

(3.4) Construction. We first compute the canonical mixed Hodge structure on the cohomology of $Y_0$ as defined by Deligne [5]. Remark that the map $f: X \to S$ satisfies the hypotheses of (2.2). We preserve the notations of (2.2). Because $Y_0$ is reduced, $e_0=1$ (if $n>0$), so $\pi: D_0 \xrightarrow{\sim} E_0$. Define $C_i=D_i \cap D_0=E_i \cap E_0$ $(i=1, \cdots, m)$. Then $C=C_1 \cup \cdots \cup C_m$ is a divisor with normal crossings on $D_0$. The cohomology of $D_0-C \cong Y_0-\{y_0\}$ can be computed in terms of the cohomology of $D_0$ and $\tilde{C}^{(i)}$, $i>0$, as in [4]. Dualizing it one obtains the cohomology groups with compact support

$H^q_c(Y_0-\{y_0\})=\tilde{H}^q(Y_0)$. This shows that one may use the following cohomological mixed Hodge complex $A^\cdot(Y_0)$ on $D_0$ to compute the mixed Hodge structure on $H^*(Y_0)$.

Let $j: D_0-C\to D_0$ be the inclusion and denote $c_p: \tilde{C}^{(p)}\to D_0$ the projection $(p>0)$.

(i) Let $A^\cdot(Y_0)_{\mathbb{Z}}=j_!\mathbb{Z}_{D_0-C}$ i.e. the constant sheaf $\mathbb{Z}$ on $D_0-C$ extended by 0 over $D_0$;

(ii) Denote $i_{j,q}: \tilde{C}^{(q+1)}\to \tilde{C}^{(q)}$ $(q>0, 0\leqq j\leqq q)$ the map which has the inclusions

$$C_{i_0}\cap\cdots\cap C_{i_q}\to C_{i_0}\cap\cdots\cap C_{i_{j-1}}\cap C_{i_{j+1}}\cap\cdots\cap C_{i_q}$$

as its components. Let $A^\cdot(Y_0)_{\mathbb{Q}}$ be the object in the filtered derived category $D^+F(D_0,\mathbb{Q})$ represented by the complex of sheaves

$$\mathbb{Q}_{D_0}\xrightarrow{d''_0}(c_1)_*\mathbb{Q}_{\tilde{C}}\xrightarrow{d''_1}(c_2)_*\mathbb{Q}_{\tilde{C}^{(2)}}\longrightarrow\cdots$$

with $d''_q=\sum_{j=0}^{q+1}(-1)^{q+j}(i_{j,q+1})^*$ and with the filtration $W$ defined by

$$W_{-q}A^\cdot(Y_0)_{\mathbb{Q}}=\sigma_{\geqq q}A^\cdot(Y_0)_{\mathbb{Q}}.$$

(iii) Let $A^{pq}(Y_0)=(c_q)_*\Omega^p_{\tilde{C}^{(q)}}$; define $d': A^{pq}\to A^{p+1,q}$ to be the differentiation in the complex $(c_q)_*\Omega^\cdot_{\tilde{C}^{(q)}}$ and define $d'': A^{pq}\to A^{p,q+1}$ by $d''=\sum_{j=0}^{q+1}(-1)^{q+j}(i_{j,q+1})^*$. In this way one obtains a double complex $A^{\cdot\cdot}(Y_0)$. Let $A^\cdot(Y_0)_{\mathbb{C}}$ be its associated single complex. One defines filtrations $F$ and $W$ on $A^\cdot(Y_0)_{\mathbb{C}}$ by

$$F^pA^\cdot(Y_0)_{\mathbb{C}}=\bigoplus_{r\geqq p}A^{r\cdot}(Y_0);$$

$$W_qA^\cdot(Y_0)_{\mathbb{C}}=\bigoplus_{s\geqq -q}A^{\cdot s}(Y_0).$$

For $q>0$ one has $Gr^W_qA^\cdot(Y_0)=0$; if $q\geqq 0$ then $Gr^W_{-q}A^\cdot(Y_0)$ is equal to $((c_q)_*\mathbb{Q}_{\tilde{C}^{(q)}}[-q], ((c_q)_*\Omega^\cdot_{\tilde{C}^{(q)}}[-q], F))$ which is indeed a cohomological Hodge complex of weight $-q$.

(3.5) Corollary. *The spectral sequence of hypercohomology for the filtered complex* $(A^\cdot(Y_0)_{\mathbb{Q}}, W)$:

$$E_1^{pq}=H^q(\tilde{C}^{(p)},\mathbb{Q})\Rightarrow H^{p+q}(Y_0,\mathbb{Q})$$

*degenerates at* $E_2$, i.e. $Gr^W_qH^{p+q}(Y_0,\mathbb{Q})=E_2^{pq}$ *is isomorphic to*

$$H(H^q(\tilde{C}^{(p-1)},\mathbb{Q})\to H^q(\tilde{C}^{(p)},\mathbb{Q})\to H^q(\tilde{C}^{(p+1)},\mathbb{Q})).$$

(3.6) Remark. The cohomological mixed Hodge complex $A^\cdot(Y_0)$ can be considered as a subcomplex of the complex $A^\cdot$ as defined in (2.8) in the following way. The inclusion $D_0 \subset D$ gives inclusions $\tilde{C}^{(i)} \subset \tilde{D}^{(i+1)}$ for all $i \leqq 0$. In this way the complex $A^{\cdot}(Y_0)_{\mathbb{Q}}$ is a direct factor of the complex

$$a_*\mathbb{Q}_{\tilde{D}} \to (a_2)_*\mathbb{Q}_{\tilde{D}^{(2)}} \to (a_3)_*\mathbb{Q}_{\tilde{D}^{(3)}} \to \cdots$$

which is a subcomplex of $A_{\mathbb{Q}}$. Moreover $A^{pq}(Y_0)$ is a direct factor of $(a_{q+1})_*\Omega^p_{\tilde{D}^{(q+1)}} = Gr^W_{q+1}\Omega^{p+q+1}_X(\log D) \subset A^{pq}$. These inclusions are compatible with the filtrations $W$ and $F$. They define a morphism of cohomological mixed Hodge complexes

$$A^{\cdot}(Y_0) \to A^{\cdot}$$

which induces morphisms of mixed Hodge structures

$$H^i(Y_0) \to H^i(X_\infty).$$

These are the same as those induced by the contraction map $Y_t \to Y_0$.

(3.7) Definition. The cohomological mixed Hodge complex $A^{\cdot}(B_\infty)$ is given by

$$A^{\cdot}(B_\infty) = A^{\cdot}/A^{\cdot}(Y_0).$$

(3.8) Theorem. *$\boldsymbol{H}^q(D, A^{\cdot}(B_\infty)_{\mathbb{Q}}) = H^q(B_\infty, \mathbb{Q})$ for $q \geqq 0$.*

Proof. We may suppose that $S$ is so small that $X_t$ intersects $\partial B$ transversally for all $t \in S$. Then $B_\infty = B \times_S H$. Denote $\tilde{g}: B_\infty \to \tilde{X}$ the natural map. Then $A^{\cdot}_{\mathbb{Q}|_{\rho^{-1}(B)\cap D}} = i^{\cdot}\boldsymbol{R}\tilde{g}_*\mathbb{Q}_{B_\infty}$, because $A^{\cdot}_{\mathbb{Q}} = i^{\cdot}\boldsymbol{R}\tilde{k}_*\mathbb{C}_{X_\infty}$. This implies that

$$\boldsymbol{H}^q(D \cap \rho^{-1}(B), A^{\cdot}_{\mathbb{Q}|_{\rho^{-1}(B)\cap D}}) = H^q(B_\infty, \mathbb{Q})$$

(compare [19], Lemmas (2.4) and (2.5)). Because $\rho^{-1}(B) \cap D/D_1 \cup \cdots \cup D_m = Y_0 \cap B$ is a cone, the restriction map $\boldsymbol{H}^q(D \cap \rho^{-1}(B), A^{\cdot}_{\mathbb{Q}|D\cap\rho^{-1}(B)}) \to \boldsymbol{H}^q(D_1 \cup \cdots \cup D_m, A^{\cdot}_{\mathbb{Q}|D_1\cup\cdots\cup D_m})$ is an isomorphism for all $q \geqq 0$. The latter is isomorphic to $\boldsymbol{H}^q(D, A^{\cdot}(B_\infty)_{\mathbb{Q}})$ because $A^{\cdot}(B_\infty)_{\mathbb{Q}}$ has support on $D_1 \cup \cdots \cup D_m$ and because $A^{\cdot}_{\mathbb{Q}|D_1\cup\cdots\cup D_m} = A^{\cdot}(B_\infty)_{\mathbb{Q}|D_1\cup\cdots\cup D_m}$.

(3.9) Corollary. *$H^n(B_\infty, \mathbb{Q})$ carries a mixed Hodge structure such that the exact sequence from* (3.3) *is an exact sequence of mixed Hodge structures.*

Proof. Use the exact sequence

$$0 \to A^{\cdot}(Y_0) \to A^{\cdot} \to A^{\cdot}(B_\infty) \to 0.$$

(3.10) Corollary. *The weight spectral sequence*

$$E_1^{-r,q+r} = \bigoplus_{k \geqq 0} H^{q-r-2k}(\tilde{D}^{(2k+r+1)})(-r-k) \qquad (r > 0)$$

$$= \bigoplus_{k \geqq 1-r} H^{q-r-2k}(\tilde{D}^{(2k+r+1)})(-r-k) \oplus H^{q+r}(\tilde{D}^{(1-r)} - \tilde{C}^{(-r)}) \qquad (r \leqq 0)$$

*abutting to $H^q(B_\infty)$, degenerates at $E_2$,* i.e. $E_2^{-r,q+r}=Gr^W_{q+r}H^q(B_\infty)$. *Moreover* $E_2^{-r,q+r}=0$ *for* $q\neq n$ *(except* $E_2^{0,0}=\mathbb{Q}$*) because* $H^q(B_\infty)=0$ *for* $q\neq 0, n$. *This enables one to compute the weights on* $H^n(B_\infty)$ *from the* $E_1$*-terms.*

(3.11) REMARK. The actions of $N=\log(T_0)$ and $\gamma_s=$ semisimple part of $T_0$ can both be lifted to actions on $A_{\mathbb{C}}^{\cdot}$ which are 0 resp. $I$ on the subcomplex $A^{\cdot}(Y_0)_{\mathbb{C}}$ (cf. [19], (4.22) for the lifting $\tilde{\nu}$ of $N$ and (2.13) for the lifting $\lambda^*$ of $\gamma_s$). Hence they act on $H^*(B_\infty)$ in such a way that the sequence of (3.3) is both $N$- and $\gamma_s$-equivariant, and such that $N$ and $\gamma_s$ are morphisms of Hodge structures of type $(-1,-1)$ and $(0,0)$ respectively. The map $T_0$ is a limit of conjugates of the monodromy $T$.

(3.12) EXAMPLE. Let $P\in\mathbb{C}[z_0,\cdots,z_n]$ be a homogeneous polynomial with an isolated singular point at 0. Let $\rho:X\to\mathbb{C}^{n+1}$ be the blowing up with center $0\in\mathbb{C}^{n+1}$. The map $f=P\rho$ satisfies our hypotheses. One has $f^{-1}(0)=E_0\cup E_1$ where $E_1\cong\mathbb{P}^n$, $E_0$ is a desingularization of $P^{-1}(0)$ and $E_0\cap E_1$ is the smooth projective hypersurface given by $P(z)=0$. Here $D_1$ is a covering of $\mathbb{P}^n$ of degree $d=$ degree of $P$, which is ramified along $E_0\cap E_1$. This implies that $D_1$ is isomorphic to the hypersurface in $\mathbb{P}^{n+1}$ with equation $P(z)-z_{n+1}^d=0$ and $\pi:D_1\to E_1$ is the projection from the point $(1,0,\cdots,0)$. One concludes that the cohomology of $B_\infty$ inherites the mixed Hodge structure of the cohomology of the affine hypersurface $D_1-D_0=\{z\in\mathbb{C}^{n+1}\mid P(z)=1\}$. A similar statement holds if $P$ is quasi-homogeneous. See [20] for a detailed description of these mixed Hodge structures.

(3.13) EXAMPLE. Families of curves. Suppose $P:\mathbb{C}^2\to\mathbb{C}$ has an isolated critical point 0. Let $\rho:X\to\mathbb{C}^2$ be a resolution of the singularities of $P$. It can be obtained by successive blowing ups of points. Let $f=P\rho$. Denote $f^{-1}(0)=E_0\cup E_1\cup\cdots\cup E_m$ with $\rho^{-1}(0)=E_1\cup\cdots\cup E_m$. Then $E_i$ is a non-singular complete rational curve for $i>0$. Denote $e_i$ its multiplicity. If $i,j>0$ and $i\neq j$, the curves $E_i$ and $E_j$ intersect in at most one point $P_{ij}$, hence if $E_i\cap E_j\neq\emptyset$, then $D_i$ and $D_j$ intersect in $m_{ij}=(e_i,e_j)$ points. Denote $m_{i0}=$ Card $E_0\cap E_i$ and $k_i=$ Card $E_i\cap\bigcup_{j\neq i}E_j$. Let $\mathring{E}_i=E_i-\bigcup_{j\neq i}E_j$, $\mathring{D}_i=D_i-\bigcup_{j\neq i}D_j$. Then $\mathring{D}_i\to\mathring{E}_i$ is an étale covering with cyclic covering group generated by $\lambda_i=\lambda_{|D_i}$ (cf. the proof of (2.13)), which is of order $e_i$. The fundamental group $\pi_1(\mathring{E}_i,*)$ is a free group of rank $k_i-1$, generated by loops $l_1,\cdots,l_{k_i}$ around the intersection points $E_i\cap\bigcup_{j\neq i}E_j$ with the relation

$$l_{k_i}l_{k_i-1}\cdots l_2l_1=1.$$

If $l_j$ is a loop going around the intersection point of $E_i$ with a curve of multiplicity $e(j)$, then the action of $l_j$ on $D_i$ is just $\lambda^{e_i/(e_i,e(j))}$. Therefore the index of the image of $\pi_1(\mathring{E}_i,*)$ in the covering group is equal to $r_i=$ $gcd(e_j\mid E_i\cap E_j\neq\emptyset)$. This is also equal to the number of connected components of $D_i$ because $D_i$ is normal.

One has the well-known relation

$$\prod_{q\geq 0} \det(I - t\lambda_i; H^q(\mathring{D}_i))^{(-1)^{q+1}} = (1-t^{e_i})^{\chi(\mathring{E}_i)}$$

so

$$\det(\lambda_i - tI; H^1(\mathring{D}_i)) = (t^{e_i}-1)^{k_i-2}(t^{r_i}-1)$$

From the action of $\lambda_i$ on the exact sequence

$$0 \to H^1(D_i) \to H^1(\mathring{D}_i) \to H^0(D_i - \mathring{D}_i)(-1) \to H^2(D_i) \to 0$$

one gets

$$\det(\lambda_i - tI; H^1(D_i)) = \frac{(t^{e_i}-1)^{k_i-2}(t^{r_i}-1)}{(t-1)^{m_{io}} \prod_{j\neq i,0} (t^{m_{ij}}-1)}$$

Remark that numerator and denominator contain an equal number of factors $t-1$. The spectral sequence (3.10) gives exact sequences

$$0 \to \mathbb{Q} \to \bigoplus_{i>0} H^0(D_i) \to \bigoplus_{i>j>0} H^0(D_i \cap D_j) \to Gr_0^W H^1(B_\infty) \to 0$$

and

$$0 \to Gr_2^W H^1(B_\infty) \to \bigoplus_{i>j\geq 0} H^0(D_i \cap D_j)(-1) \to \bigoplus_{i>0} H^2(D_i) \to 0;$$

Moreover $Gr_1^W H^1(B_\infty) = \bigoplus_{i>0} H^1(D_i)$. So the characteristic polynomial of $\lambda^*$ on $Gr_r^W H^1(B_\infty)$ is given by

$$(t-1) \prod_{i>j>0} (t^{m_{ij}}-1) \Big/ \prod_{i>0} (t^{r_i}-1) \quad \text{if} \quad r=0;$$

$$\prod_{i>0} (t^{e_i}-1)^{k_i-2}(t^{r_i}-1)^2 \Big/ (t-1)^{\Sigma_{i>0} m_{io}} \prod_{i>j>0} (t^{m_{ij}}-1)^2 \quad \text{if} \quad r=1;$$

$$(t-1)^{\Sigma_{i>0} m_{io}} \prod_{i>j>0} (t^{m_{ij}}-1) \Big/ \prod_{i>0} (t^{r_i}-1) \quad \text{if} \quad r=2.$$

Multiplication of these gives the characteristic polynomial of the monodromy and the Milnor number:

$$\Delta(t) = (t-1) \prod_{i>0} (t^{e_i}-1)^{k_i-2};$$

$$\mu = 1 + \Sigma_{i>0} e_i(k_i-2).$$

These are the formulas which follow from A'Campo's formula for the zeta function of the monodromy [1], which can also be proved using (3.10).

Durfee [6] has introduced the notion of the number of cycles in the fiber $X_0$ and uses this to give a criterion for finiteness of the monodromy. This number is equal to dim $Gr_0^W H^1(B_\infty)$. In particular the monodromy has finite order if and only if $W_0 H^1(B_\infty) = 0$. In theorem (4.4) we will give a generalization of this fact to higher dimensions.

The following lemma is useful to determine the relation between the Hodge filtration on $H^1(B_\infty)$ (or $H^n(B_\infty)$ in higher dimensions) and the eigenvalues of $\lambda^*$.

(3.14) LEMMA. *With notations as before, the action of $\lambda_j$ leads to a splitting*

$$\pi_*\mathcal{O}_{D_j} = \bigoplus_{k=0}^{e_j-1} L_k$$

*where $\lambda_j$ acts on $L_k$ by multiplication with* $\exp(2\pi ik/e_j)$. *One has*

$$L_k \cong \mathcal{O}_{E_j}\left(\Sigma_{j\neq i}\left(-\frac{ke_i}{e_j}+\left[\frac{ke_i}{e_j}\right]\right)E_i \cap E_j\right).$$

PROOF. One uses remark (2.3) and the fact that $\mathcal{O}_X(\Sigma_{i=0}^m e_i E_i)$ is isomorphic to $\mathcal{O}_X$ (by multiplication with a parameter on $S$). The divisor associated to the normal bundle of $E_j$ in $X$ is $-E_j \cdot E_j \sim \Sigma_{j\neq i}(e_i/e_j)E_j \cdot E_i$, hence the latter indeed defines an element of $\mathrm{Pic}(E_j)$ and not merely an element of $\mathrm{Pic}(E_j) \otimes \mathbb{Q}$.

(3.15) EXAMPLE. $P(x, y, z) = x^4 + y^2 + z^2$. Twice blowing up a point gives the fiber $E = E_0 \cup E_1 \cup E_2$; $E_0$ is the non-compact component, $e_1 = 4$ and $e_2 = 2$; $E_1 = \mathbb{P}^2$, $E_2$ is isomorphic to $\mathbb{P}^2$ with one point blown up, $E_1 \cap E_2$ is the exceptional curve in $E_2$, $E_0 \cap E_1$ is a quadric in $E_1$ and $E_0 \cap E_2$ is the disjoint union of two lines, each intersecting $E_1 \cap E_2$ in one point. Hence $D_1$ is the disjoint union of two quadrics in $\mathbb{P}^3$, $D_2$ is isomorphic to a quadric in $\mathbb{P}^3$ with 2 points blown up and $D_1 \cap D_2$ is the union of the two exceptional curves on $D_2$. One obtains that $H^2(B_\infty) = \mathbb{Q}(-1)^2$ is purely of type (1, 1).

(3.16) EXAMPLE. $P(x, y, z) = x^8 + y^8 + z^8 + (xyz)^2$ (Malgrange) We will determine the following data:

(i) the dimensions of the spaces $H^{pq} = Gr_F^p Gr_{p+q}^W H^2(B_\infty)$;

(ii) the eigenvalues of $\lambda^*$ on each of these.

We first describe a resolution for $P$, following A'Campo [1]. Blow up $0 \in \mathbb{C}^3$. The exceptional divisor intersects the strict transform of $P^{-1}(0)$ in three double lines which are in general position. After blowing up each of these lines one obtains a resolution $\rho: X \to \mathbb{C}^3$ for $P$. Here $f^{-1}(0) = \bigcup_{i=0}^4 E_i$ with $e_1 = 6$, $e_2 = e_3 = e_4 = 8$. One has $E_1 = \mathbb{P}^2$, $E_4$ is a $\mathbb{P}^1$-bundle over $\mathbb{P}^1$ whose zero section $E_1 \cap E_4 = a$ has self-intersection $-4$ and $E_2$ and $E_3$ are obtained form $E_4$ by blowing up 2 resp. 1 points which are not on the zero section. This implies that $D_2$ and $D_3$ are obtained from $D_4$ by blowing up 16 resp. 8 points. $D_1$ is the disjoint union of two components, each of which is a 3-fold covering of $\mathbb{P}^2$, ramified along three lines in general position. Hence $H^i(D_1) = 0$ for $i = 1, 3$ and $H^2(D_1) = \mathbb{Q}(-1)^2$. Moreover $E_0 \cap E_1 = \emptyset$, $E_0 \cap E_i$ is a curve of genus 3 for $i = 2, 3, 4$ and $E_i \cap E_j$ is a rational curve if $1 \leqq i < j \leqq 4$.

We first show that $H^1(D_4, \mathbb{Q})=0$. The composition of $\pi$ with the projection map $p: E_4 \to \mathbb{P}^1$ gives a flat projective morphism $\tau: D_4 \to \mathbb{P}^1$. Each fiber $\tau^{-1}(z)$ is an 8-fold covering of $p^{-1}(z)=\mathbb{P}^1$, ramified above $p^{-1}(z)\cap(E_0\cup E_1)$. The critical points of $\tau$ correspond to the ramification points of the map $p_{|E_0\cap E_4}$, which is a double covering of $\mathbb{P}^1$ with 8 ramification points. Hence $\tau$ has 8 singular fibers, each of which is the union of two rational singular curves which in their intersection point are of type $u^6-x^8$ and which are smooth everywhere else. Denote $C_{\lambda_i}$, $i=1,\cdots,8$ these singular fibres and let $U=D_4-\bigcup_{i=1}^8 C_{\lambda_i}$. Let $z$ be a regular value of $\tau$. The local monodromy around each of the $\lambda_i$ acts on $H^1(C_z)$ with eigenvalues $\neq 1$. In particular $H^0(\mathbb{P}^1-\{\lambda_1,\cdots,\lambda_8\}, R^1(\tau_{|U})_*\mathbb{Q})=0$. It follows from [4], (4.1.1) that $H^1(U,\mathbb{Q})=H^1(\mathbb{P}^1-\{\lambda_1,\cdots,\lambda_8\},\mathbb{Q})=\mathbb{Q}(-1)^7$ is purely of weight 2, hence $H^1(D_4,\mathbb{Q})=Gr_1^W H^1(U,\mathbb{Q})=0$. Consequently $H^1(D_4,\mathcal{O}_{D_4})=0$. Denote $b\in H^2(E_4,\mathbb{Z})$ the cohomology class of $E_2\cap E_4$. Then $a$ and $b$ form a basis for $H^2(E_4,\mathbb{Z})$ and $a^2=-4$, $ab=1$ and $b^2=0$. The canonical divisor on $D_4$ is $-2a-6b$ and the cohomology class of $E_0\cap E_4$ is $2a+8b$. Using lemma (3.14) one obtains for the cohomology class of $L_k$ $(k=0,\cdots,7)$: $[L_0]=0, [L_1]=-a-b, [L_2]=-a-2b, [L_3]=-a-3b, [L_4]=-a-4b, [L_5]=-2a-5b, [L_6]=-2a-6b, [L_7]=-2a-7b$. The Riemann-Roch theorem on $E_4$ gives that $\chi(\mathcal{O}_{E_4}(ka+lb))=(k+1)(l+1-2k)$ so $\chi(L_0)=1=\dim H^0(E_4,\mathcal{O}_{E_4})$, $\chi(L_k)=0$ for $i=1,\cdots,5$, $\chi(L_6)=1$ and $\chi(L_7)=2$. This implies that $\dim H^2(D_4,\mathcal{O}_{D_4})=3$ and that $\lambda^*$ acts on it with the eigenvalues $\xi^6,\xi^7,\xi^7$ where $\xi=\exp(\pi i/4)$. Consequently one has the following picture:

| space | dimension | eigenvalues of $\lambda$ | multiplicity |
|---|---|---|---|
| $H^{0,0}$ | 1 | $-1$ | 1 |
| $H^{0,1}$ | 9 | $\xi^5,\xi^6,\xi^7$ | 3 |
| $H^{0,2}$ | 9 | $\xi^6$ | 3 |
| | | $\xi^7$ | 6 |
| $H^{1,1}$ | 137 | 1 | 4 |
| | | $-1$ | 25 |
| | | $\xi,\xi^7$ | 15 |
| | | $\xi^2,\xi^6$ | 18 |
| | | $\xi^3,\xi^5$ | 21 |
| $H^{2,2}$ | 5 | 1 | 4 |
| | | $-1$ | 1 |
| $H^{1,2}$ | 18 | $\xi^5,\xi^6,\xi^7$ | 3 |
| | | 1 | 9 |

To show this one uses again (3.10). The eigenvalues of $\lambda^*$ on the other spaces $H^{pq}$ can be deduced from $H^{pq} = \bar{H}^{qp}$ and the fact that $\lambda^*$ is defined over $\mathbb{Q}$.

In (4.4) we will show that these data give the minimal polynomial of $T_0$: in this example it equals $(t^8-1)^2(t+1)$ while the characteristic polynomial of $T_0$ equals $(t^8-1)^{27}(t-1)^{-1}$.

## 4. Applications

In this chapter we investigate the relations between the mixed Hodge structure on the vanishing cohomology, the monodromy, the intersection form and the local cohomology groups of an isolated hypersurface singularity. We keep the notations of the preceding chapter.

(4.1) THEOREM. *The map* $N = \log \tilde{T}_0 : H^n(B_\infty) \to H^n(B_\infty)$ *is a morphism of mixed Hodge structures of type* $(-1,-1)$. *The map* $\gamma_s = \lambda^* : H^n(B_\infty) \to H^n(B_\infty)$ *is an automorphism of mixed Hodge structures.*

PROOF. See (3.11). Remark that it is no longer true in general that $N^r : Gr^W_{n+r} H^n(B_\infty) \to Gr^W_{n-r} H^n(B_\infty)$ is an isomorphism.

(4.2) LEMMA. *If* $r \geqq 1$ *the map* $Gr^W_{n+r} H^n(X_\infty) \to Gr^W_{n+r} H^n(B_\infty)$ *is injective; for* $r \geqq 2$ *it is an isomorphism.*

PROOF. This follows from the fact that $Gr^W_{n+r} H^n(Y_0) = 0$ for $r \geqq 1$ and $Gr^W_{n+r} H^{n+1}(Y_0) = 0$ for $r \geqq 2$ because $Y_0$ is complete, and from corollary (3.9).

(4.3) For $d \in \mathbb{N}$ let $\Phi_d(t)$ be the $d$th cyclotomic polynomial and

$$H^n(X_\infty)_d = \{x \in H^n(X_\infty) \mid \Phi_d(\gamma_s)(x) = 0\};$$
$$H^n(B_\infty)_d = \{x \in H^n(B_\infty) \mid \Phi_d(\gamma_s)(x) = 0\}.$$

These are the subspaces on which the monodromy acts with primitive $d$th roots of unity as eigenvalues. Then clearly $H^n(X_\infty)_d = H^n(B_\infty)_d$ for all $d > 1$. Because $N$ and $\gamma_s$ commute, one has

$$N^r : Gr^W_{n+r} H^n(B_\infty)_d \xrightarrow{\sim} Gr^W_{n-r} H^n(B_\infty)_d \quad \text{for} \quad r \geqq 0,\ d > 1.$$

Consequently for $d > 1$ the exponent of $\Phi_d(t)$ as a factor of the minimal polynomial $\delta(t)$ of the monodromy equals

$$k_d = 1 + \max(r \mid Gr^W_{n+r} H^n(B_\infty)_d \neq 0) \quad \text{if} \quad H^n(B_\infty)_d \neq 0;$$
$$0 \quad \text{if} \quad H^n(B_\infty)_d = 0.$$

To determine $k_1$ we use the exact sequence

$$0 \longrightarrow H^n(Y_0) \longrightarrow H^n(X_\infty)_1 \xrightarrow{u} H^n(B_\infty)_1 \xrightarrow{v} H^{n+1}(Y_0) \longrightarrow H^{n+1}(X_\infty) \longrightarrow 0.$$

Suppose that $H^n(B_\infty)_1 \neq 0$. Claim: $\delta(t)$ contains $t-1$ as a factor of multiplicity $k_1 = \max(1, \max(r \mid Gr^W_{n+r} H^n(B_\infty)_1 \neq 0))$.

PROOF. If $Gr^W_{n+r} H^n(B_\infty)_1 = 0$ for all $r > 0$, then $Gr^W_{n+r} H^n(X_\infty)_1 = 0$ for all $r > 0$ by lemma (4.3). Consequently $T_0 = I$ on $H^n(X_\infty)_1$ hence $H^n(Y_0) = H^n(X_\infty)_1$ by the invariant cycle theorem (cf. [19], (5.12)). So $v$ is injective and $T_0 = I$ on $H^n(B_\infty)_1$. Hence $k_1 = 1$. If $r$ is maximal such that $Gr^W_{n+r} H^n(B_\infty)_1 \neq 0$ and $r > 0$, then let $x \in Gr^W_{n+r} H^n(B_\infty)_1$ with $x \neq 0$. Then $Nv(x) = 0 = vN(x)$ so $N(x) = u(y)$ for some $y \in Gr^W_{n+r-2} H^n(X_\infty)_1$. If $r = 1$ then again lemma (4.3) says that $Gr^W_{n+k} H^n(X_\infty)_1 = 0$ for $k \geqq 2$ so $Gr^W_{n-1} H^n(Y_0) = Gr^W_{n-1} H^n(X_\infty)_1$ by the invariant cycle theorem. Hence $N(x) = u(y) = 0$. If $r > 1$ then $r$ is also maximal such that $Gr^W_{n+r} H^n(X_\infty)_1 \neq 0$. Write $x = u(z)$ for $z \in Gr^W_{n+r} H^n(X_\infty)_1$. Then $N^r(z) \in Gr^W_{n-r} H^n(X_\infty)_1 = Gr^W_{n-r} H^n(Y_0)$ so $N^r(x) = uN^r(z) = 0$. Moreover $N^{r-1}(z) \notin Gr^W_{n-r+2} H^n(Y_0)$ so $N^{r-1}(x) \neq 0$. Putting this all together we find:

(4.4) THEOREM. *With notations as in* (4.3) *we have*

$$\delta(t) = \prod_{d \geqq 1} \Phi_d(t)^{k_d}.$$

(4.5) EXAMPLE. Consider $P(x, y, z) = x^8 + y^8 + z^8 + (xyz)^2$. It follows from (3.16) that $k_1 = 2$, $k_2 = 3$, $k_4 = k_8 = 2$ and $k_i = 0$ if $i \neq 1, 2, 4, 8$. Hence $\delta(t) = (t+1)(t^8 - 1)^2$ which shows that the monodromy is not quasi-unipotent of degree 2, i.e. $N^2 \neq 0$.

(4.6) EXAMPLE. Suppose that we know that the monodromy has finite order. Then $H^n(B_\infty)_d$ is purely of weight $n$ if $d \geqq 2$ and is of weight $\leqq n+1$ if $d = 1$. If moreover $n = 1$, we even know that $H^1(B_\infty)_1$ is purely of weight 2 because $H^2(Y_0)$ is purely of weight 2. Hence in the case $n = 1$ the double factors of $\delta(t)$ are precisely the $(t - \nu)^2$ where $\nu$ is an eigenvalue of $\gamma_s$ acting on $W_0 H^1(B_\infty)$.

We can treat the general case too because of

(4.7) LEMMA. $H^{n+1}(Y_0)$ *is purely of weight* $n+1$.

PROOF. In the case $n = 2$, one may use the fact that the curve $C_1 \cup \cdots \cup C_m$ on $D_0$ can be blown down to a point, hence, by a criterion of Grauert [8], the intersection matrix $(C_i \cdot C_j)$ is negative definite. In particular the cohomology classes $[C_1], \cdots, [C_m] \in H^2(D_0)$ are linearly independent, which implies that the map $H^2(D_0) \to H^2(\tilde{C})$ is surjective.

In the general case the collapsing map $X_0 = E_0 \cup \cdots \cup E_m \to Y_0$ induces an exact sequence of mixed Hodge structures.

$$\cdots \to H^k(Y_0) \to H^k(X_0) \to H^k(E_1 \cup \cdots \cup E_m) \to H^{k+1}(Y_0) \to \cdots.$$

Because $E_1 \cup \cdots \cup E_m$ can be blown down to a point in $\mathbb{P}^{n+1}$ which is smooth, $H^k(E_1 \cup \cdots \cup E_m)$ is purely of weight $k$ and $H^k(X_0) \to H^k(E_1 \cup \cdots \cup E_m)$ is surjective for all $k \geqq 0$.

Hence it suffices to show that $H^{n+1}(X_0)$ is purely of weight $n+1$. We have the exact sequence

$$\longrightarrow H^{n+1}_{X_0}(X) \longrightarrow H^{n+1}(X_0) \longrightarrow H^{n+1}(X_\infty) \xrightarrow{N} H^{n+1}(X_\infty).$$

Because the monodromy on $H^{n+1}(X_\infty)$ is the identity, $H^{n+1}(X_\infty)$ is purely of weight $n+1$. Moreover $Gr_k^W H^{n+1}_{X_0}(X) = 0$ for $k \leqq n$, hence $H^{n+1}(X_0)$ is purely of weight $n+1$.

(4.8) Corollary. *If $T_0$ has finite order on $H^n(B_\infty)$, then $H^n(B_\infty)_1$ is purely of weight $n+1$.*

(4.9) Corollary. *$N^k$ defines for all $k \geqq 0$ an isomorphism between $Gr^W_{n+1+k} H^n(B_\infty)_1$ and $Gr^W_{n+1-k} H^n(B_\infty)_1$.*

Proof. Use (4.7), (2.13)(2) and the invariant cycle theorem.

(4.10) We look for relations between the intersection form on $H^n_c(B_\infty)$ and the mixed Hodge structure. The tools are the exact sequence (3.3), theorem (2.13) and theorem (2.18). We consider $H^n_c(B_\infty)$ as the dual space of $H^n(B_\infty)$, more precisely there is a perfect pairing $\langle\, , \rangle$:

$$H^n_c(B_\infty) \otimes_Q H^n(B_\infty) \to \mathbb{Q}(-n)$$

and we give $H^n_c(B_\infty)$ the mixed Hodge structure of $\mathrm{Hom}_{\mathbb{Q}}(H^n(B_\infty), \mathbb{Q}(-n))$. Dual to the restriction map $k: H^n(X_\infty) \to H^n(B_\infty)$ we have ${}^tk: H^n_c(B_\infty) \to H^n(X_\infty)$ and for $\omega \in H^n_c(B_\infty)$, $\omega' \in H^n(B_\infty)$ one has

$$\langle \omega, \omega' \rangle = \langle {}^tk(\omega), k(\omega') \rangle.$$

Denote $j = k\,{}^tk: H^n_c(B_\infty) \to H^n(B_\infty)$. The intersection form $S$ on $H^n_c(B_\infty)$ is defined by $S(x, y) = \langle x, j(y) \rangle$. We want to express its invariants as a real bilinear form in terms of the Hodge numbers $h^{pq} = \dim_{\mathbb{C}} H^{pq}(B_\infty)$. Let $h^{pq}_1 = \dim_{\mathbb{C}} H^{pq}(B_\infty) \cap H^n(B_\infty)_1$ and let $h^{pq}_{\neq 1} = \dim_{\mathbb{C}} H^{pq}(B_\infty) \cap \bigoplus_{d>1} H^n(B_\infty)_d$. Then $h^{pq} = h^{pq}_1 + h^{pq}_{\neq 1}$. Let $\mu$ be the Milnor number of $P$. Let $\mu_0$ be the dimension of the null-space of $S$. The rank of $S$ is the number $\mu - \mu_0$. This is the only real invariant besides $\mu$ if $n$ is odd, because in that case $S$ is antisymmetric. If $n$ is even, $S$ is symmetric and we can diagonalize $S$. Let $\mu_+$ and $\mu_-$ be the number of positive resp. negative diagonal entries of $S$. Then the numbers $\mu_0$, $\mu_+$, and $\mu_-$ form a complete set of invariants for $S$ as a real form. Particularly important is the signature $\mu_+ - \mu_-$.

(4.11) THEOREM. *With notations as above one has*:

(i)

$$\mu_0 = \sum_{p+q \leqq n+1} h_1^{pq} - \sum_{p+q \geqq n+3} h_1^{pq};$$

(ii) If $n$ is even then

$$\mu_+ = \sum_{\substack{p+q=n+2 \\ q \text{ even}}} h_1^{pq} + 2 \sum_{\substack{p+q \geqq n+3 \\ q \text{ even}}} h_1^{pq} + \sum_{q \text{ even}} h_{\neq 1}^{pq};$$

$$\mu_- = \sum_{\substack{p+q=n+2 \\ q \text{ odd}}} h_1^{pq} + 2 \sum_{\substack{p+q \geqq n+3 \\ q \text{ odd}}} h_1^{pq} + \sum_{q \text{ odd}} h_{\neq 1}^{pq}.$$

PROOF. Because the sequence (3.3) is monodromy equivariant, the map $k$ induces an isomorphism $H^n(X_\infty)_{\neq 1} \cong H^n(B_\infty)_{\neq 1} \cong H_c^n(B_\infty)_{\neq 1}$. Hence the intersection form on $H_c^n(B_\infty)_{\neq 1}$ is non-degenerate. Because the monodromy acts trivially on $H^q(X_\infty)$ for $q \neq n$, one knows that $H^n(X_\infty)_{\neq 1}$ is a direct factor of the primitive cohomology $P^n(X_\infty)$. This implies that for $n$ even, the numbers $\mu_+$ and $\mu_-$ for the restriction of $S$ to $H_c^n(B_\infty)$ equal $\sum_{q \text{ even}} h_{\neq 1}^{pq}$ resp. $\sum_{q \text{ odd}} h_{\neq 1}^{pq}$, as follows from theorem (2.18).

Next consider the restriction of $S$ to $H_c^n(B_\infty)_1$. Clearly Ker $(j)$ is contained in its null-space. Denote $U = \operatorname{Im}({}^t k) \cap H^n(X_\infty)_1$ and $V = \operatorname{Ker}(k) \cap U$. Then $H_c^n(B_\infty)_1/\operatorname{Ker}(j) = U/V$. Moreover using cup product on $H^n(X_\infty)_1$ one has $V = U \cap \operatorname{Ker}(N)$ and $U = \operatorname{Im}(N)$ by the invariant cycle theorem and its dual version. Hence $V = U \cap U^\perp$ and cup product induces a non-degenerate bilinear form on $U/V$. This shows that the null-space of $S$ coincides with Ker $(j)$. We calculate $\mu_0$ by computing the Hodge numbers of $U/V$. Clearly $N^r : Gr_{n+r}^W(U/V) \to Gr_{n-r}^W(U/V)$ is an isomorphism for all $r \geqq 0$. Then $Gr_{n+r}^W(\operatorname{Ker}(N) \cap \operatorname{Im}(N)) = 0$ so $Gr_{n+r}^W(U/V) \cong Gr_{n+r}^W(\operatorname{Im}(N)_1) \cong Gr_{n+r+2}^W H^n(X_\infty)_1$. This gives $h^{pq}(U/V) = h_1^{p+1,q+1}$ for $p+q \geqq n$ in view of lemma (4.2). One deduces from these the numbers $h^{pq}(U/V)$ for all $p, q$, using $h^{n-p,n-q}(U/V) = h^{pq}(U/V)$. Hence $\dim U/V = \sum_{p+q=n+2} h_1^{pq} + 2\sum_{p+q \geqq n+3} h_1^{pq}$. One obtains the formula for $\mu_0$ by the relation $\dim \operatorname{Ker}(j) = \dim H_c^n(B_\infty)_1 - \dim(U/V)$. Suppose that $n$ is even. Denote $b^{pq}$ the Hodge numbers of $P^n(X_\infty)$ and $a^{pq}$ those of $P^n(X_\infty) \cap H^n(Y_0)$. Then the Hodge numbers $c^{pq}$ of $P^n(Y_t)$ $(t \neq 0)$ with its pure Hodge structure are given by $c^{pq} = 0$ for $p+q \neq n$ and $c^{p,n-p} = \sum_q b^{pq}$. Hence the invariants of the intersection form on $P^n(X_\infty)$ are $\mu_+(P^n(X_\infty)) = \sum_{q \text{ even}} b^{pq}$ and $\mu_-(P^n(X_\infty)) = \sum_{q \text{ odd}} b^{pq}$. For its subspace $A = P^n(X_\infty) \cap H^n(Y_0)$ these are $\mu_0(A) = \sum_{p+q<n} a^{pq}$, $\mu_+(A) = \sum_{q \text{ even}} a^{n-q,q}$ and $\mu_-(A) = \sum_{q \text{ odd}} a^{n-q,q}$. Further

$H^n_c(B_\infty)_1/\mathrm{Ker}\ (j)=(A+A^\perp)/A$ and $\dim\ (A+A^\perp)=\sum b^{pq}-\sum_{p+q<n}\ a^{pq}$. Consequently:

$$\mu_+=\sum_{q\text{ even}} b^{pq}-\sum_{q\text{ even}} a^{n-q,q}-\sum_{p+q<n} a^{pq};$$

$$\mu_-=\sum_{q\text{ odd}} b^{pq}-\sum_{q\text{ odd}} a^{n-q,q}-\sum_{p+q<n} a^{pq}.$$

Moreover:

$$\sum_{q\text{ even}} b^{pq}=\sum_{\substack{p+q=n\\ q\text{ even}}} b_1^{pq}+2\sum_{\substack{p+q=n+1\\ q\text{ even}}} h_1^{pq}+2\sum_{\substack{p+q=n+2\\ q\text{ even}}} h_1^{pq}+2\sum_{\substack{p+q\geqq n+3\\ q\text{ even}}} h_1^{pq}+\sum_{q\text{ even}} h_{\neq}^{pq};$$

$$\sum_{q\text{ even}} a^{n-q,q}=\sum_{\substack{p+q=n\\ q\text{ even}}} b_1^{pq}-\sum_{\substack{p+q=n+2\\ q\text{ odd}}} h_1^{pq}\ \text{because}$$

$$a^{n-q,q}=b_1^{n-q,q}-b_1^{n-q+1,q+1}\quad \text{and in the same way:}$$

$$\sum_{p+q<n} a^{pq}=\sum_{p+q<n} b_1^{pq}-\sum_{p+q<n-2} b_1^{pq}=\sum_{p+q=n+1} h_1^{pq}+\sum_{p+q=n+2} h_1^{pq}.$$

The formula for $\mu_+$ follows, for $\sum_{p+q=n+1} h_1^{pq}=2\sum_{\substack{p+q=n+1\\ q\text{ odd}}} h_1^{pq}$ because $n$ is even and $h_1^{pq}=h_1^{qp}$ for all $p,q$. The formula for $\mu_-$ is proved analogously.

(4.12) EXAMPLE. The singularities $x^p+y^q+z^r+\lambda xyz(\lambda\neq 0, (1/p)+(1/q)+(1/r)<1)$ are called the hyperbolic singularities. The intersection form on $H^2_c(B_\infty)$ has been calculated by Gabrielov [7]. One obtains $\mu_0=1$, $\mu_+=1$ and $\mu=p+q+r-1$. Hence the only non-zero Hodge numbers are $h_1^{2,2}=h_1^{1,1}=1$ and $h_{\neq 1}^{1,1}=p+q+r-3$. This implies that the minimal polynomial of $T_0$ has the factor $(t-1)^2$ which implies that the monodromy has infinite order.

(4.13) EXAMPLE. For the singularity $x^8+y^8+z^8+(xyz)^2$ the Hodge numbers of $H^2(B_\infty)_1$ and $H^2(B_\infty)_{\neq 1}$ are given by the matrices

$$\begin{pmatrix}0&0&0\\0&4&9\\0&9&4\end{pmatrix}\quad\text{resp.}\quad\begin{pmatrix}1&9&9\\9&133&9\\9&9&1\end{pmatrix}$$

Hence $\mu_0=22$, $\mu_+=42$, $\mu_-=151$ so the signature equals $-109$.

(4.14) PROPOSITION. *If $n$ is odd then $\mu-\mu_0$ is even;*

*if $n\equiv 2 \bmod 4$ then $\mu-\mu_-$ is even;*

*if $n\equiv 0 \bmod 4$ then $\mu-\mu_+$ is even.*

PROOF. First remark that it follows from (4.3) and (4.9) that $h^{pq}_{\neq 1}=h^{qp}_{\neq 1}=h^{n-p,n-q}_{\neq 1}$ and $h^{pq}_1=h^{qp}_1=h^{n+1-p,n+1-q}_1$ for all $p,q\geqq 0$. This implies that for $n$ odd we have

$$\mu-\mu_0=\sum_{p+q\neq n,n+1} h^{pq}_1+\Sigma h^{pq}_{\neq 1}$$

$$=2\left(\sum_{p+q\geqq n+2} h^{pq}_1+\sum_{p+q\geqq n+1} h^{pq}_{\neq 1}+\sum_{\substack{p+q=n\\ p<q}}(h^{pq}_{\neq 1}-h^{pq}_1)\right).$$

The other statements are proved analogously. This proves the conjecture of V.I. Arnol'd [2] that $\mu_++\mu_0$ is even if $n=2$.

(4.15) EXAMPLE. Suppose that $n=2$ and that $S$ is negative semi-definite. Then $\mu_+=0$ so either $H^2(B_\infty)_1=0$ and $S$ is negative definite (this gives the simple singularities $A_k, D_k, E_6, E_7, E_8$) in which case $H^2(B_\infty)$ is purely of type (1, 1), or $H^2(B_\infty)$ is of mixed type $\{(1,1),(1,2),(2,1)\}$ with $\mu_0=2h^{1,2}>0$ (this gives the simple elliptic or parabolic singularities $\tilde{E}_6$, $\tilde{E}_7$, and $\tilde{E}_8$).

(4.16) EXAMPLE. Suppose that $n=2$, the monodromy has finite order and $\mu_0=0$. This occurs in the case of Arnol'd's exceptional singularities. Then $h^{pq}_1=0$ for all $p,q$ and $h^{pq}=0$ if $p+q\neq 2$. Then $\mu_+=2h^{2,0}$ and $\mu_-=h^{1,1}$.

(4.17) Local cohomology

The local cohomology groups of an isolated singular point $x_0$ of a projective variety $Y$ carry a mixed Hodge structure ([5], Example (8.3.8)) namely

$$H^q_{\{x_0\}}(Y)=H^q(Y \bmod Y-\{x_0\}).$$

We want to compute these for a hypersurface singularity $x_0$ and relate them to the mixed Hodge structure on $H^n(B_\infty)$. Let $Y$ be a projective variety with only one singular point $x_0$. Let $\rho:\tilde{Y}\to Y$ be a resolution of singularities for $Y$ and let $\rho^{-1}(x_0)=C=C_1\cup\cdots\cup C_m$ be a union of smooth divisors $C_i$ with normal crossings on $\tilde{Y}$.

The mixed Hodge structure on $H^*(Y-\{x_0\})$ is computed using the logarithmic De Rham complex $\Omega^{\cdot}_{\tilde{Y}}(\log C)$ and the restriction map $H^*(Y)\to H^*(Y-\{x_0\})$ is induced by the inclusion

$$\Omega^{\cdot}_{\tilde{Y}}=W_0\Omega^{\cdot}_{\tilde{Y}}(\log C)\subset\Omega^{\cdot}_{\tilde{Y}}(\log C).$$

The mixed Hodge structure on $H^*(Y)$ has been computed in (3.4) and the pull-back map $H^*(Y)\to H^*(\tilde{Y})$ is induced by the surjective map $A^\cdot(Y)_{\mathbb{C}}\to A^\cdot(Y)_{\mathbb{C}}/W_{-1}A^\cdot(Y)_{\mathbb{C}}=\Omega^\cdot_{\tilde{Y}}$. Hence we have a morphism $\phi$ of cohomological mixed Hodge complexes which on $\mathbb{C}$-level is the composed map $A^\cdot(Y)_{\mathbb{C}}\to \Omega^\cdot_{\tilde{Y}}\to\Omega^\cdot_{\tilde{Y}}(\log C)$.

The mixed Hodge structure on $H^*_{\{x_0\}}(Y)$ is obtained by taking the hypercohomology of the mapping cone of $\phi$. One has the exact sequence of cohomological mixed Hodge complexes, which on $\mathbb{C}$-level is

$$0\to\Omega^\cdot_{\tilde{Y}}(\log C)[-1]\to A^\cdot_{\{x_0\}}(Y)_{\mathbb{C}}\to A^\cdot(Y)_{\mathbb{C}}\to 0.$$

It gives the exact sequence of mixed Hodge structures

$$\cdots\to H^k_{\{x_0\}}(Y)\to H^k(Y)\to H^k(Y-\{x_0\})\to H^{k+1}_{\{x_0\}}(Y)\to\cdots$$

dual to the classical sequence of relative homology. Let's go back to the case where $Y$ is the singular fiber $Y_0$ of a one-parameter family as before. One has the isomorphism

$$H^k_{\{x_0\}}(Y_0)\to H^{k-1}(K)$$

where $K=Y_0\cap\partial B$ is the common boundary of all fibers $F$ of the Milnor fibration. The well-known exact sequence

$$0\to H_n(K)\to H_n(\bar{F})\to H_n(\bar{F},K)\to H_{n-1}(K)\to 0$$

gives an exact sequence of mixed Hodge structures

$$0\longrightarrow H^n_{\{x_0\}}(Y_0)\longrightarrow H^n_c(B_\infty)\xrightarrow{\ j\ } H^n(B_\infty)\longrightarrow H^{n+1}_{\{x_0\}}(Y_0)\longrightarrow 0$$

and $H^k_{\{x_0\}}(Y_0)=0$ for $k\neq n, n+1, 2n$. The mixed Hodge structures on $H^n_{\{x_0\}}(Y_0)$ and $\mathrm{Hom}\,(H^{n+1}_{\{x_0\}}(Y_0),\mathbb{Q}(-n))$ are isomorphic.

(4.18) Lemma. $Gr^W_n H^n(Y_0)\cong Gr^W_n H^n(Y_0-\{x_0\})$.

Proof. Cup product on $H^n(X_\infty)$ induces a non-degenerate bilinear form on $Gr^W_n H^n(Y_0)=H^n(Y_0)/H^n(Y_0)\cap H^n(Y_0)^\perp$. This implies that $Gr^W_n H^n(Y_0)$ is isomorphic to its dual $Gr^W_n H^n(Y_0-\{x_0\})$.

(4.19) Example. Let $P(x,y,z)=x^4+y^4+z^4+xyz$. Once blowing up $0\in\mathbb{C}^3$ resolves the singularity of $Y_0$ (not the one of $P$). One obtains $\rho^{-1}(0)=C_1\cup C_2\cup C_3$, a union of 3 lines in $\mathbb{P}^2$ in general position. Here $H^2_{\{x_0\}}(Y_0)=\mathbb{Q}$, $H^3_{\{x_0\}}(Y_0)=\mathbb{Q}(-2)$ hence $H^1(Y_0-\{x_0\})=0$.

## 5. Open problems and conjectures

(5.1) '*Thom-Sebastiani*'. Let $f\in\mathbb{C}[z_0,\cdots,z_n]$ and $g\in\mathbb{C}[z_{n+1},\cdots,z_{n+m+1}]$ have an isolated singularity at the origin. Then the polynomial $f+g\in$

$\mathbb{C}[z_0, \cdots, z_{n+m+1}]$ also has an isolated singularity and, denoting $B_{\infty,f}$ and $B_{\infty,g}$ the respective Milnor fibers and $T_f, T_g$ the respective monodromy transformations, on has [17]:

$$H^{n+m+1}(B_{\infty,f+g}) \cong H^n(B_{\infty,f}) \otimes H^m(B_{\infty,g});$$
$$T_{f+g} = T_f \otimes T_g.$$

(5.2) PROBLEM. What is the relation between the mixed Hodge structures on $H^n(B_{\infty,f})$, $H^m(B_{\infty,g})$ and $H^{n+m+1}(B_{\infty,f+g})$?

(5.3) The quasi-homogeneous case suggests the following possibility. To the mixed Hodge structure on $H^n(B_{\infty,f})$ together with its automorphism of finite order $\gamma_s$ we associate a sequence $(u_1, w_1), \cdots, (u_\mu, w_\mu)$ of pairs of rational numbers as follows: for $\lambda \in \mathbb{C}^*$ and $q \in \mathbb{Z}$ let $\ell_q(\lambda) \in \mathbb{C}$ be determined by $\exp 2\pi i \ell_q(\lambda) = \lambda$ and $q \leqq \operatorname{Re} \ell_q(\lambda) < q+1$; to any eigenvalue $\lambda$ of $\gamma_s$ acting on $H^{p,q}(B_\infty)$ associate the pair $(\ell_q(\lambda), p+q)$ if $\lambda \neq 1$ and $(q, p+q-1)$ if $\lambda = 1$. Doing this with all eigenvalues of $\gamma_s$ on all $H^{p,q}$ $(B_\infty)$, one obtains an unordered $\mu$-tuple $(u_i, w_i)$, $i = 1, \cdots, \mu$, which we shall call the *characteristic pairs* of $H^n(B_\infty)$.

Conversely one recovers the discrete invariants $h_1^{pq}$, $h_{\neq 1}^{pq}$ and the eigenvalues of $\gamma_s$ on each $H^{pq}(B_\infty)$ from the characteristic pairs in the obvious way.

(5.4) CONJECTURE. Let $f$, $g$ have isolated singularities at 0. Let $(u_i, w_i)$, $i = 1, \cdots, \mu_f$ and $(u'_j, w'_j)$, $j = 1, \cdots, \mu_g$ be the characteristic pairs of $H^n(B_{\infty,f})$ resp. $H^m(B_{\infty,g})$. Then the characteristic pairs for

$$H^{n+m+1}(B_{\infty,f+g}) \quad \text{are} \quad (u_i + u'_j,\ w_i + w'_j + 1),\ i = 1, \cdots, \mu_f,\ j = 1, \cdots, \mu_g.$$

This conjecture is true if $f$ and $g$ (and hence $f+g$) are quasi-homogeneous, as follows from the explicit calculations in [20].

(5.5) EXAMPLE. Let $f(x, y) = x^5 + x^2y^2 + y^4$, $g(z) = z^2$. The characteristic pairs for $f$ are:

$$(\tfrac{1}{2}, 0), (\tfrac{7}{10}, 1), (\tfrac{9}{10}, 1), (\tfrac{3}{4}, 1), (1, 1), (1, 1), (\tfrac{5}{4}, 1), (\tfrac{11}{10}, 1), (\tfrac{13}{10}, 1), (\tfrac{3}{2}, 2)$$

as one computes by resolving $f$ and applying (3.14). The characteristic pair of $g$ is $(\tfrac{1}{2}, 0)$. According to the conjecture the characteristic pairs for $f+g$ would be

$$(1, 1), (\tfrac{6}{5}, 2), (\tfrac{7}{5}, 2), (\tfrac{5}{4}, 2), (\tfrac{3}{2}, 2), (\tfrac{3}{2}, 2), (\tfrac{7}{4}, 2), (\tfrac{8}{5}, 2), (\tfrac{9}{5}, 2), (2, 3)$$

hence $H^2(B_{\infty,f+g})$ has $h_{\neq 1}^{pq} = 0$ unless $(p, q) = (1, 1)$, $h_{\neq 1}^{11} = 8$, $h_1^{1,1} = h_1^{2,2} = 1$. This is correct because $f+g$ is equivalent to a hyperbolic germ $x^4 + y^5 + z^2 + \lambda xyz$ (see example (4.12)).

More evidence that the conjecture should be true is obtained from the conjectural method of computing the characteristic pairs from the Newton diagram. See (5.8).

(5.6) Mixed Hodge structure and Newton diagram

We refer to [21] for the notions of Newton diagrams and non-degenerate principal parts of polynomials. Let $f \in \mathbb{C}[z_0, \cdots, z_n]$ be a polynomial with a non-degenerate principle part. Denote $A$ the graded ring associated to the Newton filtration on $\mathbb{C}[z_0, \cdots, z_n]$. Denote $f_0, F_0, \cdots, F_n$ the principle parts of $f, z_0\partial f/\partial z_0, \cdots, z_n\partial f/\partial z_n$. The degree on $A$ takes rational values and is normalized in such a way that

$$\deg(f_0) = \deg(F_0) = \cdots = \deg(F_n) = 1.$$

We use the method of Poincaré series to deduce the characteristic pairs of $H^n(B_{\infty,f})$ from the Newton diagram $\Gamma$ of $f$. Write $\tau \leqq \sigma$ for '$\tau$ is a face of $\sigma$'.

For $\sigma$ a face of $\Gamma$ denote $A_\sigma$ the corresponding graded subring of $A$. We consider $\{0\}$ as a common face of every face of $\Gamma$ and put $A_{\{0\}} = \mathbb{C}$, with degree 0. The Poincaré series of any $A_\sigma$ is defined by

$$p_{A_\sigma}(t) = \sum_{\alpha \in \mathbb{Q}} \dim (A_\sigma)_\alpha \cdot t^\alpha.$$

It follows from [21], p. 15 that

$$p_{A_\sigma/(F_{0,\sigma}, \cdots, F_{n,\sigma})}(t) = (1-t)^{d(\sigma)} p_{A_\sigma}(t)$$

where $d(\sigma) = \dim A_\sigma = \dim \text{Cone}(\sigma)$.

Let $q_\sigma(t)$ be the Poincaré polynomial for the 'interior' of $A_\sigma/(F_{0,\sigma}, \cdots, F_{n,\sigma})$ i.e.

$$q_\sigma(t) = \sum_{\tau \leqq \sigma} (-1)^{d(\sigma)-d(\tau)} (1-t)^{d(\tau)} p_{A_\tau}(t).$$

Then clearly

$$(1-t)^{d(\sigma)} p_{A_\sigma}(t) = \sum_{\tau \leqq \sigma} q_\sigma(t).$$

For $\sigma$ a face of $\Gamma$, define $k(\sigma) = \min \{k \in \mathbb{Z} \mid \exists i_1, \cdots, i_k \in \{0, \cdots, n\}$ such that $\sigma \subset \mathbb{R}\, e_{i_1} + \cdots + \mathbb{R}\, e_{i_k}\}$. Then [21], Prop. (2.6) implies:

$$p_A(t) = \sum_{\sigma : k(\sigma) = n+1} (-1)^{n+1-d(\sigma)} p_{A_\sigma}(t).$$

In fact we are interested in the Poincaré polynomial of

$$H = (z_0 \cdots z_n)A/(F_0, \cdots, F_n)A \cong Gr\mathbb{C}[[z_0, \cdots, z_n]]/(\partial f/\partial z_0, \cdots, \partial f/\partial z_n)$$

where $\mathbb{C}[[z_0, \cdots, z_n]]$ has been filtered such that the monomial $z_0^{a_0} \cdots z_n^{a_n}$ gets the same degree as its image $z_0^{a_0+1} \cdots z_n^{a_n+1}$ in $A$. This can be computed by calculating the Poincaré polynomials of $A/(F_0, \cdots, F_n)$ and of all its quotients corresponding to intersections of coordinate hyperplanes in $\mathbb{R}_+^{n+1}$.

By [21], Theorem 2.8 one obtains

$$P_{A/(F_0, \cdots, F_n)}(t) = (1-t)^{n+1} p_A(t)$$

hence

$$p_H(t) = \sum_\sigma (-1)^{n+1-d(\sigma)} \cdot (1-t)^{k(\sigma)} p_{A_\sigma}(t)$$

so we have proved:

(5.7) Theorem. $p_H(t) = \sum_{\tau \leqq \sigma} (-1)^{n+1-d(\sigma)} (1-t)^{k(\sigma)-d(\sigma)} q_\sigma(t)$.

Let $G$ be the free abelian group on pairs $(u, v)$ with $u, v \in \mathbb{Q}$ as generators. The characteristic pairs $(u_i, w_i)$ of $f$ correspond to the element $cp(f) = \sum_{i=1}^{\mu} (u_i, w_i)$ of $G$.

(5.8) Conjecture. For any simplex $\sigma$ write $q_\sigma(t) = \sum_{\alpha \in \mathbb{Q}} \varepsilon_{\sigma,\alpha} t^\alpha$ (because $f_0$ is non-degenerate this is a finite sum); then the characteristic pairs of $H^n(B_{\infty,f})$ are represented by the element $cp(f)$ of $G$ where

$$cp(f) = \sum_{\substack{\tau \leqq \sigma \\ \alpha \in \mathbb{Q}}} \Sigma_{j=0}^{k(\tau)-d(\tau)} (-1)^{n+1-d(\tau)+j} \binom{k(\tau)-d(\tau)}{j} \varepsilon_{\sigma,\alpha} \cdot (\alpha + j, 2k(\sigma) - d(\sigma) - 2j - 1).$$

(5.9) Example. Let $f(x, y) = x^5 + x^2 y^2 + y^4$. Then the faces of $\Gamma$ are $\sigma_1 = \{0\}$, $\sigma_2 = \langle(5, 0)\rangle$, $\sigma_3 = \langle(0, 4)\rangle$, $\sigma_4 = \langle(2, 2)\rangle$, $\sigma_5 = \langle(5, 0), (2, 2)\rangle$ and $\sigma_6 = \langle(2, 2), (0, 4)\rangle$.

Thus

$$q_{\sigma_1} = 1;$$
$$q_{\sigma_2} = t^{\frac{1}{5}} + t^{\frac{2}{5}} + t^{\frac{3}{5}} + t^{\frac{4}{5}};$$
$$q_{\sigma_3} = t^{\frac{1}{4}} + t^{\frac{1}{2}} + t^{\frac{3}{4}};$$
$$q_{\sigma_4} = t^{\frac{1}{2}};$$
$$q_{\sigma_5} = t^{\frac{7}{10}} + t^{\frac{9}{10}} + t^{\frac{11}{10}} + t^{\frac{13}{10}};$$
$$q_{\sigma_6} = t^{\frac{3}{4}} + t + t^{\frac{5}{4}}.$$

Moreover $p_H(t) = t \cdot q_1 + (1+t) q_{\sigma_4} + q_{\sigma_5} + q_{\sigma_6}$, so applying the conjecture one obtains precisely the pairs as listed in example (5.5).

(5.10) Remark. If we know a priori that the monodromy has finite order, the characteristic pairs are all of the form $(u, n)$; hence if

$$p_H(t) = \sum_{\alpha \in \mathbb{Q}} c_\alpha t^\alpha, \quad \text{then} \quad cp(f) = \Sigma_\alpha c_\alpha (\alpha, n).$$

In general we first have to split up $p_H(t)$ as in (5.7) in order to determine the weights.

(5.11) EXAMPLE. If $f$ is a quasi-homogeneous germ with weights $w_0, \cdots, w_n$, and isolated singularity at 0, i.e. $f=\Sigma_\alpha a_\alpha z^\alpha$ and $a_\alpha \neq 0 \Rightarrow \Sigma_{i=0}^n w_i\alpha_i = 1$, then the above computation shows that

$$p_H(t)=\prod_{i=0}^{n}\frac{t^{w_i}-t}{1-t^{w_i}}.$$

In particular the Hodge numbers only depend on the weights.

(5.12) EXAMPLE. $f(x, y)=x^2y+y^4$; we have $w_0=\frac{3}{8}$, $w_1=\frac{1}{4}$,

$$\begin{aligned} p_H(t) &= (t^{\frac{3}{8}}-t)(t^{\frac{1}{4}}-t)(1-t^{\frac{3}{8}})^{-1}(1-t^{\frac{1}{4}})^{-1} \\ &= t^{\frac{5}{8}}+t^{\frac{7}{8}}+t+t^{\frac{9}{8}}+t^{\frac{11}{8}}. \end{aligned}$$

(5.13) REMARK. Assuming (5.8), one can use (5.7) and (5.8) to show that conjecture (5.4) is true if $f$ and $g$ both have non-degenerate principle parts.

(5.14) We shall sketch a proof that conjecture (5.8) is true in the case $n=1$.

Let $f\in\mathbb{C}[z_0, z_1]$ be a non-degenerate function. Let $(a_0, b_0), \cdots, (a_k, b_k)$ be the vertices of its Newton diagram $\Gamma$, ordered in such a way that $a_ib_{i+1}-a_{i+1}b_i>0$ for $i=0, \cdots, k-1$. Moreover we suppose that $b_0=a_k=0$.

One obtains a resolution of $f$ as follows. Let $\Gamma^*$ be the dual diagram of $\Gamma$ and let $\Sigma$ be a subdivision of $\Gamma^*$ such that the corresponding torus-embedding $X_\Sigma$ is smooth. Let $(0, 1)=(\alpha_0, \beta_0), (\alpha_1, \beta_1), \cdots, (\alpha_r, \beta_r)=(1, 0)$ be the vertices of $\Sigma$, ordered such that $\alpha_i\beta_{i+1}-\alpha_{i+1}\beta_i=-1$ for $i=0, 1, \cdots, r$. There exists a unique increasing function $p:\{1, \cdots, k\}\to\{1, \cdots, r-1\}$ with the property that

$$\alpha_{p(i)}(a_i-a_{i+1})+\beta_{p(i)}(b_i-b_{i+1})=0.$$

One can show the following facts:

(i) $E=E_0\cup E_1\cup\cdots\cup E_{r-1}$ where $E_i$ is the canonical divisor of $X_\Sigma$ corresponding to $(\alpha_i, \beta_i)$ for $i=1, \cdots, r-1$ and $E_0$ is the union of all components of $E$ with multiplicity 1;

(ii) $D_i$ is a curve of genus $>0 \Leftrightarrow i=p(j)$ for some $j\in\{1, \cdots, k\}$; in that case the multiplicity $e_i$ of $E_i$ equals $\alpha_ia_j+\beta_ib_j$;

(iii) Moreover $E_i$ intersects the curves $E_{i-1}$ of multiplicity

$$\alpha_{i-1}a_{j+1}+\beta_{i-1}b_{j+1}=e_{i-1}, \quad E_{i+1} \quad \text{with} \quad e_{i+1}=\alpha_{i+1}a_j+\beta_{i+1}b_j$$

and precisely $m_i=(a_jb_{j+1}-a_{j+1}b_j)/e_i$ components of $E_0$;

(iv) The eigenvalues of $\gamma_s$ on $H^{0,1}(B_\infty)=\bigoplus_{j=1}^k H^1(D_{p(j)}, \mathcal{O}_{D_{p(j)}})$ are computed using (3.14).

One obtains if $i = p(j)$:

$$\pi_* \mathcal{O}_{D_i} = \bigoplus_{s=0}^{e_i-1} \mathcal{O}_{E_i}\left(-\frac{se_{i+1}}{e_i} - \frac{se_{i-1}}{e_i} + \left[\frac{se_{i+1}}{e_i}\right] + \left[\frac{se_{i-1}}{e_i}\right] - \frac{m_i s}{e_i}\right) = \bigoplus_{s=0}^{e_i-1} \mathcal{O}_{E_i}(n_{s,i})$$

where

$$n_{s,i} = \left[\frac{se_{i+1}}{e_i}\right] + \left[\frac{se_{i-1}}{e_i}\right] - \frac{s(m_i + e_{i-1} + e_{i+1})}{e_i} \leqq 0.$$

Hence the divisor of characteristic pairs for $H^{0,1}(B_\infty)$ is

$$\sum_{j=1}^{k} \sum_{s=1}^{e_{p(j)}-1} (-1 - n_{s,p(j)})(s/e_{p(j)}, 1).$$

The divisor associated to $H^{1,0}(B_\infty)$ is obtained by duality:

$$\sum_{j=1}^{k} \sum_{s=1}^{e_{p(j)}-1} (-1 - n_{s,p(j)})(2 - s/e_{p(j)}, 1).$$

The remaining eigenvalues of $\gamma_s$ on $H^1(B_\infty)$ either are equal to 1, giving the pair (1, 1), or different from 1; in that case they occur in pairs, corresponding to $H^{0,0}(B_\infty)_{\neq 1} \cong H^{1,1}(B_\infty)_{\neq 1}$.

Thus the computation above gives complete information on the characteristic pairs.

Remains to prove that if $\sigma$ is the simplex spanned by $(a_j, b_j)$ and $(a_{j+1}, b_{j+1})$ in $\Gamma$, then for $i = p(j)$:

$$q_\sigma(t) = \sum_{s=1}^{e_i-1} n_{s,i}(t^{s/e_i} + t^{2-s/e_i}) + m_i t.$$

This is left to the reader as an exercise.

Remark that in this case the monodromy has finite order if and only if $gcd(a_j, b_j) = 1$ for $j = 1, \cdots, k-1$.

(5.15) Behavior under deformations

One may ask for a relation between the mixed Hodge structures on $H^n(B_{\infty,f})$ and $H^n(B_{\infty,g})$ if $g$ is a germ in the universal unfolding of $f$. One possible question is how the discrete invariants of the mixed Hodge structures are related.

Suppose that the numbers $h^{pq}$ remain constant under some deformation of $f$. Then the deformation gives rise to a 'variation of mixed Hodge structure' on the parameter space. This occurs for example when the deformation is equisingular in the following sense: there exists a simultaneous resolution of all functions in the deformation such that all components of the exceptional divisor and all their intersections are smooth over the parameter space. In that case one should compare the moduli of the singularity with the moduli of the triple consisting of $H^n(B_{\infty,f})$ with its mixed Hodge structure (or its associated graded Hodge structure) and its endomorphisms $N$ and $\gamma_s$. In the quasi-homogeneous case one may consider the residue of the Gauss-Manin connection instead of $\gamma_s$.

(5.16) EXAMPLE. Let $P(x, y) = x^2 - y^3$. Because an elliptic curve with an automorphism of order 6 is rigid, the Hodge structure on $H^1(B_\infty)$ has no moduli in the sense of (5.15). In the same way the polynomial $x^3 + y^3$ gives rise to the Hodge structure of weight 1 with an automorphism of order 4, hence it is also rigid.

(5.17) EXAMPLE. Let $P(x, y, z) = x^3 + y^3 + z^3 + \lambda xyz$. The moduli of $P$ are reflected by the Hodge structure $Gr_3^W H^2(B_{\infty,P}) = H^1(E_P)(-1)$ with $E_P$ the curve in $\mathbb{P}^2$ with homogeneous equation $P = 0$.

## REFERENCES


[1] A'CAMPO, N., La fonction zêta d'une mondromie. Comment. Math. Helvetici *50*, 233–248 (1975).

[2] ARNOL'D, V. I., Classification of unimodal critical points of functions. Functional analysis and its applications (English translation) *7*, 75–76 (1973).

[3] BRIESKORN, E., Die Monodromie der isolierten Singularitäten von Hyperflächen. Manuscripta math. *2*, 103–160 (1970).

[4] DELIGNE, P., Théorie de Hodge II. Publ. Math. I.H.E.S. *40*, 5–58 (1971).

[5] DELIGNE, P., Théorie de Hodge III. Publ. Math. I.H.E.S. *44*, 5–77 (1972).

[6] DURFEE, A. H., The monodromy of a degenerating family of curves. Inventiones math. *28*, 231–241 (1975).

[7] GABRIELOV, A. M., Dynkin diagrams for unimodal singularities. Functional analysis and its applications (English translation) *8*, 192–197 (1974).

[8] GRAUERT, H., Über Modifikationen und exzeptionelle analytische Mengen. Math. Annalen *146*, 331–368 (1962).

[9] GRAUERT, H. and RIEMENSCHNEIDER, O., Verschwindingssätze für analytische Kohomologiegruppen auf komplexen Räumen. Inventiones math. *11*, 263–292 (1970).

[10] GROTHENDIECK, A., Local cohomology. Lecture Notes in Math. *41*, Berlin-Heidelberg-New York: Springer 1967.

[11] GROTHENDIECK, A. and DIEUDONNÉ, J., Eléments de géométrie algébrique II. Publ. Math. I.H.E.S. *8* (1961).

[12] HOCHSTER, M. and ROBERTS, J. L., Rings of invariants of reductive groups acting on regular rings are Cohen-Macaulay. Advances in Mathematics *13*, 115–175 (1974).

[13] MILNOR, J., Singular points of complex hypersurfaces. Annals of Math. Studies *61*, Princeton University Press 1968.

[14] MUMFORD, D., Abelian varieties. Oxford University Press 1970.

[15] PRILL, D., Local classification of quotients of complex manifolds by discontinuous groups. Duke Math. Journal *34*, 375–386 (1967).

[16] SCHMID, W., Variation of Hodge structure: the singularities of the period mapping. Inventiones math. *22*, 211–320 (1973).

[17] SEBASTIANI, M. and THOM, R., Un résultat sur la monodromie. Inventiones math. *13*, 90–96 (1971).

[18] SERRE, J.-P., Algèbre locale-Multiplicités. Lecture Notes in Math. *11*, 3$^{me}$ édition. Berlin-Heidelberg-New York: Springer 1965.

[19] STEENBRINK, J. H. M., Limits of Hodge structures. Inventiones math. *31*, 229–257 (1976).

[20] STEENBRINK, J. H. M., Intersection form for quasi-homogeneous singularities. Compositio Math. *34*, 211–223 (1977).

[21] KOUCHNIRENKO, A. G., Polyèdres de Newton et nombres de Milnor. Inventiones math. *32*, 1–31 (1976).

J. H. M. STEENBRINK
Universiteit van Amsterdam
Mathematisch Instituut
Roetersstraat 15, Amsterdam-1004
The Netherlands

Nordic Summer School/NAVF
Symposium in Mathematics
Oslo, August 5–25, 1976

# THE HUNTING OF INVARIANTS IN THE GEOMETRY OF DISCRIMINANTS

## Five lectures at the Nordic Summer School

Bernard Teissier

## Table of contents

## Introduction

From the table of contents, the reader might have gathered the impression that what is presented here is a survey, a medley, or even a motley of results on singularities.

However, I hope that those who read the text will agree that on the contrary, it is entirely devoted to the illustration of a single idea:

'Primitive invariants of the discriminant $D$ of a map yield rather subtle invariants of the fibre of the map.'

For the purposes at hand, let us agree to a primitive invariant of a germ of a complex hypersurface $(D, 0) \subset (\mathbb{C}^{m+1}, 0)$, being the Newton polyhedron in some coordinates of the restriction of an equation $\delta(v, t_1 \cdots t_m) = 0$ of $D$ to an $i$-plane in $(\mathbb{C}^{m+1}, 0)$ through the origin. The simplest primitive invariant in this sense is the multiplicity of $D$, corresponding to $i = 1$ and a line transversal to $D$. The next simplest is the *Newton polygon* of a plane section of $D$ (see §3) and this is what we study here in the case where $D$ is the discriminant of a miniversal deformation of a hypersurface with isolated singularity $(X_0, 0) \subset (\mathbb{C}^N, 0)$, (i.e. our map is in particular stable).

In fact, it has been an open problem for some time to understand to what extent the geometry of the discriminant determines the geometry of $(X_0, 0)$.

The real-analytic version of this problem is important in the theory of catastrophes of Thom, and problems of a related nature appear in the Jung-Zariski-Abhyankar approach to resolution of singularities (by resolving the discriminant of a projection) (see [1]) and in Zariski's theory of equisingularity. ([5]).

This problem is at least partly solved here in the special case where $(X_0, 0)$ is a plane branch since it is shown, using a theorem of Merle, that the Newton polygon of a general vertical plane section of $D$ (see 5.5.7) is a *complete invariant* of the equisingularity class of $(X_0, 0)$, i.e. in this case, of its topological type. In particular, one can compute the Puiseux characteristic exponents of $(X_0, 0)$ from the inclinations of the edges of this polygon. Of course this case is only the basic test for any theory of invariants of equisingularity, and in the general case the conclusion is not so clear-cut, but we show that one can read from the Newton polygon of a general vertical plane section of $D$ such interesting invariants of $(X_0, 0)$ as the smallest integer $a$ such that if $f(z_1, \cdots, z_N)=0$ defines $X_0$, any $g \in \mathbb{C}\{z_1, \cdots, z_N\}$ such that $g-f \in (z_1, \cdots, z_N)^{a+1}$ defines, by $g=0$, a hypersurface equisingular with $X_0$, 0),—or the diminution of class which the presence of a singularity isomorphic to $(X_0, 0)$ would impose on a projective hypersurface. Furthermore, this Newton polygon is an invariant of equisingularity, as we define it (see below).

It might seem that all this is not of great practical utility since the equations of discriminants are very hard to compute, but our method also yields a way of computing these Newton polygons which I have found quite usable in practice.

From a geometric viewpoint, what is done here is to take a dynamic view of Morse theory: the Newton polygon mentioned above can also be deemed to describe the various 'speeds' with which the (coordinates of the) quadratic critical points in a generic morsification $v=f(z_1, \cdots, z_N)+u(\sum_1^N \alpha_i z_i)(\alpha_i \in \mathbb{C})$ of an equation of $(X_0, 0)$ vanish to 0 with the parameter of morsification $u$. It is perhaps a pleasant surprise that in the case of a plane branch these speeds suffice to completely determine the topology of the function $f(z_1, z_2)$, and conversely.

From a formal viewpoint, what we do is this: we introduce on the set of germs of hypersurfaces an equivalence relation: (c)-cosécance (see 2.19), which is our working definition of equisingularity. Two (c)-cosécant germs of hypersurfaces with isolated singularity are topologically equivalent, as well as all their general plane sections. Then, on the set $\mathcal{I}_s$ of (c)-cosécance classes of isolated singularities of hypersurfaces we introduce an operation $[X_0], [X_1] \rightsquigarrow [X_0] \perp [X_1]$, which is induced by the Thom-Sebastiani operation: if $X_0$ (resp. $X_1$) is defined by $f_0(z_1, \cdots, z_N)=0$ (resp. $f_1(w_1, \cdots, w_M)=0$) then $[X_0] \perp [X_1]$ is the (c)-cosécance class of $f(z_1, \cdots, z_N)-g(w_1, \cdots, w_M)=0$. The Milnor number gives us a map $\mu : \mathcal{I}_s \to \mathbb{Z}_0$ (where the subscript 0 indicates the non-negative part) satisfying $\mu([X_0] \perp [X_1]) = \mu([X_0]) \cdot \mu([X_1])$.

What we do here is to construct a new ring $\underline{\mathrm{N}}_S$, the *ring of special Newton polygons* and to factor this map $\mu$ by a 'Jacobian Newton polygon map' $\boldsymbol{\nu}_j$:

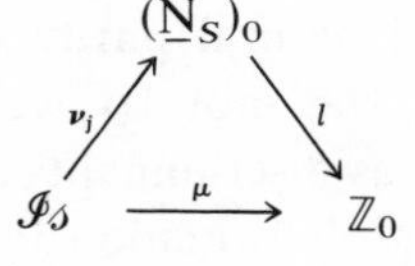


where $l$ is the length of horizontal projection of a Newton polygon (see 3.6) and $\boldsymbol{\nu}_j$ satisfies

$$\boldsymbol{\nu}_j([X_0] \perp [X_1]) = \boldsymbol{\nu}_j([X_0]) * \boldsymbol{\nu}_j([X_1])$$

where $*$ is the product of Newton polygons defined in 3.6, and the datum of $\boldsymbol{\nu}_j$ is equivalent to the datum of the Newton polygon of a general vertical plane section of the discriminant $D$ of a miniversal deformation of $(X_0, 0)$ (see 5.17).

We remark that $\mu$ and $\boldsymbol{\nu}_j$ are given by the two simplest primitive invariants of $D$, since $\mu$ is the multiplicity of $D$ (see 5.5.2). What happens for the higher-dimensional primitive invariants of $D$ is, as yet, unknown to me.

These notes therefore contain material which I believe to be essentially new, but since their aim is partly pedagogical, I decided to include not only the notions and results which I had found necessary to understand what I was doing, but also some illustration of them, to help their assimilation.

For example, to compute invariants from sections of the discriminant, it is indispensable to have a definition of the discriminant compatible with base change. Since the discriminant is – by definition – the image of the critical subspace, we are led towards a definition of the image of a finite map which is compatible with base change. In other words, finding a procedure of *elimination* stable enough to be computable. This is the subject of §1. The discriminant $D$ has the structure of an envelope, in the sense of [4]. Since this fact, although not of direct use to us, is of importance in the study of the geometry of $D$, we go a little into this in §2, and this gives the idea of our method of exposition. This method has the disadvantage – apart from the lengthening of the text – of provoking brutal changes in the level of exposition, of which I must warn the reader.

**Acknowledgements**

The proposition on developments in the first part of §2 is a version of a result first proved by Mr. A. Nobile (Some properties of the Nash blowing up, Pacific Journal of Math. Vol. 60 (1975) pp. 297–305). Also, when the definition of the image of a finite complex map given here in §1 was breezily introduced in ([7] Chap. III), I thought it 'well known' since it is so natural, I was mostly happy to have a definition of the discriminant compatible with base change, and an easy and natural purity theorem. Since then, it has appeared that the definition was not so well known or employed by geometers at all, except, as far as discriminants are concerned by D. S. Rim, especially in his papers 'Formal deformation theory' (S.G.A. 7, I, Springer Lecture Notes No. 288 and 'Torsion differentials and deformations' (Transactions A.M.S. Vol. 169 (1972)) where he had also defined the discriminant by a Fitting ideal and proved the purity theorem. Anyway, after some experimentation (partly reproduced in §3) with the Fitting ideal definition of the image of a mapping $f: X \to Y$ such that $R^i f_* \mathcal{O}_X = 0$ (e.g. $f$ finite), I became convinced that it is both a very informative and a very computable elimination process. A part of 3.3 is a maturation of an old result in common with Monique Lejeune-Jalabert and Lê Dũng Tráng (Sur un critère d'équisingularité, Note aux C. R. Acad. Sc. Paris t. 271 (1970) p. 1065). I wish to thank them for their comments on this part, and also Monique Lejeune-Jalabert for discussions on the formal study of an additive monoid of Newton polygons. I also thank P. Barril, M. Merle, and D. Trotman for suggestions which influenced the form and content of this text, and Geir Ellingsrud for writing up notes of the lectures which I used in §§1 and 2. My thanks to to the Matematisk Institutt, Oslo for typing §§1 and 2 first part, and also to Marie-Jo Lécuyer at the Centre de Mathématiques for doing a beautiful and fast typing of the rest of the text. Finally I wish to thank the organizers and participants of the Summer School for creating a very pleasant working atmosphere, and above all Per Holm for his patience with a lecturer who talked extremely fast and wrote the notes extremely slowly!

## REFERENCES FOR THE INTRODUCTION

[1] J. Lipman, Introduction to resolution of singularities, Proceedings of Symposia in Pure Mathematics, Vol. XXIX, Algebraic Geometry Arcata (A.M.S. 1975) pp. 187–230.

[2] H. Hironaka, Characteristic polyhedra of singularities, Journal of Maths. Kyoto University, Vol. 7, No. 3 (1968).

[3] A treatise of the method of fluxions and infinite series, with its application to the geometry of curve lines by the inventor Sir Isaac Newton (London 1671), Johnson Reprint Corporation.

[4] R. Thom, Sur la théorie des enveloppes, Journal de Maths. Pures et Appliquées, Tome XLI, Fasc. 2 (1962).

[5] O. Zariski, Some open questions in the theory of singularities, Bull. A.M.S. t. 77, No. 4 (July 1971) pp. 481–491.

## §1. Fitting ideals

In this section we will give the definition and elementary properties of Fitting ideals, which we will use later to give a definition of the image, as a complex analytic *space*, of a finite map between complex analytic spaces. We then give a definition of the resultant ideal of two polynomials as a Fitting ideal, and as an application give a proof of Bezout's theorem.

Let $A$ be a ring, and let $M$ be an $A$-module of finite presentation, that is, $M$ is the cokernel of an $A$-linear map between two free $A$-modules of finite rank, or if you prefer, there is an exact sequence, called a presentation of $M$:

$$A^q \xrightarrow{\psi} A^p \rightarrow M \rightarrow 0$$

where $p, q \in \mathbb{N}$. For each integer $j$ we associate to $M$ the ideal $F_j(M)$ of $A$ generated by the $(p-j)\times(p-j)$ minors of the matrix (with entries in $A$) representing $\psi$. Here we need the convention that if there are no $(p-j)\times(p-j)$ minors because $j$ is too large, i.e. $j \geqq p$, then $F_j(M)=A$ (the empty determinant is equal to 1) and if, at the other extreme, $p-j>q$, set $F_j(M)=0$ (the ideal generated by the empty set is 0).

A theorem of Fitting (for a proof see [1] p. 5 where the $F_j(M)$ are called $\sigma_j'(\psi)$) asserts that $F_j(M)$ depends only on the $A$-module $M$ and not on the choice of a presentation. We call it the $j$th Fitting ideal of $M$.

More generally, if $(X, \mathcal{O}_X)$ is a ringed space, and $\mathcal{M}$ a coherent sheaf of $\mathcal{O}_X$-modules, we can define a sheaf of ideals $\mathcal{F}_j(\mathcal{M})$ of $\mathcal{O}_X$, by defining $\mathcal{F}_j(\mathcal{M})$ locally as above, and then by uniqueness the ideals found locally patch up into a sheaf of ideals. Remark also that since $\mathcal{F}_j(\mathcal{M})$ is locally finitely generated, $\mathcal{F}_j(\mathcal{M})$ will be a coherent sheaf of ideals as soon as $\mathcal{O}_X$ is coherent, e.g. for a complex analytic space by Oka's theorem.

One important fact about Fitting ideals is that 'their formation commutes with base change' as one says. The idea is that if you have $(X, \mathcal{O}_X)$ and a coherent $\mathcal{O}_X$-module $\mathcal{M}$, then for any map $f:(Y, \mathcal{O}_Y)\rightarrow(X, \mathcal{O}_X)$ we have that $F_j(f^*\mathcal{M})=\mathcal{F}_j(\mathcal{M})\mathcal{O}_Y$, where this last expression means, here and in the sequel, the ideal in $\mathcal{O}_Y$ generated by the image of the canonical map, from $f^*(\mathcal{F}_j(M))$ to $\mathcal{O}_Y=f^*\mathcal{O}_X$ coming from the inclusion $\mathcal{F}_j(M)\subset\mathcal{O}_X$. Algebraically, it means that for any $A$-module $M$ of finite presentation, and any ring-homomorphism $g: A\rightarrow B$, we have $F_j(M\otimes_A B)=$ ideal generated by $g(F_j(M))$ in $B$, which is denoted by $F_j(M)B$. (this is immediate from the fact that tensoring a presentation of $M$ by $B$ gives a presentation of $M\otimes_A B$). In particular, for any maximal ideal $m$ in $A$ we have that $F_j(M)\subset m$ if and only if $\dim_{A/m} M\otimes_A A/m>j$ since the $j$th Fitting ideal of $M\otimes_A A/m$ is either 0 on $A/m$ depending upon whether the map $\bar{\psi}$ of $A/m$-vector spaces in the exact sequence: $(A/m)^q \xrightarrow{\bar{\psi}} (A/m)^p \rightarrow M\otimes_A A/m\rightarrow 0$ (obtained by tensoring a presentation of $M$ by $A/m$) is of rank $<p-j$ or not, i.e. $M\otimes_A A/m$ is of dimension $>j$ or not.

Geometrically, we think of this as follows: a coherent sheaf of modules $\mathcal{M}$ on a complex-analytic space $(X, \mathcal{O}_X)$ is the sheaf of sections of a mapping of complex analytic spaces $p: L(\mathcal{M})\rightarrow X$ such that for any $x\in X$, $p^{-1}(x)\simeq \mathcal{M}_{x_{\mathcal{O}_{X,x}}}\otimes\mathcal{O}_{X,x}/m_{X,x}$ is a finite dimensional vector space over $\mathcal{O}_{X,x}/m_{X,x}=\mathbb{C}$, and a section is a holomorphic map $\sigma: X\rightarrow L(\mathcal{M})$ such that $p\circ\sigma=id_X$.

Note that if $\mathcal{M}$ is not locally free, the dimension of $p^{-1}(x)$ as a vector space can vary with $x$, and since $F_j(\mathcal{M})_x = F_j(\mathcal{M}_x)$ (as an $\mathcal{O}_{X,x}$-module) we have that the set underlying the analytic subspace $V(\mathcal{F}_j(\mathcal{M}))$ of $X$ defined by the coherent sheaf of ideals $\mathcal{F}_j(\mathcal{M})$ is:

$$|V(F_j(\mathcal{M}))| = \{x \in X / \dim_{\mathbb{C}} p^{-1}(x) > j\}.$$

Let now $f:(X, \mathcal{O}_X) \to (Y, \mathcal{O}_Y)$ be a map of complex analytic spaces. We would like to define the image of $f$ as a complex analytic *subspace* of $(Y, \mathcal{O}_Y)$. This is not always possible, and in particular if one hopes to get a closed complex subspace of $Y$ it is better to assume $f$ is proper, and here we will consider only the case where $f$ is *finite* (= proper and with finite fibres).

By theorems of Grauert, the direct image sheaf $f_*\mathcal{O}_X$ is then a coherent sheaf of $\mathcal{O}_Y$-modules, and its formation commutes with base change, i.e. for any complex analytic map $h: Y' \to Y$

$$\begin{array}{ccc} X' = X_Y \times Y' & \xrightarrow{k} & X \\ \Big\downarrow{\scriptstyle f'} & \square & \Big\downarrow{\scriptstyle f} \\ Y' & \xrightarrow{h} & Y \end{array}$$

we have, in the above cartesian diagram of base change, that $f'_*\mathcal{O}_{X'} = h^* f_* \mathcal{O}_X$ (see [Cartan Seminar] 60–61, p. 15, cor. 1.6).

Now a basic requirement for the definition of the image is that again its formation should commute with base change, i.e.

$$\operatorname{im} f' = h^{-1}(\operatorname{im}(f)) \qquad \text{as complex } spaces$$

The first sheaf of ideals that comes to mind as a candidate to define $f(X)$ is the sheaf of functions $g$ on $Y$ such that $g \circ f = 0$ on $X$, i.e. the annihilator sheaf

$$\boldsymbol{Ann}_{\mathcal{O}_Y}(f_*\mathcal{O}_X) = \text{sheaf } \{\text{functions } g \text{ on } Y \text{ such that } g \cdot f_*\mathcal{O}_X = 0\}.$$

This is *not* a good choice because its formation does not commute with base extension, as we will show by an example below.

The second try is the 0th Fitting ideal of $f_*\mathcal{O}_X$, which set theoretically describes also the image of $f$, since the subspace of $Y$ defined by it, as a set is $\{y \in Y \mid \dim_{\mathbb{C}} (f_*\mathcal{O}_X)(y) > 0\} = \{y \in Y \mid (f_*\mathcal{O}_X)_y \neq 0\}$.

Now since, as we have seen, both the formation of direct images and the formation of Fitting ideals commute with base change, in any case we know that the formation of the image, with this definition will also. So we set:

DEFINITION. Let $f: X \to Y$ be a finite morphism of complex analytic spaces. The image of $f$ is the subspace of $Y$ defined by the coherent sheaf of ideals $\mathscr{F}_0(f_*\mathcal{O}_X)$.

LEMMA. *The formation of* im $(f)$ *commutes with base change. Proof:*

$$\mathscr{F}_0(f'_*\mathcal{O}_{X'}) = \mathscr{F}_0(h^* f_*\mathcal{O}_X) = \mathscr{F}_0(f_*\mathcal{O}_X)\mathcal{O}_{Y'}.$$

REMARK. Cramer's rule tells us that always $\mathbf{Ann}\, f_*\mathcal{O}_X \supset \mathscr{F}_0(f_*\mathcal{O}_X)$,

EXAMPLE. Let $f: (\mathbb{C}, 0) \to (\mathbb{C}^2, 0)$ be given by $x = t^{2k}$, $y = t^{3k}$ for some integer $k$. Clearly the set-theoretic image of $f$ is the curve $y^2 - x^3 = 0$. However, we wish to obtain an ideal defining a space supported on that curve, but possibly with nilpotent functions. To compute $\mathscr{F}_0(f_*(\mathcal{O}_\mathbb{C})$ we will look locally near 0, and therefore compute $\mathscr{F}_0(f_* O_\mathbb{C})_0$ as the 0th Fitting ideal of $\mathbb{C}\{t\}$ considered as $\mathbb{C}\{x, y\}$-module via the map of rings $\mathbb{C}\{x, y\} \to \mathbb{C}\{t\}$ sending $x$ to $t^{2k}$ and $y$ to $t^{3k}$. We must therefore write a presentation of $\mathbb{C}\{t\}$ as $\mathbb{C}\{x, y\}$-module. Let $e_0 = 1$, $e_1 = t, \cdots, e_{2k-1} = t^{2k-1}$. It is easily seen that they form a system of generators of $\mathbb{C}\{t\}$ as $\mathbb{C}\{x, y\}$-module, and that between them we have the following $2k$ relations:

$$\begin{array}{ll} xe_k - ye_0 = 0 & x^2 e_0 - ye_k = 0 \\ xe_{k+1} - ye_1 = 0 & x^2 e_1 - ye_{k+1} = 0 \\ \vdots & \vdots \\ xe_{2k-1} - ye_{k-1} = 0 & x^2 e_{k-1} - ye_{2k-1} = 0 \end{array}$$

which are obviously independent.

Hence we have a sequence of $\mathbb{C}\{x, y\}$-modules.

$$0 \longrightarrow \bigoplus_{i=0}^{2k-1} \mathbb{C}\{x, y\} e_i \xrightarrow{\psi} \bigoplus_{i=0}^{2k-1} \mathbb{C}\{x, y\} e_i \xrightarrow{\varphi} \mathbb{C}\{t\} \longrightarrow 0$$

with $\varphi(e_i) = t^i$, and where $\psi$ is given by the matrix

$$\underline{\psi} = \left[\begin{array}{ccccc|ccccc} \multicolumn{5}{c|}{\overbrace{\hphantom{-y \quad 0 \quad 0 \quad \cdot \quad \cdot \quad 0}}^{k}} & \multicolumn{5}{c}{\overbrace{\hphantom{x \quad 0 \quad \cdot \quad \cdot \quad \cdot \quad 0}}^{k}} \\ -y & 0 & 0 & \cdot\ \cdot & 0 & x & 0 & \cdot & \cdot\ \cdot & 0 \\ 0 & -y & 0 & & & 0 & x & & & \cdot \\ \cdot & & \cdot & & & \cdot & & \cdot & & \cdot \\ \cdot & & & \cdot & & \cdot & & & \cdot & \cdot \\ 0 & & & & -y & & & & & x \\ x^2 & 0 & \cdot & \cdot & & -y & & & & 0 \\ 0 & x^2 & & & & & \cdot & & & \\ \cdot & & \cdot & & & & & \cdot & & \cdot \\ \cdot & & & \cdot & & & & & & \\ & & & & \cdot & & & & \cdot & 0 \\ 0 & & & & x^2 & & & & & -y \end{array}\right] \begin{array}{l} \\ \left.\vphantom{\begin{array}{c}1\\1\\1\\1\\1\end{array}}\right\}k \\ \left.\vphantom{\begin{array}{c}1\\1\\1\\1\\1\\1\end{array}}\right\}k \end{array}$$

We give it as an exercise to check that the sequence is exact, for there is a general reason why $\mathbb{C}\{t\}$ must have such a resolution of length 1 as $\mathbb{C}\{x, y\}$-module (see §3, 3.5).

By permuting rows and columns of $\psi$ one checks that $\det \underline{\psi} = (y^2 - x^3)^k$ i.e. we have shown that

$$F_0(f_* \mathcal{O}_{\mathbb{C}})_0 = (y^2 - x^3)^k \mathbb{C}\{x, y\}.$$

Let us now calculate $\boldsymbol{Ann}_{\mathbb{C}\{x,y\}} \mathbb{C}\{t\}$. Since $1 \in \mathbb{C}\{t\}$, the annihilator is just the kernel of the map $\mathbb{C}\{x, y\} \to \mathbb{C}\{t\}$, which is the ideal generated by $(y^2 - x^3)$, certainly different from our Fitting ideal if $k > 1$.

Let us now make a base change by restricting our map over the $x$-axis, i.e. by the inclusion $\{y = 0\} \subset (\mathbb{C}^2, 0)$ or algebraically by $\mathbb{C}\{x, y\} \to \mathbb{C}\{x\}$ sending $y$ to 0. Then $\mathbb{C}\{t\} \otimes_{\mathbb{C}\{x,y\}} \mathbb{C}\{x\} = \mathbb{C}\{t\}/(t^{3k})$ viewed as $\mathbb{C}\{x\}$-module via the map sending $x$ to $t^{2k}$. Then, the annihilator of this $\mathbb{C}\{x\}$-module is $(x^2)\mathbb{C}\{x\}$ while the image in $\mathbb{C}\{x\}$ of $(y^2 - x^3)\mathbb{C}\{x, y\}$ is $(x^3)\mathbb{C}\{x\}$. This shows that the formation of the annihilator does not commute with base change.

We will now construct the resultant of two polynomials as a Fitting ideal. As an application we will prove Bezout's theorem.

Let $A$ be a ring and let $P, Q \in A[X]$ be two polynomials:

$$P = p_0 + p_1 X + \cdots + p_n X^n$$
$$Q = q_0 + q_1 X + \cdots + q_m X^m$$

We will assume that $p_n$ and $q_m$ are units in $A$. The resultant $R(P, Q)$ will be an element in $A$, satisfying the following property:

For any field $K$ and any homomorphism

$$\varphi : A \to K, \quad \text{the polynomials} \quad \varphi(P), \varphi(Q) \in K[X]$$

have a common root in an extension of $K$ if and only if $\varphi(R(P, Q)) = 0$.

Here $\varphi(P)$ denotes the polynomial $\varphi(p_0) + \varphi(p_1)X + \cdots \varphi(p_n)X^n \in K[X]$. Remark first that the $A$-module $M = A[X, Y]/(P, Q)$ satisfies $M \otimes_A K = 0$ if and only if $\varphi(P)$ and $\varphi(Q)$ have no common root in any extension of $K$. In fact, they do not have a common root if and only if we can find an extension $K'$ of $K$ and polynomials $S, T \in K'[X]$ with

$$1 = S\varphi(P) + T\varphi(Q)$$

which clearly is equivalent to $A[X]/(P, Q) \otimes_A K = 0$.

Note. The existence of such an element $R(P, Q) \in A$ is not at all clear a priori, and anyway if it exists the properties we ask of it will also be satisfied by $UR(P, Q)$, where $U$ is any invertible element of $A$. Therefore what we can see a priori, is that we can hope to define a resultant *ideal* in $A$, and then prove it is a principal ideal, which is what we do below.

The geometric meaning of the resultant is the following. Let $A$ correspond to a space $S$, i.e. let $A$ be the ring of germs of holomorphic functions at a point $s \in S$. Then $A[X]$ corresponds to the space $S \times \mathbb{C}$. Denote by $\pi : S \times \mathbb{C} \to S$ the projection.

The polynomials $P$, $Q$ define hypersurfaces $V(P)$ and $V(Q)$ in $S \times \mathbb{C}$. Let $m \subset A$ be the maximal ideal, and let $\varphi : A \to A/m$ be the canonical map. Then $\varphi(P)$ and $\varphi(Q)$ have a common root if and only if $V(P) \cap V(Q)$ intersects the fibre $\pi^{-1}(s)$, that is $s \in \pi(V(P) \cap V(Q))$. Thus by the property above the resultant should be a defining equation for the *image* of $V(P) \cap V(Q)$ by $\pi$. This motivates the following definition.

DEFINITION. Let $M$ be the $A$-module $M = A[X]/(P, Q)$. We define the resultant ideal of $P$ and $Q$ as $F_0(M) \subset A$.

LEMMA. *$F_0(M)$ is a principal ideal.*
(In fact we will by abuse of language call any generator of $F_0(M)$ the resultant of $P$ and $Q$ and write it $R(P, Q)$.)

PROOF. We will write down a presentation of $M$ as an $A$-module, namely

$$A[X]/(P) \xrightarrow{\psi_1} A[X]/(P) \longrightarrow A[X]/(P, Q) \longrightarrow 0$$

Where $\psi_1$ is multiplication by $Q$, that is $\psi_1(\bar{a}) = \overline{Qa}$, where the bar means reduction modulo $(P)$. Now as the highest-coefficient of $P$ was invertible, $A[X]/(P)$ is a free $A$-module of rank $n$. Thus we have a finite presentation and moreover $\psi_1$ is represented by a square matrix so $F_0(M)$ is generated by its determinant.

We could use other presentations too, for example:

$$A[X]/(X^n) \oplus A[X]/(X^m) \to A[X]/(P \cdot Q) \to A[X]/(P \cdot Q) \to 0$$
$$(\bar{a}, \bar{b}) \to \overline{aQ + bP}$$

or the presentation coming from the 'chinese exact sequence' (i.e. Chinese remainder theorem)

$$A[X]/(P \cdot Q) \to A[X]/(P) \oplus A[X]/(Q) \to A[X]/(P, Q) \to 0$$
$$\bar{a} \to (\bar{a}, \bar{a}) \qquad\qquad (\bar{a}, \bar{b}) \to \bar{a} - \bar{b}$$

where the bar indicates reduction modulo the ideal in question.

The natural ring in which to treat the resultant is the following

$$\mathcal{A} = \mathbb{Z}\left[p_0, \cdots, p_n, q_0, \cdots, q_m, \frac{1}{p_n}, \frac{1}{q_m}\right]$$

where the $p_i$ and $q_j$ are indeterminates.

In the ring $\mathcal{A}[X]$ we have the two polynomials $\mathcal{P}$ and $\mathcal{Q}$ of degree $n$ and $m$ respectively given by

$$\mathcal{P} = p_0 + p_1X + \cdots + p_nX^n$$
$$\mathcal{Q} = q_0 + q_1X + \cdots + q_mX^m$$

They have the following universal property: Take any ring $A$ and any two polynomials $P, Q \in A[X]$ of degree $n$ and $m$, and suppose that their highest-degree coefficients are invertible in $A$. Then there exists a unique homomorphism $\varphi: \mathcal{A} \to A$ with $\varphi(\mathcal{P}) = P$ and $\varphi(\mathcal{Q}) = Q$. It is given by sending $p_i$ to the coefficient of $X^i$ in $P$ and $q_j$ to the coefficient of $X^j$ in $Q$.

The ring $\mathcal{A}$ has a grading given by degree $p_i = n - i$ and degree $q_j = m - j$. If we give $X$ the degree 1, the polynomials $\mathcal{P}$ and $\mathcal{Q}$ are homogeneous of degree $m$ and $n$ respectively for the corresponding grading of $\mathcal{A}[X]$. Moreover if the ring $A$ above is graded and $P$ and $Q$ are homogeneous, the homomorphism $\varphi$ will be homogeneous of degree 0.

LEMMA 0. *In the situation described above, the resultant $R(\mathcal{P}, \mathcal{Q}) \in \mathcal{A}$ is homogeneous of degree $mn$ (i.e. the resultant ideal is homogeneous and generated by an element of degree $mn$).*

To prove this we need a lemma on graded modules. Let $A$ be any graded ring and let $\nu$ be an integer. We define a graded $A$-module $A(\nu)$. As an $A$-module it is just $A$ itself, we change only the grading by giving 1 the degree $\nu$. That is, for any homogeneous element $x \in A$, its degree in $A(\nu)$ is $\nu +$ (its degree in $A$).

LEMMA 1. *For any homogeneous homomorphism of degree* 0:

$$\psi: \bigoplus_{i=1}^{q} A(e_i) \to \bigoplus_{j=1}^{q} A(f_j).$$

[This represents just a graded homomorphism between two free $A$-modules, graded as indicated] the Fitting ideals $F_k$ (coker $\psi$) are homogeneous.

Moreover,

$$\deg(\det \psi) = \sum_{i=1}^{q} e_i - \sum_{j=1}^{q} f_j.$$

PROOF. Let $\psi$ be represented by the matrix $(\psi_{ij})$. By writing what is happening to a basis element in $\oplus_{i=1}^{q} A(e_i)$ it is easily verified that $\psi_{ij}$ is homogeneous of degree $e_i - f_j$. In the expansion of $\det(\psi)$ each term is a product $\psi_{i_1j_1} \cdot \psi_{i_2j_2} \cdot \psi_{i_3j_3} \cdot \cdots \cdot \psi_{i_qj_k}$ where each $i \in \{0, \cdots, g\}$ and each $j \in \{0, \cdots, q\}$ appears exactly once. Hence it is homogeneous of degree $\sum_{\nu=1}^{q} e_{i_\nu} - \sum_{\nu=1}^{q} f_{j_\nu} = \sum_{i=1}^{q} e_i - \sum_{j=1}^{q} f_j$. Thus we have proved the part of the lemma concerning $\det \psi$.

As $F_k$ (coker $\psi$) is generated by minors of $\psi$, the same argument shows that it is homogeneous.

To calculate deg $R(\mathcal{P}, \mathcal{Q})$ we go back to our presentation:

$$\mathcal{A}[X]/(\mathcal{P}) \xrightarrow{\cdot \mathcal{Q}} \mathcal{A}[X]/(\mathcal{P}) \longrightarrow \mathcal{A}[X]/(\mathcal{P}, \mathcal{Q}) \longrightarrow 0$$

The free $\mathcal{A}$-module $A[X]/(\mathcal{P})$ has a basis $1, X, X^2, \cdots, X^n$. We will assign degrees to this such that the multiplication by $\mathcal{Q}$ becomes homogeneous of degree 0, that is, as $X^i\mathcal{Q}$ has degree $m+i$, we must give $X^i$ the degree $m+i$. Hence $e_i = m+i$ and $f_j = j$, which gives $\deg(R(\mathcal{P}, \mathcal{Q})) = \sum_{i=1}^n (m+i) - \sum_{i=1}^n i = nm$.

EXERCISE. Check this computation using the other presentations of $\mathcal{A}[X]/(\mathcal{P}, \mathcal{Q})$.

Now we will, as an application, give a proof of Bezout's theorem.

THEOREM. *Let $C_1, C_2 \subseteq \mathbb{P}^2(\mathbb{C})$ be two curves defined by homogeneous polynomials $P$ and $Q$ of degree $n$ and $m$. Suppose they have no common component. Then*

$$mn = \sum_{y \in \mathbb{P}^2} \dim_{\mathbb{C}} (\mathcal{O}_{\mathbb{P}^2,y}/(P, Q)\mathcal{O}_{\mathbb{P}^2,y}).$$

In the proof we will need another lemma, very similar in nature to Lemma 1 above:

LEMMA 2. *Let $A = \mathbb{C}\{t\}$ and let $v$ be its valuation. Suppose that $\psi: A^p \to A^p$ is an homomorphism whose cokernel is of finite length, i.e. a finite dimensional vector space over $\mathbb{C}$. Then*

$$v(\det \psi) = \dim_{\mathbb{C}} (\operatorname{coker} \psi).$$

PROOF. By the main theorem on principal ideal domains we can find bases for $A^p$ such that the matrix representing $\psi$ is a diagonal matrix:

$$\psi = \begin{bmatrix} a_1 & & & & & 0 \\ & \cdot & & & & \\ & & \cdot & & & \\ & & & \cdot & & \\ & & & & \cdot & \\ & & & & & \cdot \\ 0 & & & & & a_p \end{bmatrix}$$

Clearly $v(\det \psi) = \sum v(a_i)$ and

$$\dim(\operatorname{coker} \psi) = \sum \dim_{\mathbb{C}} (A/a_i A)$$

Hence we may assume $p = 1$. Then $\psi$ is just multiplication by an element $a$. If $v(a) = n$, we can write $a = ut^n$ with $u$ a unit. One easily checks that $1, t, t^2, \cdots, t^{n-1}$ is a $\mathbb{C}$-basis for $A/aA = \mathbb{C}\{t\}/(ut^n)\mathbb{C}\{t\}$. Hence $v(a) = \dim_{\mathbb{C}} A/aA$.

We will now prove Bezout's theorem. Let $A = \mathbb{C}[x_1, x_2]$ and let $P, Q \in A[x_0]$ be the two homogeneous polynomials defining the curves $C_1$ and $C_2$. We can write

$$P = \sum_{i=1}^{n} p_i(x_1, x_2) x_0^i \qquad \deg p_i(x_1, x_2) = n - i$$

$$Q = \sum_{j=1}^{m} q_j(x_1, x_2) x_0^j \qquad \deg q_j(x_1, x_2) = m - j$$

After a change of coordinates, we may assume that $p_n(x_1, x_2)$ and $q_m(x_1, x_2)$ are non-zero, that is invertible in $A$. Geometrically this means that the point $(1, 0, 0)$ is not on any of the curves $C_1$ and $C_2$. There exists then a homogeneous homomorphism of degree $0$, $\varphi: \mathcal{A} \to A$ with $\varphi(\mathcal{P}) = P$ and $\varphi(\mathcal{Q}) = Q$. Clearly $\varphi(R(\mathcal{P}, \mathcal{Q}) = R(P, Q)$, and we obtain: If $R(P, Q) \not\equiv 0$, then $\deg(R(P, Q)) = mn$. (as a polynomial in $x_1, x_2$).

I leave it as an exercise to check, using the factoriality of polynomial rings over $\mathbb{C}$, that if $R(P, Q) \equiv 0$ then $P = 0$ and $Q = 0$ have an irreducible component in common.

But the fundamental theorem of algebra can be subsumed by the following formula:

Let $R \in \mathbb{C}[x_1, x_2]$ be a homogeneous polynomial of degree $d$, and for every point $x = (\bar{x}_1, \bar{x}_2) \in \mathbb{P}^1(\mathbb{C})$, let us denote by $\mathcal{O}_{\mathbb{P}^1, x}$ the local ring of $\mathbb{P}^1(\mathbb{C})$ at $x$, i.e. the ring of all fractions $S(x_1, x_2)/T(x_1, x_2)$ with $T$ and $S$ homogeneous polynomials and $T(\bar{x}_1, \bar{x}_2) \neq 0$. This is a discrete valuation ring, and we denote by $v_x(R)$ the valuation of the image of $R$ in this ring, i.e. $R/1$. Then:

$$d = \sum_{x \in \mathbb{P}^1(\mathbb{C})} v_x(R)$$

where of course the sum on the right is finite since $v_x(R) \neq 0$ only if $R(\bar{x}_1, \bar{x}_2) = 0$.

Let us now consider the projection $\pi: P^2(\mathbb{C}) \setminus \{(1, 0, 0)\} \to \mathbb{P}^1(\mathbb{C})$ given by $(x_0, x_1, x_2) \to (x_1, x_2)$. For each point $x \in \mathbb{P}^1$ there are finitely many points $y \in C_1 \cap C_2$ such that $\pi(y) = x$, and since $R(P, Q)$ is the ideal $\mathcal{F}_0(\mathcal{O}_{\mathbb{P}^2}/(P, Q))$

it follows by localization that

$$R(P, Q)\mathcal{O}_{\mathbb{P}^1,x} = F_0\left(\bigoplus_{\pi(y)=x} \mathcal{O}_{\mathbb{P}^2,y}/(P, Q)\mathcal{O}_{\mathbb{P}^2,y}\right)$$

so that by our Lemma 2 above

$$v_x(R(P, Q)) = \sum_{\pi(y)=x} \dim_{\mathbb{C}} \mathcal{O}_{\mathbb{P}^2,y}/(P, Q)\mathcal{O}_{\mathbb{P}^2,y}$$

and by summing up and using the fundamental theorem of algebra, and Lemma 1 above, we obtain Bezout's theorem. Remark that in Bezout's theorem the right-hand side is a sum of *local* terms, called the intersection multiplicity of $C_1$ and $C_2$ at $y \in \mathbb{P}^2$, and usually noted $(C_1, C_2)_y$ or $i(C_1, C_2, \mathbb{P}^2, y)$. Bezout's theorem provided the first and most basic examples of computation of local and global intersection multiplicities. Remark also that we deduced Bezout's formula from its analogue in $\mathbb{P}^1$, which is: $\deg R = \sum_{x\in\mathbb{P}^1} v_x(R)$, only by using a good definition of the image. Anyway, the formula can easily be generalized by these methods to the case of $n$ hypersurfaces in $\mathbb{P}^n$.

## Exercises on images

EXERCISE 1. Check that if $Y \xrightarrow{i} X$ is a closed immersion, defined by a sheaf of ideals $I$, then the image of $i$ is $Y$.

EXERCISE 2. Let $Y_1$ and $Y_2$ be two closed subspaces of $X$ defined respectively by $I_1$ and $I_2$. Then the image of the natural map $p: Y_1 \amalg Y_2 \to X$ is the subspace defined by $I_1I_2$, and therefore in general different from $Y_1 \cup Y_2$, which is defined by $I_1 \cap I_2$ (which is the annihilator ideal of $p_*\mathcal{O}_{Y_1\cup Y_2}$).

REMARK. If you are surprised by this, think of mapping the two points in $\mathbb{C}^2$ with coordinates $(x=0, y=1)$ $(x=0, y=2)$ to the $x$-axis. The image should be defined by $(x^2)\mathbb{C}\{x\}$ since as soon as you move the points a little, you get two distinct images.

## Addendum

In what follows, I shall use the definition of the image of a finite map $f: X \to Y$ of complex analytic spaces only in the case where $f$ is not surjective. In general, my definition of the image does not have the property that if $g: Y \to Z$ is another finite map, then $im(g \circ f)$ is the image by $g$ of $im(f)$. To see this, let us consider a finite map $h: X \to Z$ such that

$F_0(h_*\mathcal{O}_X) \neq \mathrm{Ann}_{\mathcal{O}_Z}(h_*\mathcal{O}_X)$. We have already seen that such maps exist. Then, let us define $Y$ to be the subspace of $Z$ defined by $\mathbf{Ann}_{\mathcal{O}_Z}(h_*\mathcal{O}_X)$: we can obviously factor $h$ as $g \circ f$ where $g: Y \to Z$ is the inclusion, and $f: X \to Y$. Since $f$ is surjective, $F_0(f_*\mathcal{O}_X) = 0$, hence $im(f) = Y$ and its image by $g$ is $Y$ in $Z$ (see exercise 1 on images, §1), but by our assumption, $Y$ is different from the image of $h = g \circ f$, which is the subspace of $Z$ defined by $F_0(h_*\mathcal{O}_X)$. Apparently, if we want a definition of the image which not only is compatible with base changes, but also behaves well under composition, the only possibility is to decide that $f_*\mathcal{O}_X$ 'is' the image, and then to remark that when $f$ is not surjective, then the Fitting ideal construction associates to $f_*\mathcal{O}_X$ a subspace of $Y$, in a natural way. In fact we have:

PROPOSITION. *Let there be given for every $A$-module of finite presentation $M$ an ideal $\mathfrak{A}_A(M)$ of $A$, this correspondence satisfying the following conditions:*

(1) *for any ring homomorphism $\varphi: A \to B$, we have*

$$\mathfrak{A}_B(M \otimes_A B) = \mathfrak{A}_A(M) \cdot B \quad \textit{(compatibility with base change)}$$

(2)
$$\sqrt{\mathfrak{A}_A(M)} = \sqrt{\mathrm{Ann}_A(M)} \quad \textit{(set-theoretically the right one)}$$

(3) *If $A$ is a discrete valuation ring*

$$v(\mathfrak{A}_A(M)) = \mathrm{lg}_A\, M \quad (\mathrm{lg} = \text{length})$$

*i.e. Lemma 2 of the proof of Bezout's theorem in §1 holds. Then $\mathfrak{A}_A(M) = F_0(M)$ for all $A$ and $M$.*

Sketch of proof: given a presentation $A^q \xrightarrow{\psi} A^p \to M \to 0$ with matrix $\psi = (\psi_{ij})$, we work in the 'universal ring' $A = \mathbb{Z}[T_{ij}]$ (or $\mathbb{C}\{T_{ij}\}$ if we really want to stay within complex analytic geometry) and consider the 'universal module' $\mathcal{M}$ = cokernel of $A^q \xrightarrow{\psi} A^p$ where the matrix of $\psi$ is $(T_{ij})$.

Then, thanks to theorems of Macaulay (algebraic theory of modular systems, Cambridge University Press 1919) and Buchsbaum-Rim (see §5) we know that the subspace of $\mathrm{Spec}\,\mathcal{A} = A^{pq}$ (or $\mathbb{C}^{pq}$ in analytic geometry) defined by the $p \times p$ minors of $\psi$ is *reduced* and all its components are of codimension $q - p + 1$. In this case, then $F_0(\mathcal{M}) = \mathrm{Ann}_{\mathcal{A}}(\mathcal{M})$. Suppose that $\mathfrak{A}_{\mathcal{A}}(\mathcal{M}) \neq F_0(\mathcal{M})$, the subspaces they define must be the same set-theoretically, and if they differ as spaces, we can already see it by restricting to arcs in $\mathbb{C}^{pq}$ (or mapping $\mathbb{Z}[T_{ij}]$ into discrete valuation rings). But condition (3)+(1) implies that $\mathfrak{A}_{\mathcal{A}}(\mathcal{M})$ and $F_0(\mathcal{M})$ cannot become different when we restrict them to arcs in $\mathbb{C}^{pq}$. Therefore they must be equal, i.e. $\mathfrak{A}_{\mathcal{A}}(\mathcal{M}) = F_0(\mathcal{M})$. Since our original $M$ is of course $\mathcal{M} \otimes_{\mathcal{A}} A$ where $\varphi: \mathcal{A} \to A$ is the morphism sending $T_{ij}$ to $\psi_{ij}$, we are done.

REFERENCES

[1] J. C. TOUGERON: Ideaux de functions différentiables. Ergebnisse der Mathematik No. 71. Springer-Verlag 1972.

[CARTAN SEMINAR] 60/61. Secretariat Mathematique, 11; rue Pierre et Marie Curie, Paris 5'. Also edited by W. A. Benjamin, New York 1967.

## §2. First part: The module of differentials

In this section, we talk about differentials and non-singularity. We will define the *critical subspace* (and not just the critical locus) and the *discriminant subspace* (and not just the branch locus) of some flat complex analytic mappings, using Fitting ideals, and we will prove a small theorem on discriminants.

2.1. There is a complex analytic space, usually denoted by $\mathbb{T}$, which is the subspace of the complex line $\mathbb{C}$ (with coordinate $v$) defined by the ideal $(v^2)$, i.e. $(v^2)\mathbb{C}\{v\}$. The underlying topological space of $\mathbb{T}$ is therefore just a point, and its 'structure sheaf' is $\mathbb{C}\{v\}/(v^2)$. It is just an infinitesimal direction, and for that reason, Mumford calls it the 'disembodied tangent vector'.

Define the Zariski tangent space $E_{X,x}$ of an analytic space $(X, \mathcal{O}_X)$ at a point $x \in X$ as the set of complex-analytic mappings $\mathbb{T} \to X$ having $x$ as (set-theoretic) image. Algebraically, it is described as the set of morphisms of $\mathbb{C}$-algebras $\mathcal{O}_{X,x} \to \mathbb{C}\{v\}/(v^2)$, which is the same as the set of $\mathbb{C}$-linear mappings $m/m^2 \to \mathbb{C}$ where $m$ is the maximal ideal of $\mathcal{O}_{X,x}$. This gives $E_{X,x}$ its natural structure of $\mathbb{C}$-vector space: $E_{X,x} = (m/m^2)^*$.

Remark that $\dim_{\mathbb{C}} E_{X,x} = \dim_{\mathbb{C}} m/m^2$ and that by Nakayama $\dim_{\mathbb{C}} m/m^2$ is the minimal number of generators of the ideal $m$, which is therefore also the smallest integer $N$ such that there exists a surjection of $\mathbb{C}$-algebras $\mathbb{C}\{x_1, \cdots, x_N\} \to \mathcal{O}_{X,x}$. Geometrically this integer is the smallest $N$ such that there exists a germ of closed imbedding $(X, x) \subset (\mathbb{C}^N, 0)$: it is called the *imbedding dimension* of $X$ at $x$.

EXERCISE. Build singular points of curves with arbitrarily large imbedding dimension.

EXERCISE. Compute the image of a morphism $p : \mathbb{T} \to X$ as a *subspace* of $X$, i.e. given a morphism of $\mathbb{C}$-algebras $\mathcal{O}_{X,x} \xrightarrow{p^*} \mathbb{C}\{v\}/(v^2)$, compute the 0th Fitting ideal of $\mathbb{C}\{v\}/(v^2)$ as $\mathcal{O}_{X,x}$-module. Hint: Use a presentation of $\mathcal{O}_{X,x}$ as quotient of a convergent power series ring as above, and distinguish two cases: $p^*$ is surjective, or not. You'll be surprised!

2.2. It turns out that there is a coherent sheaf of modules on $X$ which has as fibre at $x \in X$ exactly $m/m^2$: it is the sheaf of differentials on $X$. I will now give the construction of this sheaf not only for a space, but also for a morphism, as we will need this later:

Let $f: X \to S$ be a morphism. For simplicity, and because we will examine only local problems, we assume $f$ separated, i.e. that the diagonal $D_X \subset X \times_S X$ is closed. Let $I$ be the sheaf of ideals describing $D_X$ as subspace of $X \times_S X$. Then, by definition, the module of differentials of $X/S$ is $I/I^2$, which can be viewed as $\mathcal{O}_{X\times X}/I$-module since it is annihilated by $I$. $I/I^2$ becomes an $\mathcal{O}_X$-module via the isomorphism $\mathcal{O}_X \xrightarrow{\sim} \mathcal{O}_{X\times X/I}$ given by $p_1^*$ where $p_1: X \times X \to X$ is the first projection.

This sheaf is denoted by $\Omega^1_{X/S}$, and simply $\Omega^1_X$ in the case where $S$ is a point. Sometimes it is also written $\Omega^1_f$.

Remark that, for any $s \in S$, $\Omega^1_{X/S} \otimes_{\mathcal{O}_X} \mathcal{O}_{X_s} = \Omega^1_{X_s}$ where $X_s = f^{-1}(s)$ is the fibre of $X \to S$ over $s$, defined by $m_{S,s}\mathcal{O}_X$. More generally, one can check that 'The formation of $\Omega^1_{X/S}$ commutes with base change' i.e. given a cartesian diagram (or fibre-product)

$$\begin{array}{ccc} X' & \xrightarrow{h'} & X \\ {\scriptstyle f'}\downarrow & \square & \uparrow{\scriptstyle f} \\ S' & \xrightarrow{h} & S \end{array} \qquad (X' \simeq X \times_S S')$$

we have:

$$h'^*\Omega^1_{X/S} = \Omega^1_{X'/S'}.$$

Also, $\Omega^1_{X/S}$ comes equipped with a natural $\underline{\mathbb{C}}$-linear map $d_{X/S}: \mathcal{O}_X \to \Omega^1_{X/S}$ (we write $d$ in the case $S$ is a point) defined as $d = pr_1^* - pr_2^* \bmod I^2$, i.e. given a function $g: U \to \mathbb{C}$ on an open subset of $X$ (i.e. $g \in \Gamma(U, \mathcal{O}_X)$) we associate to $g$ the function on $U \times_S U$ which takes at $(x, x') \in U \times_S U$ the value $g(x) - g(x')$, and we remark that this function belongs to $I$. $dg$ is its class in $I/I^2$. Remark that $d(g \cdot g') = g \cdot dg' + g' \cdot dg$.

Now let us consider the following construction: Let $\mathcal{M}$ be a coherent $\mathcal{O}_X$-module. Then for any $x \in X$, there is an open neighborhood $U$ of $x$ in $X$ such that we have an exact sequence

$$\mathcal{O}_U^q \to \mathcal{O}_U^p \to \mathcal{M}|_U \to 0$$

So the symmetric algebra $\mathrm{Sym}_{\mathcal{O}_U} \mathcal{M}|_U$ is a quotient of $\mathcal{O}_U[T_1, \cdots, T_p]$ by an ideal generated by elements which are linear in the $T_j$, and finite in number. These elements describe a 'linear' subspace of $U \times \mathbb{C}^p$ (linear in the $\mathbb{C}^p$ coordinates) which is therefore a *relative vector space* over $U$, which we call $\mathrm{Specan}_U \mathrm{Sym}_{\mathcal{O}_U} \mathcal{M}|_U$. This local construction glues up naturally, and therefore we can define a relative vector space $\mathrm{Specan}_X \mathrm{Sym}_{\mathcal{O}_X} \mathcal{M} \to X$ over $X$. (It is the $L(\mathcal{M})$ of §1.) To give ourselves a section of this is, by the universal property of the symmetric algebra, to give ourselves a map $\mathcal{M} \to \mathcal{O}_X$ of $\mathcal{O}_X$-modules.

DEFINITION. The Zariski tangent space of $(X, \mathcal{O}_X)$ is the relative vector space $T_X = \text{Specan}_X \, \text{Sym}_{\mathcal{O}_X} \Omega^1_X \to X$.

REMARK. We like to think of a holomorphic vector field on $X$ as a section of the Zariski tangent space $T_X \to X$. Algebraically, it means an $\mathcal{O}_X$-linear map $\Omega^1_X \xrightarrow{\nabla} \mathcal{O}_X$. By composing $\nabla$ with $d: \mathcal{O}_X \to \Omega^1_X$, we get a $\mathbb{C}$-linear map $D: \mathcal{O}_X \to \mathcal{O}_X$ satisfying $D(g \cdot g') = g'Dg + g \cdot Dg'$. i.e. a *derivation* of $\mathcal{O}_X$ into itself. More generally, one can easily check that any $\mathbb{C}$-linear map $D: \mathcal{O}_X \to \mathcal{M}$ satisfying $D(g \cdot g') = g \cdot Dg' + g' \cdot Dg$ ($\mathcal{M}$ an $\mathcal{O}_X$-module) is obtained as $\nabla \cdot d$ where $\nabla$ is an $\mathcal{O}_X$-linear map $\nabla: \Omega^1_X \to \mathcal{M}$. [This is the universal property of the module of differentials $\Omega^1_X$, or more precisely of the differential $d: \mathcal{O}_X \to \Omega^1_X$, it is the 'universal derivation']. In particular, $\Omega^1_X$ is generated as $\mathcal{O}_X$-module by the $dg$, $g \in \mathcal{O}_X$. All this enables us to identify the holomorphic vector fields on $X$ (i.e. sections of $T_X \to X$) with derivations of $\mathcal{O}_X$ into itself. Also, we remark that the fibre $T_X(x)$ of $T_X \to X$ over $x \in X$ is in fact $\text{Hom}_{\mathbb{C}}(m/m^2, \mathbb{C})$ i.e. the Zariski tangent space $E_{X,x}$.

EXERCISE. Check that the datum of an holomorphic vector field on $X$ is the same thing as the datum of a complex-mapping $X \times \mathbb{T} \to X$ which induces the identity of $X$ on $X \times \{0\} \subset X \times \mathbb{T}$.

2.3. Exact sequences of modules of differentials. When we have a map $f: X \to S$, we expect to be able to define a tangent map $Tf: T_X \to T_S$, or more precisely: $Tf: T_X \to T_S \times_S X$. With our definition of $T_X$, and the fact that Spec is contravariant this means we want to describe the tangent map by: $\partial f: f^*\Omega^1_S \to \Omega^1_X$ and indeed, there is an exact sequence of sheaves of $\mathcal{O}_X$-modules

$$f^*\Omega^1_S \xrightarrow{\partial f} \Omega^1_X \longrightarrow \Omega^1_{X/S} \longrightarrow 0$$

where $\partial f$ is defined as follows: by linearity, it is sufficient to define $\partial f$ on elements of the form $\xi = dg \otimes 1|_V \in \Gamma(V, f^*\Omega^1_S)$, where $g \in \Gamma(U, \mathcal{O}_S)$ for some open $U$ of $S$. (This is because the $dg$'s generate $\Omega^1_S$ as $\mathcal{O}_S$-module.) Then we define $\partial f(\xi)$ as $d(g \cdot f|_V) \in \Gamma(V, \Omega^1_X)$.

We note that $\partial f$ is not in general injective. The fact that the sequence above is exact is often used as a definition of $\Omega^1_{X/S}$, and it is not very hard to check that it coincides with the definition given above.

Another important exact sequence is the following:

Suppose we have been able to imbed our mapping $f: X \to S$ locally around a point $x \in X$. That is, we have, after restricting $(X, \mathcal{O}_X)$ to some open neighborhood of $x$ in $X$ which we still write $X$ for simplicity, a diagram

$$\begin{array}{ccc} (X, x) \subset (S \times C^N, s \times 0) \\ \downarrow \quad \swarrow \\ (S, s) \end{array}$$

where $X$ is defined in $S \times C^N$ (locally near $x$) by an ideal $J = (f_1, \cdots, f_k)\mathcal{O}_{S\times\mathbb{C}^N}$, $f_i \in \mathcal{O}_S\{z_1, \cdots, z_N\}$.

Then, we have an exact sequence as follows: (of $\mathcal{O}_X$-modules)

$$J/_{J^2} \xrightarrow{(d)} \Omega^1_{S\times\mathbb{C}^N/S}|_X \longrightarrow \Omega^1_{X/S} \longrightarrow 0$$

where the map $(d)$ is defined as follows: Let $g = \sum_1^k a_i f_i \in J$ then, $dg = \sum a_i \cdot df_i + \sum da_i \cdot f_i$ in $\Omega^1_{S\times\mathbb{C}^N/S}$ ($d$ is actually $d_{S\times\mathbb{C}^N/S}$ here, i.e. we differentiate only with respect to coordinates on $\mathbb{C}^N$) is equal to $\sum a_i\, df_i$ modulo $J\Omega^1_{S\times\mathbb{C}^N/S}$ and in particular is in $J\Omega^1_{S\times\mathbb{C}^N/S}$ if $a_i \in J$, i.e. $g \in J^2$. Now $\Omega^1_{S\times\mathbb{C}^N/S|X} = \Omega^1_{S\times\mathbb{C}^N/S}/J\Omega^1_{S\times\mathbb{C}^N/S}$ so our map $(d)$ is well defined.

If we choose coordinates $z_1, \cdots, z_N$ on $\mathbb{C}^N$, then $\Omega^1_{S\times\mathbb{C}^N/S}$ is the free $\mathcal{O}_{S\times\mathbb{C}^N}$-module generated by the $dz_j$ ($1 \leqq j \leqq N$), i.e.

$$\Omega^1_{S\times\mathbb{C}^N/S} = \sum_{j=1}^N \mathcal{O}_{S\times\mathbb{C}^N} \cdot dz_j$$

and the exactness of the sequence means that $\Omega^1_{X/S}$ is just the quotient of $\Omega^1_{S\times\mathbb{C}^N/S} \mid_X$ by the submodule generated by the images of the $df_k$, which is easy to check using the universal property of the module of relative differentials $\Omega^1_{X/S}$, as above (for derivations of $\mathcal{O}_X$ into $\mathcal{M}$ which are 0 on $\mathcal{O}_S$). The sheaf $J/J^2$ is often called the conormal sheaf of $X$ in $S \times \mathbb{C}^N$ and the above exact sequence is in fact a special case of a general exact sequence, where we have $Z$ instead of $S \times \mathbb{C}^N$

$$\mathcal{N} \to \Omega^1_{Z/S} \mid X \to \Omega^1_{X/S} \to 0$$

which is commonly called the exact sequence of the normal bundle: after dualization, in the case $S$ = point, and $X$ and $Z$ smooth, it gives the usual sequence of sheaves coming from the sequence of vector bundles:

$$0 \to T_X \to T_Z \mid_X \to N^*_{Z,X} \to 0$$

where $N^*_{Z,X}$ is the normal bundle of $X$ in $Z$.

This aspect has been generalized to normal *cones*, which replace the normal bundle in the singular case, in [Lejeune-Teissier, Normal cones and sheaves of relative jets. Compositio Mathematica *28*, 1974].

2.4. Implicit function theorem, and simplicity theorem

Intuitively, we expect that if the Zariski tangent space $T_X \to X$ is actually a bundle, i.e. the dimension of its fibres is constant, then $X$ will be non-singular. There are actually several versions of this result, according to whether we assume $X$ reduced or not to start with, and whether we like to have a relative theorem or not. Let me start with the relative theorem:

SIMPLICITY THEOREM. *Let $f: X \to S$ be a* flat *morphism of complex spaces, and $x \in X$. The following are equivalent*
(i) *there is an $S$-isomorphism of germs $(X, x) \simeq (S \times \mathbb{C}^d, s \times 0)$ where $s = f(x)$*
(ii) *$f^{-1}(f(x))$ is non-singular of dimension $d$ at $x$*
(iii) *$\Omega^1_{X/S}$ is locally free of rank $d$ at $x$.*

(i.e. $T_{X_s} \to X_s$ is a vector bundle of rank $d$ for all $s \in S$). In particular, taking $S =$ a point, we have that: $X$ is non-singular at $x \Leftrightarrow \mathcal{O}_{X,x} \simeq \mathbb{C}\{z_1, \cdots, z_d\} \Leftrightarrow \Omega^1_X$ is locally free, generated by $dz_1, \cdots, dz_d$ at $x$.

We will see below that (i)$\Leftrightarrow$(ii) can be best interpreted by saying that a non-singular germ of complex space is *rigid* in the sense that any flat deformation of it (such as our $f$, as deformation of $f^{-1}(s)$) is actually locally a product.

Anyway, let me now describe an avatar of the implicit function simplicity theorem; first, why implicit function: suppose we have a map $f: (\mathbb{C}^N, 0) \to (\mathbb{C}^P, 0)$. The implicit function theorem says: if the tangent map at 0 is surjective, then $f$ is simple, in the sense of the above result. This breaks down into two parts

① $f$ is flat.
② $\Omega^1_{\mathbb{C}^N/\mathbb{C}^P}$ is locally free.

① is a very general fact: For $f: X \to S$ and $x \in X$, $s = f(x)$ if the tangent map $C_{X,x} \to C_{S,s}$ (of tangent cones) is flat, then $f$ is flat at $x$. If $X$ and $S$ are smooth, then $C_{X,x}$ and $C_{S,s}$ are just the usual tangent maps, and for linear maps, flat is equivalent to surjective.
② comes from the sequence

$$f^*\Omega^1_{\mathbb{C}^P} \xrightarrow{\partial f} \Omega^1_{\mathbb{C}^N} \longrightarrow \Omega^1_{\mathbb{C}^N/\mathbb{C}^P} \longrightarrow 0$$

and the surjectivity of $Tf$ is equivalent (via the usual Jacobian condition on the minors of the matrix describing $\partial f$ as a map between locally free modules, which is the Jacobian matrix) to the fact that $\partial f$ is everywhere of rank $P$, so that $\Omega^1_{\mathbb{C}^N/\mathbb{C}^P}$ has to be locally free of rank $N-P$.

Now here is the avatar of the simplicity theorem:

PROPOSITION. *Let $X$ be a reduced complex space of pure dimension $d$. Suppose that $\Omega^1_X$ has a locally free quotient of rank $d$. Then $X$ is non-singular (and hence $\Omega^1_X$ is in fact locally free).*

Before going into the proof, let us give the geometric interpretation of this:

Even though $\Omega^1_X$ is not the sheaf of sections of a vector bundle, we can define, after Grothendieck, an associated Grassmannian space over $X$, $G = \mathrm{Grass}_d\, \Omega^1_X \to X$ which has the property that for any $x \in X$, the fibre is the Grassmannian of $d$-dimensional subspaces of $E_{X,x}$. The characteristic property of $g: G \to X$ is that for any map $h: T \to X$ it is equivalent to give oneself a locally free quotient of rank $d$ of $h^*\Omega^1_X$ or to give oneself a factorization of $h$ through $G$.

$$\begin{array}{ccc} & & G = \mathrm{Grass}_d\, \Omega^1_X \\ & {\scriptstyle h'} \nearrow & \downarrow {\scriptstyle g} \\ T & \xrightarrow{h} & X \end{array} \qquad \Leftrightarrow h^*\Omega^1_X \to L \to 0$$

$L$ locally free rank $d$.

[This is exactly right if you think of a locally free rank $d$ quotient of $\Omega^1_X$ as a sub-relative vector space of $T_X \to X$ which is actually a vector bundle of rank $d$, i.e. picks a $d$-dimensional vector space in each fibre, in an 'analytic way'.] In particular, $g^*\Omega^1_X$ has a universal locally free quotient of rank $d$, corresponding to the tautological bundle on the grassmannian.

Now let $X^0 \subset X$ be the non-singular part of $X$. Of course, $\Omega^1_X|_{X^0} = \Omega^1_{X^0}$ is locally free of rank $d$ since $X$ is purely of dimension $d$. Hence we get a section $\sigma^0: X^0 \to G$ of $g: G \to X$. By Cartan's theorem, we can check that the closure in $G$ of the image of $\sigma^0$ is a complex analytic reduced subspace of $G$, which we call $X_1$

$$X_1 = \overline{\sigma^0(X^0)}.$$

$g$ induces a map $d: X_1 \to X$.

PROPOSITION + DEFINITION. *The induced map $d: X_1 \to X$ is a proper modification of $X$, which is surjective. We will call it the 'development' of $X$.*

REMARK. For us 'proper modification' means a proper map, which is an isomorphism over an open dense subset (here $X^0$, of course). $d$ is proper since $X_1$ is closed in $G$ and $g$ is proper of course. This implies that $d$ is surjective, since its image is closed and dense in $X$.

REMARK. A local version of $d$, presented in another manner, is often called 'Nash blowing up'. The reason why I chose another terminology is that $d$, as globally defined, is *not* the blowing up of a coherent sheaf of ideals in general, and also that $d$ is the opposite operation of taking an envelope, which seems significant to me. This construction applied to the module of *relative* differentials, has been used by Hironaka in his lectures, and he called it '$\partial f$-modification'.

REMARK. The fibre $d^{-1}(x)$ is best thought of as the set of limit directions of tangent spaces to $X$ at non-singular points near $x$. (This is the meaning of our closure operation.)

CLAIM. The assertion of the proposition is equivalent to.

$$d \text{ is an isomorphism} \Leftrightarrow X \text{ is non-singular.}$$

In effect, to say that $d$ is an isomorphism, is equivalent to saying that $\sigma^0 : X^0 \to G$ extends to a section of $g : \sigma : X \to G$ having then necessarily $X_1$ as its image. But by the universal property of the Grassmannian $G$, this is equivalent to saying that $\Omega^1_X$ has a locally free quotient of rank $d$.

Remark now that when $X$ is a small representative of a germ of an analytically irreducible curve, $d$ is in fact always an homeomorphism, (the limit of tangents at non-singular points is well defined and unique) so there is in effect something to prove. The proof is by induction on $d$, and uses the following:

Integration of vector fields (as taught by Zariski in: Studies in equisingularity I, II Amer. Journal of Maths. *87*, 1965). Let $\mathcal{O}$ be a complete local ring containing a field $k$ of characteristic zero. Let $D : \mathcal{O} \to \mathcal{O}$ be a $k$-derivation of $\mathcal{O}$ such that $D\mathcal{O} \not\subset m$ where $m$ is the maximal ideal of $\mathcal{O}$. Then, there exists $x \in \mathcal{O}$ and a subring $\mathcal{O}_1$ of $\mathcal{O}$ containing $k$, such that $\mathcal{O} = \mathcal{O}_1[[x]]$. [Translation: If you have an holomorphic vector field which is not 0 at the origin, you can integrate it to get an isomorphism $(X, x) \simeq (X_1 \times \mathbb{C}, x)$, at least formally.] In fact, we could do it in the convergent case, by extending to a non-singular ambient space and using the existence theorems of differential equations, but I think it is more informative to do it in the following way, as Zariski does:

If $D\mathcal{O} \not\subset m$, we can suppose that there exists $x \in \mathcal{O}$ such that $Dx = 1$, after multiplying $D$ by an invertible element of $\mathcal{O}$. Let us consider the operator $e^{-xD} : \mathcal{O} \to \mathcal{O}$ defined by

$$e^{-xD}(h) = h - xDh + \frac{x^2}{2!} D^2 h + \cdots + (-1)^i \frac{x^i}{i!} D^i h + \cdots$$

which is in $\mathcal{O}$ since $\mathcal{O}$ is complete for the $m$-adic topology. One checks that the image of $e^{-xD}$ is a subring $\mathcal{O}_1$ of $\mathcal{O}$ containing $k$ and that since $Dx = 1$, we have $D_{|\mathcal{O}_1} = 0$. Furthermore, the kernel of $e^{-xD}$ is $x \cdot \mathcal{O}$, so we have an isomorphism $\mathcal{O}/_{x \cdot \mathcal{O}} \simeq \mathcal{O}_1$; now, to check that the natural injection $\mathcal{O}_1[[x]] \subset \mathcal{O}$ is surjective is easy: take $h \in \mathcal{O}$, and define inductively $h_i$ by: $h_0 = h$ and

$h_{i-1} - e^{-xD}(h_{i-1}) = x \cdot h_i$. Then, setting $\bar{h}_i = e^{-xD}(h_i) \in \mathcal{O}_1$ we can see formally that

$$h = \bar{h}_0 + x\bar{h}_1 + x^2\bar{h}_2 + \cdots \in \mathcal{O}_1[[x]].$$

Now back to our proposition: The assertion is local, and to prove that an analytic algebra is a convergent power series ring, it is enough to prove that its completion is a formal power series ring. Thus we can reduce to the case of a complete local $\mathbb{C}$-algebra, with a module of differentials having a free quotient

$$\mathcal{O} \quad \text{and} \quad \Omega^1_{\mathcal{O}/\mathbb{C}} \to \mathcal{O}^d \to 0$$

Now let us choose a map $\mathcal{O}^d \to \mathcal{O}$ mapping the basis vector $(1, 0 \cdots 0)$ to 1 and all others to 0. There is an element $x \in \mathcal{O}$ such that the image of $dx$ in $\mathcal{O}$ is invertible, so this gives us a derivation $D: \mathcal{O} \to \mathcal{O}$ and $x \in \mathcal{O}$ with $Dx = 1$. By the lemma, $\mathcal{O} = \mathcal{O}_1[[x]]$ but now we remark, using for example the geometric interpretation, that $\mathcal{O}_1$ satisfies exactly the same hypothesis as $\mathcal{O}$, and $\dim \mathcal{O}_1 = \dim \mathcal{O} - 1$. (e.g. use the fact that $d \times id_{\mathbb{C}}: X_1 \times \mathbb{C} \to X \times \mathbb{C}$ is the development of $X \times \mathbb{C}$ if $d: X_1 \to X$ is the development of $X$). So by induction on the dimension, we can assume that $\mathcal{O}_1$ is a formal power series ring $C[[x_1, \cdots, x_{d-1}]]$, and get the proposition, provided we can prove it for $d = 0$, but here we use the fact that $\mathcal{O}$ (hence $\mathcal{O}_1$) is reduced. A reduced analytic algebra of dimension 0 is $\mathbb{C}$.

QUESTION. If we iterate the development

$$X = X_0 \xleftarrow{d} X_1 \xleftarrow{d} X_2 \longleftarrow X_3 \longleftarrow \cdots$$

does $d: X_{i+1} \to X_i$ become an isomorphism for some $i$?

2.5. I hope I have now given enough motivation to define the singular locus of $X$ as the set of those points of $X$ where the dimension of the fibre $\Omega^1_X(x)$ is greater than it should be, i.e. where $\Omega^1_X$ has no chance of being locally free. More precisely:

DEFINITION. Let $X$ be a reduced equidimensional complex space of dimension $d$. We define the *singular subspace* of $X$ by the coherent sheaf of ideals $F_d(\Omega^1_X)$.

And similarly, the simplicity theorem for a map suggests:

DEFINITION. Let $f: X \to S$ be a *flat* map of complex analytic spaces, the fibres of which are of pure dimension $d$. We define the *critical subspace* $C$ of $f$ by the coherent sheaf of ideals $F_d(\Omega^1_{X/S})$. Let me immediately give two examples:

EXAMPLE 1. Let $f \in \mathbb{C}\{z_0, \cdots, z_n\}$ be such that $f=0$ defines a reduced hypersurface, $(X, 0) \subset (\mathbb{C}^{n+1}, 0)$. Using the second exact sequence of the module of differentials, we get

$$(f)/(f^2) \xrightarrow{(d)} \Omega^1_{\mathbb{C}^{n+1}}|_X \longrightarrow \Omega^1_X \longrightarrow 0$$

and since $\Omega^1_{\mathbb{C}^{n+1}}$ is (locally) freely generated by the $dz_i$, and $(f)/(f^2)$ is free of rank 1; this is a presentation of $\Omega^1_X$ exactly as we need to compute a Fitting ideal. Here the matrix of $(d)$ is obviously $(\partial f/\partial z_0, \cdots, \partial f/\partial z_n)$ and therefore our $F_n(\Omega^1_X)_0$ is $(\partial f/\partial z_0, \cdots, \partial f/\partial z_n)\mathcal{O}_{X,0}$, as one would have wished!

EXAMPLE 2. Consider now $f$ as a map $\mathbb{C}^{n+1} \to \mathbb{C}$, necessarily flat. Then, the 1st exact sequence of the module of differentials gives us, if we take a coordinate $v$ on $\mathbb{C}$:

$$\mathcal{O}_{n+1} \cdot dv \xrightarrow{\psi} \sum \mathcal{O}_{n+1} \cdot dz_i \longrightarrow \Omega^1_{\mathbb{C}^{n+1}/\mathbb{C}} \longrightarrow 0$$

since clearly $\Omega^1_{\mathbb{C}} = \mathcal{O}_1\, dv$ and $f^*\Omega^1_{\mathbb{C}} = \mathcal{O}_{n+1}\, dv$ (here $\mathcal{O}_1 = \mathbb{C}\{v\}$, $\mathcal{O}_{n+1} = \mathbb{C}\{z_0, \cdots, z_n\}$).

The map $\psi$ sends $dv$ to $d(v \circ f) = \sum_{i=0}^n (\partial f/\partial z_i)\, dz_i$ hence again the matrix of $\psi$ is $(\partial f/\partial z_0, \cdots, \partial f/\partial z_n)$ and we find that the *critical subspace* of $f$ is defined by $(\partial f/\partial z_0, \cdots, \partial f/\partial z_n)\mathbb{C}\{z_0, \cdots, z_n\}$, again as one would have wished.

EXERCISE. Generalize both examples to the fibre of a *flat* map $F: \mathbb{C}^{n+1} \to \mathbb{C}^k$, and then to the map itself (i.e. compute the singular subspace of $F^{-1}(0)$ and the critical subspace of $F$) you will find the ideal generated by the minors of the Jacobian matrix, of course. [To say that $F$ is flat is the same as to say that $F^{-1}(0)$ is a complete intersection, in this case, i.e. that it is defined by a regular sequence in $\mathcal{O}_{n+1} = \mathbb{C}\{z_0, \cdots, z_n\}$.]

2.6. Given a flat map $f: X \to S$, and since we know how to define its critical subspace $C$, we can, in the case where $f \mid C: C \to S$ is a finite map, define its image, which we will call the *discriminant subspace* of $f$, after §1.

DEFINITION. Let $f: X \to S$ be a flat map such that the restriction of $f$ to the critical subspace $C$ of $f$ is finite. The discriminant subspace of $f$ is the subspace of $S$ defined by $F_0(f_*\mathcal{O}_C)$ (0th-Fitting ideal of $f_*\mathcal{O}_C$).

As an example, let us compute the discriminant of $f: (\mathbb{C}^{n+1}, 0) \to (\mathbb{C}, 0)$ in the case where it is defined, i.e. where the critical subspace defined by $(\partial f/\partial z_0, \cdots, \partial f/\partial z_n)$ is finite over $\mathbb{C}$, which means, looking locally around 0, that 0 is an isolated critical point of $f$. Now we have to compute the 0th Fitting ideal of $M = \mathbb{C}\{z_0, \cdots, z_n\}/(\partial f/\partial z_0, \cdots, \partial f/\partial z_n)$ as $\mathbb{C}\{v\}$-module. Here, we have no God-given resolution, but we can use Hilbert's Syzygy theorem, which tells us that $M$ has a resolution of length 1 because $\mathbb{C}\{v\}$ is regular of dimension 1: we get an exact sequence of $\mathbb{C}\{v\}$-modules

$$0 \longrightarrow \mathbb{C}\{v\}^q \xrightarrow{\psi} \mathbb{C}\{v\}^p \longrightarrow M \longrightarrow 0$$

hence $q \leqq p$. But the support of $M$ is the origin, so necessarily $q \geqq p$ otherwise the support of $M$ would be $\mathbb{C}$. Hence $q = p$ and $F_0(M) = (\det \psi) \subset \mathbb{C}\{v\}$. But remember Lemma 2 in the proof of Bezout's theorem in §1: we have $(\det \psi) = (v^\mu)$ where $\mu = \dim_{\mathbb{C}} M$. Hence we have:

PROPOSITION. *The discriminant of* $f:(\mathbb{C}^{n+1}, 0) \to (\mathbb{C}, 0)$ *is the subspace of* $(\mathbb{C}, 0)$ *defined by* $(v^\mu)\mathbb{C}\{v\}$ *where*

$$\mu = \dim_{\mathbb{C}} \mathbb{C}\{z_0, \cdots, z_n\}/(\partial f/\partial z_0, \cdots, \partial f/\partial z_n)$$

*is the Milnor number of f.* (See Orlik's lectures).

EXERCISE. Let now $(X, 0) \subset (\mathbb{C}^{n+1}, 0)$ be a complete intersection with isolated singularity, defined by $f_1, \cdots, f_{k-1}$. Let $f_k : (X, 0) \to (\mathbb{C}, 0)$ be such that the fibre $f_k^{-1}(0)$ still has an isolated singularity. Use the same argument as above to show that the discriminant of $f_k$ is the subspace of $\mathbb{C}$ defined by

$$(v^\Delta)\mathbb{C}\{v\} \quad \text{where}$$

$$\Delta = \dim_{\mathbb{C}} \mathcal{O}_X \Big/ \left( \left\{ \frac{\partial(f_1, \cdots, f_k)}{\partial(z_{i_1}, \cdots, z_{i_k})} \right\} (i_1, \cdots, i_k) \subset \{0, \cdots, n\} \right)$$

## §2. Second part: The idealistic Bertini theorem

2.7. In this part, we give the definition and characterization of integral dependence of ideals, which is a powerful tool to translate certain transcendental conditions on complex analytic spaces into geometric, or purely algebraic, conditions. We use this notion, and our space-theoretic viewpoint on singularities, to prove a result which is algebraically much stronger than the Bertini-Sard theorem, and then to show some connections between this result and some equisingularity and incidence conditions, such as Thom's $A_F$ conditions.

One statement of the (second) Bertini theorem, which is a complex analytic version of Sard's theorem on the critical values of a $C^\infty$ mapping (see K. Ueno: Classification theory of algebraic varieties, Springer Lect. Notes No. 439, chap. 1 §4) is as follows:

Let $f: X \to Y$ be a morphism of complex analytic spaces with $X$ reduced and $Y$ non-singular (this is not essential). Then, there exists a nowhere dense closed complex subspace $B$ of $X$ such that:

(i) $f(B)$ has measure 0 in $Y$

(ii) $|\mathrm{Sing}_x X_{f(x)}| = |\mathrm{Sing}_x X| \cap X_{f(x)}$ for all $x \in X - B$, where $X_{f(x)} = f^{-1}(f(x))$ is the fibre of $f$ through $x$, and $|\mathrm{Sing}_x X|$ is the germ at $x$ of the singular *locus* $|\mathrm{Sing}\ X|$.

Now that we have defined singular *subspaces*, we can ask whether this statement continues to hold true with singular subspaces instead of just loci. It turns out that it does not, as will be shown by an example below, but still, considerably more than the above statement is true, which can be formulated thanks to the notion of integral dependence on ideals, which I now summarize:

*Integral dependence on ideals* (for proofs see [1], [2]). Let $X$ be a reduced complex analytic space, and let $I$ be a coherent sheaf of ideals on $X$. Then, there exists a coherent sheaf of ideals $\bar{I}$ on $X$, which has the following property (and is characterized by it):

For any point $x \in X$, let $(g_1, \cdots, g_m)$ be generators for the ideal $I_x \subset \mathcal{O}_{X,x}$. Then, let us call $\bar{I}_x$ the ideal of elements $h \in \mathcal{O}_{X,x}$ satisfying an integral dependence relation:

$$h^k + a_1 \cdot h^{k-1} + \cdots + a_k = 0 \quad \text{with} \quad a_i \in I_x^i.$$

Then we have:

① $(\bar{I})_x = \bar{I}_x$

② $h \in \bar{I}_x$ if and only if there exists an open neighborhood $U$ of $x$ in $X$ and a constant $C \in \mathbb{R}_+$ such that $h$ and the $g_i$ converge in $U$ (i.e. come from elements in $\Gamma(U, \mathcal{O}_X)$, denoted by the same letters) and that: $|h(x')| \leqq C.$ $\mathrm{Sup}\,|g_j(x')|$ for all $x' \in U$.

③ [*Arcwise condition of integral dependence*]
$h \in \bar{I}_x$ if and only if for *any* mapping of $\mathbb{C}$-algebras $\varphi^*: \mathcal{O}_{X,x} \to \mathbb{C}\{t\}$ we have $v(\varphi^*(h)) \geqq \min_j v(\varphi^*(g_j))$ where $v(\ )$ denotes the order in $t$ of an element of $\mathbb{C}\{t\}$, also known as its valuation.

④ Given $U \overset{\circ}{\subset} X$ and $h \in \Gamma(U, \mathcal{O}_X)$, we have $h \in \Gamma(U, \bar{I})$ if and only if there exists a proper modification $p: Z \to U$ of complex spaces such that $(h \circ p)_z \in (I \cdot \mathcal{O}_Z)_z$ for all $z \in Z$, where $I \cdot \mathcal{O}_Z$ is the image in $\mathcal{O}_Z$ of the natural map $p^*(I_{|U}) \to p^*\mathcal{O}_U = \mathcal{O}_Z$ coming from the inclusion $I \subset \mathcal{O}_X$.

⑤ $h \in \bar{I}_x$ if and only if $h$ 'asymptotically belongs to $I_x$' in the sense that there exists a $\nu_0$ such that $h \cdot I_x^\nu \subset I_x^{\nu+1}$ for $\nu \geqq \nu_0$.

Conditions ②–⑤ can be seen to be different ways of giving a meaning to the idea that $h$ 'almost' belongs to $I$, which turn out to coincide. $\bar{I}$ (resp. $\bar{I}_x$)

is called the integral closure of the sheaf of ideals $I$ in $\mathcal{O}_X$ (resp. of $I_x$ in $\mathcal{O}_{X,x}$). The elements $h \in \bar{I}_x$ are said to be integrally dependent on $I_x$. We see that $I_x \subset \bar{I}_x \subset \sqrt{I_x}$ but $\bar{I}_x$ retains a lot of information from $I_x$ which is lost when we look at $\sqrt{I_x}$.

EXERCISE. (1) The integral closure of the ideal $(z_0^a, \cdots, z_n^a)$ in $\mathbb{C}\{z_0, \cdots, z_n\}$ is $(z_0, \cdots, z_n)^a$.

(2) If a ring $A$ is reduced and integrally closed in its total ring of fractions, and $g \in A$ is not a zero-divisor, then $\overline{(g)} = (g)$, (for example: if $A = \mathcal{O}_{X,x}$ and $X$ is normal at $x$, and $g$ is a local equation of a divisor).

(3) Check that for *any* $f \in \mathbb{C}\{z_0, \cdots, z_n\}$ belonging to the maximal ideal, we have: $f \in \overline{(z_0 \cdot (\partial f/\partial z_0), \cdots, z_n \cdot (\partial f/\partial z_n))}$ in $\mathbb{C}\{z_0, \cdots, z_n\}$, and hence, in particular (since $I \subset J \Rightarrow \bar{I} \subset \bar{J}$) we have $f \in \overline{((\partial f/\partial z_0), \cdots, (\partial f/\partial z_n))}$ always (Hint: use ③). Recall that if $f = 0$ has isolated singularity, $f \in ((\partial f/\partial z_0), \cdots, (\partial f/\partial z_n))$ is equivalent to the fact that $U \cdot f$ is quasi-homogeneous in some coordinate system, with some invertible $U$ (K. Saito: Inventiones Math. *14*, p. 123 (1971)).

REMARK. $I_1 \subset I_2 \Rightarrow \bar{I}_1 \subset \bar{I}_2$; $\bar{\bar{I}} = \bar{I}$.

2.8. Idealistic Bertini theorem

Let $f: X \to Y$ be a flat morphism of complex analytic spaces, $X$ being reduced and $Y$ non-singular. Assume the fibres of $f$ are all of pure dimension $d$. Set $n = \dim X = d + \dim Y$. Then, we have: There exists a nowhere dense closed complex subspace $B$ of $X$ such that:

(i) $f(B)$ has measure $0$ in $Y$.
(ii) $\overline{F_n(\Omega_X^1)_x} = \overline{F_d(\Omega_{X/Y}^1)_x}$ for all $x \in X - B$.

In words, the germ at $x$ of the singular subspace of $X$ and of the critical subspace of $f$ are defined by ideals having the same integral closure, outside $B$.

Now let us remark that if we induce $F_d(\Omega_{X/Y}^1)$ to $\mathcal{O}_{X_{f(x)},x}$ we obtain the ideal defining the singular subspace of $X_{f(x)}$ i.e. $F_d(\Omega_{X_{f(x)}}^1)$. Since anyway, as one can check by using the exact sequences, $F_d(\Omega_{X/Y}^1) \subset F_n(\Omega_X^1)$, and since an integral dependence relation can be restricted to a closed subspace, we see that we have

COROLLARY. $\overline{F_n(\Omega_X^1) \cdot \mathcal{O}_{X_{f(x)},x}} = \overline{F_d(\Omega_{X_{f(x)}}^1)_x}$ *for all* $x \in X - B$, *whereas Bertini's theorem quoted above can be translated by:*

$$\sqrt{F_n(\Omega_X^1) \cdot \mathcal{O}_{X_{f(x)},x}} = \sqrt{F_d(\Omega_{X_{f(x)}}^1)_x}$$

*which is a much weaker statement, algebraically speaking.*

PROOF. First, remark that the assertion is local on $X$, so we can replace $X$ by an open subset, still written $X$, so small that we have a commutative diagram:

$$\begin{array}{ccc} X & \xhookrightarrow{i} & Y\times\mathbb{C}^N \quad \text{(closed immersion)} \\ f\downarrow & \swarrow_{pr_1} & \\ Y & & \end{array}$$

where $Y\times\mathbb{C}^N$ also stands for an open subset in $Y_y\in\mathbb{C}^N$, and $i$ is a closed immersion ($Y$ also stands for an open set in $Y$) [recall, or admit, that a flat morphism is open]. Now let us choose coordinates $y_1,\cdots,y_k$ on $Y$, $z_1,\cdots,z_N$ on $\mathbb{C}^N$, and let $F_j(y,z)$ $1\leqq j\leqq m$ be generators for the ideal defining $X$ in $Y\times\mathbb{C}^N$. Here $m\geqq N-d=$ codimension of $X$ in $Y\times\mathbb{C}^N$. As we know from the exercises in §2, $F_n(\Omega^1_X)$ is the sheaf of ideals in $\mathcal{O}_X$ generated by the Jacobian determinants $(\partial(F_{j_1},\cdots,F_{j_{N-d}}))/(\partial(y_{i_1},\cdots,y_{i_l},z_{i_{l+1}},\cdots,z_{i_{N-d}}))$, $(j_1,\cdots,j_{N-d})\subset\{1,\cdots,m\},(i_1,\cdots,i_l)\subset\{1,\cdots,k\},(i_{l+1},\cdots,i_{N-d})\subset\{1,\cdots,N\}$ with $0\leqq l\leqq N-d$, and $F_d(\Omega^1_{X/Y})$ is generated in $\mathcal{O}_X$ by those among the above determinants where no $\partial/\partial y_i$ takes place, i.e. $l=0$.

Let us now consider the normalized blowing up of $F_d(\Omega^1_{X/Y})$ in $X$, i.e. the composed map:

$$\pi: X' \xrightarrow{n} X_1 \xrightarrow{\pi_0} X$$

where $\pi_0$ is the blowing up of $F_d(\Omega^1_{X/Y})$ and $n$ is the normalization of $X_1$, which is reduced since $X$ is so. Since $F_d(\Omega^1_{X/Y})\neq 0$ and $X$ is reduced, $\pi$ is a proper modification of $X$, and let us consider the exceptional divisor, i.e. the subspace $D$ of $X'$ defined by the sheaf of ideals $F_d(\Omega^1_{X/Y})\cdot\mathcal{O}_{X'}$, which is now invertible, (i.e. locally on $X'$ generated by only one element, non-zero divisor) by the fundamental property of blowing up. By further restricting $X$ and $Y$, we may assume that $D$ has only a finite number of irreducible components and write $D=\bigcup_{\alpha\in A}D_\alpha$. Let us write $A=A_0\cup A_1$ where

$$\alpha\in A_0\Leftrightarrow p(D_\alpha)\neq Y$$
$$\alpha\in A_1\Leftrightarrow p(D_\alpha)=Y$$

where $p: X'\to Y$ is $f\circ\pi$.

Now set $B=\bigcup_{\alpha\in A_0}\pi(D_\alpha)$. It is a closed complex subspace of $X$ since $\pi$ is proper, and contained in $|\mathrm{Sing}\, f|$. Furthermore, its image in $Y$ is of measure 0.

Let us now take a point $x\in X-B$, and change our coordinate system so that $x$ is the origin of $\mathbb{C}^k\times\mathbb{C}^N$, (i.e. $y_l$ and $z_i$ all vanish at $x$). Since $X'$ is normal, hence non-singular in codimension 1, we have that for each open neighborhood $U$ of $x$ in $X$, we can find a dense open analytic subspace $V_\alpha$ of $D_\alpha \cap \pi^{-1}(U)(\alpha \in A_1)$ such that if $x' \in V_\alpha$, we have:

(i) the germ of $D$ at $x'$ is equal to the germ of $D_\alpha$ at $x'$ and $p \mid D_{\alpha,\text{red}} : D_{\alpha,\text{red}} \to Y$ is a submersion of non-singular spaces, at $x'$.

(ii) $X'$ is non-singular at $x'$, and hence by (i), $p$ is a submersion of non-singular spaces at $x'$.

By the implicit function theorem we can choose local coordinates on $X'$ centered at $x'$, $(y'_1, \cdots, y'_k, v, u_2, \cdots, u_N)$ such that: $\mathcal{O}_{X',x'} \simeq \mathbb{C}\{y'_1, \cdots, y'_k, v, u_2, \cdots, u_N\}$ and:

$$F_d(\Omega^1_{X/Y}) \cdot \mathcal{O}_{X',x'} = v^\nu \cdot \mathcal{O}_{X',x'}. \tag{1}$$

$$(y_i \circ \pi)_{x'} = y'_i. \tag{2}$$

Since $(F_j \circ \pi)_{x'} \equiv 0 \; (1 \leqq j \leqq m)$ we have:

$$\frac{\partial}{\partial y'_l}(F_j \circ \pi)_{x'} \equiv 0 \equiv \left( \sum_{i=1}^{N} \frac{\partial F_j}{\partial z_i} \circ \pi \cdot \frac{\partial (z_i \circ \pi)}{\partial y'_l} + \frac{\partial F_j}{\partial y_l} \circ \pi \right)_{x'}$$

and hence

$$\left( \frac{\partial F_j}{\partial y_l} \circ \pi \right)_{x'} = -\left( \sum_{i=1}^{N} \frac{\partial F_j}{\partial z_i} \circ \pi \cdot \frac{\partial (z_i \circ \pi)}{\partial y'_l} \right)_{x'} \quad \text{for all } j \text{ and } l.$$

The multilinearity of determinants now implies immediately that we have:

$$F_n(\Omega^1_X) \cdot \mathcal{O}_{X',x'} = F_d(\Omega^1_{X/Y}) \cdot \mathcal{O}_{X',x'} \qquad (x' \in V_\alpha).$$

I claim that this implies in fact that we have $F_n(\Omega^1_X) \cdot \mathcal{O}_{\pi^{-1}(U)} = F_d(\Omega^1_{X/Y}) \cdot \mathcal{O}_{\pi^{-1}(U)}$ ($U$ open neighborhood of $x$). This because anyway $F_d(\Omega^1_{X/Y}) \subset F_n(\Omega^1_X)$ on the one hand, and $F_d(\Omega^1_{X/Y}) \cdot \mathcal{O}_{X'}$ is invertible by construction on the other hand. Therefore to check equality is to check that certain meromorphic functions generating $(F_d(\Omega^1_{X/Y}) \cdot \mathcal{O}_{\pi^{-1}(U)})^{-1} \cdot (F_n(\Omega^1_X) \cdot \mathcal{O}_{\pi^{-1}(U)})$ are in fact holomorphic on $\pi^{-1}(U)$:

this will then imply that $F_n(\Omega^1_X) \cdot \mathcal{O}_{\pi^{-1}(U)} \subseteq F_d(\Omega^1_{X/Y}) \cdot \mathcal{O}_{\pi^{-1}(U)}$, and we already know the other inclusion. The meromorphic functions in question are locally the $h_i/g$ where $g$ is a generator of $F_d(\Omega^1_{X/Y}) \cdot \mathcal{O}_{X',x'}$ and the $h_i$ are generators of $F_n(\Omega^1_X) \cdot \mathcal{O}_{X',x'}$.

But now we can argue as follows: the polar subspace of these meromorphic functions, if it is not empty, is of codimension 1 in $X'$, by a classical result on normal spaces (see R. Narasimhan: Introduction to the theory of analytic spaces, Springer Lecture Notes No. 25, 1966, p. 89). On the other hand, this polar subspace is certainly contained in the subspace defined by $F_d(\Omega^1_{X/Y}) \cdot \mathcal{O}_{X'}$, which is $D \cap \pi^{-1}(U)$, and we have just seen that each component of $D$ contains a dense open analytic subset at each point of which the meromorphic functions in question are holomorphic. Hence the polar subspace in question is empty, and therefore we have $F_n(\Omega_X^1)\cdot\mathcal{O}_{\pi^{-1}(U)} = F_d(\Omega_{X/Y}^1)\cdot\mathcal{O}_{\pi^{-1}(U)}$, which implies that $\overline{F_n(\Omega_X^1)_x} = \overline{F_d(\Omega_{X/Y}^1)_x}$ at every $x\in X-B$.

Remark. If we assume $f$ proper, then by a theorem of Frisch (Inventiones Math. *4* (1967), pp. 118–138) there exists a complex-analytic nowhere dense closed subspace $F\subset X$ such that $f(F)$ is a nowhere dense closed complex subspace of $Y$ and $f\,|\,X-F$ is flat, so we may drop the assumption of flatness on $f$ in the theorem, and in this case the conclusion of the theorem is that there exists a nowhere dense closed complex subspace $Z\subset Y$ such that for any $y\in Y-Z$ we have:

$$\overline{F_n(\Omega_X^1)_x} = \overline{F_d(\Omega_{X/Y}^1)_x} \quad \text{for all} \quad x\in f^{-1}(y)$$

and hence for $y\in Y-Z$, the equality of sheaves of ideals of $\mathcal{O}_{X_y}$:

$$\overline{F_n(\Omega_X^1)\cdot\mathcal{O}_{X_y}} = \overline{F_d(\Omega_{X/Y}^1)\cdot\mathcal{O}_{X_y}} = \overline{F_d(\Omega_{X_y}^1)}.$$

Remark. I have assumed known that the image of a *nowhere surjective* analytic mapping is of measure 0. (Cover the source by countably many semi-analytic compact subsets and use the description of subanalytic sets.)

2.9. Example. Let $f: X\to\mathbb{P}^2$ be the restriction of the natural projection $\mathbb{P}^2\times\mathbb{C}^3\to\mathbb{P}^2$ to the subspace $X$ defined by the ideal $(az_1+bz_2+cz_3,\ z_1^6+z_2^6+z_3^6)$ where $(a:b:c)$ are homogeneous coordinates on $\mathbb{P}^2$ and $z_1$, $z_2$, $z_3$ are coordinates on $\mathbb{C}^3$. $f$ is the family of plane sections of the surface in $\mathbb{C}^3$ defined by $z_1^6+z_2^6+z_3^6=0$. The Jacobian matrix of $X$ is

$$\begin{bmatrix} 6z_1^5 & 6z_2^5 & 6z_3^5 & 0 & 0 & 0 \\ a & b & c & z_1 & z_2 & z_3 \end{bmatrix}$$

and therefore $F_1(\Omega_{X/\mathbb{P}^2}^1)$ is generated by:

$$(bz_1^5-az_2^5,\ cz_2^5-bz_3^5,\ az_3^5-cz_1^5)$$

while $F_3(\Omega_X^1) = F_1(\Omega_{X/\mathbb{P}^2}^1) + (z_1^6,\ z_1z_2^5,\ z_1z_3^5,\ z_2z_1^5,\ z_2^6,\ z_2z_3^5,\ z_3z_1^5,\ z_3z_2^5,\ z_3^6)$. Now I claim that $F_1(\Omega_{X/\mathbb{P}^2}^1)\cdot\mathcal{O}_{X_p} \not\subseteq \overline{F_3(\Omega_X^1)\cdot\mathcal{O}_{X_p}}$ for a general point $p\in\mathbb{P}^2$. We work in the affine open subset of $\mathbb{P}^2$ where $abc\neq 0$ and can therefore assume $c=-1$ and replace $z_3$ by $az_1+bz_2$ on $X$. Using the homogeneity, we see that the remaining equation of $X$ in $\mathbb{C}^2\times\mathbb{C}^2$ (with coordinates $a$, $b$, $z_1$, $z_2$) already belongs to the ideal generated by $(bz_1^5-az_2^5,\ z_2^5+b(az_1+bz_2)^5)$, which is an ideal on $\mathbb{C}^2\times\mathbb{C}^2$ inducing $F_1(\Omega_{X/\mathbb{P}^2}^1)$ on $X$. Using the homogeneity again, we conclude therefore that the equality $\overline{F_3(\Omega_X^1)\cdot\mathcal{O}_{X_p}} = \overline{F_1(\Omega_{X/\mathbb{P}^2}^1)\cdot\mathcal{O}_{X_p}}$

if it were true, would imply that there exist complex numbers $\lambda_1, \lambda_2, \mu_1, \mu_2$ such that we have, for example:

$$z_1 \cdot z_2^5 = (\lambda_1 z_1 + \lambda_2 z_2)(bz_1^5 - az_2^5) + (\mu_1 z_1 + \mu_2 z_2)(z_2^5 + b(az_1 + bz_2)^5)$$

in the ring $\mathbb{C}[z_1, z_2]$. Looking at the coefficient of each monomial, we obtain 7 linear equations for $\lambda_1, \lambda_2, \mu_1, \mu_2$, and it can be checked that there are no solutions, provided $p$ is outside a curve in $\mathbb{P}^2$. This shows that when $p$ is outside this curve, $F_1(\Omega^1_{X/\mathbb{P}^2}) \cdot \mathcal{O}_{X_p} \not\subseteq F_3(\Omega^1_X) \cdot \mathcal{O}_{X_p}$.

However, using the arcwise condition of integral dependence I will now show that when the coordinates of $p$ satisfy, in addition to $abc \neq 0$, the condition $a^{\frac{6}{5}} + b^{\frac{6}{5}} + c^{\frac{6}{5}} \neq 0$, we have equality of the integral closure of the two ideals. To see this, set again $c = -1$, and check that if $1 + a^{\frac{6}{5}} + b^{\frac{6}{5}} \neq 0$, whenever we take an analytic arc $z_1(t), z_2(t)$ *in the fibre* $X_p$, the lowest order terms in $t$ cannot cancel simultaneously in the two expressions $bz_1(t)^5 - az_2(t)^5$, $z_2(t)^5 + b(az_1(t) + bz_2(t))^5$, therefore the ideal they generate in $\mathbb{C}\{t\}$ is $(t^\nu)$, where $\nu = 5 \cdot \min(v(z_1(t)), v(z_2(t)))$, and we see immediately that all the monomials appearing in the expression of $F_3(\Omega^1_X)$ give a greater valuation in $t$, which shows that $F_3(\Omega^1_X) \cdot \mathcal{O}_{X_p}$ is integrally dependent on $F_1(\Omega^1_{X/\mathbb{P}^2}) \cdot \mathcal{O}_{X_p}$.

Remark. If we denote by $\mathcal{J}_X$ and $\mathcal{J}_{X/Y}$ respectively the ideals *in* $\mathcal{O}_{Y \times \mathbb{C}^N}$ generated by the Jacobian determinants which when restricted to $X$ generate $F_n(\Omega^1_X)$ and $F_d(\Omega^1_{X/Y})$ respectively, it is *not* true in general that for many points $y \in Y$ we have for any $x \in f^{-1}(y)$ that $\overline{J_{X,x}} = \overline{J_{X/Y,x}}$. In the above example, if we did not require that the arc $z_1(t), z_2(t)$ lies on $X_p$, (or merely in $az_1 + bz_2 + cz_3 = 0$) we could always arrange to have cancellation of the lowest order terms in the two expressions in question, the ideal they generate in $\mathbb{C}\{t\}$ then being $(t^\nu)$ with $\nu \gg 5 \cdot \min(v(z_1(t)), v(z_2(t))$, and it is no longer true that, for example, $v(z_1(t)^5 \cdot z_2(t)) \geqq \nu$, so that $z_1^5 \cdot z_2$ is not integrally dependent over $\mathcal{J}_{X/Y}$ in $\mathcal{O}_{Y \times \mathbb{C}^N, x}$. However, when $X$ is a *hypersurface* in $Y \times \mathbb{C}^N$, this phenomenon does not occur, and since this fact is of importance in the theory of equisingularity. I will now prove two results in this direction, which are results of integral dependence of the $\partial F/\partial y_l$ over the ideal $((\partial F/\partial z_1, \cdots, (\partial F/\partial z_N))$ in the ambient space $Y \times \mathbb{C}^N$, and not just on the hypersurface $X$ defined by $F = 0$.

2.10. Proposition 1. Let $F(y_1, \cdots, y_k, z_1, \cdots, z_N)$ be convergent in (an open subset of) $Y \times \mathbb{C}^N$, and such that all $\partial F/\partial z_i$ do not vanish identically. Then, there exists a nowhere dense closed complex subspace $B'$ of $Y \times \mathbb{C}^N$ such that:

(i) the image of $B'$ in $Y$ by the projection has measure 0,

(ii) for all $x \in Y \times \mathbb{C}^N - B'$ we have:

$$\left(\frac{\partial F}{\partial y_l}\right)_x \in \overline{\left(\frac{\partial F}{\partial z_1}, \cdots, \frac{\partial F}{\partial z_N}\right)_x}, \quad \text{in } \mathcal{O}_{Y\times\mathbb{C}^N,x} \qquad (1 \leqq l \leqq k).$$

PROOF. The idea of the proof is the same as for the idealistic Bertini theorem, and I shall therefore give only the main features: let $\pi : Z \to Y \times \mathbb{C}^N$ be the normalized blowing up of the ideal $\mathcal{J} = ((\partial F/\partial z_1), \cdots, (\partial F/\partial z_N)) \cdot \mathcal{O}_{Y\times\mathbb{C}^N}$, let $D = \bigcup_{\alpha\in A} D_\alpha$ be the exceptional divisor, defined by $\mathcal{J} \cdot \mathcal{O}_Z$, and $D_\alpha$ its irreducible components. Note that $\pi$ is a proper modification of $Y \times \mathbb{C}^N$ since $J \neq 0$. Now set $A = A_0 \amalg A_1$

$$\alpha \in A_0 \Leftrightarrow p(D_\alpha) \neq Y$$

$$\alpha \in A_1 \Leftrightarrow p(D_\alpha) = Y$$

where $p : Z \to Y$ is $pr_1 \circ \pi$, and let $B' = \bigcup_{\alpha\in A_0} \pi(D_\alpha)$, nowhere dense closed complex subspace of $Y \times \mathbb{C}^N$. I claim that if $x \in Y \times \mathbb{C}^N - B'$, $(\partial F/\partial y_l)_x \in \bar{J}_x$: let $U$ be an open neighborhood of $x$ in $Y \times \mathbb{C}^N$. Each $D_\alpha \cap \pi^{-1}(U)$ contains a dense open analytic subspace $V_\alpha$ at each point $z$ of which the following hold:

(i) the germ of $D$ at $z$ is equal to the germ of $D_\alpha$ at $z$, and $p \mid D_{\alpha,\text{red}} : D_{\alpha,\text{red}} \to Y$ is a submersion of non-singular spaces at $z$,

(ii) $Z$ is non-singular at $z$ and hence by (i), $p : Z \to Y$ is a submersion of non-singular spaces at $z$,

(iii) the strict transform by $\pi$ of the hypersurface $X$ defined by $F = 0$ in $Y \times \mathbb{C}^N$ is empty near $z$.

[Recall that by definition, the strict transform $X'$ is $\overline{\pi^{-1}(X - F)}$ where $F$ is the subspace defined by $\mathcal{J}$, so that $X' \cap D_\alpha$ is nowhere dense in each $D_\alpha$.]

Now by the implicit function theorem, we can choose local coordinates on $Z$ centered at $z$, $(y'_1, \cdots, y'_k, v, u_2, \cdots, u_N)$ such that:

$$\mathcal{O}_{Z,z} = \mathbb{C}\{y'_1, \cdots, y'_k, v, u_2, \cdots, u_N\} \quad \text{and:}$$

(1) $\mathcal{J} \cdot \mathcal{O}_{Z,z} = (v^\mu) \cdot \mathcal{O}_{Z,z} \qquad \mu \in \quad ,$

(2) $(y_l \circ \pi)_z = y'_l \ (1 \leqq l \leqq k)$,

(3) $(F \circ \pi)_z = A \cdot v^\nu$ where $A$ is invertible in $\mathcal{O}_{Z,z}$ (this is the translation of condition (iii)).

Then we have:

$$\frac{\partial(F \circ \pi)}{\partial y'_l} = \sum_{i=1}^{N} \frac{\partial F}{\partial z_i} \circ \pi \cdot \frac{\partial(z_i \circ \pi)}{\partial y'_l} + \frac{\partial F}{\partial y_l} \quad \text{in } \mathcal{O}_{Z,z} \qquad (*)$$

and

$$\frac{\partial(F \circ \pi)}{\partial y'_l} = \frac{\partial A}{\partial y'_l} \cdot v^\nu.$$

Therefore, to check that $((\partial F/\partial y_l)\circ\pi)_z\in\mathcal{J}\cdot\mathcal{O}_{Z,z}$, it is sufficient to show that we have $\nu\geqq\mu$. To show this, let us restrict the equality (*) to the non-singular subspace $W$ of $Z$ near $z$ defined by $y'_1=\cdots=y'_k=0$. By the exercises on integral dependence, we know that $F(0, z_1,\cdots,z_N)$ is integrally dependent over $((\partial F/\partial z_1)(0, z),\cdots,(\partial F/\partial z_N)(0, z))$ in $\mathbb{C}\{z_1,\cdots,z_N\}$. Since $W=p^{-1}(0)_z$, and since an integral dependence relation can be lifted to $W=\pi^{-1}(0\times\mathbb{C}^N)$ and then localized at $z$, we see that $A(0, v, u_2,\cdots,u_N)\cdot v^\nu$ is integrally dependent over $(v^\mu)$ in $\mathbb{C}\{v, u_2,\cdots,u_N\}=\mathcal{O}_{W,z}$, and since $\mathcal{O}_{W,z}$ is integrally closed in its field of fractions, by the exercises on integral dependence it means $Av^\nu\in(v^\mu)$ but now we use the fact that $A$ is invertible in $\mathcal{O}_{Z,z}$, hence its image in $\mathcal{O}_{W,z}$ is also invertible, and finally $\nu\leqq\mu$.

The end of the proof is as before: $\mathcal{J}\cdot\mathcal{O}_{\pi^{-1}(U)}$ is invertible in a normal space by construction, and the argument above shows that the meromorphic functions $(J\cdot\mathcal{O}_{\pi^{-1}(U)})^{-1}\cdot((\partial F/\partial y_l)\circ\pi)$ are holomorphic on an open dense subset of each component of $D\cap\pi^{-1}(U)$, hence they are holomorphic everywhere on $\pi^{-1}(U)$, hence $(\partial F/\partial y_l)\cdot\mathcal{O}_{\pi^{-1}(U)}\subset\mathcal{J}\cdot\mathcal{O}_{\pi^{-1}(U)}$, hence $((\partial F/\partial y_l))_x\in\bar{J}_x$ $(1\leqq l\leqq k)$.

Remark. Proposition 1 obviously implies the idealistic Bertini theorem in the case where $X$ is a hypersurface in $Y\times\mathbb{C}^N$, by taking $B=B'\cap X$, since in this case, $F_n(\Omega^1_X)$ is generated by the $(\partial F/\partial z_i)$ and $(\partial F/\partial y_l)$, and $F_d(\Omega^1_{X/Y})$ by the $(\partial F/\partial z_i)$.

The theory of (c)-equisingularity uses the following result:

2.11. Proposition 2. *Let $F(y_1,\cdots,y_k, z_1,\cdots,z_N)$ be convergent in (an open of) $Y\times\mathbb{C}^N$, such that all $(\partial F/\partial z_i)$ do not vanish identically, and that $F(y_1,\cdots,y_k,\mathcal{O})\equiv 0$ [i.e. we are now given a section $\sigma$ $Y\times\mathbb{C}^N\overset{\sigma}{\longleftrightarrow}Y$ such that $\sigma(Y)\subset X$]. Then there exists a nowhere dense closed complex subspace $B_0\subset\sigma(Y)$ $(=Y\times\{0\})$ such that at any point $x\in\sigma(Y)-B_0$ we have:*

$$\left(\frac{\partial F}{\partial y_l}\right)_x\in\overline{\left(z_1\cdot\frac{\partial F}{\partial z_1},\cdots,z_N\cdot\frac{\partial F}{\partial z_N}\right)_x}\quad\text{in}\quad\mathcal{O}_{Y\times\mathbb{C}^N,x}\qquad(1\leqq l\leqq k).$$

Proof. It is essentially the same as that of proposition 1 above, except that we take the normalized blowing up $\pi$ of the ideal generated by $(z_1\cdot(\partial F/\partial z_1),\cdots,z_N\cdot(\partial F/\partial z_N))$ in $\mathcal{O}_{Y\times\mathbb{C}^N}$. Then $B_0$ is the union of the $\pi(D_\alpha)\cap\sigma(Y)$ for those $\alpha$ such that $\pi(D_\alpha)$ does not contain $\sigma(Y)$, where $D=UD_\alpha$ is the exceptional divisor of $\pi$. Furthermore, the open analytic $V_\alpha$ which we take in each $D_\alpha$ has to satisfy, in addition to the conditions (i), (ii), (iii) appearing in the proof of proposition 1, the condition (which is also satisfied on an open dense analytic subset):

(iv) the strict transform by $\pi$ of each of the hyperplanes $z_i=0$ $(1\leqq i\leqq N)$ is empty near $z\in V_\alpha$.

Then, the argument is the same as in the proof of proposition 1, using the equality (*), remarking that condition iv) implies that $(z_i \circ \pi)_z = \zeta_i \cdot v^{\mu_i}$ with $\zeta_i$ invertible in $\mathcal{O}_{Z,z}$, and hence $(\partial/\partial y'_l)(z_i \circ \pi) = (\partial \zeta_i/\partial y'_l) \cdot v^{\mu_i}$ is a multiple of $(z_i \circ \pi)_z$ in $\mathcal{O}_{Z,z}$ and this time using the full strength of the inclusion $f(z_0, \cdots, z_n) \in \overline{(z_0 \cdot (\partial f/\partial z_0), \cdots, z_n \cdot (\partial f/\partial z_n))}$ given in the exercises on integral dependence, to prove that $\nu \geqq \mu$, where $\mu$ is the order in $v$ of the ideal $(z_1 \cdot (\partial F/\partial z_1), \cdots, z_N \cdot (\partial F/\partial z_N)) \cdot \mathcal{O}_{Z,z}$.

The details of the adaptation of the proof of proposition 1 are given as an exercise to the reader.

REMARK. My original statement of proposition 2 was that $(\partial F/\partial y_l)_x \in \overline{(z_1, \cdots, z_N) \cdot ((\partial F/\partial z_1), \cdots, (\partial/\partial z_N))_x}$ and $I$ must thank J. P. G. Henry for remarking that my proof actually gave the statement above, which is slightly stronger, and also easier to use. Remark that the ideal $\overline{(z_1 \cdot (\partial F/\partial z_1), \cdots, z_N \cdot (\partial F/\partial z_N))}$ depends on the choice of coordinates $z_1, \cdots, z_N$, and that proposition 2 holds for any choice. The ideal $\overline{(z_1, \cdots, z_N) \cdot ((\partial F/\partial z_1), \cdots, (\partial F/\partial z_N))_x}$ does not depend upon the choice of coordinates, but presumably we have the equality $\overline{(z_1 \cdot (\partial F/\partial z_1), \cdots, z_N \cdot (\partial F/\partial z_N))_x} = \overline{(z_1, \cdots, z_N) \cdot ((\partial F/\partial z_1), \cdots, (\partial F/\partial z_N))_x}$ at a general point $x$ of $\sigma(Y)$ and for a 'sufficiently general' choice of coordinates $z_1, \cdots, z_N$. We are now ready to move into some equisingularity conditions:

2.12. DEFINITION. Let $f : X \overset{\sigma}{\leftrightarrows} Y$ be a family of germs of hypersurfaces, which means that $X$ can be locally around $0 \in X$ imbedded as a hypersurface in $Y \times \mathbb{C}^N$, $\sigma$ is a section of $f$, and we may even assume $\sigma(Y) = Y \times \{0\}$. We still assume that $Y$ is non-singular. We say that $X$ is (c)-equisingular along $\sigma(Y)$ at $x \in \sigma(Y)$ if there exists such an imbedding $X \subset Y \times \mathbb{C}^N$ described by $F(y_1, \cdots, y_k, z_1, \cdots, z_N) = 0$ and such that:

$$\left(\frac{\partial F}{\partial y_l}\right)_x \in \overline{(z_1, \cdots, z_N) \cdot \left(\frac{\partial F}{\partial z_1}, \cdots, \frac{\partial F}{\partial z_N}\right)_x} \quad \text{in} \quad \mathbb{C}\{y, z\} = \mathcal{O}_{Y \times \mathbb{C}^N, x} \cdot (1 \leqq l \leqq k)$$

EXERCISE. (1) Check that this condition depends only on $f$ and $\sigma$, i.e. is independent of the choice of the equation $F$ and the coordinates. Hint: use the fact that

$$F \in \overline{(y_1 \cdot (\partial F/\partial y_1), \cdots, y_k \cdot (\partial F/\partial y_k), z_1 \cdot (\partial F/\partial z_1), \cdots, z_N \cdot (\partial F/\partial z_N))_x}.$$

(2) Check that in fact this condition depends only upon $\sigma(Y) \subset X$ and not upon the choice of a retraction $Y \times \mathbb{C}^N \to \sigma(Y)$.

2.13. COROLLARY 1. *Given $f : X \overset{\sigma}{\leftrightarrows} Y$, $\sigma$ a section of $f$, and $f$ a family of hypersurfaces, then the set of points $x$ of $\sigma(Y)$ such that $X$ is (c)-equisingular* *along $\sigma(Y)$ at $x$ is the complement of a nowhere dense closed complex subspace of $\sigma(Y)$. This is an immediate consequence of proposition 2 above, and the fact that the set of points where we have an integral dependence relation is open and analytic in any case (possibly empty) because of the coherence of $\bar{I}$.*

2.14. COROLLARY 2. *If $f: X \overset{\sigma}{\leftrightarrows} Y$ is (c)-equisingular at $x \in \sigma(Y)$, then for any local imbedding $X \subset Y \times \mathbb{C}^N$ near $x$, and for each $i$, $1 \leqq i \leqq N$, there exists a Zariski open dense $U^{(i)}$ of the Grassmannian of $i$-planes in $(\mathbb{C}^N, 0)$ such that for any $H_0 \in U^{(i)}$ we have, setting $H = Y \times H_0$,*

(a) $\overline{j_{Y\times\mathbb{C}^N/Y}(F)\cdot \mathcal{O}_{H,x}} = \overline{j_{Y\times H_0/Y}(F\cdot\mathcal{O}_{H,x})}$ *in* $\mathcal{O}_{H,x}$ $\quad [F\cdot\mathcal{O}_{H,x} = (F \mid H)_x]$

(b) *$X \cap H$ is (c)-equisingular along $\sigma(Y)$ at $x$, [where $j_{Y\times\mathbb{C}^N/Y}(F)$ is the ideal in $\mathcal{O}_{Y\times\mathbb{C}^N,x}$ generated by $((\partial F/\partial z_1), \cdots, (\partial F/\partial z_N))$].*

PROOF. The idea is to apply proposition 2 above to the family of sections of $X$ by such $Y \times H_0$, as follows: We can restrict ourselves to an open subset of the Grassmannian of $i$-planes, of course, and therefore describe the family of sections $X \cap H$, where $H = Y \times H_0$, $H_0$ an $i$-plane in $\mathbb{C}^N$, as follows:

$$F_a = F\Big(y_1, \cdots, y_k, z_1, \cdots, z_i, \sum_{1\leqq j\leqq i} a_{i+1,j}z_j, \cdots, \sum_{1\leqq j\leqq i} a_{N,j}z_j\Big) = 0$$

$$F_a \in \mathbb{C}\{y, z_1, \cdots, z_1, (a_{p,j})\} \quad (\text{where } i+1 \leqq p \leqq N, 1 \leqq j \leqq i).$$

Given any function $g(y_1, \cdots, y_k, z_1, \cdots, z_N)$, we will write $g_a$ or $(g)_a$ for the function

$$g\Big(y_1, \cdots, y_k, z_1, \cdots, z_i, \sum_{1\leqq j\leqq i} a_{i+1,j}z_j, \cdots, \sum_{1\leqq j\leqq i} a_{N,j}z_j\Big)$$

in $\mathbb{C}\{y, z_1, \cdots, z_i, (a_{p,j})\}$.

Now, by proposition 2, we have that there exists a dense open analytic subset $V^{(i)}$ in the space $\mathbb{C}^{i(N-i)}$ of the coefficients $a = (a_{p,j})$ such that if $a \in V^{(i)}$, then we have:

$$\frac{\partial}{\partial a_{p,j}} F_a \in \overline{\Big(y_i \cdot \frac{\partial F_a}{\partial y_1}, \cdots, y_k \cdot \frac{\partial F_a}{\partial y_k}, z_1 \cdot \frac{\partial}{\partial z_1} F_a, \cdots, z_i \cdot \frac{\partial}{\partial z_i} F_a\Big)}$$

in $\mathcal{O}_{\mathbb{C}^{i(N-i)}\times(Y\times\mathbb{C}^i),a\times\{0\}}$

which can be explicited by:

$$z_j\Big(\frac{\partial F}{\partial z_p}\Big)_a \in \overline{\Big(y_1\Big(\frac{\partial F}{\partial y_1}\Big)_a, \cdots, y_k\Big(\frac{\partial F}{\partial y_k}\Big)_a, z_1\Big\{\Big(\frac{\partial F}{\partial y_1}\Big)_a + \sum_p a_{p,1}\Big(\frac{\partial F}{\partial z_p}\Big)_a\Big\}, \cdots,} \qquad (*)$$

$$\overline{z_i\Big\{\Big(\frac{\partial F}{\partial z_i}\Big)_a + \sum_p a_{p,i}\Big(\frac{\partial F}{\partial z_p}\Big)_a\Big\}\Big)}$$

for all $1 \leqq j \leqq i, i+1 \leqq p \leqq N$. Our assumption that the original family is (c)-equisingular implies:

$$\left(\frac{\partial F}{\partial y_l}\right)_a \in (z_1, \cdots, z_i)\overline{\left(\left(\frac{\partial F}{\partial z_1}\right)_a, \cdots, \left(\frac{\partial F}{\partial z_N}\right)_a\right)} \quad \text{in} \quad \mathcal{O}_{Y \times \mathbb{C}^i \times \mathbb{C}^{i(N-i)}, 0 \times \{a\}}. \qquad (**)$$

We can assume, after a change of the coordinates $z_p$, that our point $a \in V^{(i)}$ is the origin of $\mathbb{C}^{i(N-i)}$. I now give as an exercise to check, using (*) and (**) that for any mapping

$$\varphi^* : \mathbb{C}\{y_1, \cdots, y_k, z_1, \cdots, z_i, (a_{p,j})\} \to \mathbb{C}\{t\}$$

we have that for any $i+1 \leqq p \leqq N$,

$$v\left(\frac{\partial F}{\partial z_p}\right)_a \geqq \min_{1 \leqq j \leqq i} \left\{ v\left(\frac{\partial F}{\partial z_j}\right)_a \right\}$$

where $v(h) =$ order in $t$ of $\varphi^*(h)$, using the arcwise condition of integral dependence, and the fact that $v(y_l) > 0$, $v(a_{p,j}) > 0$. (Hint: use reductio ab absurdum.) From this, we deduce that

$$\left(\frac{\partial F}{\partial z_p}\right)_{H,x} \in \overline{j_{Y \times H_0/Y}(F \cdot O_{H,x})}, \; (i+1 \leqq p \leqq N), \qquad (***)$$

where $H_0$ is the plane corresponding to $a \in V^{(i)}$, and from this inclusion, the equality (a) of the proposition follows immediately.

Assertion (b) of the proposition also follows immediately from (**) and (***), in view of assertion (a).

REMARKS. (i) We have not even used the fact that the hypersurface defined by $F = 0$ is reduced. The only assumption needed is $(\partial F/\partial z_j) \not\equiv 0$.

(ii) Even when $Y$ is a point, the statement above is not a triviality. Then, condition (c) is automatically satisfied, and we have:

2.15. COROLLARY 3. *Given $F \in \mathbb{C}\{z_1, \cdots, z_N\}$ such that $F(0) = 0$, $F \not\equiv 0$, for each $i$, $1 \leqq i \leqq N$, there exists a Zariski open dense subset $U^{(i)}$ in the Grassmannian of i-planes through 0 in $\mathbb{C}^N$ such that for any $H \in U^{(i)}$ we have, denoting as above $F \cdot \mathcal{O}_{H,0}$ the germ at 0 of the restriction of $F$ to $H$:*

$$\overline{j(F) \cdot \mathcal{O}_{H,0}} = \overline{j(F \cdot \mathcal{O}_{H,0})} \quad \text{in} \quad \mathcal{O}_{H,0}$$

*where $j(F)$ is the ideal generated in the ring of functions of the ambient space by the partial derivatives of $F$. [Equivalently, since $F \in \overline{j(F)}$, we can say $\overline{(F, j(F)) \cdot O_{H,0}} = \overline{(F \cdot O_{H,0}, j(F \cdot O_{H,0})}$.]*

EXERCISE. (1) Compare corollary 3 with the idealistic Bertini theorem.

(2) Find an example where one cannot remove the bars above the ideals in corollary 3. Hint: $F = z_1^3 + z_2^3 + z_3^3$ will do. If you have done exercise 1,

you know why such a simple example will not work for the idealistic Bertini theorem.

We now turn to some geometric consequences of condition (c).

2.16. DEFINITION. Let $T_1$ and $T_2$ be two vector subspaces of $\mathbb{C}^N$ considered as hermitian space with the form $((z_i), (z'_i)) \to \sum_i z_i \bar{z}'_i$ and assume $\dim T_1 \geqq \dim T_2 \geqq 1$. The distance from $T_1$ to $T_2$ (in that order) is defined as

$$d(T_1, T_2) = \operatorname*{Sup}_{\substack{t_1 \in T_1^\perp - \{0\} \\ t_2 \in T_2 - \{0\}}} \left\{ \frac{|(t_1, t_2)|}{\|t_1\| \cdot \|t_2\|} \right\}$$

where $T_1^\perp$ is the vector space of vectors orthogonal to $T_1$ with respect to the hermitian form.

REMARK. By translating to the origin, this gives a distance between *directions* of linear subspaces.

Let now $X \subset Y \times \mathbb{C}^N$ be a hypersurface, defined locally near $x$ by $F(y_1, \cdots, y_k, z_1, \cdots, z_N) = 0$ where we have identified $Y$ with $\mathbb{C}^k$ locally as usual, and $x$ with $0 \in \mathbb{C}^k \times \mathbb{C}^N$, and $F(y, 0) \equiv 0$. Given a point $p \in Y \times \mathbb{C}^N$, let us denote by $L_p$ the level hypersurface through $p$ of the mapping $Y \times \mathbb{C}^N \xrightarrow{F} \mathbb{C}$, i.e. $F^{-1}(F(p)) = L_p$. Whenever $p$ is not a singular point of $L_p$, we can define the distance between the directions of the tangent hyperplane to $L_p$ at $p$ and $T_{Y,0}$, which we have identified with $\mathbb{C}^k$.

$T^\perp_{L_p,p}$ is generated by the vector: $(\overline{(\partial F/\partial y_1)(p)}, \cdots, \overline{(\partial F/\partial y_k)(p)}, \overline{(\partial F/\partial z_1)(p)}, \cdots, \overline{(\partial F/\partial z_N)(p)})$ and hence

$$d(T_{L_p,p}, T_{Y,0}) = \operatorname*{Sup}_{\xi \in \mathbb{C}^k - \{0\}} \left\{ \frac{\left| \sum_{l=1}^{k} \overline{\frac{\partial F}{\partial y_l}(p)} \cdot \xi_l \right|}{\left( \sum_{l=1}^{k} \left| \frac{\partial F}{\partial y_l}(p) \right|^2 + \sum_{i=1}^{N} \left| \frac{\partial F}{\partial z_i}(p) \right|^2 \right)^{\frac{1}{2}} \cdot \left( \sum_{l=1}^{k} |\xi_l|^2 \right)^{\frac{1}{2}}} \right\}.$$

2.17. PROPOSITION 3. *If $X \subset Y \times \mathbb{C}^N$ is a hypersurface satisfying condition (c) along $Y \times \{0\}$ at 0, then there exists a neighborhood $V$ of $0 \in Y \times \mathbb{C}^N$, and $C \in \mathbb{R}_+$ such that for any point $p \in V$ such that the level hypersurface $L_p$ is non-singular at $p$, we have*

$$d(T_{L_p,p}, T_{Y,0}) \leqq C \cdot \operatorname{dist}(p, Y),$$

[*where dist $(p, Y)$ is the distance from the point $p$ to $Y = \mathbb{C}^k \times \{0\}$ in $Y \times \mathbb{C}^N$*], *and conversely, this inequality implies condition (c).*

PROOF. It follows easily from the above expression of the distance from

$T_{L_p,p}$ to $T_{Y,0}$, and the criterion ② of integral dependence, if we remark that ② allows us to express condition (c) by:

$$\left|\frac{\partial F}{\partial y_l}(p)\right| \leqq K_l \cdot \operatorname{Sup}_i |z_i(p)| \cdot \operatorname{Sup}_j \left|\frac{\partial F}{\partial z_j}(p)\right| \qquad (1 \leqq l \leqq k,\ K_l \in \mathbb{R}_+)$$

i.e.

$$\left|\frac{\partial F}{\partial y_l}(p)\right| \leqq K_l' \cdot \operatorname{dist}(p, Y) \cdot \operatorname{Sup}_j \left|\frac{\partial F}{\partial z_j}(p)\right|.$$

Corollary 1. *Let $X^0$ denote the open analytic set of non-singular points of $X$, and assume $X$ reduced, so that $\overline{X^0} = X$. Set $Y_1 = Y \times \{0\} \subset X$. If $X$ is (c)-equisingular along $Y_1$ at $x \in Y_1$, we have:*

(1) *the pairs of strata $(Y \times \mathbb{C}^N - Y_1, Y_1)$ and $(X^0 - Y_1, Y_1)$ both satisfy Thom's $A_F$ condition at x, where $F: Y \times \mathbb{C}^N \to \mathbb{C}$ is the map given by F. (See Hironaka's lectures* §3, def. 3).

(2) *The pair strata $(X^0 - Y_1, Y_1)$ satisfies the Whitney condition at x (see Hironaka's lectures* §3, *def.* 2).

[Remark. In practice, most often $Y_1 \subset \operatorname{Sing} X$, so that $X^0 - Y_1 = X^0$.]

(3) *Assuming furthermore that $X^0 = X - Y_1$, we have:*
*Any holomorphic vector field on $Y_1$ can be extended to a vector field on $Y \times \mathbb{C}^N$ (in a neighborhood of x) which is:*

(i) *real-analytic outside $Y_1$,*

(ii) *tangent to the level hypersurfaces $L_p$,*

(iii) *'rugose' in the sense of Verdier [Stratifications de Whitney et théorème de Bertini-Sard, Inventiones Math. 36 (1976)] with respect to the Whitney stratification: $(Y \times \mathbb{C}^N - Y_1, Y_1)$ of $Y \times \mathbb{C}^N$, or, what amounts to the same, with respect to $(Y \times \mathbb{C}^N - X, X - Y_1, Y_1)$. [rugosity is a Lipschitz-like condition, but relative to a stratification].*

Proof. (1) and (2) follow from proposition 3 and the definitions. As for (3), we think of vector fields as derivations: a (germ of) vector field on $Y$ is given by a derivation $\partial$ of $\mathcal{O}_{Y,0} = \mathbb{C}\{y_1, \cdots, y_k\}$, and the pull-back of this vector field to $Y \times \mathbb{C}^N$ by the projection $Y \times \mathbb{C}^N \to Y$ is given by the derivations $\tilde{\partial}$ of $\mathcal{O}_{Y \times \mathbb{C}^N,0} = \mathbb{C}\{y_1, \cdots, y_k, z_1, \cdots, z_N\}$ described by $\tilde{\partial} y_l = \partial y_l$, $\tilde{\partial} z_i = 0$.

Now the projection of this vector field, at each point $p \in Y \times \mathbb{C}^N$, on the tangent space $T_{L_p,p}$ is described by the derivation:

$$D = \tilde{\partial} - \sum_{i=1}^{N} \frac{\tilde{\partial} F \cdot \overline{\dfrac{\partial F}{\partial z_i}}}{\sum_1^N \left|\dfrac{\partial F}{\partial z_i}\right|^2} \frac{\partial}{\partial z_i} \quad \text{of} \quad \mathcal{O}_{Y \times \mathbb{C}^N}$$

and now of course $DF=0$, which means it is tangent to $L_p$ at each $p \in Y \times \mathbb{C}^N - Y_1$, and to check its rugosity is just a matter of applying proposition 3 to the definition.

Since a rugose vector field is always integrable (Verdier, loc. cit.), we obtain by extending the constant vector fields on $Y_1$ a topological trivialization of $X$ viewed as a family of hypersurfaces with isolated singularity parametrized by $Y$, and in fact better than topological triviality, a 'rugose' triviality, namely a homeomorphism of pairs

$$(Y \times \mathbb{C}^N, X) \cong (Y \times \mathbb{C}^N, Y \times X_0)$$

compatible with projection to $Y$, and 'rugose' with respect to the stratifications described above.

Now that we are dealing with isolated singularities, we can ask what happens to the Jacobian ideal $j(F_y)$ associated to the singularities of hypersurfaces $(X_y, y \times \{0\}$ in our family. First we need a definition:

Definition. Let $(X_0, 0) \subset (\mathbb{C}^N, 0)$ be a germ of hypersurface with isolated singularity. For each $0 \leqq i \leqq N$ we define $\mu^{(i)}(X_0, 0) = \min_H \mu(X_0 \cap H, 0)$ $H$ running through the Grassmannian of $i$-planes in $\mathbb{C}^N$. It is not difficult to check that in fact the set of those $H$ such that $\mu(X_0 \cap H, 0) = \mu^{(i)}(X_0, 0)$ is a dense Zariski open subset of the Grassmannian. Remark that $\mu^{(N)}(X_0, 0)$ is the usual Milnor number, that $\mu^{(1)}(X_0, 0)$ is the multiplicity (= order of equation) minus none, and $\mu^{(0)}(X_0, 0) = 1$.

2.18. Theorem. *Let $X \subset Y \times \mathbb{C}^N$ ($Y$ non-singular, as always) be a hypersurface such that $|\mathrm{Sing}\, X| = Y_1 = Y \times \{0\}$. For each $y \in Y$ set $(X_y, 0) = (X \cap (\{y\} \times \mathbb{C}^N), 0)$. Then the following are equivalent*

(1) *$M(X_y, 0) = \sum_{i=0}^N \binom{N}{i} \mu^{(i)}(X_y, 0)$ is independent of $y \in Y$ (in a neighborhood of $0 \in Y$),*

(2) *for each $i$, $\mu^{(i)}(X_y, 0)$ is independent of $y \in Y$,*

(3) *$X$ satisfies the condition* (c) *along $Y_1$ at $0 \times 0$.*

I will not give the proof, referring to my 'Introduction to equisingularity problems' (Proceedings A.M.S., Conference on Algebraic Geometry, Arcata 1974, A.M.S. Pub., Providence Rhode-Island), and to the forthcoming notes of a course at the Collège de France, Spring 1976, but I will say this: the 'easy part' is to check that condition (c) implies that the Milnor number of the fibres is constant. Then by cor. 2 to prop. 2, all $\mu^{(i)}$ are constant. The converse is more delicate, and uses a connection between integral dependence and the multiplicity of ideals in the sense of algebraic geometers, which implies that equi-multiplicity conditions such as (1) or (2) have the consequence that a condition like (3), when true generically (which we know by corollary 1 to prop. 2 above) remains true at the special point, [and the fact that $M(X_y, 0)$ is the multiplicity of $m \cdot j(F_y)$]. This is the 'principle of specialization of integral dependence' [Appendix 1 to 'Sur diverses conditions numériques d'équisingularité des familles de courbes ...', preprint No M208.0675, Centre de Mathématiques, Ecole Polytechnique, 91128 Palaiseau Cedex, France, see also 'Cycles évanescents, sections planes et conditions de Whitney', ch. II, §3, No 3.3, Astérisque No 7–8, Société Math. France, 1973].

REMARK. One can generalize condition (c) to the incidence of any pair of strata $(M, N)$ with $\bar{M} \supset N$ in a complex-analytic space (in our case, $M = X^0, N = Y_1$) and prove that any complex space has a locally finite stratification such that any two distinct strata $(M, N)$ with $N \cap \bar{M} \neq \emptyset$ satisfy: $N \subset \bar{M}$ and $(M, N)$ satisfies condition (c) at every point of $N$.

REMARK. It is known [see 'Cycles évanescents ···' quoted above] that the Milnor number $\mu^{(N)}(X_y)$ of a hypersurface with isolated singularity depends only on the topological type of the *imbedded* germ $(X_y, 0) \subset (\mathbb{C}^N, 0)$. This is not true for the other $\mu^{(i)}$ (it is conjectured by Zariski for $\mu^{(1)}$ but the answer is not yet known), as is shown by the examples found by Briançon et Speder: $X$ is the family of surfaces in $\mathbb{C}^3$ with isolated singularity defined by $z_2^3 + yz_2z_1^3 + z_1^4z_3 + z_3^9 = 0$ the topological type of the fibres is independent of $y$, hence also $\mu^{(3)} = 56$, but $\mu^{(2)}(X_0) = 8$, $\mu^{(2)}(X_y) = 7$ for $y \neq 0$. This shows that the topological type of a germ of hypersurface does not determine the topological type of its generic hyperplane section (through the singular point), and that $M(X_y, 0)$ is not an invariant of the topological type of $(X_y, 0)$.

2.19. DEFINITION. Two germs of hypersurfaces $(X_0, 0)$ and $(X_1, 0)$ are (c)-cosécant if there exists a 1 parameter family $f: X \overset{\sigma}{\leftrightarrows} \mathbb{D}$ of germs of hypersurfaces everywhere (c)-equisingular along $\sigma(\mathbb{D})$, and having one fibre isomorphic to $(X_0, 0)$ and another to $(X_1, 0)$.

We have just seen that the $\mu^{(i)}$ are invariants of (c)-cosécance, and in fact (c)-cosécance implies not only that our hypersurfaces have the same topological type, but all their generic plane sections too. (c)-cosécance will be our working definition of equisingularity, and we will see what it means for plane curves in the next paragraph.

REFERENCES

[1] H. HIRONAKA, Introduction to the theory of infinitely near singular points, Memorias de Matemática del Instituto Jorge Juan, No 28, Madrid 1974.

[2] M. Lejeune-Jalabert et B. Teissier, Fermeture intégrale des idéaux et équisingularité, Séminaire Ecole Polytechnique 1973–74, to appear (chapter I available from Dept. of Math. Univ. of Grenoble, 38402 St. Martin d'Hères, France).

## §3. Families of curves

In this section, we introduce invariants attached to singularities of curves, and study families of curves where some of these invariants remain constant.

3.1. We will mostly study *germs of families of curves*, i.e. germs of *flat* mappings $f:(X,0)\to(Y,0)$ such that $(f^{-1}(0),0)$ is a germ of curve, i.e. purely 1-dimensional analytic space. We will study also (germs of) *families of germs of curves*, which means that we have also given ourselves a section $\sigma$ of $f$, so that for any small representative of $f$, we can speak of the germs $(f^{-1}(y),\sigma(y))$ as the members of our family of germs. Occasionally, we will consider mappings $f:X\to Y$, such that $X_0=F^{-1}(0)$ is a projective or affine curve. In all cases, we will say that $f$ is a family, or a deformation of its special fibre $X_0=f^{-1}(0)$ (or the germ $(X_0,0)=(f^{-1}(0),0)$).

Let us start by fixing what we mean by the datum of a germ of curve $(X_0,0)$. Abstractly, as we have said, it is a germ of a purely 1-dimensional analytic space, hence it is described by an analytic algebra $\mathcal{O}_0$ purely of dimension 1. Geometrically, $(X_0,0)$ can be effectively given in two ways:

① *By equations.* By giving an ideal $I=(F_1,\cdots,F_m)$ in $\mathbb{C}\{z_1,\cdots,z_N\}$ such that $\mathcal{O}_0=\mathbb{C}\{z_1,\cdots,z_N\}/I$. Saying that $\mathcal{O}_0$ is purely one-dimensional means that the ideal (0) has a primary decomposition $(0)=\mathfrak{q}_1\cap\cdots\cap\mathfrak{q}_r$ where $\sqrt{\mathfrak{q}_j}=\mathfrak{p}_j$ is a minimal prime ideal in $\mathcal{O}_0$, and $\dim\mathcal{O}_0/\mathfrak{p}_j=1$. All this is easily translated in terms of the primary decomposition of $I$ in $\mathbb{C}\{z_1,\cdots,z_N\}$. The rings $\mathcal{O}_0/\mathfrak{q}_j$ (resp. $\mathcal{O}_0/\mathfrak{p}_j$) correspond to germs of irreducible (resp. irreducible and reduced) analytic curves, called the irreducible components (resp. branches) of the curve. We will mostly study *reduced* curves, which means that $\mathcal{O}_0$ is a reduced ring, or equivalently that $\mathfrak{q}_j=\mathfrak{p}_j$ for all $j$. In this case, it is well known that the integral closure $\bar{\mathcal{O}}_0$ of $\mathcal{O}_0$ in its total ring of quotients is isomorphic to $\prod_{j=1}^r\mathbb{C}\{t_j\}$, and if we choose generators $z_1,\cdots,z_N$ of the maximal ideal $\mathfrak{m}$ of $\mathcal{O}_0$, the injection $\mathcal{O}_0\subset\prod_{j=1}^r\mathbb{C}\{t_j\}$ is described by the datum of the $N$ elements $z_i=(z_i(t_j))\in\prod_{j=1}^r\mathbb{C}\{t_j\}$, and such a datum is usually called a *parametrization* of the germ of curve $(X,0)$, which brings us to the second way of giving a curve.

② *By a parametrization.* By giving ourselves a germ of a finite map $p:\coprod_{j=1}^r(\mathbb{C},0)\to(\mathbb{C}^N,0)$.

Here one has to be very careful: except when $N=2$, it is not true, even if $r=1$, that the image of this mapping in the sense of §1 is a curve: it will have 'imbedded components' concentrated at the singular points, as will be shown in the examples at the end of this section. More precisely, if we give ourselves a reduced germ of curve $(X_0, 0) \hookrightarrow (\mathbb{C}^N, 0)$, then normalize it, as explained above, the normalization $n:(\bar{X}_0, n^{-1}(0)) \to (X_0, 0)$ is isomorphic with $\coprod_{j=1}^r (\mathbb{C}, 0) \to (X_0, 0)$ where $r$ is the number of irreducible components of $X_0$. Then it is not true, if $N > 2$, and $X_0$ is singular, that the image in the sense of §1 of $p = i \circ n: \coprod_{j=1}^r (\mathbb{C}, 0) \to (\mathbb{C}^N, 0)$ is $X_0$: $im(p)$ has some extra 0-dimensional components (imbedded components). However, it is true that the *root* of $F_0(p_* \mathcal{O}_{\bar{X}_0, n^{-1}(0)})$ gives an ideal defining $(X_0, 0) \subset (\mathbb{C}^N, 0)$, (which is also $\mathrm{Ann}\,(p_* \mathcal{O}_{\bar{X}_0, n^{-1}(0)})$. This is another instance of the phenomenon seen in the addendum to §1, and is the price we have to pay for the stability of our images. Anyway, $F_0(p_* \mathcal{O}_{\bar{X}_0, n^{-1}(0)})$ contains more information than is needed to recover $(X_0, 0)$.

3.2. Since a reduced curve is normal outside its singular point we see that the quotient sheaf $\bar{\mathcal{O}}_{X_0}/\mathcal{O}_{X_0}$ is concentrated at the singular points, and hence $\bar{\mathcal{O}}_0/\mathcal{O}_0$ is a finite dimensional vector space, and this will give our first invariant:

DEFINITION. Let $(X_0, 0)$ be a germ of reduced curve (resp. $X_0$ be a reduced affine algebraic curve). Then the 'invariant $\delta$' of $(X_0, 0)$ is defined by $\delta(X_0, 0) = \delta(\mathcal{O}_0) = \dim_{\mathbb{C}} \bar{\mathcal{O}}_0/\mathcal{O}_0$ (resp. $\delta(X_0) = \dim_{\mathbb{C}} \bar{A}/A$ where $A = \Gamma(X_0, \mathcal{O}_{X_0})$).

REMARK. We have: $\delta(X_0) = \sum_{x \in X_0} \delta(X_0, x)$, remarking that the sum on the right is finite, since $\delta$ is nonzero only at singular points.

This invariant has to do with the following problem: each of the above descriptions of a germ of a curve suggests a description of what a germ of a family of curves is: abstractly, anyway, we have defined it as a germ of a flat map $f:(X, 0) \to (Y, 0)$. Assume that $Y = \mathbb{C}$ with parameter $v$, i.e. that we have a 1-parameter family.

We can try to describe our family:

① By a family of equations: Let $f^{-1}(0) = X_0$ be described in $\mathbb{C}^N$ by $(F_1, \cdots, F_m)$ as above. Can we describe $(X, 0)$ as a subspace of $(\mathbb{C} \times \mathbb{C}^N, 0)$ with coordinates $(v, z_1, \cdots, z_N)$, defined by an ideal generated by $(F_1 + v \cdot G_1, \cdots, F_m + v \cdot G_m)$ in $\mathcal{O}_{\mathbb{C} \times \mathbb{C}^N, 0}$, where $G_i \in \mathbb{C}\{v, z_1, \cdots, z_N\} \in \mathcal{O}_{\mathbb{C} \times \mathbb{C}^N, 0}$? [there is a $v$ in front to mark that the perturbation of the equations $F_i$ must vanish for $v = 0$].

② By a family of parametrizations: Can we describe $(X, 0)$ as the *reduced* image of a complex-analytic mapping: $(\mathbb{C}, 0) \times \coprod_{j=1}^r (\mathbb{C}, 0) \to (\mathbb{C} \times \mathbb{C}^N, 0)$ i.e. by giving $N$ elements $z_i = (z_i(v, t_j)) \in \prod_{j=1}^r \mathbb{C}\{v, t_j\}$ where $z_i(v, t_j) = z_i(t_j) + v \cdot \zeta_i(v, t_j)$, with $\zeta_i(v, t_j) \in \mathbb{C}\{v, t_j\}$ and the $z_i(t_j)$ describe a parametrization of $(X_0, 0)$ as above and where the induced map $(\mathbb{C}, 0) \times \coprod_{j=1}^r (\mathbb{C}, 0) \to$

$(X, 0)$ is the normalization and for each fixed value of $t$ induces the normalization of the corresponding fibre?

In fact, any germ of mapping $f:(X,0)\to(Y,0)$ can be described as in ①, i.e. $X\subset Y\times\mathbb{C}^N$ defined by $\tilde{F}_i\in\mathcal{O}_{Y,0}\{z_1,\cdots,z_N\}$: it is just a matter of remarking that $\mathfrak{m}_{Y,0}\cdot\mathcal{O}_{X,0}+(z_1,\cdots,z_N)=\mathfrak{m}_{X,0}$ ($\mathfrak{m}$ = maximal ideal) and hence we have a surjection of $\mathbb{C}$-algebras $\mathcal{O}_{Y,0}\{z_1,\cdots,z_N\}\to 0_{X,0}$. We are going to see that, on the contrary, a family of reduced curves can be described by a family of parametrizations if and only if it satisfies a numerical condition, essentially that the $\delta$-invariant of its fibres is constant.

3.3. Let $f:(X,0)\to(\mathbb{C},0)$ be a germ of a flat morphism of complex analytic spaces, such that its fibre is a reduced 1-dimensional analytic space (i.e. $f$ is a germ of a family of reduced curves).

Let $n:\bar{X}\to X$ be the normalization of the surface $X$ and let $p=f\circ n:(\bar{X},n^{-1}(0))\to(\mathbb{C},0)$.

Let us agree to denote $p^{-1}(0)$ by $(\bar{X})_0$ and to set $\delta((\bar{X})_0)=\sum_{x\in n^{-1}(0)}\delta(p^{-1}(0),x)$. Let us also agree to set $\delta(X_0)=\delta(f^{-1}(0))$ and $\delta(X_y)=\delta(f^{-1}(y))$ for $y\in\mathbb{C}-\{0\}$ in a *small enough representative of* $f$ so that all the singular points of $X_y=f^{-1}(y)$ tend to 0 as $y\to 0$, and 0 is the only singular point of $X_0=f^{-1}(0)$. Then we have:

PROPOSITION. (1) $p=f\circ n$ *is a (multi-germ of a) flat mapping;* (2) $\delta((\bar{X})_0)=\delta(X_0)-\delta(X_y)(y\in\mathbb{C}-\{0\}$, *but near* 0). *In words, the invariant* $\delta$ *of the fibre over* $0\in\mathbb{C}$ *of the normalization of the surface which is the total space of our family of curves, is equal to the difference of the invariants* $\delta$ *of the 'special' and 'generic' fibres of our family.*

PROOF. The proof is entirely algebraic, and has little to do with complex-analytic geometry, or even characteristic zero for that matter: set $R=\mathcal{O}_{Y,0}=\mathbb{C}\{v\}$, $A=\mathcal{O}_{X,0}$, a reduced $R$-algebra (by the map $f^*:R\to A$ corresponding to $f$) and let $\bar{A}$ be the integral closure of $A$ in its total ring of quotients. Thus, $\bar{A}=\mathcal{O}_{\bar{X},n^{-1}(0)}$. Then, since $\bar{A}$ is a Krull domain, we can apply the results of [Bourbaki, Algèbre commutative, VII, §1.6 Prop. 10 and V, §2.1, Cor. 2]: since by our assumptions $\mathcal{O}_{X_0,0}=A/v\cdot A$ is a reduced 1-dimensional ring, we have that $v\cdot A=\mathfrak{p}_1\cap\cdots\cap\mathfrak{p}_r$ with $\mathfrak{p}_i$ prime ideals such that $\dim A/\mathfrak{p}_i=1$. Furthermore, for each $\mathfrak{p}_i$, there exists a prime ideal $\mathfrak{p}'_i$ in $\bar{A}$ such that $A/\mathfrak{p}'_i$ is of dimension 1 and $\mathfrak{p}'_i\cap A=\mathfrak{p}_i$, and furthermore that $v\cdot\bar{A}\subset\mathfrak{p}'_1\cap\cdots\cap\mathfrak{p}'_r$. Hence we see that $v\cdot\bar{A}\cap A\subseteq v\cdot A$ and since the other inclusion is obvious, we have $v\cdot\bar{A}\cap A=v\cdot A$. Now we need:

LEMMA (Universal property of the normalization). *Let A be a reduced ring and let* $\bar{A}$ *be its integral closure in its total ring of fractions. Assume that the* conductor *of* $\bar{A}$ *in A,* $\mathfrak{c}=\{d\in\bar{A}\mid d\cdot\bar{A}\subseteq A\}$ *is not zero. Then, for a mapping*

*$\varphi: A \to B$ where B is a reduced ring integrally closed in its total quotient ring, and $\varphi$ such that $\varphi(\mathfrak{c})$ contains a nonzero divisor of B, there exists a unique extension $\bar{\varphi}$ of $\varphi$ to $\bar{A}$.*

$$\begin{array}{ccc} & \bar{A} & \\ \nearrow & 0 & \searrow^{\bar{\varphi}} \\ A & \xrightarrow[\varphi]{} & B \end{array}$$

PROOF. Choose $d \in \mathfrak{c}$ such that $\varphi(d)$ is not a zero-divisor in Tot $(B)$, and set $\bar{\varphi}(a) = (\varphi(d \cdot a)/(\varphi(d)) \in \text{Tot}(B)$, which has a meaning, since $1 \in \bar{A} \Rightarrow \mathfrak{c} \subset A$, and $d \cdot a \in A$.

$\bar{\varphi}(a)$ is integral over $\varphi(A)$ since $a$ is integral over $A$, hence $\bar{\varphi}(a)$ is integral over $B$, so it is in $B$ by our assumption. Uniqueness is obvious.

Let us apply this to the composed map

$$A \longrightarrow A/v \cdot A \longrightarrow \overline{A/v \cdot A}.$$

We have that $\bar{A}/A$, which is supported by the singular locus of our surface $X$, is an $R$-module of finite type, since by our assumptions, the singular locus of $X$ is finite over $(\mathbb{C}, 0) = Y$ in view of the simplicity theorem of §2, because $f$ is flat, and with non-singular fibre at every point of $X_0 - \{0\}$, for a small enough representative. It then follows from the Weierstrass preparation theorem (see Hironaka's lectures) that any module supported by $|\text{Sing}\, X|$ is of finite type as an $R$-module. So $\bar{A}/A$ and also $A/\mathfrak{c}$ are $R$-modules of finite type.

From this we deduce that it is impossible that $\mathfrak{c} \subset \mathfrak{p}_i$ for some $i$, otherwise we would have a surjection $A/\mathfrak{c} \to A/\mathfrak{p}_i$ and $\bar{A}/\mathfrak{p}_i$ is definitely not an $R$-module of finite type since its fibre over 0 is itself because $v \in \mathfrak{p}_i$, and it is of dimension 1, hence certainly not a finite dimensional vector space, in view of Hilbert's nullstellensatz. By a useful lemma [cf. J. P. Serre, Algèbre locale et multiplicités, chap. I, Prop. 2 (old edition: Springer Lect. Notes No. 11)] we have that $\mathfrak{c} \not\subset \mathfrak{p}_1 \cup \cdots \cup \mathfrak{p}_r$, so the image of $\mathfrak{c}$ in $A/v \cdot A$ or $\bar{A}/v \cdot \bar{A}$ contains a nonzero divisor, and we conclude that the map $A \to \overline{A/v \cdot A}$ factors through $\bar{A}$, and since $v$ goes to zero, it factors in fact through $\bar{A}/v \cdot \bar{A}$ so we have a commutative diagram:

$$\begin{array}{ccccc} & & \bar{A}/v \cdot \bar{A} & & \\ & {}^{i}\nearrow & & \searrow^{\bar{\varphi}} & \\ A \longrightarrow & A/v \cdot A & \hookrightarrow & & \overline{A/v \cdot A} \end{array}$$

We know that $i$ is injective, since its injectivity amounts to: $v \cdot \bar{A} \cap A = v \cdot A$. Now let us show that $\bar{\varphi}$ is injective: to construct our factorization according to the above lemma, we have chosen an element $d \in \mathfrak{c} - \bigcup_{i=1}^{r} \mathfrak{p}_i$.

An element $a \in \bar{A}$ such that $\bar{\varphi}(a) = 0$ must be such that $d \cdot a \in v \cdot A$, but since $d \notin \cup \mathfrak{p}_i$, it means $a \in \mathfrak{p}'_1 \cap \cdots \cap \mathfrak{p}'_r$ hence $a \in v \cdot \bar{A}$, which shows that $\bar{\varphi}$ is injective. Hence we obtain the equality:

$$\dim_{\mathbb{C}} \overline{A/v \cdot A}/A/v \cdot A = \dim_{\mathbb{C}} \overline{A/v \cdot A}/\bar{A}/v \cdot \bar{A} + \dim_{\mathbb{C}} \bar{A}/v \cdot \bar{A}/A/v \cdot A$$

since $A/v \cdot A = \mathcal{O}_{X_0,0}$, the left-hand side is $\delta(X_0, 0)$ and now we remark that $v \cdot \bar{A} \cap A = v \cdot A$ implies that $\bar{A}/A$ is an $R$-module without torsion. Since $R$ is a discrete valuation ring, this implies that $\bar{A}/A$ is a (locally) free $R$-module of finite type, and its rank is necessarily $\delta(X_y)$ since for $y \neq 0$ small enough, $(\bar{X})_y = p^{-1}(y)$ is non singular [by Bertini's theorem because $\bar{X}$, being a normal surface, has only isolated singularities], and $(\bar{X})_y \to X_y$ is the normalization.

Since $\bar{A}/A$ is a locally free $R$-module of rank $\delta(X_y)$, we have

$$\dim_{\mathbb{C}} \bar{A}/v \cdot \bar{A}/A/v \cdot A = \dim_{\mathbb{C}} \bar{A}/A \bigotimes_R R/v \cdot R = \delta(X_y)$$

where of course $R/v \cdot R = \mathbb{C}$, and finally, $\overline{A/v \cdot A}$ is also the integral closure of $\bar{A}/v \cdot \bar{A}$ in its total ring of fractions, so that

$$\dim_{\mathbb{C}} \overline{A/v \cdot A}/\bar{A}/v \cdot \bar{A} = \delta((\bar{X})_0).$$

This shows the equality $\delta((\bar{X})_0) = \delta(X_0) - \delta(X_y)$ and on the way we have seen that $\bar{A}/A$ is flat over $R$, and since $A$ is by assumption flat over $R$, this is in fact equivalent to the flatness of $\bar{A}$ as an $R$-module, as follows: We have the following exact sequence:

$$\mathrm{Tor}_1^R(\bar{A}, \mathbb{C}) \xrightarrow{i} \mathrm{Tor}_1^R(\bar{A}/A, \mathbb{C}) \longrightarrow A \bigotimes_R \mathbb{C} \xrightarrow{j} \bar{A} \bigotimes_R \mathbb{C} \longrightarrow \bar{A}/A \bigotimes_R \mathbb{C} \longrightarrow 0$$

coming from the exact sequence $0 \to A \to \bar{A} \to \bar{A}/A \to 0$ and since $v \cdot \bar{A} \cap A = v \cdot A$, $j$ is injective so $i$ is surjective and (see Hironaka's lectures: appendix prop. 2) we see that $\bar{A}/A$ is a flat $R$-module $\Leftrightarrow \bar{A}$ is a flat $R$-module.

COROLLARY 1. *Let* $f: (X, 0) \to (\mathbb{C}, 0)$ *be a germ of* flat *mapping, the fibre of which is reduced of dimension* 1. *The following are equivalent:*

(i) *$f$ can be described by a deformation of a parametrization of its fibre* $(X_0, 0) = (f^{-1}(0), 0)$, *in the sense explained at the beginning of this paragraph;*

(ii) *the normalization $\bar{X}$ of $X$ is non singular for a small enough representative of f, and the composite map $\bar{X} \to X \to \mathbb{C}$ is a submersion of non-singular spaces. Furthermore, for any $y \in \mathbb{C}$ (i.e. in a small disk around* 0) *the induced map* $(\bar{X})_y \to X_y$ *is the normalization of the curve* $X_y$;

(iii) *for any small enough representative of f we have* $\delta(X_y) = \delta(X_0)$ *for any* $y \in \mathbb{C}$, *i.e. the family of curves has '$\delta$ constant'.*

PROOF. (i)$\Leftrightarrow$(ii) has just been seen. (ii)$\Leftrightarrow$(iii) follows from the proposition in view of the fact that $\bar{X}\to\mathbb{C}$ is flat, and its fibre is non-singular if and only if $\delta((\bar{X})_0)=0$, (use the simplicity theorem of §2).

COROLLARY 2. *Let $X\subset\mathbb{C}\times\mathbb{P}^N$ be a closed complex subspace such that the projection $pr_1:\mathbb{C}\times\mathbb{P}^N\to\mathbb{C}$ induces on $X$ a flat mapping $f:X\to\mathbb{C}$, the fibres of which are reduced projective connected curves in $\mathbb{P}^N$ (and $\mathbb{C}$ stands for a representative of $(\mathbb{C},0)$, as usual). Then the following conditions are equivalent:*

*(i) the family has 'simultaneous normalization', i.e. again the normalization $\bar{X}$ of $X$ is a non-singular surface, and the composed mapping $\bar{X}\to X\to\mathbb{C}$ is a differentiable fibration (since it is proper and submersive) the fibre $(\bar{X})_y$ of which over $y\in\mathbb{C}$ is the normalization of $X_y$;*

*(ii) $g(X_y)-(r_y-1)$ is constant ($y\in C$), where $g(X_y)$ is the geometric genus of $X_y$ and $r_y$ is the number of its irreducible components – (equivalently: the topological Euler characteristic $\chi_{\text{top}}(\bar{X}_y)$ of $\bar{X}_y$ is constant).*

The geometric (or effective) genus of a reduced projective curve is the genus of its normalization, which is a non-singular projective curve. It is a birational invariant, and also a topological invariant of the normalization, since $2g(\bar{X}_y)=b_1(\bar{X}_y)$. Since $r_y=b_0(\bar{X}_y)$ is also a topological invariant, it is clear that (i)$\Rightarrow$(ii).

To prove that (ii)$\Rightarrow$(i), we use the general genus formula (see H. Hironaka: On the arithmetic genera and the effective genera of algebraic curves, Memoirs College of Science Univ. of Kyoto, series A vol. XXX No 2 (1957) and J. P. Serre: Groupes algébriques et corps de classe, chap. IV §1, 2 Hermann éditeur, Paris (1959). The genus formula states that

$$p_a(X_y)=g(X_y)-(r_y-1)+\sum_{x\in X_y}\delta(X_y,x)$$

where $p_a(X_y)$ is the arithmetic genus of $X_y$, which is defined by $1-p_a(X_y)=\chi(X_y)$ where $\chi(X_y)$ is the constant term of the Hilbert polynomial of $X_y\subset\mathbb{P}^N$, which is also the Euler characteristic $\sum_{i=0}^{2}(-1)^i\dim H^i(X_y,0_{X_y})=\chi(0_{X_y})$(here $H^i(X_y,0_{X_y})=0$, $i>2$). This last definition has the consequence that $p_a(X_y)$ is constant in a flat family of curves (it is perhaps easier to see that the Hilbert polynomial of the fibres of a flat family of projective varieties is constant). We obtain in this way that the assumption of (ii) implies that $\sum_{x\in X_y}\delta(X_y,x)$ is constant. We can now localize near each one of the singular points of the special fibre $X_0$ say, and then using the semi-continuity of the invariant $\delta$ which is implied by the proposition, we obtain that the assumption of (ii) implies that each singular point of $X_0$ has a neighborhood in $X$ which satisfies the equivalent conditions of corollary 1, whence we get simultaneous normalization, since it is a local property of $X$.

REMARK. A proof in abstract algebraic geometry (characteristic $\neq 0$ for example) would require that we find a hyperplane $H$ in $\mathbb{P}^N$ such that all the singular points of the fibers $X_y$ lie outside $H$, hence in an affine open subset of $X_y$, and then repeating the argument of the proof of the proposition in this relatively affine situation (of course such a hyperplane always exists).

3.4. We are now ready to give our geometric interpretation of the invariant $\delta(X_0)$ of a *plane* curve as the maximum number of singular points which one can pile up in the same fibre of a deformation of $X_0$, meaning of course an arbitrarily small deformation. We consider $X_0$ either as a sufficiently small representative of a germ of plane curve, or as an affine plane curve, but since all our constructions and invariants are of a local nature on $X_0$, it is sufficient to treat the first case.

Let $n:\bar{X}_0 \to X_0$ be the normalization of the reduced curve $X_0$, and consider the closed complex subspace $\bar{X}_0 \times_{X_0} \bar{X}_0 \subset \bar{X}_0 \times \bar{X}_0$, defined by a sheaf of ideals $\mathcal{I}_0$ on the non-singular surface $\bar{X}_0 \times \bar{X}_0$. We will denote by $I_0$ the multi-germ of $\mathcal{I}_0$ along the finite set $|n^{-1}(0) \times n^{-1}(0)|$.

Let $\delta'(X_0, 0)$ be the maximum number of singular points which can occur in a fibre of an arbitrary small representative of a deformation $f:(X, 0) \to (Y, 0)$ of $X_0$ [$f$ is flat and $(f^{-1}(0), 0) = (X_0, 0)$]. Remark that if $X_y$ is such a fibre having $\delta'(X_0, 0)$ singular points, since the invariant $\delta$ is an integer, $\delta'(X_0, 0) \leqq \delta(X_y)$, and it follows from the proposition above that $\delta(X_y) \leqq \delta(X_0, 0)$, so that we know that $\delta'(X_0, 0) \leqq \delta(X_0, 0)$. We are now going to prove two statements at once:

COROLLARY 3. *For a germ of a reduced plane curve* $(X_0, 0)$:

(1) $\delta'(X_0, 0) = \delta(X_0, 0)$;

(2) $I_0 = K_0 \cdot \mathfrak{N}_0$ *where* $K_0$ *is invertible and* $\mathfrak{N}_0$ *defines a germ of a subspace of* $\bar{X}_0 \times \bar{X}_0$ *contained in* $|n^{-1}(0) \times n^{-1}(0)|$, *and:*

(3) $e(\mathfrak{N}_0) = \dim_{\mathbb{C}} (\mathcal{O}_{\bar{X}_0 \times \bar{X}_0, n^{-1}(0) \times n^{-1}(0)}/\mathfrak{N}_0) = 2 \cdot \delta(X_0, 0)$.

Here $e(\mathfrak{N}_0)$ is the multiplicity in the sense of Samuel of the ideal $\mathfrak{N}_0$ in the semi-local algebra in which it lives and the first equality in (3) is due to the fact that $\mathfrak{N}_0$ is generated by a regular sequence [see J. P. Serre, Algèbre locale, Multiplicités, Springer Lect. Notes, No. 11].

COMMENT. We will see that $K_0$ is in fact the ideal defining the diagonal $\bar{X}_0 \subset \bar{X}_0 \times \bar{X}_0$, and $e(\mathfrak{N}_0)$ in a way measures the difference between $\bar{X}_0 \times_{X_0} \bar{X}_0$ and this diagonal.

PROOF. We call $z_1$ and $z_2$ generators of the maximal ideal of $\mathcal{O}_0 = \mathcal{O}_{X_0,0}$, and the normalization is described algebraically by $\mathcal{O}_0 \subset \prod_{j=1}^r \mathbb{C}\{t_j\}$ given by

$(z_1(t_j))$ and $(z_2(t_j))$ as explained above. We have $\mathcal{O}_{\bar{X}_0 \times \bar{X}_0, n^{-1}(0) \times n^{-1}(0)} = \prod_{i,j} \mathbb{C}\{t_i, t_j'\}$ and to describe an ideal in such a product it is sufficient to describe the ideal it induces in the $(i, j)$th factor $\mathbb{C}\{t_i, t_j'\}$. For example, we see that our ideal $I_0$, in view of the definition of the fibre products, is described by the:

$$I_{0,i,j} = (z_1(t_i) - z_1(t_j'), z_2(t_i) - z_2(t_j'))\mathbb{C}\{t_i, t_j'\}.$$

We can define $K_0$ and $\mathfrak{N}_0$ by:

$$K_{0,i,i} = (t_i - t_i')\mathbb{C}\{t_i, t_i'\} \quad \text{and if} \quad i \neq j,\ K_{0,i,j} = \mathbb{C}\{t_i, t_j'\},$$

$$\mathfrak{N}_{0,i,i} = \left(\frac{z_1(t_i) - z_1(t_i')}{t_i - t_i'}, \frac{z_2(t_i) - z_2(t_i')}{t_i - t_i'}\right)\mathbb{C}\{t_i, t_i'\} \quad \text{and if} \quad i \neq j$$

$$\mathfrak{N}_{0,i,j} = (z_1(t_i) - z_1(t_j'), z_2(t_i) - z_2(t_j'))\mathbb{C}\{t_i, t_j'\}.$$

Clearly $K_0$ is invertible, and $I_0 = K_0 \cdot \mathfrak{N}_0$. Now we have the:

LEMMA. *Let $(X_0, 0) \subset (\mathbb{C}^2, 0)$ be a germ of reduced plane curve described parametrically by $(z_1(t_j))$ and $(z_2(t_j)) \in \prod_{j=1}^r \mathbb{C}\{t_j\}$. There exists a nowhere dense closed complex surface $B \subset (\mathbb{C}^2)^r$ such that if $((\alpha_j, \beta_j)\ 1 \leqq j \leqq r) \in (\mathbb{C}^2)^r - B$, the plane curve $X$ described parametrically by $(z_1(t_j) - \alpha_j v t_j,\ z_2(t_j) - \beta_j v t_j)$ has only ordinary double points as singularities (in a neighborhood of $0 \in \mathbb{C}^2$) for $v \neq 0$ sufficiently small.*

PROOF. Consider the graph $\Gamma \subset (\coprod_{j=1}^r (\mathbb{C}, 0)) \times (\mathbb{C}^2, 0)$ of the map $\coprod_{j=1}^r (\mathbb{C}, 0) \to (\mathbb{C}^2, 0)$ of which $X_0$ is the image. $\Gamma$ is a non-singular curve, and the parametric description above is nothing but that of the projection of $\Gamma$ to $\mathbb{C}^2$ parallel to the multi-direction with vectors $(1, \alpha_j v, \beta_j v)$. It is a well-known result that a general plane projection of a non-singular curve has only ordinary double points as singularities, 'general' meaning: for all directions of projection except those in a nowhere dense closed subspace of the space of directions of projection.

This proves the lemma.

Remark that the lemma in fact provides as with a 1-parameter deformation of $(X_0, 0)$ (see 3.5.4 below) such that all the fibres except $(X_0, 0)$ have only ordinary double points. Furthermore, this deformation is – by construction – obtained by a deformation of the parametrization, hence by corollary 1, we have $\delta(X_v) = \delta(X_0, 0)$. Since, as is readily checked, the invariant $\delta$ of an ordinary double point is 1, we see that $X_v$ must have exactly $\delta(X_0, 0)$ ordinary double points. Hence $\delta'(X_0, 0) \geqq \delta(X_0, 0)$ and since we know the other inequality, we get the equality.

Let us now see how $\mathfrak{N}_0$ varies in this deformation: we proceed as follows: consider the ideal $I$ in $\bar{X} \times_{\mathbb{C}} \bar{X}$ defining the closed subspace $\bar{X} \times_X \bar{X} \subset \bar{X} \underset{\mathbb{C}}{\times} \bar{X}$, where $(X, 0)\to(\mathbb{C}, 0)$ is the family constructed in the lemma and $n:\bar{X}\to X$ is the normalization. By corollary 1, we know that for any $v$, $(\bar{X})_v = \bar{X}_v$, so $I$ is the family of the ideals $I_v$ corresponding to the fibre $X_v$. We look at the multi-germs along $|n^{-1}(0)\times n^{-1}(0)|$ and get:

$$\mathcal{O}_{\bar{X}\times_{\mathbb{C}}\bar{X},n^{-1}(0)\times n^{-1}(0)} = \prod_{i,j}\mathbb{C}\{v, t_i, t_j'\}$$

and again $I = K\cdot\mathfrak{N}$ where

$$\mathfrak{N}_{i,i} = \left(\frac{z_1(t_i)-z_1(t_i')}{t_i-t_i'} - \alpha_i v, \frac{z_2(t_i)-z_2(t_i')}{t_i-t_i'} - \beta_i v\right)\mathbb{C}\{v, t_i, t_i'\}$$

and if $j\neq i$

$$\mathfrak{N}_{i,j} = (z_1(t_i)-z_1(t_j')-(\alpha_i t_i-\alpha_j t_j')v,\ z_2(t_i)-z_2(t_j')-(\beta_i t_i-\beta_j t_j')v)\mathbb{C}\{v, t_i, t_j'\}.$$

It is easily checked that the natural injection

$$\mathbb{C}\{v\}\to\prod_{i,j}\mathbb{C}\{v, t_i, t_j'\}/\mathfrak{N}$$

makes the quotient a $\mathbb{C}\{v\}$-module without torsion, i.e. flat (essentially because $\mathfrak{N}$ is generated by a regular sequence) and hence the dimension of the fibres is independent of $v$ (Bourbaki, Algèbre Commutative, chap. II) and thus, taking small representatives of $X$, $\bar{X}$ etc. we find that $e(\mathfrak{N}_v) = e(\mathfrak{N}_0)$ for any sufficiently small value of $v$. Now, we remark that $e(\mathfrak{N}_v)$ is the sum of the $e(\mathfrak{N}_v, x_i)$ corresponding to the various singular points of $X_v$, and we give the:

Exercise. Describe parametrically an ordinary double point of plane curve, and check that in this case, $e(\mathfrak{N}_0) = 2$.

Since $X_v$ has, for $v\neq 0$, $\delta(X_0, 0)$ ordinary points, we obtain $e(\mathfrak{N}_0) = e(\mathfrak{N}_v) = 2\cdot\delta(X_0, 0)$.

Remarks. (1) The lemma above is a parametric version of the Morse lemma which states that if $f:(\mathbb{C}^{n+1}, 0)\to(\mathbb{C}, 0)$ has an isolated critical point at 0, by an arbitrary small generic linear perturbation $f+\sum_0^n \alpha_i z_i$ of $f$, we obtain a new function which has only $\mu$ quadratic non-degenerate critical points with distinct critical values, where

$$\mu = \dim_{\mathbb{C}}\mathbb{C}\{z_0, \cdots, z_n\}\Big/\left(\frac{\partial f}{\partial z_0}, \cdots, \frac{\partial f}{\partial z_n}\right).$$

For $n = 1$, i.e. plane curves, we see that, while $\mu$ is the maximum number of critical points which one can *spread out* by a small perturbation of $f$, $\delta$ is the maximum number of critical points which one can *pile up* (in the same fibre) by a small perturbation of $f$. We shall see more about this below.

(2) The consideration of the ideal $\mathfrak{N}_0$ defined above, together with its normalized blowing up, and its deformations given by the lemma, provides a wealth of information about the numerical invariants of $X_0$. [See: F. Pham and B. Teissier, Fractions lipschitziennes d'une algèbre analytique complexe, Centre de Maths. de l'Ecole Polytechnique, Juin 1969, also: F. Pham in Congrès International des Mathématiciens, Nice 1970]. In particular, one can use it to prove algebraically a well-known relation between $\mu$ and $\delta$

$$\underline{2\delta = \mu + r - 1}$$

($r$ = number of irreducible components, or branches), given by Milnor in his book 'Singular points of complex hypersurfaces'. I will not give the proof here, referring to Milnor's book or J. J. Risler: Sur l'idéal jacobien d'une courbe plane, Bull. Soc. Math. Fr. (Risler gives an algebraic proof) but this equality will be used below.

3.5. On the images of parametrizations: the Fitting ideal as a prophet

Let $A$ be a regular local ring (for example $\mathbb{C}\{z_1, \cdots, z_N\}$) and let $M$ be an $A$-module of finite presentation. By the Hilbert Syzygy theorem, $M$ has a resolution of finite length (in fact of length $\leqq \dim A$), i.e. there exists an exact sequence of $A$-modules:

$$0 \longrightarrow L_p \longrightarrow \cdots \longrightarrow L_2 \longrightarrow L_1 \xrightarrow{\Psi_1} L_0 \longrightarrow M \longrightarrow 0$$

$p \leqq \dim A$, where each $L_i$ is a *free* $A$-module of finite type. The smallest integer $p$ such that there exists such a sequence is called the homological dimension of $M$ over $A$, and denoted $dh_A(M)$. Let us now define an $M$-sequence of elements of $A$ as a sequence $(a_1, \cdots, a_k)$ of elements of the maximal ideal of $A$ such that for any $j$, $1 \leqq j \leqq k$, $a_j$ is not a zero-divisor in the module $M/(a_1, \cdots, a_{j-1}) \cdot M$ where $(a_1, \cdots, a_{j-1}) \cdot M$ is the submodule $\sum_{k=1}^{j-1} a_k \cdot M$ (and 0 if $j = 1$). It turns out (see Serre's book on Local algebra) that all the maximal (in the obvious sense) $M$-sequences have the same length, called the *depth* of $M$, and that:

$$\underline{dh_A(M) + \operatorname{depth}_A(M) = \dim A.}$$

Application. Suppose we can check that $\operatorname{depth}_A M \geqq \dim A - 1$ and that $\operatorname{Ann}_A M \neq 0$. Then $F_0(M)$ is generated by 1 element, which is not zero.

Proof. By the above equality, $dh_A M \leqq 1$, hence we can find a resolution

$$0 \longrightarrow A^q \xrightarrow{\Psi} A^p \longrightarrow M \longrightarrow 0.$$

We see that $q \leqq p$, but since $\operatorname{Ann}_A M \neq 0$ we have $M \bigotimes_A K = 0$ where $A \to K$

is the field of fractions of $A$. This implies that $\Psi \bigotimes 1_K : K^q \to K^p$ is surjective, hence $q \geqq p$. So we must have $p = q$ and $F_0(M) = (\det \underline{\Psi}) \cdot A$, where $\underline{\Psi}$ is a matrix representing $\Psi$.

3.5.1. Suppose now that we parametrize a curve in $(\mathbb{C}^2, 0)$ by $x(t)$, $y(t)$, both $\neq 0$, thus making $\mathbb{C}\{t\}$ a $\mathbb{C}\{x, y\}$-module of finite representation. Clearly, $x(t)$ (for example) is not a zero-divisor in $\mathbb{C}\{t\}$, so the depth of $\mathbb{C}\{t\}$ as $\mathbb{C}\{x, y\}$-module is $\geqq 1$. On the other hand, we know our curve has an equation, so that $\operatorname{Ann} \mathbb{C}\{t\} \neq 0$, hence we know that $F_0(\mathbb{C}\{t\})$ is a principal ideal in $\mathbb{C}\{x, y\}$, and a generator is what we call an equation of the image curve. If we have arranged that for a given 'general' point $(x_0, y_0)$ on the image curve, the equations $x(t) = x_0$, $y(t) = y_0$ have only one solution then we know even our equation will be reduced, that is, will be a prime element in $\mathbb{C}\{x, y\}$. (Compare all this with the example given in §1.)

3.5.2. Suppose now that we consider the same plane curve but *as lying in* $\mathbb{C}^3$, that is, the curve in $\mathbb{C}^3$ given by $x = x(t)$, $y = y(t)$, $z = 0$. Then, let us compute the Fitting ideal of $\mathbb{C}\{t\}$ as $\mathbb{C}\{x, y, z\}$-module. Take for example $x(t) = t^3$, $y(t) = t^4$. Then $\mathbb{C}\{t\}$ is generated as $\mathbb{C}\{x, y, z\}$-module by $e_0 = 1$, $e_1 = t$, $e_2 = t^2$ and it is not difficult to see that the relations are described by the matrix

$$\underline{\Psi} = \begin{bmatrix} y & -x & 0 \\ 0 & y & -x \\ x^2 & 0 & -y \\ z & 0 & 0 \\ 0 & z & 0 \\ 0 & 0 & z \end{bmatrix}$$

i.e. $\underline{\Psi}$ is the matrix of a presentation

$$\mathbb{C}\{x, y, z\}^6 \xrightarrow{\Psi} \mathbb{C}\{x, y, z\}^3 \longrightarrow \mathbb{C}\{t\} \longrightarrow 0$$

here $F_0(\mathbb{C}\{t\}) = (y^3 - x^4, z^3, zx^2, zy^2, z^2y, z^2x)$ so that the image of our curve computed by $F_0(\mathbb{C}\{t\})$ consists of the curve $y^3 - x^4 = 0$, $z = 0$, plus an extra 0-dimensional component sticking out of the $(x, y)$-plane. Of course, $\sqrt{F_0(\mathbb{C}\{t\})} = (y^3 - x^4, z)\mathbb{C}\{x, y, z\}$. If we had computed $F_0(\mathbb{C}\{t\})$ for a curve which really lies in $\mathbb{C}^3$, such as $x = t^4$, $y = t^6$, $z = t^7$ which is a complete intersection with ideal: $(y^2 - x^3, z^2 - x^2y)\mathbb{C}\{x, y, z\}$, we would similarly have found $F_0(\mathbb{C}\{t\}) = ((y^2 - x^3)\varphi_1, (z^2 - x^2y)\varphi_2)\mathbb{C}\{x, y, z\}$ where $\sqrt{\varphi_1} = \sqrt{\varphi_2} = (x, y, z)\mathbb{C}\{x, y, z\}$. In general, given a morphism $\mathbb{C}\{z_1, \cdots, z_N\} \to \mathbb{C}\{t\}$ corresponding to a parametric representation of a curve, $F_0(\mathbb{C}\{t\})$ will define a curve, in the sense that $\mathbb{C}\{z_1, \cdots, z_N\}/F_0(\mathbb{C}\{t\})$ is purely 1-dimensional, only if $N=2$ or if $\min v(z_i(t))=1$, which is the case where our germ of curve is non-singular.

3.5.3. Let us now turn to the following problem: given again a parametrization of a curve in $\mathbb{C}^N$, let us consider the generalization of the construction made above for plane curves, namely, for general values of $(\alpha_i)\in\mathbb{C}^N$, the algebra $A=\mathbb{C}\{v, z_1(t)+\alpha_1 vt, \cdots, z_N(t)+\alpha_N vt\}\subset\mathbb{C}\{v, t\}$. It is not difficult to see that if $N\geqq 3$, for general values of $\alpha_i$, the curve thus described for each value of the parameter $v$ is *non-singular* for $v\neq 0$. This might seem to contradict the proposition proved above: we apparently have a family of curves given by a deformation of the parametrization, such that $\delta(X_0)>0$ (we have of course chosen our curve to be singular) but $\delta(X_v)=0$ for $v\neq 0$. What happens here is that $A/v\cdot A$ is not a purely 1-dimensional ring if $N\geqq 3$, again it has an imbedded component. Setting $\bar{A}=\mathbb{C}\{v, t\}$, it amounts to the fact that we do *not* have $v\cdot\bar{A}\cap A=v\cdot A$. Therefore $A$ does *not* describe a deformation of our original curve, if $N\geqq 3$, and the proposition above is not contradicted. [Remark that to say that $v\cdot\bar{A}\cap A=v\cdot A$ amounts to saying that $A/v\cdot A$ can be computed by setting $v=0$ in $A$.]
In order to see this, take for example $A=\mathbb{C}\{v, t^4+\alpha vt, t^6+\beta vt, t^7+\gamma vt\}$ in $\bar{A}=\mathbb{C}\{v, t\}$, and look at the element $(t^6+\beta vt)^2-(t^4+\alpha vt)^3$: it is in $v\cdot\bar{A}\cap A$ but *not* in $v\cdot A$. The choice of this element was dictated by the fact that $y^2-x^3$ lies in the annihilator of $\mathbb{C}\{t\}$ (as module over $\mathbb{C}\{x, y, z\}$ via $x=t^4$, $y=t^6$, $z=t^7$) but does not lie in the Fitting ideal $F_0(\mathbb{C}\{t\})$. In fact, the Fitting ideal *predicts*, by its imbedded components, before any deformation is made, that such 'accidents' will happen when deforming the parametrization.

3.5.4. When $N=2$, however, we can apply the results of the beginning of 3.5, and check that $F_0(\mathbb{C}\{v, t\})$ is a principal ideal. By the compatibility of the Fitting ideal with base change, the fibre over $v=0$ of the hypersurface in $\mathbb{C}\times\mathbb{C}^2$ thus described is precisely our original plane curve by 3.5.1. Hence in this case $\mathbb{C}\{v, x(t)+\alpha vt, y(t)+\beta vt\}$ does describe a flat family of curves and the argument used above in the proof of corollary 3 is indeed valid.

### 3.6. The Newton polygon of a plane curve

We do not leave the subject of parametrization, or prophecy for that matter, since the Newton polygon construction gives information in reverse of the Fitting ideal: given an equation $f(x, y)=0$ for a plane curve, it allows us to predict at least the *ratio* of the smallest exponents appearing in a parametric representation of our curve, that is, which ratios $v(x(t_j))/v(y(t_j))$ we get in the parametric representation $(x(t_j), y(t_j))$ ($v$ is the order in $t_j$).

Take $f \in \mathbb{C}\{x, y\}$, assume for simplicity $f(0, y) \not\equiv$ and $f(x, 0) \not\equiv 0$. Write $f = \Sigma c_{ij} x^i y^j$ and set $\Delta(f) = \{(i, j) \subset \mathbb{N}^2 / c_{ij} \neq 0\}$. The Newton polygon $\mathfrak{N}(f)$ (in the coordinates $x$, $y$) is the convex hull of $\Delta(f)$ in the following sense: $\mathfrak{N}(f)$ is a convex polygon with two sides of infinite length and a finite number of edges of finite length, and a line $l$ in $\mathbb{R}^2$ contains an edge of finite length of $\Delta(f)$ if and only if contains at least 2 points of $\Delta(f)$, and there are no points of $\Delta(f)$ on the same side of $l$ as $0 \in \mathbb{R}^2$.

Since we have assumed $f(0, y) \not\equiv 0$ we can, by the Weierstrass preparation theorem, assume, up to multiplication by an invertible element $U \in \mathbb{C}\{x, y\}$, which does not change the Newton polygon, that $f = y^m + a_{m-1}(x) y^{m-1} + \cdots + \cdots + a_0(x)$. It is clear that $\mathfrak{N}(f)$ is also the convex hull of the finite set $\{(v(a_k(x)), k)\} \subset \mathbb{N}^2$. Let us consider $f$ as an element of $\mathbb{C}\{\{x\}\}[y]$, where $\mathbb{C}\{\{x\}\}$ is the field of 'meromorphic functions', valued by the order in $x$ (whether $\geqq 0$ or $<0$). I think it is clear what a valued extension of a valued field is: $(K, v) \subset (L, w)$ means $w \mid K = v$. (Note that $w$ can take values in a subgroup of $\mathbb{Q}$ *isomorphic* to $\mathbb{Z}$, such as $1/k \cdot \mathbb{Z}$.)

3.6.1. We now describe a formalism on Newton polygons which will simplify matters later. We think of a Newton polygon as the following picture

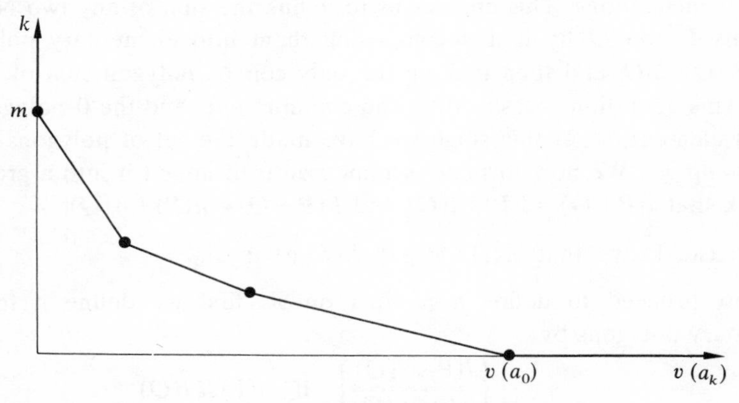


Figure 1

the coordinates of the vertices being integers. We define an *elementary* polygon as one having only one edge of finite length. An elementary polygon $P$ has a height $h(P)$, length $l(P)$ and inclination $i(P)$ as shown in Fig. 2.

We count the polygon consisting of the two coordinate axis as the 0-polygon. Its inclination is not defined. For any Newton polygon, we can still define $h(P)$ and $l(P)$. We now define the *sum* of two elementary polygons

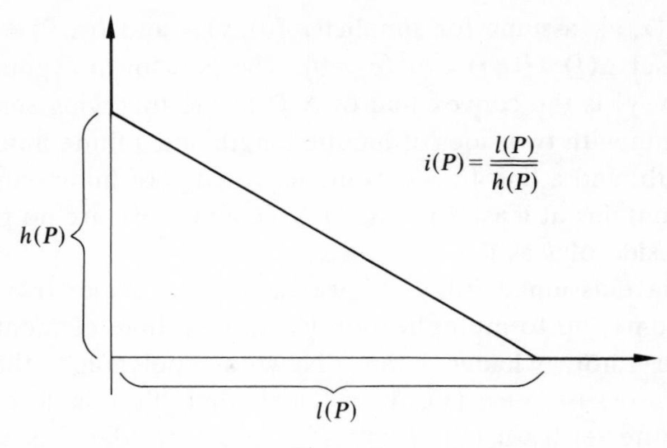


Figure 2

$P$, $Q$ as the only *convex* polygon having an edge of inclination $i(P)$ and length (of horizontal projection) $l(P)$, and an edge of inclination $i(Q)$ and length (of horizontal projection) $l(Q)$. If $i(P)=i(Q)$ then $P+Q$ is again elementary, of length $l(P)+l(Q)$ and height $h(P)+h(Q)$. We need a notation for an elementary polygon, and I propose $P=\{l(P)//h(P)\}$. We see that any Newton polygon has a decomposition as a (convex) sum of elementary polygons, and this decomposition is unique if we require that they all have different inclinations. This enables us to define the sum of any two Newton polygons $P$ and $Q$ by first decomposing them into elementary polygons, $P=\Sigma P_i$, $Q=\Sigma Q_j$ and then making the only convex polygon sum of all the $P_i$, $Q_j$. This operation is associative and commutative, and the 0-polygon is a neutral element, so at this stage we have made the set of polygons into a *semi-group* $\underline{\mathfrak{N}}_0$. We now formally symmetrize it to imbed it into a group $\underline{\mathfrak{N}}$. Remark that $l(P+Q)=l(P)+l(Q)$ and $h(P+Q)=h(P)+h(Q)$.

EXERCISE. Prove that $\mathfrak{N}(f)+\mathfrak{N}(g)=\mathfrak{N}(f\cdot g)$ in $\mathfrak{N}_0$.

We now proceed to define a *product* on $\underline{\mathfrak{N}}$: first we define it for two elementary polygons by

$$P*Q=\begin{cases}\left\{\dfrac{l(P)\cdot l(Q)}{l(P)\cdot h(Q)}\right\} & \text{if} \quad i(P)\leqq i(Q)\\[2ex] \left\{\dfrac{l(Q)\cdot l(P)}{l(Q)\cdot h(P)}\right\} & \text{if} \quad i(P)\geqq i(Q).\end{cases}$$

We remark that for elementary polygons, $i(P*Q)=\max(i(P), i(Q))$ and we now extend $*$ to all polygons by requiring distributivity, i.e. decomposing $P=\Sigma P_i$, $Q=\Sigma Q_j$ into elementary polygons, define:

$$P*Q=\sum_{i,j} P_i*Q_j$$

it is easy to see that $*$ is commutative and associative. Furthermore, define the set of *special Newton polygons* $\underline{\mathfrak{N}}_S \subset \underline{\mathfrak{N}}$ as those having only edges of inclination $\geqq 1$. $\underline{\mathfrak{N}}_S$ is stable under sum and $*$, and defining the unit Newton polygon to be $\mathbb{1} = \{1//1\}$, we have:

$$P \in \underline{\mathfrak{N}}_S \Leftrightarrow P * \mathbb{1} = P.$$

So that we have now endowed our symmetrized semi-group of Newton polygons $\underline{\mathfrak{N}}$ with a *commutative ring structure* and $\underline{\mathfrak{N}}_S \subset \underline{\mathfrak{N}}$ is even a commutative ring with unit, which we call the *ring of special Newton polygons.*

There are two other useful operations on $\underline{\mathfrak{N}}$:

(1) the symmetry $\sigma$ defined by

$$\sigma\left(\sum_i \left\{\frac{l(P_i)}{h(P_i)}\right\}\right) = \sum_i \left\{\frac{h(P_i)}{l(P_i)}\right\}.$$

(2) The horizontal expansion $\varepsilon$ defined by

$$\varepsilon\left(\sum_i \left\{\frac{l(P_i)}{h(P_i)}\right\}\right) = \sum_i \left\{\frac{l(P_i)+h(P_i)}{h(P_i)}\right\}.$$

Exercise. (1) Show that for any Newton polygons:

$$l(P * Q) = l(P) \cdot l(Q), \qquad h(P * P) = 2 \cdot S(P), \qquad l(P * P) = l(P)^2,$$

where $S(P)$ is the *area* between the polygon and the two axis.

(2) The application $\mathbb{Z} \to \underline{\mathfrak{N}}_S$ defined by $n \to \{n//n\}$ identifies $\mathbb{Z}$ with a subring of $\underline{\mathfrak{N}}_S$.

We can now state:

Theorem (Newton-Puiseux) of decomposition of a polynomial along the sides of its Newton polygon [see R. J. Walker, Algebraic curves, Dover books, and J. Dieudonné, Calcul infinitésimal, Hermann, Paris].

(1) *Let* $(L, w)$ *be a finite valued extension of the valued field* $(\mathbb{C}\{\{x\}\}, v)$ [*where* $v$ *is the order in* $x$, *whether* $\geqq 0$ *or* $<0$ *of an element of the field of fractions* $\mathbb{C}\{\{x\}\}$ *of* $\mathbb{C}\{x\}$, *and* $w$ *is a valuation of* $L$ *with values in a subgroup of* $\mathbb{Q}$ *isomorphic to* $\mathbb{Z}$ (i.e. $(1/d)\,\mathbb{Z}$ *for some* $d$) *and* $w \mid \mathbb{C}\{\{x\}\} = v$] *assume that* $f \in \mathbb{C}\{\{x\}\}[y]$ *has all its roots in* $L$, *call them* $y_1, \cdots, y_m$. *Then*

$$\underline{\mathfrak{N}(f) = \sum_{\rho \in \mathbb{Q}} \left\{\frac{m_\rho \cdot \rho}{m_\rho}\right\}}$$

*where* $m_\rho$ *is the number of roots* $y_k$ *of* $f$ *having valuation* $w(y_k) = \rho$. *Of course except for a finite number of values of* $\rho$, *the number so found is zero, so the sum on the right is finite.*

(2) *Any finite extension of* $\mathbb{C}\{\{x\}\}$ *is isomorphic with* $\mathbb{C}\{\{x^{1/d}\}\}$ *where d is the degree of the extension, and in particular, the algebraic closure of* $\mathbb{C}\{\{x\}\}$ *is* $\bigcup_{d\geq 1}\mathbb{C}\{\{x^{1/d}\}\}$.

(3) *Let* $\rho_1, \cdots, \rho_k$ *be the rational numbers such that* $m_{\rho_i}\neq 0$. *Then there is in* $\mathbb{C}\{\{x\}\}[y]$ *a decomposition* $f = f_{\rho_1}\cdots f_{\rho_k}$ *where each* $f_{\rho_i}$ *is of degree* $m_{\rho_i}$ *in y, all its roots have the same valuation* $\rho_i$, *and* $m_{\rho_i}\cdot\rho_i$ *is the order in x (or the valuation) of the constant term (in y) of* $f_{\rho_i}$.

TRANSLATION. Assume $f$ to be irreducible. Then all its roots in $y$ lie in $\mathbb{C}\{\{x^{1/m}\}\}$ where $m$ is the degree of $f$ in $y$ which means that the roots of $f(x, y)=0$ are convergent series in $x^{1/m}$ or equivalently that we can describe the curve $f(x, y)$ parametrically by

$$\begin{cases} x = t^m \\ y = \varphi(t) \quad \varphi \text{ a convergent power series.}\end{cases}$$

Finally we have that the valuation of $y$, which is $(1/m)\times$(order in $t$ of $\varphi(t)$) is uniquely determined by the equation

$$\left\{\frac{m\cdot\rho}{m}\right\} = \left\{\frac{e}{m}\right\}$$

where $\{e/\!/m\}$ is the Newton polygon of $f$, which is elementary in view of 3 above. So finally, we see that if $f$ is irreducible, its Newton polygon has only one edge, hence is elementary, say $\{e/m\}$ and from this we can predict that the parametric representation of our curve will be

$$\begin{cases} x = t^m \\ y = u(t)(t^e + c_{e+1}t^{e+1} + \cdots) \quad \text{with} \quad u(0)\neq 0.\end{cases}$$

Of course if $\mathfrak{N}(f)$ is elementary, we *cannot* conclude that $f$ is irreducible, but only that the ratios $e_k/m_k$ occurring in the parametrizations of the various branches of the curve $f(x, y)=0$ are equal.

EXERCISE. Convince yourself that at least the valuation $w(y_k)$ of a root of $f(x, y)=0$ in a valued extension is the inclination of an edge of $\mathfrak{N}(f)$.

3.6.2. One can also use the Newton polygon construction *in reverse;* namely, knowing a parametric representation of a plane curve, say $(x(t_j), y(t_j))$ in $\prod_{j=1}^r \mathbb{C}\{t_j\}$ we can immediately say something non-trivial about the equation $f=0$ of our plane curve, without having to compute the Fitting ideal;

$$\mathfrak{N}(f) = \sum_{j=1}^{r}\left\{\frac{v(y(t_j))}{v(x(t_j))}\right\} \quad (\text{where } v = \text{order in } t_j).$$

REMARKS. (1) The Newton Polygon depends very much upon the choice of the coordinates $(x, y)$ in which we expand $f$ and if we take a 'generic choice' of coordinates $(x, y)$, $\mathfrak{N}(f)=\{m//m\}$ where $m$ is order of $f$ at 0, so that $\mathfrak{N}(f)$ contains very little information about $f$. There are two ways in which the Newton polygon can be made to yield information: by taking coordinates 'as special as possible', basically asking that the inclination of $\mathfrak{N}(f)$ in these coordinates should be as large as possible: this is the theory of maximal contact of Hironaka, or alternately by looking only at those functions which are *well represented* by their Newton polygons in given coordinates: this approach is that of A. Kushnirenko (Inv. Math. *32*, 1 (1976)). Both theories were actually developed in arbitrary dimensions, and in Hironaka's theory of infinitely near points, the behaviour of Newton Polyhedra and polygons under some special modifications (permissible blowing ups) plays an important role.

Anyway, given a family of equations $f_v=\sum c_{ij}(v)x^iy^j$ with $c_{ij}(v)\in\mathbb{C}\{v\}$ we can observe that $\mathfrak{N}(f_0)$ lies *above* $\mathfrak{N}(f_v)$ simply because the coefficients $c_{ij}(v)$ can only vanish when $v=0$, and not suddenly become $\neq 0$, hence $\Delta(f_0)\subseteq\Delta(f_v)$, and the condition that $\mathfrak{N}(f_v)$ should be independent of $v$, say for some coordinate system, is a rather weak condition on the family of curves described by $f=0$, unless we assume that they are 'well represented' by their Newton polygon. However, this condition, as explained above, does contain some interesting information about the parametrizations of the curves in our family. We now turn to much stronger conditions.

### 3.7. A short summary of equisingularity for plane curves (after [6], [1])

One of the things we try to do here is to describe pertinent numerical invariants of the geometry 'up to equisingularity' of isolated singularities of hypersurfaces. The model is what happens for plane curves, where the situation is very agreeable. Let us first consider the case of a germ of an irreducible plane curve $(X_0, 0)$, defined by $f(x, y)=0$. Then, as we have seen we can obtain a parametric representation of our curve by $x=t^n$, $y=\varphi(t)$ and if we have chosen our coordinates so that of $f$ (which is the multiplicity of $X_0$), and that the $x$-axis is in the tangent cone of $X_0$, then $v(\varphi(t))>n$ and the topology of $(X_0, 0)$ *as an imbedded germ* in $(\mathbb{C}^2, 0)$ (i.e. up to a germ of a homeomorphism extending to $\mathbb{C}^2$) is completely determined by the integer $n$ and the values $v(\varphi(t)-\varphi(\omega t))$, $\omega$ running through the $n$th roots of unity. Call these values $\beta_1<\cdots<\beta_g$; then our curve is *topologically* equivalent to the curve described parametrically by

$$\begin{cases} x=t^n \\ y=t^{\beta_1}+t^{\beta_2}+\cdots+t^{\beta_g}. \end{cases}$$

The set of integers $(n, \beta_1, \cdots, \beta_g)$ is called the *characteristic* of the branch $(X_0, 0)$.

Define integers $l_i$ by $l_0 = n$, $l_1 = (n, \beta_1), \cdots, l_i = (l_{i-1}, \beta_i)$ and remark that since our branch is not multiple, $l_g = 1$. Now define $n_i$ by: $l_{i-1} = n_i \cdot l_i$: we have $n = n_1 \cdots n_g$, and finally define $m_i$ by $\beta_i = m_i \cdot l_i$. Then $(m_i, n_i) = 1$ $(1 \leqq i \leqq g)$ and we can check that any branch with characteristic $(n, \beta_1, \cdots, \beta_g)$ can be described by a *Puiseux expansion*

$$y = \sum_{i=0}^{k_0} a_{0,i} x^i + \sum_{i=0}^{k_1} a_{1,i} x^{(m_1 + i/n_1)} + \cdots + \sum_{i=0}^{k_2} a_{p,i} x^{(m_p + i/n_1 \cdots n_q)} + \\ + \cdots + \sum_{i=0}^{\infty} a_{g,i} x^{(m_g + i/n_1 \cdots n_g)}.$$

Another way to obtain the characteristics is this: consider $\mathcal{O}_0 = \mathbb{C}\{x(t), y(t)\} \subset \mathbb{C}\{t\}$. Then, the valuations of the elements of $\mathcal{O}_0$ form a semi-group $\Gamma \subset \mathbb{N}$, and $\Gamma = \mathbb{N} \Leftrightarrow 1 \in \Gamma \Leftrightarrow \mathcal{O}_0 = \mathbb{C}\{t\}$ i.e. $X_0$ is non-singular. One constructs a minimal system of generators $\bar{\beta}_0, \cdots, \bar{\beta}_{g'}$, (i.e. $\bar{\beta}_i \in \Gamma$ does not belong to the semi-group $\langle \bar{\beta}_0, \cdots, \bar{\beta}_{i-1} \rangle$ generated by the previous ones, and is the smallest non-zero element of $\Gamma$ with this property). Then, by a theorem (see [7]) one has that $g' = g$, the number of characteristic exponents, and furthermore, the $\beta_q$'s and $\bar{\beta}_q$'s are linked by the following relation: $\bar{\beta}_0 = \beta_0$, $\bar{\beta}_1 = \beta_1$

$$\bar{\beta}_q = n_{q-1} \cdot \bar{\beta}_{q-1} - \beta_{q-1} + \beta_q \qquad (1 < q \leqq g).$$

And so we see that the datum of $(n, \beta_1, \cdots, \beta_g)$ is equivalent to the datum of $\Gamma = \langle \bar{\beta}_0, \cdots, \bar{\beta}_g \rangle$.

Let us now consider a reducible, but reduced, (i.e. without nilpotent functions) plane curve, $(X_0, 0)$ with equation $f = 0$. Let us decompose $f$ in $\mathbb{C}\{x, y\}$, $f = f_1, \cdots, f_r$, where each $f_i$ is a prime element in $\mathbb{C}\{x, y\}$. Then each of the branches $X_{0,i}$ (defined by $f_i = 0$) has its own characteristic, and in addition we consider the intersection numbers $(X_{0,i}, X_{0,j}) = \dim_{\mathbb{C}} \mathbb{C}\{x, y\}/(f_i, f_j)$ (compare §1).

Let us now consider a *germ of family* of reduced plane curves, $(X, 0) \subset (Y \times \mathbb{C}^2, 0)$ defined by $(F)\mathcal{O}_{Y \times \mathbb{C}^2, 0}$ (where $(Y, 0) \simeq (\mathbb{C}^k, 0)$). By the Weierstrass preparation theorem, at the price of a linear change of the coordinates $(x, y)$, we can assume

$$F = y^n + a_{n-1}(\mathscr{y}, x) y^{n-1} + \cdots + a_0(\mathscr{y}, x) \quad \text{where} \quad a_i \in \mathbb{C}\{y_1, \cdots, y_k, x\},$$

and we can define the *discriminant* of the projection $\pi$ of $X$ to $Y \times \mathbb{C}$ $(\mathcal{O}_{Y \times \mathbb{C}, 0} = \mathbb{C}\{\mathscr{y}, x\})$ it is, as we know $F_0(\mathcal{O}_{Y \times \mathbb{C}^2}/(F, (\partial F/\partial y)))$, $F_0$ being the Fitting ideal as $\mathcal{O}_{Y \times \mathbb{C}, 0}$-module, of course. The critical locus of $\pi$ is finite over $Y \times \mathbb{C}$ because $\pi^{-1}(0)$, being defined by $(y^n)\mathbb{C}\{y\}$, has an isolated singularity. This discriminant is a hypersurface $\Delta_\pi$ in $Y \times \mathbb{C}_1$ defined by the resultant of $F$ and

$(\partial F/\partial y)$. We say $\Delta_\pi$ is *trivial* at 0 if $\Delta_{\pi,\mathrm{red}}$ is a non-singular hypersurface at 0. (This implies that $\Delta_\pi$ is itself analytically trivial along $\Delta_{\pi,\mathrm{red}}$, i.e.: $(\Delta_\pi, 0) = (Z_0 \times \Delta_{\pi,\mathrm{red}}, 0)$ where $Z_0$ is isomorphic to the subspace of $\mathbb{C}$ defined by $(x)^\Delta$ for some integer $\Delta$.) Then we have:

THEOREM OF EQUISINGULARITY FOR PLANE CURVES (see [6], [1], [2], [5]):

I. *For a family* $(X, 0) \subset (Y \times \mathbb{C}^2, 0)$ *as above, assuming that* $X \supset Y_1 = Y \times \{0\}$, *we have equivalence of the following conditions:*

(1) *All the fibres* $(X_y, 0)$ *have the same topological type.*

(2) *All the fibres* $(X_y, 0)$ *have the same Milnor number*, i.e. $\mu^{(2)}(X_y, 0)$ *is independent of* $y \in Y$.

(3) *The invariant* $\delta(X_y, 0)$ and *the number of branches* $r_y$ *of each fibre are independent of* $y \in Y$.

(4) *The composed map* $p: \bar{X} \xrightarrow{n} X \longrightarrow Y$ *is a submersion of non-singular spaces in a neighborhood of* $n^{-1}(0)$, $n$ *induces the normalization* $(\bar{X})_y = \bar{X}_y \to X_y$ *in each fibre, and the map induced by* $p: (n^{-1}(Y_1))_{\mathrm{red}} \to Y$ *is a (trivial) covering of degree* $r$ (= *the number of branches of the fibres*).

(5) *There exists a projection* $\pi: X \to Y \times \mathbb{C}$ *such that the discriminant is trivial (hence* $(\Delta_{\pi,\mathrm{red}}, 0) \simeq (Y \times \{0\}, 0)$).

(6) *There exists a projection* $\pi$ *such that the multiplicity of the discriminant* $\Delta_\pi$ *is constant along* $Y \times \{0\} \subset \Delta_\pi$.

(7) *The sum* $\mu^{(2)}(X_y) + m(X_y) - 1 = \mu^{(2)}(X_y) + \mu^{(1)}(X_y)$ *is independent of* $y \in Y$.

(8) *For any projection* $\pi: X \to Y \times \mathbb{C}$ *such that the multiplicity of* $\pi^{-1}(0)$ *at* 0 *is equal to the multiplicity of* $X$ *at* 0, *the discriminant* $\Delta_\pi$ *is trivial.*

(9) *Any holomorphic vector field on* $Y_1 = Y \times \{0\}$ *can be locally extended to a vector field which is* Lipschitz *and* meromorphic *on* $X$ *(hence extends to a holomorphic vector field on the normalization* $\bar{X}$, *see condition* 4).

(10) *For any two values* $y_1, y_2$ *of* $y \in Y$ *(near* 0) *there is a bijection* $b$ *from the set of branches of* $(X_{y_1}, 0)$ *to the set of branches of* $(X_{y_2}, 0)$ *such that* $b(X_{y_1,i})$ *has the same characteristic as* $X_{y_1,i}$ $(1 \leqq i \leqq r)$ *and we have equality of intersection numbers:*

$$(b(X_{y_1,i}), b(X_{y_1,j}))_0 = (X_{y_1,i}, X_{y_1,j})_0.$$

(11) $(X, 0)$ *satisfies the condition* (c) *of* §2 (2*nd part*) *along* $Y_1 = Y \times \{0\}$ *at* 0.

(12) $\bar{X} \xrightarrow{n} X \longrightarrow Y$ *is a submersion of non-singular spaces and* $j_{Y \times \mathbb{C}^2/Y}(F) \cdot \mathcal{O}_{\bar{X}, n^{-1}(0)}$ *is an invertible ideal.*

II. *Given two germs of plane curves* $(X_1, 0), (X_2, 0)$ *having the same topological type, or satisfying the numerical condition of* (10), *there exists a*

1-*parameter family of germs of reduced plane curves*

$$\begin{array}{c} X \subset Y \times \mathbb{C}^2 \\ {\scriptstyle\sigma}\nwarrow\searrow \quad \swarrow \\ Y \end{array}$$

*and* $y_1, y_2 \in Y(\dim Y = 1)$ *such that:*

(i) *the family satisfies the* 12 *equivalent conditions at every point of* $\sigma(Y)$.

(ii) $(X_{y_1}, \sigma(y_1)) \cong (X_1, 0)$, $(X_{y_2}, \sigma(y_2)) \cong (X_2, 0)$, *where* $\cong$ *means analytic isomorphism.*

This theorem summarizes results, mostly due to Zariski [6], also to Lê Dũng Tráng and C. P. Ramanujam [2], Pham and Teissier [3] and Teissier [4]. See [1]. Complete detailed proofs will appear in [5].

We will see examples of deformations of curves in the next section.

Exercise. (1) Check in 12 different ways that the family of curves defined by $y^2 - x^3 + v \cdot x^2 = 0$ has constant invariant $\delta$ but is not equisingular. Check that $y^2 - x^3 = 0$ has no equisingular deformation which is not analytically a product. Check that the same holds for $y^2 - x^2 = 0$, $y^3 - x^3 = 0$.

(2) Check that the family $y^4 - x^4 + v \cdot x^2y^2 = 0$ is equisingular at 0, but not analytically trivial.

REFERENCES

[1] B. Teissier, Introduction to equisingularity problems, Proc. A.M.S. Conference Algebraic Geometry, Arcata 1974 (A.M.S. 1975).

[2] Lê Dũng Tráng and C. P. Ramanujam, The invariance of the Milnor number implies topological triviality.

[3] F. Pham and B. Teissier, Fractions lipschitziennes d'une algèbre analytique complexe et saturation de Zariski. (See Pham in Congrès International des mathématiciens, Nice 1970).

[4] B. Teissier, Sur diverses conditions numériques d'équisingularité des familles de courbes, · · · , publication du Centre de Maths. de l'Ecole Polytechnique, F-91128 Palaiseau Cedex (June 75).

[5] B. Teissier, Forthcoming notes of a 'Cours Peccot', Collège de France, Spring 1976.

[6] O. Zariski, Studies in equisingularity I, II (Amer. J. Math. *87* (1965) and III (ibid. *90* (1968)).

[7] O. Zariski, Le problème des modules pour les branches planes, Cours, Centre de Maths. de l'Ecole Polytechnique, F-91128 Palaiseau Cedex (1973).

## §4. Unfoldings and deformations

In this section, we give the basic definitions of the theory of unfoldings and deformations. Our purpose is not to prove the existence of versal unfoldings, but rather to illustrate the definitions by examples and above all to emphasize the close connection between the complex-analytic avatar of the differential geometers' theory of unfoldings (Thom-Mather) and the algebraic geometers' theory of deformations (Schlessinger-Tjurina-Grauert). In fact, the local theory of unfoldings becomes simpler in the complex-analytic frame, and we have thought that the best way to show the connection mentioned above was to give a proof of the existence theorem of versal deformations for those singularities which can be presented as the fibres of a morphism $f:(\mathbb{C}^N, 0) \to (\mathbb{C}^p, 0)$ of finite singularity type in the sense of Mather, a class which includes complete intersections with isolated singularities and finite analytic spaces. This proof uses the existence of versal unfoldings and the existence of a local flattener for a map of complex-analytic spaces.

We use a transcription in complex-analytic geometry of the theory of unfoldings as found in the notes of Mather in these proceedings, and in the notes of Martinet and of Mather in 'Singularités d'applications différentiables', Springer Lecture Notes No. 535.

4.1. Definition 1. Let $f: X \to Y$ be a morphism of complex analytic spaces. An unfolding of $f$, with base a germ of complex space $(S, 0)$, is a morphism:

$$F: X \times S \to Y \times S$$

commuting with the natural projections to $S$ and such that $F \mid X \times \{0\} = f$.

We immediately remark that the datum of such an $F$ is equivalent to the datum of $\mathbf{F} = pr_1 \circ F: X \times S \to Y$ (and $\mathbf{F} \mid X \times \{0\} = f$).

The first example is that of an *infinitesimal* unfolding of $f$: it is the case where $S = \mathbb{T}$(see §2). Then, we can view $\mathbf{F}: X \times \mathbb{T} \to Y$ as a vector field on $Y$ parametrized by $X$ (vector field in the sense of §2, i.e. of Zariski tangent vectors). Namely, for every $x \in X$, $\mathbf{F}$ gives a vector tangent to $Y$ at $f(x)$, varying analytically with $x$. In this sense, as we saw in an exercise in §2, a vector field on $X$ is nothing but an infinitesimal unfolding of the identity of $X$.

Definition 2. Two unfoldings $F$ and $G$ of $f: X \to Y$ with the same base are said to be $\mathscr{A}$-isomorphic if there exists a commutative diagram

$$\begin{array}{ccc} X \times S & \xrightarrow{F} & Y \times S \\ \xi \downarrow & & \downarrow \eta \\ X \times S & \xrightarrow{G} & Y \times S \end{array}$$

where $\xi$ and $\eta$ are isomorphisms, unfoldings of the identity of $X$ and $Y$ respectively.

An unfolding of $f$ is said to be $\mathscr{A}$-*trivial* if it is $\mathscr{A}$-isomorphic to the morphism $f\times id_S: X\times S\to Y\times S$.

Let us see what it means for our infinitesimal unfolding to be $\mathscr{A}$-trivial; in order to simplify notations, we will write our unfoldings as $\mathbf{F}: X\times S\to Y$.

We can make many infinitesimal unfoldings of $f$ as shown in the following diagram:

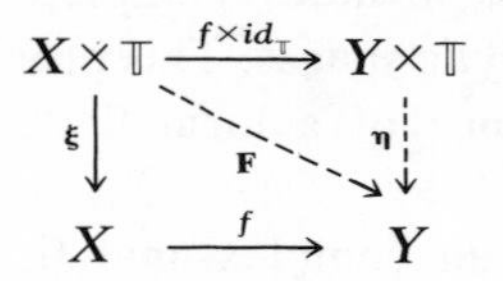


Take a vector field $\boldsymbol{\xi}$ on $X$ (resp. $\boldsymbol{\eta}$ on $Y$) and take $\mathbf{F}=f\circ\boldsymbol{\xi}$ (resp. $\mathbf{F}=\boldsymbol{\eta}\circ(f\times id_{\mathbb{T}})$) and you get an infinitesimal unfolding of $f$. However, by our definition, all such unfoldings are trivial.

Remark. In the case where $X$ and $Y$ are non-singular, $\boldsymbol{\xi}\mapsto f\circ\boldsymbol{\xi}$ is Mather's *tf*, and $\boldsymbol{\eta}\to\boldsymbol{\eta}\circ(f\times id_{\mathbb{T}})$ is his $\omega f$.

We now remark (see §2) that to give such an $F$ is the same as to give a derivation of $f^{-1}\mathcal{O}_Y$ in $\mathcal{O}_X$ and that therefore, the set of such $F$ is canonically endowed with a structure of $\mathcal{O}_X$-module. [Essentially, $\mathbf{F}$ corresponds to a map $\mathbf{F}^*:\mathcal{O}_Y\to\mathcal{O}_X[\varepsilon]$ such that the composed map $\mathcal{O}_Y\xrightarrow{\mathbf{F}^*}\mathcal{O}_X[\varepsilon]\longrightarrow\mathcal{O}_X[\varepsilon]/(\varepsilon)=\mathcal{O}_X$ is the map 'composition with $f$', and the derivation is defined by: $\Delta(h)=$ coefficient of $\varepsilon$ in $F^*(h)$].

Definition 3. $f$ is infinitesimally stable if every infinitesimal unfolding of $f$ is $A$-trivial that is, if any $\mathbf{F}: X\times\mathbb{T}\to Y$ such that $\mathbf{F}\mid X\times\{0\}=f$ can be written as

$$\mathbf{F}=f\circ\boldsymbol{\xi}+\boldsymbol{\eta}\circ(f\times id_{\mathbb{T}})$$

where $\boldsymbol{\xi}$ (resp. $\boldsymbol{\eta}$) is a vector field on $X$ (resp. $Y$).

Let us now consider *germs* of mappings $f:(X,0)\to(Y,0)$. Then the set of infinitesimal unfoldings of $f$ can be identified with the set of $\mathbb{C}$-derivations of $\mathcal{O}_{Y,0}$ in $\mathcal{O}_{X,0}$ where $\mathcal{O}_{X,0}$ is considered as $\mathcal{O}_{Y,0}$-module via the map $f^*:\mathcal{O}_{y,0}\to\mathcal{O}_{X,0}$ 'composition with $f$', which is a map of $\mathbb{C}$-algebras. The vector fields on $(X,0)$ (resp. $(Y,0)$) then correspond to the $\mathbb{C}$-derivations of $\mathcal{O}_{X,0}$ (resp. $\mathcal{O}_{Y,0}$) in itself. Let then $D\in\mathrm{Der}_{\mathbb{C}}(\mathcal{O}_{Y,0},\mathcal{O}_{Y,0})$ be a derivation of $\mathcal{O}_{Y,0}$ into itself corresponding to $\boldsymbol{\eta}$, and let $D'\in\mathrm{Der}_{\mathbb{C}}(\mathcal{O}_{X,0},\mathcal{O}_{X,0})$ correspond similarly to $\boldsymbol{\xi}$. Then, $f\circ\boldsymbol{\xi}$ corresponds to the derivation $D'\circ f^*$ of $\mathcal{O}_{Y,0}$ in $\mathcal{O}_{X,0}$ and

$\eta \circ (f \times id_{\mathbb{T}})$ corresponds to $f^* \circ D$. Assume that $(X, 0)$ and $(Y, 0)$ are non-singular, so that we can set $\mathcal{O}_{X,0} = \mathcal{O}_{\mathbb{C}^N,0} = \mathbb{C}\{z_1, \cdots, z_N\}$ and $\mathcal{O}_{Y,0} = \mathcal{O}_{\mathbb{C}^p,0} = \mathbb{C}\{y_1, \cdots, y_p\}$, with $f^*$ described by

$$f^*(y_j) = f_j(z_1, \cdots, z_N) \in \mathbb{C}\{z_1, \cdots, z_N\} \qquad (1 \leqq j \leqq p).$$

We can see that a derivation $\Delta \in \mathrm{Der}_{\mathbb{C}}(\mathcal{O}_{\mathbb{C}^p,0}, \mathcal{O}_{\mathbb{C}^N,0})$ is described by the datum of the $\Delta y_j \in \mathbb{C}\{z_1, \cdots, z_N\}$, so that, having made a choice of coordinates, we can identify the set of $\mathbb{C}$-derivations of $\mathcal{O}_{\mathbb{C}^N,0}$ as $\mathcal{O}_{X,0}$-module with $\mathcal{O}^p_{\mathbb{C}^N,0}$, while the 'trivial' ones coming from derivation of $\mathcal{O}_{Y,0}$ into itself are identified with the sub $\mathcal{O}_{Y,0}$-module $(f^*(\mathcal{O}_{\mathbb{C}^p,0}))^p$ of $\mathcal{O}^p_{\mathbb{C}^N,0}$ and the 'trivial' ones coming from derivations of $\mathcal{O}_{\mathbb{C}^N,0}$ into itself are identified with the sub $\mathcal{O}_{\mathbb{C}^N,0}$-module of $\mathcal{O}^p_{\mathbb{C}^N,0}$ generated by the $N$ elements $(\partial \vec{f}/\partial z_1), \cdots, (\partial \vec{f}/\partial z_N)$ where $\vec{f} = (f_1, \cdots, f_p)$. [N.B. This is clear because it means that we can write $\Delta y_k = D'f_k = \sum_{i=1}^N (\partial f_k/\partial z_i)\, D'z_i$ if and only if the element $(\Delta y_1, \cdots, \Delta y_p)$ of $\mathcal{O}_{\mathbb{C}^N,0}$ belongs to that submodule.]

We will write the submodule of $\mathcal{O}^p_{\mathbb{C}^N,0}$ generated by $((\partial \vec{f}/\partial z_1), \cdots, (\partial \vec{f}/\partial z_N))$ by: $((\partial \vec{f}/\partial z_1), \cdots, (\partial \vec{f}/\partial z_N))$ when no confusion is to be feared.

Finally we see that the infinitesimal unfoldings of $f$ modulo the $A$-trivial ones, which we will call the non(-$\mathcal{A}$-)trivial infinitesimal unfoldings of $f$ can be identified with the quotient:

$$A^1_f = \mathcal{O}^p_{\mathbb{C}^N,0}/(f^*(\mathcal{O}_{\mathbb{C}^p,0}))^p + ((\partial \vec{f}/\partial z_1), \cdots, (\partial \vec{f}/\partial z_N)).$$

It is very easy to check, using Hilbert's Nullstellensatz that if 0 is an isolated critical point of $f: (\mathbb{C}^N, 0) \to (\mathbb{C}^p, 0)$ then $A^1_f$ is a finite dimensional vector space.

We also remark that $A^1_f = 0 \Leftrightarrow f$ is infinitesimally stable.

In the case where $(X, 0)$ and $(Y, 0)$ are allowed to have singularities, the situation is more delicate and I refer the reader to the work of Mount and Villamayor (Publ. Math. IHES No 43, P.U.F. 1974).

4.2. Now that we have studied infinitesimal unfoldings of a map $f: (\mathbb{C}^N, 0) \to (\mathbb{C}^p, 0)$, let us turn to arbitrary unfoldings of such an $f$: Remark that we can always describe an unfolding $F: (\mathbb{C}^N \times S, 0) \to (\mathbb{C}^p \times S, 0)$ of $f$ as follows: if we describe $f$ by $f^*(y_j) = f_j(z_1, \cdots, z_N)(1 \leqq j \leqq p)$, then $F$ can be described by:

$$F^*(y_j) = f_j(z) + \sum_{A \in \mathbb{N}^N} h_{j,A} z^A,$$

where $h_{j,A} \in \mathfrak{m}_{S,0}$, maximal ideal of $\mathcal{O}_{S,0}$.

In this way, we can clearly see what the unfolding $F': (\mathbb{C}^N \times S', 0) \to (\mathbb{C}^p \times S', 0)$, obtained from $F$ by a base change $\varphi: (S', 0) \to (S, 0)$, is: letting

$\varphi^*: \mathcal{O}_{S,0} \to \mathcal{O}_{S',0}$ be the map of algebras corresponding to $\varphi$, $F'$ is described by

$$F'^*(y_j) = f_j(z) + \sum_{A \in \mathbb{N}^N} \varphi^*(h_{j,A}) z^A.$$

DEFINITION 4. An unfolding $F: (\mathbb{C}^N \times S, 0) \to (\mathbb{C}^p \times S, 0)$ of $f: (\mathbb{C}^N, 0) \to (\mathbb{C}^p, 0)$ is said to be $\mathcal{A}$-versal (resp. $\mathcal{A}$-miniversal) if any other unfolding $H: (\mathbb{C}^N \times S', 0) \to (\mathbb{C}^p \times S', 0)$ of $f$ can be obtained, up to $\mathcal{A}$-isomorphism, from $F$ by a base change $\varphi: (S', 0) \to (S, 0)$ (resp. and $S$ has the smallest possible dimension for this property to hold).

THEOREM (Mather). *An unfolding $F: (\mathbb{C}^N \times \mathbb{C}^s, 0) \to (\mathbb{C}^p \times \mathbb{C}^s, 0)$ of $f$ is $\mathcal{A}$-versal (resp. $\mathcal{A}$-miniversal) if and only if, when we describe it by*

$$F^*(y_j) = f_j(z) + \sum_{A \in \mathbb{N}^N} h_{j,A}(u) z^A$$

*$[u = (u_1, \cdots, u_s)$, coordinates on $(\mathbb{C}^s, 0)]$; we have that setting $\vec{F} = (F^*(y_1), \cdots, F^*(y_p))$, the elements*

$$\left.\frac{\partial \vec{F}}{\partial u_1}\right|_{u=0}, \cdots, \left.\frac{\partial \vec{F}}{\partial u_s}\right|_{u=0} \quad \text{in} \quad \mathbb{C}\{z_1, \cdots, z_N\}^p$$

*have images in $A_f^1$ which generate it as a $\mathbb{C}$-vector space (resp. form a basis of it). All A-miniversal unfolding of f are obtained from one another, up to A-ismorphism, by base change by (non-canonical) isomorphisms $(\mathbb{C}^s, 0) \simeq (\mathbb{C}^s, 0)$.*

In particular, we see that $f$ has an $\mathcal{A}$-miniversal unfolding $F: (\mathbb{C}^N \times \mathbb{C}^s, 0) \to (\mathbb{C}^p \times \mathbb{C}^s, 0)$ if and only if $A_f^1$ is a finite-dimensional vector space, and then $s = \dim_{\mathbb{C}} A_f^1$ and $F$ is the unfolding of $f$ constructed as follows: one takes in $\mathbb{C}\{z_1, \cdots, z_N\}^p$ $s$ elements $\vec{h}_1, \cdots, \vec{h}_s$ such that their images in $A_f^1$ form a basis of it over $\mathbb{C}$. Then $F$ is described by:

$$F^*(y_j) = f_j(z) + \sum_{i=1}^{s} u_i \cdot h_{i,j}(z) \qquad (1 \leqq j \leqq p)$$

where $h_{ij}(z)$ is the $j$th component of $\vec{h}_i$.

4.2.1. REMARK 1. Strictly speaking, the proof given in differential geometry, once transcribed in analytic geometry, proves only the versality of such an $F$ with respect to unfoldings having a non-singular base, while we have in our definitions allowed arbitrary bases, e.g. a *finite* local analytic space: the analytic spectrum of an Artinian analytic algebra (a.k.a* 'thick point'). This difficulty, however, is inessential because if we begin a theory of

* a.k.a. = also known as.

unfoldings over such bases in the spirit of Schlessinger ('Functors of Artin rings', Trans. A.M.S. 130 (1968) 208–22) we see immediately that the functor of unfoldings is *unobstructed*, having the property that given $(S_1, 0)$ and $(S_2, 0)$ closed in $(S, 0)$ and unfoldings of base $(S_1, 0)$ and $(S_2, 0)$ which coincide on $(S_1 \cap S_2, 0)$, they can be extended to an unfolding over $(S, 0)$. This follows from the explicit description given at the beginning of 4.2, and implies that the base of the $A$-miniversal unfolding is non-singular, as implied by Mather's theorem above.

4.3. Until now, we have been considering the set of mappings $\mathbb{C}^N \to \mathbb{C}^p$ modulo the group action of $A = \operatorname{Aut} \mathbb{C}^N \times \operatorname{Aut} \mathbb{C}^p : f \mapsto \eta \circ f \circ \xi^{-1}$, and the sub $\mathcal{O}_{\mathbb{C}^p,0}$-module of $\mathcal{O}^p_{\mathbb{C}^N,0}$, $(f^*(\mathcal{O}_{\mathbb{C}^p,0}))^p + ((\partial \vec{f}/\partial z_1), \cdots, (\partial \vec{f}/\partial z_N)) \subset \mathcal{O}^p_{\mathbb{C}^N,0}$ which we saw above is deemed to be the 'tangent space at $f$ of the orbit $A \circ f$ of $f$', the vector space $A^1_f$ then being its 'supplementary' in the tangent space at $f$ to the space of functions such as $f$.

We now look at another group action, corresponding to an interest in the geometry of the *fibres* of mappings more than in the mappings themselves. It is the group of *contact transformations* in the terminology of Mather, or $V$-isomorphisms in that of Martinet. The idea is that we allow more than the usual automorphisms in the target space $\mathbb{C}^p$, namely we consider $f$ and $f'$ as equivalent if $(f^{-1}(0), 0)$ and $(f'^{-1}(0), 0)$ are isomorphic, or, what amounts to the same, if there exists a complex map $\mathbb{C}^N \xrightarrow{M} GL(p, \mathbb{C})$ such that $f'(x) = M(x) \cdot f(x)$ [which simply means that the ideals generated by $(f_1, \cdots, f_p)$ and $(f'_1, \cdots, f'_p)$ in $\mathbb{C}\{z_1, \cdots, z_N\}$ are equal]. In this spirit, we have:

DEFINITION 5. Two unfoldings $\mathbf{F}, \mathbf{F}' : (X \times S, 0) \to (Y, 0)$ of $f : (X, 0) \to (Y, 0)$ are $\mathcal{K}$-isomorphic if there exists a germ of an analytic isomorphism $\Phi : X \times S \xrightarrow{\sim} X \times S$, which is an unfolding of the identity of $X$, and such that $((F' \circ \Phi)^{-1}(0), 0)$ and $(F^{-1}(0), 0)$ are isomorphic as germs of complex-analytic spaces.

Let us now see what it means for an infinitesimal unfolding to be $\mathcal{K}$-trivial, i.e. $\mathcal{K}$-isomorphic to the trivial unfolding $X \times S \to Y$ which is $f \circ pr_1$.

We are given an unfolding $\mathbf{F} : (X \times \mathbb{T}, 0) \to (Y, 0)$ viewed as a derivation $D$ of $\mathcal{O}_{Y,0}$ into $\mathcal{O}_{X,0}$. We assume $(X, 0)$ and $(Y, 0)$ non-singular again, and set $\mathcal{O}_{X,0} = \mathbb{C}\{z_1, \cdots, z_N\}$, $\mathcal{O}_{Y,0} = \mathbb{C}\{y_1, \cdots, y_p\}$. To say that $F$ is $\mathcal{K}$-trivial is to say that there exists an infinitesimal unfolding of $id_X$, i.e. a vector field on $X$, i.e. a derivation $D' : \mathcal{O}_{\mathbb{C}^N,0} \to \mathcal{O}_{\mathbb{C}^N,0}$, such that

$$Dy_j - D'f_j \in (f_1, \cdots, f_p) \cdot \mathcal{O}_{\mathbb{C}^N,0} \qquad (1 \leqq j \leqq p)$$

where $f_j = f_j(z) \in \mathcal{O}_{\mathbb{C}^N,0} = \mathbb{C}\{z_1, \cdots, z_N\}$ describe $f$ (i.e. $f^*(y_j) = f_j$). We remark that $(f_1, \cdots, f_p) \cdot \mathcal{O}_{\mathbb{C}^N,0}$ is the ideal generated in $\mathcal{O}_{\mathbb{C}^N,0}$ by the image by

$f^*:\mathcal{O}_{\mathbb{C}^p,0}\to\mathcal{O}_{\mathbb{C}^N,0}$ of the maximal ideal $\mathfrak{m}_y$ of $\mathcal{O}_{\mathbb{C}^p,0}$, i.e. what we usually write simply by $\mathfrak{m}_y\cdot\mathcal{O}_{\mathbb{C}^N,0}$.

Therefore, we see that the 'tangent space to the $\mathcal{K}$ orbit' of $f$ is exactly the sub-$\mathcal{O}_{\mathbb{C}^N,0}$-module:

$$\mathfrak{m}_y\cdot\mathcal{O}^p_{X,0}+\left(\frac{\partial\vec{f}}{\partial z_1},\cdots,\frac{\partial\vec{f}}{\partial z_N}\right)\subset\mathcal{O}^p_{\mathbb{C}^N,0}$$

because to say that $D$ describes a $\mathcal{K}$-trivial infinitesimal unfolding is exactly to say that $(Dy_1,\cdots,Dy_p)\in\mathcal{O}^p_{\mathbb{C}^N,0}$ belongs to this submodule. We shall write:

$$K^1_f=\mathcal{O}^p_{\mathbb{C}^N,0}/\mathfrak{m}_y\cdot\mathcal{O}^p_{\mathbb{C}^N,0}+\left(\frac{\partial\vec{f}}{\partial z_1},\cdots,\frac{\partial\vec{f}}{\partial z_N}\right)$$

and remark that, since the fibre $(X_0,0)$ of $f:(\mathbb{C}^N,0)\to(\mathbb{C}^p,0)$ is the subspace defined by $\mathfrak{m}_y\cdot\mathcal{O}_{\mathbb{C}^N,0}$, we have also:

$$K^1_f=\mathcal{O}^p_{X_0,0}/\sum_{i=1}^{N}\mathcal{O}_{X_0,0}\cdot\frac{\partial\vec{f}}{\partial z_i}.$$

DEFINITION. Those maps $f:(\mathbb{C}^N,0)\to(\mathbb{C}^p,0)$ such that $\dim_{\mathbb{C}}K^1_f<\infty$ are called 'of finite singularity type' or 'T.S.F.' by Mather.

EXERCISE 1. Check that when $f$ is finite, $f$ is T.S.F.

EXERCISE 2. Check that when $f$ is *flat* and its fiber $(X_0,0)$ has an isolated singularity, $f$ is T.S.F. In general, $f$ is T.S.F. if and only if the critical locus of $f$ is finite over $\mathbb{C}^p$.

EXERCISE 3. Consider the map $f:(\mathbb{C}^4,0)\to(\mathbb{C}^4,0)$ given by $(z_1,z_2,z_3,z_4)\mapsto(z_1\cdot z_3,z_2\cdot z_3,z_1\cdot z_4,z_2\cdot z_4)$ and check that it is *not* T.S.F. although its fibre $(X_0,0)$, the union of two 2-dimensional planes in $\mathbb{C}^4$ meeting in one point, has an isolated singularity. Deduce that no map $f:(\mathbb{C}^N,0)\to(\mathbb{C}^p,0)$ having $(X_0,0)$ as fibre is T.S.F.

DEFINITION 6. An unfolding $F:(\mathbb{C}^N\times S,0)\to(\mathbb{C}^p\times S,0)$ of $f:(\mathbb{C}^N,0)\to(\mathbb{C}^p,0)$ is $\mathcal{K}$-versal (resp. $\mathcal{K}$-miniversal) if any unfolding of $f$, say $H:(\mathbb{C}^N\times S',0)\to(\mathbb{C}^p\times S',0)$, is $\mathcal{K}$-isomorphic to an unfolding obtained from $F$ by a base change $\varphi:(S',0)\to(S,0)$ (resp. and $S$ has the smallest possible dimension for this property to hold).

THEOREM (Mather). *An unfolding* $F:(\mathbb{C}^N\times\mathbb{C}^t,0)\to(\mathbb{C}^p\times\mathbb{C}^t,0)$ *is* $\mathcal{K}$*-versal (resp.* $\mathcal{K}$*-miniversal) if and only if, when we describe it by:*

$$y_j\circ F=F^*(y_j)=f_j(z)+\sum_{A\subset\mathbb{N}^N}k_{j,A}(v)z^A$$

[$v=(v_1,\cdots,v_t)$, *coordinates on* $(\mathbb{C}^t,0)$] *we have that, setting* $\vec{F}=(F^*(y_1),\cdots,F^*(y_p))\in\mathcal{O}^p_{\mathbb{C}^N\times\mathbb{C}^t,0}$, *the elements* $(\partial\vec{F}/\partial v_1)|_{v=0},\cdots,(\partial\vec{F}/\partial v_t)|_{v=0}$ *in*

*$\mathcal{O}^p_{\mathbb{C}^N,0}$ have images in $K^1_f$ which generate it as $\mathbb{C}$-vector space (resp. form a basis of it).*

*All $\mathcal{K}$-miniversal unfoldings of f can be obtained from one another, up to $\mathcal{K}$-isomorphism, by base change by (non-canonical) isomorphisms $(\mathbb{C}^t, 0) \simeq (\mathbb{C}^t, 0)$.*

*In particular, f has a $\mathcal{K}$-miniversal unfolding if and only if $K^1_f$ is a finite-dimensional vector space, and it is built in the manner analogous to that explained for $\mathcal{A}$-miniversal unfoldings. In particular, it has the form*

$$F^*(y_j) = f_j(z) + \sum_{j=1}^{t} v_j h_{i,j}(z) \qquad (1 \leqq j \leqq p)$$

*where the elements $\vec{h}_i = (h_{i,j})$ have images in $K^1_f$ which form a basis of it.*

4.4. We now turn to the theory of deformations; here we (temporarily) forget mappings, and think only of spaces. Let $(X_0, 0)$ be a germ of a complex analytic space:

DEFINITION 7. A deformation of $(X_0, 0)$ is a Cartesian diagram of germs:

$$\begin{array}{ccc} (X_0, 0) & \overset{i}{\hookrightarrow} & (X, 0) \\ \downarrow & \square & \downarrow G \\ \{0\} & \hookrightarrow & (S, 0) \end{array}$$

where $G$ is a *flat* map. [Cartesian diagram means in this case that we are given an isomorphism of $(G^{-1}(0), 0)$ with $(X_0, 0)$.] A morphism of deformations is a morphism of squares inducing the identity on $(X_0, 0)$. A deformation is said to be *trivial* if it is isomorphic with the product deformation $pr_2: (X_0 \times S, 0) \to (S, 0)$, or equivalently if there exists a commutative diagram

$$\begin{array}{ccc} (X, 0) & \xrightarrow{\Psi} & (X_0 \times S, 0) \\ & {\scriptstyle G}\searrow \quad 0 \quad \swarrow{\scriptstyle pr_2} & \\ & (S, 0) & \end{array}$$

with $\Psi$ an isomorphism inducing the identity on $(X_0, 0)$.

DEFINITION 8. A deformation $G: (X, 0) \to (S, 0)$ of $(X_0, 0)$ is said to be versal (resp. miniversal) if any deformation $H: (X', 0) \to (S', 0)$ of $(X_0, 0)$ is isomorphic to a deformation obtained from $G$ by a base change $\varphi: (S', 0) \to (S, 0)$ i.e. $H$ is isomorphic to $G^\varphi$:

$$\begin{array}{ccc} (X \times S', 0) & \xrightarrow{pr_1} & (X, 0) \\ {\scriptstyle G^\varphi}\downarrow & \square & \downarrow{\scriptstyle G} \\ (S', 0) & \xrightarrow[\varphi]{} & (S, 0) \end{array}$$

(resp. and $S$ has the smallest possible dimension for this property to hold).

4.5. Now let us recall that by definition, any germ of complex analytic space $(X_0, 0)$ can be presented as the fibre of a germ of complex analytic map $f:(\mathbb{C}^N, 0)\to(\mathbb{C}^p, 0)$. Namely, the map is described by generators of the ideal defining $(X_0, 0)\subset(\mathbb{C}^N, 0)$. Recall also (see Hironaka's lectures), that $(X_0, 0)$ is a *complete intersection* if and only if it can be presented with a *flat* map $f$.

Finally recall that, given any complex-analytic map-germ $h:(X, 0)\to(Y, 0)$, there exists a germ of a complex subspace $(Z_h, 0)\subset(Y, 0)$, the *flattener* of $h$, which satisfies the following universal property:

For any complex-analytic map $\varphi:(Y', 0)\to(Y, 0)$, the map obtained by base change by $\varphi$:

$$\begin{array}{ccc} (X \times_Y Y', 0) & \xrightarrow{pr_1} & (X, 0) \\ \downarrow{\scriptstyle h^\varphi} & \square & \downarrow{\scriptstyle h} \\ (Y', 0) & \xrightarrow{\varphi} & (Y, 0) \end{array}$$

is flat (at 0, of course) if and only if $\varphi$ factors through the subspace $(Z_\varphi, 0)\subset(Y, 0)$.

(This was originally proved in [1] and is proved by a different method in Hironaka's lectures, §3).

Now the theory of unfoldings and the theory of deformations meet in the:

4.5.1. Proposition. *Let $(X_0, 0)$ be a germ of complex space with isolated singularity. Suppose there is a presentation of $(X_0, 0)$ as the fibre of an $f:(\mathbb{C}^N, 0)\to(\mathbb{C}^p, 0)$ such that $\dim_{\mathbb{C}} K_f^1<\infty$, let $F:(\mathbb{C}^N\times\mathbb{C}^t, 0)\to(\mathbb{C}^p\times\mathbb{C}^t, 0)$ be a $\mathcal{K}$-miniversal unfolding of $f$. Consider the subspace $(X', 0)=(F^{-1}(0\times\mathbb{C}^t), 0)$ and the induced map $F':(X', 0)\to(\mathbb{C}^t, 0)$. Let now $(S, 0)\hookrightarrow(\mathbb{C}^t, 0)$ be the flattener of $F'$ and $G:(X, 0)\to(S, 0)$ be the map obtained from $F'$ by base change by the inclusion $(S, 0)\hookrightarrow(\mathbb{C}^t, 0)$, i.e. restriction of $F'$ over $(S, 0)$. Then, $G$ is a* miniversal deformation *of $(X_0, 0)$. The construction is summarized in the diagram:*

$$\begin{array}{ccccccccc} (X_0, 0) & \hookrightarrow & (X, 0) & \hookrightarrow & (X', 0) & \hookrightarrow & (\mathbb{C}^N\times\mathbb{C}^t, 0) & \hookleftarrow & (\mathbb{C}^N\times 0, 0) \\ \downarrow & \square & \downarrow{\scriptstyle G} & \square & \downarrow{\scriptstyle F'} & & \downarrow{\scriptstyle F} & & \downarrow{\scriptstyle f} \\ \{0\} & \hookrightarrow & (S, 0) & \underset{\text{flattener}}{\hookrightarrow} & (0\times\mathbb{C}^t, 0) & \hookrightarrow & (\mathbb{C}^p\times\mathbb{C}^t, 0) & \hookleftarrow & (\mathbb{C}^p\times 0, 0) \end{array}$$

Proof of the proposition. First, remark that any deformation of $(X_0, 0)$ can be represented by an unfolding: let $f_j\in\mathbb{C}\{z_1, \cdots, z_N\}(1\leq j\leq p)$ be the equations of $(X_0, 0)\subset(\mathbb{C}^N, 0)$. As we saw in §3, any deformation

$H:(X',0)\to(S',0)$ of $(X_0,0)$ can be put in a commutative diagram:

$$\begin{array}{ccc} (X',0) & \hookrightarrow & (S'\times\mathbb{C}^N,0) \\ & {\scriptstyle H}\searrow \quad 0 \quad \swarrow{\scriptstyle pr_3} & \\ & (S',0) & \end{array}$$

where $(X',0)\subset(S'\times\mathbb{C}^N,0)$ is defined by an ideal generated by $F_j\in\mathcal{O}_{S',0}\{z_1,\cdots,z_N\}(1\leqq j\leqq p)$. These $F_j$ of course also describe an unfolding $\boldsymbol{F}:(\mathbb{C}^N\times S',0)\to(\mathbb{C}^p,0)$ of the map $f:(\mathbb{C}^N,0)\to(\mathbb{C}^p,0)$ described by the $f_j$. Now by definition of a $\mathcal{K}$-miniversal unfolding, there exists a base change $\varphi:(S',0)\to(\mathbb{C}^t,0)$ such that $\boldsymbol{F}$ is $\mathcal{K}$-isomorphic to the unfolding obtained from $F$ by the base change $\varphi$, and by the definition of $\mathcal{K}$-isomorphism, this means exactly that our deformation is obtained from the map $F'$ in the diagram above by the base change $\varphi$. Since $H$ is a flat map by definition, $\varphi$ must factor through the flattener $(S,0)\subset(\mathbb{C}^t,0)$, and this shows that $G$ is a *versal* deformation of $(X_0,0)$. To check that $G$ is in fact miniversal, we first stop to examine the Zariski tangent space to the base of the miniversal deformation of $(X_0,0)$, which of course coincides with the set of infinitesimal deformations of $(X_0,0)$ modulo isomorphism of deformations: suppose again $(X_0,0)\subset(\mathbb{C}^N,0)$ given by the ideal $I_0=(f_1,\cdots,f_p)\subset\mathbb{C}\{z_1,\cdots,z_N\}$. Then, any deformation of $(X_0,0)$ with base $\mathbb{T}$ can be described in $\mathbb{C}^N\times\mathbb{T}$ by an ideal $I=(f_1+\varepsilon\cdot g_1,\cdots,f_p+\varepsilon\cdot g_p)\subset\mathbb{C}\{\varepsilon,z_1,\cdots,z_N\}$, with $\varepsilon^2=0$. It is an exercise on flatness (use the appendix to Hironaka's lectures) to check that $X\subset\mathbb{C}^N\times\mathbb{T}$ described by such an $I$ is flat over $\mathbb{T}$ (by the restriction to $X$ of $pr_2:\mathbb{C}^N\times\mathbb{T}\to\mathbb{T}$) if and only if the following condition is satisfied:

(*) For any relation $\sum_1^p a_i(z)\cdot f_i(z)=0$ between the $f_i$ in $\mathbb{C}\{z_1,\cdots,z_N\}$, we have that:

$$\sum_1^p a_i(z)\cdot g_i(z)\in I_0=(f_1,\cdots,f_p).$$

Let us now remark that the datum of a set of $(g_i)$ is exactly the datum of a map $\mathbb{C}\{z_1,\cdots,z_N\}^p\to\mathbb{C}\{z_1,\cdots,z_N\}$ of $\mathbb{C}\{z_1,\cdots,z_N\}$-modules (by sending the $i$th base element to $g_i$) and (*) is equivalent to the fact that the composed map:

$$\mathbb{C}\{z_1,\cdots,z_N\}^p\to\mathbb{C}\{z_1,\cdots,z_N\}\to\mathbb{C}\{z_1,\cdots,z_N\}/I_0=\mathcal{O}_{X_0,0}$$

factors through the map

$$\mathbb{C}\{z_1,\cdots,z_N\}^p\longrightarrow I_0\hookrightarrow\mathbb{C}\{z_1,\cdots,z_N\}$$

sending the $i$th base element to $f_i$.

Finally, we can identify the set of infinitesimal deformations of $(X_0,0)$ with $\mathrm{Hom}_{\mathcal{O}_{\mathbb{C}^N,0}}(I_0,\mathcal{O}_{X_0,0})$. Now those infinitesimal deformations which are trivial are those such that the ideal generated by the $(f_i+\varepsilon g_i)$ in $\mathbb{C}\{\varepsilon, z_1, \cdots, z_N\}$ becomes equal, after a change of variables, to the ideal generated by the $(f_i)$, and since a change of variables adds to each $g_i$ an element of the form $\sum_{j=1}^N (\partial f_i/\partial z_j)h_j$ it is clear that the deformation is trivial if and only if, setting $\vec{g}=(g_1, \cdots, g_p)\in \mathcal{O}^p_{\mathbb{C}^N,0}$, we have:

$$\vec{g}\in I_0\cdot \mathcal{O}^p_{\mathbb{C}^N,0}+\left(\left(\frac{\partial \vec{f}}{\partial z_1}\right), \cdots, \left(\frac{\partial \vec{f}}{\partial z_N}\right)\right)$$

where $((\partial \vec{f}/\partial z_1), \cdots, (\partial \vec{f}/\partial z_N))=\sum_1^N \mathcal{O}_{\mathbb{C}^N,0}\cdot(\partial \vec{f}/\partial z_i)$ (as submodule of $\mathcal{O}^p_{\mathbb{C}^N,0}$). Compare this with 4.3, after remarking that if we view $\mathcal{O}_{\mathbb{C}^N,0}$ as $\mathcal{O}_{\mathbb{C}^p,0}$-module via $f^*$, we have $\mathfrak{m}_y\cdot \mathcal{O}^p_{\mathbb{C}^N,0}=I_0\cdot \mathcal{O}^p_{\mathbb{C}^N,0}$.

Exercise. The natural map of $\mathcal{O}_{\mathbb{C}^N,0}$-modules: $I_0\to \Omega^1_{\mathbb{C}^N,0}\underset{\mathcal{O}_{\mathbb{C}^N,0}}{\otimes}\mathcal{O}_{X_0,0}$ given by $h\longmapsto dh\otimes 1$ induces a map

$$d^*:\mathrm{Hom}_{\mathcal{O}_{\mathbb{C}^N,0}}(\Omega^1_{\mathbb{C}^N,0}\underset{\mathcal{O}_{\mathbb{C}^N,0}}{\otimes}\mathcal{O}_{X_0,0}, \mathcal{O}_{X_0,0})\to \mathrm{Hom}_{\mathcal{O}_{\mathbb{C}^N,0}}(I_0, \mathcal{O}_{X_0,0}).$$

Check that the trivial infinitesimal deformations of $(X_0, 0)$ correspond exactly to those elements $\vec{g}$ inducing elements of $\mathrm{Hom}_{\mathcal{O}_{\mathbb{C}^N,0}}(I_0, \mathcal{O}_{X_0,0})$ which are in the image $\mathrm{Im}\, d^*$ of $d^*$.

Therefore, the Zariski tangent space to the base of the miniversal deformation of $(X_0, 0)$ can be naturally identified with the $\mathbb{C}$-vector space:

$$T^1_{X_0,0}=\mathrm{Hom}_{\mathcal{O}_{\mathbb{C}^N,0}}(I_0, \mathcal{O}_{X_0,0})/\mathrm{Im}\, d^*=\mathrm{Hom}_{\mathcal{O}_{X_0,0}}(I_0/I_0^2, \mathcal{O}_{X_0,0})/\mathrm{Im}\, d^*$$

[which is a finite-dimensional vector space if $(X_0, 0)$ has an isolated singularity].

I claim that $T^1_{X_0,0}$ is precisely the Zariski tangent space to the flattener of the map $F'$ of the proposition. The reason is very simple: an infinitesimal unfolding of $f$ is given by $(f_i+\varepsilon g_i)$ with no condition on the $g_i$. To say that it is $\mathcal{K}$-trivial is to say, as we saw in 4.3, that

$$\vec{g}\in I_0\cdot \mathcal{O}^p_{\mathbb{C}^N,0}+((\partial \vec{f})/(\partial z_1), \cdots, (\partial \vec{f})/(\partial z_N)) \quad \text{in} \quad \mathcal{O}^p_{\mathbb{C}^N,0},$$

and therefore the Zariski tangent space to the $\mathbb{C}^t$ parametrizing the $\mathcal{K}$-miniversal unfolding is naturally identified with

$$K^1_f=\mathcal{O}^p_{X_0,0}\Big/\left(\frac{\partial \vec{f}}{\partial z_1}, \cdots, \frac{\partial \vec{f}}{\partial z_N}\right)\cdot \mathcal{O}^p_{X_0,0}$$

again as we saw in 4.3. To prove the claim is to check that $T^1_{X_0,0}$ is precisely the subset (in fact vector subspace) of $K^1_f$ corresponding to those infinitesimal unfoldings of $f$ which give infinitesimal deformations of $(X_0, 0)$, i.e. to check that in the natural map

$$\Psi : \mathrm{Hom}_{\mathcal{O}_{\mathbb{C}^N,0}}(I_0, O_{X_0,0}) \to \mathrm{Hom}_{\mathcal{O}_{\mathbb{C}^N,0}}(\mathcal{O}^p_{\mathbb{C}^N,0}, \mathcal{O}_{X_0,0}) \cong O^p_{X_0,0}$$

coming from the map $\mathcal{O}^p_{\mathbb{C}^N,0} \to I_0$ sending the $i$th base element to $f_i$, we have that $\Psi^{-1}((\partial \vec{f})/(\partial z_1), \cdots, (\partial \vec{f})/(\partial z_N)) \cdot \mathcal{O}^p_{X_0,0} = \mathrm{Im}\, d^* =$ those elements in $\mathrm{Hom}_{\mathcal{O}_{\mathbb{C}^N,0}}(I_0, \mathcal{O}_{X_0,0})$ corresponding to trivial infinitesimal deformations of $(X_0, 0)$.

But this is exactly what was checked above, hence the claim.

This shows that the flattener $(S, 0)$ of $F'$ is a versal deformation of $(X_0, 0)$ having as its Zariski tangent space $T^1_{X_0,0}$, and which therefore has the smallest possible Zariski tangent space: it must be a miniversal deformation of $(X_0, 0)$.

4.5.2. Remark 1. If $(X_0, 0)$ is a complete intersection, then $F$ is flat, hence $F'$ is also flat since flatness is preserved by base change, and in this case $F'$ itself is the miniversal deformation of $(X_0, 0)$: we recover the fact that the miniversal deformation of a complete intersection has a non-singular base – and source –. In fact, from the viewpoint we take here, we see that 'all the obstruction comes from the flatness requirement', and this is quite different from the way algebraic deformation theory constructs the obstruction. (See M. Schlessinger's papers [2].)

4.5.3. Remark 2. It is a theorem of Grauert (Inventiones Math. *15*, 3 (1972)) that any isolated singularity has a miniversal deformation. Since such an isolated singularity cannot always be presented as the fibre of a T.S.F. map $f$, (see exercise 3 in 4.3) it motivates the construction of a theory of unfoldings with an infinite-dimensional base, and the extension of the construction of the flattener to mappings between such infinite-dimensional spaces, so that hopefully the base of the miniversal deformation of an isolated singularity would appear as the (finite-dimensional) flattener of a map of infinite-dimensional spaces. (See Astérisque No 16, Soc. Math. Fr. 1974). Our viewpoint is always, given a deformation problem, to embed it in a 'bigger' problem which is unobstructed, [instead of directly constructing a prorepresentable hull of our original problem] and then to seek the base of our original problem as a subspace of the base of that bigger problem. Here of course, the 'bigger problem' is the $\mathcal{K}$-miniversal unfolding of the mapping having our space as fibre. We will see other instances of this in §5.

4.6. Examples of miniversal deformations

4.6.1. Exercise. Check that if $(X_0, 0)$ is *non-singular*, then the canonical map $(X_0, 0) \to \{0\}$ is the miniversal deformation of $(X_0, 0)$: in other words any flat map having non-singular fibre is locally a product: this is the simplicity theorem of §2.

4.6.2. EXERCISE. (1) Let $(X_0, 0) \subset (\mathbb{C}, 0)$ be defined by the ideal $(z^{n+1})$. The miniversal deformation of $(X_0, 0)$ is the map $(X, 0) \to (\mathbb{C}^n, 0)$ induced by the projection $(\mathbb{C} \times \mathbb{C}^n, 0) \to (\mathbb{C}^n, 0)$ on the hypersurface $(X, 0) \subset (\mathbb{C} \times \mathbb{C}^n, 0)$ defined by:

$$z^{n+1} + t_1 z^{n-1} + \cdots + t_n = 0,$$

(where $t_1, \cdots, t_n$ are coordinates on $(\mathbb{C}^n, 0)$).

(2) Check that in characteristic zero, where one can remove by a change of variables the term in $z^n$ of a polynomial of degree $n+1$ in $z$, the statement of the existence of a miniversal deformation of this $(X_0, 0)$ is equivalent to the statement of the classical Weierstrass preparation theorem.

4.6.3. (Taken from the appendix to [6]). Fix an integer $s$, and consider the curve in $\mathbb{C}^3$ (with coordinates $z_1, z_2, z_3$) defined by $(z_1^2 - z_0^3, z_2^2 - z_0^{s+2} z_1)$. (It is the curve parametrized by $z_0 = t^4$, $z_1 = t^6$, $z_2 = t^{2s+7}$). The miniversal deformation of this curve is the restriction of the natural projection $\mathbb{C}^3 \times \mathbb{C}^{2s+10} \to \mathbb{C}^{2s+10}$ to the subspace $X$ of $\mathbb{C}^3 \times \mathbb{C}^{2s+10}$ defined by the ideal generated by $(F_1, F_2)$ in $\mathbb{C}\{z_1, z_2, z_3, v_1, \cdots, v_{2s+10}\}$ where

$$\begin{aligned}\begin{pmatrix} F_1 \\ F_2 \end{pmatrix} = \begin{pmatrix} z_1^2 - z_0^3 \\ z_2^2 - z_0^{s+2} \cdot z_1 \end{pmatrix} &+ v_1 \begin{pmatrix} 1 \\ 0 \end{pmatrix} + v_2 \begin{pmatrix} 0 \\ 1 \end{pmatrix} + v_3 \begin{pmatrix} z_0 \\ 0 \end{pmatrix} + v_4 \begin{pmatrix} 0 \\ z_0 \end{pmatrix} + v_5 \begin{pmatrix} z_1 \\ 0 \end{pmatrix} \\ &+ v_6 \begin{pmatrix} 0 \\ z_1 \end{pmatrix} + v_7 \begin{pmatrix} z_2 \\ 0 \end{pmatrix} \\ &+ v_8 \begin{pmatrix} z_0^2 \\ 0 \end{pmatrix} + v_9 \begin{pmatrix} z_0 \cdot z_1 \\ 0 \end{pmatrix} + v_{10} \begin{pmatrix} z_0 \cdot z_2 \\ 0 \end{pmatrix} + v_{11} \begin{pmatrix} 0 \\ z_0^2 \end{pmatrix} \\ &+ \sum_{j=3}^{s+1} v_{9+j} \begin{pmatrix} 0 \\ z_0^j \end{pmatrix} + \sum_{j'=1}^{s} v_{10+s+j'} \begin{pmatrix} 0 \\ z_0^{j'} \cdot z_1 \end{pmatrix}.\end{aligned}$$

EXERCISE. Check that in this example all the curves corresponding to points in $\mathbb{C}^{2s+10}$ with $v_k = 0$ for $k \neq 7$, 10 and $v_7 \neq 0$ are isomorphic to the plane curve with equation $(z_1^2 - z_0^3)^2 - z_0^{s+2} \cdot z_1 = 0$, and that all the germs of curves corresponding to points with $v_k = 0$ for $k \neq 10$, and $v_{10} \neq 0$ are isomorphic to one another, but *not* isomorphic to the special curve $(X_0, 0)$.

Check that, given $\tilde{v}_7 \in \mathbb{C} - \{0\}$, the mapping $F: (\mathbb{C}^3 \times \mathbb{C}, 0) \to (\mathbb{C}^2 \times \mathbb{C}, 0)$ described by

$$\begin{cases} y_1 \circ F = z_1^2 - z_0^3 + \tilde{v}_7 \cdot z_2 + v_{10} \cdot z_1 z_2 \\ y_2 \circ F = z_2^2 - z_0^{s+2} \cdot z_1 \\ v_{10} \circ F = v_{10} \end{cases}$$

and considered as an unfolding of the map:

$$f:\mathbb{C}^3\to\mathbb{C}^2$$

described by

$$y_1\circ f=z_1^2-z_0^3+\tilde{v}_7 z_2$$
$$y_2\circ f=z_2^2-z_0^{s+2}z_1$$

is $\mathcal{K}$-trivial but not $\mathcal{A}$-trivial.

4.7. In the special case of *functions* $f:(\mathbb{C}^{n+1},0)\to(\mathbb{C},0)$, there is another useful notion of equivalence, known as $\mathcal{R}$-equivalence (for Right-equivalence). Two unfoldings $F, F':(\mathbb{C}^N\times S)\to(\mathbb{C}\times S,0)$, with the same base $(S,0)$ of a function $f:(\mathbb{C}^N,0)\to(\mathbb{C},0)$ are Right-equivalent if there exists an $S$-isomorphism $\Psi:(\mathbb{C}^N\times S,0)\to(\mathbb{C}^N\times S,0)$, unfolding of the identity of $\mathbb{C}^N$, such that $F'\circ\Psi=F$.

By the same methods as used above, we can check that an infinitesimal unfolding $f(z_1,\cdots,z_N)+\varepsilon g(z_1,\cdots,z_N)$ is $\mathcal{R}$-trivial if and only if $g\in((\partial f/\partial z_1),\cdots,(\partial f/\partial z_N))\cdot\mathbb{C}\{z_1,\cdots,z_N\}$ so that the space 'transversal to the $\mathcal{R}$-orbit' is

$$R_f^1=\mathbb{C}\{z_1,\cdots,z_N\}/((\partial f/\partial z_1),\cdots,(\partial f/\partial z_N))$$

which is finite-dimensional if and only if $(X_0,0)=(f^{-1}(0),0)$ has an isolated singularity, and then we have:

$$\dim_{\mathbb{C}} R_f^1=\mu^{(N)}(X_0,0)$$

where $\mu^{(N)}(X_0,0)$ is the Milnor number of $(X_0,0)$, which we have already met in §2 as a discriminant and also (2.18) in equisingularity conditions.

REMARK. The reason why one does not consider $\mathcal{R}$-equivalence when $p>1$ is that the corresponding $R_f^1$ would then be infinite dimensional, except when it is zero, which is the case where $f$ is a germ of submersion. To see this, please do the:

EXERCISE. Generalize the notion of $\mathcal{R}$-equivalence to unfoldings of an $f:(\mathbb{C}^N,0)\to(\mathbb{C}^p,0)$, and check that the corresponding $R_f^1$ is:

$$R_f^1=\mathbb{C}\{z_1,\cdots,z_N\}^p/((\partial\vec{f}/\partial z_1),\cdots,(\partial\vec{f}/\partial z_N)).$$

This shows that $R_f^1$ is the cokernel of a map $\Psi:\mathcal{O}^N_{\mathbb{C}^N,0}\to\mathcal{O}^p_{\mathbb{C}^N,0}$. Using the fact (easy consequence of what has been recalled in the addendum to §1) that the non-empty components of the subspace of $\mathbb{C}^N$ defined by the Fitting ideal of the cokernel of such a map $\Psi$ are of codimension $\leqq N-p+1$, check that if $p>1$, $\dim_{\mathbb{C}} R_f^1<\infty \Leftrightarrow R_f^1=(0)\Leftrightarrow f$ is a submersion.

4.7.1. There is a notion of $\mathcal{R}$-miniversal unfolding of a germ of function $f:(\mathbb{C}^N,0)\to(\mathbb{C},0)$: it is the function described by

$$y=f+\sum_1^{\mu} t_i\cdot s_i(z_1,\cdots,z_N)\qquad (s_i\in\mathbb{C}\{z_1,\cdots,z_N\})$$

$$\text{i.e.: } \mathbb{C}^N\times\mathbb{C}^{\mu}\to\mathbb{C}\times\mathbb{C}^{\mu}\qquad [(\mathbb{C}^{\mu},0) \text{ with coordinates } t_1,\cdots,t_{\mu})]$$

where the images of the $s_i$ form a basis of $R_f^1$.

However, with this definition, one can always take one of the $s_i$ to be 1, and the corresponding $t_i$ changes the function only by a translation.

The custom therefore is to enlarge the Right-equivalence by allowing translations in the target space, and if we call $\tilde{\mathcal{R}}$-equivalence the corresponding notion, we have of course that

$$\tilde{R}_f^1=(z_1,\cdots,z_N)\cdot\mathbb{C}\{z_1,\cdots,z_N\}/((\partial f/\partial z_1),\cdots,(\partial f/\partial z_N))$$

so that $\dim\tilde{R}_f^1=\mu^{(N)}(X_0,0)-1$ and an $\tilde{\mathcal{R}}$-miniversal unfolding of $f$ is a map $F;(\mathbb{C}^N\times\mathbb{C}^{\mu-1},0)\to(\mathbb{C}\times\mathbb{C}^{\mu-1},0)$ where $\mu=\mu^{(N)}(X_0,0)$ described by

$$t_0\circ F=f(z_1,\cdots,z_N)+\sum_1^{\mu-1} t_i\cdot s_i(z_1,\cdots,z_N)\qquad (s_i\in\mathbb{C}\{z_1,\cdots,z_N\})$$

$$t_j\circ F=t_j \quad\text{if}\quad 0<j\leqq\mu-1$$

where the images of $s_i$ form a basis of $\tilde{R}_f^1$.

This should be compared with the miniversal *deformation* of the hypersurface $(f^{-1}(0),0)=(X_0,0)\subset(\mathbb{C}^N,0)$: from what we saw above, this miniversal deformation $G$ appears in a diagram

$$\begin{array}{ccc} & (X,0)\subset(\mathbb{C}^N\times\mathbb{C}^t,0) & \\ {}_G\searrow & & \swarrow{}_{pr_2} \\ & (\mathbb{C}^t,0) & \end{array}$$

where $X$ is defined by

$$f(z_1,\cdots,z_N)+\sum_1^{t} v_i\cdot g_i(z_1,\cdots,z_N)=0\qquad (g_i\in\mathbb{C}\{z_1,\cdots,z_N\})$$

and the images of the $g_i$ form a basis of

$$K_f^1=\mathbb{C}\{z_1,\cdots,z_N\}\bigg/\left(f,\frac{\partial f}{\partial z_1},\cdots,\frac{\partial f}{\partial z_N}\right).$$

Again, one of the $g_i$ must be invertible, so can be chosen to be $-1$. Also, tradition imposes that in this case we write $\tau(X_0,0)$ instead of $t$ for $\dim_{\mathbb{C}} K_f^1$.

Finally, setting $m=\tau(X_0,0)-1$, the miniversal deformation of $(X_0,0)$ is isomorphic to the map:

$$G:(\mathbb{C}^N\times\mathbb{C}^m,0)\to(\mathbb{C}\times\mathbb{C}^m,0)$$

given by

$$\begin{cases} v_0 \circ G = f(z_1, \cdots, z_N) + \sum_1^m v_i \cdot g_i(z_1, \cdots, z_N) \\ v_j \circ G = v_j \qquad 0 < j \leqq m \end{cases}$$

where $(\mathbb{C}^m, 0)$ has coordinates $v_1, \cdots, v_m$.

EXERCISE. Write similarly the miniversal deformation of a complete intersection $(X_0, 0) \subset (\mathbb{C}^N, 0)$ given by $p$ equations as a map $(\mathbb{C}^N \times \mathbb{C}^m, 0) \to (\mathbb{C}^p \times \mathbb{C}^m, 0)$ where $m = \dim_{\mathbb{C}} K_f^1 - p$ (and $f: (\mathbb{C}^N, 0) \to (\mathbb{C}^p, 0)$ has $(X_0, 0)$ as fibre).

### 4.8. Basic results on the openness and economy of unfoldings and deformations

4.8.1. Let $F: (\mathbb{C}^N \times S, 0) \to (\mathbb{C}^p \times S, 0)$ be an unfolding of a map-germ $f: (\mathbb{C}^N, 0) \to (\mathbb{C}^p, 0)$ which is of finite singularity type. Let $\mathfrak{m}_{\mathscr{Y}}$ be the ideal defining $0 \times S$ in $\mathbb{C}^p \times S$, i.e. the ideal in $\mathcal{O}_{\mathbb{C}^p \times S, 0}$ generated by coordinates $(y_1, \cdots, y_p)$ on $(\mathbb{C}^p, 0)$. We can consider

$$A^1_{F/S} = \mathcal{O}^p_{\mathbb{C}^N \times S, 0} \Big/ \mathcal{O}^p_{\mathbb{C}^p \times S, 0} + \left(\frac{\partial \vec{F}}{\partial z_1}, \cdots, \frac{\partial \vec{F}}{\partial z_N}\right)$$

where $\mathcal{O}^p_{\mathbb{C}^p \times S, 0}$ designates the sub-$\mathcal{O}_{\mathbb{C}^p \times S, 0}$-module of $\mathcal{O}_{\mathbb{C}^N \times S, 0}$ generated by the basis elements and $\vec{F} = (y_1 \circ F, \cdots, y_p \circ F)$ and similarly:

$$K^1_{F/S} = \mathcal{O}^p_{\mathbb{C}^N \times S, 0} \Big/ \mathfrak{m}_{\mathscr{Y}} \cdot \mathcal{O}_{\mathbb{C}^N \times S, 0} + \left(\frac{\partial \vec{F}}{\partial z_1}, \cdots, \frac{\partial F}{\partial z_N}\right)$$

$$= \mathcal{O}^p_{X, 0} \Big/ \left(\frac{\partial \vec{F}}{\partial z_1}, \cdots, \frac{\partial \vec{F}}{\partial z_N}\right)$$

and in the case $p = 1$, we consider:

$$\tilde{R}^1_{F/S} = (z_1, \cdots, z_N) \cdot \mathcal{O}_{\mathbb{C}^N \times S, 0} \Big/ \left(\frac{\partial F}{\partial z_1}, \cdots, \frac{\partial F}{\partial z_N}\right).$$

Consider now the map of $\mathcal{O}_{S,0}$-modules

$$\Omega^{1V}_{S,0} \xrightarrow{\theta_B} B^1_{F/S} \quad \text{where} \quad B = A, K \text{ or } \tilde{R}$$

defined as follows: take $D \in \Omega^{1V}_{S,0} = \mathrm{Hom}_{\mathcal{O}_{S,0}}(\Omega^1_{S,0}, \mathcal{O}_{S,0})$ and extend it to a derivation $\tilde{D}$ of $\mathcal{O}_{\mathbb{C}^N \times S, 0}$ by setting $\tilde{D} z_i = 0$. Then define $\theta_B(D)$ to be the residue class in $B^1_{F/S}$ of $\tilde{D}\vec{F} = (\tilde{D}(y_1 \circ F), \cdots, \tilde{D}(y_p \circ F))$, and the kernel of $\theta_B(0)$ corresponds to elements of $\Omega^{1V}_{S,0}(0) = E_{S,0}$ which give $\mathscr{B}$-trivial infinitesimal unfoldings and $F$ is a $\mathscr{B}$-miniversal (resp. versal) unfolding if and only if $\theta_B(0)$ is an isomorphism (resp. is onto), which is just another formulation of what we saw above.

Now as soon as $B^1_{F/S}$ has a support which is finite over $S$ (by $H = pr_2 \circ F$) $\theta_B$ can be sheafified in a map of coherent $\mathcal{O}_S$-modules:

$$\Omega_S^{1\vee} \xrightarrow{\theta_B} H_* B^1_{F/S}$$

(this finiteness will occur if $f$ is T.S.F.) and if $F$ is miniversal at 0, $\theta_B(0): E_{S,0} \to B^1_f$ is an isomorphism, and by Nakayama's lemma this implies that $\theta_B(s)$ is *onto* for all $s \in S$ sufficiently near 0. This is the source of results of *openness of versality*, of which I will now quote only what I will use:

4.8.2. THEOREM (Product decomposition theorem, see [5] chap. III §1). *Let $(X_0, 0)$ be a germ of complete intersection with isolated singularity. Any sufficiently small representative $G: (X, 0) \to (\mathbb{C}^t, 0)$ of a miniversal deformation of $(X_0, 0)$ has the following property: for any $s \in \mathbb{C}^\tau$, if the fibre $X_s = G^{-1}(s)$ has $l(=l(s))$ singular points $x_i(s) (1 \leqq i \leqq l)$ there is a (non-canonical) decomposition of $\mathbb{C}$ in the neighborhood of $s$: $\mathbb{C}^\tau \simeq S_1 \times \cdots \times S_l \times \mathbb{C}^r$ where $r = \tau(X_0, 0) - \sum_{i=1}^{l} \tau(X_s, x_i)$ such that in a neighborhood of $x_i(s)$, $G$ is isomorphic as a deformation to a map:*

$$id_{S_1} \times \cdots \times id_{S_{i-1}} \times G_i \times id_{S_{i+1}} \times \cdots \times id_{S_l} \times \mathbb{C}^r:$$

$$S_1 \times \cdots \times S_{i-1} \times X_i \times S_{i+1} \times \cdots \times S_l \times \mathbb{C}^r \to S_1 \times \cdots \times S_{i-1} \times S_i \times S_{i+1} \times \cdots \times S_l \times \mathbb{C}^r$$

*where $G_i: X_i \to S_i$ is the miniversal deformation of the isolated singularity of complete intersection $(X_s, x_i(s))$ [I have omitted marked points for simplicity of notation and we have $S_i \cong \mathbb{C}^{\tau(X_s, x_i(s))}$.]*

REMARK. The theorem implies in particular that $G$ remains a *versal* deformation of $X_s$ at every point of $X_s$ near 0, hence the terminology 'openness of versality' but it is stronger than this and has in particular the:

4.8.3. COROLLARY. *Let $\Sigma \subset X$ be a closed complex subspace of $X$ (for a small enough representative) defined by conditions concentrated at each singular point of a fibre of $G$. Then, setting $\Delta = G(\Sigma)$ we have in a neighborhood of every points $s \in \mathbb{C}^t$ (and in particular of course if $s \in \Delta$) a decomposition according to the singular points $(x_i(s), 1 \leqq i \leqq l(s))$ of $G^{-1}(s)$:*

$$\Delta = \bigcup_{i=1}^{l} \tilde{\Delta}_i$$

*where* $$\tilde{\Delta}_i = S_1 \times \cdots \times S_{i-1} \times \Delta_i \times S_{i+1} \times \cdots \times S_l \times \mathbb{C}^r$$

*and $\Delta_i \subset S_i$ is $G_i(\Sigma_i)$, where $\Sigma_i \subset X_i$ is the subspace of $X_i$ defined by the conditions which define $\Sigma \subset X$.*

In particular this shows that near every $s \in \Delta$, $\Delta$ is a union of subspaces *in general position* of $\mathbb{C}^\tau$, which are in 1–1 correspondence with the singular points of $G^{-1}(s)$.

If we recall the way we built the miniversal deformation $G:(X,0)\to(S,0)$ (where $S=\mathbb{C}^{\tau(X_0,0)}$) of an isolated singularity of complete intersection, we see that we can define a coherent $\mathcal{O}_S$-module $G_* C^1_{F/S}$ with the property that for any $s \in S$

$$\dim_{\mathbb{C}} G_* C^1_{F/S}(s) = \sum_{i=1}^{l} \tau(X_s, x_i(s))$$

where $x_i(s)(1 \leqq i \leqq l)$ are the singular points of $X_s$. From this, one can deduce the existence of a (locally) finite partition of $S$, $S = \bigcup S_t$ into locally closed complex subspaces, such that

$$s \in S_t \Leftrightarrow \sum_{i=1}^{l} \tau(X_s, x_i(s)) = t$$

$S_{\tau(X_0,0)}$ is the subspace containing 0, and is a closed complex subspace of $S$. We remark that if $s \in S_{\tau(X_0,0)}$ and if $X_s$ has only one singular point, $x(s)$, then the restriction of $G$ to a neighborhood of $x(s)$ is a miniversal deformation of $(X_s, x(s))$.

4.8.4. Theorem (of economy of miniversal deformations, see [$T_1$] exp. 1, §1). *Let $(X_0, 0)$ be a germ of a complete intersection with isolated singularity. Any sufficiently small representative $G:(X,0)\to(\mathbb{C}^\tau,0)$ of a miniversal deformation of $(X_0,0)$ has the following property:*

*the set of points $x \in X$ such that the fibre of $G$ through $x$, $(X_s, x)$ where $s = G(x)$, is analytically isomorphic to $(X_0, 0)$, is reduced to $\{0\}$.*

The meaning of this theorem can be seen as follows:

Corollary 1 (proved by Seidenberg when $(X_0, 0)$ is a plane curve). *For any deformation $H:(Z,0)\to(Y,0)$ of $(X_0,0)$ where $(Y,0)$ is reduced, the following conditions are equivalent:*

(1) *there exists a representative of $H$ such that each fibre $Z$ has a point $z(y)$ such that $(Z_y, z(y))$ is analytically isomorphic to $(X_0, 0)$.*

(2) *There exists a germ of a section $\sigma:(Y,0)\to(Z,0)$ of $H$ and $H$ is isomorphic (over $(Y,0)$) to the trivial deformation $(X_0\times Y,0)\xrightarrow{pr_2}(Y,0)$ in such a way that $\sigma(Y)$ is sent to $0\times Y$.*

*In words: if all the fibres are isomorphic, the deformation is (locally) trivial.*

Corollary 2 (See my appendix to [6]):

A. *Same situation as above, but we only assume that there exists a nowhere dense closed subspace $(F,0)\subset(Y,0)$ such that for any $y \in Y-F$, the fibre $Z_y$*

*has a singular point* $z(y)$ *such that: given any other* $y' \in Y-F$, $(Z_y, z(y))$ *is isomorphic to* $(Z_{y'}, z(y'))$ *(but not necessarily to* $(X_0, 0)$ *since* $F \ni 0$.
*Then we have*: *If* $(Z_y, z(y))$ *is not isomorphic to* $(X_0, 0)$ *for* $y \in Y-F$

$$\tau(X_0, 0) > \tau(Z_y, z(y)) \qquad (y \in Y-F).$$

B. *Let* $S^1 \subset S_{\tau(X_0,0)}$ *be a complex subspace of the 'τ constant stratum of* 0' *constructed above, such that for any* $s \in S^1$, $X_s$ *has only one singular point* $x(s)$. *Then the 'analytic type of the fibres varies continuously' on* $S^1$ *meaning that any* $s \in S^1$ *has a neighborhood* $V_s$ *such that* $(X_{s'}, x(s'))$ *is not isomorphic to* $(X_s, x(s))$, *for all* $s' \in V_s$.

REMARK 1. We have nowhere been explicit about the uniqueness of the base change through which a given unfolding or deformation comes from the miniversal unfolding (or deformation): it is only the *Zariski tangent map* to this base change which is uniquely determined.

REMARK 2. In the proof of all the theorems above, the integration of holomorphic vector fields is often used, a procedure which is very far from being algebraic. Renée Elkik has proved the 'algebraicity' of the construction of miniversal deformations of isolated singularities (Ann. Sc. E. N. S. 4ème série, t. 6 (1973) 553–604) and Bruce Bennett has given a beautiful new proof, without 'integration', of the product decomposition theorem (Normalization theorems for certain modular discriminantal loci, Compositio Math. *32* (1976) 13–32).

EXERCISE. Let $f(z_1, \cdots, z_N) = 0$, $f \in \mathbb{C}\{z_1, \cdots, z_N\}$ define a germ of complex hypersurface $(X_0, 0) \subset (\mathbb{C}^N, 0)$ with isolated singularity. Write a miniversal *deformation* of $(X_0, 0)$ as

$$G: (\mathbb{C}^N \times \mathbb{C}^m, 0) \to (\mathbb{C} \times \mathbb{C}^m, 0) \quad \text{where} \quad m = \tau(X_0, 0) - 1$$

and an $\tilde{\mathcal{R}}$-miniversal unfolding of $f$ as

$$F: (\mathbb{C}^N \times \mathbb{C}^{\mu-1}, 0) \to (\mathbb{C} \times \mathbb{C}^{\mu-1}, 0) \quad \text{where} \quad \mu = \mu^{(N)}(X_0, 0).$$

Show that there is a germ of submersion

$$\varphi: (\mathbb{C} \times \mathbb{C}^{\mu-1}, 0) \to (\mathbb{C} \times \mathbb{C}^m, 0)$$

such that $F$ is obtained up to *isomorphism of deformations* from $G$ by pull back, and that the restriction of $G$ to $\varphi^{-1}(0)$ is trivial as deformation, but not as unfolding.

EXERCISE. Define the '$\mu$ constant stratum' as a closed complex subspace of the target space of the $\tilde{\mathcal{R}}$-miniversal unfolding, in analogy with the '$\tau$ constant stratum' defined above, and show that the dimension of the '$\tau$ constant stratum' in the base of the miniversal deformation is *not* upper semi-continuous.

HINT. Use the plane curve $(z_1^2-z_0^3)^2+z_0^5z_1=0$. Use what you have seen about it in an exercise above to prove that its $\tau$-constant stratum is $\{0\}$. However, for any $v_0\neq 0$, the curve $(z_1^2-z_0^3)^2+z_0^5z_1+v_0z_0^4)=0$ has a one-dimensional $\tau$-constant stratum in the base of its miniversal deformation. Check that this phenomenon does not occur for the $\mu$-constant stratum in the base of the $\tilde{\mathcal{R}}$-miniversal *unfolding* of $f=(z_1^2-z_0^3)^2+z_0^5z_1$. For more details on the $\mu$-constant stratum in the base of the $\tilde{\mathcal{R}}$-miniversal unfolding, see Arnold's article in the Proceedings, International Congress of Mathematicians, Vancouver 1974.

EXERCISE. Let $F:(\mathbb{C}^A,0)\to(\mathbb{C}^B,0)$ be a *flat* map. Show that it is infinitesimally stable if and only if it is a versal deformation of $(F^{-1}(0),0)$.

HINT. Use Nakayama's lemma and the Weierstrass preparation theorem.

REFERENCES

[1] H. HIRONAKA, M. LEJEUNE-JALABERT, B. TEISSIER, Platificateur local en géométrie analytique, et aplatissement local, Singularités à Cargèse 1972, Astérisque No 7–8 (S.M.F.).

[2] M. SCHLESSINGER, Rigidity of quotient singularities, Inventiones Math. 14, (1971) 17–26.

[3] B. TEISSIER, Déformations à type topologique constant, in Séminaire Douady-Verdier 1971–72, Astérisque No 16 (S.M.F. 1974).

[4] B. TEISSIER, Cycles évanescents, sections planes et conditions de Whitney, Singularités à Cargèse 1972, Astérisque No 7–8 (S.M.F. 1973).

[5] G. N. TJURINA, Locally semi-universal flat deformations, Math. of the U.S.S.R. Isvestija *3*, No 5 (1969).

[6] O. ZARISKI, Cours sur les modules de branches planes, Centre de Mathématiques, Ecole Polytechnique, 91128 Palaiseau Cedex France, (1973).

## §5. Discriminants

In this section, we define and study the discriminant of versal deformations $G:(X,0)\to(S,0)$ (with $S$ non-singular) of complete intersections with isolated singularities, i.e. equivalently, of stable and flat maps between non-singular spaces. In the case of hypersurfaces, we go into more detail, and finally reach the goal of these notes, which is to show connections between naive invariants of the discriminant, on one hand, and invariants of the geometry *up to* (c)-*cosécance* (§ 2) of the hypersurface, on the other hand.

On the way, we meet some naive invariants of the discriminant which are *not* invariants of (c)-cosécance of the fibre, and we emphasize the structure of the discriminant as an envelope.

5.1. To study versal deformations with a non-singular base of a complete intersection, it is sufficient to study the miniversal ones, since a versal deformation is the product of a miniversal one by the identity of some space.

Let $G:(X, 0) \to (S, 0)$ be a miniversal deformation of a germ of complete intersection with isolated singularity $(X_0, 0) = (G^{-1}(0), 0)$. As we saw in §4, $G$ is an infinitesimally stable and flat map between non-singular spaces (and is stable, by a theorem of Mather), and $G$ can be described as the restriction over $0 \times \mathbb{C}^\tau$ of a $\mathcal{K}$-miniversal unfolding $F:(\mathbb{C}^N \times \mathbb{C}^\tau, 0) \to (\mathbb{C}^p \times \mathbb{C}^\tau, 0)$ of a flat map $f:(\mathbb{C}^N, 0) \to (\mathbb{C}^p, 0)$ having $(X_0, 0)$ as fiber.

Tradition imposes that we write $T^1_{X/S}$ for the $\mathcal{O}_X$-module $K^1_{F/S}$ of §4 (here $(S, 0) = (\mathbb{C}^\tau, 0)$) and $\tau = \tau(X_0, 0)$ for $\dim_{\mathbb{C}} K^1_f$, which was denoted by $t$ in §4. We also saw in §4 that $G$ itself can be described as an unfolding of $f$, namely $G:(\mathbb{C}^N \times \mathbb{C}^m, 0) \to (\mathbb{C}^p \times \mathbb{C}^m, 0)$ commuting to projections to $\mathbb{C}^m$, where $m = \tau(X_0, 0)$-$p$.

We wish to study the *critical subspace* $C$ of $G$ in the sense of §2: it is the subspace of $X$ defined by $F_{N-p}(\Omega^1_{X/S})$.

EXERCISE. Show that $F_{N-p}(\Omega^1_{X/S}) = F_0(T^1_{X/S})$ and that $T^1_{X/S}$ has a presentation:

$$\mathcal{O}_X^N \xrightarrow{\Psi} \mathcal{O}_X^p \longrightarrow T^1_{X/S} \longrightarrow 0.$$

Since $G$ is flat and $(X_0, 0)$ has an isolated singularity, we see by using the simplicity theorem of §2 (at the non-singular points of $(X_0, 0)$) and the Weierstrass preparation theorem for a sufficiently small representative of $G$, the critical subspace $C$ will be finite over $S = \mathbb{C}^\tau$, i.e. $G \mid C: C \to \mathbb{C}^\tau$ is a finite map. By §1, $G_*(C) = \operatorname{im}(G \mid C)$ is therefore a subspace of $\mathbb{C}^\tau$, and by the theorem of Bertini, (see §2, second part) since $X$ is non-singular, $G(C)$ is a *strict* closed subspace of $\mathbb{C}^\tau$, so that $\operatorname{Ann}_{\mathcal{O}_{\mathbb{C}^\tau}}(G_*\mathcal{O}_C) \neq 0$.

5.1.1. CLAIM. $\operatorname{depth}_{\mathcal{O}_{\mathbb{C}^\tau,0}}(G_*\mathcal{O}_C) \geqq \tau - 1$.

The proof is as follows: by an easy computation of local algebra, (see A. Grothendieck: E.G.A. IV, 0.16.4.8) $\operatorname{depth}_{\mathcal{O}_{\mathbb{C}^\tau,0}}(G_*\mathcal{O}_C) = \operatorname{depth}_{\mathcal{O}_{X,0}}(\mathcal{O}_{C,0})$ since $\mathcal{O}_C$ is a finite $\mathcal{O}_{\mathbb{C}^\tau,0}$-module (see above). Now we use a result of Buschsbaum-Rim ([5] cor. 2.7) (already used in §1) to the effect that since $T^1_{X/S}$ has a presentation as above, the maximum length of a sequence of elements of $F_0(T^1_{X/S})$ which is a regular sequence for $\mathcal{O}_{X,0}$ (called the $F_0(T^1_{X/S})$-depth of $\mathcal{O}_{X,0}$ is $\leqq N - p + 1$, and also that if it is equal to $N - p + 1$, then $\operatorname{dh}_{\mathcal{O}_{X,0}}(\mathcal{O}_{C,0})$ i.e., $\operatorname{depth}_{\mathcal{O}_{X,0}}(\mathcal{O}_{X,0}/F_0(T^1_{X/S}))$, is also equal to $N - p + 1$. Therefore, if we can prove that the $F_0(T^1_{X/S})$-depth of $\mathcal{O}_{X,0}$ is at least $N - p + 1$, we will obtain $\operatorname{dh}_{\mathcal{O}_{X,0}}(\mathcal{O}_{C,0}) = N - p + 1$ and then by the equality quoted in 3.5: $\operatorname{depth}(\mathcal{O}_{C,0}) = m + p - 1 = \tau - 1$, and we win. Now, a basic property of a Cohen-Macaulay local ring $A$ is that for any proper ideal $I$ in $A$, we have $I$-depth $A = \dim A - \dim A/I$. Since $\mathcal{O}_{X_0,0}$ is regular, it is Cohen-Macaulay, and hence, setting $d = F_0(T^1_{X/S})$-depth of $\mathcal{O}_{X,0}$, we have:

$$\dim \mathcal{O}_{X,0}/F_0(T^1_{X/S}) = \dim \mathcal{O}_{X,0} - d$$

and hence:

$$d = \dim \mathcal{O}_{X,0} - \dim C$$

but we have already seen that $\dim C = \dim G_*(C) \leqq m + p - 1$ by Bertini's theorem, hence $d \geqq N + m - (m + p - 1) = N - p + 1$ which proves the claim.

Remark. If you wish to understand the geometry behind this kind of computation, as well as that in the addendum to §1, and more, I suggest reading the beautiful papers of G. Kempf ([15] and [16]). The above presentation partly follows suggestions of Patrick Barril.

Our goal is to deduce from this:

5.2. Theorem (of purity of discriminants. See [7]). *Let $G : (X, 0) \to (S, 0)$ be a miniversal deformation for a germ of a complete intersection with isolated singularity. Then the discriminant $D = G_*(C)$ of $G$, as defined in §2, (i.e. by $F_0(G_*\mathcal{O}_C)$ is a reduced and irreducible hypersurface germ in the non-singular space $(S, 0) = (\mathbb{C}^\tau, 0)$. Furthermore, $C$ is normal and $G \mid C : C \to D$ is the normalization map.*

Proof. Since $D$ is defined as the image in the sense of §1 of $C$ by $G \mid C$, the fact that it is a hypersurface follows immediately from the claim above and 3.5. To prove the rest of the assertions, we stop a while to give a typical application of the stability of $G$ (which we know thanks to Mather's theorem) and the product decomposition theorem of §4:

*The Thom-Boardman strata of G.* (See [4], [23]).

Given a stable $G$ such as above, for every sufficiently small representation there are non-singular complex subspaces $\Sigma^I(G)$ of $X$, indexed by sequences of integers $I = (i_1, \cdots, i_k)$ and defined inductively as follows: $\Sigma^\varnothing = X$, and

$$\Sigma^{I,j}(G) = \{x \in \Sigma^I / \dim \operatorname{Ker}(T_x(G \mid \Sigma^I)) = j\}$$

in words: points where the dimension of the kernel of the tangent map of $G \mid \Sigma^I$ is exactly $j$.

In fact these can be defined for any map $G$ as *subspaces* using Fitting ideals, and they can also be defined as coming from a stratification of the jet-space $J^\infty(X, S)$, and in this way one sees that if $G$ is stable, $\Sigma^I(G)$ is non-singular.

5.2.1. Exercise. Show that in the case of hypersurfaces ($p = 1$) one has $C = \Sigma^N(G)$, and that the set of points of $C$ which are of multiplicity $a = m_0(X_0, 0)$ (the multiplicity in the usual sense of $(X_0, 0)$) in their fibre (of $G$) is $\Sigma^{N, \cdots, N}(G)$ when $N$ occurs $a - 1$ times.

Boardman also gave a formula for the codimension of his strata $\Sigma^I(G)$ as follows: if $\Sigma^I(G)$ is not empty, with $I=(i_1, \cdots, i_k)$, then

($\alpha$) $i_1 \geqq i_2 \geqq \cdots \geqq i_k$

($\beta$) $N-p \leqq i_1 \leqq N+m$ $\quad (m = \dim K_f^1 - p)$

($\gamma$) if $i_1 = N-p$, $i_2 = \cdots = i_k = N-p$,

and then the codimension of $\Sigma^I(G)$ in $X$ is

$$\nu_I = (p-N+i_1)\mu_{i_1,\ldots,i_k} - (i_1-i_2)\mu_{i_2,\ldots,i_k} - \cdots - (i_{k-1}-i_k)\mu_{i_k}$$

where $\mu_{i_1,\ldots,i_k}$ is the number of non-increasing sequences $(j_1, \cdots, j_k)$ with $j_1 \neq 0$ and $j_l \leqq i_l$.

We will see these Thom-Boardman strata again later, but for the moment we use them to prove the theorem of purity: first, remark that it is an easy consequence of 4.8.3 that the mapping $G \mid \Sigma^I : \Sigma^I \to G(\Sigma^I)$ is generically one-one, just because $G(\Sigma^I)$ has to be locally irreducible outside of a nowhere dense closed subspace of itself. If $i_1 > N-p$, then $\Sigma^I \subset C$ and hence $G \mid \Sigma^I$ is finite. If we denote by $S^I$ the closure of $\Sigma^I$ in $X$, which is a closed complex subspace of $X$, we have:

5.2.2. Proposition. *If* $i_1 > N-p$, $G \mid S^I : S^I \to G_*(S^I)$ *is a* proper modification *of* $G_*(S^I) = \overline{G_*(\Sigma^I)}$, *and in particular* $G_*(S^I)$ *is reduced.* (*See* §2 *for 'proper modification'*): *this follows from the above remark and the definition of the* $\Sigma^I$ *by rank conditions.*

*End of the proof of the purity theorem.* $\mathcal{O}_C$ is of depth (as $\mathcal{O}_{\mathbf{C}^\tau,0}$-module) equal to its dimension $\tau-1$ (5.1.1) hence all its irreducible components at 0 are of the same dimension $\tau-1$, which is also the dimension of $\Sigma^{N-p+1}(G)$, and of no other Boardman stratum, by the codimension formulas. Hence, $C = S^{N-p+1}(G)$, and hence $D$ is reduced. Moreover, the singular locus of $C$ is contained in the union of $\Sigma^{I'}$ where $I' > (N-p+1, 0 \cdots 0)$ in the lexicographic order, and this is of codimension at least two in $C$. Since we already know that $C$ has depth $\tau-1$ at each of its points, it follows from the criterion of Serre ([30]) that $C$ is *normal* hence it is locally analytically irreducible, and hence its image $D$ is also locally analytically irreducible at 0, and finally since the induced map $G \mid C : C \to D$ is a proper modification of $D$ (5.2.2), it has to be the normalization.

Remarks. (1) As we saw in 5.2.1, in the case $p=1$, $C$ is even non-singular, so that $G \mid C$ is a resolution of singularities of the hypersurface $D$. However, there is a deeper fact about this map $C \to D$: it is in fact also the *development* of $D$ in the sense of §2, as we shall see below.

(2) I want to illustrate the proof above in the case $p=1$ as follows: how do we prove in this case that the discriminant $D$ is reduced? as follows: the

Boardman stratum $\Sigma^N(G)$ is dense in $C$, and $x \in \Sigma^N(G)$ if and only if the singularity at $x$ of the fibre of $G$ through $x$ is isomorphic to the ordinary quadratic singularity $A_1$ with equation $z_1^2 + \cdots + z_N^2 = 0$. Now $D$ is of course locally irreducible outside a nowhere dense closed subspace $B$, so by 4.8.3 there is an homeomorphism $C - n^{-1}(B) \to D - B$. At every point of $(C - n^{-1}(B)) \cap \Sigma^N(G)$, which is dense in $C$, we can apply the openness of versality of §4 (4.8.2) and see that our whole situation is locally isomorphic to a cylinder over the situation for the singularity $A_1$: it remains to check that for this singularity, the map from critical subspace to discriminant subspace is an isomorphism, which is obvious since the versal deformation is given by $t_0 \circ F = z_1^2 + \cdots + z_N^2$: critical subspace and discriminant are both reduced points (here $\mu^{(N)} = 1$).

### 5.3. Examples of discriminants

The purpose of these examples is to convince the reader that it is much more convenient to give oneself the discriminants *parametrically*, i.e. by $C \xrightarrow{n} D$ ($n = G \mid C$) (more precisely, as image of $G \mid C : C \to \mathbb{C}^p \times \mathbb{C}^m$) than to actually compute the equation of $D$ in $\mathbb{C}^p \times \mathbb{C}^m$. Here is what happens in the simplest cases: for the versal deformations of the singularities defined in $\mathbb{C}$ by $(z^N) \cdot C\{z\}$: the versal deformation is the restriction of $(\mathbb{C} \times \mathbb{C}^{N-1}, 0) \xrightarrow{pr_2} (\mathbb{C}^{N-1}, 0)$ to the hypersurface $X$ defined in $\mathbb{C} \times \mathbb{C}^{N-1}$ by

$$z^N + t_{N-2} z^{N-2} + \cdots + t_0 = 0$$

and the equation of the discriminant in $\mathbb{C}^{N-1}$ is of course just the $z$-resultant of this equation and its derivative with respect to $z$.

Here is what you get for the equation of the discriminant:

$N = 1$: non singular, discriminant empty.

$N = 2$: $t_0 = 0$.

$N = 3$: $27t_0^2 + 4t_1^3 = 0$.

$N = 4$: $256t_0^3 - 27t_1^4 - 128t_0^2t_2^2 + 144t_0t_1^2t_2 + 16t_0t_2^4 - 4t_1^2t_2^3 = 0$.

$N = 5$: $3.125t_0^4 - 3.750t_0^3t_2t_3 + 2.250t_0^2t_1t_2^2 + 825t_0^2t_2^2t_3^2 + 108t_0^2t_3^5$
$- 900t_0^2t_1t_3^3 + 2.000t_0^2t_1^2t_3 - 630t_0t_1t_2^3t_3 + 16t_0t_2^3t_3^3 + 108t_0t_2^5$
$- 72t_0t_1t_2t_3^4 + 560t_0t_1^2t_2t_3^2 - 1.600t_0t_1^3t_2 + 256t_1^5 - 128t_1^4t_3^2$
$+ 16t_1^3t_3^4 + 144t_1^3t_2^2t_3 - 27t_1^2t_2^4 - 4t_1^2t_2^2t_3^3 = 0$.

$N = 6$: the equation has 76 monomials (available on request).

(I thank Gérard Lejeune for programming this for me.)

The real part of the discriminant for $N=4$ is well known (not as discriminant, but as bifurcation set, see below) under the name *swallowtail:*

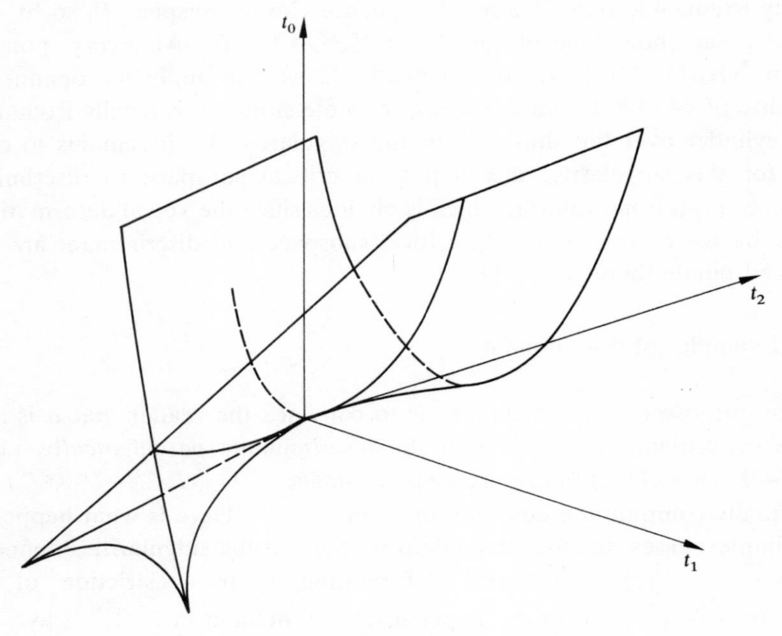


Figure 3

N.B. You can find pictures of sections of the (real part of the) discriminant also for $N=5$ (again considered as bifurcation set) in [38].

Indeed, in the case of the discriminant of the general unitary polynomial in one variable, there is a very pleasant way of building the discriminant geometrically, due to Dominique Thillaud (unpublished):

Consider $P(z)=z^N+t_{N-1}z^{N-1}+\cdots+t_0=0$.

Notice that here we have *not* removed the term in $z^{N-1}$, so this is in fact a versal deformation of $(z^N)\cdot C\{z\}$. Now the set of points where $P(z)=0$ has an $N$-uple root is a curve $M_N$ in $\mathbb{C}^N\,(t_0,\cdots,t_{N-1})$ given parametrically by

$$t_i=(-1)^{N-i}\binom{N}{i}u^{N-i}$$

where $u$ is the value of the root.

Proposition (Thillaud). *The set $M_p$ of points in $\mathbb{C}^N$ where $P$ has at least a* $p$uple *root is the* $(N-p)$th developable variety $D_{N-p}$ *of the curve $M_N$, where* *the* $(N-p)$th *developable variety of* $M_N$ *is the closure in* $\mathbb{C}^N$ *of the set* $\bigcup_{t\in M_N^*} T_{N-p}(M_N, t)$, *where* $T_{N-p}(M_N, t)$ *is the osculating space of dimension* $N-p$ *to the curve* $M_N$ *at* $t$, *meaning the affine subspace of* $\mathbb{C}^N$ *made of elements of the form* $\vec{M}_N(v)+\sum_{i=1}^{N-p} \lambda_i \vec{M}^{(i)}(u)$, $\underline{\lambda} \in \mathbb{C}^{N-p}$. $M_N^*$ *designates the points* $t$ *of* $M_N$ *where the dimension of* $T_{N-p}(M_N, t)$ *is actually* $N-p$. *Remark that* $\dim M_p = N-p+1$.

Corollary. *The discriminant of* $P(z)$ *is the* $(N-2)$th *developable variety of the curve* $M_N$.

Proof. Given as an exercise.

Question. Take a non-singular curve $M$ in projective $N$-space, and consider its $(N-2)$th developable, $D_{N-2}$ which is a hypersurface on $\mathbb{P}^N$ containing $M$. Is it true that there exists a Zariski dense set of points $U \subset M$ such that if $p \in U$, $D_{N-2}$ is locally at $p$ analytically isomorphic to the discriminant of $z^N + t_{N-1} z^{N-1} + \cdots + t_0$?

(For $N=3$ it is well known that a curve is a cuspidal edge for its developable surface.) It would then follow that the section of $D_{N-2}$ by a generic hypersurface through $p$ is locally at $p$ analytically isomorphic to the $N$-swallowtail, i.e. the discriminant of $z^N + t_{N-2} z^{N-2} + \cdots + t_0$.

5.4. We now come back to the general discriminants, with the following easy consequence of the product decomposition theorem of §4:

Remark. Given a discriminant $(D, 0) \subset (\mathbb{C}^p \times \mathbb{C}^m, 0)$ as above, for each integer $k$, $1 \leqq k \leqq m+p$, there exists for any sufficiently small representative of $D \subset \mathbb{C}^p \times \mathbb{C}^m$ a closed complex subspace $B_k \subset D$, of codimension at least $k$ in $D$, such that at every point $p \in D - B_k$, $D$ is locally analytically isomorphic to a finite union $\bigcup_i \tilde{D}_i$ of cylinders $\tilde{D}_i$ in general position over discriminants $D_i \subset \mathbb{C}^{\tau_i}$ of miniversal deformations of isolated singularities of complete intersections with $\sum_i \tau_i \leqq k$.

Corollary. *For each* $k$, *there exists a Zariski open dense subset* $U$ *in the Grassmannian* $G(m+p, k)$ *of directions of* $k$*-planes in* $\mathbb{C}^{m+p}$, *such that given* $H_0 \in U$, *for a sufficiently small representative of* $D \subset \mathbb{C}^p \times \mathbb{C}^m$, *there exists an open analytic dense subset* $V$ *of the space of affine* $k$*-planes in* $\mathbb{C}^p \times \mathbb{C}^m$ *having direction* $H_0$ *such that: If* $H \in V$, $D \cap H$ *is locally near each of its points isomorphic to a finite union* $\bigcup_i \tilde{\tilde{D}}_i$ *of cylinders, in general position in* $H$, *over discriminants* $D_i \subset \mathbb{C}^{\tau_i}$ *with* $\sum \tau_i \leqq k$.

In particular:

5.4.1. Exercise (from [7]). Show that for any sufficiently small representative of a discriminant $D \subset \mathbb{C}^p \times \mathbb{C}^m$ as above, for 'almost every' affine

2-plane $H$ (in the sense made precise above), $H \cap D$ is a plane curve having as singularities only nodes and cusps.

HINT. Use the local irreducibility of discriminants at their origin, the general position coming from the product decomposition theorem, and the fact that the only singularities of a complete intersection $(X_0, 0)$ such that $\tau(X_0, 0) = 2$ is, up to isomorphism, the singularities called $A_2$, or 'suspension' of $z^3 = 0$, namely $\sum_{i=1}^{N-1} z_i^2 + z_N^3 = 0$, where $N = \dim X_0 + 1$.

By an easy flatness argument, one sees that in fact the number $d$ of nodes and the number $k$ of cusps of $D \cap H$ are the same for 'almost all $H$' in the sense made precise above. These numbers $k$ and $d$ depend only upon the geometry of $D$, i.e. of $G$. They are mentioned here because they are related to the geometry of certain plane sections of the discriminant $D$ *through the origin*, and these plane sections, in the case $p = 1$, will be one of our main points of interest.

5.4.2. We therefore have the following picture in mind: a non-singular point $p$ in the discriminant corresponds to a singularity in $G^{-1}(p)$ at which $G^{-1}(p)$ is locally isomorphic to an ordinary quadratic singularity: $z_1^2 + \cdots + z_{N-p+1}^2 = 0$ (of the right dimension). And it is the only singularity of $G^{-1}(p)$.

A point $p$ at which the discriminant is cusp-like (i.e. cylinder over a cusp) corresponds to a point in the fiber $G^{-1}(p)$ at which it is isomorphic to $z_1^2 + \cdots + z_{N-p}^2 + z_{N-p+1}^3 = 0$ (i.e. again cusp-like) a point at which the discriminant is node-like corresponds to a fibre $G^{-1}(p)$ which has *two* ordinary quadratic singularities, and there are no points more complicated than that outside a subspace of $D$ of codimension 2.

5.5. We will from now on consider any versal deformations of germs of *hypersurfaces* with isolated singularities, i.e. we restrict to the case $p = 1$.

5.5.1. THEOREM. *The map induced by* $G:(X, 0) \to (S, 0)$ *from the critical subspace* $C$ *to the discriminant* $D = G_*(C)$ *is the development of* $D$, *in the sense of* §2. *Since* $C$ *is non-singular, we see that the singularities of* $D$ *are resolved by* one *development (and in this case the development turns out to coincide with the normalization).*

PROOF. We need the following

LEMMA. *Any* $f \in \mathbb{C}\{z_1, \cdots, z_N\}$ *can be written after a change of variables:*

$$f(z_1, \cdots, z_N) = z_1^2 + \cdots + z_p^2 + \tilde{f}(z_{p+1}, \cdots, z_N)$$

*where* $\tilde{f} \in (z_{p+1}, \cdots, z_N)^3$ *is uniquely determined, and has an isolated critical point if such was the case for* $f$.

The proof is immediate by using the classical form of the Weierstrass preparation theorem and the removal of the term in $z$ in a polynomial of degree 2 in $z$.

Next we make the

REMARK. The bases of miniversal deformations for $f$ and $\tilde{f}$ are isomorphic, and the discriminants are also isomorphic.

PROOF. A miniversal deformation for $f=0$ is described by

$$G:(\mathbb{C}^N\times\mathbb{C}^m,0)\to(\mathbb{C}\times\mathbb{C}^m,0)\quad(\text{coordinates } v, t_1,\cdots,t_m)\text{ on }\mathbb{C}\times\mathbb{C}^m)$$

$$v\circ G=f(z_1,\cdots,z_N)+\sum_{i=1}^m t_ig_i(z_1,\cdots,z_N)$$

$$t_i\circ G=t_i\quad(1\leqq i\leqq m)$$

where the images of 1 and the $g_i$ in $C\{z_1,\cdots,z_N\}/(f,(\partial f/\partial z_1,\cdots,(\partial f/\partial z_N))$ form a basis over $\mathbb{C}$. Clearly

$$\mathbb{C}\{z_1,\cdots,z_N\}\Big/\left(f,\frac{\partial f}{\partial z_1},\cdots,\frac{\partial f}{\partial z_N}\right)=\mathbb{C}\{z_{p+1},\cdots,z_N\}\Big/\left(\tilde{f},\frac{\partial\tilde{f}}{\partial z_{p+1}},\cdots,\frac{\partial\tilde{f}}{\partial z_N}\right)$$

and therefore we can assume that the $g_i$ depend only upon $z_{p+1},\cdots,z_N$. Hence

$$\tilde{G}:(\mathbb{C}^{N-p}\times\mathbb{C}^m,0)\to(\mathbb{C}\times\mathbb{C}^m,0)$$

$$v\circ\tilde{G}=\tilde{f}(z_{p+1},\cdots,t_N)+\sum_1^m t_ig_i(z_{p+1},\cdots,z_N)$$

$$t_i\circ\tilde{G}=t_i$$

is a miniversal deformation of $\tilde{f}(z_{p+1},\cdots,z_N)=0$, and has the same discriminant as $G$.

APPLICATION. For the study of $f$ from the discriminant of its miniversal deformation, we can always assume that $f\in(z_1,\cdots,z_N)^3$, and we will do so. It implies that we can choose $g_1,\cdots,g_N$ equal to $(z_1,\cdots,z_N)$, and from now on we will write our miniversal deformations by

$$v\circ G=F(z,t)\overset{\text{def}}{=}f(z_1,\cdots,z_N)+t_1z_1+\cdots+t_Nz_N+t_{N+1}g_{N+1}(z)+\cdots+t_mg_m(z),$$

where $g_k\in(z_1,\cdots,z_N)^2$ if $k\geqq N+1$.

Let now $\delta(v,t_1,\cdots,t_m)=0$ be an equation for our discriminant $(D,0)\subset(\mathbb{C}\times\mathbb{C}^m,0)$.

PROPOSITION. *We have for all* $1 \leqq j \leqq m$

$$\frac{\partial \delta}{\partial t_j} \circ G \bigg|_C = -\frac{\partial \delta}{\partial v} \circ G \bigg|_C \cdot \frac{\partial F}{\partial t_j} \bigg|_C.$$

PROOF. $C$ is defined in $\mathbb{C}^N \times \mathbb{C}^m$ by the ideal $((\partial F/\partial z_1), \cdots, (\partial F/\partial z_N))$ [which incidentally shows again that $C$ is non-singular $(\partial F/\partial z_i)$ begins with $t_i$] since $D = G_* C$, we have that $\delta \circ G$ vanishes on $C$ (i.e. $\delta \in \mathrm{Ann} \cdot G_* \mathcal{O}_C$: in fact $(\delta)\mathcal{O}_{\mathbb{C} \times \mathbb{C}^m} = \mathbf{Ann}\ G_* \mathcal{O}_C$ since $D$ is reduced, see §1).

Since $C$ is reduced (being non-singular!) it implies that

$$\delta \circ G = \sum_{i=1}^{N} A_i(z, t) \frac{\partial F}{\partial z_i}(z, t)$$

$$\frac{\partial(\delta \circ G)}{\partial t_j} = \sum_{i=1}^{N} A_i \cdot \frac{\partial^2 F}{\partial z_i \partial t_j} \bmod \left(\frac{\partial F}{\partial z_1}, \cdots, \frac{\partial F}{\partial z_N}\right)$$

and since $\delta$ comes from $\mathbb{C} \times \mathbb{C}^m$:

$$0 = \frac{\partial(\delta \circ G)}{\partial z_i} = \sum_{i=1}^{N} \sum_{j=1}^{N} A_i(z, t) \frac{\partial^2 F}{\partial z_i \partial z_j}(z, t) \bmod \left(\frac{\partial F}{\partial z_1}, \cdots, \frac{\partial F}{\partial z_N}\right)$$

but as we saw, on an open-analytic dense subset of $C$, the hessian determinant det $((\partial^2 F/\partial z_i \partial z_j))$ is different from 0. Hence $A_i(z, t)$ vanishes on $C$, hence $A_i \in ((\partial F/\partial z_1), \cdots, (\partial F/\partial z_N))$ and therefore on $C$ we have:

$$0 = \frac{\partial(\delta \circ G)}{\partial t_j}\bigg|_C = \frac{\partial \delta}{\partial v} \circ G \bigg|_C \cdot \frac{\partial(v \circ G)}{\partial t_j}\bigg|_C + \frac{\partial \delta}{\partial t_j} \circ G \bigg|_C$$

$$= \frac{\partial \delta}{\partial v} \circ G \bigg|_C \cdot \frac{\partial F}{\partial t_j}\bigg|_C + \frac{\partial \delta}{\partial t_j} \circ G \bigg|_C$$

which proves the proposition.

In particular, we can write

$$\frac{\partial \delta}{\partial t_i} \cdot \mathcal{O}_{C,0} = -\frac{\partial \delta}{\partial v} \cdot \frac{\partial F}{\partial t_i} \cdot \mathcal{O}_{C,0}$$

which shows that the restriction to $D$ of the Jacobian ideal of $\delta$ becomes invertible on $C$:

$$j(\delta) \cdot \mathcal{O}_{C,0} = \frac{\partial \delta}{\partial v} \cdot \mathcal{O}_{C,0}.$$

Since for a hypersurface, the development is isomorphic to the blowing up of the restriction to the hypersurface of the Jacobian ideal, as is easily checked (see §2), we see that $C$ must dominate the development $D_1$ of $D$, i.e. we have a factorization for $n = G \mid C$:

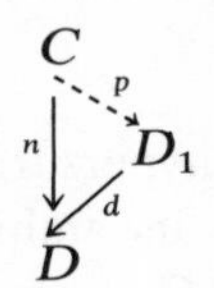


In particular, $d^{-1}(0)$ has only one point, and to prove that $p$ is a local isomorphism at 0, it is sufficient to prove that $\mathfrak{m}_1 \cdot \mathcal{O}_{C,0} = \mathfrak{m}_{C,0}$ where $\mathfrak{m}_1$ is the maximal ideal of $D_1$ at $d^{-1}(0)$, and $\mathfrak{m}_{C,0}$ is the maximal ideal of $C$ at 0. Because of the equations described above for $C$ in $\mathbb{C}^N \times \mathbb{C}^m$, we see that $\mathfrak{m}_{C,0}$ is generated by the restrictions to $C$ of $z_1, \cdots, z_N, t_{N+1}, \cdots, t_m$. Since $t_{N+1}, \cdots, t_m$ give elements which are already in $\mathfrak{m}_{D,0}$ hence a fortiori in $\mathfrak{m}_{D_1,0}$, it is sufficient to check that $z_i \in \mathfrak{m}_{D_1,0}$. But by definition of a blowing up, the ratios $(\partial\delta/\partial t_i)/(\partial\delta/\partial v)$ which are meromorphic on $D$ and tend to 0 since they do so when lifted to $C$ by the proposition above, become already holomorphic on $D_1$, which means that $(\partial F/\partial t_j)|_C$, a priori meromorphic on $D_1$ is in fact in $\mathfrak{m}_{D_1,0}$ $(1 \leqq j \leqq m)$. But for $1 \leqq j \leqq N$, $(\partial F/\partial t_j) = z_j$, which shows that $z_j \in \mathfrak{m}_{D_1,0}$ hence $D_1 \simeq C$ and this concludes the proof of the theorem.

Remark 1. The fact that $(C, 0) \to (D, 0)$ is the development seems to me to lie deeper than the fact it is the normalization. Are there similar results for the other $G_*(S^I)$?

Remark 2. This result enables one to identify $C$ with the 'projectivized conormal bundle to $D$' in the 'projectivized cotangent bundle to $\mathbb{C} \times \mathbb{C}^m$' and Pham has shown to me that the real-analytic version of this result (which I had proved only for aesthetic reasons) was of great use in the theory of caustics. It is not my purpose to go into this here, and I refer to [29], [1].

Remark 3. It follows from the theorem and the fact that there is only one point of $C$ lying over 0, that there is only one limit position at 0 of tangent hypersurfaces to $D$, and since the $(\partial F/\partial t_j)$ tend to 0 with $(z, t)$, this limit position is the 'horizontal hyperplane' $dv = 0$. This can be expressed by saying that the discriminant 'flattens' on the hyperplane $0 \times \mathbb{C}^m$ (a flattening very different from that used in §4!). We shall see more about this below.

Exercise. Show that the tangent cone to $D$ at 0 is set-theoretically given by $v = 0$ [strictly speaking one would say $V = 0$ where $V$ is the initial form of $v$ in the associated graded ring $gr_{\mathfrak{m}}\mathcal{O}_{\mathbb{C}\times\mathbb{C}^m,0}$].

Hint. Use the fact that the tangent cone to $D$ must be contained in the image of the restriction to $T_{C,0}$ of the tangent map to $G$, which can be written explicitly.

5.5.2. PROPOSITION. *The multiplicity of $D$ at 0 is equal to the Milnor number $\mu^{(N)}(X_0, 0)$, where $(X_0, 0)$ is the hypersurface with isolated singularity, fibre of $G$.*

PROOF. The multiplicity of a hypersurface at 0 can be computed as its intersection multiplicity at 0 (in its ambiant space) with any non-singular curve $C$ such that $T_{C,0} \not\subset C_{D,0}$ ($C_{D,0}$ = tangent cone). It follows from the exercise above that the line $t_1 = \cdots = t_m = 0$ (i.e. the $v$-axis) satisfies this. Therefore we have to compute the multiplicity of the restriction of our discriminant to this $v$-axis. But *since our definition of the discriminant is compatible with base change*, this is exactly what was computed in 2.6.

REMARK. Consider now the restriction $\pi : D \to \mathbb{C}^m$ to $D$ of the natural projection $\mathbb{C} \times \mathbb{C}^m \to \mathbb{C}^m$: it has a discriminant which we will denote by $(B, 0) \subset (\mathbb{C}^m, 0)$. $B_{\text{red}}$ is called the *bifurcation locus* of $G$: it is exactly the set of values $t \in \mathbb{C}^m$ such that the corresponding function $v = f(z) + \sum_1^m t_i\, g_i(z)$ is not an excellent Morse function (near the origin) i.e. a function having only ordinary quadratic singularities giving distinct critical values.

5.5.3. We remark that these notions can be defined just as well in the real-analytic case, and then the discriminant and bifurcation locus will be *semi-analytic hypersurfaces* in their respective ambient spaces $\mathbb{R} \times \mathbb{R}^m$ and $\mathbb{R}^m$, and a fundamental object of study in Thom's theory of elementary catastrophes is the bifurcation locus of an $\tilde{R}$-miniversal unfolding of a function $f : (\mathbb{R}^N, 0) \to (\mathbb{R}, 0)$ having an (algebraically) isolated critical point. The fact that these sets, defined as images, are semi-analytic, follows from Galbiati's theorem (see Hironaka's lectures) in this case, because they are images of maps having a finite (hence proper) complexification. The same remark applies to all the images occurring in a stratification of a stable map-germ in the real-analytic case (see below). In fact, using finite determinacy, Mather even made them semi-algebraic. Anyway, going back to the complex case, if we take $t \in \mathbb{C}^m - B$ (for a small representative of $G$), the line $\mathbb{C} \times \{t\}$ will meet $D$ in $\mu^{(N)}(X_0, 0)$ non-singular points of $D$, and transversally. It means the function $v = f + \sum t_i g_i$ will then have $\mu^{(N)}(X_0, 0)$ non-degenerate critical points.

5.5.4. One should not confuse $B \subset \mathbb{C}^m$ with the discriminant of the composed map $C \xrightarrow{n} D \xrightarrow{\pi} \mathbb{C}^m$ which is easily seen to be flat (it is just that $t_1, \cdots, t_m$ lift to regular sequence on $C$) and to be a ramified covering of degree $\mu^{(N)}(X_0, 0)$ in view of 5.5.2. The discriminant $\Delta$ of this composed map $p : C \to \mathbb{C}^m$, set-theoretically, is the image by $\pi$ of the locus of cusp-like points of $D$, but does not take into account the node-like points of $D$.

We will use very much from now on the following:

5.5.5. PROPOSITION (See [7]). *Let* $(X_0, 0) \subset (\mathbb{C}^N, 0)$ *be a germ of hypersurface with isolated singularity, and let* $l:(\mathbb{C}^N, 0) \to (\mathbb{C}, 0)$ *be a linear function (i.e. a coordinate function if we want). Assume that* $l^{-1}(0) \cap X_0$ *still has an isolated singularity. Then the discriminant of* $l \mid X_0:(X_0, 0) \to (\mathbb{C}, 0)$ *is the origin counted with a multiplicity equal to* $\mu^{(N)}(X_0, 0) + \mu^{(N-1)}(X_0 \cap H, 0)$, *where H is the hyperplane* $l^{-1}(0)$. *Furthermore, if* $f(z_1, \cdots, z_N) = 0$ *is an equation for* $(X_0, 0) \subset (\mathbb{C}^N, 0)$, *the ideal generated by the coefficients of the 2-form* $df \wedge dl$ *defines a curve* $S_H$ *in* $(\mathbb{C}^N, 0)$ (*the* polar curve *of f with respect to l*) *and we have*

$$\underline{(X_0, S_H)_0 = \mu^{(N)}(X_0, 0) + \mu^{(N-1)}(X_0 \cap H, 0)}$$

*where* $(\ ,\ )_0$ *is the intersection number at* 0 *in* $\mathbb{C}^N$.

*Finally, when l ranges through the* $\mathbb{P}^{N-1}$ *of linear maps,* $\mu^{(N-1)}(X_0 \cap H, 0)$ *takes its minimal value if and only if* $l^{-1}(0) = H$ is not *a limit direction of tangent hyperplanes to* $X_0$ *at non-singular points near* 0 (*in the sense of the first part of* §2). *This minimal value is the* $\mu^{(N-1)}(X_0, 0)$ *defined at the end of* 2.17.

REMARK 1. In view of the last exercise in 2.6, we have, taking $l = z_1$, and defining $H$ by $z_1 = 0$:

$$\mu^{(N)}(X_0, 0) + \mu^{(N-1)}(X_0 \cap H, 0) = \dim_{\mathbb{C}} \mathbb{C}\{z_1, \cdots, z_N\} \Big/ \left(f, \frac{\partial f}{\partial z_2}, \cdots, \frac{\partial f}{\partial z_N}\right).$$

We will meet this curve $S_H$ defined by $((\partial f/\partial z_2), \cdots, (\partial f/\partial z_N))$ again. Remark that because $(X_0, 0)$ has an isolated singularity, the $(\partial f/\partial z_i)$ form a regular sequence, so that $S_H$ is a complete intersection.

REMARK 2. The numerical part of the above proposition has been shown to be a special case of a nice general formula, proved topologically by Lê ([17]) and more algebraically by Greuel ([8]): Assume we know the Milnor number of an isolated singularity of complete intersection is. Then if $(X_0, 0)$ is such an isolated singularity, and $h:(X_0, 0) \to (\mathbb{C}, 0)$ any function such that $(h^{-1}(0), 0)$ again has an isolated singularity, than taking a coordinate $v$ on $\mathbb{C}$, the discriminant of $h$ is $(v^{\Delta})\mathbb{C}\{v\}$ where $\Delta = \mu(X_0, 0) + \mu(h^{-1}(0), 0)$.

The 'vanishing cycles' aspect of this proposition and the interpretation of $\mu^{(N)}(X_0, 0) + \mu^{(N-1)}(X_0 \cap H, 0)$ as an intersection number have also been generalized to arbitrary singularities of hypersurfaces by Lê ([18]).

Anyway, please do the

EXERCISE 1. Let $(X_0, 0) \subset (\mathbb{C}^2, 0)$ be a germ of reduced plane curve. Let $\pi:(X_0, 0) \to (\mathbb{C}, 0)$ be a projection parallel to a line $H$ in $(\mathbb{C}^2, 0)$. Show that the discriminant of $\pi$ is $(v^{\Delta})\mathbb{C}\{v\}$ ($v$: coordinate on the target of $\pi$) where

$$\Delta = \mu^{(2)}(X_0, 0) + (X_0, H)_0 - 1.$$

In particular, if $H$ is not tangent to $(X_0, 0)$, then $\Delta = \mu^{(2)}(X_0, 0) + m_0(X_0, 0) - 1$ where $m_0(\ )$ is the multiplicity (= order of the equation).

In particular, show that if $(X_0, 0)$ is a cusp (resp. a node) then if $H$ is not tangent, $\Delta = 3$ (resp. 2).

EXERCISE 2 (from [7]). Let $D \subset \mathbb{C} \times \mathbb{C}^m$ be the discriminant of a miniversal deformation of an isolated singularity of a hypersurface, and let $(B, 0) \subset (\mathbb{C}^m, 0)$ be the corresponding bifurcation subspace discriminant of the projection $\pi: (D, 0) \to (\mathbb{C}^m, 0)$. Show that

5.5.6. $$\underline{m_0(B, 0) = 2d + 3k = \tilde{\mu} + \mu - 1.}$$

Where $\mu = \mu^{(N)}(X_0, 0)$ and where $\tilde{\mu}$ is the Milnor number of the plane curve $(D \cap (\mathbb{C} \times H), 0)$, $(H, 0) \subset (\mathbb{C}^m, 0)$ being a 'generic' line (in fact, a line not in the tangent cone to $B$ at 0).

(We remark that $B$ is not reduced, so that we have to use again that we have a good definition of the discriminant.)

HINT. Use the fact that if you move $H$ away from the origin, it will meet $B$ 'transversally' in a number of points which (counted with multiplicities) is $m_0(B, 0)$.

5.5.7. DEFINITION. I shall call a section of the discriminant $D$ by an $i$-plane of the form $\mathbb{C} \times H$, where $H \subset \mathbb{C}^m$ is an $(i-1)$-plane, a *vertical* section. We will be very interested in general vertical plane sections of $D$.

5.5.8. *The invariant $\delta$ of curves and the geometry of the discriminant: first movement.*

In fact, there is another relation between $d$, $k$ and the general vertical plane sections of $D$, basically that found by Lê and Iversen [20], but which here we shall deduce from the results of §3: take a 2-plane $(H, 0)$ in $\mathbb{C}^m$, and write it $(H_0 \times H_1, 0)$. If $H$ is sufficiently general, for every $t \in H_1 \setminus \{0\}$, the line $H_t = H_0 \times \{t\}$ in $\mathbb{C}^m$ will be such that $D \cap (\mathbb{C} \times H_t)$ has as singularities only $k$ cusps and $d$ nodes. On the other hand, we can view $D \cap (\mathbb{C} \times H)$ as a flat family of plane curves parametrized by $H_1$. Furthermore, $n^{-1}(D \cap (\mathbb{C} \times H))$ is a surface in $C$ which is Cohen-Macaulay since the two coordinates on $H$ form a regular sequence in $\mathcal{O}_C$ (see the beginning of this paragraph) and is non-singular in codimension 1 (by Bertini's theorem) hence is normal, and is the normalization of $D \cap (C \times H)$. Let therefore $\Gamma$ denote the curve $(D \cap$

$(\mathbb{C}\times H_0), 0)$ and $\Gamma'$ the curve $(n^{-1}(\Gamma), 0)\subset(C, 0)$. It follows from 3.3 and the fact that the invariant $\delta$ of a cusp or a node is equal to 1 that:

5.5.9.
$$\delta(\Gamma', 0)=\delta(\Gamma)-(d+k),\quad \text{i.e.}$$
$$\underline{d+k=\delta(\Gamma)-\delta(\Gamma')}.$$

REMARK 3. If we assume known the fact, recently proved by Marc Giusti [9], that the formula $2\delta=\mu+r-1$ of 3.4 remains valid for reduced curves which are complete intersections, we deduce from 5.5.9 the equality:

5.5.10.
$$k=\mu'+\mu-1$$

where $\mu'$ is the Milnor number of the complete intersection $\Gamma'\subset(C, 0)$.

EXERCISE. Give another proof of the equality 5.5.10 using the formula of Lê and Greuel quoted in remark 2 above.

HINT. Remark that $k$ is precisely the multiplicity at 0 of the discriminant $\Delta$ of the $\mu$-fold branched cover $(C, 0)\xrightarrow{p}(\mathbb{C}^m, 0)$, and use the fact that $(p^{-1}(0), 0)$ being a 0-dimensional complete intersection, its Milnor number is equal to its multiplicity minus 1. Remark that $\Gamma'=p^{-1}(H_0)$.

REMARK. The computation of $k$ and $d$ for the discriminants of versal deformations of complete intersections with isolated singularity has been done by Lê and Greuel ([19]). The new feature is that the discriminant non longer 'flattens' so that 5.5.6 (for a generic projection of $D$ to a hyperplane) becomes $\mathfrak{m}_0(B, 0)=2d+3k+\tau$ where $\tau$ is a number of 'vertical tangents' which has to be evaluated.

REMARK. One can seek estimates for $k$ and $d$ in terms of $\mu=\mu^{(N)}(X_0, 0)$ only. It follows from 5.5.6 that ([7] chap. III).

($\alpha$) $\underline{2d+3k\geqq\mu^2-1}$

and it follows from 5.5.9 and a well-known formula for the behavior of $\delta$ by blowing up (see [10]) that

($\beta$) $\underline{d+k\geqq\dfrac{\mu(\mu-1)}{2}}$

of course ($\alpha$) follows from ($\beta$) and 5.5.10.

5.6. The invariant $\delta$ and the geometry of the discriminant: second movement

5.6.1. Let $(D, 0)\subset(\mathbb{C}^{m+1}, 0)$ be a germ of hypersurface. For any representative of $D$, define $\mathrm{Cr}_D(k)$ to be the set of points $s\in D$ at which $D$ is locally analytically isomorphic to the union of $k$ non-singular hypersurfaces in general position in $\mathbb{C}^{m+1}$. (Cr is for Cross). $\mathrm{Cr}_D(k)$ is a locally closed complex analytic subspace of $D$, of codimension $k$ in $\mathbb{C}^{m+1}$, if it is not empty.

5.6.2. Definition. Let $(X_0, 0) \subset (\mathbb{C}^N, 0)$ be a germ of a hypersurface with isolated singularity. Define $\delta(X_0, 0)$ to be the maximum number of singular points which one can pile up in the same fibre of an arbitrarily small deformation of $(X_0, 0)$.

Exercise 1 (from [36]). Show that, if $D$ is the discriminant of a versal deformation of $(X_0, 0)$, one has:

$$\delta(X_0, 0) = \text{Max}\,\{k/0 \in \overline{\text{Cr}_D\,(k)}\}.$$

Hint. Use 4.8.2.

Exercise 2. Check that when $(X_0, 0)$ is a plane curve, this definition agrees with that given in §3.

5.6.3. Remark. $\delta(X_0, 0)$, maximum number of critical points of a function nearby the function $f:(\mathbb{C}^N, 0) \to (\mathbb{C}, 0)$ having $(X_0, 0)$ as fibre which one can *pile up* in the same level variety of this function, should be compared with $\mu^{(N)}(X_0, 0)$, which is the maximum number of critical points of such a function which one can *spread out* (in the sense of critical values). In a way, $\delta$ and $\mu$ correspond to the same preoccupation, but $\delta$ is to the geometer, interested in spaces, i.e. fibres, what $\mu$ is to the function-theorist. Anyway, Mr. I. N. Iomdin has communicated to me the following results:

5.6.4. Proposition (I. N. Iomdin [12]). For an isolated singularity of hypersurface $(X_0, 0)$, one has the following inequalities:

$$\frac{\mu^{(N)}(X_0, 0)}{\mu^{(N-1)}(X_0, 0)} \leqq 2 \cdot \delta(X_0, 0) \leqq \mu^{(N)}(X_0, 0) + \mu^{(N-1)}(X_0, 0).$$

Proof. The upper bound for $\delta$ is obtained as follows: it is a theorem in ([7], chap. II) that the multiplicity in the sense of algebraic geometry of the Jacobian ideal *on the hypersurface* i.e. of $j(f) \cdot \mathcal{O}_{X_0,0}$, is equal to $\mu^{(N)}(X_0, 0) + \mu^{(N-1)}(X_0, 0)$. Suppose we have a deformation of $(X_0, 0)$ where the general fibre has $\delta(X_0, 0)$ singular points. By exercise 1 above, they have to be all ordinary quadratic singularities, and by the upper semi-continuity of multiplicities we must have:

$$\mu^{(N)}(X_0, 0) + \mu^{(N-1)}(X_0, 0) \geqq a \cdot \delta(X_0, 0)$$

where a is the value of $\mu^{(N)} + \mu^{(N-1)}$ for an ordinary quadratic singularity, which is obviously equal to 2.

5.6.5. To prove the other inequality, we must go back to the theory of polar curves according to ([14] §3, [35]). Anyway, this theory will be of essential use below: the polar curve of $f(z_1, \cdots, z_N)=0$ with respect to a hyperplane $H$ is the curve $S_H$ defined in 5.5.5. It is not difficult to convince oneself [and it is proved in [35] with the help of generic simultaneous normalization as in §3] that there exists a Zariski open dense subset $V \subset \mathbb{P}^{N-1}$ such that if the hyperplane $H \in V$, if we choose coordinates so that $H$ is given by $z_1=0$, the polar curve $S_H$, which is now $(\partial f/\partial z_2)=\cdots=(\partial f/\partial z_N)=0$, is reduced, has a number of irreducible components 1 in its decomposition $S_H=\bigcup_{q=1}^{1} \Gamma_q$ which is independent of $H \in V$ and furthermore, setting $m=m_0(\Gamma_q)$, we can define integers $e_q \geqq 0$ by $e_q+m_q=(X_0, \Gamma_q)_0$ (intersection multiplicity in $(\mathbb{C}^N, 0)$) and the sequence of integers $(e_q, m_q)$ is independent of $H \in V$. Furthermore, $e_q$ is equal to the intersection multiplicity of $\Gamma_q$ with the hypersurface defined by $(\partial f/\partial z_1)=0$, and we have $(H \in V)$

5.6.6.
$$\begin{cases} \sum_{q=1}^{1} e_q = \mu^{(N)}(X_0, 0) \\ \sum_{q=1}^{1} m_q = \mu^{(N-1)}(X_0, 0) = m_0(S_H) = (S_H, H)_0. \end{cases}$$

This means in particular that no component of $S_H$ has its reduced tangent cone (a line) contained in $H$, and therefore we can think of the branches $\Gamma_q$ as given parametrically by:

$$\Gamma_q \begin{cases} z_1 = t_q^{m_q} \\ z_i = t_q^{k_{q,i}} + \cdots \end{cases} \qquad k_{q,i} \geqq m_q \quad (2 \leqq i \leqq N).$$

While $f_{|\Gamma_q}$ has an expansion

$$f_{|\Gamma_q} = \gamma_q \cdot t_q^{e_q+m_q} + \cdots \qquad (\gamma_q \in \mathbb{C}^*)$$

and

$$\left.\frac{\partial f}{\partial z_1}\right|_{\Gamma_q} = \zeta_q \cdot t^{e_q} + \cdots \qquad (\zeta_q \in \mathbb{C}^*)$$

(by the equivalence of the several definitions of intersection multiplicities).

(Remark that the $(e_q, m_q)$ thus defined are independent of the choice of coordinates and equation as follows from [35] th. 2.)

Now Iomdin argues as follows: take $a=[\mathrm{Sup}_q\,[(e_q/m_q)]$ and a point $p$ on a component $\Gamma_q$ on which this sup is attained. Then we can find a polynomial

$P_p(z_1)$ depending on $p$ such that $f_{|\Gamma_q} - P_{p|\Gamma_q}$ vanishes with multiplicity $a+1$ at $p$, and furthermore, as $a \leqq \text{Sup}\,(e_q/m_q)$, $P_p$ tends to zero as $p \to 0$ on $\Gamma_q$.

The polar curve of $f(z_1, \cdots, z_N) - P_p(z_1) = 0$, which is $\Gamma_q$ near $p$, is non-singular and intersects it with intersection multiplicity $a+1$. Hence $f(z) - P_p(z_1) = 0$ has at $p$ a singularity of type $A_a$ i.e. locally isomorphic to $z_1^{a+1} + z_2^2 + \cdots + z_N^2 = 0$. (Recall that $\mu^{(N-1)} = 1$ implies that the general hyperplane section has only a quadratic singularity.) Therefore we have proved:

LEMMA (Iomdin). *Arbitrarily close to a singularity* $(X_0, 0)$ *with invariants* $(e_q, m_q)$ *as constructed above, there are singularities of type* $A_a$ *where* $a = [\text{Sup}_q\,(e_q/m_q)]$.

Now there only remains to prove that for $A_a$ one can find a nearby fibre with $[(a+1/2)]$ quadratic singularities. But according to exercise 1 above, this depends only upon the geometry of the discriminant and hence by the lemma in 5.5.1 we are reduced to $z_1^{a+1} = 0$, where it is obvious. All that remains now is to observe that

$$\text{Sup}_q \frac{e_q}{m_q} \geqq \frac{\mu^{(N)}(X_0, 0)}{\mu^{(N-1)}(X_0, 0)} \quad \text{by} \quad (5.6.6).$$

5.6.7. REMARK. The method used in §3 when $N=2$ to realize $\delta$ will not work in higher dimensions: take a resolution of singularities $\tilde{X}_0 \to X_0$ and perturb the composed map $\tilde{X}_0 \to X_0 \to \mathbb{C}^N$: the image will have generalized pinch-points!

5.6.8. QUESTION. How many points of $\text{Cr}_D\,(k)$ are there in a sufficiently general section of $D$ by an affine $k$-plane near the origin? (for $k=2$, it is the computation of $d$).

5.7. REMARK. The map $G \mid C : (C, 0) \to (\mathbb{C} \times \mathbb{C}^m, 0)$ is *not* sufficiently general for its image $(D, 0) \subset (\mathbb{C} \times \mathbb{C}^m, 0)$ to be *weakly normal* in the sense of [2], which means that every complex map-germ which is *continuous* on $D$ and becomes holomorphic on $C$ (i.e. when composed with $n$) is already holomorphic on $D$. (Weak normality would imply no cusps in codimension 1) but it is sufficiently general for $D$ to be *Lipschitz saturated* as was remarked in [7 chap. III, 5.18] which means that every *locally Lipschitz* function on $D$ which becomes holomorphic on $C$ is already holomorphic on $D$.

The question above is linked to what happens when one makes a generic perturbation of the map $G \mid C : C \to \mathbb{C} \times \mathbb{C}^m$.

5.8. Stratifications (see Hironaka's and Mather's lectures in these proceedings).

5.8.1. Let $G:(\mathbb{C}^N\times\mathbb{C}^m, 0)\to(\mathbb{C}\times\mathbb{C}^m, 0)$ be a versal deformation of a germ of hypersurface with isolated singularity $(X_0, 0)\subset(\mathbb{C}^N, 0)$. We want to decompose the source and target of any small representative of $G$ in a *finite* number of locally closed non-singular subspaces,

$$\mathbb{C}^N\times\mathbb{C}^m=\bigcup_\alpha Z_\alpha,\qquad \mathbb{C}\times\mathbb{C}^m=\bigcup_\beta S_\beta$$

such that

(i) for each $\alpha$ there exists a $\beta$ such that $G$ induces a subsmersive surjection $Z_\alpha\to S_\beta$

(ii) some interesting feature of $G$ remains constant along each 'stratum' $Z_\alpha$.

We shall be interested in the following three features of $G$:

(1) the local topological type at $x\in\mathbb{C}^N\times\mathbb{C}^m$ of the fibre $(G^{-1}(G(x)), x)$ of $G$ through $x$ (again a hypersurface with isolated singularity).

(2) The (c)-cosécance class (2.19) of this same fibre.

(3) The topological type of the germ at $x$ of the map $G$:

$$G:(\mathbb{C}^N\times\mathbb{C}^m, x)\to(\mathbb{C}\times\mathbb{C}^m, s).\qquad (s=G(x))$$

5.8.2. It is a basic fact of life in these problems that if we require nothing about frontier conditions and such, the constancy of (1) along a $Z_\alpha$ is in general *strictly weaker* than the constancy of (2) if $N\geqq 3$, and the constancy of (2) is in general *strictly weaker* than the constancy of (3), even for $N=2$, (i.e. plane curves). (See below 5.12.1 and 5.12.4).

5.8.3. Anyway, a process for building Thom stratifications of $G$ [i.e. of the source and target of a small enough representative of $G$] along each stratum of which (3) is constant, has been described in the lectures of Hironaka and Mather. Basically one finds Whitney stratifications of $\mathbb{C}^N\times\mathbb{C}^m$ and $\mathbb{C}\times\mathbb{C}^m$ such that (i) is satisfied and also Thom's condition $A_G$: then one has an isotopy lemma to prove the topological triviality of $G$ along the strata (by the integration of controlled vector fields), a fact which is even stronger than the constancy of 3). The point is that the finiteness of $C$ over $\mathbb{C}\times\mathbb{C}^m$ implies the 'no blowing up' condition of Thom.

Remark 1. One can prove exactly the same results but with the generalized condition (c) (2.18, Remark) instead of Whitney conditions. One then integrates rugose vector fields, which gives a stronger trivialization.

REMARK 2. Of course, $G^{-1}(\mathbb{C}\times\mathbb{C}^m - D)$ will be one stratum, and $G^{-1}(D) - C$ will be a union of strata. I wish to remark that for any stratum $Z_\alpha \subset C$, $G \mid Z_\alpha$ will in fact have to be a *local isomorphism* onto a stratum $S_\beta \subset D$. Furthermore, one can build at the same time a stratification of the *parameter space* $\mathbb{C}^m$, say $\mathbb{C}^m = \bigcup T_\gamma$ such that if $t \in T_\gamma$, then the topological type of the maps $F_t : \mathbb{C}^N \to \mathbb{C}$ are the same.

Furthermore, denoting by $S_0$ (resp. $T_0$) the stratum of the origin in the Thom stratification in $\mathbb{C}\times\mathbb{C}^m$ (resp. $\mathbb{C}^m$) the projection $\mathbb{C}\times\mathbb{C}^m \to \mathbb{C}^m$ induces a local isomorphism $S_0 \xrightarrow{\sim} T_0$: This is simply because by the properties of our stratifications, as long as $t \in T_0$, $F_t$ has only one critical value.

Finally, all this remains true in the real analytic case, with semi-analytic (even semi-algebraic) strata.

We will be mostly interested in the constancy of (2), but let us stop a while to sketch an application of the idea of the invariant $\delta$ above:

5.9. A real interlude. The catastrophic version of the Gibbs phase rule (after [32], [36]).

In the theory of elementary catastrophes of Thom, a family of physical systems depending on parameters is 'represented' by a family of functions $F_t : \mathbb{R}^N \to \mathbb{R}$, representing the family of potentials which govern the evolution of the systems. Stability conditions imply that we can think of this family of functions as an $\mathfrak{R}$-miniversal unfolding (see §4) of a given function $f = F_0 : (\mathbb{R}^N, 0) \to (\mathbb{R}, 0)$ with an algebraically isolated critical point. Our family of functions is then represented by

$$F : (\mathbb{R}^N \times \mathbb{R}^{\mu-1}, 0) \to (\mathbb{R} \times \mathbb{R}^{\mu-1}, 0).$$

Now take a Thom stratification of any small representative of this map (what follows is true for any Thom stratification, but Mather has proved the existence of a canonical one in his lectures) and let $T_0 \subset \mathbb{R}^{\mu-1}$ be the image of the stratum of the origin $S_0 \subset \mathbb{R} \times \mathbb{R}^{\mu-1}$ by the cannonical projection. Recall that $pr_2 \mid S_0 : S_0 \to T_0$ is a local isomorphism, and that $F$ induces a local isomorphism $Z_0 \to S_0$ where $Z_0$ is the stratum of 0 in $\mathbb{R}^N \times \mathbb{R}^{\mu-1}$.

Since the morphology of our family of functions does not vary along $T_0$, the true number of parameters of our systems is the dimension of the supplementary subspace to $T_0$ i.e. $\operatorname{codim}_{\mathbb{R}^{\mu-1}} T_0 = \operatorname{codim}_{\mathbb{R}\times\mathbb{R}^{\mu-1}} S_0 - 1$ and on the other hand, the *stable states* of the physical system represented by $F_t$ correspond to the ordinary minima of the function $F_t$ on $\mathbb{R}^N$ (i.e. of course in a ball around 0 in $\mathbb{R}^N$): therefore, the maximum number $\nu(F)$ of minima which a function $F_t$ in (an arbitrary small representative of) our family can present, will represent the maximum number of *phases* which can coexist in a system of our family (if we think of our systems as chemical or thermodynamical), and the well-known Gibbs phase rule to the effect that the number of phases is at most the number of parameters plus one, becomes a purely geometric statement:

$$\nu(F) \leqq \operatorname{codim}_{\mathbb{R}^{\mu-1}} T_0 + 1$$

that is

$$\underline{\nu(F) \leqq \operatorname{codim}_{\mathbb{R}\times\mathbb{R}^{\mu-1}} S_0.} \qquad 5.9.1$$

One proves this in two steps: first, one defines the real analogue $\delta_{\mathbb{R}}(X_0, 0)$ of the $\delta(X_0, 0)$ of 5.6 and since one has a real analogue of the product decomposition theorem, one proves that $\delta_{\mathbb{R}}(X_0, 0) = \max\{k/0 \in \overline{\mathrm{Cr}_D(k)}\}$ where $\mathrm{Cr}_D(k)$ is the $k$-cross locus of the real discriminant $D$ of $F$. Since a homeomorphism of ambient spaces respecting $D$ cannot help to respect $\mathrm{Cr}_D(k)$ for all $k$, we see that $S_0 \subset \overline{\mathrm{Cr}_D(\delta_{\mathbb{R}}(X_0, 0))}$ and hence $\operatorname{codim}_{\mathbb{R}\times\mathbb{R}^{\mu-1}}(S_0) \geqq \delta_{\mathbb{R}}(X_0, 0)$. On the other hand, following ideas of Smale, since all minima of a function have the same index as real critical points, there is no obstruction to putting them all at the same level, i.e. in the same fibre of $F$: this shows that $\delta_{\mathbb{R}}(X_0, 0) \geqq \nu(F)$ and proves 5.9.1.

5.10. The partition of the discriminant $D$ according to multiplicity (a.k.a. its Samuel stratification. See [33])

The fact that the Milnor number $\mu^{(N)}(X_0, 0)$ of a germ of hypersurface with isolated singularity is an invariant of its topological type (as embedded germ in $(\mathbb{C}^N, 0)$) is easily deduced from the material in Milnor's book. There is a partial converse, harder to prove:

THEOREM (Lê Dũng Tráng and C. P. Ramanujam [21]). *In an analytic family of germs of hypersurfaces with isolated singularity* (*in the sense of* 2.12), *the constancy of the Milnor number of the fibres implies the constancy of their topological type, if* $N \neq 3$.

REMARK. The restriction $N \neq 3$ comes from the fact that the proof uses the $h$-cobordism theorem. The author has recently given a purely algebraic proof of a stronger result in the case $N = 2$ (i.e. for plane curves). See [35].

5.10.1. Anyway, from our point of view this theorem and 5.5.2 show that the most primitive invariant of $D$ at 0, namely its multiplicity $\mathfrak{m}_0(D)$, already contains some good information about $(X_0, 0)$. This fact led me to introduce in [33] the so-called Samuel stratum $D_\mu$ of the origin in $D$ as a candidate for the base of a deformation 'miniversal for deformations where the topological type of the fibres remains constant'.

We must first make precise what we mean by Samuel stratum, since we can define it either as a reduced subspace of $D$ or not. Let us go back to the notations of 5.5.1 and write

$$v \circ G = F(z,t) = f(z) + t_1 z_1 + \cdots + t_N z_N + t_{N+1} g_{N+1}(z) + \cdots + t_m g_m(z)$$

assuming $f \in (z_1, \cdots, z_N)^3$.

I will also write an equation $\delta(v, t_1, \cdots, t_m)$ for $D$ in $\mathbb{C}\times\mathbb{C}^m$ as follows:

$$\delta = v^\mu + a_{\mu-1}(t)v^{\mu-1} + \cdots + a_0(t) = 0 \qquad (t = (t_1, \cdots, t_m)).$$

5.10.2. Definition. The (non-reduced) Samuel stratum $D_\mu$ of the origin in $D$ is the subspace of $\mathbb{C}\times\mathbb{C}^m$ defined by the ideal generated by the

$$\left(\frac{\partial^\alpha \delta}{\partial v^i \partial t^\gamma} \in \mathcal{O}_{\mathbb{C}\times\mathbb{C}^m,0}\right) \text{ where } (i, \gamma) = \alpha, \quad |\alpha| < \mu = \mu^{(N)}(X_0, 0).$$

Of course $(D_\mu)_{\text{red}} = \{(v, t) \in D / m_{(v,t)}(D) = \mu\}$.

5.11. I now want to make a list, with references, of positive and negative facts known about $(D_\mu)_{\text{red}}$.

5.11.1. ([33], [7], using results on the monodromy proved by Lê, Lazzeri, Gabrielov.)

Above any point $(v,t) \in D_\mu$, there is only one point of $C$, which is given by an analytic section described by:

$$z_i = \frac{-1}{\mu} \frac{\partial a_{\mu-1}(t)}{\partial t_i} \qquad (1 \leq i \leq N)$$

(recall that $z_1, \cdots, z_N$ and $t_{N+1}, \cdots, t_m$ form a system of coordinates on $C$). This shows that $n^{-1}((D_\mu)_{\text{red}})$ is the image of a section $\sigma : (D_\mu)_{\text{red}} \to C$. This is the 'non splitting' principle, (see [14]).

5.11.2. In particular, after 4.8.2 and 5.2, $D$ remains locally analytically irreducible at every point of $(D_\mu)_{\text{red}}$ and in particular, its tangent cone remains a $\mu$-fold hyperplane, the hyperplane in question being given by the tangent space at the point $(v,t)$ to the non-singular subvariety $W$ of $\mathbb{C}\times\mathbb{C}^m$ with equation

$$W : v = -\frac{1}{\mu} a_{\mu-1}(t_1, \cdots, t_m).$$

(Remark that the flattening property of $D$ of 5.5.1 implies that $a_{\mu-i} \in (t_1, \cdots, t_m)^{i+1}$.)

Remark. In the language of the theory of contact of Hironaka [11], $W$ has the *maximal contact* at every point of $(D_\mu)_{\text{red}}$ (in fact $W \supset D_\mu$) and if $s \in D_\mu$, $T_{W,s}$ in one case coincides with the *strict tangent space* to $D$ at $s$.

This is important for us because it will give us a *natural frame in which to take Newton polygons* (namely the $v$ axis and $T_{W,s}$) at every point $s$ of $D_\mu$.

5.11.3. The problem of the non-singularity of the $\mu$-constant stratum was as far as I know raised in [33], where it was shown that:

THEOREM. *If $D$ is the discriminant of the miniversal deformation of a germ of reduced plane curve, $(D_\mu)_{\text{red}}$ is non-singular.*

REMARK 1. In that first proof, the hard work had been done by J. Wahl who had shown in his thesis [37] the existence of a (formal) miniversal Zariski equisingular deformation for a germ of plane curve, and had shown it was (formally) non-singular. All I had to do was to identify the completion at 0 of the local ring of $(D_\mu)_{\text{red}}$ at 0 with that constructed by J. Wahl, thanks to the fact that $\mu$-constant $\Leftrightarrow$ equisingularity in the sense of Zariski (see 3.7).

Recently however, in the appendix to [39], I succeeded in giving a completely different proof, at least in the case of branches: the idea, which is new I believe, is to make any branch with a given semi-group of values $\Gamma$ (see 3.7) appear as a deformation of the *monomial* curve $C^\Gamma$ which has the same semi-group. Of course one has to go out of the paradise of plane curves: if $\Gamma = \langle \bar{\beta}_0, \cdots, \bar{\beta}_g \rangle$, the monomial curve $C^\Gamma$, which is given parametrically by $v_i = t^{\bar{\beta}_i}$ $(0 \leqq i \leqq g)$ has imbedding dimension $g+1$. What one wins is that it has a $\mathbb{C}^*$-action, and if it is a complete intersection, which is the case if $\Gamma$ is the semi-group of a plane curve, then it is easy to see that it has a *miniversal constant-semi-group deformation* which is a *linear* subspace in the base of a $\mathbb{C}^*$-equivariant miniversal deformation of $C^\Gamma$. Now for plane branches, equisingularity is equivalent to 'same semi-group' so that by 4.8.2, since our original plane branch $(X_0, 0)$ appears as fibre of a deformation (arbitrary small) of $C^\Gamma$, the $\mu$-constant stratum of its discriminant, multiplied by a non-singular subspace, is non-singular: therefore this $\mu$-constant stratum is non-singular (the example 4.6.3 is lifted from that theory). It seems to be easy to pass from this result to an arbitrary reduced plane curve, but I think the generalization of the ideas of this proof to higher dimensions (where the question is still open) is quite an intriguing problem (see below), and is for me the main motivation for the study of complete intersections and/or quasi-homogeneous singularities. Another form of the question, which is stronger and therefore perhaps easier to answer, is the following:

QUESTION. Is the Samuel stratum $D_\mu$ a *resolved* space in the sense of Hironaka? (see [11])

(A resolved space is a (non-reduced) space which cannot be improved by permissible blowing up: its underlying reduced space is non-singular and it is normally flat along it.)

REMARK 2. The non-singularity of $(D_\mu)_{red}$ of course would imply the non-singularity of the $\mu$-constant stratum in a $\tilde{\mathbb{R}}$-miniversal unfolding of $f$, after the first exercise in 4.8.4.

5.12. We now come to the negative facts concerning $(D_\mu)_{red}$, which are extremely important.

5.12.1. THEOREM (Pham [26], [3]). *Even when $N=2$, the topological type of $G:\mathbb{C}^N\times\mathbb{C}^m\to\mathbb{C}\times\mathbb{C}^m$ can vary along $\sigma((D_\mu)_{red})$ where $\sigma$ is the section built in 5.11.1. The example given by Pham is the miniversal deformation of the reduced plane curve $y^3+x^9=0$. The miniversal deformation is described by:*

$$v\circ G=y^3+a(x)\cdot y+b(x)$$

*where*

$$a(x)=a_0+a_1\cdot x+\cdots+a_7\cdot x^7$$
$$b(x)=b_1\cdot x+\cdots+b_7\cdot x^7+x^9.$$

*The $\mu$-constant stratum is given in this case by $v=0$, $b_j=0$ $(1\leqq j\leqq 7)$, $a_i=0$ $(0\leqq i\leqq 5)$ and the corresponding family is given by*

$$y^3+(a_6x^6+a_7x^7)y+x^9=0.$$

EXERCISE. Check that it is an equisingular family, using 3.7. Now Pham shows that near every point of $(D_\mu)_{red}$ with $a_6=0$, there is a point of the discriminant such that the corresponding fibre has 2 singular points of multiplicity 3 which have in suitable coordinates Newton polygons of smallest inclination $\geqq 4/3$ and $\geqq 5/3$ respectively, and that this is *not* the case if $a_6\neq 0$. Since the maximum over the set of coordinate choices of the inclination of the first side of the Newton polygon is a topological invariant, [11], this shows that the topological type of $G$ cannot be the same if $a_6=0$ or $a_6\neq 0$ (near a point where $a_6=0$, $a_7\neq 0$, say).

5.12.2. Furthermore, the counterexamples of Briançon and Speder (2.18, see [6]) show that even if $(D_\mu)_{red}$ is non-singular, or if we restrict to a non-singular subspace of it, it is not true that the corresponding fibres will be equisingular in any strong sense, as soon as $N\geqq 3$. (However, they will have the same topological type, at least if $N\neq 3$, by the theorem of Lê-Ramanujam).

My reaction to this was to abandon 'constant topological type' as a definition of equisingularity in high dimension, in favour of (c)-cosécance, since in my opinion the topology of the general sections is an important part of the geometry of a singularity (see for example 5.20.2 below). Furthermore (c)-cosécance can be easily generalized to the non-hypersurface case, coincides with Zariski's definition of equisingularity when $N=2$, and in the case of hypersurfaces implies that the *functions* having our hypersurfaces as fibres are topologically equivalent, a fact often deemed useful.

One could think that in the definition of equisingularity conditions 'the stronger the better' provided the condition is always satisfied for 'almost all' fibres in a family. However, I do not go all the way to the really strong possible definitions of equisingularity, such as 'miniversal deformation topologically constant', which is called also 'universal equisingularity' or '$\tau$ constant', which both satisfy the openness condition mentioned above (see also [14] §1). The reason is that with these definitions, some deformations which I like, such as the specialization of a plane branch to the monomial curve with the same semi-group, mentioned above, would not be equisingular.

5.12.3. By the results in [14] §2, all the $\mu^{(i)}(X_y, 0)$ are analytically upper semi-continuous in a family of hypersurfaces. Hence we can define a closed complex subspace of the $\mu$-constant stratum $D_\mu$, the $\mu^*$-constant stratum $D_{\mu^*}$, at least as a reduced subspace. Of course, when $N=2, D_{\mu^*}=(D_\mu)_{\text{red}}$. It seems that the extensions of the methods of proof of the non-singularity of $(D_\mu)_{\text{red}}$ for plane branches are better geared to answer the

QUESTION. Is $D_{\mu^*}$ non-singular?

5.12.4. The last negative fact on $D_\mu$ (or $D_{\mu^*}$), also due to Pham, is quite important for us:

THEOREM (Pham). *Even when* $N=2$, *the topology of a general plane section of the discriminant* $D\subset\mathbb{C}\times\mathbb{C}^m$ *of the miniversal deformation of* $(X_0, 0)\subset(\mathbb{C}^N, 0)$, *through the point* $s\in D$, *can vary when* $s$ *varies as the* $\mu$*-constant stratum* $D_\mu$. *As a consequence* (see 5.5.6) *the number* $k$ *of cusps in a general plane section of* $D$ *near* $s$ (see 5.4.1) *can vary as* $s$ *varies on* $D_\mu$.

5.12.5. IMPORTANT REMARK. I want to emphasize that after the counterexamples of Briançon and Speder (5.12.2) one is aware of the fact that a hypersurface (such as our $D$) can be topologically trivial along a non-singular subspace $S$ (such as $(D_\mu)_{\text{red}}$) while the topological type of its general $i$-plane section ($i=2$ for example) through $s\in S$ *varies* with $s$. There is no example where the topology of the discriminant varies along $(D_\mu)_{\text{red}}$ as far as I know. See [22].

5.12.6. We shall analyze the phenomenon in 5.12.4 in a way different from Pham's. The idea is that there is a very close connection between a general plane section of $D$ through the origin and the *polar curve* we saw in 5.6.5 and 5.6.6. This is seen as follows:

5.13. PROPOSITION. *The topological type (indeed, the (c)-cosécance class 2.19), of a general* vertical *plane section of the discriminant $D$ (5.5.7) is the same as that of a general plane section of $D$.*

PROOF. Since the set of 2-planes $H$ in $(\mathbb{C}\times\mathbb{C}^m, 0)$ giving the general topological type for $D\cap H$ is Zariski open and dense in the Grassmannian $Gr(m+1, 2)$, it is sufficient to check that the family of plane sections of $D$ through 0 satisfies an equisingularity condition near a plane section of the form:

$$v=\alpha_0 x, \qquad t_i=\beta_i y \qquad (1\leqq i\leqq m)$$

where $\alpha_0\neq 0$.

Let us then consider the family of plane curves $X$ defined in $\mathbb{C}^{2m+2}\times\mathbb{C}^2$ by

$$F(\alpha, \beta, x, y)=\delta(\alpha_0 x+\beta_0 y, \alpha_1 x+\beta_1 y, \cdots, \alpha_m x+\beta_m y)=0$$

where $\delta(v, t_1, \cdots, t_m)=0$ is the equation of the discriminant $D$. We see that

$$\frac{\partial F}{\partial \alpha_0}=x\cdot\frac{\partial \delta}{\partial v}, \qquad \frac{\partial F}{\partial \beta_0}=y\cdot\frac{\partial \delta}{\partial v}$$

$$\frac{\partial F}{\partial \alpha_i}=x\cdot\frac{\partial \delta}{\partial t_i}, \qquad \frac{\partial F}{\partial \beta_i}=y\cdot\frac{\partial \delta}{\partial t_i}$$

$$\frac{\partial F}{\partial x}=\alpha_0\frac{\partial \delta}{\partial v}+\sum \alpha_i\frac{\partial \delta}{\partial t_i}$$

$$\frac{\partial F}{\partial y}=\beta_0\frac{\partial \delta}{\partial v}+\sum \beta_i\frac{\partial \delta}{\partial t_i}.$$

CLAIM. Near a point $p:(\alpha, \beta, 0)$ where $\alpha_0\neq 0$ we have

$$\frac{\partial F}{\partial \alpha_i}\cdot\mathcal{O}_{X,p}\in\overline{(x, y)\cdot\left(\frac{\partial F}{\partial x}, \frac{\partial F}{\partial y}\right)\cdot\mathcal{O}_{X,p}} \qquad (0\leqq i\leqq m)$$

$$\frac{\partial F}{\partial \beta_j}\cdot\mathcal{O}_{X,p}\in\overline{(x, y)\cdot\left(\frac{\partial F}{\partial x}, \frac{\partial F}{\partial y}\right)\cdot\mathcal{O}_{X,p}} \qquad (0\leqq i\leqq m).$$

PROOF. Since an integral dependence relation can be pulled back by a map such as the map $X\to D$ induced from the map $\varphi:\mathbb{C}^{2m+2}\times\mathbb{C}^2\to\mathbb{C}\times\mathbb{C}^m$ given by $v=\alpha_0 x+\beta_0 y$, $t_1=\alpha_i x+\beta_i y$ $(1\leqq i\leqq m)$, on $X=\Phi^{-1}(D)$ we see that as soon as $\alpha_0$ or $\beta_0$ is different from 0, since $(\partial\delta/\partial t_i)\cdot\mathcal{O}_{D,0}$ is integrally dependent on the ideal $(\partial\delta/\partial v)\cdot\mathcal{O}_{D,0}$ for all $i$, as follows from the proof of 5.5.1, and 4 of 2.7, we see that if $\alpha_0\neq 0$ the ideal $((\partial F/\partial x), (\partial F/\partial y))\,\mathcal{O}_{X,p}$ is generated by $(\partial\delta/\partial v)\cdot\mathcal{O}_{X,p}$ and all $(\partial\delta/\partial t_i)\cdot\mathcal{O}_{X,p}$ are integrally dependent on it. The claim follows from this. Now we remark that the claim is nothing but the restriction to $X$ of the integral dependence required for $X$ to satisfy condition (c) along $\mathbb{C}^{2m+2}\times\{0\}$ near $p$. Of course, condition (c) does not follow directly from this, but a direct computation shows that the pull-back by $\varphi$ of the map $G\mid C: C\to\mathbb{C}\times\mathbb{C}^m$ is non-singular and is the normalization of $X$, provided $\alpha_0\neq 0$. Then one can apply the criterion (12) of 3.7 since $j(F)\cdot\mathcal{O}_{\bar{x},n^{-1}(p)}$ is then generated by $(\partial\delta/\partial v)\cdot\mathcal{O}_{\bar{x},n^{-1}(p)}$.

Remark. The same argument shows that general vertical sections of any dimension give the general (c)-cosécance class.

5.14. Proposition. *The Newton polygon (in the coordinates given by the decomposition $H=\mathbb{C}\times H_1$) of the intersection of $D$ with a general vertical plane is equal to the Newton polygon of the intersection of $D$ with the plane $\mathbb{C}\times H_1$ where $H_1$ is the line $t_2=\cdots=t_m=0$ in $\mathbb{C}^m$, provided $z_1=0$ is a sufficiently general hyperplane in the sense of* 5.6.5. *Furthermore, this section also has the same topological type ((c)-cosécance) as a general plane section of $D$.*

Proof. A vertical plane in $\mathbb{C}\times\mathbb{C}^m$ can be given parametrically by:

$$\begin{cases} v=v \\ t_i=\alpha_i u \qquad \alpha_i\in\mathbb{C} \qquad (1\leqq i\leqq m)\end{cases}$$

and if we write the corresponding functions, with the conventions of 5.5, we find

$$v=f(z_1,\cdots,z_N)+\sum_{i=1}^{N}\alpha_i u z_i+\sum_{i=N+1}^{m}\alpha_i u g_i(z_1,\cdots,z_N)$$

which, after a linear change of coordinates, becomes (if our vertical plane is general in the sense that $\alpha\neq 0$)

$$v=f(z_1,\cdots,z_N)+u\left(z_1+\sum_{i=N+1}^{m}\alpha_i g_i(z_1,\cdots,z_N)\right)$$

with $g_i\in(z_1,\cdots,z_N)^2$.

After a new change of coordinates, we can write it as

$$\underline{v=f(z_1,\cdots,z_N)+uz_1.}$$

Now since the formation of the discriminant commutes with base change, $D\cap H$ is just the discriminant of the map $(\mathbb{C}^N\times\mathbb{C},0)\to(\mathbb{C}\times\mathbb{C},0)$ thus described. The *critical subspace* is defined by the ideal

$$\left(\frac{\partial f}{\partial z_1}+u, \frac{\partial f}{\partial z_2}, \cdots, \frac{\partial f}{\partial z_N}\right)\cdot \mathcal{O}_{\mathbb{C}^N\times\mathbb{C}}$$

which is nothing but the intersection of the surface $S_H\times\mathbb{C}$, where $S_H$ is our *polar curve* with respect to $z_1=0$, with the hypersurface $(\partial f/\partial z_1)+u=0$.

As we have already assumed $z_1=0$ to be general, we are ready to obtain a parametric representation of our general vertical plane section of $D$, as follows:

Take the parametric representation of the branches $\Gamma_q$ of $S_H$ as in 5.6.6:

$$\Gamma_q \begin{cases} z_1 = t_q^{m_q} \\ z_i = a_{q,i}t^{k_{q,i}}+\cdots \end{cases} \qquad (2\leqq i\leqq N) \qquad k_{q,i}\geqq m_q.$$

We have now a parametric representation of the branches $C_q$ of the critical subspace above (which is a reduced curve) as follows:

$$C_q \begin{cases} u = -\partial f/\partial z_1|_{\Gamma_q} = -\zeta_q t_q^{e_q}+\cdots & (\zeta_q\in\mathbb{C}^*) \\ z_1 = t_q^{m_q} \\ z_i = a_{q,i}t^{k_{q,i}}+\cdots \end{cases}$$

and therefore also a parametric representation of the branches of $D\cap H$, say $D_q$, as follows: computing $v=f+uz_{1\Gamma_q}$ gives: (see 5.6.6 and remark that differentiating gives $\gamma_q=\zeta_q\cdot(m_q)/(e_q+m_q)$)

$$D_q \begin{cases} u = -\zeta_q t_q^{e_q}+\cdots \\ v = \eta_q t_q^{e_q+m}+\cdots \end{cases} \qquad \left(\eta_q = -\frac{e_q}{e_q+m_q}\,\zeta_q\in\mathbb{C}^*\right).$$

Hence we have proved:

THEOREM. *The number of branches of a general vertical plane section $D\cap H$ of $D$ is equal to that of a general polar curve of the function $f(z_1,\cdots,z_N)$, and its Newton polygon in the natural coordinates $(u,v)$ on $H=\mathbb{C}\times H_1$ is*

$$\mathfrak{N}(\delta\mid H)=\sum_{q=1}^{1}\left\{\frac{e_q+m_q}{e_q}\right\} \qquad \text{(notations of 3.6).}$$

This now follows immediately from 3.6.2.

5.15. We remark that since $k_{q,i}\geqq m_q$, it is reasonable to say that the ratios $m_q/e_q$ describe the vanishing rates of the critical points of the Morse function $f+uz_1$ *as functions of* $u$ since for example, on $C_q$:

$$z_1=(-\zeta_q)^{(1/e_q)}u^{(m_q/e_q)}+\cdots$$

and that the $(e_q+m_q)/(e_q)$ describe the vanishing rates in function of $u$ of the critical *values* of this function.

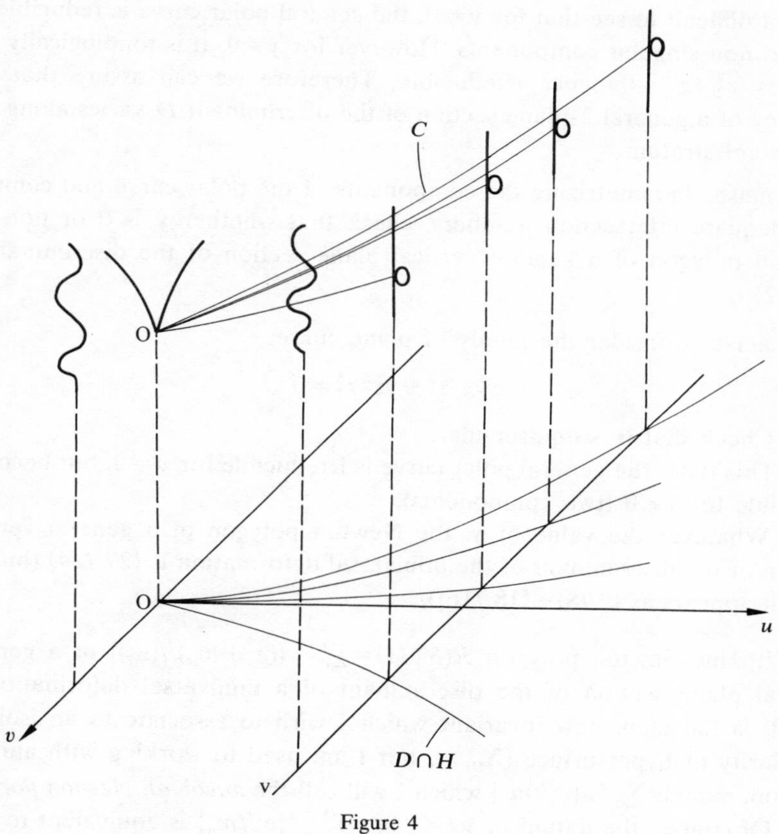


Figure 4

5.16. The point is that while it is quite hard to compute the discriminant, it is relatively easy to compute the polar curve (in practice, of course, we have to decompose it into its irreducible components, which is not so easy, but usually manageable by using the following trick: a general plane projection, and then Newton polygon again!) For example, to prove Pham's theorem in 5.12, all we have to do is to find an equisingular (i.e. $\mu$-constant) family of plane curves such that the number of irreducible components of a general polar curve does not remain constant:

Consider for example the family of plane curves given by

$$z_2^3+z_1^7+yz_1^5z_2=0 \quad \text{in} \quad \mathbb{C}\times\mathbb{C}^2$$

since it is cubic in $z_2$ it is easy to see by Zariski's discriminant criterion that it is equisingular along $\mathbb{C}\times\{0\}$. However, consider a general polar curve:

$$\lambda(3z_2^2+yz_1^5)+\mu(7z_1^6+5yz_1^4z_2)=0 \qquad (\lambda:\mu)\in\mathbb{P}^1 \text{ 'general'}.$$

It is not difficult to see that for $y=0$, the general polar curve is reducible: it has two non-singular components. However for $y\neq 0$, it is topologically the same as $z_2^2+z_1^5=0$ hence irreducible. Therefore we can assure that the topology of a general 2-plane section of the discriminant $D$ varies along the $\mu$-constant stratum.

EXERCISE. Parametrizing the components of the polar curve and computing adequate intersection numbers, check that whether $y$ is 0 or not, the Newton polygon of a general vertical plane section of the discriminant is $\{14//12\}$.

EXERCISE. Consider the family of plane curves

$$z_2^4+z_1^9+yz_2^2z_1^5=0$$

(1) Check that it is equisingular.

(2) This time, the general polar curve is irreducible for $y=0$, but becomes reducible for $y\neq 0$ (two components).

(3) Whatever the value of $y$, the Newton polygon of a general vertical section of the discriminant of the miniversal deformation is $\{27//24\}$ (but for $y\neq 0$ it appears as $\{9//8\}+\{18//16\}$).

5.17. This Newton polygon $\mathfrak{N}(\delta\mid H)=\sum_{q=1}^{1}\{(e_q+m_q)//(e_q)\}$ of a general vertical plane section of the discriminant of a miniversal deformation of $(X_0,0)$ is the main new invariant which I wish to associate to an isolated singularity of hypersurface $(X_0,0)$, but I am used to working with another polygon, namely $\sum_{q=1}^{1}\{e_q//m_q\}$ which I will call the *Jacobian Newton polygon* of $X$. Of course, the datum of $\boldsymbol{\nu}_j(X_0,0)=\sum_{q=1}^{1}\{e_q//m_q\}$ is equivalent to that of $\mathfrak{N}(\delta\mid H)$. In fact with the notations of 3.6, we have

$$\mathfrak{N}(\delta\mid H)=\varepsilon(\sigma(\boldsymbol{\nu}_j(X_0,0))).$$

We also remark that since $e_q$ is the intersection number of $\Gamma_q$ with a hypersurface, $e_q\geqq m_q$ so that $\boldsymbol{\nu}_j(X_0,0)$ is a *special* Newton polygon. Remark also that

$$\underline{l(\boldsymbol{\nu}_j(X_0,0))=\mu^{(N)}(X_0,0), \qquad h(\boldsymbol{\nu}_j(X_0,0))=\mu^{(N-1)}(X_0,0).}$$

5.18. I now quote results concerning $\boldsymbol{\nu}_j(X_0,0)$. First, it is important to remark that the analytic type of the hypersurface defined by $f(z_1,\cdots,z_N)-g(w_1,\cdots,w_M)=0$ does *not* depend only upon the analytic type of the hypersurfaces $(X_0,0)$ and $(X_1,0)$ defined respectively by $f=0$ and $g=0$ (e.g. multiply $f$ by a unit). However, we have

PROPOSITION. *The (c)-cosécance class of* $f(z)-g(w)=0$ *depends only upon the (c)-cosécance classes of* $(X_0, 0)$ *and* $(X_1, 0)$.

The proof is given as an exercise on (c)-cosécance. Therefore we have defined an operation on the set $\mathcal{I}_s$ of (c)-cosécance classes of isolated singularities of hypersurfaces. This operation will be denoted by $\perp$: the (c)-cosécance class of $f(z)-g(w)=0$ will be denoted by $[X_0]\perp[X_1]$, where $[X]$ is the (c)-cosécance class of the germ of hypersurface $(X, 0)$. Now we have

5.19. THEOREM ([35] th. 4). Assuming that $(X_0, 0)$ and $(X_1, 0)$ (given by $f(z)=0$, $g(w)=0$) have isolated singularities, the Jacobian Newton polygon of $f(z)-g(w)=0$ is given by

$$\boldsymbol{\nu}_j(f(z)-g(u)=0, 0)=\boldsymbol{\nu}_j(X_0, 0) * \boldsymbol{\nu}_j(X_1, 0).$$

This theorem is not stated in this way in [35], but the proof of theorem 4 there actually proves this. ($*$ is the product defined in 3.6).

5.20. THEOREM ([35] th. 6'). *The Jacobian Newton polygon is an invariant of (c)-cosécance.*

The idea of the proof is as follows, and relies on Theorem 2.18. First, one checks that if a family $F(y, z_1, \cdots, z_N)=0$ of hypersurfaces is (c)-equisingular, then so is the *double* family $F(y, z_1, \cdots, z_N)+F(y, w_1, \cdots, w_N)=0$. After Theorem 2.18, the Milnor number of the generic hyperplane section of the fibres of this family is constant (independent of $y$). But by an exercise in 3.6, and 5.19, this number is *twice the area* under the Jacobian Newton polygon $\boldsymbol{\nu}_j(X_y, 0)$. Since a Newton polygon can only rise under specialization we have that $\boldsymbol{\nu}_j(X_0, 0)$ is above $\boldsymbol{\nu}_j(X_y, 0)$ $(y\neq 0)$ and encloses the same area: therefore they are equal.

5.21. Now we have defined our Jacobian Newton polygon map $\boldsymbol{\nu}_j: \mathcal{I}_s \to \underline{\mathrm{N}}_S$ and shown that $\boldsymbol{\nu}_j([X_0]\perp[X_1])=\boldsymbol{\nu}_j([X_0])*\boldsymbol{\nu}_j([X_1])$.

REMARK. Since for plane curves a deformation is (c)-equisingular if and only if it is equisingular in the sense of Zariski (3.6), the two exercises given above are nothing but verifications of 5.20. However, they show that 5.20 is a rather fine phenomenon since the number of branches of the polar curve varies.

5.21.1. REMARK. I want to emphasize that 5.20 has also the following geometric meaning: Along the $\mu^*$-constant stratum $D_{\mu^*}$ in the discriminant $D\subset\mathbb{C}\times\mathbb{C}^m$ of a miniversal deformation of an isolated singularity of hypersurface $(X_0, 0)$, the Newton polygon of a general vertical plane section (which has a meaning in view of 5.11.2) is constant.

For the reader familiar with the theory of contact of Hironaka ([11]) this implies in particular that the *contact exponent* $\delta_s(W, D)$ of the discriminant $D$ with the non-singular hypersurface $W$ of 5.11.2 (which has maximal contact) is constant (and equal to $1+(1/\eta)$ where $\eta = \max_q (e_q/m_q)$) along $D_{\mu^*}$.

5.21.2. Furthermore, we have also that the intersection multiplicity at $s \in D_{\mu^*}$ of $D$ with a general non-singular curve contained in $W$ is also constant along $D_{\mu^*}$ and equal to $\mu^{(N)}(X_s, 0) + \mu^{(N-1)}(X_s, 0)$ where $X_s = G^{-1}(s) \subset \mathbb{C}^N$ is the fibre of the miniversal deformation $G$ over $s$. It is rather intriguing that this number, which is the *diminution of class* due to the presence of the singularity $(X_s, 0)$ on a projective hypersurface (i.e. what comes in the Plücker formula, see Kleiman's lectures) is also very closely linked with the contact of the discriminant $D$ with the limit direction at $s \in D_{\mu^*}$ of tangent hyperplanes, at nearby non-singular points, namely $T_{W,s}$.

I will end by quoting from [35] theorems showing that $\boldsymbol{\nu}_j(X_0, 0)$, or $\mathfrak{N}(\delta \mid H)$, contains real information on the geometry of the singularity $(X_0, 0)$ up to (c)-cosécance:

THEOREM. *A necessary and sufficient condition for* $(X_0, 0)$ *to be* (c)-*cosécant to a hypersurface defined by an equation 'with one variable separated' i.e. of the form* $f(z_2, \cdots, z_N) + z_1^{a+1}$ *where* $z_1 = 0$ *is a general hyperplane section, is that* $\boldsymbol{\nu}_j(X_0, 0)$ *is of the form* $\{a \cdot m // m\}$ *where* $a$ *is an integer.*

THEOREM. *Let* $f \in \mathbb{C}\{z_1, \cdots, z_N\}$ *define a function with an isolated critical point at* 0, *and* $(X_0, 0) = (f^{-1}(0), 0)$. *For an integer, the following conditions are equivalent*

(i) $a > \operatorname{Sup}(e_q/m_q)$ where, $(e_q/m_q)$ *are those associated to* $f$ *as in* 5.6.5.

(ii) *Any function* $g \in \mathbb{C}\{z_1, \cdots, z_N\}$ *such that* $g - f \in (z_1, \cdots, z_N)^{a+1}$ *defines by* $g = 0$ *a hypersurface having the same topological type as* $(X_0, 0)$.

(iii) *Any function* $g \in \mathbb{C}\{z_1, \cdots, z_N\}$ *such that* $g - f \in (z_1, \cdots, z_N)^{a+1}$ *defines by* $g = 0$ *a hypersurface* (c)-*cosécant to* $(X_0, 0)$ (*and hence* $g$ *has the same topological type as* $f$, *as a function*).

Finally, the test of any invariant is that it recovers, for plane branches, the classical complete systems of invariants to the geometry up to (c)-cosécance namely the characteristic $(\beta_0, \cdots, \beta_g)$ or the semi-group $\Gamma = \langle \bar{\beta}_0, \cdots, \bar{\beta}_g \rangle$: the translation in the language presented here of a theorem of M. Merle [24] gives a complete answer:

THEOREM (M. Merle). *Let* $(X_0, 0) \subset (\mathbb{C}^2, 0)$ *be a plane branch. Then (notations of* 3.6 *and* 3.7*) we have:*

$$\boldsymbol{\nu}_j(X_0, 0) = \sum_{q=1}^{g} \left\{\frac{(n_q-1)\bar{\beta}_q - n_1 \cdots n_{q-1}(n_q-1)}{n_1 \cdots n_{q-1}(n_q-1)}\right\}.$$

COROLLARY. *Given two plane branches* $(X_0, 0)$, $(X_1, 0)$, *letting* $\mathfrak{N}(\delta_0 \mid H)$ (resp. $\mathfrak{N}(\delta_1 \mid H)$) *denote the Newton polygon of a general vertical plane section of the discriminant of a miniversal deformation of* $(X_0, 0)$, *resp.* $(X_1, 0)$. *The following are equivalent:*

(1) $(X_0, 0)$ *and* $(X_1, 0)$ *have the same topological type.*

(2) $(X_0, 0)$ *and* $(X_1, 0)$ *are (c)-cosécant.*

(3) $\boldsymbol{\nu}_j(X_0, 0) = \boldsymbol{\nu}_j(X_1, 0)$.

(4) $\mathfrak{N}(\delta_0 \mid H) = \mathfrak{N}(\delta_1 \mid H)$.

EXERCISE. Using the construction in Appendix II of [34] show that $\boldsymbol{\nu}_j(X_0, 0)$ plays with respect to the local Plücker formula exactly the role that $\mathfrak{N}(\delta \mid H)$ plays with respect to Milnor's formula: it is a dynamic version. (See also [14] §3.)

EXERCISE. Let $(X_0, 0) \subset (\mathbb{C}^3, 0)$ and $(X_1, 0) \subset (\mathbb{C}^3, 0)$ be defined by $Z_2^3 + Z_1^\beta Z_3 + Z_3^{3\alpha} = 0$ and $(X_1, 0) \subset (\mathbb{C}^3, 0)$ be defined by $Z_2^3 + Z_1^\alpha Z_2 + Z_1^\beta Z_3 + Z_3^{3\alpha}$ where $3\alpha = 2\beta + 1$ and $\alpha \geqq 3$ (see [6]).

Show that

$$\boldsymbol{\nu}_j(X_0, 0) = \left\{\frac{2\beta}{2}\right\} + \left\{\frac{2\beta(2\beta-2)}{2\beta-2}\right\}$$

$$\boldsymbol{\nu}_j(X_1, 0) = \left\{\frac{2\beta(2\beta-1)}{2\beta-1}\right\}.$$

HINT. Use 5.19 and 5.20.

## REFERENCES


[1] V. I. ARNOL'D, Critical points of smooth functions, Proceedings of the International Congress of Mathematicians, Vancouver 1974.

[2] A. ANDREOTTI and P. HOLM, Parametric Spaces, this volume.

[3] P. BERTHELOT, Classification topologique universelle des singularités d'après F. Pham, Séminaire Douady-Verdier 1971–72, Astérisque No 16, S.M.F. 1974.

[4] J. M. BOARDMAN, Singularities of differentiable maps, Publ. Math. I.H.E.S. No 33 (1967) pp. 21–57.

[5] D. A. BUCHSBAUM and D. S. RIM, A generalized Koszul complex II, Trans. A.M.S. 111 (1964) pp. 197–224.

[6] J. BRIANÇON and J. P. SPEDER, La trivialité topologique n'implique pas les conditions de Whitney, Note aux C. R. Acad. Sc. Paris t. 280 Série A, (1975) p. 365.

[7] B. Teissier, Cycles évanescents, sections planes, et conditions de Whitney, Singularités à Cargèse 1972, Astérisque No 7–8, S.M.F. (1973), pp. 285–362.

[8] G. M. Greuel, Der Gauss-Manin Zusammenhang isolierter singularitäten..., Math. Ann. *214* (1975) pp. 235–266.

[9] M. Giusti, Classification des germes d'intersection complète simples (to appear).

[10] H. Hironaka, On the arithmetic genera and effective genera of algebraic curves, Memoirs Coll. of Sc. Univ. of Kyoto *30* (1967).

[11] H. Hironaka, Bimeromorphic smoothing of complex analytic spaces. Report University of Warwick 1971. Available at Centre de Mathématiques, Ecole Polytechnique, 91128 Palaiseau Cedex France.

[12] I. N. Iomdin, Letter to the author, Aug. 1976.

[13] I. N. Iomdin, in Funktional'nyi Analiz i ego prilogenia, t. 10 Fasc. 3 (1976) pp. 80–81.

[14] B. Teissier, Introduction to equisingularity problems. Proceedings A.M.S. Summer School in Algebraic Geometry, Arcata 1976, Proceedings Symposia in Pure Math. Vol. 29 A.M.S. Providence R.I. (1975).

[15] G. Kempf, On the geometry of a theorem of Riemann, Annals of Math. 98 (1973) pp. 178–185.

[16] G. Kempf, On the collapsing of homogeneous bundles, Inventiones Math. *37*, 3 (1976) pp. 229–239.

[17] Lê D. T., Calcul du nombre de Milnor d'une singularité isolée d'intersection complète, Funktsional'nyi Anal. i ego pril. *8* (1974) pp. 45–49.

[18] Lê D. T., Calcul du nombre de cycles évanouissants d'une hypersurface complexe, Ann. Inst. Fourier, t. XXIII Fasc. 4 (1973).

[19] Lê D. T. and G. M. Greuel, Spitzen, Doppelpunkte und vertikale tangenten..., Math. Ann. 222 (1976) pp. 71–88.

[20] Lê D. T. and B. Iversen, Calcul du nombre de cusps..., Bull. Soc. Math. Fr. 102 (1974) pp. 99–107.

[21] Lê D. T. and C. P. Ramanujam, The invariance of Milnor's number implies the invariance of topological type, Amer. J. Math. *98*, 1 (1976) pp. 67–78.

[22] E. Looijenga, On the semi-universal deformation of a simple-elliptic singularity, Parts I, II, Mathematisch Instituut, Katholieke Universiteit, Toernooiveld, Nijmegen, Netherlands.

[23] J. Mather, On Thom-Boardman singularities, Proceedings Dynamical systems conference, Academic Press 1973.

[24] M. Merle, Invariants polaires des courbes planes, to appear in Inventiones Math.

[25] J. Milnor, Singular points of complex hypersurfaces, Ann. of Math. Studies No. 61 (1968), Princeton U.P.

[26] F. Pham, Remarques sur l'équisingularité universelle, Preprint Université de Nice 1970.

[27] F. Pham, Classification des singularités, Preprint, Université de Nice 1971.

[28] F. Pham, Singularités planes d'ordre 3, in Singularités à Cargèse 1972, Astérisque No 7–8 S.M.F. (1973).

[29] F. Pham, Caustiques, phase stationnaire et microfonctions, Preprint Université de Nice (1976).

[30] J. P. Serre, Algèbre locale, multiplicités, Springer Lecture Notes No. 11.

[31] R. Thom, Ensembles et morphismes stratifiés, Bull. A.M.S. *75* (1969).

[32] R. Thom, Modèles mathématiques de la morphogénèse, Coll. 10/18 No 887, U.G.E. Paris (Chap. VI).

[33] B. Teissier, Déformations à type topologique constant I, II, Séminaire Douady-Verdier 1971–72, Astérisque No 16, S.M.F.

[34] B. Teissier, Sur diverses conditions numériques d'équisingularité des familles de courbes et un principe de spécialisation de la dépendance intégrale, Centre de Mathématiques de l'Ecole Polytechnique, 91128 Palaiseau Cedex France, No M208.0675 (1975).

[35] B. Teissier, Variétés polaires I, Invariants polaires des singularités d'hypersurfaces, to appear in Inventiones Mathematicae.

[36] B. Teissier, Sur la version catastrophique de la règle des phases de Gibbs···, in Singularités à Cargèse 1975, Dépt. de Mathématiques de l'Université de Nice.

[37] J. Wahl, Equisingular deformations of plane algebroid curves, Transactions of the A.M.S. Vol. 193 (1974) pp. 143–170.

[38] A. E. R. Woodcock and T. Poston, A geometrical study of the elementary catastrophes, Springer Lect. Notes No. 373.

[39] O. Zariski, Les problèmes des modules pour les branches planes, Cours au Centre de Mathématiques de l'Ecole Polytechnique 91128 Palaiseau Cedex France.

Bernard Teissier
Centre de Mathématiques de l'Ecole Polytechnique,
Plateau de Palaiseau
91128 Palaiseau Cedex, France

# NOTES ADDED AFTER PROOFREADING

p.567 line -5 of text : the $\underline{N}_S$ should be an $\underline{\mathfrak{N}}_S$ as on p. 619 line 2.

p.571 in the diagram : $X' = X \times_Y Y'$ .

p.592 line 4 of text : $Y \times \mathbb{C}^N$ instead of $X_y \in \mathbb{C}^N$.

p.603 line -18 : $\mu^{(1)}(X_o,0)$ instead of $\mu^{(\ell)}(X_o,0)$.

p.618 in the picture : $P = \left\{ \dfrac{\ell(P)}{h(P)} \right\}$ instead of $i(P) = \dfrac{\ell(P)}{h(P)}$ .

p.644 line -8 : replace $\text{depth}_{\mathcal{O}_{X,0}}(\ )$ by $dh_{\mathcal{O}_{X_o}}(\ )$.

p.660 line 13 : read $[(a+1)1/2]$ instead of $[(a+1/2)]$.

Addendum : The facts stated in Corollary 3 of §3 were apparently known to various people for different purposes. I want to quote two papers giving beautiful pictures and applications to monodromy of the deformation of the Lemma :

N. A'Campo : Le groupe de monodromie du déploiement des singularités isolées de courbes planes, Math. Ann. 213 (1975) pp. 1-22.

S. M. Gusein-Zade : Intersection matrices of some singularities of functions of two variables, Funktional. Anal. i. Prilozen. 8 (1974) No 1 pp. 11-15.

Also, this Corollary 3 shows geometrically that the conductor gives the right structure for the double-point scheme of a map from a curve to a non-singular surface. (See Kleiman's lectures, chap. 5, and the paper quoted p. 614.)

-----

Nordic Summer School/NAVF
Symposium in Mathematics
Oslo, August 5–25, 1976

# COUNTEREXAMPLES IN STRATIFICATION THEORY: TWO DISCORDANT HORNS*

D. J. A. Trotman

One of the useful properties of Whitney's (a)-regularity condition (as defined in [13]) is that the set of mappings transverse to the strata of an (a)-regular stratification is open and dense. That this set is open has often been justified by remarking that (a)-regularity implies that a submanifold transverse to a stratum at a given point is transverse to all other strata in some neighborhood of the point, a condition I have called (t)-regularity in [10]. Our first example shows that this reasoning is wrong: transversality to a (t)-regular stratification need not be open. However we verify directly that transversality to an (a)-regular stratification is open.

Our second example is that of a pair of real semialgebraic strata which are (b)-regular (as defined in [13]) but which fail Kuo's ratio test ([4], where Kuo proved that no such example exists when the smaller stratum has dimension one), and hence do not satisfy the property (w) used by Verdier in [12], where it was remarked that such an example was not known.

## 1. (a)-regularity and transversality

Let $X$, $Y$ be $C^1$ submanifolds of $\mathbb{R}^n$ and let $0 \in Y \subset \bar{X} - X$. Consider the following regularity conditions for the pair $(X, Y)$ at 0.

(a) Given $x_i$ in $X$ tending to 0, if $T_{x_i}X$ tends to $\tau$, then $T_0 Y \subset \tau$.

(t) Given a $C^1$ submanifold $S$ meeting $Y$ transversely at 0, then there is a neighborhood $U$ of 0 in $\mathbb{R}^n$ such that $S$ is transverse to $X$ within $U$.

Call a stratification *(a)-regular* if each pair of strata $(X, Y)$ satisfies (a) at each point of $Y$. Similarly for a *(t)-regular* stratification.

NOTE 1.1. That (a) implies (t) is immediate.

NOTE 1.2. It is *not* a consequence of 1.1 that mappings transverse to each of the strata of an (a)-regular stratification form an open set, as suggested for example in [8], [9], [10], [11]. It is in fact a direct consequence of (a)-regularity.

* The title was suggested by Tony Iarrobino.

PROPOSITION 1.3. *Let $N$, $P$ be $C^\infty$ manifolds. Let $P$ contain a closed subset $Q$ partitioned into a locally finite union of submanifolds forming an* (a)-*regular stratification $\mathcal{S}$, i.e. if $X$, $Y$ are strata of $\mathcal{S}$, then at each point of $Y \cap \bar{X}$, condition* (a) *is satisfied. Then $T_{\mathcal{S}} = \{f \in C^\infty(N, P) : f$ is transverse to each stratum of $\mathcal{S}\}$ is open in $C^\infty(N, P)$ with the Whitney $C^1$ topology (and hence with the Whitney $C^\infty$ topology).*

PROOF*. Suppose that $T_{\mathcal{S}}$ is not open, so that there exists $f$ in $T_{\mathcal{S}}$, a sequence $\{g_i\}$ tending to $f$ in $C^\infty(N, P)$ with $g_i \notin T_{\mathcal{S}}$, a stratum $X$, and a sequence $\{a_i\}$ tending to $a_0$ in $N$ such that $g_i$ is not transverse to $X$ at $a_i$. It is clear that $f(a_0) \notin X$, since $X$ is a smooth submanifold. So let $Y$ be the stratum containing $f(a_0)$. $(df)_{a_0}(T_{a_0}N)$ and $T_{f(a_0)}Y$ span $T_{f(a_0)}P$, and so for $i$ sufficiently large $(dg_i)_{a_i}(T_{a_i}N)$ and $T_{g_i(a_i)}X$ span $T_{g_i(a_i)}P$, by (a) and the assumption that $g_i$ tends to $f$. This gives a contradiction, proving the proposition.

NOTE 1.4. (i) When $W$ is a submanifold of $P$, it is a corollary of Thom's Transversality Theorem that $T_W = \{f \in C^\infty(N, P) : f$ is transverse to $W\}$ is dense in $C^\infty(N, P)$ with the Whitney $C^\infty$ topology. (See for example [2].) Hence $T_{\mathcal{S}}$ is both open and dense in $C^\infty(N, P)$ with the Whitney $C^\infty$ topology.

(ii) If $W$ is closed, $T_W$ is open, as proved in [2], but here the strata of $\mathcal{S}$ are not assumed to be closed.

(iii) It is easily verified that $\mathcal{S}$ is (a)-regular if and only if the set of jets transverse to $\mathcal{S}$ is open. This observation is due to C. T. C. Wall.

In [10] the curve selection lemma is used to prove that (t) implies (a) if $X$ is semianalytic. It is equally true if $X$ is subanalytic for Hironaka proved a curve selection lemma for subanalytic sets in [3] (proposition 3.9. See [5] for a proof for semialgebraic sets). Hence if the strata are subanalytic the transversal mappings to a (t)-regular stratification do form an open set.

In the next section we shall give an example of a finite (t)-regular stratification for which the set of transversal mappings is not open, and so in particular it is not (a)-regular. This is an explicit version of an example mentioned in [10].

I stress this point at length because I had mistakenly thought that proposition 1.3 was true with (a) replaced by (t). Thus in [11] (t) is used in the definition of stratification given in chapter 8. There the strata are semialgebraic (corollary 3.6 of [11]), so we could use the result of [10] mentioned above to give (a), and then apply proposition 1.3. Alternatively

* A detailed proof appears as Proposition 3.6 in E. A. Feldman, The geometry of immersions, I, Trans. Amer. Math. Soc. 120 (1965), 185–224.

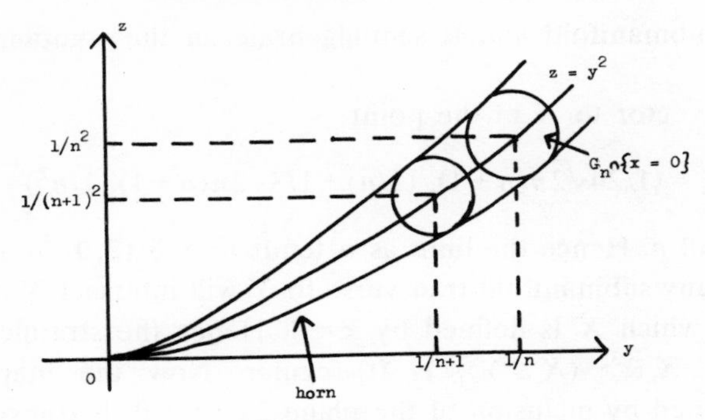

Figure 1. $x = 0$.

one can use the following elementary formulation suggested by E. C. Zeeman.

PROPOSITION 1.5. *Let X, Y be $C^1$ submanifolds of a $C^1$ manifold P, and suppose that $\phi: M \to C^1(N, P)$ is continuous, M is a topological space, N is a $C^1$ manifold, $Y = \phi(m)(N)$ for some $m \in M$, and for all open sets $U \subset M$ containing m, there is an $m' \in U$ such that $\phi(m')(N) \subset X$. Then the pair (X, Y) satisfies* (a) *at each point of Y.*

The proof is left as an exercise.

## 2. The first horn

Let $(x, y, z)$ be coordinates in $\mathbb{R}^3$. Take $Y$ to be the $y$-axis, and let $X = (\bigcup_{n=1}^{\infty} \{f_n = 0, g_n \leqq 0\}) \cup (\bigcap_{n=1}^{\infty} \{x = 0, g_n \geqq 0\})$, where $g_n \leqq 0$ defines the cylinder $G_n$ of radius $1/3n(n+1)$ with axis the line $y = 1/n$, $z = 1/n^2$, and where $f_n = 0$ defines the surface $F_n$ obtained from $x = (y^2 + z^2)^2 - (y^2 + z^2) + \frac{1}{4}$ by translating the origin to $(0, 1/n, 1/n^2)$ and reducing by a factor of $3n(n+1)/\sqrt{2}$ so that $F_n$ intersects $\partial G_n$ exactly where $x = 0$ is tangent to $F_n$. See figures 1 and 2.

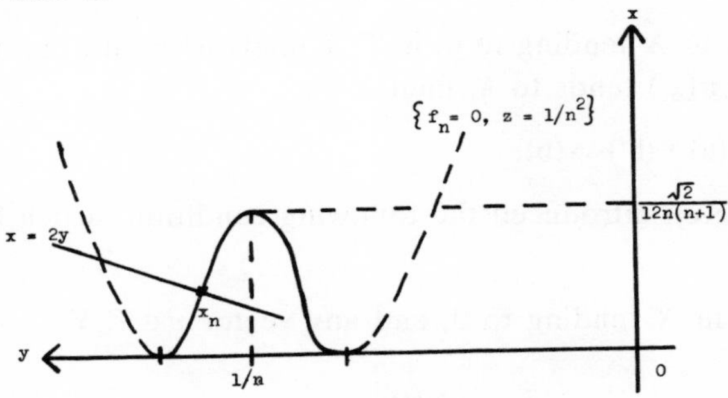

Figure 2. $z = 1/n^2$.

$X$ is a $C^1$ submanifold and is semialgebraic on the complement of the origin.

The normal vector to $X$ at the point

$$x_n = (1/24\sqrt{2}n(n+1),\ (1/n)+1/3\sqrt{2}n(n+1),\ 1/n^2)$$

is $(2, 1, 0)$ for all $n$. Hence the limit as $n$ tends to $\infty$ is $(2, 1, 0)$ and (a) fails. (t) holds since any submanifold transverse to $Y$ will intersect $X$ near $Y$ only at points near which $X$ is defined by $x = 0$. Hence the stratification of $\mathbb{R}^3$ defined by $\{Y, X, \mathbb{R}^3-(X\cup Y)\}$ is (t)-regular. Now the mapping $h$ in $C^\infty(\mathbb{R}^2, \mathbb{R}^3)$, defined by inclusion of the plane $2x+y=0$, is transverse to the stratification, but for each $n$ the mapping $h_n$ defined by inclusion of the plane

$$2x+y=(5+12\sqrt{2}(n+1))/(12\sqrt{2}n(n+1))$$

is not transverse to $X$ at $x_n$. Since $h_n$ tends to $h$ as $n$ tends to $\infty$, mappings transverse to the stratification are not open in $C^\infty(\mathbb{R}^2, \mathbb{R}^3)$.

Note that by smoothing near each circle $\{x=0, g_n=0\}$, $X$ can be made into a $C^\infty$ submanifold of $\mathbb{R}^3$, with normal vector at $x_n$ as before, for each $n$. Hence proposition 1.3 with (t) replacing (a) is false.

## 3. (b)-regularity and the ratio test

Let $X$ be a $C^1$ submanifold and a semianalytic (or subanalytic) set in $\mathbb{R}^n$. Let $Y\subset \bar{X}-X$ be an analytic submanifold of $\mathbb{R}^n$. The pair $(X, Y)$ are *(b)-regular* at $0\in Y$ if,

(b) Given $x_i$ in $X$ and $y_i$ in $Y$ tending to 0, if $T_{x_i}X$ tends to $\tau$, and the unit vector in the direction $x_iy_i$ tends to $\lambda$, then $\lambda\subset\tau$.

Apply a local analytic isomorphism at 0 to $\mathbb{R}^n$ so that, near 0, $Y$ becomes affine. Let $\pi$ denote orthogonal projection onto $Y$ and define,

(b′) Given $x_i$ in $X$ tending to 0, if $T_{x_i}X$ tends to $\tau$, and the unit vector in the direction $x_i\pi(x_i)$ tends to $\lambda$, then $\lambda\subset\tau$.

LEMMA 3.1. (a) + (b′) $\Leftrightarrow$ (b).

In [4] T.-C. Kuo introduced the following condition, which he called the *ratio test.*

(r) Given $x_i$ in $X$ tending to 0, and any vector $v\in T_0Y$,

$$\lim_{i\to\infty}\frac{|\pi_i(v)|\cdot|x_i|}{|x_i-\pi(x_i)|}=0.$$

Here $\pi_i$ denotes orthogonal projection onto the normal space to $X$ at $x_i$. Kuo proved two theorems in [4]:

THEOREM 3.2. (r)$\Rightarrow$(b).

THEOREM 3.3. (b)$\Rightarrow$(r) *if* $Y$ *is* 1-*dimensional.*

In each case the proof uses the curve selection lemma with the assumption that $X$ is a semianalytic set. As remarked in §1, by [3] we know that the same proof can be used if $X$ is a subanalytic set.

In the next section we give an example with $Y$ 2-dimensional where (b) holds and (r) fails to hold. $X$ will be a semialgebraic $C^1$ submanifold of dimension 3 in $\mathbb{R}^4$. I do not know of such an example where $X$ is the smooth part of an algebraic variety. In the special case of a family of complex hypersurfaces with isolated singularity parametrized by $Y$ it is known that (b) and (r) are equivalent, for $Y$ of arbitrary dimension. This is because (r) is a trivial consequence of (c)-cosécance as defined by Teissier in [7] and discussed by him in this volume. It follows from [1] and [6] that (b) implies (c)-cosécance.

Verdier has introduced the following condition in [12],

(w) There is a constant $C>0$ and a neighborhood $U$ of 0 in $\mathbb{R}^n$ such that if $x\in U\cap X$, and $y\in U\cap Y$, $d(T_xX, T_yY)\leqq Cd(x, y)$.

This is just (c)-cosécance restricted to $X$, so that it makes sense when $X$ is not a variety. (w) trivially implies (r), hence (b) does not imply (w), when the dimension of $Y$ is greater than 1, by the example in the next section. Even when $Y$ is 1-dimensional, (b) can hold and yet (w) fail: in $\mathbb{R}^3$ let $X$ be $\{x=0, z>0, z^2\leqq y^2\}\cup\{z^5x^2=(y^2-z^2)^4, x\geqq 0, z>0, z^2\geqq y^2\}$, let $Y$ be $\{x=z=0\}$, and consider the curve $X\cap\{z^2=3y^2\}$. Thus (w) is strictly stronger than (r) by theorem 3.3.

## 4. The second horn

Let $(x, y, z, w)$ be coordinates in $\mathbb{R}^4$, and let $Y$ be the plane $z=w=0$. Define the semialgebraic set,

$$X=\{w=0, 2(x^2+(z-y^p)^2)\geqq y^{2p}, z>0\}$$

$$\cup\{y^q w=(x^2+(z-y^p)^2)^2-y^{2p}(x^2+(z-y^p)^2)+y^{4p}/4,$$

$$2(x^2+(z-y^p)^2)\leqq y^{2p}, z>0\}$$

where $p$ and $q$ are positive integers satisfying,

$$2p<q<3p. \tag{4.1}$$

For example let $p=2$, $q=5$.

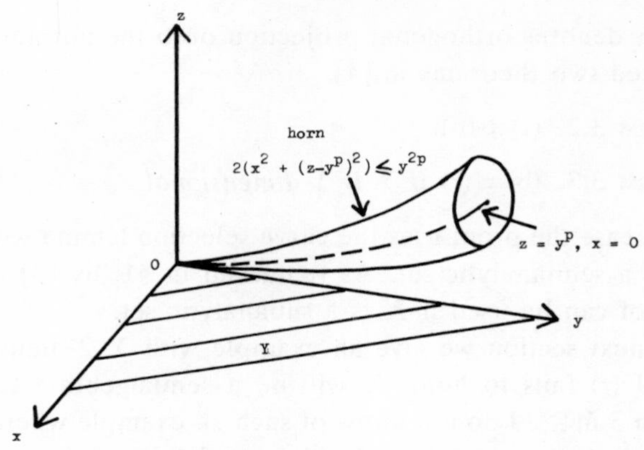


Figure 3. $w = 0$.

Observe that because the algebraic variety defined by the equality in the second part of the expression for $X$ has $w = 0$ as tangent space at every point of its intersection with $2(x^2 + (z - y^p)^2) = y^{2p}$, $X$ is a $C^1$ submanifold of $\mathbb{R}^4$.

ASSERTION 4.2. (b) holds.

PROOF. We show that there is a single limiting tangent 3-plane for sequences on $X$ tending to 0, namely $w = 0$. It suffices to consider the points on $y^q w = x^4 - y^{2p}x^2 + y^{4p}/4$ (with $y$ fixed) where $d^2w/dx^2 = 0$, since at these points the normal is furthest from the $w$-direction.

$d^2w/dx^2 = 0$ when $6x^2 = y^{2p}$, and the normal vector is $(\pm(\frac{4}{3}\sqrt{6})y^{3p}, -y^q)$ which tends to $(0, 1)$ as $y$ tends to 0 since $q < 3p$ by (4.1). Hence $w = 0$ is the unique limiting tangent plane.

At the points on $X$ where the sécant vector defined by orthogonal projection to $Y$ is furthest from the $z$-direction, the sécant vector is contained in the tangent space to $X$. Hence $0z$ is the unique limit of sécant

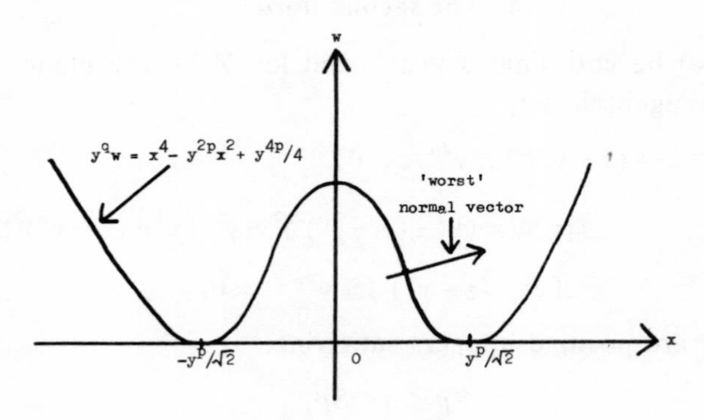


Figure 4. $z = y^p$, $y$ fixed.

vectors, and (b′) holds. (a) holds (since $\{w=0, z=0\} \subset \{w=0\}$), so we can apply lemma 3.1 to show that (b) holds, proving the assertion.

ASSERTION 4.3. (r) fails to hold.

PROOF. Consider the curve $\gamma(t)=(t^p/6, t, t^p, t^{4p-q}/9)$ which lies on $X$. The normal vector to $X$ at $\gamma(t)$ is,

$$((\tfrac{4}{3}\sqrt{6})t^{3p}, ((2p/3)-(q/9))t^{4p-1}, 0, -t^q).$$

Let $\pi_t$ denote projection onto this normal space. Then,

$$|\pi_t(0x)| \sim \frac{t^{3p}}{|(t^{3p}, t^{4p-1}, 0, t^q)|} \sim \frac{t^{3p}}{t^q},$$

since by (4.1) $q<3p$.

$$\frac{|\gamma(t)|}{|\gamma(t)-\pi(\gamma(t))|}=\frac{|(t^p/\sqrt{6}, t, t^p, t^{4p-q}/9)|}{|(0, 0, t^p, t^{4p-q}/9)|} \sim \frac{t}{t^p}.$$

Hence the ratio (as in the definition of (r)) becomes $t^{2p-q+1}$, which does not tend to zero since $2p<q$ by (4.1). This proves assertion 4.3.

Finally we check that Verdier's condition (w) fails to hold.

$$d(T_{\gamma(t)}X, T_{\pi(\gamma(t))}Y) \sim t^{3p-q},$$
$$d(\gamma(t), \pi(\gamma(t))) \sim t^p,$$

hence (w) fails exactly when $2p<q$.

NOTE 4.4. Basing the construction on $w=x^{4k}-x^{2k}+\frac{1}{4}$, $1<k<\infty$ (instead of $k=1$ as here), we can build similar examples with $X$ a $C^k$ submanifold.

REFERENCES

[1] J. BRIANÇON and J.-P. SPEDER, Les conditions de Whitney impliquent $\mu^*$-constant, *Annales de l'Institut Fourier de Grenoble 26(2)* (1976), 153–163.

[2] M. GOLUBITSKY and V. GUILLEMIN, *Stable mappings and their singularities*, Springer-Verlag, New York, 1974.

[3] H. HIRONAKA, Subanalytic sets, *Number Theory, Algebraic Geometry and Commutative Algebra, volume in honour of Y. Akizuki*, Kinokuniya, Tokyo, 1973, 453–493.

[4] T.-C. KUO, The ratio test for analytic Whitney stratifications, *Liverpool Singularities Symposium* I, Springer Lecture Notes 192, Berlin (1971), 141–149.

[5] J. MILNOR, *Singular points of complex hypersurfaces*, Ann. of Math. Studies No. 61, Princeton (1968).

[6] B. TEISSIER, Introduction to equisingularity problems, *A.M.S. Algebraic Geometry Symposium, Arcata 1974*, Providence (1975), 593–632.

[7] B. TEISSIER, Variétés polaires I: Invariants polaires des singularités d'hypersurfaces, Publication M242.0376 du *Centre de Math. de l'Ecole Polytechnique*, 1976.

[8] R. THOM, Propriétés differentielles locales des ensembles analytiques, *Séminaire Bourbaki 1964–65*, exposé no. 281.

[9] R. THOM, Ensembles et morphismes stratifiés, *Bull. Am. Math. Soc. 75* (1969), 240–284.

[10] D. J. A. TROTMAN, A transversality property weaker than Whitney (A)-regularity, *Bull. Lond. Math. Soc. 8* (1976), 225–228.

[11] D. J. A. TROTMAN and E. C. ZEEMAN, Classification of elementary catastrophes of codimension ≦5, *Structural stability, the theory of catastrophes, and applications in the sciences*, Springer Lecture Notes No. 525 (1976), 263–327.

[12] J.-L. VERDIER, Stratifications de Whitney et théorème de Bertini-Sard, *Inventiones Math.* (1976), *36*, 295–312.

[13] H. WHITNEY, Local properties of analytic varieties, *Diff. and Comb. Topology*, edited by S. Cairns, Princeton (1965), 205–244.

D. J. A. TROTMAN
Université de Paris-Sud
Faculté des Sciences
91405 Orsay, France